Treatise on Natural Philosophy

Two Volumes in One

LORD WILLIAM THOMSON KELVIN
& PETER GUTHRIE TAIT

COSIMO CLASSICS

NEW YORK

Treatise on Natural Philosophy (Two Volumes in One). First published in 1895.
Current edition published by Cosimo Classics in 2011.

ISBN: 978-1-61640-554-0

Cosimo aims to publish books that inspire, inform, and engage readers worldwide. We use innovative print-on-demand technology that enables books to be printed based on specific customer needs. This approach eliminates an artificial scarcity of publications and allows us to distribute books in the most efficient and environmentally sustainable manner. Cosimo also works with printers and paper manufacturers who practice and encourage sustainable forest management, using paper that has been certified by the FSC, SFI, and PEFC whenever possible.

Ordering Information:
Cosimo publications are available at online bookstores. They may also be purchased for educational, business, or promotional use:
Bulk orders: Special discounts are available on bulk orders for reading groups, organizations, businesses, and others.
Custom-label orders: We offer selected books with your customized cover or logo of choice.

For more information, contact us at:

Cosimo, Inc.
P.O. Box 416, Old Chelsea Station
New York, NY 10011

info@cosimobooks.com

or visit us at:
www.cosimobooks.com

Treatise on Natural Philosophy

Vol. I

PREFACE.

Les causes primordiales ne nous sont point connues; mais elles sont assujetties à des lois simples et constantes, que l'on peut découvrir par l'observation, et dont l'étude est l'objet de la philosophie naturelle.—FOURIER.

THE term Natural Philosophy was used by NEWTON, and is still used in British Universities, to denote the investigation of laws in the material world, and the deduction of results not directly observed. Observation, classification, and description of phenomena necessarily precede Natural Philosophy in every department of natural science. The earlier stage is, in some branches, commonly called Natural History; and it might with equal propriety be so called in all others.

Our object is twofold: to give a tolerably complete account of what is now known of Natural Philosophy, in language adapted to the non-mathematical reader; and to furnish, to those who have the privilege which high mathematical acquirements confer, a connected outline of the analytical processes by which the greater part of that knowledge has been extended into regions as yet unexplored by experiment.

We commence with a chapter on *Motion*, a subject totally independent of the existence of *Matter* and *Force*. In this we are naturally led to the consideration of the curvature and tortuosity of curves, the curvature of surfaces, distortions or strains, and various other purely geometrical subjects.

The *Laws of Motion*, the *Law of Gravitation and of Electric and Magnetic Attractions*, *Hooke's Law*, and other fundamental principles derived directly from experiment, lead by mathematical processes to interesting and useful results, for the full testing of which our most delicate experimental methods are as yet totally insufficient. A large part of the present volume is devoted to these deductions; which, though not immediately proved by experiment, are as certainly true as the elementary laws from which mathematical analysis has evolved them.

The analytical processes which we have employed are, as a rule, such as lead most directly to the results aimed at, and are therefore in great part unsuited to the general reader.

We adopt the suggestion of AMPÈRE, and use the term *Kinematics* for the purely geometrical science of motion in the abstract. Keeping in view the proprieties of language, and following the example of the most logical writers, we employ the term *Dynamics* in its true sense as the science which treats of the action of *force*, whether it maintains relative rest, or produces acceleration of relative motion. The two corresponding divisions of Dynamics are thus conveniently entitled *Statics* and *Kinetics*.

One object which we have constantly kept in view is the grand principle of the *Conservation of Energy*. According to modern experimental results, especially those of JOULE, Energy is as real and as indestructible as Matter. It is satisfactory to find that NEWTON anticipated, so far as the state of experimental science in his time permitted him, this magnificent modern generalization.

We desire it to be remarked that in much of our work, where we may appear to have rashly and needlessly interfered with methods and systems of proof in the present day generally accepted, we take the position of Restorers, and not of Innovators.

In our introductory chapter on Kinematics, the consideration of Harmonic Motion naturally leads us to *Fourier's Theorem*, one of the most important of all analytical results as regards usefulness in physical science. In the Appendices to that chapter we have introduced an extension of *Green's Theorem,* and a treatise on the remarkable functions known as *Laplace's Coefficients.* There can be but one opinion as to the beauty and utility of this analysis of Laplace; but the manner in which it has been hitherto presented has seemed repulsive to the ablest mathematicians, and difficult to ordinary mathematical students. In the simplified and symmetrical form in which we give it, it will be found quite within the reach of readers moderately familiar with modern mathematical methods.

In the second chapter we give NEWTON'S Laws of Motion in his own words, and with some of his own comments—every attempt that has yet been made to supersede them having ended in utter failure. Perhaps nothing so simple, and at the same time so comprehensive, has ever been given as the foundation of a system in any of the sciences. The dynamical use of the *Generalized Cöordinates* of LAGRANGE, and the *Varying Action* of HAMILTON, with kindred matter, complete the chapter.

The third chapter, "Experience," treats briefly of Observation and Experiment as the basis of Natural Philosophy.

The fourth chapter deals with the fundamental Units, and the chief Instruments used for the measurement of Time, Space, and Force.

Thus closes the First Division of the work, which is strictly preliminary, and to which we have limited the present issue.

This new edition has been thoroughly revised, and very considerably extended. The more important additions are to be found in the Appendices to the first chapter, especially that devoted to *Laplace's Coefficients;* also at the end of the second chapter, where a very full investigation of the "*cycloidal motion*" of systems is now given; and in Appendix B′, which describes a number of continuous calculating machines invented and constructed since the publication of our first edition. A great improvement has been made in the treatment of *Lagrange's Generalized Equations of Motion.*

We believe that the mathematical reader will especially profit by a perusal of the large type portion of this volume; as he will thus be forced to think out for himself what he has been too often accustomed to reach by a mere mechanical application of analysis. Nothing can be more fatal to progress than a too confident reliance on mathematical symbols; for the student is only too apt to take the easier course, and consider the *formula* and not the *fact* as the physical reality.

In issuing this new edition, of a work which has been for several years out of print, we recognise with legitimate satisfaction the very great improvement which has recently taken place in the more elementary works on Dynamics published in this country, and which we cannot but attribute, in great part, to our having effectually recalled to its deserved position Newton's system of elementary definitions, and Laws of Motion.

We are much indebted to Mr BURNSIDE and Prof. CHRYSTAL for the pains they have taken in reading proofs and verifying formulas; and we confidently hope that few erratums of serious consequence will now be found in the work.

W. THOMSON.
P. G. TAIT.

CONTENTS.

DIVISION I.—PRELIMINARY.

CHAPTER I.—KINEMATICS.

CHAPTER II.—DYNAMICAL LAWS AND PRINCIPLES.

SECTIONS

SECTIONS

DIVISION I.

PRELIMINARY.

CHAPTER I.—KINEMATICS.

1. THERE are many properties of motion, displacement, and deformation, which may be considered altogether independently of such physical ideas as force, mass, elasticity, temperature, magnetism, electricity. The preliminary consideration of such properties in the abstract is of very great use for Natural Philosophy, and we devote to it, accordingly, the whole of this our first chapter; which will form, as it were, the Geometry of our subject, embracing what can be observed or concluded with regard to actual motions, as long as the *cause* is not sought.

2. In this category we shall take up first the free motion of a point, then the motion of a point attached to an inextensible cord, then the motions and displacements of rigid systems—and finally, the deformations of surfaces and of solid or fluid bodies. Incidentally, we shall be led to introduce a good deal of elementary geometrical matter connected with the curvature of lines and surfaces.

3. When a point moves from one position to another it must evidently describe a *continuous* line, which may be curved or straight, or even made up of portions of curved and straight lines meeting each other at any angles. If the motion be that of a *material particle*, however, there cannot generally be any such abrupt changes of direction, since (as we shall afterwards see) this would imply the action of an *infinite* force, except in the case in which the velocity becomes zero at the angle. It is useful to consider at the outset various theorems connected **Motion of a point.**

Motion of a point.

with the geometrical notion of the path described by a moving point, and these we shall now take up, deferring the consideration of Velocity to a future section, as being more closely connected with physical ideas.

4. The *direction* of motion of a moving point is at each instant the tangent drawn to its path, if the path be a curve, or the path itself if a straight line.

Curvature of a plane curve.

5. If the path be not straight the direction of motion changes from point to point, and the *rate* of this change, per unit of length of the curve $\left(\frac{d\theta}{ds}\text{ according to the notation below}\right)$, is called the *curvature*. To exemplify this, suppose two tangents drawn to a circle, and radii to the points of contact. The angle between the tangents is the change of direction required, and the rate of change is to be measured by the relation between this angle and the length of the circular arc. Let I be the angle, c the arc, and ρ the radius. We see at once that (as the angle between the radii is equal to the angle between the tangents)

$$\rho I = c,$$

and therefore $\frac{I}{c} = \frac{1}{\rho}$. Hence the curvature of a circle is inversely as its radius, and, measured in terms of the proper unit of curvature, is simply the reciprocal of the radius.

6. Any small portion of a curve may be approximately taken as a circular arc, the approximation being closer and closer to the truth, as the assumed arc is smaller. The curvature is then the reciprocal of the radius of this circle.

If $\delta\theta$ be the angle between two tangents at points of a curve distant by an arc δs, the definition of curvature gives us at once as its measure, the limit of $\frac{\delta\theta}{\delta s}$ when δs is diminished without limit; or, according to the notation of the differential calculus, $\frac{d\theta}{ds}$. But we have

$$\tan\theta = \frac{dy}{dx},$$

if, the curve being a plane curve, we refer it to two rectangular axes OX, OY, according to the Cartesian method, and if θ denote the inclination of its tangent, at any point x, y, to OX. Hence Curvature of a plane curve

$$\theta = \tan^{-1}\frac{dy}{dx};$$

and, by differentiation with reference to any independent variable t, we have

$$d\theta = \frac{d\left(\frac{dy}{dx}\right)}{1+\left(\frac{dy}{dx}\right)^2} = \frac{dx\,d^2y - dy\,d^2x}{dx^2+dy^2}.$$

Also, $$ds = (dx^2 + dy^2)^{\frac{1}{2}}.$$

Hence, if ρ denote the radius of curvature, so that

$$\frac{1}{\rho} = \frac{d\theta}{ds} \qquad\qquad (1),$$

we conclude $$\frac{1}{\rho} = \frac{dx\,d^2y - dy\,d^2x}{(dx^2+dy^2)^{\frac{3}{2}}} \qquad\qquad (2).$$

Although it is generally convenient, in kinematical and kinetic formulæ, to regard time as the independent variable, and all the changing geometrical elements as functions of it, there are cases in which it is useful to regard the length of the arc or path described by a point as the independent variable. On this supposition we have

$$0 = d(ds^2) = d(dx^2 + dy^2) = 2\,(dx\,d_s^2x + dy\,d_s^2y),$$

where we denote by the suffix to the letter d, the independent variable understood in the differentiation. Hence

$$\frac{dx}{d_s^2y} = -\frac{dy}{d_s^2x} = -\frac{(dx^2+dy^2)^{\frac{1}{2}}}{\{(d_s^2y)^2+(d_s^2x)^2\}^{\frac{1}{2}}};$$

and using these, with $ds^2 = dx^2 + dy^2$, to eliminate dx and dy from (2), we have

$$\frac{1}{\rho} = \frac{\{(d_s^2y)^2 + (d_s^2x)^2\}^{\frac{1}{2}}}{ds^2};$$

or, according to the usual short, although not quite complete, notation,

$$\frac{1}{\rho} = \left\{\left(\frac{d^2y}{ds^2}\right)^2 + \left(\frac{d^2x}{ds^2}\right)^2\right\}^{\frac{1}{2}}.$$

7. If all points of the curve lie in one plane, it is called a *plane curve*, and in the same way we speak of a *plane* polygon or broken line. If various points of the line do not lie in one plane, we have in one case what is called a *curve of double* Tortuous curve.

Tortuous curve. *curvature*, in the other a *gauche polygon*. The term 'curve of double curvature' is very bad, and, though in very general use, is, we hope, not ineradicable. The fact is, that there are not two curvatures, but only a curvature (as above defined), of which the plane is continuously changing, or twisting, round the tangent line; thus exhibiting a torsion. The course of such a curve is, in common language, well called 'tortuous;' and the measure of the corresponding property is conveniently called *Tortuosity*.

8. The nature of this will be best understood by considering the curve as a polygon whose sides are indefinitely small. Any two consecutive sides, of course, lie in a plane—and in that plane the curvature is measured as above, but in a curve which is not plane the third side of the polygon will not be in the same plane with the first two, and, therefore, the new plane in which the curvature is to be measured is different from the old one. The plane of the curvature on each side of any point of a tortuous curve is sometimes called the *Osculating Plane* of the curve at that point. As two successive positions of it contain the second side of the polygon above mentioned, it is evident that the osculating plane passes from one position to the next by revolving about the tangent to the curve.

Curvature and tortuosity. 9. Thus, as we proceed along such a curve, the curvature in general varies; and, at the same time, the plane in which the curvature lies is turning about the tangent to the curve. The tortuosity is therefore to be measured by the rate at which the osculating plane turns about the tangent, per unit length of the curve.

To express the radius of curvature, the direction cosines of the osculating plane, and the tortuosity, of a curve not in one plane, in terms of Cartesian triple co-ordinates, let, as before, $\delta\theta$ be the angle between the tangents at two points at a distance δs from one another along the curve, and let $\delta\phi$ be the angle between the osculating planes at these points. Thus, denoting by ρ the radius of curvature, and τ the tortuosity, we have

$$\frac{1}{\rho} = \frac{d\theta}{ds},$$

$$\tau = \frac{d\phi}{ds},$$

according to the regular notation for the limiting values of $\frac{\delta\theta}{\delta s}$, and $\frac{\delta\phi}{\delta s}$, when δs is diminished without limit. Let OL, OL' be lines drawn through any fixed point O parallel to any two successive positions of a moving line PT, each in the directions indicated by the order of the letters. Draw OS perpendicular to their plane in the direction from O, such that OL, OL', OS lie in the same relative order in space as the positive axes of co-ordinates, OX, OY, OZ. Let OQ bisect LOL', and let OR bisect the angle between OL' and LO produced through O. Curvature and tortuosity.

Let the direction cosines of

OL	be	a, b, c;
OL'	„	a', b', c';
OQ	„	l, m, n;
OR	„	α, β, γ;
OS	„	λ, μ, ν:

and let $\delta\theta$ denote the angle LOL'. We have, by the elements of analytical geometry,

$$\cos\delta\theta = aa' + bb' + cc' \qquad (3);$$

$$l = \frac{\frac{1}{2}(a+a')}{\cos\frac{1}{2}\delta\theta}, \quad m = \frac{\frac{1}{2}(b+b')}{\cos\frac{1}{2}\delta\theta}, \quad n = \frac{\frac{1}{2}(c+c')}{\cos\frac{1}{2}\delta\theta} \qquad (4);$$

$$\alpha = \frac{a'-a}{2\sin\frac{1}{2}\delta\theta}, \quad \beta = \frac{b'-b}{2\sin\frac{1}{2}\delta\theta}, \quad \gamma = \frac{c'-c}{2\sin\frac{1}{2}\delta\theta} \qquad (5);$$

$$\lambda = \frac{bc'-b'c}{\sin\delta\theta}, \quad \mu = \frac{ca'-c'a}{\sin\delta\theta}, \quad \nu = \frac{ab'-a'b}{\sin\delta\theta} \qquad (6).$$

Now let the two successive positions of PT be tangents to a curve at points separated by an arc of length δs. We have

$$\frac{1}{\rho} = \frac{\delta\theta}{\delta s} = \frac{2\sin\frac{1}{2}\delta\theta}{\delta s} = \frac{\sin\delta\theta}{\delta s} \qquad (7)$$

when δs is infinitely small; and in the same limit

$$l = \frac{dx}{ds}, \quad m = \frac{dy}{ds}, \quad n = \frac{dz}{ds};$$

$$a'-a = d\frac{dx}{ds}, \quad b'-b = d\frac{dy}{ds}, \quad c'-c = d\frac{dz}{ds} \qquad (8);$$

$$bc'-b'c = \frac{dy}{ds}d\frac{dz}{ds} - \frac{dz}{ds}d\frac{dy}{ds}, \text{ \&c.} \qquad (9);$$

Curvature and tortuosity.

and α, β, γ become the direction cosines of the normal, PC, drawn towards the centre of curvature, C; and λ, μ, ν those of the perpendicular to the osculating plane drawn in the direction relatively to PT and PC, corresponding to that of OZ relatively to OX and OY. Then, using (8) and (9), with (7), in (5) and (6) respectively, we have

$$\alpha = \frac{d\frac{dx}{ds}}{\rho^{-1}ds}, \quad \beta = \frac{d\frac{dy}{ds}}{\rho^{-1}ds}, \quad \gamma = \frac{d\frac{dz}{ds}}{\rho^{-1}ds} \text{......... } (10);$$

$$\lambda = \frac{\frac{dy}{ds}d\frac{dz}{ds} - \frac{dz}{ds}d\frac{dy}{ds}}{\rho^{-1}ds}, \quad \mu = \frac{\frac{dz}{ds}d\frac{dx}{ds} - \frac{dx}{ds}d\frac{dz}{ds}}{\rho^{-1}ds}, \quad \nu = \frac{\frac{dx}{ds}d\frac{dy}{ds} - \frac{dy}{ds}d\frac{dx}{ds}}{\rho^{-1}ds} \quad (11).$$

The simplest expression for the curvature, with choice of independent variable left arbitrary, is the following, taken from (10):

$$\frac{1}{\rho} = \frac{\sqrt{\left\{\left(d\frac{dx}{ds}\right)^2 + \left(d\frac{dy}{ds}\right)^2 + \left(d\frac{dz}{ds}\right)^2\right\}}}{ds} \text{.........} (12).$$

This, modified by differentiation, and application of the formula

$$ds\,d^2s = dx\,d^2x + dy\,d^2y + dz\,d^2z \text{............} (13),$$

becomes

$$\frac{1}{\rho} = \frac{\sqrt{\{(d^2x)^2 + (d^2y)^2 + (d^2z)^2 - (d^2s)^2\}}}{ds^2} \text{............} (14).$$

Another formula for $\frac{1}{\rho}$ is obtained immediately from equations (11); but these equations may be put into the following simpler form, by differentiation, &c.,

$$\lambda = \frac{dy\,d^2z - dz\,d^2y}{\rho^{-1}ds^3}, \quad \mu = \frac{dz\,d^2x - dx\,d^2z}{\rho^{-1}ds^3}, \quad \nu = \frac{dx\,d^2y - dy\,d^2x}{\rho^{-1}ds^3} \quad (15);$$

from which we find

$$\rho^{-1} = \frac{\{(dy\,d^2z - dz\,d^2y)^2 + (dz\,d^2x - dx\,d^2z)^2 + (dx\,d^2y - dy\,d^2x)^2\}^{\frac{1}{2}}}{ds^3} \quad (16).$$

Each of these several expressions for the curvature, and for the directions of the relative lines, we shall find has its own special significance in the kinetics of a particle, and the statics of a flexible cord.

To find the tortuosity, $\frac{d\phi}{ds}$, we have only to apply the general equation above, with λ, μ, ν substituted for l, m, n, and $\frac{1}{\tau}\frac{d\lambda}{ds}$, $\frac{1}{\tau}\frac{d\mu}{ds}$, $\frac{1}{\tau}\frac{d\nu}{ds}$ for α, β, γ. Thus we have $\tau^2 = \left(\frac{d\lambda}{ds}\right)^2 + \left(\frac{d\mu}{ds}\right)^2 + \left(\frac{d\nu}{ds}\right)^2$,

Curvature and tortuosity

or $$\tau = \left\{ \left(\mu \frac{d\nu}{ds} - \nu \frac{d\mu}{ds} \right)^2 + \left(\nu \frac{d\lambda}{ds} - \lambda \frac{d\nu}{ds} \right)^2 + \left(\lambda \frac{d\mu}{ds} - \mu \frac{d\lambda}{ds} \right)^2 \right\}^{\frac{1}{2}}.$$

where λ, μ, ν, denote the direction cosines of the osculating plane, given by the preceding formulæ.

Integral curvature of a curve (compare § 136).

10. The *integral curvature*, or *whole change of direction* of an arc of a plane curve, is the angle through which the tangent has turned as we pass from one extremity to the other. The *average curvature* of any portion is its whole curvature divided by its length. Suppose a line, drawn from a fixed point, to move so as always to be parallel to the direction of motion of a point describing the curve: the angle through which this turns during the motion of the point exhibits what we have thus defined as the integral curvature. In estimating this, we must of course take the enlarged modern meaning of an angle, including angles greater than two right angles, and also negative angles. Thus the integral curvature of any closed curve, whether everywhere concave to the interior or not, is four right angles, provided it does not cut itself. That of a Lemniscate, or figure of 8, is *zero*. That of the Epicycloid is eight right angles; and so on.

11. The definition in last section may evidently be extended to a plane polygon, and the integral change of direction, or the angle between the first and last sides, is then the sum of its exterior angles, all the sides being produced each in the direction in which the moving point describes it while passing round the figure. This is true whether the polygon be closed or not. If closed, then, as long as it is not crossed, this sum is four right angles,—an extension of the result in Euclid, where all *re-entrant* polygons are excluded. In the case of the star-shaped figure, it is ten right angles, wanting the sum of the five acute angles of the figure; that is, eight right angles.

12. The *integral curvature* and the *average curvature* of a curve which is not plane, may be defined as follows:—Let successive lines be drawn from a fixed point, parallel to tangents at successive points of the curve. These lines will form a conical surface. Suppose this to be cut by a sphere of unit radius having its centre at the fixed point. The *length* of the

Integral curvature of a curve (compare § 136).

curve of intersection measures the *integral curvature* of the given curve. The *average curvature* is, as in the case of a plane curve, the integral curvature divided by the length of the curve. For a tortuous curve approximately plane, the integral curvature thus defined, approximates (not to the integral curvature according to the proper definition, § 10, for a plane curve, but) to the sum of the integral curvatures of all the parts of an approximately coincident plane curve, each taken as positive. Consider, for examples, varieties of James Bernouilli's plane elastic curve, § 611, and approximately coincident tortuous curves of fine steel piano-forte wire. Take particularly the plane lemniscate and an approximately coincident tortuous closed curve.

13. Two consecutive tangents lie in the osculating plane. This plane is therefore parallel to the tangent plane to the cone described in the preceding section. Thus the tortuosity may be measured by the help of the spherical curve which we have just used for defining integral curvature. We cannot as yet complete the explanation, as it depends on the theory of rolling, which will be treated afterwards (§§ 110—137). But it is enough at present to remark, that if a plane roll on the sphere, along the spherical curve, turning always round an instantaneous axis tangential to the sphere, the integral curvature of the curve of contact or trace of the rolling on the plane, is a proper measure of the *whole torsion*, or integral of tortuosity. From this and § 12 it follows that the curvature of this plane curve at any point, or, which is the same, the projection of the curvature of the spherical curve on a tangent plane of the spherical surface, is equal to the tortuosity divided by the curvature of the given curve.

Let $\frac{1}{\rho}$ be the curvature and τ the tortuosity of the given curve, and ds an element of its length. Then $\int\frac{ds}{\rho}$ and $\int\tau ds$, each integral extended over any stated length, l, of the curve, are respectively the integral curvature and the integral tortuosity. The mean curvature and the mean tortuosity are respectively

$$\frac{1}{l}\int\frac{ds}{\rho} \text{ and } \frac{1}{l}\int\tau ds.$$

Infinite tortuosity will be easily understood, by considering a helix, of inclination α, described on a right circular cylinder of radius r. The curvature in a circular section being $\frac{1}{r}$, that of the helix is, of course, $\frac{\cos^2\alpha}{r}$. The tortuosity is $\frac{\sin\alpha\cos\alpha}{r}$, or $\tan\alpha\times$ curvature. Hence, if $\alpha=\frac{\pi}{4}$ the curvature and tortuosity are equal. Integral curvature of a curve (compare § 136).

Let the curvature be denoted by $\frac{1}{\rho}$, so that $\cos^2\alpha=\frac{r}{\rho}$. Let ρ remain finite, and let r diminish without limit. The *step* of the helix being $2\pi r\tan\alpha=2\pi\sqrt{r}\left(1-\frac{r}{\rho}\right)^{\frac{1}{2}}$, is, in the limit, $2\pi\sqrt{\rho r}$, which is infinitely small. Thus the motion of a point in the curve, though infinitely nearly in a straight line (the path being always at the infinitely small distance r from the fixed straight line, the axis of the cylinder), will have finite curvature $\frac{1}{\rho}$. The tortuosity, being $\frac{1}{\rho}\tan\alpha$ or $\frac{1}{\sqrt{\rho r}}\left(1-\frac{r}{\rho}\right)^{\frac{1}{2}}$, will in the limit be a mean proportional between the curvature of the circular section of the cylinder and the finite curvature of the curve.

The acceleration (or force) required to produce such a motion of a point (or material particle) will be afterwards investigated (§ 35 *d.*).

14. A chain, cord, or fine wire, or a fine fibre, filament, or hair, may suggest what is not to be found among natural or artificial productions, a perfectly *flexible and inextensible line.* The elementary kinematics of this subject require no investigation. The mathematical condition to be expressed in any case of it is simply that the distance measured along the line from any one point to any other, remains constant, however the line be bent. Flexible line.

15. The use of a cord in mechanism presents us with many practical applications of this theory, which are in general extremely simple; although curious, and not always very easy, geometrical problems occur in connexion with it. We shall say nothing here about the theory of knots, knitting, weaving,

Flexible line. plaiting, etc., but we intend to return to the subject, under vortex-motion in Hydrokinetics.

16. In the mechanical tracing of curves, a flexible and inextensible cord is often supposed. Thus, in drawing an ellipse, the focal property of the curve shows us that by fixing the ends of such a cord to the foci and keeping it stretched by a pencil, the pencil will trace the curve.

By a ruler moveable about one focus, and a string attached to a point in the ruler and to the other focus, the hyperbola may be described by the help of its analogous focal property; and so on.

Evolute. **17.** But the consideration of evolutes is of some importance in Natural Philosophy, especially in certain dynamical and optical questions, and we shall therefore devote a section or two to this application of kinematics.

Def. If a flexible and inextensible string be fixed at one point of a plane curve, and stretched along the curve, and be then unwound in the plane of the curve, its extremity will describe an *Involute* of the curve. The original curve, considered with reference to the other, is called the *Evolute.*

18. It will be observed that we speak of *an* involute, and of *the* evolute, of a curve. In fact, as will be easily seen, a curve can have but one evolute, but it has an infinite number of involutes. For all that we have to do to vary an involute, is to change the point of the curve from which the tracing point starts, or consider the involutes described by different points of the string, and these will, in general, be different curves. The following section shows that there is but one evolute.

19. Let AB be any curve, PQ a portion of an involute, pP, qQ positions of the free part of the string. It will be seen at once that these must be tangents to the arc AB at p and q. Also (see § 90), the string at any stage, as pP, revolves about p. Hence pP is *normal* to the curve PQ. And thus the evolute of PQ is a definite curve, viz., the envelope of the normals drawn at every point of PQ,

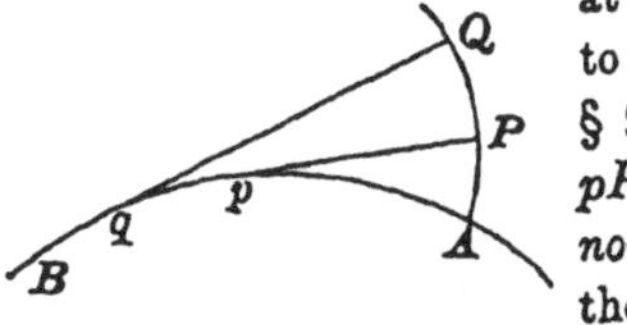

or, which is the same thing, the locus of the centres of curvature of the curve PQ. And we may merely mention, as an obvious result of the mode of tracing, that the arc pq is equal to the difference of qQ and pP, or that the arc pA is equal to pP. Evolute.

20. The rate of motion of a point, or its rate of change of position, is called its *Velocity*. It is greater or less as the space passed over in a given time is greater or less: and it may be *uniform*, *i.e.*, the same at every instant; or it may be *variable*. Velocity.

Uniform velocity is measured by the space passed over in unit of time, and is, in general, expressed in feet per second; if very great, as in the case of light, it is sometimes popularly reckoned in miles per second. It is to be observed, that time is here used in the abstract sense of a uniformly increasing quantity—what in the differential calculus is called an independent variable. Its physical definition is given in the next chapter.

21. Thus a point, which moves uniformly with velocity v, describes a space of v feet each second, and therefore vt feet in t seconds, t being any number whatever. Putting s for the space described in t seconds, we have

$$s = vt.$$

Thus with unit velocity a point describes unit of space in unit of time.

22. It is well to observe here, that since, by our formula, we have generally

$$v = \frac{s}{t};$$

and since nothing has been said as to the magnitudes of s and t, we may take these as small as we choose. Thus *we get the same result whether we derive v from the space described in a million seconds, or from that described in a millionth of a second.* This idea is very useful, as it makes our results intelligible when a variable velocity has to be measured, and we find ourselves obliged to approximate to its value by considering the space described in an interval so short, that during its lapse the velocity does not sensibly alter in value.

Velocity. **23.** When the point does not move uniformly, the velocity is variable, or different at different successive instants; but we define the *average* velocity during any time as the space described in that time, divided by the time, and, the less the interval is, the more nearly does the average velocity coincide with the actual velocity at any instant of the interval. Or again, we define the exact velocity at any instant as the space which the point would have described in one second, if for one second its velocity remained unchanged. That there is at every instant a definite value of the velocity of any moving body, is evident to all, and is matter of everyday conversation. Thus, a railway train, after starting, gradually increases its speed, and every one understands what is meant by saying that at a particular instant it moves at the rate of ten or of fifty miles an hour,—although, in the course of an hour, it may not have moved a mile altogether. Indeed, we may imagine, at any instant during the motion, the steam to be so adjusted as to keep the train running for some time at a perfectly uniform velocity. This would be the velocity which the train had at the instant in question. Without supposing any such definite adjustment of the driving power to be made, we can evidently obtain an approximation to this instantaneous velocity by considering the motion for so short a time, that during it the actual variation of speed may be small enough to be neglected.

24. In fact, if v be the velocity at either beginning or end, or at any instant of the interval, and s the space actually described in time t, the equation $v=\frac{s}{t}$ is more and more nearly true, as the velocity is more nearly uniform during the interval t; so that if we take the interval small enough the equation may be made as nearly exact as we choose. Thus the set of values—

Space described in one second,
Ten times the space described in the first tenth of a second,
A hundred " " " hundredth "

and so on, give nearer and nearer approximations to the velocity at the beginning of the first second. The whole foundation of the differential calculus is, in fact, contained in this simple question, "What is the rate at which the space described increases?" *i.e.*, What is the velocity of the moving point? Newton's notation for the velocity, *i.e.* the rate at which s increases, or the *fluxion* of s, is $\dot{s}$. This notation is very convenient, as it saves the introduction of a second letter. Velocity

Let a point which has described a space s in time t proceed to describe an additional space δs in time δt, and let v_1 be the greatest, and v_2 the least, velocity which it has during the interval δt. Then, evidently,

$$\delta s < v_1 \delta t, \quad \delta s > v_2 \delta t,$$

$$i.e., \quad \frac{\delta s}{\delta t} < v_1, \quad \frac{\delta s}{\delta t} > v_2.$$

But as δt diminishes, the values of v_1 and v_2 become more and more nearly equal, and in the limit, each is equal to the velocity at time t. Hence

$$v = \frac{ds}{dt}.$$

25. The preceding definition of velocity is equally applicable whether the point move in a straight or curved line; but, since in the latter case the direction of motion continually changes, the mere amount of the velocity is not sufficient completely to describe the motion, and we must have in every such case additional data to remove the uncertainty. Resolution of velocity.

In such cases as this the method commonly employed, whether we deal with velocities, or as we shall do farther on with accelerations and forces, consists mainly in studying, not the velocity, acceleration, or force, *directly*, but its components parallel to any three assumed directions at right angles to each other. Thus, for a train moving up an incline in a NE direction, we may have given the whole velocity and the steepness of the incline, or we may express the same ideas thus—the train is moving simultaneously northward, eastward, and upward—and the motion as to amount and direction will be completely known if we know separately the northward, eastward, and upward velocities—these being called the *components* of the whole velocity in the three mutually perpendicular directions N, E, and up.

Resolution of velocity.

In general the velocity of a point at x, y, z, is (as we have seen) $\frac{ds}{dt}$, or, which is the same, $\left\{\left(\frac{dx}{dt}\right)^2 + \left(\frac{dy}{dt}\right)^2 + \left(\frac{dz}{dt}\right)^2\right\}^{\frac{1}{2}}$.

Now denoting by u the rate at which x increases, or the velocity parallel to the axis of x, and so by v, w, for the other two; we have $u = \frac{dx}{dt}$, $v = \frac{dy}{dt}$, $w = \frac{dz}{dt}$. Hence, calling α, β, γ the angles which the direction of motion makes with the axes, and putting $q = \frac{ds}{dt}$, we have

$$\cos\alpha = \frac{dx}{ds} = \frac{\frac{dx}{dt}}{\frac{ds}{dt}} = \frac{u}{q}.$$

Hence $u = q\cos\alpha$, and therefore

26. A velocity in any direction may be resolved in, and perpendicular to, any other direction. The first component is found by multiplying the velocity by the cosine of the angle between the two directions—the second by using as factor the sine of the same angle. Or, it may be resolved into components in any three rectangular directions, each component being formed by multiplying the whole velocity by the cosine of the angle between its direction and that of the component.

It is useful to remark that if the axes of x, y, z are not rectangular, $\frac{dx}{dt}$, $\frac{dy}{dt}$, $\frac{dz}{dt}$ will still be the velocities parallel to the axes, but we shall no longer have

$$\left(\frac{ds}{dt}\right)^2 = \left(\frac{dx}{dt}\right)^2 + \left(\frac{dy}{dt}\right)^2 + \left(\frac{dz}{dt}\right)^2.$$

We leave as an exercise for the student the determination of the correct expression for the whole velocity in terms of its components.

If we resolve the velocity along a line whose inclinations to the axes are λ, μ, ν, and which makes an angle θ with the direction of motion, we find the two expressions below (which must of course be equal) according as we resolve q directly or by its components, u, v, w,

$$q\cos\theta = u\cos\lambda + v\cos\mu + w\cos\nu.$$

Substitute in this equation the values of u, v, w already given, § 25, and we have the well-known geometrical theorem for the angle between two straight lines which make given angles with the axes, Resolution of velocity.

$$\cos\theta = \cos\alpha\cos\lambda + \cos\beta\cos\mu + \cos\gamma\cos\nu.$$

From the above expression we see at once that

27. The velocity resolved in any direction is the sum of the components (in that direction) of the three rectangular components of the whole velocity. And, if we consider motion in one plane, this is still true, only we have but *two* rectangular components. These propositions are virtually equivalent to the following obvious geometrical construction:— Composition of velocities.

To compound any two velocities as OA, OB in the figure; from A draw AC parallel and equal to OB. Join OC:—then OC is the resultant velocity in magnitude and direction.

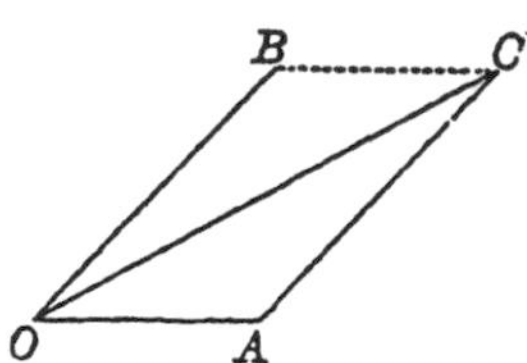

OC is evidently the diagonal of the parallelogram two of whose sides are OA, OB.

Hence the resultant of velocities represented by the sides of any closed polygon whatever, whether in one plane or not, taken all in the same order, is zero.

Hence also the resultant of velocities represented by all the sides of a polygon but one, taken in order, is represented by that one taken in the opposite direction.

When there are two velocities or three velocities in two or in three rectangular directions, the resultant is the square root of the sum of their squares—and the cosines of the inclination of its direction to the given directions are the ratios of the components to the resultant.

It is easy to see that as δs in the limit may be resolved into δr and $r\delta\theta$, where r and θ are polar co-ordinates of a plane curve, $\frac{dr}{dt}$ and $r\frac{d\theta}{dt}$ are the resolved parts of the velocity along, and perpendicular to, the radius vector. We may obtain the same result thus, $x = r\cos\theta,\quad y = r\sin\theta.$

Composition of velocities.

Hence $\frac{dx}{dt} = \frac{dr}{dt}\cos\theta - r\sin\theta\frac{d\theta}{dt}$, $\frac{dy}{dt} = \frac{dr}{dt}\sin\theta + r\cos\theta\frac{d\theta}{dt}$.

But by § 26 the whole velocity along r is $\frac{dx}{dt}\cos\theta + \frac{dy}{dt}\sin\theta$, *i.e.*, by the above values, $\frac{dr}{dt}$. Similarly the transverse velocity is

$$\frac{dy}{dt}\cos\theta - \frac{dx}{dt}\sin\theta, \text{ or } r\frac{d\theta}{dt}.$$

Acceleration.

28. The velocity of a point is said to be accelerated or retarded according as it increases or diminishes, but the word *acceleration* is generally used in either sense, on the understanding that we may regard its quantity as either positive or negative. Acceleration of velocity may of course be either uniform or variable. It is said to be uniform when the velocity receives equal increments in equal times, and is then measured by the actual increase of velocity per unit of time. If we choose as the unit of acceleration that which adds a unit of velocity per unit of time to the velocity of a point, an acceleration measured by a will add a units of velocity in unit of time—and, therefore, at units of velocity in t units of time. Hence if V be the change in the velocity during the interval t,

$$V = at, \text{ or } a = \frac{V}{t}.$$

29. Acceleration is variable when the point's velocity does not receive equal increments in successive equal periods of time. It is then measured by the increment of velocity, which would have been generated in a unit of time had the acceleration remained throughout that interval the same as at its commencement. The *average* acceleration during any time is the whole velocity gained during that time, divided by the time. In Newton's notation $\dot{v}$ is used to express the acceleration in the direction of motion; and, if $v = \dot{s}$, as in § 24, we have

$$a = \dot{v} = \ddot{s}.$$

Let v be the velocity at time t, δv its change in the interval δt, a_1 and a_2 the greatest and least values of the acceleration during the interval δt. Then, evidently,

$$\delta v < a.\delta t, \quad \delta v > a_2\delta t,$$

$$\text{or} \quad \frac{\delta v}{\delta t} < a_1, \quad \frac{\delta v}{\delta t} > a_2.$$ Acceleration.

As δt is taken smaller and smaller, the values of a_1 and a_2 approximate infinitely to each other, and to that of a the required acceleration at time t. Hence

$$\frac{dv}{dt} = a.$$

It is useful to observe that we may also write (by changing the independent variable)

$$a = \frac{dv}{ds}\frac{ds}{dt} = v\frac{dv}{ds}.$$

Since $v = \frac{ds}{dt}$, we have $a = \frac{d^2s}{dt^2}$, and it is evident from similar reasoning that the component accelerations parallel to the axes are $\frac{d^2x}{dt^2}$, $\frac{d^2y}{dt^2}$, $\frac{d^2z}{dt^2}$. But it is to be carefully observed that $\frac{d^2s}{dt^2}$ is *not* generally the resultant of the three component accelerations, but is so only when either the curvature of the path, or the velocity is zero; for [§ 9 (14)] we have

$$\left(\frac{d^2s}{dt^2}\right)^2 = \left(\frac{d^2x}{dt^2}\right)^2 + \left(\frac{d^2y}{dt^2}\right)^2 + \left(\frac{d^2z}{dt^2}\right)^2 - \left(\frac{1}{\rho}\frac{ds^2}{dt^2}\right)^2.$$

The direction cosines of the tangent to the path at any point x, y, z are

$$\frac{1}{v}\frac{dx}{dt}, \quad \frac{1}{v}\frac{dy}{dt}, \quad \frac{1}{v}\frac{dz}{dt}.$$

Those of the line of resultant acceleration are

$$\frac{1}{f}\frac{d^2x}{dt^2}, \quad \frac{1}{f}\frac{d^2y}{dt^2}, \quad \frac{1}{f}\frac{d^2z}{dt^2},$$

where, for brevity, we denote by f the resultant acceleration. Hence the direction cosines of the plane of these two lines are

$$\frac{dyd^2z - dzd^2y}{\{(dyd^2z - dzd^2y)^2 + (dzd^2x - dxd^2z)^2 + (dxd^2y - dyd^2x)^2\}^{\frac{1}{2}}}, \text{ etc.}$$

These (§ 9) show that this plane is the osculating plane of the curve. Again, if θ denote the angle between the two lines, we have

$$\sin\theta = \frac{\{(dyd^2z - dzd^2y)^2 + (dzd^2x - dxd^2z)^2 + (dxd^2y - dyd^2x)^2\}^{\frac{1}{2}}}{vfdt^3},$$

Accelera-tion.

or, according to the expression for the curvature (§ 9),

$$\sin\theta = \frac{ds^3}{\rho v f dt^3} = \frac{v^2}{f\rho}.$$

Hence $$f\sin\theta = \frac{v^2}{\rho}.$$

Again, $$\cos\theta = \frac{1}{vf}\left(\frac{dx}{dt}\frac{d^2x}{dt^2} + \frac{dy}{dt}\frac{d^2y}{dt^2} + \frac{dz}{dt}\frac{d^2z}{dt^2}\right) = \frac{ds\,d^2s}{vf dt^3} = \frac{d^2s}{f dt^2}.$$

Hence $f\cos\theta = \dfrac{d^2s}{dt^2}$, and therefore

Resolution and composition of accelerations.

30. The whole acceleration in any direction is the sum of the components (in that direction) of the accelerations parallel to any three rectangular axes—each component acceleration being found by the same rule as component velocities, that is, by multiplying by the cosine of the angle between the direction of the acceleration and the line along which it is to be resolved.

31. When a point moves in a curve the whole acceleration may be resolved into two parts, one in the direction of the motion and equal to the acceleration of the velocity—the other towards the centre of curvature (perpendicular therefore to the direction of motion), whose magnitude is proportional to the square of the velocity and also to the curvature of the path. The former of these changes the velocity, the other affects only the form of the path, or the direction of motion. Hence if a moving point be subject to an acceleration, constant or not, whose direction is continually perpendicular to the direction of motion, the velocity will not be altered—and the only effect of the acceleration will be to make the point move in a curve whose curvature is proportional to the acceleration at each instant.

32. In other words, if a point move in a curve, whether with a uniform or a varying velocity, its change of direction is to be regarded as constituting an acceleration towards the centre of curvature, equal in amount to the square of the velocity divided by the radius of curvature. The whole acceleration will, in every case, be the resultant of the acceleration, thus measuring change of direction, and the acceleration of actual velocity along the curve. Resolution and composition of accelerations.

We may take another mode of resolving acceleration for a plane curve, which is sometimes useful; along, and perpendicular to, the radius-vector. By a method similar to that employed in § 27, we easily find for the component along the radius-vector

$$\frac{d^2r}{dt^2}-r\left(\frac{d\theta}{dt}\right)^2,$$

and for that perpendicular to the radius-vector

$$\frac{1}{r}\frac{d}{dt}\left(r^2\frac{d\theta}{dt}\right).$$

33. If for any case of motion of a point we have given the whole velocity and its direction, or simply the components of the velocity in three rectangular directions, at any *time*, or, as is most commonly the case, for any *position*, the determination of the form of the path described, and of other circumstances of the motion, is a question of pure mathematics, and in all cases is capable, if not of an exact solution, at all events of a solution to any degree of approximation that may be desired. Determination of the motion from given velocity or acceleration.

The same is true if the total acceleration and its direction at every instant, or simply its rectangular components, be given, provided the velocity and direction of motion, as well as the position, of the point at any one instant, be given.

For we have in the first case

$$\frac{dx}{dt}=u=q\cos a, \text{ etc.},$$

three simultaneous equations which can contain only x, y, z, and t, and which therefore suffice when integrated to determine x, y, and z in terms of t. By eliminating t among these equations, we obtain two equations among x, y, and z—each of which represents a surface on which lies the path described, and whose intersection therefore completely determines it.

In the second case we have

$$\frac{d^2x}{dt^2}=\alpha, \quad \frac{d^2y}{dt^2}=\beta, \quad \frac{d^2z}{dt^2}=\gamma;$$

to which equations the same remarks apply, except that here each has to be twice integrated.

Determination of the motion from given velocity or acceleration.

The arbitrary constants introduced by integration are determined at once if we know the co-ordinates, and the components of the velocity, of the point at a given epoch.

Examples of velocity.

34. From the principles already laid down, a great many interesting results may be deduced, of which we enunciate a few of the most important.

a. If the velocity of a moving point be uniform, and if its direction revolve uniformly in a plane, the path described is a circle.

Let a be the velocity, and α the angle through which its direction turns in unit of time; then, by properly choosing the axes, we have

$$\frac{dx}{dt} = -a \sin \alpha t, \quad \frac{dy}{dt} = a \cos \alpha t,$$

whence

$$(x - A)^2 + (y - B)^2 = \frac{a^2}{\alpha^2}.$$

b. If a point moves in a plane, and if its component velocity parallel to each of two rectangular axes is proportional to its distance from that axis, the path is an ellipse or hyperbola whose principal diameters coincide with those axes; and the acceleration is directed to or from the origin at every instant.

$$\frac{dx}{dt} = \mu y, \quad \frac{dy}{dt} = \nu x.$$

Hence $\frac{d^2x}{dt^2} = \mu\nu x$, $\frac{d^2y}{dt^2} = \mu\nu y$, and the whole acceleration is towards or from O.

Also $\frac{dy}{dx} = \frac{\nu}{\mu}\frac{x}{y}$, from which $\mu y^2 - \nu x^2 = C$, an ellipse or hyperbola referred to its principal axes. (Compare § 65.)

c. When the velocity is uniform, but in direction revolving uniformly in a right circular cone, the motion of the point is in a circular helix whose axis is parallel to that of the cone.

Examples of acceleration.

35. *a.* When a point moves uniformly in a circle of radius R, with velocity V, the whole acceleration is directed towards the centre, and has the constant value $\frac{V^2}{R}$. See § 31.

b. With uniform acceleration in the direction of motion, a point describes spaces proportional to the squares of the times elapsed since the commencement of the motion. Examples of acceleration.

In this case the space described in any interval is that which would be described in the same time by a point moving uniformly with a velocity equal to that at the middle of the interval. In other words, the average velocity (when the acceleration is uniform) is, during any interval, the arithmetical mean of the initial and final velocities. This is the case of a stone falling vertically.

For if the acceleration be parallel to x, we have

$$\frac{d^2x}{dt^2} = a, \text{ therefore } \frac{dx}{dt} = v = at, \text{ and } x = \tfrac{1}{2}at^2.$$

And we may write the equation (§ 29) $v\dfrac{dv}{dx} = a$, whence $\dfrac{v^2}{2} = ax$.

If at time $t = 0$ the velocity was V, these equations become at once

$$v = V + at, \quad x = Vt + \tfrac{1}{2}at^2, \text{ and } \frac{v^2}{2} = \frac{V^2}{2} + ax.$$

$$\begin{aligned} \text{And initial velocity} &= V, \\ \text{final} \quad ,, \quad &= V + at; \\ \text{Arithmetical mean} &= V + \tfrac{1}{2}at, \\ &= \frac{x}{t}, \end{aligned}$$

whence the second part of the above statement.

c. When there is uniform acceleration in a constant direction, the path described is a parabola, whose axis is parallel to that direction. This is the case of a projectile moving in vacuum.

For if the axis of y be parallel to the acceleration a, and if the plane of xy be that of motion at any time,

$$\frac{d^2z}{dt^2} = 0, \quad \frac{dz}{dt} = 0, \quad z = 0,$$

and therefore the motion is wholly in the plane of xy.

Then
$$\frac{d^2x}{dt^2} = 0, \quad \frac{d^2y}{dt^2} = a;$$

Examples of acceleration.

and by integration

$$x = Ut + a, \quad y = \tfrac{1}{2}at^2 + Vt + b,$$

where U, V, a, b are constants.

The elimination of t gives the equation of a parabola of which the axis is parallel to y, parameter $\frac{2U^2}{a}$, and vertex the point whose co-ordinates are

$$x = a - \frac{UV}{a}, \quad y = b - \frac{V^2}{2a}.$$

d. As an illustration of acceleration in a tortuous curve, we take the case of § 13, or of § 34, *c*.

Let a point move in a circle of radius r with uniform angular velocity ω (about the centre), and let this circle move perpendicular to its plane with velocity V. The point describes a helix on a cylinder of radius r, and the inclination α is given by

$$\tan\alpha = \frac{V}{r\omega}.$$

The curvature of the path is $\frac{1}{r}\frac{r^2\omega^2}{V^2 + r^2\omega^2}$ or $\frac{r\omega^2}{V^2 + r^2\omega^2}$, and the tortuosity $\frac{\omega}{V}\frac{V^2}{V^2 + r^2\omega^2} = \frac{V\omega}{V^2 + r^2\omega^2}$.

The acceleration is $r\omega^2$, directed perpendicularly towards the axis of the cylinder.—Call this A.

$$\text{Curvature} = \frac{A}{V^2 + Ar} = \frac{A}{V^2 + \frac{A^2}{\omega^2}}.$$

$$\text{Tortuosity} = \frac{V}{\sqrt{Ar}}\frac{A}{V^2 + Ar} = \frac{V\omega}{V^2 + \frac{A^2}{\omega^2}}.$$

Let A be finite, r indefinitely small, and therefore ω indefinitely great.

$$\text{Curvature (in the limit)} = \frac{A}{V^2}.$$

$$\text{Tortuosity (\quad ,, \quad)} = \frac{\omega}{V}.$$

Thus, if we have a material particle moving in the manner specified, and if we consider the force (see Chap. II.) required to produce the acceleration, we find that a finite force perpendicular to the line of motion, in a direction revolving with an infinitely great angular velocity, maintains constant infinitely small deflection (in a direction opposite to its own) from the line of undisturbed motion, *finite* curvature, and infinite tortuosity. Examples of acceleration.

e. When the acceleration is perpendicular to a given plane and proportional to the distance from it, the path is a plane curve, which is the harmonic curve if the acceleration be *towards* the plane, and a more or less fore-shortened catenary (§ 580) if *from* the plane.

As in case *c*, $\frac{d^2z}{dt^2}=0$, $\frac{dz}{dt}=0$, and $z=0$, if the axis of z be perpendicular to the acceleration and to the direction of motion at any instant. Also, if we choose the origin *in* the plane,

$$\frac{d^2x}{dt^2}=0, \quad \frac{d^2y}{dt^2}=\mu y.$$

Hence $$\frac{dx}{dt}=\text{const.}=a \text{ (suppose)},$$

and $$\frac{d^2y}{dx^2}=\frac{\mu}{a^2}y=\mp\frac{y}{b^2}.$$

This gives, if μ is negative,

$$y=P\cos\left(\frac{x}{b}+Q\right), \text{ the harmonic curve, or curve of sines.}$$

If μ be positive, $y=P\epsilon^{\frac{x}{b}}+Q\epsilon^{-\frac{x}{b}}$;

and by shifting the origin along the axis of x this can be put in the form

$$y=R\left(\epsilon^{\frac{x}{b}}+\epsilon^{-\frac{x}{b}}\right):$$

which is the catenary if $2R=b$; otherwise it is the catenary stretched or fore-shortened in the direction of y.

36. [Compare §§ 233—236 below.] *a*. When the acceleration is directed to a fixed point, the path is in a plane passing through that point; and in this plane the areas traced out by the radius-vector are proportional to the times employed. This includes the case of a satellite or planet revolving about its primary. Acceleration directed to a fixed centre.

Evidently there is no acceleration perpendicular to the plane containing the fixed and moving points and the direction

Acceleration directed to a fixed centre.

of motion of the second at any instant; and, there being no velocity perpendicular to this plane at starting, there is therefore none throughout the motion; thus the point moves in the plane. And had there been no acceleration, the point would have described a straight line with uniform velocity, so that in this case the areas described by the radius-vector would have been proportional to the times. Also, the area actually described in any instant depends on the length of the radius-vector and the velocity perpendicular to it, and is shown below to be unaffected by an acceleration parallel to the radius-vector. Hence the second part of the proposition.

We have $\frac{d^2x}{dt^2} = P\frac{x}{r}, \quad \frac{d^2y}{dt^2} = P\frac{y}{r}, \quad \frac{d^2z}{dt^2} = P\frac{z}{r},$

the fixed point being the origin, and P being some function of x, y, z; in *nature* a function of r only.

Hence $$x\frac{d^2y}{dt^2} - y\frac{d^2x}{dt^2} = 0, \text{ etc.},$$

which give on integration

$$y\frac{dz}{dt} - z\frac{dy}{dt} = C_1, \quad z\frac{dx}{dt} - x\frac{dz}{dt} = C_2, \quad x\frac{dy}{dt} - y\frac{dx}{dt} = C_3.$$

Hence at once $C_1x + C_2y + C_3z = 0$, or the motion is in a plane through the origin. Take this as the plane of xy, then we have only the one equation

$$x\frac{dy}{dt} - y\frac{dx}{dt} = C_3 = h \text{ (suppose)}.$$

In polar co-ordinates this is

$$h = r^2\frac{d\theta}{dt} = 2\frac{dA}{dt}$$

if A be the area intercepted by the curve, a fixed radius-vector, and the radius-vector of the moving point. Hence the area increases uniformly with the time.

b. In the same case the velocity at any point is inversely as the perpendicular from the fixed point upon the tangent to the path, the momentary direction of motion.

For evidently the product of this perpendicular and the velocity gives double the area described in one second about the fixed point.

Or thus—if p be the perpendicular on the tangent,

Acceleration directed to a fixed centre.

$$p = x\frac{dy}{ds} - y\frac{dx}{ds},$$

and therefore
$$p\frac{ds}{dt} = x\frac{dy}{dt} - y\frac{dx}{dt} = h.$$

If we refer the motion to co-ordinates in its own plane, we have only the equations

$$\frac{d^2x}{dt^2} = \frac{Px}{r}, \quad \frac{d^2y}{dt^2} = \frac{Py}{r},$$

whence, as before,
$$r^2\frac{d\theta}{dt} = h.$$

If, by the help of this last equation, we eliminate t from $\frac{d^2x}{dt^2} = \frac{Px}{r}$, substituting polar for rectangular co-ordinates, we arrive at the polar differential equation of the path.

For variety, we may derive it from the formulæ of § 32.

They give
$$\frac{d^2r}{dt^2} - r\left(\frac{d\theta}{dt}\right)^2 = P, \quad r^2\frac{d\theta}{dt} = h.$$

Putting $\frac{1}{r} = u$, we have

$$\frac{d^2\left(\frac{1}{u}\right)}{dt^2} - \frac{1}{u}\left(\frac{d\theta}{dt}\right)^2 = P, \text{ and } \frac{d\theta}{dt} = hu^2.$$

But $\frac{d\left(\frac{1}{u}\right)}{dt} = hu^2\frac{d\left(\frac{1}{u}\right)}{d\theta} = -h\frac{du}{d\theta}$, therefore $\frac{d^2\left(\frac{1}{u}\right)}{dt^2} = -h^2u^2\frac{d^2u}{d\theta^2}$.

Also $\frac{1}{u}\left(\frac{d\theta}{dt}\right)^2 = h^2u^3$, the substitution of which values gives us

$$\frac{d^2u}{d\theta^2} + u = -\frac{P}{h^2u^2} \quad \text{..........................(1)},$$

the equation required. The integral of this equation involves *two* arbitrary constants besides h, and the remaining constant belonging to the two differential equations of the second order above is to be introduced on the farther integration of

$$\frac{d\theta}{dt} = hu^2 \quad \text{................................(2)},$$

when the value of u in terms of θ is substituted from the equation of the path.

Other examples of these principles will be met with in the chapters on Kinetics.

Hodograph. 37. If from any fixed point, lines be drawn at every instant, representing in magnitude and direction the velocity of a point describing any path in any manner, the extremities of these lines form a curve which is called the *Hodograph*. The invention of this construction is due to Sir W. R. Hamilton. One of the most beautiful of the many remarkable theorems to which it led him is that of § 38.

Since the radius-vector of the hodograph represents the velocity at each instant, it is evident (§ 27) that an elementary arc represents the velocity which must be compounded with the velocity at the beginning of the corresponding interval of time, to find the velocity at its end. Hence the velocity in the hodograph is equal to the acceleration in the path; and the tangent to the hodograph is parallel to the direction of the acceleration in the path.

If x, y, z be the co-ordinates of the moving point, ξ, η, ζ those of the corresponding point of the hodograph, then evidently

$$\xi = \frac{dx}{dt}, \quad \eta = \frac{dy}{dt}, \quad \zeta = \frac{dz}{dt},$$

and therefore

$$\frac{d\xi}{\frac{d^2x}{dt^2}} = \frac{d\eta}{\frac{d^2y}{dt^2}} = \frac{d\zeta}{\frac{d^2z}{dt^2}},$$

or the tangent to the hodograph is parallel to the acceleration in the orbit. Also, if σ be the arc of the hodograph,

$$\frac{d\sigma}{dt} = \sqrt{\left(\frac{d\xi}{dt}\right)^2 + \left(\frac{d\eta}{dt}\right)^2 + \left(\frac{d\zeta}{dt}\right)^2}$$

$$= \sqrt{\left(\frac{d^2x}{dt^2}\right)^2 + \left(\frac{d^2y}{dt^2}\right)^2 + \left(\frac{d^2z}{dt^2}\right)^2},$$

or the velocity in the hodograph is equal to the rate of acceleration in the path.

Hodograph of planet or comet, deduced from Kepler's laws. 38. *The hodograph for the motion of a planet or comet is always a circle, whatever be the form and dimensions of the orbit.* In the motion of a planet or comet, the acceleration is directed towards the sun's centre. Hence (§ 36, *b*) the velocity is inversely as the perpendicular from that point upon the tangent to the orbit. The orbit we assume to be a conic section, whose focus is the sun's centre. But we know that the intersection of the perpendicular with the tangent lies in the circle whose diameter is the major axis, if the orbit be an ellipse or hyperbola; in the tangent at the vertex if a parabola. Measure off on the perpendicular a third proportional to its own length and any constant line; this portion will thus represent the velocity in magnitude and in a direction perpendicular to its own—so that the locus of the new points in each perpendicular will be the hodograph turned through a right angle. But we see by geometry* that the locus of these points is always a circle. Hence the proposition. The hodograph surrounds its origin if the orbit be an ellipse, passes through it if a parabola, and the origin is without the hodograph if the orbit is a hyperbola.

Hodograph of planet or comet, deduced from Kepler's laws.

For a projectile unresisted by the air, it will be shewn in Kinetics that we have the equations (assumed in § 35, c)

$$\frac{d^2x}{dt^2}=0,\quad \frac{d^2y}{dt^2}=-g,$$

if the axis of y be taken vertically upwards.

Hence for the hodograph

$$\frac{d\xi}{dt}=0,\quad \frac{d\eta}{dt}=-g,$$

or $\xi=C$, $\eta=C'-gt$, and the hodograph is a vertical straight line along which the describing point moves uniformly.

For the case of a planet or comet, instead of assuming as above that the orbit is a conic with the sun in one focus, assume (Newton's deduction from that and the law of areas) that the acceleration is in the direction of the radius-vector, and varies inversely as the square of the distance. We have obviously

Hodograph for planet or comet, deduced from Newton's law of force.

$$\frac{d^2x}{dt^2}=\frac{\mu x}{r^3},\quad \frac{d^2y}{dt^2}=\frac{\mu y}{r^3},$$

where $$r^2=x^2+y^2.$$

Hence, as in § 36, $$x\frac{dy}{dt}-y\frac{dx}{dt}=h\text{..........................}(1),$$

* See our smaller work, § 51.

Hodograph for planet or comet, deduced from Newton's law of force.

and therefore

$$\frac{d^2x}{dt^2} = \frac{\mu x}{h}\,\frac{x\dfrac{dy}{dt} - y\dfrac{dx}{dt}}{r^3},$$

$$= \frac{\mu}{h}\,\frac{(x^2+y^2)\dfrac{dy}{dt} - y\left(x\dfrac{dx}{dt} + y\dfrac{dy}{dt}\right)}{r^3}, \quad = \frac{\mu}{h}\,\frac{r^2\dfrac{dy}{dt} - yr\dfrac{dr}{dt}}{r^3}.$$

Hence $$\frac{dx}{dt} + A = \frac{\mu}{h}\frac{y}{r} \quad\text{..........................(2).}$$

Similarly $$\frac{dy}{dt} + B = -\frac{\mu}{h}\frac{x}{r} \quad\text{..........................(3).}$$

Hence for the hodograph

$$(\xi + A)^2 + (\eta + B)^2 = \frac{\mu^2}{h^2},$$

the circle as before stated.

We may merely mention that the equation of the orbit will be found at once by eliminating $\frac{dx}{dt}$ and $\frac{dy}{dt}$ among the three first integrals (1), (2), (3) above. We thus get

$$-h + Ay - Bx = \frac{\mu}{h}r,$$

a conic section of which the origin is a focus.

Applications of the hodograph.

39. The intensity of heat and light emanating from a point, or from an uniformly radiating spherical surface, diminishes with increasing distance according to the same law as gravitation. Hence the amount of heat and light, which a planet receives from the sun during any interval, is proportional to the time integral of the acceleration during that interval, *i.e.* (§ 37) to the corresponding arc of the hodograph. From this it is easy to see, for example, that if a comet move in a parabola, the amount of heat it receives from the sun in any interval is proportional to the angle through which its direction of motion turns during that interval. There is a corresponding theorem for a planet moving in an ellipse, but somewhat more complicated.

Curves of pursuit.

40. If two points move, each with a definite uniform velocity, one in a given curve, the other at every instant directing its course towards the first describes a path which is called a

Curve of Pursuit. The idea is said to have been suggested by the old rule of steering a privateer always directly for the vessel pursued. (Bouguer, *Mém. de l'Acad.* 1732.) It is the curve described by a dog running to its master. Curves of pursuit.

The simplest cases are of course those in which the first point moves in a straight line, and of these there are three, for the velocity of the first point may be greater than, equal to, or less than, that of the second. The figures in the text below represent the curves in these cases, the velocities of the pursuer being $\frac{4}{3}$, 1, and $\frac{1}{2}$ of those of the pursued, respectively. In the second and third cases the second point can never overtake the first, and consequently the line of motion of the first is an asymptote. In the first case the second point overtakes the first, and the curve at that point touches the line of motion of the first. The remainder of the curve satisfies a modified form of statement of the original question, and is called the *Curve of Flight.*

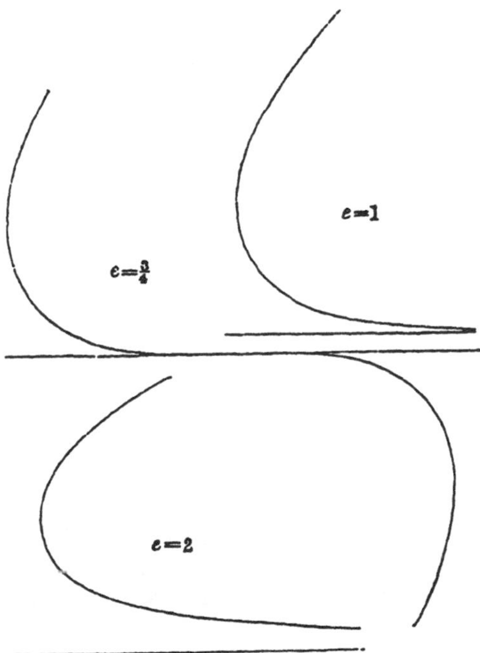

Curves of pursuit.

We will merely form the differential equation of the curve, and give its integrated form, leaving the work to the student.

Suppose Ox to be the line of motion of the first point, whose velocity is v, AP the curve of pursuit, in which the velocity is u, then the tangent at P always passes through Q, the instantaneous position of the first point. It will be evident, on a moment's consideration, that the curve AP must have a tangent perpendicular to Ox. Take this as the axis of y, and let $OA = a$. Then, if $OQ = \xi$, $AP = s$, and if x, y be the coordinates of P, we have

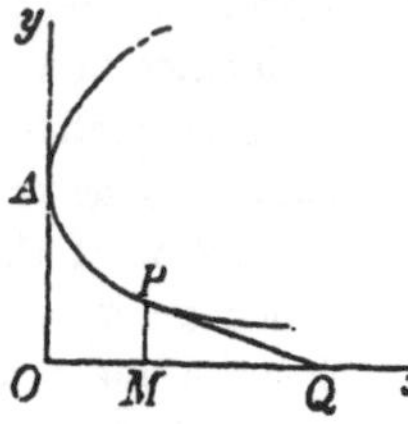

$$\frac{AP}{u} = \frac{OQ}{v},$$

because A, O and P, Q are pairs of simultaneous positions of the two points.

This gives $$\frac{v}{u}s = es = x - y\frac{dx}{dy}.$$

From this we find, unless $e = 1$,

$$2\left(x + \frac{ae}{e^2-1}\right) = \frac{y^{e+1}}{a^e(e+1)} + \frac{a^e}{y^{e-1}(e-1)};$$

and if $e = 1$, $$2\left(x + \frac{a}{4}\right) = \frac{y^2}{2a} - a\log_e\frac{y}{a},$$

the only case in which we do not get an algebraic curve. The axis of x is easily seen to be an asymptote if $e \nless 1$.

Angular velocity

41. When a point moves in any manner, the line joining it with a fixed point generally changes its direction. If, for simplicity, we consider the motion as confined to a plane passing through the fixed point, the angle which the joining line makes with a fixed line in the plane is continually altering, and its rate of alteration at any instant is called the *Angular Velocity* of the first point about the second. If uniform, it is of course measured by the angle described in unit of time; if variable, by the angle which would have been described in unit of time if the angular velocity at the instant in question were maintained constant for so long. In this respect, the process is precisely similar to that which we have already explained for the measurement of velocity and acceleration.

Unit of angular velocity is that of a point which describes, or would describe, unit angle about a fixed point in unit of time. The usual unit angle is (as explained in treatises on plane trigonometry) that which subtends at the centre of a circle an arc whose length is equal to the radius; being an angle of Angular velocity.

$$\frac{180^\circ}{\pi} = 57^\circ.29578\ldots = 57^\circ\ 17'\ 44''.8 \text{ nearly.}$$

For brevity we shall call this angle a radian.

42. The rate of increase or diminution of the angular velocity when variable is called the *angular acceleration*, and is measured in the same way and by the same unit. Angular acceleration.

By methods precisely similar to those employed for linear velocity and acceleration we see that if θ be the angle-vector of a point moving in a plane—the

$$\text{Angular velocity is } \omega = \frac{d\theta}{dt}\text{, and the}$$

$$\text{Angular acceleration is } \frac{d\omega}{dt} = \frac{d^2\theta}{dt^2} = \omega\frac{d\omega}{d\theta}.$$

Since (§ 27) $r\dfrac{d\theta}{dt}$ is the velocity perpendicular to the radius-vector, we see that

The angular velocity of a point in a plane is found by dividing the velocity perpendicular to the radius-vector by the length of the radius-vector.

43. When one point describes uniformly a circle about another, the time of describing a complete circumference being T, we have the angle 2π described uniformly in T; and, therefore, the angular velocity is $\dfrac{2\pi}{T}$. Even when the angular velocity is not uniform, as in a planet's motion, it is useful to introduce the quantity $\dfrac{2\pi}{T}$, which is then called the *mean* angular velocity. Angular velocity.

When a point moves uniformly in a straight line its angular velocity evidently diminishes as it recedes from the point about which the angles are measured.

Angular velocity.

The polar equation of a straight line is

$$r = a \sec \theta.$$

But the length of the line between the limiting angles 0 and θ is $a \tan \theta$, and this increases with uniform velocity v. Hence

$$v = \frac{d}{dt}(a \tan \theta) = a \sec^2 \theta \frac{d\theta}{dt} = \frac{r^2}{a}\frac{d\theta}{dt}.$$

Hence $\frac{d\theta}{dt} = \frac{av}{r^2}$, and is therefore inversely as the square of the radius-vector.

Similarly for the angular acceleration, we have by a second differentiation,

$$\frac{d^2\theta}{dt^2} + 2 \tan \theta \left(\frac{d\theta}{dt}\right)^2 = 0,$$

i.e., $\frac{d^2\theta}{dt^2} = -\frac{2av^2}{r^3}\left(1 - \frac{a^2}{r^2}\right)^{\frac{1}{2}}$, and ultimately varies inversely as the third power of the radius-vector.

Angular velocity of a plane.

44. We may also talk of the angular velocity of a moving plane with respect to a fixed one, as the rate of increase of the angle contained by them—but unless their line of intersection remain fixed, or at all events parallel to itself, a somewhat more laboured statement is required to give definite information. This will be supplied in a subsequent section.

Relative motion.

45. All motion that we are, or can be, acquainted with, is *Relative* merely. We can calculate from astronomical data for any instant the direction in which, and the velocity with which we are moving on account of the earth's diurnal rotation. We may compound this with the similarly calculable velocity of the earth in its orbit. This resultant again we may compound with the (roughly known) velocity of the sun relatively to the so-called fixed stars; but, even if all these elements were accurately known, it could not be said that we had attained any idea of an *absolute* velocity; for it is only the sun's relative motion among the stars that we can observe; and, in all probability, sun and stars are moving on (possibly with very great rapidity) relatively to other bodies in space. We must therefore consider how, from the actual motions of a set of points, we may find their relative motions with regard to any one of them;

and how, having given the relative motions of all but one with regard to the latter, and the actual motion of the latter, we may find the actual motions of all. The question is very easily answered. Consider for a moment a number of passengers walking on the deck of a steamer. Their relative motions with regard to the deck are what we immediately observe, but if we compound with these the velocity of the steamer itself we get evidently their actual motion relatively to the earth. Again, in order to get the relative motion of all with regard to the deck, we *abstract our ideas from* the motion of the steamer altogether—that is, we alter the velocity of each by compounding it with the actual velocity of the vessel taken in a reversed direction. Relative motion.

Hence to find the relative motions of any set of points with regard to one of their number, imagine, impressed upon each in composition with its own velocity, a velocity equal and opposite to the velocity of that one; it will be reduced to rest, and the motions of the others will be the same with regard to it as before.

Thus, to take a very simple example, two trains are running in opposite directions, say north and south, one with a velocity of fifty, the other of thirty, miles an hour. The relative velocity of the second with regard to the first is to be found by impressing on both a southward velocity of fifty miles an hour; the effect of this being to bring the first to rest, and to give the second a southward velocity of eighty miles an hour, which is the required relative motion.

Or, given one train moving north at the rate of thirty miles an hour, and another moving west at the rate of forty miles an hour. The motion of the second relatively to the first is at the rate of fifty miles an hour, in a south-westerly direction inclined to the due west direction at an angle of $\tan^{-1}\frac{3}{4}$. It is needless to multiply such examples, as they must occur to every one.

46. Exactly the same remarks apply to relative as compared with absolute acceleration, as indeed we may see at once, since accelerations are in all cases resolved and compounded by the same law as velocities.

Relative motion.

If x, y, z, and x', y', z', be the co-ordinates of two points referred to axes regarded as fixed; and ξ, η, ζ their relative co-ordinates—we have

$$\xi = x' - x, \quad \eta = y' - y, \quad \zeta = z' - z.$$

and, differentiating,

$$\frac{d\xi}{dt} = \frac{dx'}{dt} - \frac{dx}{dt}, \text{ etc.,}$$

which give the relative, in terms of the absolute, velocities; and

$$\frac{d^2\xi}{dt^2} = \frac{d^2x'}{dt^2} - \frac{d^2x}{dt^2}, \text{ etc.,}$$

proving our assertion about relative and absolute accelerations.

The corresponding expressions in polar co-ordinates in a plane are somewhat complicated, and by no means convenient. The student can easily write them down for himself.

47. The following proposition in relative motion is of considerable importance:—

Any two moving points describe similar paths relatively to each other, or relatively to any point which divides in a constant ratio the line joining them.

Let A and B be any simultaneous positions of the points. Take G or G' in AB such that the ratio $\frac{GA}{GB}$ or $\frac{G'A}{G'B}$ has a constant value. Then

G' A G B

as the form of the relative path depends only upon the *length* and *direction* of the line joining the two points at any instant, it is obvious that these will be the same for A with regard to B, as for B with regard to A, saving only the inversion of the direction of the joining line. Hence B's path about A, is A's about B turned through two right angles. And with regard to G and G' it is evident that the directions remain the same, while the lengths are altered in a given ratio; but this is the definition of similar curves.

48. As a good example of relative motion, let us consider that of the two points involved in our definition of the curve of pursuit, § 40. Since, to find the relative position and motion of the pursuer with regard to the pursued, we must impress on both a velocity equal and opposite to that of the latter, we see

at once that the problem becomes the same as the following. A boat crossing a stream is impelled by the oars with uniform velocity relatively to the water, and always towards a fixed point in the opposite bank; but it is also carried down stream at a uniform rate; determine the path described and the time of crossing. Here, as in the former problem, there are three cases, figured below. In the first, the boat, moving faster than the current, reaches the desired point; in the second, the velocities of boat and stream being equal, the boat gets across only after an infinite time—describing a parabola—but does not land at the desired point, which is indeed the focus of the parabola, while the landing point is the vertex. In the third case, its proper velocity being less than that of the water, it never reaches the other bank, and is carried indefinitely down stream. The comparison of the figures in § 40 with those in the present section cannot fail to be instructive. They are drawn to the same scale, and for the same relative velocities. The horizontal lines represent the farther bank of the river, and the vertical lines the path of the boat if there were no current.

Relative motion.

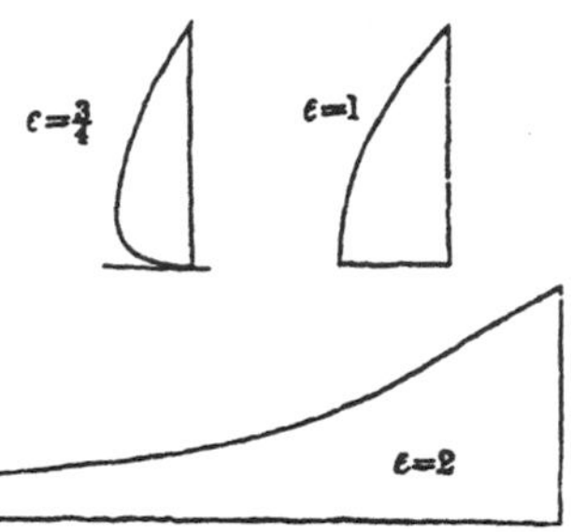

We leave the solution of this question as an exercise, merely noting that the equation of the curve is

$$\frac{y^{1+e}}{a^e} = \sqrt{x^2+y^2} - x,$$

in one or other of the three cases, according as e is $>$, $=$, or <1.

When $e=1$ this becomes

$$y^2 = a^2 - 2ax, \text{ the parabola.}$$

The time of crossing is

$$\frac{a}{u(1-e^2)},$$

which is finite only for $e<1$, because of course a negative value is inadmissible.

Relative motion.

49. Another excellent example of the transformation of relative into absolute motion is afforded by the family of cycloids. We shall in a future section consider their mechanical description, by the *rolling* of a circle on a fixed straight line or circle. In the mean time, we take a different form of enunciation, which, however, leads to precisely the same result.

Find the actual path of a point which revolves uniformly in a circle about another point—the latter moving uniformly in a straight line or circle in the same plane.

Take the former case first: let a be the radius of the relative circular orbit, and ω the angular velocity in it, v being the velocity of its centre along the straight line.

The relative co-ordinates of the point in the circle are $a\cos\omega t$ and $a\sin\omega t$, and the actual co-ordinates of the centre are vt and 0. Hence for the actual path

$$\xi = vt + a\cos\omega t,\quad \eta = a\sin\omega t.$$

Hence $\xi = \dfrac{v}{\omega}\sin^{-1}\dfrac{\eta}{a} + \sqrt{a^2-\eta^2}$, an equation which, by giving different values to v and ω, may be made to represent the cycloid itself, or either form of trochoid. See § 92.

For the epicycloids, let b be the radius of the circle which B describes about A, ω_1 the angular velocity; a the radius of A's path, ω the angular velocity.

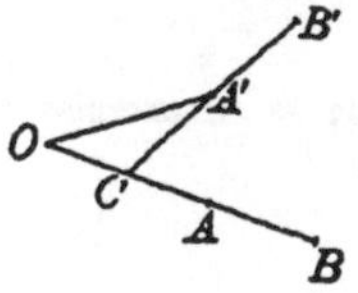

Also at time $t=0$, let B be in the radius OA of A's path. Then at time t, if A', B' be the positions, we see at once that

$$\angle AOA' = \omega t,\quad \angle B'CA = \omega_1 t.$$

Hence, taking OA as axis of x,

$$x = a\cos\omega t + b\cos\omega_1 t,\quad y = a\sin\omega t + b\sin\omega_1 t,$$

which, by the elimination of t, give an algebraic equation between x and y whenever ω and ω_1 are commensurable.

Thus, for $\omega_1 = 2\omega$, suppose $\omega t = \theta$, and we have

$$x = a\cos\theta + b\cos 2\theta,\quad y = a\sin\theta + b\sin 2\theta,$$

or, by an easy reduction,

$$(x^2+y^2-b^2)^2 = a^2\{(x+b)^2+y^2\}.$$

Put $x - b$ for x, *i.e.*, change the origin to a distance AB to the left of O, the equation becomes Relative motion.

$$a^2(x^2+y^2) = (x^2+y^2-2bx)^2,$$

or, in polar co-ordinates,

$$a^2 = (r - 2b\cos\theta)^2, \quad r = a + 2b\cos\theta,$$

and when $2b = a$, $r = a(1+\cos\theta)$, the cardioid. (See § 94.)

50. As an additional illustration of this part of our subject, we may define as follows:— Resultant motion.

If one point A executes any motion whatever with reference to a second point B; if B executes any other motion with reference to a third point C; and so on—the first is said to execute, with reference to the last, a movement which is the resultant of these several movements.

The relative position, velocity, and acceleration are in such a case the geometrical resultants of the various components combined according to preceding rules.

51. The following practical methods of effecting such a combination in the simple case of the movements of two points are useful in scientific illustrations and in certain mechanical arrangements. Let two moving points be joined by an elastic string; the middle point of this string will evidently execute a movement which is *half* the resultant of the motions of the two points. But for drawing, or engraving, or for other mechanical applications, the following method is preferable:—

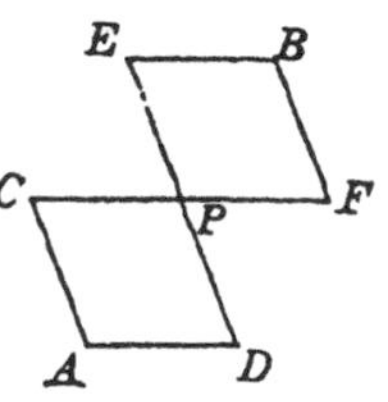

CF and ED are rods of equal length moving freely round a pivot at P, which passes through the middle point of each—CA, AD, EB, and BF, are rods of half the length of the two former, and so pivoted to them as to form a pair of equal rhombi CD, EF, whose angles can be altered at will. Whatever motions, whether in a plane, or in space of three dimensions, be given to A and B, P will evidently be subjected to half their resultant.

52. Amongst the most important classes of motions which we have to consider in Natural Philosophy, there is one, namely, *Harmonic Motion*, which is of such immense use, not only in Harmonic motion.

Harmonic motion.

ordinary kinetics, but in the theories of sound, light, heat, etc., that we make no apology for entering here into considerable detail regarding it.

Simple harmonic motion.

53. *Def.* When a point Q moves uniformly in a circle, the perpendicular QP drawn from its position at any instant to a fixed diameter AA' of the circle, intersects the diameter in a point P, whose position changes by a *simple harmonic motion.*

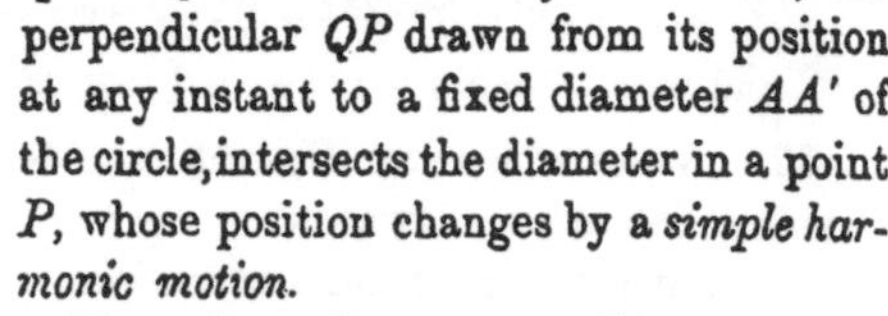

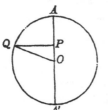

Thus, if a planet or satellite, or one of the constituents of a double star, supposed to move uniformly in a circular orbit about its primary, be viewed from a very distant position in the plane of its orbit, it will appear to move backwards and forwards in a straight line, with a simple harmonic motion. This is nearly the case with such bodies as the satellites of Jupiter when seen from the earth.

Physically, the interest of such motions consists in the fact of their being approximately those of the simplest vibrations of sounding bodies, such as a tuning-fork or pianoforte wire; whence their name; and of the various media in which waves of sound, light, heat, etc., are propagated.

54. The *Amplitude* of a simple harmonic motion is the range on one side or the other of the middle point of the course, *i.e.*, OA or OA' in the figure.

An arc of the circle referred to, measured from any fixed point to the uniformly moving point Q, is the *Argument* of the harmonic motion.

The distance of a point, performing a simple harmonic motion, from the middle of its course or range, is a *simple harmonic function of the time.* The *argument* of this function is what we have defined as the argument of the motion.

The *Epoch* in a simple harmonic motion is the interval of time which elapses from the era of reckoning till the moving point first comes to its greatest elongation in the direction reckoned as positive, from its mean position or the middle of its range. Epoch in angular measure is the angle described on the circle of reference in the period of time defined as the epoch.

The *Period* of a simple harmonic motion is the time which elapses from any instant until the moving point again moves in the same direction through the same position. Simple harmonic motion.

The *Phase* of a simple harmonic motion at any instant is the fraction of the whole period which has elapsed since the moving point last passed through its middle position in the positive direction.

55. Those common kinds of mechanism, for producing rectilineal from circular motion, or *vice versa*, in which a crank moving in a circle works in a straight slot belonging to a body which can only move in a straight line, fulfil strictly the definition of a simple harmonic motion in the part of which the motion is rectilineal, if the motion of the rotating part is uniform. Simple harmonic motion in mechanism.

The motion of the treadle in a spinning-wheel approximates to the same condition when the wheel moves uniformly; the approximation being the closer, the smaller is the angular motion of the treadle and of the connecting string. It is also approximated to more or less closely in the motion of the piston of a steam-engine connected, by any of the several methods in use, with the crank, provided always the rotatory motion of the crank be uniform.

56. The velocity of a point executing a simple harmonic motion is a simple harmonic function of the time, a quarter of a period earlier in phase than the displacement, and having its maximum value equal to the velocity in the circular motion by which the given function is defined. Velocity in S. H. motion.

For, in the fig. of § 53, if V be the velocity in the circle, it may be represented by OQ in a direction perpendicular to its own, and therefore by OP and PQ in directions perpendicular to those lines. That is, the velocity of P in the simple harmonic motion is $\frac{V}{OQ}PQ$; which, when P is at O, becomes V.

57. The acceleration of a point executing a simple harmonic motion is at any time simply proportional to the displacement from the middle point, but in opposite direction, or always towards the middle point. Its maximum value is that with which a velocity equal to that of the circular motion would Acceleration in S. H. motion.

Acceleration in S. H. motion.

be acquired in the time in which an arc equal to the radius is described.

For, in the fig. of § 53, the acceleration of Q (by § 35, a) is $\frac{V^2}{QO}$ along QO. Supposing, for a moment, QO to represent the magnitude of this acceleration, we may resolve it in QP, PO. The acceleration of P is therefore represented on the same scale by PO. Its magnitude is therefore $\frac{V^2}{QO}\cdot\frac{PO}{QO}=\frac{V^2}{QO^2}PO$, which is proportional to PO, and has at A its maximum value, $\frac{V^2}{QO}$, an acceleration under which the velocity V would be acquired in the time $\frac{QO}{V}$ as stated.

Let a be the amplitude, ϵ the epoch, and T the period, of a simple harmonic motion. Then if s be the displacement from middle position at time t, we have

$$s = a\cos\left(\frac{2\pi t}{T}-\epsilon\right).$$

Hence, for velocity, we have

$$v=\frac{ds}{dt}=-\frac{2\pi a}{T}\sin\left(\frac{2\pi t}{T}-\epsilon\right).$$

Hence V, the maximum value, is $\frac{2\pi a}{T}$, as above stated (§ 56).

Again, for acceleration,

$$\frac{dv}{dt}=-\frac{4\pi^2 a}{T^2}\cos\left(\frac{2\pi t}{T}-\epsilon\right)=-\frac{4\pi^2}{T^2}s. \quad \text{(See § 57.)}$$

Lastly, for the maximum value of the acceleration,

$$\frac{4\pi^2 a}{T^2}=\frac{V}{\frac{T}{2\pi}},$$

where, it may be remarked, $\frac{T}{2\pi}$ is the time of describing an arc equal to radius in the relative circular motion.

Composition of S. H. M. in one line.

58. Any two simple harmonic motions in one line, and of one period, give, when compounded, a single simple harmonic motion; of the same period; of amplitude equal to the diagonal of a parallelogram described on lengths equal to their amplitudes measured on lines meeting at an angle equal to their difference

of epochs; and of epoch differing from their epochs by angles equal to those which this diagonal makes with the two sides of the parallelogram. Let P and P be two points executing simple harmonic motions of one period, and in one line $B'BCAA'$. Let Q and Q be the uniformly moving points in the relative circles. On CQ and CQ' describe a parallelogram $SQCQ'$; and through S draw SR perpendicular to $B'A'$ produced. We have obviously $P'R=CP$ (being projections of the equal and parallel lines $Q'S, CQ$, on CR). Hence $CR=CP+CP'$; and therefore the point R executes the resultant of the motions P and P'. But CS, the diagonal of the parallelogram, is constant, and therefore the resultant motion is simple harmonic, of amplitude CS, and of epoch exceeding that of the motion of P, and falling short of that of the motion of P', by the angles QCS and SCQ' respectively.

Composition of S. H. M. in one line.

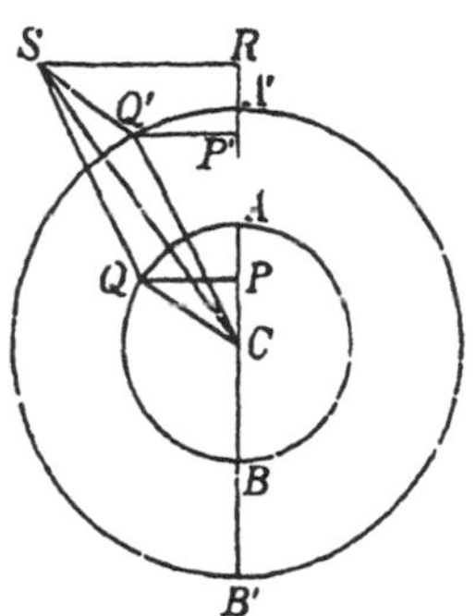

This geometrical construction has been usefully applied by the tidal committee of the British Association for a mechanical tide-indicator (compare § 60, below). An arm CQ' turning round C carries an arm $Q'S$ turning round Q'. Toothed wheels, one of them fixed with its axis through C, and the others pivoted on a framework carried by CQ', are so arranged that $Q'S$ turns very approximately at the rate of once round in 12 mean lunar hours, if CQ' be turned uniformly at the rate of once round in 12 mean solar hours. Days and half-days are marked by a counter geared to CQ. The distance of S from a fixed line through C shows the deviation from mean sea-level due to the sum of mean solar and mean lunar tides for the time of day and year marked by CQ' and the counter.

An analytical proof of the same proposition is useful, being as follows:—

$$a\cos\left(\frac{2\pi t}{T}-\epsilon\right)+a'\cos\left(\frac{2\pi t}{T}-\epsilon'\right)$$

$$=(a\cos\epsilon+a'\cos\epsilon')\cos\frac{2\pi t}{T}+(a\sin\epsilon+a'\sin\epsilon')\sin\frac{2\pi t}{T}-r\cos\left(\frac{2\pi t}{T}-\theta\right),$$

Composition of S. H. M. in one line.

where $$r = \{(a\cos\epsilon + a'\cos\epsilon')^2 + (a\sin\epsilon + a'\sin\epsilon')^2\}^{\frac{1}{2}}$$
$$= \{a^2 + a'^2 + 2aa'\cos(\epsilon - \epsilon')\}^{\frac{1}{2}},$$

and $$\tan\theta = \frac{a\sin\epsilon + a'\sin\epsilon'}{a\cos\epsilon + a'\cos\epsilon'}.$$

59. The construction described in the preceding section exhibits the resultant of two simple harmonic motions, whether of the same period or not. Only, if they are not of the same period, the diagonal of the parallelogram will not be constant, but will diminish from a maximum value, the sum of the component amplitudes, which it has at the instant when the phases of the component motions agree; to a minimum, the difference of those amplitudes, which is its value when the phases differ by half a period. Its direction, which always must be nearer to the greater than to the less of the two radii constituting the sides of the parallelogram, will oscillate on each side of the greater radius to a maximum deviation amounting on either side to the angle whose sine is the less radius divided by the greater, and reached when the less radius deviates more than this by a quarter circumference from the greater. The full period of this oscillation is the time in which either radius gains a full turn on the other. The resultant motion is therefore not simple harmonic, but is, as it were, simple harmonic with periodically increasing and diminishing amplitude, and with periodical acceleration and retardation of phase. This view is particularly appropriate for the case in which the periods of the two component motions are nearly equal, but the amplitude of one of them much greater than that of the other.

To express the resultant motion, let s be the displacement at time t; and let a be the greater of the two component half-amplitudes.

$$s = a\cos(nt - \epsilon) + a'\cos(n't - \epsilon')$$
$$= a\cos(nt - \epsilon) + a'\cos(nt - \epsilon + \phi)$$
$$= (a + a'\cos\phi)\cos(nt - \epsilon) - a'\sin\phi\sin(nt - \epsilon),$$

if $$\phi = (n't - \epsilon') - (nt - \epsilon);$$

or, finally, $$s = r\cos(nt - \epsilon + \theta),$$

Composition of S. H. M. in one line.

if $$r = (a^2 + 2aa' \cos\phi + a'^2)^{\frac{1}{2}}$$

and $$\tan\theta = \frac{a\sin\phi}{a + a'\cos\phi}.$$

The maximum value of $\tan\theta$ in the last of these equations is found by making $\phi = \frac{\pi}{2} + \sin^{-1}\frac{a'}{a}$, and is equal to $\frac{a'}{(a^2 - a'^2)^{\frac{1}{2}}}$, and hence the maximum value of θ itself is $\sin^{-1}\frac{a'}{a}$. The geometrical methods indicated above (§ 58) lead to this conclusion by the following very simple construction.

To find the time and the amount of the maximum acceleration or retardation of phase, let CA be the greater half-amplitude. From A as centre, with the less half-amplitude as radius, describe a circle. CB touching this circle is the generating radius of the most deviated resultant. Hence CBA is a right angle; and

$$\sin BCA = \frac{AB}{CA}.$$

60. A most interesting application of this case of the composition of harmonic motions is to the lunar and solar tides; which, except in tidal rivers, or long channels, or deep bays, follow each very nearly the simple harmonic law, and produce, as the actual result, a variation of level equal to the sum of variations that would be produced by the two causes separately.

Examples of composition of S. H. M in one line.

The amount of the lunar equilibrium-tide (§ 812) is about 2·1 times that of the solar. Hence, if the actual tides conformed to the equilibrium theory, the spring tides would be 3·1, and the neap tides only 1·1, each reckoned in terms of the solar tide; and at spring and neap tides the hour of high water is that of the lunar tide alone. The greatest deviation of the actual tide from the phases (high, low, or mean water) of the lunar tide alone, would be about ·95 of a lunar hour, that is, ·98 of a solar hour (being the same part of 12 lunar hours that $28^0\ 26'$, or the angle whose sine is $\frac{1}{2\cdot 1}$, is of 360^0). This maximum deviation would be in advance or in arrear according as the crown of the solar tide precedes or follows the crown of the lunar tide; and it would be exactly reached when the interval of phase between

Examples of composition of S. H. M. in one line.

the two component tides is 3·95 lunar hours. That is to say, there would be maximum advance of the time of high water $4\frac{1}{2}$ days after, and maximum retardation the same number of days before, spring tides (compare § 811).

61. We may consider next the case of equal amplitudes in the two given motions. If their periods are equal, their resultant is a simple harmonic motion, whose phase is at every instant the mean of their phases, and whose amplitude is equal to twice the amplitude of either multiplied by the cosine of half the difference of their phases. The resultant is of course nothing when their phases differ by half the period, and is a motion of double amplitude and of phase the same as theirs when they are of the same phase.

When their periods are very nearly, but not quite, equal (their amplitudes being still supposed equal), the motion passes very slowly from the former (zero, or no motion at all) to the latter, and back, in a time equal to that in which the faster has gone once oftener through its period than the slower has.

In practice we meet with many excellent examples of this case, which will, however, be more conveniently treated of when we come to apply kinetic principles to various subjects in acoustics, physical optics, and practical mechanics; such as the sympathy of pendulums or tuning-forks, the rolling of a turret ship at sea, the marching of troops over a suspension bridge, etc.

Mechanism for compounding S. H. motions in one line.

62. If any number of pulleys be so placed that a cord passing from a fixed point half round each of them has its free parts all in parallel lines, and if their centres be moved with simple harmonic motions of any ranges and any periods in lines parallel to those lines, the unattached end of the cord moves with a complex harmonic motion equal to twice the sum of the given simple harmonic motions. This is the principle of Sir W. Thomson's tide-predicting machine, constructed by the British Association, and ordered to be placed in South Kensington Museum, availably for general use in calculating beforehand for any port or other place on the sea for which the simple harmonic constituents of the tide have been determined by the "harmonic analysis" applied to

previous observations*. We may exhibit, graphically, any case of single or compound simple harmonic motion in one line by curves in which the abscissæ represent intervals of time, and the

Graphical representation of harmonic motions in one line.

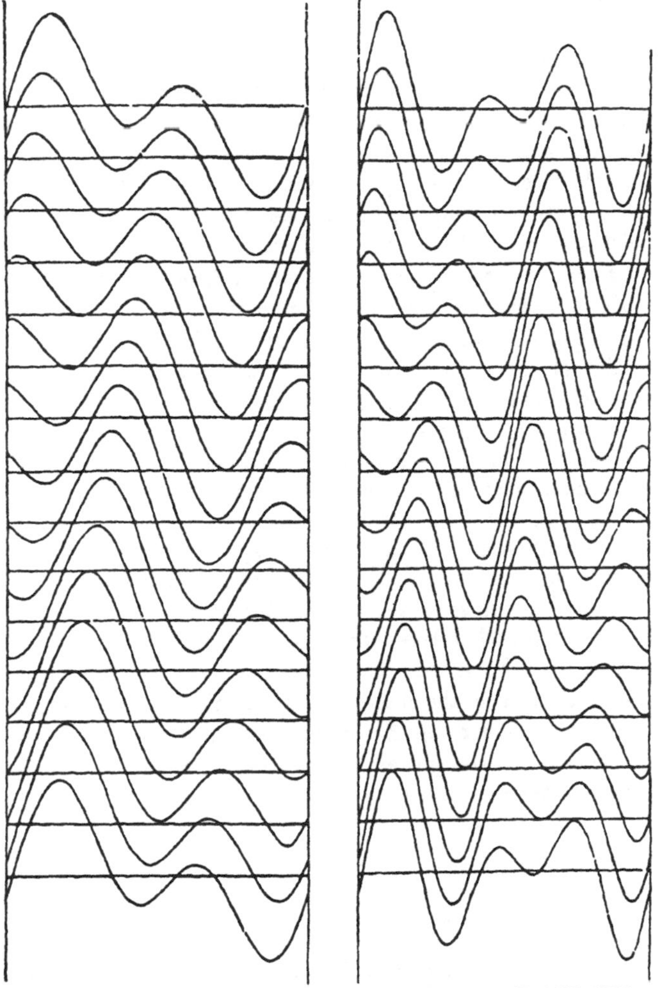

* See British Association Tidal Committee's Report, 1868, 1872, 1875: or *Lecture on Tides*, by Sir W. Thomson, "Popular Lectures and Addresses," vol. III. p. 178.

Graphical representation of harmonic motions in one line.

ordinates the corresponding distances of the moving point from its mean position. In the case of a single simple harmonic motion, the corresponding curve would be that described by the point P in § 53, if, while Q maintained its uniform circular motion, the circle were to move with uniform velocity in any direction perpendicular to OA. This construction gives the harmonic curve, or curves of sines, in which the ordinates are proportional to the sines of the abscissæ, the straight line in which O moves being the axis of abscissæ. It is the simplest possible form assumed by a vibrating string. When the harmonic motion is complex, but in one line, as is the case for any point in a violin-, harp-, or pianoforte-string (differing, as these do, from one another in their motions on account of the different modes of excitation used), a similar construction may be made. Investigation regarding complex harmonic functions has led to results of the highest importance, having their most general expression in *Fourier's Theorem*, to which we will presently devote several pages. We give, on page 45, graphic representations of the composition of two simple harmonic motions in one line, of *equal* amplitudes and of periods which are as 1 : 2 and as 2 : 3, for differences of epoch corresponding to 0, 1, 2, etc., sixteenths of a circumference. In each case the epoch of the component of greater period is a quarter of its own period. In the first, second, third, etc., of each series respectively, the epoch of the component of shorter period is less than a quarter-period by 0, 1, 2, etc., sixteenths of the period. The successive horizontal lines are the axes of abscissæ of the successive curves; the vertical line to the left of each series being the common axis of ordinates. In each of the first set the graver motion goes through one complete period, in the second it goes through two periods.

1 : 2 (Octave)	2 : 3 (Fifth)
$y = \sin x + \sin\left(2x + \frac{n\pi}{8}\right).$	$y = \sin 2x + \sin\left(3x + \frac{n\pi}{8}\right).$

Both, from $x = 0$ to $x = 2\pi$; and for $n = 0, 1, 2 \ldots\ldots 15$, in succession.

These, and similar cases, when the periodic times are not commensurable, will be again treated of under Acoustics.

S H motions in different directions.

63. We have next to consider the composition of simple harmonic motions in different directions. In the first place, we see that any number of simple harmonic motions of one period, and of the same phase, superimposed, produce a single simple harmonic motion of the same phase. For, the displacement at any instant being, according to the principle of the composition of motions, the geometrical resultant (see above, § 50) of the displacements due to the component motions separately, these component displacements, in the case supposed, all vary in simple proportion to one another, and are in constant directions. Hence the resultant displacement will vary in simple proportion to each of them, and will be in a constant direction.

But if, while their periods are the same, the phases of the several component motions do not agree, the resultant motion will generally be elliptic, with equal areas described in equal times by the radius-vector from the centre; although in particular cases it may be uniform circular, or, on the other hand, rectilineal and simple harmonic.

64. To prove this, we may first consider the case in which we have two equal simple harmonic motions given, and these in perpendicular lines, and differing in phase by a quarter period. Their resultant is a uniform circular motion. For, let BA, $B'A'$ be their ranges; and from O, their common middle point, as centre, describe a circle through $AA'BB'$. The given motion of P in BA will be (§ 53) defined by the motion of a point Q round the circumference of this circle; and the same point, if moving in the direction indicated by the arrow, will give a simple harmonic motion of P', in $B'A'$, a quarter of a period behind that of the motion of P in BA. But, since $A'OA$, QPO, and $QP'O$ are right angles, the figure $QP'OP$ is a parallelogram, and therefore Q is in the position of the displacement compounded of OP and OP'. Hence two equal simple harmonic motions in perpendicular lines, of phases differing by a quarter period, are equivalent to a uniform circular motion of radius equal to the maximum displacement of either singly, and in the direction from the positive end of the range of

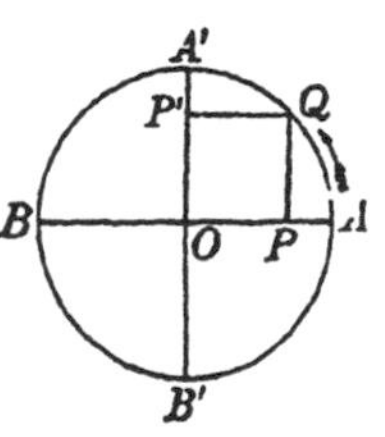

S. H. motions in different directions.

the component in advance of the other towards the positive end of the range of this latter.

65. Now, orthogonal projections of simple harmonic motions are clearly simple harmonic with unchanged phase. Hence, if we project the case of § 64 on any plane, we get motion in an ellipse, of which the projections of the two component ranges are conjugate diameters, and in which the radius-vector from the centre describes equal areas (being the projections of the areas described by the radius of the circle) in equal times. But the plane and position of the circle of which this projection is taken may clearly be found so as to fulfil the condition of having the projections of the ranges coincident with any two given mutually bisecting lines. Hence any two given simple harmonic motions, equal or unequal in range, and oblique or at right angles to one another in direction, provided only they differ by a quarter period in phase, produce elliptic motion, having their ranges for conjugate axes, and describing, by the radius-vector from the centre, equal areas in equal times (compare § 34, *b*).

66. Returning to the composition of any number of simple harmonic motions of one period, in lines in all directions and of all phases: each component simple harmonic motion may be determinately resolved into two in the same line, differing in phase by a quarter period, and one of them having any given epoch. We may therefore reduce the given motions to two sets, differing in phase by a quarter period, those of one set agreeing in phase with any one of the given, or with any other simple harmonic motion we please to choose (*i.e.*, having their epoch anything we please).

All of each set may (§ 58) be compounded into one simple harmonic motion of the same phase, of determinate amplitude, in a determinate line; and thus the whole system is reduced to two simple fully determined harmonic motions differing from one another in phase by a quarter period.

Now the resultant of two simple harmonic motions, one a quarter of a period in advance of the other, in different lines, has been proved (§ 65) to be motion in an ellipse of which the ranges of the component motions are conjugate axes, and in which equal areas are described by the radius-vector from the centre in equal times. Hence the general proposition of § 63. S. H. motions in different directions

Let
$$\left.\begin{aligned} x_1 &= l_1 a_1 \cos(\omega t - \epsilon_1), \\ y_1 &= m_1 a_1 \cos(\omega t - \epsilon_1), \\ z_1 &= n_1 a_1 \cos(\omega t - \epsilon_1), \end{aligned}\right\} \quad \ldots\ldots\ldots\ldots\ldots\ldots (1)$$

be the Cartesian specification of the first of the given motions; and so with varied suffixes for the others;

l, m, n denoting the direction cosines,

a ,, ,, half amplitude,

ϵ ,, ,, epoch,

the proper suffix being attached to each letter to apply it to each case, and ω denoting the common relative angular velocity. The resultant motion, specified by x, y, z without suffixes, is

$$x = \Sigma l_1 a_1 \cos(\omega t - \epsilon_1) = \cos \omega t \Sigma l_1 a_1 \cos \epsilon_1 + \sin \omega t \Sigma l_1 a_1 \sin \epsilon_1,$$
$$y = \text{etc.}; \quad z = \text{etc.};$$

or, as we may write for brevity,

$$\left.\begin{aligned} x &= P\cos\omega t + P'\sin\omega t, \\ y &= Q\cos\omega t + Q'\sin\omega t, \\ z &= R\cos\omega t + R'\sin\omega t, \end{aligned}\right\} \quad \ldots\ldots\ldots\ldots\ldots\ldots (2)$$

where
$$\left.\begin{aligned} P &= \Sigma\, l_1 a_1 \cos\epsilon_1, & P' &= \Sigma\, l_1 a_1 \sin\epsilon_1, \\ Q &= \Sigma m_1 a_1 \cos\epsilon_1, & Q' &= \Sigma m_1 a_1 \sin\epsilon_1, \\ R &= \Sigma\, n_1 a_1 \cos\epsilon_1, & R' &= \Sigma\, n_1 a_1 \sin\epsilon_1. \end{aligned}\right\} \quad \ldots\ldots\ldots\ldots (3)$$

The resultant motion thus specified, in terms of six component simple harmonic motions, may be reduced to two by compounding P, Q, R, and P', Q', R', in the elementary way. Thus if

$$\left.\begin{aligned} &\zeta = (P^2 + Q^2 + R^2)^{\frac{1}{2}}, \\ &\lambda = \frac{P}{\zeta}, \quad \mu = \frac{Q}{\zeta}, \quad \nu = \frac{R}{\zeta}, \\ &\zeta' = (P'^2 + Q'^2 + R'^2)^{\frac{1}{2}}, \\ &\lambda' = \frac{P'}{\zeta'}, \quad \mu' = \frac{Q'}{\zeta'}, \quad \nu' = \frac{R'}{\zeta'}, \end{aligned}\right\} \quad \ldots\ldots\ldots\ldots\ldots\ldots (4)$$

the required motion will be the resultant of $\zeta\cos\omega t$ in the line (λ, μ, ν), and $\zeta'\sin\omega t$ in the line (λ', μ', ν'). It is therefore motion in an ellipse, of which 2ζ and $2\zeta'$ in those directions are

S. H. motions in different directions.

conjugate diameters; with radius-vector from centre tracing equal areas in equal times; and of period $\frac{2\pi}{\omega}$.

H. motions of different kinds and in different lines.

67. We must next take the case of the composition of simple harmonic motions of *different* periods and in different lines. In general, whether these lines be in one plane or not, the line of motion returns into itself if the periods are commensurable; and if not, not. This is evident without proof.

If a be the amplitude, ϵ the epoch, and n the angular velocity in the relative circular motion, for a component in a line whose direction cosines are λ, μ, ν—and if ξ, η, ζ be the co-ordinates in the resultant motion,

$$\xi = \Sigma . \lambda_1 a_1 \cos(n_1 t - \epsilon_1), \quad \eta = \Sigma . \mu_1 a_1 \cos(n_1 t - \epsilon_1), \quad \zeta = \Sigma . \nu_1 a_1 \cos(n_1 t - \epsilon_1).$$

Now it is evident that at time $t + T$ the values of ξ, η, ζ will recur as soon as $n_1 T$, $n_2 T$, etc., are multiples of 2π, that is, when T is the least common multiple of $\frac{2\pi}{n_1}$, $\frac{2\pi}{n_2}$, etc.

If there be such a common multiple, the trigonometrical functions may be eliminated, and the equations (or equation, if the motion is in one plane) to the path are algebraic. If not, they are transcendental.

68. From the above we see generally that the composition of any number of simple harmonic motions in any directions and of any periods, may be effected by compounding, according to previously explained methods, their resolved parts in each of any three rectangular directions, and then compounding the final resultants in these directions.

S. H. motions in two rectangular directions.

69. By far the most interesting case, and the simplest, is that of *two* simple harmonic motions of any periods, whose directions must of course be in one plane.

Mechanical methods of obtaining such combinations will be afterwards described, as well as cases of their occurrence in Optics and Acoustics.

We may suppose, for simplicity, the two component motions to take place in perpendicular directions. Also, as we can only have a re-entering curve when their periods are commensurable, it will be advisable to commence with such a case.

The following figures represent the paths produced by the combination of simple harmonic motions of *equal* amplitude in two rectangular directions, the periods of the components being as 1 : 2, and the epochs differing successively by 0, 1, 2, etc., sixteenths of a circumference. S. H. motions in two rectangular directions

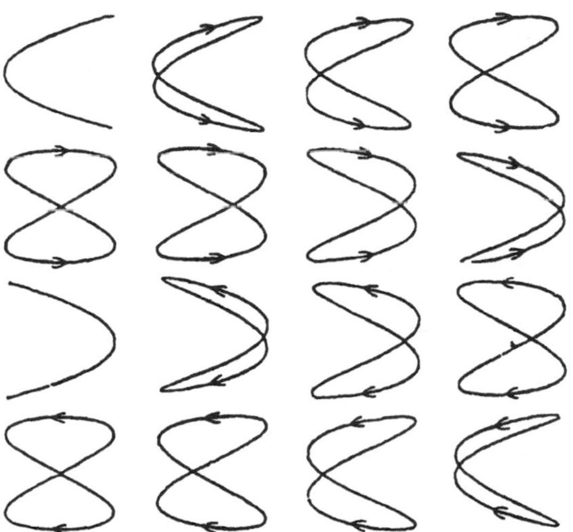

In the case of epochs equal, or differing by a multiple of π, the curve is a portion of a parabola, and is gone over twice in opposite directions by the moving point in each complete period.

For the case figured above,

$$x = a\cos(2nt - \epsilon), \quad y = a\cos nt.$$

Hence $$x = a\{\cos 2nt \cos\epsilon + \sin 2nt \sin\epsilon\}$$

$$= a\left\{\left(\frac{2y^2}{a^2} - 1\right)\cos\epsilon + 2\frac{y}{a}\sqrt{1 - \frac{y^2}{a^2}}\sin\epsilon\right\},$$

which for any given value of ϵ is the equation of the corresponding curve. Thus for $\epsilon = 0$,

$$\frac{x}{a} = \frac{2y^2}{a^2} - 1, \text{ or } y^2 = \frac{a}{2}(x + a), \text{ the parabola as above.}$$

S. H. motions in two rectangular directions.

For $\epsilon = \frac{\pi}{2}$ we have $\frac{x}{a} = 2\frac{y}{a}\sqrt{1 - \frac{y^2}{a^2}}$, or $a^2x^2 = 4y^2(a^2 - y^2)$,

the equation of the 5th and 13th of the above curves.

In general

$$x = a \cos(nt + \epsilon), \quad y = a \cos(n_1 t + \epsilon_1),$$

from which t is to be eliminated to find the Cartesian equation of the curve.

Composition of two uniform circular motions.

70. Another very important case is that of two groups of two simple harmonic motions in one plane, such that the resultant of each group is uniform circular motion.

If their periods are equal, we have a case belonging to those already treated (§ 63), and conclude that the resultant is, in general, motion in an ellipse, equal areas being described in equal times about the centre. As particular cases we may have simple harmonic, or uniform circular, motion. (Compare § 91.)

If the circular motions are in the *same* direction, the resultant is evidently circular motion in the same direction. This is the case of the motion of S in § 58, and requires no further comment, as its amplitude, epoch, etc., are seen at once from the figure.

71. If the periods of the two are very nearly equal, the resultant motion will be at any moment very nearly the circular motion given by the preceding construction. Or we may regard it as rigorously a motion in a circle with a varying radius decreasing from a maximum value, the sum of the radii of the two component motions, to a minimum, their difference, and increasing again, alternately; the direction of the resultant radius oscillating on each side of that of the greater component (as in corresponding case, § 59, above). Hence the angular velocity of the resultant motion is periodically variable. In the case of equal radii, next considered, it is constant.

72. When the radii of the two component motions are equal, we have the very interesting and important case figured below. Here the resultant radius bisects the angle between the component radii. The resultant angular velocity is the arithmetical mean of its components. We will explain in a future section

(§ 94) how this epitrochoid is traced by the rolling of one circle on another. (The particular case above delineated is that of a non-reëntrant curve.)

Composition of two uniform circular motions

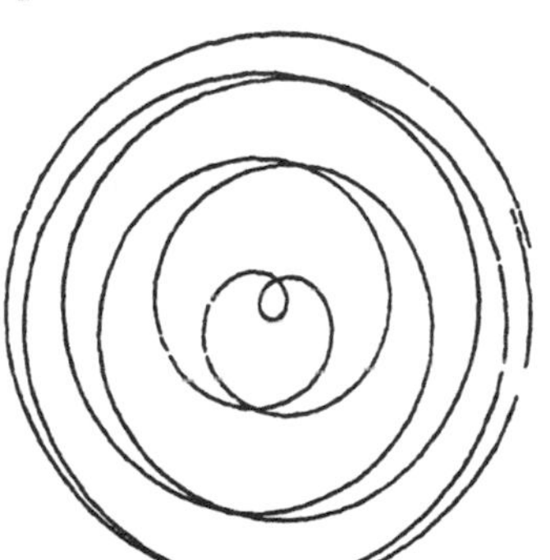

73. Let the uniform circular motions be in *opposite* directions; then, if the periods are equal, we may easily see, as before, § 66, that the resultant is in general elliptic motion, including the particular cases of uniform circular, and simple harmonic, motion.

If the periods are very nearly equal, the resultant will be easily found, as in the case of § 59.

74. If the radii of the component motions are equal, we have cases of *very* great importance in modern physics, one of which is figured below (like the preceding, a non-reëntrant curve).

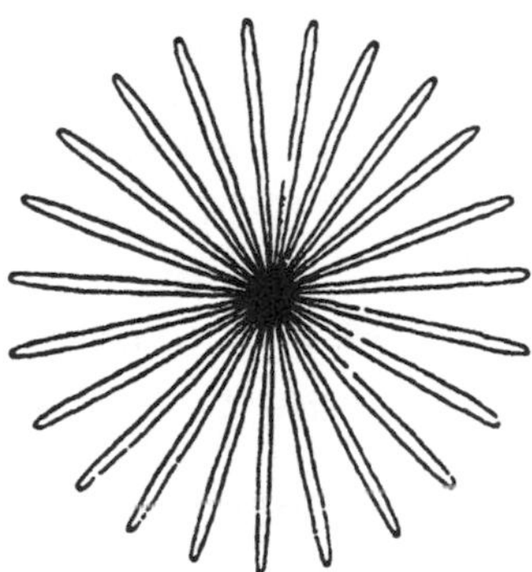

Composition of two uniform circular motions.

This is intimately connected with the explanation of two sets of important phenomena,—the rotation of the plane of polarization of light, by quartz and certain fluids on the one hand, and by transparent bodies under magnetic forces on the other. It is a case of the hypotrochoid, and its corresponding mode of description will be described in a future section. It will also appear in kinetics as the path of a pendulum-bob which contains a gyroscope in rapid rotation.

Fourier's Theorem.

75. Before leaving for a time the subject of the composition of harmonic motions, we must, as promised in § 62, devote some pages to the consideration of Fourier's Theorem, which is not only one of the most beautiful results of modern analysis, but may be said to furnish an indispensable instrument in the treatment of nearly every recondite question in modern physics. To mention only sonorous vibrations, the propagation of electric signals along a telegraph wire, and the conduction of heat by the earth's crust, as subjects in their generality intractable without it, is to give but a feeble idea of its importance. The following seems to be the most intelligible form in which it can be presented to the general reader :—

THEOREM.—*A complex harmonic function, with a constant term added, is the proper expression, in mathematical language, for any arbitrary periodic function; and consequently can express any function whatever between definite values of the variable.*

76. Any arbitrary periodic function whatever being given, the amplitudes and epochs of the terms of a complex harmonic function which shall be equal to it for every value of the independent variable, may be investigated by the "method of indeterminate coefficients."

Assume equation (14) below. Multiply both members first by $\cos \dfrac{2i\pi\xi}{p}\,d\xi$ and integrate from 0 to p: then multiply by $\sin \dfrac{2i\pi\xi}{p}\,d\xi$ and integrate between same limits. Thus instantly you find (13).

This investigation is sufficient as a solution of the problem,—to find a complex harmonic function expressing a given arbitrary periodic function,—when once we are assured that the problem is possible; and when we have this assurance, it proves that the resolution is determinate; that is to say, that no other complex harmonic function than the one we have found can satisfy the conditions. Fourier's Theorem

For description of an integrating machine by which the coefficients A_i, B_i in the Fourier expression (14) for any given arbitrary function may be obtained with exceedingly little labour, and with all the accuracy practically needed for the harmonic analysis of tidal and meteorological observations, see Proceedings of the Royal Society, Feb. 1876, or Chap. v. below.

77. The full theory of the expression investigated in § 76 will be made more intelligible by an investigation from a different point of view.

Let $F(x)$ be any periodic function, of period p. That is to say, let $F(x)$ be any function fulfilling the condition

$$F(x+ip) = F(x) \qquad (1),$$

where i denotes any positive or negative integer. Consider the integral

$$\int_{c'}^{c} \frac{F(x)\,dx}{a^2+x^2},$$

where a, c, c' denote any three given quantities. Its value is less than $F(z)\int_{c'}^{c}\frac{dx}{a^2+x^2}$, and greater than $F(z')\int_{c'}^{c}\frac{dx}{a^2+x^2}$, if z and z' denote the values of x, either equal to or intermediate between the limits c and c', for which $F(x)$ is greatest and least respectively. But

$$\int_{c'}^{c}\frac{dx}{a^2+x^2} = \frac{1}{a}\left(\tan^{-1}\frac{c}{a} - \tan^{-1}\frac{c'}{a}\right) \qquad (2),$$

and therefore

$$\left.\begin{aligned} &\int_{c'}^{c}\frac{F(x)\,a\,dx}{a^2+x^2} < F(z)\left(\tan^{-1}\frac{c}{a} - \tan^{-1}\frac{c'}{a}\right),\\ \text{and}\qquad &\qquad\text{,,}\qquad > F(z')\left(\tan^{-1}\frac{c}{a} - \tan^{-1}\frac{c'}{a}\right). \end{aligned}\right\} \qquad (3)$$

Fourier's Theorem.

Hence if A be the greatest of all the values of $F(x)$, and B the least,

$$\left.\begin{aligned}\int_c^{\infty}\frac{F(x)\,adx}{a^2+x^2} &< A\left(\frac{\pi}{2}-\tan^{-1}\frac{c}{a}\right),\\ \text{and}\qquad\qquad ,, &> B\left(\frac{\pi}{2}-\tan^{-1}\frac{c}{a}\right).\end{aligned}\right\}\quad\text{.................(4)}$$

Also, similarly,

$$\left.\begin{aligned}\int_{-\infty}^{c'}\frac{F(x)adx}{a^2+x^2} &< A\left(\tan^{-1}\frac{c'}{a}+\frac{\pi}{2}\right),\\ \text{and}\qquad\qquad ,, &> B\left(\tan^{-1}\frac{c'}{a}+\frac{\pi}{2}\right).\end{aligned}\right\}\quad\text{...........(5)}$$

Adding the first members of (3), (4), and (5), and comparing with the corresponding sums of the second members, we find

$$\left.\begin{aligned}\int_{-\infty}^{\infty}\frac{F(x)adx}{a^2-x^2} &< F(z)\left(\tan^{-1}\frac{c}{a}-\tan^{-1}\frac{c'}{a}\right)+A\left(\pi-\tan^{-1}\frac{c}{a}+\tan^{-1}\frac{c'}{a}\right),\\ \text{and}\quad ,, &> F(z')\left(\tan^{-1}\frac{c}{a}-\tan^{-1}\frac{c'}{a}\right)+B\left(\pi-\tan^{-1}\frac{c}{a}+\tan^{-1}\frac{c'}{a}\right).\end{aligned}\right\}(6)$$

But, by (1),

$$\int_{-\infty}^{\infty}\frac{F(x)dx}{a^2+x^2}-\int_0^p F(x)dx\left\{\Sigma_{i=-\infty}^{i=\infty}\left(\frac{1}{a^2+(x+ip)^2}\right)\right\}\quad\text{..........(7).}$$

Now if we denote $\sqrt{-1}$ by υ,

$$\frac{1}{a^2+(x+ip)^2}=\frac{1}{2a\upsilon}\left(\frac{1}{x+ip-a\upsilon}-\frac{1}{x+ip+a\upsilon}\right),$$

and therefore, taking the terms corresponding to positive and equal negative values of i together, and the terms for $i=0$ separately, we have

$$\begin{aligned}\Sigma_{i=-\infty}^{i=\infty}\left(\frac{1}{a^2+(x+ip)^2}\right) &= \frac{1}{2a\upsilon}\left\{\frac{1}{x-a\upsilon}-2\Sigma_{i=1}^{i=\infty}\frac{x-a\upsilon}{i^2p^2-(x-a\upsilon)^2}\right.\\ &\qquad\left.-\frac{1}{x+a\upsilon}+2\Sigma_{i=1}^{i=\infty}\frac{x+a\upsilon}{i^2p^2-(x+a\upsilon)^2}\right\}\\ &=\frac{\pi}{2ap\upsilon}\left\{\cot\frac{\pi(x-a\upsilon)}{p}-\cot\frac{\pi(x+a\upsilon)}{p}\right\}\\ &=\frac{\frac{\pi}{2ap\upsilon}\sin\frac{2\pi a\upsilon}{p}}{\cos^2\frac{\pi a\upsilon}{p}-\cos^2\frac{\pi x}{p}}=\frac{\frac{\pi}{ap\upsilon}\sin\frac{2\pi a\upsilon}{p}}{\cos\frac{2\pi a\upsilon}{p}-\cos\frac{2\pi x}{p}}\\ &=\frac{\pi}{ap}\,\frac{\epsilon^{\frac{2\pi a}{p}}-\epsilon^{-\frac{2\pi a}{p}}}{\epsilon^{\frac{2\pi a}{p}}-2\cos\frac{2\pi x}{p}+\epsilon^{-\frac{2\pi a}{p}}}.\end{aligned}$$

Hence, Fourier's Theorem.

$$\int_{-\infty}^{\infty}\frac{F(x)\,dx}{a^2+x^2}=\frac{\pi}{ap}\left(\epsilon^{\frac{2\pi a}{p}}-\epsilon^{-\frac{2\pi a}{p}}\right)\int_0^p\frac{F(x)\,dx}{\epsilon^{\frac{2\pi a}{p}}-2\cos\frac{2\pi x}{p}+\epsilon^{-\frac{2\pi a}{p}}}\ \ldots\ldots (8).$$

Next, denoting temporarily, for brevity, $\epsilon^{\frac{2\pi x\nu}{p}}$ by ζ, and putting

$$\epsilon^{-\frac{2\pi a}{p}}\quad e\ \ldots\ldots\ldots\ldots\ldots (9),$$

we have $$\frac{1}{\epsilon^{\frac{2\pi a}{p}}-2\cos\frac{2\pi x}{p}+\epsilon^{-\frac{2\pi a}{p}}}=\frac{e}{1-e(\zeta+\zeta^{-1})+e^2}$$

$$=\frac{e}{1-e^2}\left(\frac{1}{1-e\zeta}+\frac{1}{1-e\zeta^{-1}}-1\right)$$

$$\frac{e}{1-e^2}\{1+e(\zeta+\zeta^{-1})+e^2(\zeta^2+\zeta^{-2})+e^3(\zeta^3+\zeta^{-3})+\text{etc.}\}$$

$$=\frac{e}{1-e^2}\left(1+2e\cos\frac{2\pi x}{p}+2e^2\cos\frac{4\pi x}{p}+2e^3\cos\frac{6\pi x}{p}+\text{etc.}\right).$$

Hence, according to (8) and (9),

$$\int_{-\infty}^{\infty}\frac{F(x)\,dx}{a^2+x^2}=\frac{\pi}{ap}\int_0^p F(x)\,dx\left(1+2e\cos\frac{2\pi x}{p}+2e^2\cos\frac{4\pi x}{p}+\text{etc.}\right)\ldots(10).$$

Hence, by (6), we infer that

$$F(z)\left(\tan^{-1}\frac{c}{a}-\tan^{-1}\frac{c'}{a}\right)+A\left(\pi-\tan^{-1}\frac{c}{a}+\tan^{-1}\frac{c'}{a}\right)>$$

and $$F(z')\left(\tan^{-1}\frac{c}{a}-\tan^{-1}\frac{c'}{a}\right)+B\left(\pi-\tan^{-1}\frac{c}{a}+\tan^{-1}\frac{c'}{a}\right)<$$

$$\frac{\pi}{p}\int_0^p F(x)\,dx\left(1+2e\cos\frac{2\pi x}{p}+\text{etc.}\right).$$

Now let $c'=-c$, and $x=\xi'-\xi$,

ξ' being a variable, and ξ constant, so far as the integration is concerned; and let

$$F(x)=\phi(x+\xi)=\phi(\xi'),$$

and therefore $$F(z)=\phi(\xi+z),$$

$$F(z')=\phi(\xi+z').$$

Fourier's Theorem.

The preceding pair of inequalities becomes

$$\left.\begin{array}{l} \phi(\xi+z)\,.\,2\tan^{-1}\dfrac{c}{a}+A\left(\pi-2\tan^{-1}\dfrac{c}{a}\right)> \\ \text{and}\quad \phi(\xi+z')\,.\,2\tan^{-1}\dfrac{c}{a}+B\left(\pi-2\tan^{-1}\dfrac{c}{a}\right)< \\ \dfrac{\pi}{p}\left\{\int_0^p\phi(\xi')d\xi'+2\Sigma_{i=1}^{i=\infty}e^i\int_0^p\phi(\xi')d\xi'\cos\dfrac{2i\pi(\xi'-\xi)}{p}\right\}, \end{array}\right\}\ldots\ldots(11)$$

where ϕ denotes any periodic function whatever, of period p.

Now let c be a very small fraction of p. In the limit, where c is infinitely small, the greatest and least values of $\phi(\xi')$ for values of ξ' between $\xi+c$ and $\xi-c$ will be infinitely nearly equal to one another and to $\phi(\xi)$; that is to say,

$$\phi(\xi+z)=\phi(\xi+z')=\phi(\xi).$$

Next, let a be an infinitely small fraction of c. In the limit

$$\tan^{-1}\frac{c}{a}=\frac{\pi}{2},$$

and

$$e=\epsilon^{-\frac{2\pi a}{p}}=1.$$

Hence the comparison (11) becomes in the limit an equation which, if we divide both members by π, gives

$$\phi(\xi)=\frac{1}{p}\left\{\int_0^p\phi(\xi')d\xi'+2\Sigma_{i=1}^{i=\infty}\int_0^p\phi(\xi')d\xi'\cos\frac{2i\pi(\xi'-\xi)}{p}\right\}\ldots(12).$$

This is the celebrated theorem discovered by Fourier* for the development of an arbitrary periodic function in a series of simple harmonic terms. A formula included in it as a particular case had been given previously by Lagrange†.

If, for $\cos\dfrac{2i\pi(\xi'-\xi)}{p}$, we take its value

$$\cos\frac{2i\pi\xi'}{p}\cos\frac{2i\pi\xi}{p}+\sin\frac{2i\pi\xi'}{p}\sin\frac{2i\pi\xi}{p}$$

and introduce the following notation:—

$$\left.\begin{array}{l} A_0=\dfrac{1}{p}\displaystyle\int_0^p\phi(\xi)\,d\xi, \\ A_i=\dfrac{2}{p}\displaystyle\int_0^p\phi(\xi)\cos\frac{2i\pi\xi}{p}\,d\xi, \\ B_i=\dfrac{2}{p}\displaystyle\int_0^p\phi(\xi)\sin\frac{2i\pi\xi}{p}\,d\xi, \end{array}\right\}\ldots\ldots\ldots\ldots(13)$$

* *Théorie analytique de la Chaleur.* Paris, 1822.

† *Anciens Mémoires de l'Académie de Turin.*

we reduce (12) to this form :— Fourier's Theorem.

$$\phi(\xi)=A_0+\Sigma_{i=1}^{i=\infty}A_i\cos\frac{2i\pi\xi}{p}+\Sigma_{i=1}^{i=\infty}B_i\sin\frac{2i\pi\xi}{p}\text{........}(14),$$

which is the general expression of an arbitrary function in terms of a series of cosines and of sines. Or if we take

$$P_i=(A_i^2+B_i^2)^{\frac{1}{2}},\quad\text{and }\tan\epsilon_i=\frac{B_i}{A_i}\text{..................}(15),$$

we have $\phi(\xi)=A_0+\Sigma_{i=1}^{i=\infty}P_i\cos\left(\frac{2i\pi\xi}{p}-\epsilon_i\right)$..................(16),

which is the general expression in a series of single simple harmonic terms of the successive multiple periods.

Each of the equations and comparisons (2), (7), (8), (10), and (11) is a true arithmetical expression, and may be verified by actual calculation of the numbers, for any particular case; provided only that $F(x)$ has no infinite value in its period. Hence, with this exception, (12) or either of its equivalents, (14), (16), is a true arithmetical expression; and the series which it involves is therefore convergent. Hence we may with perfect rigour conclude that even the extreme case in which the arbitrary function experiences an abrupt finite change in its value when the independent variable, increasing continuously, passes through some particular value or values, is included in the general theorem. In such a case, if any value be given to the independent variable differing however little from one which corresponds to an abrupt change in the value of the function, the series must, as we may infer from the preceding investigation, converge and give a definite value for the function. But if exactly the critical value is assigned to the independent variable, the series cannot converge to any definite value. The consideration of the limiting values shown in the comparison (11) does away with all difficulty in understanding how the series (12) gives definite values having a finite difference for two particular values of the independent variable on the two sides of a critical value, but differing infinitely little from one another. Convergency of Fourier's series.

If the differential coefficient $\frac{d\phi(\xi)}{d\xi}$ is finite for every value of ξ within the period, it too is arithmetically expressible by a series of harmonic terms, which cannot be other than the series obtained by differentiating the series for $\phi(\xi)$. Hence

Convergency of Fourier's series.

$$\frac{d\phi(\xi)}{d\xi} = -\frac{2\pi}{p}\Sigma_{i=1}^{i=\infty} iP_i \sin\left(\frac{2i\pi\xi}{p} - \epsilon_i\right) \text{..................}(17),$$

and this series is convergent; and we may therefore conclude that the series for $\phi(\xi)$ is more convergent than a harmonic series with

$$1, \ \tfrac{1}{2}, \ \tfrac{1}{3}, \ \tfrac{1}{4}, \ \text{etc.},$$

for its coefficients. If $\frac{d^2\phi(\xi)}{d\xi^2}$ has no infinite values within the period, we may differentiate both members of (17) and still have an equation arithmetically true; and so on. We conclude that if the nth differential coefficient of $\phi(\xi)$ has no infinite values, the harmonic series for $\phi(\xi)$ must converge more rapidly than a harmonic series with

$$1, \ \frac{1}{2^n}, \ \frac{1}{3^n}, \ \frac{1}{4^n}, \ \text{etc.},$$

for its coefficients.

Displacement of a rigid body.

78. We now pass to the consideration of the displacement of a rigid body or group of points whose relative positions are unalterable. The simplest case we can consider is that of the motion of a plane figure in its own plane, and this, as far as kinematics is concerned, is entirely summed up in the result of the next section.

Displacements of a plane figure in its plane.

79. If a plane figure be displaced in any way in its own plane, there is always (with an exception treated in § 81) one point of it common to any two positions; that is, it may be moved from any one position to any other by rotation in its own plane about one point held fixed.

To prove this, let A, B be any two points of the plane figure in its first position, A', B' the positions of the same two after a displacement. The lines AA', BB' will not be parallel, except in one case to be presently considered. Hence the line equidistant from A and A' will meet that equidistant from B and B' in some point O. Join OA, OB, OA', OB'. Then, evidently, because $OA' = OA$, $OB' = OB$ and $A'B' = AB$, the triangles $OA'B'$ and OAB are equal and similar. Hence O is similarly situated with regard to $A'B'$ and AB, and is therefore one and

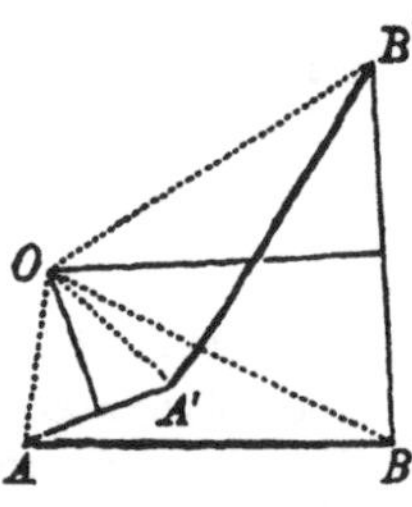

the same point of the plane figure in its two positions. If, for the sake of illustration, we actually trace the triangle OAB upon the plane, it becomes $OA'B'$ in the second position of the figure.

Displacements of a plane figure in its plane.

80. If from the equal angles $A'OB'$, AOB of these similar triangles we take the common part $A'OB$, we have the remaining angles AOA', BOB' equal, and each of them is clearly equal to the angle through which the figure must have turned round the point O to bring it from the first to the second position.

The preceding simple construction therefore enables us not only to demonstrate the general proposition, § 79, but also to determine from the two positions of one terminated line AB, $A'B'$ of the figure the common centre and the amount of the angle of rotation.

81. The lines equidistant from A and A', and from B and B', are parallel if AB is parallel to $A'B'$; and therefore the construction fails, the point O being infinitely distant, and the theorem becomes nugatory. In this case the motion is in fact a simple translation of the figure in its own plane without rotation—since, AB being parallel and equal to $A'B'$, we have AA' parallel and equal to BB'; and instead of there being one point of the figure common to both positions, the lines joining the two successive positions of all points in the figure are equal and parallel.

82. It is not necessary to suppose the figure to be a mere flat disc or plane—for the preceding statements apply to any one of a set of parallel planes in a rigid body, moving in any way subject to the condition that the points of any one plane in it remain always in a fixed plane in space.

83. There is yet a case in which the construction in § 79 is nugatory—that is when AA' is parallel to BB', but the lines of AB and $A'B'$ intersect. In this case, however, the point of intersection is the point O required, although the former method would not have enabled us to find it.

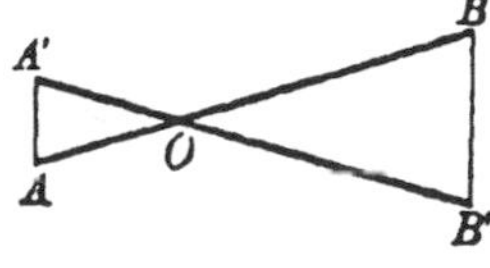

Examples of displacement in one plane.

84. Very many interesting applications of this principle may be made, of which, however, few belong strictly to our subject, and we shall therefore give only an example or two. Thus we know that if a line of given length AB move with its extremities always in two fixed lines OA, OB, any point in it as P describes an ellipse. It is required to find the direction of motion of P at any instant, *i.e.*, to draw a tangent to the ellipse. BA will pass to its next position by rotating about the point Q; found by the method of § 79 by drawing perpendiculars to OA and OB at A and B. Hence P for the instant revolves about Q, and thus its direction of motion, or the tangent to the ellipse, is perpendicular to QP. Also AB in its motion always touches a curve (called in geometry its envelop); and the same principle enables us to find the point of the envelop which lies in AB, for the motion of that point must evidently be ultimately (that is for a very small displacement) along AB, and the only point which so moves is the intersection of AB with the perpendicular to it from Q. Thus our construction would enable us to trace the envelop by points. (For more on this subject see § 91.)

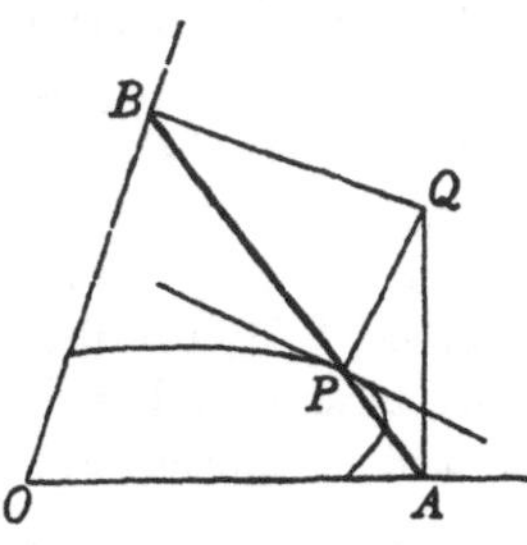

85. Again, suppose AB to be the beam of a stationary engine having a reciprocating motion about A, and by a link BD turning a crank CD about C. Determine the relation between the angular velocities of AB and CD in any position. Evidently the instantaneous direction of motion of B is transverse to AB, and of D transverse to CD—hence if AB, CD produced meet in O, the motion of BD is for an instant as if it turned about O. From this it may be easily seen that if the angular velocity of AB be ω, that of CD is $\frac{AB}{OB}\frac{OD}{CD}\omega$. A similar process is of course applicable to any combination of machinery, and we shall find it

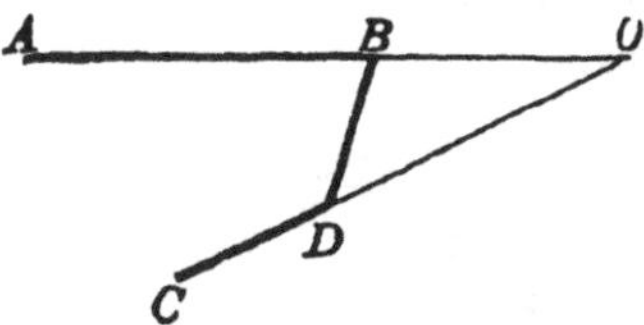

very convenient when we come to consider various dynamical problems connected with virtual velocities.

Examples of displacement in one plane.

Composition of rotations about parallel axes.

86. Since in general any movement of a plane figure in its plane may be considered as a rotation about one point, it is evident that two such rotations may in general be compounded into one; and therefore, of course, the same may be done with any number of rotations. Thus let A and B be the points of the figure about which in succession the rotations are to take place. By a rotation about A, B is brought say to B', and by a rotation about B', A is brought to A'. The construction of § 79 gives us at once the point O and the amount of rotation about it which singly gives the same effect as those about A and B in succession. But there is one case of exception, viz., when the rotations about A and B are of equal amount and in opposite directions. In this case $A'B'$ is evidently parallel to AB, and therefore the compound result is a *translation* only. That is, if a body revolve in succession through equal angles, but in opposite directions, about two parallel axes, it finally takes a position to which it could have been brought by a simple translation perpendicular to the lines of the body in its initial or final position, which were successively made axes of rotation; and inclined to their plane at an angle equal to half the supplement of the common angle of rotation.

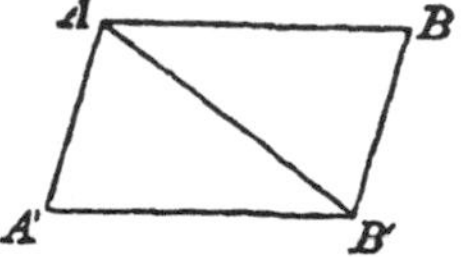

Composition of rotations and translations in one plane.

87. Hence to compound into an equivalent rotation a rotation and a translation, the latter being effected parallel to the plane of the former, we may decompose the translation into two rotations of equal amount and opposite direction, compound one of them with the given rotation by § 86, and then compound the other with the resultant rotation by the same process. Or we may adopt the following far simpler method. Let OA be the translation common to all points in the plane, and let BOC be the angle of rotation about O, BO being drawn so that OA bisects the exterior angle COB'. Take

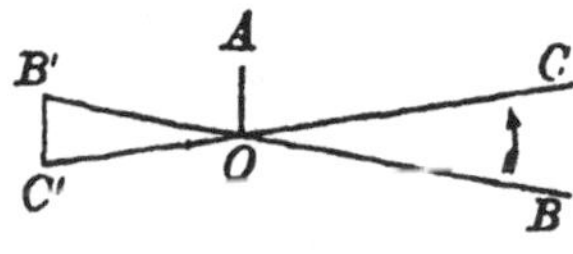

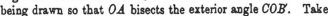

Composition of rotations and translations in one plane. the point B' in BO produced, such that $B'C'$, the space through which the rotation carries it, is equal and opposite to OA. This point retains its former position after the performance of the compound operation; so that a rotation and a translation in one plane can be compounded into an equal rotation about a different axis.

In general, if the origin be taken as the point about which rotation takes place in the plane of xy, and if it be through an angle θ, a point whose co-ordinates were originally x, y will have them changed to

$$\xi = x \cos\theta - y \sin\theta, \quad \eta = x \sin\theta + y \cos\theta,$$

or, if the rotation be very small,

$$\xi = x - y\theta, \quad \eta = y + x\theta.$$

Omission of the second and higher orders of small quantities. **88.** In considering the composition of angular velocities about different axes, and other similar cases, we may deal with infinitely small displacements only; and it results at once from the principles of the differential calculus, that if these displacements be of the *first* order of small quantities, any point whose displacement is of the *second* order of small quantities is to be considered as rigorously at rest. Hence, for instance, if a body revolve through an angle of the first order of small quantities about an axis (belonging to the body) which during the revolution is displaced through an angle or space, also of the first order, the displacement of any point of the body is rigorously what it would have been had the axis been fixed during the rotation about it, and its own displacement made either before or after this rotation. Hence in any case of motion of a rigid system the angular velocities about a system of axes moving *with* the system are the same at any instant as those about a system fixed in space, provided only that the latter coincide at the instant in question with the moveable ones.

Superposition of small motions. **89.** From similar considerations follows also the general principle of *Superposition of small motions*. It asserts that if several causes act *simultaneously* on the same particle or rigid body, and if the effect produced by each is of the first order of small quantities, the joint effect will be obtained if we consider the causes to act *successively*, each taking the point or system in the position in which the preceding one left it. It is evident at once that this is an immediate deduction from the fact that the second order of infinitely small quantities may be with rigorous accuracy neglected. This principle is of very great use, as we shall find in the sequel; its applications are of constant occurrence. **Superposition of small motions.**

A plane figure has given angular velocities about given axes perpendicular to its plane, find the resultant.

Let there be an angular velocity ω about an axis passing through the point a, b.

The consequent motion of the point x, y in the time δt is, as we have just seen (§ 87),

$$-(y-b)\omega\delta t \text{ parallel to } x, \quad \text{and } (x-a)\omega\delta t \text{ parallel to } y.$$

Hence, by the superposition of small motions, the whole motion parallel to x is

$$-(y\Sigma\omega - \Sigma b\omega)\delta t,$$

and that parallel to y $\quad (x\Sigma\omega - \Sigma a\omega)\delta t$.

Hence the point whose co-ordinates are

$$x' = \frac{\Sigma a\omega}{\Sigma\omega} \quad \text{and } y' = \frac{\Sigma b\omega}{\Sigma\omega}$$

is at rest, and the resultant axis passes through it. Any other point x, y moves through spaces

$$-(y\Sigma\omega - \Sigma b\omega)\delta t, \quad (x\Sigma\omega - \Sigma a\omega)\delta t.$$

But if the whole had turned about x', y' with velocity Ω, we should have had for the displacements of x, y,

$$-(y-y')\Omega\delta t, \quad (x-x')\Omega\delta t.$$

Comparing, we find $\Omega = \Sigma\omega$.

Hence if the sum of the angular velocities be zero, there is no rotation, and indeed the above formulæ show that there is then merely translation,

$$\Sigma(b\omega)\delta t \text{ parallel to } x, \quad \text{and } -\Sigma(a\omega)\delta t \text{ parallel to } y.$$

These formulæ suffice for the consideration of any problem on the subject.

90. Any motion whatever of a plane figure in its own plane might be produced by the rolling of a curve fixed to the figure upon a curve fixed in the plane. **Rolling of curve on curve.**

For we may consider the whole motion as made up of successive elementary displacements, each of which corresponds, as we have seen, to an elementary rotation about some point in

Rolling of curve on curve.

the plane. Let o_1, o_2, o_3, etc., be the successive points of the moving figure about which the rotations take place, O_1, O_2, O_3, etc., the positions of these points when each is the instantaneous centre of rotation. Then the figure rotates about o_1 (or O_1, which coincides with it) till o_2 coincides with O_2, then about the latter till o_3 coincides with O_3, and so on. Hence, if we join o_1, o_2, o_3, etc., in the plane of the figure, and O_1, O_2, O_3, etc., in the fixed plane, the motion will be the same as if the polygon $o_1o_2o_3$, etc., rolled upon the fixed polygon $O_1O_2O_3$, etc. By supposing the successive displacements small enough the sides of these polygons gradually diminish, and the polygons finally become continuous curves Hence the theorem.

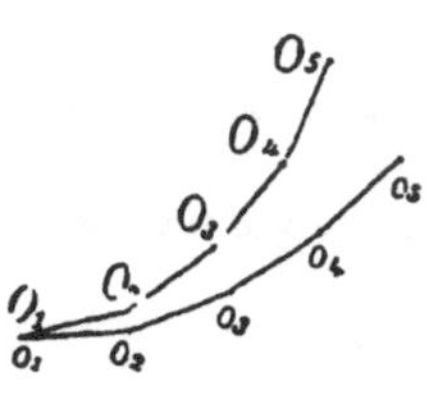

From this it immediately follows, that any displacement of a rigid solid, which is in directions wholly perpendicular to a fixed line, may be produced by the rolling of a cylinder fixed in the solid on another cylinder fixed in space, the axes of the cylinders being parallel to the fixed line.

91. As an interesting example of this theorem, let us recur to the case of § 84:—A circle may evidently be circumscribed about $OBQA$; and it must be of invariable magnitude, since in it a chord of given length AB subtends a given angle O at the circumference. Also OQ is a diameter of this circle, and is therefore constant. Hence, as Q is momentarily at rest, the motion of the circle circumscribing $OBQA$ is one of internal rolling on a circle of double its diameter. Hence if a circle roll internally on another of twice its diameter, any point in its circumference describes a diameter of the fixed circle, any other point in its plane an ellipse. This is precisely the same proposition as that of § 70, although the ways of arriving at it are very different. As it presents us with a particular case of the Hypocycloid, it warns us to return to the consideration of these and kindred curves, which give good instances of kinematical theorems, but which besides are of great use in physics generally.

Cycloids and Trochoids.

92. When a circle rolls upon a straight line, a point in its circumference describes a Cycloid; an internal point describes a

Prolate, an external one a Curtate, Cycloid. The two latter varieties are sometimes called Trochoids. Cycloids and Trochoids.

The general form of these curves will be seen in the annexed figures; and in what follows we shall confine our remarks to the cycloid itself, as of immensely greater consequence than the others. The next section contains a simple investigation of those properties of the cycloid which are most useful in our subject.

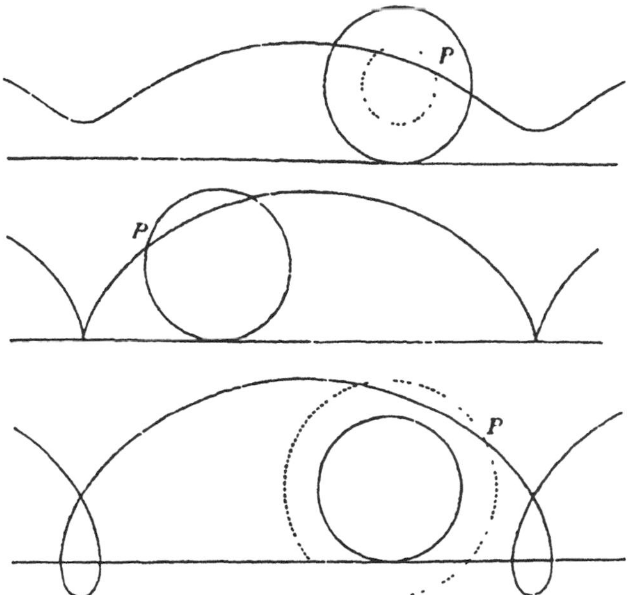

93. Let AB be a diameter of the generating (or rolling) circle, BC the line on which it rolls. The points A and B describe similar and equal cycloids, of which AQC and BS are portions. If PQR be any subsequent position of the generating circle, Q and S the new positions of A and B, $\angle QPS$ is of course a right angle. If, therefore, QR be drawn parallel to PS, PR is a diameter Properties of the cycloid

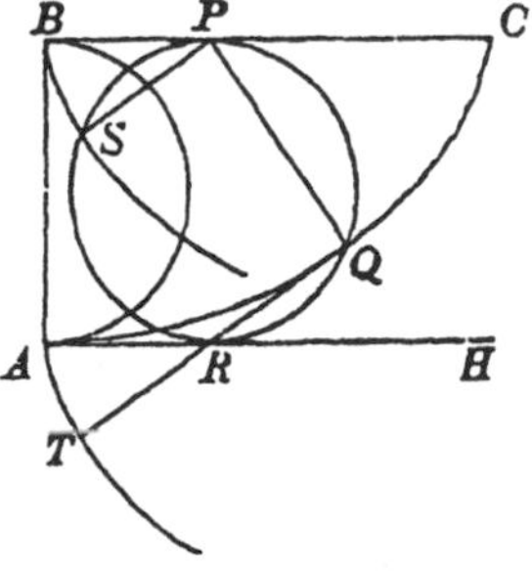

Properties of the cycloid.

of the rolling circle. Produce QR to T, making $RT = QR = PS$. Evidently the curve AT, which is the locus of T, is similar and equal to BS, and is therefore a cycloid similar and equal to AC. But QR is perpendicular to PQ, and is therefore the instantaneous direction of motion of Q, or is the tangent to the cycloid AQC. Similarly, PS is perpendicular to the cycloid BS at S, and so is therefore TQ to AT at T. Hence (§ 19) AQC is the evolute of AT, and arc $AQ = QT = 2QR$.

Epicycloids, Hypocycloids, etc.

94. When the circle rolls upon another circle, the curve described by a point in its circumference is called an Epicycloid, or a Hypocycloid, as the rolling circle is without or within the fixed circle; and when the tracing point is not in the circumference, we have Epitrochoids and Hypotrochoids. Of the latter we have already met with examples, §§ 70, 91, and others will be presently mentioned. Of the former, we have in the first of the appended figures the case of a circle rolling externally on another of equal size. The curve in this case is called the Cardioid (§ 49).

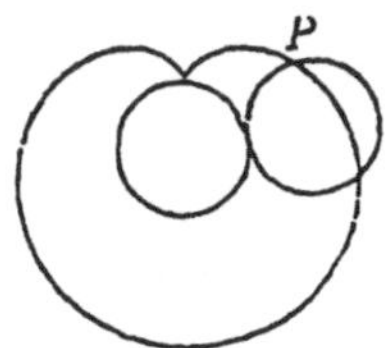

In the second diagram, a circle rolls externally on another of twice its radius. The epicycloid so described is of importance in Optics, and will, with others, be referred to when we consider the subject of Caustics by reflexion.

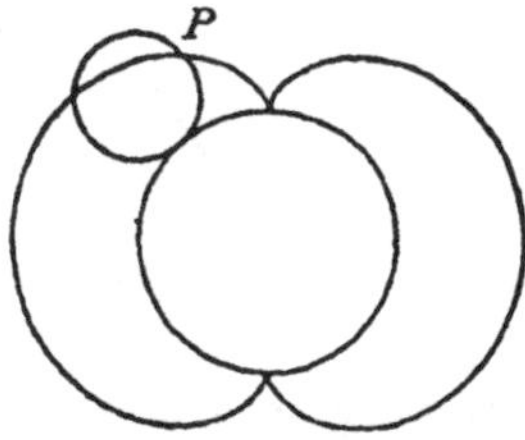

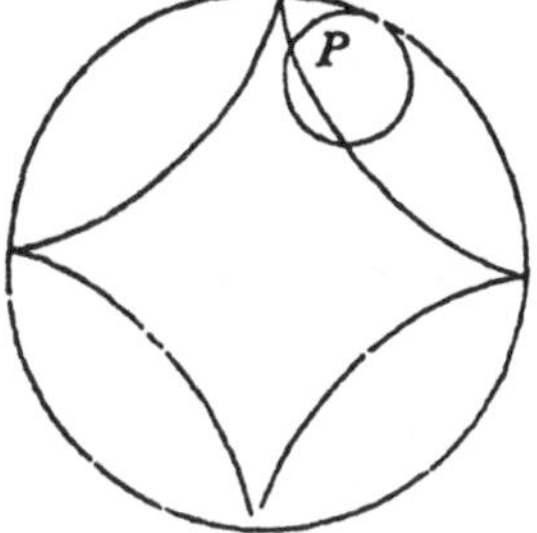

In the third diagram, we have a hypocycloid traced by the rolling of one circle internally on another of four times its radius.

The curve figured in § 72 is an epitrochoid described by a point in the plane of a large circular disc which rolls upon a circular cylinder of small diameter, so that the point passes through the axis of the cylinder. **Epicycloids, Hypocycloids, etc.**

That of § 74 is a hypotrochoid described by a point in the plane of a circle which rolls internally on another of rather more than twice its diameter, the tracing point passing through the centre of the fixed circle. Had the diameters of the circles been exactly as 1 : 2, § 72 or § 91 shows that this curve would have been reduced to a single straight line.

The general equations of this class of curves are

$$x = (a+b)\cos\theta - eb\cos\frac{a+b}{b}\theta,$$

$$y = (a+b)\sin\theta - eb\sin\frac{a+b}{b}\theta,$$

where a is the radius of the fixed, b of the rolling circle; and eb is the distance of the tracing point from the centre of the latter.

95. If a rigid solid body move in any way whatever, subject only to the condition that one of its points remains fixed, there is always (without exception) one line of it through this point common to the body in any two positions. This most important theorem is due to Euler. To prove it, consider a spherical surface within the body, with its centre at the fixed point C. All points of this sphere attached to the body will move on a sphere fixed in space. Hence the construction of § 79 may be made, but with great circles instead of straight lines; and the same reasoning will apply to prove that the point O thus obtained is common to the body in its two positions. Hence every point of the body in the line OC, joining O with the fixed point, must be common to it in the two positions. Hence the body may pass from any one position to any other by rotating through a definite angle about a definite axis. Hence any position of the body may be specified by specifying the axis, and the angle, of rotation by which it may be brought to that position from a fixed position of reference, an idea due to Euler, and revived by Rodrigues. **Motion about a fixed point. Euler's theorem.**

Rodrigues' co-ordinates.

Let OX, OY, OZ be any three fixed axes through the fixed point O round which the body turns. Let λ, μ, ν be the direction cosines, referred to these axes, of the axis OI round which the body must turn, and χ the angle through which it must turn round this axis, to bring it from some zero position to any other position. This other position, being specified by the four co-ordinates λ, μ, ν, χ (reducible, of course, to three by the relation $\lambda^2+\mu^2+\nu^2=1$), will be called for brevity $(\lambda, \mu, \nu, \chi)$. Let OA, OB, OC be three rectangular lines moving with the body, which in the "zero" position coincide respectively with OX, OY, OZ; and put

$$(XA),\ (YA),\ (ZA),\ (XB),\ (YB),\ (ZB),\ (XC),\ (YC),\ (ZC),$$

for the nine direction cosines of OA, OB, OC, each referred to OX, OY, OZ. These nine direction cosines are of course reducible to three independent co-ordinates by the well-known six relations. Let it be required now to express these nine direction cosines in terms of Rodrigues' co-ordinates λ, μ, ν, χ.

Let the lengths OX, ..., OA, ..., OI be equal, and call each unity: and describe from O as centre a spherical surface of unit radius; so that X, Y, Z, A, B, C, I shall be points on this surface. Let XA, YA, ... XB, denote arcs, and XAY, AXB, ... angles between arcs, in the spherical diagram thus obtained. We have $IA=IX=\cos^{-1}\lambda$, and $XIA=\chi$. Hence by the isosceles spherical triangle XIA,

$$\cos XA = \cos^2 IX + \sin^2 IX \cos\chi,$$

or

$$(XA) = \lambda^2 + (1-\lambda^2)\cos\chi \qquad (1).$$

And by the spherical triangle XIB,

$$\cos XB = \cos IX \cos IB + \sin IX \sin IB \cos XIB$$
$$= \lambda\mu + \sqrt{(1-\lambda^2)(1-\mu^2)}\cos XIB \qquad (2).$$

Now $XIB = XIY + YIB = XIY + \chi$; and by the spherical triangle XIY we have

$$\cos XY = 0 = \cos IX \cos IY + \sin IX \sin IY \cos XIY$$
$$= \lambda\mu + \sqrt{(1-\lambda^2)(1-\mu^2)}\cos XIY.$$

Hence $$\sqrt{(1-\lambda^2)(1-\mu^2)}\cos XIY = -\lambda\mu,$$

and $$\sqrt{(1-\lambda^2)(1-\mu^2)}\sin XIY = \sqrt{(1-\lambda^2-\mu^2)} =$$

by which we have

$$\sqrt{(1-\lambda^2)(1-\mu^2)}\cos(XIY+\chi) = -\lambda\mu\cos\chi - \nu\sin\chi;$$

and using this in (2), **Rodrigues' co-ordinates.**

$$\cos XB = \lambda\mu(1-\cos\chi) - \nu\sin\chi \qquad (3).$$

Similarly we find

$$\cos AY = \lambda\mu(1-\cos\chi) + \nu\sin\chi \qquad (4).$$

The other six formulæ may be written out by symmetry from (1), (3), and (4); and thus for the nine direction cosines we find

$$\left.\begin{aligned}(XA) &= \lambda^2 + (1-\lambda^2)\cos\chi;\ (XB) = \lambda\mu(1-\cos\chi) - \nu\sin\chi;\ (YA) = \lambda\mu(1-\cos\chi) + \nu\sin\chi;\\ (YB) &= \mu^2 + (1-\mu^2)\cos\chi;\ (YC) = \mu\nu(1-\cos\chi) - \lambda\sin\chi;\ (ZB) = \mu\nu(1-\cos\chi) + \lambda\sin\chi;\\ (ZC) &= \nu^2 + (1-\nu^2)\cos\chi;\ (ZA) = \nu\lambda(1-\cos\chi) - \mu\sin\chi;\ (XC) = \nu\lambda(1-\cos\chi) + \mu\sin\chi.\end{aligned}\right\}(5).$$

Adding the three first equations of these three lines, and remembering that

$$\lambda^2 + \mu^2 + \nu^2 = 1 \qquad (6),$$

we deduce

$$\cos\chi = \tfrac{1}{2}[(XA) + (YB) + (ZC) - 1] \qquad (7);$$

and then, by the three equations separately,

$$\left.\begin{aligned}\lambda^2 &= \frac{1+(XA)-(YB)-(ZC)}{3-(XA)-(YB)-(ZC)},\\ \mu^2 &= \frac{1-(XA)+(YB)-(ZC)}{3-(XA)-(YB)-(ZC)},\\ \nu^2 &= \frac{1-(XA)-(YB)+(ZC)}{3-(XA)-(YB)-(ZC)}.\end{aligned}\right\} \qquad (8)$$

These formulæ, (8) and (7), express, in terms of (XA), (YB), (ZC), three out of the nine direction cosines (XA), ..., the direction cosines of the axis round which the body must turn, and the cosine of the angle through which it must turn round this axis, to bring it from the zero position to the position specified by those three direction cosines.

By aid of Euler's theorem above, successive or simultaneous rotations about any number of axes through the fixed point may be compounded into a rotation about one axis. Doing this for infinitely small rotations we find the law of composition of angular velocities. **Composition of rotations.**

Let OA, OB be two axes about which a body revolves with angular velocities ϖ, ρ respectively. **Composition of angular velocities.**

With radius unity describe the arc AB, and in it take any

Composition of angular velocities.

point I. Draw Ia, $I\beta$ perpendicular to OA, OB respectively.

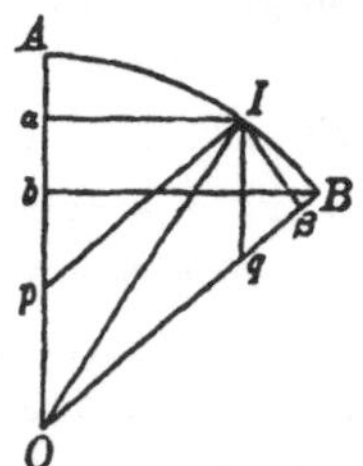

Let the rotations about the two axes be such that that about OB tends to *raise* I above the plane of the paper, and that about OA to depress it. In an infinitely short interval of time τ, the amounts of these displacements will be $\rho I\beta.\tau$ and $-\varpi Ia.\tau$. The point I, and therefore every point in the line OI, will be at rest during the interval τ if the sum of these displacements is zero, that is if $\rho.I\beta = \varpi.Ia$. Hence the line OI is instantaneously at rest, or the *two* rotations about OA and OB may be compounded into *one* about OI. Draw Ip, Iq, parallel to OB, OA respectively. Then, expressing in two ways the area of the parallelogram $IpOq$, we have

$$Oq.I\beta = Op.Ia,$$

$$Oq : Op :: \rho : \varpi.$$

Hence, if along the axes OA, OB, we measure off from O lines Op, Oq, proportional respectively to the angular velocities about these axes—the diagonal of the parallelogram of which these are contiguous sides is the resultant axis.

Again, if Bb be drawn perpendicular to OA, and if Ω be the angular velocity about OI, the whole displacement of B may evidently be represented either by $\varpi.Bb$ or $\Omega.I\beta$.
Hence

$$\Omega : \varpi :: Bb : I\beta :: \sin BOA : \sin IOB :: \sin IpO : \sin pIO,$$
$$:: OI : Op.$$

Thus it is proved that,—

Parallelogram of angular velocities.

If lengths proportional to the respective angular velocities about them be measured off on the component and resultant axes, the lines so determined will be the sides and diagonal of a parallelogram.

Composition of angular velocities about axes meeting in a point.

96. Hence the single angular velocity equivalent to three co-existent angular velocities about three mutually perpendicular axes, is determined in magnitude, and the direction of its axis is found (§ 27), as follows:—The square of the resultant angular velocity is the sum of the squares of its components,

and the ratios of the three components to the resultant are the direction cosines of the axis.

Composition of angular velocities about axes meeting in a point.

Hence simultaneous rotations about any number of axes meeting in a point may be compounded thus:—Let ω be the angular velocity about one of them whose direction cosines are l, m, n; Ω the angular velocity and λ, μ, ν the direction cosines of the resultant.

$$\lambda\Omega = \Sigma(l\omega),\quad \mu\Omega = \Sigma(m\omega),\quad \nu\Omega = \Sigma(n\omega),$$

whence
$$\Omega^2 = \Sigma^2(l\omega) + \Sigma^2(m\omega) + \Sigma^2(n\omega),$$

and
$$\lambda = \frac{\Sigma(l\omega)}{\Omega},\quad \mu = \frac{\Sigma(m\omega)}{\Omega},\quad \nu = \frac{\Sigma(n\omega)}{\Omega}.$$

Hence also, an angular velocity about any line may be resolved into three about any set of rectangular lines, the resolution in each case being (like that of simple velocities) effected by multiplying by the cosine of the angle between the directions.

Hence, just as in § 31 a uniform acceleration, perpendicular to the direction of motion of a point, produces a change in the *direction* of motion, but does not influence the *velocity;* so, if a body be rotating about an axis, and be subjected to an action tending to produce rotation about a perpendicular axis, the result will be a change of *direction* of the axis about which the body revolves, but no change in the *angular velocity*. On this kinematical principle is founded the dynamical explanation of the Precession of the Equinoxes (§ 107) and of some of the seemingly marvellous performances of gyroscopes and gyrostats.

The following method of treating the subject is useful in connexion with the ordinary methods of co-ordinate geometry. It contains also, as will be seen, an independent demonstration of the parallelogram of angular velocities:—

Angular velocities ϖ, ρ, σ about the axes of x, y, and z respectively, produce in time δt displacements of the point at x, y, z (§§ 87, 89),

$$(\rho z - \sigma y)\,\delta t \parallel x,\quad (\sigma x - \varpi z)\,\delta t \parallel y,\quad (\varpi y - \rho x)\,\delta t \parallel z.$$

Hence points for which

$$\frac{x}{\varpi} = \frac{y}{\rho} = \frac{z}{\sigma}$$

are not displaced. These are therefore the equations of the axis

Composition of angular velocities about axes meeting in a point.

Now the perpendicular from any point x, y, z to this line is, by co-ordinate geometry,

$$\left[x^2+y^2+z^2-\frac{(\varpi x+\rho y+\sigma z)^2}{\varpi^2+\rho^2+\sigma^2}\right]^{\frac{1}{2}}$$

$$=\frac{1}{\sqrt{\varpi^2+\rho^2+\sigma^2}}\sqrt{(\rho z-\sigma y)^2+(\sigma x-\varpi z)^2+(\varpi y-\rho x)^2}$$

$$=\frac{\text{whole displacement of } x, y, z}{\sqrt{\varpi^2+\rho^2+\sigma^2}\,\delta t}.$$

The actual displacement of x, y, z is therefore the same as would have been produced in time δt by a single angular velocity, $\Omega=\sqrt{\varpi^2+\rho^2+\sigma^2}$, about the axis determined by the preceding equations.

Composition of successive finite rotations.

97. We give next a few useful theorems relating to the composition of successive *finite* rotations.

If a pyramid or cone of any form roll on a heterochirally similar* pyramid (the image in a plane mirror of the first position of the first) all round, it clearly comes back to its primitive position. This (as all rolling of cones) is conveniently exhibited by taking the intersection of each with a spherical surface. Thus we see that if a spherical polygon turns about its angular points in succession, always keeping on the spherical surface, and if the angle through which it turns about each point is twice the supplement of the angle of the polygon, or, which will come to the same thing, if it be in the other direction, but equal to twice the angle itself of the polygon, it will be brought to its original position.

The polar theorem (compare § 134, below) to this is, that a body, after successive rotations, represented by the doubles of the successive sides of a spherical polygon taken in order, is restored to its original position; which also is self-evident.

98. Another theorem is the following;—

If a pyramid rolls over all its sides on a plane, it leaves its track behind it as one plane angle, equal to the sum of the plane angles at its vertex.

* The similarity of a right-hand and a left-hand is called heterochiral: that of two right-hands, homochiral. Any object and its image in a plane mirror are heterochirally similar (Thomson, *Proc. R. S. Edinburgh*, 1873).

Otherwise:—in a spherical surface, a spherical polygon having rolled over all its sides along a great circle, is found in the same position as if the side first lying along that circle had been simply shifted along it through an arc equal to the polygon's periphery. The polar theorem is:—if a body be made to take successive rotations, represented by the sides of a spherical polygon taken in order, it will finally be as if it had revolved about the axis through the first angular point of the polygon through an angle equal to the spherical excess (§ 134) or area of the polygon.

Composition of successive finite rotations.

99. The investigation of § 90 also applies to this case; and it is thus easy to show that the most general motion of a spherical figure on a fixed spherical surface is obtained by the rolling of a curve fixed in the figure on a curve fixed on the sphere. Hence as at each instant the line joining C and O contains a set of points of the body which are momentarily at rest, the most general motion of a rigid body of which one point is fixed consists in the rolling of a cone fixed in the body upon a cone fixed in space—the vertices of both being at the fixed point.

Motion about a fixed point. Rolling cones.

100. Given at each instant the angular velocities of the body about three rectangular axes attached to it, determine its position in space at any time.

Position of the body due to given rotations.

From the given angular velocities about OA, OB, OC, we know, § 95, the position of the instantaneous axis OI with reference to the body at every instant. Hence we know the conical surface in the body which rolls on the cone fixed in space. The data are sufficient also for the determination of this other cone; and these cones being known, and the lines of them which are in contact at any given instant being determined, the position of the moving body is completely determined.

If λ, μ, ν be the direction cosines of OI referred to OA, OB, OC; ϖ, ρ, σ the angular velocities, and ω their resultant:

$$\frac{\lambda}{\varpi}=\frac{\mu}{\rho}=\frac{\nu}{\sigma}=\frac{1}{\omega},$$

by § 95. These equations, in which ϖ, ρ, σ, ω are given functions of t, express explicitly the position of OI relatively to OA, OB,

Position of the body due to given rotations.

OC, and therefore determine the cone fixed in the body. For the cone fixed in space: if r be the radius of curvature of its intersection with the unit sphere, r' the same for the rolling cone, we find from § 105 below, that if s be the length of the arc of either spherical curve from a common initial point,

$$\omega r' = \frac{1}{r}\frac{ds}{dt}\sin(\sin^{-1}r + \sin^{-1}r') = \frac{1}{r}\frac{ds}{dt}(r\sqrt{1-r'^2}+r'\sqrt{1-r^2}),$$

which, as s, r' and ω are known in terms of t, gives r in terms of t, or of s, as we please. Hence, by a single quadrature, the "intrinsic" equation of the fixed cone.

101. An unsymmetrical system of angular co-ordinates ψ, θ, ϕ, for specifying the position of a rigid body by aid of a line OB and a plane AOB moving with it, and a line OY and a plane YOX fixed in space, which is essentially proper for many physical problems, such as the Precession of the Equinoxes and the spinning of a top, the motion of a gyroscope and its gimbals, the motion of a compass-card and of its bowl and gimbals, is convenient for many others, and has been used by the greatest mathematicians often even when symmetrical methods would have been more convenient, must now be described.

ON being the intersection of the two planes, let $YON = \psi$, and $NOB = \phi$; and let θ be the angle from the fixed plane, produced through ON, to the portion NOB of the moveable plane. (Example, θ the "obliquity of the ecliptic," ψ the longitude of the autumnal equinox reckoned from OY, a fixed line in the plane of the earth's orbit supposed fixed; ϕ the hour-angle of the autumnal equinox; B being in the earth's equator and in the meridian of Greenwich: thus ψ, θ, ϕ are angular co-ordinates of the earth.) To show the relation of this to the symmetrical system, let OA be perpendicular to OB, and draw OC perpendicular to both; OX perpendicular to OY, and draw OZ perpendicular to OY and OX; so that OA, OB, OC are three rectangular axes fixed relatively to the body, and OX, OY, OZ fixed in space. The annexed diagram shows ψ, θ, ϕ in angles and arc, and in arcs and angles, on a spherical surface of unit radius with centre at O.

To illustrate the meaning of these angular co-ordinates, suppose A, B, C initially to coincide with X, Y, Z respectively.

Then, to bring the body into the position specified by θ, ϕ, ψ, rotate it round OZ through an angle equal to $\psi+\phi$, thus

Position of the body due to given rotations.

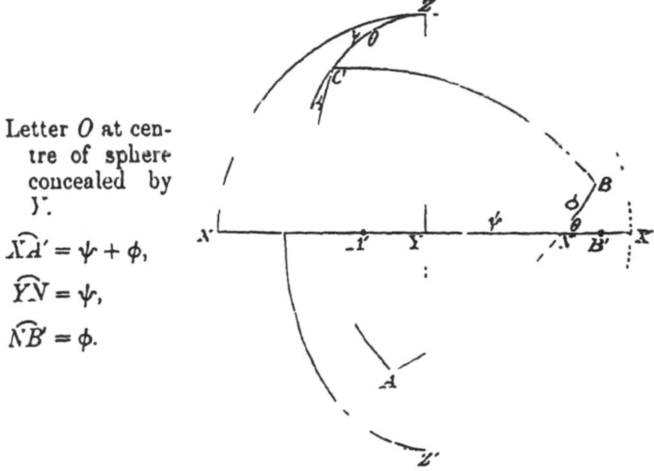

bringing A and B from X and Y to A' and B' respectively; and, (taking $\widehat{YN}=\psi$,) rotate the body round ON through an angle equal to θ, thus bringing A, B, and C from the positions A', B', and Z respectively, to the positions marked A, B, C in the diagram. Or rotate first round ON through θ, so bringing C from Z to the position marked C, and then rotate round OC through $\psi+\phi$. Or, while OC is turning from OZ to the position shown on the diagram, let the body turn round OC relatively to the plane $ZCZ'O$ through an angle equal to ϕ. It will be in the position specified by these three angles.

Let $\angle XZC=\psi$, $\angle ZCA=\pi-\phi$, and $ZC=\theta$, and ϖ, ρ, σ mean the same as in § 100. By considering in succession instantaneous motions of C along and perpendicular to ZC, and the motion of AB in its own plane, we have

$$\frac{d\theta}{dt}=\varpi\sin\phi+\rho\cos\phi, \qquad \sin\theta\frac{d\psi}{dt}=\rho\sin\phi-\varpi\cos\phi,$$

and

$$\frac{d\psi}{dt}\cos\theta+\frac{d\phi}{dt}=\sigma.$$

The nine direction cosines (XA), (YB), &c., according to the notation of § 95, are given at once by the spherical triangles

Position of the body due to given rotations.

XNA, YNB, &c.; each having N for one angular point, with θ, or its supplement or its complement, for the angle at this point. Thus, by the solution in each case for the cosine of one side in terms of the cosine of the opposite angle, and the cosines and sines of the two other sides, we find

$$
\begin{aligned}
(XA) &= \cos\theta\cos\psi\cos\phi - \sin\psi\sin\phi, \\
(XB) &= -\cos\theta\cos\psi\sin\phi - \sin\psi\cos\phi, \\
(YA) &= \cos\theta\sin\psi\cos\phi + \cos\psi\sin\phi.
\end{aligned}
$$

$$
\begin{aligned}
(YB) &= -\cos\theta\sin\psi\sin\phi + \cos\psi\cos\phi, \\
(YC) &= \sin\theta\sin\psi, \\
(ZB) &= \sin\theta\sin\phi.
\end{aligned}
$$

$$
\begin{aligned}
(ZC) &= \cos\theta, \\
(ZA) &= -\sin\theta\cos\phi, \\
(XC) &= \sin\theta\cos\psi.
\end{aligned}
$$

General motion of a rigid body.

102. We shall next consider the most general possible motion of a rigid body of which no point is fixed—and first we must prove the following theorem. There is one set of parallel planes in a rigid body which are parallel to each other in any two positions of the body. The parallel lines of the body perpendicular to these planes are of course parallel to each other in the two positions.

Let C and C' be any point of the body in its first and second positions. Move the body without rotation from its second position to a third in which the point at C' in the second position shall occupy its original position C. The preceding demonstration shows that there is a line CO common to the body in its first and third positions. Hence a line $C'O'$ of the body in its second position is parallel to the same line CO in the first position. This of course clearly applies to every line of the body parallel to CO, and the planes perpendicular to these lines also remain parallel.

Let S denote a plane of the body, the two positions of which are parallel. Move the body from its first position, without rotation, in a direction perpendicular to S, till S comes into the plane of its second position. Then to get the body into its actual position, such a motion as is treated in § 79 is farther required. But by § 79 this may be effected by rotation about a certain axis perpendicular to the plane S, unless the motion required belongs to the exceptional case of pure translation. Hence [this case excepted] the body may be brought from the first position to the second by translation through a determinate distance perpendicular to a given plane, and rotation through a determinate angle about a determinate axis perpendicular to that plane. This is precisely the motion of a screw in its nut. General motion of a rigid body

103. In the excepted case the whole motion consists of two translations, which can of course be compounded into a single one; and thus, in this case, there is no rotation at all, or every plane of it fulfils the specified condition for S of § 102.

104. Returning to the motion of a rigid body with one point fixed, let us consider the case in which the guiding cones, § 99, are both circular. The motion in this case may be called *Precessional Rotation.* Precessional Rotation.

The plane through the instantaneous axis and the axis of the fixed cone passes through the axis of the rolling cone. This plane turns round the axis of the fixed cone with an angular velocity Ω (see § 105 below), which must clearly bear a constant ratio to the angular velocity ω of the rigid body about its instantaneous axis.

105. The motion of the plane containing these axes is called the *precession* in any such case. What we have denoted by Ω is the angular velocity of the precession, or, as it is sometimes called, the rate of precession.

The angular motions ω, Ω are to one another inversely as the distances of a point in the axis of the rolling cone from the instantaneous axis and from the axis of the fixed cone.

For, let OA be the axis of the fixed cone, OB that of the rolling cone, and OI the instantaneous axis. From any point P in OB draw PN perpendicular to OI, and PQ perpendicular to OA. Then we perceive that P moves always in the circle whose centre is Q, radius PQ, and plane perpendicular to OA. Hence

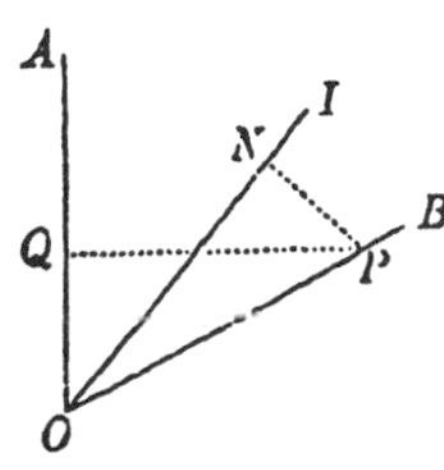

Precessional Rotation

the actual velocity of the point P is ΩQP. But, by the principles explained above, § 99, the velocity of P is the same as that of a point moving in a circle whose centre is N, plane perpendicular to ON, and radius NP, which, as this radius revolves with angular velocity ω, is $\omega . NP$. Hence

$$\Omega . QP = \omega . NP, \text{ or } \omega : \Omega :: QP : NP.$$

Let α be the semivertical angle of the fixed, β of the rolling, cone. Each of these may be supposed for simplicity to be acute, and their sum or difference less than a right angle—though, of course, the formulæ so obtained are (like all trigonometrical results) applicable to every possible case. We have the following three cases:—

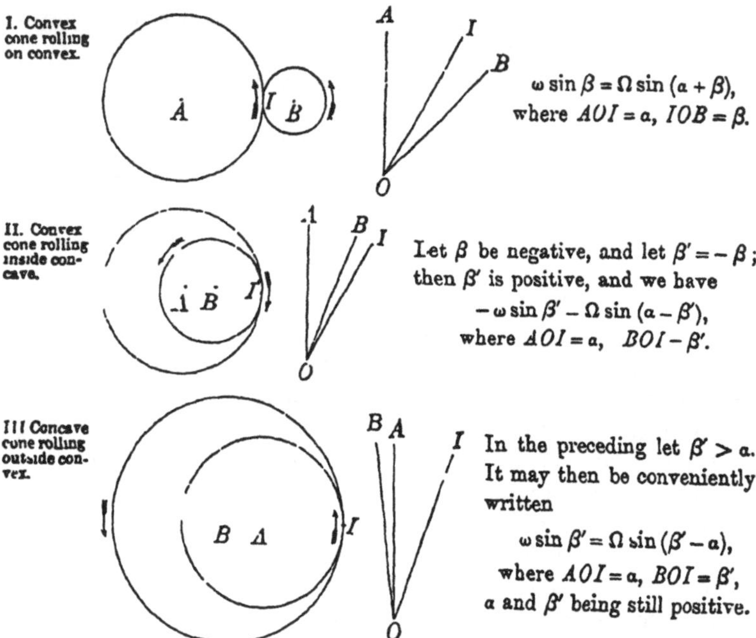

I. Convex cone rolling on convex.
A
I
B
A
I
B
O
$\omega \sin \beta = \Omega \sin (\alpha + \beta)$,
where $AOI = \alpha$, $IOB = \beta$.
II. Convex cone rolling inside concave.
A B I
A
B
I
O
Let β be negative, and let $\beta' = -\beta$;
then β' is positive, and we have
$-\omega \sin \beta' - \Omega \sin (\alpha - \beta')$,
where $AOI = \alpha$, $BOI - \beta'$.
III Concave cone rolling outside convex.
B A I
B A
I
O
In the preceding let $\beta' > \alpha$.
It may then be conveniently written
$\omega \sin \beta' = \Omega \sin (\beta' - \alpha)$,
where $AOI = \alpha$, $BOI = \beta'$,
α and β' being still positive.

Cases of precessional rotation.

106. If, as illustrated by the first of these diagrams, the case is one of a convex cone rolling on a convex cone, the precessional motion, viewed on a hemispherical surface having A for its pole and O for its centre, is in a similar direction to that of the angular rotation about the instantaneous axis. This we shall call *positive* precessional rotation. It is the case of a common spinning-top (peery), spinning on a very fine point which remains at rest in a hollow or hole bored by itself; not sleeping upright, nor nodding, but sweeping its axis round in a circular cone whose axis is vertical. In Case III. also we have *positive* precession. A good example of this occurs in the case of a coin spinning on a table when its plane is nearly horizontal.

Cases of precessional rotation.

107. Case II., that of a convex cone rolling inside a concave one, gives an example of *negative* precession: for when viewed as before on the hemispherical surface the direction of angular rotation of the instantaneous axis is opposite to that of the rolling cone. This is the case of a symmetrical cup (or figure of revolution) supported on a point, and stable when balanced, *i.e.*, having its centre of gravity below the pivot; when inclined and set spinning non-nutationally. For instance, if a Troughton's top be placed on its pivot in any inclined position, and then spun off with very great angular velocity about its axis of figure, the nutation will be insensible; but there will be slow precession.

To this case also belongs the precessional motion of the earth's axis; for which the angle $\alpha = 23^\circ\ 27'\ 28''$, the period of the rotation ω the sidereal day; that of Ω is 25,868 years. If the second diagram represent a portion of the earth's surface round the pole, the arc AI = 8,552,000 feet, and therefore the circumference of the circle in which I moves = 52,240,000 feet. Imagine this circle to be the in-

Model illustrating Precessional Equinoxes.

Precession of the equinoxes.

ner edge of a fixed ring in space (directionally fixed, that is to say, but having the same translational motion as the earth's centre), and imagine a circular post or pivot of radius BI to be fixed to the earth with its centre at B. This ideal pivot rolling on the inner edge of the fixed ring travels once round the 52,240,000 feet-circumference in 25,868 years, and therefore its own circumference must be 5·53 feet. Hence $BI = 0{\cdot}88$ feet; and angle BOI, or β, $= 0''{\cdot}00867$.

Free rotation of a body kinetically symmetrical about an axis.

108. Very interesting examples of Cases I. and III. are furnished by projectiles of different forms rotating about any axis. Thus the gyrations of an oval body or a rod or bar flung into the air belong to Class I. (the body having one axis of less moment of inertia than the other two, equal); and the seemingly irregular evolutions of an ill-thrown quoit belong to Class III. (the quoit having one axis of greater moment of inertia than the other two, which are equal). Case III. has therefore the following very interesting and important application.

If by a geological convulsion (or by the transference of a few million tons of matter from one part of the world to another) the earth's instantaneous axis OI (diagram III., § 105) were at any time brought to non-coincidence with its principal axis of greatest moment of inertia, which (§§ 825, 285) is an axis of approximate kinetic symmetry, the instantaneous axis will, and the fixed axis OA will, relatively to the solid, travel round the solid's axis of greatest moment of inertia in a period of about 306 days [this number being the reciprocal of the most probable value of $\frac{C-A}{C}$ (§ 828)]; and the motion is represented by the diagram of Case III. with $BI = 306 \times AI$. Thus in a very little less than a day (less by $\frac{1}{306}$ when BOI is a small angle) I revolves round A. It is OA, as has been remarked by Maxwell, that is found as the direction of the celestial pole by observations of the meridional zenith distances of stars, and this line being the resultant axis of the earth's moment of momentum (§ 267), would remain invariable in space did no external influence such as that of the moon and sun disturb the earth's rotation. When we neglect precession and nutation, the polar distances of the stars are constant notwithstanding the ideal motion of the fixed axis which we are now considering; and the effect of this motion will be to make a periodic variation of the latitude of every place on the earth's surface having for range on each side of its mean value the angle *BOA*, and for its period 306 days or thereabouts. Maxwell* examined a four years series of Greenwich observations of Polaris (1851-2-3-4), and concluded that there was during those years no variation exceeding half a second of angle on each side of mean range, but that the evidence did not disprove a variation of that amount, but on the contrary gave a very slight indication of a minimum latitude of Greenwich belonging to the set of months Mar. '51, Feb. '52, Dec. '52, Nov. '53, Sept. '54.

Free rotation of a body kinetically symmetrical about an axis.

"This result, however, is to be regarded as very doubtful...... "and more observations would be required to establish the "existence of so small a variation at all.

"I therefore conclude that the earth has been for a long time "revolving about an axis very near to the axis of figure, if not "coinciding with it. The cause of this near coincidence is "either the original softness of the earth, or the present fluidity "of its interior [or the existence of water on its surface]. "The axes of the earth are so nearly equal that a con-"siderable elevation of a tract of country might produce a "deviation of the principal axis within the limits of observa-"tion, and the only cause which would restore the uniform "motion, would be the action of a fluid which would gradually "diminish the oscillations of latitude. The permanence of "latitude essentially depends on the inequality of the earth's "axes, for if they had all been equal, any alteration in the "crust of the earth would have produced new principal axes, "and the axis of rotation would travel about those axes, alter-

* On a Dynamical Top, *Trans. R. S. E.*, 1857, p. 559.

Free rotation of a body kinetically symmetrical about an axis.

"ing the latitudes of all places, and yet not in the least altering "the position of the axis of rotation among the stars."

Perhaps by a more extensive "search and analysis of the "observations of different observatories, the nature of the "periodic variation of latitude, if it exist, may be determined. "I am not aware* of any calculations having been made to prove "its non-existence, although, on dynamical grounds, we have "every reason to look for some very small variation having the "periodic time of 325·6 days nearly" [more nearly 306 days], "a period which is clearly distinguished from any other astro-"nomical cycle, and therefore easily recognised†."

The periodic variation of the earth's instantaneous axis thus anticipated by Maxwell must, if it exists, give rise to a tide of 306 days period (§ 801). The amount of this tide at the equator would be a rise and fall amounting only to $5\frac{1}{2}$ centimetres above and below mean for a deviation of the instantaneous axis amounting to 1″ from its mean position OB, or for a deviation BI on the earth's surface amounting to 31 metres. This, although discoverable by elaborate analysis of long-continued and accurate tidal observations, would be less easily discovered than the periodic change of latitude by astronomical observations according to Maxwell's method‡.

* (Written twenty years ago).

† Maxwell; *Transactions of the Royal Society of Edinburgh*, 20th April, 1857.

‡ Prof. Maxwell now refers us to Peters (*Recherches sur la parallaxe des étoiles fixes*, St Petersburgh Observatory Papers, Vol. I., 1853), who seems to have been the first to raise this interesting and important question. He found from the Pulkova observations of Polaris from March 11, 1842 till April 30, 1843 an angular radius of 0″·079 (probable error 0″·017), for the circle round its mean position described by the instantaneous axis, and for the time, within that interval, when the latitude of Pulkova was a maximum, Nov. 16, 1842. The period (calculated from the dynamical theory) which Peters assumed was 304 mean solar days: the rate therefore 1·201 turns per annum, or, nearly enough, 12 turns per ten years. Thus if Peters' result were genuine, and remained constant for ten years, the latitude of Pulkova would be a maximum about the 16th of Nov. again in 1852, and Pulkova being in 30° East longitude from Greenwich, the latitude of Greenwich would be a maximum $\frac{1}{12}$ of the period, or about 25 days earlier, that is to say about Oct. 22, 1852. But Maxwell's examination of observations seemed to indicate more nearly the minimum latitude of Greenwich about the same time. This discrepance is altogether in accordance with a continuation of Peters' investigation by Dr Nyrén of the Pulkova Ob-

109. In various illustrations and arrangements of apparatus useful in Natural Philosophy, as well as in Mechanics, it is required to connect two bodies, so that when either turns about a certain axis, the other shall turn with an equal angular velocity about another axis in the same plane with the former, but inclined to it at any angle. This is accomplished in mechanism by means of equal and similar bevelled wheels, or rolling cones; when the mutual inclination of two axes is not to be varied. It is approximately accomplished by means of Hooke's joint, when the two axes are nearly in the same line, but are required to be free to vary in their mutual inclination. A chain of an infinitely great number of Hooke's joints may be imagined as constituting a perfectly flexible, untwistable cord, which, if its end-links are rigidly attached to the two bodies, connects them so as to fulfil the condition rigorously without the restriction that the two axes remain in one plane. If we imagine an infinitely short length of such a chain (still, however, having an infinitely great number of links) to have its ends attached to two bodies, it will fulfil rigorously the condition stated, and at the same time keep a definite point of one body infinitely near a definite point of the other; that is to say, it will accomplish precisely for every angle of inclination what Hooke's joint does approximately for small inclinations.

Communication of angular velocity equally between inclined axes.

Hooke's joint.

Flexible but untwistable cord.

Universal flexure joint.

The same is dynamically accomplished with perfect accuracy for every angle, by a short, naturally straight, elastic wire of

Elastic universal flexure joint.

servatory, in which, by a careful scrutiny of several series of Pulkova observations between the years 1842...1872, he concluded that there is no constancy of magnitude or phase in the deviation sought for. A similar negative conclusion was arrived at by Professor Newcomb of the United States Naval Observatory, Washington, who at our request kindly undertook an investigation of the ten-month period of latitude from the Washington Prime Vertical Observations from 1862 to 1867. His results, as did those of Peters and Nyren and Maxwell, seemed to indicate real variations of the earth's instantaneous axis amounting to possibly as much as ½″ or ¼″ from its mean position, but altogether irregular both in amount and direction; in fact, just such as might be expected from irregular heapings up of the oceans by winds in different localities of the earth.

We intend to return to this subject and to consider cognate questions regarding irregularities of the earth as a timekeeper, and variations of its figure and of the distribution of matter within it, of the ocean on its surface, and of the atmosphere surrounding it, in §§ 267, 276, 405, 406, 830, 832, 845, 846.

Elastic universal flexure joint.

truly circular section, provided the forces giving rise to any resistance to equality of angular velocity between the two bodies are infinitely small. In many practical cases this mode of connexion is useful, and permits very little deviation from the conditions of a true universal flexure joint. It is used, for instance, in the suspension of the gyroscopic pendulum (§ 74) with perfect success. The dentist's tooth-mill is an interesting illustration of the elastic universal flexure joint. In it a long spiral spring of steel wire takes the place of the naturally straight wire suggested above.

Moving body attached to a fixed object by a universal flexure joint.

Of two bodies connected by a universal flexure joint, let one be held fixed. The motion of the other, as long as the angle of inclination of the axes remains constant, will be exactly that figured in Case I., § 105, above, with the angles α and β made equal. Let O be the joint; AO the axis of the fixed body; OB the axis of the moveable body. The supplement of the angle AOB is the mutual inclination of the axes; and the angle AOB itself is bisected by the instantaneous axis of the moving body. The diagram shows a case of this motion, in which the mutual inclination, θ, of the axes is acute. According to the formulæ of Case I., § 105, we have

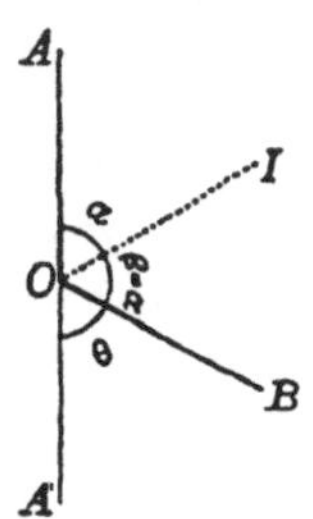

$$\omega \sin \alpha = \Omega \sin 2\alpha,$$

or

$$\omega = 2\Omega \cos \alpha = 2\Omega \sin \frac{\theta}{2},$$

where ω is the angular velocity of the moving body about its instantaneous axis, OI, and Ω is the angular velocity of its precession; that is to say, the angular velocity of the plane through the fixed axis AA', and the moving axis OB of the moving body.

Two degrees of freedom to move enjoyed by a body thus suspended.

Besides this motion, the moving body may clearly have any angular velocity whatever about an axis through O perpendicular to the plane AOB, which, compounded with ω round OI, gives the resultant angular velocity and instantaneous axis.

Two co-ordinates, $\theta = A'OB$, and ϕ measured in a plane perpendicular to AO, from a fixed plane of reference to the plane

AOB, fully specify the position of the moveable body in this case.

110. Suppose a rigid body bounded by any curved surface to be touched at any point by another such body. Any motion of one on the other must be of one or more of the forms *sliding*, *rolling*, or *spinning*. The consideration of the first is so simple as to require no comment.

General motion of one rigid body touching another.

Any motion in which there is no slipping at the point of contact must be rolling or spinning separately, or combined.

Let one of the bodies rotate about successive instantaneous axes, all lying in the common tangent plane at the point of instantaneous contact, and each passing through this point—the other body being fixed. This motion is what we call rolling, or simple rolling, of the moveable body on the fixed.

On the other hand, let the instantaneous axis of the moving body be the common normal at the point of contact. This is pure spinning, and does not change the point of contact.

Let the moving body move, so that its instantaneous axis, still passing through the point of contact, is neither in, nor perpendicular to, the tangent plane. This motion is combined rolling and spinning.

111. When a body rolls and spins on another body, the *trace of either on the other* is the curved or straight line along which it is successively touched. If the instantaneous axis is in the normal plane perpendicular to the traces, the rolling is called *direct* If not direct, the rolling may be resolved into a direct rolling, and a rotation or twisting round the tangent line to the traces.

Traces of rolling.

Direct rolling.

When there is *no spinning* the projections of the two traces on the common tangent plane at the point of contact of the two surfaces have equal and same-way directed curvature: or they have "contact of the second order." When there *is spinning*, the two projections still touch one another, but with contact of the first order only: their curvatures differ by a quantity equal to the angular velocity of spinning divided by the velocity of the point of contact. This last we see by noticing that the rate of change of direction along the pro-

Direct rolling.

jection of the fixed trace must be equal to the rate of change of direction along the projection of the moving trace if held fixed plus the angular velocity of the spinning.

At any instant let $2z = Ax^2 + 2Cxy + By^2$ ....(1)

and $2z' = A'x^2 + 2C'xy + B'y^2$ ....(2)

be the equations of the fixed and moveable surfaces S and S' infinitely near the point of contact O, referred to axes OX, OY in their common tangent plane, and OZ perpendicular to it: let ϖ, ρ, σ be the three components of the instantaneous angular velocity of S'; and let x, y, be co-ordinates of P, the point of contact at an infinitely small time t, later: the third co-ordinate, z, is given by (1).

Let P' be the point of S' which at this later time coincides with P. The co-ordinates of P' at the first instant are $x + \sigma yt$, $y - \sigma xt$; and the corresponding value of z' is given by (2). This point is infinitely near to (x, y, z'), and therefore at the first instant the direction cosines of the normal to S' through it differ but infinitely little from

$$-(A'x + C'y),\quad -(C'x + B'y),\quad 1.$$

But at time t the normal to S' at P' coincides with the normal to S at P, and therefore its direction cosines change from the preceding values, to

$$-(Ax + Cy),\quad -(Cx + By),\quad 1:$$

that is to say, it rotates through angles

$$(C' - C)\,x + (B' - B)\,y \text{ round } OX,$$

and $$-\{(A' - A)x + (C' - C)\,y\} \quad ,, \quad OY.$$

Hence
$$\left.\begin{aligned} \varpi t &= (C' - C)\,x + (B' - B)\,y \\ \rho t &= -\{(A' - A)\,x + (C' - C)\,y\} \end{aligned}\right\} \text{.............. (3)},$$

or
$$\left.\begin{aligned} \varpi &= (C' - C)\,\dot{x} + (B' - B)\,\dot{y} \\ \rho &= -\{(A' - A)\,\dot{x} + (C' - C)\,\dot{y}\} \end{aligned}\right\} \text{.................... (4)},$$

if $\dot{x}$, $\dot{y}$ denote the component velocities of the point of contact.

Put $q = \sqrt{(\dot{x}^2 + \dot{y}^2)}$ ....(5),

and take components of ϖ and ρ round the tangent to the traces and the perpendicular to it in the common tangent plane of the two surfaces, thus:

(twisting component)...... $\frac{\dot{x}}{q}\varpi + \frac{\dot{y}}{q}\rho$

$$= (C' - C)\frac{\dot{x}^2 - \dot{y}^2}{q} + [(B' - B) - (A' - A)]\frac{\dot{x}\dot{y}}{q} \text{ ......(6)},$$

and

Direct rolling.

$$\text{(direct-rolling component)}\ldots\ldots\frac{\dot{y}}{q}\varpi-\frac{\dot{x}}{q}\rho$$

$$=\frac{1}{q}\left[(A'-A)\dot{x}^2+2(C'-C)\dot{x}\dot{y}+(B'-B)\dot{y}^2\right]\ldots\ldots(7).$$

Choose OX, OY so that $C-C'=0$, and put $A'-A=\alpha$, $B'-B=\beta$ (6) and (7) become

$$\text{(twisting component)}\ \ldots\ldots\ldots\ldots\frac{\dot{x}}{q}\varpi+\frac{\dot{y}}{q}\rho=(\beta-\alpha)\frac{\dot{x}\dot{y}}{q}\ldots\ldots\ldots(8),$$

$$\text{(direct-rolling component)}\ldots\ldots\frac{\dot{y}}{q}\varpi-\frac{\dot{x}}{q}\rho=\frac{1}{q}(\alpha\dot{x}^2+\beta\dot{y}^2)\ldots\ldots(9).$$

[Compare below, § 124 (2) and (1).]

And for σ, the angular velocity of spinning, the obvious proposition stated in the preceding large print gives

$$\sigma=q\left(\frac{1}{\gamma}-\frac{1}{\gamma'}\right)\ldots\ldots\ldots\ldots\ldots\ldots\ldots\ldots(10),$$

if $\frac{1}{\gamma}$ and $\frac{1}{\gamma'}$ be the curvatures of the projections on the tangent plane of the fixed and moveable traces. [Compare below, § 124 (3).]

From (1) and (2) it follows that

When one of the surfaces is a plane, and the trace on the other is a line of curvature (§ 130), the rolling is direct.

When the trace on each body is a line of curvature, the rolling is direct. *Generally*, the rolling is direct when the twists of infinitely narrow bands (§ 120) of the two surfaces, along the traces, are equal and in the same direction.

112. Imagine the traces constructed of rigid matter, and all the rest of each body removed. We may repeat the motion with these curves alone. The difference of the circumstances now supposed will only be experienced if we vary the direction of the instantaneous axis. In the former case, we can only do this by introducing more or less of spinning, and if we do so we *alter the trace* on each body. In the latter, we have always the same moveable curve rolling on the same fixed curve; and therefore a determinate line perpendicular to their common tangent for one component of the rotation; but along with this we may give arbitrarily any velocity of twisting round the common tangent. The consideration of this case is very in-

Curve rolling on curve.

structive. It may be roughly imitated in practice by two stiff wires bent into the forms of the given curves, and prevented from crossing each other by a short piece of elastic tube clasping them together.

First, let them be both plane curves, and kept in one plane. We have then *rolling*, as of one cylinder on another.

Let ρ' be the radius of curvature of the rolling, ρ of the fixed, cylinder; ω the angular velocity of the former, V the linear velocity of the point of contact. We have

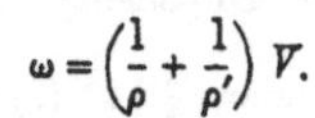

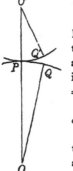

$$\omega = \left(\frac{1}{\rho} + \frac{1}{\rho'}\right) V.$$

For, in the figure, suppose P to be at any time the point of contact, and Q and Q' the points which are to be in contact after an infinitely small interval t; O, O' the centres of curvature; $POQ = \theta$, $PO'Q' = \theta'$.

Then $PQ = PQ' =$ space described by point of contact. In symbols $\rho\theta = \rho'\theta' = Vt$.

Also, before $O'Q'$ and OQ can coincide in direction, the former must evidently turn through an angle $\theta + \theta'$.

Therefore $\omega t = \theta + \theta'$; and by eliminating θ and θ', and dividing by t, we get the above result.

It is to be understood, that as the radii of curvature have been considered positive here when both surfaces are convex, the negative sign must be introduced for either radius when the corresponding curve is concave.

Angular velocity of rolling in a plane.

Hence the angular velocity of the rolling curve is in this case equal to the product of the linear velocity of the point of contact by the sum or difference of the curvatures, according as the curves are both convex, or one concave and the other convex.

Plane curves not in same plane.

113. When the curves are both plane, but in different planes, the plane in which the rolling takes place divides the angle between the plane of one of the curves, and that of the other produced through the common tangent line, into parts whose sines are inversely as the curvatures in them respectively; and the angular velocity is equal to the linear velocity

of the point of contact multiplied by the difference of the projections of the two curvatures on this plane. The projections of the circles of the two curvatures on the plane of the common tangent and of the instantaneous axis coincide. Plane curves not in same plane.

For, let PQ, Pp be equal arcs of the two curves as before, and let PR be taken in the common tangent (*i.e.*, the intersection of the planes of the curves) equal to each. Then QR, pR are ultimately perpendicular to PR.

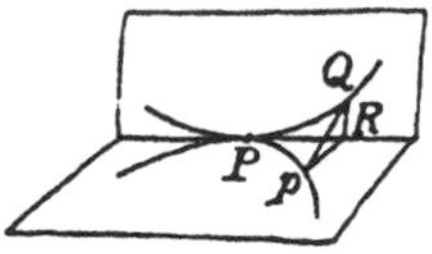

Hence
$$pR = \frac{PR^2}{2\sigma},$$
$$QR = \frac{PR^2}{2\rho}.$$

Also, $\angle QRp = a$, the angle between the planes of the curves.

We have
$$Qp^2 = \frac{PR^4}{4}\left(\frac{1}{\sigma^2} + \frac{1}{\rho^2} - \frac{2}{\sigma\rho}\cos a\right).$$

Therefore if ω be the velocity of rotation as before,
$$\omega = V\sqrt{\frac{1}{\sigma^2} + \frac{1}{\rho^2} - \frac{2\cos a}{\sigma\rho}}.$$

Also the instantaneous axis is evidently perpendicular, and therefore the plane of rotation parallel, to Qp. Whence the above. In the case of $a = \pi$, this agrees with the result of § 112.

A good example of this is the case of a coin spinning on a table (mixed rolling and spinning motion), as its plane becomes gradually horizontal. In this case the curvatures become more and more nearly equal, and the angle between the planes of the curves smaller and smaller. Thus the resultant angular velocity becomes exceedingly small, and the motion of the point of contact very great compared with it.

114. The preceding results are, of course, applicable to tortuous as well as to plane curves; it is merely requisite to substitute the osculating plane of the former for the plane of the latter. Curve rolling on curve: two degrees of freedom.

115. We come next to the case of a curve rolling, with or without spinning, on a surface. Curve rolling on surface: three degrees of freedom.

It may, of course, roll on any curve traced on the surface. When this curve is given, the moving curve may, while rolling along it, revolve arbitrarily round the tangent. But the com-

Curve rolling on surface: three degrees of freedom.

ponent instantaneous axis perpendicular to the common tangent, that is, the axis of the direct rolling of one curve on the other, is determinate, § 113. If this axis does not lie in the surface, there is spinning. Hence, when the trace on the surface is given, there are two independent variables in the motion; the space traversed by the point of contact, and the inclination of the moving curve's osculating plane to the tangent plane of the fixed surface.

Trace prescribed and no spinning permitted.

116. If the trace is given, and it be prescribed as a condition that there shall be no spinning, the angular position of the rolling curve round the tangent at the point of contact is determinate. For in this case the instantaneous axis must be in the tangent plane to the surface. Hence, if we resolve the rotation into components round the tangent line, and round an axis perpendicular to it, the latter must be in the tangent plane. Thus the rolling, as of curve on curve, must be in a normal plane to the surface; and therefore (§§ 114, 113) the rolling curve must be always so situated relatively to its trace on the surface that the projections of the two curves on the tangent plane may be of coincident curvature.

Two degrees of freedom.

The curve, as it rolls on, must continually revolve about the tangent line to it at the point of contact with the surface, so as in every position to fulfil this condition.

Let α denote the inclination of the plane of curvature of the trace, to the normal to the surface at any point, α' the same for the plane of the rolling curve; $\frac{1}{\rho}$, $\frac{1}{\rho'}$ their curvatures. We reckon α as obtuse, and α' acute, when the two curves lie on opposite sides of the tangent plane. Then

$$\frac{1}{\rho'}\sin\alpha' = \frac{1}{\rho}\sin\alpha,$$

which fixes α' or the position of the rolling curve when the point of contact is given.

Angular velocity of direct rolling.

Let ω be the angular velocity of rolling about an axis perpendicular to the tangent, ϖ that of twisting about the tangent, and let V be the linear velocity of the point of contact. Then, since $\frac{1}{\rho'}\cos\alpha'$

and $-\frac{1}{\rho}\cos\alpha$ (each positive when the curves lie on opposite sides of the tangent plane) are the projections of the two curvatures on a plane through the normal to the surface containing their common tangent, we have, by § 112,

Angular velocity of direct rolling.

$$\omega = V\left(\frac{1}{\rho'}\cos\alpha' - \frac{1}{\rho}\cos\alpha\right),$$

α' being determined by the preceding equation. Let τ and τ' denote the tortuosities of the trace, and of the rolling curve, respectively. Then, first, if the curves were both plane, we see that one rolling on the other about an axis always perpendicular to their common tangent could never change the inclination of their planes. Hence, secondly, if they are both tortuous, such rolling will alter the inclination of their osculating planes by an indefinitely small amount $(\tau-\tau')\,ds$ during rolling which shifts the point of contact over an arc ds. Now α is a known function of s if the trace is given, and therefore so also is α'. But $\alpha-\alpha'$ is the inclination of the osculating planes, hence

Angular velocity round tangent.

$$V\left\{\frac{d(\alpha-\alpha')}{ds} - (\tau-\tau')\right\} = \varpi.$$

117. Next, for one surface rolling and spinning on another. First, if the trace on each is given, we have the case of § 113 or § 115, one curve rolling on another, with this farther condition, that the former must *revolve* round the tangent to the two curves so as to keep the tangent planes of the two surfaces coincident.

Surface on surface.

It is well to observe that when the points in contact, and the two traces, are given, the position of the moveable surface is quite determinate, being found thus:—Place it in contact with the fixed surface, the given points together, and *spin* it about the common normal till the tangent lines to the traces coincide.

Both traces prescribed: one degree of freedom.

Hence when both the traces are given the condition of no spinning cannot be imposed. During the rolling there must in general be spinning, such as to keep the tangents to the two traces coincident. The rolling along the trace is due to rotation round the line perpendicular to it in the tangent plane. The whole rolling is the resultant of this rotation and a rotation about the tangent line required to keep the two tangent planes coincident.

Surface on surface, both traces prescribed; one degree of freedom.

In this case, then, there is but one independent variable—the space passed over by the point of contact: and when the velocity of the point of contact is given, the resultant angular velocity, and the direction of the instantaneous axis of the rolling body are determinate. We have thus a sufficiently clear view of the general character of the motion in question, but it is right that we consider it more closely, as it introduces us very naturally to an important question, the measurement of the *twist* of a rod, wire, or narrow plate, a quantity wholly distinct from the *tortuosity* of its axis (§ 7).

118. Suppose all of each surface cut away except an infinitely narrow strip, including the trace of the rolling. Then we have the rolling of one of these strips upon the other, each having at every point a definite curvature, tortuosity, and twist.

Twist.

119. Suppose a flat bar of small section to have been bent (the requisite amount of stretching and contraction of its edges being admissible) so that its axis assumes the form of any plane or tortuous curve. If it be unbent without twisting, *i.e.*, if the curvature of each element of the bar be removed by bending it through the requisite angle in the osculating plane, and it be found untwisted when thus rendered straight, it had no *twist* in its original form. This case is, of course, included in the general theory of *twist*, which is the subject of the following sections.

Axis and transverse.

120. A bent or straight rod of circular or any other form of section being given, a line through the centres, or any other chosen points of its sections, may be called its *axis*. Mark a line on its side all along its length, such that it shall be a straight line parallel to the axis when the rod is unbent and untwisted. A line drawn from any point of the axis perpendicular to this side line of reference, is called the *transverse* of the rod at this point.

The whole twist of any length of a straight rod is the angle between the transverses of its ends. The average twist is the integral twist divided by the length. The twist at any point is the average twist in an infinitely short length through this point; in other words, it is the rate of rotation of its transverse per unit of length along it.

The twist of a curved, plane or tortuous, rod at any point is the rate of component rotation of its transverse round its tangent line, per unit of length along it. Twist.

If t be the twist at any point, $\int t ds$ over any length is the integral twist in this length.

121. Integral twist in a curved rod, although readily defined, as above, in the language of the integral calculus, cannot be exhibited as the angle between any two lines readily constructible. The following considerations show how it is to be reckoned, and lead to a geometrical construction exhibiting it in a spherical diagram, for a rod bent and twisted in any manner:—

122. If the axis of the rod forms a plane curve lying in one plane, the integral twist is clearly the difference between the inclinations of the transverse at its ends to its plane. For if it be simply unbent, without altering the twist in any part, the inclination of each transverse to the plane in which its curvature lay will remain unchanged; and as the axis of the rod now has become a straight line in this plane, the mutual inclination of the transverses at any two points of it has become equal to the difference of their inclinations to the plane. Estimation of integral twist: In a plane curve;

123. No simple application of this rule can be made to a tortuous curve, in consequence of the change of the plane of curvature from point to point along it; but, instead, we may proceed thus:—

First, Let us suppose the plane of curvature of the axis of the wire to remain constant through finite portions of the curve, and to change abruptly by finite angles from one such portion to the next (a supposition which involves no angular points, that is to say, no infinite curvature, in the curve). Let planes parallel to the planes of curvature of three successive portions, PQ, QR, RS (not shown in the diagram), cut a spherical surface in the great circles GAG', ACA', CE. The radii of the sphere parallel to the tangents at the points Q and R of the curve where its curvature changes will cut its surface in A and C, the intersections of these circles. In a curve consisting of plane portions in different planes.

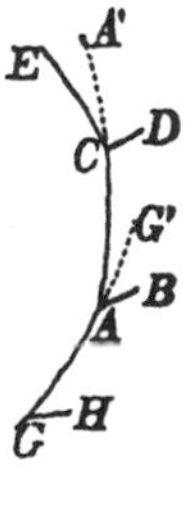

Let G be the point in which the radius of the sphere parallel to the tangent at P cuts the surface; and let GH, AB, CD (lines necessarily in tangent planes to the spherical surface), be parallels to the transverses of the bar drawn from the points P, Q, R of its axis. Then (§ 122) the twist from P to Q is equal to the difference of the angles HGA and BAG'; and the twist from Q to R is equal to the difference between BAC and DCA'. Hence the whole twist from P to R is equal to

Estimation of integral twist: in a curve consisting of plane portions in different planes.

$$HGA - BAG' + BAC - DCA',$$

or, which is the same thing,

$$A'CE + G'AC - (DCE - HGA).$$

Continuing thus through any length of rod, made up of portions curved in different planes, we infer that the integral twist between any two points of it is equal to the sum of the exterior angles in the spherical diagram, wanting the excess of the inclination of the transverse at the second point to the plane of curvature at the second point above the inclination at the first point to the plane of curvature at the first point. The sum of those exterior angles is what is defined below as the "change of direction in the spherical surface" from the first to the last side of the polygon of great circles. When the polygon is closed, and the sum includes all its exterior angles, it is (§ 134) equal to 2π wanting the area enclosed if the radius of the spherical surface be unity. The construction we have made obviously holds in the limiting case, when the lengths of the plane portions are infinitely small, and is therefore applicable to a wire forming a tortuous curve with continuously varying plane of curvature, for which it gives the following conclusion:—

In a continuously tortuous curve.

Let a point move uniformly along the axis of the bar: and, parallel to the tangent at every instant, draw a radius of a sphere cutting the spherical surface in a curve, the hodograph of the moving point. From points of this hodograph draw parallels to the transverses of the corresponding points of the bar. The excess of the change of direction (§ 135) from any point to another of the hodograph, above the increase of its inclination to the transverse, is equal to the twist in the corresponding part of the bar.

The annexed diagram, showing the hodograph and the parallels to the transverses, illustrates this rule. Thus, for instance, the excess of the change of direction in the spherical surface along the hodograph from A to C, above $DCS—BAT$, is equal to the twist in the bar between the points of it to which A and C correspond. Or, again, if we consider a portion of the bar from any point of it, to another point at which the tangent to its axis is parallel to the tangent at its first point, we shall have a closed curve as the spherical hodograph; and if A be the point of the hodograph corresponding to them, and AB and AB' the parallels to the transverses, the whole twist in the included part of the bar will be equal to the change of direction all round the hodograph, wanting the excess of the exterior angle $B'AT$ above the angle BAT; that is to say, the whole twist will be equal to the excess of the angle BAB' above the area enclosed by the hodograph.

Estimation of integral twist: in a continuously tortuous curve.

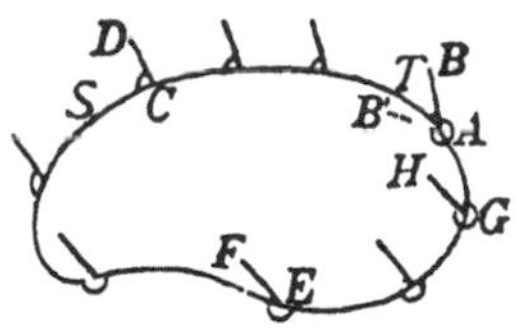

The principles of twist thus developed are of vital importance in the theory of rope-making, especially the construction and the dynamics of wire ropes and submarine cables, elastic bars, and spiral springs.

For example: take a piece of steel pianoforte-wire carefully straightened, so that when free from stress it is straight: bend it into a circle and join the ends securely so that there can be no turning of one relatively to the other. Do this first without torsion: then twist the ring into a figure of 8, and tie the two parts together at the crossing. The area of the spherical hodograph is zero, and therefore there is one full turn (2π) of twist; which (§ 600 below) is uniformly distributed throughout the length of the wire. The form of the wire, (which is not in a plane,) will be investigated in § 610. Meantime we can see that the "torsional couples" in the normal sections farthest from the crossing give rise to forces by which the tie at the crossing is pulled in opposite directions perpendicular to the plane of the crossing. Thus if the tie is cut the wire springs back into the circular form. Now do the same thing again,

Dynamics of twist in kinks.

Dynamics of twist in kinks.

beginning with a straight wire, but giving it one full turn (2π) of twist before bending it into the circle. The wire will stay in the S form without any pull on the tie. Whether the circular or the S form is stable or unstable depends on the relations between torsional and flexural rigidity. If the torsional rigidity is small in comparison with the flexural rigidity [as (§§ 703, 704, 705, 709) would be the case if, instead of round wire, a rod of + shaped section were used], the circular form would be stable, the 8 unstable.

Lastly, suppose any degree of twist, either more or less than 2π, to be given before bending into the circle. The circular form, which is always a figure of free equilibrium, may be stable or unstable, according as the ratio of torsional to flexural rigidity is more or less than a certain value depending on the actual degree of twist. The tortuous 8 form is not (except in the case of whole twist $= 2\pi$, when it becomes the plane elastic lemniscate of Fig. 4, § 610,) a continuous figure of free equilibrium, but involves a positive pressure of the two crossing parts on one another when the twist $> 2\pi$, and a negative pressure (or a pull on the tie) between them when twist $< 2\pi$: and with this force it is a figure of stable equilibrium.

Surface rolling on surface; both traces given.

124. Returning to the motion of one surface rolling and spinning on another, the trace on each being given, we may consider that, of each, the curvature (§ 6), the tortuosity (§ 7), and the twist reckoned according to transverses in the tangent plane of the surface, are known; and the subject is fully specified in § 117 above.

Let $\frac{1}{\rho'}$ and $\frac{1}{\rho}$ be the curvatures of the traces on the rolling and fixed surfaces respectively; α' and α the inclinations of their planes of curvature to the normal to the tangent plane, reckoned as in § 116; τ' and τ their tortuosities; t' and t their twists; and q the velocity of the point of contact. All these being known, it is required to find:—

ω the angular velocity of rotation about the transverse of the traces; that is to say, the line in the tangent plane perpendicular to their tangent line,

ϖ the angular velocity of rotation about the tangent line, and

σ ,, ,, of spinning.

We have

Surface rolling on surface, both traces given.

$$\omega = q\left(\frac{1}{\rho'}\cos\alpha' - \frac{1}{\rho}\cos\alpha\right) \qquad (1),$$

$$\varpi = q(t - t') = q\left\{\frac{d(\alpha - \alpha')}{ds} - (\tau - \tau')\right\} \qquad (2),$$

and
$$\sigma = q\left(\frac{1}{\rho'}\sin\alpha' - \frac{1}{\rho}\sin\alpha\right) \qquad (3).$$

These three formulas are respectively equivalent to (9), (8), and (10) of § 111.

125. In the same case, suppose the trace on *one* only of the surfaces to be given. We may evidently impose the condition of no spinning, and then the trace on the other is determinate. This case of motion is thoroughly examined in § 137, below.

Surface rolling on surface without spinning.

The condition is that the projections of the curvatures of the two traces on the common tangent plane must coincide.

If $\frac{1}{r'}$ and $\frac{1}{r}$ be the curvatures of the rolling and stationary surfaces in a normal section of each through the tangent line to the trace, and if α, α', ρ, ρ' have their meanings of § 124,

$\rho' = r'\cos\alpha'$, $\rho = r\cos\alpha$ (Meunier's Theorem, § 129, below).

But $\frac{1}{\rho'}\sin\alpha' = \frac{1}{\rho}\sin\alpha$, hence $\tan\alpha' = \frac{r'}{r}\tan\alpha$, the condition required.

126. If a straight rod with a straight line marked on one side of it be bent along any curve on a spherical surface, so that the marked line is laid in contact with the spherical surface, it acquires no twist in the operation. For if it is laid so along any finite arc of a small circle there will clearly be no twist. And no twist is produced in continuing from any point along another small circle having a common tangent with the first at this point.

Examples of tortuosity and twist.

If a rod be bent round a cylinder so that a line marked along one side of it may lie in contact with the cylinder, or if, what presents somewhat more readily the view now de-

Examples of tortuosity and twist.

sired, we wind a straight ribbon spirally on a cylinder, the axis of bending is parallel to that of the cylinder, and therefore oblique to the axis of the rod or ribbon. We may therefore resolve the instantaneous rotation which constitutes the bending at any instant into two components, one round a line perpendicular to the axis of the rod, which is pure bending, and the other round the axis of the rod, which is pure twist.

The twist at any point in a rod or ribbon, so wound on a circular cylinder, and constituting a uniform helix, is

$$\frac{\cos a \sin a}{r},$$

if r be the radius of the cylinder and a the inclination of the spiral. For if V be the velocity at which the bend proceeds along the previously straight wire or ribbon, $\frac{V\cos a}{r}$ will be the angular velocity of the instantaneous rotation round the line of bending (parallel to the axis), and therefore

$$\frac{V\cos a}{r}\sin a \text{ and } \frac{V\cos a}{r}\cos a$$

are the angular velocities of twisting and of pure bending respectively.

From the latter component we may infer that the curvature of the helix is

$$\frac{\cos^2 a}{r},$$

a known result, which agrees with the expression used above (§ 13).

127. The hodograph in this case is a small circle of the sphere. If the specified condition as to the mode of laying on of the rod on the cylinder is fulfilled, the transverses of the spiral rod will be parallel at points along it separated by one or more whole turns. Hence the integral twist in a single turn is equal to the excess of four right angles above the spherical area enclosed by the hodograph. If a be the inclination of the spiral, $\frac{1}{2}\pi - a$ will be the arc-radius of the hodograph, and therefore its area is $2\pi(1-\sin a)$. Hence the integral twist in a turn of the spiral is $2\pi \sin a$, which agrees with the result previously obtained (§ 126).

128. As a preliminary to the further consideration of the rolling of one surface on another, and as useful in various parts of our subject, we may now take up a few points connected with the curvature of surfaces. Curvature of surface.

The tangent plane at any point of a surface may or may not cut it at that point. In the former case, the surface bends away from the tangent plane partly towards one side of it, and partly towards the other, and has thus, in some of its normal sections, curvatures oppositely directed to those in others. In the latter case, the surface on every side of the point bends away from the same side of its tangent plane, and the curvatures of all normal sections are similarly directed. Thus we may divide curved surfaces into *Anticlastic* and *Synclastic*. A saddle gives a good example of the former class; a ball of the latter. Curvatures in opposite directions, with reference to the tangent plane, have of course different signs. The outer portion of an anchor-ring is synclastic, the inner anticlastic. Synclastic and anticlastic surfaces.

129. *Meunier's Theorem.*—The curvature of an oblique section of a surface is equal to that of the normal section through the same tangent line multiplied by the secant of the inclination of the planes of the sections. This is evident from the most elementary considerations regarding projections. Curvature of oblique sections.

130. *Euler's Theorem.*—There are at every point of a synclastic surface two normal sections, in one of which the curvature is a maximum, in the other a minimum; and these are at right angles to each other. Principal curvatures.

In an anticlastic surface there is maximum curvature (but in opposite directions) in the two normal sections whose planes bisect the angles between the lines in which the surface cuts its tangent plane. On account of the difference of sign, these may be considered as a maximum and a minimum.

Generally the sum of the curvatures at a point, in any two normal planes at right angles to each other, is independent of the position of these planes. Sum of curvatures in normal sections at right angles to each other.

Taking the tangent plane as that of x, y, and the origin at the point of contact, and putting

$$\left(\frac{d^2z}{dx^2}\right)_0 = A,\ \left(\frac{d^2z}{dxdy}\right)_0 = B,\ \left(\frac{d^2z}{dy^2}\right)_0 = C;$$

we have $$z = \frac{1}{2}(Ax^2 + 2Bxy + Cy^2) + \text{etc.} \qquad (1)$$

The curvature of the normal section which passes through the point x, y, z is (in the limit)

$$\frac{1}{r} = \frac{2z}{x^2+y^2} = \frac{Ax^2 + 2Bxy + Cy^2}{x^2+y^2}.$$

If the section be inclined at an angle θ to the plane of XZ, this becomes

$$\frac{1}{r} = A\cos^2\theta + 2B\sin\theta\cos\theta + C\sin^2\theta. \qquad (2)$$

Hence, if $\frac{1}{r}$ and $\frac{1}{s}$ be curvatures in normal sections at right angles to each other,

$$\frac{1}{r} + \frac{1}{s} = A + C = \text{constant}.$$

(2) may be written

$$\frac{1}{r} = \frac{1}{2}\{A(1+\cos 2\theta) + 2B\sin 2\theta + C(1-\cos 2\theta)\}$$

$$= \frac{1}{2}\{\overline{A+C} + \overline{A-C}\cos 2\theta + 2B\sin 2\theta\},$$

or if $$\frac{1}{2}(A - C) = R\cos 2\alpha,\ B = R\sin 2\alpha,$$

that is $$R = \sqrt{\left\{\frac{1}{4}(A-C)^2 + B^2\right\}},\ \text{and}\ \tan 2\alpha = \frac{2B}{A-C},$$

we have $$\frac{1}{r} = \frac{1}{2}(A+C) + \sqrt{\left\{\frac{1}{4}(A-C)^2 + B^2\right\}}\cos 2(\theta - \alpha)$$

Principal normal sections. The maximum and minimum curvatures are therefore those in normal places at right angles to each other for which $\theta = \alpha$ and $\theta = \alpha + \frac{\pi}{2}$, and are respectively

$$\frac{1}{2}(A+C) \pm \sqrt{\left\{\frac{1}{4}(A-C)^2 + B^2\right\}}.$$

Hence their product is $AC - B^2$.

If this be positive we have a synclastic, if negative an anticlastic, surface. If it be zero we have one curvature only, and the surface is *cylindrical* at the point considered. It is demonstrated

(§ 152, below) that if this condition is fulfilled at every point, the surface is "developable" (§ 139, below). Principal normal sections.

By (1) a plane parallel to the tangent plane and very near it cuts the surface in an ellipse, hyperbola, or two parallel straight lines, in the three cases respectively. This section, whose nature informs us as to whether the curvature be synclastic, anticlastic, or cylindrical, at any point, was called by Dupin the *Indicatrix*.

A line of curvature of a surface is a line which at every point is cotangential with normal section of maximum or minimum curvature. Definition of Line of Curvature.

131. Let P, p be two points of a surface infinitely near to each other, and let r be the radius of curvature of a normal section passing through them. Then the radius of curvature of an oblique section through the same points, inclined to the former at an angle a, is (§ 129) $r \cos a$. Also the length along the normal section, from P to p, is less than that along the oblique section—since a given chord cuts off an arc from a circle, longer the less the radius of that circle. Shortest line between two points on a surface.

If a be the length of the chord Pp, we have

$$\text{Distance } Pp \text{ along normal section} = 2r \sin^{-1}\frac{a}{2r} = a\left(1 + \frac{a^2}{24r^2}\right),$$

$$\text{,,} \quad \text{,,} \quad \text{oblique section} = a\left(1 + \frac{a^2}{24r^2\cos^2 a}\right).$$

132. Hence, if the shortest possible line be drawn from one point of a surface to another, its plane of curvature is everywhere perpendicular to the surface.

Such a curve is called a *Geodetic* line. And it is easy to see that it is the line in which a flexible and inextensible string would touch the surface if stretched between those points, the surface being supposed smooth. Geodetic Lines.

133. If an infinitely narrow ribbon be laid on a surface along a geodetic line, its twist is equal to the tortuosity of its axis at each point. We have seen (§ 125) that when one body rolls on another without spinning, the projections of the traces on the common tangent plane agree in curvature at the point

Shortest line between two points on a surface.

of contact. Hence, if one of the surfaces be a plane, and the trace on the other be a geodetic line, the trace on the plane is a straight line. Conversely, if the trace on the plane be a straight line, that on the surface is a geodetic line.

And, quite generally, if the given trace be a geodetic line, the other trace is also a geodetic line.

Spherical excess.

134. The area of a spherical triangle (on a sphere of unit radius) is known to be equal to the "spherical excess," *i.e.*, the excess of the sum of its angles over two right angles, or the excess of four right angles over the sum of its exterior angles.

Area of spherical polygon.

The area of a spherical polygon whose n sides are portions of great circles—*i.e.*, geodetic lines—is to that of the hemisphere as the excess of four right angles over the sum of its exterior angles is to four right angles. (We may call this the "spherical excess" of the polygon.)

For the area of a spherical triangle is known to be equal to

$$A + B + C - \pi.$$

Divide the polygon into n such triangles, with a common vertex, the angles about which, of course, amount to 2π.

$$\begin{aligned} \text{Area} &= \text{sum of interior angles of triangles} - n\pi \\ &= 2\pi + \text{sum of interior angles of polygon} - n\pi \\ &= 2\pi - \text{sum of exterior angle of polygon.} \end{aligned}$$

Reciprocal polars on a sphere.

Given an open or closed spherical polygon, or line on the surface of a sphere composed of consecutive arcs of great circles. Take either pole of the first of these arcs, and the corresponding poles of all the others (all the poles to be on the right hand, or all on the left, of a traveller advancing along the given great circle arcs in order). Draw great circle arcs from the first of these poles to the second, the second to the third, and so on in order. Another closed or open polygon, constituting what is called the polar diagram to the given polygon, is thus obtained. The sides of the second polygon are evidently equal to the exterior angles in the first; and the exterior angles of the second are equal to the sides of the first. Hence the relation between the two diagrams is reciprocal, or each is polar to the other. The polar figure to any continuous curve on a spherical surface is the locus of the ultimate intersections of great circles equatorial to points taken infinitely near each other along it. Reciprocal polars on a sphere.

The area of a closed spherical figure is, consequently, according to what we have just seen, equal to the excess of 2π above the periphery of its polar, if the radius of the sphere be unity.

135. If a point move on a surface along a figure whose sides are geodetic lines, the sum of the exterior angles of this polygon is defined to be the *integral change of the direction in the surface.* Integral change of direction in a surface.

In great circle sailing, unless a vessel sail on the equator, or on a meridian, her course, as indicated by points of the compass (true, not magnetic, for the latter change even on a meridian), perpetually changes. Yet just as we say her direction does not change if she sail in a meridian, or in the equator, so we ought to say her direction does not change if she moves in *any* great circle. Now, the great circle is the geodetic line on the sphere, and by extending these remarks to other curved surfaces, we see the connexion of the above definition with that in the case of a plane polygon (§ 10).

Note.—We cannot define integral change of direction here by any angle directly constructible from the first and last tangents to the path, as was done (§ 10) in the case of a plane curve or polygon; but from §§ 125 and 133 we have the following statement:—The whole change of direction in a curved surface, from one end to another of any arc of a curve traced on it, is equal to the change of direction from end to end of the trace of this arc on a plane by pure rolling. Change of direction in a surface, of any arc traced on it.

136. *Def.* The excess of four right angles above the integral change of direction from one side to the same side next time in going round a closed polygon of geodetic lines on a curved surface, is the *integral curvature* of the enclosed portion of surface. This excess is zero in the case of a polygon traced on a plane. We shall presently see that this corresponds exactly to what Gauss has called the *curvatura integra.* Integral curvature.

Def. (Gauss.) The *curvatura integra* of any given portion of a curved surface, is the area enclosed on a spherical surface Curvatura integra.

of unit radius by a straight line drawn from its centre, parallel to a normal to the surface, the normal being carried round the boundary of the given portion.

Horograph. The curve thus traced on the sphere is called the *Horograph* of the given portion of curved surface.

The *average curvature* of any portion of a curved surface is the integral curvature divided by the area. The *specific curvature* of a curved surface at any point is the average curvature of an infinitely small area of it round that point.

Change of direction round the boundary in the surface, together with area of the horograph, equals four rightangles: or "Integral Curvature" equals "Curvatura Integra."

137. The excess of 2π above the change of direction, in a surface, of a point moving round any closed curve on it, is equal to the area of the horograph of the enclosed portion of surface.

Let a tangent plane roll without spinning on the surface over every point of the bounding line. (Its instantaneous axis will always lie in it, and pass through the point of contact, but will not, as we have seen, be at right angles to the given bounding curve, except when the twist of a narrow ribbon of the surface along this curve is nothing.) Considering the auxiliary sphere of unit radius, used in Gauss's definition, and the moving line through its centre, we perceive that the motion of this line is, at each instant, in a plane perpendicular to the instantaneous axis of the tangent plane to the given surface. The direction of motion of the point which cuts out the area on the spherical surface is therefore perpendicular to this instantaneous axis. Hence, if we roll a tangent plane on the spherical surface also, making it keep time with the other, the trace on this tangent plane will be a curve always perpendicular to the instantaneous axis of each tangent plane. The change of direction, in the spherical surface, of the point moving round and cutting out the area, being equal to the change of direction in its own trace on its own tangent plane (§ 135), is therefore equal to the change of direction of the instantaneous axis in the tangent plane to the given surface reckoned from a line fixed relatively to this plane. But having rolled all round, and being in position to roll round again, the instantaneous axis of the fresh start must be inclined to the trace at the same angle as in the beginning. Hence the change of direction of the instantaneous axis in either tangent plane is equal to the change of direction, in the given surface, of

Curvatura integra, and horograph.

a point going all round the boundary of the given portion of it (§ 135); to which, therefore, the change of direction, in the spherical surface, of the point going all round the spherical area is equal. But, by the well known theorem (§ 134) of the "spherical excess," this change of direction subtracted from 2π leaves the spherical area. Hence the spherical area, called by Gauss the *curvatura integra*, is equal to 2π wanting the change of direction in going round the boundary. *Curvatura integra, and horograph*

It will be perceived that when the two rollings we have considered are each complete, each tangent plane will have come back to be parallel to its original position, but any fixed line in it will have changed direction through an angle equal to the equal changes of direction just considered.

Note.—The two rolling tangent planes are at each instant parallel to one another, and a fixed line relatively to one drawn at any time parallel to a fixed line relatively to the other, remains parallel to the last-mentioned line.

If, instead of the closed curve, we have a closed polygon of geodetic lines on the given surface, the trace of the rolling of its tangent plane will be an unclosed rectilineal polygon. If each geodetic were a plane curve (which could only be if the given surface were spherical), the instantaneous axis would be always perpendicular to the particular side of this polygon which is rolled on at the instant; and, of course, the spherical area on the auxiliary sphere would be a similar polygon to the given one. But the given surface being other than spherical, there must (except in the particular case of some of the geodetics being lines of curvature) be tortuosity in every geodetic of the closed polygon; or, which is the same thing, twist in the corresponding ribbons of the surface. Hence the portion of the whole trace on the second rolling tangent plane which corresponds to any one side of the given geodetic polygon, must in general be a curve; and as there will generally be finite angles in the second rolling corresponding to (but not equal to) those in the first, the trace of the second on its tangent plane will be an unclosed polygon of curves. The trace of the same rolling on the spherical surface in which it takes place will generally be a spherical polygon, not of great circle arcs, but of other curves. The sum of the exterior angles of this polygon, and of the changes of direction from one end to the other of each of its sides, is the whole change of direction considered, and is, by the proper

Curvatura integra, and horograph.

application of the theorem of § 134, equal to 2π wanting the spherical area enclosed.

Or again, if, instead of a geodetic polygon as the given curve, we have a polygon of curves, each fulfilling the condition that the normal to the surface through any point of it is parallel to a fixed plane; one plane for the first curve, another for the second, and so on; then the figure on the auxiliary spherical surface will be a polygon of arcs of great circles; its trace on its tangent plane will be an unclosed rectilineal polygon; and the trace of the given curve on the tangent plane of the first rolling will be an unclosed polygon of curves. The sum of changes of direction in these curves, and of exterior angles in passing from one to another of them, is of course equal to the change of direction in the given surface, in going round the given polygon of curves on it. The change of direction in the other will be simply the sum of the exterior angles of the spherical polygon, or of its rectilineal trace. Remark that in this case the instantaneous axis of the first rolling, being always perpendicular to that plane to which the normals are all parallel, remains parallel to one line, fixed with reference to the tangent plane, during rolling along each curved side, and also remains parallel to a fixed line in space.

Lastly, remark that although the whole change of direction of the trace in one tangent plane is equal to that in the trace on the other, when the rolling is completed round the given circuit; the changes of direction in the two are generally unequal in any part of the circuit. They may be equal for particular parts of the circuit, viz., between those points, if any, at which the instantaneous axis is equally inclined to the direction of the trace on the first tangent plane.

Any difficulty which may have been felt in reading this Section will be removed if the following exercises on the subject be performed.

(1) Find the horograph of an infinitely small circular area of any continuous curved surface. It is an ellipse or a hyperbola according as the surface is synclastic or anticlastic (§ 128). Find the axes of the ellipse or hyperbola in either case.

(2) Find the horograph of the area cut off a synclastic surface by a plane parallel to the tangent plane at any given point of it, and infinitely near this point. Find and interpret the corresponding result for the case in which the surface is anticlastic in the neighbourhood of the given point.

(3) Let a tangent plane roll without spinning over the boundary of a given closed curve or geodetic polygon on any curved surface. Show that the points of the trace in the tangent plane which successively touch the same point of the given surface are at equal distances successively on the circumference of a circle, the angular values of the intermediate arcs being each $2\pi - K$ if taken in the direction in which the trace is actually described, and K if taken in the contrary direction, K being the "integral curvature" of the portion of the curved surface enclosed by the given curve or geodetic polygon. Hence if K be commensurable with π the trace on the tangent plane, however complicatedly autotomic it may be, is a finite closed curve or polygon. *Curvatura integra*, and horograph.

(4) The trace by a tangent plane rolling successively over three principal quadrants bounding an eighth part of the circumference of an ellipsoid is represented in the accompanying diagram, the whole of which is traced when the tangent plane is

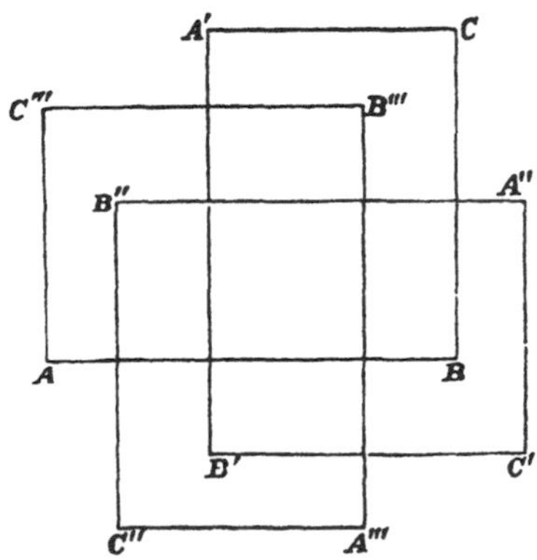

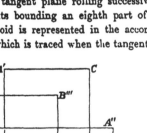

rolled four times over the stated boundary. A, B, C; A', B', C', &c. represent the points of the tangent plane touched in order by ends of the mean principal axis (A), the greatest principal axis (B), and least principal axis (C), and AB, BC, CA' are the lengths of the three principal quadrants.

138. It appears from what precedes, that the same equality or identity subsists between "whole curvature" in a plane arc and the excess of π above the angle between the terminal Analogy between lines and surfaces as regards curvature.

Analogy between lines and surfaces as regards curvature.

tangents, as between "whole curvature" and excess of 2π above change of direction along the bounding line in the surface for any portion of a curved surface.

Or, according to Gauss, whereas the whole curvature in a plane arc is the angle between two lines parallel to the terminal normals, the whole curvature of a portion of curve surface is the solid angle of a cone formed by drawing lines from a point parallel to all normals through its boundary.

Again, average curvature in a plane curve is $\frac{\text{change of direction}}{\text{length}}$; and specific curvature, or, as it is commonly called, curvature, at any point of it $= \frac{\text{change of direction in infinitely small length}}{\text{length}}$.

Thus average curvature and specific curvature are for surfaces analogous to the corresponding terms for a plane curve.

Lastly, in a plane arc of uniform curvature, *i.e.*, in a circular arc, $\frac{\text{change of direction}}{\text{length}} = \frac{1}{\rho}$. And it is easily proved (as below) that, in a surface throughout which the specific curvature is uniform, $\frac{2\pi - \text{change of direction}}{\text{area}}$, or $\frac{\text{integral curvature}}{\text{area}}$, $= \frac{1}{\rho\rho'}$, where ρ and ρ' are the principal radii of curvature. Hence in a surface, whether of uniform or non-uniform specific curvature, the specific curvature at any point is equal to $\frac{1}{\rho\rho'}$. In geometry of three dimensions, $\rho\rho'$ (an area) is clearly analogous to ρ in a curve and plane.

Horograph.

Consider a portion S, of a surface of any curvature, bounded by a given closed curve. Let there be a spherical surface, radius r, and centre C. Draw a radius CQ, parallel to the normal at any point P of S. If this be done for every point of the boundary, the line so obtained encloses the spherical area used in Gauss's definition. Now let there be an infinitely small rectangle on S, at P, having for its sides arcs of angles ζ and ζ', on the normal sections of greatest and least curvature, and let their radii of curvature be denoted by ρ and ρ'. The lengths of these sides will be $\rho\zeta$ and $\rho'\zeta'$ respectively. Its area will therefore be $\rho\rho'\zeta\zeta'$. The corresponding figure at Q on the spherical surface will be bounded by arcs of angles equal to those, and therefore of lengths $r\zeta$ and $r\zeta'$ respectively, and its area will be $r^2\zeta\zeta'$. Hence if $d\sigma$ denote this area, the area of the infinitely small portion of the given surface will be $\frac{\rho\rho' d\sigma}{r^2}$. In a surface for which $\rho\rho'$ is constant, the area is therefore $= \frac{\rho\rho'}{r^2}\iint d\sigma = \rho\rho' \times$ integral curvature. Area of the horograph.

139. A perfectly flexible but inextensible surface is suggested, although not realized, by paper, thin sheet metal, or cloth, when the surface is plane; and by sheaths of pods, seed vessels, or the like, when it is not capable of being stretched flat without tearing. The process of changing the form of a surface by bending is called "*developing*." But the term "*Developable Surface*" is commonly restricted to such inextensible surfaces as can be developed into a plane, or, in common language, "smoothed flat." Flexible and inextensible surface.

140. The geometry or kinematics of this subject is a great contrast to that of the flexible line (§ 14), and, in its merest elements, presents ideas not very easily apprehended, and subjects of investigation that have exercised, and perhaps even overtasked, the powers of some of the greatest mathematicians.

141. Some care is required to form a correct conception of what is a perfectly flexible inextensible surface. First let us consider a plane sheet of paper. It is very flexible, and we can easily form the conception from it of a sheet of ideal matter perfectly flexible. It is very inextensible; that is to say, it yields very little to any application of force tending to pull or stretch it in any direction, up to the strongest it can bear without tearing. It does, of course, stretch a little. It is easy to test that it stretches when under the influence of force, and that it contracts again when the force is removed, although not always to its original dimensions, as it may and generally does remain to some sensible extent permanently stretched. Also, flexure stretches one side and condenses the other temporarily; and, to a less extent, permanently. Under elasticity (§§ 717, 718, 719) we shall return to this. In the meantime, in considering illustrations of our kinematical propositions, it is necessary to anticipate such physical circumstances.

Surface inextensible in two directions.

142. Cloth woven in the simple common way, very fine muslin for instance, illustrates a surface perfectly inextensible in two directions (those of the warp and the woof), but susceptible of any amount of extension from 1 up to $\sqrt{2}$ along one diagonal, with contraction from 1 to 0 (each degree of extension along one diagonal having a corresponding determinate degree of contraction along the other, the relation being $e^2+e'^2=2$, where $1:e$ and $1:e'$ are the ratios of elongation, which will be contraction in the case in which e or e' is < 1) in the other.

"Elastic finish" of muslin goods.

143. The flexure of a surface fulfilling any case of the geometrical condition just stated, presents an interesting subject for investigation, which we are reluctantly obliged to forego. The moist paper drapery that Albert Dürer used on his little lay figures must hang very differently from cloth. Perhaps the stiffness of the drapery in his pictures may be to some extent owing to the fact that he used the moist paper in preference to cloth on account of its superior flexibility, while unaware of the great distinction between them as regards extensibility. Fine muslin, prepared with starch or gum, is, during the process of drying, kept moving by a machine, which, by producing a to-and-fro relative angular motion of warp and woof, stretches and contracts the diagonals of its structure alternately, and thus prevents the parallelograms from becoming stiffened into rectangles.

Flexure of inextensible developable.

144. The flexure of an inextensible surface which can be plane, is a subject which has been well worked by geometrical investigators and writers, and, in its elements at least, presents little difficulty. The first elementary conception to be formed is, that such a surface (if perfectly flexible), taken plane in the first place, may be bent about any straight line ruled on it, so that the two plane parts may make any angle with one another.

Such a line is called a "generating line" of the surface to be formed.

Next, we may bend one of these plane parts about any other line which does not (within the limits of the sheet) intersect the former; and so on. If these lines are infinite in number, and the angles of bending infinitely small, but such that their sum may be finite, we have our plane surface bent into a curved surface, which is of course "developable" (§ 139). Flexure of inextensible developable.

145. Lift a square of paper, free from folds, creases, or ragged edges, gently by one corner, or otherwise, without crushing or forcing it, or very gently by two points. It will hang in a form which is very rigorously a developable surface; for although it is not absolutely inextensible, yet the forces which tend to stretch or tear it, when it is treated as above described, are small enough to produce no sensible stretching. Indeed the greatest stretching it can experience without tearing, in any direction, is not such as can affect the form of the surface much when sharp flexures, singular points, etc., are kept clear of.

146. Prisms and cylinders (when the lines of bending, § 144, are parallel, and finite in number with finite angles, or infinite in number with infinitely small angles), and pyramids and cones (the lines of bending meeting in a point if produced), are clearly included.

147. If the generating lines, or line-edges of the angles of bending, are not parallel, they must meet, since they are in a plane when the surface is plane. If they do not meet all in one point, they must meet in several points: in general, each one meets its predecessor and its successor in different points.

148. There is still no difficulty in understanding the form of, say a square, or circle, of the plane surface when bent as explained above, provided it does not include any of these points of intersection. When the number is infinite, and the surface finitely curved, the developable lines will in general be tangents to a curve (the locus of the points of intersection when the number is infinite). This curve is called the *edge of regression*. The surface must clearly, when *complete* (according to mathematical ideas), consist of two sheets meeting in this edge

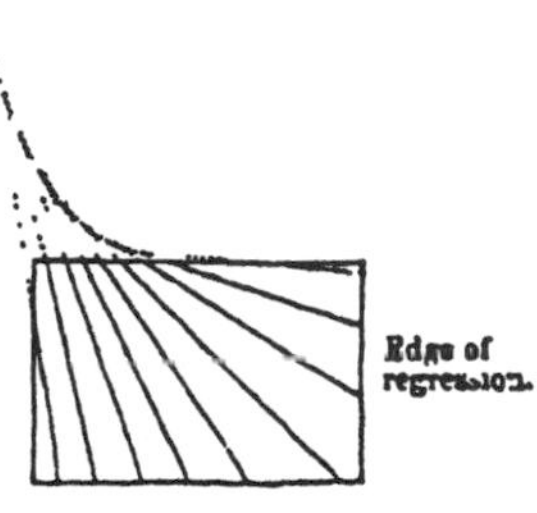

Edge of regression.

Edge of regression. of regression (just as a cone consists of two sheets meeting in the vertex), because each tangent may be produced beyond the point of contact, instead of stopping at it, as in the annexed diagram.

Practical construction of a developable from its edge. 149. To construct a complete developable surface in two sheets from its edge of regression—

Lay one piece of perfectly flat, unwrinkled, smooth-cut paper on the top of another. Trace any curve on the upper,

and let it have no point of inflection, but everywhere finite curvature. Cut the two papers along the curve and remove the convex portions. If the curve traced is closed, it must be cut open (see second diagram).

Attach the two sheets together by very slight paper or muslin clamps gummed to them along the common curved

edge. These must be so slight as not to interfere sensibly with the flexure of the two sheets. Take hold of one corner of one sheet and lift the whole. The two will open out into the two sheets of a developable surface, of which the curve, bending into a curve of double curvature, is the edge of regression. The tangent to the curve drawn in one direction from the point of contact, will always lie in one of the sheets, and its continuation on the other side in the other sheet. Of course a double-sheeted developable polyhedron can be constructed by this process, by starting from a polygon instead of a curve.

General property of inextensible surface. 150. A flexible but perfectly inextensible surface, altered in form in any way possible for it, must keep any line traced on it unchanged in length; and hence any two intersecting lines unchanged in mutual inclination. Hence, also, geodetic lines must remain geodetic lines. Hence "the change of direction" in a surface, of a point going round any portion of it, must be the same, however this portion is bent. Hence (§ 136) the integral curvature remains the same in any and every portion however the surface is bent. Hence (§ 138,

Gauss's Theorem) the product of the principal radii of curvature at each point remains unchanged. General property of inextensible surface.

151. The general statement of a converse proposition, expressing the condition that two given areas of curved surfaces may be bent one to fit the other, involves essentially some mode of specifying corresponding points on the two. A full investigation of the circumstances would be out of place here.

152. In one case, however, a statement in the simplest possible terms is applicable. Any two surfaces, in each of which the specific curvature is the same at all points, and equal to that of the other, may be bent one to fit the other. Thus any surface of uniform positive specific curvature (*i.e.*, wholly convex one side, and concave the other) may be bent to fit a sphere whose radius is a mean proportional between its principal radii of curvature at any point. A surface of uniform negative, or anticlastic, curvature would fit an imaginary sphere, but the interpretation of this is not understood in the present condition of science. But practically, of any two surfaces of uniform anticlastic curvature, either may be bent to fit the other. Surface of constant specific curvature.

153. It is to be remarked, that geodetic trigonometry on any surface of uniform positive, or synclastic, curvature, is identical with spherical trigonometry. Geodetic triangles on such a surface.

If $a=\frac{s}{\sqrt{\rho\rho'}}$, $b=\frac{t}{\sqrt{\rho\rho'}}$, $c=\frac{u}{\sqrt{\rho\rho'}}$, where s, t, u are the lengths of three geodetic lines joining three points on the surface, and if A, B, C denote the angles between the tangents to the geodetic lines at these points; we have six quantities which agree perfectly with the three sides and the three angles of a certain spherical triangle. A corresponding anticlastic trigonometry exists, although we are not aware that it has hitherto been noticed, for any surface of uniform anticlastic curvature. In a geodetic triangle on an anticlastic surface, the sum of the three angles is of course less than three right angles, and the difference, or "anticlastic defect" (like the "spherical excess"), is equal to the area divided by $\rho \times -\rho'$, where ρ and $-\rho'$ are positive.

154. We have now to consider the very important kinematical conditions presented by the changes of volume or figure Strain.

Strain. experienced by a solid or liquid mass, or by a group of points whose positions with regard to each other are subject to known conditions. Any such definite alteration of form or dimensions is called a *Strain*.

Thus a rod which becomes longer or shorter is strained. Water, when compressed, is strained. A stone, beam, or mass of metal, in a building or in a piece of framework, if condensed or dilated in any direction, or bent, twisted, or distorted in any way, is said to experience a strain. A ship is said to "strain" if, in launching, or when working in a heavy sea, the different parts of it experience relative motions.

Definition of homogeneous strain. 155. If, when the matter occupying any space is strained in any way, all pairs of points of its substance which are initially at equal distances from one another in parallel lines remain equidistant, it may be at an altered distance; and in parallel lines, altered, it may be, from their initial direction; the strain is said to be homogeneous.

Properties of homogeneous strain. 156. Hence if any straight line be drawn through the body in its initial state, the portion of the body cut by it will continue to be a straight line when the body is homogeneously strained. For, if ABC be any such line, AB and BC, being parallel to one line in the initial, remain parallel to one line in the altered, state; and therefore remain in the same straight line with one another. Thus it follows that a plane remains a plane, a parallelogram a parallelogram, and a parallelepiped a parallelepiped.

157. Hence, also, similar figures, whether constituted by actual portions of the substance, or mere geometrical surfaces, or straight or curved lines passing through or joining certain portions or points of the substance, similarly situated (*i.e.*, having corresponding parameters parallel) when altered according to the altered condition of the body, remain similar and similarly situated among one another.

158. The lengths of parallel lines of the body remain in the same proportion to one another, and hence all are altered in the same proportion. Hence, and from § 156, we infer that any plane figure becomes altered to another plane figure which

is a diminished or magnified orthographic projection of the first on some plane. For example, if an ellipse be altered into a circle, its principal axes become radii at right angles to one another. Properties of homogeneous strain.

The elongation of the body along any line is the proportion which the addition to the distance between any two points in that line bears to their primitive distance.

159. Every orthogonal projection of an ellipse is an ellipse (the case of a circle being included). Hence, and from § 158, we see that an ellipse remains an ellipse; and an ellipsoid remains a surface of which every plane section is an ellipse; that is, remains an ellipsoid.

A plane curve remains (§ 156) a plane curve. A system of two or of three straight lines of reference (Cartesian) remains a rectilineal system of lines of reference; but, in general, a rectangular system becomes oblique.

Let
$$\frac{x^2}{a^2}+\frac{y^2}{b^2}=1$$
be the equation of an ellipse referred to any rectilineal conjugate axes, in the substance, of the body in its initial state. Let α and β be the proportions in which lines respectively parallel to OX and OY are altered. Thus, if we call ξ and η the altered values of x and y, we have
$$\xi=\alpha x,\ \eta=\beta y.$$
Hence
$$\frac{\xi^2}{(\alpha a)^2}+\frac{\eta^2}{(\beta b)^2}=1,$$
which also is the equation of an ellipse, referred to oblique axes at, it may be, a different angle to one another from that of the given axes, in the initial condition of the body.

Or again, let
$$\frac{x^2}{a^2}+\frac{y^2}{b^2}+\frac{z^2}{c^2}=1$$
be the equation of an ellipsoid referred to three conjugate diametral planes, as oblique or rectangular planes of reference, in the initial condition of the body. Let α, β, γ be the proportion in which lines parallel to OX, OY, OZ are altered; so that if ξ, η, ζ be the altered values of x, y, z, we have
$$\xi=\alpha x,\ \eta=\beta y,\ \zeta=\gamma z.$$
Thus
$$\frac{\xi^2}{(\alpha a)^2}+\frac{\eta^2}{(\beta b)^2}+\frac{\zeta^2}{(\gamma c)^2}=1,$$

Properties of homogeneous strain.

which is the equation of an ellipsoid, referred to conjugate diametral planes, altered it may be in mutual inclination from those of the given planes of reference in the initial condition of the body.

Strain ellipsoid.

160. The ellipsoid which any surface of the body initially spherical becomes in the altered condition, may, to avoid circumlocutions, be called the strain ellipsoid.

161. In any absolutely unrestricted homogeneous strain there are three directions (the three principal axes of the strain ellipsoid), at right angles to one another, which remain at right angles to one another in the altered condition of the body (§ 158). Along one of these the elongation is greater, and along another less, than along any other direction in the body. Along the remaining one, the elongation is less than in any other line in the plane of itself and the first mentioned, and greater than along any other line in the plane of itself and the second.

Note.—Contraction is to be reckoned as a negative elongation: the maximum elongation of the preceding enunciation may be a minimum contraction: the minimum elongation may be a maximum contraction.

162. The ellipsoid into which a sphere becomes altered may be an ellipsoid of revolution, or, as it is called, a spheroid, prolate, or oblate. There is thus a maximum or minimum elongation along the axis, and equal minimum or maximum elongation along all lines perpendicular to the axis.

Or it may be a sphere; in which case the elongations are equal in all directions. The effect is, in this case, merely an alteration of dimensions without change of figure of any part.

Change of volume.

The original volume (sphere) is to the new (ellipsoid) evidently as $1 : \alpha\beta\gamma$.

Axes of a Strain.

163. The principal axes of a strain are the principal axes of the ellipsoid into which it converts a sphere. The principal elongations of a strain are the elongations in the direction of its principal axes.

164. When the position of the principal axes, and the magnitudes of the principal elongations of a strain are given, the elongation of any line of the body, and the alteration of angle between any two lines, may be obviously determined by a simple geometrical construction, Elongation and change of direction of any line of the body.

Analytically thus:—let $\alpha - 1$, $\beta - 1$, $\gamma - 1$ denote the principal elongations, so that α, β, γ may be now the ratios of alteration along the three principal axes, as we used them formerly for the ratios for any three oblique or rectangular lines. Let l, m, n be the direction cosines of any line, with reference to the three principal axes. Thus,

$$lr,\ mr,\ nr$$

being the three initial co-ordinates of a point P, at a distance $OP = r$, from the origin in the direction l, m, n; the co-ordinates of the same point of the body, with reference to the same rectangular axes, become, in the altered state,

$$\alpha lr,\ \beta mr,\ \gamma nr.$$

Hence the altered length of OP is

$$(\alpha^2 l^2 + \beta^2 m^2 + \gamma^2 n^2)^{\frac{1}{2}} r,$$

and therefore the "elongation" of the body in that direction is

$$(\alpha^2 l^2 + \beta^2 m^2 + \gamma^2 n^2)^{\frac{1}{2}} - 1.$$

For brevity, let this be denoted by $\zeta - 1$, *i.e.*

let $$\zeta = (\alpha^2 l^2 + \beta^2 m^2 + \gamma^2 n^2)^{\frac{1}{2}}.$$

The direction cosines of OP in its altered position are

$$\frac{\alpha l}{\zeta},\ \frac{\beta m}{\zeta},\ \frac{\gamma n}{\zeta};$$

and therefore the angles XOP, YOP, ZOP are altered to having their cosines of these values respectively, from having them of the values l, m, n.

The cosine of the angle between any two lines OP and OP', specified in the initial condition of the body by the direction cosines l', m', n', is

$$ll' + mm' + nn',$$

in the initial condition of the body, and becomes

$$\frac{\alpha^2 ll' + \beta^2 mm' + \gamma^2 nn'}{(\alpha^2 l^2 + \beta^2 m^2 + \gamma^2 n^2)^{\frac{1}{2}}\,(\alpha^2 l'^2 + \beta^2 m'^2 + \gamma^2 n'^2)^{\frac{1}{2}}}$$

in the altered condition.

Change of plane in the body.

165. With the same data the alteration of angle between any two planes of the body may also be easily determined, either geometrically or analytically.

Let l, m, n be the cosines of the angles which a plane makes with the planes YOZ, ZOX, XOY, respectively, in the initial condition of the body. The effects of the change being the same on all parallel planes, we may suppose the plane in question to pass through O; and therefore its equation will be

$$lx + my + nz = 0.$$

In the altered condition of the body we shall have, as before,

$$\xi = \alpha x, \quad \eta = \beta y, \quad \zeta = \gamma z,$$

for the altered co-ordinates of any point initially x, y, z. Hence the equation of the altered plane is

$$\frac{l\xi}{\alpha} + \frac{m\eta}{\beta} + \frac{n\zeta}{\gamma} = 0.$$

But the planes of reference are still rectangular, according to our present supposition. Hence the cosines of the inclinations of the plane in question, to YOZ, ZOX, XOY, in the altered condition of the body, are altered from l, m, n to

$$\frac{l}{\alpha\vartheta}, \quad \frac{m}{\beta\vartheta}, \quad \frac{n}{\gamma\vartheta},$$

respectively, where for brevity

$$\vartheta = \left(\frac{l^2}{\alpha^2} + \frac{m^2}{\beta^2} + \frac{n^2}{\gamma^2}\right)^{\frac{1}{2}}.$$

If we have a second plane similarly specified by l', m', n', in the initial condition of the body, the cosine of the angle between the two planes, which is

$$ll' + mm' + nn'$$

in the initial condition, becomes altered to

$$\frac{\dfrac{ll'}{\alpha^2} + \dfrac{mm'}{\beta^2} + \dfrac{nn'}{\gamma^2}}{\left(\dfrac{l^2}{\alpha^2} + \dfrac{m^2}{\beta^2} + \dfrac{n^2}{\gamma^2}\right)^{\frac{1}{2}} \left(\dfrac{l'^2}{\alpha^2} + \dfrac{m'^2}{\beta^2} + \dfrac{n'^2}{\gamma^2}\right)^{\frac{1}{2}}}.$$

Conical surface of equal elongation.

166. Returning to elongations, and considering that these are generally different in different directions, we perceive that all lines through any point, in which the elongations have any one

value intermediate between the greatest and least, must lie on a determinate conical surface. This is easily proved to be in general a cone of the second degree.

Conical surface of equal elongation.

For, in a direction denoted by direction cosines l, m, n, we have

$$\alpha^2 l^2 + \beta^2 m^2 + \gamma^2 n^2 = \zeta^2,$$

where ζ denotes the ratio of elongation, intermediate between α the greatest and γ the least. This is the equation of a cone of the second degree, l, m, n being the direction cosines of a generating line.

167. In one particular case this cone becomes two planes, the planes of the circular sections of the strain ellipsoid.

Two planes of no distortion,

Let $\zeta = \beta$. The preceding equation becomes

$$\alpha^2 l^2 + \gamma^2 n^2 - \beta^2 (1 - m^2) = 0,$$

or, since

$$1 - m^2 = l^2 + n^2,$$

$$(\alpha^2 - \beta^2)\, l^2 - (\beta^2 - \gamma^2)\, n^2 = 0.$$

The first member being the product of two factors, the equation is satisfied by putting either $= 0$, and therefore the equation represents the two planes whose equations are

$$l\,(\alpha^2 - \beta^2)^{\frac{1}{2}} + n\,(\beta^2 - \gamma^2)^{\frac{1}{2}} = 0,$$

and

$$l\,(\alpha^2 - \beta^2)^{\frac{1}{2}} - n\,(\beta^2 - \gamma^2)^{\frac{1}{2}} = 0,$$

respectively.

This is the case in which the given elongation is equal to that along the mean principal axis of the strain ellipsoid. The two planes are planes through the mean principal axis of the ellipsoid, equally inclined on the two sides of either of the other axes. The lines along which the elongation is equal to the mean principal elongation, all lie in, or parallel to, either of these two planes. This is easily proved as follows, without any analytical investigation.

being the circular sections of the strain ellipsoid.

168. Let the ellipse of the annexed diagram represent the section of the strain ellipsoid through the greatest and least principal axes. Let $S'OS$, $T'OT$ be the two diameters of this ellipse, which are equal to the mean principal axis of the ellipsoid. Every plane through O, perpendicular to the plane of the diagram, cuts the ellipsoid in an ellipse of which

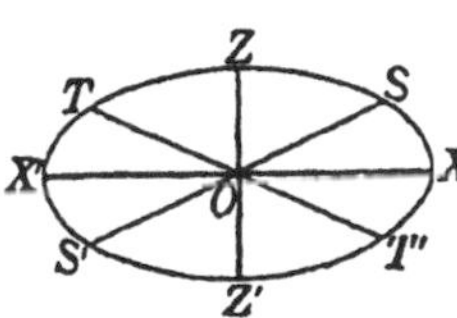

Two planes of no distortion, being the circular sections of the strain ellipsoid.

one principal axis is the diameter in which it cuts the ellipse of the diagram, and the other, the mean principal diameter of the ellipsoid. Hence a plane through either SS', or TT', perpendicular to the plane of the diagram, cuts the ellipsoid in an ellipse of which the two principal axes are equal, that is to say, in a circle. Hence the elongations along all lines in either of these planes are equal to the elongation along the mean principal axis of the strain ellipsoid.

Distortion in parallel planes without change of volume.

169. The consideration of the circular sections of the strain ellipsoid is highly instructive, and leads to important views with reference to the analysis of the most general character of a strain. First, let us suppose there to be no alteration of volume on the whole, and neither elongation nor contraction along the mean principal axis. That is to say, let $\beta = 1$, and $\gamma = \frac{1}{\alpha}$ (§ 162).

Let OX and OZ be the directions of elongation $\alpha - 1$ and contraction $1 - \frac{1}{\alpha}$ respectively. Let A be any point of the body in its primitive condition, and $A_{,}$ the same point of the altered body, so that $OA_{,} = \alpha OA$.

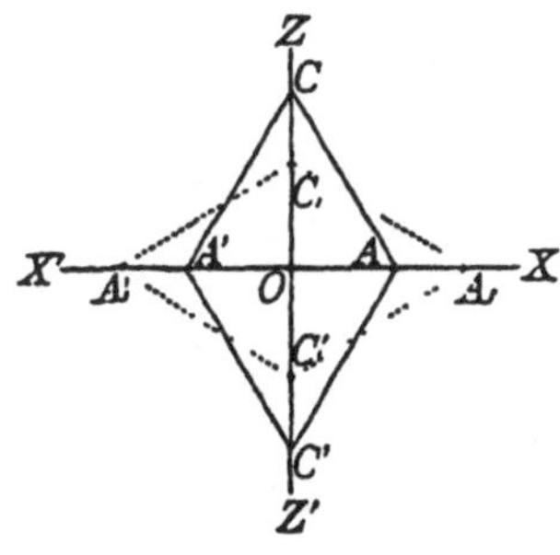

Now, if we take $OC = OA_{,}$, and if $C_{,}$ be the position of that point of the body which was in the position C initially, we shall have $OC_{,} = \frac{1}{\alpha} OC$, and therefore $OC_{,} = OA$. Hence the two triangles COA and $C_{,}OA_{,}$ are equal and similar.

Initial and altered position of lines of no elongation.

Hence CA experiences no alteration of length, but takes the altered position $C_{,}A_{,}$ in the altered position of the body. Similarly, if we measure on XO produced, OA' and $OA_{,}'$ equal respectively to OA and $OA_{,}$, we find that the line CA' experiences no alteration in length, but takes the altered position $C_{,}A_{,}'$.

Consider now a plane of the body initially through CA perpendicular to the plane of the diagram, which will be altered into a plane through $C_{,}A_{,}$, also perpendicular to the plane of the diagram. All lines initially perpendicular to the plane of the diagram remain so, and remain unaltered in length. AC has just been proved to remain unaltered in length. Hence (§ 158) all lines in the plane we have just drawn remain unaltered in length and in mutual inclination. Similarly we see that all lines in a plane through CA', perpendicular to the plane of the diagram, altering to a plane through $C_{,}A_{,}'$, perpendicular to the plane of the diagram, remain unaltered in length and in mutual inclination.

Initial and altered position of lines of no elongation.

170. The precise character of the strain we have now under consideration will be elucidated by the following:—Produce CO, and take OC' and $OC_{,}'$ respectively equal to OC and $OC_{,}$. Join $C'A$, $C'A'$, $C_{,}'A_{,,}$ and $C_{,}'A_{,}'$, by plain and dotted lines as in the diagram. Then we see that the rhombus $CAC'A'$ (plain lines) of the body in its initial state becomes the rhombus $C_{,}A_{,}C_{,}'A_{,}'$ (dotted) in the altered condition. Now imagine the body thus strained to be moved as a rigid body (*i.e.*, with its state of strain kept unchanged) till $A_{,}$ coincides with A, and $C_{,}'$ with C', keeping all the lines of the diagram still in the same plane. $A_{,}'C_{,}$ will take a position in CA' produced, as shown in the new diagram, and the original and the altered parallelogram will be on the same base AC', and between the same parallels AC' and $CA_{,}'$, and their other sides will be equally inclined on the two sides of a perpendicular to these parallels. Hence, irrespectively of any rotation, or other absolute motion of the body not involving change of form or dimensions, the strain under consideration may be produced by holding fast and unaltered the plane of the body through AC' perpendicular to the plane of the diagram, and making every plane parallel to it slide, keeping the same distance, through a space proportional to this distance (*i.e.*, different planes parallel to the fixed plane slide through spaces proportional to their distances).

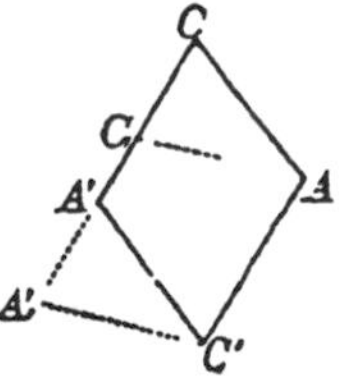

171. This kind of strain is called a *simple shear*. The plane of a shear is a plane perpendicular to the undistorted planes, and parallel to the lines of their relative motion. It

Simple shear.

Simple shear.

has (1) the property that one set of parallel planes remain each unaltered in itself; (2) that another set of parallel planes remain each unaltered in itself. This other set is found when the first set and the degree or amount of shear are given, thus:—Let $CC_{,}$ be the motion of one point of one plane, relative to a plane KL held fixed—the diagram being in a plane of the shear. Bisect $CC_{,}$ in N. Draw NA perpendicular to it. A plane perpendicular to the plane of the diagram, initially through AC, and finally through $AC_{,}$, remains unaltered in its dimensions.

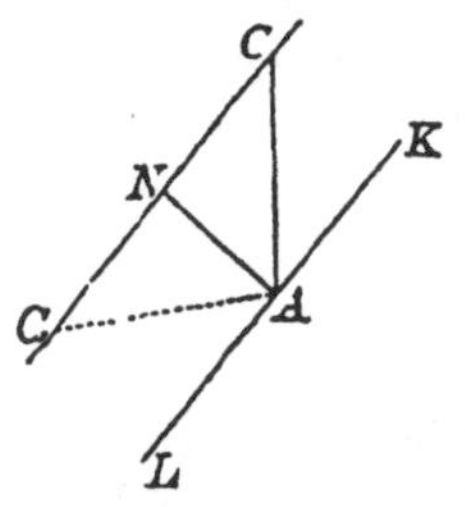

172. One set of parallel undistorted planes, and the amount of their relative parallel shifting having been given, we have just seen how to find the other set. The shear may be otherwise viewed, and considered as a shifting of this second set of parallel planes, relative to any one of them. The amount of this relative shifting is of course equal to that of the first set, relatively to one of them.

Axes of a shear.

173. The principal axes of a shear are the lines of maximum elongation and of maximum contraction respectively. They may be found from the preceding construction (§ 171), thus:—In the plane of the shear bisect the obtuse and acute angles between the planes destined not to become deformed. The former bisecting line is the principal axis of elongation, and the latter is the principal axis of contraction, in their initial positions. The former angle (obtuse) becomes equal to the latter, its supplement (acute), in the altered condition of the body, and the lines bisecting the altered angles are the principal axes of the strain in the altered body.

Otherwise, taking a plane of shear for the plane of the diagram, let AB be a line in which it is cut by one of either set of parallel planes of no distortion. On any portion AB of this as diameter, describe a semicircle. Through C, its middle point, draw, by the preceding construction, CD the initial, and CE

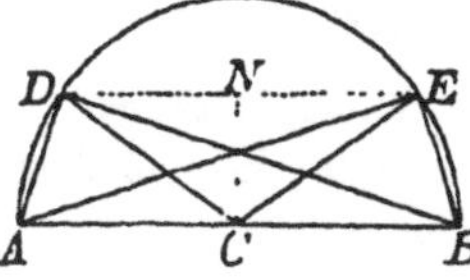

the final, position of an unstretched line. Join DA, DB, EA, EB. DA, DB are the initial, and EA, EB the final, positions of the principal axes. Axes of a shear.

174. The ratio of a shear is the ratio of elongation or contraction of its principal axes. Thus if one principal axis is elongated in the ratio $1:a$, and the other therefore (§ 169) contracted in the ratio $a:1$, a is called the ratio of the shear. It will be convenient generally to reckon this as the ratio of elongation; that is to say, to make its numerical measure greater than unity. Measure of a shear.

In the diagram of § 173, the ratio of DB to EB, or of EA to DA, is the ratio of the shear.

175. The amount of a shear is the amount of relative motion per unit distance between planes of no distortion.

It is easily proved that this is equal to the excess of the ratio of the shear above its reciprocal.

Since $DCA = 2DBA$, and $\tan DBA = \frac{1}{a}$ we have $\tan DCA = \frac{2a}{a^2-1}$.

But $DE = 2CN \tan DCN = 2CN \cot DCA$.

Hence $\frac{DE}{CN} = 2\frac{a^2-1}{2a} = a - \frac{1}{a}$.

176. The planes of no distortion in a simple shear are clearly the circular sections of the strain ellipsoid. In the ellipsoid of this case, be it remembered, the mean axis remains unaltered, and is a mean proportional between the greatest and the least axis. Ellipsoidal specification of a shear.

177. If we now suppose all lines perpendicular to the plane of the shear to be elongated or contracted in any proportion, without altering lengths or angles in the plane of the shear, and if, lastly, we suppose every line in the body to be elongated or contracted in some other fixed ratio, we have clearly (§ 161) the most general possible kind of strain. Thus if s be the ratio of the simple shear, for which case s, 1, $\frac{1}{s}$ are the three principal ratios, and if we elongate lines perpendicular to its plane in the Shear, simple elongation, and expansion combined.

Shear, simple elongation, and expansion, combined. ratio $1 : m$, without any other change, we have a strain of which the principal ratios are

$$s, m, \frac{1}{s}.$$

If, lastly, we elongate all lines in the ratio $1 : n$, we have a strain in which the principal ratios are

$$ns, nm, \frac{n}{s},$$

where it is clear that ns, nm, and $\frac{n}{s}$ may have any values whatever. It is of course not necessary that nm be the mean principal ratio. Whatever they are, if we call them α, β, γ respectively, we have

$$s = \sqrt{\frac{\alpha}{\gamma}}; \; n = \sqrt{\alpha\gamma}; \text{ and } m = \frac{\beta}{\sqrt{\alpha\gamma}}.$$

Analysis of a strain. **178.** Hence any strain (, β, γ) whatever may be viewed as compounded of a uniform dilatation in all directions, of linear ratio $\sqrt{\alpha\gamma}$, superimposed on a simple elongation $\frac{\beta}{\sqrt{\alpha\gamma}}$ in the direction of the principal axis to which β refers, superimposed on a simple shear, of ratio $\sqrt{\frac{\alpha}{\gamma}}$ $\left(\text{or of amount } \sqrt{\frac{\alpha}{\gamma}} - \sqrt{\frac{\gamma}{\alpha}}\right)$ in the plane of the two other principal axes.

179. It is clear that these three elementary component strains may be applied in any other order as well as that stated. Thus, if the simple elongation is made first, the body thus altered must get just the same shear in planes perpendicular to the line of elongation, as the originally unaltered body gets when the order first stated is followed. Or the dilatation may be first, then the elongation, and finally the shear, and so on.

Displacement of a body, rigid or not, one point of which is held fixed. **180.** In the preceding sections on strains, we have considered the alterations of lengths of lines of the body, and of angles between lines and planes of it; and we have, in particular cases, founded on particular suppositions (the principal axes of the strain remaining fixed in direction, § 169, or one of either set of undistorted planes in a simple shear remaining fixed, § 170), considered the actual displacements of parts of the body from their original positions. But to complete the kinematics of a non-rigid solid, it is necessary to take a more general view of the relation between displacements and strains. It will be sufficient for us to suppose one point of the body to remain fixed, as it is easy to see the effect of superimposing upon any motion with one point fixed, a motion of translation without strain or rotation.

Displacement of a body, rigid or not, one point of which is held fixed.

181. Let us therefore suppose one point of a body to be held fixed, and any displacement whatever given to any point or points of it, subject to the condition that the whole substance if strained at all is homogeneously strained.

Let OX, OY, OZ be any three rectangular axes, fixed with reference to the initial position and condition of the body. Let x, y, z be the initial co-ordinates of any point of the body, and x_1, y_1, z_1 be the co-ordinates of the same point of the altered body, with reference to those axes unchanged. The condition that the strain is homogeneous throughout is expressed by the following equations :—

$$\left.\begin{aligned} x_1 &= [Xx]x + [Xy]y + [Xz]z, \\ y_1 &= [Yx]x + [Yy]y + [Yz]z, \\ z_1 &= [Zx]x + [Zy]y + [Zz]z, \end{aligned}\right\} \qquad (1)$$

where $[Xx]$, $[Xy]$, etc., are nine quantities, of absolutely arbitrary values, the same for all values of x, y, z.

$[Xx]$, $[Yx]$, $[Zx]$ denote the three final co-ordinates of a point originally at unit distance along OX, from O. They are, of course, proportional to the direction-cosines of the altered position of the line primitively coinciding with OX. Similarly for $[Xy]$, $[Yy]$, $[Zy]$, etc.

Let it be required to find, if possible, a line of the body which remains unaltered in direction, during the change specified by $[Xx]$, etc. Let x, y, z, and x_1, y_1, z_1, be the co-ordinates of the primitive and altered position of a point in such a line. We must have $\frac{x_1}{x} = \frac{y_1}{y} = \frac{z_1}{z} = 1 + \epsilon$, where ϵ is the elongation of the line in question.

Displacement of a body, rigid or not, one point of which is held fixed.

Thus we have $x_1 = (1+\epsilon)x$, etc., and therefore if $\eta = 1 + \epsilon$

$$\left.\begin{array}{l}\{[Xx]-\eta\}x \quad + [Xy]y \quad + [Xz]z = 0,\\ \quad [Yx]x + \{[Yy]-\eta\}y \quad + [Yz]z = 0,\\ \quad [Zx]x \quad + [Zy]y + \{[Zz]-\eta\}z = 0.\end{array}\right\} \qquad (2)$$

From these equations, by eliminating the ratios $x:y:z$ according to the well-known algebraic process, we find

$$\begin{array}{l}([Xx]-\eta)([Yy]-\eta)([Zz]-\eta)\\ -[Yz][Zy]([Xx]-\eta)-[Zx][Xz]([Yy]-\eta)-[Xy][Yx]([Zz]-\eta)\\ \qquad +[Xz][Yx][Zy]+[Xy][Yz][Zx]=0.\end{array}$$

This cubic equation is necessarily satisfied by at least one real value of η, and the two others are either both real or both imaginary. Each real value of η gives a real solution of the problem, since any two of the preceding three equations with it, in place of η, determine real values of the ratios $x:y:z$. If the body is rigid (*i.e.*, if the displacements are subject to the condition of producing no strain), we know (*ante*, § 95) that there is just one line common to the body in its two positions, the axis round which it must turn to pass from one to the other, except in the peculiar cases of *no* rotation, and of rotation through *two* right angles, which are treated below. Hence, in this case, the cubic equation has only one real root, and therefore it has two imaginary roots. The equations just formed solve the problem of finding the axis of rotation when the data are the actual displacements of the points primitively lying in three given fixed axes of reference, OX, OY, OZ; and it is worthy of remark, that the practical solution of this problem is founded on the one real root of a cubic which has two imaginary roots.

Again, on the other hand, let the given displacements be made so as to produce a strain of the body with no angular displacement of the principal axes of the strain. Thus three lines of the body remain unchanged. Hence there must be three real roots of the equation in η, one for each such axis; and the three lines determined by them are necessarily at right angles to one another.

But if neither of these conditions holds, we may have three real solutions and three oblique lines of directional identity; or we may have only one real root and only one line of directional identity.

An analytical proof of these conclusions may easily be given; thus we may write the cubic in the form— *Displacement of a body, rigid or not, one point of which is held fixed.*

$$\begin{vmatrix} [Xx], & [Xy], & [Xz] \\ [Yx], & [Yy], & [Yz] \\ [Zx], & [Zy], & [Zz] \end{vmatrix} - \eta\left\{\begin{vmatrix} [Yy], & [Yz] \\ [Zy], & [Zz] \end{vmatrix} + \begin{vmatrix} [Zz], & [Zx] \\ [Xz], & [Xx] \end{vmatrix} + \begin{vmatrix} [Xx], & [Xy] \\ [Yx], & [Yy] \end{vmatrix}\right\} + \eta^2\{[Xx]+[Yy]+[Zz]\} - \eta^3 = 0 \ldots\ldots\ldots\ldots (3)$$

In the particular case of no strain, since $[Xx]$, etc., are then *equal*, not merely *proportional*, to the direction cosines of three mutually perpendicular lines, we have by well-known geometrical theorems

$$\begin{vmatrix} [Xx], & [Xy], & [Xz] \\ [Yx], & [Yy], & [Yz] \\ [Zx], & [Zy], & [Zz] \end{vmatrix} = 1, \text{ and } \begin{vmatrix} [Yy], & [Yz] \\ [Zy], & [Zz] \end{vmatrix} = [Xx], \text{ etc.}$$

Hence the cubic becomes

$$1 - (\eta - \eta^2)\{[Xx] + [Yy] + [Zz]\} - \eta^3 = 0,$$

of which one root is evidently $\eta = 1$. This leads to the above explained rotational solution, the line determined by the value 1 of η being the axis of rotation. Dividing out the factor $1 - \eta$, we get for the two remaining roots the equation

$$1 + (1 - [Xx] - [Yy] - [Zz])\,\eta + \eta^2 = 0,$$

whose roots are imaginary if the coefficient of η lies between $+2$ and -2. Now -2 is evidently its *least* value, and for that case the roots are real, each being unity. Here there is no rotation. Also $+2$ is its *greatest* value, and this gives us a pair of values each $= -1$, of which the interpretation is, that there is rotation through two right angles. In this case, as in general, one line (the axis of rotation) is determined by the equations (2) with the value $+1$ for η; but with $\eta = -1$ these equations are satisfied by any line perpendicular to the former.

The limiting case of two equal roots, when there is strain, is an interesting subject which may be left as an exercise. It separates the cases in which there is only one axis of directional identity from those in which there are three.

Let it next be proposed to find those lines of the body whose elongations are greatest or least. For this purpose we must find the equations expressing that $x_1^2 + y_1^2 + z_1^2$ is a maximum, when $x^2 + y^2 + z^2 = r^2$, a constant. First, we have

$$x_1^2 + y_1^2 + z_1^2 = Ax^2 + By^2 + Cz^2 + 2(ayz + bzx + cxy) \ldots\ldots\ldots (4),$$

Displacement of a body, rigid or not, one point of which is held fixed.

where

$$\left.\begin{aligned} A &= [Xx]^2 + [Yx]^2 + [Zx]^2 \\ B &= [Xy]^2 + [Yy]^2 + [Zy]^2 \\ C &= [Xz]^2 + [Yz]^2 + [Zz]^2 \\ a &= [Xy][Xz] + [Yy][Yz] + [Zy][Zz] \\ b &= [Xz][Xx] + [Yz][Yx] + [Zz][Zx] \\ c &= [Xx][Xy] + [Yx][Yy] + [Zx][Zy] \end{aligned}\right\} \quad \text{..............(5).}$$

The equation

$$Ax^2 + By^2 + Cz^2 + 2(ayz + bzx + cxy) = r_1^2 \text{..........(6),}$$

where r_1 is any constant, represents clearly the ellipsoid which a spherical surface, radius r_1, of the altered body, would become if the body were restored to its primitive condition. The problem of making r_1 a maximum when r is a given constant, leads to the following equations:—

$$x^2 + y^2 + z^2 = r^2 \text{..................................(7),}$$

$$\left.\begin{aligned} xdx + ydy + zdz &= 0, \\ (Ax + cy + bz)dx + (cx + By + az)dy + (bx + ay + Cz)dz &= 0. \end{aligned}\right\} \quad (8)$$

On the other hand, the problem of making r a maximum or minimum when r_1 is given, that is to say, the problem of finding maximum and minimum diameters, or principal axes, of the ellipsoid (6), leads to these same two differential equations (8), and only differs in having equation (6) instead of (7) to complete the determination of the absolute values of x, y, and z. Hence the ratios $x : y : z$ will be the same in one problem as in the other; and therefore the *directions* determined are those of the principal axes of the ellipsoid (6). We know, therefore, by the properties of the ellipsoid, that there are three real solutions, and that the directions of the three radii so determined are mutually rectangular. The ordinary method (Lagrange's) for dealing with the differential equations, being to multiply one of them by an arbitrary multiplier, then add, and equate the coefficients of the separate differentials to zero, gives, if we take $-\eta$ as the arbitrary multiplier, and the first of the two equations the one multiplied by it,

$$\left.\begin{aligned} (A-\eta)x + cy + bz &= 0, \\ cx + (B-\eta)y + az &= 0, \\ bx + ay + (C-\eta)z &= 0. \end{aligned}\right\} \quad (9)$$

We may find what η means if we multiply the first of these by x, the second by y, and the third by z, and add; because we thus obtain

Displacement of a body, rigid or not, one point of which is held fixed.

$$Ax^2 + By^2 + Cz^2 + 2(ayz + bzx + cxy) - \eta(x^2 + y^2 + z^2) = 0,$$

or $$r_1^2 - \eta r^2 = 0,$$

which gives $$\eta = \left(\frac{r_1}{r}\right)^2 \quad \ldots\ldots\ldots\ldots\ldots\ldots\ldots\ldots (10).$$

Eliminating the ratios $x:y:z$ from (9), by the usual method, we have the well-known determinant cubic

$$(A-\eta)(B-\eta)(C-\eta) - a^2(A-\eta) - b^2(B-\eta) - c^2(C-\eta) + 2abc = 0 \ldots (11),$$

of which the three roots are known to be all real. Any one of the three roots if used for η, in (9), harmonizes these three equations for the true ratios $x:y:z$; and, making the coefficients of x, y, z in them all known, allows us to determine the required ratios by any two of the equations, or symmetrically from the three, by the proper algebraic processes. Thus we have only to determine the absolute magnitudes of x, y, and z, which (7) enables us to do when their ratios are known.

It is to be remarked, that when $[Yz] = [Zy]$, $[Zx] = [Xz]$, and $[Xy] = [Yx]$, equation (3) becomes a cubic, the squares of whose roots are the roots of (11), and that the three lines determined by (2) in this case are identical with those determined by (9). The reader will find it a good analytical exercise to prove this directly from the equations. It is a necessary consequence of § 183, below.

We have precisely the same problem to solve when the question proposed is, to find what radii of a sphere remain perpendicular to the surface of the altered figure. This is obvious when viewed geometrically. The tangent plane is perpendicular to the radius when the radius is a maximum or minimum. Therefore, every plane of the body parallel to such tangent plane is perpendicular to the radius in the altered, as it was in the initial condition.

The analytical investigation of the problem, presented in the second way, is as follows:—

Let $$l_1x_1 + m_1y_1 + n_1z_1 = 0 \quad \ldots\ldots\ldots\ldots\ldots\ldots\ldots (12)$$

be the equation of any plane of the altered substance, through the origin of co-ordinates, the axes of co-ordinates being the same fixed axes, OX, OY, OZ, which we have used of late. The direction cosines of a perpendicular to it are, of course, proportional to l_1, m_1, n_1. If, now, for x_1, y_1, z_1, we substitute their

Displacement of a body, rigid or not, one point of which is held fixed.

values, as in (1), in terms of the co-ordinates which the same point of the substance had initially, we find the equation of the same plane of the body in its initial position, which, when the terms are grouped properly, is this—

$$\{l_1[Xx] + m_1[Yx] + n_1[Zx]\}x + \{l_1[Xy] + m_1[Yy] + n_1[Zy]\}y + \{l_1[Xz] + m_1[Yz] + n_1[Zz]\}z = 0 \quad\text{.................}(13).$$

The direction cosines of the perpendicular to the plane are proportional to the co-efficients of x, y, z. Now these are to be the direction cosines of the same line of the substance as was altered into the line $l_1 : m_1 : n_1$. Hence, if $l : m : n$ are quantities proportional to the direction cosines of this line in its initial position, we must have

$$\left.\begin{array}{l} l_1[Xx] + m_1[Yx] + n_1[Zx] = \eta l \\ l_1[Xy] + m_1[Yy] + n_1[Zy] = \eta m \\ l_1[Xz] + m_1[Yz] + n_1[Zz] = \eta n \end{array}\right\} \quad\text{...............}(14),$$

where η is arbitrary. Suppose, to fix the ideas, that l_1, m_1, n_1 are the co-ordinates of a certain point of the substance in its altered state, and that l, m, n are proportional to the initial co-ordinates of the same point of the substance. Then we shall have, by the fundamental equations, the expressions for l_1, m_1, n_1 in terms of l, m, n. Using these in the first members of (14), and taking advantage of the abbreviated notation (5), we have precisely the same equations for l, m, n as (9) for x, y, z above.

Analysis of a strain into distortion and rotation.

182. From the preceding analysis it follows that any homogeneous strain whatever applied to a body generally changes a sphere of the body into an ellipsoid, and causes the latter to rotate about a definite axis through a definite angle. In particular cases the sphere may remain a sphere. Also there may be no rotation. In the general case, when there is no rotation, there are three directions in the body (the axes of the ellipsoid) which remain fixed; when there *is* rotation, there are generally three such directions, but not rectangular. Sometimes, however, there is but one.

Pure strain.

183. When the axes of the ellipsoid are lines of the body whose directions do not change, the strain is said to be *pure*, or unaccompanied by rotation. The strains we have already considered were more general than this, being pure strains accompanied by rotation. We proceed to find the analytical conditions of the existence of a pure strain. Pure strain.

Let $O\Xi$, $O\Xi'$, $O\Xi''$ be the three principal axes of the strain, and let

$$l, m, n, \quad l', m', n', \quad l'', m'', n'',$$

be their direction cosines. Let a, a', a'' be the principal elongations. Then, if ξ, ξ', ξ'' be the position of a point of the unaltered body, with reference to $O\Xi$, $O\Xi'$, $O\Xi''$, its position in the body when altered will be $a\xi$, $a'\xi'$, $a''\xi''$. But if x, y, z be its initial, and x_1, y_1, z_1 its final, positions with reference to OX, OY, OZ, we have

$$\xi = lx + my + nz, \quad \xi' = \text{etc.}, \quad \xi'' = \text{etc.} \quad \ldots\ldots\ldots\ldots (15),$$

and $\quad x_1 = la\xi + l'a'\xi' + l''a''\xi'', \quad y_1 = \text{etc.}, \quad z_1 = \text{etc.}$

For ξ, ξ', ξ'' substitute their values (15), and we have x_1, y_1, z_1 in terms of x, y, z, expressed by the following equations :—

$$\left.\begin{aligned} x_1 &= (al^2 + a'l'^2 + a''l''^2)x + (alm + a'l'm' + a''l''m'')y + (aln + a'l'n' + a''l''n'')z \\ y_1 &= (aml + a'm'l' + a''m''l'')x + (am^2 + a'm'^2 + a''m''^2)y + (amn + a'm'n' + a''m''n'')z \\ z_1 &= (anl + a'n'l' + a''n''l'')x + (anm + a'n'm' + a''n''m'')y + (an^2 + a'n'^2 + a''n''^2)z \end{aligned}\right\} . (16).$$

Hence, comparing with (1) of § 181, we have

$$\left.\begin{aligned} [Xx] &= al^2 + a'l'^2 + a''l''^2, \text{ etc.}; \\ [Zy] &= [Yz] = amn + a'm'n' + a''m''n'', \text{ etc.} \end{aligned}\right\} \ldots\ldots\ldots (17).$$

In these equations, l, l', l'', m, m', m'', n, n', n'', are deducible from three independent elements, the three angular co-ordinates (§ 100, above) of a rigid body, of which one point is held fixed; and therefore, along with a, a', a'', constituting in all six independent elements, may be determined so as to make the six members of these equations have any six prescribed values. Hence the conditions necessary and sufficient to insure no rotation are

$$[Zy] = [Yz], \quad [Xz] = [Zx], \quad [Xy] = [Yx] \ldots\ldots\ldots\ldots (18).$$

184. If a body experience a succession of strains, each un- Composition of pure strains. accompanied by rotation, its resulting condition will generally be producible by a strain and a rotation. From this follows the remarkable corollary that three pure strains produced one after another, in any piece of matter, each without rotation, may be so adjusted as to leave the body unstrained, but rotated through some angle about some axis. We shall have, later, most important and interesting applications to fluid motion,

Composition of pure strains. which (Vol. II.) will be proved to be instantaneously, or differentially, irrotational; but which may result in leaving a whole fluid mass merely turned round from its primitive position, as if it had been a rigid body. The following elementary geometrical investigation, though not bringing out a thoroughly comprehensive view of the subject, affords a rigorous demonstration of the proposition, by proving it for a particular case.

Let us consider, as above (§ 171), a simple shearing motion. A point O being held fixed, suppose the matter of the body in a plane, cutting that of the diagram perpendicularly in CD, to move in this plane from right to left parallel to DC; and in other planes parallel to it let there be motions proportional to their distances from O. Consider first a shear from P to P_1; then from P_1 on to P_2; and let O be taken in a line through P_1, perpendicular to CD. During the shear from P to P_1 a point Q moves of course to Q_1 through a distance

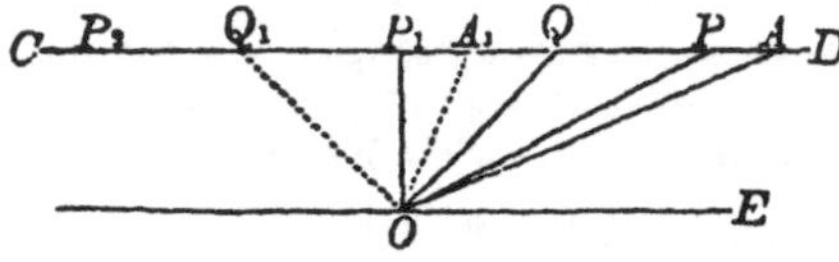

$QQ_1 = PP_1$. Choose Q midway between P and P_1, so that $P_1Q = QP = \frac{1}{2}P_1P$. Now, as we have seen above (§ 152), the line of the body, which is the principal axis of contraction in the shear from Q to Q_1, is OA, bisecting the angle QOE at the beginning, and OA_1, bisecting Q_1OE at the end, of the whole motion considered. The angle between these two lines is half the angle Q_1OQ, that is to say, is equal to P_1OQ. Hence, if the plane CD is rotated through an angle equal to P_1OQ, in the plane of the diagram, in the same way as the hands of a watch, during the shear from Q to Q_1, or, which is the same thing, the shear from P to P_1, this shear will be effected without final rotation of its principal axes. (Imagine the diagram turned round till OA_1 lies along OA. The actual and the newly imagined position of CD will show how this plane of the body has moved during such non-rotational shear.)

Now, let the second step, P_1 to P_2, be made so as to complete the whole shear, P to P_2, which we have proposed to consider. Such second partial shear may be made by the common shearing process parallel to the new position (imagined in the preceding parenthesis) of CD, and to make itself also non-rotational, as its predecessor has been made, we must turn further round, in the same direction, through an angle equal to Q_1OP_1. Thus in these two steps, each made non-rotational, we have turned the plane CD round through an angle equal to Q_1OQ. But now, we have a whole shear PP_2; and to make this as one non-rotational shear, we must turn CD through an angle P_1OP only, which is less than Q_1OQ by the excess of P_1OQ above QOP. Hence the resultant of the two shears, PP_1, P_1P_2, each separately deprived of rotation, is a single shear PP_2, and a rotation of its principal axes, in the direction of the hands of a watch, through an angle equal to $QOP_1 - POQ$. Composition of pure strains.

185. Make the two partial shears each non-rotationally. Return from their resultant in a single non-rotational shear: we conclude with the body unstrained, but turned through the angle $QOP_1 - POQ$, in the same direction as the hands of a watch.

$$\begin{aligned} x_1 &= Ax + cy + bz \\ y_1 &= cx + By + az \\ z_1 &= bx + ay + Cz \end{aligned}$$

is (§ 183) the most general possible expression for the displacement of any point of a body of which one point is held fixed, strained according to any three lines at right angles to one another, as principal axes, which are kept fixed in direction, relatively to the lines of reference OX, OY, OZ.

Similarly, if the body thus strained be again non-rotationally strained, the most general possible expressions for x_2, y_2, z_2, the co-ordinates of the position to which x_1, y_1, z_1, will be brought, are

$$\begin{aligned} x_2 &= A_1x_1 + c_1y_1 + b_1z_1 \\ y_2 &= c_1x_1 + B_1y_1 + a_1z_1 \\ z_2 &= b_1x_1 + a_1y_1 + C_1z_1. \end{aligned}$$

Substituting in these, for x_1, y_1, z_1, their preceding expressions, in terms of the primitive co-ordinates, x, y, z, we have the following expressions for the co-ordinates of the position to which the point in question is brought by the two strains:—

$$\begin{aligned} x_2 &= (A_1A + c_1c + b_1b)\,x + (A_1c + c_1B + b_1a)\,y + (A_1b + c_1a + b_1C)\,z \\ y_2 &= (c_1A + B_1c + a_1b)\,x + (c_1c + B_1B + a_1a)\,y + (c_1b + B_1a + a_1C)\,z \\ z_2 &= (b_1A + a_1c + C_1b)\,x + (b_1c + a_1B + C_1a)\,y + (b_1b + a_1a + C_1C)\,z. \end{aligned}$$

Composition of pure strains.

The resultant displacement thus represented is not generally of the non-rotational character, the conditions (18) of § 183 not being fulfilled, as we see immediately. Thus, for instance, we see that the coefficient of y in the expression for x_2 is not necessarily equal to the coefficient of x in the expression for y_2.

Cor.—If both strains are infinitely small, the resultant displacement is a pure strain without rotation. For A, B, C, A_1, B_1, C_1 are each infinitely nearly unity, and a, b, etc., each infinitely small. Hence, neglecting the products of these infinitely small quantities among one another, and of any of them with the differences between the former and unity, we have a resultant displacement

$$x_2 = A_1Ax + (c + c_1)y + (b + b_1)z$$

$$y_2 = (c_1 + c)x + B_1By + (a + a_1)z$$

$$z_2 = (b_1 + b)x + (a_1 + a)y + C_1Cz,$$

which represents a pure strain unaccompanied by rotation.

Displacement of a curve.

186. The measurement of rotation in a strained elastic solid, or in a moving fluid, is much facilitated by considering separately the displacement of any line of the substance. We are therefore led now to a short digression on the displacement of a curve, which may either belong to a continuous solid or fluid mass, or may be an elastic cord, given in any position. The propositions at which we shall arrive are, of course, applicable to a flexible but inextensible cord (§ 14, above) as a particular case.

It must be remarked, that the displacements to be considered do not depend merely on the curves occupied by the given line in its successive positions, but on the corresponding points of these curves.

Tangential displacement.

What we shall call tangential displacement is to be thus reckoned:—Divide the undisplaced curve into an infinite number of infinitely small equal parts. The sum of the tangential components of the displacements from all the points of division, multiplied by the length of each of the infinitely small parts, is *the entire tangential displacement of the curve reckoned along the undisplaced curve.* The same reckoning carried out in the displaced curve is *the entire tangential displacement reckoned on the displaced curve.*

187. The whole tangential displacement of a curve reckoned along the displaced curve, exceeds the whole tangential displacement reckoned along the undisplaced curve by half the rectangle under the sum and difference of the absolute terminal displacements, taken as positive when the displacement of the end towards which the tangential components are if positive exceeds that at the other. This theorem may be proved by a geometrical demonstration which the reader may easily supply. Two reckonings of tangential displacement compared.

Analytically thus:—Let x, y, z be the co-ordinates of any point, P, in the undisplaced curve; x_1, y_1, z_1, those of P_1 the point to which the same point of the curve is displaced. Let dx, dy, dz be the increments of the three co-ordinates corresponding to any infinitely small arc, ds, of the first; so that

$$ds = (dx^2 + dy^2 + dz^2)^{\frac{1}{2}},$$

and let corresponding notation apply to the corresponding element of the displaced curve. Let θ denote the angle between the line PP_1 and the tangent to the undisplaced curve through P; so that we have

$$\cos\theta = \frac{x_1 - x}{D}\frac{dx}{ds} + \frac{y_1 - y}{D}\frac{dy}{ds} + \frac{z_1 - z}{D}\frac{dz}{ds},$$

where for brevity

$$D = \{(x_1 - x)^2 + (y_1 - y)^2 + (z_1 - z)^2\}^{\frac{1}{2}},$$

being the absolute space of displacement. Hence

$$D\cos\theta ds = (x_1 - x)dx + (y_1 - y)dy + (z_1 - z)dz.$$

Similarly we have

$$D\cos\theta_1 ds_1 = (x_1 - x)\,dx_1 + (y_1 - y)\,dy_1 + (z_1 - z)\,dz_1,$$

and therefore

$$D\cos\theta_1 ds_1 - D\cos\theta ds = (x_1 - x)d(x_1 - x) + (y_1 - y)d(y_1 - y) + (z_1 - z)d(z_1 - z),$$

or

$$D\cos\theta_1 ds_1 - D\cos\theta ds = \tfrac{1}{2}\,d(D^2).$$

To find the difference of the tangential displacements reckoned the two ways, we have only to integrate this expression. Thus we obtain

$$\int D\cos\theta_1 ds_1 - \int D\cos\theta ds = \tfrac{1}{2}(D''^2 - D'^2) = \tfrac{1}{2}(D'' + D')(D'' - D'),$$

where D'' and D' denote the displacements of the two ends.

Tangential displacement of a closed curve.

188. The entire tangential displacement of a closed curve is the same whether reckoned along the undisplaced or the displaced curve.

189. The entire tangential displacement from one to another of two conterminous arcs, is the same reckoned along either as along the other.

Rotation of a rigid closed curve.

190. The entire tangential displacement of a rigid closed curve when rotated through any angle about any axis, is equal to twice the area of its projection on a plane perpendicular to the axis, multiplied by the sine of the angle.

Tangential displacement in a solid, in terms of components of strain.

(*a*) *Prop.*—The entire tangential displacement round a closed curve of a homogeneously strained solid, is equal to

$$2(P\varpi + Q\rho + R\sigma),$$

where P, Q, R denote, for its initial position, the areas of its projections on the planes YOZ, ZOX, XOY respectively, and ϖ, ρ, σ are as follows :—

$$\varpi = \tfrac{1}{2}\{[Zy] - [Yz]\}$$
$$\rho = \tfrac{1}{2}\{[Xz] - [Zx]\}$$
$$\sigma = \tfrac{1}{2}\{[Yx] - [Xy]\}.$$

To prove this, let, farther,

$$a = \tfrac{1}{2}\{[Zy] + [Yz]\}$$
$$b = \tfrac{1}{2}\{[Xz] + [Zx]\}$$
$$c = \tfrac{1}{2}\{[Yx] + [Xy]\}.$$

Thus we have

$$x_1 = Ax + cy + bz + \sigma y - \rho z$$
$$y_1 = cx + By + az + \varpi z - \sigma x$$
$$z_1 = bx + ay + Cz + \rho x - \varpi y.$$

Hence, according to the previously investigated expression, we have, for the tangential displacement, reckoned along the undisplaced curve,

$$\int\{(x_1 - x)\,dx + (y_1 - y)\,dy + (z_1 - z)\,dz\}$$
$$= \int[\tfrac{1}{2}d\{(A-1)\,x^2 + (B-1)\,y^2 + (C-1)\,z^2 + 2\,(ayz + bzx + cxy)\}$$
$$+ \varpi\,(ydz - zdy) + \rho\,(zdx - xdz) + \sigma\,(xdy - ydx)].$$

The first part, $\int\tfrac{1}{2}d\{\ \}$, vanishes for a closed curve.

The remainder of the expression is

Tangential displacement in a solid, in terms of components of strain.

$$\varpi\int(ydz - zdy) + \rho\int(zdx - xdz) + \sigma\int(xdy - ydx),$$

which, according to the formulæ for projection of areas, is equal to

$$2P\varpi + 2Q\rho + 2R\sigma.$$

For, as in § 36 (*a*), we have in the plane of xy

$$\int(xdy - ydx) = \int r^2 d\theta,$$

double the area of the orthogonal projection of the curve on that plane; and similarly for the other integrals.

(*b*) From this and § 190, it follows that if the body is rigid, and therefore only rotationally displaced, if at all, $[Zy] - [Yz]$ is equal to twice the sine of the angle of rotation multiplied by the cosine of the inclination of the axis of rotation to the line of reference OX.

(*c*) And in general $[Zy] - [Yz]$ measures the entire tangential displacement, divided by the area on ZOY, of any closed curve given, if a plane curve, in the plane YOZ, or, if a tortuous curve, given so as to have zero area projections on ZOX and XOY. The entire tangential displacement of any closed curve given in a plane, A, perpendicular to a line whose direction cosines are proportional to ϖ, ρ, σ, is equal to twice its area multiplied by $\sqrt{(\varpi^2 + \rho^2 + \sigma^2)}$. And the entire tangential displacement of any closed curve whatever is equal to twice the area of its projection on A, multiplied by $\sqrt{(\varpi^2 + \rho^2 + \sigma^2)}$.

In the transformation of co-ordinates, ϖ, ρ, σ transform by the elementary cosine law, and of course $\varpi^2 + \rho^2 + \sigma^2$ is an invariant; that is to say, its value is unchanged by transformation from one set of rectangular axes to another.

(*d*) In non-rotational homogeneous strain, the entire tangential displacement along any curve from the fixed point to (x, y, z), reckoned along the undisplaced curve, is equal to

$$\tfrac{1}{2}\{(A-1)x^2 + (B-1)y^2 + (C-1)z^2 + 2(ayz + bzx + cxy)\}.$$

Reckoned along displaced curve, it is, from this and § 187,

$$\tfrac{1}{2}\{(A-1)x^2 + (B-1)y^2 + (C-1)z^2 + 2(ayz + bzx + cxy)\}$$
$$+ \tfrac{1}{2}\{[(A-1)x + cy + bz]^2 + [cx + (B-1)y + az]^2$$
$$+ [bx + ay + (C-1)z]^2\}.$$

And the entire tangential displacement from one point along any curve to another point, is independent of the curve, *i.e.*, is the same along any number of conterminous curves, this of course whether reckoned in each case along the undisplaced or along the displaced curve.

Hetero-geneous strain.

(e) Given the absolute displacement of every point, to find the strain. Let a, β, γ, be the components, relative to fixed axes, OX, OY, OZ, of the displacement of a particle, P, initially in the position x, y, z. That is to say, let $x+a$, $y+\beta$, $z+\gamma$ be the co-ordinates, in the strained body, of the point of it which was initially at x, y, z.

Consider the matter all round this point in its first and second positions. Taking this point P as moveable origin, let ξ, η, ζ be the initial co-ordinates of any other point near it, and ξ_1, η_1, ζ_1 the final co-ordinates of the same.

The initial and final co-ordinates of the last-mentioned point, with reference to the fixed axes OX, OY, OZ, will be

$$x+\xi,\ y+\eta,\ z+\zeta,$$

and $$x+a+\xi_1,\ y+\beta+\eta_1,\ z+\gamma+\zeta_1,$$

respectively; that is to say,

$$a+\xi_1-\xi,\ \beta+\eta_1-\eta,\ \gamma+\zeta_1-\zeta$$

are the components of the displacement of the point which had initially the co-ordinates $x+\xi$, $y+\eta$, $z+\zeta$, or, which is the same thing, are the values of a, β, γ, when x, y, z are changed into

$$x+\xi,\ y+\eta,\ z+\zeta.$$

Hence, by Taylor's theorem,

$$\xi_1-\xi=\frac{da}{dx}\xi+\frac{da}{dy}\eta+\frac{da}{dz}\zeta$$

$$\eta_1-\eta=\frac{d\beta}{dx}\xi+\frac{d\beta}{dy}\eta+\frac{d\beta}{dz}\zeta$$

$$\zeta_1-\zeta=\frac{d\gamma}{dx}\xi+\frac{d\gamma}{dy}\eta+\frac{d\gamma}{dz}\zeta,$$

the higher powers and products of ξ, η, ζ being neglected. Comparing these expressions with (1) of § 181, we see that they express the changes in the co-ordinates of any displaced point of a body relatively to three rectangular axes in fixed directions through one point of it, when all other points of it are displaced relatively to this one, in any manner subject only to the condition of giving a homogeneous strain. Hence we perceive that at distances all round any point, so small that the first terms only of the expressions by Taylor's theorem for the differences of displacement are sensible, the strain is sensibly homogeneous, and we conclude that the directions of the principal axes of the strain at any point (x, y, z), and the amounts of the elongations of the matter along them, and the tangential displacements in closed curves, are to be found according to the general methods described above, by taking **Heterogeneous strain.**

$$[Xx] = \frac{d\alpha}{dx} + 1, \quad [Xy] = \frac{d\alpha}{dy}, \quad [Xz] = \frac{d\alpha}{dz},$$

$$[Yx] = \frac{d\beta}{dx}, \quad [Yy] = \frac{d\beta}{dy} + 1, \quad [Yz] = \frac{d\beta}{dz},$$

$$[Zx] = \frac{d\gamma}{dx}, \quad [Zy] = \frac{d\gamma}{dy}, \quad [Zz] = \frac{d\gamma}{dz} + 1.$$

If each of these nine quantities is constant (*i.e.*, the same for all values of x, y, z), the strain is homogeneous: not unless. **Homogeneous strain.**

(*f*) The condition that the strain may be infinitely small is that **Infinitely small strain.**

$$\frac{d\alpha}{dx}, \frac{d\alpha}{dy}, \frac{d\alpha}{dz},$$

$$\frac{d\beta}{dx}, \frac{d\beta}{dy}, \frac{d\beta}{dz},$$

$$\frac{d\gamma}{dx}, \frac{d\gamma}{dy}, \frac{d\gamma}{dz},$$

must be each infinitely small.

(*g*) These formulæ apply to the most general possible motion of any substance, and they may be considered as the fundamental equations of kinematics. If we introduce time as independent variable, we have for component velocities u, v, w, parallel to the fixed axes OX, OY, OZ, the following expressions; x, y, z, t being independent variables, and α, β, γ functions of them:— **Most general motion of matter.**

$$u = \frac{d\alpha}{dt}, \quad v = \frac{d\beta}{dt}, \quad w = \frac{d\gamma}{dt}.$$

(*h*) If we introduce the condition that no line of the body experiences any elongation, we have the general equations for the kinematics of a rigid body, of which, however, we have had enough already. The equations of condition to express this will be six in number, among the nine quantities $\frac{d\alpha}{dx}$, etc., which (*g*) are, in this case, each constant relatively to x, y, z. There are left three independent arbitrary elements to express any angular motion of a rigid body. **Change of position of a rigid body**

Non-rotational strain.

(*i*) If the disturbed condition is so related to the initial condition that every portion of the body can pass from its initial to its disturbed position and strain, by a translation and a strain without rotation; *i.e.*, if the three principal axes of the strain at any point are lines of the substance which retain their parallelism, we must have, § 183 (18),

$$\frac{d\beta}{dz}=\frac{d\gamma}{dy},\quad \frac{d\gamma}{dx}=\frac{d\alpha}{dz},\quad \frac{d\alpha}{dy}=\frac{d\beta}{dx};$$

and if these equations are fulfilled, the strain is non-rotational, as specified. But these three equations express neither more nor less than that

$$\alpha dx+\beta dy+\gamma dz$$

is the differential of a function of three independent variables. Hence we have the remarkable proposition, and its converse, that if $F(x, y, z)$ denote any function of the co-ordinates of any point of a body, and if every such point be displaced from its given position (x, y, z) to the point whose co-ordinates are

$$x_1=x+\frac{dF}{dx},\quad y_1=y+\frac{dF}{dy},\quad z_1=z+\frac{dF}{dz}\quad\ldots\ldots\ldots\ldots (1),$$

the principal axes of the strain at every point are lines of the substance which have retained their parallelism. The displacement back from (x_1, y_1, z_1) to (x, y, z) fulfils the same condition, and therefore we must have

Theorem of pure analysis.

$$x=x_1+\frac{dF_1}{dx_1},\quad y=y_1+\frac{dF_1}{dy_1},\quad z=z_1+\frac{dF_1}{dz_1}\ldots\ \ldots\ldots (2),$$

where F_1 denotes a function of x_1, y_1, z_1, and $\frac{dF_1}{dx_1}$, etc., its partial differential coefficients with reference to this system of variables. The relation between F and F_1 is clearly

$$F+F_1=-\tfrac{1}{2}D^2\ldots\ldots\ldots\ldots\ldots\ldots\ldots\ldots(3),$$

$$\text{where } D^2=\frac{dF^2}{dx^2}+\frac{dF^2}{dy^2}+\frac{dF^2}{dz^2}=\frac{dF_1^2}{dx_1^2}+\frac{dF_1^2}{dy_1^2}+\frac{dF_1^2}{dz_1^2}\ldots\ldots\ldots (4).$$

This, of course, may be proved by ordinary analytical methods, applied to find x, y, z in terms of x_1, y_1, z_1, when the latter are given by (1) in terms of the former.

(*j*) Let α, β, γ be any three functions of x, y, z. Let dS be any element of a surface; l, m, n the direction cosines of its normal.

Then $\iint dS\left\{l\left(\frac{d\gamma}{dy}-\frac{d\beta}{dz}\right)+m\left(\frac{d\alpha}{dz}-\frac{d\gamma}{dx}\right)+n\left(\frac{d\beta}{dx}-\frac{d\alpha}{dy}\right)\right\}$ — Hetero-geneous strain.

$$=\int(\alpha dx+\beta dy+\gamma dz)\ldots\ldots\ldots\ldots(5),$$

the former integral being over any curvilinear area bounded by a closed curve; and the latter, which may be written

$$\int ds\left(\alpha\frac{dx}{ds}+\beta\frac{dy}{ds}+\gamma\frac{dz}{ds}\right),$$

being round the periphery of this curve line*. To demonstrate this, begin with the part of the first member of (5) depending on α; that is

$$\iint dS\left(m\frac{d\alpha}{dz}-n\frac{d\alpha}{dy}\right);$$

and to evaluate it divide S into bands by planes parallel to ZOY, and each of these bands into rectangles. The breadth at x, y, z, of the band between the planes $x-\frac{1}{2}dx$ and $x+\frac{1}{2}dx$ is $\frac{dx}{\sin\theta}$, if θ denote the inclination of the tangent plane of S to the plane x. Hence if ds denote an element of the curve in which the plane x cuts the surface S, we may take

$$dS=\frac{1}{\sin\theta}dx\,ds.$$

And we have $l=\cos\theta$, and therefore may put

$$m=\sin\theta\cos\phi,\quad n=\sin\theta\sin\phi.$$

Hence

$$\iint dS\left(m\frac{d\alpha}{dz}-n\frac{d\alpha}{dy}\right)=\iint dx\,ds\left(\cos\phi\frac{d\alpha}{dz}-\sin\phi\frac{d\alpha}{dy}\right)$$

$$=\iint dx\,ds\frac{d\alpha}{ds}=\int\alpha dx.$$

The limits of the s integration being properly attended to we see that the remaining integration, $\int\alpha dx$, must be performed round the periphery of the curve bounding S. By this, and corresponding evaluations of the parts of the first member of (5) depending on β and γ, the equation is proved.

* This theorem was given by Stokes in his Smith's Prize paper for 1854 (*Cambridge University Calendar*, 1854). The demonstration in the text is an expansion of that indicated in our first edition. A more synthetical proof is given in § 69 (q) of Sir W. Thomson's paper on "Vortex Motion," *Trans. R. S. E.* 1869. A thoroughly analytical proof is given by Prof. Clerk Maxwell in his *Electricity and Magnetism* (§ 24).

Hetero-geneous strain.

(*k*) It is remarkable that

$$\iint dS\left\{l\left(\frac{d\gamma}{dy}-\frac{d\beta}{dz}\right)+m\left(\frac{d\alpha}{dz}-\frac{d\gamma}{dx}\right)+n\left(\frac{d\beta}{dx}-\frac{d\alpha}{dy}\right)\right\}$$

is the same for all surfaces having common curvilinear boundary; and when α, β, γ are the components of a displacement from x, y, z, it is the entire tangential displacement round the said curvilinear boundary, being a closed curve. It is therefore this that is nothing when the displacement of every part is non-rotational. And when it is not nothing, we see by the above propositions and corollaries precisely what the measure of the rotation is.

Displacement function.

(*l*) *Lastly*, We see what the meaning, for the case of no rotation, of $\int(\alpha dx+\beta dy+\gamma dz)$, or, as it has been called, "the displacement function," is. It is, the entire tangential displacement along any curve from the fixed point O, to the point $P\ (x, y, z)$. And the entire tangential displacement, being in this case the same along all different curves proceeding from one to another of any two points, is equal to the difference of the values of the displacement functions at those points.

"Equation of continuity."

191. As there can be neither annihilation nor generation of matter in any natural motion or action, the whole quantity of a fluid within any space at any time must be equal to the quantity originally in that space, increased by the whole quantity that has entered it and diminished by the whole quantity that has left it. This idea when expressed in a perfectly comprehensive manner for every portion of a fluid in motion constitutes what is called the "*equation of continuity*," an unhappily chosen expression.

Integral equation of continuity.

192. Two ways of proceeding to express this idea present themselves, each affording instructive views regarding the properties of fluids. In one we consider a definite portion of the fluid; follow it in its motions; and declare that the average density of the substance varies inversely as its volume. We thus obtain the equation of continuity in an integral form.

Let a, b, c be the co-ordinates of any point of a moving fluid, at a particular era of reckoning, and let x, y, z be the co-ordinates of the position it has reached at any time t from that era. To specify completely the motion, is to give each of these three varying co-ordinates as a function of a, b, c, t.

Integral equation of continuity.

Let δa, δb, δc denote the edges, parallel to the axes of co-ordinates, of a very small rectangular parallelepiped of the fluid, when $t=0$. Any portion of the fluid, if only small enough in all its dimensions, must (§ 190, *e*), in the motion, approximately fulfil the condition of a body uniformly strained throughout its volume. Hence if δa, δb, δc are taken infinitely small, the corresponding portion of fluid must (§ 156) remain a parallelepiped during the motion.

If a, b, c be the initial co-ordinates of one angular point of this parallelepiped: and $a+\delta a, b, c$; $a, b+\delta b, c$; $a, b, c+\delta c$; those of the other extremities of the three edges that meet in it: the co-ordinates of the same points of the fluid at time t, will be

$$x,\ y,\ z;$$

$$x-\frac{dx}{da}\delta a,\quad y+\frac{dy}{da}\delta a,\quad z+\frac{dz}{da}\delta a;$$

$$x+\frac{dx}{db}\delta b,\quad y+\frac{dy}{db}\delta b,\quad z+\frac{dz}{db}\delta b;$$

$$x+\frac{dx}{dc}\delta c,\quad y+\frac{dy}{dc}\delta c,\quad z+\frac{dz}{dc}\delta c.$$

Hence the lengths and direction cosines of the edges are respectively—

$$\left(\frac{dx^2}{da^2}+\frac{dy^2}{da^2}+\frac{dz^2}{da^2}\right)^{\frac{1}{2}}\delta a,\qquad \frac{\frac{dx}{da}}{\left(\frac{dx^2}{da^2}+\frac{dy^2}{da^2}+\frac{dz^2}{da^2}\right)^{\frac{1}{2}}},\ \text{etc.}$$

$$\left(\frac{dx^2}{db^2}+\frac{dy^2}{db^2}+\frac{dz^2}{db^2}\right)^{\frac{1}{2}}\delta b,\qquad \frac{\frac{dx}{db}}{\left(\frac{dx^2}{db^2}+\frac{dy^2}{db^2}+\frac{dz^2}{db^2}\right)^{\frac{1}{2}}},\ \text{etc.}$$

$$\left(\frac{dx^2}{dc^2}+\frac{dy^2}{dc^2}+\frac{dz^2}{dc^2}\right)^{\frac{1}{2}}\delta c,\qquad \frac{\frac{dx}{dc}}{\left(\frac{dx^2}{dc^2}+\frac{dy^2}{dc^2}+\frac{dz^2}{dc^2}\right)^{\frac{1}{2}}},\ \text{etc.}$$

The volume of this parallelepiped is therefore

$$\left(\frac{dx}{da}\frac{dy}{db}\frac{dz}{dc}-\frac{dx}{da}\frac{dy}{dc}\frac{dz}{db}+\frac{dx}{db}\frac{dy}{dc}\frac{dz}{da}-\frac{dx}{db}\frac{dy}{da}\frac{dz}{dc}+\frac{dx}{dc}\frac{dy}{da}\frac{dz}{db}-\frac{dx}{dc}\frac{dy}{db}\frac{dz}{da}\right)\delta a\delta b\delta c$$

Integral equation of continuity.

or, as it is now usually written,

$$\begin{vmatrix} \frac{dx}{da}, & \frac{dy}{da}, & \frac{dz}{da} \\ \frac{dx}{db}, & \frac{dy}{db}, & \frac{dz}{db} \\ \frac{dx}{dc}, & \frac{dy}{dc}, & \frac{dz}{dc} \end{vmatrix} \delta a \delta b \delta c.$$

Now as there can be neither increase nor diminution of the quantity of matter in any portion of the fluid, the density, or the quantity of matter per unit of volume, in the infinitely small portion we have been considering, must vary inversely as its volume if this varies. Hence, if ρ denote the density of the fluid in the neighbourhood of (x, y, z) at time t, and ρ_0 the initial density, we have

$$\rho \begin{vmatrix} \frac{dx}{da}, & \frac{dy}{da}, & \frac{dz}{da} \\ \frac{dx}{db}, & \frac{dy}{db}, & \frac{dz}{db} \\ \frac{dx}{dc}, & \frac{dy}{dc}, & \frac{dz}{dc} \end{vmatrix} = \rho_0 \quad \ldots\ldots\ldots\ldots\ldots\ldots (1),$$

which is the integral "equation of continuity."

Differential equation of continuity.

193. The form under which the equation of continuity is most commonly given, or the *differential equation of continuity*, as we may call it, expresses that the rate of diminution of the density bears to the density, at any instant, the same ratio as the rate of increase of the volume of an infinitely small portion bears to the volume of this portion at the same instant.

To find it, let a, b, c denote the co-ordinates, not when $t = 0$, but at any time $t - dt$, of the point of fluid whose co-ordinates are x, y, z at t; so that we have

$$x - a = \frac{dx}{dt} dt, \quad y - b = \frac{dy}{dt} dt, \quad z - c = \frac{dz}{dt} dt,$$

according to the ordinary notation for partial differential coefficients; or, if we denote by u, v, w, the components of the velocity of this point of the fluid, parallel to the axes of co-ordinates,

$$x - a = u dt, \quad y - b = v dt, \quad z - c = w dt.$$

Differential equation of continuity.

Hence

$$\frac{dx}{da} = 1 + \frac{du}{da}dt, \quad \frac{dy}{da} = \frac{dv}{da}dt, \quad \frac{dz}{da} = \frac{dw}{da}dt;$$

$$\frac{dx}{db} = \frac{du}{db}dt, \quad \frac{dy}{db} = 1 + \frac{dv}{db}dt, \quad \frac{dz}{db} = \frac{dw}{db}dt;$$

$$\frac{dx}{dc} = \frac{du}{dc}dt, \quad \frac{dy}{dc} = \frac{dv}{dc}dt, \quad \frac{dz}{dc} = 1 + \frac{dw}{dc}dt;$$

and, as we must reject all terms involving higher powers of dt than the first, the determinant becomes simply

$$1 + \left(\frac{du}{da} + \frac{dv}{db} + \frac{dw}{dc}\right)dt.$$

This therefore expresses the ratio in which the volume is augmented in time dt. The corresponding ratio of variation of density is

$$1 + \frac{D\rho}{\rho}$$

if $D\rho$ denote the differential of ρ, the density of one and the same portion of fluid as it moves from the position (a, b, c) to (x, y, z) in the interval of time from $t - dt$ to t. Hence

$$\frac{1}{\rho}\frac{D\rho}{dt} + \frac{du}{da} + \frac{dv}{db} + \frac{dw}{dc} = 0. \quad \ldots\ldots\ldots\ldots\ldots\ (1).$$

Here ρ, u, v, w are regarded as functions of a, b, c, and t, and the variation of ρ implied in $\frac{D\rho}{dt}$ is the rate of the actual variation of the density of an indefinitely small portion of the fluid as it moves away from a fixed position (a, b, c). If we alter the principle of the notation, and consider ρ as the density of whatever portion of the fluid is at time t in the neighbourhood of the fixed point (a, b, c), and u, v, w the component velocities of the fluid passing the same point at the same time, we shall have

$$\frac{D\rho}{dt} = \frac{d_t\rho}{dt} + u\frac{d_a\rho}{da} + v\frac{d_b\rho}{db} + w\frac{d_c\rho}{dc} \quad \ldots\ldots\ldots\ldots\ldots\ (2).$$

Omitting again the suffixes, according to the usual imperfect notation for partial differential co-efficients, which on our new understanding can cause no embarrassment, we thus have, in virtue of the preceding equation,

$$\frac{1}{\rho}\left(\frac{d\rho}{dt} + u\frac{d\rho}{da} + v\frac{d\rho}{db} + w\frac{d\rho}{dc}\right) + \frac{du}{da} + \frac{dv}{db} + \frac{dw}{dc} = 0;$$

or,

$$\frac{d\rho}{dt} + \frac{d(\rho u)}{da} + \frac{d(\rho v)}{db} + \frac{d(\rho w)}{dc} = 0 \quad \ldots\ldots\ldots\ (3),$$

Differential equation of continuity.

which is the differential equation of continuity, in the form in which it is most commonly given.

194. The other way referred to above (§ 192) leads immediately to the differential equation of continuity.

Imagine a space fixed in the interior of a fluid, and consider the fluid which flows into this space, and the fluid which flows out of it, across different parts of its bounding surface, in any time. If the fluid is of the same density and incompressible, the whole quantity of matter in the space in question must remain constant at all times, and therefore the quantity flowing in must be equal to the quantity flowing out in any time. If, on the contrary, during any period of motion, more fluid enters than leaves the fixed space, there will be condensation of matter in that space; or if more fluid leaves than enters, there will be dilatation. The rate of augmentation of the average density of the fluid, per unit of time, in the fixed space in question, bears to the actual density, at any instant, the same ratio that the rate of acquisition of matter into that space bears to the whole matter in that space.

Let the space S be an infinitely small parallelepiped, of which the edges α, β, γ are parallel to the axes of co-ordinates, and let x, y, z be the co-ordinates of its centre; so that $x \pm \frac{1}{2}\alpha$, $y \pm \frac{1}{2}\beta$, $z \pm \frac{1}{2}\gamma$ are the co-ordinates of its angular points. Let ρ be the density of the fluid at (x, y, z), or the mean density through the space S, at the time t. The density at the time $t + dt$ will be $\rho + \frac{d\rho}{dt}dt$; and hence the quantities of fluid contained in the space S, at the times t, and $t + dt$, are respectively $\rho\alpha\beta\gamma$ and $\left(\rho + \frac{d\rho}{dt}dt\right)\alpha\beta\gamma$. Hence the quantity of fluid lost (there will of course be an absolute gain if $\frac{d\rho}{dt}$ be positive) in the time dt is

$$-\frac{d\rho}{dt}\alpha\beta\gamma dt \quad\quad (a).$$

"Steady motion" defined.

Now let u, v, w be the three components of the velocity of the fluid (or of a fluid particle) at P. These quantities will be functions of x, y, z (involving also t, except in the case of "steady motion"), and will in general vary gradually from point to point of the fluid; although the analysis which follows is not restricted by this consideration, but holds even in cases where in certain places of the fluid there are abrupt transitions in the velocity, as may be seen by considering them as limiting cases of motions in which there are very sudden continuous transitions of velocity. If ω be a small plane area, perpendicular to the axis of x, and having its centre of gravity at P, the volume of fluid which flows across it in the time dt will be equal to $u\omega dt$, and the mass or quantity will be $\rho u\omega dt$. If we substitute $\beta\gamma$ for ω, the quantity which flows across either of the faces β, γ of the parallelepiped S, will differ from this only on account of the variation in the value of ρu; and therefore the quantities which flow across the two sides $\beta\gamma$ are respectively

$$\left\{\rho u - \tfrac{1}{2}\alpha\frac{d(\rho u)}{dx}\right\}\beta\gamma dt,$$

and

$$\left\{\rho u + \tfrac{1}{2}\alpha\frac{d(\rho u)}{dx}\right\}\beta\gamma dt.$$

Hence $\alpha\dfrac{d(\rho u)}{dx}\beta\gamma dt$, or $\dfrac{d(\rho u)}{dx}\alpha\beta\gamma dt$, is the excess of the quantity of fluid which leaves the parallelepiped across one of the faces $\beta\gamma$ above that which enters it across the other. By considering in addition the effect of the motion across the other faces of the parallelepiped, we find for the total quantity of fluid lost from the space S, in the time dt,

$$\left\{\frac{d(\rho u)}{dx} + \frac{d(\rho v)}{dy} + \frac{d(\rho w)}{dz}\right\}\alpha\beta\gamma dt \text{ .... } (b).$$

Equating this to the expression (a), previously found, we have

$$\left\{\frac{d(\rho u)}{dx} + \frac{d(\rho v)}{dy} + \frac{d(\rho w)}{dz}\right\}\alpha\beta\gamma dt = -\frac{d\rho}{dt}\alpha\beta\gamma dt;$$

and we deduce

$$\frac{d(\rho u)}{dx} + \frac{d(\rho v)}{dy} + \frac{d(\rho w)}{dz} + \frac{d\rho}{dt} = 0 \text{ .... } (4),$$

which is the required equation.

195. Several references have been made in preceding sections to the number of independent variables in a displacement, or to the degrees of *freedom* or *constraint* under which the displacement takes place. It may be well, therefore, to take a general view of this part of the subject by itself. **Freedom and constraint.**

Of a point. **196.** A free point has *three* degrees of freedom, inasmuch as the most general displacement which it can take is resolvable into three, parallel respectively to any three directions, and independent of each other. It is generally convenient to choose these three directions of resolution at right angles to one another.

If the point be constrained to remain always on a given surface, *one* degree of constraint is introduced, or there are left but *two* degrees of freedom. For we may take the normal to the surface as one of three rectangular directions of resolution. No displacement can be effected parallel to it: and the other two displacements, at right angles to each other, in the tangent plane to the surface, are independent.

If the point be constrained to remain on *each* of two surfaces, it loses two degrees of freedom, and there is left but one. In fact, it is constrained to remain on the curve which is common to both surfaces, and along a curve there is at each point but one direction of displacement.

Of a rigid body. **197.** Taking next the case of a free rigid body, we have evidently *six* degrees of freedom to consider—*three* independent translations in rectangular directions as a point has, and three independent rotations about three mutually rectangular axes.

Freedom and constraint of a rigid body. If it have one point fixed, it loses *three* degrees of freedom; in fact, it has now only the rotations just mentioned.

If a second point be fixed, the body loses *two* more degrees of freedom, and keeps only one freedom to rotate about the line joining the two fixed points. See § 102 above.

If a third point, not in a line with the other two, be fixed, the body is fixed.

198. If a rigid body is forced to touch a smooth surface, *one* degree of freedom is lost; there remain *five*, two displacements parallel to the tangent plane to the surface, and three rotations. As a degree of freedom is lost by a constraint of the body to touch a smooth surface, *six* such conditions completely determine the position of the body. Thus if six points on the barrel and stock of a rifle rest on six convex portions of the surface of a fixed rigid body, the rifle may be placed, and replaced any number of times, in precisely the same position, and always left quite free to recoil when fired, for the purpose of testing its accuracy. Freedom and constraint of a rigid body.

A fixed V under the barrel near the muzzle, and another under the swell of the stock close in front of the trigger-guard, give four of the contacts, bearing the weight of the rifle. A fifth (the one to be broken by the recoil) is supplied by a nearly vertical fixed plane close behind the second V, to be touched by the trigger-guard, the rifle being pressed forward in its V's as far as this obstruction allows it to go. This contact may be dispensed with and nothing sensible of accuracy lost, by having a mark on the second V, and a corresponding mark on barrel or stock, and sliding the barrel backwards or forwards in the V's till the two marks are, as nearly as can be judged by eye, in the same plane perpendicular to the barrel's axis. The sixth contact may be dispensed with by adjusting two marks on the heel and toe of the butt to be as nearly as need be in one vertical plane judged by aid of a plummet. This method requires less of costly apparatus, and is no doubt more accurate and trustworthy, and more quickly and easily executed, than the ordinary method of clamping the rifle in a massive metal cradle set on a heavy mechanical slide.

A geometrical clamp is a means of applying and maintaining six mutual pressures between two bodies touching one another at six points. Geometrical clamp.

A "geometrical slide" is any arrangement to apply five degrees of constraint, and leave one degree of freedom, to the relative motion of two rigid bodies by keeping them pressed together at just five points of their surfaces. Geometrical slide.

Ex. 1. The transit instrument would be an instance if one end of one pivot, made slightly convex, were pressed against a fixed vertical end-plate, by a spring pushing at the other end of the axis. The other four guiding points are the points, or small areas, of contact of the pivots on the Y's. Examples of geometrical slide.

Ex. 2. Let two rounded ends of legs of a three-legged stool rest in a straight, smooth, V-shaped canal, and the third on a smooth horizontal plane*. Gravity maintains positive determinate pressures on the five bearing points; and there is a determinate distribution and amount of friction to be overcome, to produce the rectilineal translational motion thus accurately provided for.

Example of geometrical clamp.

Ex. 3. Let only one of the feet rest in a V canal, and let another rest in a trihedral hollow† in line with the canal, the third still resting on a horizontal plane. There are thus six bearing points, one on the horizontal plane, two on the sides of the canal, and three on the sides of the trihedral hollow: and the stool is fixed in a determinate position as long as all these six contacts are unbroken. Substitute for gravity a spring, or a screw and nut (of not infinitely rigid material), binding the stool to the rigid body to which these six planes belong. Thus we have a "geometrical clamp," which clamps two bodies together with perfect firmness in a perfectly definite position,

* Thomson's reprint of *Electrostatics and Magnetism*, § 346.

† A conical hollow is more easily made (as it can be bored out at once by an ordinary drill), and fulfils nearly enough for most practical applications the geometrical principle. A conical, or otherwise rounded, hollow is touched at three points by knobs or ribs projecting from a round foot resting in it, and thus again the geometrical principle is rigorously fulfilled. The virtue of the geometrical principle is well illustrated by its possible violation in this very case. Suppose the hollow to have been drilled out not quite "true," and instead of being a circular cone to have slightly elliptic horizontal sections:—A hemispherical foot will not rest steadily in it, but will be liable to a slight horizontal displacement in the direction parallel to the major axes of the elliptic sections, besides the legitimate rotation round any axis through the centre of the hemispherical surface: in fact, on this supposition there are just two points of contact of the foot in the hollow instead of three. When the foot and hollow are large enough in any particular case to allow the possibility of this defect to be of moment, it is to be obviated, not by any vain attempt to turn the hollow and the foot each perfectly "true:"—even if this could be done the desired result would be lost by the smallest particle of matter such as a chip of wood, or a fragment of paper, or a hair, getting into the hollow when, at any time in the use of the instrument, the foot is taken out and put in again. On the contrary, the true geometrical method, (of which the general principle was taught to one of us by the late Professor Willis thirty years ago,) is to alter one or other of the two surfaces so as to render it manifestly not a figure of revolution, thus:—Roughly file three round notches in the hollow so as to render it something between a trihedral pyramid and a circular cone, leaving the foot approximately round; or else roughly file at three places of the rounded foot so that horizontal sections through and a little above and below the points of contact may be (roughly) equilateral triangles with rounded corners.

without the aid of friction (except in the screw, if a screw is used); and in various practical applications gives very readily and conveniently a more securely firm connexion by one screw slightly pressed, than a clamp such as those commonly made hitherto by mechanicians can give with three strong screws forced to the utmost. **Example of geometrical clamp.**

Do away with the canal and let two feet (instead of only one) rest on the plane, the other still resting in the conical hollow. The number of contacts is thus reduced to five (three in the hollow and two on the plane), and instead of a "clamp" we have again a slide. This form of slide,—a three-legged stool with two feet resting on a plane and one in a hollow,—will be found very useful in a large variety of applications, in which motion about an axis is desired when a material axis is not conveniently attainable. Its first application was to the "azimuth mirror," an instrument placed on the glass cover of a mariner's compass and used for taking azimuths of sun or stars to correct the compass, or of landmarks or other terrestrial objects to find the ship's position. It has also been applied to the "Deflector," an adjustible magnet laid on the glass of the compass bowl and used, according to a principle first we believe given by Sir Edward Sabine, to discover the "semicircular" error produced by the ship's iron. The movement may be made very frictionless when the plane is horizontal, by weighting the moveable body so that its centre of gravity is very nearly over the foot that rests in the hollow. One or two guard feet, not to touch the plane except in case of accident, ought to be added to give a broad enough base for safety. **Example of geometrical slide.**

The geometrical slide and the geometrical clamp have both been found very useful in electrometers, in the "siphon recorder," and in an instrument recently brought into use for automatic signalling through submarine cables. An infinite variety of forms may be given to the geometrical slide to suit varieties of application of the general principle on which its definition is founded.

An old form of the geometrical clamp, with the six pressures produced by gravity, is the three V grooves on a stone slab bearing the three legs of an astronomical or magnetic instru-

Example of geometrical slide.

ment. It is not generally however so "well-conditioned" as the trihedral hole, the V groove, and the horizontal plane contact, described above.

For investigation of the pressures on the contact surfaces of a geometrical slide or a geometrical clamp, see § 551, below.

There is much room for improvement by the introduction of geometrical slides and geometrical clamps, in the mechanism of mathematical, optical, geodetic, and astronomical instruments: which as made at present are remarkable for disregard of geometrical and dynamical principles in their slides, micrometer screws, and clamps. Good workmanship cannot compensate for bad design, whether in the safety-valve of an iron-clad, or the movements and adjustments of a theodolite.

199. If one point be constrained to remain in a curve, there remain four degrees of freedom.

If two points be constrained to remain in given curves, there are four degrees of constraint, and we have left two degrees of freedom. One of these may be regarded as being a simple rotation about the line joining the constrained points, a motion which, it is clear, the body is free to receive. It may be shown that the other possible motion is of the most general character for one degree of freedom; that is to say, translation and rotation in any fixed proportions as of the nut of a screw.

If one line of a rigid system be constrained to remain parallel to itself, as, for instance, if the body be a three-legged stool standing on a perfectly smooth board fixed to a common window, sliding in its frame with perfect freedom, there remain *three* translations and one rotation.

But we need not further pursue this subject, as the number of combinations that might be considered is endless; and those already given suffice to show how simple is the determination of the degrees of freedom or constraint in any case that may present itself.

One degree of constraint of the most general character.

200. One degree of constraint, of the most general character, is not producible by constraining one point of the body to a curve surface; but it consists in stopping one line of the body from longitudinal motion, except accompanied by rotation round this line, in fixed proportion to the longitudinal motion, and leaving unimpeded every other motion: that is to say, free rotation about any axis perpendicular to this line (two degrees of freedom); and translation in any direction perpendicular to the same line (two degrees of freedom). These four, with the one degree of freedom to screw, constitute the five degrees of freedom, which, with one degree of constraint, make up the six elements. Remark that it is only in case (*b*) below (§ 201) that there is any point of the body which cannot move in every direction.

201. Let a screw be cut on one shaft, *A*, of a Hooke's joint, and let the other shaft, *L*, be joined to a fixed shaft, *B*, by a second Hooke's joint. A nut, *N*, turning on *A*, has the most general kind of motion admitted by one degree of constraint; or it is subjected to just one degree of constraint of the most general character. It has five degrees of freedom; for it may move, 1*st*, by screwing on *A*, the two Hooke's joints being at rest; 2*d*, it may rotate about either axis of the first Hooke's joint, or any axis in their plane (two more degrees of freedom: being freedom to rotate about two axes through one point); 3*d*, it may, by the two Hooke's joints, each bending, have irrotational translation in any direction perpendicular to the link, *L*, which connects the joints (two more degrees of freedom). But it cannot have a translation parallel to the line of the shafts and link without a definite proportion of rotation round this line; nor can it have rotation round this line without a definite proportion of translation parallel to it. The same statements apply to the motion of *B* if *N* is held fixed; but it is now a fixed axis, not as before a moveable one round which the screwing takes place. Mechanical illustration.

No simpler mechanism can be easily imagined for producing one degree of constraint, of the most general kind.

Particular case (*a*).—Step of screw infinite (straight rifling), *i.e.*, the nut may slide freely, but cannot turn. Thus the one degree of constraint is, that there shall be no rotation about a certain axis, a fixed axis if we take the case of *N* fixed and *B* moveable. This is the kind and degree of freedom enjoyed by the outer ring of a gyroscope with its fly-wheel revolving infinitely fast. The outer ring, supposed taken off its stand, and held in the hand, cannot revolve about an axis perpen-

Mechanical illustration. dicular to the plane of the inner ring*, but it may revolve freely about either of two axes at right angles to this, namely, the axis of the fly-wheel, and the axis of the inner ring relative to the outer; and it is of course perfectly free to translation in any direction.

Particular case (*b*).—Step of the screw = 0. In this case the nut may run round freely, but cannot move along the axis of the shaft. Hence the constraint is simply that the body can have no translation parallel to the line of shafts, but may have every other motion. This is the same as if any point of the body in this line were held to a fixed surface. This constraint may be produced less frictionally by not using a guiding surface, but the link and second Hooke's joint of the present arrangement, the first Hooke's joint being removed, and by pivoting one point of the body in a cup on the end of the link. Otherwise, let the end of the link be a continuous surface, and let a continuous surface of the body press on it, rolling or spinning when required, but not permitted to slide.

One degree of constraint expressed analytically. A single degree of constraint is expressed by a single equation among the six co-ordinates specifying the position of one rigid body, relatively to another considered fixed. The effect of this on the body in any particular position is to prevent it from getting out of this position, except by means of component velocities (or infinitely small motions) fulfilling a certain linear equation among themselves.

Thus if ϖ_1, ϖ_2, ϖ_3, ϖ_4, ϖ_5, ϖ_6, be the six co-ordinates, and $F(\varpi_1 \ldots\ldots\ldots) = 0$ the condition; then

$$\frac{dF}{d\varpi_1}\delta\varpi_1 + \ldots\ldots\ldots\ldots = 0$$

is the linear equation which guides the motion through any particular position, the special values of ϖ_1, ϖ_2, ϖ_3, etc., for the particular position, being used in $\frac{dF}{d\varpi_1}$, $\frac{dF}{d\varpi_2}$, &c.

Now, whatever may be the co-ordinate system adopted, we may, if we please, reduce this equation to one between three velocities of translation u, v, w, and three angular velocities ϖ, ρ, σ.

* "The plane of the inner ring" is the plane of the axis of the fly-wheel and of the axis of the inner ring by which it is pivoted on the outer ring.

Let this equation be

One degree of constraint expressed analytically.

$$Au + Bv + Cw + A'\varpi + B'\rho + C'\sigma = 0.$$

This is equivalent to the following:—

$$q + a\omega = 0,$$

if q denote the component velocity along or parallel to the line whose direction cosines are proportional to

$$A,\ B,\ C,$$

ω the component angular velocity round an axis through the origin and in the direction whose direction cosines are proportional to

$$A',\ B',\ C',$$

and lastly,

$$a = \sqrt{\frac{A'^2 + B'^2 + C'^2}{A^2 + B^2 + C^2}}.$$

It might be supposed that by altering the origin of co-ordinates we could do away with the angular velocities, and leave only a linear equation among the components of translational velocity. It is not so; for let the origin be shifted to a point whose co-ordinates are ξ, η, ζ. The angular velocities about the new axes, parallel to the old, will be unchanged; but the linear velocities which, in composition with these angular velocities about the new axes, give ϖ, ρ, σ, u, v, w, with reference to the old, are (§ 89)

$$u - \sigma\eta + \rho\zeta = u',$$
$$v - \varpi\zeta + \sigma\xi = v',$$
$$w - \rho\xi + \varpi\eta = w'.$$

Hence the equation of constraint becomes

$$Au' + Bv' + Cw' + (A' + B\zeta - C\eta)\,\varpi + \text{etc.} = 0.$$

Now we cannot generally determine ξ, η, ζ, so as to make ϖ, etc., disappear, because this would require three conditions, whereas their coefficients, as functions of ξ, η, ζ, are not independent, since there exists the relation

$$A\,(B\zeta - C\eta) + B\,(C\xi - A\zeta) + C\,(A\eta - B\xi) = 0.$$

The simplest form we can reduce to is

$$lu' + mv' + nw' + a\,(l\varpi + m\rho + n\sigma) = 0,$$

that is to say, every longitudinal motion of a certain axis must be accompanied by a definite proportion of rotation about it.

202. These principles constitute in reality part of the general theory of "co-ordinates" in geometry. The three co-ordinates

Generalised co-ordinates. Of a point. of either of the ordinary systems, rectangular or polar, required to specify the position of a point, correspond to the three degrees of freedom enjoyed by an unconstrained point. The most general system of co-ordinates of a point consists of three sets of surfaces, on one of each of which it lies. When one of these surfaces only is given, the point may be anywhere on it, or, in the language we have been using above, it enjoys two degrees of freedom. If a second and a third surface, on each of which also it must lie, it has, as we have seen, no freedom left: in other words, its position is completely specified, being the point in which the three surfaces meet. The analytical ambiguities, and their interpretation, in cases in which the specifying surfaces meet in more than one point, need not occupy us here.

To express this analytically, let $\psi=\alpha$, $\phi=\beta$, $\theta=\gamma$, where ψ, ϕ, θ are functions of the position of the point, and α, β, γ constants, be the equations of the three sets of surfaces, different values of each constant giving the different surfaces of the corresponding set. Any one value, for instance, of α, will determine one surface of the first set, and so for the others: and three particular values of the three constants specify a particular point, P, being the intersection of the three surfaces which they determine. Thus α, β, γ are the "co-ordinates" of P; which may be referred to as "the point (α, β, γ)." The form of the co-ordinate surfaces of the (ψ, ϕ, θ) system is defined in terms of co-ordinates (x, y, z) on any other system, plane rectangular co-ordinates for instance, if ψ, ϕ, θ are given each as a function of (x, y, z).

Origin of the differential calculus. **203.** Component velocities of a moving point, parallel to the three axes of co-ordinates of the ordinary plane rectangular system, are, as we have seen, the rates of augmentation of the corresponding co-ordinates. These, according to the Newtonian fluxional notation, are written $\dot{x}$, $\dot{y}$, $\dot{z}$; or, according to Leibnitz's notation, which we have used above, $\frac{dx}{dt}, \frac{dy}{dt}, \frac{dz}{dt}$. Lagrange has combined the two notations with admirable skill and taste in the first edition* of his *Mécanique Analytique*, as we shall

* In later editions the Newtonian notation is very unhappily altered by the

see in Chap. II. In specifying the motion of a point according to the generalized system of co-ordinates, ψ, ϕ, θ must be considered as varying with the time: $\dot{\psi}$, $\dot{\phi}$, $\dot{\theta}$, or $\frac{d\psi}{dt}$, $\frac{d\phi}{dt}$, $\frac{d\theta}{dt}$, will then be the generalized components of velocity: and $\ddot{\psi}$, $\ddot{\phi}$, $\ddot{\theta}$, or $\frac{d\dot{\psi}}{dt}$, $\frac{d\dot{\phi}}{dt}$, $\frac{d\dot{\theta}}{dt}$, or $\frac{d^2\psi}{dt^2}$, $\frac{d^2\phi}{dt^2}$, $\frac{d^2\theta}{dt^2}$, will be the generalized components of acceleration.

204. On precisely the same principles we may arrange sets of co-ordinates for specifying the position and motion of a material system consisting of any finite number of rigid bodies, or material points, connected together in any way. Thus if ψ, ϕ, θ, etc., denote any number of elements, independently variable, which, when all given, fully specify its position and configuration, being of course equal in number to the degrees of freedom to move enjoyed by the system, these elements are its *co-ordinates*. When it is actually moving, their rates of variation per unit of time, or $\dot{\psi}$, $\dot{\phi}$, etc., express what we shall call its generalized component velocities; and the rates at which $\dot{\psi}$, $\dot{\phi}$, etc., augment per unit of time, or $\ddot{\psi}$, $\ddot{\phi}$, etc., its component accelerations. Thus, for example, if the system consists of a single rigid body quite free, ψ, ϕ, etc, in number six, may be three common co-ordinates of one point of the body, and three angular co-ordinates (§ 101, above) fixing its position relatively to axes in a given direction through this point. Then $\dot{\psi}$, $\dot{\phi}$, etc., will be the three components of the velocity of this point, and the velocities of the three angular motions explained in § 101, as corresponding to variations in the angular co-ordinates. Or, again, the system may consist of one rigid body supported on a fixed axis; a second, on an axis fixed relatively to the first; a third, on an axis fixed relatively to the second, and so on. There will be in this case only as many co-ordinates as there are of rigid bodies. These co-ordinates might be, for instance, the angle between a plane of the first body and a fixed plane, through the first axis; the angle between planes through the

Co-ordinates of any system.

Generalized components of velocity.

Examples.

substitution of accents, ′ and ″, for the · and ·· signifying velocities and accelerations.

Generalized components of velocity. Examples. second axis, fixed relatively to the first and second bodies, and so on; and the component velocities, $\dot{\psi}$, $\dot{\phi}$, etc. would then be the angular velocity of the first body relatively to directions fixed in space; the angular velocity of the second body relatively to the first; of the third relatively to the second, and so on. Or if the system be a set, i in number, of material points perfectly free, one of its $3i$ co-ordinates may be the sum of the squares of their distances from a certain point, either fixed or moving in any way relatively to the system, and the remaining $3i-1$ may be angles, or may be mere ratios of distances between individual points of the system. But it is needless to multiply examples here. We shall have illustrations enough of the principle of generalized co-ordinates, by actual use of it in Chap. II., and other parts of this book.

APPENDIX TO CHAPTER I.

A_0.—Expression in Generalized Co-ordinates for Poisson's extension of Laplace's equation.

(*a*) In § 491 (*c*) below is to be found Poisson's extension of Laplace's equation, expressed in rectilineal rectangular co-ordinates; and in § 492 an equivalent in a form quite independent of the particular kind of co-ordinates chosen: all with reference to the theory of attraction according to the Newtonian law. The same analysis is largely applicable through a great range of physical mathematics, including hydro-kinematics (the "equation of continuity" § 192), the equilibrium of elastic solids (§ 734), the vibrations of elastic solids and fluids (Vol. II.), Fourier's theory of heat, &c. Hence detaching the analytical subject from particular physical applications, consider the equation

$$\frac{d^2U}{dx^2}+\frac{d^2U}{dy^2}+\frac{d^2U}{dz^2}=-4\pi\rho\ldots\ldots\ldots\ldots\ldots(1)$$

where ρ is a given function of x, y, z, (arbitrary and discontinuous it may be). Let it be required to express in terms of generalized co-ordinates ξ, ξ', ξ'', the property of U which this equation expresses in terms of rectangular rectilinear co-ordinates. This may be done of course directly [§ (m) below] by analytical transformation, finding the expression in terms of ξ, ξ', ξ'', for the operation $\frac{d^2}{dx^2} + \frac{d^2}{dy^2} + \frac{d^2}{dz^2}$. But it is done in the form most convenient for physical applications much more easily as follows, by taking advantage of the formula of § 492 which expresses the same property of U independently of any particular system of co-ordinates. This expression is

Laplace's equation in generalized co-ordinates.

$$\iint \delta U dS = -4\pi \iiint \rho dB \text{..........................} (2),$$

where $\iint dS$ denotes integration over the whole of a closed surface S, $\iiint dB$ integration throughout the volume B enclosed by it, and δU the rate of variation of U at any point of S, per unit of length in the direction of the normal outwards.

(b) For B take an infinitely small curvilineal parallelepiped having its centre at (ξ, ξ', ξ''), and angular points at

$$(\xi \pm \tfrac{1}{2}\delta\xi, \ \xi' \pm \tfrac{1}{2}\delta\xi', \ \xi'' \pm \tfrac{1}{2}\delta\xi'').$$

Let $R\delta\xi$, $R'\delta\xi'$, $R''\delta\xi''$ be the lengths of the edges of the parallelepiped, and a, a', a'' the angles between them in order of symmetry, so that $R'R'' \sin a\, \delta\xi'\delta\xi''$, &c., are the areas of its faces.

Let DU, $D'U$, $D''U$ denote the rates of variation of U, per unit of length, perpendicular to the three surfaces $\xi = \text{const.}$, $\xi' = \text{const.}$, $\xi'' = \text{const.}$, intersecting in (ξ, ξ', ξ'') the centre of the parallelepiped. The value of $\iint \delta U dS$ for a section of the parallelepiped by the surface $\xi = \text{const.}$ through (ξ, ξ', ξ'') will be

$$R'R'' \sin a\, \delta\xi'\, \delta\xi''\, DU.$$

Hence the values of $\iint \delta U\, dS$ for the two corresponding sides of the parallelepiped are

$$R'R'' \sin a\, \delta\xi'\, \delta\xi''\, DU \pm \frac{d}{d\xi}(R'R'' \sin a\, \delta\xi'\, \delta\xi''\, DU) \,.\, \tfrac{1}{2}\delta\xi.$$

Hence the value of $\iint \delta U\, dS$ for the pair of sides is

$$\frac{d}{d\xi}(R'R'' \sin a\, \delta\xi'\, \delta\xi''\, DU) \,.\, \delta\xi,$$

or
$$\frac{d}{d\xi}(R'R'' \sin a\, DU)\, \delta\xi\, \delta\xi'\, \delta\xi''.$$

Dealing similarly with the two other pairs of sides of the parallelepiped and adding we find the first member of (2). Its

Laplace's equation in generalized co-ordinates.

second member is $-4\pi\rho \,.\, Q \,.\, RR'R''\,\delta\xi\,\delta\xi'\,\delta\xi''$, if Q denote the ratio of the bulk of the parallelepiped to a rectangular one of equal edges. Hence equating and dividing both sides by the bulk of the parallelepiped we find

$$\frac{1}{QRR'R''}\left\{\frac{d}{d\xi}(R'R''\sin a\, DU)+\frac{d}{d\xi'}(R''R\sin a'\, D'U)\right.$$
$$\left.+\frac{d}{d\xi''}(RR'\sin a''\, D''U)\right\}=-4\pi\rho \ldots (3).$$

(c) It remains to express DU, $D'U$, $D''U$ in terms of the co-ordinates ξ, ξ', ξ''.

Denote by K, L the two points (ξ, ξ', ξ'') and $(\xi+\delta\xi, \xi', \xi'')$. From L (not shown in the diagram) draw LM perpendicular to the surface $\xi=$ const. through K. Taking an infinitely small portion of this surface for the plane of our diagram, let $K\Xi'$, $K\Xi''$ be the lines in which it is cut respectively by the surfaces $\xi''=$ const. and $\xi'=$ const. through K. Draw MN parallel to $\Xi''K$, and MG perpendicular to $K\Xi'$.

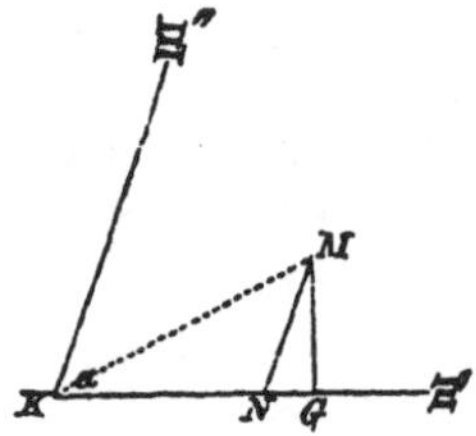

Let now p denote the angle LKM,
A' „ „ „ LGM.

We have

$ML=KL\sin p=R\sin p\,\delta\xi$,

$NM=GM\operatorname{cosec} a=ML\operatorname{cosec} a\cot A'=R\sin p\operatorname{cosec} a\cot A'\,\delta\xi$.

Similarly $\quad KN=R\sin p\operatorname{cosec} a\cot A''\,\delta\xi$,

if A'' denotes an angle corresponding to A'; so that A' and A'' are respectively the angles at which the surfaces $\xi''=$ const. and $\xi'=$ const. cut the plane of the diagram in the lines $K\Xi'$ and $K\Xi''$.

Now the difference of values of ξ' for K and N is $\frac{KN}{R'}$,

and „ „ „ „ „ ξ'' „ N „ M „ $\frac{MN}{R''}$.

Hence if $U(K)$, $U(M)$, $U(L)$ denote the values of U respectively at the points K, M, L, we have

$$U(M)=U(K)+\frac{dU}{d\xi'}\cdot\frac{KN}{R'}+\frac{dU}{d\xi''}\cdot\frac{NM}{R''},$$

and $$U(L)=U(K)+\frac{dU}{d\xi}\delta\xi.$$

But
$$DU = \frac{U(L) - U(M)}{ML},$$
Laplace's equation in generalized co-ordinates.

and so using the preceding expressions in the terms involved we find

$$DU = \frac{1}{R \sin p} \frac{dU}{d\xi} - \frac{1}{R' \sin a \tan A''} \frac{dU}{d\xi'} - \frac{1}{R'' \sin a \tan A'} \frac{dU}{d\xi''} \ldots (4).$$

Using this and the symmetrical expressions for $D'U$ and $D''U$, in (3), we have the required equation.

(*d*) It is to be remarked that a, a', a'' are the three sides of a spherical triangle of which A, A', A'' are the angles, and p the perpendicular from the angle A to the opposite side.

Hence by spherical trigonometry

$$\cos A = \frac{\cos a - \cos a \cos a'}{\sin a \sin a'};$$

$$\sin A = \frac{\sqrt{(1 - \cos^2 a - \cos^2 a' - \cos^2 a'' + 2 \cos a \cos a' \cos a'')}}{\sin a \sin a'} \ldots (5):$$

$$\sin p = \sin A' \sin a''$$

$$= \frac{\sqrt{(1 - \cos^2 a - \cos^2 a' - \cos^2 a'' + 2 \cos a \cos a' \cos a'')}}{\sin a} \ldots\ldots\ldots (6).$$

To find Q remark that the volume of the parallelepiped is equal to $f \sin p \,.\, gh \sin a$ if f, g, h be its edges: therefore

$$Q = \sin p \sin a \ldots\ldots\ldots (7),$$

whence by (6)

$$Q = \sqrt{(1 - \cos^2 a - \cos^2 a' - \cos^2 a'' + 2 \cos a \cos a' \cos a'')} \ldots\ldots (8).$$

Lastly by (5) and (8) we have

$$\tan A = \frac{Q}{\cos a - \cos a' \cos a''} \ldots\ldots\ldots (9).$$

(*e*) Using these in (4) we find

$$DU = \frac{1}{Q \sin a} \left(\frac{\sin^2 a}{R} \frac{dU}{d\xi} - \frac{\cos a'' - \cos a \cos a'}{R'} \frac{dU}{d\xi'} \right.$$
$$\left. - \frac{\cos a' - \cos a \cos a''}{R''} \frac{dU}{d\xi''} \right) \ldots\ldots (10).$$

Using this and the two symmetrical expressions in (3) and adopting a common notation [App. B (*g*), § 491 (*c*), &c. &c.], according to which Poisson's equation is written

$$\nabla^2 U = -4\pi\rho \ldots\ldots\ldots (11),$$

Laplace's equation in generalized co-ordinates.

we find for the symbol ∇^2 in terms of the generalized co-ordinates ξ, ξ', ξ'',

$$\nabla^2 = \frac{1}{QRR'R''}\left\{\frac{d}{d\xi}\frac{1}{Q}\left[\frac{R'R''\sin^2 a}{R}\frac{d}{d\xi} + R''(\cos a\cos a' - \cos a'')\frac{d}{d\xi'} + R'(\cos a''\cos a - \cos a')\frac{d}{d\xi''}\right]\right.$$

$$+\frac{d}{d\xi'}\frac{1}{Q}\left[\frac{R''R\sin^2 a'}{R'}\frac{d}{d\xi'} + R(\cos a'\cos a'' - \cos a)\frac{d}{d\xi''} + R''(\cos a\cos a' - \cos a'')\frac{d}{d\xi}\right],$$

$$\left. + \frac{d}{d\xi''}\frac{1}{Q}\left[\frac{RR'\sin^2 a''}{R''}\frac{d}{d\xi''} + R'(\cos a''\cos a - \cos a')\frac{d}{d\xi} + R(\cos a'\cos a'' - \cos a)\frac{d}{d\xi'}\right]\right\} \ldots(12),$$

where for Q, its value by (8) in terms of a, a', a'' is to be used, and a, a', a'', R, R', R'' are all known functions of ξ, ξ', ξ'' when the system of co-ordinates is completely defined.

Case of rectangular co-ordinates, curved or plane.

(*f*) For the case of rectangular co-ordinates whether plane or curved $a = a' = a'' = A = A' = A'' = 90°$ and $Q = 1$, and therefore we have

$$\nabla^2 = \frac{1}{RR'R''}\left\{\frac{d}{d\xi}\left(\frac{R'R''}{R}\frac{d}{d\xi}\right) + \frac{d}{d\xi'}\left(\frac{R''R}{R'}\frac{d}{d\xi'}\right) + \frac{d}{d\xi''}\left(\frac{RR'}{R''}\frac{d}{d\xi''}\right)\right\} \ldots(13),$$

which is the formula originally given by Lamé for expressing in terms of his orthogonal curved co-ordinate system the Fourier equations of the conduction of heat. The proof of the more general formula (12) given above is an extension, in purely analytical form, of a demonstration of Lamé's formula (13) which was given in terms relating to thermal conduction in an article "On the equations of Motion of Heat referred to curvilinear co-ordinates" in the *Cambridge Mathematical Journal* (1843).

(*g*) For the particular case of polar co-ordinates, r, θ, ϕ, considering the rectangular parallelepiped corresponding to δr, $\delta\theta$, $\delta\phi$ we see in a moment that the lengths of its edges are δr, $r\delta\theta$, $r\sin\theta\delta\phi$. Hence in the preceding notation $R = 1$, $R' = r$, $R'' = r\sin\theta$, and Lamé's formula (13) gives

$$\nabla^2 = \frac{1}{r^2\sin\theta}\left\{\sin\theta\frac{d}{dr}\left(r^2\frac{d}{dr}\right) + \frac{d}{d\theta}\left(\sin\theta\frac{d}{d\theta}\right) + \frac{1}{\sin\theta}\frac{d^2}{d\phi^2}\right\} \ldots(14).$$

(h) Again let the co-ordinates be of the kind which has been called "columnar"; that is to say, distance from an axis (r), angle from a plane of reference through this axis to a plane through the axis and the specified point (ϕ), and distance from a plane of reference perpendicular to the axis (z). The co-ordinate surfaces here are

Laplace's equation in columnar co-ordinates.

coaxal circular cylinders (r = const.),

planes through the axis (ϕ = const.),

planes perpendicular to the axis (z = const.).

The three edges of the infinitesimal rectangular parallelepiped are now dr, $rd\phi$, and dz. Hence $R = 1$, $R' = r$, $R'' = 1$, and Lamé's formula gives

$$\nabla^2 = \frac{1}{r}\frac{d}{dr}\left(r\frac{d}{dr}\right) + \left(\frac{d}{rd\phi}\right)^2 + \left(\frac{d}{dz}\right)^2 \;\ldots\ldots\ldots\ldots\ldots(15),$$

which is very useful for many physical problems, such as the conduction of heat in a solid circular column, the magnetization of a round bar or wire, the vibrations of air in a closed circular cylinder, the vibrations of a vortex column, &c. &c.

(i) For plane rectangular co-ordinates we have $R = R' = R''$; so in this case (13) becomes (with x, y, z for ξ, ξ', ξ''),

Algebraic transformation from plane rectangular to generalized co-ordinates

$$\nabla^2 = \frac{d^2}{dx^2} + \frac{d^2}{dy^2} + \frac{d^2}{dz^2} \;\ldots\ldots\ldots\ldots\ldots\ldots\ldots(16),$$

which is Laplace's and Fourier's original form.

(j) Suppose now it be desired to pass from plane rectangular co-ordinates to the generalized co-ordinates.

Let x, y, z be expressed as functions of ξ, ξ', ξ''; then putting for brevity

$$\frac{dx}{d\xi} = X,\; \frac{dy}{d\xi} = Y,\; \frac{dz}{d\xi} = Z;\; \frac{dx}{d\xi'} = X', \text{ \&c.};\; \frac{dx}{d\xi''} = X'', \text{ \&c.} \ldots(17);$$

we have

$$\left.\begin{aligned} \delta x &= X\delta\xi + X'\delta\xi' + X''\delta\xi'', \\ \delta y &= Y\delta\xi + Y'\delta\xi' + Y''\delta\xi'', \\ \delta z &= Z\delta\xi + Z'\delta\xi' + Z''\delta\xi'', \end{aligned}\right\} \;\ldots\ldots\ldots\ldots(18);$$

whence

$$R = \sqrt{(X^2 + Y^2 + Z^2)},\qquad R' = \sqrt{(X'^2 + Y'^2 + Z'^2)},$$
$$R'' = \sqrt{(X''^2 + Y''^2 + Z''^2)} \ldots\ldots\ldots\ldots\ldots\ldots(19),$$

Algebraic transformation from plane rectangular to generalized co-ordinates.

and the direction cosines of the three edges of the infinitesimal parallelepiped corresponding to $\delta\xi$, $\delta\xi'$, $\delta\xi''$ are

$$\left(\frac{X}{R}, \frac{Y}{R}, \frac{Z}{R}\right), \left(\frac{X'}{R'}, \frac{Y'}{R'}, \frac{Z'}{R'}\right), \left(\frac{X''}{R''}, \frac{Y''}{R''}, \frac{Z''}{R''}\right) \ldots\ldots(20).$$

Hence

$$\cos a = \frac{X'X'' + Y'Y'' + Z'Z''}{R'R''}, \quad \cos a' = \frac{X''X + Y''Y + Z''Z}{R''R},$$

$$\cos a'' = \frac{XX' + YY' + ZZ'}{RR'} \ldots\ldots\ldots\ldots\ldots\ldots(21).$$

(k) It is important to remark that when these expressions for $\cos a$, $\cos a'$, $\cos a''$, R, R', R'', in terms of X, &c. are used in (8), Q^2 becomes a complete square, so that $QRR'R''$ is a rational homogeneous function of the 3rd degree of X, Y, Z, X', &c.

For the ordinary process of finding from the direction cosines (20) of three lines, the sine of the angle between one of them and the plane of the other two gives

$$\sin p = \begin{vmatrix} X, & Y, & Z \\ X', & Y', & Z' \\ X'', & Y'', & Z'' \end{vmatrix} \div RR'R'' \sin a \ldots\ldots\ldots(21);$$

from this and (7) we see that $QRR'R''$ is equal to the determinant. From this and (8) we see that

Square of a determinant.

$$(X^2 + Y^2 + Z^2)(X'^2 + Y'^2 + Z'^2)(X''^2 + Y''^2 + Z''^2)$$
$$-(X^2+Y^2+Z^2)(X'X''+Y'Y''+Z'Z'')^2-(X'^2+Y'^2+Z'^2)(X''X+Y''Y+Z''Z)^2$$
$$-(X''^2 + Y''^2 + Z''^2)(XX' + YY' + ZZ')^2$$
$$+ 2(X'X''+Y'Y''+Z'Z'')(X''X+Y''Y+Z''Z)(XX'+YY'+ZZ')$$
$$= \begin{vmatrix} X, & Y, & Z, \\ X', & Y', & Z', \\ X'', & Y'', & Z'', \end{vmatrix}^2 \ldots\ldots\ldots\ldots\ldots\ldots(22),$$

an algebraic identity which may be verified by expanding both members and comparing.

(l) Denoting now by T the complete determinant, we have

$$Q = \frac{T}{RR'R''} \ldots\ldots\ldots\ldots\ldots\ldots\ldots\ldots(23),$$

and using this for Q in (12) we have a formula for ∇^2 in which only rational functions of X, Y, Z, X', &c. appear, and which is readily verified by comparing with the following derived from (16) by direct transformation.

Algebraic transformation from plane rectangular to generalized co-ordinates.

(m) Go back to (18) and resolve for $\delta\xi$, $\delta\xi'$, $\delta\xi''$. We find

$$\delta\xi = \frac{L}{T}\delta x + \frac{M}{T}\delta y + \frac{N}{T}\delta z, \quad \delta\xi' = \&c., \quad \delta\xi'' = \&c.,$$

where

$$\left.\begin{array}{lll} L = Y'Z'' - Y''Z', & M = Z'X'' - Z''X', & N = X'Y'' - X''Y', \\ L' = Y''Z - YZ'', & M' = Z''X - ZX'', & N' = X''Y - XY'', \\ L'' = YZ' - Y'Z, & M'' = ZX' - Z'X, & N'' = XY' - X'Y, \end{array}\right\} \ldots\ldots (24).$$

Hence

$$\frac{d}{dx} = \frac{L}{T}\frac{d}{d\xi} + \frac{L'}{T}\frac{d}{d\xi'} + \frac{L''}{T}\frac{d}{d\xi''}, \quad \frac{d}{dy} = \&c., \quad \frac{d}{dz} = \&c.,$$

and thus we have

$$\nabla^2 = \left(\frac{L}{T}\frac{d}{d\xi} + \frac{L'}{T}\frac{d}{d\xi'} + \frac{L''}{T}\frac{d}{d\xi''}\right)^2 + \left(\frac{M}{T}\frac{d}{d\xi} + \frac{M'}{T}\frac{d}{d\xi'} + \frac{M''}{T}\frac{d}{d\xi''}\right)^2$$
$$+ \left(\frac{N}{T}\frac{d}{d\xi} + \frac{N'}{T}\frac{d}{d\xi'} + \frac{N''}{T}\frac{d}{d\xi''}\right)^2 \ldots\ldots\ldots\ldots (25).$$

Expanding this and comparing the coefficients of $\frac{d^2}{d\xi^2}$, $\frac{d^2}{d\xi d\xi'}$, $\frac{d}{d\xi}$, &c. with those of the corresponding terms of (12) with (21) and (23) we find the two formulas, (12) and (25), identical.

A.—Extension of Green's Theorem.

It is convenient that we should here give the demonstration of a few theorems of pure analysis, of which we shall have many and most important applications, not only in the subject of spherical harmonics, which follows immediately, but in the general theories of attraction, of fluid motion, and of the conduction of heat, and in the most practical investigations regarding electricity, and magnetic and electro-magnetic force.

(a) Let U and U' denote two functions of three independent variables, x, y, z, which we may conveniently regard as rectangular co-ordinates of a point P, and let α denote a quantity which may be either constant, or any arbitrary function of the variables. Let $\iiint dxdydz$ denote integration throughout a finite *singly continuous* space bounded by a close surface S; let $\iint dS$ denote integration over the whole surface S; and let δ, prefixed to any function, denote its rate of variation at any point of S, per unit of length in the direction perpendicular to S outwards.

Then

a constant gives a theorem of Green's.

$$\iiint a^2\left(\frac{dU}{dx}\frac{dU'}{dx}+\frac{dU}{dy}\frac{dU'}{dy}+\frac{dU}{dz}\frac{dU'}{dz}\right)dxdydz$$

$$\left.\begin{aligned}&=\iint dS.U'a^2\delta U-\iiint U'\left\{\frac{d\left(a^2\frac{dU}{dx}\right)}{dx}+\frac{d\left(a^2\frac{dU}{dy}\right)}{dy}+\frac{d\left(a^2\frac{dU}{dz}\right)}{dz}\right\}dxdydz\\&=\iint dS.Ua^2\delta U'-\iiint U\left\{\frac{d\left(a^2\frac{dU'}{dx}\right)}{dx}+\frac{d\left(a^2\frac{dU'}{dy}\right)}{dy}+\frac{d\left(a^2\frac{dU'}{dz}\right)}{dz}\right\}dxdydz\end{aligned}\right\}\quad\ldots\ldots\ldots\ldots\ldots(1).$$

For, taking one term of the first member alone, and integrating "by parts," we have

$$\iiint a^2\frac{dU}{dx}\frac{dU'}{dx}dxdydz=\iint U'a^2\frac{dU}{dx}dydz-\iiint U'\frac{d\left(a^2\frac{dU}{dx}\right)}{dx}dxdydz,$$

the first integral being between limits corresponding to the surface S; that is to say, being from the negative to the positive end of the portion within S, or of each portion within S, of the line x through the point $(0, y, z)$. Now if A_2 and A_1 denote the inclination of the outward normal of the surface to this line, at points where it enters and emerges from S respectively, and if dS_2 and dS_1 denote the elements of the surface in which it is cut at these points by the rectangular prism standing on $dydz$, we have

$$dydz=-\cos A_2 dS_2=\cos A_1 dS_1.$$

Thus the first integral, between the proper limits, involves the elements $U'a^2\frac{dU}{dx}\cos A_1 dS_1$, and $-U'a^2\frac{dU}{dx}\cos A_2 dS_2$; the latter of which, as corresponding to the lower limit, is subtracted. Hence, there being in the whole of S an element dS_2 for each element dS_1, the first integral is simply

$$\iint U'a^2\frac{dU}{dx}\cos A\, dS,$$

for the whole surface. Adding the corresponding terms for y and z, and remarking that

a constant gives a theorem of Green's.

$$\frac{dU}{dx}\cos A + \frac{dU}{dy}\cos B + \frac{dU}{dz}\cos C = \delta U,$$

where B and C denote the inclinations of the outward normal through dS to lines drawn through dS in the positive directions parallel to y and z respectively, we perceive the truth of (1).

(b) Again, let U and U' denote two functions of x, y, z, which have equal values at every point of S, and of which the first fulfils the equation

Equation of the conduction of heat.

$$\frac{d\left(a^2\frac{dU}{dx}\right)}{dx} + \frac{d\left(a^2\frac{dU}{dy}\right)}{dy} + \frac{d\left(a^2\frac{dU}{dz}\right)}{dz} = 0 \quad\quad (2),$$

for every point within S.

Then if $U' - U = u$, we have

$$\iiint\left\{\left(a\frac{dU'}{dx}\right)^2 + \left(a\frac{dU'}{dy}\right)^2 + \left(a\frac{dU'}{dz}\right)^2\right\}dxdydz$$

$$= \iiint\left\{\left(a\frac{dU}{dx}\right)^2 + \left(a\frac{dU}{dy}\right)^2 + \left(a\frac{dU}{dz}\right)^2\right\}dxdydz$$

$$+ \iiint\left\{\left(a\frac{du}{dx}\right)^2 + \left(a\frac{du}{dy}\right)^2 + \left(a\frac{du}{dz}\right)^2\right\}dxdydz \quad\quad (3).$$

For the first member is equal identically to the second member with the addition of

$$2\iiint a^2\left(\frac{dU}{dx}\frac{du}{dx} + \frac{dU}{dy}\frac{du}{dy} + \frac{dU}{dz}\frac{du}{dz}\right)dxdydz.$$

But, by (1), this is equal to

$$2\iint dS . ua^2\delta U - 2\iiint u\left\{\frac{d\left(a^2\frac{dU}{dx}\right)}{dx} + \frac{d\left(a^2\frac{dU}{dy}\right)}{dy} + \frac{d\left(a^2\frac{dU}{dz}\right)}{dz}\right\}dxdydz,$$

of which each term vanishes; the first, or the double integral, because, by hypothesis, u is equal to nothing at every point of S, and the second, or the triple integral, because of (2).

(c) The second term of the second member of (3) is essentially positive, provided a has a real value, whether positive, zero, or negative, for every point (x, y, z) within S. Hence the first member of (3) necessarily exceeds the first term of the second member. But the sole characteristic of U is that it satisfies (2). Hence U' cannot also satisfy (2). That is to say, U being any

Property of solution with U given over S.

Solution proved to

be determinate; one solution of (2), there can be no other solution agreeing with it at every point of S, but differing from it for some part of the space within S.

proved to be possible. (*d*) One solution of (2) exists, satisfying the condition that U has an arbitrary value for every point of the surface S. For let U denote any function whatever which has the given arbitrary value at each point of S; let u be any function whatever which is equal to nothing at each point of S, and which is of any real finite or infinitely small value, of the same sign as the value of

$$\frac{d\left(a^2\frac{dU}{dx}\right)}{dx}+\frac{d\left(a^2\frac{dU}{dy}\right)}{dy}+\frac{d\left(a^2\frac{dU}{dz}\right)}{dz}$$

at each internal point, and therefore, of course, equal to nothing at every internal point, if any, for which the value of this expression is nothing; and let $U'=U+\theta u$, where θ denotes any constant. Then, using the formulæ of (*b*), modified to suit the altered circumstances, and taking Q and Q' for brevity to denote

$$\iiint\left\{\left(a\frac{dU}{dx}\right)^2+\left(a\frac{dU}{dy}\right)^2+\left(a\frac{dU}{dz}\right)^2\right\}dxdydz,$$

and the corresponding integral for U', we have

$$Q'=Q-2\theta\iiint u\left\{\frac{d}{dx}\left(a^2\frac{dU}{dx}\right)+\frac{d}{dy}\left(a^2\frac{dU}{dy}\right)+\frac{d}{dz}\left(a^2\frac{dU}{dz}\right)\right\}dxdydz$$
$$+\theta^2\iiint\left\{\left(a\frac{du}{dx}\right)^2+\left(a\frac{du}{dy}\right)^2+\left(a\frac{du}{dz}\right)^2\right\}dxdydz.$$

The coefficient of -2θ here is essentially positive, in consequence of the condition under which u is chosen, unless (2) is satisfied, in which case it is nothing; and the coefficient of θ^2 is essentially positive, if not zero, because all the quantities involved are real. Hence the equation may be written thus:—

$$Q'=Q-m\theta(n-\theta),$$

where m and n are each positive. This shows that if any positive value less than n is assigned to θ, Q' is made smaller than Q; that is to say, unless (2) is satisfied, a function, having the same value at S as U, may be found which shall make the Q integral smaller than for U. In other words, a function U, which, having any prescribed value over the surface S, makes the integral Q for the interior as small as possible, must satisfy equation (2). But the Q integral is essentially positive, and therefore there is a limit than which it cannot be made smaller.

Hence there is a solution of (2) subject to the prescribed surface condition. **Solution proved possible.**

(*e*) We have seen (*c*) that there is, if one, only one, solution of (2) subject to the prescribed surface condition, and now we see that there is one. To recapitulate,—we conclude that, if the value of U be given arbitrarily at every point of any closed surface, the equation **a constant gives "Green's theorem."**

$$\frac{d}{dx}\left(a^2\frac{dU}{dx}\right)+\frac{d}{dy}\left(a^2\frac{dU}{dy}\right)+\frac{d}{dz}\left(a^2\frac{dU}{dz}\right)=0$$

determines its value without ambiguity for every point within that surface. That this important proposition holds also for the whole infinite space without the surface S, follows from the preceding demonstration, with only the precaution, that the different functions dealt with must be so taken as to render all the triple integrals convergent. S need not be merely a single closed surface, but it may be any number of surfaces enclosing isolated portions of space. The extreme case, too, of S, or any detached part of S, an open shell, that is a finite unclosed surface, is clearly included. Or lastly, S, or any detached part of S, may be an infinitely extended surface, provided the value of U arbitrarily assigned over it be so assigned as to render the triple and double integrals involved all convergent.

B.—Spherical Harmonic Analysis.

The mathematical method which has been commonly referred to by English writers as that of "Laplace's Coefficients," but which is here called *spherical harmonic analysis*, has for its object the expression of an arbitrary periodic function of two independent variables in the proper form for a large class of physical problems involving arbitrary data over a spherical surface, and the deduction of solutions for every point of space. **Object of spherical harmonic analysis.**

(*a*) A *spherical harmonic function* is defined as a homogeneous function, V, of x, y, z, which satisfies the equation **Definition of spherical harmonic functions.**

$$\frac{d^2V}{dx^2}+\frac{d^2V}{dy^2}+\frac{d^2V}{dz^2}=0\ldots\ldots\ldots\ldots\ldots\ldots(4).$$

Its degree may be any positive or negative integer; or it may be fractional; or it may be imaginary

Examples of spherical harmonics.

EXAMPLES. The functions written below are spherical harmonics of the degrees noted; r representing $(x^2+y^2+z^2)^{\frac{1}{2}}$:—

Degree Zero.

I. $$1; \qquad \log\frac{r+z}{r-z}.$$
$$\tan^{-1}\frac{y}{x}; \quad \tan^{-1}\frac{y}{x}\log\frac{r+z}{r-z}; \quad \frac{rz(x^2-y^2)}{(x^2+y^2)^2}; \quad \frac{2rzxy}{(x^2+y^2)^2}.$$

II. Generally, in virtue of (*g*) (15) and (13) below,
$$r\frac{dV_0}{dx}, \quad r\frac{dV_0}{dy}, \quad r\frac{dV_0}{dz},$$
if V_0 denote any harmonic of degree 0: for instance, group III. below.

III. $$\frac{rx}{x^2+y^2}; \quad \frac{zx}{x^2+y^2}; \quad \frac{x}{r+z}\left(=\frac{rx-zx}{x^2+y^2}\right); \quad \frac{2zy}{x^2+y^2}\tan^{-1}\frac{y}{x}-\frac{xr}{x^2+y^2}\log\frac{r+z}{r-z}.$$
$$\frac{ry}{x^2+y^2}; \quad \frac{zy}{x^2+y^2}; \quad \frac{y}{r+z}; \qquad \frac{2zx}{x^2+y^2}\tan^{-1}\frac{y}{x}+\frac{yr}{x^2+y^2}\log\frac{r+z}{r-z}.$$

IV. Generally, in virtue of (*g*) (15), (13), below,
$$\delta_{n-j-1}(r^{2(n-j)-1}\delta_n V_j),$$
where V_j denotes any spherical harmonic of integral degree, j, and δ_n, δ_{n-j-1} homogeneous integral functions of $\frac{d}{dx}, \frac{d}{dy}, \frac{d}{dz}$, of degrees n and $n-j-1$ respectively: for instance, some of group II. above, and groups V. and VI. below.

V. $$\frac{d^{n-1}(r^{2n-1})}{dz^{n-1}} \cdot \frac{d^n\tan^{-1}\frac{y}{x}}{dx^n};$$
$$\frac{d^{n-1}(r^{2n-1})}{dz^{n-1}} \cdot \frac{d^n\tan^{-1}\frac{y}{x}}{dy^n}.$$
Remark that
$$\tan^{-1}\frac{y}{x}=\frac{1}{2\sqrt{-1}}\log\frac{x+y\sqrt{-1}}{x-y\sqrt{-1}},$$
and therefore
$$\frac{d^n\tan^{-1}\frac{y}{x}}{dx^n}=(-1)^n 1\,.\,2\ldots(n-1)\frac{\sin n\phi}{(x^2+y^2)^{\frac{n}{2}}};$$

Examples of spherical harmonics.

so the preceding yields

$$\frac{d^{n-1}(r^{2n-1})}{dz^{n-1}} \frac{\begin{smallmatrix}\sin\\ \cos\end{smallmatrix} n\phi}{(x^2+y^2)^{\frac{n}{2}}},$$

where ϕ denotes $\tan^{-1}\frac{y}{x}$.

Taking, in IV., $j=-1$,

$$V_j = \frac{1}{r}\log\frac{r+z}{r-z},$$

$$\delta_n = \tfrac{1}{2}\left\{\left(\frac{d}{dx}+\frac{d}{dy}\sqrt{-1}\right)^n + \left(\frac{d}{dx}-\frac{d}{dy}\sqrt{-1}\right)^n\right\},$$

or $$\delta_n = \frac{1}{2\sqrt{-1}}\left\{\left(\frac{d}{dx}+\frac{d}{dy}\sqrt{-1}\right)^n - \left(\frac{d}{dx}-\frac{d}{dy}\sqrt{-1}\right)^n\right\},$$

VI. $$\delta_{n-j-1} = \left(\frac{d}{dz}\right)^n,$$

we find

$$\left(\frac{d}{dz}\right)^n\left[r^{2n+1}\left(\frac{d}{rdr}\right)^n\left(\frac{1}{r}\log\frac{r+z}{r-z}\right)\right](x^2+y^2)^{\frac{n}{2}} \begin{smallmatrix}\cos\\ \sin\end{smallmatrix} n\phi,$$

where $\frac{d}{dr}$ denotes differentiation with reference to r on the supposition of z constant, and $\frac{d}{dz}$ differentiation with reference to z on supposition of x and y constant.

Degree $-i-1$, *or* $+i$, *and type* $H\{z, \sqrt{(x^2+y^2)}\} \begin{smallmatrix}\sin\\ \cos\end{smallmatrix} n\phi$.

H denoting a homogeneous function; n any integer; and i any positive integer.

Let $U_0^{(n)}$ and $V_0^{(n)}$ denote functions yielded by V. and VI. preceding. The following are the two* distinct functions of the degrees and types now sought, and found in virtue of (g) (15) below:—

$$U_{-i-1}^{(n)} = \frac{d^{i+1}}{dz^{i+1}} U_0^{(n)}, \quad V_{-i-1}^{(n)} = \frac{d^{i+1}}{dz^{i+1}} V_0;$$

* See § (l) below.

Examples of spherical harmonics.

or explicitly

I. $$\begin{cases} U^{(n)}_{-i-1} = \dfrac{d^{n+i}(r^{2n-1})}{dz^{n+i}} \dfrac{\begin{matrix}\cos\\ \sin\end{matrix} n\phi}{(x^2+y^2)^{\frac{n}{2}}}, \\[2ex] V^{(n)}_{-i-1} = \left(\dfrac{d}{dz}\right)^{n+i+1} \left[r^{2n+1} \left(\dfrac{d}{rdr}\right)^n \left(\dfrac{1}{r} \log \dfrac{r+z}{r-z}\right)\right] (x^2+y^2)^{\frac{n}{2}} \begin{matrix}\cos\\ \sin\end{matrix} n\phi. \end{cases}$$

In the particular case of $n=0$, these two are not distinct. Either of them yields

II. $$U^{(0)}_{-i-1} = \frac{d^i\left(\dfrac{1}{r}\right)}{dz^i}.$$

The other harmonic of the same degree and type is

$$V^{(0)}_{-i-1} = \frac{d^i\left(\dfrac{1}{r}\log\dfrac{r+z}{r-z}\right)}{dz^i}.$$

III. To obtain the harmonics of the same types, but of degree i, multiply each of the preceding groups I. and II. by r^{2i+1}, in virtue of (g) (13) below.

Degree -1.

I. Generally, in virtue of (g) (13) below, any of the preceding functions of degree zero divided by r; or, in virtue of (g) (15), the differential coefficient of any of them with reference to x, or y, or z. For instance,

II. $\dfrac{1}{r}$;

III. $$\frac{1}{r}\tan^{-1}\frac{y}{x};\quad \frac{1}{r}\log\frac{r+z}{r-z};\quad \frac{1}{r}\tan^{-1}\frac{y}{x}\log\frac{r+z}{r-z}.$$

IV. $$\begin{cases} \dfrac{x}{x^2+y^2};\quad \dfrac{xz}{r(x^2+y^2)};\quad \dfrac{x}{r(r+z)}; \\[2ex] \dfrac{y}{x^2+y^2};\quad \dfrac{yz}{r(x^2+y^2)};\quad \dfrac{y}{r(r+z)}. \end{cases}$$

Examples of spherical harmonics

Degrees -2 *and* $+1$.

I. $\left\{\frac{x}{r^3},\ \frac{y}{r^3},\ \frac{z}{r^3};\quad x,\ y,\ z.\right.$

II. $\left\{\frac{z\tan^{-1}\frac{y}{x}}{r^3};\quad z\tan^{-1}\frac{y}{x}.\right.$

III. $\left\{\frac{z}{r^3}\log\frac{r+z}{r-z}-\frac{2}{r^2};\quad z\log\frac{r+z}{r-z}-2r.\right.$

IV. $\left\{\begin{array}{l}\frac{x^2-y^2}{(x^2+y^2)^2},\ \frac{2xy}{(x^2+y^2)^2};\ \frac{r^3(x^2-y^2)}{(x^2+y^2)^2},\ \frac{2r^3xy}{(x^2+y^2)^2};\\ \text{or}\ \frac{\cos 2\phi}{x^2+y^2},\ \frac{\sin 2\phi}{x^2+y^2};\ \frac{r^3\cos 2\phi}{x^2+y^2},\ \frac{r^3\sin 2\phi}{x^2+y^2}.\end{array}\right.$

V. $\left\{\frac{1}{r^3}\left(\log\frac{r+z}{r-z}+\frac{2rz}{x^2+y^2}\right)x;\quad\left(\log\frac{r+z}{r-z}+\frac{2rz}{x^2+y^2}\right)x\right.$

(the former being $\frac{d}{dx}$ of III. 2 dogree -1, and the latter being $-\int dz$ of VI. degree 0 with $n=1$).

The Rational Integral Harmonics of Degree 2.

I. Five distinct functions, for instance,

$$2z^2-x^2-y^2;\ x^2-y^2;\ yz;\ xz;\ xy.$$

Or one function with five arbitrary constants.

II. $\left\{\begin{array}{ll} & ax^2+by^2+cz^2+eyz+fzx+gxy,\\ \text{where} & a+b+c=0.\end{array}\right.$

Degrees $-n-1$, *and* $+n$ (*n any integer*).

With same notation and same references for proof as above for Degree 0, group IV.

I. $\delta_{n+1}V_0,\quad \delta_n V_{-1},\quad \text{or}\quad \delta_{n+j}V_{j-1}.$

II. $\delta_{n+i-j+1}(r^{2(i-j)+1}\delta_i V_{j-1})$, and $r^{2n+1}\delta_{n+i-j+1}(r^{2(i-j)+1}\delta_i V_{j-1})$.

Examples of spherical harmonics.

Degrees $e + vf$, *and* $-e-1-vf$.

(v denoting $\sqrt{-1}$, and e and f any real quantities.)

I.

$$\tfrac{1}{2}\left[(x+vy)^{e+vf}+(x-vy)^{e+vf}\right]; \quad \frac{1}{2v}\left[(x+vy)^{e+vf}-(x-vy)^{e+vf}\right]:$$

or $$q^{e+vf}\cos\left[(e+vf)\phi\right]; \quad q^{e+vf}\sin\left[(e+vf)\phi\right],$$

where $$q=\sqrt{(x^2+y^2)} \text{ and } \phi=\tan^{-1}\frac{y}{x}:$$

or $$\tfrac{1}{2}q^{e+vf}\left[\epsilon^{v(e+vf)\phi}+\epsilon^{-v(e+vf)\phi}\right]; \quad \frac{1}{2v}q^{e+vf}\left[\epsilon^{v(e+vf)\phi}-\epsilon^{-v(e+vf)\phi}\right]:$$

or $$\tfrac{1}{2}q^e\{\epsilon^{f\phi}\left[\cos(f\log q-e\phi)+v\sin(f\log q-e\phi)\right] + \epsilon^{-f\phi}\left[\cos(f\log q+e\phi)+v\sin(f\log q+e\phi)\right]\};$$

II. the same with $\frac{3\pi}{2}+e\phi$ instead of $e\phi$.

III.

$$\frac{\tfrac{1}{2}\left[(x+vy)^{e+vf}+(x-vy)^{e+vf}\right]}{r^{2(e+vf)+1}};$$

or $$\tfrac{1}{2}r^{-2e-1}q^e\left[\epsilon^{f\phi}\epsilon^{v(f\log q-2f\log r-e\phi)}+\epsilon^{-f\phi}\epsilon^{v(f\log q-2f\log r+e\phi)}\right];$$

or $$\tfrac{1}{2}r^{-2e-1}q^e\left\{\epsilon^{f\phi}\left[\cos\left(f\log\frac{q}{r^2}-e\phi\right)+v\sin\left(f\log\frac{q}{r^2}-e\phi\right)\right] + \epsilon^{-f\phi}\left[\cos\left(f\log\frac{q}{r^2}+e\phi\right)+v\sin\left(f\log\frac{q}{r^2}+e\phi\right)\right]\right\}.$$

(*b*) A *spherical surface harmonic* is the function of two angular co-ordinates, or spherical surface co-ordinates, which a spherical harmonic becomes at any spherical surface described from O, the origin of co-ordinates, as centre. Sometimes a function which, according to the definition (*a*), is simply a spherical harmonic, will be called a *spherical solid harmonic*, when it is desired to call attention to its not being confined to a spherical surface.

(*c*) A *complete spherical harmonic* is one which is finite and of single value for all finite values of the co-ordinates.

Partial Harmonics.

A *partial harmonic* is a spherical harmonic which either does not continuously satisfy the fundamental equation (4) for space completely surrounding the centre, or does not return to the same value in going once round every closed curve. The "partial" harmonic is as it were a harmonic for a part of the spherical surface: but it may be for a part which is greater than the whole, or a part of which portions jointly and independently occupy the same space.

(d) It will be shown, later, § (h), that a complete spherical harmonic is necessarily either a rational integral function of the co-ordinates, or reducible to one by a factor of the form

Algebraic quality of complete harmonics.

$$(x^2+y^2+z^2)^{\frac{m}{2}},$$

m being an integer.

(e) The general problem of finding harmonic functions is most concisely stated thus :—

Differential equations of a harmonic of degree n.

To find the most general integral of the equation

$$\frac{d^2u}{dx^2}+\frac{d^2u}{dy^2}+\frac{d^2u}{dz^2}=0 \qquad (4')$$

subject to the condition

$$x\frac{du}{dx}+y\frac{du}{dy}+z\frac{du}{dz}=nu \qquad (5),$$

the second of these equations being merely the analytical expression of the condition that u is a homogeneous function of x, y, z of the degree n, which may be any whole number positive or negative, any fraction, or any imaginary quantity.

Let $P+vQ$ be a harmonic of degree $e+vf$, P, Q, e, f being real. We have

Differential equation for real constituents of a homogeneous function of imaginary degree.

$$\left(x\frac{d}{dx}+y\frac{d}{dy}+z\frac{d}{dz}\right)(P+vQ)=(e+vf)(P+vQ);$$

and therefore

$$\left.\begin{aligned} x\frac{dP}{dx}+y\frac{dP}{dy}+z\frac{dP}{dz}&=eP-fQ \\ x\frac{dQ}{dx}+y\frac{dQ}{dy}+z\frac{dQ}{dz}&=fP+eQ \end{aligned}\right\} \qquad (5');$$

whence

$$\left.\begin{aligned} &\left[\left(x\frac{d}{dx}+y\frac{d}{dy}+z\frac{d}{dz}-e\right)^2+f^2\right]P=0, \\ \text{and}\quad &\left[\left(x\frac{d}{dx}+y\frac{d}{dy}+z\frac{d}{dz}-e\right)^2+f^2\right]Q=0, \end{aligned}\right\} \qquad (5'').$$

(f) Analytical expressions in various forms for an absolutely general integration of these equations, may be found without much difficulty; but with us the only value or interest which any such investigation can have, depends on the availability of

Value of general symbolical expressions.

its results for solutions fulfilling the conditions at bounding surfaces presented by physical problems. In a very large and most important class of physical problems regarding space bounded by a complete spherical surface, or by two complete concentric spherical surfaces, or by closed surfaces differing very little from spherical surfaces, the case of n any positive or negative integer, integrated particularly under the restriction stated in (*d*), is of paramount importance. It will be worked out thoroughly below.

Use of complete spherical harmonics in physical problems.

Again, in similar problems regarding sections cut out of spherical spaces by two diametral planes making any angle with one another *not a sub-multiple of two right angles*, or regarding spaces bounded by two circular cones having a common vertex and axis, and by the included portion of two spherical surfaces described from their vertex as centre, solutions for cases of fractional and imaginary values of n are useful. Lastly, when the subject is a solid or fluid, shaped as a section cut from the last-mentioned spaces by two planes through the axis of the cones, inclined to one another at any angle, whether a submultiple of π or not, we meet with the case of n either integral or not, but to be integrated under a restriction differing from that specified in (*d*). We shall accordingly, after investigating general expressions for complete spherical harmonics, give some indications as to the determination of the incomplete harmonics, whether of fractional, of imaginary, or of integral degrees, which are required for the solution of problems regarding such portions of spherical spaces as we have just described.

Uses of incomplete spherical harmonics in physical problems.

A few formulæ, which will be of constant use in what follows, are brought together in the first place.

Working formulæ.

(*g*) Calling O the origin of co-ordinates, and P the point x, y, z, let $OP=r$, so that $x^2+y^2+z^2=r^2$. Let δ, prefixed to any function, denote its rate of variation per unit of space in the direction OP; so that

$$\delta=\frac{x}{r}\frac{d}{dx}+\frac{y}{r}\frac{d}{dy}+\frac{z}{r}\frac{d}{dz} \quad \text{.......................(6).}$$

If H_n denote any homogeneous function of x, y, z of order n, we have clearly

$$\delta H_n=\frac{n}{r}H_n \quad \text{........................ (7);}$$

whence
$$x\frac{dH_n}{dx}+y\frac{dH_n}{dy}+z\frac{dH_n}{dz}=nH_n \quad \text{......... (5) or (8),}$$

the well-known differential equation of a homogeneous function; in which, of course, n may have any value, positive, integral, negative, fractional, or imaginary. Again, denoting, for brevity, Working formulæ.

$\frac{d^2}{dx^2}+\frac{d^2}{dy^2}+\frac{d^2}{dz^2}$ by ∇^2, we have, by differentiation,

$$\nabla^2(r^m)=m(m+1)r^{m-2} \quad\ldots\ldots(9).$$

Also, if u, u' denote any two functions,

$$\nabla^2(uu')=u'\nabla^2u+2\left(\frac{du}{dx}\frac{du'}{dx}+\frac{du}{dy}\frac{du'}{dy}+\frac{du}{dz}\frac{du'}{dz}\right)+u\nabla^2u' \ldots\ldots(10);$$

whence, if u and u' are both solutions of (4),

$$\nabla^2(uu')=2\left(\frac{du}{dx}\frac{du'}{dx}+\frac{du}{dy}\frac{du'}{dy}+\frac{du}{dz}\frac{du'}{dz}\right)\ldots\ldots(11);$$

or, by taking $u=V_n$, a harmonic of degree n, and $u'=r^m$,

$$\nabla^2(r^mV_n)=2mr^{m-2}\left(x\frac{dV_n}{dx}+y\frac{dV_n}{dy}+z\frac{dV_n}{dz}\right)+V_n\nabla^2(r^m),$$

or, by (8) and (9),

$$\nabla^2(r^mV_n)=m(2n+m+1)r^{m-2}V_n\ldots\ldots(12).$$

From this last it follows that $r^{-2n-1}V_n$ is a harmonic; which, being of degree $-n-1$, may be denoted by V_{-n-1}, so that we have

$$\left.\begin{array}{ll} & V_{-n-1}=r^{-2n-1}V_n \\ \text{or} & \dfrac{V_{n'}}{r^{n'}}=\dfrac{V_n}{r^n} \\ \text{if} & n+n'=-1 \end{array}\right\}\ldots\ldots(13),$$

a formula showing a reciprocal relation between two solid harmonics which give the same form of surface harmonic at any spherical surface described from O as centre. Again, by taking $m=-1$, in (9), we have

$$\nabla^2\frac{1}{r}=0\ldots\ldots(14).$$

Hence $\frac{1}{r}$ is a harmonic of degree -1. We shall see later § (h), that it is the only *complete harmonic*, of this degree.

If u be any solution of the equation $\nabla^2u=0$, we have also

$$\nabla^2\frac{du}{dx}=0,$$

Working formulæ.

and so on for any number of differentiations. Hence if V_ι is a harmonic of any degree ι, $\frac{d^{j+k+l}V_\iota}{dx^j dy^k dz^l}$ is a harmonic of degree $\iota-j-k-l$; or, as we may write it,

$$\frac{d^{j+k+l}V_\iota}{dx^j dy^k dz^l}=V_{\iota-j-k-l} \quad\text{.......................(15).}$$

Theorem due to Laplace.

Again, we have a most important theorem expressed by the following equation:—

$$\iint S_i S_{i'} d\varpi = 0 \quad\text{..............................(16),}$$

where $d\varpi$ denotes an element of a spherical surface, described from O as centre with radius unity; $\iint$ an integration over the whole of this surface; and S_i, $S_{i'}$ two complete surface harmonics, of which the degrees, i and i', are neither equal to one another, nor such that $i+i'=-1$. For, denoting the solid harmonics $r^i S_i$ and $r^{i'} S_{i'}$ by V_i and $V_{i'}$ for any point (x, y, z), we have, by the general theorem (1) of A (a), above, applied to the space between any two spherical surfaces having O for their common centre, and a and a_1 their radii;—

$$\iiint\left(\frac{dV_i}{dx}\frac{dV_{i'}}{dx}+\frac{dV_i}{dy}\frac{dV_{i'}}{dy}+\frac{dV_i}{dz}\frac{dV_{i'}}{dz}\right)dxdydz$$
$$=\iint V_i \delta V_{i'} d\sigma=\iint V_{i'} \delta V_i d\sigma.$$

But, according to (7), $\delta V_{i'}=\frac{i'}{r}V_{i'}$, and $\delta V_i=\frac{i}{r}V_i$. And for the portions of the bounding surface constituted by the two spherical surfaces respectively, $d\sigma=a^2 d\varpi$, and $d\sigma=a_1{}^2 d\varpi$. Hence the two last equal members of the preceding double equations become

$$i(a^{i+i'+1}-a_1{}^{i+i'+1})\iint S_i S_{i'} d\varpi = i'(a^{i+i'+1}-a_1{}^{i+i'+1})\iint S_i S_{i'} d\varpi,$$

to satisfy which, when i differs from i', and $a^{i+i'+1}$ from $a_1{}^{i+i'+1}$, (16) must hold.

The corresponding theorem for partial harmonics is this:—

Extension of theorem of Laplace to partial harmonics.

Let S_i, $S_{i'}$ denote any two different partial surface harmonics of degrees i, i', having their sum different from -1; and further, fulfilling the condition that, at every point of the boundary of some one part of the spherical surface either each of them vanishes, or the rate of variation of each of them perpendicular to this boundary vanishes, and that each is finite and single in its value at every point of the enclosed portion of surface; then, with the integration $\iint$ limited to the portion of surface in question, equation (16) holds. The proof differs from the preceding only in this, that instead of taking the whole space between two concentric spherical surfaces, we must now take only the part of it enclosed by the cone having O for vertex, and containing the boundary of the spherical area considered.

Extension of theorem of Laplace to partial harmonics.

(h) Proceeding now to the investigation of complete harmonics, we shall first prove that every such function is either rational and integral in terms of the co-ordinates x, y, z, or is made so by a factor of the form r^m.

Investigation of complete harmonics.

Let V be any function of x, y, z, satisfying the equation

$$\nabla^2 V = 0 \qquad (17)$$

at every point within a spherical surface, S, described from O as centre, with any radius a. Its value at this surface, if a known function of any arbitrary character, may be expanded according to the general theorem of § 51, below, in the following series:—

$$(r = a), \quad V = S_0 + S_1 + S_2 + \ldots\ldots + S_i + \text{etc.} \qquad (18)$$

where S_1, S_2,...S_i denote the surface values of solid spherical harmonics of degrees 1, 2,...i, each a rational integral function for every point within S. But

$$S_0 + S_1\frac{r}{a} + S_2\frac{r^2}{a^2} + \ldots + S_i\frac{r^i}{a^i} + \text{etc.} \qquad (19)$$

Harmonic solution of Green's problem for the space within a spherical surface.

is a function fulfilling these conditions, and therefore, as was proved above, A (c), V cannot differ from it. Now, as a particular case, let V be a harmonic function of positive degree ι, which may be denoted by $S_\iota\frac{r^\iota}{a^\iota}$: we must have

$$S_\iota\frac{r^\iota}{a^\iota} = S_0 + S_1\frac{r}{a} + S_2\frac{r^2}{a^2} + \ldots + S_i\frac{r^i}{a^i} + \text{etc.}$$

This cannot be unless $\iota = i$, $S_\iota = S_i$, and all the other functions S_0, S_1, S_2, etc., vanish. Hence there can be no complete spherical harmonic of positive degree, which is not, as $S_i\frac{r^i}{a^i}$, of integral degree and an integral rational function of the co-ordinates.

Complete harmonics of positive degrees, proved rational and integral.

Again, let V be any function satisfying (17) for every point without the spherical surface S, and vanishing at an infinite distance in every direction; and let, as before, (18) express its surface value at S. We similarly prove that it cannot differ from

$$\frac{aS_0}{r} + \frac{a^2S_1}{r^2} + \frac{a^3S_2}{r^3} + \ldots\ldots + \frac{a^{i+1}S_i}{r^{i+1}} + \text{etc.} \qquad (20).$$

Harmonic solution of Green's problem for space external to a spherical surface.

Harmonic solution of Green's problem for space external to a spherical surface.

Hence if, as a particular case, V be any complete harmonic $\frac{r^\kappa S_\kappa}{a^\kappa}$, of negative degree κ, we must have, for all points outside S,

$$\frac{r^\kappa S_\kappa}{a^\kappa} = \frac{aS_0}{r} + \frac{a^2 S_1}{r^2} + \frac{a^3 S_2}{r^3} + \ldots\ldots + \frac{a^{i+1} S_i}{r^{i+1}} + \text{etc.},$$

which requires that $\kappa = -(i+1)$, $S_\kappa = S_i$, and that all the other functions S_0, S_1, S_2, etc., vanish. Hence a complete spherical harmonic of negative degree cannot be other than $\frac{a^{i+1} S_i}{r^{i+1}}$, or $\frac{a^{i+1}}{r^{2i+1}} S_i r^i$, where $S_i r^i$ is not only a rational integral function of the co-ordinates, as asserted in the enunciation, but is itself a spherical harmonic.

Complete harmonics of negative degree.

(*i*) Thus we have proved that a complete spherical harmonic, if of positive, is necessarily of integral, degree, and is, besides, a rational integral function of the co-ordinates, or if of negative degree, $-(i+1)$, is necessarily of the form $\frac{V_i}{r^{2i+1}}$, where V_i is a harmonic of positive degree, i. We shall therefore call the *order* of a complete spherical harmonic of negative degree, the *degree or order* of the complete harmonic of positive degree allied to it; and we shall call the *order* of a surface harmonic, the degree or order of the solid harmonic of positive degree, or the order of the solid harmonic of negative degree, which agrees with it at the spherical surface.

Orders and degrees of complete harmonics.

General expressions for complete harmonics.

(*j*) To obtain general expressions for complete spherical harmonics of all orders, we may first remark that, inasmuch as a constant is the only rational integral function of degree 0, a complete harmonic of degree 0 is necessarily constant. Hence, by what we have just seen, a complete harmonic of the degree -1 is necessarily of the form $\frac{A}{r}$. That this function is a harmonic we knew before, by (14).

By differentiation of harmonic of degree -1.

Hence, by (15), we see that

$$\left.\begin{aligned} V_{-i-1} &= \frac{d^{j+k+l}}{dx^j dy^k dz^l} \frac{1}{(x^2+y^2+z^2)^{\frac{1}{2}}} \\ \text{if} \qquad j+k+l &= i \end{aligned}\right\} \ldots\ldots\ldots\ldots\ldots\ldots(21),$$

where V_{-i-1} denotes a harmonic, which is clearly a complete harmonic, of degree $-(i+1)$. The differential coefficient here indicated, when worked out, is easily found to be a fraction, of which the numerator is a rational integral function of degree i, and the denominator is r^{2i+1}. By what we have just seen, the numerator must be a harmonic; and, denoting it by V_i, we thus have **By differentiation of harmonic of degree −1.**

$$V_i = r^{2i+1}\frac{d^{j+k+l}}{dx^j dy^k dz^l}\frac{1}{r} \text{ ....(22).}$$

The number of independent harmonics of order i, which we can thus derive by differentiation from $\frac{1}{r}$, is $2i+1$. For, although there are $\frac{(i+2)(i+1)}{2}$ differential coefficients $\frac{d^{j+k+l}}{dx^j dy^k dz^l}$, for which $j+k+l=i$, only $2i+1$ of these are independent when $\frac{1}{r}$ is the subject of differentiation, inasmuch as **Number of independent harmonics of any order.**

$$\left(\frac{d^2}{dx^2}+\frac{d^2}{dy^2}+\frac{d^2}{dz^2}\right)\frac{1}{r}=0 \text{ ....(14),}$$

which gives
$$\frac{d^{2n}}{dz^{2n}}\frac{1}{r}=(-1)^n\left(\frac{d^2}{dx^2}+\frac{d^2}{dy^2}\right)^n\frac{1}{r} \text{ ....(23),}$$
Relation between differential coefficients of harmonics.

n being any integer, and shows that

$$\left.\begin{aligned}\frac{d^{j+k+l}}{dx^j dy^k dz^l}\frac{1}{r} &= (-1)^{\frac{l}{2}}\frac{d^{j+k}}{dx^j dy^k}\left(\frac{d^2}{dx^2}+\frac{d^2}{dy^2}\right)^{\frac{l}{2}}\frac{1}{r}, \text{ if } l \text{ is even,}\\ \text{or}\quad &= (-1)^{\frac{l-1}{2}}\frac{d^j d^k}{dx^j dy^k}\left(\frac{d^2}{dx^2}+\frac{d^2}{dy^2}\right)^{\frac{l-1}{2}}\frac{d}{dz}\frac{1}{r}, \text{ if } l \text{ is odd.}\end{aligned}\right\}\text{...(24).}$$

Hence, by taking $l=0$, and $j+k=i$, in the first place, we have $i+1$ differential coefficients $\frac{d^{j+k}}{dx^j dy^k}$; and by taking next $l=1$, and $j+k=i-1$, we have i varieties of $\frac{d^{j+k}}{dx^j dy^k}$; that is to say, we have in all $2i+1$ varieties, and no more, when $\frac{1}{r}$ is the subject. It is easily seen that these $2i+1$ varieties are in reality independent. We need not stop at present to show this, as it will be apparent in the actual expansions given below.

Now if $H_i(x, y, z)$ denote any rational integral function of x, y, z of degree i, $\nabla^2 H_i(x, y, z)$ is of degree $i-2$. Hence since in H_i there are $\frac{(i+2)(i+1)}{2}$ terms, in $\nabla^2 H_i$ there are $\frac{i(i-1)}{2}$.

Complete harmonic of any degree investigated algebraically.

Hence if $\nabla^2 H_i = 0$, we have $\frac{i(i-1)}{2}$ equations among the constant coefficients, and the number of independent constants remaining is $\frac{(i+2)(i+1)}{2} - \frac{i(i-1)}{2}$, or $2i+1$; that is to say, there are $2i+1$ constants in the general rational integral harmonic of degree i. But we have seen that there are $2i+1$ distinct varieties of differential coefficients of $\frac{1}{r}$ of order i, and that the numerator of each is a harmonic of degree i. Hence every complete harmonic of order i is expressible in terms of differential coefficients of $\frac{1}{r}$. It is impossible to form $2i+1$ functions symmetrically among three variables, except when $2i+1$ is divisible by 3; that is to say, when $i = 3n+1$, n being any integer. This class of cases does not seem particularly interesting or important, but here are two examples of it.

Example 1. $i = 1$, $2i+1 = 3$.

The harmonics are obviously

$$\frac{d}{dx}\frac{1}{r}, \quad \frac{d}{dy}\frac{1}{r}, \quad \frac{d}{dz}\frac{1}{r}.$$

Formula (25) involves z singularly, and x and y symmetrically, for every value of i greater than unity, but for the case of $i = 1$ it is essentially symmetrical in respect to x, y, and z, as in this case it becomes

$$\frac{V_1}{r^3} = \left(A_0 \frac{d}{dx} + A_1 \frac{d}{dy} + B_0 \frac{d}{dz}\right)\frac{1}{r}.$$

Example 2. $i = 4$, $2i+1 = 9$.

Looking first for three differential coefficients of the 4th order, singular with respect to x, and symmetrical with respect to y and z; and thence changing cyclically to yzx and zxy, we find

$$\frac{d^4}{dy^2 dz^2}, \quad \frac{d^4}{dx\,dy^3}, \quad \frac{d^4}{dx\,dz^3},$$

$$\frac{d^4}{dz^2 dx^2}, \quad \frac{d^4}{dy\,dz^3}, \quad \frac{d^4}{dy\,dx^3},$$

$$\frac{d^4}{dx^2 dy^2}, \quad \frac{d^4}{dz\,dx^3}, \quad \frac{d^4}{dz\,dy^3}.$$

These nine differentiations of $\frac{1}{r}$ are essentially distinct and give us therefore nine distinct harmonics of the 4th order formed symmetrically among x, y, z. By putting in them for $\frac{d^2}{dz^2}$, wherever it occurs, its equivalent $-\left(\frac{d^2}{dx^2}+\frac{d^2}{dy^2}\right)$, considering that it is $\frac{1}{r}$ which is differentiated, and for $\frac{d^3}{dz^3}$, its equivalent $-\frac{d}{dz}\left(\frac{d^2}{dx^2}+\frac{d^2}{dy^2}\right)$, we may pass from them to (25).

Complete harmonic of any degree investigated algebraically.

But for every value of i the general harmonic may be exhibited as a function, with $2i+1$ constants, involving two out of the three variables symmetrically. This may be done in a variety of ways, of which we choose the two following, as being the most useful :—First,

General expression for complete harmonic of order i.

$$\left.\begin{aligned}\underset{2i+1}{V_i}=&\left\{A_0\left(\frac{d}{dx}\right)^i+A_1\left(\frac{d}{dx}\right)^{i-1}\frac{d}{dy}+A_2\left(\frac{d}{dx}\right)^{i-2}\left(\frac{d}{dy}\right)^2+\ldots+A_i\left(\frac{d}{dy}\right)^i\right\}\frac{1}{r}\\&+\left\{B_0\left(\frac{d}{dx}\right)^{i-1}+B_1\left(\frac{d}{dx}\right)^{i-2}\frac{d}{dy}+B_2\left(\frac{d}{dx}\right)^{i-3}\left(\frac{d}{dy}\right)^2+\ldots\right.\\&\qquad\qquad\left.+B_{i-1}\left(\frac{d}{dy}\right)^{i-1}\right\}\frac{d}{dz}\frac{1}{r}\end{aligned}\right\}\ldots(25).$$

Secondly, let $x+yv=\xi,\quad x-yv=\eta$ .... (26),

where, as formerly, v is taken to denote $\sqrt{-1}$.

This gives

$$\left.\begin{aligned}&x=\tfrac{1}{2}(\xi+\eta),\quad y=\frac{1}{2v}(\xi-\eta),\\&\frac{1}{r}=\frac{1}{(\xi\eta+z^2)^{\frac{1}{2}}},\end{aligned}\right\}\ldots\ldots\ldots\ldots\ldots(27);$$

$$\left.\begin{aligned}&\frac{d}{dx}[x,y]=\left(\frac{d}{d\xi}+\frac{d}{d\eta}\right)[\xi,\eta],\quad\frac{d}{dy}[x,y]=v\left(\frac{d}{d\xi}-\frac{d}{d\eta}\right)[\xi,\eta],\\&\frac{d}{d\xi}[\xi,\eta]=\tfrac{1}{2}\left(\frac{d}{dx}-v\frac{d}{dy}\right)[x,y],\quad\frac{d}{d\eta}[\xi,\eta]=\tfrac{1}{2}\left(\frac{d}{dx}+v\frac{d}{dy}\right)[x,y],\end{aligned}\right\}\ldots(28),$$

Imaginary linear transformation.

where $[x, y]$ and $[\xi, \eta]$ denote the same quantity, expressed in terms of x, y, and of ξ, η respectively. From these we have, further,

Imaginary linear transformation.

$$\left.\begin{array}{c}\left(\frac{d^2}{dx^2}+\frac{d^2}{dy^2}+\frac{d^2}{dz^2}\right)[x,y,z]=\left(4\frac{d^2}{d\xi d\eta}+\frac{d^2}{dz^2}\right)[\xi,\eta,z],\\ \text{or, according to our abbreviated notation,}\\ \nabla^2=4\frac{d^2}{d\xi d\eta}+\frac{d^2}{dz^2}.\end{array}\right\}\ldots(29).$$

Hence, as $\nabla^2 V=0$, if V denote $\frac{1}{r}$ or any other solid harmonic,

$$\frac{d^2}{dz^2}V=-4\frac{d^2}{d\xi d\eta}V\ldots\ldots(29').$$

Using (28) in (25) and taking $\mathfrak{A}_0$, $\mathfrak{A}_1$, $\mathfrak{B}_0$, $\mathfrak{B}_1$, ... to denote another set of coefficients readily expressible in terms of A_0, A_1, B_0, B_1, ... we find

$$\left.\begin{array}{l}\frac{V_i}{r^{2i+1}}=\left\{\mathfrak{A}_0\left(\frac{d}{d\xi}\right)^i+\mathfrak{A}_1\left(\frac{d}{d\xi}\right)^{i-1}\frac{d}{d\eta}+\mathfrak{A}_2\left(\frac{d}{d\xi}\right)^{i-2}\left(\frac{d}{d\eta}\right)^2+\ldots+\mathfrak{A}_i\left(\frac{d}{d\eta}\right)^i\right\}\frac{1}{r}\\ \quad+\left\{\mathfrak{B}_0\left(\frac{d}{d\xi}\right)^{i-1}+\mathfrak{B}_1\left(\frac{d}{d\xi}\right)^{i-2}\frac{d}{d\eta}+\mathfrak{B}_2\left(\frac{d}{d\xi}\right)^{i-3}\left(\frac{d}{d\eta}\right)^2+\ldots+\mathfrak{B}_{i-1}\left(\frac{d}{d\eta}\right)^{i-1}\right\}\frac{d}{dz}\frac{1}{r}\end{array}\right\}\ldots(30).$$

Expansion of elementary term.

The differentiations here are performed with great ease, by the aid of Leibnitz's theorem. Thus we have

$$\left.\begin{array}{l}r^{2(m+n)+1}\frac{d^{m+n}}{d\xi^m d\eta^n}\frac{1}{r}=(-)^{m+n}\tfrac{1}{2}.\tfrac{3}{2}.\tfrac{5}{2}\ldots(m+n-\tfrac{1}{2})\\ \left[\eta^m\xi^n-\frac{mn}{1.(m+n-\frac{1}{2})}\eta^{m-1}\xi^{n-1}r^2+\frac{m(m-1).n(n-1)}{1.2.(m+n-\frac{1}{2})(m+n-\frac{3}{2})}\eta^{m-2}\xi^{n-2}r^4-\text{etc.}\right]\\ \text{and}\\ r^{2(m+n)+3}\frac{d^{m+n+1}}{d\xi^m d\eta^n dz}\frac{1}{r}=(-)^{m+n+1}\tfrac{1}{2}.\tfrac{3}{2}.\tfrac{5}{2}\ldots(m+n+\tfrac{1}{2}).2z\\ \left[\eta^m\xi^n-\frac{mn}{1.(m+n+\frac{1}{2})}\eta^{m-1}\xi^{n-1}r^2+\frac{m(m-1).n(n-1)}{1.2.(m+n+\frac{1}{2})(m+n-\frac{1}{2})}\eta^{m-2}\xi^{n-2}r^4-\text{etc.}\right].\end{array}\right\}\ldots(31)$$

Polar transformation.

This expression leads at once to a real development, in terms of polar co-ordinates, thus:—Let

$$z=r\cos\theta,\quad x=r\sin\theta\cos\phi,\quad y=r\sin\theta\sin\phi\ldots\ldots(32);$$

so that $\xi=r\sin\theta\epsilon^{v\phi}$, $\eta=r\sin\theta\epsilon^{-v\phi}$ ...... (33).

Then, since $\xi\eta=x^2+y^2=r^2\sin^2\theta$,

and

$$\xi^n\eta^m=(\xi\eta)^m\xi^s=(\xi\eta)^m(r\sin\theta)^s(\cos\phi+v\sin\phi)^s=(r\sin\theta)^{m+n}(\cos s\phi+v\sin s\phi),$$

where $s=n-m$; and if, further, we take

$$\left.\begin{array}{ll}\mathfrak{A}_n+\mathfrak{A}_m=\mathbf{A}_s, & (\mathfrak{A}_n-\mathfrak{A}_m)v=\mathbf{A}_s',\\ \mathfrak{B}_n+\mathfrak{B}_m=\mathbf{B}_s, & (\mathfrak{B}_n-\mathfrak{B}_m)v=\mathbf{B}_s',\end{array}\right\}\ldots\ldots(34)$$

Trigonometrical expansions.

we have

$$\left.\begin{aligned}
&\mathfrak{A}_n\frac{d^{m+n}}{d\xi^m d\eta^n}\frac{1}{r}+\mathfrak{A}_m\frac{d^{m+n}}{d\xi^n d\eta^m}\frac{1}{r}\\
&=(-)^{m+n}\tfrac{1}{2}\cdot\tfrac{3}{2}\cdot\tfrac{5}{2}\ldots(m+n-\tfrac{1}{2})r^{-(m+n+1)}(\mathbf{A}_s\cos s\phi+\mathbf{A}_s'\sin s\phi)\\
&\left[\sin^{m+n}\theta-\frac{mn}{1.(m+n-\frac{1}{2})}\sin^{m+n-2}\theta+\frac{m(m-1).n(n-1)}{1.2.(m+n-\frac{1}{2})(m+n-\frac{3}{2})}\right.\\
&\left.\qquad\qquad\sin^{m+n-4}\theta-\text{etc.}\right]\\
&\mathfrak{B}_n\frac{d^{m+n+1}}{d\xi^m d\eta^n dz}\frac{1}{r}+\mathfrak{B}_m\frac{d^{m+n+1}}{d\xi^n d\eta^m dz}\frac{1}{r}\\
&=(-)^{m+n+1}\tfrac{1}{2}\cdot\tfrac{3}{2}\cdot\tfrac{5}{2}\ldots(m+n+\tfrac{1}{2})r^{-(m+n+2)}(\mathbf{B}_s\cos s\phi+\mathbf{B}_s'\sin s\phi)2\cos\theta\\
&\left[\sin^{m+n}\theta-\frac{mn}{1.(m+n+\frac{1}{2})}\sin^{m+n-2}\theta+\frac{m(m-1).n(n-1)}{1.2.(m+n+\frac{1}{2})(m+n-\frac{1}{2})}\right.\\
&\left.\qquad\qquad\sin^{m+n-4}\theta-\text{etc.}\right].
\end{aligned}\right\}\quad(35)$$

Setting aside now constant factors, which have been retained hitherto to show the relations of the expressions we have investigated, to differential coefficients of $\frac{1}{r}$; taking Σ to denote summation with respect to the arbitrary constants, $\mathbf{A}$, $\mathbf{A}'$, $\mathbf{B}$, $\mathbf{B}'$; and putting $\sin\theta=\nu$, $\cos\theta=\mu$; we have the following perfectly general expression for a complete surface harmonic of order i:—

$$S_i=\sum^{m+n=i}(\mathbf{A}_s\cos s\phi+\mathbf{A}_s'\sin s\phi)\Theta_{(m,n)}+\sum^{m+n+1=i}(\mathbf{B}_s\cos s\phi+\mathbf{B}_s'\sin s\phi)\mu Z_{(m,n)}\ldots(36)$$

where $s=m\sim n$, and

$$\Theta_{(m,n)}=\nu^{m+n}-\frac{mn}{1.(m+n-\frac{1}{2})}\nu^{m+n-2}+\frac{m(m-1).n(n-1)}{1.2.(m+n-\frac{1}{2})(m+n-\frac{3}{2})}\nu^{m+n-4}-\text{etc.}$$

while $Z_{(m,n)}$ differs from $\Theta_{(m,n)}$ only in having $m+n+1$ in place of $m+n$, in the denominators.

The formula most commonly given for a spherical harmonic of order i (Laplace, *Mécanique Celeste*, livre III. chap. II., or Murphy's *Electricity*, Preliminary Prop. xi.) is somewhat simpler, being as follows:—

$$S_i=\sum_{s=0}^{s=i}(\mathbf{A}_s\cos s\phi+\mathbf{B}_s\sin s\phi)\Theta_i^{(s)}\ \ldots\ldots\ldots\ldots\ldots\ldots\ (37).$$

$$\Theta_i^{(s)}=\nu^s\left[\mu^{i-s}-\frac{(i-s)(i-s-1)}{2.(2i-1)}.\mu^{i-s-2}+\frac{(i-s)(i-s-1)(i-s-2)(i-s-3)}{2.4.(2i-1)(2i-3)}\mu^{i-s-4}-\text{etc.}\right]$$

$$\ldots\ldots\ldots\ldots\ldots\ldots\ldots\ldots(38),$$

Trigonometrical expansions.

where it may be remarked that $\Theta_i^{(s)}$ means the same as $(-1)^{\frac{i-s}{2}}\Theta_{(m,n)}$ if $m+n=i$ and $m\sim n=s$, or as $(-1)^{\frac{i-s-1}{2}}\mu Z_{(m,n)}$ if $m+n+1=i$ and $m\sim n=s$. Formula (38) may be derived algebraically from (36) by putting $\surd(1-\mu^2)$ for ν in $\Theta_{(m,n)}\div\nu^s$ and in $Z_{(m,n)}\div\nu^s\mu$: or it may be obtained directly by the method of differentiation followed above, varied suitably. But it may also be obtained by assuming (with a_s and b_s as arbitrary constants)

$$V_i = S_i r^i = \Sigma(a_s\xi^s + b_s\eta^s)(z^{i-s} + pr^2z^{i-s-2} + qr^4z^{i-s-4} + \text{etc.}),$$

which is obviously a proper form; and determining p, q, etc., by the differential equation $\nabla^2 V_i = 0$, with (29).

Another form may be obtained with even greater ease, thus: Assuming

$$V_i = \Sigma(a_s\xi^s + b_s\eta^s)(z^{i-s} + p_1z^{i-s-2}\xi\eta + p_2z^{i-s-4}\xi^2\eta^2 + \text{etc.}),$$

and determining p_1, p_2, etc., by the differential equation, we have

$$\left.\begin{aligned}V_i = \Sigma(a_s\xi^s+\beta_s\eta^s)\Big[z^{i-s}-\frac{(i-s)(i-s-1)}{4.(s+1).1}z^{i-s-2}\xi\eta \\ +\frac{(i-s)(i-s-1)(i-s-2)(i-s-3)}{4^2.(s+1)(s+2).1.2}z^{i-s-4}\xi^2\eta^2-\text{etc.}\Big]\end{aligned}\right\}(39)$$

which might also have been found easily by the differentiation of $\frac{1}{r}$. Hence, eliminating imaginary symbols, and retaining the notation of (37) and (38), we have

$$\left.\begin{aligned}\Theta_i^{(s)} = C\Big[\sin^s\theta\,\mu^{i-s}-\frac{(i-s)(i-s-1)}{4.(s+1).1}\mu^{i-s-2}\nu^2 \\ +\frac{(i-s)(i-s-1)(i-s-2)(i-s-3)}{4^2.(s+1)(s+2).1.2}\mu^{i-s-4}\nu^4-\text{etc.}\Big], \\ \text{where}\qquad C=\frac{(2s+1)(2s+2)\dots(i+s)}{(2s+1)(2s+3)\dots(2i-1)}.\end{aligned}\right\}(40)$$

This value of C is found by comparing with (35). Thus we see that C must be equal to the numerical coefficient of the last term of (35), irrespectively of sign. Or C is found by comparing (40) with (38): it is equal to the coefficient of the last term of (38) divided by the coefficient of the last term within the brackets of (40). Or it is found directly (that is to say, independently of other equivalent formulas) thus :—We have, by (29′),

Trigonometrical expansions.

$$\left.\begin{aligned}\frac{d^i}{dz^{i-s}d\eta^s}\frac{1}{r} &= (-)^{\frac{i-s}{2}}2^{i-s}\frac{d^i}{d\xi^{\frac{i-s}{2}}d\eta^{\frac{i+s}{2}}}\frac{1}{r}, \text{ if } i-s \text{ is even,}\\ \text{or} \quad &= (-)^{\frac{i-s-1}{2}}2^{i-s-1}\frac{d^i}{dz\,d\xi^{\frac{i-s-1}{2}}d\eta^{\frac{i+s-1}{2}}}\frac{1}{r}, \text{ if } i-s \text{ is odd.}\end{aligned}\right\}\text{(41)}$$

Expanding the first member in terms of z, ξ, η, by successive differentiation, with reference first to η, s times, and then z, $i-s$ times, we find

$$(-)^i\tfrac{1}{2}\cdot\tfrac{3}{2}\ldots(s-\tfrac{1}{2})(2s+1)(2s+2)(2s+3)\ldots(i+s)z^{i-s}\xi^s \ldots\text{(42)},$$

for a term in its numerator: comparing this with (39) and (40), and the second number of (41) with (35), we find C.

(*k*) It is very important to remark, first, that

Important properties of elementary terms and auxiliary functions.

$$\iint U_i U_{i'}' d\sigma = 0 \ldots\ldots\ldots\ldots\ldots\text{(43)},$$

where U_i and $U_{i'}'$ denote any two of the elements of which V is composed in one of the preceding expressions; and secondly, that

$$\int_0^\pi \Theta_i^{(s)}\Theta_{i'}^{(s)}\sin\theta d\theta = 0 \ldots\ldots\ldots\ldots\text{(44)},$$

the case of $i=i'$ being of course excluded. For, taking $r=a$, the radius of the spherical surface; and $d\sigma = a^2 d\varpi$, as above; we have $d\varpi = \sin\theta d\theta d\phi$, etc., the limits of θ and ϕ, in the integration for the whole spherical surface, being 0 to π, and 0 to 2π, respectively. Thus, since $\int_0^{2\pi}\cos s\phi\cos s'\phi\, d\phi = 0$, we see the truth of the first remark; and from (16) and (36) we infer the second, which the reader may verify algebraically, as an exercise.

(*l*) Each one of the preceding series may be taken by either end, and used with i or s, either or both of them negative or fractional or imaginary. Whether finite or infinite in its number of terms, any series thus obtained expresses when multiplied by r^i a harmonic of degree i; since it is of degree i, and satisfies $\nabla^2 V_i = 0$. In any case in which one of the preceding series is not finite, the formula taken by one end gives a converging series; taken by the other end a diverging series. Thus (40) taken in the order shown above, converges when θ is between 0 and 45°, or between 135° and 180°: and taken with the last term of that order first it converges when $\theta > 45°$ and

Expansions of partial harmonics for cones and wedges.

Dismissal of series which diverge when θ is real.

$<135°$. Thus, again, $\Theta_{(m,n)}$ and $Z_{(m,n)}$ of (36), being each of a finite number of terms when either m or n is a positive integer, become when neither is so, infinite series, which diverge when $\nu<1$ and converge when $\nu>1$. These two series, whether both infinite or one finite and the other infinite, when convergent are so related that

$$\mu Z_{(m-\frac{1}{2},\, n-\frac{1}{2})}=\sqrt{-1}\,\Theta_{(m,\,n)}\ldots\ldots\ldots\ldots\ldots\ldots(36'),$$

as is easily verified for a few terms by multiplying $Z_{(m-\frac{1}{2},\, n-\frac{1}{2})}$ by the expansion of $\left(1-\frac{1}{\nu^2}\right)^{\frac{1}{2}}$ in ascending powers of $\frac{1}{\nu^2}$. But expansions in ascending powers of $\frac{1}{\nu^2}$ are of comparatively little interest, as they are divergent for real values of θ, and therefore not available for the proposed physical applications. To find expansions which converge when $\nu<1$ take the last terms of (36) first. Thus, if we put

Finding of convergent expansions.

$$K=(-)^n\frac{m(m-1)\ldots(m-n+2)(m-n+1).n(n-1)\ldots2.1}{1.2\ldots.(n-1)n.(m+n-\frac{1}{2})(m+n-\frac{3}{2})\ldots(m+\frac{3}{2})(m+\frac{1}{2})}\cdot\ \ldots(36'');$$

supposing m to be $>n$, and n to be a positive integer, we find

$$\Theta_{(m,n)}=K.\nu^{m-n}\left[1-\frac{n(m+\frac{1}{2})}{(m-n+1).1}\nu^2+\frac{n(n-1).(m+\frac{1}{2})(m+\frac{3}{2})}{(m-n+1)(m-n+2).1.2}\nu^4-\text{etc.}\right]\ldots(36''').$$

Writing down the corresponding expression for $Z_{(m-\frac{1}{2},\, n-\frac{1}{2})}$ from (36), and using (36′), we find

$$\Theta_{(m,n)}=K\mu\nu^{m-n}\left[1-\frac{(n-\frac{1}{2})(m+1)}{(m-n+1).1}\nu^2+\frac{(n-\frac{1}{2})(n-\frac{3}{2}).(m+1)(m+2)}{(m-n+1)(m-n+2).1.2}\nu^4+\&\text{c.}\right]\ldots(36^{\text{iv}}).$$

This expansion of $\Theta_{(m,n)}$ is derivable algebraically from (36‴) by multiplying the second member of (36‴) by

$$\mu\left(1+\tfrac{1}{2}\nu^2+\frac{1.3}{2.4}\nu^4+\text{etc.}\right)$$

(which is equal to unity). Both expansions converge when $\nu^2<1$, or, for all real values of θ; just failing when $\theta=\frac{1}{2}\pi$. In choosing between the two expansions (36‴) and (36^{iv}), prefer (36^{iv}) when n differs by less than $\frac{1}{4}$ from zero or some positive integer, otherwise choose (36‴); but it is chiefly important to have them both, because (36^{iv}) is finite, but (36‴) infinite, when $n=\frac{2j-1}{2}$; and (36‴) is finite, but (36^{iv}) infinite, when $n=j-1$; j being any positive integer.

Put now $m+n=i, \quad m-n=s,$ or $m=\frac{1}{2}(i+s), \quad n=\frac{1}{2}(i-s),$ ....(36^{v}) and denote by $Ku_i^{(s)}$

Complete expressions for spherical harmonics of any tesseral type;

what the second members of (36‴) and (36^{iv}) become with these values for m and n. Again, put

$m+n=i, \quad n-m=s,$ or $m=\frac{1}{2}(i-s), \quad n=\frac{1}{2}(i+s),$ .... (36^{vi}) and denote by $Kv_i^{(s)}$

what the second members of (36‴) and (36^{iv}) become with these values for m and n. We thus have two equal convergent series for $u_i^{(s)}$ and two equal convergent series for $v_i^{(s)}$, and $u_i^{(s)}$, $v_i^{(s)}$ are functions of ν (or of θ) such that

$$\left.\begin{array}{l} u_i^{(s)}(A\cos s\phi + B\sin s\phi) \\ \text{and} \quad v_i^{(s)}(A\cos s\phi + B\sin s\phi) \end{array}\right\} \text{.................} (36^{vii})$$

in ascending powers of ν.

are surface harmonics of order i.

The first terms of $u_i^{(s)}$ and $v_i^{(s)}$ are ν^s and ν^{-s}, or $\mu\nu^s$ and $\mu\nu^{-s}$, according as they are taken from (36‴) or (36^{iv}), and in general $u_i^{(s)}$ and $v_i^{(s)}$ are distinct from one another.

Two distinct solutions are clearly needed for the physical problems. But in the particular case of s an integer, $u_i^{(s)}$ and $v_i^{(s)}$ are not distinct. For in this case each term of $v_i^{(s)}$ after the first s terms has the infinite factor $\dfrac{1}{s-s}$; thus if C_j denote the coefficient of the $(j+1)^{\text{th}}$ term of $v_i^{(s)}$, the first s terms of $\dfrac{v_i^{(s)}}{C_s}$ vanish when s is an integer, and those that follow constitute the same series as that expressing $u_i^{(s)}$, whether we take (36‴) or (36^{iv}). For the case of s an integer the wanting solution is to be found by putting

$$w_i^{(j)} = \frac{u_i^{(j+\sigma)} - \dfrac{v_i^{(j+\sigma)}}{C_j}}{2\sigma}, \text{ when } \sigma = 0: \quad \text{........} (36^{viii})$$

$w_i^{(s)}$ thus found is such that

$$w_i^{(s)}(A\cos s\phi + B\sin s\phi)$$

Complete expressions for spherical harmonics of any tesseral type;

is a surface harmonic of order i distinct from $u_i^{(s)}$. The first term of $w_i^{(s)}$, according to (36'''), is $\nu^s \log \nu$, or $\mu\nu^s \log \nu$ according to (36^{iv}), and subsequent terms are of the form $(a + b \log \nu)\nu^{2j}$, or $(a + b \log \nu)\mu\nu^{2j}$, j being an integer. The circumstances belong to a well-known class of cases in the solution of linear differential equations of the second order (see § (f') below).

Again, lastly, remark that (38), unless it is finite (which it is if and only if $i - s$ is a positive integer), diverges when $\mu < 1$ and converges when $\mu > 1$, if taken in the order in which it is given above. To obtain series which converge when $\mu < 1$ (that is to say, for real values of θ), reverse the order of (38) for the case of $i - s$ a positive integer. Thus, according as $i - s$ is even or odd, we find

in ascending powers of μ.

$$\Theta_i^{(s)} = H\nu^s \left\{1 - \frac{(i-s).(i+s+1)}{1.2}\mu^2 + \frac{(i-s)(i-s-2).(i+s+1)(i+s+3)}{1.2.3.4}\mu^4 - \text{etc.}\right.$$

where, $i - s$ being even,

$$H = (-)^{\frac{1}{2}(i-s)} \frac{(i-s)(i-s-1)(i-s-2)(i-s-3)\ldots4.3.2.1}{2.4\ldots(i-s-2)(i-s).(2i-1)(2i-3)\ldots(i+s+3)(i+s+1)} \quad \ldots(38')\text{,}$$

and

$$\Theta_i^{(s)} = H'\nu^s \left\{\mu - \frac{(i-s-1)(i+s+2)}{2.3}\mu^3 + \frac{(i-s-1)(i-s-3).(i+s+2)(i+s+4)}{2.3.4.5}\mu^5 - \text{etc.}\right.$$

where, $i - s$ being odd,

$$H' = (-)^{\frac{1}{2}(i-s-1)} \frac{(i-s)(i-s-1)(i-s-2)(i-s-3)\ldots5.4.3.2}{2.4\ldots(i-s-3)(i-s-1).(2i-1)(2i-3)\ldots(i+s+4)(i+s+2)} \quad (38'').$$

Then, whatever be $i - s$, or i, or s, integral or fractional, positive or negative, real or imaginary, the formulas within the brackets { } are convergent series when they are not finite integral functions of μ. Hence we see that if we put

$$p_i^{(s)} = 1 - \frac{(i-s).(i+s+1)}{1.2}\mu^2 + \frac{(i-s)(i-s-2).(i+s+1)(i+s+3)}{1.2.3.4}\mu^4 - \text{etc.}$$

and

$$q_i^{(s)} = \mu - \frac{(i-s-1).(i+s+2)}{2.3}\mu^3 + \frac{(i-s-1)(i-s-3).(i+s+2)(i+s+4)}{2.3.4.5}\mu^5 - \text{etc.} \quad (38''')$$

or $\quad p_i^{(s)} = A_0 + A_2\mu^2 + A_4\mu^4 + \text{\&c.},$

and $\quad q_i^{(s)} = A_1\mu + A_3\mu^3 + A_5\mu^5 + \text{\&c.},$

where $A_0 = 1$, $A_1 = 1$, and $A_{n+2} = \dfrac{(n-i+s)(n+1+i+s)}{(n+1)(n+2)}A_n$, $\quad$ (38^{iv}), the functions $p_i^{(s)}$, $q_i^{(s)}$ thus expressed, whether they be algebraic or transcendental, are such that

Complete expressions for spherical harmonics of any tesseral type in ascending powers of μ.

$$\left.\begin{array}{ll} & p_i^{(s)}(A\cos s\phi + B\sin s\phi)\nu^s, \\ \text{and} & q_i^{(s)}(A\cos s\phi + B\sin s\phi)\nu^s, \end{array}\right\}\quad\text{.................}(38^{\text{v}})$$

are the two surface harmonics of order i, and of the form $f(\theta)\frac{\sin}{\cos}s\phi$. For example, if $i-s$ be an even integer, $p_i^{(s)}$ is the finite function with which we are familiar as giving a rational integral solution of the form (38^{v}), and $q_i^{(s)}$ gives the solution of the same form which is not integral or rational. And if $i-s$ is odd, $q_i^{(s)}$ gives the familiar rational integral solution, and $p_i^{(s)}$ the other solution of the same form but not integral or rational.

Corresponding solid harmonics.

The corresponding solid harmonics of degrees i and $-i-1$ are obtained by multiplying (38^{v}) by r^i and r^{-i-1}. Reducing the latter from polar to rectangular co-ordinates, we find them of the form

$$\left.\begin{array}{l} \left[r^{-i-s-1} - \dfrac{(i-s)(i+s+1)}{1\,.\,2}r^{-i-s-3}z^2 + \text{etc.}\right]H_s(x,y) \\ \text{and} \\ \left[r^{-i-s-2}z - \text{etc.}\right]H_s(x,y) \end{array}\right\}\text{.....}\ (38^{\text{vi}}),$$

where H_s denotes a homogeneous function of degree s. Now (15) $\frac{d}{dz}$ of any solid harmonic of degree $-i$ is a solid harmonic of degree $-i-1$. Hence

Successive derivation from lower orders.

$$r^{i+1}r^s\nu^s\frac{\sin}{\cos}s\phi\,\frac{d}{dz}\left[r^{-i-s}q_{i-1}^{(s)}\right],$$

and

$$r^{i+1}r^s\nu^s\frac{\sin}{\cos}s\phi\,\frac{d}{dz}\left[r^{-i-s}p_{i-1}^{(s)}\right],$$

are surface harmonics of order i, and they are clearly of the first and second forms of (38^{v}). Hence, putting into the forms shown in (38^{vi}) and performing the indicated differentiation for the first term of the q function and the first and second terms of the p function, so as to find the numerical coefficients of r^{-i-s-1} and $r^{-i-s-2}z$ in the immediate results of the differentiation, and then putting μr for z, we find

Successive derivation from lower orders.

$$\left.\begin{aligned} r^{-i-s-1}p_i^{(s)} &= \frac{d}{dz}\left[r^{-i-s}q_{i-1}^{(s)}\right] \\ \text{and} \qquad r^{-i-s-1}q_i^{(s)} &= -\frac{1}{i^2-s^2}\frac{d}{dz}\left[r^{-i-s}p_{i-1}^{(s)}\right] \end{aligned}\right\} \text{.........}(38^{vii}).$$

To reduce back to polar co-ordinates put for a moment $x^2+y^2=a^2$. Then we have

$$r=\frac{a}{\sqrt{(1-\mu^2)}}=\frac{a}{\nu},$$

$$z=\frac{a\mu}{\sqrt{(1-\mu^2)}}=\frac{a\mu}{\nu},$$

and

$$dz=\frac{ad\mu}{(1-\mu^2)^{\frac{3}{2}}}=\frac{ad\mu}{\nu^3}.$$

Hence, instead of (38^{vii}), we have

$$\left.\begin{aligned} p_i^{(s)} &= \nu^{-i-s+2}\frac{d}{d\mu}\{\nu^{i+s}q_{i-1}^{(s)}\}, \\ \text{and} \qquad q_i^{(s)} &= -\frac{1}{i^2-s^2}\nu^{-i-s+2}\frac{d}{d\mu}\{\nu^{i+s}p_{i-1}^{(s)}\} \end{aligned}\right\} \text{....} (38^{viii}).$$

[Compare § 782 (5) below.]

Supposing now s and i to be real quantities, and going back to (38^{iv}), to investigate the convergency of the series for $p_i^{(s)}$ and $q_i^{(s)}$, we see that, when n is infinitely great,

$$\frac{A_{n+2}}{A_n}=1+\frac{2(s-1)}{n}.$$

Now if $\qquad (1-\mu^2)^{-\kappa}=\Sigma B_n\mu^n,$

we have, by the binomial theorem,

$$B_0=1,\quad B_1=0,\quad \text{and}\ \frac{B_{n+2}}{B_n}=1+\frac{2(\kappa-1)}{n}.$$

Hence, when $\mu=\pm(1-e)$, where e is an infinitely small positive quantity,

$$\left.\begin{aligned} &p_i^{(s)}\nu^{2\kappa}=0 \ \text{ or } =\infty, \\ \text{and} \qquad &q_i^{(s)}\nu^{2\kappa}=0 \ \text{ or } =\infty, \\ \text{according as} \qquad &\kappa>s \ \text{ or } \ \kappa<s. \end{aligned}\right\} \text{......................}(38^{x}).$$

Hence if $i>s$, the quantities within the brackets under $\frac{d}{d\mu}$ in (38^{viii}) vanish when $\mu=\pm 1$; and as they vary continuously, and within finite limits, when μ is continuously increased from -1 to $+1$, it follows that $p_i^{(s)}$ vanishes one time more than does $q_{i-1}^{(s)}$, and $q_i^{(s)}$ one time more than does $p_{i-1}^{(s)}$. Now looking to (38‴), and supposing (as we clearly may without loss of generality) that s is positive, we see that every term of $p_{i-1}^{(s)}$ is positive if $i < s+1$. Hence if i is any quantity between s and $s+1$, $\nu^{i+s} p_{i-1}^{(s)}$ vanishes when $\mu = \pm 1$, and is finite and positive for every intermediate value of μ.

Acquisition of roots with rise of order.

The rootless form of lowest order.

Hence and from the second formula of (38^{viii}), $q_i^{(s)}$ vanishes just once as μ is increased continuously from -1 to $+1$: thence and from the first of (38^{viii}), $p_{i+1}^{(s)}$ vanishes twice: hence and from the second again, $q_{i+2}^{(s)}$ vanishes thrice, and so on. Again, as the coefficient of every term of the series (38‴) for $q_i^{(s)}$ is positive when $i < s+1$, this is the case for $q_{i-1}^{(s)}$, and therefore this function vanishes only for $\mu = 0$, as μ is increased from -1 to $+1$. Hence $p_i^{(s)}$ vanishes twice; and, then, continuing alternate applications of the second and first of (38″), we see that $q_{i+1}^{(s)}$ vanishes thrice, $p_{i+2}^{(s)}$ four times, and so on. Thus, putting all together, we see that $q_{i+j-1}^{(s)}$ has j or $j+1$ roots, and $p_{i+j-1}^{(s)}$ has $j+1$ or j roots, according as j is odd or even; j being any integer and i, as defined above, any quantity between s and $s+1$. In other words, the number of roots of $p_i^{(s)}$ is the even number next above $i-s$; and the number of roots of $q_i^{(s)}$ is the odd number next above $i-s$. Farther, from (38^{viii}) we see that the roots of $p_i^{(s)}$ lie in order between those of $q_{i-1}^{(s)}$, and the roots of $q_i^{(s)}$ between those of $p_{i-1}^{(s)}$. [Compare § (*p*) below.] These properties of the p and q functions are of paramount importance, not only in the theory of the development of arbitrary functions by aid of them, but in the physical applications of the fractional harmonic analysis. In each case of physical application they belong to the foundation of the theory of the simple and nodal modes of the action investigated. They afford the principles for the determination of values of $i-s$, which shall make $\Theta_i^{(s)}$ or $\frac{d}{d\theta}\Theta_i^{(s)}$ vanish for each of two stated

The other form of lowest order has one root,—zero.

Census of roots of tesseral harmonics of any order.

Electric induction, motion of water, etc., in space between two coaxal cones.

values of θ. This is an analytical problem of high interest in connexion with these extensions of spherical harmonic analysis: it is essentially involved in the physical application referred to above where the spaces concerned are bounded partly by coaxal cones. When the boundary is completed by the intercepted portions of two concentric spherical surfaces, functions of the class described in (o) below also enter into the solution. When prepared to take advantage of physical applications we shall return to the subject; but it is necessary at present to restrict ourselves to these few observations.

Electric induction, motion of water, etc., in space between spherical surface and two planes meeting in a diameter.

(m) If, in physical problems such as those already referred to, the space considered is bounded by two planes meeting, at any angle $\frac{\pi}{s}$, in a diameter, and the portion of spherical surface in the angle between them (the case of $s<1$, that is to say, the case of angle exceeding two right angles, not being excluded) the harmonics required are all of fractional degrees, but each a finite algebraic function of the co-ordinates ξ, η, z if s is any incommensurable number. Thus, for instance, if the problem be to find the internal temperature at any point of a solid of the shape in question, when each point of the curved portion of its surface is maintained permanently at any arbitrarily given temperature, and its plane sides at one constant temperature, the forms and the degrees of the harmonics referred to are as follows:—

Degree.	Harmonic.	Degree.	Harmonic.	Degree.	Harmonic.
s,	ξ^s	$2s$,	ξ^{2s}	$3s$,	ξ^{3s}
$s+1$,	$r^{2s+3}\frac{d}{dz}\frac{\xi^s}{r^{2s+1}}$	$2s+1$,	$r^{4s+3}\frac{d}{dz}\frac{\xi^{2s}}{r^{4s+1}}$	$3s+1$,	$r^{6s+1}\frac{d}{dz}\frac{\xi^{3s}}{r^{6s+1}}$
$s+2$,	$r^{2s+5}\frac{d^2}{dz^2}\frac{\xi^s}{r^{2s+1}}$	$2s+2$,	$r^{4s+5}\frac{d^2}{dz^2}\frac{\xi^{2s}}{r^{4s+1}}$	.....	....
$s+3$,	$r^{2s+7}\frac{d^3}{dz^3}\frac{\xi^s}{r^{2s+1}}$	$2s+3$,	$r^{4s+7}\frac{d^3}{dz^3}\frac{\xi^{2s}}{r^{4s+1}}$	.....	....
......	....	......	....	......	....
......	....	.....	....	......	....

These harmonics are expressed, by various formulæ (36)...(40), etc., in terms of real co-ordinates, in what precedes.

(n) It is worthy of remark that these, and every other spherical harmonic, of whatever degree, integral, real but fractional, or imaginary, are derivable by a general form of process, which includes differentiation as a particular case. Thus if $\left(\frac{d}{d\eta}\right)^s$ denotes an operation which, when s is an integer, constitutes taking the sth differential coefficient, we have clearly

Harmonic functions of all degrees derived from that of degree −1 by generalized differentiation.

$$\xi^s = r^{2s+1} P_s \left(\frac{d}{d\eta}\right)^s \frac{1}{(\xi\eta + z^2)^{\frac{1}{2}}},$$

where P_s denotes a function of s, which, when s is a real integer, becomes $(-)^s \frac{1}{2} \cdot \frac{3}{2} \cdot \frac{5}{2} \ldots (s - \frac{1}{2})$.
The investigation of this generalized differentiation presents difficulties which are confined to the evaluation of P_s, and which have formed the subject of highly interesting mathematical investigations by Liouville, Gregory, Kelland, and others.

If we set aside the factor P_s, and satisfy ourselves with determinations of *forms* of spherical harmonics, we have only to apply Leibnitz's and other obvious formulæ for differentiation with any fractional or imaginary number as index, to see that the equivalent expressions above given for a complete spherical harmonic of any degree, are derivable from $\frac{1}{r}$ by the process of generalized differentiation now indicated, so as to include every possible partial harmonic, of whatever degree, whether integral, or fractional and real, or imaginary. But, as stated above, those expressions may be used, in the manner explained, for partial harmonics, whether finite algebraic functions of ξ, η, z, or transcendents expressed by converging infinite series; quite irrespectively of the manner of derivation now remarked.

Expansions of partial harmonics obtained by common formulæ, with generalized indices.

(*o*) To illustrate the use of spherical harmonics of imaginary degrees, the problem regarding the conduction of heat specified above may be varied thus:—Let the solid be bounded by two concentric spherical surfaces, of radii a and a', and by two cones or planes, and let every point of each of these flat or conical sides be maintained with any arbitrarily given distribution of temperature, and the whole spherical portion of the boundary at one constant temperature. Harmonics will enter into the solution, of degree

Imaginary degrees useful when arbitrary functions of r are to be expressed.

$$-\frac{1}{2} + \frac{j\pi\sqrt{-1}}{\log \frac{a}{a'}},$$

where j denotes any integer. [Compare § (d') below.] Converging series for these and the others required for the solution are included in our general formulas (36)...(40), etc.

Derivation of any harmonic from that of degree −1 indicates the character and number of its nodes.

(p) The method of finding complete spherical harmonics by the differentiation of $\frac{1}{r}$, investigated above, has this great advantage, that it shows immediately very important properties which they possess with reference to the values of the variables for which they vanish. Thus, inasmuch as $\frac{1}{r}$ and all its differential coefficients vanish for $x = \pm\infty$, and for $y = \pm\infty$, and for $z = \pm\infty$, it follows that

$$\frac{d^{j+k+l}}{dx^j dy^k dz^l} \frac{1}{r}$$

vanishes j times when x is increased from $-\infty$ to $+\infty$

" k " y " " " "

and " l " z " " " "

[Compare with the investigation of the roots of $p_i^{(s)}$ and $q_i^{(s)}$ in § (l) above.]

The reader who is not familiar with Fourier's theory of equations will have no difficulty in verifying for himself the present application of the principles developed in that admirable work. Its interpretation for fractional or imaginary values of j, k, l is wonderfully interesting, and of obvious value for the physical applications of partial harmonics.

Expression of an arbitrary function in a series of surface harmonics.

Thus it appears that spherical harmonics of large real degrees, integral or fractional, or of imaginary degrees with large real parts ($\alpha + \beta\sqrt{-1}$, with α large), belong to the general class, to which Sir William R. Hamilton has applied the designation "Fluctuating Functions." This property is essentially involved in their capacity for expressing arbitrary functions, to the demonstration of which for the case of complete harmonics we now proceed, in conclusion.

Preliminary proposition.

(r) Let C be the centre and a the radius of a spherical surface, which we shall denote by S. Let P be any external or internal point, and let f denote its distance from C. Let $d\sigma$ denote an element of S, at a point E, and let $EP = D$. Then, $\iint$ denoting an integration extended over S, it is easily proved that

Preliminary proposition.

$$\left.\begin{array}{ll} & \iint\frac{d\sigma}{D^3}=\frac{a}{f}\frac{4\pi a}{f^2-a^2} \text{ when } P \text{ is external to } S \\ \text{and} & \iint\frac{d\sigma}{D^3}=\frac{4\pi a}{a^2-f^2} \text{ when } P \text{ is within } S \end{array}\right\} \ldots\ldots\ldots\ldots(45).$$

This is merely a particular case of a very general theorem of Green's, included in that of A (a), above, as will be shown when we shall be particularly occupied, later, with the general theory of Attraction: a geometrical proof of a special theorem, of which it is a case, (§ 474, fig. 2, with P infinitely distant,) will occur in connexion with elementary investigations regarding the distribution of electricity on spherical conductors: and, in the meantime, the following direct evaluation of the integral itself is given, in order that no part of the important investigation with which we are now engaged may be even temporarily incomplete.

Choosing polar co-ordinates, $\theta = ECP$, and ϕ the angle between the plane of ECP and a fixed plane through CP, we have

$$d\sigma = a^2 \sin\theta\, d\theta\, d\phi.$$

Hence, by integration from $\phi = 0$ to $\phi = 2\pi$,

$$\iint\frac{d\sigma}{D^3} = 2\pi a^2\int_0^\pi \frac{\sin\theta d\theta}{D^3}.$$

But
$$D^2 = a^2 - 2af\cos\theta + f^2;$$

and therefore
$$\sin\theta d\theta = \frac{DdD}{af};$$

the limiting values of D in the integral being

$$f-a,\ f+a, \text{ when } f>a,$$

and
$$a-f,\ a+f, \text{ when } f<a.$$

Hence we have

$$\iint\frac{d\sigma}{D^3} = \frac{2\pi a}{f}\left(\frac{1}{f-a}-\frac{1}{f+a}\right), \text{ or} = \frac{2\pi a}{f}\left(\frac{1}{a-f}-\frac{1}{a+f}\right),$$

in the two cases respectively, which proves (45).

(s) Let now $F(E)$ denote any arbitrary function of the position of E on S, and let

Solution of Green's problem for case of spherical surface, expressed by definite integral.

$$u = \iint\frac{(f^2-a^2)F(E)d\sigma}{D^3} \ldots\ldots\ldots\ldots\ldots(46).$$

When f is infinitely nearly equal to a, every element of this integral will vanish except those for which D is infinitely small.

Green's problem for case of spherical surface, expressed by definite integral.

Hence the integral will have the same value as it would have if $F(E)$ had everywhere the same value as it has at the part of S nearest to P; and, therefore, denoting this value of the arbitrary function by $F(P)$, we have

$$u = F(P) \iint \frac{(f^2 \sim a^2)\,d\sigma}{D^3},$$

when f differs infinitely little from a; or, by (45),

$$u = 4\pi a F(P) \quad \text{.................................}(46').$$

Its expansion in harmonic series.

Now, if e denote any positive quantity less than unity, we have, by expansion in a convergent series,

$$\frac{1}{(1 - 2e\cos\theta + e^2)^{\frac{1}{2}}} = 1 + Q_1 e + Q_2 e^2 + \text{etc.} \quad \text{............}(47),$$

Q_1, Q_2, etc., denoting functions of θ, for which expressions will be investigated below. Each of them is equal to $+1$, when $\theta = 0$, and they are alternately equal to -1 and $+1$, when $\theta = \pi$. It is easily proved that each is > -1 and $< +1$, for all values of θ between 0 and π. Hence the series, which becomes the geometrical series $1 \pm e + e^2 \pm \text{etc.}$, in the extreme cases, converges more rapidly than the geometrical series, except in those extreme cases of $\theta = 0$ and $\theta = \pi$.

$$\left.\begin{aligned} &\text{Hence} \quad \frac{1}{D} = \frac{1}{f}\left(1 + \frac{Q_1 a}{f} + \frac{Q_2 a^2}{f^2} + \text{etc.}\right) \text{ when } f > a \\ &\text{and} \quad \frac{1}{D} = \frac{1}{a}\left(1 + \frac{Q_1 f}{a} + \frac{Q_2 f^2}{a^2} + \text{etc.}\right) \text{ when } f < a \end{aligned}\right\} \text{...........}(48).$$

Now we have

$$\frac{d\frac{1}{D}}{df} = \frac{a\cos\theta - f}{D^3},$$

and therefore

$$\frac{f^2 - a^2}{D^3} = -\left(2f\frac{d\frac{1}{D}}{df} + \frac{1}{D}\right).$$

Hence by (48),

$$\left.\begin{aligned} &\frac{f^2 - a^2}{D^3} = \frac{1}{f}\left(1 + \frac{3Q_1 a}{f} + \frac{5Q_2 a^2}{f^2} + \text{.........}\right) \text{ when } f > a \\ &\text{and } \frac{a^2 - f^2}{D^3} = \frac{1}{a}\left(1 + \frac{3Q_1 f}{a} + \frac{5Q_2 f^2}{a^2} + \text{.........}\right) \text{ when } f < a \end{aligned}\right\} \quad (49).$$

Hence, for u (46), we have the following expansions:—

Green's problem for case of spherical surface, solved explicitly in harmonic series.

$$u=\frac{1}{f}\left\{\iint F(E)d\sigma+\frac{3a}{f}\iint Q_1F(E)d\sigma+\frac{5a^2}{f^2}\iint Q_2F(E)d\sigma+\ldots\right\},\text{ when } f>a,$$

and

$$u=\frac{1}{a}\left\{\iint F(E)d\sigma+\frac{3f}{a}\iint Q_1F(E)d\sigma+\frac{5f^2}{a^2}\iint Q_2F(E)d\sigma+\ldots\right\},\text{ when } f<a$$

.........(51).

These series being clearly convergent, except in the case of $f=a$, and, in this limiting case, the unexpanded value of u having been proved (46′) to be finite and equal to $4\pi aF(P)$, it follows that the sum of each series approaches more and more nearly to this value when f approaches to equality with a. Hence, in the limit,

Laplace's spherical harmonic expansion of an arbitrary function.

$$F(P)=\frac{1}{4\pi a^2}\left\{\iint F(E)d\sigma+3\iint Q_1F(E)d\sigma+5\iint Q_2F(E)d\sigma+\text{etc.},\right\}\ldots(52),$$

which is the celebrated development of an arbitrary function in a series of "Laplace's coefficients," or, as we now call them, *spherical harmonics*.

(*t*) The preceding investigation shows that when there is one determinate value of the arbitrary function F for every point of S, the series (52) converges to the value of this function at P. The same reason shows that when there is an abrupt transition in the value of F, across any line on S, the series cannot converge when P is *exactly on*, but must still converge, however near it may be to, this line. [Compare with last two paragraphs of § 77 above.] The degree of non-convergence is so slight that, as we see from (51), the introduction of factors e, e^2, e^3, &c. to the successive terms e being <1 by a very small difference, produces decided convergence for every position of P, and the value of the series differs very little from $F(P)$, passing very rapidly through the finite difference when P is moved across the line of abrupt change in the value of $F(P)$.

Convergence of series never lost except at abrupt changes in value of the function expressed.

(*u*) In the development (47) of

$$\frac{1}{(1-2e\cos\theta+e^2)^{\frac{1}{2}}},$$

the coefficients of e, e^2,... e^i, are clearly rational integral functions of $\cos\theta$, of degrees 1, 2...i, respectively. They are given explicitly below in (60) and (61), with $\theta'=0$. But, if x, y, z and

Expansion of $\frac{1}{D}$ in symmetrical harmonic functions of the co-ordinates of the two points.

x', y', z' denote rectangular co-ordinates of P and of E respectively, we have

$$\cos\theta = \frac{xx' + yy' + zz'}{rr'},$$

where $r = (x^2 + y^2 + z^2)^{\frac{1}{2}}$, and $r' = (x'^2 + y'^2 + z'^2)^{\frac{1}{2}}$. Hence, denoting, as above, by Q_i the coefficient of e^i in the development, we have

$$Q_i = \frac{H_i[(x, y, z), (x', y', z')]}{r^i r'^i} \quad\text{.....................(53)},$$

$H_i[(x, y, z), (x', y', z')]$ denoting a symmetrical function of (x, y, z) and (x', y', z'), which is homogeneous with reference to either set alone. An explicit expression for this function is of course found from the expression for Q_i in terms of $\cos\theta$.

Biaxal harmonic.

Viewed as a function of (x, y, z), $Q_i r^i r'^i$ is symmetrical round OE; and as a function of (x', y', z') it is symmetrical round OP. We shall therefore call it the biaxal harmonic of (x, y, z) (x', y', z') of degree i; and Q_i the biaxal surface harmonic of order i.

Expansion of $\frac{1}{D}$ by Taylor's theorem.

(v) But it is important to remark, that the coefficient of any term, such as $x'^j y'^k z'^l$, in it may be obtained alone, by means of Taylor's theorem, applied to a function of three variables, thus:—

$$\frac{1}{(1-2e\cos\theta+e^2)^{\frac{1}{2}}} = \frac{r}{(r^2-2rr'\cos\theta+r'^2)^{\frac{1}{2}}} = \frac{r}{[(x-x')^2+(y-y')^2+(z-z')^2]^{\frac{1}{2}}}$$

Now if $F(x, y, z)$ denote any function of x, y, and z, we have

$$F(x+f, y+g, z+h) = \sum_{j=0}^{j=\infty}\sum_{k=0}^{k=\infty}\sum_{l=0}^{l=\infty} \frac{f^j g^k h^l}{1.2\ldots j.1.2\ldots k.1.2\ldots l}\,\frac{d^{j+k+l}F(x,y,z)}{dx^j dy^k dz^l};$$

where it must be remarked that the interpretation of $1.2\ldots j$, when $j=0$, is unity, and so for k and l also. Hence, by taking $F(x, y, z) = \frac{1}{(x^2+y^2+z^2)^{\frac{1}{2}}}$, we have

$$\frac{1}{[(x-x')^2+(y-y')^2+(z-z')^2]^{\frac{1}{2}}}$$
$$= \Sigma\Sigma\Sigma\,\frac{(-1)^{j+k+l}x'^j y'^k z'^l}{1.2\ldots j.1.2\ldots k.1.2\ldots l}\,\frac{d^{j+k+l}}{dx^j dy^k dz^l}\,\frac{1}{(x^2+y^2+z^2)^{\frac{1}{2}}},$$

a development which, by comparing it with (48), above, we see to be convergent whenever

$$x'^2 + y'^2 + z'^2 < x^2 + y^2 + z^2.$$

Expression for biaxal harmonic deduced.

Hence

$$(rr')^i Q_i = r^{2i+1} \sum\sum\sum^{(j+k+l=i)} \frac{(-1)^{j+k+l} x'^j y'^k z'^l}{1.2\ldots j.1.2\ldots k.1.2\ldots l} \frac{d^{j+k+l}}{dx^j dy^k dz^l} \frac{1}{(x^2+y^2+z^2)^{\frac{1}{2}}} \ldots (54),$$

the summation including all terms which fulfil the indicated condition $(j+k+l=i)$. It is easy to verify that the second member is not only integral and homogeneous of the degree i, in x, y, z, as it is expressly in x', y', z'; but that it is symmetrical with reference to these two sets of variables. Arriving thus at the conclusion expressed above by (53), we have now, for the function there indicated, an explicit expression in terms of differential coefficients, which, further, may be immediately expanded into an algebraic form with ease.

(v') In the particular case of $x'=0$ and $y'=0$, (54) becomes reduced to a single term, a function of x, y, z symmetrical about the axis OZ; and, dividing each member by r'^i, or its equal, z'^i, we have

Axial harmonic of order i.

$$r^i Q_i = \frac{(-1)^i r^{2i+1}}{1.2.3\ldots i} \frac{d^i}{dz^i} \frac{1}{(x^2+y^2+z^2)^{\frac{1}{2}}} \ldots\ldots\ldots\ldots\ldots (55).$$

Axial harmonic with its co-ordinates transformed becomes biaxal.

By actual differentiation it is easy to find the law of successive derivation of the numerators; and thus we find, with about equal ease, either of the expansions (31), (40), or (41), above, for the case $m=n$, or the trigonometrical formulæ, which are of course obtained by putting $z = r\cos\theta$ and $x^2+y^2 = r^2\sin^2\theta$.

(w) If now we put in these, $\cos\theta = \dfrac{xx'+yy'+zz'}{rr'}$, introducing again, as in (u) above, the notation (x, y, z), (x', y', z'), we arrive at expansions of Q_i in the terms indicated in (53).

Expansions of the biaxal harmonic, of order i.

(x) Some of the most useful expansions of Q_i are very readily obtained by introducing, as before, the imaginary co-ordinates (ξ, η) instead of (x, y), according to equations (26) of (j), and similarly, (ξ', η') instead of (x', y'). Thus we have

$$D^2 = (\xi-\xi')(\eta-\eta') + (z-z')^2.$$

Hence, as above,

$$\frac{1}{[(\xi-\xi')(\eta-\eta')+(z-z')^2]^{\frac{1}{2}}}$$

$$= \sum\sum\sum \frac{(-1)^{j+k+l}\xi'^j\eta'^k z'^l}{1.2\ldots j.1.2\ldots k.1.2\ldots l} \frac{d^{j+k+l}}{d\xi^j d\eta^k dz^l} \frac{1}{(\xi\eta+z^2)^{\frac{1}{2}}}.$$

Expansions of the biaxal harmonic, of order i.

Hence

$$(rr')^i Q_i = r'^{2i+1} \sum^{(j+k+l=i)} \sum \sum \frac{(-1)^{j+k+l} \xi'^j \eta'^k z'^l}{1.2\ldots j.1.2\ldots k.1.2\ldots l} \frac{d^{j+k+l}}{d\xi^j d\eta^k dz^l} \frac{1}{(\xi\eta + z^2)^{\frac{1}{2}}} \ldots (56).$$

Of course we have in this case

$$r^2 = \xi\eta + z^2, \quad r'^2 = \xi'\eta' + z'^2,$$

and $$\cos\theta = \frac{\xi\eta' + \xi'\eta + zz'}{rr'}.$$

And, just as above, we see that this expression, obviously a homogeneous function of ξ', η', z', of degree i, and also of η, ξ, z, involves these two systems of variables symmetrically.

Now, as we have seen above, all the ith differential coefficients of $\frac{1}{r}$ are reducible to the $2i+1$ independent forms

$$\left(\frac{d}{dz}\right)^i \frac{1}{r}, \quad \begin{matrix} \left(\frac{d}{dz}\right)^{i-1} \frac{d}{d\eta}\frac{1}{r}, & \left(\frac{d}{dz}\right)^{i-2} \left(\frac{d}{d\eta}\right)^2 \frac{1}{r}, & \ldots & \left(\frac{d}{d\eta}\right)^i \frac{1}{r}, \\ \left(\frac{d}{dz}\right)^{i-1} \frac{d}{d\xi}\frac{1}{r}, & \left(\frac{d}{dz}\right)^{i-2} \left(\frac{d}{d\xi}\right)^2 \frac{1}{r}, & \ldots & \left(\frac{d}{d\xi}\right)^i \frac{1}{r}. \end{matrix}$$

Hence $r^i Q_i$, viewed as a function of z, ξ, η, is expressed by these $2i+1$ terms, each with a coefficient involving z', ξ', η'. And because of the symmetry we see that this coefficient must be the same function of z', η', ξ', into some factor involving none of these variables (z, ξ, η), (z', η', ξ'). Also, by the symmetry with reference to ξ, η' and η, ξ', we see that the numerical factor must be the same for the terms similarly involving ξ, η' on the one hand, and η, ξ' on the other. Hence,

Biaxal harmonic expressed in symmetrical series of differential coefficients.

$$\left.\begin{aligned} Q_i = (rr')^{i+1} \Bigg[& E_0 \left(\frac{d}{dz'}\right)^i \frac{1}{r'} \left(\frac{d}{dz}\right)^i \frac{1}{r} \\ & + \sum_{s=1}^{s=i} E_i^{(s)} \left\{ \frac{d^i}{dz'^{i-s} d\xi'^s} \frac{1}{r'} \frac{d^i}{dz^{i-s} d\eta^s} \frac{1}{r} + \frac{d^i}{dz'^{i-s} d\eta'^s} \frac{1}{r'} \frac{d^i}{dz^{i-s} d\xi^s} \frac{1}{r} \right\} \Bigg] \\ \text{where} \qquad & \\ E_i^{(s)} = \frac{1}{1.2\ldots s.1.2\ldots(i-s).\frac{1}{2}.\frac{3}{2}\ldots(s-\frac{1}{2}).(2s+1)(2s+2)\ldots(i+s)} & \end{aligned}\right\} \ldots (57).$$

The value of $E_i^{(s)}$ is obtained thus:—Comparing the coefficient of the term $(zz')^{i-s}(\xi\eta')^s$ in the numerator of the expression which (56) becomes when the differential coefficient is expanded, with the coefficient of the same term in (57), we have

$$\frac{(-)^i M}{1.2\ldots(i-s).1.2\ldots s} = E_i^{(s)} M^2 \ldots\ldots\ldots (58),$$

where M denotes the coefficient of $z^{i-s}\xi^s$ in $r^{2i+1}\frac{d^i}{dz^{i-s}d\eta^s}\frac{1}{r}$, or, which is the same, the coefficient of $z'^{i-s}\eta'^s$ in $r'^{2i+1}\frac{d^i}{dz'^{i-s}d\xi'^s}\frac{1}{r'}$. From this, with the value (42) for M, we find $E_i^{(s)}$ as above.

Biaxal harmonic expressed in symmetrical series of differential coefficients.

(y) We are now ready to reduce the expansion of Q_i to a real trigonometrical form. First, we have, by (33),

$$(\xi\eta')^s + (\xi'\eta)^s = 2(rr'\sin\theta\sin\theta')^s\cos s(\phi-\phi') \quad \text{.........} (59).$$

Let now

$$\vartheta_i^{(s)} = \sin^s\theta\left[\cos^{i-s}\theta - \frac{(i-s)(i-s-1)}{4(s+1).1}\cos^{i-s-2}\theta\sin^2\theta + \frac{(i-s)(i-s-1)(i-s-2)(i-s-3)}{4^2(s+1)(s+2).1.2}\cos^{i-s-4}\theta\sin^4\theta - \text{etc.}\right] \text{...}(60);$$

(that is to say, $C\vartheta_i^{(s)} = \Theta_i^{(s)}$, in accordance with the notation of 40,) and let the corresponding notation with accents apply to θ'. Then, by the aid of (57), (58), and (59), we have

$$Q_i = 2\sum_{s=0}^{s=i}\frac{\frac{1}{2}.\frac{3}{2}\text{...}(s-\frac{1}{2})}{1.2\text{...}s}\cdot\frac{(2s+1)(2s+2)\text{...}(2s+i-s)}{1.2\text{...}(i-s)}\cos s(\phi-\phi')\vartheta_i^{(s)}\vartheta_i'^{(s)} \text{...}(61),$$

Trigonometrical expansion of biaxal surface harmonic.

of which, however, the first term (that for which $s=0$) must be halved.

(z) As a supplement to the fundamental proposition $\iint S_i S_{i'} d\varpi = 0$, (16) of (g), and the corresponding propositions, (43) and (44), regarding elementary terms of harmonics, we are now prepared to evaluate $\iint S_i^2 d\varpi$.

First, using the general expression (37) investigated above for S_i, and modifying the arbitrary constants to suit our present notation, we have

Fundamental definite integral investigated.

$$S_i = \sum_{s=0}^{s=i} A_s\cos(s\phi + \alpha_s)\vartheta_i^{(s)} \quad \text{..........................} (62).$$

Hence

$$\iint S_i^2\, d\varpi = \pi\sum_0^i A_s^2\int_0^\pi (\vartheta_i^{(s)})^2\sin\theta\, d\theta \quad \text{..................} (63).$$

To evaluate the definite integral in the second member, we have only to apply the general theorem (52) for expansion, in terms of surface harmonics, to the particular case in which the arbitrary function $F(E)$ is itself the harmonic, $\cos s\phi\vartheta_i^{(s)}$. Thus, remembering (16), we have

$$\cos s\phi\,\vartheta_i^{(s)} = \frac{2i+1}{4\pi}\int_0^\pi \sin\theta' d\theta'\int_0^{2\pi} d\phi'\cos s\phi'\vartheta_i'^{(s)} Q_i \quad \text{.........} (64).$$

Fundamental definite integral evaluated.

Using here for Q_i its trigonometrical expansion just investigated, and performing the integration for ϕ' between the stated limits, we find that $\cos s\phi\, \vartheta_i^{(s)}$ may be divided out, and (omitting the accents in the residual definite integral) we conclude,

$$\int_0^\pi \sin\theta (\vartheta_i^{(s)})^2 d\theta = \frac{2}{2i+1} \cdot \frac{1.2\ldots s}{\frac{1}{2}.\frac{3}{2}\ldots(s-\frac{1}{2})} \cdot \frac{1.2\ldots(i-s)}{(2s+1)(2s+2)\ldots(2s+i-s)} \ldots(65).$$

This holds without exception for the case $s=0$, in which the second member becomes $\frac{2}{2i+1}$. It is convenient here to recal equation (44), which, when expressed in terms of $\vartheta_i^{(s)}$ instead of $\Theta_{(m,n)}$, becomes

$$\int_0^\pi \sin\theta\, \vartheta_i^{(s)} \vartheta_{i'}^{(s)} d\theta = 0 \ldots\ldots\ldots\ldots\ldots\ldots (66),$$

where i and i' must be different. The properties expressed by these two equations, (65) and (66), may be verified by direct integration, from the explicit expression (60) for $\vartheta_i^{(s)}$; and to do so will be a good analytical exercise on the subject.

(a') Denote for brevity the second member of (65) by (i, s), so that

$$\int_0^\pi \sin\theta (\vartheta_i^{(s)})^2 d\theta = (i, s) \ldots\ldots\ldots\ldots\ldots\ldots (67).$$

Suppose the co-ordinates θ, ϕ to be used in (52); so that a, θ, ϕ are the three co-ordinates of P, and we may take $d\sigma = a^2 \sin\theta d\theta d\phi$. Working out by aid of (61), (65), the processes indicated symbolically in (52), we find

Spherical harmonic synthesis of arbitrary function concluded.

$$F(\theta, \phi) = \sum_{i=0}^{i=\infty} \{A_i \vartheta_i^{(0)} + \sum_{s=1}^{s=i} (A_i^{(s)} \cos s\phi + B_i^{(s)} \sin s\phi) \vartheta_i^{(s)}\} \ldots\ldots\ldots(68),$$

where

$$\left.\begin{aligned} A_i &= \frac{2i+1}{4\pi} \int_0^\pi \vartheta_i^{(0)} \sin\theta d\theta \int_0^{2\pi} F(\theta, \phi) d\phi \\ A_i^{(s)} &= \frac{1}{(i, s)\pi} \int_0^\pi \vartheta_i^{(s)} \sin\theta d\theta \int_0^{2\pi} \cos s\phi\, F(\theta, \phi) d\phi \\ B_i^{(s)} &= \frac{1}{(i, s)\pi} \int_0^\pi \vartheta_i^{(s)} \sin\theta d\theta \int_0^{2\pi} \sin s\phi\, F(\theta, \phi) d\phi \end{aligned}\right\} \ldots\ldots (69),$$

which is the explicit form most convenient for general use, of the expansion of an arbitrary function of the co-ordinates θ, ϕ in spherical surface harmonics. It is most easily proved, [when once the general theorem expressed by (66) and (65) has been in any way established,] by assuming the form of expansion (68), and then determining the coefficients by multiplying both members by $S_i^{(s)} \cos s\phi \sin\theta d\theta d\phi$, and again by $S_i^{(s)} \sin s\phi \sin\theta d\theta d\phi$, and integrating in each case over the whole spherical surface.

Spherical harmonic analysis of arbitrary function.

(b') In what precedes the expansions of surface harmonics, whether complete or not, have been obtained solely by the differentiation of $\frac{1}{r}$ with reference to rectilineal rectangular co-ordinates x, y, z. The expansions of the complete harmonics have been found simply as expressions for differential coefficients, or for linear functions of differential coefficients of $\frac{1}{r}$. The expansions of harmonics of fractional and imaginary orders have been inferred from the expansions of the complete harmonics merely by generalizing their algebraic forms. The properties of the harmonics have been investigated solely from the differential equation

Review of preceding expansions and investigations of properties.

$$\frac{d^2 V}{dx^2} + \frac{d^2 V}{dy^2} + \frac{d^2 V}{dz^2} = 0 \ldots\ldots\ldots\ldots\ldots\ldots\ldots(70),$$

in terms of the rectilineal rectangular co-ordinates. The original investigations of Laplace, on the other hand, were founded exclusively on the transformation of this equation into polar co-ordinates. In our first edition this transformation was not given—we now supply the omission, not only on account of the historical interest attached to "Laplace's equation" in terms of polar co-ordinates, but also because in this form it leads directly by the ordinary methods of treating differential equations, to every possible expansion of surface harmonics in polar co-ordinates.

(c') By App. A₀(g)(14) we find for Laplace's equation (70) transformed to polar co-ordinates,

$$\frac{d}{dr}\left(\frac{r^2 dV}{dr}\right) + \frac{1}{\sin\theta}\frac{d}{d\theta}\left(\sin\theta\frac{dV}{d\theta}\right) + \frac{1}{\sin^2\theta}\frac{d^2 V}{d\phi^2} = 0 \ldots\ldots\ldots (71).$$

In this put

$$V = S_i r^i, \text{ or } V = S_i r^{-i-1} \ldots\ldots\ldots\ldots\ldots\ldots\ldots(72).$$

We find

$$i(i+1)S_i + \frac{1}{\sin\theta}\frac{d}{d\theta}\left(\sin\theta\frac{dS_i}{d\theta}\right) + \frac{1}{\sin^2\theta}\frac{d^2 S_i}{d\phi^2} = 0 \ldots\ldots\ldots(73),$$

Laplace's equation for surface harmonic in polar coordinates.

which is the celebrated formula commonly known in England as "Laplace's Equation" for determining S_i, the "Laplace's coefficient" of order i; i being an integer, and the solutions admitted or sought for being restricted to rational integral functions of $\cos\theta$, $\sin\theta\cos\phi$ and $\sin\theta\sin\phi$.

(d') Doing away now with all such restrictions, suppose i to be any number, integral or fractional, real or imaginary, only if imaginary let it be such as to make $i(i+1)$ real [compare § (o)] above. On the supposition that S_i is a rational integral function of $\cos\theta$, $\sin\theta\cos\phi$ and $\sin\theta\sin\phi$, it would be the sum of terms such as $\Theta_i^{(s)} \begin{smallmatrix}\sin\\ \cos\end{smallmatrix} s\phi$. Now, allowing s to have any value integral or fractional, real or imaginary, assume

$$S = \Theta_i^{(s)} \begin{smallmatrix}\sin\\ \cos\end{smallmatrix} s\phi \quad\text{.......................(74).}$$

This will be a form of particular solution adapted for application to problems such as those referred to in §§ (l), (m) above; and (73) gives, for the determination of $\Theta_i^{(s)}$,

$$\frac{1}{\sin\theta}\frac{d}{d\theta}\left(\sin\theta\frac{d\Theta_i^{(s)}}{d\theta}\right) + \left[\frac{-s^2}{\sin^2\theta} + i(i+1)\right]\Theta_i^{(s)} = 0 \quad\text{.........(75).}$$

Definition of "Laplace's functions."

(e') When i and s are both integers we know from § (h) above, and we shall verify presently, by regular treatment of it in its present form, that the differential equation (75) has for one solution a rational integral function of $\sin\theta$ and $\cos\theta$. It is this solution that gives the "Laplace's Function," or the "complete surface harmonic" of the form $\Theta_i^{(s)} \begin{smallmatrix}\sin\\ \cos\end{smallmatrix} s\phi$. But being a differential equation of the second order, (75) must have another distinct solution, and from § (h) above it follows that this second solution cannot be a rational integral function of $\sin\theta$, $\cos\theta$. It may of course be found by quadratures from the rational integral solution according to the regular process for finding the second particular solution of a differential equation of the second order when one particular solution is known. Thus denoting by $\Theta_i^{(s)}$ any solution, as for example the known rational integral solution expressed by equation (38), or (36) or (40) above, or § 782 (e) or (f) with (5) below, we have for the complete solution,

$$\Theta'^{(s)}_i = \Theta^{(s)}_i \int \frac{d\mu}{(1-\mu^2)\left[\Theta^{(s)}_i\right]^2} \quad \text{.................}(76).$$

Definition of "Laplace's functions."

For a direct investigation of the complete solution in finite terms for the case $i-s$ a positive integer, see below § (n'), Example 2; and for the case i an integer, and s either not an integer or not $<i$, see § (o') (111).

The rational integral solution alone can enter, and it alone suffices, when the problem deals with the complete spherical surface. When there are boundaries, whether by two planes meeting in a diameter at an angle equal to a submultiple of four right angles, or by coaxal cones corresponding to certain particular values of θ, or by planes and cones, both the rational integral solution and the other are required. But when there are coaxal cones for boundaries, the values of i required by the boundary conditions [§ (l')] are not generally integral, and it is only when $i-s$ is integral that either solution is a rational and integral function of $\sin\theta$ and $\cos\theta$. Hence, in general, for the class of problems referred to, two solutions are required and neither is a rational integral function of $\sin\theta$ and $\cos\theta$.

(f') The ordinary process for the solution of linear differential equations in series of powers of the independent variable when the multipliers of the differential coefficients are rational algebraic functions of the independent variable leads easily from the equation (75) to any of the forms of rational integral solutions referred to above, as well as to the second solution in a form corresponding to each of them, when i and s are integers; and, quite generally, to the two particular solutions in every case, whether i and s be integral or fractional, real or imaginary. Thus, putting as above, § (k),

$$\cos\theta = \mu, \quad \sin\theta = \nu \quad \text{.....................}(77),$$

make μ the independent variable in the first place, in order to find expansions in powers of μ: thus (75) becomes

Differential equation with ν independent variable omitted here for brevity.

$$\frac{d}{d\mu}\left[(1-\mu^2)\frac{d\Theta^{(s)}_i}{d\mu}\right] + \left[\frac{-s^2}{1-\mu^2} + i(i+1)\right]\Theta^{(s)}_i = 0 \quad \text{......}(78).$$

This is the form in which "Laplace's equation" has been most commonly presented. To avoid the appearance of supposing

Commonest form of "Laplace's equation"

i and s to be integers or even real, put

$$\Theta_i^{(s)} = w, \quad i(i+1) = a, \quad s^2 = b, \ldots\ldots\ldots\ldots\ldots(79).$$

Using this notation, and multiplying both members by $(1-\mu^2)$, we have, instead of (78),

generalized.

$$(1-\mu^2)\frac{d}{d\mu}\left[(1-\mu^2)\frac{dw}{d\mu}\right] + [a(1-\mu^2) - b]\, w = 0 \ldots\ldots\ldots(80).$$

To integrate this equation, assume

$$w = \Sigma K_n \mu^n,$$

Obvious solution in ascending powers of μ;

and in the series so found for its first member equate to zero the coefficient of μ^n. Thus we find

$$(n+1)(n+2)K_{n+2} = [2n^2 - a + b]K_n - [(n-1)(n-2) - a]K_{n-2}, \ldots(81).$$

The first member of this vanishes for $n=-1$, and for $n=-2$, if K_1 and K_0 be finite. Hence, we may put $K_n = 0$ for all negative values of n, give arbitrary values to K_0 and K_1, and then find K_2, K_3, K_4, &c., by applications of (81) with $n=0$, $n=1$, $n=2$,... successively. Thus if we first put $K_0 = 1$, and $K_1 = 0$; then again $K_0 = 0$, $K_1 = 1$; we find two series of the forms

$$1 + K_2\mu^2 + K_4\mu^4 + \&c.$$

and

$$\mu + K_3\mu^3 + K_5\mu^5 + \&c.,$$

each of which satisfies (80); and therefore the complete solution is

$$w = C(1 + K_2\mu^2 + K_4\mu^4 + \&c.) + C'(\mu + K_3\mu^3 + K_5\mu^5 + \&c.) \ldots(82).$$

From the form of (81) we see that for very great values of n we have

$$K_{n+2} = 2K_n - K_{n-2} \text{ approximately,}$$

and therefore

$$K_{n+2} - K_n = K_n - K_{n-2} \text{ approximately.}$$

Hence each of the series in (82) converges for every value of μ less than unity.

why dismissed.

(g') But this is a very unsatisfactory form of solution. It gives in the form of an infinite series $1 + K_2\mu^2 + K_4\mu^4 + \&c.$ or $\mu + K_3\mu^3 + K_5\mu^5 + \&c.$, the finite solution which we know exists in the form

$$(1-\mu^2)^{\frac{s}{2}}(A_0+A_2\mu^2+\ldots A_{i-s}\mu^{i-s})$$

or $$(1-\mu^2)^{\frac{s}{2}}(A_1\mu+A_3\mu^3+\ldots A_{i-s}\mu^{i-s}),$$

Geometrical antecedents suggest modified form of solution;

when b is the square of an odd integer (s), and when $a=i(i+1)$, i being an odd integer or an even integer; and, a minor defect, but still a serious one, it does not show without elaborate verification that one or other of its constituents $1+K_2\mu^2+$ &c. or $\mu+K_3\mu^3+$ &c. consists of a finite number, $\frac{1}{2}i$ or $\frac{1}{2}(i+1)$, of terms when b is the square of an even integer and $a=i(i+1)$, i being an even integer or an odd integer.

(h') A form of solution which turns out to be much simpler in every case is suggested by our primary knowledge [§ (j) above] of integral solutions. Put

$$w=(1-\mu^2)^{\frac{\sqrt{b}}{2}}v \;\ldots\ldots\ldots\ldots\ldots\ldots\ldots\ldots(83),$$

in (80) and divide the first member by $(1-\mu^2)^{\frac{\sqrt{b}}{2}}$. Thus we find

correspondingly modified differential equation:

$$(1-\mu^2)\frac{d^2v}{d\mu^2}-2(\sqrt{b}+1)\mu\frac{dv}{d\mu}+[a-\sqrt{b}(\sqrt{b}+1)]v=0\ldots\ldots(84).$$

Assume now

$$v=\Sigma A_n\mu^n\ldots\;\ldots\ldots\ldots\ldots\ldots\ldots\ldots(85);$$

equating to zero the coefficient of μ^n in the first member of (84) gives

its solution in powers of μ:

$$(n+1)(n+2)A_{n+2}-[(n-1)n+2(\sqrt{b}+1)n-a+\sqrt{b}(\sqrt{b}+1)]A_n=0\ldots(86),$$

or $$(n+1)(n+2)A_{n+2}=(n+\tfrac{1}{2}+s+\alpha)(n+\tfrac{1}{2}+s-\alpha)A_n\;\ldots..(87),$$

if we put $$\alpha=\sqrt{(a+\tfrac{1}{4})},\qquad s=\sqrt{b}\;\ldots\ldots\ldots\ldots\ldots\ldots\ldots(88),$$

and with this notation (84) becomes

$$(1-\mu^2)\frac{d^2v}{d\mu^2}-2(s+1)\mu\frac{dv}{d\mu}+[\alpha^2-(s+\tfrac{1}{2})^2]v=0\;\ldots\ldots\;(84').$$

The second member of (87) shows that if the series (85) is in descending powers of μ its first term must have either

$$n=-\tfrac{1}{2}-s+\alpha,\quad\text{or } n=-\tfrac{1}{2}-s-\alpha:$$

descending order dismissed, ascending chosen.

the expansion thus obtained would, if not finite, be convergent when $\mu>1$ and divergent when $\mu<1$, and it is therefore not suited for the physical applications. On the other hand, the first member of (87) shows that if the series (85) is in ascending powers of μ, its first term must have either $n=0$ or

Complete solution for tesseral harmonics.

$n=1$: the expansions thus obtained are necessarily convergent when $\mu<1$, and it is therefore these that are suited for our purposes. Taking then $A_0=1$ and $A_1=0$, and denoting by p the series so found, and again $A_0=0$ and $A_1=1$, and q the series; so that we have

$$\left.\begin{aligned} p &= 1+A_2\mu^2+A_4\mu^4+\text{etc.} \\ \text{and}\qquad q &= \mu+A_3\mu^3+A_5\mu^5+\text{etc.}\end{aligned}\right\}\text{..................(89)},$$

A_2, A_4, etc. and A_3, A_5, etc. being found by two sets of successive applications of (87); then the complete solution of (84) is

$$v=Cp+C'q\text{ .... (90).}$$

This solution is identical with (38^{iv}) of § (l) above, as we see by (88) and (79), which give

$$a=i+\tfrac{1}{2}\text{ ....(91).}$$

Alternative solution by changing sign of s:

(i') The sign of either a or s may be changed, in virtue of (88). No variation however is made in the solution by changing the sign of a [which corresponds to changing i into $-i-1$, and verifies (13) (g) above]: but a very remarkable variation is made by changing the sign of s, from which, looking to (88), (83), (87), we infer that if $\mathfrak{p}$ and $\mathfrak{q}$ denote what p and q become when $-s$ is substituted for s in (89), we have

$$\left.\begin{aligned} \mathfrak{p} &= (1-\mu^2)^s p \\ \text{and}\qquad \mathfrak{q} &= (1-\mu^2)^s q\end{aligned}\right\}\text{.............................. (92);}$$

and the prescribed modification of (89) gives

$$\left.\begin{aligned} \mathfrak{p} &= 1+\mathfrak{A}_2\mu^2+\mathfrak{A}_4\mu^4+\text{etc.} \\ \mathfrak{q} &= \mu+\mathfrak{A}_3\mu^3+\mathfrak{A}_5\mu^5+\text{etc.}\end{aligned}\right\}\text{.................. (93),}$$

$\mathfrak{A}_2$, $\mathfrak{A}_4$, etc., and $\mathfrak{A}_3$, $\mathfrak{A}_5$, etc. being found by successive applications of

$$\mathfrak{A}_{n+2}=\frac{(n+\frac{1}{2}-s+a)(n+\frac{1}{2}-s-a)}{(n+1)(n+2)}\mathfrak{A}_n\text{ .... (94).}$$

(j') In the case of "complete harmonics" s is zero or an integer, and the $\mathfrak{p}$ or $\mathfrak{q}$ solution expressing the result of multiplying the already finite and integral p or q solution by the integral polynomial $(1-\mu^2)^s$, is only interesting on account of the way of obtaining it from (87), etc. in virtue of (88). But when either $a-\frac{1}{2}$ or s is not an integer, the possession of the alternative solutions, p or $\mathfrak{p}$, q or $\mathfrak{q}$ may come to be of great intrinsic importance, in respect to obtaining results in finite form. For, supposing a and s to be both positive, it is impossible that both p and q can be finite polynomials, but one or both of $\mathfrak{p}$ and $\mathfrak{q}$ may be so; or

its importance.

one of the p, q forms and the other of the $\mathfrak{p}$, $\mathfrak{q}$ forms may be finite. This we see from (87) and (94), which show as follows:— Cases of finite algebraic solution.

1. If $\frac{1}{2}+s-a$ is positive, p and q must each be an infinite series; but $\mathfrak{p}$ *or* $\mathfrak{q}$ will be finite if either $\frac{1}{2}+s-a$ or $\frac{1}{2}+s+a$ is a positive integer*; and *both* $\mathfrak{p}$ and $\mathfrak{q}$ will be finite if $\frac{1}{2}+s-a$ and $\frac{1}{2}+s+a$ are positive integers differing by unity or any odd number.

2. If $a \overline{\gtrless} s+\frac{1}{2}$, one of the two series $\mathfrak{p}$, $\mathfrak{q}$ must be infinite; and if $a-s-\frac{1}{2}$ is zero or a positive integer, one of the two series p, q is finite. If, lastly, $a+s-\frac{1}{2}$ is zero or a positive integer, one of the two $\mathfrak{p}$, $\mathfrak{q}$ is finite. It is p that is finite if $a-s-\frac{1}{2}$ is zero or even, q if it is odd: and $\mathfrak{p}$ that is finite if $a+s-\frac{1}{2}$ is zero or even, $\mathfrak{q}$ if it is odd. Hence it is p and $\mathfrak{p}$, or q and $\mathfrak{q}$ that are finite if $2s$ be zero or even; but it is p and $\mathfrak{q}$, or q and $\mathfrak{p}$ that are finite if $2s$ be odd. Hence in this latter case the complete solution is a finite algebraic function of μ.

(*k'*) Remembering that by a and s we denote the positive values of the square roots indicated in (88), we collect from (*j'*) 1 and 2, that, if $\mathbf{F}$ denote a rational integral function of μ and $(1-\mu^2)^{\pm\frac{1}{2}s}$, the character of the solution of (80) is as follows in the several cases indicated:—

I. { **A**; $a < s+\frac{1}{2}$; if s and $a-\frac{1}{2}$ are integers.
B; $a \overline{\gtrless} s+\frac{1}{2}$; if $s+\frac{1}{2}$ and a are integers.
The complete solution is $\mathbf{F}$.

II. { **A**; $a < s+\frac{1}{2}$; if $s \pm (a-\frac{1}{2})$ is an integer, but $a-\frac{1}{2}$ not an integer.
B; $a \overline{\gtrless} s+\frac{1}{2}$; if $a-\frac{1}{2} \pm s$ is an integer, but $s+\frac{1}{2}$ not an integer.
A particular solution *is* $\mathbf{F}$; but the complete solution *is not* $\mathbf{F}$.

(*l'*) "Complete Spherical Harmonics," or "Laplace's Coefficients," are included in the particular solution $\mathbf{F}$ of Case II. **B**.

(*m'*) Differentiate (84') and put

$$\frac{dv}{d\mu} = u \qquad\qquad (95).$$

* Unity being understood as included in the class of "positive integers."

Three modes of derivation with change of type.

We find immediately

$$(1-\mu^2)\frac{d^2u}{d\mu^2}-2(s+2)\mu\frac{du}{d\mu}+[a^2-(s+\tfrac{3}{2})^2]u=0 \quad\text{.........(96).}$$

Let $$u'=\mu\frac{dv}{d\mu}+(\pm a+s+\tfrac{1}{2})v \quad\text{....................(97).}$$

We have, as will be proved presently,

$$(1-\mu^2)\frac{d^2u'}{d\mu^2}-2(s+2)\mu\frac{du'}{d\mu}+[(a\pm 1)^2-(s+\tfrac{3}{2})^2]u'=0 \quad\text{...(98).}$$

Lastly, let $$u''=(1-\mu^2)\frac{dv}{d\mu}-(\pm a+s+\tfrac{1}{2})\mu v \quad\text{..............(99).}$$

We have, as will be proved presently,

$$(1-\mu^2)\frac{d^2u''}{d\mu^2}-2(s+1)\mu\frac{du''}{d\mu}+[(a\pm 1)^2-(s+\tfrac{1}{2})^2]u''=0 \quad\text{...(100).}$$

The operation $\frac{d}{dz}$ performed on a solid harmonic of degree $-a-\frac{1}{2}$, and type $H\{z, \sqrt{(x^2+y^2)}\}\,{}^{\sin}_{\cos}\,s\phi$, and transformed to polar co-ordinates r, μ, ϕ, with attention to (83), gives the transition from v to u'', as expressed in (99), and thus (100) is proved by (g) (15).

Similarly the operation

$$\left(\frac{d}{dx}+v\frac{d}{dy}\right)H[z,\sqrt{(x^2+y^2)}](x+vy)^s+\left(\frac{d}{dx}-v\frac{d}{dy}\right)H[z,\sqrt{(x^2+y^2)}](x-vy)^s,$$

transformed to co-ordinates r, μ, ϕ, gives (97), and thus (98) is proved by (g) (15).

Thus it was that (97) (98), and (99) (100) were found. But, assuming (97) and (99) arbitrarily as it were, we prove (98) and (100) most easily as follows. Let

$$u'=\Sigma B'_n\mu^n, \quad\text{and}\quad u''=\Sigma B''_n\mu^n \quad\text{................(101).}$$

Then, by (97) and (99), with (85), we find

$$\left.\begin{aligned} B'_{n+2}&=(n+2\pm a+s+\tfrac{1}{2})A_{n+2}\\ \text{and}\quad B''_{n+2}&=(\mp a+s-\tfrac{1}{2})\frac{n+1+\frac{1}{2}+s\pm a}{n+2}A_{n+1}\end{aligned}\right\}\quad\text{.........(102).}$$

Lastly, applying (87), we find that the corresponding equation is satisfied by $B'_{n+2}\div B'_n$, with $a\pm 1$ and $s+1$ instead of a and s; and by $B''_{n+2}\div B''_n$, with $a\pm 1$ instead of a, but with s unchanged.

As to (95) and (96), they merely express for the generalized surface harmonics the transition from s to $s+1$ without change of i shown for complete harmonics by Murphy's formula, § 782 (6) below.

Examples of derivation.

(n') *Examples of* (95) (96), *and* (99) (100).

Example 1. Let $\alpha = s + \frac{1}{2}$.

(84′) becomes $(1-\mu^2)\dfrac{d^2v}{d\mu^2} - 2(s+1)\mu\dfrac{dv}{d\mu} = 0,$

of which the complete solution is

$$v = C\int\frac{d\mu}{(1-\mu^2)^{s+1}} + C', \qquad \ldots\ldots\ldots\ldots(103).$$

Tesserals from sectorial by increase of s, with order i unchanged.

By (95) (96) we find

$$u = C\frac{d^{n-1}[(1-\mu^2)^{-s-1}]}{d\mu^{n-1}}$$

as a solution of

$$(1-\mu^2)\frac{d^2u}{d\mu^2} - 2(n+s+1)\mu\frac{du}{d\mu} - n(n+2s+1)v = 0 \qquad \ldots\ldots\ldots(104).$$

This is the particular finite solution indicated in § (k') II. **A.**

The liberty we now have to let α be negative as well as positive allows us now to include in our formula for u the cases represented by the double sign $\pm$ in II. **A** of (k').

Example 2. By m successive applications of (99) (100), with the upper sign, to v of (103), we find for the complete integral of

$$\left.\begin{aligned}(1-\mu^2)\frac{d^2u'}{d\mu^2} - 2(s+1)\mu\frac{du'}{d\mu} + m(m+2s+1)u' = 0 \\ u' = C\left\{f(\mu)\int\frac{d\mu}{(1-\mu^2)^{s+1}} + F(\mu)\right\} + C'\mathbf{F}(\mu)\end{aligned}\right\}\ldots\ (105),$$

Tesserals from sectorial by increase of i, with s unchanged.

where $f(\mu)$, $F(\mu)$, $\mathbf{F}(\mu)$ denote rational integral algebraic functions of μ.

Of this solution the part $C'\mathbf{F}(\mu)$ is the particular finite solution indicated in § (k') II. **B.** We now see that the complete solution involves no other transcendent than $\int\frac{d\mu}{(1-\mu^2)^{s+1}}$. When s is an integer, this is reducible to the form

$$a\log\frac{1+\mu}{1-\mu} + \mathfrak{f}(\mu),$$

Examples of derivation continued.

a being a constant and $f(\mu)$ a rational integral algebraic function of μ. In this case, remembering that (105) is what (84') becomes when $m+s+\frac{1}{2}$ is put for a, we may recur to our notation of §§ (g) (j), by putting i for $m+s$, which is now an integer: and going back, by (83) to (80) or (78), put

$$w = (1-\mu^2)^s u' \quad\text{..........................(83')};$$

thus (105) is equivalent to

$$\frac{d}{d\mu}\left[(1-\mu^2)\frac{dw}{d\mu}\right] + \left[\frac{-s^2}{1-\mu^2} + i(i+1)\right] w = 0 \quad\text{............ (78')}.$$

The process of Example 2, § (n'), gives the complete integral of this equation when $i-s$ is a positive integer. When also s, and therefore also i, is an integer, the transcendent involved becomes $\log\frac{1+\mu}{1-\mu}$: in this case the algebraic part of the solution [or $C'\mathbf{F}(\mu)(1-\mu^2)^{\frac{s}{2}}$ according to the notation of (105) and (78')] is the ordinary "Laplace's Function" of order and type (i, s); the $\Theta_s^{(i)}$, $\vartheta_i^{(s)}$, &c. of our previous notations of §§ (j), (y). It is interesting to know that the other particular solution which we now have, completing the solution of the differential equation for these functions, involves nothing of transcendent but $\log\frac{1+\mu}{1-\mu}$.

(o') *Examples of* (99) (100), *and* (95) (96) *continued.*

Algebraic case of last example.

Example 3. Returning to (n'), Example 2, let $s+\frac{1}{2}$ be an integer: the integral $\int\frac{d\mu}{(1-\mu^2)^{s+1}}$ is algebraic. Thus we have the case of (k') I. **B**, in which the complete solution is algebraic.

(p') Returning to (n'), Example 1: let $a=\frac{1}{2}$ and $s=0$, (103) becomes

Zonal of order zero:

$$\left.\begin{aligned} &(1-\mu^2)\frac{d^2v}{d\mu^2} - 2\mu\frac{dv}{d\mu} = 0, \\ &\text{of which the complete integral is} \\ &v = \tfrac{1}{2}C\log\frac{1+\mu}{1-\mu} + C' \end{aligned}\right\} \quad\text{........... (103')}.$$

growing into one sectorial by augmentation of s, with order still zero:

As before, apply (95) (96) n times successively: we find

$$u_1 = \tfrac{1}{2}.1.2\ldots(n-1)C\left[\left(\frac{1}{1-\mu}\right)^n - \left(\frac{-1}{1+\mu}\right)^n\right] \quad\text{...........(106)}$$

as one solution of

$$(1-\mu^2)\frac{d^2u}{d\mu^2}-2(n+1)\mu\frac{du}{d\mu}-n(n+1)u=0\text{..........}(96').$$

the other derived from this by lowering order to -1, equivalent to zero.

To find the other: treat (106) by (99) (100) with the lower sign; the effect is to diminish a from $\frac{1}{2}$ to $-\frac{1}{2}$, and therefore to make no change in the differential equation, but to derive from (106) another particular solution, which is as follows:

$$u_s=\tfrac{1}{2}.1.2\dots(n-1).n.C\left[\left(\frac{1}{1-\mu}\right)^n+\left(\frac{-1}{1+\mu}\right)^n\right]\text{......}(106').$$

Giving any different values to C in (106) and (106′), and, using K, K' to denote two arbitrary constants, adding we have the complete solution of (96′), which we may write as follows:

Complete solution for tesserals of order zero:

$$u=\frac{K}{(1-\mu)^n}+\frac{K'}{(1+\mu)^n}\text{..................}(107).$$

(q') That (107) is the solution of (96′) we verify in a moment by trial, and in so doing we see farther that it is the complete solution, whether n be integral or not.

(r') Example 4. Apply (99) (100) with upper sign i times to (107) and successive results. We get thus the complete solution of (84′) for $a-\frac{1}{2}=i$ *any* integer, if n is *not* an integer. But if n is an integer we get the complete solution only provided $i<n$: this is case I. A of § (k'). If we take $i=n-1$, the result, algebraic as it is, may be proved to be expressible in the form

derivation from it of both tesserals of every integral order;

$$\left.\begin{array}{l} u=\dfrac{C+C'\int d\mu(1-\mu^2)^{n-1}}{(1-\mu^2)^n}, \\ \text{which is therefore for } n \text{ an integer the complete} \\ \text{integral of} \\ (1-\mu^2)\dfrac{d^2u}{d\mu^2}-2(n+1)\mu\dfrac{du}{d\mu}-2nu=0, \end{array}\right\}\text{..........}(108):$$

except case of s an integer and $i\overline{>}s$, when only Laplace's function is found.

being the case of (84′) for which $a=s-\frac{1}{2}$, and $s=n$ an integer: applying to this (99) (100) with upper sign, the constant C disappears, and we find $u'=C'$ as a solution of

$$(1-\mu^2)\frac{d^2u'}{d\mu^2}-2(n+1)\mu\frac{du'}{d\mu}=0\text{............}(109).$$

Hence, for $i\overline{>}n$ one solution is lost. The other, found by

Examples of derivation continued.

continued applications of (99) (100) with upper sign, is the regular "Laplace's function" growing from $C' \sin^n \theta \begin{matrix}\sin\\ \cos\end{matrix} n\phi$, which is the case represented by $u' = C'$ in (109). But in this continuation we are only doing for the case of n an integer, part of what was done in § (n'), Example 2, where the other part, from the other part of the solution of (109) now lost, gives the other part of the complete solution of Laplace's equation subject to the limitation $i-n$ (or $i-s$) a positive integer, but not to the limitation of i an integer or n an integer.

(s') Returning to the commencement of § (r'), with s put for n, we find a complete solution growing in the form

$$\frac{Kf_i(\mu)}{(1-\mu)^s} + (-)^i \frac{K'f_i(-\mu)}{(1+\mu)^s} \quad \text{.....................} \quad (110);$$

which may be immediately reduced to

$$\frac{Kf_i(\mu)(1+\mu)^s + (-)^i K'f_i(-\mu)(1-\mu)^s}{(1-\mu^2)^s} \quad \text{............} \quad (110');$$

f_i denoting an integral algebraic function of the ith degree, readily found by the proper successive applications of (99) (100). Hence, by (83) (79), we have

$$w = \frac{Kf_i(\mu)(1+\mu)^s + (-)^i K'f_i(-\mu)(1-\mu)^s}{(1-\mu^2)^{\frac{s}{2}}} \quad \text{..........} \quad (111),$$

as the complete solution of Laplace's equation

$$\frac{d}{d\mu}\left[(1-\mu^2)\frac{dw}{d\mu}\right] + \left[\frac{-s^2}{1-\mu^2} + i(i+1)\right] w = 0 \quad \text{.........} (112),$$

Finite algebraic expression of complete solution for tesseral harmonics of *integral order*.

for the case of i an integer without any restriction as to the value of s, which may be integral or fractional, real or imaginary, with no failure except the case of s an integer and $i > s$, of which the complete treatment is included in § (m'), Example 2, above.

CHAPTER II.

DYNAMICAL LAWS AND PRINCIPLES.

205. In the preceding chapter we considered as a subject of pure geometry the motion of points, lines, surfaces, and volumes, whether taking place with or without change of dimensions and form; and the results we there arrived at are of course altogether independent of the idea of *matter*, and of the *forces* which matter exerts. We have heretofore assumed the *existence* merely of motion, distortion, etc.; we now come to the consideration, not of how we *might* consider such motions, etc., to be produced, but of the *actual* causes which in the material world *do* produce them. The axioms of the present chapter must therefore be considered to be due to actual experience, in the shape either of observation or experiment. How this experience is to be obtained will form the subject of a subsequent chapter. **Ideas of matter and force introduced.**

206. We cannot do better, at all events in commencing, than follow Newton somewhat closely. Indeed the introduction to the *Principia* contains in a most lucid form the general foundations of Dynamics. The *Definitiones* and *Axiomata sive Leges Motûs*, there laid down, require only a few amplifications and additional illustrations, suggested by subsequent developments, to suit them to the present state of science, and to make a much better introduction to dynamics than we find in even some of the best modern treatises.

207. We cannot, of course, give a definition of *Matter* which will satisfy the metaphysician, but the naturalist may be content to know matter as *that which can be perceived by the senses*, or as *that which can be acted upon by, or can exert, force.* The **Matter.**

latter, and indeed the former also, of these definitions involves the idea of *Force*, which, in point of fact, is a direct object of sense; probably of all our senses, and certainly of the "muscular sense." To our chapter on Properties of Matter we must refer for further discussion of the question, *What is matter?* And we shall then be in a position to discuss the question of the subjectivity of *Force*.

Force.

208. *The Quantity of Matter* in a body, or, as we now call it, the *Mass* of a body, is proportional, according to Newton, to the *Volume* and the *Density* conjointly. In reality, the definition gives us the meaning of density rather than of mass; for it shows us that if twice the original quantity of matter, air for example, be forced into a vessel of given capacity, the density will be doubled, and so on. But it also shows us that, of matter of uniform density, the mass or quantity is proportional to the volume or space it occupies.

Mass. Density.

Let M be the mass, ρ the density, and V the volume, of a homogeneous body. Then

$$M = V\rho;$$

if we so take our units that unit of mass is that of unit volume of a body of unit density.

If the density vary from point to point of the body, we have evidently, by the above formula and the elementary notation of the integral calculus,

$$M = \iiint \rho \, dxdydz,$$

where ρ is supposed to be a known function of x, y, z, and the integration extends to the whole space occupied by the matter of the body whether this be continuous or not.

It is worthy of particular notice that, in this definition, Newton says, if there be anything which *freely* pervades the interstices of all bodies, this is *not* taken account of in estimating their Mass or Density.

Measurement of mass

209. Newton further states, that a practical measure of the mass of a body is its *Weight*. His experiments on pendulums, by which he establishes this most important result, will be described later, in our chapter on Properties of Matter.

As will be presently explained, the unit mass most convenient for British measurements is an imperial pound of matter.

210. The *Quantity of Motion*, or the *Momentum*, of a rigid body moving without rotation is proportional to its mass and velocity conjointly. The whole motion is the sum of the motions of its several parts. Thus a doubled mass, or a doubled velocity, would correspond to a double quantity of motion; and so on. **Momentum.**

Hence, if we take as unit of momentum the momentum of a unit of matter moving with unit velocity, the momentum of a mass M moving with velocity v is Mv.

211. *Change of Quantity of Motion*, or *Change of Momentum*, is proportional to the mass moving and the change of its velocity conjointly. **Change of momentum**

Change of velocity is to be understood in the general sense of § 27. Thus, in the figure of that section, if a velocity represented by OA be changed to another represented by OC, the change of velocity is represented in magnitude and direction by AC.

212. *Rate of Change of Momentum* is proportional to the mass moving and the acceleration of its velocity conjointly. Thus (§ 35, *b*) the rate of change of momentum of a falling body is constant, and in the vertical direction. Again (§ 35, *a*) the rate of change of momentum of a mass M, describing a circle of radius R, with uniform velocity V, is $\frac{MV^2}{R}$, and is directed to the centre of the circle; that is to say, it is a change of direction, not a change of speed, of the motion. Hence if the mass be compelled to keep in the circle by a cord attached to it and held fixed at the centre of the circle, the force with which the cord is stretched is equal to $\frac{MV^2}{R}$: this is called the centrifugal force of the mass M moving with velocity V in a circle of radius R. **Rate of change of momentum.**

Generally (§ 29), for a body of mass M moving anyhow in space there is change of momentum, at the rate, $M\frac{d^2s}{dt^2}$ in the direc-

Rate of change of momentum.

tion of motion, and $M\frac{v^2}{\rho}$ towards the centre of curvature of the path; and, if we choose, we may exhibit the whole acceleration of momentum by its three rectangular components $M\frac{d^2x}{dt^2}$, $M\frac{d^2y}{dt^2}$, $M\frac{d^2z}{dt^2}$, or, according to the Newtonian notation, $M\ddot{x}$, $M\ddot{y}$, $M\ddot{z}$.

Kinetic energy.

213. The *Vis Viva*, or *Kinetic Energy*, of a moving body is proportional to the mass and the square of the velocity, conjointly. If we adopt the same units of mass and velocity as before, there is particular advantage in defining kinetic energy as *half* the product of the mass and the square of its velocity.

214. *Rate of Change of Kinetic Energy* (when defined as above) is the product of the velocity into the component of rate of change of momentum in the direction of motion.

For $$\frac{d}{dt}\left(\frac{Mv^2}{2}\right) = v\,\frac{d(Mv)}{dt}.$$

Particle and point.

215. It is to be observed that, in what precedes, with the exception of the definition of mass, we have taken no account of the dimensions of the moving body. This is of no consequence so long as it does not rotate, and so long as its parts preserve the same relative positions amongst one another. In this case we may suppose the whole of the matter in it to be condensed in one point or particle. We thus speak of a *material particle*, as distinguished from a *geometrical point*. If the body rotate, or if its parts change their relative positions, then we cannot choose any one point by whose motions alone we may determine those of the other points. In such cases the momentum and change of momentum of the whole body in any direction are, the sums of the momenta, and of the changes of momentum, of its parts, in these directions; while the kinetic energy of the whole, being non-directional, is simply the sum of the kinetic energies of the several parts or particles.

Inertia.

216. Matter has an innate power of resisting external influences, so that every body, as far as it can, remains at rest, or moves uniformly in a straight line.

This, the *Inertia* of matter, is proportional to the quantity of matter in the body. And it follows that some *cause* is requisite to disturb a body's uniformity of motion, or to change its direction from the natural rectilinear path. Inertia.

217. Force is any cause which tends to alter a body's natural state of rest, or of uniform motion in a straight line. Force.

Force is wholly expended in the *Action* it produces; and the body, after the force ceases to act, retains by its inertia the direction of motion and the velocity which were given to it. Force may be of divers kinds, as pressure, or gravity, or friction, or any of the attractive or repulsive actions of electricity, magnetism, etc.

218. The three elements specifying a force, or the three elements which must be known, before a clear notion of the force under consideration can be formed, are, its place of application, its direction, and its magnitude. Specification of a force.

(*a*) The place of application of a force. The first case to be considered is that in which the place of application is a point. It has been shown already in what sense the term "point" is to be taken, and, therefore, in what way a force may be imagined as acting at a point. In reality, however, the place of application of a force is always either a surface or a space of three dimensions occupied by matter. The point of the finest needle, or the edge of the sharpest knife, is still a surface, and acts by pressing over a finite area on bodies to which it may be applied. Even the most rigid substances, when brought together, do not touch at a point merely, but mould each other so as to produce a surface of application. On the other hand, gravity is a force of which the place of application is the whole matter of the body whose weight is considered; and the smallest particle of matter that has weight occupies some finite portion of space. Thus it is to be remarked, that there are two kinds of force, distinguishable by their place of application—force, whose place of application is a surface, and force, whose place of application is a solid. When a heavy body rests on the ground, or on a table, force of the second character, acting downwards, is balanced by force of the first character acting upwards. Place of application.

Direction. (*b*) The second element in the specification of a force is its direction. The direction of a force is the line in which it acts. If the place of application of a force be regarded as a point, a line through that point, in the direction in which the force tends to move the body, is the direction of the force. In the case of a force distributed over a surface, it is frequently possible and convenient to assume a single point and a single line, such that a certain force acting at that point in that line would produce sensibly the same effect as is really produced.

Magnitude. (*c*) The third element in the specification of a force is its magnitude. This involves a consideration of the method followed in dynamics for measuring forces. Before measuring anything, it is necessary to have a unit of measurement, or a standard to which to refer, and a principle of numerical specification, or a mode of referring to the standard. These will be supplied presently. See also § 258, below.

Accelerative effect. 219. The *Accelerative Effect of a Force* is proportional to the velocity which it produces in a given time, and is measured by that which is, or would be, produced in unit of time; in other words, the *rate of change of velocity* which it produces. This is simply what we have already defined as acceleration, § 28.

Measure of force. 220. The *Measure of a Force* is the quantity of motion which it produces per unit of time.

The reader, who has been accustomed to speak of a force of so many pounds, or so many tons, may be startled when he finds that such expressions are not definite unless it be specified at what part of the earth's surface the pound, or other definite quantity of matter named, is to be weighed; for the *heaviness* or *gravity* of a given quantity of matter differs in different latitudes. But the force required to produce a stated quantity of motion in a given time is perfectly definite, and independent of locality. Thus, let W be the mass of a body, g the velocity it would acquire in falling freely for a second, and P the force of gravity upon it, measured in kinetic or absolute units. We have

$$P = Wg.$$

221. According to the system commonly followed in mathematical treatises on dynamics till fourteen years ago, when a small instalment of the first edition of the present work was issued for the use of our students, the unit of mass was g times the mass of the standard or unit weight. This definition, giving a varying and a very unnatural unit of mass, was exceedingly inconvenient. By taking the gravity of a constant mass for the unit of force it makes the unit of force greater in high than in low latitudes. In reality, standards of weight are *masses*, not *forces*. They are employed primarily in commerce for the purpose of measuring out a definite *quantity* of matter; not an amount of matter which shall be attracted by the earth with a given force.

Inconvenient system of modern treatises.

Standards of weight are *masses*, and not primarily intended for measurement of force.

A merchant, with a balance and a set of standard weights, would give his customers the same quantity of the same kind of matter however the earth's attraction might vary, depending as he does upon *weights* for his measurement; another, using a spring-balance, would defraud his customers in high latitudes, and himself in low, if his instrument (which depends on constant forces and not on the gravity of constant masses) were correctly adjusted in London.

It is a secondary application of our standards of weight to employ them for the measurement of *forces*, such as steam pressures, muscular power, etc. In all cases where great accuracy is required, the results obtained by such a method have to be reduced to what they would have been if the measurements of force had been made by means of a perfect spring-balance, graduated so as to indicate the forces of gravity on the standard weights in some conventional locality.

It is therefore very much simpler and better to take the imperial pound, or other national or international standard weight, as, for instance, the gramme (see the chapter on Measures and Instruments), as the unit of mass, and to derive from it, according to Newton's definition above, the unit of force. This is the method which Gauss has adopted in his great improvement (§ 223 below) of the system of measurement of forces.

Clairault's formula for the amount of gravity.

222. The formula, deduced by Clairault from observation, and a certain theory regarding the figure and density of the earth, may be employed to calculate the most probable value of the apparent force of gravity, being the resultant of true gravitation and centrifugal force, in any locality where no pendulum observation of sufficient accuracy has been made. This formula, with the two coefficients which it involves, corrected according to the best modern pendulum observations (Airy, *Encyc. Metropolitana, Figure of the Earth*), is as follows:—

Let G be the apparent force of gravity on a unit mass at the equator, and g that in any latitude λ; then

$$g = G\,(1 + \cdot 005133 \sin^2 \lambda).$$

The value of G, in terms of the British absolute unit, to be explained immediately, is

$$32\cdot 088.$$

According to this formula, therefore, polar gravity will be

$$g = 32\cdot 088 \times 1\cdot 005133 = 32\cdot 2527.$$

223. Gravity having failed to furnish a definite standard, independent of locality, recourse must be had to something else. The principle of measurement indicated as above by Newton, but first introduced practically by Gauss, furnishes us with what we want. According to this principle, the unit force is that force which, acting on a national standard unit of matter during the unit of time, generates the unity of velocity.

Gauss's absolute Unit of Force.

This is known as Gauss's absolute unit; absolute, because it furnishes a standard force independent of the differing amounts of gravity at different localities. It is however terrestrial and inconstant if the unit of time depends on the earth's rotation, as it does in our present system of chronometry. The period of vibration of a piece of quartz crystal of specified shape and size and at a stated temperature (a tuning-fork, or bar, as one of the bars of glass used in the "musical glasses") gives us a unit of time which is constant through all space and all time, and independent of the earth. A unit of force founded on such a unit of time would be better entitled to the designation *abso-*

lute than is the "absolute unit" now generally adopted, which is founded on the *mean solar second*. But this depends essentially on one particular piece of matter, and is therefore liable to all the accidents, etc. which affect so-called National Standards however carefully they may be preserved, as well as to the almost insuperable practical difficulties which are experienced when we attempt to make exact copies of them. Still, in the present state of science, we are really confined to such approximations. The recent discoveries due to the Kinetic theory of gases and to Spectrum analysis (especially when it is applied to the light of the heavenly bodies) indicate to us *natural standard* pieces of matter such as atoms of hydrogen, or sodium, ready made in infinite numbers, all absolutely alike in every physical property. The time of vibration of a sodium particle corresponding to any one of its modes of vibration, is known to be absolutely independent of its position in the universe, and it will probably remain the same so long as the particle itself exists. The wave-length for that particular ray, *i.e.* the space through which light is propagated *in vacuo* during the time of one complete vibration of this period, gives a perfectly invariable unit of length; and it is possible that at some not very distant day the mass of such a sodium particle may be employed as a natural standard for the remaining fundamental unit. This, the latest improvement made upon our original suggestion of a *Perennial Spring* (First edition, § 406), is due to Clerk Maxwell*; who has also communicated to us another very important and interesting suggestion for founding the unit of time upon physical properties of a substance without the necessity of specifying any particular quantity of it. It is this, water being chosen as the substance of all others known to us which is most easily obtained in perfect purity and in perfectly definite physical condition.—Call the standard density of water the maximum density of the liquid when under the pressure of its own vapour alone. The time of revolution of an infinitesimal satellite close to the surface of a globe of water at standard density (or of any kind of matter at the same density) may be taken as the unit of time; for it is independent of the size of the globe. This has

Maxwell's two suggestions for Absolute Unit of Time.

* *Electricity and Magnetism*, 1872.

Third suggestion for Absolute Unit of Time.

suggested to us still another unit, founded, however, still upon the same physical principle. The time of the gravest simple harmonic infinitesimal vibration of a globe of liquid, water at standard density, or of other perfect liquids at the same density, may be taken as the unit of time; for the time of the simple harmonic vibration of any one of the fundamental modes of a liquid sphere is independent of the size of the sphere.

Let f be the force of gravitational attraction between two units of matter at unit distance. The force of gravity at the surface of a globe of radius r, and density ρ, is $\frac{4\pi}{3}f\rho r$. Hence if ω be the angular velocity of an infinitesimal satellite, we have, by the equilibrium of centrifugal force and gravity (§§ 212, 477),

$$\omega^2 r = \frac{4\pi}{3} f\rho r.$$

Hence $$\omega = \sqrt{\frac{4\pi f\rho}{3}},$$

and therefore if T be the satellite's period,

$$T = 2\pi\sqrt{\frac{3}{4\pi f\rho}}$$

(which is equal to the period of a simple pendulum whose length is the globe's radius, and weighted end infinitely near the surface of the globe). And it has been proved* that if a globe of liquid be distorted infinitesimally according to a spherical harmonic of order i, and left at rest, it will perform simple harmonic oscillations in a period equal to

$$2\pi\sqrt{\left\{\frac{3}{4\pi f\rho}\cdot\frac{2i+1}{2i(i-1)}\right\}}.$$

Hence if T' denote the period of the gravest, that, namely, for which $i=2$, we have

$$T' = T\sqrt{\frac{5}{4}}.$$

The semi-period of an infinitesimal satellite round the earth is equal, reckoned in seconds, to the square root of the number of metres in the earth's radius, the metre being very approximately

* "Dynamical Problems regarding Elastic Spheroidal Shells and Spheroids of Incompressible Liquid" (W. Thomson), *Phil. Trans.* Nov. 27, 1862.

the length of the seconds pendulum, whose period is two seconds. Hence taking the earth's radius as 6,370,000 metres, and its density as $5\frac{1}{2}$ times that of our standard globe,

Suggestions for Absolute Unit of Time.

$$T = 3 \text{ h. } 17 \text{ m.}$$
$$T' = 3 \text{ h. } 40 \text{ m.}$$

224. The absolute unit depends on the unit of matter, the unit of time, and the unit of velocity; and as the unit of velocity depends on the unit of space and the unit of time, there is, in the definition, a single reference to mass and space, but a *double* reference to time; and this is a point that must be particularly attended to.

225. The unit of mass may be the British imperial pound; the unit of space the British standard foot; and, accurately enough for practical purposes for a few thousand years, the unit of time may be the mean solar second.

We accordingly define the British absolute unit force as "the force which, acting on one pound of matter for one second, generates a velocity of one foot per second." Prof. James Thomson has suggested the name "Poundal" for this unit of force.

British absolute unit.

226. To illustrate the reckoning of force in "absolute measure," find how many absolute units will produce, in any particular locality, the same effect as the force of gravity on a given mass. To do this, measure the effect of gravity in producing acceleration on a body unresisted in any way. The most accurate method is indirect, by means of the pendulum. The result of pendulum experiments made at Leith Fort, by Captain Kater, is, that the velocity which would be acquired by a body falling unresisted for one second is at that place 32·207 feet per second. The preceding formula gives exactly 32·2, for the latitude 55° 33′, which is approximately that of Edinburgh. The variation in the force of gravity for one degree of difference of latitude about the latitude of Edinburgh is only 0000832 of its own amount. It is nearly the same, though somewhat more, for every degree of latitude southwards, as far as the southern limits of the British Isles. On the other hand, the variation per degree is sensibly less, as far north as the Orkney and Shetland Isles. Hence

Comparison with gravity.

Gravity of Unit weight or mass in terms of Kinetic Unit. the augmentation of gravity per degree from south to north throughout the British Isles is at most about $\frac{1}{12000}$ of its whole amount in any locality. The average for the whole of Great Britain and Ireland differs certainly but little from 32·2. Our present application is, that the force of gravity at Edinburgh is 32·2 times the force which, acting on a pound for a second, would generate a velocity of one foot per second; in other words, 32·2 is the number of absolute units which measures the weight of a pound in this latitude. Thus, approximately, the poundal is equal to the gravity of about half an ounce.

227. Forces (since they involve only direction and magnitude) may be represented, as velocities are, by straight lines in their directions, and of lengths proportional to their magnitudes, respectively.

Also the laws of composition and resolution of any number of forces acting at the same point, are, as we shall show later (§ 255), the same as those which we have already proved to hold for velocities; so that with the substitution of force for velocity, §§ 26, 27, are still true.

Effective component of a force. 228. In rectangular resolution the *Component* of a force in any direction, (sometimes called the *Effective Component* in that direction,) is therefore found by multiplying the magnitude of the force by the cosine of the angle between the directions of the force and the component. The remaining component in this case is perpendicular to the other.

It is very generally convenient to resolve forces into components parallel to three lines at right angles to each other; each such resolution being effected by multiplying by the cosine of the angle concerned.

Geometrical Theorem preliminary to definition of centre of inertia. 229. The point whose distances from three planes at right angles to one another are respectively equal to the mean distances of any group of points from these planes, is at a distance from any plane whatever, equal to the mean distance of the group from the same plane. Hence of course, if it is in motion, its velocity perpendicular to that plane is the mean of the velocities of the several points, in the same direction.

Let (x_1, y_1, z_1), etc., be the points of the group in number i; and $\bar{x}$, $\bar{y}$, $\bar{z}$ be the co-ordinates of a point at distances respectively equal to their mean distances from the planes of reference; that is to say, let Geometrical Theorem preliminary to definition of centre of inertia.

$$\bar{x}=\frac{x_1+x_2+\text{etc.}}{i},\ \bar{y}=\frac{y_1+y_2+\text{etc.}}{i},\ \bar{z}=\frac{z_1+z_2+\text{etc.}}{i}.$$

Thus, if p_1, p_2, etc., and p, denote the distances of the points in question from any plane at a distance a from the origin of co-ordinates, perpendicular to the direction (l, m, n), the sum of a and p_1 will make up the projection of the broken line x_1, y_1, z_1 on (l, m, n), and therefore

$$p_1 = lx_1 + my_1 + nz_1 - a, \text{ etc.};$$

and similarly, $\quad p = l\bar{x} + m\bar{y} + n\bar{z} - a.$

Substituting in this last the expressions for $\bar{x}$, $\bar{y}$, $\bar{z}$, we find

$$p=\frac{p_1+p_2+\text{etc.}}{i},$$

which is the theorem to be proved. Hence, of course,

$$\frac{dp}{dt}=\frac{1}{i}\left(\frac{dp_1}{dt}+\frac{dp_2}{dt}+\text{etc.}\right).$$

230. The *Centre of Inertia* of a system of equal material points (whether connected with one another or not) is the point whose distance is equal to their average distance from any plane whatever (§ 229). Centre of inertia.

A group of material points of unequal masses may always be imagined as composed of a greater number of equal material points, because we may imagine the given material points divided into different numbers of very small parts. In any case in which the magnitudes of the given masses are incommensurable, we may approach as near as we please to a rigorous fulfilment of the preceding statement, by making the parts into which we divide them sufficiently small.

On this understanding the preceding definition may be applied to define the centre of inertia of a system of material points, whether given equal or not. The result is equivalent to this:—

Centre of Inertia.

The centre of inertia of any system of material points whatever (whether rigidly connected with one another, or connected in any way, or quite detached), is a point whose distance from any plane is equal to the sum of the products of each mass into its distance from the same plane divided by the sum of the masses.

We also see, from the proposition stated above, that a point whose distance from three rectangular planes fulfils this condition, must fulfil this condition also for every other plane.

The co-ordinates of the centre of inertia, of masses w_1, w_2, etc., at points (x_1, y_1, z_1), (x_2, y_2, z_2), etc., are given by the following formulæ:—

$$\bar{x} = \frac{w_1x_1 + w_2x_2 + \text{etc.}}{w_1 + w_2 + \text{etc.}} = \frac{\Sigma wx}{\Sigma w}, \quad \bar{y} = \frac{\Sigma wy}{\Sigma w}, \quad \bar{z} = \frac{\Sigma wz}{\Sigma w}$$

These formulæ are perfectly general, and can easily be put into the particular shape required for any given case. Thus, suppose that, instead of a set of detached material points, we have a continuous distribution of matter through certain definite portions of space; the density at x, y, z being ρ, the elementary principles of the integral calculus give us at once

$$\bar{x} = \frac{\iiint \rho x dx dy dz}{\iiint \rho dx dy dz}, \text{ etc.},$$

where the integrals extend through all the space occupied by the mass in question, in which ρ has a value different from zero.

The Centre of Inertia or Mass is thus a perfectly definite point in every body, or group of bodies. The term *Centre of Gravity* is often very inconveniently used for it. The theory of the resultant action of gravity which will be given under Abstract Dynamics shows that, except in a definite class of distributions of matter, there is no one fixed point which can properly be called the Centre of Gravity of a rigid body. In ordinary cases of terrestrial gravitation, however, an approximate solution is available, according to which, in common parlance, the term "*Centre of Gravity*" may be used as equivalent to *Centre of Inertia;* but it must be carefully remembered that the fundamental ideas involved in the two definitions are essentially different.

The second proposition in § 229 may now evidently be stated thus:—The sum of the momenta of the parts of the system in any direction is equal to the momentum in the same direction of a mass equal to the sum of the masses moving with a velocity equal to the velocity of the centre of inertia. Centre of Inertia.

231. The *Moment* of any physical agency is the numerical measure of its importance. Thus, the moment of a force round a point or round a line, signifies the measure of its importance as regards producing or balancing rotation round that point or round that line. Moment.

232. The *Moment* of a force about a point is defined as the product of the force into its perpendicular distance from the point. It is numerically double the area of the triangle whose vertex is the point, and whose base is a line representing the force in magnitude and direction. It is often convenient to represent it by a line numerically equal to it, drawn through the vertex of the triangle perpendicular to its plane, through the front of a watch held in the plane with its centre at the point, and facing so that the force tends to turn round this point in a direction opposite to the hands. The moment of a force round any axis is the moment of its component in any plane perpendicular to the axis, round the point in which the plane is cut by the axis. Here we imagine the force resolved into two components, one parallel to the axis, which is ineffective so far as rotation round the axis is concerned; the other perpendicular to the axis (that is to say, having its line in any plane perpendicular to the axis). This latter component may be called the effective component of the force, with reference to rotation round the axis. And its moment round the axis may be defined as its moment round the nearest point of the axis, which is equivalent to the preceding definition. It is clear that the moment of a force round any axis, is equal to the area of the projection on any plane perpendicular to the axis, of the figure representing its moment round any point of the axis. Moment of a force about a point. Moment of a force about an axis.

233. The projection of an area, plane or curved, on any plane, is the area included in the projection of its bounding line. Digression on projection of areas.

Digression on projection of areas.

If we imagine an area divided into any number of parts, the projections of these parts on any plane make up the projection of the whole. But in this statement it must be understood that the areas of partial projections are to be reckoned as positive if particular sides, which, for brevity, we may call the outside of the projected area and the front of the plane of projection, face the same way, and negative if they face oppositely.

Of course if the projected surface, or any part of it, be a plane area at right angles to the plane of projection, the projection vanishes. The projections of any two shells having a common edge, on any plane, are equal, but with the same, or opposite, signs as the case may be. Hence, by taking two such shells facing opposite ways, we see that the projection of a closed surface (or a shell with evanescent edge), on any plane, is nothing.

Equal areas in one plane, or in parallel planes, have equal projections on any plane, whatever may be their figures.

Hence the projection of any plane figure, or of any shell, edged by a plane figure, on another plane, is equal to its area, multiplied by the cosine of the angle at which its plane is inclined to the plane of projection. This angle is acute or obtuse, according as the outside of the projected area, and the front of plane of projection, face on the whole towards the same parts, or oppositely. Hence lines representing, as above described, moments about a point in different planes, are to be compounded as forces are.—See an analogous theorem in § 96.

Couple.

234. A *Couple* is a pair of equal forces acting in dissimilar directions in parallel lines. The *Moment* of a couple is the sum of the moments of its forces about any point in their plane, and is therefore equal to the product of either force into the shortest distance between their directions. This distance is called the *Arm* of the couple.

The *Axis of a Couple* is a line drawn from any chosen point of reference perpendicular to the plane of the couple, of such magnitude and in such direction as to represent the magnitude of the moment, and to indicate the direction in which the couple tends to turn. The most convenient rule for fulfilling the latter condition is this:—Hold a watch with its centre at the point of reference, and with its plane parallel to the plane of the couple. Then, according as the motion of the hands is contrary to or along with the direction in which the couple tends to turn, draw the axis of the couple through the face or through the back of the watch, *from* its centre. Thus a couple is completely represented by its axis; and couples are to be resolved and compounded by the same geometrical constructions performed with reference to their axes as forces or velocities, with reference to the lines directly representing them.

Couple.

235. If we substitute, for the force in § 232, a velocity, we have the moment of a velocity about a point; and by introducing the mass of the moving body as a factor, we have an important element of dynamical science, the *Moment of Momentum*. The laws of composition and resolution are the same as those already explained; but for the sake of some simple applications we give an elementary investigation.

Moment of velocity.

Moment of momentum.

The moment of a rectilineal motion is the product of its length into the distance of its line from the point.

Moment of a rectilineal displacement.

The moment of the resultant velocity of a particle about any point in the plane of the components is equal to the algebraic sum of the moments of the components, the proper sign of each moment being determined as above, § 233. The same is of course true of moments of displacements, of moments of forces and of moments of momentum.

First, consider two component motions, AB and AC, and let AD be their resultant (§ 27). Their half moments round the point O are respectively the areas OAB, OCA. Now OCA, together with half the area of the parallelogram $CABD$, is equal to OBD. Hence the sum of the two half moments together with half the area of the parallelogram, is equal to AOB together with BOD, that is to say, to the area of the whole figure $OABD$. But ABD, a part of this figure, is equal to half the area of the parallelogram; and therefore the remainder, OAD, is equal to the sum of the two half moments. But OAD is half the moment of the resultant velocity round the point O. Hence the moment of the

For two forces, motions, velocities, or moments, in one plane, the sum of their moments proved equal to the moment of their resultant round any point in that plane.

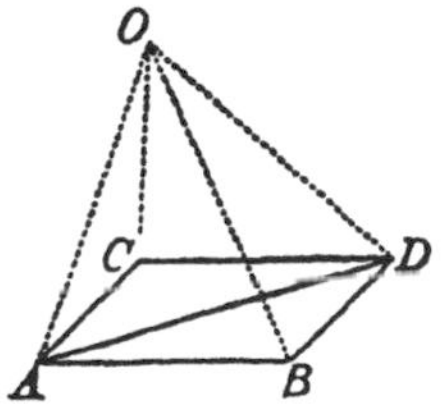

O
C
D
A
B

resultant is equal to the sum of the moments of the two components.

If there are any number of component rectilineal motions in one plane, we may compound them in order, any two taken together first, then a third, and so on; and it follows that the sum of their moments is equal to the moment of their resultant. It follows, of course, that the sum of the moments of any number of component velocities, all in one plane, into which the velocity of any point may be resolved, is equal to the moment of their resultant, round any point in their plane. It follows also, that if velocities, in different directions all in one plane, be successively given to a moving point, so that at any time its velocity is their resultant, the moment of its velocity at any time is the sum of the moments of all the velocities which have been successively given to it.

Any number of moments in one plane compounded by addition.

Cor.—If one of the components always passes through the point, its moment vanishes. This is the case of a motion in which the acceleration is directed to a fixed point, and we thus reproduce the theorem of § 36, *a*, that in this case the areas described by the radius-vector are proportional to the times; for, as we have seen, the moment of velocity is double the area traced out by the radius-vector in unit of time.

Moment round an axis.

236. The moment of the velocity of a point round any axis is the moment of the velocity of its projection on a plane perpendicular to the axis, round the point in which the plane is cut by the axis.

Moment of a whole motion, round an axis.

The moment of the whole motion of a point during any time, round any axis, is twice the area described in that time by the radius-vector of its projection on a plane perpendicular to that axis.

If we consider the conical area traced by the radius-vector drawn from any fixed point to a moving point whose motion is not confined to one plane, we see that the projection of this area on any plane through the fixed point is half of what we have just defined as the moment of the whole motion round an axis perpendicular to it through the fixed point. Of all these planes, there is one on which the projection of the area is greater than on any other; and the projection of the conical area on any plane perpendicular to this plane, is equal to nothing, the proper interpretation of positive and negative projections being used. Moment of a whole motion, round an axis.

If any number of moving points are given, we may similarly consider the conical surface described by the radius-vector of each drawn from one fixed point. The same statement applies to the projection of the many-sheeted conical surface, thus presented. The resultant axis of the whole motion in any finite time, round the fixed point of the motions of all the moving points, is a line through the fixed point perpendicular to the plane on which the area of the whole projection is greater than on any other plane; and the moment of the whole motion round the resultant axis, is twice the area of this projection. Resultant axis.

The resultant axis and moment of velocity, of any number of moving points, relatively to any fixed point, are respectively the resultant axis of the whole motion during an infinitely short time, and its moment, divided by the time.

The moment of the whole motion round any axis, of the motion of any number of points during any time, is equal to the moment of the whole motion round the resultant axis through any point of the former axis, multiplied into the cosine of the angle between the two axes.

The resultant axis, relatively to any fixed point, of the whole motion of any number of moving points, and the moment of the whole motion round it, are deduced by the same elementary constructions from the resultant axes and moments of the individual points, or partial groups of points of the system, as the direction and magnitude of a resultant displacement are deduced from any given lines and magnitudes of component displacements. Moment of momentum.

Corresponding statements apply, of course, to the moments of velocity and of momentum.

237. If the point of application of a force be displaced through a small space, the resolved part of the displacement in the direction of the force has been called its *Virtual Velocity.* Virtual velocity.

Virtual velocity. This is positive or negative according as the virtual velocity is in the same, or in the opposite, direction to that of the force.

The product of the force, into the virtual velocity of its point of application, has been called the *Virtual Moment* of the force. These terms we have introduced since they stand in the history and developments of the science; but, as we shall show further on, they are inferior substitutes for a far more useful set of ideas clearly laid down by Newton.

Work. 238. A force is said to *do work* if its place of application has a positive component motion in its direction; and the work done by it is measured by the product of its amount into this component motion.

Thus, in lifting coals from a pit, the amount of work done is proportional to the weight of the coals lifted; that is, to the force overcome in raising them; and also to the height through which they are raised. Practical unit. The unit for the measurement of work adopted in practice by British engineers, is that required to overcome a force equal to the gravity of a pound through the space of a foot; and is called a *Foot-Pound.*

Scientific unit. In purely scientific measurements, the unit of work is not the foot-pound, but the kinetic unit force (§ 225) acting through unit of space. Thus, for example, as we shall show further on, this unit is adopted in measuring the work done by an electric current, the units for electric and magnetic measurements being founded upon the kinetic unit force.

If the weight be raised obliquely, as, for instance, along a smooth inclined plane, the space through which the force has to be overcome is increased in the ratio of the length to the height of the plane; but the force to be overcome is not the whole gravity of the weight, but only the component of the gravity parallel to the plane; and this is less than the gravity in the ratio of the height of the plane to its length. By Work of a force. multiplying these two expressions together, we find, as we might expect, that the amount of work required is unchanged by the substitution of the oblique for the vertical path.

239. Generally, for any force, the work done during an infinitely small displacement of the point of application is the

virtual moment of the force (§ 237), or is the product of the resolved part of the force in the direction of the displacement into the displacement.

Work of a force.

From this it appears, that if the motion of the point of application be always perpendicular to the direction in which a force acts, such a force does no work. Thus the mutual normal pressure between a fixed and moving body, as the tension of the cord to which a pendulum bob is attached, or the attraction of the sun on a planet if the planet describe a circle with the sun in the centre, is a case in which no work is done by the force.

240. The work done by a force, or by a couple, upon a body turning about an axis, is the product of the moment of the force or couple into the angle (in radians, or fraction of a radian) through which the body acted on turns, if the moment remains the same in all positions of the body. If the moment be variable, the statement is only valid for infinitely small displacements, but may be made accurate by employing the proper *average* moment of the force or of the couple. The proof is obvious.

Work of a couple.

If Q be the moment of the force or couple for a position of the body given by the angle θ, $Q(\theta_1-\theta_0)$ if Q is constant, or $\int_{\theta_0}^{\theta_1} Q d\theta = q(\theta_1-\theta_0)$ where q is the proper average value of Q when variable, is the work done by the couple during the rotation from θ_0 to θ_1.

241. Work done on a body by a force is always shown by a corresponding increase of vis viva, or kinetic energy, if no other forces act on the body which can do work or have work done against them. If work be done against any forces, the increase of kinetic energy is less than in the former case by the amount of work so done. In virtue of this, however, the body possesses an equivalent in the form of *Potential Energy* (§ 273), if its physical conditions are such that these forces will act equally, and in the same directions, if the motion of the system is reversed. Thus there may be no change of kinetic energy pro-

Transformation of work.

Potential energy.

Potential energy. duced, and the work done may be wholly stored up as potential energy.

Thus a weight requires work to raise it to a height, a spring requires work to bend it, air requires work to compress it, etc.; but a raised weight, a bent spring, compressed air, etc., are *stores* of energy which can be made use of at pleasure.

Newton's Laws of Motion. **242.** In what precedes we have given some of Newton's *Definitiones* nearly in his own words; others have been enunciated in a form more suitable to modern methods; and some terms have been introduced which were invented subsequent to the publication of the *Principia*. But the *Axiomata, sive Leges Motûs*, to which we now proceed, are given in Newton's own words; the two centuries which have nearly elapsed since he first gave them have not shown a necessity for any addition or modification. The first two, indeed, were discovered by Galileo, and the third, in some of its many forms, was known to Hooke, Huyghens, Wallis, Wren, and others; before the publication of the *Principia*. Of late there has been a tendency to split the second law into two, called respectively the second and third, and to ignore the third entirely, though using it *directly* in every dynamical problem; but all who have done so have been forced *indirectly* to acknowledge the completeness of Newton's system, by introducing as an axiom what is called D'Alembert's principle, which is really Newton's rejected third law in another form. Newton's own interpretation of his third law directly points out not only D'Alembert's principle, but also the modern principles of Work and Energy.

Axiom. **243.** An Axiom is a proposition, the truth of which must be admitted as soon as the terms in which it is expressed are clearly understood. But, as we shall show in our chapter on "Experience," physical axioms are axiomatic to those only who have sufficient knowledge of the action of physical causes to enable them to see their truth. Without further remark we shall give Newton's Three Laws; it being remembered that, as the properties of matter *might* have been such as to render a totally different set of laws axiomatic, these laws must be considered as resting on convictions drawn from observation and experiment, *not* on intuitive perception.

244. Lex I. *Corpus omne perseverare in statu suo quiescendi vel movendi uniformiter in directum, nisi quatenus illud à viribus impressis cogitur statum suum mutare.* Newton's first law.

Every body continues in its state of rest or of uniform motion in a straight line, except in so far as it may be compelled by force to change that state.

245. The meaning of the term *Rest*, in physical science is essentially relative. Absolute rest is undefinable. If the universe of matter were finite, its centre of inertia might fairly be considered as absolutely at rest; or it might be imagined to be moving with any uniform velocity in any direction whatever through infinite space. But it is remarkable that the first law of motion enables us (§ 249, below) to explain what may be called *directional* rest. As will soon be shown, § 267, the plane in which the moment of momentum of the universe (if finite) round its centre of inertia is the greatest, which is clearly determinable from the actual motions at any instant, is fixed in direction in space. Rest.

246. We may logically convert the assertion of the first law of motion as to velocity into the following statements:—

The times during which any particular body, not compelled by force to alter the speed of its motion, passes through equal spaces, are equal. And, again—Every other body in the universe, not compelled by force to alter the speed of its motion, moves over equal spaces in successive intervals, during which the particular chosen body moves over equal spaces.

247. The first part merely expresses the convention universally adopted for the measurement of *Time*. The earth, in its rotation about its axis, presents us with a case of motion in which the condition, of not being compelled by force to alter its speed, is more nearly fulfilled than in any other which we can easily or accurately observe. And the numerical measurement of time practically rests on defining *equal intervals of time, as times during which the earth turns through equal* Time.

angles. This is, of course, a mere convention, and not a law of nature; and, as we now see it, is a part of Newton's first law.

Examples of the law.

248. The remainder of the law is not a convention, but a great truth of nature, which we may illustrate by referring to small and trivial cases as well as to the grandest phenomena we can conceive.

A curling-stone, projected along a horizontal surface of ice, travels equal distances, except in so far as it is retarded by friction and by the resistance of the air, in successive intervals of time during which the earth turns through equal angles. The sun moves through equal portions of interstellar space in times during which the earth turns through equal angles, except in so far as the resistance of interstellar matter, and the attraction of other bodies in the universe, alter his speed and that of the earth's rotation.

Directional fixedness.

249. If two material points be projected from one position, A, at the same instant with any velocities in any directions, and each left to move uninfluenced by force, the line joining them will be always parallel to a fixed direction. For the law asserts, as we have seen, that $AP : AP' :: AQ : AQ'$, if P, Q, and again P', Q' are simultaneous positions; and therefore PQ is parallel to $P'Q'$. Hence if four material points O, P, Q, R are all projected at one instant from one position, OP, OQ, OR are fixed directions of reference ever after. But, practically, the determination of fixed directions in space, § 267, is made to depend upon the rotation of groups of particles exerting forces on each other, and thus involves the Third Law of Motion.

The "Invariable Plane" of the solar system.

250. The whole law is singularly at variance with the tenets of the ancient philosophers who maintained that circular motion is perfect.

The last clause, "*nisi quatenus*," etc., admirably prepares for the introduction of the second law, by conveying the idea that *it is force alone which can produce a change of motion.* How, we naturally inquire, does the change of motion produced depend on the magnitude and direction of the force which produces it? And the answer is—

251. Lex II. *Mutationem motûs proportionalem esse vi motrici impressæ, et fieri secundum lineam rectam quâ vis illa imprimitur.* Newton's second law.

Change of motion is proportional to force applied, and takes place in the direction of the straight line in which the force acts.

252. If any force generates motion, a double force will generate double motion, and so on, whether simultaneously or successively, instantaneously, or gradually applied. And this motion, if the body was moving beforehand, is either added to the previous motion if directly conspiring with it; or is subtracted if directly opposed; or is geometrically compounded with it, according to the kinematical principles already explained, if the line of previous motion and the direction of the force are inclined to each other at an angle. (This is a paraphrase of Newton's own comments on the second law.)

253. In Chapter I. we have considered change of velocity, or acceleration, as a purely geometrical element, and have seen how it may be at once inferred from the given initial and final velocities of a body. By the definition of quantity of motion (§ 210), we see that, if we multiply the change of velocity, thus geometrically determined, by the mass of the body, we have the change of motion referred to in Newton's law as the measure of the force which produces it.

It is to be particularly noticed, that in this statement there is nothing said about the actual motion of the body before it was acted on by the force: it is only the *change* of motion that concerns us. Thus the same force will produce precisely the same change of motion in a body, whether the body be at rest, or in motion with any velocity whatever.

254. Again, it is to be noticed that nothing is said as to the body being under the action of *one* force only; so that we may logically put a part of the second law in the following (apparently) amplified form:—

When any forces whatever act on a body, then, whether the body be originally at rest or moving with any velocity and in any direction, each force produces in the body the exact change of motion which it would have produced if it had acted singly on the body originally at rest.

Composition of forces. **255.** A remarkable consequence follows immediately from this view of the second law. Since forces are measured by the changes of motion they produce, and their directions assigned by the directions in which these changes are produced; and since the changes of motion of one and the same body are in the directions of, and proportional to, the changes of velocity—a single force, measured by the resultant change of velocity, and in its direction, will be the equivalent of any number of simultaneously acting forces. Hence

The resultant of any number of forces (applied at one point) is to be found by the same geometrical process as the resultant of any number of simultaneous velocities.

256. From this follows at once (§ 27) the construction of the *Parallelogram of Forces* for finding the resultant of two forces, and the *Polygon of Forces* for the resultant of any number of forces, in lines all through one point.

The case of the equilibrium of a number of forces acting at one point, is evidently deducible at once from this; for if we introduce one other force equal and opposite to their resultant, this will produce a change of motion equal and opposite to the resultant change of motion produced by the given forces; that is to say, will produce a condition in which the point experiences no change of motion, which, as we have already seen, is the only kind of rest of which we can ever be conscious.

257. Though Newton perceived that the Parallelogram of Forces, or the fundamental principle of Statics, is essentially involved in the second law of motion, and gave a proof which is virtually the same as the preceding, subsequent writers on Statics (especially in this country) have very generally ignored the fact; and the consequence has been the introduction of various unnecessary Dynamical Axioms, more or less obvious, but in reality included in or dependent upon Newton's laws of motion. We have retained Newton's method, not only on account of its admirable simplicity, but because we believe it contains the most philosophical foundation for the static as well as for the kinetic branch of the dynamic science.

258. But the second law gives us the means of measuring force, and also of measuring the mass of a body. Measurement of force and mass.

For, if we consider the actions of various forces upon the same body for equal times, we evidently have changes of velocity produced which are *proportional to* the forces. The changes of velocity, then, give us in this case the means of comparing the magnitudes of different forces. Thus the velocities acquired in one second by the same mass (falling freely) at different parts of the earth's surface, give us the relative amounts of the earth's attraction at these places.

Again, if equal forces be exerted on different bodies, the changes of velocity produced in equal times must be *inversely* as the masses of the various bodies. This is approximately the case, for instance, with trains of various lengths started by the same locomotive: it is exactly realized in such cases as the action of an electrified body on a number of solid or hollow spheres of the same external diameter, and of different metals or of different thicknesses.

Again, if we find a case in which different bodies, each acted on by a force, acquire in the same time the same changes of velocity, the forces must be proportional to the masses of the bodies. This, when the resistance of the air is removed, is the case of falling bodies; and from it we conclude that the weight of a body in any given locality, or the force with which the earth attracts it, is proportional to its mass; a most important physical truth, which will be treated of more carefully in the chapter devoted to "Properties of Matter."

259. It appears, lastly, from this law, that every theorem of Kinematics connected with acceleration has its counterpart in Kinetics. Translations from the kinematics of a point.

For instance, suppose X, Y, Z to be the components, parallel to fixed axes of x, y, z respectively, of the whole force acting on a particle of mass M. We see by § 212 that

$$M\frac{d^2x}{dt^2} = X, \quad M\frac{d^2y}{dt^2} = Y, \quad M\frac{d^2z}{dt^2} = Z;$$

or

$$M\ddot{x} = X, \quad M\ddot{y} = Y, \quad M\ddot{z} = Z.$$

Translations from the kinematics of a point.

Also, from these, we may evidently write,

$$M\ddot{s} = X\frac{dx}{ds} + Y\frac{dy}{ds} + Z\frac{dz}{ds} = X\frac{\dot{x}}{\dot{s}} + Y\frac{\dot{y}}{\dot{s}} + Z\frac{\dot{z}}{\dot{s}},$$

$$0 = X\frac{\dot{y}\ddot{z} - \dot{z}\ddot{y}}{\rho^{-1}\dot{s}^3} + Y\frac{\dot{z}\ddot{x} - \dot{x}\ddot{z}}{\rho^{-1}\dot{s}^3} + Z\frac{\dot{x}\ddot{y} - \dot{y}\ddot{x}}{\rho^{-1}\dot{s}^3},$$

$$\frac{M\dot{s}^2}{\rho} = X\frac{\dot{s}\ddot{x} - \dot{x}\ddot{s}}{\rho^{-1}\dot{s}^3} + Y\frac{\dot{s}\ddot{y} - \dot{y}\ddot{s}}{\rho^{-1}\dot{s}^3} + Z\frac{\dot{s}\ddot{z} - \dot{z}\ddot{s}}{\rho^{-1}\dot{s}^3}.$$

The second members of these equations are respectively the components of the impressed force, along the tangent (§ 9), perpendicular to the osculating plane (§ 9), and towards the centre of curvature, of the path described.

Measurement of force and mass.

260. We have, by means of the first two laws, arrived at a *definition* and a *measure* of force; and have also found how to compound, and therefore also how to resolve, forces; and also how to investigate the motion of a single particle subjected to given forces. But more is required before we can completely understand the more complex cases of motion, especially those in which we have mutual actions between or amongst two or more bodies; such as, for instance, attractions, or pressures, or transference of energy in any form. This is perfectly supplied by

Newton's third law.

261. LEX III. *Actioni contrariam semper et æqualem esse reactionem: sive corporum duorum actiones in se mutuò semper esse æquales et in partes contrarias dirigi.*

To every action there is always an equal and contrary reaction: or, the mutual actions of any two bodies are always equal and oppositely directed.

262. If one body presses or draws another, it is pressed or drawn by this other with an equal force in the opposite direction. If any one presses a stone with his finger, his finger is pressed with the same force in the opposite direction by the stone. A horse towing a boat on a canal is dragged backwards by a force equal to that which he impresses on the towing-rope forwards. By whatever amount, and in whatever direction, one body has its motion changed by impact upon another, this other body has its motion changed by the same amount in the opposite direction; for at each instant during the impact the force between them was equal and opposite on the two. When neither of the two bodies has any rotation, whether before or after impact, the changes of velocity which they experience are inversely as their masses. Newton's third law.

When one body attracts another from a distance, this other attracts it with an equal and opposite force. This law holds not only for the attraction of gravitation, but also, as Newton himself remarked and verified by experiment, for magnetic attractions: also for electric forces, as tested by Otto-Guericke.

263. What precedes is founded upon Newton's own comments on the third law, and the actions and reactions contemplated are simple forces. In the scholium appended, he makes the following remarkable statement, introducing another description of actions and reactions subject to his third law, the full meaning of which seems to have escaped the notice of commentators:—

Si æstimetur agentis actio ex ejus vi et velocitate conjunctim; et similiter resistentis reactio æstimetur conjunctim ex ejus partium singularum velocitatibus et viribus resistendi ab earum attritione, cohæsione, pondere, et acceleratione oriundis; erunt actio et reactio, in omni instrumentorum usu, sibi invicem semper æquales.

In a previous discussion Newton has shown what is to be understood by the velocity of a force or resistance; *i.e.*, that it is the velocity of the point of application of the force *resolved in the direction of the force.* Bearing this in mind, we may read the above statement as follows:—

If the Activity of an agent be measured by its amount and its velocity conjointly; and if, similarly, the Counter-activity of the resistance be measured by the velocities of its several parts and their several amounts conjointly, whether these arise from friction, cohesion, weight, or acceleration;—Activity and Counter-activity, in all combinations of machines, will be equal and opposite.*

Farther on (§§ 264, 293) we shall give an account of the

* We translate Newton's word "*Actio*" here by "Activity" to avoid confusion with the word "Action" so universally used in modern dynamical treatises, according to the definition of § 326 below, in relation to Maupertuis' principle of "Least Action."

splendid dynamical theory founded by D'Alembert and Lagrange on this most important remark.

D'Alembert's principle.

264. Newton, in the passage just quoted, points out that forces of resistance against acceleration are to be reckoned as reactions equal and opposite to the actions by which the acceleration is produced. Thus, if we consider any one material point of a system, its reaction against acceleration must be equal and opposite to the resultant of the forces which that point experiences, whether by the actions of other parts of the system upon it, or by the influence of matter not belonging to the system. In other words, it must be in equilibrium with these forces. Hence Newton's view amounts to this, that all the forces of the system, with the reactions against acceleration of the material points composing it, form groups of equilibrating systems for these points considered individually. Hence, by the principle of superposition of forces in equilibrium, all the forces acting on points of the system form, with the reactions against acceleration, an equilibrating set of forces on the whole system. This is the celebrated principle first explicitly stated, and very usefully applied, by D'Alembert in 1742, and still known by his name. We have seen, however, that it is very distinctly implied in Newton's own interpretation of his third law of motion. As it is usual to investigate the general equations or conditions of equilibrium, in dynamical treatises, before entering in detail on the kinetic branch of the subject, this principle is found practically most useful in showing how we may write down at once the equations of motion for any system for which the equations of equilibrium have been investigated.

Mutual forces between particles of a rigid body.

265. Every rigid body may be imagined to be divided into indefinitely small parts. Now, in whatever form we may eventually find a *physical* explanation of the origin of the forces which act between these parts, it is certain that each such small part may be considered to be held in its position relatively to the others by mutual forces in lines joining them.

266. From this we have, as immediate consequences of the second and third laws, and of the preceding theorems relating

to Centre of Inertia and Moment of Momentum, a number of important propositions such as the following:—

(*a*) The centre of inertia of a rigid body moving in any manner, but free from external forces, moves uniformly in a straight line. Motion of centre of inertia of a rigid body.

(*b*) When any forces whatever act on the body, the motion of the centre of inertia is the same as it would have been had these forces been applied with their proper magnitudes and directions at that point itself.

(*c*) Since the moment of a force acting on a particle is the same as the moment of momentum it produces in unit of time, the changes of moment of momentum in any two parts of a rigid body due to their mutual action are equal and opposite. Hence the moment of momentum of a rigid body, about any axis which is fixed in direction, and passes through a point which is either fixed in space or moves uniformly in a straight line, is unaltered by the mutual actions of the parts of the body. Moment of momentum of a rigid body.

(*d*) The rate of increase of moment of momentum, when the body is acted on by external forces, is the sum of the moments of these forces about the axis.

267. We shall for the present take for granted, that the mutual action between two rigid bodies may in every case be imagined as composed of pairs of equal and opposite forces in straight lines. From this it follows that the sum of the quantities of motion, parallel to any fixed direction, of two rigid bodies influencing one another in any possible way, remains unchanged by their mutual action; also that the sum of the moments of momentum of all the particles of the two bodies, round any line in a fixed direction in space, and passing through any point moving uniformly in a straight line in any direction, remains constant. From the first of these propositions we infer that the centre of inertia of any number of mutually influencing bodies, if in motion, continues moving uniformly in a straight line, unless in so far as the direction or velocity of its motion is changed by forces acting mutually between them and some other matter not belonging to them; also that the centre of inertia of any body or system of bodies moves Conservation of momentum, and of moment of momentum.

The "Invariable Plane" is a plane through the centre of inertia, perpendicular to the resultant axis.

just as all their matter, if concentrated in a point, would move under the influence of forces equal and parallel to the forces really acting on its different parts. From the second we infer that the axis of resultant rotation through the centre of inertia of any system of bodies, or through any point either at rest or moving uniformly in a straight line, remains unchanged in direction, and the sum of moments of momenta round it remains constant if the system experiences no force from without. This principle used to be called *Conservation of Areas*, a very ill-considered designation. From this principle it follows that if by internal action such as geological upheavals or subsidences, or pressure of the winds on the water, or by evaporation and rain- or snow-fall, or by any influence not depending on the attraction of sun or moon (even though dependent on solar heat), the disposition of land and water becomes altered, the component round any fixed axis of the moment of momentum of the earth's rotation remains constant.

Terrestrial application.

Rate of doing work.

268. The foundation of the abstract theory of energy is laid by Newton in an admirably distinct and compact manner in the sentence of his scholium already quoted (§ 263), in which he points out its application to mechanics*. The *actio agentis*, as he defines it, which is evidently equivalent to the product of the effective component of the force, into the velocity of the point on which it acts, is simply, in modern English phraseology, the rate at which the agent works. The subject for measurement here is precisely the same as that for which Watt, a hundred years later, introduced the practical unit of a "*Horse-power*," or the rate at which an agent works when overcoming 33,000 times the weight of a pound through the space of a foot in a minute; that is, producing 550 foot-pounds of work per second. The unit, however, which is most generally convenient is that which Newton's definition implies, namely, the rate of doing work in which the unit of energy is produced in the unit of time.

Horse-power.

* The reader will remember that we use the word "mechanics" in its true classical sense, the science of machines, the sense in which Newton himself used it, when he dismissed the further consideration of it by saying (in the scholium referred to), *Cæterum mechanicam tractare non est hujus instituti.*

269. Looking at Newton's words (§ 263) in this light, we see that they may be logically converted into the following form:— Energy in abstract dynamics

Work done on any system of bodies (in Newton's statement, the parts of any machine) has its equivalent in work done against friction, molecular forces, or gravity, if there be no acceleration; but if there be acceleration, part of the work is expended in overcoming the resistance to acceleration, and the additional kinetic energy developed is equivalent to the work so spent. This is evident from § 214.

When part of the work is done against molecular forces, as in bending a spring; or against gravity, as in raising a weight; the recoil of the spring, and the fall of the weight, are capable at any future time, of reproducing the work originally expended (§ 241). But in Newton's day, and long afterwards, it was supposed that work was *absolutely lost* by friction; and, indeed, this statement is still to be found even in recent authoritative treatises. But we must defer the examination of this point till we consider in its modern form the principle of *Conservation of Energy*.

270. If a system of bodies, given either at rest or in motion, be influenced by no forces from without, the sum of the kinetic energies of all its parts is augmented in any time by an amount equal to the whole work done in that time by the mutual forces, which we may imagine as acting between its points. When the lines in which these forces act remain all unchanged in length, the forces do no work, and the sum of the kinetic energies of the whole system remains constant. If, on the other hand, one of these lines varies in length during the motion, the mutual forces in it will do work, or will consume work, according as the distance varies with or against them.

271. A limited system of bodies is said to be *dynamically conservative* (or simply *conservative*, when force is understood to be the subject), if the mutual forces between its parts always perform, or always consume, the same amount of work during any motion whatever, by which it can pass from one particular configuration to another. Conservative system.

Foundation of the theory of energy.

272. The whole theory of energy in physical science is founded on the following proposition:—

If the mutual forces between the parts of a material system are independent of their velocities, whether relative to one another, or relative to any external matter, the system must be dynamically conservative.

Physical axiom that "the Perpetual Motion is impossible" introduced.

For if more work is done by the mutual forces on the different parts of the system in passing from one particular configuration to another, by one set of paths than by another set of paths, let the system be directed, by frictionless constraint, to pass from the first configuration to the second by one set of paths and return by the other, over and over again for ever. It will be a continual source of energy without any consumption of materials, which is impossible.

Potential energy of conservative system.

273. The *potential energy* of a conservative system, in the configuration which it has at any instant, is the amount of work required to bring it to that configuration against its mutual forces during the passage of the system from any one chosen configuration to the configuration at the time referred to. It is generally, but not always, convenient to fix the particular configuration chosen for the zero of reckoning of potential energy, so that the potential energy, in every other configuration practically considered, shall be positive.

274. The potential energy of a conservative system, at any instant, depends solely on its configuration at that instant, being, according to definition, the same at all times when the system is brought again and again to the same configuration. It is therefore, in mathematical language, said to be a function of the co-ordinates by which the positions of the different parts of the system are specified. If, for example, we have a conservative system consisting of two material points; or two rigid bodies, acting upon one another with force dependent only on the relative position of a point belonging to one of them, and a point belonging to the other; the potential energy of the system depends upon the co-ordinates of one of these points relatively to lines of reference in fixed directions through the other. It will therefore, in general, depend on three indepen-

dent co-ordinates, which we may conveniently take as the distance between the two points, and two angles specifying the absolute direction of the line joining them. Thus, for example, let the bodies be two uniform metal globes, electrified with any given quantities of electricity, and placed in an insulating medium such as air, in a region of space under the influence of a vast distant electrified body. The mutual action between these two spheres will depend solely on the relative position of their centres. It will consist partly of gravitation, depending solely on the distance between their centres, and of electric force, which will depend on the distance between them, but also, in virtue of the inductive action of the distant body, will depend on the absolute direction of the line joining their centres. In our divisions devoted to gravitation and electricity respectively, we shall investigate the portions of the mutual potential energy of the two bodies depending on these two agencies separately. The former we shall find to be the product of their masses divided by the distance between their centres; the latter a somewhat complicated function of the distance between the centres and the angle which this line makes with the direction of the resultant electric force of the distant electrified body. Or again, if the system consist of two balls of soft iron, in any locality of the earth's surface, their mutual action will be partly gravitation, and partly due to the magnetism induced in them by terrestrial magnetic force. The portion of the mutual potential energy depending on the latter cause, will be a function of the distance between their centres and the inclination of this line to the direction of the terrestrial magnetic force. It will agree in mathematical expression with the potential energy of electric action in the preceding case, so far as the inclination is concerned, but the law of variation with the distance will be less easily determined.

Potential energy of conservative system

275. In nature the hypothetical condition of § 271 is *apparently violated* in all circumstances of motion. A material system can never be brought through any returning cycle of motion without spending more work against the mutual forces of its parts than is gained from these forces, because no relative motion can take place without meeting with frictional or

Inevitable loss of energy of visible motions.

Inevitable loss of energy of visible motions. other forms of resistance; among which are included (1) mutual friction between solids sliding upon one another; (2) resistances due to the viscosity of fluids, or imperfect elasticity of solids; (3) resistances due to the induction of electric currents; (4) resistances due to varying magnetization under the influence of imperfect magnetic retentiveness. No motion in nature can take place without meeting resistance due to some, if not to all, of these influences. It is matter of every day experience that friction and imperfect elasticity of solids impede the action of all artificial mechanisms; and that even when bodies are detached, and left to move freely in the air, as falling bodies, or as projectiles, they experience resistance owing to the viscosity of the air.

The greater masses, planets and comets, moving in a less resisting medium, show less indications of resistance*. Indeed it cannot be said that observation upon any one of these bodies, with the exception of Encke's comet, has demonstrated resistance. But the analogies of nature, and the ascertained facts of physical science, forbid us to doubt that every one of them, every star, and every body of any kind moving in any part of space, has its relative motion impeded by the air, gas, vapour, medium, or whatever we choose to call the substance occupying the space immediately round it; just as the motion of a rifle bullet is impeded by the resistance of the air.

Effect of tidal friction. 276. There are also indirect resistances, owing to friction impeding the tidal motions, on all bodies (like the earth) partially or wholly covered by liquid, which, as long as these bodies move relatively to neighbouring bodies, must keep drawing off energy from their relative motions. Thus, if we consider, in the first place, the action of the moon alone, on the earth with its oceans, lakes, and rivers, we perceive that it must tend to equalize the periods of the earth's rotation about its axis, and of the revolution of the two bodies about their centre of inertia; because as long as these periods differ, the tidal action on the

* Newton, *Principia*. (Remarks on the first law of motion.) "Majora autem Planetarum et Cometarum corpora motus suos et progressivos et circulares, in spatiis minus resistentibus factos, conservant diutius."

earth's surface must keep subtracting energy from their motions. To view the subject more in detail, and, at the same time, to avoid unnecessary complications, let us suppose the moon to be a uniform spherical body. The mutual action and reaction of gravitation between her mass and the earth's, will be equivalent to a single force in some line through her centre; and must be such as to impede the earth's rotation as long as this is performed in a shorter period than the moon's motion round the earth. It must therefore lie in some such direction as the line *MQ* in the diagram, which represents, necessarily with enormous exaggeration, its deviation, *OQ*, from the earth's centre. Now the actual force on the moon in the line *MQ*, may be regarded as consisting of a force in the line *MO* towards the earth's centre, sensibly equal in amount to the whole force, and a comparatively very small force in the line *MT* perpendicular to *MO*. This latter is very nearly tangential to the moon's path, and is in the direction *with* her motion. Such a force, if suddenly commencing to act, would, in the first place, increase the moon's velocity; but after a certain time she would have moved so much farther from the earth, in virtue of this acceleration, as to have lost, by moving against the earth's attraction, as much velocity as she had gained by the tangential accelerating force. The effect of a continued tangential force, acting with the motion, but so small in amount as to make only a small deviation at any moment from the circular form of the orbit, is to gradually increase the distance from the central body, and to cause as much again as its own amount of work to be done against the attraction of the central mass, by the kinetic energy of motion lost. The circumstances will be readily understood, by considering this motion round the central body in a very gradual spiral path tending outwards. Provided the law of the central force is the inverse square of the distance, the tangential component of the central force against the motion will be twice as great as the disturbing tangential force in the direction with the motion; and therefore one-half of the amount of work done

Effect of tidal friction.

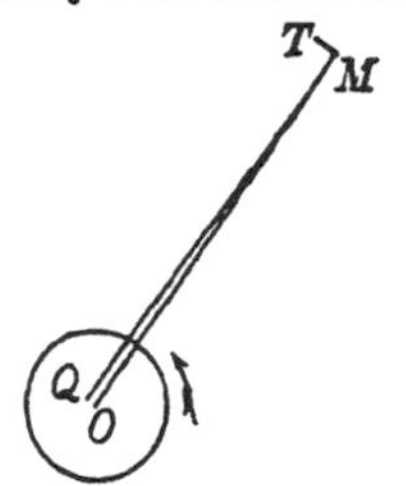

Inevitable loss of energy of visible motions. Tidal friction.

against the former, is done by the latter, and the other half by kinetic energy taken from the motion. The integral effect on the moon's motion, of the particular disturbing cause now under consideration, is most easily found by using the principle of moments of momenta. Thus we see that as much moment of momentum is gained in any time by the motions of the centres of inertia of the moon and earth relatively to their common centre of inertia, as is lost by the earth's rotation about its axis. The sum of the moments of momentum of the centres of inertia of the moon and earth as moving at present, is about 4·45 times the present moment of momentum of the earth's rotation. The average plane of the former is the ecliptic; and therefore the axes of the two momenta are inclined to one another at the average angle of 23° 27½′, which, as we are neglecting the sun's influence on the plane of the moon's motion, may be taken as the actual inclination of the two axes at present. The resultant, or whole moment of momentum, is therefore 5·38 times that of the earth's present rotation, and its axis is inclined 19° 13′ to the axis of the earth. Hence the ultimate tendency of the tides is, to reduce the earth and moon to a simple uniform rotation with this resultant moment round this resultant axis, as if they were two parts of one rigid body: in which condition the moon's distance would be increased (approximately) in the ratio 1 : 1·46, being the ratio of the square of the present moment of momentum of the centres of inertia to the square of the whole moment of momentum; and the period of revolution in the ratio 1 : 1·77, being that of the cubes of the same quantities. The distance would therefore be increased to 347,100 miles, and the period lengthened to 48·36 days. Were there no other body in the universe but the earth and the moon, these two bodies might go on moving thus for ever, in circular orbits round their common centre of inertia, and the earth rotating about its axis in the same period, so as always to turn the same face to the moon, and therefore to have all the liquids at its surface at rest relatively to the solid. But the existence of the sun would prevent any such state of things from being permanent. There would be solar tides—twice high water and twice low water—in the period of the earth's revolution relatively to the sun (that is to say, twice in the solar day, or, which would be the same thing, the month). This could not go on without loss of energy by fluid friction. It is easy to trace the whole course of the disturbance in the earth's and moon's motions which this cause would produce*: its first effect must be to bring the moon to fall in to the earth, with compensation for loss of moment of momentum of the two round their centre of inertia in increase of its distance from the sun, and then to reduce the very rapid rotation of the compound body, Earth-and-Moon, after the collision, and farther increase its distance from the Sun till ultimately, (corresponding action on liquid matter on the Sun having *its* effect also, and it being for our illustration supposed that there are no other planets,) the two bodies shall rotate round their common centre of inertia, like parts of one rigid body. It is remarkable that the whole frictional effect of the lunar and solar tides should be, first to augment the moon's distance from the earth to a maximum, and then to diminish it, till ultimately the moon falls in to the earth: and first to diminish, after that to increase, and lastly to diminish the earth's rotational velocity. We hope to return to the subject later, and to consider the general problem of the motion of any number of rigid bodies or material points acting on one another with mutual forces, under any actual physical law, and therefore, as we shall see, necessarily subject to loss of energy as long as any of their mutual distances vary; that is to say, until all subside into a state of motion in circles round an axis passing through their centre of inertia, like parts of one rigid body. It is probable

Inevitable loss of energy of visible motions. Tidal friction.

* The friction of these solar tides on the earth would cause the earth to rotate still slower; and then the moon's influence, tending to keep the earth rotating with always the same face towards herself, would resist this further reduction in the speed of the rotation. Thus (as explained above with reference to the moon) there would be from the sun a force opposing the earth's rotation, and from the moon a force promoting it. Hence according to the preceding explanation applied to the altered circumstances, the line of the earth's attraction on the moon passes now as before, not through the centre of inertia of the earth, but now in a line slightly *behind* it (instead of *before*, as formerly). It therefore now resists the moon's motion of revolution. The combined effect of this resistance and of the earth's attraction on the moon is, like that of a resisting medium, to cause the moon to fall in towards the earth in a spiral path with gradually increasing velocity.

Inevitable loss of energy of visible motions. Tidal friction.

that the moon, in ancient times liquid or viscous in its outer layer if not throughout, was thus brought to turn always the same face to the earth.

Ultimate tendency of the solar system.

277. We have no data in the present state of science for estimating the relative importance of tidal friction, and of the resistance of the resisting medium through which the earth and moon move; but whatever it may be, there can be but one ultimate result for such a system as that of the sun and planets, if continuing long enough under existing laws, and not disturbed by meeting with other moving masses in space. That result is the falling together of all into one mass, which, although rotating for a time, must in the end come to rest relatively to the surrounding medium.

Conservation of energy.

278. The theory of energy cannot be completed until we are able to examine the physical influences which accompany loss of energy in each of the classes of resistance mentioned above, § 275. We shall then see that in every case in which energy is lost by resistance, heat is generated; and we shall learn from Joule's investigations that the quantity of heat so generated is a perfectly definite equivalent for the energy lost. Also that in no natural action is there ever a development of energy which cannot be accounted for by the disappearance of an equal amount elsewhere by means of some known physical agency. Thus we shall conclude, that if any limited portion of the material universe could be perfectly isolated, so as to be prevented from either giving energy to, or taking energy from, matter external to it, the sum of its potential and kinetic energies would be the same at all times: in other words, that every material system subject to no other forces than actions and reactions between its parts, is a dynamically conservative system, as defined above, § 271. But it is only when the inscrutably minute motions among small parts, possibly the ultimate molecules of matter, which constitute light, heat, and magnetism; and the intermolecular forces of chemical affinity; are taken into account, along with the palpable motions and measurable forces of which we become cognizant by direct observation, that we can recognise

the universally conservative character of all natural dynamic action, and perceive the bearing of the principle of reversibility on the whole class of natural actions involving resistance, which seem to violate it. In the meantime, in our studies of abstract dynamics, it will be sufficient to introduce a special reckoning for energy lost in working against, or gained from work done by, forces not belonging palpably to the conservative class.

Conservation of energy.

279. As of great importance in farther developments, we prove a few propositions intimately connected with energy.

280. The kinetic energy of any system is equal to the sum of the kinetic energies of a mass equal to the sum of the masses of the system, moving with a velocity equal to that of its centre of inertia, and of the motions of the separate parts relatively to the centre of inertia.

Kinetic energy of a system.

For if x, y, z be the co-ordinates of any particle, m, of the system; ξ, η, ζ its co-ordinates relative to the centre of inertia; and $\bar{x}$, $\bar{y}$, $\bar{z}$, the co-ordinates of the centre of inertia itself; we have for the whole kinetic energy

$$\tfrac{1}{2}\Sigma m\left\{\left(\frac{dx}{dt}\right)^2+\left(\frac{dy}{dt}\right)^2+\left(\frac{dz}{dt}\right)^2\right\}=\tfrac{1}{2}\Sigma m\left\{\left(\frac{d(\bar{x}+\xi)}{dt}\right)^2+\left(\frac{d(\bar{y}+\eta)}{dt}\right)^2+\left(\frac{d(\bar{z}+\zeta)}{dt}\right)^2\right\}.$$

But by the properties of the centre of inertia, we have

$$\Sigma m\frac{d\bar{x}}{dt}\frac{d\xi}{dt}=\frac{d\bar{x}}{dt}\Sigma m\frac{d\xi}{dt}=0, \text{ etc. etc.}$$

Hence the preceding is equal to

$$\tfrac{1}{2}\Sigma m\left\{\left(\frac{d\bar{x}}{dt}\right)^2+\left(\frac{d\bar{y}}{dt}\right)^2+\left(\frac{d\bar{z}}{dt}\right)^2\right\}+\tfrac{1}{2}\Sigma m\left\{\left(\frac{d\xi}{dt}\right)^2+\left(\frac{d\eta}{dt}\right)^2+\left(\frac{d\zeta}{dt}\right)^2\right\},$$

which proves the proposition.

281. The kinetic energy of rotation of a rigid system about any axis is (§ 95) expressed by $\frac{1}{2}\Sigma mr^2\omega^2$, where m is the mass of any part, r its distance from the axis, and ω the angular velocity of rotation. It may evidently be written in the form $\frac{1}{2}\omega^2\Sigma mr^2$. The factor Σmr^2 is of very great importance in kinetic investigations, and has been called the *Moment of Inertia* of the system about the axis in question. The moment of inertia about any axis is therefore found by summing the

Moment of inertia.

Moment of inertia. products of the masses of all the particles each into the square of its distance from the axis.

Moment of momentum of a rotating rigid body. It is important to notice that the moment of momentum of any rigid system about an axis, being $\Sigma mvr = \Sigma mr^2\omega$, is the product of the angular velocity into the moment of inertia.

If we take a quantity k, such that

$$k^2\Sigma m = \Sigma mr^2$$

Radius of gyration. k is called the *Radius of Gyration* about the axis from which r is measured. The radius of gyration about any axis is therefore the distance from that axis at which, if the whole mass were placed, it would have the same moment of inertia as before. Fly-wheel. In a fly-wheel, where it is desirable to have as great a moment of inertia with as small a mass as possible, within certain limits of dimensions, the greater part of the mass is formed into a ring of the largest admissible diameter, and the radius of this ring is then approximately the radius of gyration of the whole.

Moment of inertia about any axis. A rigid body being referred to rectangular axes passing through any point, it is required to find the moment of inertia about an axis through the origin making given angles with the co-ordinate axes.

Let λ, μ, ν be its direction-cosines. Then the distance (r) of the point x, y, z from it is, by § 95,

$$r^2 = (\mu z - \nu y)^2 + (\nu x - \lambda z)^2 + (\lambda y - \mu x)^2,$$

and therefore

$$Mk^2 = \Sigma mr^2 = \Sigma m\left[\lambda^2(y^2+z^2)+\mu^2(z^2+x^2)+\nu^2(x^2+y^2)-2\mu\nu yz-2\nu\lambda zx-2\lambda\mu xy\right]$$

which may be written

$$A\lambda^2 + B\mu^2 + C\nu^2 - 2\alpha\mu\nu - 2\beta\nu\lambda - 2\gamma\lambda\mu,$$

where A, B, C are the moments of inertia about the axes, and $\alpha = \Sigma myz$, $\beta = \Sigma mzx$, $\gamma = \Sigma mxy$. From its derivation we see that this quantity is *essentially positive*. Hence when, by a proper linear transformation, it is deprived of the terms containing the products of λ, μ, ν, it will be brought to the form

$$Mk^2 = A\lambda^2 + B\mu^2 + C\nu^2 = Q,$$

where A, B, C are essentially positive. They are evidently the moments of inertia about the new rectangular axes of co-ordinates, and λ, μ, ν the corresponding direction-cosines of the axis round which the moment of inertia is to be found. Moment of inertia about any axis.

Let $A > B > C$, if they are unequal. Then

$$A\lambda^2 + B\mu^2 + C\nu^2 = Q(\lambda^2 + \mu^2 + \nu^2)$$

shows that Q cannot be greater than A, nor less than C. Also, if A, B, C be equal, Q is equal to each.

If a, b, c be the radii of gyration about the new axes of x, y, z,

$$A = Ma^2, \quad B = Mb^2, \quad C = Mc^2,$$

and the above equation gives

$$k^2 = a^2\lambda^2 + b^2\mu^2 + c^2\nu^2.$$

But if x, y, z be any point in the line whose direction-cosines are λ, μ, ν, and r its distance from the origin, we have

$$\frac{x}{\lambda} = \frac{y}{\mu} = \frac{z}{\nu} = r, \text{ and therefore}$$

$$k^2r^2 = a^2x^2 + b^2y^2 + c^2z^2.$$

If, therefore, we consider the ellipsoid whose equation is

$$a^2x^2 + b^2y^2 + c^2z^2 = \epsilon^4,$$

we see that it intercepts on the line whose direction-cosines are λ, μ, ν—and about which the radius of gyration is k, a length r which is given by the equation

$$k^2r^2 = \epsilon^4;$$

or the rectangle under any radius-vector of this ellipsoid and the radius of gyration about it is constant. Its semi-axes are evidently $\frac{\epsilon^2}{a}$, $\frac{\epsilon^2}{b}$, $\frac{\epsilon^2}{c}$ where ϵ may have any value we may assign. Thus it is evident that

282. For every rigid body there may be described about any point as centre, an ellipsoid (called *Poinsot's Momental Ellipsoid**) which is such that the length of any radius-vector is Momental ellipsoid.

* The definition is not Poinsot's, but ours. The momental ellipsoid as we define it is fairly called Poinsot's, because of the splendid use he has made of it in his well-known kinematic representation of the solution of the problem —to find the motion of a rigid body with one point held fixed but otherwise influenced by no forces—which, with Sylvester's beautiful theorem completing it so as to give a purely kinematical mechanism to show the time which the body takes to attain any particular position, we reluctantly keep back for our Second Volume.

Momental ellipsoid. inversely proportional to the radius of gyration of the body about that radius-vector as axis.

Principal axes. The axes of this ellipsoid are, and might be defined as, the *Principal Axes* of inertia of the body for the point in question: but the best definition of principal axes of inertia is given below. First take two preliminary lemmas:—

Equilibration of Centrifugal Forces. (1) If a rigid body rotate round any axis, the centrifugal forces are reducible to a single force perpendicular to the axis of rotation, and to a couple (§ 234 above) having its axis parallel to the line of this force.

(2) But in particular cases the couple may vanish, or both couple and force may vanish and the centrifugal forces be in equilibrium. The force vanishes if, and only if, the axis of rotation passes through the body's centre of inertia.

Definition of Principal Axes of Inertia. DEF. (1). Any axis is called a principal axis of a body's inertia, or simply a principal axis of the body, if when the body rotates round it the centrifugal forces either balance or are reducible to a single force.

DEF. (2). A principal axis not through the centre of inertia is called a principal axis of inertia for the point of itself through which the resultant of centrifugal forces passes.

DEF. (3). A principal axis which passes through the centre of inertia is a principal axis for every point of itself.

The proofs of the lemmas may be safely left to the student as exercises on § 559 below; and from the proof the identification of the principal axes as now defined with the principal axes of Poinsot's momental ellipsoid is seen immediately by aid of the analysis of § 281.

283. The proposition of § 280 shows that the moment of inertia of a rigid body about any axis is equal to that which the mass, if collected at the centre of inertia, would have about this axis, together with that of the body about a parallel axis through its centre of inertia. It leads us naturally to investigate the relation between principal axes for any point and principal axes for the centre of inertia. The following investigation proves the remarkable theorem of § 284, which was first given in 1811 by Binet in the *Journal de l'École Polytechnique*.

Let the origin, O, be the centre of inertia, and the axes the principal axes at that point. Then, by §§ 280, 281, we have for the moment of inertia about a line through the point P (ξ, η, ζ), whose direction-cosines are λ, μ, ν; Principal axes.

$$Q = A\lambda^2 + B\mu^2 + C\nu^2 + M\{(\mu\zeta - \nu\eta)^2 + (\nu\xi - \lambda\zeta)^2 + (\lambda\eta - \mu\xi)^2\}$$
$$= \{A + M(\eta^2 + \zeta^2)\}\lambda^2 + \{B + M(\zeta^2 + \xi^2)\}\mu^2 + \{C + M(\xi^2 + \eta^2)\}\nu^2$$
$$- 2M(\mu\nu\eta\zeta + \nu\lambda\zeta\xi + \lambda\mu\xi\eta).$$

Substituting for Q, A, B, C their values, and dividing by M, we have

$$k^2 = (a^2 + \eta^2 + \zeta^2)\lambda^2 + (b^2 + \zeta^2 + \xi^2)\mu^2 + (c^2 + \xi^2 + \eta^2)\nu^2$$
$$- 2(\eta\zeta\mu\nu + \zeta\xi\nu\lambda + \xi\eta\lambda\mu).$$

Let it be required to find λ, μ, ν so that the direction specified by them may be a principal axis. Let $s = \lambda\xi + \mu\eta + \nu\zeta$, *i.e.* let s represent the projection of OP on the axis sought.

The axes of the ellipsoid

$$(a^2 + \eta^2 + \zeta^2)x^2 + \ldots\ldots - 2(\eta\zeta yz + \ldots\ldots) = H \;\ldots\ldots(a),$$

are found by means of the equations

$$\left.\begin{array}{r} (a^2 + \eta^2 + \zeta^2 - p)\lambda - \xi\eta\mu - \zeta\xi\nu = 0 \\ -\xi\eta\lambda + (b^2 + \zeta^2 + \xi^2 - p)\mu - \eta\zeta\nu = 0 \\ -\zeta\xi\lambda - \eta\zeta\mu + (c^2 + \xi^2 + \eta^2 - p)\nu = 0 \end{array}\right\}\ldots\ldots\ldots\ldots\ldots(b).$$

If, now, we take f to denote OP, or $(\xi^2 + \eta^2 + \zeta^2)^{\frac{1}{2}}$, these equations, where p is clearly the square of the radius of gyration about the axis to be found, may be written

$$(a^2 + f^2 - p)\lambda - \xi(\xi\lambda + \eta\mu + \zeta\nu) = 0,$$
$$\text{etc.} = \text{etc.},$$

or

$$(a^2 + f^2 - p)\lambda - \xi s = 0,$$
$$\text{etc.} = \text{etc.},$$

or

$$\left.\begin{array}{r} (a^2 - K)\lambda - \xi s = 0 \\ (b^2 - K)\mu - \eta s = 0 \\ (c^2 - K)\nu - \zeta s = 0 \end{array}\right\}\ldots\ldots\ldots\ldots\ldots\ldots(c)$$

where $K = p - f^2$. Hence

$$\lambda = \frac{\xi s}{a^2 - K}, \text{ etc.}$$

Multiply, in order, by ξ, η, ζ, add, and divide by s, and we get

$$\frac{\xi^2}{a^2 - K} + \frac{\eta^2}{b^2 - K} + \frac{\zeta^2}{c^2 - K} = 1 \;\ldots\ldots\ldots\ldots\ldots\ldots(d).$$

Principal axes.

By (c) we see that (λ, μ, ν) is the direction of the normal through the point P, (ξ, η, ζ) of the surface represented by the equation

$$\frac{x^2}{a^2-K}+\frac{y^2}{b^2-K}+\frac{z^2}{c^2-K}=1 \quad\text{..................}(e),$$

which is obviously a surface of the second degree confocal with the ellipsoid

$$\frac{x^2}{a^2}+\frac{y^2}{b^2}+\frac{z^2}{c^2}=1 \quad\text{......................}(f),$$

and passing through P in virtue of (d), which determines K accordingly. The three roots of this cubic are clearly all real; one of them is less than the least of a^2, b^2, c^2, and positive or negative according as P is within or without the ellipsoid (f). And if $a>b>c$, the two others are between c^2 and b^2, and between b^2 and a^2, respectively. The addition of f^2 to each gives the square of the radius of gyration round the corresponding principal axis. Hence

Binet's Theorem.

284. The principal axes for any point of a rigid body are normals to the three surfaces of the second order through that point, confocal with the ellipsoid, which has its centre at the centre of inertia, and its three principal diameters co-incident with the three principal axes for that point, and equal respectively to the doubles of the radii of gyration round them. This ellipsoid is called the *Central Ellipsoid.*

Central ellipsoid.

Kinetic symmetry round a point:

285. A rigid body is said to be kinetically symmetrical about its centre of inertia when its moments of inertia about three principal axes through that point are equal; and therefore necessarily the moments of inertia about *all* axes through that point equal, § 281, and all these axes principal axes. About it uniform spheres, cubes, and in general any complete crystalline solid of the first system (see chapter on Properties of Matter), are kinetically symmetrical.

round an axis.

A rigid body is kinetically symmetrical about an *axis* when this axis is one of the principal axes through the centre of inertia, and the moments of inertia about the other two, and therefore about any line in their plane, are equal. A spheroid, a square or equilateral triangular prism or plate, a circular ring, disc, or cylinder, or any complete crystal of the second or fourth system, is kinetically symmetrical about its axis.

Energy in abstract dynamics.

286. The only actions and reactions between the parts of a system, not belonging palpably to the conservative class, which we shall consider in abstract dynamics, are those of friction between solids sliding on solids, except in a few instances in which we shall consider the general character and ultimate results of effects produced by viscosity of fluids, imperfect elasticity of solids, imperfect electric conduction, or imperfect magnetic retentiveness. We shall also, in abstract dynamics, consider forces as applied to parts of a limited system arbitrarily from without. These we shall call, for brevity, the applied forces.

287. The law of energy may then, in abstract dynamics, be expressed as follows:—

The whole work done in any time, on any limited material system, by applied forces, is equal to the whole effect in the forms of potential and kinetic energy produced in the system, together with the work lost in friction.

288. This principle may be regarded as comprehending the whole of abstract dynamics, because, as we now proceed to show, the conditions of equilibrium and of motion, in every possible case, may be immediately derived from it.

Equilibrium.

289. A material system, whose relative motions are unresisted by friction, is in equilibrium in any particular configuration if, and is not in equilibrium unless, the work done by the applied forces is equal to the potential energy gained, in any possible infinitely small displacement from that configuration. This is the celebrated principle of "virtual velocities" which Lagrange made the basis of his *Mécanique Analytique*. The ill-chosen name "virtual velocities" is now falling into disuse.

Principle of virtual velocities.

290. To prove it, we have first to remark that the system cannot possibly move away from any particular configuration except by work being done upon it by the forces to which it is subject: it is therefore in equilibrium if the stated condition is fulfilled. To ascertain that nothing less than this condition can secure its equilibrium, let us first consider a system having only one degree of freedom to move. Whatever forces act on the whole system, we may always hold it in equilibrium by a single force applied to any one point of the system in its line

Principle of virtual velocities. of motion, opposite to the direction in which it tends to move, and of such magnitude that, in any infinitely small motion in either direction, it shall resist, or shall do, as much work as the other forces, whether applied or internal, altogether do or resist. Now, by the principle of superposition of forces in equilibrium, we might, without altering their effect, apply to any one point of the system such a force as we have just seen would hold the system in equilibrium, and another force equal and opposite to it. All the other forces being balanced by one of these two, they and it might again, by the principle of superposition of forces in equilibrium, be removed; and therefore the whole set of given forces would produce the same effect, whether for equilibrium or for motion, as the single force which is left acting alone. This single force, since it is in a line in which the point of its application is free to move, must move the system. Hence the given forces, to which this single force has been proved equivalent, cannot possibly be in equilibrium unless their whole work for an infinitely small motion is nothing, in which case the single equivalent force is reduced to nothing. But whatever amount of freedom to move the whole system may have, we may always, by the application of frictionless constraint, limit it to one degree of freedom only; —and this may be freedom to execute any particular motion whatever, possible under the given conditions of the system. If, therefore, in any such infinitely small motion, there is variation of potential energy uncompensated by work of the applied forces, constraint limiting the freedom of the system to only this motion will bring us to the case in which we have just demonstrated there cannot be equilibrium. But the application of constraints limiting motion cannot possibly disturb equilibrium, and therefore the given system under the actual conditions cannot be in equilibrium in any particular configuration if there is more work done than resisted in any possible infinitely small motion from that configuration by all the forces to which it is subject.

Neutral equilibrium. **291.** If a material system, under the influence of internal and applied forces, varying according to some definite law, is balanced by them in any position in which it may be placed, its equilibrium is said to be neutral. This is the case with any spherical body of uniform material resting on a horizontal plane. A right cylinder or cone, bounded by plane ends perpendicular to the axis, is also in neutral equilibrium on a horizontal plane. Practically, any mass of moderate dimensions is in neutral equilibrium when its centre of inertia only is fixed, since, when its longest dimension is small in comparison with the earth's radius, gravity is, as we shall see, approximately equivalent to a single force through this point. **Neutral equilibrium.**

But if, when displaced infinitely little in any direction from a particular position of equilibrium, and left to itself, it commences and continues vibrating, without ever experiencing more than infinitely small deviation in any of its parts, from the position of equilibrium, the equilibrium in this position is said to be stable. A weight suspended by a string, a uniform sphere in a hollow bowl, a loaded sphere resting on a horizontal plane with the loaded side lowest, an oblate body resting with one end of its shortest diameter on a horizontal plane, a plank, whose thickness is small compared with its length and breadth, floating on water, etc. etc., are all cases of stable equilibrium; if we neglect the motions of rotation about a vertical axis in the second, third, and fourth cases, and horizontal motion in general, in the fifth, for all of which the equilibrium is neutral. **Stable equilibrium.**

If, on the other hand, the system can be displaced in any way from a position of equilibrium, so that when left to itself it will not vibrate within infinitely small limits about the position of equilibrium, but will move farther and farther away from it, the equilibrium in this position is said to be unstable. Thus a loaded sphere resting on a horizontal plane with its load as high as possible, an egg-shaped body standing on one end, a board floating edgeways in water, etc. etc., would present, if they could be realised in practice, cases of unstable equilibrium. **Unstable equilibrium.**

When, as in many cases, the nature of the equilibrium varies with the direction of displacement, if unstable for any possible displacement it is practically unstable on the whole. Thus a coin standing on its edge, though in neutral equilibrium for displacements in its plane, yet being in unstable equilibrium

Unstable equilibrium. for those perpendicular to its plane, is practically unstable. A sphere resting in equilibrium on a saddle presents a case in which there is stable, neutral, or unstable equilibrium, according to the direction in which it may be displaced by rolling, but, practically, it would be unstable.

Test of the nature of equilibrium. 292. The theory of energy shows a very clear and simple test for discriminating these characters, or determining whether the equilibrium is neutral, stable, or unstable, in any case. If there is just as much work resisted as performed by the applied and internal forces in any possible displacement the equilibrium is neutral, but not unless. If in every possible infinitely small displacement from a position of equilibrium they do less work among them than they resist, the equilibrium is thoroughly stable, and not unless. If in any or in every infinitely small displacement from a position of equilibrium they do more work than they resist, the equilibrium is unstable. It follows that if the system is influenced only by internal forces, or if the applied forces follow the law of doing always the same amount of work upon the system passing from one configuration to another by all possible paths, the whole potential energy must be constant, in all positions, for neutral equilibrium; must be a minimum for positions of thoroughly stable equilibrium; must be either an absolute maximum, or a maximum for some displacements and a minimum for others when there is unstable equilibrium.

Deduction of the equations of motion of any system. 293. We have seen that, according to D'Alembert's principle, as explained above (§ 264), forces acting on the different points of a material system, and their reactions against the accelerations which they actually experience in any case of motion, are in equilibrium with one another. Hence in any actual case of motion, not only is the actual work done by the forces equal to the kinetic energy produced in any infinitely small time, in virtue of the actual accelerations; but so also is the work which would be done by the forces, in any infinitely small time, if the velocities of the points constituting the system, were at any instant changed to any possible infinitely small velocities, and the accelerations unchanged. This statement, when put in the concise language of mathematical analysis, constitutes Lagrange's application of the "principle of virtual velocities" to express the conditions of D'Alembert's equilibrium between the forces acting, and the resistances of the masses to acceleration. It comprehends, as we have seen, every possible condition of every case of motion. The "equations of motion" in any particular case are, as Lagrange has shown, deduced from it with great ease.

Deduction of the equations of motion of any system.

Let m be the mass of any one of the material points of the system; x, y, z its rectangular co-ordinates at time t, relatively to axes fixed in direction (§ 249) through a point reckoned as fixed (§ 245); and X, Y, Z the components, parallel to the same axes, of the whole force acting on it. Thus $-m\dfrac{d^2x}{dt^2}$, $-m\dfrac{d^2y}{dt^2}$, $-m\dfrac{d^2z}{dt^2}$ are the components of the reaction against acceleration. And these, with X, Y, Z, for the whole system, must fulfil the conditions of equilibrium. Hence if $\delta x, \delta y, \delta z$ denote any arbitrary variations of x, y, z consistent with the conditions of the system, we have

$$\Sigma\left\{\left(X - m\frac{d^2x}{dt^2}\right)\delta x + \left(Y - m\frac{d^2y}{dt^2}\right)\delta y + \left(Z - m\frac{d^2z}{dt^2}\right)\delta z\right\} = 0 \;..(1),$$

Indeterminate equation of motion of any system.

where Σ denotes summation to include all the particles of the system. This may be called the indeterminate, or the variational, equation of motion. Lagrange used it as the foundation of his whole kinetic system, deriving from it all the common equations of motion, and his own remarkable equations in generalized co-ordinates (presently to be given). We may write it otherwise as follows:

$$\Sigma m\,(\ddot{x}\delta x + \ddot{y}\delta y + \ddot{z}\delta z) = \Sigma\,(X\delta x + Y\delta y + Z\delta z) \;......(2),$$

where the first member denotes the work done by forces equal to those required to produce the real accelerations, acting through the spaces of the arbitrary displacements; and the second member the work done by the actual forces through these imagined spaces.

If the moving bodies constitute a conservative system, and if V denote its potential energy in the configuration specified by (x, y, z, etc.), we have of course (§§ 241, 273)

$$\delta V = -\Sigma\,(X\delta x + Y\delta y + Z\delta z) \;.................(3),$$

and therefore the indeterminate equation of motion becomes

$$\Sigma m(\ddot{x}\delta x + \ddot{y}\delta y + \ddot{z}\delta z) = -\delta V \qquad (4),$$

Of conservative system.

where δV denotes the excess of the potential energy in the configuration ($x+\delta x$, $y+\delta y$, $z+\delta z$, etc.) above that in the configuration (x, y, z, etc.).

One immediate particular result must of course be the common equation of energy, which must be obtained by supposing δx, δy, δz, etc., to be the actual variations of the co-ordinates in an infinitely small time δt. Thus if we take $\delta x = \dot{x}\delta t$, etc., and divide both members by δt, we have

$$\Sigma(X\dot{x} + Y\dot{y} + Z\dot{z}) = \Sigma m(\ddot{x}\dot{x} + \ddot{y}\dot{y} + \ddot{z}\dot{z}) \qquad (5).$$

Equation of energy.

Here the first member is composed of Newton's *Actiones Agentium*; with his *Reactiones Resistentium* so far as friction, gravity, and molecular forces are concerned, subtracted: and the second consists of the portion of the *Reactiones* due to acceleration. As we have seen above (§ 214), the second member is the rate of increase of $\Sigma\frac{1}{2}m(\dot{x}^2 + \dot{y}^2 + \dot{z}^2)$ per unit of time. Hence, denoting by v the velocity of one of the particles, and by W the integral of the first member multiplied by dt, that is to say, the integral work done by the working and resisting forces in any time, we have

$$\Sigma\tfrac{1}{2}mv^2 = W + E_0 \qquad (6),$$

E_0 being the initial kinetic energy. This is the integral equation of energy. In the particular case of a conservative system, W is a function of the co-ordinates, irrespectively of the time, or of the paths which have been followed. According to the previous notation, with besides V_0 to denote the potential energy of the system in its initial configuration, we have $W = V_0 - V$, and the integral equation of energy becomes

$$\Sigma\tfrac{1}{2}mv^2 = V_0 - V + E_0,$$

or, if E denote the sum of the potential and kinetic energies, a constant,

$$\Sigma\tfrac{1}{2}mv^2 = E - V \qquad (7).$$

The general indeterminate equation gives immediately, for the motion of a system of free particles,

$$m_1\ddot{x}_1 = X_1,\ m_1\ddot{y}_1 = Y_1,\ m_1\ddot{z}_1 = Z_1,\ m_2\ddot{x}_2 = X_2, \text{ etc.}$$

Of these equations the three for each particle may of course be treated separately if there is no mutual influence between the particles: but when they exert force on one another, X_1, Y_1, etc., will each in general be a function of all the co-ordinates.

From the indeterminate equation (1) Lagrange, by his method of multipliers, deduces the requisite number of equations for determining the motion of a rigid body, or of any system of connected particles or rigid bodies, thus:—Let the number of the particles be i, and let the connexions between them be expressed by n equations, Constraint introduced into the indeterminate equation.

$$\left.\begin{array}{c} F(x_1, y_1, z_1, x_2, \ldots) = 0 \\ F_{\prime}(x_1, y_1, z_1, x_2, \ldots) = 0 \\ \text{etc.} \qquad \text{etc.} \end{array}\right\} \qquad (8)$$

being the *kinematical equations* of the system. By taking the variations of these we find that every possible infinitely small displacement δx_1, δy_1, δz_1, δx_2, ... must satisfy the n linear equations

$$\frac{dF}{dx_1}\delta x_1 + \frac{dF}{dy_1}\delta y_1 + \text{etc.} = 0, \quad \frac{dF_{\prime}}{dx_1}\delta x_1 + \frac{dF_{\prime}}{dy_1}\delta y_1 + \text{etc.} = 0, \quad \text{etc.} \ldots (9).$$

Multiplying the first of these by λ, the second by $\lambda_{\prime}$, etc., adding to the indeterminate equation, and then equating the coefficients of δx_1, δy_1, etc., each to zero, we have

$$\left.\begin{array}{c} \lambda\frac{dF}{dx_1} + \lambda_{\prime}\frac{dF_{\prime}}{dx_1} + \ldots + X_1 - m_1\frac{d^2x_1}{dt^2} = 0 \\ \lambda\frac{dF}{dy_1} + \lambda_{\prime}\frac{dF_{\prime}}{dy_1} + \ldots + Y_1 - m_1\frac{d^2y_1}{dt^2} = 0 \\ \text{etc.} \qquad \text{etc.} \end{array}\right\} \qquad (10).$$

These are in all $3i$ equations to determine the n unknown quantities λ, $\lambda_{\prime}$, ..., and the $3i-n$ independent variables to which x_1, y_1, ... are reduced by the kinematical equations (8). The same equations may be found synthetically in the following manner, by which also we are helped to understand the precise meaning of the terms containing the multipliers λ, $\lambda_{\prime}$, etc. Determinate equations of motion deduced.

First let the particles be free from constraint, but acted on both by the given forces X_1, Y_1, etc., and by forces depending on mutual distances between the particles and upon their positions relatively to fixed objects subject to the law of conservation, and having for their potential energy

$$-\tfrac{1}{2}(kF^2 + k_{\prime}F_{\prime}^2 + \text{etc.}),$$

so that components of the forces actually experienced by the different particles shall be

Determinate equations of motion deduced.

$$X_1 + kF\frac{dF}{dx} + k_,F_,\frac{dF_,}{dx_1} + \text{etc.} + \tfrac{1}{2}\left(F^2\frac{dk}{dx_1} + F_,^2\frac{dk_,}{dx_1} + \text{etc.}\right)$$

etc., etc.

Hence the equations of motion are

$$\left.\begin{aligned} m_1\frac{d^2x_1}{dt^2} &= X_1 + kF\frac{dF}{dx_1} + k_,F_,\frac{dF_,}{dx_1} + \text{etc.} + \tfrac{1}{2}\left(F^2\frac{dk}{dx_1} + F_,^2\frac{dk_,}{dx_1} + \text{etc.}\right) \\ m_1\frac{d^2y_1}{dt^2} &= \text{etc.} \\ \text{etc.}, &\qquad \text{etc.} \end{aligned}\right\} \quad (11).$$

Now suppose k, $k_,$, etc. to be infinitely great:—in order that the *forces on the particles* may not be infinitely great, we must have

$$F = 0, \quad F_, = 0, \quad \text{etc.},$$

that is to say, the equations of condition (8) must be fulfilled; and the last groups of terms in the second members of (11) now disappear because they contain the squares of the infinitely small quantities F, $F_,$, etc. Put now $kF = \lambda$, $k_,F_, = \lambda_,$, etc., and we have equations (10). This second mode of proving Lagrange's equations of motion of a constrained system corresponds precisely to the imperfect approach to the ideal case which can be made by real mechanism. The levers and bars and guide-surfaces cannot be infinitely rigid. Suppose then k, $k_,$, etc. to be finite but very great quantities, and to be some functions of the co-ordinates depending on the elastic qualities of the materials of which the guiding mechanism is composed:—equations (11) will express the motion, and by supposing k, $k_,$, etc. to be greater and greater we approach more and more nearly to the ideal case of absolutely rigid mechanism constraining the precise fulfilment of equations (8).

The problem of finding the motion of a system subject to any *unvarying* kinematical conditions whatever, under the action of any given forces, is thus reduced to a question of pure analysis. In the still more general problem of determining the motion when certain parts of the system are constrained to move in a specified manner, the equations of condition (8) involve not only the co-ordinates, but also t, the time. It is easily seen however that the equations (10) still hold, and with (8) fully determine the motion. For:—consider the equations of equilibrium of the particles acted on by any forces X_1', Y_1', etc., and constrained by proper mechanism to fulfil the equations of condition (8) with the actual values of the parameters for any particular value of t. The equations of equilibrium will be uninfluenced by the fact that some of the parameters of the conditions (8) have different values at different times. Hence, with $X_1 - m_1 \frac{d^2x_1}{dt^2}$, $Y_1 - m_1 \frac{d^2y_1}{dt^2}$, instead of X_1', Y_1', etc., according to D'Alembert's principle, the equations of motion will still be (8), (9), and (10) quite independently of whether the parameters of (8) are all constant, or have values varying in any arbitrary manner with the time.

Determinate equations of motion deduced.

To find the equation of energy multiply the first of equations (10) by $\dot{x}_1$, the second by $\dot{y}_1$, etc., and add. Then remarking that in virtue of (8) we have

Equation of energy.

$$\frac{dF}{dx_1}\dot{x}_1 + \frac{dF}{dy_1}\dot{y}_1 + \text{etc.} + \left(\frac{dF}{dt}\right) = 0,$$

$$\frac{dF_{\prime}}{dx_1}\dot{x}_1 + \frac{dF_{\prime}}{dy_1}\dot{y}_1 + \text{etc.} + \left(\frac{dF_{\prime}}{dt}\right) = 0,$$

partial differential coefficients of F, $F_{\prime}$, etc. with reference to t being denoted by $\left(\frac{dF}{dt}\right)$, $\left(\frac{dF_{\prime}}{dt}\right)$, etc.; and denoting by T the kinetic energy or $\frac{1}{2}\Sigma m(\dot{x}^2 + \dot{y}^2 + \dot{z}^2)$, we find

$$\frac{dT}{dt} = \Sigma(X\dot{x} + Y\dot{y} + Z\dot{z}) - \lambda\left(\frac{dF}{dt}\right) - \lambda_{\prime}\left(\frac{dF_{\prime}}{dt}\right) - \text{etc.} = 0 \text{....(12)}.$$

When the kinematic conditions are "*unvarying*," that is to say, when the equations of condition are equations among the co-ordinates with constant parameters, we have

$$\left(\frac{dF}{dt}\right) = 0, \quad \left(\frac{dF_{\prime}}{dt}\right) = 0, \text{ etc.},$$

and the equation of energy becomes

$$\frac{dT}{dt} = \Sigma(X\dot{x} + Y\dot{y} + Z\dot{z}) \text{ ....(13)},$$

showing that in this case the fulfilment of the equations of condition involves neither gain nor loss of energy. On the other hand, equation (12) shows how to find the work performed or consumed in the fulfilment of the kinematical conditions when they are not unvarying.

Equation of energy.

As a simple example of varying constraint, which will be very easily worked out by equations (8) and (10), perfectly illustrating the general principle, the student may take the case of a particle acted on by any given forces and free to move anywhere in a plane which is kept moving with any given uniform or varying angular velocity round a fixed axis.

Gauss's principle of least constraint.

When there are connexions between any parts of a system, the motion is in general not the same as if all were free. If we consider any particle during any infinitely small time of the motion, and call the product of its mass into the square of the distance between its positions at the end of this time, on the two suppositions, the *constraint*: the sum of the constraints is a minimum. This follows easily from (1).

Impact.

294. When two bodies, in relative motion, come into contact, pressure begins to act between them to prevent any parts of them from jointly occupying the same space. This force commences from nothing at the first point of collision, and gradually increases per unit of area on a gradually increasing surface of contact. If, as is always the case in nature, each body possesses some degree of elasticity, and if they are not kept together after the impact by cohesion, or by some artificial appliance, the mutual pressure between them will reach a maximum, will begin to diminish, and in the end will come to nothing, by gradually diminishing in amount per unit of area on a gradually diminishing surface of contact. The whole process would occupy not greatly more or less than an hour if the bodies were of such dimensions as the earth, and such degrees of rigidity as copper, steel, or glass. It is finished, probably, within a thousandth of a second if they are globes of any of these substances not exceeding a yard in diameter.

295. The whole amount, and the direction, of the *"Impact"* experienced by either body in any such case, are reckoned according to the "change of momentum" which it experiences. The amount of the impact is measured by the amount, and its direction by the direction, of the change of momentum which is produced. The component of an impact in a direction parallel to any fixed line is similarly reckoned according to the component change of momentum in that direction.

296. If we imagine the whole time of an impact divided into a very great number of equal intervals, each so short that the force does not vary sensibly during it, the component change of momentum in any direction during any one of these intervals will (§ 220) be equal to the force multiplied by the measure of the interval. Hence the component of the impact is equal to the sum of the forces in all the intervals, multiplied by the length of each interval. Impact.

Let P be the component force in any direction at any instant, τ, of the interval, and let I be the amount of the corresponding component of the whole impact. Then

$$I = \int P d\tau.$$

297. Any force in a constant direction acting in any circumstances, for any time great or small, may be reckoned on the same principle; so that what we may call its whole amount during any time, or its "*time-integral,*" will measure, or be measured by, the whole momentum which it generates in the time in question. But this reckoning is not often convenient or useful except when the whole operation considered is over before the position of the body, or configuration of the system of bodies, involved, has altered to such a degree as to bring any other forces into play, or alter forces previously acting, to such an extent as to produce any sensible effect on the momentum measured. Thus if a person presses gently with his hand, during a few seconds, upon a mass suspended by a cord or chain, he produces an effect which, if we know the degree of the force at each instant, may be thoroughly calculated on elementary principles. No approximation to a full determination of the motion, or to answering such a partial question as "how great will be the whole deflection produced?" can be founded on a knowledge of the "*time-integral*" alone. If, for instance, the force be at first very great and gradually diminish, the effect will be very different from what it would be if the force were to increase very gradually and to cease suddenly, even although the time-integral were the same in the two cases. But if the same body is "struck a blow," in a horizontal direction, either by the hand, or by a mallet or other somewhat Time-integral.

Time-integral.

hard mass, the action of the force is finished before the suspending cord has experienced any sensible deflection from the vertical. Neither gravity nor any other force sensibly alters the effect of the blow. And therefore the whole momentum at the end of the blow is sensibly equal to the "amount of the impact," which is, in this case, simply the time-integral.

Ballistic pendulum.

298. Such is the case of Robins' *Ballistic Pendulum*, a massive cylindrical block of wood cased in a cylindrical sheath of iron closed at one end and moveable about a horizontal axis at a considerable distance above it—employed to measure the velocity of a cannon or musket-shot. The shot is fired into the block in a horizontal direction along the axis of the block and perpendicular to the axis of suspension. The impulsive penetration is so nearly instantaneous, and the inertia of the block so large compared with the momentum of the shot, that the ball and pendulum are moving on as one mass before the pendulum has been sensibly deflected from the vertical. This is essential to the regular use of the apparatus. The iron sheath with its flat end must be strong enough to guard against splinters of wood flying sidewise, and to keep in the bullet.

299. Other illustrations of the cases in which the time-integral gives us the complete solution of the problem may be given without limit. They include all cases in which the direction of the force is always coincident with the direction of motion of the moving body, and those special cases in which the time of action of the force is so short that the body's motion does not, during its lapse, sensibly alter its relation to the direction of the force, or the action of any other forces to which it may be subject. Thus, in the vertical fall of a body, the time-integral gives us at once the change of momentum; and the same rule applies in most cases of forces of brief duration, as in a "drive" in cricket or golf.

Direct impact of spheres.

300. The simplest case which we can consider, and the one usually treated as an introduction to the subject, is that of the collision of two smooth spherical bodies whose centres before collision were moving in the same straight line. The force between them at each instant must be in this line, because of

Direct impact of spheres.

the symmetry of circumstances round it; and by the third law it must be equal in amount on the two bodies. Hence (LEX II.) they must experience changes of motion at equal rates in contrary directions; and at any instant of the impact the integral amounts of these changes of motion must be equal. Let us suppose, to fix the ideas, the two bodies to be moving both before and after impact in the same direction in one line: one of them gaining on the other before impact, and either following it at a less speed, or moving along with it, as the case may be, after the impact is completed. Cases in which the former is driven backwards by the force of the collision, or in which the two moving in opposite directions meet in collision, are easily reduced to dependence on the same formula by the ordinary algebraic convention with regard to positive and negative signs.

In the standard case, then, the quantity of motion lost, up to any instant of the impact, by one of the bodies, is equal to that gained by the other. Hence at the instant when their velocities are equalized they move as one mass with a momentum equal to the sum of the momenta of the two before impact. That is to say, if v denote the common velocity at this instant, we have

$$(M + M')\,v = MV + M'V',$$

or

$$v = \frac{MV + M'V'}{M + M'},$$

if M, M' denote the masses of the two bodies, and V, V' their velocities before impact.

During this first period of the impact the bodies have been, on the whole, coming into closer contact with one another, through a compression or deformation experienced by each, and resulting, as remarked above, in a fitting together of the two surfaces over a finite area. No body in nature is perfectly inelastic; and hence, at the instant of closest approximation, the mutual force called into action between the two bodies continues, and tends to separate them. Unless prevented by natural surface cohesion or welding (such as is always found, as we shall see later in our chapter on Properties of Matter, however hard and well polished the surfaces may

Direct impact of spheres.

be), or by artificial appliances (such as a coating of wax, applied in one of the common illustrative experiments; or the coupling applied between two railway carriages when run together so as to push in the springs, according to the usual practice at railway stations), the two bodies are actually separated by this force, and move away from one another. Newton found that, *provided the impact is not so violent as to make any sensib'e permanent indentation in either body*, the relative velocity of separation after the impact bears a proportion to their previous relative velocity of approach, which is constant for the same two bodies. This proportion, always less than unity, approaches more and more nearly to it the harder the bodies are. Thus with balls of compressed wool he found it $\frac{5}{9}$, iron nearly the same, glass $\frac{15}{16}$. The results of more recent experiments on the same subject have confirmed Newton's law. These will be described later. In any case of the collision of two balls, let e denote this proportion, to which we give the name *Coefficient of Restitution*;* and, with previous notation, let in addition U, U' denote the velocities of the two bodies after the conclusion of the impact; in the standard case each being positive, but $U' > U$. Then we have

Effect of elasticity.

Newton's experiments.

$$U' - U = e(V - V')$$

and, as before, since one has lost as much momentum as the other has gained,

$$MU + M'U' = MV + M'V'.$$

From these equations we find

$$(M + M')\,U = MV + M'V' - eM'(V - V'),$$

with a similar expression for U'.

Also we have, as above,

$$(M + M')\,v = MV + M'V'.$$

Hence, by subtraction,

$$(M + M')(v - U) = eM'(V - V') = e\{M'V - (M + M')\,v + MV'\}$$

* In most modern treatises this is called a "coefficient of elasticity," which is clearly a mistake; suggested, it may be, by Newton's words, but inconsistent with his facts, and utterly at variance with modern language and modern knowledge regarding elasticity.

and therefore

Direct impact of spheres.

$$v - U = e(V - v).$$

Of course we have also

$$U' - v = e(v - V').$$

These results may be put in words thus:—The *relative* velocity of either of the bodies with regard to the centre of inertia of the two is, after the completion of the impact, reversed in direction, and diminished in the ratio $e : 1$.

301. Hence the loss of kinetic energy, being, according to §§ 267, 280, due only to change of kinetic energy relative to the centre of inertia, is to this part of the whole as $1 - e^2 : 1$.

Thus

Initial kinetic energy $= \frac{1}{2}(M + M')v^2 + \frac{1}{2}M(V - v)^2 + \frac{1}{2}M'(v - V')^2.$

Final " " $= \frac{1}{2}(M + M')v^2 + \frac{1}{2}M(v - U)^2 + \frac{1}{2}M'(U' - v)^2.$

Loss $= \frac{1}{2}(1 - e^2)\{M(V - v)^2 + M'(v - V')^2\}.$

302. When two elastic bodies, the two balls supposed above for instance, impinge, some portion of their previous kinetic energy will always remain in them as vibrations. A *portion* of the loss of energy (miscalled the effect of imperfect elasticity) is necessarily due to this cause in every real case.

Distribution of energy after impact.

Later, in our chapter on Properties of Matter, it will be shown as a result of experiment, that forces of elasticity are, to a very close degree of accuracy, simply proportional to the strains (§ 154), within the limits of elasticity, in elastic solids which, like metals, glass, etc., bear but small deformations without permanent change. Hence when two such bodies come into collision, sometimes with greater and sometimes with less mutual velocity, but with all other circumstances similar, the velocities of all particles of either body, at corresponding times of the impacts, will be always in the same proportion. Hence the velocity of separation of the centres of inertia after impact will bear a constant proportion to the previous velocity of approach; which agrees with the Newtonian Law. It is therefore probable that a very sensible portion, if not the whole, of the loss of energy in the visible motions of two elastic bodies, after impact, experimented on by Newton, may have been due

Newton's experimental law consistent with *perfect* elasticity.

Distribution of energy after impact.

to vibrations; but unless some other cause also was largely operative, it is difficult to see how the loss was so much greater with iron balls than with glass.

303. In certain definite extreme cases, imaginable although not realizable, no energy will be spent in vibrations, and the two bodies will separate, each moving simply as a rigid body, and having in this simple motion the whole energy of work done on it by elastic force during the collision. For instance, let the two bodies be cylinders, or prismatic bars with flat ends, of the same kind of substance, and of equal and similar transverse sections; and let this substance have the property of compressibility with perfect elasticity, in the direction of the length of the bar, and of absolute resistance to change in every transverse dimension. Before impact, let the two bodies be placed with their lengths in one line, and their transverse sections (if not circular) similarly situated, and let one or both be set in motion in this line. The result, as regards the motions of the two bodies after the collision, will be sensibly the same if they are of any real ordinary elastic solid material, provided the greatest transverse diameter of each is very small in comparison with its length. Then, if the lengths of the two be equal, they will separate after impact with the same relative velocity as that with which they approached, and neither will retain any vibratory motion after the end of the collision.

304. If the two bars are of unequal length, the shorter will, after the impact, be exactly in the same state as if it had struck another of its own length, and it therefore will move as a rigid body after the collision. But the other will, along with a motion of its centre of gravity, calculable from the principle that its whole momentum must (§ 267) be changed by an amount equal exactly to the momentum gained or lost by the first, have also a vibratory motion, of which the whole kinetic and potential energy will make up the deficiency of energy which we shall presently calculate in the motions of the centres of inertia. For simplicity, let the longer body be supposed to be at rest before the collision. Then the shorter on striking it will be left at rest; this being clearly the result in the case of

$e=1$ in the preceding formulæ (§ 300) applied to the impact of one body striking another of equal mass previously at rest. The longer bar will move away with the same momentum, and therefore with less velocity of its centre of inertia, and less kinetic energy of this motion, than the other body had before impact, in the ratio of the smaller to the greater mass. It will also have a very remarkable vibratory motion, which, when its length is more than double of that of the other, will consist of a wave running backwards and forwards through its length, and causing the motion of its ends, and, in fact, of every particle of it, to take place by "fits and starts," not continuously. The full analysis of these circumstances, though very simple, must be reserved until we are especially occupied with waves, and the kinetics of elastic solids. It is sufficient at present to remark, that the motions of the centres of inertia of the two bodies after impact, whatever they may have been previously, are given by the preceding formulæ with for e the value $\frac{M'}{M}$, where M' and M are the smaller and the larger mass respectively.

Distribution of energy after impact.

305. The mathematical theory of the vibrations of solid elastic spheres has not yet been worked out; and its application to the case of the vibrations produced by impact presents considerable difficulty. Experiment, however, renders it certain, that but a small part of the whole kinetic energy of the previous motions can remain in the form of vibrations after the impact of two equal spheres of glass or of ivory. This is proved, for instance, by the common observation, that one of them remains nearly motionless after striking the other previously at rest; since, the velocity of the common centre of inertia of the two being necessarily unchanged by the impact, we infer that the second ball acquires a velocity nearly equal to that which the first had before striking it. But it is to be expected that unequal balls of the same substance coming into collision will, by impact, convert a very sensible proportion of the kinetic energy of their previous motions into energy of vibrations; and generally, that the same will be the case when equal or unequal masses of different substances come into colli-

Distribution of energy after impact.

sion; although for one particular proportion of their diameters, depending on their densities and elastic qualities, this effect will be a minimum, and possibly not much more sensible than it is when the substances are the same and the diameters equal.

306. It need scarcely be said that in such cases of impact as that of the tongue of a bell, or of a clock-hammer striking its bell (or spiral spring as in the American clocks), or of piano-forte hammers striking the strings, or of a drum struck with the proper implement, a large part of the kinetic energy of the blow is spent in generating vibrations.

Moment of an impact about an axis.

307. The *Moment of an impact* about any axis is derived from the line and amount of the impact in the same way as the moment of a velocity or force is determined from the line and amount of the velocity or force, §§ 235, 236. If a body is struck, the change of its moment of momentum about any axis is equal to the moment of the impact round that axis. But, without considering the measure of the impact, we see (§ 267) that the moment of momentum round any axis, lost by one body in striking another, is, as in every case of mutual action, equal to that gained by the other.

Ballistic pendulum.

Thus, to recur to the ballistic pendulum—the line of motion of the bullet at impact may be in any direction whatever, but the only part which is effective is the component in a plane perpendicular to the axis. We may therefore, for simplicity, consider the motion to be in a line perpendicular to the axis, though not necessarily horizontal. Let m be the mass of the bullet, v its velocity, and p the distance of its line of motion from the axis. Let M be the mass of the pendulum with the bullet lodged in it, and k its radius of gyration. Then if ω be the angular velocity of the pendulum when the impact is complete,

$$mvp = Mk^2\omega,$$

from which the solution of the question is easily determined.

For the kinetic energy after impact is changed (§ 241) into its equivalent in potential energy when the pendulum reaches its position of greatest deflection. Let this be given by the angle θ: then the height to which the centre of inertia is raised is $h(1-\cos\theta)$ if h be its distance from the axis. Thus

$$Mgh(1-\cos\theta) = \tfrac{1}{2}Mk^2\omega^2 = \tfrac{1}{2}\frac{m^2v^2p^2}{Mk^2},$$ Ballistic pendulum.

or $$2\sin\frac{\theta}{2} = \frac{mvp}{Mk\sqrt{gh}},$$

an expression for the chord of the angle of deflection. In practice the chord of the angle θ is measured by means of a light tape or cord attached to a point of the pendulum, and slipping with small friction through a clip fixed close to the position occupied by that point when the pendulum hangs at rest.

308. *Work done by an impact* is, in general, the product of the impact into half the sum of the initial and final velocities of the point at which it is applied, resolved in the direction of the impact. In the case of direct impact, such as that treated in § 300, the initial kinetic energy of the body is $\frac{1}{2}MV^2$, the final $\frac{1}{2}MU^2$, and therefore the gain, by the impact, is Work done by impact.

$$\tfrac{1}{2}M(U^2 - V^2),$$

or, which is the same,

$$M(U-V)\,.\,\tfrac{1}{2}(U+V).$$

But $M(U-V)$ is (§ 295) equal to the amount of the impact. Hence the proposition: the extension of which to the most general circumstances is easily seen.

Let ι be the amount of the impulse up to time τ, and I the whole amount, up to the end, T. Thus,—

$$\iota = \int_0^\tau Pd\tau,\quad I = \int_0^T Pd\tau;\ \text{also } P = \frac{d\iota}{d\tau}.$$

Whatever may be the conditions to which the body struck is subjected, the change of velocity in the point struck is proportional to the amount of the impulse up to any part of its whole time, so that, if $\mathfrak{M}$ be a constant depending on the masses and conditions of constraint involved, and if U, v, V denote the component velocities of the point struck, in the direction of the impulse, at the beginning, at the time τ, and at the end, respectively, we have

$$v = U + \frac{\iota}{\mathfrak{M}},\quad V = U + \frac{I}{\mathfrak{M}}.$$

Hence, for the rate of the doing of work by the force P, at the instant t, we have

$$Pv = PU + \frac{\iota P}{\mathfrak{M}}.$$

Work done by impact.

Hence for the whole work (W) done by it,

$$W = \int_0^T \left(PU + \frac{\iota P}{\mathfrak{M}}\right) d\tau$$

$$= UI + \frac{1}{\mathfrak{M}} \int_0^I \iota d\iota = UI + \frac{1}{2} \frac{I^2}{\mathfrak{M}}$$

$$= UI + \tfrac{1}{2} I (V - U) = I \cdot \tfrac{1}{2} (U + V).$$

309. It is worthy of remark, that if any number of impacts be applied to a body, their whole effect will be the same whether they be applied together or successively (provided that the whole time occupied by them be infinitely short), although the work done by each particular impact is in general different according to the order in which the several impacts are applied. The whole amount of work is the sum of the products obtained by multiplying each impact by half the sum of the components of the initial and final velocities of the point to which it is applied.

Equations of impulsive motion.

310. The effect of any stated impulses, applied to a rigid body, or to a system of material points or rigid bodies connected in any way, is to be found most readily by the aid of D'Alembert's principle; according to which the given impulses, and the impulsive reaction against the generation of motion, measured in amount by the momenta generated, are in equilibrium; and are therefore to be dealt with mathematically by applying to them the equations of equilibrium of the system.

Let P_1, Q_1, R_1 be the component impulses on the first particle, m_1, and let $\dot{x}_1$, $\dot{y}_1$, $\dot{z}_1$ be the components of the velocity instantaneously acquired by this particle. Component forces equal to $(P_1 - m_1\dot{x}_1)$, $(Q_1 - m_1\dot{y}_1)$, ... must equilibrate the system, and therefore we have (§ 290)

$$\Sigma \{(P - m\dot{x})\, \delta x + (Q - m\dot{y})\, \delta y + (R - m\dot{z})\, \delta z\} = 0 \ldots\ldots\ldots (a)$$

where δx_1, δy_1, ... denote the components of any infinitely small displacements of the particles possible under the conditions of the system. Or, which amounts to the same thing, since any possible infinitely small displacements are simply proportional to any possible velocities in the same directions,

$$\Sigma \{(P - m\dot{x})\, u + (Q - m\dot{y})\, v + (Q - m\dot{z})\, w\} = 0 \ldots\ldots\ldots\ldots (b)$$

where u_1, v_1, w_1 denote any possible component velocities of the first particle, etc. **Equations of impulsive motion.**

One particular case of this equation is of course had by supposing u_1, v_1, ... to be equal to the velocities $\dot{x}_1$, $\dot{y}_1$, ... actually acquired; and, by halving, etc., we find

$$\Sigma\left(P\,.\,\tfrac{1}{2}\dot{x} + Q\,.\,\tfrac{1}{2}\dot{y} + R\,.\,\tfrac{1}{2}\dot{z}\right) = \tfrac{1}{2}\Sigma m\left(\dot{x}^2 + \dot{y}^2 + \dot{z}^2\right)\ldots\ldots\ldots(c).$$

This agrees with § 308 above.

311. Euler discovered that the kinetic energy acquired from rest by a rigid body in virtue of an impulse fulfils a maximum-minimum condition. Lagrange* extended this proposition to a system of bodies connected by any invariable kinematic relations, and struck with any impulses. Delaunay found that it is really always a maximum *when the impulses are given, and when different motions possible under the conditions of the system, and fulfilling the law of energy* [§ 310 (c)], *are considered.* Farther, Bertrand shows that the energy actually acquired is not merely a "maximum," but exceeds the energy of any other motion fulfilling these conditions; and that the amount of the excess is equal to the energy of the motion which must be compounded with either to produce the other. **Theorem of Euler, extended by Lagrange.** **Equation of impulsive motion.**

Let $\dot{x}_1'$, $\dot{y}_1'$ ... be the component velocities of any motion whatever fulfilling the equation (c), which becomes

$$\tfrac{1}{2}\Sigma\left(P\dot{x}' + Q\dot{y}' + R\dot{z}'\right) = \tfrac{1}{2}\Sigma m\left(\dot{x}'^2 + \dot{y}'^2 + \dot{z}'^2\right) = T'\ldots\ldots\ldots(d).$$

If, then, we take $\dot{x}_1' - \dot{x}_1 = u_1$, $\dot{y}_1' - \dot{y}_1 = v_1$, etc., we have

$$T' - T = \tfrac{1}{2}\Sigma m\left\{(2\dot{x} + u)\,u + (2\dot{y} + v)\,v + (2\dot{z} + w)\,w\right\}$$
$$= \Sigma m\left(\dot{x}u + \dot{y}v + \dot{z}w\right) + \tfrac{1}{2}\Sigma m\left(u^2 + v^2 + w^2\right)\ldots\ldots(e).$$

But, by (b),

$$\Sigma m\left(\dot{x}u + \dot{y}v + \dot{z}w\right) = \Sigma\left(Pu + Qv + Rw\right)\ldots\ldots\ldots(f);$$

and, by (c) and (d),

$$\Sigma\left(Pu + Qv + Rw\right) = 2T' - 2T\ldots\ldots\ldots\ldots\ldots(g).$$

Hence (e) becomes

$$T' - T = 2\left(T' - T\right) + \tfrac{1}{2}\Sigma m\left(u^2 + v^2 + w^2\right),$$

whence

$$T - T' = \tfrac{1}{2}\Sigma m\left(u^2 + v^2 + w^2\right)\ldots\ldots\ldots\ldots\ldots(h),$$

which is Bertrand's result.

* *Mécanique Analytique*, 2nde partie, 3me section, § 37.

Liquid set in motion impulsively. **312.** The energy of the motion generated suddenly in a mass of incompressible liquid given at rest completely filling a vessel of any shape. when the vessel is suddenly set in motion, or when it is suddenly bent out of shape in any way whatever, subject to the condition of not changing its volume, *is less than the energy of any other motion it can have with the same motion of its bounding surface.* The consideration of this theorem, which, so far as we know, was first published in the *Cambridge and Dublin Mathematical Journal* [Feb. 1849], has led us to a general *minimum* property regarding motion acquired by any system when *any prescribed velocities* are generated suddenly in any of its parts; announced in the *Proceedings of the Royal Society of Edinburgh* for April, 1863. It is, that provided impulsive forces are applied to the system only at places where the velocities to be produced are prescribed, the kinetic energy is *less* in the actual motion than in any other motion which the system can take, and which has the same values for the prescribed velocities. The excess of the energy of any possible motion above that of the actual motion is (as in Bertrand's theorem) equal to the energy of the motion which must be compounded with either to produce the other. The proof is easy:—here it is :—

Equations (d), (e), and (f) hold as in § (311). But now each velocity component, u_1, v_1, w_1, u_2, etc. vanishes for which the component impulse P_1, Q_1, R_1, P_2, etc. does not vanish (because $\dot{x}_1 + u_1$, $\dot{y}_1 + v_1$, etc. fulfil the prescribed velocity conditions). Hence every product P_1u_1, Q_1v_1, etc. vanishes. Hence now instead of (g) and (h) we have

$$\Sigma(\dot{x}u + \dot{y}v + \dot{z}w) = 0 \quad \ldots\ldots\ldots\ldots (g'),$$

and

$$T' - T = \tfrac{1}{2}\Sigma m(u^2 + v^2 + w^2) \quad \ldots\ldots\ldots\ldots (h').$$

We return to the subject in §§ 316, 317 as an illustration of the use of Lagrange's generalized co-ordinates; to the introduction of which into Dynamics we now proceed.

Impulsive motion referred to generalized co-ordinates. **313.** The method of generalized co-ordinates explained above (§ 204) is extremely useful in its application to the dynamics of a system; whether for expressing and working out the details of any particular case in which there is any finite number of degrees of freedom, or for proving general principles applicable even to cases, such as that of a liquid, as described in the preceding section, in which there may be an infinite number of degrees of freedom. It leads us to generalize the measure of inertia, and the resolution and composition of forces, impulses, and momenta, on dynamical principles corresponding with the kinematical principles explained in § 204, which gave us generalized component velocities: and, as we shall see later, the generalized equations of continuous motion are not only very convenient for the solution of problems, but most *instructive* as to the nature of relations, however complicated, between the motions of different parts of a system. In the meantime we shall consider the generalized expressions for the impulsive generation of motion. We have seen above (§ 308) that the kinetic energy acquired by a system given at rest and struck with any given impulses, is equal to half the sum of the products of the component forces multiplied each into the corresponding component of the velocity acquired by its point of application, when the ordinary system of rectangular co-ordinates is used. Precisely the same statement holds on the generalized system, and if stated as the convention agreed upon, it suffices to define the generalized components of impulse, those of velocity having been fixed on kinematical principles (§ 204). Generalized components of momentum of any specified motion are, of course, equal to the generalized components of the impulse by which it could be generated from rest.

Impulsive motion referred to generalized co-ordinates.

Generalized components of impulse or momentum.

(*a*) Let ψ, ϕ, θ, ... be the generalized co-ordinates of a material system at any time; and let $\dot\psi$, $\dot\phi$, $\dot\theta$, ... be the corresponding generalized velocity-components, that is to say, the rates at which ψ, ϕ, θ, ... increase per unit of time, at any instant, in the actual motion. If x_1, y_1, z_1 denote the common rectangular co-ordinates of one particle of the system, and $\dot x_1$, $\dot y_1$, $\dot z_1$ its component velocities, we have

$$\left.\begin{array}{l} \dot x_1 = \dfrac{dx_1}{d\psi}\dot\psi + \dfrac{dx_1}{d\phi}\dot\phi + \text{etc.} \\ \dot y_1 = \dfrac{dy_1}{d\psi}\dot\psi + \dfrac{dy_1}{d\phi}\dot\phi + \text{etc.} \\ \quad\text{etc.}\qquad\quad \text{etc.} \end{array}\right\} \quad\ldots\ldots\ldots\ldots\ldots\ldots(1).$$

Hence the kinetic energy, which is $\Sigma\frac{1}{2}m(\dot{x}^2+\dot{y}^2+\dot{z}^2)$, in terms of rectangular co-ordinates, becomes a quadratic function of $\dot{\psi}$, $\dot{\phi}$, etc., when expressed in terms of generalized co-ordinates, so that if we denote it by T we have

Generalized expression for kinetic energy.

$$T=\tfrac{1}{2}\{(\psi,\psi)\dot{\psi}^2+(\phi,\phi)\dot{\phi}^2+\ldots+2(\psi,\phi)\dot{\psi}\dot{\phi}+\ldots\}\ldots\ldots(2),$$

where (ψ,ψ), (ϕ,ϕ), (ψ,ϕ), etc., denote various functions of the co-ordinates, determinable according to the conditions of the system. The only condition essentially fulfilled by these coefficients is, that they must give a finite positive value to T for all values of the variables.

(b) Again let (X_1, Y_1, Z_1), (X_2, Y_2, Z_2), etc., denote component forces on the particles (x_1, y_1, z_1), (x_2, y_2, z_2), etc., respectively; and let $(\delta x_1, \delta y_1, \delta z_1)$, etc., denote the components of any infinitely small motions possible without breaking the conditions of the system. The work done by those forces, upon the system when so displaced, will be

$$\Sigma(X\delta x+Y\delta y+Z\delta z)\ldots\ldots\ldots\ldots\ldots\ldots(3).$$

To transform this into an expression in terms of generalized co-ordinates, we have

$$\left.\begin{array}{l}\delta x_1=\dfrac{dx_1}{d\psi}\delta\psi+\dfrac{dx_1}{d\phi}\delta\phi+\text{etc.}\\[2mm] \delta y_1=\dfrac{dy_1}{d\psi}\delta\psi+\dfrac{dy_1}{d\phi}\delta\phi+\text{etc.}\\ \qquad\text{etc.}\qquad\quad\text{etc.}\end{array}\right\}\ldots\ldots\ldots\ldots\ldots(4),$$

and it becomes

$$\Psi\delta\psi+\Phi\delta\phi+\text{etc.}\ldots\ldots\ldots\ldots\ldots\ldots\ldots\ldots(5),$$

where

Generalized components of force,

$$\left.\begin{array}{l}\Psi=\Sigma\left(X\dfrac{dx}{d\psi}+Y\dfrac{dy}{d\psi}+Z\dfrac{dz}{d\psi}\right)\\[2mm] \Phi=\Sigma\left(X\dfrac{dx}{d\phi}+Y\dfrac{dy}{d\phi}+Z\dfrac{dz}{d\phi}\right)\\ \qquad\text{etc.}\qquad\quad\text{etc.}\end{array}\right\}\ldots\ldots\ldots\ldots\ldots(6).$$

These quantities, Ψ, Φ, etc., are clearly *the generalized components of the force on the system.*

Let $\underset{\cdot}{\Psi}$, $\underset{\cdot}{\Phi}$, etc. denote component impulses, generalized on the same principle; that is to say, let

of impulse.

$$\underset{\cdot}{\Psi}=\int_0^\tau\Psi dt,\quad \underset{\cdot}{\Phi}=\int_0^\tau\Phi dt,\ \text{etc.},$$

where Ψ, Φ, ... denote generalized components of the continuous force acting at any instant of the infinitely short time τ, within which the impulse is completed.

If this impulse is applied to the system, previously in motion in the manner specified above, and if $\delta\dot{\psi}$, $\delta\dot{\phi}$, ... denote the resulting augmentations of the components of velocity, the means of the component velocities before and after the impulse will be

Impulsive generation of motion referred to generalized co-ordinates.

$$\dot{\psi} + \tfrac{1}{2}\delta\dot{\psi},\quad \dot{\phi} + \tfrac{1}{2}\delta\dot{\phi},\ \ldots\ldots$$

Hence, according to the general principle explained above for calculating the work done by an impulse, the whole work done in this case is

$$\underset{\cdot}{\Psi}(\dot{\psi} + \tfrac{1}{2}\delta\dot{\psi}) + \underset{\cdot}{\Phi}(\dot{\phi} + \tfrac{1}{2}\delta\dot{\phi}) + \text{etc.}$$

To avoid unnecessary complications, let us suppose $\delta\dot{\psi}$, $\delta\dot{\phi}$, etc., to be each infinitely small. The preceding expression for the work done becomes

$$\underset{\cdot}{\Psi}\dot{\psi} + \underset{\cdot}{\Phi}\dot{\phi} + \text{etc.};$$

and, as the effect produced by this work is augmentation of kinetic energy from T to $T + \delta T$, we must have

$$\delta T = \underset{\cdot}{\Psi}\dot{\psi} + \underset{\cdot}{\Phi}\dot{\phi} + \text{etc.}$$

Now let the impulses be such as to augment $\dot{\psi}$ to $\dot{\psi} + \delta\dot{\psi}$, and to leave the other component velocities unchanged. We shall have

$$\underset{\cdot}{\Psi}\dot{\psi} + \underset{\cdot}{\Phi}\dot{\phi} + \text{etc.} = \frac{dT}{d\dot{\psi}}\delta\dot{\psi}.$$

Dividing both members by $\delta\dot{\psi}$, and observing that $\dfrac{dT}{d\dot{\psi}}$ is a linear function of $\dot{\psi}$, $\dot{\phi}$, etc., we see that $\dfrac{\underset{\cdot}{\Psi}}{\delta\dot{\psi}}$, $\dfrac{\underset{\cdot}{\Phi}}{\delta\dot{\psi}}$, etc., must be equal to the coefficients of $\dot{\psi}$, $\dot{\phi}$, ... respectively in $\dfrac{dT}{d\dot{\psi}}$.

(*c*) From this we see, further, that the impulse required to produce the component velocity $\dot{\psi}$ from rest, or to generate it in the system moving with any other possible velocity, has for its components

$$(\psi, \psi)\dot{\psi},\quad (\psi, \phi)\dot{\psi},\quad (\psi, \theta)\dot{\psi},\ \text{etc.}$$

Hence we conclude that to generate the whole resultant velocity $(\dot{\psi}, \dot{\phi}, \ldots)$ from rest, requires an impulse, of which the components, if denoted by ξ, η, ζ, ..., are expressed as follows :—

Momenta in terms of velocities.

$$\left.\begin{aligned}\xi &= (\psi, \psi)\dot\psi + (\phi, \psi)\dot\phi + (\theta, \psi)\dot\theta + \ldots \\ \eta &= (\psi, \phi)\dot\psi + (\phi, \phi)\dot\phi + (\theta, \phi)\dot\theta + \ldots \\ \zeta &= (\psi, \theta)\dot\psi + (\phi, \theta)\dot\phi + (\theta, \theta)\dot\theta + \ldots \\ &\text{etc.}\end{aligned}\right\} \ldots\ldots\ldots\ldots\ldots(7),$$

where it must be remembered that, as seen in the original expression for T, from which they are derived, (ϕ, ψ) means the same thing as (ψ, ϕ), and so on. The preceding expressions are the differential coefficients of T with reference to the velocities; that is to say,

$$\xi = \frac{dT}{d\dot\psi}, \quad \eta = \frac{dT}{d\dot\phi}, \quad \zeta = \frac{dT}{d\dot\theta} \ldots\ldots\ldots\ldots\ldots\ldots\ldots (8).$$

(d) The second members of these equations being linear functions of $\dot\psi$, $\dot\phi$, ..., we may, by ordinary elimination, find $\dot\psi$, $\dot\phi$, etc., in terms of ξ, η, etc., and the expressions so obtained are of course linear functions of the last-named elements. And, since T is a quadratic function of $\dot\psi$, $\dot\phi$, etc., we have

$$2T = \xi\dot\psi + \eta\dot\phi + \zeta\dot\theta + \text{etc.} \ldots\ldots\ldots\ldots\ldots\ldots\ldots(9).$$

Kinetic energy in terms of momentums and velocities.

From this, on the supposition that T, $\dot\psi$, $\dot\phi$, ... are expressed in terms of ξ, η, ..., we have by differentiation

$$2\frac{dT}{d\xi} = \dot\psi + \xi\frac{d\dot\psi}{d\xi} + \eta\frac{d\dot\phi}{d\xi} + \zeta\frac{d\dot\theta}{d\xi} + \text{etc.}$$

Now the algebraic process by which $\dot\psi$, $\dot\phi$, etc., are obtained in terms of ξ, η, etc., shows that, inasmuch as the coefficient of $\dot\phi$ in the expression, (7), for ξ, is equal to the coefficient of $\dot\psi$, in the expression for η, and so on; the coefficient of η in the expression for $\dot\psi$ must be equal to the coefficient of ξ in the expression for $\dot\phi$, and so on; that is to say,

$$\frac{d\dot\psi}{d\eta} = \frac{d\dot\phi}{d\xi}, \quad \frac{d\dot\psi}{d\zeta} = \frac{d\dot\theta}{d\xi}, \text{ etc.}$$

Hence the preceding expression becomes

$$2\frac{dT}{d\xi} = \dot\psi + \xi\frac{d\dot\psi}{d\xi} + \eta\frac{d\dot\psi}{d\eta} + \zeta\frac{d\dot\psi}{d\zeta} + \ldots = 2\dot\psi,$$

and therefore

Velocities in terms of momentums.

$$\left.\begin{aligned}&\dot\psi = \frac{dT}{d\xi}. \\ \text{Similarly} \quad &\dot\phi = \frac{dT}{d\eta}, \text{ etc.}\end{aligned}\right\} \ldots\ldots\ldots\ldots\ldots\ldots\ldots\ldots(10).$$

These expressions solve the direct problem,—to find the velocity produced by a given impulse ($\xi, \eta, \ldots$), when we have the kinetic energy, T, expressed as a quadratic function of the components of the impulse. Velocities in terms of momentums.

(*e*) If we consider the motion simply, without reference to the impulse required to generate it from rest, or to stop it, the quantities $\xi, \eta, \ldots$ are clearly to be regarded as the components of the momentum of the motion, according to the system of generalized co-ordinates.

(*f*) The following algebraic relation will be useful:— Reciprocal relation between momentums and velocities in two motions.

$$\xi_{,}\dot{\psi} + \eta_{,}\dot{\phi} + \zeta_{,}\dot{\theta} + \text{etc.} = \xi\dot{\psi}_{,} + \eta\dot{\phi}_{,} + \zeta\dot{\theta}_{,} + \text{etc} \ldots\ldots\ldots\ldots (11),$$

where, $\xi, \eta, \dot{\psi}, \dot{\phi}$, etc., having the same signification as before, $\xi_{,}, \eta_{,}, \zeta_{,}$, etc., denote the impulse-components corresponding to any other values, $\dot{\psi}_{,}, \dot{\phi}_{,}, \dot{\theta}_{,}$, etc., of the velocity-components. It is proved by observing that each member of the equation becomes a symmetrical function of $\dot{\psi}, \dot{\psi}_{,}; \dot{\phi}, \dot{\phi}_{,}$; etc.; when for $\xi_{,}, \eta_{,}$, etc., their values in terms of $\dot{\psi}_{,}, \dot{\phi}_{,}$, etc., and for ξ, η, etc., their values in terms of $\dot{\psi}, \dot{\phi}$, etc., are substituted.

314. A material system of any kind, given at rest, and subjected to an impulse in any specified direction, and of any given magnitude, moves off so as to take the greatest amount of kinetic energy which the specified impulse can give it, subject to § 308 or § 309 (*c*). Application of generalized co-ordinates to theorems of § 311.

Let $\xi, \eta, \ldots$ be the components of the given impulse, and $\dot{\psi}, \dot{\phi}, \ldots$ the components of the actual motion produced by it, which are determined by the equations (10) above. Now let us suppose the system be guided, by means of merely directive constraint, to take, from rest, under the influence of the given impulse, some motion ($\dot{\psi}_{,}, \dot{\phi}_{,}, \ldots$) different from the actual motion; and let $\xi_{,}, \eta_{,}, \ldots$ be the impulse which, with this constraint removed, would produce the motion ($\dot{\psi}_{,}, \dot{\phi}_{,}, \ldots$). We shall have, for this case, as above,

$$T_{,} = \tfrac{1}{2}(\xi_{,}\dot{\psi}_{,} + \eta_{,}\dot{\phi}_{,} + \ldots).$$

But $\xi_{,} - \xi, \ \eta_{,} - \eta \ldots$ are the components of the impulse experienced in virtue of the constraint we have supposed introduced. They neither perform nor consume work on the system when moving as directed by this constraint; that is to say,

$$(\xi_{,} - \xi)\dot{\psi}_{,} + (\eta_{,} - \eta)\dot{\phi}_{,} + (\zeta_{,} - \zeta)\dot{\theta}_{,} + \text{etc.} = 0 \ldots\ldots\ldots\ldots (12);$$

Application of generalized co-ordinates to theorems of § 311.

and therefore

$$2T_, = \xi\dot{\psi}_, + \eta\dot{\phi}_, + \zeta\dot{\theta}_, + \text{etc.} \quad \text{.................(13).}$$

Hence we have

$$2(T - T_,) = \xi(\dot{\psi} - \dot{\psi}_,) + \eta(\dot{\phi} - \dot{\phi}_,) + \text{etc.}$$
$$= (\xi - \xi_,)(\dot{\psi} - \dot{\psi}_,) + (\eta - \eta_,)(\dot{\phi} - \dot{\phi}_,) + \text{etc.}$$
$$+ \xi_,(\dot{\psi} - \dot{\psi}_,) + \eta_,(\dot{\phi} - \dot{\phi}_,) + \text{etc.}$$

But, by (11) and (12) above, we have

$$\xi_,(\dot{\psi} - \dot{\psi}_,) + \eta_,(\dot{\phi} - \dot{\phi}_,) + \text{etc.} = (\xi - \xi_,)\dot{\psi}_, + (\eta - \eta_,)\dot{\phi}_, + \text{etc.} = 0,$$

and therefore we have finally

$$2(T - T_,) = (\xi - \xi_,)(\dot{\psi} - \dot{\psi}_,) + (\eta - \eta_,)(\dot{\phi} - \dot{\phi}_,) + \text{etc.} \quad \text{...(14).}$$

Theorems of § 311 in terms of generalized co-ordinates.

that is to say, T exceeds $T_,$ by the amount of the kinetic energy that would be generated by an impulse ($\xi - \xi_,$, $\eta - \eta_,$, $\zeta - \zeta_,$, etc.) applied simply to the system, which is essentially positive. In other words,

315. If the system is guided to take, under the action of a given impulse, any motion ($\dot{\psi}_,$, $\dot{\phi}_,$, ...) different from the natural motion ($\dot{\psi}$, $\dot{\phi}$, ...), it will have less kinetic energy than that of the natural motion, by a difference equal to the kinetic energy of the motion ($\dot{\psi} - \dot{\psi}_,$, $\dot{\phi} - \dot{\phi}_,$, ...).

Cor. If a set of material points are struck independently by impulses each given in amount, more kinetic energy is generated if the points are perfectly free to move each independently of all the others, than if they are connected in any way. And the deficiency of energy in the latter case is equal to the amount of the kinetic energy of the motion which geometrically compounded with the motion of either case would give that of the other.

Problems whose data involve impulses and velocities.

(*a*) Hitherto we have either supposed the motion to be fully given, and the impulses required to produce them, to be to be found; or the impulses to be given and the motions produced by them to be to be found. A not less important class of problems is presented by supposing as many linear equations of condition between the impulses and components of motion to be given as there are degrees of freedom of the system to move (or independent co-ordinates). These equations, and as many more supplied by (8) or their equivalents (10), suffice for the complete solution of the problem, to determine the impulses and the motion.

(*b*) A very important case of this class is presented by prescribing, among the velocities alone, a number of linear equations with constant terms, and supposing the impulses to be so directed and related as to do no work on any velocities satisfying another prescribed set of linear equations with no constant terms; the whole number of equations of course being equal to the number of independent co-ordinates of the system. The equations for solving this problem need not be written down, as they are obvious; but the following reduction is useful, as affording the easiest proof of the *minimum* property stated below. Problems whose data involve impulses and velocities.

(*c*) The given equations among the velocities may be reduced to a set, each homogeneous, except one equation with a constant term. Those homogeneous equations diminish the number of degrees of freedom; and we may transform the co-ordinates so as to have the number of independent co-ordinates diminished accordingly. Farther, we may choose the new co-ordinates, so that the linear function of the velocities in the single equation with a constant term may be one of the new velocity-components; and the linear functions of the velocities appearing in the equation connected with the prescribed conditions as to the impulses may be the remaining velocity-components. Thus the impulse will fulfil the condition of doing no work on any other component velocity than the one which is given, and the general problem—

316. Given any material system at rest: let any parts of it be set in motion suddenly with any specified velocities, possible according to the conditions of the system; and let its other parts be influenced only by its connexions with those; required the motion: General problem (compare § 312).

takes the following very simple form:—An impulse of the character specified as a particular component, according to the generalized method of co-ordinates, acts on a material system; its amount being such as to produce a given velocity-component of the corresponding type. It is required to find the motion.

The solution of course is to be found from the equations

$$\dot{\psi} = A, \qquad \eta = 0, \qquad \zeta = 0 \ldots\ldots\ldots\ldots\ldots\ldots\ldots(15)$$

(which are the special equations of condition of the problem) and the general kinetic equations (7), or (10). Choosing the latter, and denoting by $[\xi, \xi]$, $[\xi, \eta]$, etc., the coefficients of $\frac{1}{2}\xi^2$, $\xi\eta$, etc.,

General problem (compare § 312).

in T, we have

$$\xi = \frac{A}{[\xi,\xi]}, \quad \dot\phi = \frac{[\xi,\eta]}{[\xi,\xi]}A, \quad \dot\theta = \frac{[\xi,\zeta]}{[\xi,\xi]}A, \text{ etc.} \qquad (16)$$

for the result.

This result possesses the remarkable property, that the kinetic energy of the motion expressed by it is less than that of any other motion which fulfils the prescribed condition as to velocity. For, if $\xi_{,}, \eta_{,}, \zeta_{,}$, etc., denote the impulses required to produce any other motion, $\dot\psi_{,}, \dot\phi_{,}, \dot\theta_{,}$, etc., and $T_{,}$ the corresponding kinetic energy, we have, by (9),

$$2T_{,} = \xi_{,}\dot\psi_{,} + \eta_{,}\dot\phi_{,} + \zeta_{,}\dot\theta_{,} + \text{etc.}$$

But by (11),

$$\xi_{,}\dot\psi + \eta_{,}\dot\phi + \zeta_{,}\dot\theta + \text{etc.} = \xi\dot\psi_{,},$$

since, by (15), we have $\eta = 0$, $\zeta = 0$, etc. Hence

$$2T_{,} = \xi\dot\psi_{,} + \xi_{,}(\dot\psi_{,} - \dot\psi) + \eta_{,}(\dot\phi_{,} - \dot\phi) + \zeta_{,}(\dot\theta_{,} - \dot\theta) + \ldots$$

Now let also this second case $(\dot\psi_{,}, \dot\phi_{,}, \ldots)$ of motion fulfil the prescribed velocity-condition $\dot\psi_{,} = A$. We shall have

$$\xi_{,}(\dot\psi_{,} - \dot\psi) + \eta_{,}(\dot\phi_{,} - \dot\phi) + \zeta_{,}(\dot\theta_{,} - \dot\theta) + \ldots$$
$$= (\xi_{,} - \xi)(\dot\psi_{,} - \dot\psi) + (\eta_{,} - \eta)(\dot\phi_{,} - \dot\phi) + (\zeta_{,} - \zeta)(\dot\theta_{,} - \dot\theta) + \ldots.$$

since $\dot\psi_{,} - \dot\psi = 0$, $\eta = 0$, $\zeta = 0, \ldots.$ Hence if $\mathfrak{T}$ denote the kinetic energy of the differential motion $(\dot\psi_{,} - \dot\psi, \dot\phi_{,} - \dot\phi, \ldots)$ we have

$$2T_{,} = 2T + 2\mathfrak{T} \qquad (17);$$

but $\mathfrak{T}$ is essentially positive and therefore $T_{,}$, the kinetic energy of any motion fulfilling the prescribed velocity-condition, but differing from the actual motion, is greater than T the kinetic energy of the actual motion; and the amount, $\mathfrak{T}$, of the difference is given by the equation

$$2\mathfrak{T} = \eta_{,}(\dot\phi_{,} - \dot\phi) + \zeta_{,}(\dot\theta_{,} - \dot\theta) + \text{etc.} \qquad (18),$$

or in words,

Kinetic energy a minimum in this case.

317. The solution of the problem is this:—The motion actually taken by the system is the motion which has less kinetic energy than any other fulfilling the prescribed velocity-conditions. And the excess of the energy of any other such motion, above that of the actual motion, is equal to the energy of the motion which must be compounded with either to produce the other.

In dealing with cases it may often happen that the use of the co-ordinate system required for the application of the solution (16) is not convenient; but in all cases, even in such as in examples (2) and (3) below, which involve an infinite number of degrees of freedom, the minimum property now proved affords an easy solution.

Kinetic energy a minimum in this case.

Example (1). Let a smooth plane, constrained to keep moving with a given normal velocity, q, come in contact with a free inelastic rigid body at rest: to find the motion produced. The velocity-condition here is, that the motion shall consist of any motion whatever giving to the point of the body which is struck a stated velocity, q, perpendicular to the impinging plane, compounded with any motion whatever giving to the same point any velocity parallel to this plane. To express this condition, let u, v, w be rectangular component linear velocities of the centre of gravity, and let ϖ, ρ, σ be component angular velocities round axes through the centre of gravity parallel to the line of reference. Thus, if x, y, z denote the co-ordinates of the point struck relatively to these axes through the centre of gravity, and if l, m, n be the direction cosines of the normal to the impinging plane, the prescribed velocity-condition becomes

Impact of a smooth rigid plane of infinite mass on a free rigid body at rest.

$$(u+\rho z-\sigma y)\,l+(v+\sigma x-\varpi z)\,m+(w+\varpi y-\rho x)\,n=-q \quad\ldots\ldots\ldots(a),$$

the negative sign being placed before q on the understanding that the motion of the impinging plane is obliquely, if not directly, *towards* the centre of gravity, when l, m, n are each positive. If, now, we suppose the rectangular axes through the centre of gravity to be principal axes of the body, and denote by Mf^2, Mg^2, Mh^2 the moments of inertia round them, we have

$$T=\tfrac{1}{2}M\left(u^2+v^2+w^2+f^2\varpi^2+g^2\rho^2+h^2\sigma^2\right)\quad\ldots\ldots\ldots\ldots\ldots(b).$$

This must be made a minimum subject to the equation of condition (a). Hence, by the ordinary method of indeterminate multipliers,

$$\left.\begin{aligned}&Mu+\lambda l=0,\ Mv+\lambda m=0,\ Mw+\lambda n=0\\ &Mf^2\varpi+\lambda(ny-mz)=0,\ Mg^2\rho+\lambda(lz-nx)=0,\ Mh^2\sigma+\lambda(mx-ly)=0\end{aligned}\right\}\ (c).$$

These six equations give each of them explicitly the value of one of the six unknown quantities u, v, w, ϖ, ρ, σ, in terms of λ and data. Using the values thus found in (a), we have an equation to determine λ; and thus the solution is completed. The first three of equations (c) show that λ, which has entered as an indeterminate multiplier, is to be interpreted as the measure of the amount of the impulse.

Generation of motion by impulse in an inextensible cord or chain.

Example (2). A stated velocity in a stated direction is communicated impulsively to each end of a flexible inextensible cord forming any curvilineal arc: it is required to find the initial motion of the whole cord.

Let x, y, z be the co-ordinates of any point P in it, and $\dot{x}$, $\dot{y}$, $\dot{z}$ the components of the required initial velocity. Let also s be the length from one end to the point P.

If the cord were extensible, the rate per unit of time of the stretching per unit of length which it would experience at P, in virtue of the motion $\dot{x}$, $\dot{y}$, $\dot{z}$, would be

$$\frac{dx}{ds}\frac{d\dot{x}}{ds}+\frac{dy}{ds}\frac{d\dot{y}}{ds}+\frac{dz}{ds}\frac{d\dot{z}}{ds}.$$

Hence, as the cord is inextensible, by hypothesis,

$$\frac{dx}{ds}\frac{d\dot{x}}{ds}+\frac{dy}{ds}\frac{d\dot{y}}{ds}+\frac{dz}{ds}\frac{d\dot{z}}{ds}=0\text{.....................}(a).$$

Subject to this, the kinematical condition of the system, and

$$\left.\begin{array}{l}\dot{x}=u\\ \dot{y}=v\\ \dot{z}=w\end{array}\right\}\text{ when } s=0,\qquad \left.\begin{array}{l}\dot{x}=u'\\ \dot{y}=v'\\ \dot{z}=w'\end{array}\right\}\text{ when } s=l,$$

l denoting the length of the cord, and (u, v, w), (u', v', w'), the components of the given velocities at its two ends: it is required to find $\dot{x}$, $\dot{y}$, $\dot{z}$ at every point, so as to make

$$\int_0^l \tfrac{1}{2}\mu\,(\dot{x}^2+\dot{y}^2+\dot{z}^2)\,ds\text{........................}(b)$$

a minimum, μ denoting the mass of the string per unit of length, at the point P, which need not be uniform from point to point; and of course

$$ds=(dx^2+dy^2+dz^2)^{\frac{1}{2}}\text{..........................}(c).$$

Multiplying (a) by λ, an indeterminate multiplier, and proceeding as usual according to the method of variations, we have

$$\int_0^l\left\{\mu(\dot{x}\delta\dot{x}+\dot{y}\delta\dot{y}+\dot{z}\delta\dot{z})+\lambda\left(\frac{dx}{ds}\frac{d\delta\dot{x}}{ds}+\frac{dy}{ds}\frac{d\delta\dot{y}}{ds}+\frac{dz}{ds}\frac{d\delta\dot{z}}{ds}\right)\right\}ds=0,$$

in which we may regard x, y, z as known functions of s, and this it is convenient we should make independent variable. Integrating "by parts" the portion of the first member which contains λ, and attending to the terminal conditions, we find, according to the regular process, for the equations containing the solution

Generation of motion by impulse in an inextensible cord or chain.

$$\mu\dot{x} = \frac{d}{ds}\left(\lambda\frac{dx}{ds}\right),\ \mu\dot{y} = \frac{d}{ds}\left(\lambda\frac{dy}{ds}\right),\ \mu\dot{z} = \frac{d}{ds}\left(\lambda\frac{dz}{ds}\right) \ldots\ldots\ldots\ldots (d).$$

These three equations with (a) suffice to determine the four unknown quantities, $\dot{x}$, $\dot{y}$, $\dot{z}$, and λ. Using (d) to eliminate $\dot{x}$, $\dot{y}$, $\dot{z}$ from (a), we have

$$0 = \frac{d\frac{1}{\mu}}{ds}\left\{\frac{dx}{ds}\frac{d}{ds}\left(\lambda\frac{dx}{ds}\right) + \ldots\right\} + \frac{1}{\mu}\left\{\frac{dx}{ds}\frac{d^2}{ds^2}\left(\lambda\frac{dx}{ds}\right) + \ldots\right\}.$$

Taking now s for independent variable, and performing the differentiation here indicated, with attention to the following relations:—

$$\frac{dx^2}{ds^2} + \ldots = 1,\ \frac{dx}{ds}\frac{d^2x}{ds^2} + \ldots = 0,$$

$$\frac{dx}{ds}\frac{d^3x}{ds^3} + \ldots + \left(\frac{d^2x}{ds^2}\right)^2 + \ldots = 0,$$

and the expression (§ 9) for ρ, the radius of curvature, we find

$$\frac{1}{\mu}\frac{d^2\lambda}{ds^2} + \frac{d\left(\frac{1}{\mu}\right)}{ds}\frac{d\lambda}{ds} - \frac{\lambda}{\mu\rho^2} = 0 \ldots\ldots\ldots\ldots (e)$$

a linear differential equation of the second order to determine λ, when μ and ρ are given functions of s.

The interpretation of (d) is very obvious. It shows that λ is the impulsive tension at the point P of the string; and that the velocity which this point acquires instantaneously is the resultant of $\frac{1}{\mu}\frac{d\lambda}{ds}$ tangential, and $\frac{\lambda}{\rho\mu}$ towards the centre of curvature. The differential equation (e) therefore shows the law of transmission of the instantaneous tension along the string, and proves that it depends solely on the mass of the cord per unit of length in each part, and the curvature from point to point, but not at all on the plane of curvature, of the initial form. Thus, for instance, it will be the same along a helix as along a circle of the same curvature.

Generation of motion by impulse in an inextensible cord or chain.

With reference to the fulfilling of the six terminal equations, a difficulty occurs inasmuch as $\dot{x}, \dot{y}, \dot{z}$ are expressed by (d) immediately, without the introduction of fresh arbitrary constants, in terms of λ, which, as the solution of a differential equation of the second degree, involves only two arbitrary constants. The explanation is, that at any point of the cord, at any instant, any velocity in any direction perpendicular to the tangent may be generated without at all altering the condition of the cord even at points infinitely near it. This, which seems clear enough without proof, may be demonstrated analytically by transforming the kinematical equation (a) thus. Let f be the component tangential velocity, q the component velocity towards the centre of curvature, and p the component velocity perpendicular to the osculating plane. Using the elementary formulas for the direction cosines of these lines (§ 9), and remembering that s is now independent variable, we have

$$\dot{x} = f\frac{dx}{ds} + q\frac{\rho d^2x}{ds^2} + p\frac{\rho\,(dzd^2y - dyd^2z)}{ds^3},\quad \dot{y} = \text{etc.}$$

Substituting these in (a) and reducing, we find

$$\frac{df}{ds} = \frac{q}{\rho} \qquad\qquad (f),$$

a form of the kinematical equation of a flexible line which will be of much use to us later.

We see, therefore, that if the tangential components of the impressed terminal velocities have any prescribed values, we may give besides, to the ends, any velocities whatever perpendicular to the tangents, without altering the motion acquired by any part of the cord. From this it is clear also, that the directions of the terminal impulses are necessarily tangential; or, in other words, that an impulse inclined to the tangent at either end, would generate an infinite transverse velocity.

To express, then, the terminal conditions, let F and F' be the tangential velocities produced at the ends, which we suppose known. We have, for any point, P, as seen above from (d),

$$f = \frac{1}{\mu}\frac{d\lambda}{ds} \qquad\qquad (g),$$

Generation of motion by impulse in an inextensible cord or chain.

and hence when

$$\left.\begin{array}{l} s=0,\ \dfrac{1}{\mu}\dfrac{d\lambda}{ds}=F \\ s=l,\ \dfrac{1}{\mu}\dfrac{d\lambda}{ds}=F' \end{array}\right\} \qquad (h),$$

and when (second line above)

which suffice to determine the constants of integration of (d). Or if the data are the tangential impulses, I, I', required at the ends to produce the motion, we have

when $s=0,\ \lambda=I,$
and when $s=l,\ \lambda=I'$ $\qquad (i)$.

Or if either end be free, we have $\lambda=0$ at it, and any prescribed condition as to impulse applied, or velocity generated, at the other end.

The solution of this problem is very interesting, as showing how rapidly the propagation of the impulse falls off with "change of direction" along the cord. The reader will have no difficulty in illustrating this by working it out in detail for the case of a cord either uniform or such that $\mu\dfrac{d\frac{1}{\mu}}{ds}$ is constant, and given in the form of a circle or helix. When μ and ρ are constant, for instance, the impulsive tension decreases in the proportion of 1 to ϵ per space along the curve equal to ρ. The results have curious, and dynamically most interesting, bearings on the motions of a whip lash, and of the rope in harpooning a whale.

Example (3). Let a mass of incompressible liquid be given at rest completely filling a closed vessel of any shape; and let, by suddenly commencing to change the shape of this vessel, any arbitrarily prescribed normal velocities be suddenly produced in the liquid at all points of its bounding surface, subject to the condition of not altering the volume: It is required to find the instantaneous velocity of any interior point of the fluid. Impulsive motion of incompressible liquid.

Let x, y, z be the co-ordinates of any point P of the space occupied by the fluid, and let u, v, w be the components of the required velocity of the fluid at this point. Then ρ being the density of the fluid, and $\iiint$ denoting integration throughout the space occupied by the fluid, we have

$$T=\iiint \tfrac{1}{2}\rho\,(u^2+v^2+w^2)\,dxdydz \qquad (a),$$

Impulsive motion of incompressible liquid.

which, subject to the kinematical condition (§ 193),

$$\frac{du}{dx}+\frac{dv}{dy}+\frac{dw}{dz}=0 \qquad (b),$$

must be the least possible, with the given surface values of the normal component velocity. By the method of variation we have

$$\iiint\left\{\rho(u\delta u+v\delta v+w\delta w)+\lambda\left(\frac{d\delta u}{dx}+\frac{d\delta v}{dy}+\frac{d\delta w}{dz}\right)\right\}dxdydz=0 \qquad (c).$$

But integrating by parts we have

$$\iiint\lambda\left(\frac{d\delta u}{dx}+\frac{d\delta v}{dy}+\frac{d\delta w}{dz}\right)dxdydz=\iint\lambda\,(\delta udydz+\delta vdzdx+\delta wdxdy)$$
$$-\iiint\left(\delta u\frac{d\lambda}{dx}+\delta v\frac{d\lambda}{dy}+\delta w\frac{d\lambda}{dz}\right)dxdydz \qquad (d),$$

and if l, m, n denote the direction cosines of the normal at any point of the surface, dS an element of the surface, and $\iint$ integration over the whole surface, we have

$$\iint\lambda\,(\delta udydz+\delta vdzdx+\delta wdxdy)=\iint\lambda\,(l\delta u+m\delta v+n\delta w)\,dS=0,$$

since the normal component of the velocity is given, which requires that $l\delta u+m\delta v+n\delta w=0$. Using this in going back with the result to (c), (d), and equating to zero the coefficients of δu, δv, δw, we find

$$\rho u=\frac{d\lambda}{dx},\quad \rho v=\frac{d\lambda}{dy},\quad \rho w=\frac{d\lambda}{dz} \qquad (e).$$

These, used to eliminate u, v, w from (b), give

$$\frac{d}{dx}\left(\frac{1}{\rho}\frac{d\lambda}{dx}\right)+\frac{d}{dy}\left(\frac{1}{\rho}\frac{d\lambda}{dy}\right)+\frac{d}{dz}\left(\frac{1}{\rho}\frac{d\lambda}{dz}\right)=0 \qquad (f),$$

an equation for the determination of λ, whence by (e) the solution is completed.

The condition to be fulfilled, besides the kinematical equation (b), amounts to this merely,—that $\rho\,(udx+vdy+wdz)$ must be a complete differential. If the fluid is homogeneous, ρ is constant, and $udx+vdy+wdz$ must be a complete differential; in other words, the motion suddenly generated must be of the "non-rotational" character [§ 190, (i)] throughout the fluid mass. The equation to determine λ becomes, in this case,

$$\frac{d^2\lambda}{dx^2}+\frac{d^2\lambda}{dy^2}+\frac{d^2\lambda}{dz^2}=0 \qquad (g).$$

Impulsive motion of incompressible liquid.

From the hydrodynamical principles explained later it will appear that λ, the function of which $\rho\,(udx + vdy + wdz)$ is the differential, is the impulsive pressure at the point (x, y, z) of the fluid. Hence we may infer that the equation (f), with the condition that λ shall have a given value at every point of a certain closed surface, has a possible and a determinate solution for every point within that surface. This is precisely the same problem as the determination of the permanent temperature at any point within a heterogeneous solid of which the surface is kept permanently with any non-uniform distribution of temperature over it, (f) being Fourier's equation for the uniform conduction of heat through a solid of which the conducting power at the point (x, y, z) is $\frac{1}{\rho}$. The possibility and the determinateness of this problem (with an exception regarding multiply continuous spaces, to be fully considered in Vol. II.) were both proved above [Chap. I. App. A, (e)] by a demonstration, the comparison of which with the present is instructive. The other case of superficial condition—that with which we have commenced here—shows that the equation (f), with $l\frac{d\lambda}{dx} + m\frac{d\lambda}{dy} + n\frac{d\lambda}{dz}$ given arbitrarily for every point of the surface, has also (with like qualification respecting multiply continuous spaces) a possible and single solution for the whole interior space. This, as we shall see in examining the mathematical theory of magnetic induction, may also be inferred from the general theorem (e) of App. A above, by supposing a to be zero for all points without the given surface, and to have the value $\frac{1}{\rho}$ for any internal point (x, y, z).

Lagrange's equations of motion in terms of generalized co-ordinates

318. The equations of continued motion of a set of free particles acted on by any forces, or of a system connected in any manner and acted on by any forces, are readily obtained in terms of Lagrange's Generalized Co-ordinates by the regular and direct process of analytical transformation, from the ordinary forms of the equations of motion in terms of Cartesian (or rectilineal rectangular) co-ordinates. It is convenient first to effect the transformation for a set of free particles acted on by any forces. The case of any system with invariable connexions, or with connexions varied in a given manner, is

then to be dealt with by supposing one or more of the generalized co-ordinates to be constant: or to be given functions of the time. Thus the generalized equations of motion are merely those for the reduced number of the co-ordinates remaining un-given; and their integration determines these co-ordinates.

deduced direct by transformation from the equations of motion in terms of Cartesian co-ordinates.

Let m_1, m_2, etc. be the masses, x_1, y_1, z_1, x_2, etc. be the co-ordinates of the particles; and X_1, Y_1, Z_1, X_2, etc. the components of the forces acting upon them. Let ψ, ϕ, etc. be other variables equal in number to the Cartesian co-ordinates, and let there be the same number of relations given between the two sets of variables; so that we may either regard ψ, ϕ, etc. as known functions of x_1, y_1, etc., or x_1, y_1, etc. as known functions of ψ, ϕ, etc. Proceeding on the latter supposition we have the equations (a), (1), of § 313; and we have equations (b), (6), of the same section for the generalized components Ψ, Φ, etc. of the force on the system.

For the Cartesian equations of motion we have

$$X_1 = m_1\frac{d^2x_1}{dt^2},\quad Y_1 = m_1\frac{d^2y_1}{dt^2},\quad Z_1 = m_1\frac{d^2z_1}{dt^2},\quad X_2 = m_2\frac{d^2x_2}{dt^2}\text{ etc.}\ldots(19).$$

Multiplying the first by $\frac{dx_1}{d\psi}$, the second by $\frac{dy_1}{d\psi}$, and so on, and adding all the products, we find by 313 (6)

$$\Psi = m_1\left(\frac{d^2x_1}{dt^2}\frac{dx_1}{d\psi} + \frac{d^2y_1}{dt^2}\frac{dy_1}{d\psi} + \frac{d^2z_1}{dt^2}\frac{dz_1}{d\psi}\right) + m_2(\text{etc.}) + \text{etc.}\ldots(20).$$

Now

$$\frac{d^2x_1}{dt^2}\frac{dx_1}{d\psi} = \frac{d}{dt}\left(\dot{x}_1\frac{dx_1}{d\psi}\right) - \dot{x}_1\frac{d}{dt}\frac{dx_1}{d\psi} = \frac{d}{dt}\left(\dot{x}_1\frac{d\dot{x}_1}{d\dot{\psi}}\right) - \dot{x}_1\frac{d\dot{x}_1}{d\psi}$$

$$= \frac{d}{dt}\left\{\tfrac{1}{2}\frac{d(\dot{x}_1^2)}{d\dot{\psi}}\right\} - \tfrac{1}{2}\frac{d(\dot{x}_1^2)}{d\psi}\ldots\ldots(21).$$

Using this and similar expressions with reference to the other co-ordinates in (20), and remarking that

$$\tfrac{1}{2}m_1(\dot{x}_1^2 + \dot{y}_1^2 + \dot{z}_1^2) + \tfrac{1}{2}m_2(\text{etc.}) + \text{etc.} = T\ldots\ldots(22),$$

if, as before, we put T for the kinetic energy of the system; we find

$$\Psi = \frac{d}{dt}\frac{dT}{d\dot{\psi}} - \frac{dT}{d\psi}\ldots\ldots(23).$$

The substitutions of $\frac{d\dot{x}_1}{d\dot{\psi}}$ for $\frac{dx_1}{d\psi}$ and of $\frac{d\dot{x}_1}{d\psi}$ for $\frac{d}{dt}\frac{dx_1}{d\psi}$ used above, suppose $\dot{x}_1$ to be a function of the co-ordinates, and of the generalized velocity-components, as shown in equations (1) of § 313. It is on this supposition [which makes T a quadratic function of the generalized velocity-components with functions of the co-ordinates as coefficients as shown in § 313 (2)] that the differentiations $\frac{d}{d\dot{\psi}}$ and $\frac{d}{d\psi}$ in (23) are performed. Proceeding similarly with reference to ϕ, etc., we find expressions similar to (23) for Φ, etc., and thus we have for the equations of motion in terms of the generalized co-ordinates

Lagrange's equations of motion in terms of generalized co-ordinates deduced direct by transformation from the equations of motion in terms of Cartesian co-ordinates.

$$\left.\begin{array}{c} \frac{d}{dt}\frac{dT}{d\dot{\psi}} - \frac{dT}{d\psi} = \Psi, \\ \frac{d}{dt}\frac{dT}{d\dot{\phi}} - \frac{dT}{d\phi} = \Phi, \\ \text{etc.} \end{array}\right\} \quad (24).$$

It is to be remarked that there is nothing in the preceding transformation which would be altered by supposing t to appear in the relations between the Cartesian and the generalized co-ordinates: thus if we suppose these relations to be

$$\left.\begin{array}{c} F(x_1, y_1, z_1, x_2, \ldots\ldots \psi, \phi, \theta, \ldots\ldots t) = 0 \\ F_1(x_1, y_1, z_1, x_2, \ldots\ldots \psi, \phi, \theta, \ldots\ldots t) = 0 \\ \text{etc.} \end{array}\right\} \quad (25),$$

we now, instead of § 313 (1), have

$$\left.\begin{array}{c} \dot{x}_1 = \left(\frac{dx_1}{dt}\right) + \frac{dx_1}{d\psi}\dot{\psi} + \frac{dx_1}{d\phi}\dot{\phi} + \text{etc.} \\ \dot{y}_1 = \left(\frac{dy_1}{dt}\right) + \frac{dy_1}{d\psi}\dot{\psi} + \frac{dy_1}{d\phi}\dot{\phi} + \text{etc.} \\ \text{etc.} \end{array}\right\} \quad (26),$$

where $\left(\frac{dx_1}{dt}\right)$ denotes what the velocity-component $\dot{x}_1$ would be if ψ, ϕ, etc. were constant; being analytically the partial differential coefficient with reference to t of the formula derived from (26) to express x_1 as a function of t, ψ, ϕ, θ, etc.

Using (26) in (22) we now find instead of a homogeneous quadratic function of $\dot{\psi}$, $\dot{\phi}$, etc., as in (2) of § 313, a mixed

Lagrange's equations of motion in terms of generalized co-ordinates deduced direct by transformation from the equations of motion in terms of Cartesian co-ordinates.

function of zero degree and first and second degrees, for the kinetic energy, as follows:—

$$T = K + (\psi)\dot{\psi} + (\phi)\dot{\phi} + \ldots + \tfrac{1}{2}\{(\psi,\psi)\dot{\psi}^2 + (\phi,\phi)\dot{\phi}^2 + \ldots 2(\psi,\phi)\dot{\psi}\dot{\phi}\ldots\} \quad \ldots(27),$$

where

$$\left.\begin{aligned} K &= \tfrac{1}{2}\Sigma m\left\{\left(\left(\frac{dx}{dt}\right)\right)^2 + \left(\left(\frac{dy}{dt}\right)\right)^2 + \left(\left(\frac{dz}{dt}\right)\right)^2\right\} \\ (\psi) &= \Sigma m\left\{\left(\frac{dx}{dt}\right)\frac{dx}{d\psi} + \left(\frac{dy}{dt}\right)\frac{dy}{d\psi} + \left(\frac{dz}{dt}\right)\frac{dz}{d\psi}\right\}, \text{ etc.} \\ (\psi,\psi) &= \Sigma m\left\{\left(\frac{dx}{d\psi}\right)^2 + \left(\frac{dy}{d\psi}\right)^2 + \left(\frac{dz}{d\psi}\right)^2\right\}, \text{ etc.} \\ (\psi,\phi) &= \Sigma m\left(\frac{dx}{d\psi}\frac{dx}{d\phi} + \frac{dy}{d\psi}\frac{dy}{d\phi} + \frac{dz}{d\psi}\frac{dz}{d\phi}\right), \text{ etc.} \\ &\text{etc.} \end{aligned}\right\} \ldots(28);$$

K, (ψ), (ϕ), (ψ, ψ), (ψ, ϕ), etc. being thus in general each a known function of t, ψ, ϕ, etc.

Equations (24) above are Lagrange's celebrated equations of motion in terms of generalized co-ordinates. It was first pointed out by Vieille* that they are applicable not only when ψ, ϕ, etc. are related to x_1, y_1, z_1, x_2, etc. by invariable relations as supposed in Lagrange's original demonstration, but also when the relations involve t in the manner shown in equations (25). Lagrange's original demonstration, to be found in the Fourth Section of the Second Part of his *Mécanique Analytique*, consisted of a transformation from Cartesian to generalized co-ordinates of the indeterminate equation of motion; and it is the same demonstration with unessential variations that has been hitherto given, so far as we know, by all subsequent writers including ourselves in our first edition (§ 329). It seems however an unnecessary complication to introduce the indeterminate variations δx, δy, etc.; and we find it much simpler to deduce Lagrange's generalized equations by direct transformation from the equations of motion (19) of a free particle.

* Sur les équations différentielles de la dynamique, *Liouville's Journal*, 1849, p. 201.

When the kinematic relations are invariable, that is to say when t does not appear in the equations of condition (25), we find from (27) and (28), Lagrange's generalized form of the equations of motion expanded.

$$T = \tfrac{1}{2}\{(\psi,\psi)\dot\psi^2 + 2(\psi,\phi)\dot\psi\dot\phi + (\phi,\phi)\dot\phi^2 + \ldots\} \ldots\ldots(29),$$

$$\left.\begin{aligned}\frac{d}{dt}\frac{dT}{d\dot\psi} &= (\psi,\psi)\ddot\psi + (\psi,\phi)\ddot\phi + \ldots\\ &\quad + \left\{\frac{d(\psi,\psi)}{d\psi}\dot\psi + \frac{d(\psi,\psi)}{d\phi}\dot\phi + \ldots\right\}\dot\psi\\ &\quad + \left\{\frac{d(\psi,\phi)}{d\psi}\dot\psi + \frac{d(\psi,\phi)}{d\phi}\dot\phi + \ldots\right\}\dot\phi\\ &\quad + \ldots\ldots\ldots\ldots\ldots\ldots\end{aligned}\right\} \ldots\ldots\ldots(29'),$$

and

$$\frac{dT}{d\psi} = \tfrac{1}{2}\left\{\frac{d(\psi,\psi)}{d\psi}\dot\psi^2 + 2\frac{d(\psi,\phi)}{d\psi}\dot\psi\dot\phi + \frac{d(\phi,\phi)}{d\psi}\dot\phi^2 + \ldots\right\} \ (29'').$$

Hence the ψ-equation of motion expanded in this, the most important class of cases, is as follows:

$$\left.\begin{aligned}&(\psi,\psi)\ddot\psi + (\psi,\phi)\ddot\phi + \ldots + Q_\psi(T) = \Psi,\\ &\text{where}\\ &Q_\psi(T) = \tfrac{1}{2}\left\{\frac{d(\psi,\psi)}{d\psi}\dot\psi^2 + 2\frac{d(\psi,\psi)}{d\phi}\dot\psi\dot\phi + \left[2\frac{d(\psi,\phi)}{d\phi} - \frac{d(\phi,\phi)}{d\psi}\right]\dot\phi^2 + \ldots\right\}\\ &\ldots\ldots\ldots\ldots\ldots\ldots(29''').\end{aligned}\right\}$$

Remark that $Q_\psi(T)$ is a quadratic function of the velocity-components derived from that which expresses the kinetic energy (T) by the process indicated in the second of these equations, in which ψ appears singularly, and the other co-ordinates symmetrically with one another.

Multiply the ψ-equation by $\dot\psi$, the ϕ-equation by $\dot\phi$, and so on; and add. In what comes from $Q_\psi(T)$ we find terms Equation of energy.

$$+2\frac{d(\psi,\psi)}{d\phi}\dot\psi\dot\phi\,.\,\dot\psi, \text{ and} -\frac{d(\psi,\psi)}{d\phi}\dot\psi^2\,.\,\dot\phi;$$

which together yield $+\dfrac{d(\psi,\psi)}{d\phi}\dot\psi^2\,.\,\dot\phi.$

With this, and the rest simply as shown in (29'''), we find

$$\begin{aligned}&[(\psi,\psi)\ddot\psi + (\psi,\phi)\ddot\phi + \ldots]\dot\psi\\ &+ [(\psi,\phi)\ddot\psi + (\phi,\phi)\ddot\phi + \ldots]\dot\phi\\ &+ \ldots\ldots\ldots\ldots\ldots\ldots\end{aligned}$$

Equation of energy.

$$+\frac{dT}{d\psi}\dot{\psi}+\frac{dT}{d\phi}\dot{\phi}+\ldots \qquad =\Psi\dot{\psi}+\Phi\dot{\phi}+\ldots\ldots\ldots\ldots(29^{iv}),$$

or

$$\frac{dT}{dt}=\Psi\dot{\psi}+\Phi\dot{\phi}+\ldots\ldots\ldots\ldots\ldots(29^{v}).$$

Hamilton's form.

When the kinematical relations are invariable, that is to say, when t does not appear in the equations of condition (25), the equations of motion may be put under a slightly different form first given by Hamilton, which is often convenient; thus:—Let T, $\dot{\psi}$, $\dot{\phi}$,..., be expressed in terms of ξ, η,..., the impulses required to produce the motion from rest at any instant [§ 313 (d)]; so that T will now be a homogeneous quadratic function, and $\dot{\psi}$, $\dot{\phi}$, ... each a linear function, of these elements, with coefficients—functions of ψ, ϕ, etc., depending on the kinematical conditions of the system, but not on the particular motion. Thus, denoting, as in § 322 (29), by ∂, partial differentiation with reference to ξ, η, ..., ψ, ϕ,..., considered as independent variables, we have [§ 313 (10)]

$$\dot{\psi}=\frac{\partial T}{d\xi}, \qquad \dot{\phi}=\frac{\partial T}{d\eta}, \ldots\ldots\ldots\ldots(30),$$

and, allowing d to denote, as in what precedes, the partial differentiations with reference to the system $\dot{\psi}$, $\dot{\phi}$, ..., ψ, ϕ, ..., we have [§ 313 (8)]

$$\xi=\frac{dT}{d\dot{\psi}}, \qquad \eta=\frac{dT}{d\dot{\phi}}, \ldots\ldots\ldots\ldots(31).$$

The two expressions for T being, as above, § 313,

$$T=\tfrac{1}{2}\{(\psi,\psi)\dot{\psi}^2+\ldots+2(\psi,\phi)\dot{\psi}\dot{\phi}+\ldots\}=\tfrac{1}{2}\{[\psi,\psi]\xi^2+\ldots+2[\psi,\phi]\xi\eta+\ldots\}(32),$$

the second of these is to be obtained from the first by substituting for $\dot{\psi}$, $\dot{\phi}$...., their expressions in terms of ξ, η, ... Hence

$$\frac{\partial T}{d\psi}=\frac{dT}{d\psi}+\frac{dT}{d\dot{\psi}}\frac{\partial\dot{\psi}}{d\psi}+\frac{dT}{d\dot{\phi}}\frac{\partial\dot{\phi}}{d\psi}+\ldots=\frac{dT}{d\psi}+\xi\frac{\partial}{d\psi}\frac{\partial T}{d\xi}+\eta\frac{\partial}{d\psi}\frac{\partial T}{d\eta}+\ldots$$

$$=\frac{dT}{d\psi}+\frac{\partial}{d\psi}\left(\xi\frac{\partial T}{d\xi}+\eta\frac{\partial T}{d\eta}+\ldots\right)=\frac{dT}{d\psi}+2\frac{\partial T}{d\psi}.$$

From this we conclude

$$\frac{\partial T}{d\psi}=-\frac{dT}{d\psi}\text{; and, similarly, }\frac{\partial T}{d\phi}=-\frac{dT}{d\phi},\text{ etc. }\ldots\ldots(33).$$

Hence Lagrange's equations become

$$\frac{d\xi}{dt}+\frac{\partial T}{d\psi}=\Psi,\text{ etc.}\ldots\ldots\ldots\ldots\ldots(34).$$

In § 327 below a purely analytical proof will be given of Lagrange's generalized equations of motion, establishing them directly as a deduction from the principle of "Least Action," independently of any expression either of this principle or of the equations of motion in terms of Cartesian co-ordinates. In their Hamiltonian form they are also deduced in § 330 (33) from the principle of Least Action ultimately, but through the beautiful "Characteristic Equation" of Hamilton. Hamilton's form.

319. Hamilton's form of Lagrange's equations of motion in terms of generalized co-ordinates expresses that what is required to prevent any one of the components of momentum from varying is a corresponding component force equal in amount to the rate of change of the kinetic energy per unit increase of the corresponding co-ordinate, with all components of momentum constant: and that whatever is the amount of the component force, its excess above this value measures the rate of increase of the component momentum.

In the case of a conservative system, the same statement takes the following form:—The rate at which any component momentum increases per unit of time is equal to the rate, per unit increase of the corresponding co-ordinate, at which the sum of the potential energy, and the kinetic energy for constant momentums, diminishes. This is the celebrated "canonical form" of the equations of motion of a system, though why it has been so called it would be hard to say.

Let V denote the potential energy, so that [§ 293 (3)] "Canonical form" of Hamilton's general equations of motion of a conservative system.

$$\Psi\delta\psi + \Phi\delta\phi + \ldots = -\delta V,$$

and therefore

$$\Psi = -\frac{dV}{d\psi}, \quad \Phi = -\frac{dV}{d\phi}, \ \ldots$$

Let now U denote the algebraic expression for the sum of the potential energy, V, in terms of the co-ordinates, ψ, ϕ..., and the kinetic energy, T, in terms of the co-ordinates and the components of momentum, ξ, η,.... Then

$$\left.\begin{aligned} \frac{d\xi}{dt} &= -\frac{\partial U}{d\psi}, \text{ etc.} \\ \text{also} \quad \frac{d\psi}{dt} &= \frac{\partial U}{d\xi}, \text{ etc.} \end{aligned}\right\} \quad \ldots\ldots\ldots\ldots(35),$$

the latter being equivalent to (30), since the potential energy does not contain ξ, η, etc.

In the following examples we shall adhere to Lagrange's form (24), as the most convenient for such applications.

Examples of the use of Lagrange's generalized equations of motion;—polar co-ordinates.

Example (A).—Motion of a single point (m) referred to polar co-ordinates (r, θ, ϕ). From the well-known geometry of this case we see that δr, $r\delta\theta$, and $r\sin\theta\delta\phi$ are the amounts of linear displacement corresponding to infinitely small increments, δr, $\delta\theta$, $\delta\phi$, of the co-ordinates: also that these displacements are respectively in the direction of r, of the arc $r\delta\theta$ (of a great circle) in the plane of r and the pole, and of the arc $r\sin\theta\delta\phi$ (of a small circle in a plane perpendicular to the axis); and that they are therefore at right angles to one another. Hence if F, G, H denote the components of the force experienced by the point, in these three rectangular directions, we have

$$F = R,\ Gr = \Theta,\ \text{and}\ Hr\sin\theta = \Phi\,;$$

R, Θ, Φ being what the generalized components of force (§ 313) become for this particular system of co-ordinates. We also see that $\dot{r}$, $r\dot{\theta}$, and $r\sin\theta\dot{\phi}$ are three components of the velocity, along the same rectangular directions. Hence

$$T = \tfrac{1}{2}m(\dot{r}^2 + r^2\dot{\theta}^2 + r^2\sin^2\theta\dot{\phi}^2).$$

From this we have

$$\frac{dT}{d\dot{r}} = m\dot{r},\ \frac{dT}{d\dot{\theta}} = mr^2\dot{\theta},\ \frac{dT}{d\dot{\phi}} = mr^2\sin^2\theta\dot{\phi}\,;$$

$$\frac{dT}{dr} = mr(\dot{\theta}^2 + \sin^2\theta\dot{\phi}^2),\ \frac{dT}{d\theta} = mr^2\sin\theta\cos\theta\dot{\phi}^2,\ \frac{dT}{d\phi} = 0.$$

Hence the equations of motion become

$$m\left\{\frac{d\dot{r}}{dt} - r(\dot{\theta}^2 + \sin^2\theta\dot{\phi}^2)\right\} = F,$$

$$m\left\{\frac{d(r^2\dot{\theta})}{dt} - r^2\sin\theta\cos\theta\dot{\phi}^2\right\} = Gr,$$

$$m\frac{d(r^2\sin^2\theta\dot{\phi})}{dt} = Hr\sin\theta\,;$$

or, according to the ordinary notation of the differential calculus,

$$m\left\{\frac{d^2r}{dt^2} - r\left(\frac{d\theta^2}{dt^2} + \sin^2\theta\frac{d\phi^2}{dt^2}\right)\right\} = F,$$

$$m\left\{\frac{d}{dt}\left(r^2\frac{d\theta}{dt}\right) - r^2\sin\theta\cos\theta\frac{d\phi^2}{dt^2}\right\} = Gr,$$

$$m\frac{d}{dt}\left(r^2\sin^2\theta\frac{d\phi}{dt}\right) = Hr\sin\theta.$$

Examples of the use of Lagrange's generalized equations of motion;—polar co-ordinates.

If the motion is confined to one plane, that of r, θ, we have $\frac{d\phi}{dt} = 0$, and therefore $H = 0$, and the two equations of motion which remain are

$$m\left(\frac{d^2r}{dt^2} - r\frac{d\theta^2}{dt^2}\right) = F, \quad m\frac{d}{dt}\left(r^2\frac{d\theta}{dt}\right) = Gr.$$

These equations might have been written down at once in terms of the second law of motion from the kinematical investigation of § 32, in which it was shown that $\frac{d^2r}{dt^2} - r\frac{d\theta^2}{dt^2}$, and $\frac{1}{r}\frac{d}{dt}\left(r^2\frac{d\theta}{dt}\right)$ are the components of acceleration along and perpendicular to the radius-vector, when the motion of a point in a plane is expressed according to polar co-ordinates, r, θ.

The same equations, with ϕ instead of θ, are obtained from the polar equations in three dimensions by putting $\theta = \frac{1}{2}\pi$, which implies that $G = 0$, and confines the motion to the plane (r, ϕ).

Example (B).—Two particles are connected by a string; one of them, m, moves in any way on a smooth horizontal plane, and the string, passing through a smooth infinitely small aperture in this plane, bears the other particle m', hanging vertically downwards, and only moving in this vertical line: (the string remaining always stretched in any practical illustration, but, in the problem, being of course supposed capable of transmitting negative tension with its two parts straight.) Let l be the whole length of the string, r that of the part of it from m to the aperture in the plane, and let θ be the angle between the direction of r and a fixed line in the plane. We have

Dynamical problem.

$$T = \tfrac{1}{2}\{m(\dot{r}^2 + r^2\dot{\theta}^2) + m'\dot{r}^2\},$$

$$\frac{dT}{d\dot{r}} = (m + m')\dot{r}, \quad \frac{dT}{d\dot{\theta}} = mr^2\dot{\theta},$$

$$\frac{dT}{dr} = mr\dot{\theta}^2, \quad \frac{dT}{d\theta} = 0.$$

Also, there being no other external force than gm', the weight of the second particle,

$$R = -gm', \quad \Theta = 0.$$

Examples of the use of Lagrange's generalized equations of motion; dynamical problem.

Hence the equations of motion are

$$(m+m')\ddot{r}-mr\dot{\theta}^2=-m'g, \qquad m\frac{d(r^2\dot{\theta})}{dt}=0.$$

The motion of m' is of course that of a particle influenced only by a force towards a fixed centre; but the law of this force, P (the tension of the string), is remarkable. To find it we have (§ 32), $P=m(-\ddot{r}+r\dot{\theta}^2)$. But, by the equations of the motion,

$$\ddot{r}-r\dot{\theta}^2=-\frac{m'}{m+m'}(g+r\dot{\theta}^2), \text{ and } \dot{\theta}=\frac{h}{mr^2},$$

where h (according to the usual notation) denotes the moment of momentum of the motion, being an arbitrary constant of integration. Hence

$$P=\frac{mm'}{m+m'}\left(g+\frac{h^2}{m^2}r^{-3}\right).$$

Case of stable equilibrium due to motion.

The particular case of projection which gives m a circular motion and leaves m' at rest is interesting, inasmuch as (§ 350, below) the motion of m is stable, and therefore m' is in stable equilibrium.

Examples continued; C (a), folding door.

Example (C).—A rigid body m is supported on a fixed axis, and another rigid body n is supported on the first, by another axis; the motion round each axis being perfectly free.

Case (a).—*The second axis parallel to the first.* At any time, t, let ϕ and ψ be the inclinations of a fixed plane through the first axis to the plane of it and the second axis, and to a plane through the second axis and the centre of inertia of the second body. These two co-ordinates, ϕ, ψ, it is clear, completely specify the configuration of the system. Now let a be the distance of the second axis from the first, and b that of the centre of inertia of the second body from the second axis. The velocity of the second axis will be $a\dot{\phi}$; and the velocity of the centre of inertia of the second body will be the resultant of two velocities

$$a\dot{\phi}, \text{ and } b\dot{\psi},$$

in lines inclined to one another at an angle equal to $\psi-\phi$, and its square will therefore be equal to

$$a^2\dot{\phi}^2+2ab\dot{\phi}\dot{\psi}\cos(\psi-\phi)+b^2\dot{\psi}^2.$$

Hence, if m and n denote the masses, j the radius of gyration of the first body about the fixed axis, and k that of the second body about a parallel axis through its centre of inertia; we have, according to §§ 280, 281, **Examples continued; C (a, folding door.**

$$T = \tfrac{1}{2}\{mj^2\dot{\phi}^2 + n[a^2\dot{\phi}^2 + 2ab\dot{\phi}\dot{\psi}\cos(\psi - \phi) + b^2\dot{\psi}^2 + k^2\dot{\psi}^2]\}.$$

Hence we have,

$$\frac{dT}{d\dot{\phi}} = mj^2\dot{\phi} + na^2\dot{\phi} + nab\cos(\psi - \phi)\dot{\psi};\ \frac{dT}{d\dot{\psi}} = nab\cos(\psi - \phi)\dot{\phi} + n(b^2 + k^2)\dot{\psi};$$

$$\frac{dT}{d\phi} = -\frac{dT}{d\psi} = nab\sin(\psi - \phi)\dot{\phi}\dot{\psi}.$$

The most general supposition we can make as to the applied forces, is equivalent to assuming a couple, Φ, to act on the first body, and a couple, Ψ, on the second, each in a plane perpendicular to the axes; and these are obviously what the generalized components of stress become in this particular co-ordinate system, ϕ, ψ. Hence the equations of motion are

$$(mj^2 + na^2)\ddot{\phi} + nab\,\frac{d[\dot{\psi}\cos(\psi - \phi)]}{dt} - nab\sin(\psi - \phi)\dot{\phi}\dot{\psi} = \Phi,$$

$$nab\,\frac{d[\dot{\phi}\cos(\psi - \phi)]}{dt} + n(b^2 + k^2)\ddot{\psi} + nab\sin(\psi - \phi)\dot{\phi}\dot{\psi} = \Psi.$$

If there is no other applied force than gravity, and if, as we may suppose without losing generality, the two axes are horizontal, the potential energy of the system will be

$$gmh(1 - \cos\phi) + gn\{a[1 - \cos(\phi + A)] + b[1 - \cos(\psi + A)]\},$$

the distance of the centre of inertia of the first body from the fixed axis being denoted by h, the inclination of the plane through the fixed axis and the centre of inertia of the first body, to the plane of the two axes, being denoted by A, and the fixed plane being so taken that $\phi = 0$ when the former plane is vertical. By differentiating this, with reference to ϕ and ψ, we therefore have

$$-\Phi = gmh\sin\phi + gna\sin(\phi + A),\quad -\Psi = gnb\sin(\psi + A).$$

We shall examine this case in some detail later, in connexion with the interference of vibrations, a subject of much importance in physical science.

When there are no applied or intrinsic working forces, we have $\Phi = 0$ and $\Psi = 0$: or, if there are mutual forces between the two bodies, but no forces applied from without, $\Phi + \Psi = 0$. In

Examples the use of Lagrange's generalized equations of motion.

either of these cases we have the following first integral:—

$$(mj^2 + na^2)\dot{\phi} + m'ab\cos(\psi-\phi)(\dot{\phi}+\dot{\psi}) + n(b^2+k^2)\dot{\psi} = C;$$

obtained by adding the two equations of motion and integrating. This, which clearly expresses the constancy of the whole moment of momentum, gives $\dot{\phi}$ and $\dot{\psi}$ in terms of $(\dot{\psi}-\dot{\phi})$ and $(\psi-\phi)$. Using these in the integral equation of energy, provided the mutual forces are functions of $\psi-\phi$, we have a single equation between $\frac{d(\psi-\phi)}{dt}$, $(\psi-\phi)$, and constants, and thus the full solution of the problem is reduced to quadratures. [It is worked out fully below, as Sub-example G_1.]

C (b). Motion of governing masses in Watt's centrifugal governor: also of gimballed compass-bowl.

Case (*b*).—*The second axis perpendicular to the first.* For simplicity suppose the pivoted axis of the second body, n, to be a principal axis relatively [§ 282 Def. (2)] to the point, N, in which it is cut by a plane perpendicular to it through the fixed axis of the first body, m. Let NE and NF be n's two other principal axes. Denote now by

h the distance from N to m's fixed axis;

k, e, f the radii of gyration of n round its three principal axes through N;

j the radius of gyration of m round its fixed axis;

θ the inclination of NE to m's fixed axis;

ψ the inclination of the plane parallel to n's pivoted axis through m's fixed axis, to a fixed plane through the latter.

Remarking that the component angular velocities of n round NE and NF are $\dot{\psi}\cos\theta$ and $\dot{\psi}\sin\theta$, we find immediately

$$T = \tfrac{1}{2}\{[mj^2 + n(h^2 + e^2\cos^2\theta + f^2\sin^2\theta)]\dot{\psi}^2 + nk^2\dot{\theta}^2\},$$

or, if we put

$$mj^2 + n(h^2+f^2) = G,\quad n(e^2-f^2) = D;$$

$$T = \tfrac{1}{2}\{(G + D\cos^2\theta)\dot{\psi}^2 + nk^2\dot{\theta}^2\}.$$

The farther working out of this case we leave as a simple but most interesting exercise for the student. We may return to it later, as its application to the theory of centrifugal chronometric regulators is very important.

Example (C′). Take the case C (b) and mount a third body M upon an axis OC fixed relatively to n in any position parallel to NE. Suppose for simplicity O to be the centre of inertia of M and OC one of its principal axes; and let OA, OB be its two other principal axes relative to O. The notation being in other respects the same as in Example C (b), denote now farther by A, B, C the moments of inertia of M round OA, OB, OC; ϕ the angle between the plane AOC and the plane through the fixed axis of m perpendicular to the pivoted axis of n; ϖ, ρ, σ the component angular velocities of M round OA, OB, OC.

Motion of a rigid body pivoted on one of its principal axes mounted on a gimballed bowl.

In the annexed diagram, taken from § 101 above, ZCZ' is a

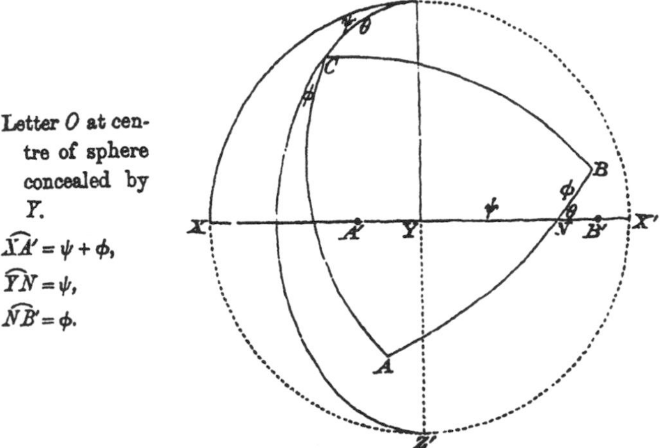

Letter O at centre of sphere concealed by Y.

$\widehat{XA'} = \psi + \phi$,

$\widehat{YN} = \psi$,

$\widehat{NB'} = \phi$.

circle of unit radius having its centre at O and its plane parallel to the fixed axis of m and perpendicular to the pivoted axis of n.

The component velocities of C in the direction of the arc ZC and perpendicular to it are $\dot{\theta}$ and $\dot{\psi}\sin\theta$; and the component angular velocity of the plane ZCZ' round OC is $\dot{\psi}\cos\theta$. Hence

$$\varpi = \dot{\theta}\sin\phi - \dot{\psi}\sin\theta\cos\phi,$$

$$\rho = \dot{\theta}\cos\phi + \dot{\psi}\sin\theta\sin\phi,$$

and

$$\sigma = \dot{\psi}\cos\theta + \dot{\phi}.$$

[Compare § 101.]

Motion of a rigid body pivoted on one of its principal axes mounted on a gimballed bowl.

The kinetic energy of the motion of M relatively to O, its centre of inertia, is (§ 281)

$$\tfrac{1}{2}(A\varpi^2 + B\rho^2 + C\sigma^2);$$

and (§ 280) its whole kinetic energy is obtained by adding the kinetic energy of a material point equal to its mass moving with the velocity of its centre of inertia. This latter part of the kinetic energy of M is most simply taken into account by supposing n to include a material point equal to M placed at O; and using the previous notation k, e, f for radii of gyration of n on the understanding that n now includes this addition. Hence for the present example, with the preceding notation G, D, we have

$$T = \tfrac{1}{2}\{(G + D\cos^2\theta)\dot\psi^2 + nk^2\dot\theta^2\}$$
$$+ A(\dot\theta\sin\phi - \dot\psi\sin\theta\cos\phi)^2 + B(\dot\theta\cos\phi + \dot\psi\sin\theta\sin\phi)^2$$
$$+ C(\dot\psi\cos\theta + \dot\phi)^2\}.$$

Rigid body rotating freely; referred to the ψ, ϕ, θ co-ordinates (§ 101).

From this the three equations of motion are easily written down.

By putting $G = 0$, $D = 0$, and $k = 0$, we have the case of the motion of a free rigid body relatively to its centre of inertia.

Gyroscopes and gyrostats.

By putting $B = A$ we fall on a case which includes gyroscopes and gyrostats of every variety; and have the following much simplified formula:

$$T = \tfrac{1}{2}\{[G + A + (D - A)\cos^2\theta]\dot\psi^2 + (nk^2 + A)\dot\theta^2 + C(\dot\psi\cos\theta + \dot\phi)^2\},$$

or

$$T = \tfrac{1}{2}\{(E + F\cos^2\theta)\dot\psi^2 + (nk^2 + A)\dot\theta^2 + C(\dot\psi\cos\theta + \dot\phi)^2\},$$

if we put $E = G + A$, and $F = D - A$.

Gyroscopic pendulum.

Example (D).—*Gyroscopic pendulum.*—A rigid body, P, is attached to one axis of a universal flexure joint (§ 109), of which the other is held fixed, and a second body, Q, is supported on P by a fixed axis, in line with, or parallel to, the first-mentioned arm of the joint. For simplicity, we shall suppose Q to be kinetically symmetrical about its bearing axis, and OB to be a principal axis of an ideal rigid body, PQ, composed of P and a mass so distributed along the bearing axis of the actual body Q as to have the same centre of inertia and the same moments of inertia round axes perpendicular to it. Let AO be the fixed arm, O the joint, OB the movable arm bearing the body P, and coinciding with, or parallel to, the axis of Q. Let $BOA' = \theta$; let ϕ be the angle which the plane AOB makes with a fixed plane of reference, through OA, chosen so as to contain a second principal axis of the imagined rigid body, PQ, when OB is placed in line with AO; and let ψ be the angle between a plane of reference in Q through its axis of symmetry and the plane of the two principal axes of PQ already mentioned. These three co-ordinates (θ, ϕ, ψ) clearly specify the configuration of the system at any time, t. Let the moments of inertia of the imagined rigid body PQ, round its principal axis OB, the other principal axis referred to above, and the remaining one, be denoted by $\mathfrak{A}$, $\mathfrak{B}$, $\mathfrak{C}$ respectively; and let $\mathfrak{A}'$ be the moment of inertia of Q round its bearing axis.

Gyroscopic pendulum.

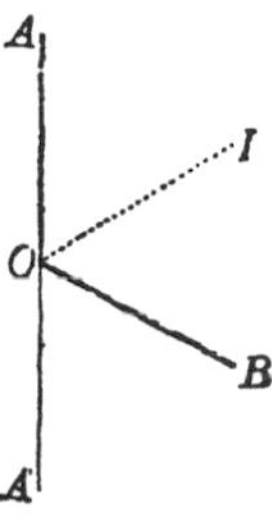

We have seen (§ 109) that, with the kind of joint we have supposed at O, every possible motion of a body rigidly connected with OB, is resolvable into a rotation round OI, the line bisecting the angle AOB, and a rotation round the line through O perpendicular to the plane AOB. The angular velocity of the latter is $\dot{\theta}$, according to our present notation. The former would give to any point in OB the same absolute velocity by rotation round OI, that it has by rotation with angular velocity $\dot{\phi}$ round AA'; and is therefore equal to

$$\frac{\sin A'OB}{\sin IOB}\dot{\phi} = \frac{\sin\theta}{\cos\frac{1}{2}\theta}\dot{\phi} = 2\dot{\phi}\sin\tfrac{1}{2}\theta.$$

This may be resolved into $2\dot{\phi}\sin^2\frac{1}{2}\theta = \dot{\phi}(1-\cos\theta)$ round OB, and $2\dot{\phi}\sin\frac{1}{2}\theta\cos\frac{1}{2}\theta = \dot{\phi}\sin\theta$ round the perpendicular to OB, in plane AOB. Again, in virtue of the symmetrical character of the joint with reference to the line OI, the angle ϕ, as defined above, will be equal to the angle between the plane of the two first-mentioned principal axes of body P, and the plane AOB. Hence the axis of the angular velocity $\dot{\phi}\sin\theta$, is inclined to the principal axis of moment $\mathfrak{B}$ at an angle equal to ϕ. Resolving therefore this angular velocity, and $\dot{\theta}$, into components round the axes of $\mathfrak{B}$ and $\mathfrak{C}$, we find, for the whole component angular velocities of the imagined rigid body PQ, round these axes, $\dot{\phi}\sin\theta\cos\phi + \dot{\theta}\sin\phi$, and $-\dot{\phi}\sin\theta\sin\phi + \dot{\theta}\cos\phi$, respectively. The whole kinetic energy, T, is composed of that of the imagined rigid body PQ, and that of Q about axes through its centre of

Gyroscopic pendulum.

inertia: we therefore have

$$2T = \mathfrak{A}(1-\cos\theta)^2\dot{\phi}^2 + \mathfrak{B}(\dot{\phi}\sin\theta\cos\phi + \dot{\theta}\sin\phi)^2 + \mathfrak{C}(\dot{\phi}\sin\theta\sin\phi - \dot{\theta}\cos\phi)^2 + \mathfrak{A}'\{\dot{\psi} - \dot{\phi}(1-\cos\theta)\}^2.$$

Hence $\dfrac{dT}{d\dot{\psi}} = \mathfrak{A}'\{\dot{\psi} - \dot{\phi}(1-\cos\theta)\}$, $\dfrac{dT}{d\psi} = 0$,

$$\frac{dT}{d\dot{\phi}} = \mathfrak{A}(1-\cos\theta)^2\dot{\phi} + \mathfrak{B}(\dot{\phi}\sin\theta\cos\phi + \dot{\theta}\sin\phi)\sin\theta\cos\phi$$
$$+ \mathfrak{C}(\dot{\phi}\sin\theta\sin\phi - \dot{\theta}\cos\phi)\sin\theta\sin\phi - \mathfrak{A}'\{\dot{\psi} - \dot{\phi}(1-\cos\theta)\}(1-\cos\theta),$$

$$\frac{dT}{d\phi} = -\mathfrak{B}(\dot{\phi}\sin\theta\cos\phi + \dot{\theta}\sin\phi)(\dot{\phi}\sin\theta\sin\phi - \dot{\theta}\cos\phi)$$
$$+ \mathfrak{C}(\dot{\phi}\sin\theta\sin\phi - \dot{\theta}\cos\phi)(\dot{\phi}\sin\theta\cos\phi + \dot{\theta}\sin\phi),$$

$$\frac{dT}{d\dot{\theta}} = \mathfrak{B}(\dot{\phi}\sin\theta\cos\phi + \dot{\theta}\sin\phi)\sin\phi - \mathfrak{C}(\dot{\phi}\sin\theta\sin\phi - \dot{\theta}\cos\phi)\cos\phi$$

and
$$\frac{dT}{d\theta} = \mathfrak{A}(1-\cos\theta)\sin\theta\dot{\phi}^2 + \mathfrak{B}\cos\theta\cos\phi\dot{\phi}(\dot{\phi}\sin\theta\cos\phi + \dot{\theta}\sin\phi)$$
$$+ \mathfrak{C}\cos\theta\sin\phi\dot{\phi}(\dot{\phi}\sin\theta\sin\phi - \dot{\theta}\cos\phi) - \mathfrak{A}'\sin\theta\dot{\phi}\{\dot{\psi} - (1-\cos\theta)\dot{\phi}\}.$$

Now let a couple, G, act on the body Q, in a plane perpendicular to its axis, and let L, M, N act on P, in the plane perpendicular to OB, in the plane $A'OB$, and in the plane through OB perpendicular to the diagram. If ψ is kept constant, and ϕ varied, the couple G will do or resist work in simple addition with L. Hence, resolving $L+G$ and N into components round OI, and perpendicular to it, rejecting the latter, and remembering that $2\sin\frac{1}{2}\theta\dot{\phi}$ is the angular velocity round OI, we have

$$\Phi = 2\sin\tfrac{1}{2}\theta\{-(L+G)\sin\tfrac{1}{2}\theta + N\cos\tfrac{1}{2}\theta\} = \{-(L+G)(1-\cos\theta) + N\sin\theta\}.$$

Also, obviously

$$\Psi = G, \quad \Theta = M.$$

Using these several expressions in Lagrange's general equations (24), we have the equations of motion of the system. They will be of great use to us later, when we shall consider several particular cases of remarkable interest and of very great importance.

Example of varying relation without constraint (rotating axes).

Example (E).—*Motion of a free particle referred to rotating axes.* Let x, y, z be the co-ordinates of a moving particle referred to axes rotating with a constant or varying angular velocity round the axis OZ. Let x_1, y_1, z, be its co-ordinates referred to the same axis, OZ, and two axes OX_1, OY_1, fixed in the plane perpendicular to it. We have

Example of varying relation without constraint (rotating axes).

$$x_1 = x \cos a - y \sin a, \quad y_1 = x \sin a + y \cos a;$$
$$\dot{x}_1 = \dot{x} \cos a - \dot{y} \sin a - (x \sin a + y \cos a)\, \dot{a}, \quad \dot{y}_1 = \text{etc.}$$

where a, the angle X_1OX, must be considered as a given function of t. Hence

$$T = \tfrac{1}{2} m \{\dot{x}^2 + \dot{y}^2 + \dot{z}^2 + 2(x\dot{y} - y\dot{x})\,\dot{a} + (x^2 + y^2)\,\dot{a}^2\},$$

$$\frac{dT}{d\dot{x}} = m(\dot{x} - y\dot{a}), \quad \frac{dT}{d\dot{y}} = m(\dot{y} + x\dot{a}), \quad \frac{dT}{d\dot{z}} = m\dot{z},$$

$$\frac{dT}{dx} = m(\dot{y}\dot{a} + x\dot{a}^2), \quad \frac{dT}{dy} = m(-\dot{x}\dot{a} + y\dot{a}^2), \quad \frac{dT}{dz} = 0.$$

Also,

$$\frac{d}{dt}\frac{dT}{d\dot{x}} = m(\ddot{x} - \dot{y}\dot{a} - y\ddot{a}), \quad \frac{d}{dt}\frac{dT}{d\dot{y}} = m(\ddot{y} + \dot{x}\dot{a} + x\ddot{a}),$$

and hence the equations of motion are

$$m(\ddot{x} - 2\dot{y}\dot{a} - x\dot{a}^2 - y\ddot{a}) = X, \quad m(\ddot{y} + 2\dot{x}\dot{a} - y\dot{a}^2 + x\ddot{a}) = Y, \quad m\ddot{z} = Z,$$

X, Y, Z denoting simply the components of the force on the particle, parallel to the moving axes at any instant. In this example t enters into the relation between fixed rectangular axes and the co-ordinate system to which the motion is referred; but there is no constraint. The next is given as an example of varying, or kinetic, constraint.

Example of varying relation due to kinetic constraint.

Example (F).—*A particle, influenced by any forces, and attached to one end of a string of which the other is moved with any constant or varying velocity in a straight line.* Let θ be the inclination of the string at time t, to the given straight line, and ϕ the angle between two planes through this line, one containing the string at any instant, and the other fixed. These two co-ordinates (θ, ϕ) specify the position, P, of the particle at any instant, the length of the string being a given constant, a, and the distance OE, of its other end E, from a fixed point, O, of the line in which it is moved, being a given function of t, which we shall denote by u. Let x, y, z be the co-ordinates of the particle referred to three fixed rectangular axes. Choosing OX as the given straight line, and YOX the fixed plane from which ϕ is measured, we have

$$x = u + a \cos\theta, \quad y = a \sin\theta \cos\phi, \quad z = a \sin\theta \sin\phi,$$
$$\dot{x} = \dot{u} - a \sin\theta\,\dot{\theta};$$

Example of varying relation due to kinetic constraint.

and for $\dot{y}$, $\dot{z}$ we have the same expressions as in Example (A). Hence

$$T = \mathfrak{T} + \tfrac{1}{2}m\,(\dot{u}^2 - 2\dot{u}\dot{\theta}a\sin\theta)$$

where $\mathfrak{T}$ denotes the same as the T of Example (A), with $\dot{r} = 0$, and $r = a$. Hence, denoting as there, by G and H the two components of the force on the particle, perpendicular to EP, respectively in the plane of θ and perpendicular to it, we find, for the two required equations of motion,

$$m\{a(\ddot{\theta} - \sin\theta\cos\theta\dot{\phi}^2) - \sin\theta\,\ddot{u}\} = G, \text{ and } ma\frac{d(\sin^2\theta\dot{\phi})}{dt} = H.$$

These show that the motion is the same as if E were fixed, and a force equal to $-m\ddot{u}$ were applied to the particle in a direction parallel to EX; a result that might have been arrived at at once by superimposing on the whole system an acceleration equal and opposite to that of E, to effect which on P the force $-m\ddot{u}$ is required.

Example (F). Any case of varying relations such that in 318 (27) the coefficients (ψ, ψ), (ψ, ϕ) ... are independent of t. Let $\mathfrak{T}$ denote the quadratic part, L the linear part, and K [as in § 318 (27)] the constant part of T in respect to the velocity components, so that

$$\left.\begin{aligned} \mathfrak{T} &= \tfrac{1}{2}\{(\psi,\psi)\dot{\psi}^2 + 2(\psi,\phi)\dot{\psi}\dot{\phi} + (\phi,\phi)\dot{\phi}^2 + \ldots\} \\ L &= (\psi)\dot{\psi} + (\phi)\dot{\phi} + \ldots \\ K &= (\psi, \phi, \theta, \ldots) \end{aligned}\right\} \ldots\ldots(a),$$

where (ψ, ψ), (ψ, ϕ), (ϕ, ϕ) ... denote functions of the co-ordinates without t, and (ψ), (ϕ), ..., $(\psi, \phi, \theta, \ldots)$ functions of the co-ordinates and, may be also, of t; and

$$T = \mathfrak{T} + L + K \ldots\ldots\ldots\ldots(b).$$

We have
$$\frac{dK}{d\dot{\psi}} = 0.$$

Hence the contribution from K to the first member of the ψ-equation of motion is simply $-\dfrac{dK}{d\psi}$. Again we have

$$\frac{dL}{d\dot{\psi}} = (\psi);$$

hence
$$\frac{d}{dt}\frac{dL}{d\dot{\psi}} = \frac{d(\psi)}{d\psi}\dot{\psi} + \frac{d(\psi)}{d\phi}\dot{\phi} + \text{etc.} + \left(\frac{d(\psi)}{dt}\right).$$

Farther we have

$$\frac{dL}{d\dot{\psi}} = \frac{d(\psi)}{d\psi}\dot{\psi} + \frac{d(\phi)}{d\psi}\dot{\phi} + \ldots.$$

Example of varying relation due to kinetic constraint.

Hence the whole contribution from L to the ψ-equation of motion is

$$\left(\frac{d(\psi)}{d\phi} - \frac{d(\phi)}{d\psi}\right)\dot{\phi} + \left(\frac{d(\psi)}{d\theta} - \frac{d(\theta)}{d\psi}\right)\dot{\theta} + \ldots + \left(\frac{d(\psi)}{dt}\right) \ldots\ldots(c).$$

Lastly, the contribution from $\mathfrak{T}$ is the same as the whole from T in § 318 (29‴); so that we have

$$\frac{d}{dt}\frac{d\mathfrak{T}}{d\dot{\psi}} - \frac{d\mathfrak{T}}{d\psi} = (\psi, \psi)\ddot{\psi} + (\psi, \phi)\ddot{\phi} + \ldots$$

$$+\tfrac{1}{2}\left\{\frac{d(\psi,\psi)}{d\psi}\dot{\psi}^2 + 2\frac{d(\psi,\psi)}{d\phi}\dot{\psi}\dot{\phi} + \left[2\frac{d(\psi,\phi)}{d\phi} - \frac{d(\phi,\phi)}{d\psi}\right]\dot{\phi}^2 + \ldots\right\} \quad (d),$$

and the completed ψ-equation of motion is

$$\frac{d}{dt}\frac{d\mathfrak{T}}{d\dot{\psi}} - \frac{d\mathfrak{T}}{d\psi} + \left(\frac{d(\psi)}{d\phi} - \frac{d(\phi)}{d\psi}\right)\dot{\phi} + \left(\frac{d(\psi)}{d\theta} - \frac{d(\theta)}{d\psi}\right)\dot{\theta} + \ldots$$

$$+ \left(\frac{d(\psi)}{dt}\right) - \frac{dK}{d\psi} = \Psi \ldots\ldots\ldots(e).$$

It is important to remark that the coefficient of $\dot{\phi}$ in this ψ-equation is equal but of opposite sign to the coefficient of $\dot{\psi}$ in the ϕ-equation. [Compare Example G (19) below.]

Proceeding as in § 318 (29ⁱᵛ) (29ᵛ), we have in respect to $\mathfrak{T}$ precisely the same formulas as there in respect to T. The terms involving first powers of the velocities simply, balance in the sum: and we find finally

Equation of energy.

$$\frac{d\mathfrak{T}}{dt} + \left(\frac{dL}{dt}\right) - \frac{d_{(\psi,\phi,\ldots)}K}{dt} = \Psi\dot{\psi} + \Phi\dot{\phi} + \ldots\ldots\ldots\ldots(f),$$

where $d_{(\psi,\phi,\ldots)}$ denotes differentiation on the supposition of $\psi, \phi, \ldots$ variable; and t constant, where it appears explicitly.

Now with this notation we have

$$\frac{dL}{dt} = \left(\frac{dL}{dt}\right) + \frac{d_{(\psi,\phi,\ldots)}L}{dt} + (\psi)\ddot{\psi} + (\phi)\ddot{\phi} + \ldots,$$

and

$$\frac{dK}{dt} = \left(\frac{dK}{dt}\right) + \frac{d_{(\psi,\phi,\ldots)}K}{dt}.$$

Hence from (f) we have

$$\frac{dT}{dt} = \frac{d(\mathfrak{T}+L+K)}{dt} = \Psi\dot{\psi} + \Phi\dot{\phi} + \ldots + \frac{d_{(\psi,\phi,\ldots)}L}{dt} + (\psi)\ddot{\psi} + (\phi)\ddot{\phi} + \ldots$$

$$+ 2\frac{d_{(\psi,\phi,\ldots)}K}{dt} + \left(\frac{dK}{dt}\right) \ldots\ldots\ldots\ldots(g).$$

Exercise for student.

Take, for illustration, Examples (E) and (F) from above; in which we have

[Example (E)] $\mathfrak{T} = \frac{1}{2} m (\dot{x}^2 + \dot{y}^2 + \dot{z}^2)$,

$L = m\dot{a} (x\dot{y} - y\dot{x})$,

$K = \frac{1}{2} m\dot{a}^2 (x^2 + y^2)$,

and [Example (F)] $\mathfrak{T} = \frac{1}{2} ma^2 (\sin^2\theta\dot{\phi}^2 + \dot{\theta}^2)$,

$L = - m\dot{u}a \sin\theta\dot{\theta}$,

$K = \frac{1}{2} m\dot{u}^2$.

Write out explicitly in each case equations (f) and (g), and verify them by direct work from the equations of motion forming the conclusions of the examples as treated above (remembering that $\dot{a}$ and $\dot{u}$ are to be regarded as given explicit functions of t).

Ignoration of co-ordinates.

Example (G).—*Preliminary to Gyrostatic connexions and to Fluid Motion.* Let there be one or more co-ordinates χ, χ', etc. which do not appear in the coefficients of velocities in the expression for T; that is to say let $\frac{dT}{d\chi} = 0$, $\frac{dT}{d\chi'} = 0$, etc. The equations corresponding to these co-ordinates become

$$\frac{d}{dt}\frac{dT}{d\dot{\chi}} = \mathrm{X}, \quad \frac{d}{dt}\frac{dT}{d\dot{\chi}'} = \mathrm{X}', \text{ etc.} \quad\text{..................(1).}$$

Farther let us suppose that the force-components X, X′, etc. corresponding to the co-ordinates χ, χ', etc. are each zero: we shall have

$$\frac{dT}{d\dot{\chi}} = C, \quad \frac{dT}{d\dot{\chi}'} = C', \text{ etc.} \quad\text{..................(2);}$$

or, expanded according to previous notation [318 (29)],

$$\left.\begin{array}{l} (\psi, \chi)\,\dot{\psi} + (\phi, \chi)\,\dot{\phi} + \ldots + (\chi, \chi)\,\dot{\chi} + (\chi, \chi')\,\dot{\chi}' + \ldots \;= C \\ (\psi, \chi')\,\dot{\psi} + (\phi, \chi')\,\dot{\phi} + \ldots + (\chi', \chi)\,\dot{\chi} + (\chi', \chi')\,\dot{\chi}' + \ldots = C' \\ \ldots\ldots\ldots\ldots\ldots\ldots\ldots\ldots\ldots\ldots\ldots\ldots \end{array}\right\} \ldots(3).$$

Hence, if we put

$$\left.\begin{array}{l} (\psi, \chi)\,\dot{\psi} + (\phi, \chi)\,\dot{\phi} + \ldots \;= P \\ (\psi, \chi')\,\dot{\psi} + (\phi, \chi')\,\dot{\phi} + \ldots = P' \\ \ldots\ldots\ldots\ldots\ldots\ldots \end{array}\right\} \ldots\ldots\ldots\ldots\ldots\ldots(4),$$

Ignoration of co-ordinates.

we have

$$\left.\begin{array}{l}(\chi,\chi)\dot\chi+(\chi,\chi')\dot\chi'+\ldots\ =C-P\\ (\chi',\chi)\dot\chi+(\chi',\chi')\dot\chi'+\ldots=C'-P'\\ \ldots\ldots\ldots\ldots\ldots\ldots\ldots\ldots\ldots\end{array}\right\}\ldots\ldots\ldots\ldots(5).$$

Resolving these for $\dot\chi$, $\dot\chi'$, ... we find

$$\dot\chi=\frac{\begin{vmatrix}(\chi',\chi'),&(\chi',\chi''),&\ldots\\(\chi'',\chi'),&(\chi'',\chi''),&\ldots\\ \ldots&\ldots&\ldots\end{vmatrix}(C-P)+\begin{vmatrix}(\chi'',\chi'),&(\chi'',\chi''),&\ldots\\(\chi''',\chi'),&(\chi''',\chi''),&\ldots\\ \ldots&\ldots&\ldots\end{vmatrix}(C'-P')+\ldots}{\begin{vmatrix}(\chi,\chi),&(\chi,\chi'),&(\chi,\chi''),&\ldots\\(\chi',\chi),&(\chi',\chi'),&(\chi',\chi''),&\ldots\\(\chi'',\chi),&(\chi'',\chi'),&(\chi'',\chi''),&\ldots\\ \ldots&\ldots&\ldots&\ldots\end{vmatrix}}\quad(6),$$

and symmetrical expressions for $\dot\chi'$, $\dot\chi''$, ..., or, as we may write them short,

$$\left.\begin{array}{l}\dot\chi=(C,C)(C-P)+(C,C')(C'-P')\ +\ldots\\ \dot\chi'=(C',C)(C-P)+(C',C')(C'-P')+\ldots\\ \ldots\ldots\ldots\ldots\ldots\ldots\ldots\ldots\ldots\ldots\end{array}\right\}\ldots\ldots\ldots(7),$$

where (C,C), (C,C'), (C',C'), ... denote functions of the retained co-ordinates ψ, ϕ, θ, ... It is to be remembered that, because $(\chi,\chi')=(\chi',\chi)$, $(\chi,\chi'')=(\chi'',\chi)$, we see from (6) that

$$(C,C')=(C',C),\ (C,C'')=(C'',C),\ (C',C'')=(C'',C'),\ \text{and so on}\ldots(8).$$

The following formulas for $\dot\chi$, $\dot\chi'$, ..., condensed in respect to C, C', C'' by aid of the notation (14) below, and expanded in respect to $\dot\psi$, $\dot\phi$, ..., by (4), will also be useful.

$$\left.\begin{array}{l}\dot\chi=\dfrac{dK}{dC}-(M\dot\psi+N\dot\phi+\ldots)\\[2ex] \dot\chi'=\dfrac{dK}{dC'}-(M'\dot\psi+N'\dot\phi+\ldots)\\[2ex] \ldots\ldots\ldots\ldots\ldots\ldots\ldots\ldots\end{array}\right\}\ldots\ldots\ldots\ldots(9),$$

where

$$\left.\begin{array}{l}M=(C,C).(\psi,\chi)+(C,C').(\psi,\chi')\ +\ldots\\ N=(C,C).(\phi,\chi)+(C,C').(\phi,\chi')+\ldots\\ \ldots\ldots\ldots\ldots\ldots\ldots\ldots\ldots\ldots\ldots\\ M'=(C',C).(\psi,\chi)+(C',C').(\psi,\chi')+\ldots\\ \ldots\ldots\ldots\ldots\ldots\ldots\ldots\ldots\ldots\ldots\end{array}\right\}\ldots\ldots(10).$$

The elimination of $\dot\chi$, $\dot\chi'$, ... from T by these expressions for

Ignoration of co-ordinates.

them is facilitated by remarking that, as it is a quadratic function of $\dot{\psi}$, $\dot{\phi}$, ... $\dot{\chi}$, $\dot{\chi}'$, ..., we have

$$T = \tfrac{1}{2}\left\{\dot{\psi}\frac{dT}{d\dot{\psi}} + \dot{\phi}\frac{dT}{d\dot{\phi}} + \ldots + \dot{\chi}\frac{dT}{d\dot{\chi}} + \dot{\chi}'\frac{dT}{d\dot{\chi}'} + \ldots\right\}.$$

Hence by (3),

$$T = \tfrac{1}{2}\left\{\dot{\psi}\frac{dT}{d\dot{\psi}} + \dot{\phi}\frac{dT}{d\dot{\phi}} + \ldots + \dot{\chi}C + \dot{\chi}'C' + \ldots\right\},$$

so that we have now only first powers of $\dot{\chi}$, $\dot{\chi}'$, ... to eliminate. Gleaning out $\dot{\chi}$, $\dot{\chi}'$, ... from the first group of terms, and denoting by T_0 the part of T not containing $\dot{\chi}$, $\dot{\chi}'$, ..., we find

$$\begin{aligned} T = T_0 + \tfrac{1}{2}\{&[(\psi, \chi)\dot{\psi} + (\phi, \chi)\dot{\phi} + \ldots + C]\dot{\chi} \\ &+ [(\psi, \chi')\dot{\psi} + (\phi, \chi')\dot{\phi} + \ldots + C']\dot{\chi}' \\ &+ \ldots\ldots\ldots\ldots\ldots\ldots\ldots\ldots\},\end{aligned}$$

or, according to the notation of (4),

$$T = T_0 + \tfrac{1}{2}\{(C + P)\dot{\chi} + (C' + P')\dot{\chi}' + \ldots\}.$$

Eliminating now $\dot{\chi}$, $\dot{\chi}'$, ... by (7) we find

$$T = T_0 + \tfrac{1}{2}\{(C, C)(C^2 - P^2) + 2(C, C')(CC' - PP') + (C', C')(C'^2 - P'^2) + \ldots\} \ldots\ldots\ldots (11).$$

It is remarkable that only second powers, and products, *not first powers*, of the velocity-components $\dot{\psi}$, $\dot{\phi}$, ... appear in this expression. We may write it thus:—

$$T = \mathfrak{T} + K \ldots\ldots\ldots\ldots\ldots\ldots\ldots\ldots\ldots (12),$$

where $\mathfrak{T}$ denotes a quadratic function of $\dot{\psi}$, $\dot{\phi}$, ..., as follows:—

$$\mathfrak{T} = T_0 - \tfrac{1}{2}\{(C, C)P^2 + 2(C, C')PP' + (C', C')P'^2 + \ldots\} \ldots (13),$$

and K a quantity independent of $\dot{\psi}$, $\dot{\phi}$, ..., as follows:—

$$K = \tfrac{1}{2}\{(C, C)C^2 + 2(C, C')CC' + (C', C')C'^2 + \ldots\} \ldots\ldots (14).$$

Next, to eliminate $\dot{\chi}$, $\dot{\chi}'$, ... from the Lagrange's equations, we have, in virtue of (12) and of the constitutions of T, $\mathfrak{T}$, and K,

$$\frac{dT}{d\dot{\psi}} + \frac{dT}{d\dot{\chi}}\frac{d\dot{\chi}}{d\dot{\psi}} + \frac{dT}{d\dot{\chi}'}\frac{d\dot{\chi}'}{d\dot{\psi}} + \text{etc.} = \frac{d\mathfrak{T}}{d\dot{\psi}} \ldots\ldots\ldots\ldots (15),$$

where $\dfrac{d\dot{\chi}}{d\dot{\psi}}$, $\dfrac{d\dot{\chi}'}{d\dot{\psi}}$, etc. are to be found by (7) or (9), and therefore are simply the coefficients of $\dot{\psi}$ in (9); so that we have

$$\frac{d\dot{\chi}}{d\dot{\psi}} = -M,\quad \frac{d\dot{\chi}'}{d\dot{\psi}} = -M' \ldots\ldots\ldots\ldots\ldots\ldots (16),$$

where M, M' are functions of ψ, ϕ, ... explicitly expressed by (10). Using (16) in (15) we find

Ignoration of co-ordinates.

$$\frac{dT}{d\psi} = \frac{d\mathfrak{T}}{d\psi} + CM + C'M' + \text{etc.} \qquad (17).$$

Again remarking that $\mathfrak{T} + K$ contains ψ, both as it appeared originally in T, and as farther introduced in the expressions (7) for $\dot{\chi}, \dot{\chi}', \ldots$, we see that

$$\frac{d}{d\psi}(\mathfrak{T} + K) = \frac{dT}{d\psi} + \frac{dT}{d\dot{\chi}}\frac{d\dot{\chi}}{d\psi} + \frac{dT}{d\dot{\chi}'}\frac{d\dot{\chi}'}{d\psi} + \ldots$$

$$= \frac{dT}{d\psi} + C\frac{d\dot{\chi}}{d\psi} + C'\frac{d\dot{\chi}'}{d\psi} + \ldots$$

And by (9) we have

$$\frac{d\dot{\chi}}{d\psi} = -\left(\dot{\psi}\frac{dM}{d\psi} + \dot{\phi}\frac{dN}{d\psi} + \ldots\right) + \frac{d}{d\psi}\frac{dK}{dC};$$

which, used in the preceding, gives

$$\frac{d}{d\psi}(\mathfrak{T} + K) = \frac{dT}{d\psi} - C\left(\dot{\psi}\frac{dM}{d\psi} + \dot{\phi}\frac{dN}{d\psi} + \ldots\right) - C'\left(\dot{\psi}\frac{dM'}{d\psi} + \dot{\phi}\frac{dN'}{d\psi} + \ldots\right) - \text{etc.} + 2\frac{dK}{d\psi}.$$

Hence

$$\frac{dT}{d\psi} = \frac{d\mathfrak{T}}{d\psi} - \frac{dK}{d\psi} + \Sigma C\left(\dot{\psi}\frac{dM}{d\psi} + \dot{\phi}\frac{dN}{d\psi} + \ldots\right) \qquad (18),$$

where Σ denotes summation with regard to the constants C, C', etc.

Using this and (17) in the Lagrange's ψ-equation, we find finally for the ψ-equation of motion in terms of the non-ignored co-ordinates alone, and conclude the symmetrical equations for ϕ, etc., as follows,

$$\left.\begin{aligned}
&\frac{d}{dt}\left(\frac{d\mathfrak{T}}{d\dot{\psi}}\right) - \frac{d\mathfrak{T}}{d\psi} + \Sigma C\left\{\left(\frac{dM}{d\phi} - \frac{dN}{d\psi}\right)\dot{\phi} + \left(\frac{dM}{d\theta} - \frac{dO}{d\psi}\right)\dot{\theta} + \ldots\right\} + \frac{dK}{d\psi} = \Psi \\
&\frac{d}{dt}\left(\frac{d\mathfrak{T}}{d\dot{\phi}}\right) - \frac{d\mathfrak{T}}{d\phi} + \Sigma C\left\{\left(\frac{dN}{d\psi} - \frac{dM}{d\phi}\right)\dot{\psi} + \left(\frac{dN}{d\theta} - \frac{dO}{d\phi}\right)\dot{\theta} + \ldots\right\} + \frac{dK}{d\phi} = \Phi \\
&\frac{d}{dt}\left(\frac{d\mathfrak{T}}{d\dot{\theta}}\right) - \frac{d\mathfrak{T}}{d\theta} + \Sigma C\left\{\left(\frac{dO}{d\psi} - \frac{dM}{d\theta}\right)\dot{\psi} + \left(\frac{dO}{d\phi} - \frac{dN}{d\theta}\right)\dot{\phi} + \ldots\right\} + \frac{dK}{d\theta} = \Theta \\
&\ldots\ldots\ldots\ldots\ldots\ldots\ldots\ldots\ldots\ldots
\end{aligned}\right\} (19).$$

[Compare Example F' (e) above. It is important to remark that in each equation of motion the first power of the related velocity-component disappears; and the coefficient of each of the other velocity-components in this equation is equal but of opposite sign to the coefficient of the velocity-component corresponding to this equation, in the equation corresponding to that other velocity-component.]

Equation of Energy.

The equation of energy, found as above [§ 318 (29^{iv}) and (29^{v})], is

$$\frac{d(\mathfrak{T}+K)}{dt} = \Psi\dot{\psi} + \Phi\dot{\phi} + \text{etc.} \quad \ldots\ldots\ldots\ldots\ldots(20).$$

The interpretation, considering (12), is obvious. The contrast with Example F′ (g) is most instructive.

Sub-Example (G_1).—Take, from above, Example C, case (*a*): and put $\phi = \psi + \theta$; also, for brevity, $mj^2 + na^2 = B$, $n(b^2+k^2) = A$, and $nab = c$. We have*

$$T = \tfrac{1}{2}\{A\dot{\psi}^2 + 2c\dot{\psi}(\dot{\psi}+\dot{\theta})\cos\theta + B(\dot{\psi}+\dot{\theta})^2\};$$

and from this find

$$\frac{dT}{d\psi} = 0, \quad \frac{dT}{d\dot{\psi}} = A\dot{\psi} + c(2\dot{\psi}+\dot{\theta})\cos\theta + B(\dot{\psi}+\dot{\theta});$$

$$\frac{dT}{d\theta} = -c\dot{\psi}(\dot{\psi}+\dot{\theta})\sin\theta, \quad \frac{dT}{d\dot{\theta}} = c\dot{\psi}\cos\theta + B(\dot{\psi}+\dot{\theta}).$$

Here the co-ordinate θ alone, and not the co-ordinate ψ, appears in the coefficients. Suppose now $\Psi = 0$ [which is the case considered at the end of C (*a*) above]. We have $\dfrac{dT}{d\dot{\psi}} = C$, and deduce

$$\dot{\psi} = \frac{C - (c\cos\theta + B)\dot{\theta}}{A + B + 2c\cos\theta},$$

$$T = \tfrac{1}{2}\left(\dot{\psi}\frac{dT}{d\dot{\psi}} + \dot{\theta}\frac{dT}{d\dot{\theta}}\right) = \tfrac{1}{2}\{\dot{\psi}C + \dot{\theta}[(c\cos\theta + B)\dot{\psi} + B\dot{\theta}]\}$$

$$= \tfrac{1}{2}\{\dot{\psi}[C + (c\cos\theta + B)\dot{\theta}] + B\dot{\theta}\}$$

$$= \tfrac{1}{2}\left\{\frac{C^2 - (c\cos\theta + B)^2\dot{\theta}^2}{A + B + 2c\cos\theta} + B\dot{\theta}^2\right\} = \tfrac{1}{2}\,\frac{C^2 + (AB - c^2\cos^2\theta)\dot{\theta}^2}{A + B + 2c\cos\theta}.$$

Hence $$\mathfrak{T} = \tfrac{1}{2}\,\frac{AB - c^2\cos^2\theta}{A + B + 2c\cos\theta}\,\dot{\theta}^2,$$

and $$K = \tfrac{1}{2}\,\frac{C^2}{A + B + 2c\cos\theta}:$$

* Remark that, according to the alteration from $\psi, \dot{\psi}, \phi, \dot{\phi}$, to $\psi, \dot{\psi}, \theta, \dot{\theta}$, as independent variables,

$$\frac{dT}{d\psi} = \left(\frac{dT}{d\psi}\right) + \left(\frac{dT}{d\phi}\right), \quad \frac{dT}{d\theta} = \left(\frac{dT}{d\phi}\right);$$

and $$\frac{dT}{d\dot{\psi}} = \left(\frac{dT}{d\dot{\psi}}\right) + \left(\frac{dT}{d\dot{\phi}}\right), \quad \frac{dT}{d\dot{\theta}} = \left(\frac{dT}{d\dot{\phi}}\right);$$

where ( ) indicates the original notation of C (*a*).

and the one equation of the motion becomes Equation Energy.

$$\frac{d}{dt}\left(\frac{AB-c^2\cos^2\theta}{A+B+2c\cos\theta}\dot{\theta}\right)-\tfrac{1}{2}\dot{\theta}^2\frac{d}{d\theta}\left(\frac{AB-c^2\cos^2\theta}{A+B+2c\cos\theta}\right)=\Theta-\frac{dK}{d\theta};$$

which is to be fully integrated first by multiplying by $d\theta$ and integrating once; and then solving for dt and integrating again with respect to θ. The first integral, being simply the equation of energy integrated, is [Example G (20)]

$$\mathfrak{T}=\int\Theta\, d\theta-K;$$

and the final integral is

$$t=\int d\theta\sqrt{\frac{AB-\cos^2\theta}{2(A+B+2c\cos\theta)(\int\Theta d\theta-K)}}.$$

In the particular case in which the motion commences from rest, or is such that it can be brought to rest by proper applications of force-components, Ψ, Φ, etc. without any of the force-components X, X′, etc., we have $C=0$, $C'=0$, etc.; and the elimination of $\dot{\chi}$, $\dot{\chi}'$, etc. by (3) renders T a homogeneous quadratic function of $\dot{\psi}$, $\dot{\phi}$, etc. without C, C', etc.; and the equations of motion become Ignoration of co-ordinates.

$$\left.\begin{array}{l}\dfrac{d}{dt}\dfrac{dT}{d\dot{\psi}}-\dfrac{dT}{d\psi}=\Psi\\ \dfrac{d}{dt}\dfrac{dT}{d\dot{\phi}}-\dfrac{dT}{d\phi}=\Phi\\ \dfrac{d}{dt}\dfrac{dT}{d\dot{\theta}}-\dfrac{dT}{d\theta}=\Theta\\ \quad\text{etc.}\qquad\text{etc.}\end{array}\right\}\quad\text{.....................}\ (21).$$

We conclude that on the suppositions made, the elimination of the velocity-components corresponding to the non-appearing co-ordinates gives an expression for the kinetic energy in terms of the remaining velocity-components and corresponding co-ordinates which may be used in the generalised equations just as if these were the sole co-ordinates. The reduced number of equations of motion thus found suffices for the determination of the co-ordinates which they involve without the necessity for knowing or finding the other co-ordinates. If the farther question be put,—to determine the ignored co-ordinates, it is to be answered by a simple integration of equations (7) with $C=0$, $C'=0$, etc.

One obvious case of application for this example is a system in which any number of fly wheels, that is to say, bodies which are

Ignoration of co-ordinates.

kinetically symmetrical round an axis (§ 285), are pivoted frictionlessly on any moveable part of the system. In this case with the particular supposition $C=0$, $C'=0$, etc., the result is simply that the motion is the same as if each fly wheel were deprived of moment of inertia round its bearing axis, that is to say reduced to a line of matter fixed in the position of this axis and having unchanged moment of inertia round any axis perpendicular to it. But if C, C', etc. be not each zero we have a case embracing a very interesting class of dynamical problems in which the motion of a system having what we may call gyrostatic links or connexions is the subject. Example (D) above is an example, in which there is just one fly wheel and one moveable body on which it is pivoted. The ignored co-ordinate is ψ; and supposing now Ψ to be zero, we have

$$\dot{\psi}-\dot{\phi}(1-\cos\theta)=C \qquad (a).$$

If we suppose $C=0$ all the terms having $\mathfrak{A}'$ for a factor vanish and the motion is the same as if the fly wheel were deprived of inertia round its bearing axis, and we had simply the motion of the "ideal rigid body PQ" to consider. But when C does not vanish we eliminate $\dot{\psi}$ from the equations by means of (a). It is important to remark that in every case of Example (G) in which $C=0$, $C'=0$, etc. the motion at each instant possesses the property (§ 312 above) of having less kinetic energy than any other motion for which the velocity-components of the non-ignored co-ordinates have the same values.

Take for another example the final form of Example C' above, putting B for C, and A for nk^2+A. We have

$$T=\tfrac{1}{2}\{(E+F\cos^2\theta)\dot{\psi}^2+B(\dot{\psi}\cos\theta+\dot{\phi})^2+A\dot{\theta}^2\} \quad \ldots(22).$$

Here neither ψ nor ϕ appears in the coefficients. Let us suppose $\Phi=0$, and eliminate $\dot{\phi}$, to let us ignore ϕ. We have

$$\frac{dT}{d\dot{\phi}}=B(\dot{\psi}\cos\theta+\dot{\phi})=C.$$

Hence
$$\dot{\phi}=\frac{C}{B}-\dot{\psi}\cos\theta \qquad \ldots(23),$$

$$\mathfrak{T}=\tfrac{1}{2}\{(E+F\cos^2\theta)\dot{\psi}^2+A\dot{\theta}^2\} \qquad \ldots(24),$$

and
$$K=\tfrac{1}{2}\frac{C^2}{B} \qquad \ldots(25).$$

The place of $\dot{\chi}$ in (9) above is now taken by $\dot{\phi}$, and comparing with (23) we find

$$M=\cos\theta,\ N=0,\ O=0.$$

Hence, and as K is constant, the equations of motion (19) become

Ignoration of co-ordinates.

$$\left.\begin{array}{l} \frac{d}{dt}\frac{d\mathfrak{T}}{d\dot{\psi}} - \frac{d\mathfrak{T}}{d\psi} - C\sin\theta\dot{\theta} = \Psi \\ \text{and} \quad \frac{d}{dt}\frac{d\mathfrak{T}}{d\dot{\theta}} - \frac{d\mathfrak{T}}{d\theta} + C\sin\theta\dot{\psi} = \Theta \end{array}\right\} \ldots\ldots\ldots\ldots\ldots (26);$$

and, using (24) and expanding,

$$\left.\begin{array}{l} \frac{d\{(E+F\cos^2\theta)\dot{\psi}\}}{dt} - C\sin\theta\dot{\theta} = \Psi \\ A\ddot{\theta} + F\sin\theta\cos\theta\dot{\psi}^2 + C\sin\theta\dot{\psi} = \Theta \end{array}\right\} \ldots\ldots\ldots\ldots (27).$$

A most important case for the "ignoration of co-ordinates" is presented by a large class of problems regarding the motion of frictionless incompressible fluid in which we can ignore the infinite number of co-ordinates of individual portions of the fluid and take into account only the co-ordinates which suffice to specify the whole boundary of the fluid, including the bounding surfaces of any rigid or flexible solids immersed in the fluid. The analytical working out of Example (G) shows in fact that when the motion is such as could be produced from rest by merely moving the boundary of the fluid without applying force to its individual particles otherwise than by the transmitted fluid pressure we have exactly the case of $C=0$, $C'=0$, etc.: and Lagrange's generalized equations with the kinetic energy expressed in terms of velocity-components completely specifying the motion of the boundary are available. Thus,

320. Problems in fluid motion of remarkable interest and importance, not hitherto attacked, are very readily solved by the aid of Lagrange's generalized equations of motion. For brevity we shall designate a mass which is absolutely incompressible, and absolutely devoid of resistance to change of shape, by the simple appellation of a *liquid*. We need scarcely say that matter perfectly satisfying this definition does not exist in nature: but we shall see (under properties of matter) how nearly it is approached by water and other common real liquids. And we shall find that much practical and interesting information regarding their true motions is obtained by deductions from the principles of abstract dynamics applied to the ideal perfect liquid of our definition. It follows from Example

Kinetics of a perfect liquid.

Kinetics of a perfect liquid.

(G) above (and several other proofs, some of them more synthetical in character, will be given in our Second Volume,) that the motion of a homogeneous liquid, whether of infinite extent, or contained in a finite closed vessel of any form, with any rigid or flexible bodies moving through it, if it has ever been at rest, is the same at each instant as that determinate motion (fulfilling, § 312, the condition of having the least possible kinetic energy) which would be impulsively produced from rest by giving instantaneously to every part of the bounding surface, and of the surface of each of the solids within it, its actual velocity at that instant. So that, for example, however long it may have been moving, if all these surfaces were suddenly or gradually brought to rest, the whole fluid mass would come to rest at the same time. Hence, if none of the surfaces is flexible, but we have one or more rigid bodies moving in any way through the liquid, under the influence of any forces, the kinetic energy of the whole motion at any instant will depend solely on the finite number of co-ordinates and component velocities, specifying the position and motion of those bodies, whatever may be the positions reached by particles of the fluid (expressible only by an infinite number of co-ordinates). And an expression for the whole kinetic energy in terms of such elements, finite in number, is precisely what is wanted, as we have seen, as the foundation of Lagrange's equations in any particular case.

It will clearly, in the hydrodynamical, as in all other cases, be a homogeneous quadratic function of the components of velocity, if referred to an invariable co-ordinate system; and the coefficients of the several terms will in general be functions of the co-ordinates, the determination of which follows immediately from the solution of the minimum problem of Example (3) § 317, in each particular case.

Example (1).—*A ball set in motion through a mass of incompressible fluid extending infinitely in all directions on one side of an infinite plane, and originally at rest.* Let x, y, z be the co-ordinates of the centre of the ball at time t, with reference to rectangular axes through a fixed point O of the bounding plane, with OX perpendicular to this plane. If at any instant either component $\dot{y}$ or $\dot{z}$ of the velocity be reversed, the kinetic energy will clearly be unchanged, and hence no terms $\dot{y}\dot{z}$, $\dot{z}\dot{x}$, or $\dot{x}\dot{y}$ can appear in the expression for the kinetic energy: which, on this account, and because of the symmetry of circumstances with reference to y and z, is

Kinetics of a perfect liquid.

$$T = \tfrac{1}{2}\{P\dot{x}^2 + Q(\dot{y}^2 + \dot{z}^2)\}.$$

Also, we see that P and Q are functions of x simply, since the circumstances are similar for all values of y and z. Hence, by differentiation,

$$\frac{dT}{d\dot{x}} = P\dot{x},\ \frac{dT}{d\dot{y}} = Q\dot{y},\ \frac{dT}{d\dot{z}} = Q\dot{z},$$

$$\frac{d}{dt}\left(\frac{dT}{d\dot{x}}\right) = P\ddot{x} + \frac{dP}{dx}\dot{x}^2,\quad \frac{d}{dt}\left(\frac{dT}{d\dot{y}}\right) = Q\ddot{y} + \frac{dQ}{dx}\dot{y}\dot{x},\ \text{etc.},$$

$$\frac{dT}{dx} = \tfrac{1}{2}\left\{\frac{dP}{dx}\dot{x}^2 + \frac{dQ}{dx}(\dot{y}^2 + \dot{z}^2)\right\},\quad \frac{dT}{dy} = 0,\ \text{etc.},$$

and the equations of motion are

$$P\ddot{x} + \tfrac{1}{2}\left\{\frac{dP}{dx}\dot{x}^2 - \frac{dQ}{dx}(\dot{y}^2 + \dot{z}^2)\right\} = X,$$

$$Q\ddot{y} + \frac{dQ}{dx}\dot{y}\dot{x} = Y,\ Q\ddot{z} + \frac{dQ}{dx}\dot{z}\dot{x} = Z.$$

Principles sufficient for a practical solution of the problem of determining P and Q will be given later. In the meantime, it is obvious that each decreases as x increases. Hence the equations of motion show that

321. A ball projected through a liquid perpendicularly *from* an infinite plane boundary, and influenced by no other forces than those of fluid pressure, experiences a gradual acceleration, quickly approximating to a limiting velocity which it sensibly reaches when its distance from the plane is many times its diameter. But if projected *parallel* to the plane, it experiences, as the resultant of fluid pressure, a resultant attraction towards the plane. The former of these results is easily proved by first considering projection *towards* the plane (in which case the motion of the ball will obviously be retarded), and by taking into account the general principle of reversibility (§ 272) which has perfect application in the ideal case of a perfect liquid. The second result is less easily foreseen without

Effect of a rigid plane on the motion of a ball through a liquid.

the aid of Lagrange's analysis; but it is an obvious consequence of the Hamiltonian form of his equations, as stated in words in § 319 above. In the precisely equivalent case, of a liquid extending infinitely in all directions, and given at rest; and two equal balls projected through it with equal velocities perpendicular to the line joining their centres—the result that the two balls will seem to attract one another is most remarkable, and very suggestive.

Seeming attraction between two ships moving side by side in the same direction.

Hydro-dynamical examples continued.

Example (2).—*A solid symmetrical round an axis, moving through a liquid so as to keep its axis always in one plane.* Let ω be the angular velocity of the body at any instant about any axis perpendicular to the fixed plane, and let u and q be the component velocities along and perpendicular to the axis of figure, of any chosen point, C, of the body in this line. By the general principle stated in § 320 (since changing the sign of u cannot alter the kinetic energy), we have

$$T = \tfrac{1}{2}(Au^2 + Bq^2 + \mu'\omega^2 + 2E\omega q) \ldots\ldots\ldots\ldots\ldots\ldots (a),$$

where A, B, μ', and E are constants depending on the figure of the body, its mass, and the density of the liquid. Now let v denote the velocity, perpendicular to the axis, of a point which we shall call the *centre of reaction*, being a point in the axis and at a distance $\frac{E}{B}$ from C, so that (§ 87) $q = v - \frac{E}{B}\omega$. Then, denoting $\mu' - \frac{E^2}{B}$ by μ, we have $T = \tfrac{1}{2}(Au^2 + Bv^2 + \mu\omega^2) \ldots\ldots (a')$.

"Centre of reaction" defined.

Let x and y be the co-ordinates of the centre of reaction relatively to any fixed rectangular axes in the plane of motion of the axis of figure, and let θ be the angle between this line and OX, at any instant, so that

$$\omega = \dot\theta,\ u = \dot x \cos\theta + \dot y \sin\theta,\ v = -\dot x \sin\theta + \dot y \cos\theta \ldots\ldots\ldots (b).$$

Substituting in T, differentiating, and retaining the notation u, v where convenient for brevity, we have

$$\left.\begin{aligned} &\frac{dT}{d\dot\theta} = \mu\dot\theta,\ \frac{dT}{d\dot x} = Au\cos\theta - Bv\sin\theta,\ \frac{dT}{d\dot y} = Au\sin\theta + Bv\cos\theta, \\ &\frac{dT}{d\theta} = (A - B)uv,\ \frac{dT}{dx} = 0,\ \frac{dT}{dy} = 0, \end{aligned}\right\} \quad (c)$$

Hence the equations of motion are Hydro-dynamical examples continued.

$$\left.\begin{array}{l}\mu\ddot{\theta}-(A-B)uv=L,\\ \dfrac{d(Au\cos\theta-Bv\sin\theta)}{dt}=X,\quad \dfrac{d(Au\sin\theta+Bv\cos\theta)}{dt}=Y\end{array}\right\}\ (d),$$

where X, Y are the component forces in lines through C parallel to OX and OY, and L the couple, applied to the body.

Denoting by λ, ξ, η the impulsive couple, and the components of impulsive force through C, required to produce the motion at any instant, we have of course [§ 313 (*c*)],

$$\lambda=\frac{dT}{d\dot{\theta}}=\mu\dot{\theta},\ \xi=\frac{dT}{d\dot{x}},\ \eta=\frac{dT}{d\dot{y}}\ \dots\dots\dots\dots\ (e),$$

and therefore by (*c*), and (*b*),

$$u=\frac{1}{A}(\xi\cos\theta+\eta\sin\theta),\ v=\frac{1}{B}(-\xi\sin\theta+\eta\cos\theta),\ \dot{\theta}=\frac{\lambda}{\mu}\dots\dots(f),$$

$$\left.\begin{array}{l}\dot{x}=\left(\dfrac{\cos^2\theta}{A}+\dfrac{\sin^2\theta}{B}\right)\xi+\left(\dfrac{1}{A}-\dfrac{1}{B}\right)\sin\theta\cos\theta\eta,\\ \dot{y}=\left(\dfrac{1}{A}-\dfrac{1}{B}\right)\sin\theta\cos\theta\xi+\left(\dfrac{\sin^2\theta}{A}+\dfrac{\cos^2\theta}{B}\right)\eta\end{array}\right\}\dots\dots\dots(g),$$

and the equations of motion become

$$\mu\frac{d^2\theta}{dt^2}-\frac{A-B}{2AB}\{(-\xi^2+\eta^2)\sin 2\theta+2\xi\eta\cos 2\theta\}=L,\ \frac{d\xi}{dt}=X,\ \frac{d\eta}{dt}=Y,\ \ (h).$$

The simple case of $X=0$, $Y=0$, $L=0$, is particularly interesting. In it ξ and η are each constant; and we may therefore choose the axes OX, OY, so that η shall vanish. Thus we have, in (*g*), two first integrals of the equations of motion; and they become

$$\dot{x}=\xi\left(\frac{\cos^2\theta}{A}+\frac{\sin^2\theta}{B}\right),\ \dot{y}=-\frac{A-B}{2AB}\xi\sin 2\theta\dots\dots\dots\dots(k):$$

and the first of equations (*h*) becomes

$$\mu\frac{d^2\theta}{dt^2}+\frac{A-B}{2AB}\xi^2\sin 2\theta=0\dots\dots\dots\dots(l).$$

In this let, for a moment, $2\theta=\phi$, and $\dfrac{A-B}{AB}\xi^2=ghW$. It becomes

$$\mu\frac{d^2\phi}{dt^2}+ghW\sin\phi=0,$$

which is the equation of motion of a common pendulum, of mass W, moment of inertia μ round its fixed axis, and length

Hydro-dynamical examples continued.

h from axis to centre of gravity; if ϕ be the angle from the position of equilibrium to the position at time t. As we shall see, under kinetics, the final integral of this equation expresses ϕ in terms of t by means of an elliptic function. By using the value thus found for θ or $\frac{1}{2}\phi$, in (k), we have equations giving x and y in terms of t by common integration; and thus the full solution of our present problem is reduced to quadratures. The detailed working out to exhibit both the actual curve described by the centre of reaction, and the position of the axis of the body at any instant, is highly interesting. It is very easily done approximately for the case of very small angular vibrations; that is to say, when either $A-B$ is positive, and ϕ always very small, or $A-B$ negative, and ϕ very nearly equal to $\frac{1}{2}\pi$. But without attending at present to the final integrals, rigorous or approximate, we see from (k) and (l) that

322. If a solid of revolution in an infinite liquid, be set in motion round any axis perpendicular to its axis of figure, or simply projected in any direction without rotation, it will move with its axis always in one plane, and every point of it moving only parallel to this plane; and the strange evolutions which it will, in general, perform, are perfectly defined by comparison with the common pendulum thus. First, for brevity, we shall call by the name of *quadrantal pendulum* (which will be further exemplified in various cases described later, under electricity and magnetism; for instance, an elongated mass of soft iron pivoted on a vertical axis, in a "uniform field of magnetic force"), a body moving about an axis, according to the same law with reference to a quadrant on each side of its position of equilibrium, as the common pendulum with reference to a half circle on each side.

Quadrantal pendulum defined.

Let now the body in question be set in motion by an impulse, ξ, in any line through the centre of reaction, and an impulsive couple λ in the plane of that line and the axis. This will (as will be proved later in the theory of statical couples) have the same effect as a simple impulse ξ (applied to a point, if not of the real body, connected with it by an imaginary infinitely light framework) in a certain fixed line, which we shall call the line of resultant impulse, or of resultant momentum, being parallel to the former line, and at a distance from it equal to $\frac{\lambda}{\xi}$. The whole momentum of the motion generated is of course (§ 295) equal to ξ. The body will move ever afterwards according to the following conditions :—(1.) The angular velocity follows the law of the quadrantal pendulum. (2.) The distance of the centre of reaction from the line of resultant impulse varies simply as the angular velocity. (3.) The velocity of the centre of reaction parallel to the line of impulse is found by dividing the excess of the whole constant energy of the motion above the part of it due to the angular velocity round the centre of reaction, by half the momentum. (4.) If A, B, and μ denote constants, depending on the mass of the solid and its distribution, the density of the liquid, and the form and dimensions of the solid, such that $\frac{\xi}{A}$, $\frac{\xi}{B}$, $\frac{\lambda}{\mu}$ are the linear velocities, and the angular velocity, respectively produced by an impulse ξ along the axis, an impulse ξ in a line through the centre of reaction perpendicular to the axis, and an impulsive couple λ in a plane through the axis; the length of the simple gravitation pendulum, whose motion would keep time with the periodic motion in question, is $\frac{g\mu AB}{\xi^2(A-B)}$, and, when the angular motion is vibratory, the vibrations will, according as $A > B$, or $A < B$, be of the axis, or of a line perpendicular to the axis, vibrating on each side of the line of impulse. The angular motion will in fact be vibratory if the distance of the line of resultant impulse from the centre of reaction is anything less than $\sqrt{\frac{(A \sim B)\mu\cos 2\alpha}{AB}}$ where α denotes the inclination of the impulse to the initial position of the axis. In this case the path of the centre of reaction will be a sinuous curve symmetrical on the two sides of the line of impulse; every time it cuts this line, the angular motion will reverse, and the maximum inclination will be attained; and every time the centre of reaction is at its greatest distance on either side, the angular velocity will be at its greatest, positive or negative, value, and the linear velocity of

Motion of a solid of revolution through a liquid.

Motion of a solid of revolution through a liquid.

the centre of reaction will be at its least. If, on the other hand, the line of the resultant impulse be at a greater distance than $\sqrt{\frac{(A-B)\mu\cos 2\alpha}{AB}}$ from the centre of reaction, the angular motion will be always in one direction, but will increase and diminish periodically, and the centre of reaction will describe a sinuous curve on one side of that line; being at its greatest and least deviations when the angular velocity is greatest and least. At the same points the curvature of the path will be greatest and least respectively, and the linear velocity of the describing point will be least and greatest.

323. At any instant the component linear velocities along and perpendicular to the axis of the solid will be $\frac{\xi\cos\theta}{A}$ and $-\frac{\xi\sin\theta}{B}$ respectively, if θ be its inclination to the line of resultant impulse; and the angular velocity will be $\frac{\xi y}{\mu}$ if y be the distance of the centre of reaction from that line. The whole kinetic energy of the motion will be

$$\frac{\xi^2\cos^2\theta}{2A}+\frac{\xi^2\sin^2\theta}{2B}+\frac{\xi^2y^2}{2\mu},$$

and the last term is what we have referred to above as the part due to rotation round the centre of reaction (defined in § 321). To stop the whole motion at any instant, a simple impulse equal and opposite to ξ in the fixed "line of resultant impulse" will suffice (or an equal and parallel impulse in any line through the body, with the proper impulsive couple, according to the principle already referred to).

324. From Lagrange's equations applied as above to the case of a solid of revolution moving through a liquid, the couple which must be kept applied to it to prevent it from turning is immediately found to be

$$uv\,(A-B),$$

Motion of a solid of revolution through a liquid.

if u and v be the component velocities along and perpendicular to the axis, or [§ 321 (f)]

$$\xi^2 \frac{(A-B)\sin 2\theta}{2AB},$$

if, as before, ξ be the generating impulse, and θ the angle between its line and the axis. The direction of this couple must be such as to prevent θ from diminishing or from increasing, according as A or B is the greater. The former will clearly be the case of a flat disc, or oblate spheroid; the latter that of an elongated, or oval-shaped body. The actual values of A and B we shall learn how to calculate (hydrodynamics) for several cases, including a body bounded by two spherical surfaces cutting one another at any angle a submultiple of two right angles; two complete spheres rigidly connected; and an oblate or a prolate spheroid.

Observed phenomena.

325. The tendency of a body to turn its flat side, or its length (as the case may be), across the direction of its motion through a liquid, to which the accelerations and retardations of rotatory motion described in § 322 are due, and of which we have now obtained the statical measure, is a remarkable illustration of the statement of § 319; and is closely connected with the dynamical explanation of many curious observations well known in practical mechanics, among which may be mentioned:—

(1) That the course of a symmetrical square-rigged ship sailing in the direction of the wind with rudder amidships is unstable, and can only be *kept* by manipulating the rudder to check infinitesimal deviations;—and that a child's toy-boat, whether "square-rigged" or "fore-and-aft rigged*," cannot be

* "Fore-and-aft" rig is any rig in which (as in "cutters" and "schooners") the chief sails come into the plane of mast or masts and keel, by the action of the wind upon the sails when the vessel's head is to wind. This position of the sails is unstable when the wind is right astern. Accordingly, in "wearing" a fore-and-aft rigged vessel (that is to say turning her round stern to wind, from sailing with the wind on one side to sailing with the wind on the other side) the mainsail must be hauled in as closely as may be towards the middle position before the wind is allowed to get on the other side of the sail from that on which it had been pressing, so that when the wind

Applications to nautical dynamics

got to sail permanently before the wind by any permanent adjustment of rudder and sails, and that (without a wind vane, or a weighted tiller, acting on the rudder to do the part of steersman) it always, after running a few yards before the wind, turns round till nearly in a direction perpendicular to the wind (either "gibing" first, or "luffing" without gibing if it is a cutter or schooner) :—

(2) That the towing rope of a canal boat, when the rudder is left straight, takes a position in a vertical plane cutting the axis *before* its middle point :—

(3) That a boat sculled rapidly across the direction of the wind, always (unless it is extraordinarily unsymmetrical in its draught of water, and in the amounts of surface exposed to the wind, towards its two ends) requires the weather oar to be worked hardest to prevent it from running up on the wind, and that for the same reason a sailing vessel generally "carries a weather helm*" or "gripes;" and that still more does so a steamer with sail even if only in the forward half of her length—griping so badly with any after canvass† that it is often impossible to steer :—

(4) That in a heavy gale it is exceedingly difficult, and often found impossible, to get a ship out of "the trough of the sea," and that it cannot be done at all without rapid motion ahead, whether by steam or sails :—

(5) That in a smooth sea with moderate wind blowing parallel to the shore, a sailing vessel heading towards the shore with not enough of sail set can only be saved from creeping ashore by setting more sail, and sailing rapidly towards the shore, or the danger that is to be avoided, so as to allow her to be steered away from it. The risk of going ashore in fulfilment

does get on the other side, and when therefore the sail dashes across through the mid-ship position to the other side, carrying massive boom and gaff with it, the range of this sudden motion, which is called "gibing," shall be as small as may be.

* The weather side of any object is the side of it towards the wind. A ship is said to "carry a weather helm" when it is necessary to hold the "helm" or "tiller" permanently on the weather side of its middle position (by which the rudder is held towards the lee side) to keep the ship on her course.

† Hence mizen masts are altogether condemned in modern war-ships by many competent nautical authorities.

of Lagrange's equations is a frequent incident of "getting under way" while lifting anchor, or even after slipping from moorings :—

(6) That an elongated rifle-bullet requires rapid rotation about its axis to keep its point foremost. And gunnery.

(7) The curious motions of a flat disc, oyster-shell, or the like, when dropped obliquely into water, resemble, no doubt, to some extent those described in § 322. But it must be remembered that the real circumstances differ greatly, because of fluid friction, from those of the abstract problem, of which we take leave for the present.

326. Maupertuis' celebrated principle of *Least Action* has been, even up to the present time, regarded rather as a curious and somewhat perplexing property of motion, than as a useful guide in kinetic investigations. We are strongly impressed with the conviction that a much more profound importance will be attached to it, not only in abstract dynamics, but in the theory of the several branches of physical science now beginning to receive dynamic explanations. As an extension of it, Sir W. R. Hamilton* has evolved his method of *Varying Action*, which undoubtedly must become a most valuable aid in future generalizations. Least action.

What is meant by "Action" in these expressions is, unfortunately, something very different from the *Actio Agentis* defined by Newton†, and, it must be admitted, is a much less judiciously chosen word. Taking it, however, as we find it, now universally used by writers on dynamics, we define the *Action of a Moving System* as proportional to the average kinetic energy, which the system has possessed during the time from any convenient epoch of reckoning, multiplied by the time. According to the unit generally adopted, the action of a system which has not varied in its kinetic energy, is twice the amount of the energy multiplied by the time from the epoch. Or if the energy has been sometimes greater and sometimes less, Action. Time average of energy.

* *Phil. Trans.* 1834—1835.

† Which, however (§ 263), we have translated "activity" to avoid confusion.

Time average of energy doubled.

the action at time t is the double of what we may call the *time-integral* of the energy, that is to say, it is what is denoted in the integral calculus by

$$2\int_0^t T d\tau,$$

where T denotes the kinetic energy at any time τ, between the epoch and t.

Let m be the mass, and v the velocity at time τ, of any one of the material points of which the system is composed. We have

$$T = \Sigma \tfrac{1}{2} m v^2 \ldots\ldots\ldots\ldots\ldots\ldots\ldots\ldots (1),$$

and therefore, if A denote the action at time t,

$$A = \int_0^t \Sigma m v^2 d\tau \ldots\ldots\ldots\ldots\ldots\ldots\ldots\ldots (2).$$

This may be put otherwise by taking ds to denote the space described by a particle in time $d\tau$, so that $v d\tau = ds$, and therefore

$$A = \int \Sigma m v ds \ldots\ldots\ldots\ldots\ldots\ldots\ldots\ldots (3),$$

or, if x, y, z be the rectangular co-ordinates of m at any time,

$$A = \int \Sigma m (\dot{x} dx + \dot{y} dy + \dot{z} dz) \ldots\ldots\ldots\ldots (4).$$

Hence we might, as many writers in fact have virtually done, define action thus:—

Space average of momentums.

The action of a system is equal to the sum of the *average momentums for the spaces* described by the particles from any era each multiplied by the length of its path.

Least action.

327. The principle of Least Action is this:—Of all the different sets of paths along which a conservative system may be guided to move from one configuration to another, with the sum of its potential and kinetic energies equal to a given constant, that one for which the action is the least is such that the system will require only to be started with the proper velocities, to move along it unguided. Consider the Problem:— Given the whole initial kinetic energy; find the initial velocities through one given configuration, which shall send the system unguided to another specified configuration. This problem is essentially determinate, but generally has multiple solutions (§ 363 below); (or only imaginary solutions.)

General Problem of *aim*. High aim and low aim for same target. Target too far off to be reached.

If there are any real solutions, there is one of them for which the action is less than for any other real solution, and less than for any constrainedly guided motion with proper sum of potential and kinetic energies. Compare §§ 346—366 below. Least action.

Let x, y, z be the co-ordinates of a particle, m, of the system, at time τ, and V the potential energy of the system in its particular configuration at this instant; and let it be required to find the way to pass from one given configuration to another with velocities at each instant satisfying the condition

$$\Sigma \tfrac{1}{2} m (\dot{x}^2 + \dot{y}^2 + \dot{z}^2) + V = E, \text{ a constant} \dots\dots\dots\dots\dots (5),$$

so that A, or

$$\int \Sigma m (\dot{x}dx + \dot{y}dy + \dot{z}dz)$$

may be the least possible.

By the method of variations we must have $\delta A = 0$, where

$$\delta A = \int \Sigma m (\dot{x}d\delta x + \dot{y}d\delta y + \dot{z}d\delta z + \delta\dot{x}dx + \delta\dot{y}dy + \delta\dot{z}dz) \dots\dots (6).$$

Taking in this $dx = \dot{x}d\tau$, $dy = \dot{y}d\tau$, $dz = \dot{z}d\tau$, and remarking that

$$\Sigma m (\dot{x}\delta\dot{x} + \dot{y}\delta\dot{y} + \dot{z}\delta\dot{z}) = \delta T \dots\dots\dots\dots\dots\dots\dots (7),$$

we have

$$\int \Sigma m (\delta\dot{x}dx + \delta\dot{y}dy + \delta\dot{z}dz) = \int_0^t \delta T d\tau \dots\dots\dots\dots (8).$$

Also by integration by parts,

$$\int \Sigma m (\dot{x}d\delta x + \dots) = \{\Sigma m (\dot{x}\delta x + \dots)\} - [\Sigma m (\dot{x}\delta x + \dots)] - \int \Sigma m (\ddot{x}\delta x + \dots) d\tau,$$

where $[\dots]$ and $\{\dots\}$ denote the values of the quantities enclosed, at the beginning and end of the motion considered, and where, further, it must be remembered that $d\dot{x} = \ddot{x}d\tau$, etc. Hence, from above,

$$\delta A = \{\Sigma m (\dot{x}\delta x + \dot{y}\delta y + \dot{z}\delta z)\} - [\Sigma m (\dot{x}\delta x + \dot{y}\delta y + \dot{z}\delta z)]$$
$$+ \int_0^t d\tau \, [\delta T - \Sigma m (\ddot{x}\delta x + \ddot{y}\delta y + \ddot{z}\delta z)] \dots\dots\dots\dots (9).$$

This, it may be observed, is a perfectly general kinematical expression, unrestricted by any terminal or kinetic conditions. Now in the present problem we suppose the initial and final positions to be invariable. Hence the terminal variations, δx, etc., must all vanish, and therefore the integrated expressions $\{\dots\}$, $[\dots]$ disappear. Also, in the present problem $\delta T = -\delta V$, by the equation of energy (5). Hence, to make $\delta A = 0$, since the intermediate variations, δx, etc., are quite arbitrary, subject only to the con-

Least action.

ditions of the system, we must have

$$\Sigma m(\ddot{x}\delta x + \ddot{y}\delta y + \ddot{z}\delta z) + \delta V = 0 \ldots\ldots\ldots\ldots\ldots\ldots\ldots (10),$$

which [(4), § 293 above] is the general variational equation of motion of a conservative system. This proves the proposition.

Principle of Least Action applied to find Lagrange's generalized equations of motion.

It is interesting and instructive as an illustration of the principle of least action, to derive directly from it, without any use of Cartesian co-ordinates, Lagrange's equations in generalized co-ordinates, of the motion of a conservative system [§ 318 (24)]. We have

$$A = \int 2T dt,$$

where T denotes the formula of § 313 (2). If now we put

$$T = \tfrac{1}{2}\frac{ds^2}{dt^2},$$

so that $$ds^2 = (\psi, \psi)\, d\psi^2 + 2\,(\psi, \phi)\, d\psi d\phi + \text{etc.},$$

we have $$A = \int \frac{ds}{dt}\, ds.$$

Hence

$$\delta A = \int \left(\delta\frac{ds}{dt}\, ds + \frac{ds\delta ds}{dt}\right) = \int dt \frac{ds}{dt}\,\delta\,\frac{ds}{dt} + \int \frac{\tfrac{1}{2}\delta\,(ds^2)}{dt}$$

$$= \int dt\delta T + \int \frac{(\psi, \psi)\, d\psi + (\psi, \phi)\, d\phi + \text{etc}}{dt}\,\delta d\psi$$

$$+ \int \frac{(\psi\ \phi)\, d\psi + (\phi, \phi)\, d\phi + \text{etc.}}{dt}\,\delta d\phi + \text{etc.} + \int dt \delta_{(\psi, \phi, \text{etc.})}\, T,$$

where $\delta_{(\psi, \phi, \text{etc.})}$ denotes variation dependent on the explicit appearance of ψ, ϕ, etc. in the coefficients of the quadratic function T. The second chief term in the formula for δA is clearly equal to $\int \frac{dT}{d\dot{\psi}}\, d\delta\psi$, and this, integrated by parts, becomes

$$\frac{dT}{d\dot{\psi}}\delta\psi - \int d\,\frac{dT}{d\dot{\psi}}\delta\psi, \quad \text{or} \quad \left[\frac{dT}{d\dot{\psi}}\,\delta\psi\right] - \int dt\,\frac{d}{dt}\frac{dT}{d\dot{\psi}}\,\delta\psi,$$

where [ ] denotes the difference of the values of the bracketed expression, at the beginning and end of the time $\int dt$. Thus we have finally

$$\delta A = \left[\frac{dT}{d\dot{\psi}}\,\delta\psi + \frac{dT}{d\dot{\phi}}\,\delta\phi + \text{etc.}\right]$$

$$+ \int dt \left\{-\left(\frac{d}{dt}\frac{dT}{d\dot{\psi}}\,\delta\psi + \frac{d}{dt}\frac{dT}{d\dot{\phi}}\,\delta\phi + \text{etc.}\right) + \delta T + \delta_{(\psi, \phi, \text{etc.})}\, T\right\} \ldots. (10)'.$$

So far we have a purely kinematical formula. Now introduce the dynamical condition [§ 293 (7)]

Principle of Least Action applied to find Lagrange's generalized equations of motion.

$$T = C - V \text{ ....} (10)''.$$

From it we find

$$\delta T = -\left(\frac{dV}{d\psi}\delta\psi + \frac{dV}{d\phi}\delta\phi + \text{etc.}\right) \text{ ....} (10)'''.$$

Again, we have

$$\delta_{(\psi,\phi,\text{etc.})} T = \frac{dT}{d\psi}\delta\psi + \frac{dT}{d\phi}\delta\phi + \text{etc. ....} (10)^{\text{iv}}.$$

Hence (10)′ becomes

$$\delta A = \left[\frac{dT}{d\dot{\psi}}\delta\psi + \frac{dT}{d\dot{\phi}}\delta\phi + \text{etc.}\right]$$

$$+ \int dt\left\{\left(-\frac{d}{dt}\frac{dT}{d\dot{\psi}} + \frac{dT}{d\psi} + \frac{dV}{d\psi}\right)\delta\psi + (\text{etc.})\delta\phi + \text{etc.}\right\} \text{...} (10)^{\text{v}}.$$

To make this a minimum we have

$$-\frac{d}{dt}\frac{dT}{d\dot{\psi}} + \frac{dT}{d\psi} + \frac{dV}{d\psi} = 0, \text{ etc. ....} (10)^{\text{vi}},$$

which are the required equations [§ 318 (24)].

From the proposition that $\delta A = 0$ implies the equations of motion, it follows that

328. In any unguided motion whatever, of a conservative system, the Action from any one stated position to any other, though not necessarily a minimum, fulfils the *stationary condition*, that is to say, the condition that the variation vanishes, which secures either a minimum or maximum, or maximum-minimum.

Why called "stationary action" by Hamilton.

This can scarcely be made intelligible without mathematical language. Let (x_1, y_1, z_1), (x_2, y_2, z_2), etc., be the co-ordinates of particles, m_1, m_2, etc., composing the system; at any time τ of the actual motion. Let V be the potential energy of the system, in this configuration; and let E denote the given value of the sum of the potential and kinetic energies. The equation of energy is—

Stationary action.

$$\tfrac{1}{2}\{m_1(\dot{x}_1^2 + \dot{y}_1^2 + \dot{z}_1^2) + m_2(\dot{x}_2^2 + \dot{y}_2^2 + \dot{z}_2^2) + \text{etc.}\} + V = E \text{...} (5) \text{ bis.}$$

Choosing any part of the motion, for instance that from time 0 to time t, we have, for the action during it,

$$A = \int_0^t (E - V)\,d\tau = Et - \int_0^t V d\tau \text{ ....} (11).$$

Stationary action.

Let now the system be guided to move in any other way possible for it, with any other velocities, from the same initial to the same final configuration as in the given motion, subject only to the condition, that the sum of the kinetic and potential energies shall still be E. Let (x_1', y_1', z_1'), etc., be the co-ordinates, and V' the corresponding potential energy; and let $(\dot{x}_1', \dot{y}_1', \dot{z}_1')$, etc., be the component velocities, at time τ in this arbitrary motion; equation (2) still holding, for the accented letters, with only E unchanged. For the action we shall have

$$A' = Et' - \int_0^{t'} V' d\tau \quad \ldots\ldots\ldots\ldots\ldots\ldots\ldots (12),$$

where t' is the time occupied by this supposed motion. Let now θ denote a small numerical quantity, and let ξ_1, η_1, etc., be finite lines such that

$$\frac{x_1' - x_1}{\xi_1} = \frac{y_1' - y_1}{\eta_1} = \frac{z_1' - z_1}{\zeta} = \frac{x_2' - x_2}{\xi_2} = \text{etc.} = \theta.$$

The "principle of stationary action" is, that $\frac{V' - V}{\theta}$ vanishes when θ is made infinitely small, for every possible deviation $(\xi_1\theta, \eta_1\theta$, etc.) from the natural way and velocities, subject only to the equation of energy and to the condition of passing through the stated initial and final configurations: and conversely, that if $\frac{V' - V}{\theta}$ vanishes with θ for every possible such deviation from a certain way and velocities, specified by (x_1, y_1, z_1), etc., as the co-ordinates at t, *this* way and *these* velocities are such that the system unguided will move accordingly if only started with proper velocities from the initial configuration.

Varying action.

329. From this principle of stationary action, founded, as we have seen, on a comparison between a natural motion, and any other motion, arbitrarily guided and subject only to the law of energy, the initial and final configurations of the system being the same in each case, Hamilton passes to the consideration of the variation of the action in a natural or unguided motion of the system produced by varying the initial and final configurations, and the sum of the potential and kinetic energies. The result is, that

330. The rate of *decrease* of the action per unit of increase of any one of the free (generalized) co-ordinates (§ 204) specifying the initial configuration, is equal to the corresponding (generalized) component momentum [§ 313, (c)] of the actual motion from that configuration: the rate of *increase* of the action per unit increase of any one of the free co-ordinates specifying the final configuration, is equal to the corresponding component momentum of the actual motion towards this second configuration: and the rate of increase of the action per unit increase of the constant sum of the potential and kinetic energies, is equal to the time occupied by the motion of which the action is reckoned.

Varying action.

To prove this we must, in our previous expression (9) for δA, now suppose the terminal co-ordinates to vary; δT to become $\delta E - \delta V$, in which δE is a constant during the motion; and each set of paths and velocities to belong to an unguided motion of the system, which requires (10) to hold. Hence

Action expressed as a function of initial and final co-ordinates and the energy;

$$\delta A = \{\Sigma m(\dot{x}\delta x + \dot{y}\delta y + \dot{z}\delta z)\} - [\Sigma m(\dot{x}\delta x + \dot{y}\delta y + \dot{z}\delta z)] + t\delta E \ldots (13).$$

If, now, in the first place, we suppose the particles constituting the system to be all free from constraint, and therefore (x, y, z) for each to be three independent variables, and if, for distinctness, we denote by (x_1', y_1', z_1') and (x_1, y_1, z_1) the co-ordinates of m_1 in its initial and final positions, and by $(\dot{x}_1', \dot{y}_1', \dot{z}_1')$, $(\dot{x}_1, \dot{y}_1, \dot{z}_1)$ the components of the velocity it has at those points, we have, from the preceding, according to the ordinary notation of partial differential coefficients,

$$\left.\begin{array}{lll} \dfrac{dA}{dx_1'} = -m_1\dot{x}_1', & \dfrac{dA}{dy_1'} = -m_1\dot{y}_1', & \dfrac{dA}{dz_1'} = -m_1\dot{z}_1', \text{ etc.} \\ \dfrac{dA}{dx_1} = m_1\dot{x}_1, & \dfrac{dA}{dy_1} = m_1\dot{y}_1, & \dfrac{dA}{dz_1} = m_1\dot{z}_1, \text{ etc.} \\ \text{and} & \dfrac{dA}{dE} = t. & \end{array}\right\} \ldots (14).$$

its differential coefficients equal respectively to initial and final momentums, and to the time from beginning to end.

In these equations we must suppose A to be expressed as a function of the initial and final co-ordinates, in all six times as many independent variables as there are of particles; and E, one more variable, the sum of the potential and kinetic energies.

If the system consist not of free particles, but of particles connected in any way forming either one rigid body or any number

Varying action.

of rigid bodies connected with one another or not, we might, it is true, be contented to regard it still as a system of free particles, by taking into account among the impressed forces, the forces necessary to compel the satisfaction of the conditions of connexion. But although this method of dealing with a system of connected particles is very simple, so far as the law of energy merely is concerned, Lagrange's methods, whether that of "equations of condition," or, what for our present purposes is much more convenient, his "generalized co-ordinates," relieve us from very troublesome interpretations when we have to consider the displacements of particles due to arbitrary variations in the configuration of a system.

Let us suppose then, for any particular configuration (x_1, y_1, z_1) $(x_2, y_2, z_2)\ldots$, the expression

$$m_1(\dot{x}_1\delta x_1 + \dot{y}_1\delta y_1 + \dot{z}_1\delta z_1) + \text{etc., to become } \xi\delta\psi + \eta\delta\phi + \zeta\delta\theta + \text{etc.} \quad (15),$$

Same propositions for generalized co-ordinates.

when transformed into terms of ψ, ϕ, $\theta\ldots$, generalized co-ordinates, as many in number as there are of degrees of freedom for the system to move [§ 313, (c)].

The same transformation applied to the kinetic energy of the system would obviously give

$$\tfrac{1}{2}m_1(\dot{x}_1^2 + \dot{y}_1^2 + \dot{z}_1^2) + \text{etc.} = \tfrac{1}{2}(\xi\dot{\psi} + \eta\dot{\phi} + \zeta\dot{\theta} + \text{etc.})\ldots\ldots(16),$$

and hence ξ, η, ζ, etc., are those linear functions of the generalized velocities which, in § 313 (e), we have designated as "generalized components of momentum;" and which, when T, the kinetic energy, is expressed as a quadratic function of the velocities (of course with, in general, functions of the co-ordinates ψ, ϕ, θ, etc., for the coefficients) are derivable from it thus:

$$\xi = \frac{dT}{d\dot{\psi}}, \quad \eta = \frac{dT}{d\dot{\phi}}, \quad \zeta = \frac{dT}{d\dot{\theta}}, \text{ etc.}\ldots\ldots\ldots\ldots\ldots(17).$$

Hence, taking as before non-accented letters for the second, and accented letters for the initial, configurations of the system respectively, we have

$$\left.\begin{array}{lll} \dfrac{dA}{d\psi'} = -\xi', & \dfrac{dA}{d\phi'} = -\eta', & \dfrac{dA}{d\theta'} = -\zeta', \text{ etc.} \\ \dfrac{dA}{d\psi} = \xi, & \dfrac{dA}{d\phi} = \eta, & \dfrac{dA}{d\theta} = \zeta, \text{ etc.} \\ \text{and, as before,} & \dfrac{dA}{dE} = t, & \end{array}\right\}\ldots\ldots\ldots\ldots(18).$$

These equations (18), including of course (14) as a particular case, express in mathematical terms the proposition stated in words above, as the *Principle of Varying Action.* Varying action.

The values of the momentums, thus, (14) and (18), expressed in terms of differential coefficients of A, must of course satisfy the equation of energy. Hence, for the case of free particles,

Hamilton's "characteristic equation" of motion in Cartesian co-ordinates.

$$\Sigma \frac{1}{m}\left(\frac{dA^2}{dx^2}+\frac{dA^2}{dy^2}+\frac{dA^2}{dz^2}\right)=2(E-V) \quad \ldots\ldots\ldots\ldots(19),$$

$$\Sigma \frac{1}{m}\left(\frac{dA^2}{dx'^2}+\frac{dA^2}{dy'^2}+\frac{dA^2}{dz'^2}\right)=2(E-V') \quad \ldots\ldots\ldots\ldots(20).$$

Or, in general, for a system of particles or rigid bodies connected in any way, we have, (16) and (18),

Hamilton's characteristic equation of motion in generalized co-ordinates.

$$\dot\psi\frac{dA}{d\psi}+\dot\phi\frac{dA}{d\phi}+\dot\theta\frac{dA}{d\theta}+\text{etc.}=2(E-V)\ldots\ldots\ldots(21),$$

$$-\left(\dot\psi'\frac{dA}{d\psi'}+\dot\phi'\frac{dA}{d\phi'}+\dot\theta'\frac{dA}{d\theta'}+\text{etc.}\right)=2(E-V')\ldots\ldots(22),$$

where $\dot\psi$, $\dot\phi$, etc., are expressible as linear functions of $\frac{dA}{d\psi}$, $\frac{dA}{d\phi}$, etc., by the solution of the equations

$$\left.\begin{array}{l}(\psi,\psi)\,\dot\psi+(\psi,\phi)\,\dot\phi+(\psi,\theta)\,\dot\theta+\text{etc.}=\xi=\dfrac{dA}{d\psi}\\ (\phi,\psi)\,\dot\psi+(\phi,\phi)\,\dot\phi+(\phi,\theta)\,\dot\theta+\text{etc.}=\eta=\dfrac{dA}{d\phi}\\ \qquad\text{etc.}\qquad\qquad\qquad\text{etc.}\end{array}\right\}\ldots\ldots(23),$$

and $\dot\psi'$, $\dot\phi'$, etc., as similar functions of $-\frac{dA}{d\psi'}$, $-\frac{dA}{d\phi'}$, etc, by

$$\left.\begin{array}{l}(\psi',\psi')\,\dot\psi'+(\psi',\phi')\,\dot\phi'+(\psi',\theta')\,\dot\theta'+\text{etc.}=\xi'=-\dfrac{dA}{d\psi'}\\ (\phi',\psi')\,\dot\psi'+(\phi',\phi')\,\dot\phi'+(\phi',\theta')\,\dot\theta'+\text{etc.}=\eta'=-\dfrac{dA}{d\phi'}\\ \qquad\text{etc.}\qquad\qquad\qquad\text{etc.}\end{array}\right\}\ldots(24),$$

where it must be remembered that (ψ, ψ), (ψ, ϕ), etc., are functions of the specifying elements, ψ, ϕ, θ, etc., depending on the kinematical nature of the co-ordinate system alone, and quite independent of the dynamical problem with which we are now concerned; being the coefficients of the half squares and the products of the generalized velocities in the expression for the

Varying action.

kinetic energy of any motion of the system; and that (ψ', ψ'), (ψ', ϕ'), etc., are the same functions with ψ', ϕ', etc., written for ψ, ϕ, θ, etc.; but, on the other hand, that A is a function of all the elements ψ, ϕ, etc., ψ', ϕ', etc. Thus the first member of (21) is a quadratic function of $\frac{dA}{d\psi}$, $\frac{dA}{d\phi}$, etc., with coefficients, known functions of ψ, ϕ, etc., depending merely on the kinematical relations of the system, and the masses of its parts, but not at all on the actual forces or motions; while the second member is a function of the co-ordinates ψ, ϕ, etc., depending on the forces in the dynamical problem, and a constant expressing the particular value given to the sum of the potential and kinetic energies in the actual motion; and so for (22), and ψ', ϕ', etc.

Proof that the characteristic equation defines the motion, for free particles.

It is remarkable that the single linear partial differential equation (19) of the first order and second degree, for the case of free particles, or its equivalent (21), is sufficient to determine a function A, such that the equations (14) or (18) express the momentums in an actual motion of the system, subject to the given forces. For, taking the case of free particles first, and differentiating (19) still on the Hamiltonian understanding that A is expressed merely as a function of initial and final co-ordinates, and of E, the sum of the potential and kinetic energies, we have

$$2\Sigma \frac{1}{m}\left(\frac{dA}{dx}\frac{d^2A}{dx_1 dx} + \frac{dA}{dy}\frac{d^2A}{dx_1 dy} + \frac{dA}{dz}\frac{d^2A}{dx_1 dz}\right) = -2\frac{dV}{dx_1}.$$

But, by (14),

$$\frac{1}{m_1}\frac{dA}{dx_1} = \dot{x}_1,\ \frac{1}{m_1}\frac{dA}{dy_1} = \dot{y}_1, \text{ etc.},$$

and therefore

$$\frac{d^2A}{dx_1^2} = m_1\frac{d\dot{x}_1}{dx_1},\ \frac{d^2A}{dx_1 dy_1} = m_1\frac{d\dot{y}_1}{dx_1} = m_1\frac{d\dot{x}_1}{dy_1},\ \frac{d^2A}{dx_1 dz_1} = m_1\frac{d\dot{z}_1}{dx_1} = m_1\frac{d\dot{x}_1}{dz_1},$$

$$\frac{d^2A}{dx_1 dx_2} = m_2\frac{d\dot{x}_2}{dx_1} = m_1\frac{d\dot{x}_1}{dx_2}, \text{ etc.}$$

Using these properly in the preceding and taking half; and writing out for two particles to avoid confusion as to the meaning of Σ, we have

$$m_1\left(\dot{x}_1\frac{d\dot{x}_1}{dx_1} + \dot{y}_1\frac{d\dot{x}_1}{dy_1} + \dot{z}_1\frac{d\dot{x}_1}{dz_1} + \dot{x}_2\frac{d\dot{x}_1}{dx_2} + \dot{y}_2\frac{d\dot{x}_1}{dy_2} + \dot{z}_2\frac{d\dot{x}_1}{dz_2} + \text{etc.}\right) = -\frac{dV}{dx_1} \quad (25).$$

Now if we multiply the first member by dt, we have clearly the change of the value of $m_1\dot{x}_1$ due to varying, still on the Hamiltonian supposition, the co-ordinates of all the points, that is to say, the configuration of the system, from what it is at any moment to what it becomes at a time dt later; and it is therefore the actual change in the value of $m\dot{x}_1$, in the natural motion, from the time, t, when the configuration is $(x_1, y_1, z_1, x_2, \ldots, E)$, to the time $t+dt$. It is therefore equal to $m_1\ddot{x}_1 dt$, and hence (25) becomes simply $m_1\ddot{x}_1 = -\dfrac{dV}{dx_1}$. Similarly we find

Varying action.

Proof that the characteristic equation defines the motion, for free particles.

$$m_1\ddot{y}_1 = -\frac{dV}{dy_1},\ m_1\ddot{z}_1 = -\frac{dV}{dz_1},\ m_2\ddot{x}_2 = -\frac{dV}{dx_2},\ \text{etc.}$$

But these are [§ 293, (4)] the elementary differential equations of the motions of a conservative system composed of free mutually influencing particles.

If next we regard x_1, y_1, z_1, x_2, etc., as constant, and go through precisely the same process with reference to x_1', y_1', z_1', x_2', etc., we have exactly the same equations among the accented letters, with only the difference that $-A$ appears in place of A; and end with $m_1\ddot{x}_1' = \dfrac{dV'}{dx_1'}$, from which we infer that, if (20) is satisfied, the motion represented by (14) is a natural motion through the configuration $(x_1', y_1', z_1', x_2', \text{etc.})$.

Hence if both (19) and (20) are satisfied, and if when $x_1 = x_1'$, $y_1 = y_1'$, $z_1 = z_1'$, $x_2 = x_2'$, etc., we have $\dfrac{dA}{dx_1} = -\dfrac{dA}{dx_1'}$, etc., the motion represented by (14) is a natural motion through the two configurations $(x_1', y_1', z_1', x_2', \text{etc.})$, and $(x_1, y_1, z_1, x_2, \text{etc.})$. Although the signs in the preceding expressions have been fixed on the supposition that the motion is *from* the former, to the latter configuration, it may clearly be from either towards the other, since whichever way it is, the reverse is also a natural motion (§ 271), according to the general property of a conservative system.

To prove the same thing for a conservative system of particles or rigid bodies connected in any way, we have, in the first place, from (18)

Same proposition for a connected system, and generalized co-ordinates.

$$\frac{d\eta}{d\psi} = \frac{d\xi}{d\phi},\ \frac{d\zeta}{d\psi} = \frac{d\xi}{d\theta},\ \text{etc.} \quad \ldots\ldots\ldots\ldots\ldots\ldots (26),$$

where, on the Hamiltonian principle, we suppose $\dot{\psi}$, $\dot{\phi}$, etc., and ξ, η, etc., to be expressed as functions of ψ, ϕ, etc., ψ', ϕ', etc.,

Varying action.

and the sum of the potential and kinetic energies. On the same supposition, differentiating (21), we have

$$\dot\psi\frac{d\xi}{d\psi}+\dot\phi\frac{d\eta}{d\psi}+\dot\theta\frac{d\zeta}{d\psi}+\text{etc.}+\xi\frac{d\dot\psi}{d\psi}+\eta\frac{d\dot\phi}{d\psi}+\zeta\frac{d\dot\theta}{d\psi}+\text{etc.}=-2\frac{dV}{d\psi}\quad\text{....(27)}.$$

But, by (26), and by the considerations above, we have

$$\dot\psi\frac{d\xi}{d\psi}+\dot\phi\frac{d\eta}{d\psi}+\dot\theta\frac{d\zeta}{d\psi}+\text{etc.}=\dot\psi\frac{d\xi}{d\psi}+\dot\phi\frac{d\xi}{d\phi}+\dot\theta\frac{d\xi}{d\theta}+\text{etc.}=\dot\xi\quad\text{....(28)},$$

where $\dot\xi$ denotes the rate of variation of ξ per unit of time in the actual motion.

Again, we have

$$\left.\begin{aligned}\frac{d\dot\psi}{d\psi}&=\frac{\partial\dot\psi}{d\xi}\frac{d\xi}{d\psi}+\frac{\partial\dot\psi}{d\eta}\frac{d\eta}{d\psi}+\text{etc.}+\frac{\partial\dot\psi}{d\psi}\\\frac{d\dot\phi}{d\psi}&=\frac{\partial\dot\phi}{d\xi}\frac{d\xi}{d\psi}+\frac{\partial\dot\phi}{d\eta}\frac{d\eta}{d\psi}+\text{etc.}+\frac{\partial\dot\phi}{d\psi}\\&\quad\text{etc.}\qquad\qquad\text{etc.}\end{aligned}\right\}\quad\text{.......... (29)},$$

if, as in Hamilton's system of canonical equations of motion, we suppose $\dot\psi$, $\dot\phi$, etc., to be expressed as linear functions of ξ, η, etc., with coefficients involving ψ, ϕ, θ, etc., and if we take ∂ to denote the partial differentiation of these functions with reference to the system ξ, η,...ψ, ϕ,..., regarded as independent variables. Let the coefficients be denoted by $[\psi, \dot\psi]$, etc., according to the plan followed above; so that, if the formula for the kinetic energy be

$$T=\tfrac{1}{2}\{[\psi,\psi]\,\xi^2+[\phi,\phi]\,\eta^2+\ldots+2\,[\psi,\phi]\,\xi\eta+\text{etc.}\}\text{.....(30)},$$

we have

$$\left.\begin{aligned}\dot\psi&=\frac{\partial T}{d\xi}=[\psi,\psi]\,\xi+[\psi,\phi]\,\eta+[\psi,\theta]\,\zeta+\text{etc.}\\\dot\phi&=\frac{\partial T}{d\eta}=[\phi,\psi]\,\xi+[\phi,\phi]\,\eta+[\phi,\theta]\,\zeta+\text{etc.}\\&\qquad\text{etc.}\qquad\qquad\text{etc.}\end{aligned}\right\}\text{......(31)},$$

where of course $[\psi, \phi]$, and $[\phi, \psi]$, mean the same.

Hence $$\frac{\partial\dot\psi}{d\xi}=[\psi,\psi],\ \frac{\partial\dot\phi}{d\xi}=[\phi,\psi],\ \ldots;$$

$$\frac{\partial\dot\psi}{d\psi}=\frac{d\,[\psi,\psi]}{d\psi}\xi+\frac{d\,[\psi,\phi]}{d\psi}\eta+\text{etc.};\ \frac{\partial\dot\phi}{d\psi}=\frac{d\,[\phi,\psi]}{d\psi}\xi+\text{etc. etc.},$$

and therefore, by (29),

$$\begin{aligned}\xi\frac{d\dot\psi}{d\psi}+\eta\frac{d\dot\phi}{d\psi}+\zeta\frac{d\dot\theta}{d\psi}+\text{etc.}&=\{[\psi,\psi]\xi+[\phi,\psi]\eta+\text{etc.}\}\frac{d\xi}{d\psi}+\{[\psi,\phi]\xi+[\phi,\phi]\eta+\text{etc.}\}\frac{d\eta}{d\psi}\\&\quad+\text{etc.}+\frac{d[\psi,\psi]}{d\psi}\xi^2+\frac{d[\phi,\phi]}{d\psi}\eta^2+\text{etc.}+2\frac{d[\psi,\phi]}{d\psi}\xi\eta+\text{etc.}\\&=\dot\psi\frac{d\xi}{d\psi}+\dot\phi\frac{d\eta}{d\psi}+\text{etc.}+2\frac{\partial T}{d\psi};\end{aligned}$$

whence, by (28), we see that

Hamiltonian form of Lagrange's generalized equations deduced from characteristic equation.

$$\xi\frac{d\dot\psi}{d\psi}+\eta\frac{d\dot\phi}{d\psi}+\zeta\frac{d\dot\theta}{d\psi}+\text{etc.}=\dot\xi+2\frac{\partial T}{d\psi}\text{...........}(32).$$

This, and (28), reduce the first member of (27) to $2\dot\xi+2\frac{\partial T}{d\psi}$, and therefore, halving, we conclude

$$\dot\xi+\frac{\partial T}{d\psi}=-\frac{dV}{d\psi},\text{ and similarly, }\dot\eta+\frac{\partial T}{d\phi}=-\frac{dV}{d\phi},\text{ etc...}(33).$$

These, in all as many differential equations as there are of variables, ψ, ϕ, etc., suffice for determining them in terms of t and twice as many arbitrary constants. But every solution of the dynamical problem, as has been demonstrated above, satisfies (21) and (23); and therefore it must satisfy these (33), which we have derived from them. These (33) are therefore *the* equations of motion, of the system referred to generalized co-ordinates, as many in number as it has of degrees of freedom. They are the Hamiltonian explicit equations of motion, of which a direct demonstration was given in § 318 above. Just as above, it appears therefore, that if (21) and (22) are satisfied, (18) expresses a natural motion of the system from one to another of the two configurations $(\psi, \phi, \theta, \ldots)$ $(\psi', \phi', \theta', \ldots)$. Hence

331. The determination of the motion of any conservative system from one to another of any two configurations, when the sum of its potential and kinetic energies is given, depends on the determination of a single function of the co-ordinates of those configurations by solution of two quadratic partial differential equations of the first order, with reference to those two sets of co-ordinates respectively, with the condition that the corresponding terms of the two differential equations become separately equal when the values of the two sets of co-ordinates agree. The function thus determined and employed to express the solution of the kinetic problem was called the *Characteristic Function* by Sir W. R. Hamilton, to whom the method is due. It is, as we have seen, the "action" from one of the configurations to the other; but its peculiarity in Hamilton's system is, that it is to be expressed as a function of the co-ordinates and a constant, the whole energy, as explained above. It is evi-

Hence proof concluded.

Characteristic function.

dently symmetrical with respect to the two configurations, changing only in sign if their co-ordinates are interchanged.

Characteristic equation of motion.

Since not only the complete solution of the problem of motion gives a solution, A, of the partial differential equation (19) or (21), but, as we have just seen [§ 330 (33), etc.], every solution of this equation corresponds to an actual problem relative to the motion, it becomes an object of mathematical analysis, which could not be satisfactorily avoided, to find what character of completeness a solution or integral of the differential equation must have in order that a complete integral of the dynamical equations may be derivable from it—a question which seems to have been first noticed by Jacobi. What is called a "complete integral" of the differential equation; that is to say, an expression,

Complete integral of characteristic equation.

$$A = A_0 + F(\psi, \phi, \theta, \ldots \alpha, \beta, \ldots) \quad \text{.................. (34)},$$

for A satisfying it and involving the same number i, let us suppose, of independent arbitrary constants, A_0, α, β,...as there are of the independent variables, ψ, ϕ, etc.; leads, as he found, to a complete final integral of the equations of motion, expressed as follows:—

$$\frac{dF}{d\alpha} = \mathfrak{A}, \; \frac{dF}{d\beta} = \mathfrak{B} \quad \text{........................ (35)},$$

and, as above,

$$\frac{dF}{dE} = t + \epsilon \quad \text{................................ (36)},$$

where ϵ is the constant depending on the epoch, or era of reckoning, chosen, and $\mathfrak{A}$, $\mathfrak{B}$,... are $i-1$ other arbitrary constants, constituting in all, with E, α, β,..., the proper number, $2i$, of arbitrary constants. This is proved by remarking that (35) are the equations of the "course" (or *paths* in the case of a system of free particles), which is obvious. For they give

$$\left.\begin{array}{l} 0 = \dfrac{d}{d\psi}\dfrac{dF}{d\alpha}d\psi + \dfrac{d}{d\phi}\dfrac{dF}{d\alpha}d\phi + \dfrac{d}{d\theta}\dfrac{dF}{d\alpha}d\theta + \ldots \\ 0 = \dfrac{d}{d\psi}\dfrac{dF}{d\beta}d\psi + \dfrac{d}{d\phi}\dfrac{dF}{d\beta}d\phi + \dfrac{d}{d\theta}\dfrac{dF}{d\beta}d\theta + \ldots \\ \qquad \text{etc.} \qquad\qquad\qquad \text{etc.} \end{array}\right\} \text{........ (37)},$$

in all $i-1$ equations to determine the ratios $d\psi : d\phi : d\theta : \ldots$ From these, and (21), we find

$$\frac{d\psi}{\dot{\psi}} = \frac{d\phi}{\dot{\phi}} = \frac{d\theta}{\dot{\theta}} \quad \text{.......................... (38)}$$

[since (37) are the same as the equations which we obtain by differentiating (21) and (23) with reference to α, β,... successively, only that they have $d\psi$, $d\phi$, $d\theta$,... in place of $\dot{\psi}$, $\dot{\phi}$, $\dot{\theta}$,...]. Complete integral of characteristic equation.

A perfectly general solution of the partial differential equation, that is to say, an expression for A including every function of ψ, ϕ, θ,... which can satisfy (21), may of course be found, by the regular process, from the complete integral (34), by eliminating A_0, α, β,... from it by means of an arbitrary equation General solution derived from complete integral.

$$f(A_0, \alpha, \beta, \ldots) = 0,$$

and the $(i-1)$ equations

$$\frac{1}{\frac{df}{dA_0}} = \frac{\frac{dF}{d\alpha}}{\frac{df}{d\alpha}} = \frac{\frac{dF}{d\beta}}{\frac{df}{d\beta}} = \ldots$$

where f denotes an arbitrary function of the i elements A_0, α, β,... now made to be variables depending on ψ, ϕ,... But the full meaning of the general solution of (21) will be better understood in connexion with the physical problem if we first go back to the Hamiltonian solution, and then from it to the general. Thus, first, let the equations (35) of the course be assumed to be satisfied for each of two sets ψ, ϕ, θ,..., and ψ', ϕ', θ',..., of the co-ordinates. They will give $2(i-1)$ equations for determining the $2(i-1)$ constants α, β,..., $\mathfrak{A}$, $\mathfrak{B}$,..., in terms of ψ, ϕ, ..., ψ', ϕ',..., to fulfil these conditions. Using the values of α, β,..., so found, and assigning A_0 so that A shall vanish when $\psi = \psi'$, $\phi = \phi'$, etc., we have the Hamiltonian expression for A in terms of ψ, ϕ, ..., ψ', ϕ', ..., and E, which is therefore equivalent to a "complete integral" of the partial differential equation (21). Now let ψ', ϕ', ..., be connected by any single arbitrary equation

$$\mathbf{f}(\psi', \phi', \ldots) = 0 \quad \ldots\ldots\ldots\ldots (39),$$

and by means of this equation and the following $(i-1)$ equations, let their values be determined in terms of ψ, ϕ, ..., and E:—

$$\frac{\frac{dA}{d\psi'}}{\frac{d\mathbf{f}}{d\psi'}} = \frac{\frac{dA}{d\phi'}}{\frac{d\mathbf{f}}{d\phi'}} = \frac{\frac{dA}{d\theta'}}{\frac{d\mathbf{f}}{d\theta'}} = \text{etc.} \quad \ldots\ldots\ldots\ldots (40).$$

Substituting the values thus found for ψ', ϕ', θ', etc., in the Hamiltonian A, we have an expression for A, which is the general

General solution derived from complete integral.

solution of (21). For we see immediately that (40) expresses that the values of A are equal for all configurations satisfying (39), that is to say, we have

$$\frac{dA}{d\psi'}d\psi' + \frac{dA}{d\phi'}d\phi' + \ldots = 0$$

when ψ', ϕ', etc., satisfy (39) and (40). Hence when, by means of these equations, ψ', ϕ', ..., are eliminated from the Hamiltonian expression for A, the complete Hamiltonian differential

$$dA = \left(\frac{dA}{d\psi}\right)d\psi + \left(\frac{dA}{d\phi}\right)d\phi + \ldots + \frac{dA}{d\psi'}d\psi' + \frac{dA}{d\phi'}d\phi' + \ldots\ldots\ldots (41)$$

becomes merely

$$dA = \left(\frac{dA}{d\psi}\right)d\psi + \left(\frac{dA}{d\phi}\right)d\phi + \ldots\ldots\ldots\ldots\ldots\ldots (42),$$

where $\left(\frac{dA}{d\psi}\right)$, etc., denote the differential coefficients in the Hamiltonian expression. Hence, A being now a function of ψ, ϕ, etc., both as these appear in the Hamiltonian expression and as they are introduced by the elimination of ψ', ϕ', etc., we have

$$\frac{dA}{d\psi} = \left(\frac{dA}{d\psi}\right),\ \frac{dA}{d\phi} = \left(\frac{dA}{d\phi}\right),\ \text{etc.}\ldots\ldots\ldots\ldots\ldots (43);$$

and therefore the new expression satisfies the partial differential equation (21). That it is a completely general solution we see, because it satisfies the condition that the action is equal for all configurations fulfilling an absolutely arbitrary equation (39).

For the case of a single free particle, the interpretation of (39) is that the point (x', y', z') is on an arbitrary surface, and of (40) that each line of motion cuts this surface at right angles. Hence

Practical interpretation of the complete solution of the characteristic equation.

332. The most general possible solution of the quadratic, partial, differential equation of the first order, which Hamilton showed to be satisfied by his Characteristic Function (either terminal configuration alone varying), when interpreted for the case of a single free particle, expresses the action up to any point (x, y, z), from some point of a certain arbitrarily given surface, from which the particle has been projected, in the direction of the normal, and with the proper velocity to make the sum of the potential and actual energies have a given value. In other

words, the physical problem solved by the most general solution of that partial differential equation, is this:—

Let free particles, not mutually influencing one another, be projected normally from all points of a certain arbitrarily given surface, each with the proper velocity to make the sum of its potential and kinetic energies have a given value. To find, for the particle which passes through a given point (x, y, z), the "action" in its course from the surface of projection to this point. The Hamiltonian principles stated above, show that the surfaces of equal action cut the paths of the particles at right angles; and give also the following remarkable properties of the motion:— Properties of surfaces of equal action.

If, from all points of an arbitrary surface, particles not mutually influencing one another be projected with the proper velocities in the directions of the normals; points which they reach with equal actions lie on a surface cutting the paths at right angles. The infinitely small thickness of the space between any two such surfaces corresponding to amounts of action differing by any infinitely small quantity, is inversely proportional to the velocity of the particle traversing it; being equal to the infinitely small difference of action divided by the whole momentum of the particle.

Let λ, μ, ν be the direction cosines of the normal to the surface of equal action through (x, y, z). We have

$$\lambda = \frac{\frac{dA}{dx}}{\left(\frac{dA^2}{dx^2} + \frac{dA^2}{dy^2} + \frac{dA^2}{dz^2}\right)^{\frac{1}{2}}}, \text{ etc. } \ldots\ldots\ldots\ldots\ldots(1).$$

But $\frac{dA}{dx} = m\dot{x}$, etc., and, if q denote the resultant velocity,

$$mq = \left(\frac{dA^2}{dx^2} + \frac{dA^2}{dy^2} + \frac{dA^2}{dz^2}\right)^{\frac{1}{2}} \ldots\ldots\ldots\ldots\ldots\ldots(2).$$

Hence $$\lambda = \frac{\dot{x}}{q}, \quad \mu = \frac{\dot{y}}{q}, \quad \nu = \frac{\dot{z}}{q},$$

which proves the first proposition. Again, if δA denote the in-

Properties of surfaces of equal action.

finitely small difference of action from (x, y, z) to any other point $(x+\delta x, y+\delta y, z+\delta z)$, we have

$$\delta A = \frac{dA}{dx}\delta x + \frac{dA}{dy}\delta y + \frac{dA}{dz}\delta z.$$

Let the second point be at an infinitely small distance, e, from the first, in the direction of the normal to the surface of equal action; that is to say, let

$$\delta x = e\lambda,\ \delta y = e\mu,\ \delta z = e\nu.$$

Hence, by (1), $$\delta A = e\left(\frac{dA^2}{dx^2} + \frac{dA^2}{dy^2} + \frac{dA^2}{dz^2}\right)^{\frac{1}{2}} \text{..............(3)};$$

whence, by (2), $$e = \frac{\delta A}{mq} \text{ ....(4)},$$

which is the second proposition.

Examples of varying action.

333. Irrespectively of methods for finding the "characteristic function" in kinetic problems, the fact that any case of motion whatever can be represented by means of a single function in the manner explained in § 331, is most remarkable, and, when geometrically interpreted, leads to highly important and interesting properties of motion, which have valuable applications in various branches of Natural Philosophy. One of the many applications of the general principle made by Hamilton* led to a general theory of optical instruments, comprehending the whole in one expression.

Some of its most direct applications; to the motions of planets, comets, etc., considered as free points, and to the celebrated problem of perturbations, known as the Problem of Three Bodies, are worked out in considerable detail by Hamilton (*Phil. Trans.*, 1834-35), and in various memoirs by Jacobi, Liouville, Bour, Donkin, Cayley, Boole, etc. The now abandoned, but still interesting, corpuscular theory of light furnishes a good and exceedingly simple illustration. In this theory light is supposed to consist of material particles not mutually influencing one another, but subject to molecular forces from the particles of bodies—not sensible at sensible distances, and therefore not causing any deviation from uniform rectilinear motion in a homogeneous medium, except within an indefinitely small dis-

* *On the Theory of Systems of Rays.* Trans. R. I. A., 1824, 1830, 1832.

tance from its boundary. The laws of reflection and of single refraction follow correctly from this hypothesis, which therefore suffices for what is called geometrical optics.

Examples of varying action.

We hope to return to this subject, with sufficient detail, in treating of Optics. At present we limit ourselves to state a theorem comprehending the known rule for measuring the magnifying power of a telescope or microscope (by comparing the diameter of the object-glass with the diameter of pencil of parallel rays emerging from the eye-piece, when a point of light is placed at a great distance in front of the object-glass), as a particular case.

Application to common optics,

334. Let any number of attracting or repelling masses, or perfectly smooth elastic objects, be fixed in space. Let two stations, O and O', be chosen. Let a shot be fired with a stated velocity, V, from O, in such a direction as to pass through O'. There may clearly be more than one natural path by which this may be done; but, generally speaking, when one such path is chosen, no other, not considerably diverging from it, can be found; and any infinitely small deviation in the line of fire from O, will cause the bullet to pass infinitely near to, but not through, O'. Now let a circle, with infinitely small radius r, be described round O as centre, in a plane perpendicular to the line of fire from this point, and let—all with infinitely nearly the same velocity, but fulfilling the condition that the sum of the potential and kinetic energies is the same as that of the shot from O—bullets be fired from all points of this circle, all directed infinitely nearly parallel to the line of fire from O, but each precisely so as to pass through O'. Let a target be held at an infinitely small distance, a', beyond O', in a plane perpendicular to the line of the shot reaching it from O. The bullets fired from the circumference of the circle round O, will, after passing through O', strike this target in the circumference of an exceedingly small ellipse, each with a velocity (corresponding of course to its position, under the law of energy) differing infinitely little from V', the common velocity with which they pass through O'. Let now a circle, equal to the former, be described round O', in the plane perpendicular to the central path through O', and let bullets be fired from points in its circumference, each

or kinetics of a single particle

Application to common optics, or kinetics of a single particle.

with the proper velocity, and in such a direction infinitely nearly parallel to the central path as to make it pass through O. These bullets, if a target is held to receive them perpendicularly at a distance $a = a'\frac{V}{V'}$, beyond O, will strike it along the circumference of an ellipse equal to the former and placed in a "corresponding" position; and the points struck by the individual bullets will correspond; according to the following law of "correspondence":—Let P and P' be points of the first and second circles, and Q and Q' the points on the first and second targets which bullets from them strike; then if P' be in a plane containing the central path through O' and the position which Q would take if its ellipse were made circular by a pure strain (§ 183); Q and Q' are similarly situated on the two ellipses.

For, let XOY, $X'O'Y'$, be planes perpendicular to the central path through O and through O'. Let A be the "action" from O to O', and ϕ the action from a point $P(x, y, z)$, in the neighbourhood of O, specified with reference to the former axes of co-ordinates, to a point $P'(x', y', z')$, in the neighbourhood of O', specified with reference to the latter.

The function $\phi - A$ vanishes, of course, when $x=0$, $y=0$, $z=0$, $x'=0$, $y'=0$, $z'=0$. Also, for the same values of the co-ordinates, its differential coefficients $\frac{d\phi}{dx}$, $\frac{d\phi}{dy}$, and $\frac{d\phi}{dx'}$, $\frac{d\phi}{dy'}$, must vanish, and $\frac{d\phi}{dz}$, $-\frac{d\phi}{dz'}$ must be respectively equal to V and V', since, for any values whatever of the co-ordinates, $\frac{d\phi}{dx}$ and $\frac{d\phi}{dy}$ are the component velocities parallel to the two lines OX, OY, of the particle passing through P, when it comes from P', and $-\frac{d\phi}{dx'}$ and $-\frac{d\phi}{dy'}$ are the components parallel to $O'X'$, $O'Y'$, of the velocity through P' directed so as to reach P. Hence by Taylor's (or Maclaurin's) theorem we have

$$\begin{aligned}\phi - A = &- V'z' + Vz \\ &+ \tfrac{1}{2}\{(X, X)x^2 + (Y, Y)y^2 + (X', X')x'^2 + (Y', Y')y'^2 + \ldots \\ &+ 2(Y, Z)yz + \ldots + 2(Y', Z')y'z' + \ldots \\ &+ 2(X, X')xx' + 2(Y, Y')yy' + 2(Z, Z')zz' \\ &+ 2(X, Y')xy' + 2(X, Z')xz' + \ldots + 2(Z, Y')zy'\} + R \ldots (1),\end{aligned}$$

where (X, X), (X, Y), etc., denote constants, viz., the values of the differential coefficients $\frac{d^2\phi}{dx^2}$, $\frac{d^2\phi}{dxdy}$, etc., when each of the six co-ordinates x, y, z, x', y', z' vanishes; and R denotes the remainder after the terms of the second degree. According to Cauchy's principles regarding the convergence of Taylor's theorem, we have a rigorous expression for $\phi - A$ in the same form, without R, if the coefficients (X, X), etc., denote the values of the differential coefficients with some variable values intermediate between 0 and the actual values of x, y, etc., substituted for these elements. Hence, provided the values of the differential coefficients are infinitely nearly the same for any infinitely small values of the co-ordinates as for the vanishing values, R becomes infinitely smaller than the terms preceding it, when x, y, etc., are each infinitely small. Hence when each of the variables x, y, z, x', y', z' is infinitely small, we may omit R in the expression (1) for $\phi - A$. Now, as in the proposition to be proved, let us suppose z and z' each to be rigorously zero: and we have

Application to common optics, or kinetics of a single particle.

$$\frac{d\phi}{dx} = (X, X)\,x + (X, Y)\,y + (X, X')\,x' + (X, Y')\,y';$$

$$\frac{d\phi}{dy} = (Y, Y)\,y + (X, Y)\,x + (Y, X')\,x' + (Y, Y')\,y'.$$

These expressions, if in them we make $x = 0$, and $y = 0$, become the component velocities parallel to OX, OY, of a particle passing through O having been projected from P'. Hence, if ξ, η, ζ denote its co-ordinates, an infinitely small time, $\frac{a}{V}$, after it passes through O, we have $\zeta = a$, and

$$\xi = \{(X, X')\,x' + (X, Y')\,y'\}\frac{a}{V},\quad \eta = \{(Y, X')\,x' + (Y, Y')\,y'\}\frac{a}{V} \ldots (2).$$

Here ξ and η are the rectangular co-ordinates of the point Q' in which, in the second case, the supposed target is struck. And by hypothesis

$$x'^2 + y'^2 = r^2 \ldots\ldots\ldots\ldots\ldots\ldots (3).$$

If we eliminate x', y' between these three equations, we have clearly an ellipse; and the former two express the relation of the "corresponding" points. Corresponding equations with x and y for x' and y'; with ξ', η' for ξ, η; and with $-(X, X')$, $-(Y, X')$, $-(X, Y')$, $-(Y, Y')$, in place of (X, X'), (X, Y'),

Application to common optics, or kinetics of a single particle.

(Y, X'), (Y, Y'), express the first case. Hence the proposition, as is most easily seen by choosing OX and $O'X'$ so that (X, Y') and (Y, X') may each be zero.

Application to common optics.

335. The most obvious optical application of this remarkable result is, that in the use of any optical apparatus whatever, if the eye and the object be interchanged without altering the position of the instrument, the magnifying power is unaltered. This is easily understood when, as in an ordinary telescope, microscope, or opera-glass (Galilean telescope), the instrument is symmetrical about an axis, and is curiously contradictory of the common idea that a telescope "diminishes" when looked through the wrong way, which no doubt is true if the telescope is simply reversed about the middle of its length, eye and object remaining fixed. But if the telescope be removed from the eye till its eye-piece is close to the object, the part of the object seen will be seen enlarged to the same extent as when viewed with the telescope held in the usual manner. This is easily verified by looking from a distance of a few yards, in through the object-glass of an opera-glass, at the eye of another person holding it to his eye in the usual way.

The more general application may be illustrated thus:—Let the points, O, O' (the centres of the two circles described in the preceding enunciation), be the optic centres of the eyes of two persons looking at one another through any set of lenses, prisms, or transparent media arranged in any way between them. If their pupils are of equal sizes in reality, they will be seen as similar ellipses of equal apparent dimensions by the two observers. Here the imagined particles of light, projected from the circumference of the pupil of either eye, are substituted for the projectiles from the circumference of either circle, and the retina of the other eye takes the place of the target receiving them, in the general kinetic statement.

Application to system of free mutually influencing particles.

336. If instead of one free particle we have a conservative system of any number of mutually influencing free particles, the same statement may be applied with reference to the initial position of one of the particles and the final position of another, or with reference to the initial positions or to the final positions

of two of the particles. It serves to show how the influence of an infinitely small change in one of those positions, on the direction of the other particle passing through the other position, is related to the influence on the direction of the former particle passing through the former position produced by an infinitely small change in the latter position. A corresponding statement, in terms of generalized co-ordinates, may of course be adapted to a system of rigid bodies or particles connected in any way. All such statements are included in the following very general proposition:—

Application to system of free mutually influencing particles, and to generalized system.

The rate of increase of any one component momentum, corresponding to any one of the co-ordinates, per unit of increase of any other co-ordinate, is equal to the rate of increase of the component momentum corresponding to the latter per unit increase or diminution of the former co-ordinate, according as the two co-ordinates chosen belong to one configuration of the system, or one of them belongs to the initial configuration and the other to the final.

Let ψ and χ be two out of the whole number of co-ordinates constituting the argument of the Hamiltonian characteristic function A; and ξ, η the corresponding momentums. We have [§ 330 (18)]

$$\frac{dA}{d\psi} = \pm \xi, \; \frac{dA}{d\chi} = \pm \eta,$$

the upper or lower sign being used according as it is a final or an initial co-ordinate that is concerned. Hence

$$\frac{d^2A}{d\psi d\chi} = \pm \frac{d\xi}{d\chi} = \pm \frac{d\eta}{d\psi},$$

and therefore

$$\frac{d\xi}{d\chi} = \frac{d\eta}{d\psi},$$

if both co-ordinates belong to one configuration, or

$$\frac{d\xi}{d\chi} = -\frac{d\eta}{d\psi},$$

if one belongs to the initial configuration, and the other to the final, which is the second proposition. The geometrical interpretation of this statement for the case of a free particle, and two co-ordinates both belonging to one position, its final position, for

Application to system of free mutually influencing particles, and to generalised system.

instance, gives merely the proposition of § 332 above, for the case of particles projected from one point, with equal velocities in all directions; or, in other words, the case of the arbitrary surface of that enunciation, being reduced to a point. To complete the set of variational equations derived from § 330 we have $\frac{dt}{d\chi} = \pm \frac{d\eta}{dE}$ which expresses another remarkable property of conservative motion.

Slightly disturbed equilibrium.

337. By the help of Lagrange's form of the equations of motion, § 318, we may now, as a preliminary to the consideration of stability of motion, investigate the motion of a system infinitely little disturbed from a position of equilibrium, and left free to move, the velocities of its parts being initially infinitely small. The resulting equations give the values of the independent co-ordinates at any future time, provided the displacements *continue* infinitely small; and the mathematical expressions for their values must of course show the nature of the equilibrium, giving at the same time an interesting example of the *coexistence of small motions*, § 89. The method consists simply in finding what the equations of motion, and their integrals, become for co-ordinates which differ infinitely little from values corresponding to a configuration of equilibrium—and for an infinitely small initial kinetic energy. The solution of these differential equations is always easy, as they are linear and have constant coefficients. If the solution indicates that these differences *remain infinitely small*, the position is one of stable equilibrium; if it shows that one or more of them may *increase indefinitely*, the result of an infinitely small displacement from or infinitely small velocity through the position of equilibrium may be a finite departure from it—and thus the equilibrium is unstable.

Since there is a position of equilibrium, the kinematic relations must be invariable. As before,

$$T = \tfrac{1}{2}\{(\psi, \psi)\dot{\psi}^2 + (\phi, \phi)\dot{\phi}^2 + 2(\psi, \phi)\dot{\psi}\dot{\phi} + \text{etc.}\ldots\}\ldots(1),$$

which cannot be negative for any values of the co-ordinates. Now, though the values of the coefficients in this expression are not generally constant, they are to be taken as constant in the approximate investigation, since their variations, depending on the infinitely small variations of ψ, ϕ, etc., can only give rise to terms of the third or higher orders of small quantities. Hence Lagrange's equations become simply

Slightly disturbed equilibrium.

$$\frac{d}{dt}\left(\frac{dT}{d\dot{\psi}}\right)=\Psi,\quad \frac{d}{dt}\left(\frac{dT}{d\dot{\phi}}\right)=\Phi,\ \text{etc.}\ldots\ldots\ldots\ldots\ldots\ldots(2),$$

and the first member of each of these equations is a linear function of $\ddot{\psi}$, $\ddot{\phi}$, etc., with constant coefficients.

Now, since we may take what origin we please for the generalized co-ordinates, it will be convenient to assume that ψ, ϕ, θ, etc., are measured from the position of equilibrium considered; and that their values are therefore always infinitely small.

Hence, infinitely small quantities of higher orders being neglected, and the forces being supposed to be independent of the velocities, we shall have linear expressions for Ψ, Φ, etc., in terms of ψ, ϕ, etc., which we may write as follows:—

$$\left.\begin{array}{l}\Psi=a\psi+b\phi+c\theta+\ldots\\ \Phi=a'\psi+b'\phi+c'\theta+\ldots\\ \quad\text{etc.}\qquad\text{etc.}\end{array}\right\}\ldots\ldots\ldots\ldots\ldots\ldots\ldots(3).$$

Equations (2) consequently become linear differential equations of the second order, with constant coefficients; as many in number as there are variables ψ, ϕ, etc., to be determined.

The regular processes explained in elementary treatises on differential equations, lead of course, independently of any particular relation between the coefficients, to a general form of solution (§ 343 below). But this form has very remarkable characteristics in the case of a conservative system; which we therefore examine particularly in the first place. In this case we have

$$\Psi=-\frac{dV}{d\psi},\quad \Phi=-\frac{dV}{d\phi},\ \text{etc.}$$

where V is, in our approximation, a homogeneous quadratic function of ψ, ϕ, ... if we take the origin, or configuration of equilibrium, as the configuration from which (§ 273) the potential energy is reckoned. Now, it is obvious*, from the theory

* For in the first place any such assumption as

Simultaneous transformation of two quadratic functions to sums of squares.

$$\begin{array}{c}\psi=A\psi_{,}+B\phi_{,}+\ldots\\ \phi=A'\psi_{,}+B'\phi_{,}+\ldots\\ \text{etc., etc.}\end{array}$$

gives equations for $\dot{\psi}$, $\dot{\phi}$, etc., in terms of $\dot{\psi}_{,}$, $\dot{\phi}_{,}$, etc., with the same coefficients, A, B, etc., if these are independent of t. Hence (the co-ordinates being i in

Slightly disturbed equilibrium.

of the transformation of quadratic functions, that we may, by a determinate linear transformation of the co-ordinates, reduce the

Simultaneous transformation of two quadratic functions to sums of squares.

number) we have i^2 quantities $A, A', A'', \ldots B, B', B'', \ldots$ etc., to be determined by i^2 equations expressing that in $2T$ the coefficients of $\dot{\psi}_{,}^2$, $\dot{\phi}_{,}^2$, etc. are each equal to unity, and of $\dot{\psi}_{,}\dot{\phi}_{,}$ etc. each vanish, and that in V the coefficients of $\psi_{,}\phi_{,}$, etc. each vanish. But, particularly in respect to our dynamical problem, the following process in two steps is instructive:—

(1) Let the quadratic expression for T in terms of $\dot{\psi}^2$, $\dot{\phi}^2$, $\dot{\psi}\dot{\phi}$, etc., be reduced to the form $\dot{\psi}_{,}^2 + \dot{\phi}_{,}^2 + \ldots$ by proper assignment of values to A, B, etc. This may be done arbitrarily, in an infinite number of ways, without the solution of any algebraic equation of degree higher than the first; as we may easily see by working out a synthetical process algebraically according to the analogy of finding first the conjugate diametral plane to any chosen diameter of an ellipsoid, and then the diameter of its elliptic section, conjugate to any chosen diameter of this ellipse. Thus, of the $\frac{i(i-1)}{2}$ equations expressing that the coefficients of the products $\dot{\psi}_{,}\dot{\phi}_{,}$, $\dot{\psi}_{,}\dot{\theta}_{,}$, $\dot{\phi}_{,}\dot{\theta}_{,}$, etc. vanish in T, take first the one expressing that the coefficient of $\dot{\psi}_{,}\dot{\phi}_{,}$ vanishes, and by it find the value of one of the B's, supposing all the A's and all the B's but one to be known. Then take the two equations expressing that the coefficients of $\dot{\psi}_{,}\dot{\theta}_{,}$ and $\dot{\phi}_{,}\dot{\theta}_{,}$ vanish, and by them find two of the C's supposing all the C's but two to be known, as are now all the A's and all the B's: and so on. Thus, in terms of all the A's, all the B's but one, all the C's but two, all the D's but three, and so on, supposed known, we find by the solution of linear equations the remaining B's, C's, D's, etc. Lastly, using the values thus found for the unassumed quantities, B, C, D, etc., and equating to unity the coefficients of $\dot{\psi}_{,}^2$, $\dot{\phi}_{,}^2$, $\dot{\theta}_{,}^2$, etc. in the transformed expression for $2T$, we have i equations among the squares and products of the $\frac{(i+1)i}{2}$ assumed quantities, (i) A's, $(i-1)$ B's, $(i-2)$ C's, etc., by which any one of the A's, any one of the B's, any one of the C's, and so on, are given immediately in terms of the $\frac{i(i-1)}{2}$ ratios of the others to them. Thus the thing is done, and $\frac{i(i-1)}{2}$ disposable ratios are left undetermined.

(2) These quantities may be determined by the $\frac{i(i-1)}{2}$ equations expressing that also in the transformed quadratic V the coefficients of $\psi_{,}\phi_{,}$, $\psi_{,}\theta_{,}$, $\phi_{,}\theta_{,}$, etc. vanish.

Or, having made the first transformation as in (1) above, with assumed values for $\frac{i(i-1)}{2}$ disposable ratios, make a second transformation determinately thus:

Generalized orthogonal transformation of co-ordinates.

—Let

$$\psi_{,} = l\psi_{,,} + m\phi_{,,} + \ldots$$
$$\phi_{,} = l'\psi_{,,} + m'\phi_{,,} + \ldots$$
$$\text{etc., etc.}$$

where the i^2 quantities $l, m, \ldots, l', m', \ldots$ satisfy the $\frac{1}{2}i(i+1)$ equations

$$ll' + mm' + \ldots = 0, \quad l'l'' + m'm'' + \ldots = 0, \text{ etc.},$$

expression for $2T$, which is essentially positive, to a sum of squares of generalized component velocities, and at the same time V to a sum of the squares of the corresponding co-ordinates, each multiplied by a constant, which may be either positive or negative, but is essentially real. [In the case of an equality or of any number of equalities among the values of these constants (α, β, etc. in the notation below), roots as they are of a determinantal equation, the linear transformation ceases to be wholly determinate; but the degree or degrees of indeterminacy which supervene is the reverse of embarrassing in respect to either the process of obtaining the solution, or the interpretation and use of it when obtained.] Hence ψ, ϕ, ... may be so chosen that

Simplified expressions for the kinetic and potential energies.

$$T = \tfrac{1}{2}(\dot{\psi}^2 + \dot{\phi}^2 + \text{etc.}) \quad\ldots\ldots\ldots\ldots(4),$$

and

$$V = \tfrac{1}{2}(\alpha\psi^2 + \beta\phi^2 + \text{etc.}) \quad\ldots\ldots\ldots\ldots(5),$$

α, β, etc., being real positive or negative constants. Hence Lagrange's equations become

$$\ddot{\psi} = -\alpha\psi, \quad \ddot{\phi} = -\beta\phi, \quad \text{etc.} \ldots\ldots\ldots\ldots(6).$$

The solutions of these equations are

$$\psi = A\cos(t\sqrt{\alpha} - e), \quad \phi = A'\cos(t\sqrt{\beta} - e'), \quad \text{etc.} \ldots\ldots(7),$$

Integrated equations of motion, expressing the fundamental modes of vibration.

A, e, A', e', etc., being the arbitrary constants of integration. Hence we conclude the motion consists of a simple harmonic variation of each co-ordinate, provided that α, β, etc., are all positive. This condition is satisfied when V is a true minimum at the configuration of equilibrium; which, as we have seen (§ 292), is necessarily the case when the equilibrium is stable. If any one or more of α, β, ... vanishes, the equilibrium might

and $\quad l^2+m^2+\ldots=1,\ l'^2+m'^2+\ldots=1$, etc.,

leaving $\frac{1}{2}i(i-1)$ disposables.

Simultaneous transformation of two quadratic functions to sums of squares.

We shall still have, obviously, the same form for $2T$, that is:—

$$2T = \dot{\psi}_{\prime\prime}^2 + \dot{\phi}_{\prime\prime}^2 + \ldots$$

And, according to the known theory of the transformation of quadratic functions, we may determine the $\frac{1}{2}i(i-1)$ disposables of l, m, ..., l', m', ... so as to make the $\frac{1}{2}i(i-1)$ products of the co-ordinates $\psi_{\prime\prime}$, $\phi_{\prime\prime}$, etc. disappear from the expression for V, and give

$$2V = \alpha\psi_{\prime\prime}^2 + \beta\phi_{\prime\prime}^2 + \ldots,$$

where α, β, γ, etc., are the roots, necessarily real, of an equation of the ith degree of which the coefficients depend on the coefficients of the squares and products in the expression for V in terms of $\psi_{\prime}$, $\phi_{\prime}$, etc. Later [(7′), (8) and (9) of § 343 *f*], a *single process* for carrying out this investigation will be worked out.

Integrated equations of motion, expressing the fundamental modes of vibration; or of falling away from configuration of unstable equilibrium.

be either stable or unstable, or neutral; but terms of higher orders in the expansion of V in ascending powers and products of the co-ordinates would have to be examined to test it; and if it were stable, the period of an infinitely small oscillation in the value of the corresponding co-ordinate or co-ordinates would be infinitely great. If any or all of α, β, γ, ... are negative, V is not a minimum, and the equilibrium is (§ 292) essentially unstable. The form (7) for the solution, for each co-ordinate for which this is the case, becomes imaginary, and is to be changed into the exponential form, thus; for instance, let $-\alpha = p$, a positive quantity. Thus

Infinitely small disturbance from unstable equilibrium.

$$\psi = C\epsilon^{+t\sqrt{p}} + K\epsilon^{-t\sqrt{p}} \quad \ldots\ldots\ldots\ldots\ldots\ldots(8),$$

which (unless the disturbance is so adjusted as to make the arbitrary constant C vanish) indicates an unlimited increase in the deviation. This form of solution expresses the approximate law of falling away from a configuration of unstable equilibrium. In general, of course, the approximation becomes less and less accurate as the deviation increases.

Potential and Kinetic energies expressed as functions of time.

We have, by (5), (4), (7) and (8),

$$\left.\begin{aligned} V &= \tfrac{1}{4}\alpha A^2\left[1 + \cos 2\left(t\sqrt{\alpha} - e\right)\right] + \text{etc.} \\ \text{or} \quad V &= -\tfrac{1}{2}p\left[2CK + C^2\epsilon^{2t\sqrt{p}} + K^2\epsilon^{-2t\sqrt{p}}\right] - \text{etc.} \end{aligned}\right\} \ldots\ldots(9),$$

$$\left.\begin{aligned} \text{and} \quad T &= \tfrac{1}{4}\alpha A^2\left[1 - \cos 2\left(t\sqrt{\alpha} - e\right)\right] + \text{etc.} \\ \text{or} \quad T &= \tfrac{1}{2}p\left[-2CK + C^2\epsilon^{2t\sqrt{p}} + K^2\epsilon^{-2t\sqrt{p}}\right] + \text{etc.} \end{aligned}\right\} \ldots\ldots(10);$$

and, verifying the constancy of the sum of potential and kinetic energies,

$$\left.\begin{aligned} T + V &= \tfrac{1}{2}\left(\alpha A^2 + \beta A'^2 + \text{etc.}\right) \\ \text{or} \quad T + V &= -2\left(pCK + qC'K' + \text{etc.}\right) \end{aligned}\right\} \ldots\ldots\ldots\ldots(11).$$

Example of fundamental modes.

One example for the present will suffice. Let a solid, immersed in an infinite *liquid* (§ 320), be prevented from any motion of rotation, and left only freedom to move parallel to a certain fixed plane, and let it be influenced by forces subject to the conservative law, which vanish in a particular position of equilibrium. Taking any point of reference in the body, choosing its position when the body is in equilibrium, as origin of rectangular co-ordinates OX, OY, and reckoning the potential energy from it, we shall have, as in general,

$$2T = A\dot{x}^2 + B\dot{y}^2 + 2C\dot{x}\dot{y}; \quad 2V = ax^2 + by^2 + 2cxy,$$

the principles stated in § 320 above, allowing us to regard the co-ordinates x and y as fully specifying the system, provided always, that if the body is given at rest, or is brought to rest, the whole liquid is at rest (§ 320) at the same time. By solving the obviously determinate problem of finding that pair of conjugate diameters which are in the same directions for the ellipse

Example of fundamental modes.

$$Ax^2 + By^2 + 2Cxy = \text{const.},$$

and the ellipse or hyperbola,

$$ax^2 + by^2 + 2cxy = \text{const.},$$

and choosing these as oblique axes of co-ordinates (x_1, y_1), we shall have

$$2T = A_1\dot{x}_1^2 + B_1\dot{y}_1^2, \text{ and } 2V = a_1x_1^2 + b_1y_1^2.$$

And, as A_1, B_1 are essentially positive, we may, to shorten our expressions, take $x_1\sqrt{A_1} = \psi$, $y_1\sqrt{B_1} = \phi$; so that we shall have

$$2T = \dot{\psi}^2 + \dot{\phi}^2, \quad 2V = \alpha\psi^2 + \beta\phi^2,$$

the normal expressions, according to the general forms shown above in (4) and (5).

The interpretation of the general solution is as follows :—

338. If a conservative system is infinitely little displaced from a configuration of stable equilibrium, it will ever after vibrate about this configuration, remaining infinitely near it; each particle of the system performing a motion which is composed of simple harmonic vibrations. If there are i degrees of freedom to move, and we consider any system (§ 202) of generalized co-ordinates specifying its position at any time, the deviation of any one of these co-ordinates from its value for the configuration of equilibrium will vary according to a complex harmonic function (§ 68), composed of i simple harmonics generally of incommensurable periods, and therefore (§ 67) the whole motion of the system will not in general recur periodically through the same series of configurations. There are, however, i distinct displacements, generally quite determinate, which we shall call *the normal displacements*, fulfilling the condition, that if any one of them be produced alone, and the system then left to itself for an instant at rest, this displacement will diminish and increase periodically according to a simple harmonic func-

General theorem of fundamental modes of infinitely small motion about a configuration of equilibrium.

Normal displacements from equilibrium.

Fundamental modes of vibration. tion of the time, and consequently every particle of the system will execute a simple harmonic movement in the same period. This result, we shall see later (Vol. II.), includes cases in which there are an infinite number of degrees of freedom; as for instance a stretched cord; a mass of air in a closed vessel; waves in water, or oscillations of water in a vessel of limited extent, or of an elastic solid; and in these applications it gives the theory of the so-called "fundamental vibration," and successive "harmonics" of a cord or organ-pipe, and of all the different possible simple modes of vibration in the other cases. In all these cases it is convenient to give the name "fundamental mode" to any one of the possible simple harmonic vibrations, and not to restrict it to the gravest simple harmonic mode, as has been hitherto usual in respect to vibrating cords and organ-pipes.

Theorem of kinetic energy; of potential energy. The whole kinetic energy of any complex motion of the system is [§ 337 (4)] equal to the sum of the kinetic energies of the fundamental constituents; and [§ 337 (5)] the potential energy of any displacement is equal to the sum of the potential energies of its normal components.

Infinitesimal motions in neighbourhood of configuration of unstable equilibrium. Corresponding theorems of normal constituents and fundamental modes of motion, and the summation of their kinetic and potential energies in complex motions and displacements, hold for motion in the neighbourhood of a configuration of *unstable* equilibrium. In this case, some or all of the constituent motions are fallings away from the position of equilibrium (according as the potential energies of the constituent normal vibrations are negative).

Case of equality among periods. 339. If, as may be in particular cases, the periods of the vibrations for two or more of the normal displacements are equal, any displacement compounded of them will also fulfil the condition of being a normal displacement. And if the system be displaced according to any one such normal displacement, and projected with velocity corresponding to another, it will execute a movement, the resultant of two simple harmonic movements in equal periods. Graphic representation. The graphic representation of the variation of the corresponding co-ordinates of the system, laid down as two rectangular co-ordinates in a plane diagram, will consequently (§ 65) be a circle or an ellipse; which will therefore, of course, be the form of the orbit of any particle of the system which has a distinct direction of motion, for two of the displacements in question. But it must be remembered that some of the principal parts [as for instance the body supported on the fixed axis, in the illustration of § 319, *Example* (C)] may have only one degree of freedom; or even that each part of the system may have only one degree of freedom, as for instance if the system is composed of a set of particles each constrained to remain on a given line, or of rigid bodies on fixed axes, mutually influencing one another by elastic cords or otherwise. In such a case as the last, no particle of the system can move otherwise than in one line; and the ellipse, circle, or other graphical representation of the composition of the harmonic motions of the system, is merely an aid to comprehension, and is not the orbit of a motion actually taking place in any part of the system.

Graphic representation.

340. In nature, as has been said above (§ 278), every system uninfluenced by matter external to it is conservative, when the ultimate molecular motions constituting heat, light, and magnetism, and the potential energy of chemical affinities, are taken into account along with the palpable motions and measurable forces. But (§ 275) practically we are obliged to admit forces of friction, and resistances of the other classes there enumerated, as causing losses of energy, to be reckoned, in abstract dynamics, without regard to the equivalents of heat or other molecular actions which they generate. Hence when such resistances are to be taken into account, forces opposed to the motions of various parts of a system must be introduced into the equations. According to the approximate knowledge which we have from experiment, these forces are independent of the velocities when due to the friction of solids: but are simply proportional to the velocities when due to fluid viscosity directly, or to electric or magnetic influences; with corrections depending on varying temperature, and on the varying configuration of the system. In consequence of the last-mentioned cause, the resistance of a *real liquid* (which is always more or less viscous) against a body moving rapidly enough through it, to leave a great deal of irregular motion, in the shape of

Dissipative systems.

Views of Stokes on resistance to a solid moving through a liquid. "eddies," in its wake, seems, when the motion of the solid has been kept long enough uniform, to be nearly in proportion to the square of the velocity; although, as Stokes has shown, at the lowest speeds the resistance is probably in simple proportion to the velocity, and for all speeds, after long enough time of one speed, may, it is probable, be approximately expressed as Stokes' probable law. the sum of two terms, one simply as the velocity, and the other as the square of the velocity. If a solid is started from rest in an incompressible fluid, the initial law of resistance is no doubt simple proportionality to velocity, (however great, if suddenly enough given;) until by the gradual growth of eddies the resistance is increased gradually till it comes to fulfil Stokes' law.

Friction of solids. **341.** The effect of friction of solids rubbing against one another is simply to render impossible the *infinitely* small vibrations with which we are now particularly concerned; and to allow any system in which it is present, to rest balanced when displaced, within certain finite limits, from a configuration of frictionless equilibrium. In mechanics it is easy to estimate its effects with sufficient accuracy when any practical case of finite oscillations is in question. But the other classes of dissipative agencies give rise to resistances simply as the velocities, Resistances varying as velocities. without the corrections referred to, when the motions are infinitely small; and can never balance the system in a configuration deviating to any extent, however small, from a configuration of equilibrium. In the theory of infinitely small vibrations, they are to be taken into account by adding to the expressions for the generalized components of force, proper (§ 343 *a*, below) linear functions of the generalized velocities, which gives us equations still remarkably amenable to rigorous mathematical treatment.

The result of the integration for the case of a single degree of freedom is very simple; and it is of extreme importance, both for the explanation of many natural phenomena, and for use in a large variety of experimental investigations in Natural Philosophy. Partial conclusions from it are as follows:—

If the resistance per unit velocity is less than a certain critical value, in any particular case, the motion is a simple harmonic oscillation, with amplitude decreasing in the same ratio in equal successive intervals of time. But if the resistance equals or exceeds the critical value, the system when displaced from its position of equilibrium, and left to itself, returns gradually towards its position of equilibrium, never oscillating through it to the other side, and only reaching it after an infinite time.

Resistances varying as velocities.

In the unresisted motion, let n^2 be the rate of acceleration, when the displacement is unity; so that (§ 57) we have $T = \frac{2\pi}{n}$: and let the rate of retardation due to the resistance corresponding to unit velocity be k. Then the motion is of the oscillatory or non-oscillatory class according as $k^2 < (2n)^2$ or $k^2 > (2n)^2$. In the first case, the period of the oscillation is increased by the resistance from T to $T \frac{n}{(n^2 - \frac{1}{4}k^2)^{\frac{1}{2}}}$, and the rate at which the Napierian logarithm of the amplitude diminishes per unit of time is $\frac{1}{2}k$. If a negative value be given to k, the case represented will be one in which the motion is assisted, instead of resisted, by force proportional to the velocity: but this case is purely ideal.

Effect of resistance varying as velocity in a simple motion.

The differential equation of motion for the case of one degree of motion is

$$\ddot{\psi} + k\dot{\psi} + n^2\psi = 0;$$

of which the complete integral is

$$\psi = \{A \sin n't + B \cos n't\}\epsilon^{-\frac{1}{2}kt}, \text{ where } n' = \sqrt{(n^2 - \tfrac{1}{4}k^2)},$$

or, which is the same,

$$\psi = (C\epsilon^{-n_{,}t} + C'\epsilon^{n_{,}t})\epsilon^{-\frac{1}{2}kt}, \text{ where } n_{,} = \sqrt{(\tfrac{1}{4}k^2 - n^2)},$$

A and B in one case, or C and C' in the other, being the arbitrary constants of integration. Hence the propositions above. In the case of $k^2 = (2n)^2$ the general solution is $\psi = (C + C't)\,\epsilon^{-\frac{1}{2}kt}$.

Case of equal roots.

342. The general solution [§ 343 *a* (2) and § 345‴] of the problem, to find the motion of a system having any number, i, of degrees of freedom, when infinitely little disturbed from a position of stable equilibrium, and left to move subject to resistances proportional to velocities, shows that the whole motion may be resolved, in general determinately, into $2i$ different motions each

Infinitely small motion of a dissipative system.

Infinitely small motion of a dissipative system.

either simple harmonic with amplitude diminishing according to the law stated above, or non-oscillatory and consisting of equi-proportionate diminutions of the components of displacement in equal successive intervals of time.

Cycloidal system defined.

Easy and instructive lecture illustration.

343. It is now convenient to cease limiting our ideas to infinitely small motions of an absolutely general system through configurations infinitely little different from a configuration of equilibrium, and to consider any motions large or small of a system so constituted that the positional* forces are proportional to displacements and the motional* to velocities, and that the kinetic energy is a quadratic function of the velocities with constant coefficients. Such a system we shall call a cycloidal† system; and we shall call its motions cycloidal motions. A good and instructive illustration is presented in the motion of one two or more weights in a vertical line, hung one from another, and the highest from a fixed point, by spiral springs.

343 *a*. If now instead of ψ, ϕ,... we denote by ψ_1, ψ_2,... the generalized co-ordinates, and if we take 11, 12, 21, 22..., ɪɪ, ɪ2, 2ɪ, 22,... to signify constant coefficients (not numbers as in the ordinary notation of arithmetic), the most general equations of motions of a cycloidal system may be written thus:

Positional and Motional Forces.

* Much trouble and verbiage is to be avoided by the introduction of these adjectives, which will henceforth be in frequent use. They tell their own meanings as clearly as any definition could.

† A single adjective is needed to avoid a sea of troubles here. The adjective 'cycloidal' is already classical in respect to any motion with one degree of freedom, curvilineal or rectilineal, lineal or angular (Coulomb-torsional, for example), following the same law as the cycloidal pendulum, that is to say:—*the displacement a simple harmonic function of the time.* The motion of a particle on a cycloid with vertex up may as properly be called cycloidal; and in it the displacement is an imaginary simple harmonic, or a real exponential, or the sum of two real exponentials of the time

$$\left(C\epsilon^{t\sqrt{\frac{g}{l}}} + C'\epsilon^{-t\sqrt{\frac{g}{l}}}\right).$$

In cycloidal motion as defined in the text, each component of displacement is proved to be a sum of exponentials $(C\epsilon^{\lambda t} + C'\epsilon^{\lambda' t} + \text{etc.})$ real or imaginary, reducible to a sum of products of real exponentials and real simple harmonics $[C\epsilon^{mt}\cos(nt-e) + C'\epsilon^{m't}\cos(n't-e') + \text{etc.}]$.

$$\left.\begin{array}{l}\frac{d}{dt}\left(\frac{dT}{d\dot{\psi}_1}\right)+11\dot{\psi}_1+12\dot{\psi}_2+\ldots+\mathrm{11}\psi_1+\mathrm{12}\psi_2+\ldots=0\\ \frac{d}{dt}\left(\frac{dT}{d\dot{\psi}_2}\right)+21\dot{\psi}_1+22\dot{\psi}_2+\ldots+\mathrm{21}\psi_1+\mathrm{22}\psi_2+\ldots=0\\ \qquad\text{etc.}\qquad\qquad\text{etc.}\end{array}\right\}\ldots\ldots(1).$$

Differential equations of complex cycloidal motion.

Positional forces of the non-conservative class are included by *not* assuming $\mathrm{12}=\mathrm{21}$, $\mathrm{13}=\mathrm{31}$, $\mathrm{23}=\mathrm{32}$, etc.

The theory of simultaneous linear differential equations with constant coefficients shows that the general solution for each co-ordinate is the sum of particular solutions, and that every particular solution is of the form

$$\psi_1=a_1\epsilon^{\lambda t},\ \psi_2=a_2\epsilon^{\lambda t}\ldots\ldots\ldots\ldots(2).$$

Their solution.

Assuming, then, this to be a solution, and substituting in the differential equations, we have

$$\left.\begin{array}{l}\lambda^2\frac{d\mathfrak{T}}{da_1}+\lambda(11a_1+12a_2+\ldots)+\mathrm{11}a_1+\mathrm{12}a_2+\ldots=0\\ \lambda^2\frac{d\mathfrak{T}}{da_2}+\lambda(21a_1+22a_2+\ldots)+\mathrm{21}a_1+\mathrm{22}a_2+\ldots=0\\ \qquad\text{etc.}\qquad\qquad\text{etc.}\end{array}\right\}\ldots\ldots(3),$$

where $\mathfrak{T}$ denotes the same homogeneous quadratic function of $a_1, a_2,\ldots$, that T is of $\dot{\psi}_1, \dot{\psi}_2,\ldots$. These equations, i in number, determine λ by the determinantal equation

$$\left|\begin{array}{l}(\mathrm{11})\lambda^2+11\lambda+\mathrm{11},\ (\mathrm{12})\lambda^2+12\lambda+\mathrm{12},\ldots\\ (\mathrm{21})\lambda^2+21\lambda+\mathrm{21},\ (\mathrm{22})\lambda^2+22\lambda+\mathrm{22},\ldots\\ \ldots\ldots\ldots\ldots\ldots\ldots\\ \ldots\ldots\ldots\ldots\ldots\ldots\end{array}\right|=0\ldots..(4),$$

where $(\mathrm{11}), (\mathrm{22}), (\mathrm{12}), (\mathrm{21})$, etc. denote the coefficients of squares and doubled products in the quadratic, $2T$; with identities

$$(\mathrm{12})=(\mathrm{21}),\ (\mathrm{13})=(\mathrm{31}),\ \text{etc}\ldots\ldots\ldots\ldots(5).$$

The equation (4) is of the degree $2i$, in λ; and if any one of its roots be used for λ in the i linear equations (3), these become harmonized and give the $i-1$ ratios a_2/a_1, a_3/a_1, etc.; and we have then, in (2), a particular solution with one arbitrary constant, a_1. Thus, from the $2i$ roots, when unequal, we have $2i$ distinct particular solutions, each with an arbitrary constant; and the addition of these solutions, as explained above, gives the general solution.

Solution of differential equations of complex cycloidal motion.

343 *b*. To show explicitly the determination of the ratios a_2/a_1, a_3/a_1, etc. put for brevity

$$(11)\lambda^2 + 11\lambda + 11 = 1\cdot 1, \quad (12)\lambda^2 + 12\lambda + 12 = 1\cdot 2, \text{ etc.},$$
$$(32)\lambda^2 + 32\lambda + 32 = 3\cdot 2, \text{ etc. } \ldots\ldots(5)';$$

and generally let $j\cdot k$ denote the coefficient of a_k in the jth equation of (3), or the kth term of the jth line of the determinant (to be called D for brevity) constituting the first member of (4).

Algebra of linear equations.

Let $M(j\cdot k)$ denote the factor of $j\cdot k$ in D so that $j\cdot k\,.\,M(j\cdot k)$ is the sum of all the terms of D which contain $j\cdot k$, and we have

$$D = \frac{1}{i}\sum_{j=1}^{j=i}\sum_{k=1}^{k=i} j\cdot k\,.\,M(j\cdot k)\ldots\ldots\ldots\ldots\ldots\ldots(5)'',$$

because in the sum $\Sigma\Sigma$ each term of D clearly occurs i times: and taking different groupings of terms, but each one only once, we have

$$\left.\begin{aligned}
D &= 1\cdot 1\,M(1\cdot 1) + 1\cdot 2\,M(1\cdot 2) + 1\cdot 3\,M(1\cdot 3) + \text{etc.}\\
&= 2\cdot 1\,M(2\cdot 1) + 2\cdot 2\,M(2\cdot 2) + 2\cdot 3\,M(2\cdot 3) + \text{etc.}\\
&= 3\cdot 1\,M(3\cdot 1) + 3\cdot 2\,M(3\cdot 2) + 3\cdot 3\,M(3\cdot 3) + \text{etc.}\\
&\ldots\ldots\ldots\ldots\ldots\ldots\ldots\ldots\ldots\ldots\ldots\ldots\ldots\ldots\\
&= 1\cdot 1\,M(1\cdot 1) + 2\cdot 1\,M(2\cdot 1) + 3\cdot 1\,M(3\cdot 1) + \text{etc.}\\
&= 1\cdot 2\,M(1\cdot 2) + 2\cdot 2\,M(2\cdot 2) + 3\cdot 2\,M(3\cdot 2) + \text{etc.}\\
&= 1\cdot 3\,M(1\cdot 3) + 2\cdot 3\,M(2\cdot 3) + 3\cdot 3\,M(3\cdot 3) + \text{etc.}\\
&\ldots\ldots\ldots\ldots\ldots\ldots\ldots\ldots\ldots\ldots\ldots\ldots\ldots\ldots
\end{aligned}\right\}\ldots(5)'''$$

in all $2i$ different expressions for D.

Minor determinants.

Farther, by the elementary law of formation of determinants we see that

$$M(j-1\cdot k-1) = (-1)^{(i-1)(j+k)}\begin{vmatrix}
j\cdot k, & j\cdot(k+1), \; j\cdot(k+2), \ldots j\cdot i, \; j\cdot 1, \; j\cdot 2, \ldots, & j\cdot(k-2)\\
(j+1)\cdot k, & \ldots\ldots\ldots\ldots\ldots\ldots\ldots\ldots & \\
(j+2)\cdot k, & \ldots\ldots\ldots\ldots\ldots\ldots\ldots\ldots & \\
\ldots\ldots & \ldots\ldots\ldots\ldots\ldots\ldots\ldots\ldots & \\
\ldots\ldots & \ldots\ldots\ldots\ldots\ldots\ldots\ldots\ldots & \\
i\cdot k, & \ldots\ldots\ldots\ldots\ldots\ldots\ldots\ldots & \\
1\cdot k, & \ldots\ldots\ldots\ldots\ldots\ldots\ldots\ldots & \\
2\cdot k, & \ldots\ldots\ldots\ldots\ldots\ldots\ldots\ldots & \\
\ldots\ldots & \ldots\ldots\ldots\ldots\ldots\ldots\ldots\ldots & \\
\ldots\ldots & \ldots\ldots\ldots\ldots\ldots\ldots\ldots\ldots & \\
\ldots\ldots & \ldots\ldots\ldots\ldots\ldots\ldots\ldots\ldots & \\
(j-2)\cdot k, & \ldots\ldots\ldots\ldots\ldots\ldots\ldots\ldots, & (j-2)\cdot(k-2)
\end{vmatrix}$$
$$\ldots\ldots\ldots\ldots\ldots\ldots\ldots\ldots(5)^{iv}.$$

The quantities $M(1\cdot 1)$, $M(1\cdot 2)$, ... $M(j\cdot k)$, thus defined are what are commonly called the first minors of the determinant D, with just this variation from ordinary usage that the proper signs are given to them by the factor

Minors of a determinant.

$$(-1)^{(i-1)(j+k)}$$

in 5^{iv} so that in the formation of D the ordinary complication of alternate positive and negative signs when i is even and all signs positive when i is odd is avoided. In terms of the notation $(5)'$ the linear equations (3) become

$$\left.\begin{array}{l} 1\cdot 1a_1 + 1\cdot 2a_2 + \ldots\ldots + 1\cdot ia_i = 0 \\ 2\cdot 1a_1 + 2\cdot 2a_2 + \ldots\ldots + 2\cdot ia_i = 0 \\ \ldots\ldots\ldots\ldots\ldots\ldots\ldots\ldots\ldots\ldots \\ i\cdot 1a_1 + i\cdot 2a_2 + \ldots\ldots + i\cdot ia_i = 0 \end{array}\right\} \ldots\ldots\ldots\ldots (5)^{v},$$

and when $D=0$, which is required to harmonize them, they may be put under any of the following i different but equivalent forms,

$$\left.\begin{array}{ll} & \dfrac{a_1}{M(1\cdot 1)} = \dfrac{a_2}{M(1\cdot 2)} = \dfrac{a_3}{M(1\cdot 3)} = \text{etc.} \\ \text{or} & \dfrac{a_1}{M(2\cdot 1)} = \dfrac{a_2}{M(2\cdot 2)} = \dfrac{a_3}{M(2\cdot 3)} = \text{etc.} \\ \text{or} & \dfrac{a_1}{M(3\cdot 1)} = \dfrac{a_2}{M(3\cdot 2)} = \dfrac{a_3}{M(3\cdot 3)} = \text{etc.} \end{array}\right\} \ldots\ldots\ldots (5)^{vi},$$

from which we find

$$\left.\begin{array}{l} \dfrac{a_2}{a_1} = \dfrac{M(1\cdot 2)}{M(1\cdot 1)} = \dfrac{M(2\cdot 2)}{M(2\cdot 1)} = \dfrac{M(3\cdot 2)}{M(3\cdot 1)} = \text{etc.} \\ \dfrac{a_3}{a_1} = \dfrac{M(1\cdot 3)}{M(1\cdot 1)} = \dfrac{M(2\cdot 3)}{M(2\cdot 1)} = \dfrac{M(3\cdot 3)}{M(3\cdot 1)} = \text{etc.} \\ \ldots\ldots\ldots\ldots\ldots\ldots\ldots\ldots\ldots\ldots\ldots\ldots \end{array}\right\} \ldots\ldots\ldots (5)^{vii}.$$

The remarkable relations here shown among the minors, due to the evanescence of the major determinant D, are well known in algebra. They are all included in the following formula,

Relations among the minors of an evanescent determinant.

$$M(j\cdot k)\,.\,M(l\cdot n) - M(j\cdot n)\,.\,M(l\cdot k) = 0 \ldots\ldots\ldots (5)^{viii},$$

which is given in Salmon's *Higher Algebra* (§ 33 Ex. 1), as a consequence of the formula

$$M(j\cdot k)\,.\,M(l\cdot n) - M(j\cdot n)\,.\,M(l\cdot k) = D\,.\,M(j, l\cdot k, n) \ldots (5)^{ix},$$

where $M(j, l\cdot k, n)$ denotes the second minor formed by suppressing the jth and l^{th} columns and the kth and n^{th} lines.

Relations among the minors of an evanescent determinant.

343 c. When there are equalities among the roots the problem has generally solutions of the form

$$\psi_1=(c_1t+b_1)\epsilon^{\lambda t},\quad \psi_2=(c_2t+b_2)\epsilon^{\lambda t},\quad \text{etc.} \qquad (6).$$

To prove this let λ, λ' be two unequal roots which become equal with some slight change of the values of some or all of the given constants (11), 11, 11, (12), 12, 12, etc.; and let

$$\psi_1=A_1'\epsilon^{\lambda' t}-A_1\epsilon^{\lambda t},\quad \psi_2=A_2'\epsilon^{\lambda' t}-A_2\epsilon^{\lambda t},\quad \text{etc.} \qquad (6)'$$

be a particular solution of (1) corresponding to these roots.

Now let

$$\left.\begin{array}{lll} c_1=A_1'(\lambda'-\lambda), & c_2=A_2'(\lambda'-\lambda), & \text{etc.} \\ b_1=A_1'-A_1, & b_2=A_2'-A_2, & \text{etc.} \end{array}\right\} \qquad (6)''.$$

and

Using these in (6)′ we find

$$\psi_1=c_1\frac{\epsilon^{\lambda' t}-\epsilon^{\lambda t}}{\lambda'-\lambda}+b_1\epsilon^{\lambda t},\quad \psi_2=c_2\frac{\epsilon^{\lambda' t}-\epsilon^{\lambda t}}{\lambda'-\lambda}+b_2\epsilon^{\lambda t},\quad \text{etc.} \ldots (6)'''.$$

To find proper equations for the relations among $b_1, b_2, \ldots c_1, c_2, \ldots$ in order that (6)‴ may be a solution of (1), proceed thus:—first write down equations (3) for the λ' solution, with constants A_1', A_2', etc.: then subtract from these the corresponding equations for the λ solution: thus, and introducing the notation (6)″, we find

$$\left.\begin{array}{l} \{(11)\lambda'^2+11\lambda'+11\}c_1+\{(12)\lambda'^2+12\lambda'+12\}c_2+\text{etc.}=0 \\ \{(21)\lambda'^2+21\lambda'+21\}c_1+\{(22)\lambda'^2+22\lambda'+22\}c_2+\text{etc.}=0 \\ \quad\text{etc.} \qquad\qquad\qquad \text{etc.} \end{array}\right\} \ldots (6)^{iv},$$

and

$$\left.\begin{array}{l} \{(11)\lambda'^2+11\lambda'+11\}b_1+\{(12)\lambda'^2+12\lambda'+12\}b_2+\text{etc.} \\ \quad =-[c_1-b_1(\lambda'-\lambda)]\{(11)(\lambda+\lambda')+11\} \\ \qquad -[c_2-b_2(\lambda'-\lambda)]\{(12)(\lambda+\lambda')+12\}-\text{etc.} \\ \{(21)\lambda'^2+21\lambda'+21\}b_1+\{(22)\lambda'^2+22\lambda'+22\}b_2+\text{etc.} \\ \quad =-[c_1-b_1(\lambda'-\lambda)]\{(21)(\lambda+\lambda')+21\} \\ \qquad -[c_2-b_2(\lambda'-\lambda)]\{(22)(\lambda+\lambda')+22\}+\text{etc.} \\ \quad\text{etc.} \qquad\qquad\qquad \text{etc.} \end{array}\right\} \ldots (6)^{v}.$$

Case of equal roots.

Equations $(6)^{iv}$ require that λ' be a root of the determinant, and $i-1$ of them determine $i-1$ of the quantities c_1, c_2, etc. in terms of one of them assumed arbitrarily. Supposing now c_1, c_2, etc. to be thus all known, the i equations $(6)^{v}$ fail to determine the i quantities b_1, b_2, etc. in terms of the right-hand members because λ' is a root of the determinant. The two sets of equations $(6)^{iv}$ and $(6)^{v}$ require that λ be also a root of the determinant: and $i-1$ of the equations $(6)^{v}$ determine $i-1$ of the quantities b_1, b_2, etc. in terms of c_1, c_2, etc. (supposed already known as above) and a properly assumed value of one of the b's. Case of equal roots.

343 *d*. When λ' is infinitely nearly equal to λ, $(6)'''$ becomes infinitely nearly the same as (6), and $(6)^{iv}$ and $(6)^{v}$ become in terms of the notation $(5)'$

$$\left.\begin{array}{l} 1\cdot 1\, c + 1\cdot 2\, c_2 + \text{etc.} = 0 \\ 2\cdot 1\, c_1 + 2\cdot 2\, c_2 + \text{etc.} = 0 \\ \quad\text{etc.} \qquad\qquad \text{etc.} \end{array}\right\} \ldots\ldots\ldots\ldots\ldots\ldots (6)^{vi},$$

$$\left.\begin{array}{l} 1\cdot 1\, b_1 + 1\cdot 2\, b_2 + \text{etc.} = -c_1 \dfrac{d1\cdot 1}{d\lambda} - c_2 \dfrac{d1\cdot 2}{d\lambda} - \text{etc.} \\ 2\cdot 1\, b_1 + 2\cdot 2\, b_2 + \text{etc.} = -c_1 \dfrac{d2\cdot 1}{d\lambda} - c_2 \dfrac{d2\cdot 2}{d\lambda} - \text{etc.} \\ \qquad\quad \text{etc.} \qquad\qquad\qquad \text{etc.} \end{array}\right\} \ldots\ldots (6)^{vii}.$$

These, $(6)^{vi}$, $(6)^{vii}$, are clearly the equations which we find simply by trying if (6) is a solution of (1). $(6)^{vi}$ requires that λ be a root of the determinant D; and they give by $(5)^{vi}$ with c substituted for a the values of $i-1$ of the quantities c_1, c_2, etc. in terms of one of them assumed arbitrarily. And by the way we have found them we know that $(6)^{vii}$ superadded to $(6)^{vi}$ shows that λ must be a dual root of the determinant. To verify this multiply the first of them by $M(1\cdot 1)$, the second by $M(2\cdot 1)$, etc., and add. The coefficients of b_2, b_3, etc. in the sum are each identically zero in virtue of the elementary constitution of determinants, and the coefficient of b_1 is the major determinant D. Thus irrespectively of the value of λ we find in the first place,

$$Db_1 = -c_1 \left\{ M(1\cdot 1)\frac{d1\cdot 1}{d\lambda} + M(2\cdot 1)\frac{d2\cdot 1}{d\lambda} + \text{etc.} \right\}$$
$$- c_2 \left\{ M(1\cdot 1)\frac{d1\cdot 2}{d\lambda} + M(2\cdot 1)\frac{d2\cdot 2}{d\lambda} + \text{etc.} \right\} - \text{etc.} \ldots (6)^{viii}.$$

Now in virtue of $(6)^{vi}$ and $(5)^{v}$ we have

$$\frac{c_2}{c_1} = \frac{a_2}{a_1},\ \ \frac{c_3}{c_1} = \frac{a_3}{a_1},\ \text{etc.}$$

Using successively the several expressions given by $(5)^{vii}$ for these ratios, in $(6)^{viii}$, and putting $D = 0$, we find

$$0 = \Sigma_1^i \Sigma_1^i M(j\cdot k)\frac{dj\cdot k}{d\lambda}, \quad \text{or } 0 = \frac{dD}{d\lambda},$$

which with $D = 0$ shows that λ is a double root.

Suppose now that one of the c's has been assumed, and the others found by $(6)^{vi}$: let one of the b's be assumed: the other

Case of equal roots.

$i-1$ b's are to be calculated by $i-1$ of the equations $(6)^{vii}$. Thus for example take $b_1=0$. In the first place use all except the first of equations $(6)^{vii}$ to determine b_2, b_3, etc.: we thus find

$$\left.\begin{array}{l} M(1\cdot1)b_2=-\left\{M(1,2\cdot1,2)\dfrac{d2\cdot1}{d\lambda}+M(1,2\cdot1,3)\dfrac{d3\cdot1}{d\lambda}+\text{etc.}\right\}c_1 \\ \qquad -\left\{M(1,2\cdot1,2)\dfrac{d2\cdot2}{d\lambda}+M(1,2\cdot1,3)\dfrac{d3\cdot2}{d\lambda}+\text{etc.}\right\}c_2-\text{etc.} \\ M(1\cdot1)b_3=\text{etc.}\quad M(1\cdot1)b_4=\text{etc.}\qquad \text{etc.}\qquad \text{etc.} \end{array}\right\}\ldots(6)^{ix}.$$

Secondly, use all except the second of $(6)^{vii}$ to find b_2, b_3, etc.: we thus find

$$M(2\cdot1)\,b_2=\text{etc.},\quad M(2\cdot1)\,b_3=\text{etc.},\quad M(2\cdot1)\,b_4=\text{etc.}\ldots\ldots(6)^{x}.$$

Thirdly, by using all of $(6)^{vii}$ except the third, fourthly, all except the fourth, and so on, we find

$$M(3\cdot1)\,b_2=\text{etc.},\quad M(3\cdot1)\,b_3=\text{etc.},\quad M(3\cdot1)\,b_4=\text{etc.}\ldots\ldots(6)^{xi}.$$

343 *e*. In certain cases of equality among the roots (343 *m*) it is found that values of the coefficients (11), 11, 11, etc. differing infinitely little from particular values which give the equality give values of a_1 and a_1', a_2 and a_2', etc., which are not infinitely nearly equal. In such cases we see by $(6)''$ that b_1, b_2, etc. are finite, and c_1, c_2, etc. vanish: and so the solution does not contain terms of the form $t\epsilon^{\lambda t}$: but the requisite number of arbitrary constants is made up by a proper degree of indeterminateness in the residuary equations for the ratios b_2/b_1, b_3/b_1, etc.

Now when $c_1=0$, $c_2=0$, etc. the second members of equations $(6)^{ix}$, $(6)^{x}$, $(6)^{xi}$, etc. all vanish, and as b_2, b_3, b_4, etc. do not all vanish, it follows that we have

$$M(1\cdot1)=0,\quad M(2\cdot1)=0,\quad M(3\cdot1)=0,\quad \text{etc.}\ldots\ldots(6)^{xii}.$$

Case of equal roots and evanescent minors:

Hence by $(5)^{vii}$ or $(5)^{viii}$ we infer that all the first minors are zero for any value of λ which is doubly a root, and which yet does not give terms of the form $t\epsilon^{\lambda t}$ in the solution. This important proposition is due to Routh*, who, escaping the errors of previous writers (§ 343 *m* below), first gave the complete theory of equal roots of the determinant in cycloidal motion.

* *Stability of Motion* (Adams Prize Essay for 1877), chap. I. § 5.

He also remarked that the factor t does not necessarily imply instability, as terms of the form $t\epsilon^{-pt}$, or $t\epsilon^{-pt}\cos(nt-e)$, when p is positive, do not give instability, but on the contrary correspond to non-oscillatory or oscillatory subsidence to equilibrium. Routh's theorem.

343 *f.* We fall back on the case of no motional forces by taking $11=0$, $12=0$, etc., which reduces the equations (3) for determining the ratios a_2/a_1, a_3/a_1, etc. to Case of no motional forces.

$$\lambda^2\frac{d\mathfrak{T}}{da_1}+11a_1+12a_2+\text{etc.}=0,\quad \lambda^2\frac{d\mathfrak{T}}{da_2}+21a_1+22a_2+\text{etc.}=0,\text{ etc. } (7),$$

or, expanded,

$$\left.\begin{array}{l}[(11)\lambda^2+11]a_1+[(12)\lambda^2+12]a_2+\text{etc.}=0\\ [(21)\lambda^2+21]a_1+[(22)\lambda^2+22]a_2+\text{etc.}=0\end{array}\right\}\ldots.(7').$$

The determinantal equation (4) to harmonize these simplified equations (7) or (7′) becomes

$$\begin{vmatrix}(11)\lambda^2+11, & (12)\lambda^2+12, \ldots\\ (21)\lambda^2+21, & (22)\lambda^2+22, \ldots\\ \ldots\ldots & \ldots\ldots\end{vmatrix}=0\ldots\ldots\ldots\ldots(8).$$

This is of degree i, in λ^2: therefore λ has i pairs of oppositely signed equal values, which we may now denote by

$$\pm\lambda,\quad \pm\lambda',\quad \pm\lambda'',\ \ldots;$$

and for each of these pairs the series of ratio-equations (7′) are the same. Hence the complete solution of the differential equations of motion may be written as follows, to show its arbitraries explicitly :—

$$\left.\begin{array}{l}\psi_1=\quad(A\epsilon^{\lambda t}+B\epsilon^{-\lambda t})+\quad(A'\epsilon^{\lambda' t}+B'\epsilon^{-\lambda' t})+\quad(A''\epsilon^{\lambda'' t}+B''\epsilon^{-\lambda'' t})+\text{etc.}\\ \psi_2=\frac{a_2}{a_1}(A\epsilon^{\lambda t}+B\epsilon^{-\lambda t})+\frac{a_2'}{a_1'}(A'\epsilon^{\lambda' t}+B'\epsilon^{-\lambda' t})+\frac{a_2''}{a_1''}(A''\epsilon^{\lambda'' t}+B''\epsilon^{-\lambda'' t})+\text{etc.}\\ \psi_3=\frac{a_3}{a_1}(A\epsilon^{\lambda t}+B\epsilon^{-\lambda t})+\frac{a_3'}{a_1'}(A'\epsilon^{\lambda' t}+B'\epsilon^{-\lambda' t})+\frac{a_3''}{a_1''}(A''\epsilon^{\lambda'' t}+B''\epsilon^{-\lambda'' t})+\text{etc.}\\ \qquad\text{etc.}\qquad\qquad\qquad\text{etc.}\qquad\qquad\qquad\text{etc.}\end{array}\right\}\ldots\ldots(9),$$

where A, B; A', B'; A'', B''; etc. denote $2i$ arbitrary constants, and

$$\frac{a_2}{a_1},\ \frac{a_3}{a_1},\ \ldots,\ \frac{a_2'}{a_1'},\ \frac{a_3'}{a_1'},\ \ldots,\ \frac{a_2''}{a_1''},\ \frac{a_3''}{a_1''},\ \ldots,\ \text{etc.},$$

are i sets of $i-1$ ratios each, the values of which, when all the i roots of the determinantal equation in λ^2 have different values, are fully determined by giving successively these i values to λ^2 in (7′).

Case of no motional forces.

343 *g*. When there are equal roots, the solution is to be completed according to § 343 *d* or *e*, as the case may be. The case of a conservative system (343 *h*) necessarily falls under § 343 *e*, as is proved in § 343 *m*. The same form, (9), still represents the complete solutions when there are equalities among the roots, but with changed conditions as to arbitrariness of the elements appearing in it. Suppose $\lambda^2 = \lambda'^2$ for example. In this case any value may be chosen arbitrarily for a_2/a_1, and the remainder of the set a_3/a_1, a_4/a_1 ... are then fully determined by (7'); again another value may be chosen for a_2'/a_1', and with it a_3'/a_1', a_4'/a_1', ... are determined by a fresh application of (7') with the same value for λ^2: and the arbitraries now are $A+A'$, $B+B'$,

$$\frac{a_2}{a_1}A+\frac{a_2'}{a_1'}A',\ \frac{a_2}{a_1}B+\frac{a_2'}{a_1'}B',\ A'',\ B'',\ A''',\ B''',\ \ldots A^{(i-1)},\ \text{and } B^{(i-1)},$$

numbering still $2i$ in all. Similarly we see how, beginning with the form (9), convenient for the general case of i different roots, we have in it also the complete solution when λ^2 is triply, or quadruply, or any number of times a root, and when any other root or roots also are double or multiple.

Cycloidal motion. Conservative positional, and no motional, forces.

343 *h*. For the case of a conservative system, that is to say, the case in which

$$12=21,\quad 13=31,\quad 23=32,\quad \text{etc., etc.} \qquad (10),$$

the differential equations of motion, (1), become

$$\frac{d}{dt}\left(\frac{dT}{d\dot\psi}\right)+\frac{dV}{d\psi}=0,\quad \frac{d}{dt}\left(\frac{dT}{d\dot\phi}\right)+\frac{dV}{d\phi}=0,\ \text{etc.} \qquad (10'),$$

and the solving linear algebraic equations, (3), become

$$\frac{d\mathfrak{T}}{da_1}+\frac{d\mathfrak{V}}{da_1}=0,\quad \frac{d\mathfrak{T}}{da_2}+\frac{d\mathfrak{V}}{da_2}=0 \qquad (10''),$$

where

$$V=\tfrac{1}{2}(11\psi_1^2+2.12\psi_1\psi_2+\text{etc.}),\ \text{and}\ \mathfrak{V}=\tfrac{1}{2}(11a_1^2+2.12a_1a_2+\text{etc.}) \qquad (10''').$$

In this case the i roots, λ^2, of the determinantal equation are the negatives of the values of α, β, ... of our first investigation; and thus in (10''), (8), and (9) we have the promised solution by one completely expressed process. From § 337 and its footnote we infer that in the present case the roots λ^2 are all real, whether negative or positive.

Cycloidal motion. Conservative positional, and no motional, forces.

In § 337 it was expressly assumed that T (as it must be in the dynamical problem) is essentially positive; but the investigation was equally valid for any case in which either of the quadratics T or V is incapable of changing sign for real values of the variables ($\dot\psi_1$, $\dot\psi_2$, etc. for T, or ψ_1, ψ_2, etc. for V). Thus we see that the roots λ^2 are all real when the relations (5) and (9) are satisfied, and when the magnitudes of the residual independent coefficients (11), (22), (12), ... and 11, 22, 12, ... are such that of the resulting quadratics, $\mathfrak{T}$, $\mathfrak{V}$, one or other is essentially positive or essentially negative. This property of the determinantal equation (7') is very remarkable. A more direct algebraic proof is to be desired. Here is one:—

343 *k*. Writing out (7') for λ^2, and for λ'^2, multiplying the first for λ^2 by $\frac{1}{2}a_1'$, the second by $\frac{1}{2}a_2'$, and so on, and adding; and again multiplying the first for λ'^2 by $\frac{1}{2}a_1$, the second by $\frac{1}{2}a_2$, and so on, and adding, we find

$$\left.\begin{aligned} &\lambda^2\mathfrak{T}(a, a') + \mathfrak{V}(a, a') = 0 \\ \text{and}\quad &\lambda'^2\mathfrak{T}(a, a') + \mathfrak{V}(a, a') = 0 \end{aligned}\right\} \quad \ldots\ldots\ldots\ldots\ldots(11),$$

where

$$\left.\begin{aligned} &\mathfrak{T}(a, a') = \tfrac{1}{2}\{(11)\,a_1a_1' + (12)(a_1a_2' + a_2a_1') + \text{etc.}\} \\ \text{and}\quad &\mathfrak{V}(a, a') = \tfrac{1}{2}\{\,11\;a_1a_1' + 12\;(a_1a_2' + a_2a_1') + \text{etc.}\} \end{aligned}\right\} \ldots(12).$$

Remark that according to this (12) notation $\mathfrak{T}$ (a, a) means the same thing as $\mathfrak{T}$ simply, according to the notation of (3) etc. above, and $\mathfrak{T}$ $(\dot\psi, \dot\psi)$ the same thing as T. Remark farther that $\mathfrak{T}$ (a, a') is a linear function of a_1, a_2, ... with coefficients each involving a_1', a_2', ... linearly; and that it is symmetrical with reference to a_1, a_1', and a_2, a_2', etc.; and that we therefore have

$$\left.\begin{aligned} &\mathfrak{T}(mp, p') = m\mathfrak{T}(p, p') = \mathfrak{T}(p, mp') \text{ and} \\ &\mathfrak{T}(mp + nq, m'p + n'q) = mm'\,\mathfrak{T}(p, p) + (mn' + m'n)\,\mathfrak{T}(p, q) + nn'\,\mathfrak{T}(q, q) \end{aligned}\right\} \ldots(13).$$

Precisely similar statements and formulas hold for $\mathfrak{V}$ (a, a').

From (11) we infer that if λ^2 and λ'^2 be unequal we must have

$$\mathfrak{T}(a, a') = 0, \text{ and } \mathfrak{V}(a, a') = 0 \ldots\ldots\ldots\ldots\ldots(14).$$

Now if there can be imaginary roots, λ^2, let $\lambda^2 = \rho + \sigma\sqrt{-1}$ and $\lambda'^2 = \rho - \sigma\sqrt{-1}$ be a pair of them, ρ and σ being real. And, p_1, q_1, p_2, q_2, etc. being all real, let $p_1 + q_1\sqrt{-1}$, $p_1 - q_1\sqrt{-1}$, be arbitrarily chosen values of a_1, a_1', and let

$$p_2 + q_2\sqrt{-1},\ p_3 + q_3\sqrt{-1},\ \ldots,\ p_2 - q_2\sqrt{-1},\ p_3 - q_3\sqrt{-1},\ \ldots$$

Cycloidal motion. Conservative positional, and no motional, forces.

be the *determinately deduced* values* of $a_2, a_3, \ldots, a_2', a_3', \ldots$ according to (7′); we have, by (13), with

$$m = m' = 1, \; n = \sqrt{-1}, \; n' = -\sqrt{-1},$$

$$\left.\begin{array}{l}\mathfrak{T}(a, a') = \mathfrak{T}(p, p) + \mathfrak{T}(q, q) \\ \mathfrak{V}(a, a') = \mathfrak{V}(p, p) + \mathfrak{V}(q, q)\end{array}\right\}\ldots\ldots\ldots\ldots\ldots\ldots(14').$$

and

Now by hypothesis either $\mathfrak{T}(x, x)$, or $\mathfrak{V}(x, x)$ is essentially of one sign for all real values of x_1, x_2, etc. Hence the second member of one or other of equations (14′) cannot be zero, because $p_1, p_2, \ldots,$ and $q_1, q_2, \ldots$ are all real. But by (14) the first member of each of the equations (14′) is zero if λ^2 and λ'^2 are unequal: hence they are equal: hence either $p_1 = 0$, $p_2 = 0$, etc., or $q_1 = 0$, $q_2 = 0$, etc., that is to say the roots λ^2 are all necessarily real, whether negative or positive.

343 *l*. Farther we now see by going back to (11):—

(*a*) if for all real values of $x_1, x_2, \ldots$ the values of $\mathfrak{T}(x, x)$ and $\mathfrak{V}(x, x)$ have the same unchanging sign, the roots λ^2 are all negative;

(*b*) if for different real values of x_1, x_2, etc., one of the two $\mathfrak{T}(x, x)$, $\mathfrak{V}(x, x)$ has different signs (the other by hypothesis having always one sign), some of the roots λ^2 are negative and some positive;

(*c*) if the values of $\mathfrak{T}$ and $\mathfrak{V}$ have essentially opposite signs (and each therefore according to hypothesis unchangeable in sign), the roots λ^2 are all positive.

The (*a*) and (*c*) of this tripartite conclusion we see by taking $\lambda'^2 = \lambda^2$ in (11), which reduces them to

$$\lambda^2 \mathfrak{T}(a, a) + \mathfrak{V}(a, a) = 0 \ldots\ldots\ldots\ldots\ldots\ldots(15),$$

and remarking that a_2, a_3, etc. are now all real if we please to give a real value to a_1. The (*b*) is proved in § 343 *o* below.

Case of equal roots.

343 *m*. From (14) we see that when two roots λ^2, λ'^2, are infinitely nearly equal there is no approach to equality between a_1 and a_1', a_2 and a_2', and therefore, when there are no motional forces, and when the positional forces are conservative, equality of roots essentially falls under the case of § 343 *e* above. This may be proved explicitly as follows:—let

$$\psi_1 = (a_1 t + b_1)\,\epsilon^{\lambda t}, \quad \psi_2 = (a_2 t + b_2)\,\epsilon^{\lambda t}, \text{ etc.} \ldots\ldots\ldots (15)'$$

* Cases of equalities among the roots are disregarded for the moment merely to avoid circumlocutions, but they obviously form no exception to the reasoning and conclusion.

be the complete solution corresponding to the root λ supposed to be a dual root. Using this in equations (1) and equating to zero in each equation so found the coefficients of $t\epsilon^{\lambda t}$ and of $\epsilon^{\lambda t}$, with the notation of (12) we find

Cycloidal motion. Conservative positional, and no motional, forces.

$$\lambda^2 \frac{d\mathfrak{T}(a,a)}{da_1} + \frac{d\mathfrak{V}(a,a)}{da_1} = 0,\ \lambda^2 \frac{d\mathfrak{T}(a,a)}{da_2} + \frac{d\mathfrak{V}(a,a)}{da_2} = 0, \text{ etc.} \ldots (15)'',$$

$$\left.\begin{aligned} &\lambda^2 \frac{d\mathfrak{T}(b,b)}{db_1} + 2\lambda \frac{d\mathfrak{T}(a,a)}{da_1} + \frac{d\mathfrak{V}(b,b)}{db_1} = 0, \\ &\lambda^2 \frac{d\mathfrak{T}(b,b)}{db_2} + 2\lambda \frac{d\mathfrak{T}(a,a)}{da_2} + \frac{d\mathfrak{V}(b,b)}{db_2} = 0, \qquad \text{etc.} \end{aligned}\right\} \ldots (15)'''.$$

Multiplying the first, second, third, etc. of (15)″ by b_1, b_2, b_3, etc. and adding we find

$$\lambda^2 \mathfrak{T}(a,b) + \mathfrak{V}(a,b) = 0 \ldots\ldots\ldots\ldots\ldots\ldots (15)^{\text{iv}};$$

and similarly from (15)‴ with multipliers a_1, a_2, etc.

$$\lambda^2 \mathfrak{T}(a,b) + \mathfrak{V}(a,b) + 2\lambda \mathfrak{T}(a,a) = 0 \ldots\ldots\ldots (15)^{\text{v}}.$$

Subtracting $(15)^{\text{iv}}$ from $(15)^{\text{v}}$ we see that $\mathfrak{T}(a,a) = 0$. Hence we must have $a_1 = 0$, $a_2 = 0$, etc., that is to say there are no terms of the form $t\epsilon^{\lambda t}$ in the solution. It is to be remarked that the inference of $a_1 = 0$, $a_2 = 0$, etc. from $\mathfrak{T}(a,a) = 0$, is not limited to real roots λ because λ^2 in the present case is essentially real, and whether it be positive or negative the ratios a_2/a_1, a_3/a_1, etc., are essentially real.

It is remarkable that both Lagrange and Laplace fell into the error of supposing that equality among roots necessarily implies terms in the solution of the form $t\epsilon^{\lambda t}$ (or $t \cos pt$), and therefore that for stability the roots must be all unequal. This we find in the *Mécanique Analytique,* Seconde Partie, section VI. Art. 7 of the second edition of 1811 published three years before Lagrange's death, and repeated without change in the posthumous edition of 1853. It occurs in the course of a general solution of the problem of the infinitely small oscillations of a system of bodies about their positions of equilibrium, with conservative forces of position and no motional forces, which from the "Avertissement" (p. vi.) prefixed to the 1811 edition seems to have been first published in the 1811 edition, and not to have appeared in the original edition of 1788*. It would be

* Since this statement was put in type, the first edition of the *Mécanique Analytique* (which had been inquired for in vain in the University libraries of Cambridge and Glasgow) has been found in the University library of Edinburgh,

Cycloidal motion. Conservative positional, and no motional, forces.

curious if such an error had remained for twenty-three years in Lagrange's mind. It could scarcely have existed even during the writing and printing of the Article for his last edition if he had been in the habit of considering particular applications of his splendid analytical work: if he had he would have seen that a proposition which asserted that the equilibrium of a particle in the bottom of a frictionless bowl is unstable if the bowl be a figure of revolution with its axis vertical, cannot be true. No such obvious illustration presents itself to suggest or prove the error as Laplace has it in the *Mécanique Céleste* (Première Partie, Livre II. Art. 57) in the course of an investigation of the secular inequalities of the planetary system. But as [by a peculiarly simple case of the process of § 345^{xiv} (54)] he has reduced his analysis of this problem virtually to the same as that of conservative oscillations about a configuration of equilibrium, the physical illustrations which abound for this case suffice to prove the error in Laplace's statement, different and comparatively recondite as its dynamical subject is. An error the converse of that of Laplace and Lagrange occurred in page 278 of our First Edition where it was said that "Cases in which "there are equal roots leave a corresponding number of degrees "of indeterminateness in the ratios $l:m$, $l:n$, etc., and so allow "the requisite number of arbitrary constants to be made up," without limiting this statement to the case of conservative positional and no motional forces, for which its truth is obvious from the nature of the problem, and for which alone it is obvious at first sight; although for the cases of adynamic oscillations, and of stable precessions, § 345^{xiv}, it is also essentially true. The correct theory of equal roots in the generalized problem of cycloidal motion has been so far as we know first given by Routh in his investigation referred to above (§ 343 *e*).

343 *n*. Returning to § 343 *l*, to make more of (*b*), and to understand the efficiency of the oppositely signed roots, λ^2, asserted in it, let $\sigma^2=-\lambda^2$ in any case in which λ^2 is negative, and let

$$\psi_1=r_1\cos(\sigma t-e),\ \ \psi_2=r_2\cos(\sigma t-e),\ \text{etc.}\ldots\ldots(16),$$

be the corresponding particular solution in fully *realized* terms,

and it does contain the problem of infinitely small oscillations, with the remarkable error referred to in the text.

as in § 337 (6) above but with somewhat different notation. By substituting in (1) and multiplying the first of the resulting equations by r_1, the second by r_2, and so on and adding, virtually as we found (15), we now find

Cycloidal motion. Conservative positional, and no motional, forces.

$$-\sigma^2 \mathfrak{T}(r, r) + \mathfrak{V}(r, r) = 0 \quad\quad (17).$$

Adopting now the notation of (9) for the real positive ones of the roots λ^2, but taking, for brevity, $a_1 = 1$, $a_1' = 1$, $a_1'' = 1$, etc., we have for the complete solution when there are both negative and positive roots of the determinantal equation (7');

$$\left.\begin{array}{l} \psi_1 = (A\epsilon^{\lambda t}+B\epsilon^{-\lambda t}) + (A'\epsilon^{\lambda' t}+B'\epsilon^{-\lambda' t}) + \text{etc.} + r_1\cos(\sigma t-e) + r_1'\cos(\sigma' t-e') + \text{etc.} \\ \psi_2 = a_2(A\epsilon^{\lambda t}+B\epsilon^{-\lambda t}) + a_2'(A'\epsilon^{\lambda' t}+B'\epsilon^{-\lambda' t}) + \text{etc.} + r_2\cos(\sigma t-e) + r_2'\cos(\sigma' t-e') + \text{etc.} \\ \psi_3 = \text{etc.}, \quad \psi_4 = \text{etc.} \quad \text{etc.} \quad \text{etc.} \end{array}\right\} \ldots(18).$$

343 *o*. Using this in the general expressions for T and V, with the notation (12), and remarking that the products $\epsilon^{\lambda t} \times \epsilon^{\lambda' t}$, etc. and $\epsilon^{\lambda t} \times \sin(\sigma t - e)$, etc., and $\sin(\sigma t - e) \times \sin(\sigma' t - e')$, etc., disappear from the terms in virtue of (11), we find

$$\left.\begin{array}{l} T = \lambda^2\mathfrak{T}(a, a)(A\epsilon^{\lambda t} - B\epsilon^{-\lambda t})^2 + \lambda'^2\mathfrak{T}(a', a')(A'\epsilon^{\lambda' t} - B'\epsilon^{-\lambda' t})^2 + \text{etc.} \\ \quad + \sigma^2\mathfrak{T}(r, r)\sin^2(\sigma t - e) + \sigma'^2\mathfrak{T}(r', r')\sin^2(\sigma' t - e') + \text{etc.} \end{array}\right\} (19),$$

and

$$\left.\begin{array}{l} V = \mathfrak{V}(a, a)(A\epsilon^{\lambda t} + B\epsilon^{-\lambda t})^2 + \mathfrak{V}(a', a')(A'\epsilon^{\lambda' t} + B'\epsilon^{-\lambda' t})^2 + \text{etc.} \\ \quad + \mathfrak{V}(r, r)\cos^2(\sigma t - e) + \mathfrak{V}(r', r')\cos^2(\sigma' t - e') + \text{etc.} \end{array}\right\} (20).$$

The factors which appear with

$$\mathfrak{T}(a, a), \mathfrak{T}(a', a'), \ldots \mathfrak{T}(r, r), \mathfrak{T}(r', r')$$

in this expression (19) for T are all essentially positive; and the same is true of $\mathfrak{V}$ in (20) for V. Now for every set of real co-ordinates and velocity-components the potential and kinetic energies are expressible by the formulas (20) and (19) because (18) is the complete solution with $2i$ arbitraries. Hence if the value of V can change sign with real values of the co-ordinates, the quantities $\mathfrak{V}(a, a)$, $\mathfrak{V}(a', a')$, etc., and $\mathfrak{V}(r, r)$, $\mathfrak{V}(r', r')$, etc., for the several roots must be some of them positive and some of them negative; and if the value of T could change sign with real values of the velocity-components, some of the quantities $\mathfrak{T}(a, a)$, $\mathfrak{T}(a', a')$, etc., and $\mathfrak{T}(r, r)$, $\mathfrak{T}(r', r')$, etc. would need to be positive and some negative. So much being learned from (20) and (19) we must now recal to mind that according to hypothesis one only of the two quadratics T and V can change sign, to conclude from (15) and (17) that there are both positive and negative roots λ^2 when either T or V can change sign. Thus (*b*) of the tripartite conclusion above is rigorously proved.

Cycloidal motion. Conservative positional, and no motional, forces.

343 *p*. A short algebraic proof of (*b*) could no doubt be easily given; but our somewhat elaborate discussion of the subject is important as showing in (15)...(20) the whole relation between the previous short algebraic investigation, conducted in terms involving quantities which are essentially imaginary for the case of oscillations about a configuration of stable equilibrium, and the fully realized solution, with formulas for the potential and kinetic energies realized both for oscillations and for fallings away from unstable equilibrium.

We now see definitively by (15) and (17) that, in *real* dynamics (that is to say T essentially positive) the factors $\mathfrak{V}(a, a)$, $\mathfrak{V}(a', a')$, etc., are all negative, and $\mathfrak{V}(r, r)$, $\mathfrak{V}(r', r')$, etc., all positive in the expression (20) for the potential energy. Adding (20) to (19) and using (15) and (17) in the sum, we find

Equation of energy in realized general solution.

$$\left.\begin{array}{l} T + V = -4AB\lambda^2\mathfrak{T}(a, a) - 4A'B'\lambda'^2\mathfrak{T}(a', a'), \text{ etc.} \\ \qquad + \sigma^2\mathfrak{T}(r, r) + \sigma'^2\mathfrak{T}(r', r') + \text{etc.} \end{array}\right\} \ldots (21).$$

It is interesting to see in this formula how the constancy of the sum of the potential and kinetic energies is attained in any solution of the form $A\epsilon^{\lambda t} + B\epsilon^{-\lambda t}$ [which, with $\lambda = \sigma\sqrt{-1}$, includes the form $r\cos(\sigma t - e)$], and to remark that for any single solution $a\epsilon^{\lambda t}$, or solution compounded of single solutions depending on unequal values of λ^2 (whether real or imaginary), the sum of the potential and kinetic energies is essentially zero.

Artificial or ideal accumulative system.

344. When the positional forces of a system violate the law of conservatism, we have seen (§ 272) that energy without limit may be drawn from it by guiding it perpetually through a returning cycle of configurations, and we have inferred that in every real system, not supplied with energy from without, the positional forces fulfil the conservative law. But it is easy to arrange a system artificially, in connexion with a source of energy, so that its positional forces shall be non-conservative; and the consideration of the kinetic effects of such an arrangement, especially of its oscillations about or motions round a configuration of equilibrium, is most instructive, by the contrasts which it presents to the phenomena of a natural system. The preceding formulas, (7)...(9) of § 343 *f* and § 343 *g*, express the general solution of the problem—to find the infinitely small motion of a cycloidal system, when, without motional forces, there is deviation from conservatism by the character of the positional forces.

In this case [(10) not fulfilled,] just as in the case of motional forces fulfilling the conservative law (10), the character of the equilibrium as to stability or instability is discriminated according to the character of the roots of an algebraic equation of degree equal to the number of degrees of freedom of the system.

Artificial or ideal accumulative system.

Criterion of stability.

If the roots (λ^2) of the determinantal equation § 343 (8) are all real and negative, the equilibrium is stable: in every other case it is unstable.

345. But although, when the equilibrium is stable, no possible infinitely small displacement and velocity given to the system can cause it, when left to itself, to go on moving farther and farther away till either a finite displacement is reached, or a finite velocity acquired; it is very remarkable that stability should be possible, considering that even in the case of stability an endless increase of velocity may, as is easily seen from § 272, be obtained merely by *constraining* the system to a particular closed course, or circuit of configurations, nowhere deviating by more than an infinitely small amount from the configuration of equilibrium, and leaving it at rest anywhere in a certain part of this circuit. This result, and the distinct peculiarities of the cases of stability and instability, will be sufficiently illustrated by the simplest possible example, that of a material particle moving in a plane.

Let the mass be unity, and the components of force parallel to two rectangular axes be $ax+by$, and $a'x+b'y$, when the position of the particle is (x, y). The equations of motion will be

$$\ddot{x} = ax + by,\quad \ddot{y} = a'x + b'y \qquad\qquad (1).$$

Let $\frac{1}{2}(a'+b) = c$, and $\frac{1}{2}(a'-b) = e$:

the components of the force become

$$ax + cy - ey,\ \text{and}\ cx + b'y + ex,$$

or

$$-\frac{dV}{dx} - ey,\ \text{and}\ -\frac{dV}{dy} + ex,$$

where

$$V = -\tfrac{1}{2}(ax^2 + b'y^2 + 2cxy).$$

The terms $-ey$ and $+ex$ are clearly the components of a force $e(x^2+y^2)^{\frac{1}{2}}$, perpendicular to the radius-vector of the particle. Hence if we turn the axes of co-ordinates through any angle, the

Artificial or ideal accumulative system.

corresponding terms in the transformed components are still $-ey$ and $+ex$. If, therefore, we choose the axes so that

$$V = \tfrac{1}{2}(\alpha x^2 + \beta y^2) \quad\text{..........................(2)},$$

the equations of motion become, without loss of generality,

$$\ddot{x} = -\alpha x - ey, \quad \ddot{y} = -\beta y + ex.$$

To integrate these, assume, as in general [§ 343 (2)],

$$x = l\epsilon^{\lambda t}, \quad y = m\epsilon^{\lambda t}.$$

Then, as before [§ 343 (7)],

$$(\lambda^2 + \alpha)\, l + em = 0, \text{ and } -el + (\lambda^2 + \beta)\, m = 0.$$

Whence $$(\lambda^2 + \alpha)(\lambda^2 + \beta) = -e^2 \quad\text{......................(3)},$$

which gives

$$\lambda^2 = -\tfrac{1}{2}(\alpha + \beta) \pm \{\tfrac{1}{4}(\alpha - \beta)^2 - e^2\}^{\frac{1}{2}}.$$

This shows that the equilibrium is stable if both $\alpha\beta + e^2$ and $\alpha + \beta$ are positive and $e^2 < \frac{1}{4}(\alpha - \beta)^2$ but unstable in every other case.

But let the particle be constrained to remain on a circle, of radius r. Denoting by θ its angle-vector from OX, and transforming (§ 27) the equations of motion, we have

$$\ddot{\theta} = -(\beta - \alpha)\sin\theta\cos\theta + e = -\tfrac{1}{2}(\beta - \alpha)\sin 2\theta + e \quad\text{.......(4)}.$$

If we had $e = 0$ (a conservative system of force) the positions of equilibrium would be at $\theta = 0$, $\theta = \frac{1}{2}\pi$, $\theta = \pi$, and $\theta = \frac{3}{2}\pi$; and the motion would be that of the quadrantal pendulum. But when e has any finite value less than $\frac{1}{2}(\beta - \alpha)$ which, for convenience, we may suppose positive, there are positions of equilibrium at

$$\theta = \vartheta, \quad \theta = \frac{\pi}{2} - \vartheta, \quad \theta = \pi + \vartheta, \text{ and } \theta = \frac{3\pi}{2} - \vartheta,$$

where ϑ is half the acute angle whose sine is $\dfrac{2e}{\beta - \alpha}$: the first and third being positions of stable, and the second and fourth of unstable, equilibrium. Thus it appears that the effect of the constant tangential force is to displace the positions of stable and unstable equilibrium forwards and backwards on the circle through angles each equal to ϑ. And, by multiplying (4) by $2\dot{\theta}dt$ and integrating, we have as the integral equation of energy

$$\dot{\theta}^2 = C + \tfrac{1}{2}(\beta - \alpha)\cos 2\theta + 2e\theta \quad\text{..................(5)}.$$

From this we see that the value of C, to make the particle just reach the position of unstable equilibrium, is

Artificial or ideal accumulative system.

$$C = -\tfrac{1}{2}(\beta - a)\cos(\pi - 2\vartheta) - e(\pi - 2\vartheta),$$

$$= \sqrt{\frac{(\beta - a)^2}{4} - e^2} - e\left(\pi - \sin^{-1}\frac{2e}{\beta - a}\right),$$

and by equating to zero the expression (5) for $\dot{\theta}^2$, with this value of C substituted, we have a transcendental equation in θ, of which the least negative root, $\theta_{,}$, gives the limit of vibrations on the side reckoned backwards from a position of stable equilibrium. If the particle be placed at rest on the circle at any distance less than $\frac{\pi}{2} - 2\vartheta$ *before* a position of stable equilibrium, or less than $\vartheta - \theta_{,}$ *behind* it, it will vibrate. But if placed anywhere beyond those limits and left either at rest or moving with any velocity in either direction, it will end by flying round and round forwards with a periodically increasing and diminishing velocity, but increasing every half turn by equal additions to its squares.

If on the other hand $e > \frac{1}{2}(\beta - a)$, the positions both of stable and unstable equilibrium are imaginary; the tangential force predominating in every position. If the particle be left at rest in any part of the circle it will fly round with continually increasing velocity, but periodically increasing and diminishing acceleration.

345^{i}. Leaving now the ideal case of positional forces violating the law of conservatism, interestingly curious as it is, and instructive in respect to the contrast it presents with the positional forces of nature which are essentially conservative, let us henceforth suppose the positional forces of our system to be conservative and let us admit infringement of conservatism only as in nature through motional forces. We shall soon see (§ 345^{vii} and ^{viii}) that we may have motional forces which do not violate the law of conservatism. At present we make no restriction upon the motional forces and no other restriction on the positional forces than that they are conservative.

Cycloidal system with conservative positional forces and unrestricted motional forces.

The differential equations of motion, taken from (1) of 343a above, with the relations (10), and with V to denote the potential energy, are,

Cycloidal system with conservative positional forces and unrestricted motional forces.

$$\left.\begin{array}{l}\frac{d}{dt}\left(\frac{dT}{d\dot{\psi}_1}\right) + 11\dot{\psi}_1 + 12\dot{\psi}_2 + \ldots + \frac{dV}{d\psi_1} = 0 \\ \frac{d}{dt}\left(\frac{dT}{d\dot{\psi}_2}\right) + 21\dot{\psi}_1 + 22\dot{\psi}_2 + \ldots + \frac{dV}{d\psi_2} = 0 \\ \text{etc.} \qquad\qquad \text{etc.}\end{array}\right\} \ldots\ldots\ldots\ldots(1).$$

Multiplying the first of these by $\dot{\psi}_1$, the second by $\dot{\psi}_2$, adding and transposing, we find

$$\frac{d(T+V)}{dt} = -Q \ldots\ldots\ldots\ldots\ldots\ldots\ldots\ldots(2),$$

where

$$Q = 11\dot{\psi}_1^2 + (12+21)\dot{\psi}_1\dot{\psi}_2 + 22\dot{\psi}_2^2 + (13+31)\dot{\psi}_1\dot{\psi}_3 + \text{etc.}\ldots\ldots(3).$$

Dissipativity defined.

345''. The quadratic function of the velocities here denoted by Q has been called by Lord Rayleigh* the Dissipation Function. We prefer to call it *Dissipativity*. It expresses the rate at which the *palpable energy* of our supposed cycloidal system is lost, not, as we now know, annihilated but (§§ 278, 340, 341, 342) dissipated away into other forms of energy. *It is essentially positive when the assumed motional forces are such as can exist in nature.* That it is equal to a quadratic function of the velocities is an interesting and important theorem.

Lord Rayleigh's theorem of Dissipativity.

Integral equation of energy.

Multiplying (2) by dt, and integrating, we find

$$T + V = E_0 - \int_0^t Q dt \ldots\ldots\ldots\ldots\ldots\ldots\ldots\ldots(4),$$

where E_0 is a constant denoting the sum of the kinetic and potential energies at the instant $t=0$. Now T and Q are each of them essentially positive except when the system is at rest, and then each of them is zero. Therefore $\int_0^t Q dt$ must increase to infinity unless the system comes more and more nearly to rest as time advances. Hence either this must be the case, or V must diminish to $-\infty$. It follows that when V is positive for all real values of the co-ordinates the system must as time advances come more and more nearly to rest in its zero-configuration, whatever may have been the initial values of the co-ordinates and velocities. Even if V is negative for some or for all values of the co-ordinates, the system may be projected from *some given*

* *Proceedings of the London Mathematical Society*, May, 1873; *Theory of Sound*, Vol. I. § 81.

configurations with such velocities that when $t=\infty$ it shall be at rest in its zero configuration: this we see by taking, as a particular solution, the terms of (9) § 345^{iv} below, for which m is negative. But this equilibrium is essentially unstable, unless V is positive for all real values of the co-ordinates. To prove this imagine the system placed in any configuration in which V is negative, and left there either at rest or with any motion of kinetic energy less than or at the most equal to $-V$: thus E_0 will be negative or zero; $T+V$ will therefore have increasing negative value as time advances; therefore V must always remain negative; and therefore the system can never reach its zero configuration. It is clear that $-V$ and T must each on the whole increase though there may be fluctuations, of T diminishing for a time, during which $-V$ must also diminish so as to make the excess $(-V)-T$ increase at the rate equal to Q per unit of time according to formula (2).

Cycloidal system with conservative positional forces and unrestricted motional forces.

345'''. To illustrate the circumstances of the several cases let $\lambda = m + n\sqrt{-1}$ be a root of the determinantal equation, m and n being both real. The corresponding realized solution of the dynamical problem is

$$\psi_1 = r_1\epsilon^{mt}\cos(nt-e_1),\ \psi_2 = r_2\epsilon^{mt}\cos(nt-e_2),\ \text{etc.} \quad \ldots\ldots\ldots(5),$$

where the differences of epochs e_2-e_1, e_3-e_1, etc. and the ratios r_2/r_1, etc., in all $2i-2$ numerics*, are determined by the $2i$ simultaneous linear equations (3) of § 343 harmonized by taking for $\lambda = m+n\sqrt{-1}$, and again $\lambda = m-n\sqrt{-1}$. Using these expressions for ψ_1, ψ_2, etc. in the expressions for V, Q, T, we find,—

$$\left.\begin{aligned} V &= \epsilon^{2mt}(C + A\cos 2nt + B\sin 2nt) \\ Q &= \epsilon^{2mt}(C' + A'\cos 2nt + B'\sin 2nt) \\ T &= \epsilon^{2mt}(C'' + A''\cos 2nt + B''\sin 2nt) \end{aligned}\right\} \ldots\ldots\ldots\ldots(6),$$

* The term numeric has been recently introduced by Professor James Thomson to denote a number, or a proper fraction, or an improper fraction, or an incommensurable ratio (such as π or ϵ). It must also to be useful in mathematical analysis include imaginary expressions such as $m+n\sqrt{-1}$, where m and n are real numerics. "Numeric" may be regarded as an abbreviation for "numerical expression." It lets us avoid the intolerable verbiage of integer or proper or improper fraction which mathematical writers hitherto are so often compelled to use; and is more appropriate for mere number or ratio than the designation "quantity," which rather implies quantity of something than the mere numerical expression by which quantities of any measurable things are reckoned in terms of the unit of quantity.

Cycloidal system with conservative positional forces and unrestricted motional forces.

where $C, A, B, C', A', B', C'', A'', B''$, are determinate constants: and in order that Q and T may be positive we have

$$C' > +\sqrt{(A'^2+B'^2)}, \text{ and } C'' > +\sqrt{(A''^2+B''^2)} \ldots\ldots\ldots\ldots(7).$$

Substituting these in (2), and equating coefficients of corresponding terms, we find

$$\left.\begin{array}{r} 2m(C+C'') = -C' \\ 2\{m(A+A'') + n(B+B'')\} = -A' \\ 2\{m(B+B'') - n(A+A'')\} = -B' \end{array}\right\} \ldots\ldots\ldots\ldots(8).$$

Real part of every root proved negative when V positive for all real co-ordinates;

The first of these shows that $C+C''$ and m must be of contrary signs. Hence if V be essentially positive [which requires that C be greater than $+\sqrt{(A^2+B^2)}$], every value of m must be negative.

345^{iv}. If V have negative values for some or all real values of the co-ordinates, m must clearly be positive for some roots, but there must still, and always, be roots for which m is negative. To prove this last clause let us instead of (5) take sums of particular solutions corresponding to different roots

positive for some roots when V has negative values;

$$\lambda = m \pm n\sqrt{-1},\ \lambda' = m' \pm n'\sqrt{-1}, \text{ etc.},$$

m and n denoting real numerics. Thus we have

$$\left.\begin{array}{l} \psi_1 = r_1\epsilon^{mt}\cos(nt-e_1) + r'_1\epsilon^{m't}\cos(n't-e'_1) + \text{etc.} \\ \psi_2 = r_2\epsilon^{mt}\cos(nt-e_2) + r'_2\epsilon^{m't}\cos(n't-e'_2) + \text{etc.} \\ \text{etc.} \end{array}\right\} \quad (9).$$

Suppose now m, m', etc. to be all positive: then for $t=-\infty$, we should have $\psi_1=0, \psi_2=0, \dot\psi_1=0, \dot\psi_2=0$, etc., and therefore $V=0, T=0$. Hence, for finite values of t, T would in virtue of (4) be less than $-V$ (which in this case is essentially positive): but we may place the system in any configuration and project it with any velocity we please, and therefore the amount of kinetic energy we may give it is unlimited. Hence, if (9) be the complete solution, it must include some negative value or values of m, and therefore of all the roots λ, λ', etc. there must be some of which the real part is negative. This conclusion is also obvious on purely algebraic grounds, because the coefficient of λ^{2i-1} in the determinant is obviously $11+22+33+\ldots$, which is essentially positive when Q is positive for all real values of the co-ordinates.

but always negative for some roots.

345^{v}. It is an important subject for investigation, interesting both in mere Algebra and in Dynamics, to find how many roots there are with m positive, or how many with m negative in any particular case or class of cases; also to find under what conditions n disappears [or the motion non-oscillatory (compare § 341)]. We hope to return to it in our second volume, and should be very glad to find it taken up and worked out fully by mathematicians in the mean time. At present it is obvious that if V be negative for all real values of ψ_1, ψ_2, etc., the motion must be non-oscillatory for every mode (or every value of λ must be real) if Q be but large enough: but as we shall see immediately with *Q not too large*, n may appear in some or in all the roots, even though V be negative for all real co-ordinates, when there are forces of the gyroscopic class [§ 319, Examp. (G) above and § 345x below). When the motional forces are wholly of the viscous class it is easily seen that n can only appear if V is positive for some or all real values of the co-ordinates: n must disappear if V is negative for all real values of the co-ordinates (again compare § 341).

Non-oscillatory subsidence to stable equilibrium, or falling away from unstable. Oscillatory subsidence to stable equilibrium, or falling away from unstable. Falling away from wholly unstable equilibrium is essentially non-oscillatory if motional forces wholly viscous.

345vi. A chief part of the substance of §§ 345ii ... 345v above may be expressed shortly without symbols thus:—When there is any dissipativity the equilibrium in the zero position is stable or unstable according as the same system with no motional forces, but with the same positional forces, is stable or unstable. The gyroscopic forces which we now proceed to consider may convert instability into stability, as in the gyrostat § 345x below, *when there is no dissipativity*:—but when there is any dissipativity gyroscopic forces may convert rapid falling away from an unstable configuration into falling by (as it were) exceedingly gradual spirals, but they cannot convert instability into stability if there be any dissipativity.

Stability of Dissipative system.

The theorem of Dissipativity [§ 345i, (2) and (3)] suggests the following notation,—

$$\left.\begin{aligned} &\tfrac{1}{2}(12+21)=[12] \text{ or } [21],\ \tfrac{1}{2}(13+31)=[13] \text{ or } [31], \text{ etc.}\\ \text{and } &\tfrac{1}{2}(12-21)=12] \text{ or } -21],\ \tfrac{1}{2}(13-31)=13] \text{ or } -31], \text{ etc.}\end{aligned}\right\}\ (10),$$

so that the symbols [12], [21], [13], etc., and 12], 21], 13], etc. denote quantities which respectively fulfil the following mutual relations,

$$\left.\begin{aligned} &[12]=[21],\ [13]=[31],\ [23]=[32], \text{ etc.}\\ &12]=-21],\ 13]=-31],\ 23]=-32], \text{ etc.}\end{aligned}\right\}\ \ldots\ldots\ldots\ (11).$$

Thus (3) of § 345i becomes

$$Q=11\dot\psi_1^2+2\,[12]\,\dot\psi_1\dot\psi_2+22\dot\psi_2^2+2\,[13]\,\dot\psi_1\dot\psi_3+\text{etc.}\ldots\ldots(12),$$

and going back to (1), with (10) and (12) we have

$$\left.\begin{array}{l}\frac{d}{dt}\frac{dT}{d\dot\psi_1}+\frac{dQ}{d\dot\psi_1}+12]\dot\psi_2+13]\dot\psi_3+\text{etc.}+\frac{dV}{d\psi_1}=0\\ \frac{d}{dt}\frac{dT}{d\dot\psi_2}+\frac{dQ}{d\dot\psi_2}+21]\dot\psi_1+23]\dot\psi_3+\text{etc.}+\frac{dV}{d\psi_2}=0\\ \frac{d}{dt}\frac{dT}{d\dot\psi_3}+\frac{dQ}{d\dot\psi_3}+31]\dot\psi_1+32]\dot\psi_2+\text{etc.}+\frac{dV}{d\psi_3}=0\\ \dots\dots\dots\dots\dots\dots\dots\dots\dots\dots\dots\end{array}\right\}\dots\dots(13).$$

Various origins of gyroscopic terms.

In these equations the terms $12]\dot\psi_2$, $21]\dot\psi_1$, $13]\dot\psi_3$, $31]\dot\psi_1$, etc. represent what we may call gyroscopic forces, because, as we have seen in § 319, Ex. G, they occur when fly-wheels each given in a state of rapid rotation form part of the system by being mounted on frictionless bearings connected through framework with other parts of the system; and because, as we have seen in § 319, Ex. F, they occur when the motion considered is motion of the given system relatively to a rigid body revolving with a constrainedly constant angular velocity round a fixed axis This last reason is especially interesting on account of Laplace's dynamical theory of the tides at the foundation of which it lies, and in which it is answerable for some of the most curious and instructive results, such as the beautiful vortex problem presented by what Laplace calls "Oscillations of the First Species*."

Equation of energy.

345vii. The gyrostatic terms disappear from the equation of energy as we see by § 345i, (2) and (3), and as we saw previously by § 319, Example G (19), and in § 319, Ex. F′ (*f*). Comparing § 319 (*f*) and (*g*), we see that in the case of motion relatively to a body revolving uniformly round a fixed axis it is not the equation of total absolute energy but the equation of energy of the *relative motion* that the gyroscopic terms disappear from, as (*f*) of § 319; and (2) and (3) of § 345i when the subject of their application is to such relative motion.

* The integrated equation for this species of tidal motions, in an ideal ocean equally deep over the whole solid rotating spheroid, is given in a form ready for numerical computation in "Note on the 'Oscillations of the First Species' in Laplace's Theory of the Tides" (W. Thomson), *Phil. Mag.* Oct. 1875.

345viii. To discover something of the character of the gyroscopic influence on the motion of a system, suppose there to be no resistances (or viscous influences), that is to say let the dissipativity, Q, be zero. The determinantal equation (4) becomes

Gyrostatic conservative system:

$$\begin{vmatrix} (11)\lambda^2 \quad\quad +11, & (12)\lambda^2+12]\lambda+12, \dots \\ (21)\lambda^2+21]\lambda+21, & (22)\lambda^2 \quad\quad +22, \dots \\ \dots\dots\dots & \dots\dots\dots \\ \dots\dots\dots & \dots\dots\dots \end{vmatrix} = 0 \dots\dots (14).$$

Now by the relations (12)=(21), etc., 12 = 21, etc., and 12] = −21], we see that if λ be changed into $-\lambda$ the determinant becomes altered merely by interchange of terms between columns and rows, and hence the value of the determinant remains unchanged. Hence the first member of (14) cannot contain odd powers of λ, and therefore its roots must be in pairs of oppositely signed equals. The condition for stability of equilibrium in the zero configuration is therefore that the roots λ^2 of the determinantal equation be each real and negative.

345ix. The equations are simplified by transforming the co-ordinates (§ 337) so as to reduce T to a sum of squares with positive coefficients and V to a sum of squares with positive or negative coefficients as the case may be, or which is the same thing to adopt for co-ordinates those displacements which would correspond to "fundamental modes" (§ 338), if the positional forces were as they are and there were no motional forces. Suppose farther the unit values of the co-ordinates to be so chosen that the coefficients of the squares of the velocities in $2T$ shall be each unity; and let us put ϖ_1, ϖ_2, ϖ_3, etc. instead of the coefficients 11, 22, 33, etc., remaining in $2V$. Thus we have

Simplification of its equations.

$$T = \tfrac{1}{2}(\dot\psi_1^2 + \dot\psi_2^2 + \text{etc.}), \text{ and } V = \tfrac{1}{2}(\varpi_1\psi_1^2 + \varpi_2\psi_2^2 + \text{etc.}) \dots\dots (15).$$

If now we omit the half brackets] as no longer needed to avoid ambiguity, and understand that 12 = −21, 13 = −31, 23 = −32, etc., the equations of motion are

$$\left.\begin{aligned} &\ddot\psi_1 + 12\dot\psi_2 + 13\dot\psi_3 + \dots\dots + \varpi_1\psi_1 = 0 \\ &\ddot\psi_2 + 21\dot\psi_1 + 23\dot\psi_3 + \dots\dots + \varpi_2\psi_2 = 0 \\ &\ddot\psi_3 + 31\dot\psi_1 + 32\dot\psi_2 + \dots\dots + \varpi_3\psi_3 = 0 \\ &\dots\dots\dots\dots\dots\dots\dots\dots \end{aligned}\right\} \dots\dots\dots (16),$$

and the determinantal equation becomes

$$\begin{vmatrix} \lambda^2+\varpi_1, & 12\lambda, & 13\lambda,\dots \\ 21\lambda, & \lambda^2+\varpi_2, & 23\lambda,\dots \\ 31\lambda, & 32\lambda, & \lambda^2+\varpi_3,\dots \\ \dots & \dots & \dots \end{vmatrix} = 0 \qquad (17).$$

Determinant of gyrostatic conservative system.

The determinant (which for brevity we shall denote by D) in this case is what has been called by Cayley a skew determinant. What it would become if zero were substituted for $\lambda^2+\varpi_1$, $\lambda^2+\varpi_2$, etc. in its principal diagonal is what is called a skew symmetric determinant. The known algebra of skew and skew symmetric determinants gives

$$\left.\begin{aligned} D = &(\lambda^2+\varpi_1)(\lambda^2+\varpi_2)\dots(\lambda^2+\varpi_i) \\ &+\lambda^2\Sigma(\lambda^2+\varpi_3)(\lambda^2+\varpi_4)\dots(\lambda^2+\varpi_i)\,12^2 \\ &+\lambda^4\Sigma(\lambda^2+\varpi_5)(\lambda^2+\varpi_6)\dots(\lambda^2+\varpi_i)(12\,.\,34+31\,.\,24+23\,.\,14)^2 \\ &+\lambda^6\Sigma(\lambda^2+\varpi_7)(\lambda^2+\varpi_8)\dots(\lambda^2+\varpi_i)(\Sigma 12\,.\,34\,.\,56)^2+\text{etc.} \\ &+\lambda^i(\Sigma 12\,.\,34\,.\,56\,.\dots.\,i-1,\ i)^2 \end{aligned}\right\} (18),$$

when i is even. For example see (30) below. When λ is odd the last term is

$$\lambda^{i-1}\Sigma(\lambda^2+\varpi_i)(\Sigma 12\,.\,34\,.\,56\dots i-2,\ i-1)^2 \qquad (18'),$$

and no other change in the formula is necessary. In each case the small Σ denotes the sum of the products obtained by making every possible permutation of the numbers in the line of factors following it, with orders chosen acccording to a proper rule to render the sign of each product positive (Salmon's *Higher Algebra*, Lesson v. Art. 40). This sum is in each case the square root of a certain corresponding skew symmetric determinant.

Square roots of skew symmetrics.

An easy rule to find other products from any one given to begin with is this:—Invert the order in any one factor, and make a simple interchange of any two numbers in different factors. Thus, in the last Σ of (18) alter $i-1$, i to i, $i-1$, and interchange $i-1$ with 3: so we find $12\,.\,i-1,\ 4\,.\,56\dots\dots i,\ 3$ for a term: similarly $12\,.\,64\,.\,53\dots i,\ i-1$, and $62\,.\,14\,.\,53\dots i-1,\ i$, for two others. The same number must not occur more than once in any one product. Two products differing only in the orders of the two numbers in factors are not admitted. If n be the number of factors in each term, the whole number of factors is clearly $1\,.\,3\,.\,5\dots(2n-1)$, and they may be found in regular

Square roots of skew symmetrics.

progression thus: Begin with a single factor and single term 12. Then apply to it the factor 34, and permute to suit 24 instead of 34, and permute the result to suit 14 instead of 24. Thirdly, apply to the sum thus found the factor 56, and permute successively from 56 to 46, from 46 to 36, from 36 to 26, and from 26 to 16. Fourthly, introduce the factor 78; and so on. Thus we find

$$\left.\begin{aligned}
&\begin{vmatrix} 0, & 12 \\ 21, & 0 \end{vmatrix} = 12 \\
&\begin{vmatrix} 0, & 12, & 13, & 14 \\ 21, & 0, & 23, & 24 \\ 31, & 32, & 0, & 34 \\ 41, & 42, & 43, & 0 \end{vmatrix} = 12 \cdot 34 + 31 \cdot 24 + 23 \cdot 14 \\
&\begin{vmatrix} 0, & 12, & 13, & 14, & 15, & 16 \\ 21, & 0, & 23, & 24, & 25, & 26 \\ 31, & 32, & 0, & 34, & 35, & 36 \\ 41, & 42, & 43, & 0, & 45, & 46 \\ 51, & 52, & 53, & 54, & 0, & 56 \\ 61, & 62, & 63, & 64, & 65, & 0 \end{vmatrix} = \begin{aligned} &(12 \cdot 34 + 31 \cdot 24 + 23 \cdot 14)56 \\ &+ (12 \cdot 53 + 13 \cdot 52 + 23 \cdot 51)46 \\ &+ (12 \cdot 45 + 41 \cdot 52 + 42 \cdot 51)36 \\ &+ (31 \cdot 45 + 41 \cdot 35 + 34 \cdot 51)26 \\ &+ (23 \cdot 45 + 24 \cdot 35 + 34 \cdot 25)16 \end{aligned}
\end{aligned}\right\} \quad (19).$$

The second member of the last of these equations is what is denoted by $\Sigma 12 \cdot 34 \cdot 56$ in (18).

345^{x}. Each term of the determinant D except

$$(\lambda^2 + \varpi_1)(\lambda^2 + \varpi_2) \ldots (\lambda^2 + \varpi_i)$$

Gyrostatic system with two freedoms.

contains λ^2 as a factor. Hence, when all are expanded in powers of λ^2, the term independent of λ is $\varpi_1 \varpi_2 \ldots \varpi_i$. If this be negative there must be at least one real positive and one real negative root λ^2. Hence for stability either must all of ϖ_1, ϖ_2, ... ϖ_i be positive or an even number of them negative. Ex.:—Two modes of motion, x and y the co-ordinates. Let the equations of motion be

$$\left.\begin{aligned} I\ddot{x} + g\dot{y} + Ex = 0 \\ J\ddot{y} - g\dot{x} + Fy = 0 \end{aligned}\right\} \ldots\ldots\ldots\ldots\ldots\ldots (20),$$

and the determinantal equation is

$$(I\lambda^2 + E)(J\lambda^2 + F) + g^2\lambda^2 = 0.$$

If we put

$$x = \xi / \sqrt{I}, \quad y = \eta / \sqrt{J} \ldots\ldots\ldots\ldots\ldots\ldots (21),$$

and

$$E = \varpi I, \quad F = \zeta J, \quad \text{and} \quad g = \gamma \sqrt{(IJ)} \ldots\ldots\ldots (22),$$

Gyrostatic system with two freedoms.

equations (20) and the determinantal equation become

$$\left.\begin{array}{l}\ddot{\xi}+\gamma\dot{\eta}+\varpi\xi=0\\ \ddot{\eta}-\gamma\dot{\xi}+\zeta\eta=0\end{array}\right\}\text{..................(23),}$$

and $$(\lambda^2+\varpi)(\lambda^2+\zeta)+\gamma^2\lambda^2=0 \text{ ....(24).}$$

The solution of this quadratic in λ^2 may be put under the following forms,—

$$\left.\begin{array}{l}-\lambda^2=\tfrac{1}{2}(\gamma^2+\varpi+\zeta)\pm\tfrac{1}{2}\{[\gamma^2+(\sqrt{\varpi}+\sqrt{\zeta})^2]\,[\gamma^2+(\sqrt{\varpi}-\sqrt{\zeta})^2]\}^{\frac{1}{2}}\\ -\lambda^2=\tfrac{1}{2}(\gamma^2+\varpi+\zeta)\pm\tfrac{1}{2}\{[\gamma^2-(\sqrt{-\varpi}+\sqrt{-\zeta})^2][\gamma^2-(\sqrt{-\varpi}-\sqrt{-\zeta})^2]\}^{\frac{1}{2}}\end{array}\right\}\text{...(25).}$$

To make both values of $-\lambda^2$ real and positive ϖ and ζ must be of the same sign. If they are both positive no farther condition is necessary. If they are both negative we must have

$$\gamma>\sqrt{-\varpi}+\sqrt{-\zeta}\text{ ....(26).}$$

These are the conditions that the zero configuration may be stable. Remark that when (as practically in all the gyrostatic illustrations) γ^2 is very great in comparison with $\sqrt{(\varpi\zeta)}$, the greater value of $-\lambda^2$ is approximately equal to γ^2, and therefore (as the product of the two roots is exactly $\varpi\zeta$), the less is approximately equal to $\varpi\zeta/\gamma^2$. Remark also that $2\pi/\sqrt{\varpi}$ and $2\pi/\sqrt{\zeta}$ are the periods of the two fundamental vibrations of a system otherwise the same as the given system, but with $\gamma=0$. Hence, using the word irrotational to refer to the system with $g=0$, and gyroscopic, or gyrostatic, or gyrostat, to refer to the actual system;

Gyrostatic influence dominant.

From the preceding analysis we have the curious and interesting result that, in a system with two freedoms, two irrotational instabilities are converted into complete gyrostatic stability (each freedom stable) by sufficiently rapid rotation; but that with one irrotational stability the gyrostat is essentially unstable, with one of its freedoms unstable and the other stable, if there be one irrotational instability. Various good illustrations of gyrostatic systems with two, three, and four freedoms (§§ 345x, xi and xii) are afforded by the several different modes of mounting shown in the accompanying sketches, applied to the ordinary gyrostat* (a rapidly rotating fly-wheel pivoted as finely as possible within a rigid case, having a convex curvilinear polygonal border, in the plane perpendicular to the axis through the centre of gravity of the whole).

Gyrostatic stability.

* *Nature*, No. 379, Vol. 15 (February 1, 1877), page 297.

Ordinary gyrostats.

[Translational motions not considered] there are two freedoms, one azimuthal the other inclinational; the first neutral the other unstable when fly-wheel not rotating: the first still neutral the second stable when fly-wheel in rapid rotation. Equations (23) with $\zeta=0$ express the problem, and (24) and (25) its solution.

Gyrostats,

on gimbals;

on universal-flexure-joint (§ 109) in place of gimbals; constituting an inverted gyroscopic pendulum (§ 319, Ex. D).

Gyrostat on knife-edge gimbals with its axis vertical. Two freedoms; each unstable without rotation of the fly-wheel; each stable when it is rotating rapidly. Neglecting inertia of the knife-edges and gimbal-ring we have $I=J$ in (20), and supposing the levels of the knife-edges to be the same, we have $E=F$. Thus its determinantal equation is $(I\lambda^2+E)^2+\mathfrak{g}^2\lambda^2=0$. A similar result, expressed by the same equations of motion, is obtained by supporting the gyrostat on a little elastic universal flexure-joint of, for example, thin steel pianoforte-wire one or two centimetres long between end clamps or solderings. A drawing is unnecessary.

on stilts;

Two freedoms, one azimuthal the other inclinational, both unstable without, both stable with, rapid rotation of the fly-wheel.

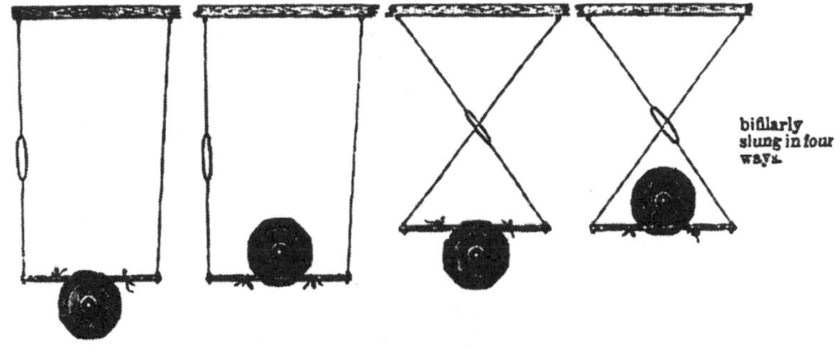

Gyrostat bifilarly slung in four ways.

Four freedoms, reducible to three if desired by a third thread in each case, diagonal in the first and second, lateral in the third and fourth, the freedom thus annulled being in each case stable and independent of the rotation of the fly-wheel. Three modes essentially involved in the gyrostatic system in each case, two inclinational and one azimuthal.

No. 1.—Azimuthally stable without rotation; with rotation all three modes stable.

No. 2.—Azimuthally stable, one inclinational mode unstable the other stable without rotation; with rotation two unstable, one stable.

No. 3.—The azimuthal mode unstable, two inclinational modes stable the other unstable, without rotation; with rotation one azimuthal mode and one inclinational mode unstable, and one inclinational mode stable.

No. 4.—Azimuthally and one inclinational mode unstable, one inclinational mode stable, without rotation; with rotation all three stable.

345^xi. Take for another example a system having three freedoms (that is to say, three independent co-ordinates ψ_1, ψ_2, ψ_3), (16) become

Gyrostatic system with three freedoms.

$$\left.\begin{array}{l}\ddot{\psi}_1 + g_3\dot{\psi}_2 - g_2\dot{\psi}_3 + \varpi_1\psi_1 = 0\\ \ddot{\psi}_2 + g_1\dot{\psi}_3 - g_3\dot{\psi}_1 + \varpi_2\psi_2 = 0\\ \ddot{\psi}_3 + g_2\dot{\psi}_1 - g_1\dot{\psi}_2 + \varpi_3\psi_3 = 0\end{array}\right\} \quad\ldots\ldots\ldots\ldots\ldots (27),$$

where g_1, g_2, g_3 denote the values of the three pairs of equals 23 or −32, 31 or −13, 12 or −21. Imagine ψ_1, ψ_2, ψ_3 to be rectangular co-ordinates of a material point, and let the co-ordinates be transformed to other axes OX, OY, OZ, so chosen that OZ coincides with the line whose direction cosines relatively to the ψ_1-, ψ_2-, ψ_3- axes are proportional to g_1, g_2, g_3. The equations become

$$\left.\begin{array}{l}\ddot{x} - 2\omega\dot{y} = X\\ \ddot{y} + 2\omega\dot{x} = Y\\ \ddot{z} \qquad\;\;\, = Z\end{array}\right\} \quad\ldots\ldots\ldots\ldots\ldots\ldots\ldots (28),$$

Gyrostatic system with three freedoms: where $\omega=\sqrt{(g_1^2+g_2^2+g_3^2)}$, and the force-components parallel to the fresh axes are denoted by X, Y, Z (instead of $-\frac{dV}{dx}$, $-\frac{dV}{dy}$, $-\frac{dV}{dz}$, because the present transformation is clearly independent of the assumption we have been making latterly that the positional forces are conservative). These (28) are simply the equations [§ 319, Ex. (E)] of the motion of a particle relatively to co-ordinates revolving with angular velocity ω round the axis OZ, if we suppose X, Y, Z to include the components of the centrifugal force due to this rotation.

reduced to a mere rotating system. Hence the influence of the gyroscopic terms however originating in any system with three freedoms (and therefore also in any system with only two freedoms) may be represented by the motion of a material particle supported by massless springs attached to a rigid body revolving uniformly round a fixed axis. It is an interesting and instructive exercise to imagine or to actually construct mechanical arrangements for the motion of a material particle to illustrate the experiments described in § 345^{x}.

345^{xii}. Consider next the case of a system with four freedoms. The equations are

$$\left.\begin{array}{l}\ddot\psi_1+12\dot\psi_2+13\dot\psi_3+14\dot\psi_4+\varpi_1\psi_1=0\\ \ddot\psi_2+21\dot\psi_1+23\dot\psi_3+24\dot\psi_4+\varpi_2\psi_2=0\\ \ddot\psi_3+31\dot\psi_1+32\dot\psi_2+34\dot\psi_4+\varpi_3\psi_3=0\\ \ddot\psi_4+41\dot\psi_1+42\dot\psi_2+43\dot\psi_3+\varpi_4\psi_4=0\end{array}\right\}\quad\ldots\ldots\ldots(29).$$

Denoting by D the determinant we have, by (18),

$$\left.\begin{array}{l}D=(\lambda^2+\varpi_1)(\lambda^2+\varpi_2)(\lambda^2+\varpi_3)(\lambda^2+\varpi_4)\\ \quad+\lambda^2\{34^2(\lambda^2+\varpi_1)(\lambda^2+\varpi_2)+12^2(\lambda^2+\varpi_3)(\lambda^2+\varpi_4)+42^2(\lambda^2+\varpi_1)(\lambda^2+\varpi_3)\\ \qquad+13^2(\lambda^2+\varpi_4)(\lambda^2+\varpi_2)+23^2(\lambda^2+\varpi_1)(\lambda^2+\varpi_4)+14^2(\lambda^2+\varpi_2)(\lambda^2+\varpi_3)\}\\ \quad+\lambda^4(12\;34+13\;42+14\;23)^2\end{array}\right\}\ldots(30).$$

If ϖ_1, ϖ_2, ϖ_3, ϖ_4 be each zero, D becomes

$$\lambda^8+(12^2+13^2+14^2+23^2+42^2+34^2)\lambda^6+(12\;34+13\;42+14\;23)^2\lambda^4.$$

This equated to zero and viewed as an equation for λ^2 has two roots each equal to 0, and two others given by the residual quadratic

Quadruply free gyrostatic system without force.

$$\lambda^4 + (12^2+13^2+14^2+23^2+24^2+34^2)\lambda^2 + (12\,34+13\,42+14\,23)^2 = 0 \ldots (31).$$

Now remarking that the solution of $z^2+pz+q^2=0$ may be written

$$-z = \tfrac{1}{2}\{p \pm \sqrt{(p+2q)(p-2q)}\} = \tfrac{1}{4}\{\sqrt{(p+2q)} \pm \sqrt{(p-2q)}\}^2,$$

we have from (31)

$$\left.\begin{aligned} -\lambda^2 &= \tfrac{1}{2}(12^2+13^2+14^2+23^2+24^2+34^2 \pm \sqrt{s}) \\ &= \tfrac{1}{4}(r \pm s)^2 \end{aligned}\right\}\ldots\ldots(32),$$

$$\left.\begin{aligned} \text{where}\quad r &= \sqrt{\{(12+34)^2+(13+42)^2+(14+23)^2\}} \\ \text{and}\quad s &= \sqrt{\{(12-34)^2+(13-42)^2+(14-23)^2\}} \end{aligned}\right\}\ldots\ldots(33).$$

As 12, 34, 13, etc. are essentially real, r and s are real, and (unless $12\,43+13\,42+14\,23=0$, when one of the values of λ^2 is zero, a case which must be considered specially, but is excluded for the present,) they are unequal. Hence the two values of $-\lambda^2$ given by (32) are real and positive. Hence two of the four freedoms are stable. The other two (corresponding to $-\lambda^2=0$) are neutral.

Excepted case of failing gyrostatic predominance.

345^{xiii}. Now suppose ϖ_1, ϖ_2, ϖ_3, ϖ_4 to be not zero, but each very small. The determinantal equation will be a biquadratic in λ^2, of which two roots (the two which vanish when ϖ_1, etc. vanish) are approximately equal to the roots of the quadratic

Quadruply free cycloidal system, gyrostatically dominated.

$$(12\,34+13\,42+14\,23)^2\lambda^4 + (12^2\varpi_3\varpi_4 + 13^2\varpi_2\varpi_4 + 14^2\varpi_2\varpi_3 + 23^2\varpi_1\varpi_4 + 24^2\varpi_1\varpi_3 + 34^2\varpi_1\varpi_2)\lambda^2 + \varpi_1\varpi_2\varpi_3\varpi_4 = 0 \ldots (34),$$

and the other two roots are approximately equal to those of the previous residual quadratic (31).

To solve equation (34), first write it thus:—

$$\left(\frac{1}{\lambda^2}\right)^2 + (12'^2+13'^2+14'^2+23'^2+24'^2+34'^2)\frac{1}{\lambda^2} + (12'34'+13'42'+14'23')^2 = 0 \qquad \ldots\ldots\ldots(35),$$

where

$$12' = \frac{12}{\sqrt{(\varpi_1\varpi_2)}},\quad 13' = \frac{13}{\sqrt{(\varpi_1\varpi_3)}},\quad 23' = \frac{23}{\sqrt{(\varpi_2\varpi_3)}},\ \text{etc.}\ldots\ldots(36).$$

Quadruply free cycloidal system, gyrostatically dominated.

Thus, taken as a quadratic for λ^{-2}, it has the same form as (31) for λ^2, and so, as before in (32) and (33), we find

$$\frac{-1}{\lambda^2} = \tfrac{1}{4}(r' \pm s')^2 \text{ .... } (37)$$

$$\left.\begin{array}{ll}\text{where} & r' = \sqrt{\{(12'+34')^2+(13'+42')^2+(14'+23')^2\}} \\ \text{and} & s' = \sqrt{\{(12'-34')^2+(13'-42')^2+(14'-23')^2\}}\end{array}\right\} \text{ ......(38).}$$

Four irrotational stabilities confirmed, four irrotational instabilities rendered stable, by gyrostatic links.

Now if $\varpi_1, \varpi_2, \varpi_3, \varpi_4$ be all four positive or all four negative, $12', 34', 13'$, etc. are all real, and therefore both the values of $-\frac{1}{\lambda^2}$ given by (37) are real and positive (the excluded case referred to at the end of § 345^xi, which makes

$$12'34' + 13'42' + 14'23' = 0,$$

and therefore the smaller value of $-\frac{1}{\lambda^2} = 0$, being still excluded). Hence the corresponding freedoms are stable. But it is not *necessary* for stability that $\varpi_1, \varpi_2, \varpi_3, \varpi_4$ be all four of one sign: it *is* necessary that their product be positive: since if it were negative the values of λ^2 given by (34) would both be real, but one only negative and the other positive. Suppose two of them, ϖ_3, ϖ_4 for example, be negative, and the other two, ϖ_1, ϖ_2, positive: this makes $\varpi_1\varpi_3, \varpi_1\varpi_4, \varpi_2\varpi_3$, and $\varpi_2\varpi_4$ negative, and therefore $13', 14', 23'$, and $24'$ imaginary. Instead of four of the six equations (36), put therefore

$$13'' = \frac{13}{\sqrt{(-\varpi_1\varpi_3)}},\ 14'' = \frac{14}{\sqrt{(-\varpi_1\varpi_4)}},\ 23'' = \frac{23}{\sqrt{(-\varpi_2\varpi_3)}},\ 24'' = \frac{24}{\sqrt{(-\varpi_2\varpi_4)}} \text{ (39).}$$

Thus $13''$ etc. are real, and $13' = 13''\sqrt{-1}$ etc., and (38) become

$$\left.\begin{array}{l} r' = \sqrt{\{(12'+34')^2-(13''+42'')^2-(14''+23'')^2\}} \\ s' = \sqrt{\{(12'-34')^2-(13''-42'')^2-(14''-23'')^2\}}\end{array}\right\} \text{ ......(40).}$$

Hence for stability it is necessary and sufficient that

$$\left.\begin{array}{ll} & (12'+34')^2 > (13''+42'')^2+(14''+23'')^2 \\ \text{and} & (12'-34')^2 > (13''-42'')^2+(14''-23'')^2\end{array}\right\} \text{ ....(41).}$$

Combined dynamic and gyrostatic sta-

If these inequalities are reversed, the stabilities due to ϖ_1, ϖ_2 and $34'$ are undone by the gyrostatic connexions $13'', 42'', 14''$ and $23''$.

345^{xiv}. Going back to (29) we see that for the particular solution $\psi_1 = a_1 \epsilon^{\lambda t}$, $\psi_2 = a_2 \epsilon^{\lambda t}$, etc., given by the first pair of roots of (32), they become approximately

bility gyrostatically counteracted.

$$\left.\begin{array}{l} \lambda a_1 + 12\, a_2 + 13\, a_3 + 14\, a_4 = 0 \\ \lambda a_2 + 21\, a_1 + 23\, a_3 + 24\, a_4 = 0 \\ \lambda a_3 + 31\, a_1 + 32\, a_2 + 34\, a_4 = 0 \\ \lambda a_4 + 41\, a_1 + 42\, a_2 + 43\, a_3 = 0 \end{array}\right\} \ldots\ldots\ldots\ldots\ldots (42);$$

being in fact the linear algebraic equations for the solution in the form $\epsilon^{\lambda t}$ of the simple simultaneous differential equations (53) below. And if we take

Completed solution.

$$\psi_1 = \frac{b_1}{\sqrt{\varpi_1}} \epsilon^{\lambda t},\quad \psi_2 = \frac{b_2}{\sqrt{\varpi_2}} \epsilon^{\lambda t}, \text{ etc.} \ldots\ldots\ldots\ldots (43),$$

for either particular approximate solution of (29) corresponding to (37), we find from (29) approximately

$$\left.\begin{array}{l} \lambda^{-1} b_1 + 12' b_2 + 13' b_3 + 14' b_4 = 0 \\ \lambda^{-1} b_2 + 21' b_1 + 23' b_3 + 24' b_4 = 0 \\ \lambda^{-1} b_3 + 31' b_1 + 32' b_2 + 34' b_4 = 0 \\ \lambda^{-1} b_4 + 41' b_1 + 42' b_2 + 43' b_3 = 0 \end{array}\right\} \ldots\ldots\ldots\ldots\ldots (44).$$

Remark that in (42) the coefficients of the first terms are imaginary and those of all the others real. Hence the ratios a_1/a_2, a_1/a_3, etc., are imaginary. To realize the equations put

$$\lambda = n\sqrt{-1},\ a_1 = p_1 + q_1\sqrt{-1},\ a_2 = p_2 + q_2\sqrt{-1}, \text{ etc.} \ldots (45),$$

and let p_1, q_1, p_2, etc. be real; we find, as equivalent to (42),

Realization of completed solution.

$$\left.\begin{array}{l} \left\{\begin{array}{l} -nq_1 + 12\, p_2 + 13\, p_3 + 14\, p_4 = 0 \\ \ \ np_1 + 12\, q_2 + 13\, q_3 + 14\, q_4 = 0 \end{array}\right. \\ \left\{\begin{array}{l} -nq_2 + 21\, p_1 + 23\, p_3 + 24\, p_4 = 0 \\ \ \ np_2 + 21\, q_1 + 23\, q_3 + 24\, q_4 = 0 \end{array}\right. \\ \qquad \text{etc.} \qquad\qquad \text{etc.} \end{array}\right\} \ldots\ldots\ldots\ldots (46).$$

Eliminating q_1, q_2, etc. from the seconds by the firsts of these pairs, we find

$$\left.\begin{array}{llll} (n^2 + 11) p_1 + & 12\, p_2 + & 13\, p_3 + & 14\, p_4 = 0 \\ 21\, p_1 + & (n^2 + 22) p_2 + & 23\, p_3 + & 24\, p_4 = 0 \\ 31\, p_1 + & 32\, p_2 + & (n^2 + 33) p_3 + & 34\, p_4 = 0 \\ 41\, p_1 + & 42\, p_2 + & 43\, p_3 + & (n^2 + 44) p_4 = 0 \end{array}\right\} (47);$$

and by eliminating p_1, p_2, etc. similarly we find similar equations

Realization of completed solution.

for the q's; with the same coefficients 11, 12, etc., given by the following formulas:—

$$\left.\begin{aligned} 11 &= 12\,21 + 13\,31 + 14\,41 \\ 12 &= 13\,32 + 14\,42 \\ 13 &= 12\,23 + 14\,43 \\ 21 &= 23\,31 + 24\,41 \\ \text{etc.} & \quad \text{etc.} \end{aligned}\right\} \quad \ldots\ldots\ldots\ldots\ldots (48).$$

Remember now that

$$12 = -21,\ 13 = -31,\ 32 = -23, \text{ etc.} \ldots\ldots\ldots\ldots\ldots (49),$$

and we see in (48) that

$$12 = 21,\ 13 = 31,\ 23 = 31, \text{ etc.} \ldots\ldots\ldots\ldots\ldots\ldots\ldots (50);$$

and farther, that 11, 12, etc. are the negatives of the coefficients of $\frac{1}{2}\alpha_1^2$, $\alpha_1\alpha_2$, etc. in the quadratic

$$\tfrac{1}{2}\{(12\,\alpha_2 + 13\,\alpha_3 + 14\,\alpha_4)^2 + (21\,\alpha_1 + 23\,\alpha_3 + 24\,\alpha_4)^2 + \text{etc.}\} \ldots (51)$$

Resultant motion reduced to motion of a conservative system with four fundamental periods equal two and two.

expanded. Hence if $G(\alpha\alpha)$ denote this quadratic, and $G(pp)$, $G(qq)$ the same of the p's and the q's, we may write (47) and the corresponding equations for the q's as follows:

$$\left.\begin{aligned} -n^2 p_1 + \frac{dG(pp)}{dp_1} = 0, \quad -n^2 p_2 + \frac{dG(pp)}{dp_2} = 0, \text{ etc.} \\ -n^2 q_1 + \frac{dG(qq)}{dq_1} = 0, \quad -n^2 q_2 + \frac{dG(qq)}{dq_2} = 0, \text{ etc.} \end{aligned}\right\} \quad (52).$$

These equations are harmonized by, and as is easily seen, only by, assigning to n^2 one or other of the two values of $-\lambda^2$ given in (32), above. Hence their determinantal equation, a biquadratic in n^2, has two pairs of equal real positive roots. We readily verify this by verifying that the square of the determinant of (42), with λ^2 replaced by $-n^2$, is equal to the determinant of (47) with 11, 12, etc. replaced by their values (48).

Algebraic theorem.

Hence (§ 343g) there is for each root an indeterminacy in the ratios p_1/p_2, p_1/p_3, p_1/p_4, according to which one of them may be assumed arbitrarily and the two others then determined by two of the equations (47); so that with two of the p's assumed arbitrarily the four are known: then the corresponding set of four q's is determined explicitly by the firsts of the pairs (46).

Details of realized solution

Similarly the other root, n'^2, of the determinantal equation gives another solution with two fresh arbitraries. Thus we have the complete solution of the four equations

$$\left.\begin{array}{l}\frac{d\psi_1}{dt} + 12\,\psi_2 + 13\,\psi_3 + 14\,\psi_4 = 0 \\ \frac{d\psi_2}{dt} + 21\,\psi_1 + 23\,\psi_3 + 24\,\psi_4 = 0 \\ \quad\text{etc.} \qquad\qquad \text{etc.}\end{array}\right\} \ldots\ldots\ldots\ldots\ldots (53),$$

Details of realized solution

with its four arbitraries. The formulas (46)...(52) are clearly the same as we should have found if we had commenced with assuming

$$\psi_1 = p_1 \sin nt + q_1 \cos nt, \quad \psi_2 = p_2 \sin nt + q_2 \cos nt, \text{ etc.}\ldots(54),$$

as a particular solution of (53).

345^{xv}. Important properties of the solution of (53) are found thus:—

Orthogonalities proved between two components of one fundamental oscillation:

(*a*) Multiply the firsts of (46) by p_1, p_2, p_3, p_4 and add: or the seconds by q_1, q_2, q_3, q_4 and add: either way we find

$$p_1q_1 + p_2q_2 + p_3q_3 + p_4q_4 = 0 \ldots\ldots\ldots\ldots\ldots\ldots\ldots (55).$$

(*b*) Multiply the firsts of (46) by q_1, q_2, q_3, q_4 and add: multiply the seconds by p_1, p_2, p_3, p_4 and add: and compare the results: we find

$$n\Sigma p^2 = n\Sigma q^2 = \Sigma 12\,(p_2q_1 - p_1q_2) \ldots\ldots\ldots\ldots\ldots (56),$$

and equality of their energies.

where Σ of the last member denotes a sum of such double terms as the sample without repetition of their equals, such as $21\,(p_1q_2 - p_2q_1)$.

(*c*) Let n^2, n'^2 denote the two values of $-\lambda^2$ given in (32), and let (54) and

Orthogonalities proved between different fundamental oscillations.

$$\psi_1 = p'_1 \sin n't + q'_1 \cos n't, \quad \psi_2 = p'_2 \sin n't + q'_2 \cos n't, \text{ etc.}\ldots(57)$$

be the two corresponding solutions of (53). Imagine (46) to be written out for n'^2 and call them (46′): multiply the firsts of (46) by p'_1, p'_2, p'_3, p'_4 and add: multiply the firsts of (46′) by p_1, p_2, p_3, p_4 and add. Proceed correspondingly with the seconds. Proceed similarly with multipliers q for the firsts and p for the seconds. By comparisons of the sums we find that when n' is not equal to n we must have

$$\left.\begin{array}{l}\Sigma p'q = 0, \quad \Sigma 12\,(p'_1p_2 - p'_2p_1) = 0 \\ \Sigma q'p = 0, \quad \Sigma 12\,(q'_1q_2 - q'_2q_1) = 0 \\ \left.\begin{array}{l}\Sigma q'q = 0 \\ \Sigma p'p = 0\end{array}\right\} \Sigma 12\,(q'_1p_2 - q'_2p_1) = 0, \quad \Sigma 12\,(p'_1q_2 - p'_2q_1) = 0\end{array}\right\} (58).$$

Case of equal periods.

345^{xvi}. The case of $n=n'$ is interesting. The equations $\Sigma q'q=0$, $\Sigma p'p=0$, $\Sigma p'q=0$, $\Sigma q'p=0$, when n differs however little from n', show (as we saw in a corresponding case in § 343m) that equality of n to n' does *not* bring into the solution terms of the form $Ct\cos nt$, and it must therefore come under § 343e. The condition to be fulfilled for the equality of the roots is seen from (32) and (33) to be

$$12=34,\ 13=42,\ \text{and}\ 14=23 \qquad (59):$$

and to give

$$n^2=12^2+13^2+14^2 \qquad (60)$$

for the common value of the roots. It is easy to verify that these relations reduce to zero each of the first minors of (42), as they must according to Routh's theorem (§ 343e), because each root, λ, of (42) is a double root. According to the same theorem all the first, second and third minors of (47) must vanish for each root, because each root, n^2, of (47) is a quadruple root: for this, as there are just four equations, it is necessary and sufficient that

$$11=22=33=44\ \text{and}\ 12=0,\ 13=0,\ 14=0,\ 23=0,\ \text{etc.} \qquad (60'),$$

which we see at once by (48) is the case when (59) are fulfilled. In fact, these relations immediately reduce (51) to

$$G(aa)=\tfrac{1}{2}(12^2+13^2+14^2)(a_1^2+a_2^2+a_3^2+a_4^2) \qquad (61).$$

In this case one particular solution is readily seen from (52) and (46) to be

$$\left.\begin{array}{llll} p_1=1, & p_2=0, & p_3=0, & p_4=0 \\ q_1=0, & q_2=-\dfrac{12}{n}, & q_3=-\dfrac{13}{n}, & q_4=-\dfrac{14}{n} \\ \psi_1=\sin nt, & \psi_2=-\dfrac{12}{n}\cos nt, & \psi_3=-\dfrac{13}{n}\cos nt, & \psi_4=-\dfrac{14}{n}\cos nt \end{array}\right\} \qquad (62).$$

Completed solution for case of equal periods.

Hence the general solution, with four arbitraries p_1, p_2, p_3, p_4, is

$$\left.\begin{array}{l} \psi_1=p_1\sin nt+\dfrac{1}{n}(12p_2+13p_3+14p_4)\cos nt \\ \psi_2=p_2\sin nt+\dfrac{1}{n}(-12p_1+14p_3-13p_4)\cos nt \\ \psi_3=p_3\sin nt+\dfrac{1}{n}(-13p_1-14p_2+12p_4)\cos nt \\ \psi_4=p_4\sin nt+\dfrac{1}{n}(-14p_1+13p_2-12p_3)\cos nt \end{array}\right\} \qquad (63).$$

It is easy to verify that this satisfies the four differential equations (53).

345[xvii]. Quite as we have dealt with (42), (45), (53), (54) in § 345[xiv], we may deal with (44) and the simple simultaneous equations for the solution of which they serve, which are

Two higher,

$$\left.\begin{array}{l} 12\dfrac{d\psi_2}{dt}+13\dfrac{d\psi_3}{dt}+14\dfrac{d\psi_4}{dt}+\varpi_1\psi_1=0,\\ 21\dfrac{d\psi_1}{dt}+23\dfrac{d\psi_3}{dt}+24\dfrac{d\psi_4}{dt}+\varpi_2\psi_2=0,\\ \quad\text{etc.}\qquad\qquad\text{etc.}\end{array}\right\}\text{............(64);}$$

and two lower, of the four fundamental oscillations,

and all the formulas which we meet in so doing are real when ϖ_1, ϖ_2, ϖ_3, ϖ_4 are all of one sign, and therefore 12′, 13′, etc., all real. In the case of some of the ϖ's negative and some positive there is no difficulty in realizing the formulas, but the consideration of the simultaneous reduction of the two quadratics,

similarly dealt with by solution of two similar quadratics.

$$\left.\begin{array}{l}\frac{1}{2}\left\{\dfrac{(12\,a_2+13\,a_3+14\,a_4)^2}{\varpi_1}+\dfrac{(21\,a_1+23\,a_3+24\,a_4)^2}{\varpi_2}+\text{etc.}\right\}\\ \text{and}\qquad\qquad \frac{1}{2}\left(\varpi_1 a_1^2+\varpi_2 a_2^2+\varpi_3 a_3^2+\varpi_4 a_4^2\right)\end{array}\right\}\ (65),$$

to which we are led when we go back from the notation 12′, etc. of (36), is not completely instructive in respect to stability, as was our previous explicit working out of the two roots of the determinantal equation in (37), (38), and (40).

345[xviii]. The conditions to be fulfilled that the system may be dominated by gyrostatic influence are that the smaller value of $-\lambda^2$ found from (31) and the greater found from (34) be respectively very great in comparison with the greatest and very small in comparison with the smallest, of the four quantities ϖ_1, ϖ_2, ϖ_3, ϖ_4 irrespectively of their signs. Supposing ϖ_1 to be the greatest and ϖ_4 the smallest, these conditions are easily proved to be fulfilled when, and only when,

provided that gyrostatic influence be fully dominant.

$$\frac{(12\,.\,34+13\,.\,42+14\,.\,23)^2}{12^2+13^2+14^2+34^2+42^2+23^2}>>\pm\varpi_1\text{..........(66),}$$

and

$$\frac{(12\,.\,34+13\,.\,42+14\,.\,23)^2}{12^2\varpi_3\varpi_4+13^2\varpi_4\varpi_2+14^2\varpi_2\varpi_3+34^2\varpi_1\varpi_2+42^2\varpi_1\varpi_3+23^2\varpi_1\varpi_4}>>\pm\varpi_4^{-1}\ (67),$$

where >> denotes "*very great in comparison with.*" When these conditions are fulfilled, let 12, 13, 23, etc., be each increased in the ratio of N to 1. The two greater values of n (or $\lambda\sqrt{-1}$) will be increased in the same ratio, N to 1; and the two smaller will be diminished each in the inverse ratio, 1 to N. Again, let $\sqrt{\pm\varpi_1}$, $\sqrt{\pm\varpi_2}$, $\sqrt{\pm\varpi_3}$, $\sqrt{\pm\varpi_4}$ be each diminished in the ratio M to 1; the two larger values of n will be sensibly unaltered; and the two smaller will be diminished in the ratio M^2 to 1.

Limits of smallest and second smallest of the four periods.

345xix. Remark that

(*a*) When (66) is satisfied the two greater values of n are each

$$\left.\begin{array}{l} < \sqrt{\{(12^2+13^2+14^2+34^2+42^2+23^2)\}} \\ \text{and} \quad > \dfrac{12\cdot 34+13\cdot 42+14\cdot 23}{\sqrt{(12^2+13^2+14^2+34^2+42^2+23^2)}} \end{array}\right\} \quad \ldots\ldots\ldots(68);$$

and that when they are very unequal the greater is approximately equal to the former limit and the less to the latter.

(*b*) When (67) is satisfied, and when the equilibrium is stable, the two smaller values of n are each

$$\left.\begin{array}{l} < \dfrac{\sqrt{(12^2\varpi_3\varpi_4+13^2\varpi_4\varpi_2+14^2\varpi_2\varpi_3+34^2\varpi_1\varpi_2+42^2\varpi_1\varpi_3+23^2\varpi_1\varpi_4)}}{12\cdot 34+13\cdot 42+14\cdot 23} \\ \text{and} \\ > \dfrac{\sqrt{(\varpi_1\varpi_2\varpi_3\varpi_4)}}{\sqrt{\{(12^2\varpi_3\varpi_4+13^2\varpi_4\varpi_2+14^2\varpi_2\varpi_3+34^2\varpi_1\varpi_2+42^2\varpi_1\varpi_3+23^2\varpi_1\varpi_4)\}}} \end{array}\right\} (69),$$

Limits of the next greatest and greatest of the four periods.

and that when they are very unequal the greater of the two is approximately equal to the former limit, and the less to the latter.

345xx. Both (66) and (67) must be satisfied in order that the four periods may be found approximately by the solution of the two quadratics (31), (34). If (66) is satisfied but not (67), the biquadratic determinant still splits into two quadratics, of which one is approximately (31) but the other is not approximately (34). Similarly, if (67) is satisfied but not (66), the biquadratic splits into two quadratics of which one is approximately (34) but the other not approximately (31).

Quadruply free cycloidal system with non-dominant gyrostatic influences.

345xxi. When neither (66) nor (67) is fulfilled there is not generally any splitting of the biquadratic into two rational quadratics; and the conditions of stability, the determination of the fundamental periods, and the working out of the complete solution depend essentially on the roots of a biquadratic equation. When ϖ_1, ϖ_2, ϖ_3, ϖ_4 are all positive it is clear from the equation of energy [345^{ii}, (4), with $Q=0$] that the motion is stable whatever be the values of the gyrostatic coefficients 12, 34, 13, etc. and therefore in this case each of the four roots λ^2 of the biquadratic is real and negative, a proposition included in the general theorem of § 345^{xxvi} below. To illustrate the interesting questions which occur when the ϖ's are not all positive put

Quadruply free cycloidal system with non-dominant gyrostatic influences.

$$12 = {}_{12}g,\ 34 = {}_{34}g,\ 13 = {}_{13}g, \text{ etc.} \quad\ldots\ldots\ldots\ldots\ldots\ldots(70),$$

where ${}_{12}$, ${}_{34}$, ${}_{13}$, etc. denote any numerics whatever subject only to the condition that they do not make zero of

$$_{12}\,.\,_{34} + {}_{13}\,.\,_{42} + {}_{14}\,.\,_{23}.$$

When ϖ_1, ϖ_2, ϖ_3, ϖ_4, are all negative each root λ^2 of the biquadratic is as we have seen in § 345^{xiii} real and negative when the gyrostatic influences dominate. It becomes an interesting question to be answered by treatment of the biquadratic, how small may g be to keep all the roots λ^2 real and negative, and how large may g be to render them other than real and positive as they are when $g=0$? Similar questions occur in connexion with the case of two of the ϖ's negative and two positive, when the gyrostatic influences are so proportioned as to fulfil 345^{xiii} (41), so that when g is infinitely great there is complete gyrostatic stability, though when $g=0$ there are two instabilities and two stabilities.

345^{xxii}. Returning now to 345^{x} and 345^{vii}, 345^{viii}, and 345^{ix}, for a gyrostatic system with any number of freedoms, we see by 345^{vi} that the roots λ^2 of the determinantal equation (14) or (17) are necessarily real and negative when ϖ_1, ϖ_2, ϖ_3, ϖ_4, etc. are all positive. This conclusion is founded on the reasoning of § 345^{ii} regarding the equation of energy (4) applied to the case $Q=0$, for which it becomes $T+V=E_0$, or the same as for the case of no motional forces. It is easy of course to eliminate dynamical considerations from the reasoning and to give a purely algebraic proof that the roots λ^2 of the determinantal equation (14) of 345^{viii} are necessarily real and negative, provided both of the two quadratic functions $(11)\,a_1^2 + 2\,(12)\,a_1a_2 + \text{etc.}$, and $11a_1^2 + 2\ 12a_1a_2 + \text{etc.}$ are positive for all real values of a_1, a_2, etc. But the equations (14) of § 343 (k), which we obtained and used in the course of the corresponding demonstration for the case of no motional forces, do not hold in our present case of gyrostatic motional forces. Still for this present case we have the con-

Gyrostatic system with any number of freedoms.

Case of equal roots with stability.

clusion of § 343 (*m*) that equality among the roots falls essentially under the case of § 343 (*e*) above. For we know from the consideration of energy, as in § 345^ii, that no particular solution can be of the form $te^{\lambda t}$ or $t \sin \sigma t$, when the potential energy is positive for all displacements: yet [though there cannot be equal roots for the gyrostatic system of two freedoms (§ 345^x) as we see from the solution (25) of the determinantal equation for this case] there obviously may be equality of roots* in a quadruply free gyrostatic system, or in one with more than four freedoms. Hence, if both the quadratic functions have the same sign for all real values of a_1, a_2, etc., all the first minors

Application of Routh's theorem.

* Examples of this may be invented *ad libitum* by commencing with pairs of equations such as (23) and altering the variables by (generalized) orthogonal transformations. For one very simple example put $\zeta = \varpi$ and take (23) as one pair of equations of motion, and as a second pair take

$$\ddot{\xi}' + \gamma\dot{\eta}' + \varpi\xi' = 0,$$
$$\ddot{\eta}' - \gamma\dot{\xi}' + \varpi\eta' = 0.$$

The second of (23) and the first of these multiplied respectively by $\cos a$ and $\sin a$, and again by $\sin a$ and $\cos a$, and added and subtracted, give

$$\ddot{\psi}_2 - \gamma \cos a\dot{\xi} + \gamma \sin a\dot{\eta}' + \varpi\psi_2 = 0,$$

and

$$\ddot{\psi}_3 + \gamma \sin a\dot{\xi} + \gamma \cos a\dot{\eta}' + \varpi\psi_3 = 0,$$

where

$$\psi_2 = \xi' \sin a + \eta \cos a,$$

and

$$\psi_3 = \xi' \cos a - \eta \sin a.$$

Eliminating ξ' and η by these last equations, from the first and fourth of the equations of motion, and for symmetry putting ψ_1 instead of ξ, and ψ_4 instead of η', and for simplicity putting $\gamma \cos a = g$, and $\gamma \sin a = h$, and collecting the equations of motion in order, we have the following,—

$$\ddot{\psi}_1 + g\dot{\psi}_2 - h\dot{\psi}_3 + \varpi\psi_1 = 0,$$
$$\ddot{\psi}_2 - g\dot{\psi}_1 + h\dot{\psi}_4 + \varpi\psi_2 = 0,$$
$$\ddot{\psi}_3 + h\dot{\psi}_1 + g\dot{\psi}_4 + \varpi\psi_3 = 0,$$
$$\ddot{\psi}_4 - h\dot{\psi}_2 - g\dot{\psi}_3 + \varpi\psi_4 = 0,$$

for the equations of motion of a quadruply free gyrostatic system having two equalities among its four fundamental periods. The two different periods are the two values of the expression

$$2\pi/\{\sqrt{(\tfrac{1}{4}g^2 + \tfrac{1}{4}h^2)} \pm \sqrt{(\tfrac{1}{4}g^2 + \tfrac{1}{4}h^2 + \varpi)}\}.$$

When these two values are unequal the equalities among the roots *do not* give rise to terms of the form $te^{\lambda t}$ or $t \cos \sigma t$ in the solution. But if $\varpi = -(\tfrac{1}{4}g^2 + \tfrac{1}{4}h^2)$, which makes these two values equal, and therefore all four roots equal, terms of the form $t \cos \sigma t$ *do appear* in the solution, and the equilibrium is unstable in the transitional case though it is stable if $-\varpi$ be less than $\tfrac{1}{4}g^2 + \tfrac{1}{4}h^2$ by ever so small a difference.

of the determinantal equation (14), § 345^{viii}, must vanish for each double, triple, or multiple root of the equation, if it has any such roots. Application of Routh's theorem.

It will be interesting to find a purely algebraic proof of this theorem, and we leave it as an exercise to the student; remarking only that, when the quadratic functions have contrary signs for some real values of a_1, a_2, etc., there may be equality among the roots without the evanescence of all the first minors; or, in dynamical language, there may be terms of the form $te^{\lambda t}$, or $t \sin \sigma t$, in the solution expressing the motion of a gyrostatic system, in transitional cases between stability and instability. It is easy to invent examples of such cases, taking for instance the quadruply free gyrostatic system, whether gyrostatically dominated as in § 345^{xiii}, but in this case with some of the four quantities negative, and some positive; or, as in § 345^{xxi}, not gyrostatically dominated, with either some or all of the quantities $\varpi_1, \varpi_2, \ldots, \varpi_i$ negative. All this we recommend to the student as interesting and instructive exercise. Equal roots with instability in transitional cases between stability and instability.

345^{xxiii}. When all the quantities $\varpi_1, \varpi_2, \ldots, \varpi_i$ are of the same sign it is easy to find the conditions that must be fulfilled in order that the system may be gyrostatically dominated. For if $\rho_1, \rho_2, \ldots, \rho_n$ are the roots of the equation Conditions of gyrostatic domination.

$$c_0 z^n + c_1 z^{n-1} + \ldots + c_{n-1} z + c_n = 0,$$

we have

$$-(\rho_1 + \rho_2 + \ldots + \rho_n) = \frac{c_1}{c_0}, \text{ and } -\left(\frac{1}{\rho_1} + \frac{1}{\rho_2} + \ldots + \frac{1}{\rho_n}\right) = \frac{c_{n-1}}{c_n}.$$

Hence if $-\rho_1, -\rho_2, \ldots -\rho_n$ be each positive, c_1/nc_0 is their arithmetic mean, and nc_n/c_{n-1} is their harmonic mean. Hence c_1/nc_0 is greater than nc_n/c_{n-1}, and the greatest of $-\rho_1, -\rho_2, \ldots, -\rho_n$ is greater than c_1/nc_0, and the least of them is less than nc_n/c_{n-1}. Take now the two following equations:

$$\lambda^i + \lambda^{i-2} \sum 12^2 + \lambda^{i-4} \sum (\Sigma 12 \cdot 34)^2 + \lambda^{i-6} \sum (\Sigma 12 \cdot 34 \cdot 56)^2 + \text{etc.} = 0 \quad \ldots\ldots\ldots (71),$$

$$\left(\frac{1}{\lambda}\right)^i + \left(\frac{1}{\lambda}\right)^{i-2} \sum 12'^2 + \left(\frac{1}{\lambda}\right)^{i-4} \sum (\Sigma 12' \cdot 34')^2 + \left(\frac{1}{\lambda}\right)^{i-6} \sum (\Sigma 12' \cdot 34' \cdot 56')^2 + \text{etc.} = 0 \quad (72),$$

where

$$12' = \frac{12}{\sqrt{(\varpi_1 \varpi_2)}},\ 13' = \frac{13}{\sqrt{(\varpi_1 \varpi_3)}},\ 34' = \frac{34}{\sqrt{(\varpi_3 \varpi_4)}}, \ldots,\ i-1, i' = \frac{i-1, i}{\sqrt{(\varpi_{i-1} \varpi_i)}} \quad (73).$$

Cycloidal motion. Conditions of gyrostatic domination.

Suppose for simplicity i to be even. All the roots λ^2 of (71) are (§ 345^{xxvi} below) essentially real and negative. So are those of (72) provided $\varpi_1, \varpi_2, \ldots, \varpi_i$ are all of one sign as we now suppose them to be. Hence the smallest root $-\lambda^2$ of (71) is less than

$$\frac{\frac{1}{2}i\Sigma(12.34.56,\ldots,i-1,i)^2}{\Sigma(\Sigma 12.34.56,\ldots,i-3,i-2)^2}\ldots\ldots\ldots\ldots(74),$$

and the greatest root $-\lambda^2$ of (72) is greater than

$$\frac{\Sigma(\Sigma 12'.34'.56',\ldots,i-3,i-2')^2}{\frac{1}{2}i\Sigma(12'.34'.56',\ldots,i-1,i')}\ldots\ldots\ldots\ldots(75).$$

Hence the conditions for gyrostatic domination are that (74) must be much greater than the greatest of the positive quantities $\pm\varpi_1$, $\pm\varpi_2, \ldots, \pm\varpi_i$, and that (75) must be very much less than the least of these positive quantities. When these conditions are fulfilled the i roots of (18) § 345^{ix} equated to zero are separable into two groups of $\frac{1}{2}i$ roots which are infinitely nearly equal to the roots of equations (71) and (72) respectively, conditions of reality of which are investigated in § 345^{xxvi} below. The interpretation leads to the following interesting conclusions:—

Gyrostatic links explained.

345^{xxiv}. Consider a cycloidal system provided with non-rotating flywheels mounted on frames so connected with the moving parts as to give infinitesimal angular motions to the axes of the flywheels proportional to the motions of the system. Let the number of freedoms of the system exclusive of the ignored co-ordinates [§ 319, Ex. (G)] of the flywheels relatively to their frames be even. Let the forces of the system be such that when the flywheels are given at rest, when the system is at rest, the equilibrium is either stable for all the freedoms, or unstable for all the freedoms. Let the number and connexions of the gyrostatic links be such as to permit gyrostatic domination (§ 345^{xxi}) when each of the flywheels is set into sufficiently rapid rotation. Now let the flywheels be set each into sufficiently rapid rotation to fulfil the conditions of gyrostatic domination (§ 345^{xxi}): the equilibrium of the system becomes stable: with half the whole number i of its modes of vibration exceedingly rapid, with frequencies equal to the roots of a certain algebraic equation of the degree $\frac{1}{2}i$; and the other half of

Gyrostatically dominated system:

its modes of vibration very slow, with frequencies given by the roots of another algebraic equation of degree $\frac{1}{2}i$. The first class of fundamental modes may be called adynamic because they are the same as if no forces were applied to the system, or acted between its moving parts, except actions and reactions in the normals between mutually pressing parts (depending on the inertias of the moving parts). The second class of fundamental modes may be called precessional because the precession of the equinoxes, and the slow precession of a rapidly spinning top supported on a very fine point, are familiar instances of it. Remark however that the obliquity of the ecliptic should be infinitely small to bring the precession of the equinoxes precisely within the scope of the equations of our "cycloidal system."

Gyrostatically dominated system: its adynamic oscillations (very rapid); and precessional oscillations (very slow).

345^xv. If the angular velocities of all the flywheels be altered in the same proportion the frequencies of the adynamic oscillations will be altered in the same proportion directly, and those of the precessional modes in the same proportion inversely. Now suppose there to be either no inertia in the system except that of the flywheels round their pivoted axes and round their equatorial diameters, or suppose the effective inertia of the connecting parts to be comparable with that of the flywheels when given without rotation. The period of each of the adynamic modes is comparable with the periods of the flywheels. And the periods of the precessional modes are comparable with a third proportional to a mean of the periods of the flywheels and a mean of the irrotational periods of the system, if the system be stable when the flywheels are deprived of rotation. For the last mentioned term of the proportion we may, in the case of irrotational *instability*, substitute the time of increasing a displacement a thousandfold, supposing the system to be falling away from its configuration of equilibrium according to one of its fundamental modes of motion ($\epsilon^{\lambda t}$). The reciprocal of this time we shall call, for brevity, the rapidity of the system, for convenience of comparison with the frequency of a vibrator or of a rotator, which is the name commonly given to the reciprocal of its period.

Comparison between adynamic frequencies, rotational frequencies of the flywheels, precessional frequencies of the system, and frequencies or rapidities of the system, with flywheels deprived of rotation.

Proof of reality of adynamic and of precessional periods when system's irrotational periods are either all real or all imaginary.

345^{xxvi}. It remains to prove that the roots λ^2 of (71), and of (72) also when $\varpi_1, \varpi_2, \ldots, \varpi_i$ are all of one sign, are essentially real and negative. (71) is the determinantal equation of § 345^{xiv} (42) with any even number of equations instead of only four. The treatment of §§ 345^{xiv} and 345^{xv} is all directly applicable without change to this extension; and it proves that the roots λ^2 are real and negative by bringing the problem to that of the orthogonal reduction of the essentially positive quadratic function

$$G(aa)=\tfrac{1}{2}\{(12a_2+13a_3+\text{etc.})^2+(21a_1+23a_3+\text{etc.})^2+(31a_1+32a_2+\text{etc.})^2+\text{etc.}\} \quad (76):$$

Algebraic theorem.

it proves also the equalities of energies of (56), § 345^{xv}, and the orthogonalities of (55), (58) § 345^{xv}: also the curious algebraic theorem that the determinantal roots of the quadratic function consist of $\frac{1}{2}i$ pairs of equals.

Inasmuch as (72) is the same as (71) with λ^{-1} put for λ and $12'$, $13'$, $23'$, etc. for 12, 13, 23, etc., all the formulas and propositions which we have proved for (71) hold correspondingly for (72) when $12'$, $13'$, $23'$, etc. are all real, as they are when $\varpi_1, \varpi_2, \ldots \varpi_i$ are all of one sign.

345^{xxvii}. Going back now to § 345^{viii}, and taking advantage of what we have learned in § 345^{ix} and the consequent treatment of the problem, particularly that in § 345^{xiv}, we see now how to simplify equations (14) of § 345^{viii} otherwise than was done in § 345^{ix}, by a new method which has the advantage of being applicable also to materially simplify the general equations (13) of § 345^{vi}. Apply orthogonal transformation of the co-ordinates to reduce to a sum of squares of simple co-ordinates, the quadratic function (76). Thus denoting by $G(\psi\psi)$ what $G(aa)$ becomes when ψ_1, ψ_2, etc. are substituted for a_1, a_2, etc.; and denoting by $n_1^2, n_2^2, \ldots, n_{\frac{1}{2}i}^2$ the values of the pairs of roots of the determinantal equation of degree i, which are simply the negative of the roots λ^2 of equation (71) of degree $\frac{1}{2}i$ in λ^2; and denoting by $\xi_1, \eta_1, \xi_2, \eta_2, \ldots \xi_{\frac{1}{2}i}\eta_{\frac{1}{2}i}$, the fresh co-ordinates, we have

$$G(\psi\psi)=\tfrac{1}{2}\{n_1^2(\xi_1^2+\eta_1^2)+n_2^2(\xi_2^2+\eta_2^2)+\ldots+n_{\frac{1}{2}i}^2(\xi_{\frac{1}{2}i}^2+\eta_{\frac{1}{2}i}^2)\}\ldots(77).$$

It is easy to see that the general equations of cycloidal motion (13) of § 345^{vi} transformed to the ξ-co-ordinates come out in $\frac{1}{2}i$ pairs as follows:

Cycloidal motion.

$$\left.\begin{array}{l}\left\{\begin{array}{l}\frac{d}{dt}\frac{dT}{d\dot{\xi}_1}+\frac{dQ}{d\dot{\xi}_1}+n_1\dot{\eta}_1+\frac{dV}{d\xi_1}=0\\ \frac{d}{dt}\frac{dT}{d\dot{\eta}_1}+\frac{dQ}{d\dot{\eta}_1}-n_1\dot{\xi}_1+\frac{dV}{d\eta_1}=0\end{array}\right.\\ \left\{\begin{array}{l}\frac{d}{dt}\frac{dT}{d\dot{\xi}_2}+\frac{dQ}{d\dot{\xi}_2}+n_2\dot{\eta}_2+\frac{dV}{d\xi_2}=0\\ \frac{d}{dt}\frac{dT}{d\dot{\eta}_2}+\frac{dQ}{d\dot{\eta}_2}-n_2\dot{\xi}_2+\frac{dV}{d\eta_2}=0\end{array}\right.\\ \dots\dots\dots\dots\dots\dots\\ \left\{\begin{array}{l}\frac{d}{dt}\frac{dT}{d\dot{\xi}_{\frac{1}{2}i}}+\frac{dQ}{d\dot{\xi}_{\frac{1}{2}i}}+n_{\frac{1}{2}i}\dot{\eta}_{\frac{1}{2}i}+\frac{dV}{d\xi_{\frac{1}{2}i}}=0\\ \frac{d}{dt}\frac{dT}{d\dot{\eta}_{\frac{1}{2}i}}+\frac{dQ}{d\dot{\eta}_{\frac{1}{2}i}}-n_{\frac{1}{2}i}\dot{\xi}_{\frac{1}{2}i}+\frac{dV}{d\eta_{\frac{1}{2}i}}=0\end{array}\right.\end{array}\right\}\dots\dots\dots\dots(78).$$

345[xxviii]. Considerations of space and time prevent us from detailed treatment at present of gyrostatic systems with odd numbers of degrees of freedom, but it is obvious from § 345[xxvii] and 345[xi] that the general equations (13) of § 345[vi] may, when i the number of freedoms is odd, by proper transformation from co-ordinates ψ_1, ψ_2, etc. to a set of co-ordinates ζ, ξ_1, η_2, ... $\xi_{\frac{1}{2}(i-1)}$, $\eta_{\frac{1}{2}(i-1)}$ be reduced to the following form:

$$\left.\begin{array}{l}\left\{\begin{array}{l}\frac{d}{dt}\frac{dT}{d\dot{\xi}_1}+\frac{dQ}{d\dot{\xi}_1}+n_1\dot{\eta}_1+\frac{dV}{d\xi_1}=0\\ \frac{d}{dt}\frac{dT}{d\dot{\eta}_1}+\frac{dQ}{d\dot{\eta}_1}-n_1\dot{\xi}_1+\frac{dV}{d\eta_1}=0\end{array}\right.\\ \left\{\begin{array}{l}\frac{d}{dt}\frac{dT}{d\dot{\xi}_2}+\frac{dQ}{d\dot{\xi}_2}+n_2\dot{\eta}_2+\frac{dV}{d\xi_2}=0\\ \frac{d}{dt}\frac{dT}{d\dot{\eta}_2}+\frac{dQ}{d\dot{\eta}_2}-n_2\dot{\xi}_2+\frac{dV}{d\eta_2}=0\end{array}\right.\\ \dots\dots\dots\dots\dots\dots\\ \left\{\begin{array}{l}\frac{d}{dt}\frac{dT}{d\dot{\xi}_{\frac{1}{2}(i-1)}}+\frac{dQ}{d\dot{\xi}_{\frac{1}{2}(i-1)}}+n_{\frac{1}{2}(i-1)}\dot{\eta}_{\frac{1}{2}(i-1)}+\frac{dV}{d\xi_{\frac{1}{2}(i-1)}}=0\\ \frac{d}{dt}\frac{dT}{d\dot{\eta}_{\frac{1}{2}(i-1)}}+\frac{dQ}{d\dot{\eta}_{\frac{1}{2}(i-1)}}-n_{\frac{1}{2}(i-1)}\dot{\xi}_{\frac{1}{2}(i-1)}+\frac{dV}{d\eta_{\frac{1}{2}(i-1)}}=0\end{array}\right.\\ \qquad\frac{d}{dt}\frac{dT}{d\dot{\zeta}}+\frac{dQ}{d\dot{\zeta}}+\frac{dV}{d\zeta}=0\end{array}\right\}\dots\dots(79).$$

Kinetic stability.

346. There is scarcely any question in dynamics more important for Natural Philosophy than the stability or instability of motion. We therefore, before concluding this chapter, propose to give some general explanations and leading principles regarding it.

Conservative disturbance.

A "conservative disturbance of motion" is a disturbance in the motion or configuration of a conservative system, not altering the sum of the potential and kinetic energies. A conservative disturbance of the motion through any particular configuration is a change in velocities, or component velocities, not altering the whole kinetic energy. Thus, for example, a conservative disturbance of the motion of a particle through any point, is a change in the direction of its motion, unaccompanied by change of speed.

Kinetic stability and instability discriminated.

347. The actual motion of a system, from any particular configuration, is said to be *stable* if every possible infinitely small conservative disturbance of its motion through that configuration may be compounded of conservative disturbances, any one of which would give rise to an alteration of motion which would bring the system again to some configuration belonging to the undisturbed path, in a finite time, and without more than an infinitely small digression. If this condition is not fulfilled, the motion is said to be *unstable*.

Examples.

348. For example, if a body, A, be supported on a fixed vertical axis; if a second, B, be supported on a parallel axis belonging to the first; a third, C, similarly supported on B, and so on; and if B, C, etc., be so placed as to have each its centre of inertia as far as possible from the fixed axis, and the whole set in motion with a common angular velocity about this axis, the motion will be stable, from every configuration, as is evident from the principles regarding the resultant centrifugal force on a rigid body, to be proved later. If, for instance, each of the bodies is a flat rectangular board hinged on one edge, it is obvious that the whole system will be kept stable by centrifugal force, when all are in one plane and as far out from the axis as possible. But if A consist partly of a shaft and crank, as a common spinning-wheel, or the fly-wheel and crank of a steam-engine, and if B be supported on the crank-pin as axis, and turned inwards (towards the fixed axis, or across the fixed axis), then, even although the centres of inertia of C, D, etc., are placed as far from the fixed axis as possible, consistent with this position of B, the motion of the system will be unstable.

Examples.

349. The rectilinear motion of an elongated body lengthwise, or of a flat disc edgewise, through a fluid is unstable. But the motion of either body, with its length or its broadside perpendicular to the direction of motion, is stable. This is demonstrated for the ideal case of a perfect liquid (§ 320), in § 321, Example (2); and the results explained in § 322 show, for a solid of revolution, the precise character of the motion consequent upon an infinitely small disturbance in the direction of the motion from being exactly along or exactly perpendicular to the axis of figure; whether the infinitely small oscillation, in a definite period of time, when the rectilineal motion is stable, or the swing round to an infinitely nearly inverted position when the rectilineal motion is unstable. Observation proves the assertion we have just made, for real fluids, air and water, and for a great variety of circumstances affecting the motion. Several illustrations have been referred to in § 325; and it is probable we shall return to the subject later, as being not only of great practical importance, but profoundly interesting although very difficult in theory.

Kinetic stability. Hydrodynamic example.

350. The motion of a single particle affords simpler and not less instructive illustrations of stability and instability. Thus if a weight, hung from a fixed point by a light inextensible cord, be set in motion so as to describe a circle about a vertical line through its position of equilibrium, its motion is stable. For, as we shall see later, if disturbed infinitely little in direction without gain or loss of energy, it will describe a sinuous path, cutting the undisturbed circle at points successively distant from one another by definite fractions of the circumference, depending upon the angle of inclination of the string to the vertical. When this angle is very small, the motion is sensibly the same as that of a particle confined to one plane and moving under the influence of an attractive

Circular simple pendulum.

force towards a fixed point, simply proportional to the distance; and the disturbed path cuts the undisturbed circle four times in a revolution. Or if a particle confined to one plane, move under the influence of a centre in this plane, attracting with a force inversely as the square of the distance, a path infinitely little disturbed from a circle will cut the circle twice in a revolution. Or if the law of central force be the nth power of the distance, and if $n+3$ be positive, the disturbed path will cut the undisturbed circular orbit at successive angular intervals, each equal to $\pi/\sqrt{n+3}$. But the motion will be unstable if n be negative, and $-n>3$.

Circular orbit.

Kinetic stability in circular orbit.

The criterion of stability is easily investigated for circular motion round a centre of force from the differential equation of the general orbit (§ 36),

$$\frac{d^2u}{d\theta^2}+u=\frac{P}{h^2u^2}.$$

Let the value of h be such that motion in a circle of radius a^{-1} satisfies this equation. That is to say, let $P/h^2u^2=u$, when $u=a$. Let now $u=a+\rho$, ρ being infinitely small. We shall have

$$u-\frac{P}{h^2u^2}=\alpha\rho,$$

if α denotes the value of $\dfrac{d}{du}\left(u-\dfrac{P}{h^2u^2}\right)$ when $u=a$: and therefore the differential equation for motion infinitely nearly circular is

$$\frac{d^2\rho}{d\theta^2}+\alpha\rho=0.$$

The integral of this is most conveniently written

$$\rho=A\sin(\theta\sqrt{\alpha}+\beta)$$

when α is positive, and

$$\rho=C'\epsilon^{\theta\sqrt{-\alpha}}+C''\epsilon^{-\theta\sqrt{-\alpha}}$$

when α is negative.

Hence we see that the circular motion is stable in the former case, and unstable in the latter.

For instance, if $P = \mu r^n = \mu u^{-n}$, we have

Kinetic stability in circular orbit.

$$\frac{d}{du}\left(u - \frac{P}{h^2u^2}\right) = 1 + (n+2)\frac{P}{h^2u^3};$$

and putting $\frac{P}{h^2u^2} = u = a$, in this we find $a = n+3$; whence the result stated above.

Or, taking Example (B) of § 319, and putting mP for P, and mh for h,

$$\frac{P}{h^2u^2} = \frac{m'}{m+m'}\left(\frac{g}{h^2}u^{-2} + u\right),$$

$$\frac{d}{du}\left(u - \frac{P}{h^2u^2}\right) = \frac{m + \frac{2m'g}{h^2u^3}}{m+m'}.$$

Hence, putting $u = a$, and making $h^2 = gm'/ma^3$ so that motion in a circle of radius a^{-1} may be possible, we find

$$\alpha = \frac{3m}{m+m'}.$$

Hence the circular motion is always stable; and the period of the variation produced by an infinitely small disturbance from it is

$$2\pi\sqrt{\frac{m+m'}{3m}}.$$

351. The case of a particle moving on a smooth fixed surface under the influence of no other force than that of the constraint, and therefore always moving along a geodetic line of the surface, affords extremely simple illustrations of stability and instability. For instance, a particle placed on the inner circle of the surface of an anchor-ring, and projected in the plane of the ring, would move perpetually in that circle, but unstably, as the smallest disturbance would clearly send it away from this path, never to return until after a digression round the outer edge. (We suppose of course that the particle is held to the surface, as if it were placed in the infinitely narrow space between a solid ring and a hollow one enclosing it.) But if a particle is placed on the outermost, or greatest,

Kinetic stability of a particle moving on a smooth surface.

Kinetic stability of a particle moving on a smooth surface.

circle of the ring, and projected in its plane, an infinitely small disturbance will cause it to describe a sinuous path cutting the circle at points round it successively distant by angles each equal to $\pi\sqrt{b/a}$, or intervals of time, $\pi\sqrt{b}/\omega\sqrt{a}$, where a denotes the radius of that circle, ω the angular velocity in it, and b the radius of the circular cross section of the ring. This is proved by remarking that an infinitely narrow band from the outermost part of the ring has, at each point, a and b for its principal radii of curvature, and therefore (§ 150) has for its geodetic lines the great circles of a sphere of radius $\sqrt{ab}$, upon which (§ 152) it may be bent.

352. In all these cases the undisturbed motion has been circular or rectilineal, and, when the motion has been stable, the effect of a disturbance has been *periodic*, or recurring with the same phases in equal successive intervals of time. An illustration of thoroughly stable motion in which the effect of a disturbance is not "periodic," is presented by a particle sliding down an inclined groove under the action of gravity. To take the simplest case, we may consider a particle sliding down along the lowest straight line of an inclined hollow cylinder. If slightly disturbed from this straight line, it will oscillate on each side of it perpetually in its descent, but not with a uniform periodic motion, though the durations of its excursions to each side of the straight line are all equal.

Kinetic stability. Incommensurable oscillations.

353. A very curious case of stable motion is presented by a particle constrained to remain on the surface of an anchor-ring fixed in a vertical plane, and projected along the great circle from any point of it, with any velocity. An infinitely small disturbance will give rise to a disturbed motion of which the path will cut the vertical circle over and over again for ever, at unequal intervals of time, and unequal angles of the circle; and obviously not recurring periodically in any cycle, except with definite particular values for the whole energy, some of which are less and an infinite number are greater than that which just suffices to bring the particle to the highest point of the ring. The full mathematical investigation of these

circumstances would afford an excellent exercise in the theory of differential equations, but it is not necessary for our present illustrations.

354. In this case, as in all of stable motion with only two degrees of freedom, which we have just considered, there has been stability throughout the motion; and an infinitely small disturbance from any point of the motion has given a disturbed path which intersects the undisturbed path over and over again at finite intervals of time. But, for the sake of simplicity at present confining our attention to two degrees of freedom, we have a *limited* stability in the motion of an unresisted projectile, which satisfies the criterion of stability only at points of its upward, not of its downward, path. Thus if $MOPQ$ be

Oscillatory kinetic stability.

Limited kinetic stability.

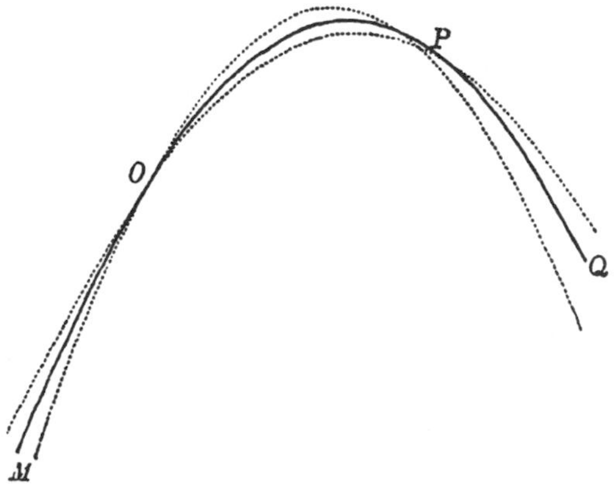

the path of a projectile, and if at O it be disturbed by an infinitely small force either way perpendicular to its instantaneous direction of motion, the disturbed path will cut the undisturbed infinitely near the point P where the direction of motion is perpendicular to that at O: as we easily see by considering that the line joining two particles projected from one point at the same instant with equal velocities in the directions of any two lines, will always remain perpendicular to the line bisecting the angle between these two lines.

Kinetic stability of a projectile.

General criterion.

355. The principle of varying action gives a mathematical criterion for stability or instability in every case of motion Thus in the first place it is obvious, and it will be proved below (§§ 358, 361), that if the action is a true minimum in the motion of a system from any one configuration to the configuration reached at any other time, however much later, the motion is thoroughly unstable. For instance, in the motion of a particle constrained to remain on a smooth fixed surface, and uninfluenced by gravity, the action is simply the length of the path, multiplied by the constant velocity. Hence in the particular case of a particle uninfluenced by gravity, moving round the inner circle in the plane of an anchor-ring considered above, the action, or length of path, is clearly a minimum from any one point to the point reached at any subsequent time. (The action is not merely a minimum, but is the smaller of two minimums, when the course is from any point of the circular path to any other, through less than half a circumference of the circle.) On the other hand, although the path from any point in the greatest circle of the ring to any other at a distance from it along the circle, less than $\pi\sqrt{ab}$, is clearly least possible if along the circumference; the path of absolutely least length is not along the circumference between two points at a greater circular distance than $\pi\sqrt{ab}$ from one another, nor is the path along the circumference between them a minimum at all in this latter case. On any surface whatever which is everywhere anticlastic, or along a geodetic of any surface which passes altogether through an anticlastic region, the motion is thoroughly unstable. For if it were stable from any point O, we should have the given undisturbed path, and the disturbed path from O cutting it at some point Q;—two different geodetic lines joining two points; which is impossible on an anticlastic surface, inasmuch as the sum of the exterior angles of any closed figure of geodetic lines exceeds four right angles (§ 136) when the integral curvature of the enclosed area is negative, which (§§ 138, 128) is the case for every portion of surface thoroughly anticlastic. But, on the other hand, it is easily proved that if we have an endless rigid band of curved surface everywhere synclastic, with a geodetic line running through its

Examples.

Motion on an anticlastic surface proved unstable.

Motion of a particle on an anticlastic surface, unstable;

middle, the motion of a particle projected along this line will be stable throughout, and an infinitely slight disturbance will give a disturbed path cutting the given undisturbed path again and again for ever at successive distances differing according to the different specific curvatures of the intermediate portions of the surface. If from any point, N, of the undisturbed path, a perpendicular be drawn to cut the infinitely near disturbed path in E, the angles OEN and NOE must (§ 138) be together greater than a right angle by an amount equal to the integral curvature of the area EON. From this the differential equation of the disturbed path may be obtained immediately.

On a synclastic surface, stable.

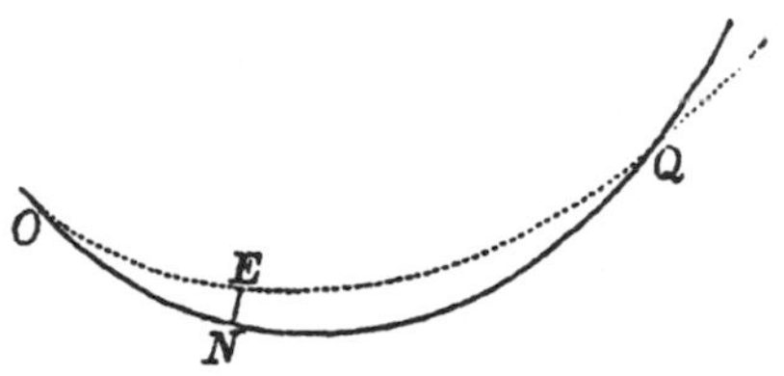

Differential equation of disturbed path.

Let $\angle EON = a$, $ON = s$, and $NE = u$; and let ϑ, a known function of s, be the specific curvature (§ 136) of the surface in the neighbourhood of N. Let also, for a moment, ϕ denote the complement of the angle OEN. We have

$$a - \phi = \int_0^s \vartheta u ds.$$

Hence $$\frac{d\phi}{ds} = -\vartheta u.$$

But, obviously, $$\phi = \frac{du}{ds};$$

hence $$\frac{d^2u}{ds^2} + \vartheta u = 0.$$

When ϑ is constant (as in the case of the equator of a surface of revolution considered above, § 351), this gives

$$u = A \cos (s\sqrt{\vartheta} + E),$$

agreeing with the result (§ 351) which we obtained by development into a spherical surface.

The case of two or more bodies supported on parallel axes in the manner explained above in § 348, and rotating with the centre of inertia of the whole at the least possible distance from the fixed axis, affords a very good illustration also of this proposition which may be safely left as an exercise to the student.

General investigation of disturbed path.

356. To investigate the effect of an infinitely small conservative disturbance produced at any instant in the motion of any conservative system, may be reduced to a practicable problem (however complicated the required work may be) of mathematical analysis, provided the undisturbed motion is thoroughly known.

General equation of motion free in two degrees.

(*a*) First, for a system having but two degrees of freedom to move, let

$$2T = P\dot{\psi}^2 + Q\dot{\phi}^2 + 2R\dot{\psi}\dot{\phi} \quad \text{.....................(1)},$$

where P, Q, R are functions of the co-ordinates not depending on the actual motion. Then

$$\left.\begin{aligned} &\frac{dT}{d\dot{\psi}} = P\dot{\psi} + R\dot{\phi}, \quad \frac{dT}{d\dot{\phi}} = Q\dot{\phi} + R\dot{\psi} \\ &\frac{d}{dt}\frac{dT}{d\dot{\psi}} = P\ddot{\psi} + R\ddot{\phi} + \frac{dP}{d\psi}\dot{\psi}^2 + \left(\frac{dP}{d\phi} + \frac{dR}{d\psi}\right)\dot{\psi}\dot{\phi} + \frac{dR}{d\phi}\dot{\phi}^2 \end{aligned}\right\} \text{...(2)};$$

and the Lagrangian equations of motion [§ 318 (24)] are

$$\left.\begin{aligned} &P\ddot{\psi} + R\ddot{\phi} + \tfrac{1}{2}\left\{\frac{dP}{d\psi}\dot{\psi}^2 + 2\frac{dP}{d\phi}\dot{\psi}\dot{\phi} + \left(2\frac{dR}{d\phi} - \frac{dQ}{d\psi}\right)\dot{\phi}^2\right\} = \Psi \\ &R\ddot{\psi} + Q\ddot{\phi} + \tfrac{1}{2}\left\{\left(2\frac{dR}{d\psi} - \frac{dP}{d\phi}\right)\dot{\psi}^2 + 2\frac{dQ}{d\psi}\dot{\psi}\dot{\phi} + \frac{dQ}{d\phi}\dot{\phi}^2\right\} = \Phi \end{aligned}\right\} \text{...(3)}.$$

We shall suppose the system of co-ordinates so chosen that none of the functions P, Q, R, nor their differential coefficients $\frac{dP}{d\phi}$, etc., can ever become infinite.

(*b*) To investigate the effects of an infinitely small disturbance, we may consider a motion in which, at any time t, the co-ordinates are $\psi + p$ and $\phi + q$, p and q being infinitely small; and, by simply taking the variations of equations (3) in the usual manner, we arrive at two simultaneous differential equations of the second degree, linear with respect to

$$p, \; q, \; \dot{p}, \; \dot{q}, \; \ddot{p}, \; \ddot{q},$$

but having variable coefficients which, when the undisturbed motion ψ, ϕ is fully known, may be supposed to be known functions of t. In these equations obviously none of the coefficients can at any time become infinite if the data correspond to a real dynamical problem, provided the system of co-ordinates is properly chosen (*a*); and the coefficients of $\ddot{p}$ and $\ddot{q}$ are the values, at the time t, of P, R, and R, Q, respectively, in the order in which they appear in (3), P, Q, R being the coefficients of a homogeneous quadratic function (1) which is essentially positive. These properties being taken into account, it may be shown that in no case can an infinitely small interval of time be the solution of the problem presented (§ 347) by the question of kinetic stability or instability, which is as follows:—

General investigation of disturbed path.

(c) The component velocities $\dot{\psi}$, $\dot{\phi}$ are at any instant changed to $\dot{\psi}+a$, $\dot{\phi}+\beta$, subject to the condition of not changing the value of T. Then, a and β being infinitely small, it is required to find the interval of time until q/p first becomes equal to $\dot{\phi}/\dot{\psi}$.

(d) The differential equations in p and q reduce this problem, and in fact the full problem of finding the disturbance in the motion when the undisturbed motion is given, to a practicable form. But, merely to prove the proposition that the disturbed course cannot meet the undisturbed course until after some finite time, and to estimate a limit which this time must exceed in any particular case, it may be simpler to proceed thus:—

(e) To eliminate t from the general equations (3), let them first be transformed so as not to have t independent variable. We must put

$$\ddot{\psi}=\frac{dtd^2\psi-d\psi d^2t}{dt^3},\quad \ddot{\phi}=\frac{dtd^2\phi-d\phi d^2t}{dt^3}\quad\text{..................(4).}$$

And by the equation of energy we have

$$dt=\frac{(Pd\psi^2+Qd\phi^2+2Rd\psi d\phi)^{\frac{1}{2}}}{\{2(E-V)\}^{\frac{1}{2}}}\quad\text{.....................(5),}$$

it being assumed that the system is conservative. Eliminating dt and d^2t between this and the two equations (3), we find a differential equation of the second degree between ψ and ϕ, which is the differential equation of the course. For simplicity, let us suppose one of the co-ordinates, ϕ for instance, to be independent variable; that is, let $d^2\phi=0$. We have, by (4),

$$d^2t=-\ddot{\phi}\frac{dt^3}{d\phi},$$

and therefore $$\ddot{\psi}dt^2=d^2\psi+\frac{d\psi}{d\phi}\ddot{\phi}dt^2,$$

General investigation of disturbed path.

and the result of the elimination becomes

$$(PQ-R^2)\frac{d^2\psi}{d\phi^2}+F\left(\frac{d\psi}{d\phi}\right)=\frac{\left(P\frac{d\psi^2}{d\phi^2}+2R\frac{d\psi}{d\phi}+Q\right)\left[\left(Q+R\frac{d\psi}{d\phi}\right)\Psi-\left(R+P\frac{d\psi}{d\phi}\right)\Phi\right]}{2(E-V)}$$

..............................(6),

$F\left(\frac{d\psi}{d\phi}\right)$ denoting a function of $\frac{d\psi}{d\phi}$ of the third degree, with variable coefficients, none of which can become infinite as long as $E-V$, the kinetic energy, is finite.

(*f*) Taking the variation of this equation on the supposition that ψ becomes $\psi+p$, where p is infinitely small, we have

$$(PQ-R^2)\frac{d^2p}{d\phi^2}+L\frac{dp}{d\phi}+Mp=0\text{..................(7)},$$

where L and M denote known functions of ϕ, neither of which has any infinitely great value. This determines the deviation, p, of the course. Inasmuch as the quadratic (1) is essentially always positive, $PQ-R^2$ must be always positive. Hence, if for a particular value of ϕ, p vanishes, and $\frac{dp}{d\phi}$ has a given value which defines the disturbance we suppose made at any instant, ϕ must increase by a finite amount (and therefore a finite time must elapse) before the value of p can be again zero; that is to say, before the disturbed course can again cut the undisturbed course.

(*g*) The same proposition consequently holds for a system having any number of degrees of freedom. For the preceding proof shows it to hold for the system subjected to any frictionless constraint, leaving it only two degrees of freedom; including that particular frictionless constraint which would not alter either the undisturbed or the disturbed course. The full general investigation of the disturbed motion, with more than two degrees of freedom, takes a necessarily complicated form, but the principles on which it is to be carried out are sufficiently indicated by what we have done.

(*h*) If for $L/PQ-R^2$ we substitute a constant $2a$, less than its least value, irrespectively of sign, and for $M/PQ-R^2$, a constant β greater algebraically than its greatest value, we have an equation

General investigation of disturbed path.

$$\frac{d^2p}{d\phi^2} + 2a\frac{dp}{d\phi} + \beta = 0 \qquad \text{.....}(8).$$

Here the value of p vanishes for values of ϕ successively exceeding one another by $\pi/\sqrt{\beta - a^2}$, which is clearly less than the increase that ϕ must have in the actual problem before p vanishes a second time. Also, we see from this that if $a^2 > \beta$ the actual motion is unstable. It might of course be unstable even if $a^2 < \beta$; and the proper analytical methods for finding either the rigorous solution of (7), or a sufficiently near practical solution, would have to be used to close the criterion of stability or instability, and to thoroughly determine the disturbance of the course.

(*i*) When the system is only a single particle, confined to a plane, the differential equation of the deviation may be put under a remarkably simple form, useful for many practical problems. Let N be the normal component of the force, per unit of the mass, at any instant, v the velocity, and ρ the radius of curvature of the path. We have (§ 259)

Differential equation of disturbed path of single particle in a plane.

$$N = \frac{v^2}{\rho}.$$

Let, in the diagram, ON be the undisturbed, and OE the disturbed path. Let EN, cutting ON at right angles, be denoted by u, and ON by s. If further we denote by ρ' the radius of curvature in the disturbed path, remembering that u is infinitely small, we easily find

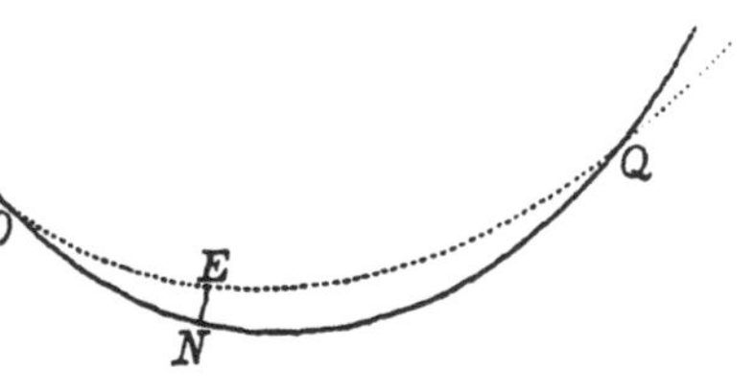

$$\frac{1}{\rho'} = \frac{1}{\rho} + \frac{d^2u}{ds^2} + \frac{u}{\rho^2} \qquad \text{.....}(9).$$

Hence, using δ to denote variations from N to E, we have

$$\delta N = \delta\frac{v^2}{\rho} = \frac{\delta(v^2)}{\rho} + v^2\left(\frac{d^2u}{ds^2} + \frac{u}{\rho^2}\right) \qquad \text{.....}(10).$$

Differential equation of disturbed path of single particle in a plane.

But, by the equation of energy,

$$v^2 = 2(E - V),$$

and therefore

$$\delta(v^2) = -2\delta V = 2Nu = \frac{2v^2}{\rho}u.$$

Hence (10) becomes

$$\frac{d^2u}{ds^2} + \frac{3u}{\rho^2} - \frac{\delta N}{v^2} = 0 \quad\text{.....................(11)},$$

or, if we denote by ζ the rate of variation of N, per unit of distance from the point N in the normal direction, so that $\delta N = \zeta u$,

$$\frac{d^2u}{ds^2} + \left(\frac{3}{\rho^2} - \frac{\zeta}{v^2}\right)u = 0 \quad\text{.....................(12)}.$$

This includes, as a particular case, the equation of deviation from a circular orbit, investigated above (§ 350).

Kinetic foci.

357. If, from any one configuration, two courses differing infinitely little from one another have again a configuration in common, this second configuration will be called a kinetic focus relatively to the first: or (because of the reversibility of the motion) these two configurations will be called conjugate kinetic foci. Optic foci, if for a moment we adopt the corpuscular theory of light, are included as a particular case of kinetic foci in general. By § 356 (*g*) we see that there must be finite intervals of space and time between two conjugate foci in every motion of every kind of system, only provided the kinetic energy does not vanish.

Theorem of minimum action.

Action never a minimum in a course including kinetic foci.

358. Now it is obvious that, provided only a sufficiently short course is considered, the *action*, in any natural motion of a system, is less than for any other course between its terminal configurations. It will be proved presently (§ 361) that the first configuration up to which the action, reckoned from a given initial configuration, ceases to be a minimum, is the first kinetic focus; and conversely, that when the first kinetic focus is passed, the action, reckoned from the initial configuration, ceases to be a minimum; and therefore of course can never again be a minimum, because a course of shorter action, deviating infinitely little from it, can be found for a part, without altering the remainder of the whole, natural course.

359. In such statements as this it will frequently be convenient to indicate particular configurations of the system by single letters, as O, P, Q, R; and any particular course, in which it moves through configurations thus indicated, will be called the course $O...P...Q...R$. The *action* in any natural course will be denoted simply by the terminal letters, taken in the order of the motion. Thus OR will denote the action from O to R; and therefore $OR = -RO$. When there are more real natural courses from O to R than one, the analytical expression for OR will have more than one real value; and it may be necessary to specify for which of these courses the action is reckoned. Thus we may have **Notation for configurations, courses, and action.**

$$OR \text{ for } O...E...R,$$
$$OR \text{ for } O...E'...R,$$
$$OR \text{ for } O...E''...R,$$

three different values of one algebraic irrational expression.

360. In terms of this notation the preceding statement (§ 358) may be expressed thus:—If, for a conservative system, moving on a certain course $O...P...O'...P'$, the first kinetic focus conjugate to O be O', the action OP', in this course, will be less than the action along any other course deviating infinitely little from it: but, on the other hand, OP' is greater than the actions in some courses from O to P' deviating infinitely little from the specified natural course $O...P...O'...P'$. **Theorem of minimum action.**

361. It must not be supposed that the action along OP is necessarily *the least possible* from O to P. There are, in fact, cases in which the action ceases to be *least of all possible*, before **Two or more courses of minimum action possible.**

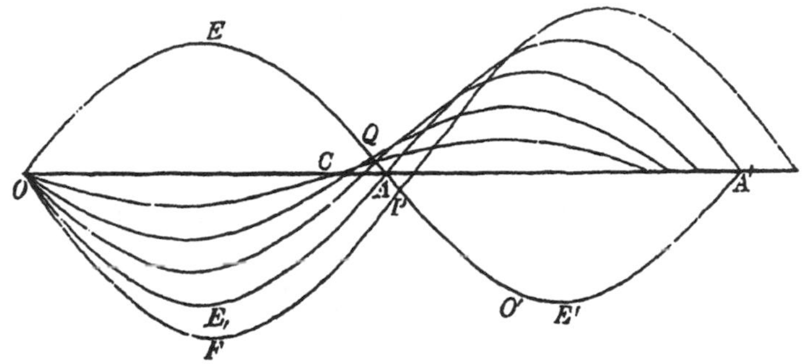

Two or more courses of minimum action possible.

a kinetic focus is reached. Thus if $OEAPO'E'A'$ be a sinuous geodetic line cutting the outer circle of an anchor-ring, or the equator of an oblate spheroid, in successive points O, A, A', it is easily seen that O', the first kinetic focus conjugate to O, must lie somewhat beyond A. But the length $OEAP$, although a *minimum* (a stable position for a stretched string), is not the shortest distance on the surface from O to P, as *this* must obviously be a line lying entirely on one side of the great circle. From O, to any point, Q, short of A, the distance along the geodetic $OEQA$ is clearly the least possible: but if Q be near enough to A (that is to say, between A and the point in which the envelope of the geodetics drawn from O, cuts OEA), there will also be two other geodetics from O to Q. The length of one of these will be a minimum, and that of the other not a minimum. If Q is moved forward to A, the former becomes $OE_,A$, equal and similar to OEA, but on the other side of the great circle: and the latter becomes the great circle from O to A. If now Q be moved on, to P, beyond A, the minimum geodetic $OEAP$ ceases to be the less of the two minimums, and the geodetic OFP lying altogether on the other side of the great circle becomes the least possible line from O to P. But until P is advanced beyond the point, O', in which it is cut by another geodetic from O lying infinitely nearly along it, the length $OEAP$ remains a minimum, according to the general proposition of § 358, which we now proceed to prove.

Case of two minimum, and one not minimum, geodetic lines between two points.

Difference between two sides and the third of a kinetic triangle.

(*a*) Referring to the notation of § 360, let $P_,$ be any configuration differing infinitely little from P, but not on the course $O...P...O'...P'$; and let S be a configuration on this course, reached at some finite time after P is passed. Let ψ, ϕ,... be the co-ordinates of P, and $\psi_,$, $\phi_,$,... those of $P_,$, and let

$$\psi_, - \psi = \delta\psi,\ \phi_, - \phi = \delta\phi, \ldots$$

Thus, by Taylor's theorem,

$$OP_, + P_,S = OS + \left\{ \frac{d\,(OP+PS)}{d\psi}\delta\psi + \frac{d\,(OP+PS)}{d\phi}\delta\phi + \ldots \right\}$$

$$+ \tfrac{1}{2}\left\{ \frac{d^2\,(OP+PS)}{d\psi^2}(\delta\psi)^2 + 2\frac{d^2\,(OP+PS)}{d\psi d\phi}\delta\psi\delta\phi + \ldots \right\}$$

$$+ \text{etc.}$$

But if ξ, η,... denote the components of momentum at P in the course $O...P$, which are the same as those at P in the continuation, $P...S$, of this course, we have [§ 330 (18)] Difference between two sides and the third of a kinetic triangle.

$$\xi = \frac{dOP}{d\psi} = -\frac{dPS}{d\psi}, \quad \eta = \frac{dOP}{d\phi} = -\frac{dPS}{d\phi}, \ldots$$

Hence the coefficients of the terms of the first degree of $\delta\psi$, $\delta\phi$, in the preceding expression vanish, and we have

$$OP_, + P_,S - OS = \tfrac{1}{2}\left\{\frac{d^2(OP+PS)}{d\psi^2}\delta\psi^2 + 2\frac{d^2(OP+PS)}{d\psi d\phi}\delta\psi\delta\phi + \ldots\right\} + \text{etc.} \quad \ldots (1).$$

(b) Now, assuming

$$\left.\begin{array}{l} x_1 = \alpha_1\delta\psi + \beta_1\delta\phi + \ldots \\ x_2 = \alpha_2\delta\psi + \beta_2\delta\phi + \ldots \\ \quad \text{etc.} \qquad \text{etc.} \end{array}\right\} \ldots\ldots\ldots\ldots\ldots (2),$$

according to the known method of linear transformations, let α_1, β_1,... α_2, β_2,... be so chosen that the preceding quadratic function be reduced to the form

$$A_1x_1^2 + A_2x_2^2 + \ldots + A_ix_i^2,$$

the whole number of degrees of freedom being i.

This may be done in an infinite variety of ways; and, towards fixing upon one particular way, we may take $\alpha_i = \dot\psi$, $\beta_i = \dot\phi$, etc.; and subject the others to the conditions

$$\dot\psi\alpha_1 + \dot\phi\beta_1 + \ldots = 0, \quad \dot\psi\alpha_2 + \dot\phi\beta_2 + \ldots = 0, \text{ etc.}$$

This will make $A_i = 0$: for if for a moment we suppose $P_,$ to be on the course $O...P...O'$, we have

$$\frac{\delta\psi}{\dot\psi} = \frac{\delta\phi}{\dot\phi} = \ldots,$$

and therefore

$$x_i = \frac{\dot\psi}{\delta\psi}(\delta\psi^2 + \delta\phi^2 + \ldots), \quad x_{i-1} = 0, \quad \ldots x_2 = 0, \quad x_1 = 0.$$

But in this case $OP_, + P_,S = OS$; and therefore the value of the quadratic must be zero; that is to say, we must have $A_i = 0$. Hence we have

$$OP_, + P_,S - OS = \tfrac{1}{2}(A_1x_1^2 + A_2x_2^2 + \ldots + A_{i-1}x_{i-1}^2) + R \quad \ldots\ldots\ldots (3)$$

where R denotes a remainder consisting of terms of the third and higher degrees in $\delta\psi$, $\delta\phi$, etc., or in x_1, x_2, etc.

Difference between two sides and the third of a kinetic triangle.

(c) Another form, which will be used below, may be given to the same expression thus:—Let $(\xi_{,}, \eta_{,}, \zeta_{,}, ...)$ and $(\xi_{,}', \eta_{,}', \zeta_{,}', ...)$ be the components of momentum at $P_{,}$, in the courses $OP_{,}$ and $P_{,}S$ respectively. By § 330 (18) we have

$$\xi_{,} = \frac{dOP'}{d\psi_{,}},$$

and therefore by Taylor's theorem

$$\xi_{,} = \frac{dOP}{d\psi} + \frac{d^2OP}{d\psi^2}\delta\psi + \frac{d^2OP}{d\psi d\phi}\delta\phi + ... + \text{etc.}$$

Similarly,

$$-\xi_{,}' = \frac{dPS}{d\psi} + \frac{d^2PS}{d\psi^2}\delta\psi + \frac{d^2PS}{d\psi d\phi}\delta\phi + ... + \text{etc.};$$

and therefore, as $\frac{dOP}{d\psi} = -\frac{dPS}{d\psi}$,

$$\xi_{,}' - \xi_{,} = -\left\{\frac{d^2(OP+PS)}{d\psi^2}\delta\psi + \frac{d^2(OP+PS)}{d\psi d\phi}\delta\phi + ...\right\} + \text{etc.} ...(4),$$

and so for $\eta_{,}' - \eta_{,}$, etc. Hence (1) is the same as

$$\left.\begin{array}{l} OP_{,} + P_{,}S - OS = -\tfrac{1}{2}\{(\xi_{,}' - \xi')\delta\psi + (\eta_{,}' - \eta_{,})\delta\phi + ...\} \\ \qquad + R \end{array}\right\} ......(5),$$

where R denotes a remainder consisting of terms of the third and higher degrees. Also the transformation from $\delta\psi$, $\delta\phi$, ... to x_1, x_2, ..., gives clearly

$$\left.\begin{array}{l} \xi_{,}' - \xi_{,} = -(A_1\alpha_1x_1 + A_2\alpha_2x_2 + ... + A_{i-1}\alpha_{i-1}x_{i-1}) \\ \eta_{,}' - \eta_{,} = -(A_1\beta_1x_1 + A_2\beta_2x_2 + ... + A_{i-1}\beta_{i-1}x_{i-1}) \\ \qquad \text{etc.} \qquad\qquad \text{etc.} \end{array}\right\} ....(6).$$

(d) Now for any infinitely small time the velocities remain sensibly constant; as also do the coefficients (ψ, ψ), (ψ, ϕ), etc., in the expression [§ 313 (2)] for T: and therefore for the action we have

$$\int 2Tdt = \sqrt{2T}\int\sqrt{2T}dt$$

$$= \sqrt{2T}\,\{(\psi, \psi)(\psi-\psi_0)^2 + 2(\psi, \phi)(\psi-\psi_0)(\phi-\phi_0) + \text{etc.}\}^{\frac{1}{2}}$$

where $(\psi_0, \phi_0, ...)$ are the co-ordinates of the configuration from which the action is reckoned. Hence, if P, P', P'' be any three configurations infinitely near one another, and if Q, with the proper differences of co-ordinates written after it, be used to denote square roots of quadratic functions such as that in the preceding expression, we have

$$\left.\begin{aligned} PP' &= \sqrt{2T}\,.\,Q\,\{(\psi - \psi'),\ (\phi - \phi'),\ \ldots\} \\ P'P'' &= \sqrt{2T}\,.\,Q\,\{(\psi' - \psi''),\ (\phi' - \phi''),\ \ldots\} \\ P''P &= \sqrt{2T}\,.\,Q\,\{(\psi'' - \psi'),\ (\phi'' - \phi'),\ \ldots\} \end{aligned}\right\} \ldots\ldots\ldots\ldots(7).$$

Difference between two sides and the third of a kinetic triangle.

In the particular case of a single free particle, these expressions become simply proportional to the distances PP', $P'P''$, $P''P$; and by Euclid we have

$$P'P + PP'' < P'P''$$

unless P is in the straight line $P'P''$.

The verification of this proposition by the preceding expressions (7) is merely its proof by co-ordinate geometry with an oblique rectilineal system of co-ordinates, and is necessarily somewhat complicated. If $(\psi, \phi) = (\phi, \theta) = (\theta, \psi) = 0$, the co-ordinates become rectangular and the algebraic proof is easy. There is no difficulty, by following the analogies of these known processes, to prove that, for any number of co-ordinates, ψ, ϕ, etc., we have

$$P'P + PP'' > P'P'',$$

unless

$$\frac{\psi - \psi'}{\psi'' - \psi'} = \frac{\phi - \phi'}{\phi'' - \phi'} = \frac{\theta - \theta'}{\theta'' - \theta'} = \ldots$$

(expressing that P is on the course from P to P''), in which case

$$P'P + PP'' = P'P'',$$

$P'P$, etc., being given by (7). And further, by the aid of (1), it is easy to find the proper expression for $P'P + PP'' - P'P''$, when P is infinitely little off the course from P' to P'': but it is quite unnecessary for us here to enter on such purely algebraic investigations.

(e) It is obvious indeed, as has been already said (§ 358), that the action along any natural course is *the least possible between its terminal configurations* if only a sufficiently short course is included. Hence for all cases in which the time from O to S is less than some particular amount, the quadratic term in the expression (3) for $OP_{,} + P_{,}S - OS$ is necessarily positive, for all values of x_1, x_2, etc.; and therefore A_1, A_2,...A_{i-1} must each be positive.

(f) Let now S be removed further and further from O, along the definite course $O\ldots P\ldots O'$, until it becomes O'. When it is O', let $P_{,}$ be taken on a natural course through O and O', de-

Actions on different courses infinitely near one another

between two conjugate kinetic foci, proved ultimately equal.

viating infinitely little from the course $OP_{,}O'$. Then, as $OP_{,}O'$ is a natural course,

$$\xi_{,}' - \xi_{,} = \eta_{,}' - \eta_{,}, \ldots = 0;$$

and therefore (5) becomes

$$OP_{,} + P_{,}O' - OO' = R,$$

which proves that the chief, or quadratic, term in the other expression (3) for the same, vanishes. Hence one at least of the coefficients A_1, A_2,... must vanish, and if one only, $A_{i-1} = 0$ for instance, we must have

$$x_1 = 0,\ x_2 = 0, \ldots x_{i-2} = 0.$$

These equations express the condition that $P_{,}$ lies on a natural course from O to O'.

If two sides, deviating infinitely little from the third, are together equal to it, they constitute an unbroken natural course.

(*g*) Conversely if one or more of the coefficients A_1, A_2, etc., vanishes, if for instance $A_{i-1} = 0$, S must be a kinetic focus. For if we take $P_{,}$ so that

$$x_1 = 0,\ x_2 = 0, \ldots x_{i-2} = 0,$$

we have, by (6),

$$\xi_{,}' - \xi_{,} = \eta_{,}' - \eta' = \ldots = 0.$$

(*h*) Thus we have proved that at a kinetic focus conjugate to O the action from O is not a minimum of the first order*, and that the last configuration, up to which the action from O *is* a minimum of the first order, is a kinetic focus conjugate to O.

(*i*) It remains to be proved that the action from O ceases to be a minimum when the first kinetic focus conjugate to O is passed. Let, as above (§ 360), $O \ldots P \ldots O' \ldots P'$ be a natural course extending beyond O', the first kinetic focus conjugate to O. Let P and P' be so near one another that there is no focus conjugate to either, between them; and let $O \ldots P_{,} \ldots O'$ be a natural course from O to O' deviating infinitely little from $O \ldots P \ldots O'$. By what we have just proved (*e*), the action OO' along $O \ldots P_{,} \ldots O'$ differs only by R, an infinitely small quantity of the third order, from the action OO' along $O \ldots P \ldots O'$, and therefore

Natural course proved not a course of minimum action, beyond a kinetic focus.

$$Ac.\,(O \ldots P \ldots O' \ldots P') = Ac.\,(O \ldots P_{,} \ldots O') + O'P' + R$$
$$= OP_{,} + P_{,}O' + O'P' + R.$$

* A maximum or minimum "of the first order" of any function of one or more variables, is one in which the differential of the first degree vanishes, but not that of the second degree.

Natural course proved not a course of minimum action, beyond a kinetic focus.

But, by a proper application of (*e*) we see that

$$P_,O' + O'P' = P_,P' + Q$$

where Q denotes an infinitely small quantity of the second order, which is essentially positive. Hence

$$Ac\,(O...P...O'...P') = OP_, + P_,P' + Q + R,$$

and therefore, as R is infinitely small in comparison with Q,

$$Ac\,(O...P...O'...P') > OP_, + P_,P'.$$

Hence the broken course $O...P_,$, $P_,...P'$ has less action than the natural course $O...P...O'...P'$, and therefore, as the two are infinitely near one another, the latter is not a minimum.

A course which includes no focus conjugate to either extremity includes no pair of conjugate foci.

362. As it has been proved that the action from any configuration ceases to be a minimum at the first conjugate kinetic focus, we see immediately that if O' be the first kinetic focus conjugate to O, reached after passing O, no two configurations on this course from O to O' can be kinetic foci to one another. For, the action from O just ceasing to be a minimum when O' is reached, the action between any two intermediate configurations of the same course is necessarily a minimum.

How many kinetic foci in any case.

363. When there are i degrees of freedom to move there are in general, on any natural course from any particular configuration, O, at least $i-1$ kinetic foci conjugate to O. Thus, for example, on the course of a ray of light emanating from a luminous point O, and passing through the centre of a convex lens held obliquely to its path, there are two kinetic foci conjugate to O, as defined above, being the points in which the line of the central ray is cut by the so-called "focal lines"* of a pencil of rays diverging from O and made convergent after passing through the lens. But some or all of these kinetic foci may be on the course previous to O; as for instance in the case of a common projectile when its course passes obliquely downwards through O. Or some or all may be lost; as when, in the optical illustration just referred to, the lens is only strong enough to produce convergence in one of the principal planes, or too weak to produce convergence in either. Thus

* In our second volume we hope to give all necessary elementary explanations on this subject.

How many kinetic foci in any case. also in the case of the undisturbed rectilineal motion of a point, or in the motion of a point uninfluenced by force, on an anticlastic surface (§ 355), there are no real kinetic foci. In the motion of a projectile (not confined to one vertical plane) there can only be one kinetic focus on each path, conjugate to one given point; though there are three degrees of freedom. Again, there may be any number more than $i-1$, of foci in one course, all conjugate to one configuration, as for instance on the course of a particle uninfluenced by force, moving round the surface of an anchor-ring, along either the outer great circle, or along a sinuous geodetic such as we have considered in § 361, in which clearly there are an infinite number of foci each conjugate to any one point of the path, at equal successive distances from one another.

Referring to the notation of § 361 (f), let S be gradually moved on until first one of the coefficients, A_{i-1} for instance, vanishes; then another, A_{i-2}, etc.; and so on. We have seen that each of these positions of S is a kinetic focus: and thus by the successive vanishing of the $i-1$ coefficients we have $i-1$ foci. If none of the coefficients can ever vanish, there are no kinetic foci. If one or more of them, after vanishing, comes to a minimum, and again vanishes, as S is moved on, there may be any number more than $i-1$ of foci each conjugate to the same configuration, O.

Theorem of maximum action. 364. If $i-1$ distinct* courses from a configuration O, each differing infinitely little from a certain natural course

$$O \ldots E \ldots O_1 \ldots O_2 \ldots O_{i-1} \ldots Q,$$

cut it in configurations O_1, O_2, O_3,...O_{i-1}, and if, besides these, there are not on it any other kinetic foci conjugate to O, between O and Q, and no focus at all, conjugate to E, between E and Q, the action in this natural course from O to Q is the maximum for all courses $O \ldots P_{\prime}$, $P_{\prime} \ldots Q$; $P_{\prime}$ being a configuration infinitely nearly agreeing with some configuration between E and O_1 of the standard course $O \ldots E \ldots O_1 \ldots O_2 \ldots O_{i-1} \ldots Q$, and $O \ldots P_{\prime}$, $P_{\prime} \ldots Q$

* Two courses are not called distinct if they differ from one another only in the absolute magnitude, not in the proportions of the components, of the deviations by which they differ from the standard course.

denoting the natural courses between O and $P_{,}$, and $P_{,}$ and Q, which deviate infinitely little from this standard course. Theorem of maximum action.

In § 361 (i), let O' be any one, O_1, of the foci O_1, O_2, ... O_{i-1}, and let $P_{,}$ be called P_1 in this case. The demonstration there given shows that

$$OQ > OP_1 + P_1Q.$$

Hence there are $i-1$ different broken courses

$$O \ldots P_1,\ P_1 \ldots Q;\ O \ldots P_2,\ P_2 \ldots Q;\ \text{etc.},$$

in each of which the action is less than in the standard course from O to Q. But whatever be the deviation of $P_{,}$, it may clearly be compounded of deviations P to P_1, P to P_2, P to P_3, ..., P to P_{i-1}, corresponding to these $i-1$ cases respectively; and it is easily seen from the analysis that

$$OP_{,} + P_{,}Q - OQ = (OP_1 + P_1Q - OQ) + (OP_2 + P_2Q - OQ) + \ldots$$

Hence $OP_{,} + P_{,}Q < OQ$, which was to be proved.

365. Considering now, for simplicity, only cases in which there are but two degrees (§§ 195, 204) of freedom to move, we see that after any infinitely small conservative disturbance of a system in passing through a certain configuration, the system will first again pass through a configuration of the undisturbed course, at the first configuration of the latter at which the action in the undisturbed motion ceases to be a minimum. For instance, in the case of a particle, confined to a surface, and subject to any conservative system of force, an infinitely small conservative disturbance of its motion through any point, O, produces a disturbed path, which cuts the undisturbed path at the first point, O', at which the action in the undisturbed path from O ceases to be a minimum. Or, if projectiles, under the influence of gravity alone, be thrown from one point, O, in all directions with equal velocities, in one vertical plane, their paths, as is easily proved, intersect one another consecutively in a parabola, of which the focus is O, and the vertex the point reached by the particle projected directly upwards. The actual course of each particle from O is the course of least possible action to any point, P, reached before the enveloping parabola, but is not a course of minimum action to any point, Q, in its path after the envelope is passed. Applications to two degrees of freedom.

Applications to two degrees of freedom.

366. Or again, if a particle slides round along the greatest circle of the smooth inner surface of a hollow anchor-ring, the "action," or simply the length of path, from point to point, will be least possible for lengths (§ 351) less than $\pi\sqrt{ab}$. Thus, if a string be tied round outside on the greatest circle of a perfectly smooth anchor-ring, it will slip off unless held in position by staples, or checks of some kind, at distances of not less than $\pi\sqrt{ab}$ from one another in succession round the circle. With reference to this example, see also § 361, above.

Or, of a particle sliding down an inclined cylindrical groove, the action from any point will be the least possible along the straight path to any other point reached in a time less than that of the vibration one way of a simple pendulum of length equal to the radius of the groove, and influenced by a force equal $g \cos i$, instead of g the whole force of gravity. But the action will not be a minimum from any point, along the straight path, to any other point reached in a longer time than this. The case in which the groove is horizontal ($i=0$) and the particle is projected along it, is particularly simple and instructive, and may be worked out in detail with great ease, without assuming any of the general theorems regarding action.

Hamilton's second form.

367. In the preceding account of the Hamiltonian principle, and of developments and applications which it has received, we have adhered to the system (§§ 328, 330) in which the initial and final co-ordinates and the constant sum of potential and kinetic energies are the elements of which the action is supposed to be a function. Another system was also given by Hamilton, according to which the action is expressed in terms of the initial and final co-ordinates and the *time prescribed for the motion;* and a set of expressions quite analogous to those with which we have worked, are established. For practical applications this method is generally less convenient than the other; and the analytical relations between the two are so obvious that we need not devote any space to them here.

Liouville's kinetic theorem.

368. We conclude by calling attention to a very novel analytical investigation of the motion of a conservative system, by Liouville (*Comptes Rendus*, June 16, 1856), which leads im-

mediately to the principle of least action, and the Hamiltonian principle with the developments by Jacobi and others; but which also establishes a very remarkable and absolutely new theorem regarding the amount of the action along any constrained course. For brevity we shall content ourselves with giving it for a singlo free particle, referring the reader to the original article for Liouville's complete investigation in terms of generalized co-ordinates, applicable to any conservative system whatever.

Liouville's kinetic theorem.

Let (x, y, z) be the co-ordinates of any point through which the particle may move: V its potential energy in this position: E the sum of the potential and kinetic energies of the motion in question: A the action, from any position (x_0, y_0, z_0) to (x, y, z) along any course arbitrarily chosen (supposing, for instance, the particle to be guided along it by a frictionless guiding tube). Then (§ 326), the mass of the particle being taken as unity,

$$A = \int v ds = \int \sqrt{2(E-V)} \sqrt{(dx^2+dy^2+dz^2)}.$$

Now let ϑ be a function of x, y, z, which satisfies the partial differential equation

$$\frac{d\vartheta^2}{dx^2} + \frac{d\vartheta^2}{dy^2} + \frac{d\vartheta^2}{dz^2} = 2(E-V).$$

Then

$$A = \int \sqrt{\left(\frac{d\vartheta^2}{dx^2} + \frac{d\vartheta^2}{dy^2} + \frac{d\vartheta^2}{dz^2}\right)(dx^2+dy^2+dz^2)}$$

$$= \int \sqrt{\left[\frac{d\vartheta}{dx}dx + \frac{d\vartheta}{dy}dy + \frac{d\vartheta}{dz}dz\right)^2 + \left(\frac{d\vartheta}{dz}dy - \frac{d\vartheta}{dy}dz\right)^2 + \left(\frac{d\vartheta}{dx}dz - \frac{d\vartheta}{dz}dx\right)^2}$$
$$+ \left(\frac{d\vartheta}{dy}dx - \frac{d\vartheta}{dx}dy\right)^2\bigg].$$

But $$\frac{d\vartheta}{dx}dx + \frac{d\vartheta}{dy}dy + \frac{d\vartheta}{dz}dz = d\vartheta,$$

and, if $\dot{x}, \dot{y}, \dot{z}$ denote the actual component velocities along the arbitrary path, and $\dot{\vartheta}$ the rate at which ϑ increases per unit of time in this motion,

$$dx = \dot{x}dt,\quad dy = \dot{y}dt,\quad dz = \dot{z}dt,\quad d\vartheta = \dot{\vartheta}dt.$$

Hence the preceding becomes

$$A = \int d\vartheta \sqrt{\left\{1 + \frac{\left(\dot{y}\frac{d\vartheta}{dz} - \dot{z}\frac{d\vartheta}{dy}\right)^2 + \left(\dot{z}\frac{d\vartheta}{dx} - \dot{x}\frac{d\vartheta}{dz}\right)^2 + \left(\dot{x}\frac{d\vartheta}{dy} - \dot{y}\frac{d\vartheta}{dx}\right)^2}{\dot{\vartheta}^2}\right\}}.$$

CHAPTER III.

EXPERIENCE.

Observation and experiment.

369. By the term Experience, in physical science, we designate, according to a suggestion of Herschel's, our means of becoming acquainted with the material universe and the laws which regulate it. In general the actions which we see ever taking place around us are *complex*, or due to the simultaneous action of many causes. When, as in astronomy, we endeavour to ascertain these causes by simply watching their effects, we *observe;* when, as in our laboratories, we interfere arbitrarily with the causes or circumstances of a phenomenon, we are said to *experiment.*

Observation.

370. For instance, supposing that we are possessed of instrumental means of measuring time and angles, we may trace out by successive observations the relative position of the sun and earth at different instants; and (the method is not susceptible of any accuracy, but is alluded to here only for the sake of illustration) from the variations in the apparent diameter of the former we may calculate the ratios of our distances from it at those instants. We have thus a set of observations involving time, angular position with reference to the sun, and ratios of distances from it: sufficient (if numerous enough) to enable us to discover the laws which connect the variations of these co-ordinates.

Similar methods may be imagined as applicable to the motion of any planet about the sun, of a satellite about its primary, or of one star about another in a binary group.

371. In general all the data of Astronomy are determined in this way, and the same may be said of such subjects as

Tides and Meteorology. Isothermal Lines, Lines of Equal Dip, Lines of Equal Intensity, Lines of Equal "Variation" (or "Declination" as it has still less happily been sometimes called), the Connexion of Solar Spots with Terrestrial Magnetism, and a host of other data and phénomena, to be explained under the proper heads in the course of the work, are thus deducible from *Observation* merely. In these cases the apparatus for the gigantic experiments is found ready arranged in Nature, and all that the philosopher has to do is to watch and measure their progress to its last details. Observation.

372. Even in the instance we have chosen above, that of the planetary motions, the observed effects are complex; because, unless possibly in the case of a double star, we have no instance of the *undisturbed* action of one heavenly body on another; but to a first approximation the motion of a planet about the sun is found to be the same as if no other bodies than these two existed; and the approximation is sufficient to indicate the probable law of mutual action, whose full confirmation is obtained when, *its* truth being assumed, the disturbing effects thus calculated are allowed for, and found to account completely for the observed deviations from the consequences of the first supposition. This may serve to give an idea of the mode of obtaining the laws of phenomena, which can only be observed in a complex form—and the method can always be directly applied when one cause is known to be pre-eminent.

373. Let us take cases of the other kind—in which the effects are so complex that we cannot deduce the causes from the observation of combinations arranged in Nature, but must endeavour to form for ourselves other combinations which may enable us to study the effects of each cause separately, or at least with only slight modification from the interference of other causes. Experiment.

374. A stone, when dropped, falls to the ground; a brick and a boulder, if dropped from the top of a cliff at the same moment, fall side by side, and reach the ground together. But a brick and a slate do not; and while the former falls in a nearly vertical direction, the latter describes a most complex

Experiment. path. A sheet of paper or a fragment of gold leaf presents even greater irregularities than the slate. But by a slight modification of the circumstances, we gain a considerable insight into the nature of the question. The paper and gold leaf, if rolled into balls, fall nearly in a vertical line. Here, then, there are evidently at least two causes at work, one which tends to make all bodies fall, and fall vertically; and another which depends on the form and substance of the body, and tends to retard its fall and alter its course from the vertical direction. How can we study the effects of the former on all bodies without sensible complication from the latter? The effects of Wind, etc., at once point out *what* the latter cause is, the air (whose existence we may indeed suppose to have been discovered by such effects); and to study the nature of the action of the former it is necessary to get rid of the complications arising from the presence of air. Hence the necessity for *Experiment*. By means of an apparatus to be afterwards described, we remove the greater part of the air from the interior of a vessel, and in *that* we try again our experiments on the fall of bodies; and now a general law, simple in the extreme, though most important in its consequences, is at once apparent—viz., that *all* bodies, of whatever size, shape, or material, if dropped side by side at the same instant, fall side by side in a space void of air. Before experiment had thus separated the phenomena, hasty philosophers had rushed to the conclusion that some bodies possess the quality of *heaviness*, others that of *lightness*, etc. Had this state of confusion remained, the law of gravitation, vigorous though its action be throughout the universe, could never have been recognised as a general principle by the human mind.

Mere observation of lightning and its effects could never have led to the discovery of their relation to the phenomena presented by rubbed amber. A modification of the course of nature, such as the collecting of atmospheric electricity in our laboratories, was necessary. Without experiment we could never even have learned the existence of terrestrial magnetism.

Rules for the conduct of experiments. 375. When a particular agent or cause is to be studied, experiments should be arranged in such a way as to lead if possible to results depending on it alone; or, if this cannot be done, they should be arranged so as to show differences produced by varying it.

Rules for the conduct of experiments.

376. Thus to determine the resistance of a wire against the conduction of electricity through it, we may measure the whole strength of current produced in it by electromotive force between its ends when the amount of this electromotive force *is given*, or can be ascertained. But when the wire is that of a submarine telegraph cable there is always an *unknown* and ever varying electromotive force between its ends, due to the earth (producing what is commonly called the "earth-current"), and to determine its resistance, the difference in the strength of the current produced by suddenly adding to or subtracting from the terrestrial electromotive force the electromotive force of a given voltaic battery, is to be *very quickly* measured; and this is to be done over and over again, to eliminate the effect of variation of the earth-current during the few seconds of time which must elapse before the electrostatic induction permits the current due to the battery to reach nearly enough its full strength to practically annul error on this score.

377. Endless patience and perseverance in designing and trying different methods for investigation are necessary for the advancement of science: and indeed, in discovery, he is the most likely to succeed who, not allowing himself to be disheartened by the non-success of one form of experiment, judiciously varies his methods, and thus interrogates in every conceivably useful manner the subject of his investigations.

378. A most important remark, due to Herschel, regards what are called *residual* phenomena. When, in an experiment, all known causes being allowed for, there remain certain unexplained effects (excessively slight it may be), these must be carefully investigated, and every conceivable variation of arrangement of apparatus, etc., tried; until, if possible, we manage so to isolate the residual phenomenon as to be able to detect its cause. It is here, perhaps, that in the present state of science we may most reasonably look for extensions of our knowledge; at all events we are warranted by the recent history of Natural Philosophy in so doing. Thus, to take only

Residual phenomena.

Residual phenomena.

a very few instances, and to say nothing of the discovery of electricity and magnetism by the ancients, the peculiar smell observed in a room in which an electrical machine is kept in action, was long ago observed, but called the "smell of electricity," and thus left unexplained. The sagacity of Schönbein led to the discovery that this is due to the formation of Ozone, a most extraordinary body, of great chemical activity; whose nature is still uncertain, though the attention of chemists has for years been directed to it.

379. Slight anomalies in the motion of Uranus led Adams and Le Verrier to the discovery of a new planet; and the fact that the oscillations of a magnetized needle about its position of equilibrium are "damped" by placing a plate of copper below it, led Arago to his beautiful experiment showing a resistance to relative motion between a magnet and a piece of copper; which was first supposed to be due to magnetism in motion, but which soon received its correct explanation from Faraday, and has since been immensely extended, and applied to most important purposes. In fact, from this accidental remark about the oscillation of a needle was evolved the grand discovery of the Induction of Electrical Currents by magnets or by other currents.

We need not enlarge upon this point, as in the following pages the proofs of the truth and usefulness of the principle will continually recur. Our object has been not so much to give applications as principles, and to show how to attack a new combination, with the view of separating and studying in detail the various causes which generally conspire to produce observed phenomena, even those which are apparently the simplest.

Unexpected agreement or discordance of results of different trials.

380. If on repetition several times, an experiment continually gives different results, it must either have been very carelessly performed, or there must be some disturbing cause not taken account of. And, on the other hand, in cases where no very great coincidence is likely on repeated trials, an unexpected degree of agreement between the results of various trials should be regarded with the utmost suspicion, as probably due to some unnoticed peculiarity of the apparatus employed. In

either of these cases, however, careful observation cannot fail to detect the cause of the discrepancies or of the unexpected agreement, and may possibly lead to discoveries in a totally unthought-of quarter. Instances of this kind may be given without limit; one or two must suffice.

Unexpected agreement or discordance of results of different trials.

381. Thus, with a *very* good achromatic telescope a star appears to have a sensible disc. But, as it is observed that the discs of all stars appear to be of equal angular diameter, we of course suspect some common error. Limiting the aperture of the object-glass *increases* the appearance in question, which, on full investigation, is found to have nothing to do with discs at all. It is, in fact, a diffraction phenomenon, and will be explained in our chapters on Light.

382. Again, in measuring the velocity of Sound by experiments conducted at night with cannon, the results at one station were never found to agree exactly with those at the other; sometimes, indeed, the differences were very considerable. But a little consideration led to the remark, that on those nights in which the discordance was greatest a strong wind was blowing nearly from one station to the other. Allowing for the obvious effect of this, or rather eliminating it altogether, the mean velocities on different evenings were found to agree very closely.

383. It may perhaps be advisable to say a few words here about the use of hypotheses, and especially those of very different gradations of value which are promulgated in the form of Mathematical Theories of different branches of Natural Philosophy.

Hypotheses.

384. Where, as in the case of the planetary motions and disturbances, the forces concerned are thoroughly known, the mathematical theory is absolutely true, and requires only analysis to work out its remotest details. It is thus, in general, far ahead of observation, and is competent to predict effects not yet even observed—as, for instance, Lunar Inequalities due to the action of Venus upon the Earth, etc. etc., to which no amount of observation, unaided by theory, could ever have enabled us to assign the true cause. It may also, in such subjects as Geometrical Optics, be carried to developments far beyond the reach

Hypotheses. of experiment; but in this science the assumed bases of the theory are only approximate; and it fails to explain in all their peculiarities even such comparatively simple phenomena as Halos and Rainbows—though it is perfectly successful for the practical purposes of the maker of microscopes and telescopes, and has enabled really scientific instrument-makers to carry the construction of optical apparatus to a degree of perfection which merely tentative processes never could have reached.

385. Another class of mathematical theories, based to some extent on experiment, is at present useful, and has even in certain cases pointed to new and important results, which experiment has subsequently verified. Such are the Dynamical Theory of Heat, the Undulatory Theory of Light, etc. etc. In the former, which is based upon the conclusion from experiment that *heat is a form of energy*, many formulæ are at present obscure and uninterpretable, because we do not know the mechanism of the motions or distortions of the particles of bodies. Results of the theory in which these are not involved, are of course experimentally verified. The same difficulties exist in the Theory of Light. But before this obscurity can be perfectly cleared up, we must know something of the ultimate, or *molecular*, constitution of the bodies, or groups of molecules, at present known to us only in the aggregate.

Deduction of most probable result from a number of observations.

386. A third class is well represented by the Mathematical Theories of Heat (Conduction), Electricity (Statical), and Magnetism (Permanent). Although we do not know *how* Heat is propagated in bodies, nor *what* Statical Electricity or Permanent Magnetism are—the laws of their fluxes and forces are as certainly known as that of Gravitation, and can therefore like it be developed to their consequences, by the application of Mathematical Analysis. The works of Fourier*, Green†, and Poisson‡ areremarkable instances of such development. Another good example is Ampère's Theory of Electro-dynamics.

* *Théorie analytique de la Chaleur.* Paris, 1822.

† *Essay on the Application of Mathematical Analysis to the Theories of Electricity and Magnetism.* Nottingham, 1828. Reprinted in Crelle's Journal.

‡ *Mémoires sur le Magnétisme.* Mém. de l'Acad. des Sciences, 1811.

387. When the most probable result is required from a number of observations of the same quantity which do not exactly agree, we must appeal to the mathematical theory of probabilities to guide us to a method of combining the results of experience, so as to eliminate from them, as far as possible, the inaccuracies of observation. Of course it is to be understood that we do not here class as *inaccuracies of observation* any errors which may affect alike every one of a series of observations, such as the inexact determination of a zero point, or of the essential units of time and space, the personal equation of the observer, etc. The process, whatever it may be, which is to be employed in the elimination of errors, is applicable even to these, but only when *several distinct series* of observations have been made, with a change of instrument, or of observer, or of both.

Deduction of most probable result from a number of observations.

388. We understand as inaccuracies of observation the whole class of errors which are as likely to lie in one direction as in another in successive trials, and which we may fairly presume would, on the average of an infinite number of repetitions, exactly balance each other in excess and defect. Moreover, we consider only errors of such a kind that their probability is the less the greater they are; so that such errors as an accidental reading of a wrong number of whole degrees on a divided circle (which, by the way, can in general be "probably" corrected by comparison with other observations) are not to be included.

389. Mathematically considered, the subject is by no means an easy one, and many high authorities have asserted that the reasoning employed by Laplace, Gauss, and others, is not well founded; although the results of their analysis have been generally accepted. As an excellent treatise on the subject has recently been published by Airy, it is not necessary for us to do more than to sketch in the most cursory manner a simple and apparently satisfactory method of arriving at what is called the *Method of Least Squares.*

390. Supposing the zero-point and the graduation of an instrument (micrometer, mural circle, thermometer, electrometer,

Deduction of most probable result from a number of observations.

galvanometer, etc.) to be *absolutely* accurate, successive readings of the value of a quantity (linear distance, altitude of a star, temperature, potential, strength of an electric current, etc.) may, and in general do, continually differ. What is most probably the true value of the observed quantity?

The most probable value, in all such cases, if the observations are all equally trustworthy, will evidently be the simple mean; or if they are not equally trustworthy, the mean found by attributing *weights* to the several observations in proportion to their presumed exactness. But if several such means have been taken, or several single observations, and if these several means or observations have been differently qualified for the determination of the sought quantity (some of them being likely to give a more exact value than others), we must assign *theoretically* the best practical method of combining them.

391. Inaccuracies of observation are, in general, as likely to be in excess as in defect. They are also (as before observed) more likely to be small than great; and (practically) large errors are not to be expected at all, as such would come under the class of *avoidable mistakes*. It follows that in any one of a series of observations of the same quantity the probability of an error of magnitude x must depend upon x^2, and must be expressed by some function whose value diminishes very rapidly as x increases. The probability that the error lies between x and $x + \delta x$, where δx is very small, must also be proportional to δx.

Hence we may assume the probability of an error of any magnitude included in the range of x to $x + \delta x$ to be

$$\phi(x^2)\,\delta x.$$

Now the error must be included between $+\infty$ and $-\infty$. Hence, as a first condition,

$$\int_{-\infty}^{+\infty} \phi(x^2)\,dx = 1 \qquad\qquad (1).$$

The consideration of a very simple case gives us the means of determining the form of the function ϕ involved in the preceding expression*.

* Compare Boole, *Trans. R.S.E.*, 1857. See also Tait, *Trans. R.S.E.*, 1864.

Suppose a stone to be let fall with the object of hitting a mark on the ground. Let two horizontal lines be drawn through the mark at right angles to one another, and take them as axes of x and y respectively. The chance of the stone falling at a distance between x and $x + \delta x$ from the axis of y is $\phi(x^2)\,\delta x$.

Deduction of most probable result from a number of observations.

Of its falling between y and $y + \delta y$ from the axis of x the chance is $\phi(y^2)\,\delta y$.

The chance of its falling on the elementary area $\delta x \delta y$, whose co-ordinates are x, y, is therefore (since these are independent events, and it is to be observed that this is the assumption on which the whole investigation depends)

$$\phi(x^2)\,\phi(y^2)\,\delta x \delta y, \text{ or } a\phi(x^2)\,\phi(y^2),$$

if a denote the indefinitely small area about the point xy.

Had we taken any other set of rectangular axes with the same origin, we should have found for the same probability the expression $a\phi(x'^2)\,\phi(y'^2)$,

x', y' being the new co-ordinates of a. Hence we must have

$$\phi(x^2)\,\phi(y^2) = \phi(x'^2)\,\phi(y'^2), \text{ if } x^2 + y^2 = x'^2 + y'^2.$$

From this functional equation we have at once

$$\phi(x^2) = A\epsilon^{mx^2},$$

where A and m are constants. We see at once that m must be negative (as the chance of a large error is very small), and we may write for it $-\frac{1}{h^2}$, so that h will indicate the degree of delicacy or coarseness of the system of measurement employed.

Substituting in (1) we have

$$A\int_{-\infty}^{+\infty} \epsilon^{-\frac{x^2}{h^2}}\,dx = 1,$$

whence $A = \frac{1}{h\sqrt{\pi}}$, and the law of error is

$$\frac{1}{\sqrt{\pi}}\,\epsilon^{-\frac{x^2}{h^2}}\,\frac{\delta x}{h}.$$

Law of error.

The law of error, as regards *distance from the mark, without reference to the direction* of error, is evidently

$$\iint \phi(x^2)\,\phi(y^2)\,dxdy,$$

taken through the space between concentric circles whose radii are r and $r + \delta r$, and is therefore

$$\frac{2}{h^2}\,\epsilon^{-\frac{r^2}{h^2}}\,r\delta r,$$

Law of error.

which is of the same form as the law of error to the right or left of a line, with the additional factor r for the greater space for error at greater distances from the centre. As a verification, we see at once that

$$\frac{2}{h^2}\int_0^\infty \epsilon^{-\frac{r^2}{h^2}} r dr = 1$$

as was to be expected.

Probable error.

392. The *Probable Error* of an observation is a numerical quantity such that the error of the observation is as likely to exceed as to fall short of it in magnitude.

If we assume the law of error just found, and call P the probable error in one trial,

$$\int_0^P \epsilon^{-\frac{x^2}{h^2}} dx = \int_P^\infty \epsilon^{-\frac{x^2}{h^2}} dx.$$

The solution of this equation by trial and error leads to the approximate result

$$P = 0{\cdot}477\, h.$$

Probable error of a sum, difference, or multiple.

393. The probable error of any given multiple of the value of an observed quantity is evidently the same multiple of the probable error of the quantity itself.

The probable error of the sum or difference of two quantities, affected by *independent* errors, is the square root of the sum of the squares of their separate probable errors.

To prove this, let us investigate the *law* of error of

$$X \pm Y = Z$$

where the laws of error of X and Y are

$$\frac{1}{\sqrt{\pi}} \epsilon^{-\frac{x^2}{a^2}} \frac{dx}{a}, \text{ and } \frac{1}{\sqrt{\pi}} \epsilon^{-\frac{y^2}{b^2}} \frac{dy}{b},$$

respectively. The chance of an error in Z, of a magnitude included between the limits z, $z + \delta z$, is evidently

$$\frac{1}{\pi ab}\int_{-\infty}^{+\infty} \epsilon^{-\frac{x^2}{a^2}} dx \int_{z-x}^{z+\delta z - x} \epsilon^{-\frac{y^2}{b^2}} dy.$$

For, whatever value is assigned to x, the value of y is given by the limits $z - x$ and $z + \delta z - x$ [or $z + x$, $z + \delta z + x$; but the chances of $\pm x$ are the same, and both are included in the limits ($\pm \infty$) of integration with respect to x].

The value of the above integral becomes, by effecting the integration with respect to y, Probable error of a sum, difference, or multiple.

$$\frac{\delta z}{\pi ab}\int_{-\infty}^{+\infty}\epsilon^{-\frac{x^2}{a^2}}\epsilon^{-\frac{(z-x)^2}{b^2}}dx,$$

and this is easily reduced to

$$\frac{1}{\sqrt{\pi}}\epsilon^{-\frac{z^2}{a^2+b^2}}\frac{\delta z}{\sqrt{a^2+b^2}}.$$

Thus the probable error is $0{\cdot}477\sqrt{a^2+b^2}$, whence the proposition. And the same theorem is evidently true for *any* number of quantities.

394. As above remarked, the principal use of this theory is in the deduction, from a large series of observations, of the values of the quantities sought in such a form as to be liable to the smallest probable error. As an instance—by the principles of physical astronomy, the place of a planet is calculated from assumed values of the elements of its orbit, and tabulated in the *Nautical Almanac*. The *observed* places do not exactly agree with the predicted places, for two reasons—first, the data for calculation are not exact (and in fact the main object of the observation is to correct their assumed values); second, each observation is in error to some unknown amount. Now the difference between the observed, and the calculated, places depends on the errors of assumed elements and of observation. The methods are applied to eliminate as far as possible the second of these, and the resulting equations give the required corrections of the elements. Practical application.

Thus if θ be the calculated R.A. of a planet: $\delta\alpha$, δe, $\delta\varpi$, etc., the corrections required for the assumed elements—the true R.A. is $\theta + A\delta a + E\delta e + \Pi\delta\varpi + \text{etc.}$, Method of least square.
where A, E, Π, etc., are approximately known. Suppose the observed R.A. to be Θ, then

$$\theta + A\delta a + E\delta e + \Pi\delta\varpi + \ldots = \Theta$$

or

$$A\delta a + E\delta e + \Pi\delta\varpi + \ldots = \Theta - \theta,$$

a known quantity, subject to error of observation. Every observation made gives us an equation of the same *form* as this, and in general the number of observations greatly exceeds that of the quantities δa, δe, $\delta\varpi$, etc., to be found. But it will be sufficient to consider the simple case where only *one* quantity is to be found.

Method of least squares.

Suppose a number of observations, of the same quantity x, lead to the following equations:—

$$x = B_1, \quad x = B_2, \quad \text{etc.},$$

and let the probable errors be E_1, E_2, ... Multiply the terms of each equation by numbers inversely proportional to E_1, E_2, ... This will make the probable errors of the second members of all the equations the same, e suppose. The equations have now the general form

$$ax = b,$$

and it is required to find a system of linear factors, by which these equations, being multiplied in order and added, shall lead to a final equation giving the value of x with the probable error a minimum. Let them be f_1, f_2, etc. Then the final equation is

$$(\Sigma af)\, x = \Sigma\,(bf)$$

and therefore

$$P^2\,(\Sigma af)^2 = e^2\Sigma\,(f^2)$$

by the theorems of § 393, if P denote the probable error of x.

Hence $\dfrac{\Sigma\,(f^2)}{(\Sigma af)^2}$ is a minimum, and its differential coefficients with respect to each separate factor f must vanish.

This gives a series of equations, whose general form is

$$f\Sigma\,(af) - a\Sigma\,(f^2) = 0,$$

which give evidently $f_1 = a_1$, $f_2 = a_2$, etc.

Hence the following rule, which may easily be seen to hold for any number of linear equations containing a smaller number of unknown quantities,

Make the probable error of the second member the same in each equation, by the employment of a proper factor; multiply each equation by the coefficient of x in it and add all, for one of the final equations; and so, with reference to y, z, etc., for the others. The probable errors of the values of x, y, etc., found from these final equations will be less than those of the values derived from any other *linear* method of combining the equations.

This process has been called the method of *Least Squares*, because the values of the unknown quantities found by it are such as to render the sum of the squares of the errors of the original equations a minimum.

That is, in the simple case taken above,

$$\Sigma\,(ax - b)^2 = \text{minimum}.$$

For it is evident that this gives, on differentiating with respect to x, Method of least squares.

$$\Sigma a (ax - b) = 0,$$

which is the law above laid down for the formation of the single equation.

395. When a series of observations of the same quantity has been made at different times, or under different circumstances, the law connecting the value of the quantity with the time, or some other variable, may be derived from the results in several ways—all more or less approximate. Two of these methods, however, are so much more extensively used than the others, that we shall devote a page or two here to a preliminary notice of them, leaving detailed instances of their application till we come to Heat, Electricity, etc. They consist in (1) a *Curve*, giving a graphic representation of the relation between the ordinate and abscissa, and (2) an *Empirical Formula* connecting the variables. Methods of representing experimental results.

396. Thus if the abscissæ represent intervals of time, and the ordinates the corresponding height of the barometer, we may construct curves which show at a glance the dependence of barometric pressure upon the time of day; and so on. Such curves may be accurately drawn by photographic processes on a sheet of sensitive paper placed behind the mercurial column, and made to move past it with a uniform horizontal velocity by clockwork. A similar process is applied to the Temperature and Electrification of the atmosphere, and to the components of terrestrial magnetism. Curves.

397. When the observations are not, as in the last section, continuous, they give us only a series of points in the curve, from which, however, we may in general approximate very closely to the result of continuous observation by drawing, *liberâ manu*, a curve passing through these points. This process, however, must be employed with great caution; because, unless the observations are sufficiently close to each other, most important fluctuations in the curve may escape notice. It is applicable, with abundant accuracy, to all cases where the quantity observed changes very slowly. Thus, for instance, weekly observations of the temperature at depths of from 6 to

Curves. 24 feet underground were found by Forbes sufficient for a very accurate approximation to the law of the phenomenon.

Interpolation and empirical formulæ. 398. As an instance of the processes employed for obtaining an empirical formula, we may mention methods of *Interpolation*, to which the problem can always be reduced. Thus from sextant observations, at known intervals, of the altitude of the sun, it is a common problem of astronomy to determine at what instant the altitude is greatest, and what is that greatest altitude. The first enables us to find the true solar time at the place; and the second, by the help of the *Nautical Almanac*, gives the latitude. The differential calculus, and the calculus of finite differences, give us formulæ for any required data; and Lagrange has shown how to obtain a very useful one by elementary algebra.

By Taylor's Theorem, if $y = f(x)$, we have

$$y = f(x_0 + \overline{x - x_0}) = f(x_0) + (x - x_0) f'(x_0) + \frac{(x - x_0)^2}{1 \cdot 2} f''(x_0) + \ldots$$
$$+ \frac{(x - x_0)^n}{1 \cdot 2 \ldots n} f^{(n)}[x_0 + \theta(x - x_0)] \ldots\ldots\ldots (1),$$

where θ is a proper fraction, and x_0 is *any* quantity whatever. This formula is useful only when the successive derived values of $f(x_0)$ diminish very rapidly.

In finite differences we have

$$f(x + h) = D^h f(x) = (1 + \Delta)^h f(x)$$
$$= f(x) + h\Delta f(x) + \frac{h(h-1)}{1 \cdot 2} \Delta^2 f(x) + \ldots\ldots\ldots\ldots\ldots (2);$$

a very useful formula when the higher differences are small.

(1) suggests the proper form for the required expression, but it is only in rare cases that $f'(x_0)$, $f''(x_0)$, etc., are derivable directly from observation. But (2) is useful, inasmuch as the successive differences, $\Delta f(x)$, $\Delta^2 f(x)$, etc., are easily calculated from the tabulated results of observation, provided these have been taken for equal successive increments of x.

If for values $x_1, x_2, \ldots x_n$ a function takes the values $y_1, y_2, y_3, \ldots y_n$, Lagrange gives for it the obvious expression

$$\left[\frac{y_1}{x - x_1} \frac{1}{(x_1 - x_2)(x_1 - x_3)\ldots(x_1 - x_n)} + \frac{y_2}{x - x_2} \frac{1}{(x_2 - x_1)(x_2 - x_3)\ldots(x_2 - x_n)} + \ldots\right](x - x_1)(x - x_2)\ldots(x - x_n).$$

Here it is of course assumed that the function required is a rational and integral one in x of the $n-1^{th}$ degree; and, in general, a similar limitation is in practice applied to the other formulæ above; for in order to find the complete expression for $f(x)$ in either, it is necessary to determine the values of $f'(x_0)$, $f''(x_0)$, ... in the first, or of $\Delta f(x)$, $\Delta^2 f(x)$, ... in the second. If n of the coefficients be required, so as to give the n chief terms of the general value of $f(x)$, we must have n observed simultaneous values of x and $f(x)$, and the expressions become determinate and of the $n-1^{th}$ degree in $x-x_0$ and h respectively.

Interpolation and empirical formulæ.

In practice it is usually sufficient to employ at most three terms of either of the first two series. Thus to express the length l of a rod of metal as depending on its temperature t, we may assume from (1)

$$l = l_0 + A(t - t_0) + B(t - t_0)^2,$$

l_0 being the measured length at any temperature t_0.

398′. These formulæ are practically useful for calculating the probable values of any observed element, for values of the independent variable lying within the range for which observation has given values of the element. But except for values of the independent variable either actually within this range, or not far beyond it in either direction, these formulæ express functions which, in general, will differ more and more widely from the truth the further their application is pushed beyond the range of observation.

Periodic functions.

In a large class of investigations the observed element is in its nature a periodic function of the independent variable. The harmonic analysis (§ 77) is suitable for all such. When the values of the independent variable for which the element has been observed are not equidifferent the coefficients, determined according to the method of least squares, are found by a process which is necessarily very laborious; but when they are equidifferent, and especially when the difference is a submultiple of the period, the equation derived from the method of least squares becomes greatly simplified. Thus, if θ denote an angle increasing in proportion to t, the time, through four right angles in the period, T, of the phenomenon; so that

$$\theta = \frac{2\pi t}{T};$$

Periodic functions.

let
$$f(\theta) = A_0 + A_1 \cos\theta + A_2 \cos 2\theta + \dots$$
$$+ B_1 \sin\theta + B_2 \sin 2\theta + \dots$$
where A_0, A_1, A_2, ... B_1, B_2, ... are unknown coefficients, to be determined so that $f(\theta)$ may express the most probable value of the element, not merely at times between observations, but through all time as long as the phenomenon is strictly periodic. By taking as many of these coefficients as there are of distinct data by observation, the formula is made to agree precisely with these data. But in most applications of the method, the periodically recurring part of the phenomenon is expressible by a small number of terms of the harmonic series, and the higher terms, calculated from a great number of data, express either irregularities of the phenomenon not likely to recur, or errors of observation. Thus a comparatively small number of terms may give values of the element even for the very times of observation, more probable than the values actually recorded as having been observed, if the observations are numerous but not minutely accurate.

The student may exercise himself in writing out the equations to determine five, or seven, or more of the coefficients according to the method of least squares; and reducing them by proper formulæ of analytical trigonometry to their simplest and most easily calculated forms where the values of θ for which $f(\theta)$ is given are equidifferent. He will thus see that when the difference is $\frac{2\pi}{i}$, i being any integer, and when the number of the data is i or any multiple of it, the equations contain each of them only one of the unknown quantities: so that the method of least squares affords the most probable values of the coefficients, by the easiest and most direct elimination.

CHAPTER IV.

MEASURES AND INSTRUMENTS.

399. HAVING seen in the preceding chapter that for the investigation of the laws of nature we must carefully watch experiments, either those gigantic ones which the universe furnishes, or others devised and executed by man for specia objects—and having seen that in all such observations accurate measurements of Time, Space, Force, etc., are absolutely necessary, we may now appropriately describe a few of the more useful of the instruments employed for these purposes, and the various standards or units which are employed in them. Necessity of accurate measurements.

400. Before going into detail we may give a rapid *résumé* of the principal Standards and Instruments to be described in this chapter. As most, if not all, of them depend on physical principles to be detailed in the course of this work—we shall assume in anticipation the establishment of such principles, giving references to the future division or chapter in which the experimental demonstrations are more particularly explained. This course will entail a slight, but unavoidable, confusion—slight, because Clocks, Balances, Screws, etc., are familiar even to those who know nothing of Natural Philosophy; unavoidable, because it is in the very nature of our subject that no one part can grow alone, each requiring for its full development the utmost resources of all the others. But if one of our departments thus borrows from others, it is satisfactory to find that it more than repays by the power which its improvement affords them.

Classes of instruments.

401. We may divide our more important and fundamental instruments into four classes—

Those for measuring Time;
 „ „ Space, linear or angular;
 „ „ Force;
 „ „ Mass.

Other instruments, adapted for special purposes such as the measurement of Temperature, Light, Electric Currents, etc., will come more naturally under the head of the particular physical energies to whose measurement they are applicable. Descriptions of self-recording instruments such as tide-gauges, and barometers, thermometers, electrometers, recording photographically or otherwise the continuously varying pressure, temperature, moisture, electric potential of the atmosphere, and magnetometers recording photographically the continuously varying direction and magnitude of the terrestrial magnetic force, must likewise be kept for their proper places in our work.

Calculating Machines.

Calculating Machines have also important uses in assisting physical research in a great variety of ways. They belong to two classes.:—

I. Purely Arithmetical, dealing with integral numbers of units. All of this class are evolved from the primitive use of calculuses or little stones for counters (from which we derived the very names *calculation* and "The Calculus"), through such mechanism as that of the Chinese Abacus, still serving its original purpose well in infant schools, up to the Arithmometer of Thomas of Colmar and the grand but partially realized conceptions of calculating machines by Babbage.

II. Continuous Calculating Machines. As these are not only useful as auxiliaries for physical research but also involve dynamical and kinematical principles belonging properly to our subject, some of them have been described in the Appendix to this Chapter, from which dynamical illustrations will be taken in our chapters on Statics and Kinetics.

402. We shall consider in order the more prominent fundamental instruments of the four classes, and some of their most important applications :— Classes of Instruments.

Clock, Chronometer, Chronoscope, Applications to Observation and to self-registering Instruments.

Vernier and Screw-Micrometer, Cathetometer, Spherometer, Dividing Engine, Theodolite, Sextant or Circle.

Common Balance, Bifilar Balance, Torsion Balance, Pendulum, Ergometer.

Among Standards we may mention—

1. *Time.*—Day, Hour, Minute, Second, sidereal and solar.
2. *Space.*—Yard and Mètre: Radian, Degree, Minute, Second.
3. *Force.*—Weight of a Pound or Kilogramme, etc., in any particular locality (gravitation unit); poundal, or dyne (kinetic unit).
4. *Mass.* Pound, Kilogramme, etc.

403. Although without instruments it is impossible to procure or apply any standard, yet, as without the standards no instrument could give us *absolute* measure, we may consider the standards first—referring to the instruments as if we already knew their principles and applications.

404. First we may notice the standards or units of angular measure : Angular measure.

Radian, or angle whose arc is equal to radius;

Degree, or ninetieth part of a right angle, and its successive subdivisions into sixtieths called *Minutes*, *Seconds*, *Thirds*, etc. The division of the right angle into 90 degrees is convenient because it makes the half-angle of an equilateral triangle ($\sin^{-1} \frac{1}{2}$) an integral number (30) of degrees. It has long been universally adopted by all Europe. The decimal division of the right angle, decreed by the French Republic when it successfully introduced other more sweeping changes, utterly and deservedly failed.

The division of the degree into 60 minutes and of the minute into 60 seconds is not convenient; and tables of the

Angular measure.

circular functions for degrees and hundredths of the degree are much to be desired. Meantime, when reckoning to tenths of a degree suffices for the accuracy desired, in any case the ordinary tables suffice, as $6'$ is $\frac{1}{10}$ of a degree.

The decimal system is exclusively followed in reckoning by radians. The value of two right angles in this reckoning is 3·14159..., or π. Thus π radians is equal to 180°. Hence $180^\circ \div \pi$ is $57^\circ{\cdot}29578\ldots$, or $57^\circ\, 17'\, 44''{\cdot}8$ is equal to one radian. In mathematical analysis, angles are uniformly reckoned in terms of the radian.

Measure of time.

405. The practical standard of time is the *Sidereal Day*, being the period, nearly constant*, of the earth's rotation about its axis (§ 247). From it is easily derived the *Mean Solar Day*, or the mean interval which elapses between successive passages of the sun across the meridian of any place. This is not so nearly as the Sidereal Day, an absolute or invariable unit:

* In our first edition it was stated in this section that Laplace had calculated from ancient observations of eclipses that the period of the earth's rotation about its axis had not altered by $\frac{1}{10000000}$ of itself since 720 B.C. In § 830 it was pointed out that this conclusion is overthrown by farther information from Physical Astronomy acquired in the interval between the printing of the two sections, in virtue of a correction which Adams had made as early as 1863 upon Laplace's dynamical investigation of an acceleration of the moon's mean motion, produced by the sun's attraction, showing that only about half of the observed acceleration of the moon's mean motion relatively to the angular velocity of the earth's rotation was accounted for by this cause. [Quoting from the first edition, § 830] "In 1859 Adams communicated to Delaunay his final result:—that at "the end of a century the moon is $5''{\cdot}7$ before the position she would have, "relatively to a meridian of the earth, according to the angular velocities of the "two motions, at the beginning of the century, and the acceleration of the "moon's motion truly calculated from the various disturbing causes then recog-"nized. Delaunay soon after verified this result: and about the beginning of "1866 suggested that the true explanation may be a retardation of the earth's "rotation by tidal friction. Using this hypothesis, and allowing for the conse-"quent retardation of the moon's mean motion by tidal reaction (§ 276), Adams, "in an estimate which he has communicated to us, founded on the rough as-"sumption that the parts of the earth's retardation due to solar and lunar tides "are as the squares of the respective tide-generating forces, finds 22^s as the "error by which the earth would in a century get behind a perfect clock rated "at the beginning of the century. If the retardation of rate giving this integral "effect were uniform (§ 35, *b*), the earth, as a timekeeper, would be going slower "by ·22 of a second per year in the middle, or ·44 of a second per year at the "end, than at the beginning of a century."

secular changes in the period of the earth's revolution about the sun affect it, though very slightly. It is divided into 24 hours, and the hour, like the degree, is subdivided into successive sixtieths, called minutes and seconds. The usual subdivision of seconds is decimal.

Measure of time.

It is well to observe that seconds and minutes of time are distinguished from those of angular measure by notation Thus we have for time $13^h\ 43^m\ 27^s{\cdot}58$, but for angular measure $13^\circ\ 43'\ 27''{\cdot}58$.

When long periods of time are to be measured, the mean solar year, consisting of 366·242203 sidereal days, or 365·242242 mean solar days, or the century consisting of 100 such years, may be conveniently employed as the unit.

Necessity for a perennial standard. A spring suggested.

406. The ultimate standard of accurate chronometry must (if the human race live on the earth for a few million years) be founded on the physical properties of some body of more constant character than the earth: for instance, a carefully arranged metallic spring, hermetically sealed in an exhausted glass vessel. The time of vibration of such a spring would be necessarily more constant from day to day than that of the balance-spring of the best possible chronometer, disturbed as this is by the train of mechanism with which it is connected: and it would almost certainly be more constant from age to age than the time of rotation of the earth (cooling and shrinking, as it certainly is, to an extent that must be very considerable in fifty million years).

Measure of length, founded on artificial metallic standards.

407. The British standard of length is the *Imperial Yard*, defined as the distance between two marks on a certain metallic bar, preserved in the Tower of London, when the whole has a temperature of 60° Fahrenheit. It was not directly derived from any fixed quantity in nature, although some important relations with such have been measured with great accuracy. It has been carefully compared with the length of a seconds pendulum vibrating at a certain station in the neighbourhood of London, so that if it should again be destroyed, as it was at the burning of the Houses of Parliament in 1834, and should all exact copies of it, of which several are preserved in various

Earth's dimensions not constant, places, be also lost, it can be restored by pendulum observations. A less accurate, but still (except in the event of earthquake disturbance) a very good, means of reproducing it exists in the measured base-lines of the Ordnance Survey, and the thence calculated distances between definite stations in the British Islands, which have been ascertained in terms of it with a degree of accuracy sometimes within an inch per mile, that is to say, within about $\frac{1}{60000}$.

408. In scientific investigations, we endeavour as much as possible to keep to one unit at a time, and the foot, which is defined to be one-third part of the yard, is, for British measurement, generally the most convenient. Unfortunately the inch, or one-twelfth of a foot, must sometimes be used. The statute mile, or 1760 yards, is most unhappily often used when great lengths are considered. The British measurements of area and volume are infinitely inconvenient and wasteful of brain-energy, and of plodding labour. Their contrast with the simple, uniform, metrical system of France, Germany, and Italy, is but little creditable to English intelligence.

409. In the French metrical system the decimal division is exclusively employed. The standard, (unhappily) called the *Mètre*, was defined originally as the ten-millionth part of the length of the quadrant of the earth's meridian from the pole to the equator; but it is now defined practically by the accurate standard metres laid up in various national repositories in Europe. It is somewhat longer than the yard, as the following Table shows:

nor easily measured with great accuracy.

Measure of length.

Inch = 25·39977 millimètres.	Centimètre = ·3937043 inch.
Foot = 3·047972 decimètres.	Mètre = 3·280869 feet.
British statute mile	Kilomètre = ·6213767 British
= 1609·329 mètres.	statute mile.

Measure of surface.

410. The unit of superficial measure is in Britain the square yard, in France the mètre carré. Of course we may use square inches, feet, or miles, as also square millimètres, kilomètres, etc., or the *Hectare* = 10,000 square mètres.

Measure of surface.

Square inch = 6·451483 square centimètres.
" foot = 9·290135 " decimètres.
" yard = 83·61121 " decimètres.
Acre = ·4046792 of a hectare.
Square British statute mile = 258·9946 hectares.
Hectare = 2·471093 acres.

411. Similar remarks apply to the cubic measure in the two countries, and we have the following Table:— Measure of volume.

Cubic inch = 16·38661 cubic centimètres.
" foot = 28·31606 " decimètres or *Litres*.
Gallon = 4·543808 litres.
" = 277·274 cubic inches, by Act of Parliament now repealed.
Litre = ·035315 cubic feet.

412. The British unit of mass is the Pound (defined by standards only); the French is the *Kilogramme*, defined originally as a litre of water at its temperature of maximum density; but now practically defined by existing standards. Measure of mass.

Grain = 64·79896 milligrammes.	Gramme = 15·43235 grains.
Pound = 453·5927 grammes.	Kilogramme = 2·20462125 lbs.

Professor W. H. Miller finds (*Phil. Trans.* 1857) that the "*kilogramme des Archives*" is equal in mass to 15432·34874 grains; and the "*kilogramme type laiton,*" deposited in the Ministère de l'Intérieure in Paris, as standard for French commerce, is 15432·344 grains.

413. The measurement of force, whether in terms of the weight of a stated mass in a stated locality, or in terms of the *absolute* or *kinetic* unit, has been explained in Chap. II. (See §§ 220—226). From the measures of force and length, we derive at once the measure of work or mechanical effect. That practically employed by engineers is founded on the gravitation measure of force. Neglecting the difference of gravity at London and Paris, we see from the above tables that the following relations exist between the London and the Parisian reckoning of work:— Measure of force.

Foot-pound = 0·13825 kilogramme-mètre.
Kilogramme-mètre = 7·2331 foot-pounds.

Clock. **414.** A *Clock* is primarily an instrument which, by means of a train of wheels, records the number of vibrations executed by a pendulum; a *Chronometer* or *Watch* performs the same duty for the oscillations of a flat spiral spring—just as the train of wheel-work in a gas-metre counts the number of revolutions of the main shaft caused by the passage of the gas through the machine. As, however, it is impossible to avoid friction, resistance of air, etc., a pendulum or spring, left to itself, would not long continue its oscillations, and, while its motion continued, would per.orm each oscillation in less and less time as the arc of vibration diminished: a continuous supply of energy is furnished by the descent of a weight, or the uncoiling of a powerful spring. This is so applied, through the train of wheels, to the pendulum or balance-wheel by means of a mechanical contrivance called an *Escapement*, that the oscillations are maintained of nearly uniform extent, and therefore of nearly uniform duration. The construction of escapements, as well as of trains of clock-wheels, is a matter of *Mechanics*, with the details of which we are not concerned, although it may easily be made the subject of mathematical investigation. The means of avoiding errors introduced by changes of temperature, which have been carried out in *Compensation* pendulums and balances, will be more properly described in our chapters on Heat. It is to be observed that there is little inconvenience if a clock lose or gain *regularly;* that can be easily and accurately allowed for: irregular rate is fatal.

Electrically controlled clocks. **415.** By means of a recent application of electricity to be afterwards described, one good clock, carefully regulated from time to time to agree with astronomical observations, may be made (without injury to its own performance) to control any number of other less-perfectly constructed clocks, so as to compel their pendulums to vibrate, beat for beat, with its own.

Chronoscope. **416.** In astronomical observations, time is estimated to tenths of a second by a practised observer, who, while watching the phenomena, counts the beats of the clock. But for the *very* accurate measurement of short intervals, many instruments have been devised. Thus if a small orifice be opened in a large and deep vessel full of mercury, and if we know by trial the weight of metal that escapes say in five minutes, a simple proportion gives the interval which elapses during the escape of any given weight. It is easy to contrive an adjustment by which a vessel may be placed under, and withdrawn from, the issuing stream at the time of occurrence of any two successive phenomena. Chronoscope.

417. Other contrivances, called Stop-watches, Chronoscopes, etc., which can be read off at rest, started on the occurrence of any phenomenon, and stopped at the occurrence of a second, then again read off; or which allow of the making (by pressing a stud) a slight mark, on a dial revolving at a given rate, at the instant of the occurrence of each phenomenon to be noted, are common enough. But, of late, these have almost entirely given place to the Electric Chronoscope, an instrument which will be fully described later, when we shall have occasion to refer to experiments in which it has been usefully employed.

418. We now come to the measurement of space, and of angles, and for these purposes the most important instruments are the *Vernier* and the *Screw*.

419. Elementary geometry, indeed, gives us the means of dividing any straight line into any assignable number of equal parts; but in practice this is by no means an accurate or reliable method. It was formerly used in the so-called Diagonal Scale, of which the construction is evident from the diagram. The reading is effected by a sliding-piece whose edge is perpendicular to the length of the scale. Suppose that it is PQ whose position on the scale is required. This can evidently cut only *one* of the transverse lines. *Its* number gives the number of tenths of an inch [4 in the figure], and the horizontal line next above the point of intersection gives evidently the number of hundredths [in the present case 4]. Hence the reading is 7·44. As an idea of the comparative uselessness of this Diagonal scale.

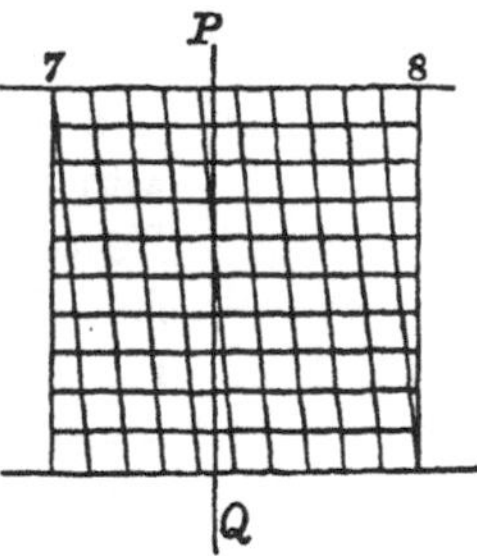

Diagonal scale. method, we may mention that a quadrant of 3 feet radius which belonged to Napier of Merchiston, and is divided on the limb by this method, reads to minutes of a degree; no higher accuracy than is now attainable by the pocket sextants made by Troughton and Simms, the radius of whose arc is virtually little more than an inch. The latter instrument is read by the help of a Vernier.

Vernier. **420.** The Vernier is commonly employed for such instruments as the Barometer, Sextant, and Cathetometer, while the Screw is micrometrically applied to the more delicate instruments, such as Astronomical Circles, and Micrometers, and the Spherometer.

421. The vernier consists of a slip of metal which slides along a divided scale, the edges of the two being coincident. Hence, when it is applied to a divided circle, its edge is circular, and it moves about an axis passing through the centre of the divided limb.

In the sketch let 0, 1, 2,...10 be the divisions on the vernier, 0, 1, 2, etc., any set of consecutive divisions on the limb or scale along whose edge it slides. If, when 0 and 0 coincide, 10 and 11 coincide also, then 10 divisions of the vernier are equal in length to 11 on the limb; and therefore each division on the vernier is $\frac{11}{10}$ths or $1\frac{1}{10}$ of a division on the limb. If, then, the vernier be moved till 1 coincides with 1, 0 will be $\frac{1}{10}$th of a division of the limb beyond 0; if 2 coincide with 2, 0 will be $\frac{2}{10}$ths beyond 0; and so on. Hence to read the vernier in any position, note first the division next to 0, and behind it on the limb. This is the *integral* number of divisions to be read. For the fractional part, see which division of the vernier is in a line with one on the limb; if it be the 4th (as in the figure), that indicates an addition to the reading of $\frac{4}{10}$ths of a division of the limb; and so on. Thus, if the figure represent a barometer scale divided into inches and tenths, the reading is $30\overset{\text{in}}{\cdot}34$, the zero line of the vernier being adjusted to the level of the mercury.

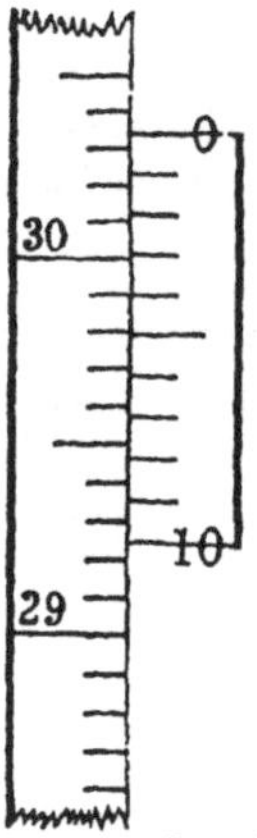

422. If the limb of a sextant be divided, as it usually is, to third parts of a degree, and the vernier be formed by dividing 21 of these into 20 equal parts, the instrument can be read to twentieths of divisions on the limb, that is, to minutes of arc. Vernier.

If no line on the vernier coincide with one on the limb, then since the divisions of the former are the longer there will be one of the latter included between the two lines of the vernier, and it is usual in practice to take the mean of the readings which would be given by a coincidence of either pair of bounding lines.

423. In the above sketch and description, the numbers on the scale and vernier have been supposed to run *opposite* ways. This is generally the case with British instruments. In some foreign ones the divisions run in the same direction on vernier and limb, and in that case it is easy to see that to read to tenths of a scale division we must have ten divisions of the vernier equal to *nine* of the scale.

In general, to read to the nth part of a scale division, n divisions of the vernier must equal $n+1$ or $n-1$ divisions on the limb, according as these run in opposite or similar directions.

424. The principle of the *Screw* has been already noticed (§ 102). It may be used in either of two ways, *i.e.*, the nut may be fixed, and the screw advance through it, or the screw may be prevented from moving longitudinally by a fixed collar, in which case the nut, if prevented by fixed guides from rotating, will move in the direction of the common axis. The advance in either case is evidently proportional to the angle through which the screw has turned about its axis, and this may be measured by means of a divided head fixed perpendicularly to the screw at one end, the divisions being read off by a pointer or vernier attached to the frame of the instrument. The nut carries with it either a tracing point (as in the dividing engine) or a wire, thread, or half the object-glass of a telescope (as in micrometers), the thread or wire, or the play of the tracing point, being at right angles to the axis of the screw. Screw.

425. Suppose it be required to divide a line into any number of equal parts. The line is placed parallel to the axis

Screw. of the screw with one end exactly under the tracing point, or under the fixed wire of a microscope carried by the nut, and the screw-head is read off. By turning the head, the tracing point or microscope wire is brought to the other extremity of the line; and the number of turns and fractions of a turn required for the whole line is thus ascertained. Dividing this by the number of equal parts required, we find at once the number of turns and fractional parts corresponding to *one* of the required divisions, and by giving that amount of rotation to the screw over and over again, drawing a line after each rotation, the required division is effected.

Screw-Micrometer. **426.** In the Micrometer, the movable wire carried by the nut is parallel to a fixed wire. By bringing them into optical contact the zero reading of the head is known; hence when another reading has been obtained, we have by subtraction the number of turns corresponding to the length of the object to be measured. The *absolute* value of a turn of the screw is determined by calculation from the number of threads in an inch, or by actually applying the micrometer to an object of known dimensions.

Spherometer. **427.** For the measurement of the thickness of a plate, or the curvature of a lens, the *Spherometer* is used. It consists of a screw nut rigidly fixed in the middle of a very rigid three-legged table, with its axis perpendicular to the plane of the three feet (or finely rounded ends of the legs), and an accurately cut screw working in this nut. The lower extremity of the screw is also finely rounded. The number of turns, whole or fractional, of the screw, is read off by a divided head and a pointer fixed to the stem. Suppose it be required to measure the thickness of a plate of glass. The three feet of the instrument are placed upon a nearly enough flat surface of a hard body, and the screw is gradually turned until its point touches and presses the surface. The muscular sense of touch perceives resistance to the turning of the screw when, after touching the hard body, it presses on it with a force somewhat exceeding the weight of the screw. The first effect of the contact is a diminution of resistance to the turning, due to the weight of the screw coming to be borne on its fine pointed end instead of on the thread of the nut. The *sudden* increase of resistance at the instant when the screw commences to bear part of the weight of the nut finds the sense prepared to perceive it with remarkable delicacy on account of its contrast with the immediately preceding diminution of resistance. The screw-head is now read off, and the screw turned backwards until room is left for the insertion, beneath its point, of the plate whose thickness is to be measured. The screw is again turned until increase of resistance is again perceived; and the screw-head is again read off. The difference of the readings of the head is equal to the thickness of the plate, reckoned in the proper unit of the screw and the division of its head. Spherometer.

428. If the curvature of a lens is to be measured, the instrument is first placed, as before, on a plane surface, and the reading for the contact is taken. The same operation is repeated on the spherical surface. The difference of the screw readings is evidently the greatest thickness of the glass which would be cut off by a plane passing through the three feet. This enables us to calculate the radius of the spherical surface (the distance from foot to foot of the instrument being known).

Let a be the distance from foot to foot, l the length of screw corresponding to the difference of the two readings, R the radius of the spherical surface; we have at once $2R = \frac{a^2}{3l} + l$, or, as l is generally very small compared with a, the diameter is, very approximately, $\frac{a^2}{3l}$.

429. The *Cathetometer* is used for the accurate determination of differences of level—for instance, in measuring the height to which a fluid rises in a capillary tube above the exterior free surface. It consists of a long divided metallic stem, turning round an axis as nearly as may be parallel to its length, on a fixed tripod stand: and, attached to the stem, a spirit-level. Upon the stem slides a metallic piece bearing a telescope of which the length is approximately enough perpendicular to the axis. The telescope tube is as nearly as may be perpendicular to the length of the stem. By levelling screws in two feet of the Cathetometer.

Catheto-meter. tripod the bubble of the spirit-level is brought to one position of its glass when the stem is turned all round its axis. This secures that the axis is vertical. In using the instrument the telescope is directed in succession to the two objects whose difference of level is to be found, and in each case moved (generally by a delicate screw) up or down the stem, until a horizontal wire in the focus of its eye-piece coincides with the image of the object. The difference of readings on the vertical stem (each taken generally by aid of a vernier sliding-piece) corresponding to the two positions of the telescope gives the required difference of level.

Balance. **430.** The common *Gravity Balance* is an instrument for testing the equality of the gravity of the masses placed in the two pans. We may note here a few of the precautions adopted in the best balances to guard against the various defects to which the instrument is liable; and the chief points to be attended to in its construction to secure delicacy, and rapidity of weighing.

The balance-beam should be very stiff, and as light as possible consistently with the requisite stiffness. For this purpose it is generally formed either of tubes, or of a sort of lattice-framework. To avoid friction, the axle consists of a knife-edge, as it is called; that is, a wedge of hard steel, which, when the balance is in use, rests on horizontal plates of polished agate. A similar contrivance is applied in very delicate balances at the points of the beam from which the scale-pans are suspended. When not in use, and just before use, the beam with its knife-edge is lifted by a lever arrangement from the agate plates. While thus secured it is loaded with weights as nearly as possible equal (this can be attained by previous trial with a coarser instrument), and the accurate determination is then readily effected. The last fraction of the required weight is determined by a rider, a very small weight, generally formed of wire, which can be worked (by a lever) from the outside of the glass case in which the balance is enclosed, and which may be placed in different positions upon one arm of the beam. This arm is graduated to tenths, etc., and thus shows at once the value of the rider in any case as depending on its moment or leverage, § 232.

431. Qualities of a balance: Balance.

1. *Stability.*—For stability of the beam alone without pans and weights, its centre of gravity must be below its bearing knife-edge. For stability with the heaviest weights the line joining the points at the ends of the beam from which the pans are hung must be below the knife-edge bearing the whole.

2. *Sensibility.*—The beam should be sensibly deflected from a horizontal position by the smallest difference between the weights in the scale-pans. The definite measure of the sensibility is the angle through which the beam is deflected by a stated difference between the loads in the pans.

3. *Quickness.*—This means rapidity of oscillation, and consequently speed in the performance of a weighing. It depends mainly upon the depth of the centre of gravity of the whole below the knife-edge and the length of the beam.

In our Chapter on Statics we shall give the investigation. The sensibility and quickness will there be calculated for any given form and dimensions of the instrument.

A fine balance should turn with about a 500,000th of the greatest load which can safely be placed in either pan. In fact few measurements of any kind are correct to more than *six* significant figures.

The process of *Double Weighing*, which consists in counterpoising a mass by shot, or sand, or pieces of fine wire, and then substituting weights for it in the same pan till equilibrium is attained, is more laborious, but more accurate, than single weighing; as it eliminates all errors arising from unequal length of the arms, etc.

Correction is required for the weights of air displaced by the two bodies weighed against one another when their difference is too large to be negligible.

432. In the *Torsion-balance*, invented and used with great effect by Coulomb, a force is measured by the torsion of a glass fibre, or of a metallic wire. The fibre or wire is fixed at its upper end, or at both ends, according to circumstances. In general it carries a very light horizontal rod or needle, to the extremities of which are attached the body on Torsion-balance.

Torsion-balance.

which is exerted the force to be measured, and a counterpoise. The upper extremity of the torsion fibre is fixed to an index passing through the centre of a divided disc, so that the angle through which that extremity moves is directly measured. If, at the same time, the angle through which the needle has turned be measured, or, more simply, if the index be always turned till the needle assumes a definite position determined by marks or sights attached to the case of the instrument—we have the amount of torsion of the fibre, and it becomes a simple statical problem to determine from the latter the force to be measured; its direction, and point of application, and the dimensions of the apparatus, being known. The force of torsion as depending on the angle of torsion was found by Coulomb to follow the law of simple proportion up to the limits of perfect elasticity—as might have been expected from Hooke's Law (see *Properties of Matter*), and it only remains that we determine the amount for a particular angle in absolute measure. This determination is in general simple enough in theory; but in practice requires considerable care and nicety. The torsion-balance, however, being chiefly used for comparative, not absolute, measure, this determination is often unnecessary. More will be said about it when we come to its applications.

433. The ordinary spiral spring-balances used for roughly comparing either small or large weights or forces, are, properly speaking, only a modified form of torsion-balance*, as they act almost entirely by the torsion of the wire, and not by longitudinal extension or by flexure. Spring-balances we believe to be capable, if carefully constructed, of rivalling the ordinary balance in accuracy, while, for some applications, they far surpass it in sensibility and convenience. They measure directly *force*, not *mass;* and therefore if used for determining masses in different parts of the earth, a correction must be applied for the varying force of gravity. The correction for temperature must not be overlooked. These corrections may be avoided by the method of double weighing.

* Binet, *Journal de l'École Polytechnique*, x. 1815: and J. Thomson, *Cambridge and Dublin Math. Journal* (1848).

434. Perhaps the most delicate of all instruments for the measurement of force is the *Pendulum*. It is proved in kinetics (see Div. II.) that for any pendulum, whether oscillating about a mean vertical position under the action of gravity, or in a horizontal plane, under the action of magnetic force, or force of torsion, the square of the number of *small* oscillations in a given time is proportional to the magnitude of the force under which these oscillations take place. Pendulum.

For the estimation of the relative amounts of gravity at different places, this is by far the most perfect instrument. The method of coincidences by which this process has been rendered so excessively delicate will be described later.

435. The *Bifilar Suspension*, an arrangement for measuring small horizontal forces, or couples in horizontal planes, in terms of the weight of the suspended body, is due originally to Sir William Snow Harris, who used it in one of his electrometers, as a substitute for the simple torsion-balance of Coulomb. It was used also by Gauss in his bifilar magnetometer for measuring the horizontal component of the terrestrial magnetic force*. In this instrument the bifilar suspension is adjusted to keep a bar-magnet in a position approximately perpendicular to the magnetic meridian. The small natural augmentations and diminutions of the horizontal component are shown by small azimuthal motions of the bar. On account of some obvious mechanical and dynamical difficulties this instrument was not found very convenient for absolute determinations, but from the time of its first practical introduction by Gauss and Weber it has been in use in all Magnetic Observatories for measuring the natural variations of the horizontal magnetic component. It is now made with a much smaller magnet than the great bar weighing twenty-five pounds originally given with it by Gauss; but the bars in actual use at the present day are still enormously too large† for their duty. The weight of the Bifilar Balance. Bifilar Magnetometer.

* Gauss, *Resultate aus den Beobachtungen des magnetischen Vereins im Jahre* 1837. Translated in Taylor's *Scientific Memoirs*, Vol. II., Article VI.

† The suspended magnets used for determining the direction and the intensity of the horizontal magnetic force in the Dublin Magnetic Observatory,

Bifilar Magnetometer.

bar with attached mirror ought not to exceed eight grammes, so that two single silk fibres may suffice for the bearing threads. The only substantial alteration, besides the diminution of its magnitude, which has been made in the instrument since Gauss and Weber's time is the addition of photographic apparatus and clockwork for automatic record of its motions. For absolute determinations of the horizontal component force, Gauss's method of deflecting a freely suspended magnet by a magnetic bar brought into proper positions in its neighbourhood, and again making an independent set of observations to determine the period of oscillation of the same deflecting bar when suspended by a fine fibre and set to vibrate through a small horizontal angle on each side of the magnetic meridian, is the method which has been uniformly in use both in magnetic observatories and in travellers' observations with small portable apparatus since it was first invented by Gauss*.

Absolute measurement of Terrestrial Magnetic Force.

Bifilar Balance.

In the bifilar balance the two threads may be of unequal lengths, the line joining their upper fixed ends need not be horizontal, and their other ends may be attached to any two points of the suspended body: but for most purposes, and particularly for regular instruments such as electrometers and magnetometers with bifilar suspension, it is convenient to have, as nearly as may be, the two threads of equal length, their fixed ends at the same level, and their other ends attached to the suspended body symmetrically with reference to its centre of gravity (as illustrated in the last set of drawings of § 345[x]). Supposing the instrument-maker to have fulfilled these conditions of symmetry as nearly as he can with reference to the four points of attachment of the threads, we have still to adjust properly the lengths of the threads. For this purpose remark that a small difference in the lengths will throw the suspended body into an unsymmetrical

as described by Dr Lloyd in his *Treatise on Magnetism* (London, 1874), are each of them 15 inches long, $\frac{7}{8}$ of an inch broad, and $\frac{1}{8}$ of an inch in thickness, and must therefore weigh about a pound each. The corresponding magnets used at the Kew Observatory are much smaller. They are each 5·4 inches long, 0·8 inch broad, and 0·1 inch thick, and therefore the weight of each is about 0·012 pound, or nearly 55 grammes.

* *Intensitas Vis Magneticae Terrestris ad Mensuram Absolutam revocuta*, Commentationes Societatis Gottingensis, 1832.

Bifilar Balance.

position, in which, particularly if its centre of gravity be very low (as it is in Sir W. Thomson's Quadrant Electrometer), much more of its weight will be borne by one thread than by the other. This will diminish very much the amount of the horizontal couple required to produce a stated azimuthal deflection in the regular use of the instrument, in other words will increase its sensibility above its proper amount, that is to say, the amount which it would have if the conditions of symmetry were fully realized. Hence the proper adjustment for equalizing the lengths of the threads in a symmetrical bifilar balance, or for giving them their right difference in an unsymmetrical arrangement, in order to make the instrument as accurate as it can be, is to alter the length of one or both of the threads, until we attain to the condition of *minimum sensibility*, that is to say minimum angle of deflection under the influence of a given amount of couple.

The great merit of the bifilar balance over the simple torsion-balance of Coulomb for such applications as that to the horizontal magnetometer in the continuous work of an observatory, is the comparative smallness of the influence it experiences from changes of temperature. The torsional rigidity of iron, copper, and brass wires is diminished about $\frac{1}{2}$ per cent. with 10° elevation of temperature, while the linear expansions of the same metals are each less than $\frac{1}{50}$ per cent. with the same elevation of temperature. Hence in the unifilar torsion-balance, if iron, copper, or brass (the only metals for which the change of torsional rigidity with change of temperature has hitherto been measured) is used for the material of the bearing fibre, the sensibility is augmented $\frac{1}{2}$ per cent. by 10° elevation of temperature.

On the other hand, in the bifilar balance, if torsional rigidity does not contribute any sensible proportion to the whole directive couple (and this condition may be realized as nearly as we please by making the bearing wires long enough and making the distance between them great enough to give the requisite amount of directive couple), the sensibility of the balance is affected only by the linear expansions of the substances concerned. If the equal distances between the two pairs of points

Bifilar Balance.

of attachment, in the normal form of bifilar balance (or that in which the two threads are vertical when the suspended body is uninfluenced by horizontal force or couple), remained constant, the sensibility would be augmented with elevation of temperature in simple proportion to the linear expansions of the bearing wires; and this small influence might, if it were worth while to make the requisite mechanical arrangements, be perfectly compensated by choosing materials for the frames or bars bearing the attachments of the wires so that the proportionate augmentation of the distance between them should be just half the elongation of either wire, because the sensibility, as shown by the mathematical formula below, is simply proportional to the length of the wires and inversely proportional to the square of the distance between them. But, even without any such compensation, the temperature-error due to linear expansions of the materials of the bifilar balance is so small that in the most accurate regular use of the instrument in magnetic observatories it may be almost neglected; and at most it is less than $\frac{1}{25}$ of the error of the unifilar torsion-balance, at all events if, as is probably the case, the changes of rigidity with changes of temperature in other metals are of similar amounts to those for the three metals on which experiments have been made. In reality the chief temperature-error of the bifilar magnetometer depends on the change of the magnetic moment of the suspended magnet with change of temperature. It seems that the magnetism of a steel magnet diminishes with rise of temperature and augments with fall of temperature, but experimental information is much wanted on this subject.

The amount of the effect is very different in different bars, and it must be experimentally determined for each bar serving in a bifilar magnetometer. The amount of the change of magnetic moment in the bar which had been most used in the Dublin Magnetic Observatory was found to be ·000029 per degree Fahrenheit or at the rate of ·000052 per degree Centigrade, being about the same amount as that of the change of torsional rigidity with temperature of the three metals referred to above.

Let a be the half length of the bar between the points of attachment of the wires, θ the angle through which the bar has been turned (in a horizontal plane) from its position of equilibrium, l the length of one of the wires, ι its inclination to the vertical. Bifilar Balance.

Then $l \cos \iota$ is the difference of levels between the ends of each wire, and evidently, by the geometry of the case,

$$\tfrac{1}{2} l \sin \iota = a \sin \tfrac{1}{2} \theta.$$

Now if Q be the couple tending to turn the bar, and W its weight, the principle of mechanical effect gives

$$\begin{aligned} Qd\theta &= -Wd(l \cos \iota) \\ &= Wl \sin \iota d\iota. \end{aligned}$$

But, by the geometrical condition above,

$$l^2 \sin \iota \cos \iota d\iota = a^2 \sin \theta d\theta.$$

Hence
$$\frac{Q}{a^2 \sin \theta} = \frac{W}{l \cos \iota},$$

or
$$Q = \frac{Wa^2}{l} \frac{\sin \theta}{\sqrt{1 - \frac{4a^2}{l^2} \sin^2 \frac{\theta}{2}}},$$

which gives the couple in terms of the deflection θ.

If the torsion of the wires be taken into account, it is sensibly equal to θ (since the greatest inclination to the vertical is small), and therefore the couple resulting from it will be $E\theta$. This must be added to the value of Q just found in order to get the whole deflecting couple.

436. Ergometers are instruments for measuring energy. Ergometers. *White's friction brake* measures the amount of work actually performed in any time by an engine or other "prime mover," by allowing it during the time of trial to waste all its work on friction. *Morin's ergometer* measures work without wasting any of it, in the course of its transmission from the prime mover to machines in which it is usefully employed. It consists of a simple arrangement of springs, measuring at every instant the *couple* with which the prime mover turns the shaft that transmits its work, and an integrating machine from which the work done by this couple during any time can be read off.

Let L be the couple at any instant, and ϕ the whole angle through which the shaft has turned from the moment at which the reckoning commences. The integrating machine shows at any moment the value of $\int L d\phi$, which (§ 240) is the whole work done.

Ergometers. **437.** White's friction brake consists of a lever clamped to the shaft, but not allowed to turn with it. The moment of the force required to prevent the lever from going round with the shaft, multiplied by the whole angle through which the shaft turns, measures the whole work done against the friction of the clamp. The same result is much more easily obtained by wrapping a rope or chain several times round the shaft, or round a cylinder or drum carried round by the shaft, and applying measured forces to its two ends in proper directions to keep it nearly steady while the shaft turns round without it. The difference of the moments of these two forces round the axis, multiplied by the angle through which the shaft turns, measures the whole work spent on friction against the rope. If we remove all other resistance to the shaft, and apply the proper amount of force at each end of the dynamimetric rope or chain (which is very easily done in practice), the prime mover is kept running at the proper speed for the test, and having its whole work thus wasted for the time and measured.

APPENDIX B'.

CONTINUOUS CALCULATING MACHINES.

I. Tide-predicting Machine.

The object is to predict the tides for any port for which the tidal constituents have been found from the harmonic analysis from tide-gauge observations; not merely to predict the times and heights of high water, but the depths of water at any and every instant, showing them by a continuous curve, for a year, or for any number of years in advance. Tide-predicting Machine.

This object requires the summation of the simple harmonic functions representing the several constituents* to be taken into account, which is performed by the machine in the following manner:—For each tidal constituent to be taken into account the machine has a shaft with an overhanging crank, which carries a pulley pivoted on a parallel axis adjustable to a greater or less distance from the shaft's axis, according to the greater or less range of the particular tidal constituent for the different ports for which the machine is to be used. The several shafts, with their axes all parallel, are geared together so that their periods are to a sufficient degree of approximation proportional to the periods of the tidal constituents. The crank on each shaft can be turned round on the shaft and clamped in any position: thus it is set to the proper position for the epoch of the particular tide which it is to produce. The axes of the several shafts are horizontal, and their vertical planes are at successive distances one from another, each equal to the diameter of one of the pulleys (the diameters of these being equal). The shafts are in two rows, an upper and a lower, and the grooves of the pulleys are all in one plane perpendicular to their axes.

Suppose, now, the axes of the pulleys to be set each at zero distance from the axis of its shaft, and let a fine wire or chain,

* See Report for 1876 of the Committee of the British Association appointed for the purpose of promoting the Extension, Improvement, and Harmonic Analysis of Tidal Observations.

Tide-predicting Machine.

with one end hanging down and carrying a weight, pass alternately over and under the pulleys in order, and vertically upwards or downwards (according as the number of pulleys is even or odd) from the last pulley to a fixed point. The weight is to be properly guided for vertical motion by a geometrical slide. Turn the machine now, and the wire will remain undisturbed with all its free parts vertical and the hanging weight unmoved. But now set the axis of any one of the pulleys to a distance $\frac{1}{2}T$ from its shaft's axis and turn the machine. If the distance of this pulley from the two on each side of it in the other row is a considerable multiple of $\frac{1}{2}T$, the hanging weight will now (if the machine is turned uniformly) move up and down with a simple harmonic motion of amplitude (or semi-range) equal to T in the period of its shaft. If, next, a second pulley is displaced to a distance $\frac{1}{2}T'$, a third to a distance $\frac{1}{2}T''$, and so on, the hanging weight will now perform a complex harmonic motion equal to the sum of the several harmonic motions, *each* in its proper period, which would be produced separately by the displacements T, T', T''. Thus, if the machine was made on a large scale, with T, T',... equal respectively to the actual semi-ranges of the several constituent tides, and if it was turned round slowly (by clockwork, for example), each shaft going once round in the actual period of the tide which it represents, the hanging weight would rise and fall exactly with the water-level as affected by the whole tidal action. This, of course, could be of no use, and is only suggested by way of illustration. The actual machine is made of such magnitude, that it can be set to give a motion to the hanging weight equal to the actual motion of the water-level reduced to any convenient scale: and provided the whole range does not exceed about 30 centimetres, the geometrical error due to the deviation from perfect parallelism in the successive free parts of the wire is not so great as to be practically objectionable. The proper order for the shafts is the order of magnitude of the constituent tides which they produce, the greatest next the hanging weight, and the least next the fixed end of the wire: this so that the greatest constituent may have only one pulley to move, the second in magnitude only two pulleys, and so on.

One machine of this kind has already been constructed for the British Association, and another (with a greater number of shafts to include a greater number of tidal constituents) is being constructured for the Indian Government. The British Association machine, which is kept available for general use, under charge of the Science and Art Department in South Kensington, has ten shafts, which taken in order, from the hanging weight, give respectively the following tidal constituents*: Tide-predicting Machine.

1. The mean lunar semi-diurnal.
2. The mean solar semi-diurnal.
3. The larger elliptic semi-diurnal.
4. The luni-solar diurnal declinational.
5. The lunar diurnal declinational.
6. The luni-solar semi-diurnal declinational.
7. The smaller elliptic semi-diurnal.
8. The solar diurnal declinational.
9. The lunar quarter-diurnal, or first shallow-water tide of mean lunar semi-diurnal.
10. The luni-solar quarter-diurnal, shallow-water tide.

The hanging weight consists of an ink-bottle with a glass tubular pen, which marks the tide level in a continuous curve on a long band of paper, moved horizontally across the line of motion of the pen, by a vertical cylinder geared to the revolving shafts of the machine. One of the five sliding points of the geometrical slide is the point of the pen sliding on the paper stretched on the cylinder, and the couple formed by the normal pressure on this point, and on another of the five, which is about four centimetres above its level and one and a half centimetres from the paper, balances the couple due to gravity of the ink-bottle and the vertical component of the pull of the bearing wire, which is in a line about a millimetre or two farther from the paper than that in which the centre of gravity moves. Thus is ensured, notwithstanding small inequalities on the paper, a pressure of the pen on the paper very approximately constant and as small as is desired.

Hour marks are made on the curve by a small horizontal movement of the ink-bottle's lateral guides, made once an hour; a somewhat greater movement, giving a deeper notch, serves to mark the noon of every day.

The machine may be turned so rapidly as to run off a year's tides for any port in about four hours.

Each crank should carry an adjustable counterpoise, to be

* See Report for 1876 of the British Association's Tidal Committee.

Tide-predicting Machine.

adjusted so that when the crank is not vertical the pulls of the approximately vertical portions of wire acting on it through the pulley which it carries shall, as exactly as may be, balance on the axis of the shaft, and the motion of the shaft should be resisted by a slight weight hanging on a thread wrapped once round it and attached at its other end to a fixed point. This part of the design, planned to secure against "lost time" or "back lash" in the gearings, and to preserve uniformity of pressure between teeth and teeth, teeth and screws, and ends of axles and "end-plates," was not carried out in the British Association machine.

II. Machine for the Solution of Simultaneous Linear Equations*.

Equation-Solver.

Let $B_1, B_2, \ldots B_n$ be n bodies each supported on a fixed axis (in practice each is to be supported on knife-edges like the beam of a balance).

Let $P_{11}, P_{21}, P_{31}, \ldots P_{n1}$ be n pulleys each pivoted on B_1;
$P_{12}, P_{22}, P_{32} \ldots P_{n2}$ " " B_2;
$P_{13}, P_{23}, P_{33} \ldots P_{n3}$ " " B_3;
....

" $C_1, C_2, C_3 \ldots C_n$ be n cords passing over the pulleys;
" $D_1, P_{11}, P_{12}, P_{13} \ldots P_{1n}, E_1$, be the course of C_1;
" $D_2, P_{21}, P_{22}, P_{23} \ldots P_{2n}, E_2$, " " C_2;
....

" $D_1, E_1, D_2, E_2, \ldots D_n, E_n$, be fixed points;

" $l_1, l_2, l_3 \ldots l_n$ be the lengths of the cords between D_1, E_1, and $D_2, E_2, \ldots$ and D_n, E_n, along the courses stated above, when $B_1, B_2, \ldots B_n$, are in particular positions which will be called their zero positions;

" $l_1+e_1, l_2+e_2, \ldots l_n+e_n$ be their lengths between the same fixed points, when $B_1, B_2 \ldots B_n$ are turned through angles $x_1, x_2, \ldots x_n$ from their zero positions;

(11), (12), (13), ... (1n),
(21), (22), (23), ... (2n),
(31), (32), (33), ... (3n),
....

* Sir W. Thomson, *Proceedings of the Royal Society*, Vol. XXVIII., 1878.

quantities such that Equation-Solver.

$$\left.\begin{array}{l}(11)\,x_1+(12)\,x_2+\ldots+(1n)\,x_n=e_1\\(21)\,x_1+(22)\,x_2+\ldots+(2n)\,x_n=e_2\\(31)\,x_1+(32)\,x_2+\ldots+(3n)\,x_n=e_3\\\ldots\ldots\ldots\ldots\ldots\ldots\ldots\ldots\\(n1)\,x_1+(n2)\,x_2+\ldots+(nn)\,x_n=e_n\end{array}\right\}\ldots\ldots\ldots\ldots\ldots\ (\text{I}).$$

We shall suppose $x_1, x_2, \ldots x_n$ to be each so small that (11), (12),...(21), etc., do not vary sensibly from the values which they have when $x_1, x_2, \ldots x_n$, are each infinitely small. In practice it will be convenient to so place the axes of $B_1, B_2, \ldots B_n$, and the mountings of the pulleys on $B_1, B_2, \ldots B_n$, and the fixed points D_1, E_1, D_2, etc., that when $x_1, x_2, \ldots x_n$ are infinitely small, the straight parts of each cord and the lines of infinitesimal motion of the centres of the pulleys round which it passes shall be all parallel. Then $\frac{1}{2}(11)$, $\frac{1}{2}(21)$, ... $\frac{1}{2}(n1)$ will be simply equal to the distances of the centres of the pulleys $P_{11}, P_{21}, \ldots P_{n1}$, from the axis of B_1; $\frac{1}{2}(12)$, $\frac{1}{2}(22)$ ... $\frac{1}{2}(n2)$ the distances of $P_{12}, P_{22}, \ldots P_{n2}$ from the axis of B_2; and so on.

In practice the mountings of the pulleys are to be adjustable by proper geometrical slides, to allow any prescribed positive or negative value to be given to each of the quantities (11), (12), ... (21), etc.

Suppose this to be done, and each of the bodies $B_1, B_2, \ldots B_n$ to be placed in its zero position and held there. Attach now the cords firmly to the fixed points $D_1, D_2, \ldots D_n$ respectively; and, passing them round their proper pulleys, bring them to the other fixed points $E_1, E_2, \ldots E_n$, and pass them through infinitely small smooth rings fixed at these points. Now hold the bodies $B_1, B_2, \ldots$ each fixed, and (in practice by weights hung on their ends, outside $E_1, E_2, \ldots E_n$) pull the cords through $E_1, E_2, \ldots E_n$ with any given tensions* $T_1, T_2, \ldots T_n$. Let $G_1, G_2, \ldots G_n$ be moments round the fixed axes of $B_1, B_2, \ldots B_n$ of the forces required to hold the bodies fixed when acted on by the cords thus

* The idea of force here first introduced is not essential, indeed is not technically admissible to the purely kinematic and algebraic part of the subject proposed. But it is not merely an ideal kinematic construction of the algebraic problem that is intended; and the design of a kinematic machine, for success in practice, essentially involves dynamical considerations. In the present case some of the most important of the purely algebraic questions concerned are very interestingly illustrated by these dynamical considerations.

Equation-Solver.

stretched. The principle of "virtual velocities," just as it came from Lagrange (or the principle of "work"), gives immediately, in virtue of (I),

$$\left.\begin{aligned} G_1 &= (11)\,T_1 + (21)\,T_2 + \ldots + (n1)\,T_n \\ G_2 &= (12)\,T_1 + (22)\,T_2 + \ldots + (n2)\,T_n \\ &\ldots\ldots\ldots\ldots\ldots\ldots\ldots\ldots\ldots\ldots \\ G_n &= (1n)\,T_1 + (2n)\,T_2 + \ldots + (nn)\,T_n \end{aligned}\right\} \ldots\ldots\ldots\ldots \text{(II)}.$$

Apply and keep applied to each of the bodies, $B_1, B_2, \ldots B_n$ (in practice by the weights of the pulleys, and by counter-pulling springs), such forces as shall have for their moments the values $G_1, G_2 \ldots G_n$, calculated from equations (II) with whatever values seem desirable for the tensions $T_1, T_2, \ldots T_3$. (In practice, the straight parts of the cords are to be approximately vertical, and the bodies B_1, B_2, are to be each balanced on its axis when the pulleys belonging to it are removed, and it is advisable to make the tensions each equal to half the weight of one of the pulleys with its adjustable frame.) The machine is now ready for use. To use it, pull the cords simultaneously or successively till lengths equal to $e_1, e_2, \ldots e_n$ are passed through the rings E_1, $E_2, \ldots E_n$, respectively.

The *pulls* required to do this may be positive or negative; in practice, they will be infinitesimal downward or upward pressures applied by hand to the stretching weights which remain permanently hanging on the cords.

Observe the angles through which the bodies $B_1, B_2, \ldots B_n$ are turned by this given movement of the cords. These angles are the required values of the unknown $x_1, x_2, \ldots x_n$, satisfying the simultaneous equations (I).

The actual construction of a practically useful machine for calculating as many as eight or ten or more of unknowns from the same number of linear equations does not promise to be either difficult or over-elaborate. A fair approximation having been found by a first application of the machine, a very moderate amount of straightforward arithmetical work (aided very advantageously by Crelle's multiplication tables) suffices to calculate the residual errors, and allow the machines (with the setting of the pulleys unchanged) to be re-applied to calculate the corrections (which may be treated decimally, for convenience): thus, 100 times the amount of the correction on each of the original unknowns may be made the new unknowns, if the magnitudes thus falling to be dealt with are convenient for the machine. There is, of course, no limit to the accuracy thus obtainable by successive approximations. The exceeding easiness of each application of the machine promises well for its real usefulness, whether for cases in which a single application suffices, or for others in which the requisite accuracy is reached after two, three, or more, of successive approximations. Equation-Solver.

The accompanying drawings represent a machine for finding six* unknowns from six equations. Fig. 1 represents in elevation and plan one of the six bodies B_1, B_2, etc. Fig. 2 shows in elevation and plan one of the thirty-six pulleys P, with its cradle on geometrical slide (§ 198). Fig. 3 shows in front-elevation the general disposition of the instrument.

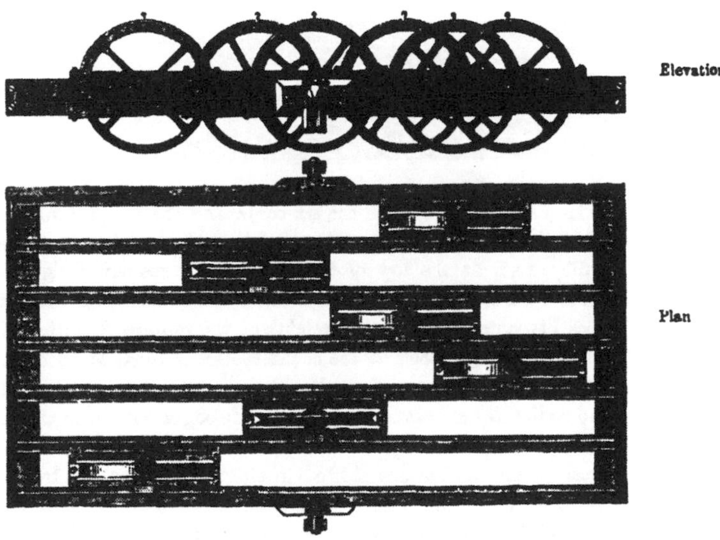

* This number has been chosen for the first practical machine to be constructed, because a chief application of the machine may be to the calculation of the corrections on approximate values already found of the six elements of the orbit of a comet or asteroid.

Equation-Solver.

FIG. 2. One of the thirty-six pulleys, P, with its sliding cradle. Full Size.

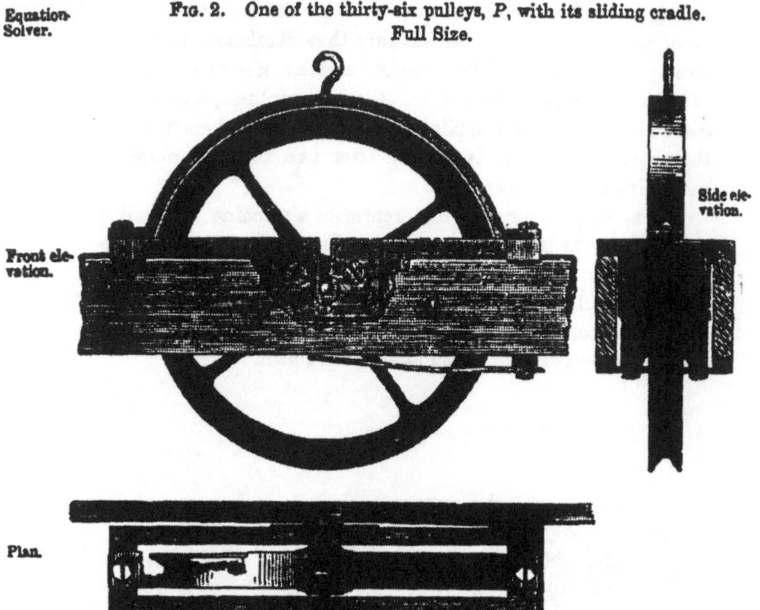

In Fig. 3 only one of the six cords, and the six pulleys over which it passes, is shown, not any of the other thirty. The three pulleys seen at the top of the sketch are three out of eighteen pivoted on immoveable bearings above the machine, for the purpose of counterpoising the weights of the pulleys P, with their sliding cradles. Each of the counterpoises is equal to twice the weight of one of the pulleys P with its sliding cradle. Thus if the bodies B are balanced on their knife-edges with each sliding cradle in its central position, they remain balanced when one or all of the cradles are shifted to either side; and the tension of each of the thirty-six essential cords is exactly equal to half the weight of one of the pulleys with its adjustable frame, as specified above (the deviations from exact verticality of all the free portions of the thirty-six essential cords and the eighteen counterpoising cords being neglected).

Equation-Solver.

FIG. 3. General disposition of machine.

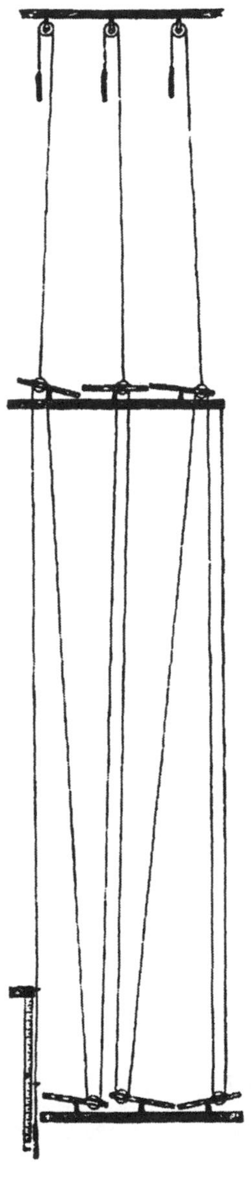

III. An Integrating Machine having a New Kinematic Principle*.

Disk-Globe-, and Cylinder-Integrating Machine.

The kinematic principle for integrating *ydx*, which is used in the instruments well known as Morin's Dynamometer† and Sang's Planimeter‡, admirable as it is in many respects, involves one element of imperfection which cannot but prevent our contemplating it with full satisfaction. This imperfection consists in the sliding action which the edge wheel or roller is required to take in conjunction with its rolling action, which alone is desirable for exact communication of motion from the disk or cone to the edge roller.

The very ingenious, simple, and practically useful instrument well known as Amsler's Polar Planimeter, although different in its main features of principle and mode of action from the instruments just referred to, ranks along with them in involving the like imperfection of requiring to have a sidewise sliding action of its edge rolling wheel, besides the desirable rolling action on the surface which imparts to it its revolving motion—a surface

* Professor James Thomson, *Proceedings of the Royal Society*, Vol. xxiv., 1876, p. 262.

† Instruments of this kind, and any others for measuring mechanical work, may better in future be called Ergometers than Dynamometers. The name "dynamometer" has been and continues to be in common use for signifying a spring instrument for measuring *force;* but an instrument for measuring *work*, being distinct in its nature and object, ought to have a different and more suitable designation. The name "dynamometer," besides, appears to be badly formed from the Greek; and for designating an instrument for *measurement of force*, I would suggest that the name may with advantage be changed to *dynamimeter*. In respect to the mode of forming words in such cases, reference may be made to Curtius's Grammar, Dr Smith's English edition, § 354, p. 220.—J. T., 26th February, 1876.

‡ Sang's Planimeter is very clearly described and figured in a paper by its inventor, in the Transactions of the Royal Scottish Society of Arts, Vol. iv. January 12, 1852.

which in this case is not a disk or cone, but is the surface of the paper, or any other plane face, on which the map or other plane diagram to be evaluated in area is drawn. Disk-, Globe-, and Cylinder-Integrating Machine.

Professor J. Clerk Maxwell, having seen Sang's Planimeter in the Great Exhibition of 1851, and having become convinced that the combination of slipping and rolling was a drawback on the perfection of the instrument, began to search for some arrangement by which the motion should be that of perfect rolling in every action of the instrument, corresponding to that of combined slipping and rolling in previous instruments. He succeeded in devising a new form of planimeter or integrating machine with a quite new and very beautiful principle of kinematic action depending on the mutual rolling of two equal spheres, each on the other. He described this in a paper submitted to the Royal Scottish Society of Arts in January 1855, which is published in Vol. IV. of the Transactions of that Society. In that paper he also offered a suggestion, which appears to be both interesting and important, proposing the attainment of the desired conditions of action by the mutual rolling of a cone and cylinder with their axes at right angles.

The idea of using pure rolling instead of combined rolling and slipping was communicated to me by Prof. Maxwell, when I had the pleasure of learning from himself some particulars as to the nature of his contrivance. Afterwards (some time between the years 1861 and 1864), while endeavouring to contrive means for the attainment in meteorological observatories of certain integrations in respect to the motions of the wind, and also in endeavouring to devise a planimeter more satisfactory in principle than either Sang's or Amsler's planimeter (even though, on grounds of practical simplicity and convenience, unlikely to turn out preferable to Amsler's in ordinary cases of taking areas from maps or other diagrams, but something that I hoped might possibly be attainable which, while having the merit of working by pure rolling contact, might be simpler than the instrument of Prof. Maxwell and preferable to it in mechanism), I succeeded in devising for the desired object a new kinematic method, which has ever since appeared to me likely sometime to prove valuable when occasion for its employment might be found. Now, within the last few days, this principle, on being suggested to my brother as perhaps capable of being usefully employed towards the development of tide-calculating machines

Disk-, Globe-, and Cylinder-Integrator.

which he had been devising, has been found by him to be capable of being introduced and combined in several ways to produce important results. On his advice, therefore, I now offer to the Royal Society a brief description of the new principle as devised by me.

The new principle consists primarily in the transmission of motion from a disk or cone to a cylinder by the intervention of a loose ball, which presses by its gravity on the disk and cylinder, or on the cone and cylinder, as the case may be, the pressure being sufficient to give the necessary frictional coherence at each point of rolling contact; and the axis of the disk or cone and that of the cylinder being both held fixed in position by bearings in stationary framework, and the arrangement of these axes being such that when the disk or the cone and the cylinder are kept steady, or, in other words, without rotation on their axes, the ball can roll along them in contact with both, so that the point of rolling contact between the ball and the cylinder shall traverse a straight line on the cylindric surface parallel necessarily to the axis of the cylinder—and so that, in the case of a disk being used, the point of rolling contact of the ball with the disk shall traverse a straight line passing through the centre of the disk—or that, in case of a cone being used, the line of rolling contact of the ball on the cone shall traverse a straight line on the conical surface, directed necessarily towards the vertex of the cone. It will thus readily be seen that, whether the cylinder and the disk or cone be at rest or revolving on their axes, the two lines of rolling contact of the ball, one on the cylindric surface and the other on the disk or cone, when both considered as lines traced out in space fixed relatively to the framing of the whole instrument, will be two parallel straight lines, and that the line of motion of the ball's centre will be straight and parallel to them. For facilitating explanations, the motion of the centre of the ball along its path parallel to the axis of the cylinder may be called the ball's longitudinal motion.

Now for the integration of ydx: the distance of the point of contact of the ball with the disk or cone from the centre of the disk or vertex of the cone in the ball's longitudinal motion is to represent y, while the angular space turned by the disk or cone from any initial position represents x; and then the angular space turned by the cylinder will, when multiplied by a suitable constant numerical coefficient, express the integral in terms of any required unit for its evaluation.

Disk-, Globe-, and Cylinder-Integrator.

The longitudinal motion may be imparted to the ball by having the framing of the whole instrument so placed that the lines of longitudinal motion of the two points of contact and of the ball's centre, which are three straight lines mutually parallel, shall be inclined to the horizontal sufficiently to make the ball tend decidedly to descend along the line of its longitudinal motion, and then regulating its motion by an abutting controller, which may have at its point of contact, where it presses on the ball, a plane face perpendicular to the line of the ball's motion. Otherwise the longitudinal motion may, for some cases, preferably be imparted to the ball by having the direction of that motion horizontal, and having two controlling flat faces acting in close contact without tightness at opposite extremities of the ball's diameter, which at any moment is in the line of the ball's motion or is parallel to the axis of the cylinder.

It is worthy of notice that, in the case of the disk-, ball-, and cylinder-integrator, no theoretical nor important practical fault in the action of the instrument would be involved in any deficiency of perfect exactitude in the practical accomplishment of the desired condition that the line of motion of the ball's point of contact with the disk should pass through the centre of the disk. The reason of this will be obvious enough on a little consideration.

The plane of the disk may suitably be placed inclined to the horizontal at some such angle as 45°; and the accompanying sketch, together with the model, which will be submitted to the Society by my brother, will aid towards the clear understanding of the explanations which have been given.

My brother has pointed out to me that an additional operation, important for some purposes, may be effected by arranging that the machine shall give a continuous record of the growth of the integral by introducing additional mechanisms suitable for continually describing a curve such that for each point of it the abscissa shall represent the value of x, and the ordinate shall represent the integral attained from $x=0$ forward to that value of x. This, he has pointed out, may be effected in practice by having a cylinder axised on the axis of the disk, a roll of paper covering this cylinder's surface, and a straight bar situated parallel to this cylinder's axis and resting with enough of pres-

Disk-, Globe-, and Cylinder-Integrator.

sure on the surface of the primary registering or *the indicating* cylinder (the one, namely, which is actuated by its contact with the ball) to make it have sufficient frictional coherence with that

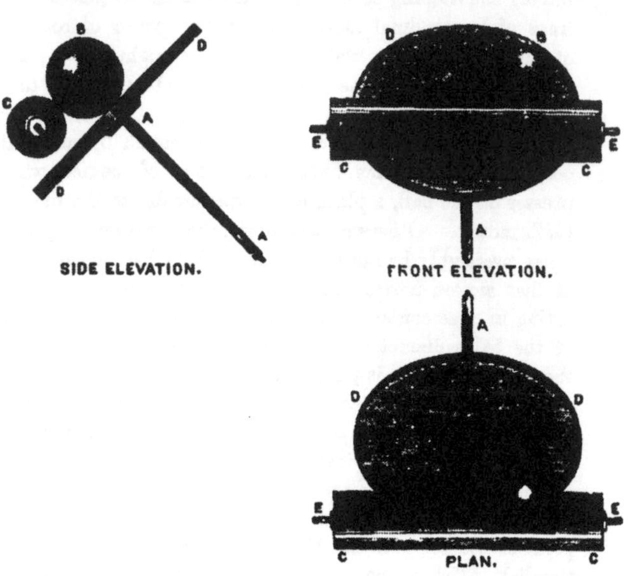

surface, and by having this bar made to carry a pencil or other tracing point which will mark the desired curve on the secondary registering or *the recording* cylinder. As, from the nature of the apparatus, the axis of the disk and of the secondary registering or recording cylinder ought to be steeply inclined to the horizontal, and as, therefore, this bar, carrying the pencil, would have the line of its length and of its motion alike steeply inclined with that axis, it seems that, to carry out this idea, it may be advisable to have a thread attached to the bar and extending off in the line of the bar to a pulley, passing over the pulley, and having suspended at its other end a weight which will be just sufficient to counteract the tendency of the rod, in virtue of gravity, to glide down along the line of its own slope, so as to leave it perfectly free to be moved up or down by the frictional coherence between itself and the moving surface of the indicating cylinder worked directly by the ball.

IV. An Instrument for calculating $\left(\int \phi(x)\psi(x)\,dx\right)$, the Integral of the Product of two given Functions*.

Machine to calculate Integral of Product of two Functions.

In consequence of the recent meeting of the British Association at Bristol, I resumed an attempt to find an instrument which should supersede the heavy arithmetical labour of calculating the integrals required to analyze a function into its simple harmonic constituents according to the method of Fourier. During many years previously it had appeared to me that the object ought to be accomplished by some simple mechanical means; but it was not until recently that I succeeded in devising an instrument approaching sufficiently to simplicity to promise practically useful results. Having arrived at this stage, I described my proposed machine a few days ago to my brother Professor James Thomson, and he described to me in return a kind of mechanical integrator which had occurred to him many years ago, but of which he had never published any description. I instantly saw that it gave me a much simpler means of attaining my special object than anything I had been able to think of previously. An account of his integrator is communicated to the Royal Society along with the present paper.

To calculate $\int \phi(x)\psi(x)\,dx$, the rotating disk is to be displaced from a zero or initial position through an angle equal to

$$\int_0^x \phi(x)\,dx,$$

while the rolling globe is moved so as always to be at a distance from its zero position equal to $\psi(x)$. This being done, the cylinder obviously turns through an angle equal to $\int_0^x \phi(x)\psi(x)\,dx$, and thus solves the problem.

One way of giving the required motions to the rotating disk and rolling globe is as follows:—

* Sir W. Thomson, *Proceedings of the Royal Society*, Vol. xxiv., 1876, p. 266.

Machine to calculate Integral of Product of two Functions.

On two pieces of paper draw the curves

$$y=\int_0^x \phi(x)\,dx, \text{ and } y=\psi(x).$$

Attach these pieces of paper to the circumference of two circular cylinders, or to different parts of the circumference of one cylinder, with the axis of x in each in the direction perpendicular to the axis of the cylinder. Let the two cylinders (if there are two) be geared together so as that their circumferences shall move with equal velocities. Attached to the framework let there be, close to the circumference of each cylinder, a slide or guide-rod to guide a moveable point, moved by the hand of an operator, so as always to touch the curve on the surface of the cylinder, while the two cylinders are moved round.

Two operators will be required, as one operator could not move the two points so as to fulfil this condition—at all events unless the motion were very slow. One of these points, by proper mechanism, gives an angular motion to the rotating disk equal to its own linear motion, the other gives a linear motion equal to its own to the centre of the rolling globe.

The machine thus described is immediately applicable to calculate the values H_1, H_2, H_3, etc. of the harmonic constituents of a function $\psi(x)$ in the splendid generalization of Fourier's simple harmonic analysis, which he initiated himself in his solutions for the conduction of heat in the sphere and the cylinder, and which was worked out so ably and beautifully by Poisson*, and by Sturm and Liouville in their memorable papers on this subject published in the first volume of Liouville's *Journal des Mathématiques*. Thus if

$$\psi(x)=H_1\phi_1(x)+H_2\phi_2(x)+H_3\phi_3(x)+\text{etc.}$$

be the expression for an arbitrary function ψx, in terms of the generalized harmonic functions $\phi_1(x)$, $\phi_2(x)$, $\phi_3(x)$, etc., these functions being such that

$$\int_0^l \phi_1(x)\,\phi_2(x)\,dx=0,\quad \int_0^l \phi_1(x)\,\phi_3(x)\,dx=0,\quad \int_0^l \phi_2(x)\,\phi_3(x)=0, \text{ etc.},$$

* His general demonstration of the reality of the roots of transcendental equations essential to this analysis (an exceedingly important step in advance from Fourier's position), which he first gave in the *Bulletin de la Société Philomathique* for 1828, is reproduced in his *Théorie Mathématique de la Chaleur*, § 90.

we have

Machine to calculate Integral of Product of two Functions.

$$H_1 = \frac{\int_0^l \phi_1(x)\,\psi(x)\,dx}{\int_0^l \{\phi_1(x)\}^2\,dx},$$

$$H_2 = \frac{\int_0^l \phi_2(x)\,\psi(x)\,dx}{\int_0^l \{\phi_2(x)\}^2\,dx},$$

etc.

In the physical applications of this theory the integrals which constitute the denominators of the formulas for H_1, H_2, etc. are always to be evaluated in finite terms by an extension of Fourier's formula for the $\int_0^X xu_i^2\,dx$ of his problem of the cylinder* made by Sturm in equation (10), § iv. of his *Mémoire sur une Classe d'Équations à différences partielles* in Liouville's *Journal*, Vol. I. (1836). The integrals in the numerators are calculated with great ease by aid of the machine worked in the manner described above.

The great practical use of this machine will be to perform the simple harmonic Fourier-analysis for tidal, meteorological, and perhaps even astronomical, observations. It is the case in which

$$\phi(x) = {\sin \atop \cos}(nx);$$

and the integration is performed through a range equal to $\frac{2i\pi}{n}$ (i any integer) that gives this application. In this case the addition of a simple crank mechanism, to give a simple harmonic angular motion to the rotating disk in the proper period $\frac{2\pi}{n}$, when the cylinder bearing the curve $y = \psi(x)$ moves uniformly, supersedes the necessity for a cylinder with the curve $y = \phi(x)$ traced on it, and an operator keeping a point always on this curve in the manner described above. Thus one operator will be enough to carry on the process; and I believe that in the application of it to the tidal harmonic analysis he will be able in an

* Fourier's *Théorie Analytique de la Chaleur*, § 319, p. 391 (Paris, 1822).

Machine to calculate Integral of Product of two Functions.

hour or two to find by aid of the machine any one of the simple harmonic elements of a year's tides recorded in curves in the usual manner by an ordinary tide-gauge—a result which hitherto has required not less than twenty hours of calculation by skilled arithmeticians. I believe this instrument will be of great value also in determining the diurnal, semi-diurnal, ter-diurnal, and quarter-diurnal constituents of the daily variations of temperature, barometric pressure, east and west components of the velocity of the wind, north and south components of the same; also of the three components of the terrestrial magnetic force; also of the electric potential of the air at the point where the stream of water breaks into drops in atmospheric electrometers, and of other subjects of ordinary meteorological or magnetic observations; also to estimate precisely the variation of terrestrial magnetism in the eleven years sun-spot period, and of sun-spots themselves in this period; also to disprove (or prove, as the case may be) supposed relations between sun-spots and planetary positions and conjunctions; also to investigate lunar influence on the height of the barometer, and on the components of the terrestrial magnetic force, and to find if lunar influence is sensible on any other meteorological phenomena—and if so, to determine precisely its character and amount.

From the description given above it will be seen that the mechanism required for the instrument is exceedingly simple and easy. Its accuracy will depend essentially on the accuracy of the circular cylinder, of the globe, and of the plane of the rotating disk used in it. For each of the three surfaces a much less elaborate application of the method of scraping than that by which Sir Joseph Whitworth has given a true plane with such marvellous accuracy will no doubt suffice for the practical requirements of the instrument now proposed.

V. Mechanical Integration of Linear Differential Equations of the Second Order with Variable Coefficients*.

Mechanical Integration of Linear Differential Equations of Second Order.

Every linear differential equation of the second order may, as is known, be reduced to the form

$$\frac{d}{dx}\left(\frac{1}{P}\frac{du}{dx}\right) = u \quad \text{..........................} (1),$$

where P is any given function of x.

On account of the great importance of this equation in mathematical physics (vibrations of a non-uniform stretched cord, of a hanging chain, of water in a canal of non-uniform breadth and depth, of air in a pipe of non-uniform sectional area, conduction of heat along a bar of non-uniform section or non-uniform conductivity, Laplace's differential equation of the tides, etc. etc.), I have long endeavoured to obtain a means of facilitating its practical solution.

Methods of calculation such as those used by Laplace himself are exceedingly valuable, but are very laborious, too laborious unless a serious object is to be attained by calculating out results with minute accuracy. A ready means of obtaining approximate results which shall show the general character of the solutions, such as those so well worked out by Sturm†, has always seemed to me a desideratum. Therefore I have made many attempts to plan a mechanical integrator which should give solutions by successive approximations. This is clearly done now, when we have the instrument for calculating $\int \phi(x)\,\psi(x)\,dx$, founded on my brother's disk-, globe-, and cylinder-integrator, and described in a previous communication to the Royal Society; for it is easily proved‡ that if

* Sir W. Thomson, *Proceedings of the Royal Society*, Vol. xxiv., 1876, p. 269.

† *Mémoire sur les équations différentielles linéaires du second ordre*, Liouville's *Journal*, Vol. I. 1836.

‡ Cambridge Senate-House Examination, Thursday afternoon, January 22nd, 1874.

Mechanical Integration of Linear Differential Equations of Second Order.

$$\left.\begin{aligned} u_2 &= \int_0^x P\left(C - \int_0^x u_1\,dx\right)dx, \\ u_3 &= \int_0^x P\left(C - \int_0^x u_2\,dx\right)dx, \\ &\text{etc.,} \end{aligned}\right\} \qquad (2)$$

where u_1 is any function of x, to begin with, as for example $u_1 = x$; then u_2, u_3, etc. are successive approximations converging to that one of the solutions of (1) which vanishes when $x = 0$.

Now let my brother's integrator be applied to find $C - \int_0^x u_1\,dx$, and let its result feed, as it were, continuously a second machine, which shall find the integral of the product of its result into Pdx. The second machine will give out continuously the value of u_2. Use again the same process with u_2 instead of u_1, and then u_3, and so on.

After thus altering, as it were, u_1 into u_2 by passing it through the machine, then u_2 into u_3 by a second passage through the machine, and so on, the thing will, as it were, become refined into a solution which will be more and more nearly rigorously correct the oftener we pass it through the machine. If u_{i+1} does not sensibly differ from u_i, then each is sensibly a solution.

So far I had gone and was satisfied, feeling I had done what I wished to do for many years. But then came a pleasing surprise. Compel agreement between the function fed into the double machine and that given out by it. This is to be done by establishing a connexion which shall cause the motion of the centre of the globe of the first integrator of the double machine to be the same as that of the surface of the second integrator's cylinder. The motion of each will thus be necessarily a solution of (1). Thus I was led to a conclusion which was quite unexpected; and it seems to me very remarkable that the general differential equation of the second order with variable coefficients may be rigorously, continuously, and in a single process solved by a machine.

Take up the whole matter *ab initio:* here it is. Take two of my brother's disk-, globe-, and cylinder-integrators, and connect the fork which guides the motion of the globe of each of the integrators, by proper mechanical means, with the circumference of the other integrator's cylinder. Then move one integrator's disk through an angle $= x$, and simultaneously move the other integrator's disk through an angle always $=\int_0^x Pdx$, a given function of x. The circumference of the second integrator's cylinder and the centre of the first integrator's globe move each of them through a space which satisfies the differential equation (1).

Mechanical Integration of Linear Differential Equations of Second Order.

To prove this, let at any time g_1, g_2 be the displacements of the centres of the two globes from the axial lines of the disks; and let dx, Pdx be infinitesimal angles turned through by the two disks. The infinitesimal motions produced in the circumferences of two cylinders will be

$$g_1 dx \text{ and } g_2 Pdx.$$

But the connexions pull the second and first globes through spaces respectively equal to those moved through by the circumferences of the first and second cylinders. Hence

$$g_1 dx = dg_2, \text{ and } g_2 Pdx = dg_1;$$

and eliminating g_2,

$$\frac{d}{dx}\left(\frac{1}{P}\frac{dg_1}{dx}\right)=g_1,$$

which shows that g_1 put for u satisfies the differential equation (1).

The machine gives the complete integral of the equation with its two arbitrary constants. For, for any particular value of x, give arbitrary values G_1, G_2. [That is to say mechanically; disconnect the forks from the cylinders, shift the forks till the globes' centres are at distances G_1, G_2 from the axial lines, then connect, and move the machine.]

We have for this value of x,

$$g_1 = G_1, \text{ and } \frac{dg_1}{dx}=G_2P;$$

that is, we secure arbitrary values for g_1 and $\frac{dg_1}{dx}$ by the arbitrariness of the two initial positions G_1, G_2 of the globes.

VI. Mechanical Integration of the general Linear Differential Equation of any Order with Variable Coefficients*.

Mechanical Integration of General Linear Differential Equation of Any Order

Take any number i of my brother's disk-, globe-, and cylinder-integrators, and make an integrating chain of them thus:—Connect the cylinder of the first so as to give a motion equal to its own† to the fork of the second. Similarly connect the cylinder of the second with the fork of the third, and so on. Let g_1, g_2, g_3 up to g_n be the positions‡ of the globes at any time. Let infinitesimal motions P_1dx, P_2dx, P_3dx, ... be given simultaneously to all the disks (dx denoting an infinitesimal motion of some part of the mechanism whose displacement it is convenient to take as independent variable). The motions ($d\kappa_1$, $d\kappa_2$, ... $d\kappa_i$) of the cylinders thus produced are

$$d\kappa_1 = g_1 P\, dx,\ d\kappa_2 = g_2 P_2 dx, \dots d\kappa_i = g_i P_i dx \dots\dots(1).$$

But, by the connexions between the cylinders and forks which move the globes, $d\kappa_1 = dg_2$, $d\kappa_2 = dg_3$, ... $d\kappa_{i-1} = dg_i$; and therefore

$$\left.\begin{array}{l} dg_2 = g_1 P_1 dx,\ dg_3 = g_2 P_2 dx, \dots\ dg_i = g_{i-1} P_{i-1} dx \\ \text{and}\quad d\kappa_1 = g_1 P_1 dx,\ d\kappa_2 = g_2 P_2 dx, \dots\ d\kappa_i = g_i P_i dx. \end{array}\right\}\dots(2).$$

Hence

$$g_1 = \frac{1}{P_1}\frac{d}{dx}\frac{1}{P_2}\frac{d}{dx}\cdots\frac{1}{P_{i-1}}\frac{d}{dx}\frac{1}{P_i}\frac{d\kappa_i}{dx}\dots\dots\dots\dots(3).$$

Suppose, now, for the moment that we couple the last cylinder with the first fork, so that their motions shall be equal—that is to say, $\kappa_i = g_1$. Then, putting u to denote the common value of these variables, we have

$$u = \frac{1}{P_1}\frac{d}{dx}\frac{1}{P_2}\frac{d}{dx}\cdots\frac{1}{P_{i-1}}\frac{d}{dx}\frac{1}{P_i}\frac{du}{dx}\dots\dots\dots\dots(4).$$

* Sir W. Thomson, *Proceedings of the Royal Society*, Vol. xxiv., 1876, p. 271.

† For brevity, the motion of the circumference of the cylinder is called the cylinder's motion.

‡ For brevity, the term "position" of any one of the globes is used to denote its distance, positive or negative, from the axial line of the rotating disk on which it presses.

Mechanical Integration of General Linear Differential Equation of Any Order.

Thus an endless chain or cycle of integrators with disks moved as specified above gives to each fork a motion fulfilling a differential equation, which for the case of the fork of the ith integrator is equation (4). The differential equations of the displacements of the second fork, third fork, ... $(i-1)$th fork may of course be written out by inspection from equation (4).

This seems to me an exceedingly interesting result; but though $P_1, P_2, P_3, \ldots P_i$ may be any given functions whatever of x, the differential equations so solved by the simple cycle of integrators cannot, except for the case of $i=2$, be regarded as the general linear equation of the order i, because, so far as I know, it has not been proved for any value of i greater than 2 that the general equation, which in its usual form is as follows,

$$Q_1\frac{d^iu}{dx^i}+Q_2\frac{d^{i-1}u}{dx^{i-1}}+\ldots Q_i\frac{du}{dx}-u=0\ldots\ldots\ldots\ldots(5),$$

can be reduced to the form (4). The general equation of the form (5), where $Q_1, Q_2, \ldots Q_i$ are any given forms of x, may be integrated mechanically by a chain of connected integrators thus:—

First take an open chain of i simple integrators as described above, and simplify the movement by taking

$$P_1=P_2=P_3=\ldots=P_2=1,$$

so that the speeds of all the disks are equal, and dx denotes an infinitesimal angular motion of each. Then by (2) we have

$$g_i=\frac{d\kappa_i}{dx},\ g_{i-1}=\frac{d^2\kappa_i}{dx^2},\ \ldots,\ g_2=\frac{d^{i-1}\kappa_i}{dx^{i-1}},\ g_1=\frac{d^i\kappa_i}{dx^i}\ldots\ldots(6).$$

Now establish connexions between the i forks and the ith cylinder, so that

$$Q_1g_1+Q_2g_2+\ldots+Q_{i-1}g_{i-1}+Q_ig_i=\kappa_i\ldots\ldots\ldots\ldots(7).$$

Putting in this for g_1, g_2, etc. their values by (6), we find an equation the same as (5), except that κ_i appears instead of u. Hence the mechanism, when moved so as to fulfil the condition (7), performs by the motion of its last cylinder an integration of the equation (5). This mechanical solution is complete; for we may give arbitrarily any initial values to $\kappa_i, g_i, g_{i-1}, \ldots g_3, g_2$; that is to say, to

$$u,\ \frac{du}{dx},\ \frac{d^2u}{dx^2},\ \ldots\ \frac{d^{i-1}u}{dx^{i-1}}.$$

Mechanical Integration of General Linear Differential Equation of Any Order.

Until it is desired actually to construct a machine for thus integrating differential equations of the third or any higher order, it is not necessary to go into details as to plans for the mechanical fulfilment of condition (7); it is enough to know that it can be fulfilled by pure mechanism working continuously in connexion with the rotating disks of the train of integrators.

ADDENDUM.

Mechanical Integration of any Differential Equation of Any Order.

The integrator may be applied to integrate any differential equation of any order. Let there be i simple integrators; let x_1, g_1, κ_1 be the displacements of disk, globe, and cylinder of the first, and so for the others. We have

$$g_1 = \frac{d\kappa_1}{dx_1}, \qquad g_2 = \frac{d\kappa_2}{dx_2}, \text{ etc.}$$

Now by proper mechanism establish such relations between

$$x_1, g_1, \kappa_1, x_2, g_2, \text{ etc.}$$

that

$$f^{(1)}(x_1, g_1, \kappa_1, x_2, \ldots) = 0,$$
$$f^{(2)}(x_1, g_1, \kappa_1, x_2, \ldots) = 0,$$
$$\ldots\ldots\ldots\ldots\ldots\ldots\ldots\ldots$$
$$f^{(2i-1)}(x_., g_1, \kappa_1, x_2, \ldots) = 0$$

($2i-1$ relations).

This will leave just one degree of freedom; and thus we have $2i-1$ simultaneous equations solved. As one particular case of relations take

$$x_1 = x_2 = \ldots (i-1 \text{ relations}),$$

and $$g_2 = \kappa_1, \quad g_3 = \kappa_2, \quad \text{etc. } (i-1 \text{ relations});$$

so that

$$g_1 = \frac{d\,\kappa_i}{dx^i}, \qquad g_2 = \frac{d^{i-1}\kappa_i}{dx^{i-1}}, \text{ etc.}$$

Thus one relation is still available. Let it be

$$f(x, g_1, g_2, \ldots g_i, \kappa_i) = 0.$$

Thus the machine solves the differential equation

$$f\left(x, \frac{d^i u}{dx^i}, \frac{d^{i-1}u}{dx^{i-1}}, \ldots \frac{du}{dx}, u\right) = 0 \text{ (putting } u \text{ for } \kappa_i).$$

Or again, take $2i$ double integrators. Let the disks of all be connected so as to move with the same speed, and let t be the displacement of any one of them from any particular position. Let

Mechanical Integration of any Differential Equation of Any Order.

$$x, y, x', y', x'', y'', \ldots x^{(i-1)}, y^{(i-1)}$$

be the displacements of the second cylinders of the several double integrators. Then (the second globe-frame of each being connected to its first cylinder) the displacements of the first globe-frames will be

$$\frac{d^2x}{dt^2}, \frac{d^2y}{dt^2}, \frac{d^2x'}{dt^2}, \frac{d^2y'}{dt^2}, \text{ etc.}$$

Let now X, Y, X', Y', etc. be each a given function of

$$x, y, x', y', x'', \text{ etc.}$$

By proper mechanism make the first globe of the first double integrator-frame move so that its displacement shall be equal to X, and so on. The machine then solves the equations

$$\frac{d^2x}{dt^2} = X, \quad \frac{d^2y}{dt^2} = Y, \quad \frac{d^2x'}{dt^2} = X', \text{ etc.}$$

For example, let

$$\begin{aligned} X = {} & (x'-x)f\{(x'-x)^2 + (y'-y)^2\} \\ & + (x''-x)f\{(x''-x)^2 + (y''-y)^2\} \\ & + \ldots\ldots\ldots\ldots\ldots\ldots \\ Y = {} & (y'-y)f\{(x'-x)^2 + (y'-y)^2\} \\ & + (y''-y)f\{(x''-x)^2 + (y''-y)^2\} \\ & + \ldots\ldots\ldots\ldots\ldots\ldots \\ X' = {} & \text{etc.}, \quad Y' = \text{etc.}, \end{aligned}$$

where f denotes any function.

Construct in (frictionless) steel the surface whose equation is

$$z = \xi f(\xi^2 + \eta^2)$$

(and repetitions of it, for practical convenience, though *one* theoretically suffices). By aid of it (used as if it were a cam, but for two independent variables) arrange that one moving auxiliary piece (an x-auxiliary I shall call it), capable of moving to and fro in a straight line, shall have displacement always equal to

$$(x'-x)f\{(x'-x)^2 + (y'-y)^2\},$$

that another (a y-auxiliary) shall have displacement always equal to

$$(y'-y)f\{(x'-x)^2 + (y'-y)^2\},$$

Mechanical Integration of any Differential Equation of Any Order.

that another (an x-auxiliary) shall have displacement equal to

$$(x''-x)f\{(x''-x)^2+(y''-y)^2\},$$

and so on.

Then connect the first globe-frame of the first double integrator, so that its displacement shall be equal to the sum of the displacements of the x-auxiliaries; that is to say, to

$$\begin{aligned}&(x'-x)f\{(x'-x)^2+(y'-y)^2\}\\&+(x''-x)f\{(x''-x)^2+(y''-y)^2\}\\&+\text{etc.}\end{aligned}$$

This may be done by a cord passing over pulleys attached to the x-auxiliaries, with one end of it fixed and the other attached to the globe-frame (as in my tide-predicting machine, or in Wheatstone's alphabetic telegraph-sending instrument).

Then, to begin with, adjust the second globe-frames and the second cylinders to have their displacements equal to the initial velocity-components and initial co-ordinates of i particles free to move in one plane. Turn the machine, and the positions of the particles at time t are shown by the second cylinders of the several double integrators, supposing them to be free particles attracting or repelling one another with forces varying according to any function of the distance.

The same may clearly be done for particles moving in three dimensions of space, since the components of force on each may be mechanically constructed by aid of a cam-surface whose equation is

$$z=\xi f(\eta)$$

and taking η for the distance between any two particles, and

$$\xi=x'-x$$

or $$=y'-y$$

or $$=x''-x,\text{ etc.}$$

Thus we have a complete mechanical integration of the problem of finding the free motions of any number of mutually influencing particles, not restricted by any of the approximate suppositions which the analytical treatment of the lunar and planetary theories requires.

VII. Harmonic Analyzer*.

This is a realization of an instrument designed rudimentarily in the author's communication to the Royal Society ("Proceedings," February 3rd, 1876), entitled "On an Instrument for Calculating ($\int\phi(x)\psi(x)\,dx$), the Integral of the Product of two given Functions." Harmonic Analyzer

It consists of five disk-, globe-, and cylinder integrators of the kind described in Professor James Thomson's paper "On an Integrating Machine having a new Kinematic Principle," of the same date, and represented in the woodcuts of Appendix B′, III.

The five disks are all in one plane, and their centres in one line. The axes of the cylinders are all in a line parallel to it. The diameters of the five cylinders are all equal, so are those of the globes; hence the centres of the globes are in a line parallel to the line of the centres of the disks, and to the line of the axes of the cylinders.

One long wooden rod, properly supported and guided, and worked by a rack and pinion, carries five forks to move the five globes and a pointer to trace the curve on the paper cylinder. The shaft of the paper cylinder carries at its two ends cranks at right angles to one another; and a toothed wheel which turns a parallel shaft, and a third shaft in line with the first, by means of three other toothed wheels. This third shaft carries at its two ends two cranks at right angles to one another.

Another toothed wheel on the shaft of the paper drum turns another parallel shaft, which, by a slightly oblique toothed wheel working on a crown wheel with slightly oblique teeth, turns one of the five disks uniformly (supposing to avoid circumlocution the paper drum to be turning uniformly). The cylinder of the integrator, of which this one is the disk, gives the continuously growing value of $\int y\,dx$.

Each of the four cranks gives a simple harmonic angular motion to one of the other four disks by means of a slide and crosshead, carrying a rack which works a sector attached to the disk. Hence, the cylinders moved by the disks, driven by the

* Sir W. Thomson, *Proceedings of the Royal Society*, Vol. XXVII., 1878, p.371.

Harmonic Analyzer.

first mentioned pair of cranks, give the continuously growing values of

$$\int y \cos \frac{2\pi x}{c}\, dx, \text{ and } \int y \sin \frac{2\pi x}{c}\, dx;$$

where c denotes the circumference of the paper drum: and the two remaining cylinders give

$$\int y \cos \frac{2\pi\omega x}{c}\, dx, \text{ and } \int y \sin \frac{2\pi\omega x}{c}\, dx;$$

where ω denotes the angular velocity of the shaft carrying the second pair of shafts, that of the first being unity.

The machine, with the toothed wheels actually mounted on it when shown to the Royal Society, gave $\omega = 2$, and was therefore adopted for the meteorological application. By removal of two of the wheels and substitution of two others, which were laid on the table of the Royal Society, the value of ω becomes $\frac{39 \times 109}{40 \times 110}$* (according to factors found by Mr E. Roberts, and supplied by him to the author, for the ratio of the mean lunar to the mean solar periods relatively to the earth's rotation). Thus, the same machine can serve for analyzing out simultaneously the mean lunar and mean solar semi-diurnal tides from a tide-gauge curve. But the dimensions of the actual machine do not allow range enough of motion for the majority of tide-gauge curves, and they are perfectly sufficient and suitable for meteorological work. The machine, with the train giving $\omega = 2$, is therefore handed over to the Meteorological Office to be brought immediately into practical work by Mr Scott (as soon as a brass cylinder of proper diameter to suit the $24h$ length of his curves is substituted for the wooden model cylinder in the machine as shown to the Royal Society): and the construction of a new machine for the tidal analysis, to have eleven disk-, globe-, and cylinder-integrators in line, and four crank shafts having their axes in line with the paper drum, according to the preceding description, in proper periods to analyse a tide curve by one process for mean level, and for the two components of each of the five chief tidal constituents—that is to say,

Tidal Harmonic Analyzer.

* The actual numbers of the teeth in the two pairs of wheels constituting the train are 78 : 80 and 109 : 110.

(1) The mean solar semi-diurnal;
(2) „ „ lunar „
(3) „ „ lunar quarter-diurnal, shallow-water tide;
(4) „ „ lunar declinational diurnal;
(5) „ „ luni-solar declinational diurnal;

Tidal Harmonic Analyzer.

is to be immediately commenced. It is hoped that it may be completed without need to apply for any addition to the grant already made by the Royal Society for harmonic analyzers.

Counterpoises are applied to the crank shafts to fulfil the condition that gravity on cranks, and sliding pieces, and sectors, is in equilibrium. Error from "back lash" or "lost time" is thus prevented simply by frictional resistance against the rotation of the uniformly rotating disk and of the tertiary shafts, and by the weights of the sectors attached to the oscillating disks.

Addition, April, 1879. The machine promised in the preceding paper has now been completed with one important modification:—Two of the eleven constituent integrators, instead of being devoted, as proposed in No. 3 of the preceding schedule, to evaluate the lunar quarter-diurnal shallow-water tide, are arranged to evaluate the solar declinational diurnal tide, this being a constituent of great practical importance in all other seas than the North Atlantic, and of very great scientific interest. For the evaluation of quarter-diurnal tides, whether lunar or solar, and of semi-diurnal tides of periods the halves of those of the diurnal tides, that is to say of all tidal constituents whose periods are the halves of those of the five main constituents for which the machine is primarily designed, an extra paper-cylinder, of half the diameter of the one used in the primary application of the machine, is constructed. By putting in this secondary cylinder and repassing the tidal curve through the machine the secondary tidal constituents (corresponding to the first "overtones" or secondary harmonic constituents of musical sounds) are to be evaluated. Similarly tertiary, quaternary, etc. tides (corresponding to the second and higher overtones in musical sounds) may be evaluated by passing the curve over cylinders of one-third and of smaller sub-multiples of the diameter of the primary cylinder. These secondary and tertiary tidal constituents are only perceptible at places where the rise and fall is influenced by a large area of sea, or a considerable length of

Secondary, tertiary, quaternary, etc. tides, due to influence of shallow water,—analogous to musical overtones.

Tidal Harmonic Analyser.

channel through which the whole amount of the rise and fall is notable in proportion to the mean depth. They are very perceptible at almost all commercial ports, except in the Mediterranean, and to them are due such curious and practically important tidal characteristics as the double high waters at Southampton and in the Solent and on the south coast of England from the Isle of Wight to Portland, and the protracted duration of high water at Havre.

END OF PART I.

Treatise on Natural Philosophy

Vol. II

To take another case: in the consideration of the
propagation of waves at the surface of a fluid, it is
impossible, not only on account of mathematical
difficulties, but on account of our ignorance of the what matters is,
and what forces its particles exert on each other, to form the
equations which would give us the separate motion of each.

—from Chapter V: "Abstract Dynamics"

PREFACE.

THE original design of the Authors in commencing this work about twenty years ago has not been carried out beyond the production of the first of a series of volumes, in which it was intended that the various branches of mathematical and experimental physics should be successively treated. The intention of proceeding with the other volumes is now definitely abandoned; but much new matter has been added to the first volume, and it has been divided into two parts, in the second edition now completed in this second part. The original first volume contained many references to the intended future volumes; and these references have been allowed to remain in the present completion of the new edition of the first volume, because the plan of treatment followed depended on the expectation of carrying out the original design.

Throughout the latter part of the book extensive use has, according to Prof. Stokes' revival of this valuable notation, been made of the "solidus" to replace the horizontal stroke in fractions; for example $\frac{a}{b}$ is printed a/b. This notation is (as is illustrated by the spacing between these lines) advantageous for the introduction of isolated analytical expressions in the midst of the text, and its use in printing complex fractional and exponential expressions permits the printer to dispense with much of the troublesome process known as "justification," and effects a considerable saving in space and expense.

An index to the *whole* of the first volume has been prepared by Mr BURNSIDE, and is placed at the end.

A schedule is also given below of all the amendments and additions (excepting purely verbal changes and corrections) made in the present edition of the first volume.

Inspection of the schedules on pages xxii. to xxv. will shew that much new matter has been imported into the present edition, both in Part I. and Part II. These additions are indicated by the word "new."

The most important part of the labour of editing Part II. has been borne by Mr G. H. DARWIN, and it will be seen from the schedule below that he has made valuable contributions to the work.

CONTENTS.

DIVISION II.—ABSTRACT DYNAMICS.

SECTIONS

SCHEDULE OF ALTERATIONS AND ADDITIONS IN PART I., VOL. I.

§ 9. Partly rewritten.
§ 12. Small addition.
§ 58. Reference to Tide Predicter added.
§ 62. Addition.
§ 76. Small addition.
§ 90. Some alteration.
§ 95. Rodrigues' co-ordinates (new).
§ 96. Addition.
§ 97. Small addition.
§ 100. Somewhat altered.
§ 101. Rewritten and enlarged.
§ 102. Rewritten.
§ 107. Addition.
§ 108. Addition—transient terrestrial nutation of 306 days (new).
§ 110. Slight alteration.
§ 111. Addition—rolling and spinning bodies (new).
§ 123. Addition—dynamics of twist in kinks (new).
§ 130. Small alteration.
§ 137. Addition—integral curvature, horograph (new).
§ 184. Some alteration.
§ 190*j*. Addition.
§ 198. Rewritten—degrees of freedom—geometrical slide (new).
§ 200. Small alterations.
§ 201. Slight alteration.
Appendix A_0. Laplace's equation in generalised co-ordinates (new).
Appendix B. Examples of spherical harmonics—rewritten and much extended (new).
§ 212. Slight addition.
§ 223. Units of length and time (new).
§ 225. Small addition.
§ 245. Part omitted.
§ 267. Small addition.
§ 276. Footnote (new).
§ 282. Definition of "Principal axes" (new).
§ 283. Reference added.
§ 289. Slight addition.
§ 293. Considerable addition (new).
§ 298. Small alterations and additions.
§ 312. Addition.

§ 314 } Slight alteration.
§ 316 }
§ 317. Small alteration.
§ 318. Old § 329 rewritten and extended.
§ 319. Old § 330—with considerable additions—ignoration of co-ordinates (new).
§ 320 to § 324. Same as old § 331 to § 335.
§ 325. Extended from old § 336—addition to observed phenomena of fluid motion.
§ 326 to § 336. Same as old § 318 to § 328, with some alterations—considerable addition, to § 319 now § 327.
§ 337. Addition including slightly disturbed equilibrium (new).
§ 338 } Some addition.
§ 340 }
§ 341. Extended to include old § 342 with addition.
§ 342. Same as non-mathematical portion of old § 343.
§ 343, *a* to *p*. On the motions of a cycloidal system rewritten and greatly extended.
§ 344. Rewritten.
§ 345, i. to xxviii. Oscillations with friction—dissipation of energy—positional and motional forces—gyrostatics—stability (new).
§ 373 and § 374. Same as old § 373.
§ 374 to § 380. Same as old § 375 to § 379, with alterations.
§ 381 and § 382. Same as old § 380.
§ 383 to § 386. Same as old §§ 381 to § 384. Old § 385 and § 386 omitted.
§ 398′. Harmonic analysis (new).
§ 401. Addition on calculating machines (new).
§ 404. Rewritten.
§ 405. Foot-note quoted from old § 830; compare with new § 830.
§ 408. Slight alteration.
§ 409 } Rewritten.
§ 427 }
§ 429. Part rewritten.
§ 431. Rewritten.
§ 435. Extended—bifilar balance (new).
Appendix B′, I. Tide-predicter (new).
,, II. Equation-solver (new).
,, III. to VI. Mechanical integrator (new).
,, VII. Harmonic analyser (new).

SCHEDULE OF ALTERATIONS AND ADDITIONS IN PART II., VOL. I.

§ 443. Part rewritten.
§ 451. Slightly altered and part omitted.
§ 452. Same as part of old § 451—old § 452 omitted.
§ 453 and § 454. Rewritten.
§ 455 } Small omission.
§ 458 }
§ 478 and § 479. Small addition.
§ 491 (*f*) and § 492. Slight alteration.
§ 493. Integral of normal attraction over a closed surface (new).
§ 494, *a* to *q*. Theory of potential—attraction of ellipsoids (new).
§ 495 (*a*), (*b*), & (*c*). Same as old §§ 493, 494, and 495.
§ 496. Small addition.
§ 501. Example added.
§ 506. Part rewritten.
§ 507. Slight alteration.
§ 519. Old § 520 rewritten, including part of § 519.
§ 520. Distribution of electricity on an ellipsoidal conductor (new).
§ 521 to § 525. Attraction of Homoeoids (new), including old § 523.
§ 526 and § 527. Attraction of ellipsoids (new), rewritten for old § 522.
§ 528 to § 530. Mathematical part of old § 519 rewritten.
§ 531. Old § 524 rewritten.
§ 532. Old § 521 rewritten.
§ 533. Same as old § 525 with small addition.
§ 534. Same as old § 526 and § 527.
§ 534 (*a*) to § 534 (*g*). Same as old § 528 to § 534.
§ 551 to § 557. Equilibrium of free and constrained rigid bodies, including Theory of Screws (new)—old § 551 omitted.
§ 558 to § 559 (*f*). Same as old § 552 to § 559, partly rewritten and slightly altered.
§ 561. Rewritten.
§ 562 to § 569. Slight alterations.
§ 572. Theory of balance—considerably altered.
§ 597. Modified.
§ 599. Proof added.
§ 609. Rewritten.
§ 628. Slight alteration.

§ 632. Slightly modified.
§ 686. Note on modulus of elasticity (new).
§ 688. Slightly modified.
§ 691, *a* to *f*. Theory of elasticity—wave propagation—resilience (new).
§ 737 (*h*). Small addition.
§ 740. Small addition.
§ 755. Note on cyclic functions (new).
§ 771. Somewhat altered—includes part of old § 772.
§ 772. Rewritten, table altered.
§ 776 and § 777. Rewritten.
§ 778 and § 778'. Old § 778 rewritten.
§ 778''. Equilibrium of rotating masses of fluid (new).
§ 797. Addition—latest results of geodesy (new).
§ 803 and § 809. Slight alterations.
§ 810. Small addition.
§ 812. Addition—lunar disturbance of gravity (new).
§ 818'. Gravitational observatories (new, G. H. D.).
§ 824, 824'. Addition—ellipticity of internal strata of earth and other planets (new, G. H. D.).
§ 830. Entirely rewritten and extended (new, G. H. D.).
§ 832'. Rigidity and strength of materials of earth (new, G. H. D.).
§ 834. Modification of analysis and correction (G. H. D.).
§ 835. Slight alteration.
§ 837. Small addition (G. H. D.).
§ 840'. Theory of elastic tides (new, G. H. D.).
§ 847 to end. Rigidity of earth deduced from tidal observations (new, G. H. D.) —old § 847 to end entirely omitted.
Appendix E. Heat of the Sun (new).
Appendix F. Size of atoms (new).
Appendix G. Tidal friction (new, G. H. D.).

DIVISION II.

ABSTRACT DYNAMICS.

CHAPTER V.

INTRODUCTORY.

438. UNTIL we know thoroughly the nature of matter and the forces which produce its motions, it will be utterly impossible to submit to mathematical reasoning the *exact* conditions of any physical question. It has been long understood, however, that approximate solutions of problems in the ordinary branches of Natural Philosophy may be obtained by a species of *abstraction*, or rather *limitation of the data*, such as enables us easily to solve the modified form of the question, while we are well assured that the circumstances (so modified) affect the result only in a superficial manner. **Approximate treatment of physical questions.**

439. Take, for instance, the very simple case of a crowbar employed to move a heavy mass. The accurate mathematical investigation of the action would involve the simultaneous treatment of the motions of every part of bar, fulcrum, and mass raised; but our ignorance of the nature of matter and molecular forces, precludes any such complete treatment of the problem.

It is a result of observation that the particles of the bar, fulcrum, and mass, separately, retain throughout the process nearly the same relative positions. Hence the idea of solving,

Approximate treatment of physical questions

instead of the complete but infinitely transcendent problem, another, in reality quite different, but which, while amply simple, obviously leads to practically the same results so far as concerns the equilibrium and motions of the bodies as a whole.

440. The new form is given at once by the experimental result of the trial. Imagine the masses involved to be *perfectly rigid*, that is, incapable of changing form or dimensions. Then the infinite series of forces, really acting, may be left out of consideration; so that the mathematical investigation deals with a finite (and generally small) number of forces instead of a practically infinite number. Our warrant for such a substitution is to be established thus.

441. The effects of the intermolecular forces could be exhibited only in alterations of the form or volume of the masses involved. But as these (practically) remain almost unchanged, the forces which produce, or tend to produce, them may be left out of consideration. Thus we are enabled to investigate the action of machinery supposed to consist of separate portions whose form and dimensions are unalterable.

Further approximations.

442. If we go a little further into the question, we find that the lever *bends*, some parts of it are extended and others compressed. This would lead us into a very serious and difficult inquiry if we had to take account of the whole circumstances. But (by experience) we find that a sufficiently accurate solution of this more formidable case of the problem may be obtained by supposing (what can *never* be realized in practice) the mass to be homogeneous, and the forces consequent on a dilatation, compression, or distortion, to be proportional in magnitude, and opposed in direction, to these deformations respectively. By this further assumption, close approximations may be made to the vibrations of rods, plates, etc., as well as to the statical effect of springs, etc.

443. We may pursue the process further. Compression, in general, produces heat, and extension, cold. The elastic forces of the material are thus rendered sensibly different from what they would be with the same changes of bulk and shape, but with no change of temperature. By introducing such considerations, we reach, without great difficulty, what may be called a *third* approximation to the solution of the physical problem considered. Further approximations.

444. We might next introduce the conduction of the heat, so produced, from point to point of the solid, with its accompanying modifications of elasticity, and so on; and we might then consider the production of thermo-electric currents, which (as we shall see) are always developed by unequal heating in a mass if it be not perfectly homogeneous. Enough, however, has been said to show, *first*, our utter ignorance as to the true and complete solution of any physical question by the only perfect method, that of the consideration of the circumstances which affect the motion of every portion, separately, of each body concerned; and, *second*, the practically sufficient manner in which practical questions may be attacked by limiting their generality, *the limitations introduced being themselves deduced from experience*, and being therefore Nature's own solution (to a less or greater degree of accuracy) of the infinite additional number of equations by which we should otherwise have been encumbered.

445. To take another case: in the consideration of the propagation of waves at the surface of a fluid, it is impossible, not only on account of mathematical difficulties, but on account of our ignorance of *what* matter is, and what forces its particles exert on each other, to form the equations which would give us the separate motion of each. Our first approximation to a solution, and one sufficient for most practical purposes, is derived from the consideration of the motion of a homogeneous, incompressible, and perfectly plastic mass; a hypothetical substance which may have no existence in nature.

446. Looking a little more closely, we find that the actual motion differs considerably from that given by the analytical solution of the restricted problem, and we introduce further considerations, such as the *compressibility* of fluids, their *internal friction*, the heat generated by the latter, and its effects in dilating the mass, etc. etc. By such successive corrections we

Further approximations. attain, at length, to a mathematical result which (at all events in the present state of experimental science) agrees, within the limits of experimental error, with observation.

447. It would be easy to give many more instances substantiating what has just been advanced, but it seems scarcely necessary to do so. We may therefore at once say that there is no question in physical science which can be *completely and accurately* investigated by mathematical reasoning, but that there are different degrees of approximation, involving assumptions more and more nearly coincident with observation, which may be arrived at in the solution of any particular question.

Object of the present division of the work. 448. *The object of the present division of this volume is to deal with the first and second of these approximations.* In it we shall suppose all solids either RIGID, *i.e.*, unchangeable in form and volume, or ELASTIC; but in the latter case, we shall assume the law, connecting a compression or a distortion with the force which causes it, to have a particular form deduced from experiment. And we shall in the latter case neglect the thermal or electric effects which compression or distortion generally cause. We shall also suppose fluids, whether liquids or gases, to be either INCOMPRESSIBLE or compressible according to certain known laws; and we shall omit considerations of fluid friction, although we admit the consideration of friction between solids. Fluids will therefore be supposed *perfect*, *i.e.*, such that any particle may be moved amongst the others by the slightest force.

449. When we come to Properties of Matter and the various forms of Energy, we shall give in detail, as far as they are yet known, the modifications which further approximations have introduced into the previous results.

Laws of friction. 450. The laws of friction between solids were very ably investigated by Coulomb; and, as we shall require them in the succeeding chapters, we give a brief summary of them here; reserving the more careful scrutiny of experimental results to our chapter on Properties of Matter.

451. To produce and to maintain sliding of one solid body on another requires a tangential force which depends—(1) upon

the nature of the bodies; (2) upon their polish, or the species and quantity of lubricant which may have been applied; (3) upon the normal pressure between them, to which it is in general directly proportional. It does not (except in some extreme cases where scratching or excessive abrasion takes place) depend sensibly upon the area of the surfaces in contact. When two bodies are pressed together without being caused to slide one on another, the force which prevents sliding is called Statical Friction. It is capable of opposing a tangential resistance to motion which may be of any amount less than or at most equal to μR; where R is the whole normal pressure between the bodies; and μ (which depends mainly upon the nature of the surfaces in contact) is what is commonly called the *coefficient of Statical Friction*. This coefficient varies greatly with the circumstances, being in some cases as low as 0·03, in others as high as 0·80. Later, we shall give a table of its values. When the applied forces are insufficient to produce motion, the whole amount of statical friction is not called into play; its amount then just reaches what is sufficient to equilibrate the other forces, and its direction is the opposite of that in which their resultant tends to produce motion. **Laws of friction.**

452. When the statical friction has been overcome, and sliding is produced, experiment shows that a force of friction continues to act, opposing the motion; that this force of *Kinetic Friction* is in most cases considerably less than the extreme force of static friction which had to be overcome before the sliding commenced; that it too is sensibly proportional to the normal pressure; and that it is approximately the same whatever be the velocity of the sliding.

453. In the following Chapters on Abstract Dynamics we confine ourselves mainly to the general principles, and the fundamental formulas and equations of the mathematics of this extensive subject; and, neither seeking nor avoiding mathematical exercitations, we enter on special problems solely with a view to possible usefulness for physical science, whether in the way of the *material* of experimental investigation, or for illustrating physical principles, or for aiding in speculations of Natural Philosophy. **Rejection of merely *curious* speculations.**

CHAPTER VI.

STATICS OF A PARTICLE.—ATTRACTION.

Objects of the chapter. 454. We naturally divide Statics into two parts—the equilibrium of a particle, and that of a rigid or elastic body or system of particles whether solid or fluid. In a very few sections we shall dispose of the first of these parts, and the rest of this chapter will be devoted to a digression on the important subject of Attraction.

Conditions of equilibrium of a particle. 455. By § 255, forces acting at the same point, or on the same material particle, are to be compounded by the same laws as velocities. Hence, evidently, the sum of their components in any direction must vanish if there is equilibrium; and there is equilibrium if the sums of the components in each of three lines not in one plane are each zero. And thence the necessary and sufficient mathematical equations of equilibrium.

Thus, for the equilibrium of a material particle, it is *necessary*, and *sufficient*, that the (algebraic) sums of the components of the applied forces, resolved in any three rectangular directions, should vanish.

Equilibrium of a particle. If P be one of the forces, l, m, n its direction-cosines, we have

$$\Sigma lP=0, \quad \Sigma mP=0, \quad \Sigma nP=0.$$

If there be not equilibrium, suppose R, with direction-cosines λ, μ, ν, to be the resultant force. If reversed in direction, it will, with the other forces, produce equilibrium. Hence

$$\Sigma lP-\lambda R=0, \quad \Sigma mP-\mu R=0, \quad \Sigma nP-\nu R=0.$$

Equilibrium of a particle

And $$R^2 = (\Sigma lP)^2 + (\Sigma mP)^2 + (\Sigma nP)^2,$$

while $$\frac{\lambda}{\Sigma lP} = \frac{\mu}{\Sigma mP} = \frac{\nu}{\Sigma nP}.$$

456. We may take one or two particular cases as examples of the general results above. Thus,

(1) If the particle rest on a frictionless curve, the component force along the curve must vanish.

If x, y, z be the co-ordinates of the point of the curve at which the particle rests, we have evidently

$$\Sigma P\left(l\frac{dx}{ds} + m\frac{dy}{ds} + n\frac{dz}{ds}\right) = 0.$$

When P, l, m, n are given in terms of x, y, z, this, with the *two* equations to the curve, determines the position of equilibrum.

(2) If the curve be frictional, the resultant force along it must be balanced by the friction.

If F be the friction, the condition is

$$\Sigma P\left(l\frac{dx}{ds} + m\frac{dy}{ds} + n\frac{dz}{ds}\right) - F = 0.$$

This gives the amount of friction which will be called into play; and equilibrium will subsist until, as a limit, the friction is μ times the normal pressure on the curve. But the normal pressure is

$$\Sigma P\left\{\left(m\frac{dz}{ds} - n\frac{dy}{ds}\right)^2 + \left(n\frac{dx}{ds} - l\frac{dz}{ds}\right)^2 + \left(l\frac{dy}{ds} - m\frac{dx}{ds}\right)^2\right\}^{\frac{1}{2}}.$$

Hence, the limiting positions, between which equilibrium is possible, are given by the two equations to the curve, combined with

$$\Sigma P\left(l\frac{dx}{ds} + m\frac{dy}{ds} + n\frac{dz}{ds}\right) \pm \mu\Sigma P\left\{\left(m\frac{dz}{ds} - n\frac{dy}{ds}\right)^2 + \left(n\frac{dx}{ds} - l\frac{dz}{ds}\right)^2 + \left(l\frac{dy}{ds} - m\frac{dx}{ds}\right)^2\right\}^{\frac{1}{2}} = 0.$$

(3) If the particle rest on a smooth surface, the resultant of the applied forces must evidently be perpendicular to the surface.

If $\phi(x, y, z) = 0$ be the equation of the surface, we must therefore have

$$\frac{\frac{d\phi}{dx}}{\Sigma lP} = \frac{\frac{d\phi}{dy}}{\Sigma mP} = \frac{\frac{d\phi}{dz}}{\Sigma nP},$$

and these three equations determine the position of equilibrium.

Equilibrium of a particle. (4) If it rest on a rough surface, friction will be called into play, resisting motion along the surface; and there will be equilibrium at any point within a certain boundary, determined by the condition that at *it* the friction is μ times the normal pressure on the surface, while within it the friction bears a less ratio to the normal pressure. When the only applied force is gravity, we have a very simple result, which is often practically useful. Let θ be the angle between the normal to the surface and the vertical at any point; the normal pressure on the surface is evidently $W\cos\theta$, where W is the weight of the particle; and the resolved part of the weight parallel to the surface, which must of course be balanced by the friction, is $W\sin\theta$. In the limiting position, when sliding is just about to commence, the greatest possible amount of statical friction is called into play, and we have

$$W\sin\theta = \mu W\cos\theta,$$

or

$$\tan\theta = \mu.$$

Angle of repose. The value of θ thus found is called the *Angle of Repose*.

Let $\phi(x, y, z) = 0$ be the surface: P, with direction-cosines l, m, n, the resultant of the applied forces. The normal pressure is

$$P\frac{l\frac{d\phi}{dx}+m\frac{d\phi}{dy}+n\frac{d\phi}{dz}}{\sqrt{\left(\frac{d\phi}{dx}\right)^2+\left(\frac{d\phi}{dy}\right)^2+\left(\frac{d\phi}{dz}\right)^2}}.$$

The resolved part of P parallel to the surface is

$$P\sqrt{\frac{\left(m\frac{d\phi}{dz}-n\frac{d\phi}{dy}\right)^2+\left(n\frac{d\phi}{dx}-l\frac{d\phi}{dz}\right)^2+\left(l\frac{d\phi}{dy}-m\frac{d\phi}{dx}\right)^2}{\left(\frac{d\phi}{dx}\right)^2+\left(\frac{d\phi}{dy}\right)^2+\left(\frac{d\phi}{dz}\right)^2}}.$$

Hence, for the boundary of the portion of the surface within which equilibrium is possible, we have the additional equation

$$\left(m\frac{d\phi}{dz}-n\frac{d\phi}{dy}\right)^2+\left(n\frac{d\phi}{dx}-l\frac{d\phi}{dz}\right)^2+\left(l\frac{d\phi}{dy}-m\frac{d\phi}{dx}\right)^2=\mu^2\left(l\frac{d\phi}{dx}+m\frac{d\phi}{dy}+n\frac{d\phi}{dz}\right)^2.$$

Attraction. **457.** A most important case of the composition of forces acting at one point is furnished by the consideration of the attraction of a body of any form upon a material particle anywhere situated. Experiment has shown that the attraction exerted by any portion of matter upon another is not modified by the proximity, or even by the interposition, of other matter; and thus the attraction of a body on a particle is the resultant of the attractions exerted by its several parts. To treatises on applied mathematics we must refer for the examination of the consequences, often very curious, of various laws of attraction; but, dealing with Natural Philosophy, we confine ourselves mainly, (and except where we give the mathematics of Laplace's beautiful and instructive and physically important, though unreal, theory of capillary attraction,) to the law of the inverse square of the distance which Newton discovered for gravitation. This, indeed, furnishes us with an ample supply of most interesting as well as useful results. Attraction

458. The law, which (as a property of matter) is to be carefully considered in the next proposed Division of this Treatise, may be thus enunciated. Universal law of attraction.

Every particle of matter in the universe attracts every other particle, with a force whose direction is that of the line joining the two, and whose magnitude is directly as the product of their masses, and inversely as the square of their distance from each other.

Experiment shows (as will be seen further on) that the same law holds for electric and magnetic attractions under properly defined conditions.

459. For the special applications of Statical principles to which we proceed, it will be convenient to use a special unit of mass, or quantity of matter, and corresponding units for the measurement of electricity and magnetism. Special unit of quantity of matter.

Thus if, in accordance with the physical law enunciated in § 458, we take as the expression for the forces exerted on each other by masses M and m, at distance D,

$$\frac{Mm}{D^2};$$

it is obvious that our *unit* force is the mutual attraction of two units of mass placed at unit of distance from each other.

Linear, surface, and volume, densities.

460. It is convenient for many applications to speak of the *density* of a distribution of matter, electricity, etc., along a line, over a surface, or through a volume.

Here line-density	= quantity of matter per unit of length.			
surface-density =	,,	,,	,,	area.
volume-density =	,,	,,	,,	volume.

Electric and magnetic reckonings of *quantity*.

461. In applying the succeeding investigations to electricity or magnetism, it is only necessary to premise that M and m stand for *quantities* of free electricity or magnetism, whatever these may be, and that here the idea of *mass* as depending on *inertia* is not necessarily involved. The formula $\frac{Mm}{D^2}$ will still represent the mutual action, if we take as unit of imaginary electric or magnetic matter, such a quantity as exerts unit force on an equal quantity at unit distance. Here, however, one or both of M, m may be negative; and, as in these applications like kinds *repel* each other, the mutual action will be attraction or repulsion, according as its sign is negative or positive. With these provisos, the following theory is applicable to any of the above-mentioned classes of forces. We commence with a few simple cases which can be completely treated by means of elementary geometry.

Positive and negative masses admitted in abstract theory of attraction.

Uniform spherical shell. Attraction on *internal* point.

462. *If the different points of a spherical surface attract equally with forces varying inversely as the squares of the distances, a particle placed within the surface is not attracted in any direction.*

Let $HIKL$ be the spherical surface, and P the particle within it. Let two lines HK, IL, intercepting very small arcs HI, KL, be drawn through P; then, on account of the similar triangles HPI, KPL, those arcs will be proportional to the distances HP, LP; and any small elements of the spherical surface at HI and KL, each bounded all round by straight lines passing through P [and very nearly coinciding with HK], will be in the duplicate ratio of those lines.

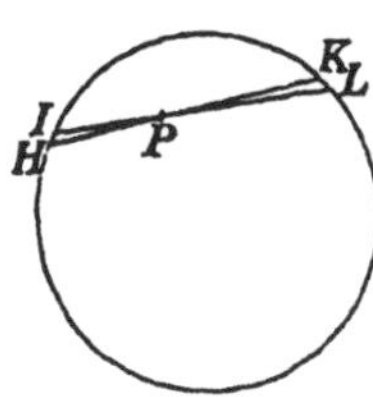

Hence the forces exercised by the matter of these elements on the particle P are equal; for they are as the quantities of matter directly, and the squares of the distances, inversely; and these two ratios compounded give that of equality. The attractions therefore, being equal and opposite, balance one another: and a similar proof shows that the attractions due to all parts of the whole spherical surface are balanced by contrary attractions. Hence the particle P is not urged in any direction by these attractions.

Uniform spherical shell. Attraction on *internal* point.

463. The division of a spherical surface into infinitely small elements will frequently occur in the investigations which follow: and Newton's method, described in the preceding demonstration, in which the division is effected in such a manner that all the parts may be taken together in *pairs of opposite elements with reference to an internal point;* besides other methods deduced from it, suitable to the special problems to be examined; will be repeatedly employed. The present digression, in which some definitions and elementary geometrical propositions regarding this subject are laid down, will simplify the subsequent demonstrations, both by enabling us, through the use of convenient terms, to avoid circumlocution, and by affording us convenient means of reference for elementary principles, regarding which repeated explanations might otherwise be necessary.

Digression on the division of surfaces into elements.

464. If a straight line which constantly passes through a fixed point be moved in any manner, it is said to describe, or generate, a *conical surface* of which the fixed point is the vertex.

Explanations and definitions regarding cones.

If the generating line be carried from a given position continuously through any series of positions, no two of which coincide, till it is brought back to the first, the entire line on the two sides of the fixed point will generate a complete conical surface, consisting of two sheets, which are called *vertical or opposite cones.* Thus the elements HI and KL, described in Newton's demonstration given above, may be considered as being cut from the spherical surface by two *opposite cones* having P for their common vertex.

The solid angle of a cone, or of a complete conical surface.

465. If any number of spheres be described from the vertex of a cone as centre, the segments cut from the concentric spherical surfaces will be similar, and their areas will be as the squares of the radii. The quotient obtained by dividing the area of one of these segments by the square of the radius of the spherical surface from which it is cut, is taken as the measure of the *solid angle of the cone.* The segments of the same spherical surfaces made by the opposite cone, are respectively equal and similar to the former (but "perverted"). Hence the solid angles of two vertical or opposite cones are equal: either may be taken as the solid angle of the complete conical surface, of which the opposite cones are the two sheets.

Sum of all the solid angles round a point $=4\pi$.

466. Since the area of a spherical surface is equal to the square of its radius multiplied by 4π, it follows that the sum of the solid angles of all the distinct cones which can be described with a given point as vertex, is equal to 4π.

Sum of the solid angles of all the complete conical surfaces $=2\pi$.

467. The solid angles of vertical or opposite cones being equal, we may infer from what precedes that the sum of the solid angles of all the complete conical surfaces which can be described without mutual intersection, with a given point as vertex, is equal to 2π.

Solid angle subtended at a point by a terminated surface.

468. The solid angle subtended at a point by a superficial area of any kind, is the solid angle of the cone generated by a straight line passing through the point, and carried entirely round the boundary of the area.

Orthogonal and oblique sections of a small cone.

469. A very small cone, that is, a cone such that any two positions of the generating line contain but a very small angle, is said to be cut at right angles, or orthogonally, by a spherical surface described from its vertex as centre, or by any surface, whether plane or curved, which touches the spherical surface at the part where the cone is cut by it.

A very small cone is said to be cut obliquely, when the section is inclined at any finite angle to an orthogonal section; and this angle of inclination is called the *obliquity of the section.*

The area of an orthogonal section of a very small cone is equal to the area of an oblique section in the same position, multiplied by the cosine of the obliquity. Orthogonal and oblique sections of a small cone.

Hence the area of an oblique section of a small cone is equal to the quotient obtained by dividing the product of the square of its distance from the vertex, into the solid angle, by the cosine of the obliquity.

470. Let E denote the area of a very small element of a spherical surface at the point E (that is to say, an element every part of which is very near the point E), let ω denote the solid angle subtended by E at any point P, and let PE, produced if necessary, meet the surface again in E': then, a denoting the radius of the spherical surface, we have Area of segment cut from spherical surface by small cone.

$$E = \frac{2a \cdot \omega \cdot PE^2}{EE'}.$$

For, the obliquity of the element E, considered as a section of the cone of which P is the vertex and the element E a section; being the angle between the given spherical surface and another described from P as centre, with PE as radius; is equal to the angle between the radii, EP and EC, of the two spheres. Hence, by considering the isosceles triangle ECE', we find that the cosine of the obliquity is equal to $\frac{\frac{1}{2}EE'}{EC}$ or to $\frac{EE'}{2a}$, and we arrive at the preceding expression for E.

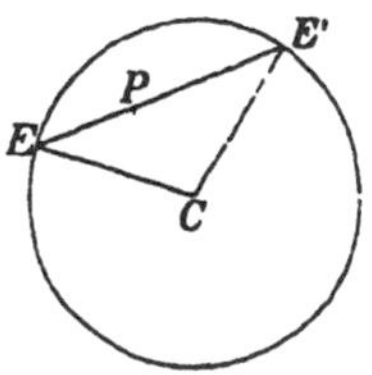

471. *The attraction of a uniform spherical surface on an external point is the same as if the whole mass were collected at the centre**. Uniform spherical shell. Attraction of external point.

* This theorem, which is more comprehensive than that of Newton in his first proposition regarding attraction on an external point (Prop. LXXI.), is fully established as a corollary to a subsequent proposition (Prop. LXXIII. cor. 2). If we had considered the proportion of the forces exerted upon two external points at different distances, instead of, as in the text, investigating the absolute force on one point, and if besides we had taken together all the pairs of elements which would constitute two narrow annular portions of the surface, in planes perpendicular to PC, the theorem and its demonstration would have coincided precisely with Prop. LXXI. of the *Principia*.

Uniform spherical shell. Attraction on *external* point.

Let P be the external point, C the centre of the sphere, and CAP a straight line cutting the spherical surface in A. Take I in CP, so that CP, CA, CI may be continual proportionals, and let the whole spherical surface be divided into *pairs of opposite elements with reference to the point I.*

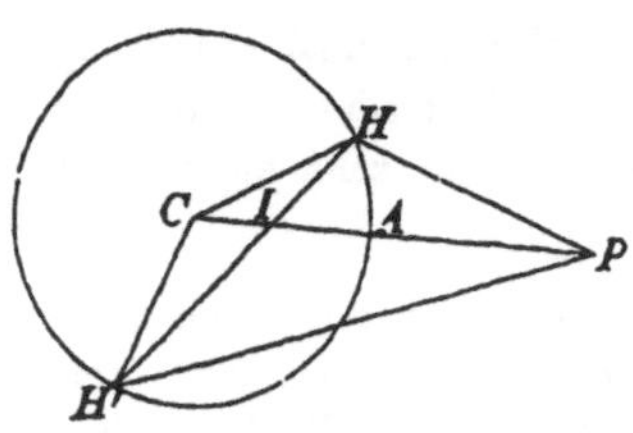

Let H and H' denote the magnitudes of a pair of such elements, situated respectively at the extremities of a chord HH'; and let ω denote the magnitude of the solid angle subtended by either of these elements at the point I.

We have (§ 469),

$$H = \frac{\omega . IH^2}{\cos CHI}, \text{ and } H' = \frac{\omega . IH'^2}{\cos CH'I}.$$

Hence, if ρ denote the density of the surface, the attractions of the two elements H and H' on P are respectively

$$\rho \frac{\omega}{\cos CHI} \cdot \frac{IH^2}{PH^2}, \text{ and } \rho \frac{\omega}{\cos CH'I} \cdot \frac{IH'^2}{PH'^2}.$$

Now the two triangles PCH, HCI have a common angle at C, and, since $PC : CH :: CH : CI$, the sides about this angle are proportional. Hence the triangles are similar; so that the angles CPH and CHI are equal, and

$$\frac{IH}{HP} = \frac{CH}{CP} = \frac{a}{CP}.$$

In the same way it may be proved, by considering the triangles PCH', $H'CI$, that the angles CPH' and $CH'I$ are equal, and that

$$\frac{IH'}{H'P} = \frac{CH'}{CP} = \frac{a}{CP}.$$

Hence the expressions for the attractions of the elements H and H' on P become

$$\rho \frac{\omega}{\cos CHI} \cdot \frac{a^2}{CP^2}, \text{ and } \rho \frac{\omega}{\cos CH'I} \cdot \frac{a^2}{CP^2},$$

which are equal, since the triangle HCH' is isosceles; and, for the same reason, the angles CPH, CPH', which have been proved to be respectively equal to the angles CHI, $CH'I$, are equal. We infer that the resultant of the forces due to the two elements is in the direction PC, and is equal to

Uniform spherical shell. Attraction on external point.

$$2\omega \cdot \rho \cdot \frac{a^2}{CP^2}.$$

To find the total force on P, we must take the sum of all the forces along PC due to the pairs of opposite elements; and, since the multiplier of ω is the same for each pair, we must add all the values of ω, and we therefore obtain (§ 467), for the required resultant,

$$\frac{4\pi\rho a^2}{CP^2}.$$

The numerator of this expression; being the product of the density, into the area of the spherical surface; is equal to the whole mass; and therefore the force on P is the same as if the whole mass were collected at C.

Cor. The force on an external point, infinitely near the surface, is equal to $4\pi\rho$, and is in the direction of a normal at the point. The force on an internal point, however near the surface, is, by a preceding proposition, *nil*.

Attraction on an element of the surface.

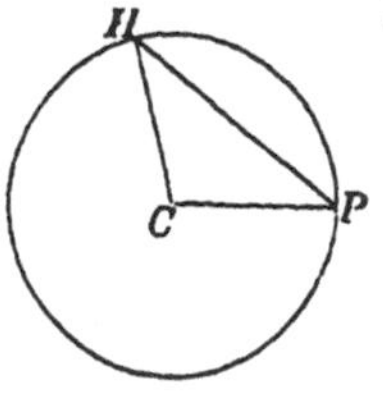

472. Let σ be the area of an infinitely small element of the surface at any point P, and at any other point H of the surface let a small element subtending a solid angle ω, at P, be taken. The area of this element will be equal to

$$\frac{\omega \cdot PH^2}{\cos CHP},$$

and therefore the attraction along HP, which it exerts on the element σ at P, will be equal to

$$\frac{\rho\omega \cdot \rho\sigma}{\cos CHP}, \text{ or } \frac{\omega}{\cos CHP}\rho^2\sigma.$$

Now the total attraction on the element at P is in the direction CP; the component in this direction of the attraction due to the element H, is

$$\omega \cdot \rho^2\sigma;$$

Attraction on an element of the surface.

and, since all the cones corresponding to the different elements of the spherical surface lie on the same side of the tangent plane at P, we deduce, for the resultant attraction on the element σ,

$$2\pi\rho^2\sigma.$$

From the corollary to the preceding proposition, it follows that this attraction is half the force which would be exerted on an external point, possessing the same quantity of matter as the element σ, and placed infinitely near the surface.

473. In some of the most important elementary problems of the theory of electricity, spherical surfaces with densities varying inversely as the cubes of distances from eccentric points occur: and it is of fundamental importance to find the attraction of such a shell on an internal or external point. This may be done synthetically as follows; the investigation being, as we shall see below, virtually the same as that of § 462, or § 471.

Attraction of a spherical surface of which the density varies inversely as the cube of the distance from a given point.

474. Let us first consider the case in which the given point S and the attracted point P are separated by the spherical surface. The two figures represent the varieties of this case in which, the point S being without the sphere, P is within; and, S being within, the attracted point is external. The same demonstration is applicable literally with reference to the two figures; but, to avoid the consideration of negative quantities, some of the expressions may be conveniently modified to suit the second figure. In such instances the two expressions are given in a double line, the upper being that which is most convenient for the first figure, and the lower for the second.

Let the radius of the sphere be denoted by a, and let f be the distance of S from C, the centre of the sphere (not represented in the figures).

Join SP and take T in this line (or its continuation) so that

$$\text{(fig. 1)} \quad SP \,.\, ST = f^2 - a^2.$$
$$\text{(fig. 2)} \quad SP \,.\, TS = a^2 - f^2.$$

Through T draw any line cutting the spherical surface at K', K. Join SK, SK', and let the lines so drawn cut the spherical surface again in E, E'.

Let the whole spherical surface be divided into pairs of opposite elements with reference to the point T. Let K and K' be a pair of such elements situated at the extremities of the chord KK', and subtending the solid angle ω at the point T; and let elements E and E' be taken subtending at S the same solid angles respectively as the elements K and K'. By this means we may divide the whole spherical surface into pairs of conjugate elements, E, E', since it is easily seen that when we have taken every pair of elements, K, K', the whole surface

Attraction of a spherical surface of which the density varies inversely as the cube of the distance from a given point.

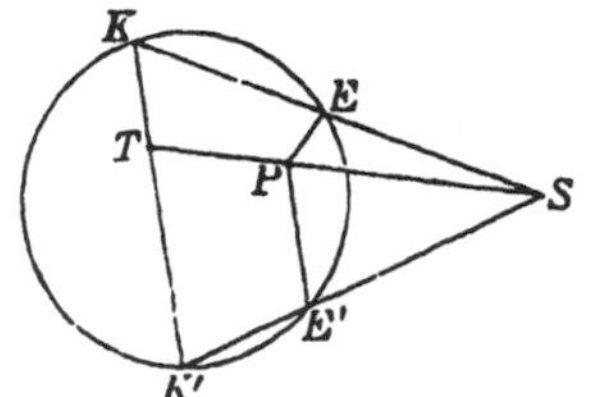

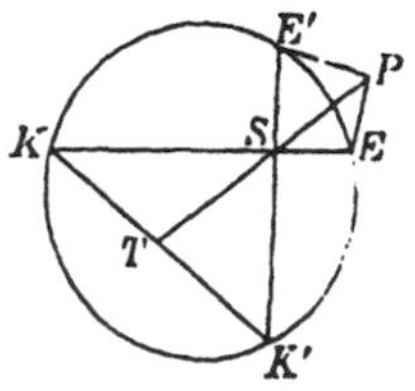

will have been exhausted, without repetition, by the deduced elements, E, E'. Hence the attraction on P will be the final resultant of the attractions of all the pairs of elements, E, E'.

Now if ρ be the surface density at E, and if F denote the attraction of the element E on P, we have

$$F = \frac{\rho . E}{EP^2}.$$

According to the given law of density we shall have

$$\rho = \frac{\lambda}{SE^3},$$

where λ is a constant. Again, since SEK is equally inclined to the spherical surface at the two points of intersection, we have

$$E = \frac{SE^2}{SK^2} . K = \frac{SE^2}{SK^2} . \frac{2a\omega . TK^2}{KK'};$$

and hence

$$F = \frac{\frac{\lambda}{SE^3} . \frac{SE^2}{SK^2} . \frac{2a\omega . TK^2}{KK'}}{EP^2} = \lambda . \frac{2a}{KK'} . \frac{TK^2}{SE . SK^2 . EP^2} \omega.$$

Attraction of a spherical surface of which the density varies inversely as the cube of the distance from a given point.

Now, by considering the great circle in which the sphere is cut by a plane through the line SK, we find that

$$\text{(fig. 1) } SK \,.\, SE = f^2 - a^2,$$

$$\text{(fig. 2) } KS \,.\, SE = a^2 - f^2,$$

and hence $SK.SE = SP \,.\, ST$, from which we infer that the triangles KST, PSE are similar; so that $TK : SK :: PE : SP$.

Hence
$$\frac{TK^2}{SK^2 \,.\, PE^2} = \frac{1}{SP^2},$$

and the expression for F becomes

$$F = \lambda \,.\, \frac{2a}{KK'} \,.\, \frac{1}{SE \,.\, SP^2} \,.\, \omega.$$

Modifying this by preceding expressions we have

$$\text{(fig. 1) } F = \lambda \,.\, \frac{2a}{KK'} \,.\, \frac{\omega}{(f^2 - a^2)\, SP^2} \,.\, SK,$$

$$\text{(fig. 2) } F = \lambda \,.\, \frac{2a}{KK'} \,.\, \frac{\omega}{(a^2 - f^2)\, SP^2} \,.\, KS.$$

Similarly, if F' denote the attraction of E' on P, we have

$$\text{(fig. 1) } F' = \lambda \frac{2a}{KK'} \,.\, \frac{\omega}{(f^2 - a^2)\, SP^2} \,.\, SK',$$

$$\text{(fig. 2) } F' = \lambda \frac{2a}{KK'} \,.\, \frac{\omega}{(a^2 - f^2)\, SP^2} \,.\, K'S.$$

Now in the triangles which have been shown to be similar, the angles TKS, EPS are equal; and the same may be proved of the angles $TK'S$, $E'PS$. Hence the two sides SK, SK' of the triangle KSK' are inclined to the third at the same angles as those between the line PS and directions PE, PE' of the two forces on the point P; and the sides SK, SK' are to one another as the forces, F, F', in the directions PE, PE'. It follows, by "the triangle of forces," that the resultant of F and F' is along PS, and that it bears to the component forces the same ratios as the side KK' of the triangle bears to the other two sides. Hence the resultant force due to the two elements E and E' on the point P, is towards S, and is equal to

$$\lambda \,.\, \frac{2a}{KK'} \,.\, \frac{\omega}{(f^2 \sim a^2) \,.\, SP^2} \,.\, KK', \text{ or } \frac{\lambda \,.\, 2a \,.\, \omega}{(f^2 \sim a^2)\, SP^2}.$$

The total resultant force will consequently be towards S; and we find, by summation (§ 467) for its magnitude,

Attraction of a spherical surface of which the density varies inversely as the cube of the distance from a given point.

$$\frac{\lambda \, . \, 4\pi a}{(f^2 \sim a^2)\, SP^2}.$$

Hence we infer that the resultant force at any point P, separated from S by the spherical surface, is the same as if a quantity of matter equal to $\frac{\lambda \, . \, 4\pi a}{f^2 \sim a^2}$ were concentrated at the point S.

475. To find the attraction when S and P are either both without or both within the spherical surface.

Take in CS, or in CS produced through S, a point S_1, such that

$$CS \, . \, CS_1 = a^2.$$

Then, by a well-known geometrical theorem, if E be any point on the spherical surface, we have

$$\frac{SE}{S_1E} = \frac{f}{a}.$$

Hence we have

$$\frac{\lambda}{SE^3} = \frac{\lambda a^3}{f^3 \, . \, S_1E^3}.$$

Hence, ρ being the surface-density at E, we have

$$\rho = \frac{\frac{\lambda a^3}{f^3}}{S_1E^3} = \frac{\lambda_1}{S_1E^3},$$

if

$$\lambda_1 = \frac{\lambda a^3}{f^3}.$$

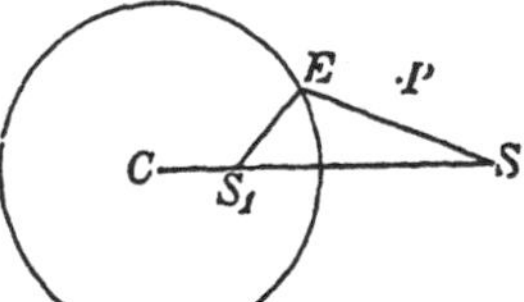

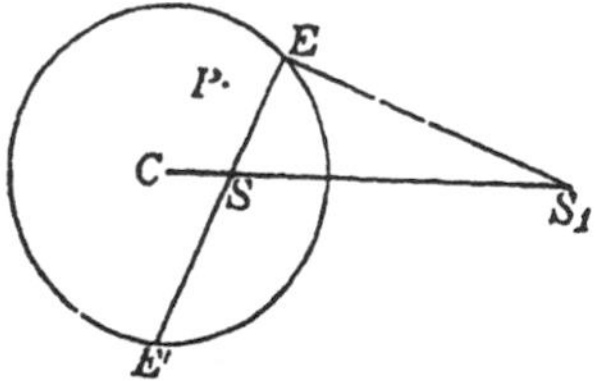

Hence, by the investigation in the preceding section, the attraction on P is towards S_1, and is the same as if a quantity

Attraction of a spherical surface of which the density varies inversely as the cube of the distance from a given point.

of matter equal to $\frac{\lambda_1 . 4\pi a}{f_1^2 - a^2}$ were concentrated at that point; f_1 being taken to denote CS_1. If for f_1 and λ_1 we substitute their values, $\frac{a^2}{f}$ and $\frac{\lambda a^3}{f^3}$, we have the modified expression

$$\frac{\lambda \frac{a}{f} . 4\pi a}{a^2 \sim f^2}$$

for the quantity of matter which we must conceive to be collected at S_1.

Uninsulated sphere under the influence of an electric point.

476. If a spherical surface be electrified in such a way that the electrical density varies inversely as the cube of the distance from an internal point S, or from the corresponding external point S_1, it will attract any external point, as if its whole electricity were concentrated at S, and any internal point, as if a quantity of electricity greater than its own in the ratio of a to f were concentrated at S_1.

Let the density at E be denoted, as before, by $\frac{\lambda}{SE^3}$. Then, if we consider two opposite elements at E and E', which subtend a solid angle ω at the point S, the areas of these elements being $\frac{\omega . 2a\, SE^2}{EE'}$ and $\frac{\omega . 2a . SE'^2}{EE'}$, the quantity of electricity which they possess will be

$$\frac{\lambda . 2a . \omega}{EE'}\left(\frac{1}{SE} + \frac{1}{SE'}\right) \text{ or } \frac{\lambda . 2a . \omega}{SE . SE'}.$$

Now $SE . SE'$ is constant (Euc. III. 35) and its value is $a^2 - f^2$. Hence, by summation, we find for the total quantity of electricity on the spherical surface

$$\frac{\lambda . 4\pi a}{a^2 - f^2}.$$

Hence, if this be denoted by m, the expressions in the preceding paragraphs, for the quantities of electricity which we must suppose to be concentrated at the point S or S_1, according as P is without or within the spherical surface, become respectively

$$m, \text{ and } \frac{a}{f} m.$$

477. The *direct* analytical solution of such problems consists in the expression, by § 455, of the three components of the whole attraction as the sums of its separate parts due to the several particles of the attracting body; the transformation, by the usual methods, of these sums into definite integrals; and the evaluation of the latter. This is, in general, inferior in elegance and simplicity to the less direct mode of solution depending upon the determination of the potential energy of the attracted particle with reference to the forces exerted upon it by the attracting body, a method which we shall presently develop with peculiar care, as being of incalculable value in the theories of Electricity and Magnetism as well as in that of Gravitation. But before we proceed to it, we give some instances of the direct method, beginning with the case of a spherical shell. **Direct analytical calculation of attractions.**

(*a*) Let P be the attracted point, O the centre of the shell. Let any plane perpendicular to OP cut it in N, and the sphere in the small circle QR. **Uniform spherical shell.**

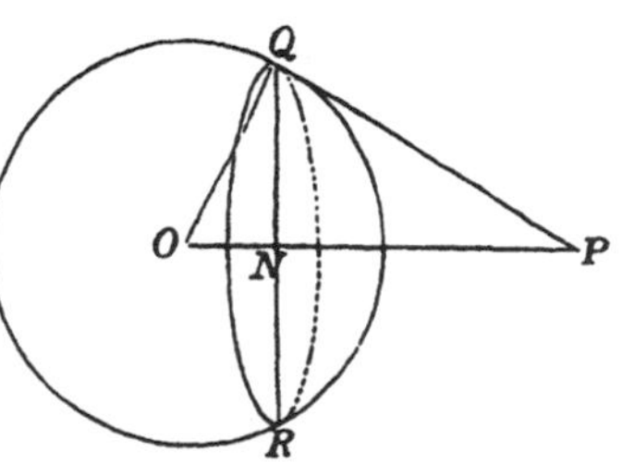

Let $QOP=\theta$, $OQ=a$, $OP=D$. Then as the whole attraction is evidently along PO, we may at once resolve the parts of it in that direction. The circular band corresponding to θ, $\theta+d\theta$ has for area $2\pi a^2 \sin\theta d\theta$. Hence if M be the mass of the shell, the component attraction of the band on P, along PO, is

$$\frac{M}{2}\sin\theta d\theta \cdot \frac{PN}{PQ^3}\text{; and } PQ^2=a^2+D^2-2aD\cos\theta.$$

Hence if $PQ=x$, $\quad xdx=aD\sin\theta d\theta.$

Also $$PN=D-a\cos\theta=\frac{x^2-a^2+D^2}{2D}\text{;}$$

hence the attraction of the band is

$$\frac{M}{4D^2}\frac{x^2-a^2+D^2}{ax^2}dx.$$

Uniform spherical shell.

This divides itself, on integration, into two cases,

(1) P external, *i.e.*, $D > a$. Here the limits of x are $D-a$ and $D+a$, and the attraction is $\frac{M}{4D^2}\left[\frac{x}{a}-\frac{D^2-a^2}{ax}\right]_{D-a}^{D+a}=\frac{M}{D^2}$, as before.

(2) P internal, *i.e.*, $D < a$. Here the limits are $a-D$ and $a+D$, and the attraction is $\frac{M}{4D^2}\left[\frac{x}{a}+\frac{a^2-D^2}{ax}\right]_{a-D}^{a+D}=0.$

Uniform circular disc, on particle in its axis.

(*b*) A useful case is that of the attraction of a circular plate of uniform surface density on a point in a line through its centre, and perpendicular to its plane.

If a be the radius of the plate, h the distance of the point from it, and M its mass, the attraction (which is evidently in a direction perpendicular to the plate) is easily seen to be

$$\frac{M}{a^2}\int_0^a \frac{2hrdr}{(h^2+r^2)^{\frac{3}{2}}}=\frac{2M}{a^2}\left\{1-\frac{h}{\sqrt{h^2+a^2}}\right\}.$$

If ρ denote the surface density of the plate, this becomes

$$2\pi\rho\left(1-\frac{h}{\sqrt{h^2+a^2}}\right);$$

which, for an infinite plate, becomes

$$2\pi\rho.$$

From the preceding formula many useful results may easily be deduced: thus,

Cylinder on particle in axis.

(*c*) A uniform *cylinder* of length l, and diameter a, attracts a point in its axis at a distance x from the nearest end with a force

$$2\pi\rho\int_x^{x+l}\left(1-\frac{h}{\sqrt{h^2+a^2}}\right)dh=2\pi\rho\left\{l-\sqrt{(x+l)^2+a^2}+\sqrt{x^2+a^2}\right\}.$$

When the cylinder is of infinite length (in one direction) the attraction is therefore

$$2\pi\rho\left(\sqrt{x^2+a^2}-x\right);$$

and, when the attracted particle is in contact with the centre of the end of the infinite cylinder, this is

$$2\pi\rho a.$$

(*d*) A right cone, of semivertical angle a, and length l, attracts a particle at its vertex. Here we have at once for the attraction, the expression

Right cone on particle at vertex.

$$2\pi\rho l\,(1-\cos a),$$

which is simply proportional to the length of the axis.

It is of course easy, when required, to find the necessarily less simple expression for the attraction on any point of the axis.

(*e*) For magnetic and electro-magnetic applications a very useful case is that of two equal discs, each perpendicular to the line joining their centres, on any point in that line—their masses (§ 461) being of opposite sign—that is, one repelling and the other attracting.

Positive and negative discs.

Let a be the radius, ρ the mass of a superficial unit, of either, c their distance, x the distance of the attracted point from the nearest disc. The whole action is evidently

$$2\pi\rho\left\{\frac{x+c}{\sqrt{(x+c)^2+a^2}}-\frac{x}{\sqrt{x^2+a^2}}\right\}.$$

In the particular case when c is diminished without limit, this becomes

$$2\pi\rho c\,\frac{a^2}{(x^2+a^2)^{\frac{3}{2}}}.$$

478. Let P and P' be two points infinitely near one another on two sides of a surface over which matter is distributed; and let ρ be the density of this distribution on the surface in the neighbourhood of these points. Then whatever be the resultant attraction, R, at P, due to all the attracting matter, whether lodging on this surface, or elsewhere, the resultant force, R', on P' is the resultant of a force equal and parallel to R, and a force equal to $4\pi\rho$, in the direction from P' perpendicularly towards the surface. For, suppose PP' to be perpendicular to the surface, which will not limit the generality of the proposition, and consider a circular disc, of the surface, having its centre in PP', and radius infinitely small in comparison with the radii of curvature of the surface but infinitely great in comparison with PP'. This disc will [§ 477, (*b*)] attract P and P' with forces, each equal to $2\pi\rho$ and opposite to one another in the line PP'. Whence the proposition. It is one of much importance in the theory of electricity.

Variation of force in crossing an attracting surface.

Uniform hemisphere attracting particle at edge.

(a) As a further example of the direct analytical process, let us find the components of the attraction exerted by a uniform *hemisphere* on a particle at its edge. Let A be the particle, AB a diameter of the base, AC the tangent to the base at A; and AD perpendicular to AC, and AB. Let RQA be a section by a plane passing through AC; AQ any radius-vector of this section; P a point in AQ. Let $AP=r$, $CAQ=\theta$, $RAB=\phi$. The volume of an element at P is

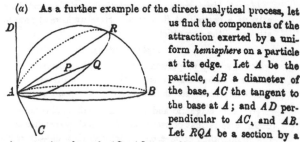

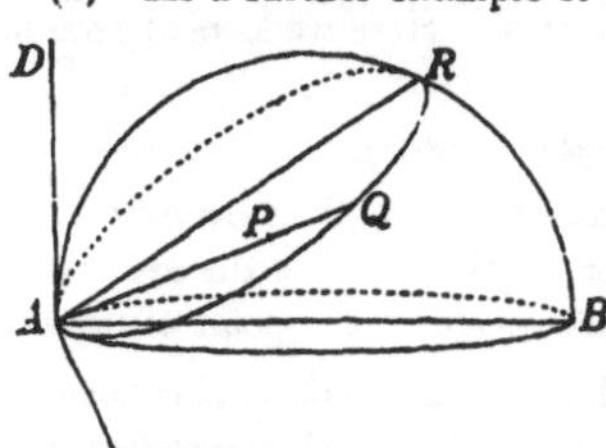

$$rd\theta\,.\,r\sin\theta d\phi\,.\,dr = r^2\sin\theta d\phi d\theta dr.$$

The resultant attraction on unit of matter at A has zero component along AC. Along AB the component is

$$\rho\iiint\sin\theta d\phi d\theta dr\cos\phi\sin\theta,$$

between proper limits. The limits of r are 0 and $2a\sin\theta\cos\phi$, those of ϕ are 0 and $\frac{\pi}{2}$, and those of θ are 0 and π. Hence, Attraction along $AB = \frac{2}{3}\pi\rho a$.

Along AD the component is

$$\rho\int_0^{+\pi}\int_0^{\frac{\pi}{2}}\int_0^{2a\sin\theta\cos\phi}\sin\theta d\theta d\phi dr\sin\phi\sin\theta = \tfrac{4}{3}\rho a.$$

Alteration of latitude; by hemispherical hill or cavity.

(b) Hence at the southern base of a hemispherical hill of radius a and density ρ, the true latitude (as measured by the aid of the plumb-line, or by reflection of starlight in a trough of mercury) is diminished by the attraction of the mountain by the angle

$$\frac{\frac{2}{3}\pi\rho a}{G-\frac{4}{3}\rho a}$$

where G is the attraction of the earth, estimated in the same units. Hence, if R be the radius and σ the mean density of the earth, the angle is

$$\frac{\frac{2}{3}\pi\rho a}{\frac{4}{3}\pi\sigma R-\frac{4}{3}\rho a},\text{ or }\frac{1}{2}\frac{\rho a}{\sigma R}\text{ approximately.}$$

Hence the latitudes of stations at the base of the hill, north and south of it, differ by $\frac{a}{R}\left(2+\frac{\rho}{\sigma}\right)$; instead of by $\frac{2a}{R}$, as they would do if the hill were removed. Alteration of latitude; by hemispherical hill or cavity.

In the same way the latitude of a place at the southern edge of a hemispherical *cavity* is increased on account of the cavity by $\frac{1}{2}\frac{\rho a}{\sigma R}$ where ρ is the density of the superficial strata.

(*c*) For mutual attraction between two segments of a homogeneous solid sphere, investigated indirectly on a hydrostatic principle, see § 753 below.

479. As a curious additional example of the class of questions considered in § 478 (*a*) (*b*), a deep crevasse, extending east and west, increases the latitude of places at its southern edge by (approximately) the angle $\frac{3}{4}\frac{\rho a}{\sigma R}$ where ρ is the density of the crust of the earth, and a is the width of the crevasse. Thus the north edge of the crevasse will have a *lower* latitude than the south edge if $\frac{3}{2}\frac{\rho}{\sigma}>1$, which might be the case, as there are rocks of density $\frac{2}{3}\times 5{\cdot}5$ or 3·67 times that of water. At a considerable depth in the crevasse, this change of latitudes is nearly *doubled,* and then the southern side has the greater latitude if the density of the crust be not less than 1·83 times that of water. The reader may exercise himself by drawing lines of equal latitude in the neighbourhood of the crevasse in this case: and by drawing meridians for the corresponding case of a crevasse running north and south. by crevasse.

480. It is interesting, and will be useful later, to consider as a particular case, the attraction of a sphere whose mass is composed of concentric layers, each of uniform density. Attraction of a sphere composed of concentric shells of uniform density.

Let R be the radius, r that of any layer, $\rho=F(r)$ its density. Then, if σ be the mean density,

$$\tfrac{4}{3}\pi\sigma R^3=4\pi\int_0^R \rho r^2 dr,$$

from which σ may be found.

The surface attraction is $\frac{4}{3}\pi\sigma R$, $=G$, suppose.

At a distance r from the centre the attraction is $\frac{4\pi}{r^2}\int_0^r \rho r^2 dr$.

Attraction of a sphere composed of concentric shells of uniform density.

If it is to be the same for all points inside the sphere

$$\int_0^r \rho r^2 dr = \frac{G}{4\pi} r^2.$$

Hence $\rho = F(r) = \frac{1}{2\pi} \cdot \frac{G}{r}$ is the requisite law of density.

If the density of the upper crust be τ, the attraction at a depth h, small compared with the radius, is

$$\tfrac{4}{3}\pi\sigma_1 (R-h) = G_1,$$

where σ_1 is the mean density of nucleus when a shell of thickness h is removed from the sphere. Also, evidently,

$$\tfrac{4}{3}\pi\sigma_1 (R-h)^3 + 4\pi\tau (R-h)^2 h = \tfrac{4}{3}\pi\sigma R^3,$$

or $$G_1 (R-h)^2 + 4\pi\tau (R-h)^2 h = GR^2,$$

whence $$G_1 = G\left(1 + \frac{2h}{R}\right) - 4\pi\tau h.$$

The attraction is therefore unaltered at a depth h if

$$\frac{G}{R} = \tfrac{4}{3}\pi\sigma = 2\pi\tau.$$

481. Some other simple cases may be added here, as their results will be of use to us subsequently.

Attraction of a uniform circular arc.

(a) The attraction of a circular arc, AB, of uniform density, on a particle at the centre, C, of the circle, lies evidently in the line CD bisecting the arc. Also the resolved part parallel to CD of the attraction of an element at P is

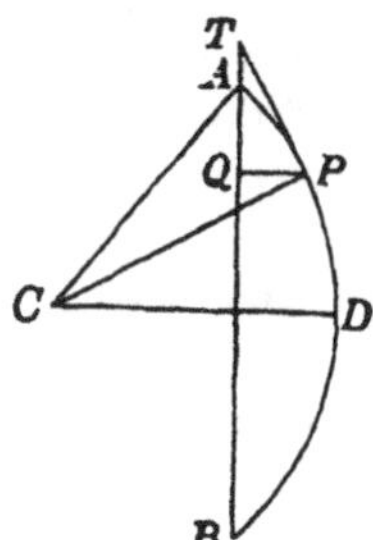

$$\frac{\text{mass of element at } P}{CD^2} \cos . P\hat{C}D.$$

Now suppose the density of the chord AB to be the same as that of the arc. Then for (mass of element at $P \times \cos P\hat{C}D$) we may put mass of projection of element on AB at Q; since, if PT be the tangent at P, $P\hat{T}Q = P\hat{C}D$.

Hence attraction along $CD = \dfrac{\text{Sum of projected elements}}{CD^2}$

$$= \frac{\rho AB}{CD^2},$$

if ρ be the density of the given arc,

Attraction of a uniform circular arc,

$$=\frac{2\rho\sin A\hat{C}D}{CD}.$$

It is therefore the same as the attraction of a mass equal to the chord, with the arc's density, concentrated at the point D.

(b) Again a limited straight line of uniform density attracts any external point in the same direction and with the same force as the corresponding arc of a circle of the same density, which has the point for centre, and touches the straight line.

straight line.

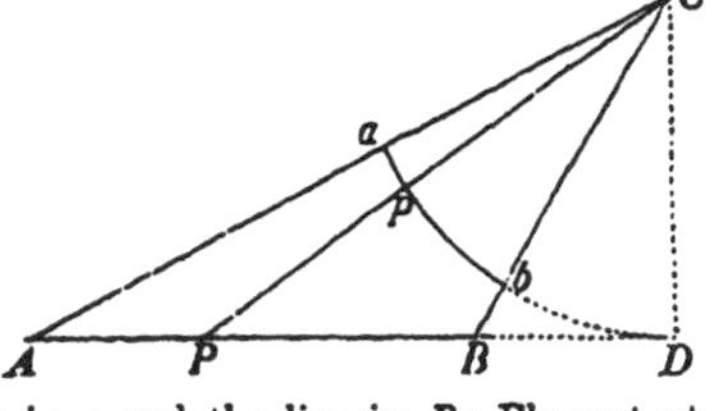

For if CpP be drawn cutting the circle in p and the line in P; Element at p : element at P :: $Cp : CP\frac{CP}{CD}$; that is, as $Cp^2 : CP^2$. Hence the attractions of these elements on C are equal and in the same line. Thus the arc ab attracts C as the line AB does; and, by the last proposition, the attraction of AB bisects the angle ACB, and is equal to

$$\frac{2\rho}{CD}\sin \tfrac{1}{2}A\hat{C}B.$$

(c) This may be put into other useful forms — thus, let CKF bisect the angle ACB, and let Aa, Bb, EF, be drawn perpendicular to CF from the ends and middle point of AB. We have

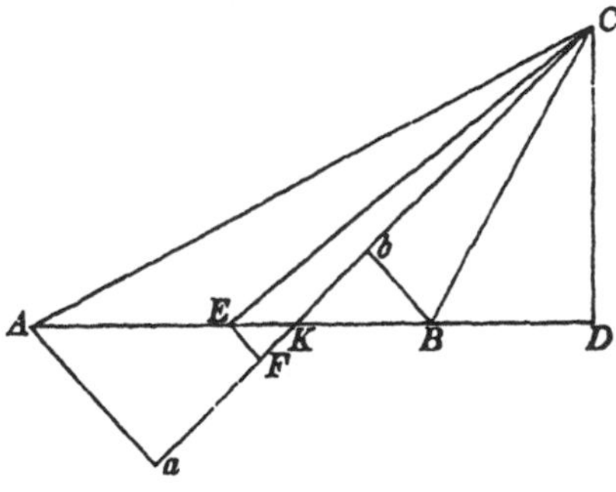

$$\sin K\hat{C}B=\frac{KB}{CB}\sin C\hat{K}D=\frac{AB}{AC+CB}\frac{CD}{CK}.$$

Attraction of a uniform straight line.

Hence the attraction, which is along CK, is

$$\frac{2\rho AB}{(AC+CB)\,CK} = \frac{\rho AB}{8\,(AC+CB)\,(\overline{AC+CB}^2 - AB^2)} \, . \, CF. \quad (1)$$

For, evidently,

$$bK : Ka :: BK : KA :: BC : CA :: bC : Ca,$$

i.e., ab is divided, externally in C, and internally in K, in the same ratio. Hence, by geometry,

$$KC \, . \, CF = aC \, . \, Cb = \tfrac{1}{4}\{\overline{AC+CB}^2 - AB^2\},$$

which gives the transformation in (1).

(*d*) CF is obviously the tangent at C to a hyperbola, passing through that point, and having A and B as foci. Hence, if in *any* plane through AB any hyperbola be described, with foci A and B, it will be a line of force as regards the attraction of the line AB; that is, as will be more fully explained later, a curve which at every point indicates the direction of attraction.

(*e*) Similarly, if a prolate spheroid be described with foci A and B, and passing through C, CF will evidently be the normal at C; thus the force on a particle at C will be perpendicular to the spheroid; and the particle would evidently rest in equilibrium on the surface, even if it were smooth. This is an instance of (what we shall presently develop at some length) a surface of equilibrium, a level surface, or an equipotential surface.

(*f*) We may further prove, by a simple application of the preceding theorem, that the lines of force due to the attraction of two infinitely long rods in the line AB produced, one of which is attractive and the other repulsive, are the series of ellipses described from the extremities, A and B, as foci, while the surfaces of equilibrium are generated by the revolution of the confocal hyperbolas.

Potential.

482. As of immense importance, in the theory not only of gravitation but of electricity, of magnetism, of fluid motion, of the conduction of heat, etc., we give here an investigation of the most important properties of the *Potential.*

483. This function was introduced for gravitation by Laplace, but the name was first given to it by Green, who may almost be said to have in 1828 created the theory, as we now have it,

Green's work was neglected till 1846, and before that time most of its important theorems had been re-discovered by Gauss, Chasles, Sturm, and Thomson. Potential.

In § 273, the *potential energy* of a conservative system in any configuration was defined. When the forces concerned are forces acting, either really or apparently, at a distance, as attraction of gravitation, or attractions or repulsions of electric or magnetic origin, it is in general most convenient to choose, for the zero configuration, infinite distance between the bodies concerned. We have thus the following definition:—

484. The mutual potential energy of two bodies in any relative position is the amount of work obtainable from their mutual repulsion, by allowing them to separate to an infinite distance asunder. When the bodies attract mutually, as for instance when no other force than gravitation is operative, their mutual potential energy, according to the convention for zero now adopted, is negative, or (§ 547 below) their *exhaustion of potential energy* is positive.

485. The *Potential* at any point, due to any attracting or repelling body, or distribution of matter, is the mutual potential energy between it and a unit of matter placed at that point. But in the case of gravitation, to avoid defining the potential as a negative quantity, it is convenient to change the sign. Thus the gravitation potential, at any point, due to any mass, is the quantity of work required to remove a unit of matter from that point to an infinite distance.

486. Hence if V be the potential at any point P, and V_1 that at a proximate point Q, it evidently follows from the above definition that $V - V_1$ is the work required to remove an independent unit of matter from P to Q; and it is useful to note that this is altogether independent of the form of the path chosen between these two points, as it gives us a preliminary idea of the power we acquire by the introduction of this mode of representation.

Suppose Q to be so near to P that the attractive forces exerted on unit of matter at these points, and therefore at any

Potential. point in the line PQ, may be assumed to be equal and parallel. Then if F represent the resolved part of this force along PQ, $F.PQ$ is the work required to transfer unit of matter from P to Q. Hence

$$V - V_1 = F.PQ,$$

or

$$F = \frac{V - V_1}{PQ},$$

Force in terms of the potential. that is, the attraction on unit of matter at P in any direction PQ, is the rate at which the potential at P increases per unit of length of PQ.

Equipotential surface. 487. A surface, at every point of which the potential has the same value, and which is therefore called an *Equipotential Surface*, is such that the attraction is everywhere in the direction of its normal. For in no direction along the surface does the potential change in value, and therefore there is no force in any such direction. Hence if the attracted particle be placed on such a surface (supposed smooth and rigid), it will rest in any position, and the surface is therefore sometimes called a *Surface of Equilibrium*. We shall see later, that the force on a particle of a liquid at the free surface is always in the direction of the normal, hence the term *Level Surface*, which is often used for the other terms above.

Relative intensities of force at different points of an equipotential surface. 488. If a series of equipotential surfaces be constructed for values of the potential increasing by equal small amounts, it is evident from § 486 that the attraction at any point is inversely proportional to the normal distance between two successive surfaces close to that point; since the numerator of the expression for F is, in this case, constant.

Line of force. 489. A line drawn from any origin, so that at every point of its length its tangent is the direction of the attraction at that point, is called a *Line of Force;* and it obviously cuts at right angles every equipotential surface which it meets.

These three last sections are true *whatever* be the law of attraction; in the next we are restricted to the law of the inverse square of the distance.

490. If, through every point of the boundary of an infinitely small portion of an equipotential surface, the corresponding lines of force be drawn, we shall evidently have a tubular surface of infinitely small section. The force in any direction, at any point within such a tube, so long as it does not cut through attracting matter, is inversely as the section of the tube made by a plane passing through the point and perpendicular to the given direction. Or, more simply, the whole force is at every point tangential to the direction of the tube, and inversely as its transverse section: from which the more general statement above is easily seen to follow.

Variation of intensity along a line of force.

This is an immediate consequence of a most important theorem, which will be proved later, § 492. *The surface integral of the attraction exerted by any distribution of matter in the direction of the normal at every point of any closed surface is $4\pi M$; where M is the amount of matter within the surface, while the attraction is considered positive or negative according as it is inwards or outwards at any point of the surface.*

For in the present case the force perpendicular to the tubular part of the surface vanishes, and we need consider the ends only. When none of the attracting mass is within the portion of the tube considered, we have at once

$$F\varpi - F'\varpi' = 0,$$

F being the force at any point of the section whose area is ϖ. This is equivalent to the celebrated equation of Laplace—App. B (*a*); and below, § 491 (*c*).

When the attracting body is symmetrical about a point, the lines of force are obviously straight lines drawn from this point. Hence the tube is in this case a cone, and, by § 469, ϖ is proportional to the square of the distance from the vertex. Hence F is inversely as the square of the distance for points external to the attracting mass.

When the mass is symmetrically disposed about an axis in infinitely long cylindrical shells, the lines of force are evidently perpendicular to the axis. Hence the tube becomes a *wedge*, whose section is proportional to the distance from the axis, and the attraction is therefore inversely as the distance from the axis.

Variation of intensity along a line of force.

When the mass is arranged in infinite parallel planes, each of uniform density, the lines of force are obviously perpendicular to these planes; the tube becomes a *cylinder;* and, since its section is constant, the force is the same at all distances.

If an infinitely small length l of the portion of the tube considered pass through matter of density ρ, and if ω be the area of the section of the tube in this part, we have

$$F\varpi - F'\varpi' = 4\pi l\omega\rho.$$

This is equivalent to Poisson's extension of Laplace's equation [§ 491 (c)].

Potential due to an attracting point.

491. In estimating work done against a force which varies inversely as the square of the distance from a fixed point, the mean force is to be reckoned as the geometrical mean between the forces at the beginning and end of the path: and, whatever may be the path followed, the effective space is to be reckoned as the difference of distances from the attracting point. Thus the work done in any course is equal to the product of the difference of distances of the extremities from the attracting point, into the geometrical mean of the forces at these distances; or, if O be the attracting point, and m its force on a unit mass at unit distance, the work done in moving a particle, of unit mass, from any position P to any other position P', is

$$(OP' - OP)\sqrt{\frac{m^2}{OP^2 OP'^2}}, \text{ or } \frac{m}{OP} - \frac{m}{OP'}.$$

To prove this it is only necessary to remark, that for any infinitely small step of the motion, the effective space is clearly the difference of distances from the centre, and the working force may be taken as the force at either end, or of any intermediate value, the geometrical mean for instance: and the preceding expression applied to each infinitely small step shows that the same rule holds for the sum making up the whole work done through any finite range, and by any path.

Hence, by § 485, it is obvious that the potential at P, of a mass m situated at O, is $\frac{m}{OP}$; and thus that the potential of any mass at a point P is to be found by adding the quotients of every portion of the mass, each divided by its distance from P.

Potential due to an attracting point.

a. For the analytical proof of these propositions, consider, first, a pair of particles, O and P, whose masses are m and unity, and co-ordinates abc, xyz. If D be their distance

Analytical investigation of the value of the potential.

$$D^2 = (x-a)^2 + (y-b)^2 + (z-c)^2.$$

The components of the mutual attraction are

$$X = m\frac{x-a}{D^3}, \quad Y = m\frac{y-b}{D^3}, \quad Z = m\frac{z-c}{D^3};$$

and therefore the work required to remove P to infinity is

$$m\int\frac{(x-a)\,dx + (y-b)\,dy + (z-c)\,dz}{D^3}$$

$$= m\int\frac{dD}{D^2}$$

which, since the superior limit is $D = \infty$, is equal to

$$\frac{m}{D}.$$

The mutual potential energy is therefore, in this case, the product of the masses divided by their mutual distance; and therefore the potential at x, y, z, due to m, is $\frac{m}{D}$.

Again, if there be more than one fixed particle m, the same investigation shows us that the potential at xyz is

$$\Sigma\frac{m}{D}.$$

And if the particles form a continuous mass, whose density at a, b, c is ρ, we have of course for the potential the expression

$$\iiint\rho\frac{da\,db\,dc}{D},$$

the limits depending on the boundaries of the mass.

If we call V the potential at any point P (x, y, z), it is evident (from the way in which we have obtained its value) that the components of the attraction on unit of matter at P are

Force at any point.

$$X = -\frac{dV}{dx}, \quad Y = -\frac{dV}{dy}, \quad Z = -\frac{dV}{dz}.$$

Force at any point.

Hence the force, resolved along any curve of which s is the arc,

$$\text{is} \quad X\frac{dx}{ds}+Y\frac{dy}{ds}+Z\frac{dz}{ds}=-\left(\frac{dV}{dx}\frac{dx}{ds}+\frac{dV}{dy}\frac{dy}{ds}+\frac{dV}{dz}\frac{dz}{ds}\right)$$

$$=-\frac{dV}{ds}.$$

All this is evidently independent of the question whether P lies within the attracting mass or not.

Force within a homogeneous sphere.

b. If the attracting mass be a sphere of density ρ, and centre a, b, c, and if P be within its surface, we have, since the exterior shell has no effect,

$$X=-\frac{dV}{dx}=\frac{4}{3}\pi\rho D^3 . \frac{x-a}{D^3}$$

$$=\frac{4}{3}\pi\rho(x-a).$$

Rate of increase of the force in any direction.

Hence
$$\frac{dX}{dx}=-\frac{d^2V}{dx^2}=\frac{4}{3}\pi\rho.$$

c. Now if

$$\nabla^2=\frac{d^2}{dx^2}+\frac{d^2}{dy^2}+\frac{d^2}{dz^2},$$

we have $\nabla^2\frac{1}{D}=0$, as was proved before, App. B *g* (14) as a particular case of *g*. The proof for this case alone is as follows:

$$\frac{d}{dx}\frac{1}{D}=-\frac{x-a}{D^3}; \quad \frac{d^2}{dx^2}\frac{1}{D}=-\frac{1}{D^3}+\frac{3(x-a)^2}{D^5}:$$

and from this, and the similar expressions for the second differentials in y and z, the theorem follows by summation.

Hence as
$$V=\iiint\rho\frac{da\,db\,dc}{D}$$

and ρ does not involve x, y, z, we see that *as long as D does not vanish within the limits of integration*, i.e., as long as P is not a point of the attracting mass

$$\nabla^2V=0;$$

or, in terms of the components of the force,

Laplace's equation.

$$\frac{dX}{dx}+\frac{dY}{dy}+\frac{dZ}{dz}=0.$$

If P be within the attracting mass, suppose a small sphere to be described so as to contain P. Divide the potential into two parts, V_1 that of the sphere, V_2 that of the rest of the body. **Laplace's equation.**

The expression above shows that

$$\nabla^2 V_2 = 0.$$

Also the expressions for $\frac{d^2V}{dx^2}$, etc., in the case of a sphere (b) give

$$\nabla^2 V_1 = -4\pi\rho,$$

where ρ is the density of the sphere.

Hence as

$$V = V_1 + V_2$$

$$\nabla^2 V = -4\pi\rho,$$

Poisson's extension of Laplace's equation.

which is the general equation of the potential, and includes the case of P being wholly external to the attracting mass, since there $\rho = 0$. In terms of the components of the force, this equation becomes

$$\frac{dX}{dx} + \frac{dY}{dy} + \frac{dZ}{dz} = 4\pi\rho.$$

d. We have already, in these most important equations, the means of verifying various former results, and also of adding new ones.

Thus, to find the attraction of a hollow sphere composed of concentric shells, each of uniform density, on an external point (by which we mean a point *not* part of the mass). In this case symmetry shows that V must depend upon the distance from the centre of the sphere alone. Let the centre of the sphere be origin, and let **Potential of matter arranged in concentric spherical shells of uniform density**

$$r^2 = x^2 + y^2 + z^2.$$

Then V is a function of r alone, and consequently

$$\frac{dV}{dx} = \frac{dV}{dr}\frac{dr}{dx} = \frac{x}{r}\frac{dV}{dr},$$

$$\frac{d^2V}{dx^2} = \frac{1}{r}\frac{dV}{dr} - \frac{x^2}{r^3}\frac{dV}{dr} + \frac{x^2}{r^2}\frac{d^2V}{dr^2},$$

and

$$\nabla^2 V = \frac{2}{r}\frac{dV}{dr} + \frac{d^2V}{dr^2}.$$

Hence, when P is outside the sphere, or in the hollow space within it,

$$\frac{2}{r}\frac{dV}{dr} + \frac{d^2V}{dr^2} = 0.$$

Potential of matter arranged in concentric spherical shells of uniform density.

A first integral of this is $r^2 \dfrac{dV}{dr} = C$.

For a point outside the shell C has a finite value, which is easily seen to be $-M$, where M is the mass of the shell.

For a point in the internal cavity $C = 0$, because evidently at the centre there is no attraction—*i.e.*, there $r = 0$, $\dfrac{dV}{dr} = 0$ together. Hence there is no attraction on *any* point in the cavity.

We need not be surprised at the apparent discontinuity of this solution. It is owing to the *discontinuity of the given distribution of matter*. Thus it appears, by § 491 *c*, that the true general equation of the potential is not what we have taken above, but

$$\frac{d^2V}{dr^2} + \frac{2}{r}\frac{dV}{dr} = -4\pi\rho,$$

where ρ, the density of the matter at distance r from the centre, is zero when $r < a$ the radius of the cavity: has a finite value σ, which for simplicity we may consider constant, when $r > a$ and $< a'$ the radius of the outer bounding surface: and is zero, again, for all values of r exceeding a'. Hence, integrating from $r = 0$, to $r = r$, any value, we have (since $r^2 \dfrac{dV}{dr} = 0$ when $r = 0$),

$$r^2 \frac{dV}{dr} = -4\pi \int_0^r \rho r^2 dr = -M_1,$$

if M_1 denote the whole amount of matter within the spherical surface of radius r; which is the discontinuous function of r specified as follows:—

From $r = 0$ to $r = a$, $\quad r = a$ to $r = a'$, $\quad r = a'$ to $r = \infty$,

$$M_1 = 0, \qquad M_1 = \frac{4\pi\sigma}{3}(r^3 - a^3), \qquad M_1 = \frac{4\pi\sigma}{3}(a'^3 - a^3).$$

The corresponding values of V are, in order,

$$V = 2\pi\sigma(a'^2 - a^2), \quad V = \frac{4\pi\sigma}{3}\left(\frac{3a'^2 - r^2}{2} - \frac{a^3}{r}\right), \quad V = \frac{4\pi\sigma}{3r}(a'^3 - a^3).$$

We have entered thus into detail in this case, because such apparent anomalies are very common in the analytical solution of physical questions. To make this still more clear, we subjoin a graphic representation of the values of V, $\dfrac{dV}{dr}$, and $\dfrac{d^2V}{dr^2}$ for this case. $ABQC$, the curve for V, is partly a straight line, and has a point of inflection at Q: but there is no discontinuity and no abrupt change of direction. $OEFD$, that for $\frac{dV}{dr}$, is continuous, but its direction twice changes abruptly. That for $\frac{d^2V}{dr^2}$ consists of three detached portions, OE, GH, KL.

Potential of matter arranged in concentric spherical shells of uniform density.

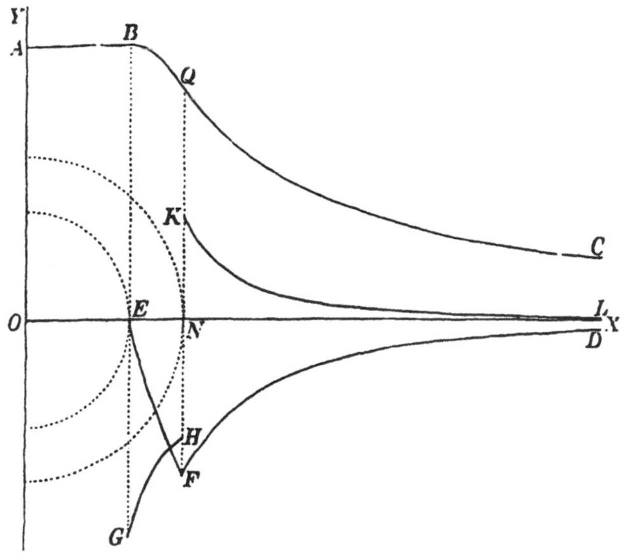

e. For a mass disposed in infinitely long concentric cylindrical shells, each of uniform density, if the axis of the cylinders be z, we must evidently have V a function of x^2+y^2 only.

Coaxal right cylinders of uniform density and infinite length.

Hence $\frac{dV}{dz}=0$, or the attraction is wholly perpendicular to the axis.

Also, $\frac{d^2V}{dz^2}=0$; and therefore by (d)

$$\nabla^2 V = \frac{d^2V}{dr^2}+\frac{1}{r}\frac{dV}{dr}=-4\pi\rho.$$

Hence
$$r\frac{dV}{dr}=C-4\pi\int\rho r\,dr,$$

from which conclusions similar to the above may be drawn.

f. If, finally, the mass be arranged in infinite parallel planes, each of uniform density, and perpendicular to the axis

Matter arranged in infinite parallel planes of uniform density.

of x; the resultant force must be parallel to this direction: that is to say, $Y=0$, $Z=0$, and therefore

$$\frac{dX}{dx}=4\pi\rho,$$

which, if ρ is known in terms of x, is completely integrable.

Outside the mass, $\rho=0$, and therefore

$$X=C,$$

or the attraction is the same at all distances, a result easily verified by the direct methods.

If within the mass the density is constant, we have

$$X=C'+4\pi\rho x;$$

and if the origin be in the middle of the lamina, we have, obviously, $C'=0$. Hence if t denote the thickness, the values of X at the two sides and in the spaces beyond are respectively $-2\pi\rho t$ and $+2\pi\rho t$. The difference of these is $4\pi\rho t$ (§ 478).

Equipotential surface.

g. Since in any case $\frac{dV}{ds}$ is the component of the attraction in the direction of the tangent to the arc s, the attraction will be perpendicular to that arc if

$$\frac{dV}{ds}=0,$$

or

$$V=C.$$

This is the equation of an *equipotential* surface.

If n be the normal to such a surface, measured outwards, the whole force at any point is evidently

$$\frac{dV}{dn},$$

and its direction is that in which V increases.

Integral of normal attraction over a closed surface.

492. Let S be any closed surface, and let O be a point, either external or internal, where a mass, m, of matter is collected. Let N be the component of the attraction of m in the direction of the normal drawn inwards from any point P, of S. Then, if $d\sigma$ denotes an element of S, and $\iint$ integration over the whole of it,

$$\iint N d\sigma=4\pi m, \text{ or } =0 \text{ ....(1)},$$

according as O is internal or external.

Case 1, *O internal.* Let $OP_1P_2P_3\ldots$ be a straight line drawn in any direction from O, cutting S in P_1, P_2, P_3, etc., and therefore passing out at P_1, in at P_2, out again at P_3, in again at P_4, and so on. Let a conical surface be described by lines through O, all infinitely near $OP_1P_2\ldots$, and let ω be its solid angle (§ 465). The portions of $\iint N d\sigma$ corresponding to the elements cut from S by this case will be clearly each equal in absolute magnitude to ωm, but will be alternately positive and negative. Hence as there is an odd number of them their sum is $+\omega m$. And the sum of these, for all solid angles round O is (§ 466) equal to $4\pi m$; that is to say, $\iint N d\sigma = 4\pi m$.

Integral of normal attraction over a closed surface.

Equivalent to Poisson's extension of Laplace's equation, § 491 c.

Case 2, *O external.* Let $OP_1P_2P_3\ldots$ be a line drawn from O passing across S, inwards at P_1, outwards at P_2, and so on. Drawing, as before, a conical surface of infinitely small solid angle, ω, we have still ωm for the absolute value of each of the portions of $\iint N d\sigma$ corresponding to the elements which it cuts from S; but their signs are alternately negative and positive: and therefore as their number is even, their sum is zero. Hence $\iint N d\sigma = 0$.

Equivalent to Laplace's equation, § 491 c.

From these results it follows immediately that if there be any distribution of matter, partly within and partly without a closed surface S, and N and $d\sigma$ be still used with the same signification, we have

$$\iint N d\sigma = 4\pi M \ldots\ldots\ldots\ldots\ldots\ldots\ldots\ldots (2)$$

if M denote the whole amount of matter within S.

This, with M eliminated from it by Poisson's theorem, § 491 *c*, is the particular case of the analytical theorem of Chap. I. App. A (*a*), found by taking $\alpha = 1$, and $U' = 1$, by which it becomes

$$0 = \iint d\sigma \partial U - \iiint \nabla^2 U dx dy dz \ldots\ldots\ldots\ldots\ldots (3).$$

For let U be the potential at (x, y, z), due to the distribution of matter in question. Then, according to the meaning of ∂, we have $\partial U = -N$. Also, let ρ be the density of the matter at (x, y, z). Then [§ 491 (*c*)] we have

$$\nabla^2 U = -4\pi\rho.$$

Hence (3) gives

$$\iint N d\sigma = 4\pi \iiint \rho dx dy dz = 4\pi M.$$

Integral of normal attraction over a closed surface.

493. If in crossing any surface K we find an abrupt change in the value of the component force perpendicular to K, it follows from (2) that there must be a condensation of matter on K, and that the surface-density of this distribution is $N/4\pi$, if N be the difference of the values of the normal component on the two sides of K; as we see by taking for our closed surface S an infinitely small rectangular parallelepiped with two of its faces parallel to K and on opposite sides of it. This result was found in § 478, in a thoroughly synthetical manner. The same result is found by the proper analytical interpretation of Poisson's equation

$$\frac{dX}{dx} + \frac{dY}{dy} + \frac{dZ}{dz} = 4\pi\rho.$$

It is to be remarked that in travelling across K abrupt change in the value of the component force along any line parallel to K is forbidden by the Conservation of Energy.

494. The theorem of Laplace and Poisson, § 492, for the present application most conveniently taken (§ 491 *c*) in its differential form

$$\rho = -\frac{1}{4\pi}\left(\frac{d^2V}{dx^2} + \frac{d^2V}{dy^2} + \frac{d^2V}{dz^2}\right) \ldots\ldots\ldots\ldots\ldots (1),$$

Inverse problem.

is explicitly the solution of the inverse problem,—*given the potential at every point of space*, or, which is virtually the same, *given the direction and magnitude of the resultant force at every point of space,—it is required to find the distribution of matter by which it is produced.*

494 *a.* Example. Let the potential be given equal to zero for all space external to a given closed surface S, and let

$$V = \phi(x, y, z) \ldots\ldots\ldots\ldots\ldots\ldots\ldots\ldots (2)$$

for all space within this surface; $\phi(x, y, z)$ being any arbitrary function subject to no other condition than that its value is zero at S, and that it has no abrupt changes of value within S. Abrupt changes in the values of differential coefficients,

$$\frac{d\phi}{dx},\ \frac{d\phi}{dy},\ \frac{d\phi}{dz},$$

are not excluded, but are subject to interpretations, as in § 493, if they occur.

494 *b*. The required distribution of matter must include a surface distribution on S, because there is abrupt change in the value of the normal component force from Inverse problem.

$$\sqrt{\left(\frac{d\phi^2}{dx^2}+\frac{d\phi^2}{dy^2}+\frac{d\phi^2}{dz^2}\right)}$$

at the inside of S to zero at the outside. Thus, by § 493, and by § 494 (1), we have for our complete solution (compare §§ 501. 505, 506, 507 below)

$$\left.\begin{aligned} &\rho=0, \text{ for space external to } S \\ &\sigma=\frac{1}{4\pi}\left(\frac{d\phi^2}{dx^2}+\frac{d\phi^2}{dy^2}+\frac{d\phi^2}{dz^2}\right)^{\frac{1}{2}} \text{ on } S, \\ \text{and} \quad &\rho=-\frac{1}{4\pi}\left(\frac{d^2\phi}{dx^2}+\frac{d^2\phi}{dy^2}+\frac{d^2\phi}{dz^2}\right) \\ &\text{for space enclosed by } S. \end{aligned}\right\} \text{........(2).}$$

494 *c*. From § 492 (2), remembering that $N=0$ outside of S, we infer that the total mass on and within S is zero, and therefore the quantity of matter condensed on S is equal and of opposite sign to the quantity enclosed by it.

494 *d*. Sub-Example. Let the potential be given equal to zero for all space external to the ellipsoidal surface

$$\frac{x^2}{a^2}+\frac{y^2}{b^2}+\frac{z^2}{c^2}=1,$$

and equal to

$$\tfrac{1}{2}\left(1-\frac{x^2}{a^2}-\frac{y^2}{b^2}-\frac{z^2}{c^2}\right) \text{.....................(3),}$$

for the space enclosed by it: in other words let the potential be zero wherever the value of (3) is negative, and equal to the value of (3) wherever it is positive.

494 *e*. The solution (2) becomes

$$\left.\begin{aligned} &\rho=0, && \text{wherever } \frac{x^2}{a^2}+\frac{y^2}{b^2}+\frac{z^2}{c^2}>1; \\ &\sigma=-\frac{1}{4\pi p}, && \text{at the surface } \frac{x^2}{a^2}+\frac{y^2}{b^2}+\frac{z^2}{c^2}=1; \\ \text{and } &\rho=\frac{1}{4\pi}\left(\frac{1}{a^2}+\frac{1}{b^2}+\frac{1}{c^2}\right) && \text{wherever } \frac{x^2}{a^2}+\frac{y^2}{b^2}+\frac{z^2}{c^2}<1. \end{aligned}\right\} \text{...(4);}$$

Inverse problem.

p denoting the perpendicular from the centre to the tangent plane of the ellipsoidal surface.

494 *f*. Let q be an infinitely small quantity. The equation

$$\frac{x^2}{a^2-q}+\frac{y^2}{b^2-q}+\frac{z^2}{c^2-q}=1 \text{..................(5)}$$

represents an ellipsoidal surface confocal with the given one, and infinitely near it. The distance between the two surfaces infinitely near any point (x, y, z) of either is easily proved to be equal to $\frac{1}{2}q/p$. Calling this t, we have, from (4),

$$\sigma=-\frac{1}{4\pi}\cdot\frac{2t}{q} \text{......................(6).}$$

We conclude from (6) and (4) and the theorem (§ 494 *c*) of masses that

Attractions of solid homogeneous ellipsoid and circumscribed focaloid of equal mass found equal.

494 *g*. The attraction of a homogeneous solid ellipsoid is the same through all external space as the attraction of a homogeneous focaloid* of equal mass coinciding with its surface.

Homoeoids and Focaloids defined.

* To avoid complexity of diction we now propose to introduce two new words, "focaloid" and "homoeoid," according to the following definitions:—

(1) A *homoeoid* is an infinitely thin shell bounded by two similar surfaces similarly oriented.

The one point which is situated similarly relatively to the two similar surfaces of a homoeoid is called the homoeoidal centre. Supposing the homoeoid to be a finite closed surface, the homoeoidal centre may be any internal or external point. In the extreme case of two equal surfaces, the homoeoidal centre is at an infinite distance. The homoeoid in this extreme case (which is interesting as representing the surface-distribution of ideal magnetic matter constituting the free polarity of a body magnetized uniformly in parallel lines) may be called a homoeoidal couple. In every case the thickness of the homoeoid is directly proportional to the perpendicular from the centre to the tangent plane at any point. When (the surface being still supposed to be finite and closed) the centre is external, the thickness is essentially negative in some places, and positive in others.

The bulk of a homoeoid is the excess of the bulk of the part where the thickness is positive above that where the thickness is negative. The bulk of a homoeoidal couple is essentially zero. Its moment and its axis are important qualities, obvious in their geometric definition, and useful in magnetism as

494 h. Take now a homogeneous solid ellipsoid and divide it into an infinite number of focaloids, numbered 1, 2, 3, ... from the surface inwards. Take the mass of No. 1 and distribute it uniformly through the space enclosed by its inner boundary. This makes no difference in the attraction through space external to the original ellipsoid. Take the infinitesimally increased mass of No. 2 and distribute it uniformly through the space enclosed by *its* inner boundary. And so on with Nos. 3, 4, &c., till instead of the given homogeneous ellipsoid we have another of the same mass and correspondingly greater density enclosed by any smaller confocal ellipsoidal surface. Proof of Maclaurin's Theorem.

494 i. We conclude that

Any two confocal homogeneous solid ellipsoids of equal masses produce equal attraction through all space external to both. Maclaurin's Theorem.

This is Maclaurin's splendid theorem. It is tantamount to the following, which presents it in a form specially interesting in some respects:

Any two thick or thin confocal focaloids of equal masses, each homogeneous, produce equal attraction through all space external to both. Equivalent in shells of Maclaurin's Theorem.

494 j. Maclaurin's theorem reduces the problem of finding the attraction of an ellipsoid* on any point in external space, (which when attempted by direct integration presents difficulties not hitherto directly surmounted,) to the problem of Digression on the attraction of an ellipsoid.

representing the magnetic moment and the magnetic axis of a piece of matter uniformly magnetized in parallel lines.

(2) An *elliptic homoeoid* is an infinitely thin shell bounded by two concentric similar ellipsoidal surfaces.

(3) A *focaloid* is an infinitely thin shell bounded by two confocal ellipsoidal surfaces.

(4) The terms "thick homoeoid" and "thick focaloid" may be used in the comparatively rare cases (see for example §§ 494 i, 519, 522) when forms satisfying the definitions (1) and (3) except that they are not infinitely thin, are considered.

* To avoid circumlocutions we call simply "an ellipsoid" a homogeneous solid ellipsoid.

Digression on the attraction of an ellipsoid.

finding the attraction of an ellipsoid on a point at its surface which, as the limiting case of the attraction of an ellipsoid on an internal point, is easily solved by direct integration, thus:

To find the potential of an ellipsoid at any interior point.

494 *k*. Divide the whole solid into pairs of vertically opposite infinitesimal cones or pyramids, having the attracted point P for common vertex.

Let $E'PE$ be any straight line through P, cut by the surface at E' and E, and let $d\sigma$ be the solid angle of the pair of cones lying along it. The potentials at P of the two are easily shown to be $\frac{1}{2} PE^2 d\sigma$ and $\frac{1}{2} PE'^2 d\sigma$, and therefore the whole contribution of potential at P by the pair is $\frac{1}{2}(PE'^2 + PE^2)\, d\sigma$.

Hence, if V denote the potential at P of the whole ellipsoid, the density being taken as unity, we have

$$V = \iint \tfrac{1}{2}(PE^2 + PE'^2)\, d\sigma \qquad (7),$$

where $\iint$ denotes integration over a hemisphere of spherical surface of unit radius.

Now if x, y, z be the co-ordinates of P relative to the principal axes of the ellipsoid; and l, m, n the direction cosines of PE, we have, by the equation of the ellipsoid,

$$\frac{(x + lPE)^2}{a^2} + \frac{(y + mPE)^2}{b^2} + \frac{(z + mPE)^2}{c^2} = 1;$$

whence

$$\left(\frac{l^2}{a^2} + \frac{m^2}{b^2} + \frac{n^2}{c^2}\right) PE^2 + 2\left(\frac{lx}{a^2} + \frac{my}{b^2} + \frac{nz}{c^2}\right) PE - \left(1 - \frac{x^2}{a^2} - \frac{y^2}{b^2} - \frac{z^2}{c^2}\right) = 0.$$

When (x, y, z) is within the ellipsoid this equation, viewed as a quadratic in PE, has its roots of opposite signs; the positive one is PE, the negative is $-PE'$.

Now if r_1, r_2 be the two roots of $gr^2 + 2fr - e = 0$, we have

$$\tfrac{1}{2}(r_1^2 + r_2^2) = (2f^2 + ge)/g^2.$$

Hence

$$\left.\begin{aligned} \tfrac{1}{2}(PE^2 + PE'^2) &= \frac{\frac{l^2}{a^2}\left(\frac{2x^2}{a^2} + e\right) + \frac{m^2}{b^2}\left(\frac{2y^2}{b^2} + e\right) + \frac{n^2}{c^2}\left(\frac{2z^2}{c^2} + e\right) + Q}{\left(\frac{l^2}{a^2} + \frac{m^2}{b^2} + \frac{n^2}{c^2}\right)^2}, \\ \text{where} \quad e &= 1 - \frac{x^2}{a^2} - \frac{y^2}{b^2} - \frac{z^2}{c^2}, \\ \text{and} \quad Q &= 4\left(\frac{mnyz}{b^2c^2} + \frac{nlzx}{c^2a^2} + \frac{lmxy}{a^2b^2}\right) \end{aligned}\right\} \quad ..(8).$$

Now in the $\iint$ integration of (7), as we see readily by taking for example one of the hemispheres into which the whole sphere round P is cut by the plane through P perpendicular to z, it is clear that Digression on the attraction of an ellipsoid.

$$\iint \frac{Q d\sigma}{\frac{l^2}{a^2}+\frac{m^2}{b^2}+\frac{n^2}{c^2}} = 0 \quad \text{..................(9)};$$

and therefore (7) and (8) give

$$V = \iint d\sigma \frac{\frac{l^2}{a^2}\left(\frac{2x^2}{a^2}+e\right)+\frac{m^2}{b^2}\left(\frac{2y^2}{b^2}+e\right)+\frac{n^2}{c^2}\left(\frac{2z^2}{c^2}+e\right)}{\left(\frac{l^2}{a^2}+\frac{m^2}{b^2}+\frac{n^2}{c^2}\right)^2} \quad \text{...(10)};$$

or

$$V = e\Phi + \frac{x^2}{a}\frac{d\Phi}{da} + \frac{y^2}{b}\frac{d\Phi}{db} + \frac{z^2}{c}\frac{d\Phi}{dc} \quad \text{..........(11)},$$

where

$$\Phi = \iint \frac{d\sigma}{\frac{l^2}{a^2}+\frac{m^2}{b^2}+\frac{n^2}{c^2}} \quad \text{......................(12)}.$$

494 l. A symmetrical evaluation of Φ not being obvious, we may be content to take

$$l = \cos\theta, \quad m = \sin\theta\cos\phi, \quad n = \sin\theta\sin\phi,$$

and

$$d\sigma = \sin\theta\, d\theta\, d\phi.$$

Using these, replacing l, and putting

$$\frac{1}{b^2} - \left(\frac{1}{b^2}-\frac{1}{a^2}\right) l^2 = H, \quad \text{and} \quad \frac{1}{c^2} - \left(\frac{1}{c^2}-\frac{1}{a^2}\right) l^2 = K,$$

we find

$$\Phi = \int_0^1 dl \int_0^{2\pi} \frac{d\phi}{H\cos^2\phi + K\sin^2\phi}.$$

$$\int_0^{2\pi} \frac{d\phi}{H\cos^2\phi + K\sin^2\phi} = 4\int_0^\infty \frac{dt}{H+Kt^2} = \frac{2\pi}{\sqrt{(HK)}}.$$

Hence

$$\Phi = 2\pi \int_0^1 \frac{dl}{\left[\frac{1}{b^2}-\left(\frac{1}{b^2}-\frac{1}{a^2}\right)l^2\right]^{\frac{1}{2}}\left[\frac{1}{c^2}-\left(\frac{1}{c^2}-\frac{1}{a^2}\right)l^2\right]^{\frac{1}{2}}} \quad \text{...(13)}.$$

By (12) we know that Φ is a symmetrical function of a, b, c.

Digression on the attraction of an ellipsoid.

To bring (12) to this form, take

$$l=\frac{a}{\sqrt{(a^2+u)}}\text{ ....(14)},$$

which reduces (13) to

$$\Phi=\pi abc\int_0^\infty\frac{du}{(a^2+u)^{\frac{1}{2}}(b^2+u)^{\frac{1}{2}}(c^2+u)^{\frac{1}{2}}}\text{ ....(15)}.$$

The expression (11) for V, with (15) for Φ, is worth preserving for its own sake and for some applications; but the following, derived from it by performing the indicated differentiations, is simpler and is generally preferable:

$$V=\pi abc\int_0^\infty\left(1-\frac{x^2}{a^2+u}-\frac{y^2}{b^2+u}-\frac{z^2}{c^2+u}\right)\frac{du}{(a^2+u)^{\frac{1}{2}}(b^2+u)^{\frac{1}{2}}(c^2+u)^{\frac{1}{2}}}\text{ ..(16)};$$

or, if M denote the mass of the ellipsoid,

$$V=\frac{3M}{4}\int_0^\infty\left(1-\frac{x^2}{a^2+u}-\frac{y^2}{b^2+u}-\frac{z^2}{c^2+u}\right)\frac{du}{(a^2+u)^{\frac{1}{2}}(b^2+u)^{\frac{1}{2}}(c^2+u)^{\frac{1}{2}}}\text{ ...(17)}.$$

This, or (16), expresses the potential at any point (x, y, z) within the ellipsoid (a, b, c) or on its surface.

494 *m*. The potential at any external point is deduced from (17) through Maclaurin's theorem [§§ 494 *i*] simply by substituting for a, b, c the semi-axes of the ellipsoid confocal with (a, b, c), and passing through x, y, z: these semi-axes are $\sqrt{(a^2+q)}$, $\sqrt{(b^2+q)}$, $\sqrt{(c^2+q)}$, where q denotes the positive root of the equation

$$\frac{x^2}{a^2+q}+\frac{y^2}{b^2+q}+\frac{z^2}{c^2+q}=1\text{ ....(18)};$$

which is a cubic in q. Thus, for an external point, we find

$$V=\frac{3M}{4}\int_0^\infty\left(1-\frac{x^2}{a^2+q+u}-\frac{y^2}{b^2+q+u}-\frac{z^2}{c^2+q+u}\right)\frac{du}{(a^2+q+u)^{\frac{1}{2}}(b^2+q+u)^{\frac{1}{2}}(c^2+q+u)^{\frac{1}{2}}}\text{ ....(19)};$$

which may be written shorter as follows:

$$V=\frac{3M}{4}\int_q^\infty\left(1-\frac{x^2}{a^2+u}-\frac{y^2}{b^2+u}-\frac{z^2}{c^2+u}\right)\frac{du}{(a^2+u)^{\frac{1}{2}}(b^2+u)^{\frac{1}{2}}(c^2+u)^{\frac{1}{2}}}\text{ ...(20)}.$$

494 *n.* These formulas, (17) and (20), are, we believe, due to Lejeune Dirichlet, who proves them (Crelle's *Journal*, 1846, Vol. XXXII.) by showing that they satisfy the equation Digression on the attraction of an ellipsoid

$$\frac{d^2V}{dx^2}+\frac{d^2V}{dy^2}+\frac{d^2V}{dz^2}=-4\pi,$$

when

$$\frac{x^2}{a^2}+\frac{y^2}{b^2}+\frac{z^2}{c^2}<1,$$

and

$$\frac{d^2V}{dx^2}+\frac{d^2V}{dy^2}+\frac{d^2V}{dz^2}=0,$$

when

$$\frac{x^2}{a^2}+\frac{y^2}{b^2}+\frac{z^2}{c^2}>1;$$

and that

$$\frac{dV}{dx},\ \frac{dV}{dy},\ \frac{dV}{dz}$$

have equal values at points infinitely near the surface

$$\frac{x^2}{a^2}+\frac{y^2}{b^2}+\frac{z^2}{c^2}=1,$$

outside and inside it. His first step towards this proof (the completion of which we leave as an exercise to our readers) is the evaluation of dV/dx, dV/dy, dV/dz. In this it is necessary to remark that, for the external point, terms depending on the variation of q as it appears in (20) vanish because of (18): and taking the results which we then get instantly by plain differentiation, and remembering that $X=-dV/dx$, &c., we have, for the principal components of the resultant force,

$$\left.\begin{aligned}X&=\frac{3Mx}{2}\int_q^\infty\frac{du}{(a^2+u)^{\frac{3}{2}}(b^2+u)^{\frac{1}{2}}(c^2+u)^{\frac{1}{2}}}\\ Y&=\frac{3My}{2}\int_q^\infty\frac{du}{(a^2+u)^{\frac{1}{2}}(b^2+u)^{\frac{3}{2}}(c^2+u)^{\frac{1}{2}}}\\ Z&=\frac{3Mz}{2}\int_q^\infty\frac{du}{(a^2+u)^{\frac{1}{2}}(b^2+u)^{\frac{1}{2}}(c^2+u)^{\frac{3}{2}}}\end{aligned}\right\}\ldots\ldots(21),$$

where $q=0$ when (x, y, z) is internal, and q is the positive root of the cubic (18), when (x, y, z) is external.

Using (21) in (20) and (17), we see that

$$V=\frac{3M}{4}\int_q^\infty\frac{du}{(a^2+u)^{\frac{1}{2}}(b^2+u)^{\frac{1}{2}}(c^2+u)^{\frac{1}{2}}}-\tfrac{1}{2}(Xx+Yy+Zz)\ldots(22).$$

Digression on the attraction of an ellipsoid.

494 *o*. For the case of an internal point or a point on the surface, by putting $q=0$, we fall back on the original expressions (16) for V, and the proper differential coefficients of it for X, Y, Z.

These results may be written as follows:

$$\left.\begin{array}{l} X=\frac{4\pi}{3}\mathfrak{A}x, \quad Y=\frac{4\pi}{3}\mathfrak{B}y, \quad Z=\frac{4\pi}{3}\mathfrak{C}z, \\ V=\Phi-\frac{2\pi}{3}(\mathfrak{A}x^2+\mathfrak{B}y^2+\mathfrak{C}z^2) \end{array}\right\} \ldots (23),$$

where Φ, $\mathfrak{A}$, $\mathfrak{B}$, $\mathfrak{C}$ are constants, of which Φ is given by (12), or (13), or (15), and the others by (21) with $q=0$; all expressed in terms of elliptic integrals.

It follows that the internal equipotential surfaces are concentric similar ellipsoids with axes proportional to $\mathfrak{A}^{-\frac{1}{2}}$, $\mathfrak{B}^{-\frac{1}{2}}$, $\mathfrak{C}^{-\frac{1}{2}}$; and that the internal surfaces of equal resultant force are concentric similar ellipsoids with axes proportional to $\mathfrak{A}^{-1}$, $\mathfrak{B}^{-1}$, $\mathfrak{C}^{-1}$.

The external equipotentials are transcendental plinthoids* of an interesting character. So are the equipotentials partly internal (where they are ellipsoidal) and external (where they are not ellipsoidal).

It is interesting, and useful in helping to draw the external equipotentials, to remark the following relations between the internal equipotentials, the external equipotentials, and the surface of the attracting ellipsoid.

(1) The external equipotential $V=C$ is the envelope of the series of ellipsoidal surfaces obtained by giving an infinite number of constant values to q in the equation

$$\int_q^\infty \left(1-\frac{x^2}{a^2+u}-\frac{y^2}{b^2+u}-\frac{z^2}{c^2+u}\right)\frac{du}{(a^2+u)^{\frac{1}{2}}(b^2+u)^{\frac{1}{2}}(c^2+u)^{\frac{1}{2}}}=\frac{4C}{3M} \ldots (\alpha).$$

(2) This envelope is cut by the ellipsoidal surface

$$\frac{x^2}{a^2+q}+\frac{y^2}{b^2+q}+\frac{z^2}{c^2+q}=1 \ldots\ldots\ldots\ldots\ldots (\beta),$$

* From πλινθοειδής, brick-like. Plinthoid, as we now use the term, denotes as it were a sea-worn brick; any figure with three rectangular axes, and surfaces everywhere convex, such as an ellipsoid, or a perfectly symmetrical bale of cotton with slightly rounded sides and rounded edges and corners. One extreme of plinthoidal figure is a rectangular parallelepiped; another extreme, just not excluded by our definition, is a figure composed of two equal and similar right rectangular pyramids fixed together base to base, that is a "regular octohedron."

for any particular value of q in the line along which it is touched by the particular one of the series of consecutive ellipsoidal surfaces (β) corresponding to this value of q.

Digression on the attraction of an ellipsoid.

(3) If the ellipsoidal surface (β) be filled with homogeneous matter, the complete equipotential for any particular value of C is composed of an interior ellipsoidal surface passing tangentially to the external plinthoidal (but not ellipsoidal) surface across the transitional line defined in (2).

It is easy to make graphic illustrations for the case of ellipsoids of revolution, by aid of § 527 below.

494 p. In the case of an elliptic cylinder, which is important in many physical investigations, replace M by $4\pi abc/3$, and put $c=\infty$.

Attraction of an infinitely long elliptic cylinder.

Thus we find

$$\left.\begin{aligned}
X&=2\pi abx\int_q^\infty\frac{du}{(a^2+u)^{\frac{3}{2}}(b^2+u)^{\frac{1}{2}}}=\frac{4\pi ab[\sqrt{(a^2+q)}-\sqrt{(b^2+q)}]x}{(a^2-b^2)\sqrt{(a^2+q)}}\\
&=\frac{4\pi abx}{\sqrt{(a^2+q)}[\sqrt{(a^2+q)}+\sqrt{(b^2+q)}]}\\
Y&=2\pi aby\int_q^\infty\frac{du}{(a^2+u)^{\frac{1}{2}}(b^2+u)^{\frac{3}{2}}}=\frac{4\pi ab[\sqrt{(a^2+q)}-\sqrt{(b^2+q)}]y}{(a^2-b^2)\sqrt{(b^2+q)}}\\
&=\frac{4\pi aby}{\sqrt{(b^2+q)}[\sqrt{(a^2+q)}+\sqrt{(b^2+q)}]}\\
&\text{where}\quad q=0,\ \text{when}\ \frac{x^2}{a^2}+\frac{y^2}{b^2}<1;\\
&\text{and } q \text{ is the positive root of the quadratic}\\
&\frac{x^2}{a^2+q}+\frac{y^2}{b^2+q}=1,\ \text{when}\ \frac{x^2}{a^2}+\frac{y^2}{b^2}>1.
\end{aligned}\right\}\ldots(24).$$

For the case of $q=0$, that is to say, the case of an internal point, (24) becomes

$$X=\frac{4\pi ab}{a+b}\frac{x}{a},\quad\text{and}\quad Y=\frac{4\pi ab}{a+b}\frac{y}{b}\ldots\ldots\ldots\ldots(25).$$

494 q. For the magnitude of the resultant force we deduce

Internal isodynamic surfaces are similar to the bounding surface.

$$R=\sqrt{(X^2+Y^2)}=\frac{4\pi ab}{a+b}\sqrt{\left(\frac{x^2}{a^2}+\frac{y^2}{b^2}\right)}\ldots\ldots(26);$$

Attraction of an infinitely long elliptic cylinder.

and it is remarkable that this is constant for all points on the surface of the elliptic cylinder $\frac{x^2}{a^2}+\frac{y^2}{b^2}=1$, and on each similar internal surface, and that its values on different ones of these surfaces are as their linear magnitudes.

Potential in free space cannot have a maximum or minimum value:

495 a. At any point of zero force, the potential is a *maximum* or a *minimum*, or a "*minimax*." Now from § 492 (2) it follows that the potential cannot be a maximum or a minimum at a point in free space. For if it were so, a closed surface could be described about the point, and indefinitely near it, so that at every point of it the value of the potential would be less than, or greater than, that at the point; so that N would be negative or positive all over the surface, and therefore $\iint N d\sigma$ would be finite, which is impossible, as the surface encloses none of the attracting mass.

is a minimax at a point of zero force in free space.

Earnshaw's theorem of unstable equilibrium.

495 b. Consider, now, a point of zero force in free space:—the potential, if it varies at all in the neighbourhood, must be a minimax at the point, because, as has just been proved, it cannot be a maximum or a minimum. Hence a material particle placed at a point of zero force under the action of any attracting bodies, and free from all constraint, is in unstable equilibrium, a result due to Earnshaw*.

495 c. If the potential be constant over a closed surface which contains none of the attracting mass, it has the same constant value throughout the interior. For if not, it must have a maximum or a minimum value somewhere within the surface, which (§ 495, a) is impossible.

Mean potential over a spherical surface equal to that at its centre.

496. The mean potential over any spherical surface, due to matter entirely without it, is equal to the potential at its centre; a theorem apparently first given by Gauss. See also *Cambridge Mathematical Journal*, Feb. 1845 (Vol. IV. p. 225). It is one of the most elementary propositions of spherical harmonic analysis, applied to potentials, found by applying App. B. (16) to the formulæ of § 539, below. But the following proof taken from the paper now referred to is noticeable as independent of the harmonic expansion.

* *Cambridge Phil. Trans.*, March, 1839.

Let, in Chap. I. App. A. (*a*), S be a spherical surface, of radius a; and let U be the potential at (x, y, z), due to matter altogether external to it; let U' be the potential of a unit of matter uniformly distributed through a smaller concentric spherical surface; so that, outside S and to some distance within it, $U' = \frac{1}{r}$; and lastly, let $a = 1$. The middle member of App. A (*a*) (1) becomes

Mean potential over a spherical surface equal to that at its centre.

$$\frac{1}{a}\iint \partial U d\sigma - \iiint U' \nabla^2 U dx dy dz,$$

which is equal to zero, since $\nabla^2 U = 0$ for the whole internal space, and (§ 492) $\iint \partial U d\sigma = 0$. Equating therefore the third member to zero we have

$$\iint d\sigma U \partial U' = \iiint U \nabla^2 U' dx dy dz.$$

Now at the surface, S, $\partial U' = -\frac{1}{a^2}$; and for all points external to the sphere of matter to which U' is due, $\nabla^2 U' = 0$, and for all internal points $\nabla^2 U' = -4\pi\rho'$, if ρ' be the density of the matter. Hence the preceding equation becomes

$$\frac{1}{a^2}\iint U d\sigma = 4\pi \iiint \rho' U dx dy dz.$$

Let now the density ρ' increase without limit, and the spherical space within which the triple integral extends, therefore become infinitely small. If we denote by U_0 the value of U at its centre, which is also the centre of S, we shall have

$$\iiint \rho' U dx dy dz = U_0 \iiint \rho' dx dy dz = U_0.$$

Hence the equation becomes

$$\frac{\iint U d\sigma}{4\pi a^2} = U_0,$$

which was to be proved.

The following more elementary proof is preferable:—imagine any quantity of matter to be uniformly distributed over the spherical surface. The mutual potential (§ 547 below) of this and the external mass is the same as if the matter were condensed from the spherical surface to its centre.

497. If the potential of any masses has a constant value, V, through any finite portion, K, of space, unoccupied by matter, it is equal to V through every part of space which can be reached

Theorem of Gauss:

Theorem of Gauss, proved. in any way without passing through any of those masses: a very remarkable proposition, due to Gauss, proved thus:—If the potential differ from V in space contiguous to K, we may, from any point C within K, as centre, in the neighbourhood of a place where the potential differs from V, describe a spherical surface not large enough to contain any part of any of the attracting masses, nor to include any of the space external to K except such as has potential all greater than V, or all less than V. But this is impossible, since we have just seen (§ 496) that the mean potential over the spherical surface must be V. Hence the supposition that the potential differs from V in any place contiguous to K and not including masses, is false.

498. Similarly we see that in any case of symmetry round an axis, if the potential is constant through a certain finite distance, however short, along the axis, it is constant throughout the whole space that can be reached from this portion of the axis, without crossing any of the masses. (See § 546, below.)

Green's problem.

499. Let S be any finite portion of a surface, or a complete closed surface, or an infinite surface; and let E be any point on S. (*a*) It is possible to distribute matter over S so as to produce, over the whole of S, potential equal to $F(E)$, any arbitrary function of the position of E. (*b*) There is only one whole quantity of matter, and one distribution of it, which can do this.

In Chap. I. App. A. (*b*) (*e*), etc., let $\alpha = 1$. By (*e*) we see that there is one, and that there is only one, solution of the equation

$$\nabla^2 U = 0$$

for all points not belonging to S, subject to the condition that U shall have a value arbitrarily given over the whole of S. Continuing to denote by U the solution of this problem, and considering first the case of S an open shell, that is to say, a finite portion of curved surface (including a plane, of course, as a particular case), let, in Chap. I. App. A. (*a*), U' be the potential at (x, y, z) due to a distribution of matter, having $\varpi(Q)$ for density at any point, Q. Let the triple integration extend throughout infinite space, exclusive of the infinitely thin shell S. Although in the investigation referred to [App. A. (a)] the triple integral extended only through the finite space contained within a closed surface, the same process shows that we have now, instead of the second and third members of (1) of that investigation, the following equated expressions:—

Green's problem:

$$\iint d\sigma U' \{[\partial U] - (\partial U)\} - \iiint dxdydz U' \nabla^2 U$$
$$= \iint d\sigma U \{[\partial U'] - (\partial U')\} - \iiint dxdydz U \nabla^2 U'$$

where $[\partial U]$ denotes the rate of variation of U on either side of S, infinitely near E, reckoned per unit of length *from* S; and (∂U) denotes the rate of variation of U infinitely near E, on the other side of S, reckoned per unit of length *towards* S; and $[\partial U']$, $(\partial U')$ denote the same for U'. Now we shall suppose the matter of which U' is the potential not to be condensed in finite quantities on any finite areas of S, which will make

$$[\partial U'] = (\partial U'):$$

and the conditions defining U and U' give, throughout the space of the triple integral,

$$\nabla^2 U = 0, \text{ and } \nabla^2 U' = -4\pi\varpi;$$

ϖ denoting the value of ϖ (Q) when Q is the point (x, y, z). Hence the preceding equation becomes

$$\iint d\sigma U' \{[\partial U] - (\partial U)\} = 4\pi \iiint dxdydz \varpi U \text{............} (1).$$

Let now the matter of which U' is the potential be equal in amount to unity and be confined to an infinitely small space round a point Q. We shall have

$$\iiint dxdydz \varpi U = U(Q) \iiint \varpi dxdydz = U(Q),$$

if we denote the value of U at (Q) by $U(Q)$:

also $$U' = \frac{1}{EQ}.$$

Hence (1) becomes

$$\iint \frac{[\partial U] - (\partial U)}{EQ} d\sigma = 4\pi U(Q) \text{..............} (2).$$

Hence a distribution of matter over S, having

reduced to the proper general solution of Laplace's equation.

$$\frac{1}{4\pi} \{[\partial U] - (\partial U)\} \text{ ....} (3)$$

for density at the point E, gives U as its potential at (x, y, z). We conclude, therefore, that it is possible to find one, but only one, distribution of matter over S which shall produce an arbi-

Green's problem;

trarily given potential, $F(E)$, over the whole of S; and in (2) we have the solution of this problem, when the problem of finding U to fulfil the conditions stated above, has been solved.

If S is any finite closed surface, any group of surfaces, open or closed, or an infinite surface, the same conclusions clearly hold. The triple integration used in the investigation must then be separately carried out through all the portions of space separated from one another by S, or by portions of S.

If the solution, ρ, of the problem has been obtained for the case in which the arbitrary function is the potential at any point of S, due to a unit of matter at any point P not belonging to S, that is to say, for the case of $F(E)=\frac{1}{EP}$, the solution of the general problem was shown by Green to be deducible from it thus:—

solved synthetically in terms of particular solution of Laplace's equation.

$$U=\iint \rho F(E)\,d\sigma \quad\ldots\ldots\ldots\ldots\ldots\ldots\ldots (4).$$

The proof is obvious: For let, for a moment, ${}_{,}\rho$ denote the superficial density required to produce U, then ${}_{,}\rho'$ denoting the value of ${}_{,}\rho$ for any other element, E', of S, we have

$$F(E)=\iint \frac{{}_{,}\rho' d\sigma'}{E'E}.$$

Hence the preceding double integral becomes

$$\iint d\sigma\rho\iint d\sigma' \frac{{}_{,}\rho'}{E'E}, \text{ or } \iint d\sigma' {}_{,}\rho' \iint d\sigma \frac{\rho}{E'E}.$$

But, by the definition of ρ,

$$\iint d\sigma \frac{\rho}{E'E}=\frac{1}{E'P}\ldots\ldots\ldots\ldots\ldots\ldots\ldots\ldots (5);$$

and therefore

$$\iint \rho F(E)\,d\sigma=\iint d\sigma' \frac{{}_{,}\rho'}{E'P}\ldots\ldots\ldots\ldots\ldots (6).$$

The second member of this is equal to U, according to the definition of ${}_{,}\rho$.

The expression (46) of App. B., from which the spherical harmonic expansion of an arbitrary function was derived, is a case of the general result (4) now proved.

Isolation of effect by closed portion of surface.

500. It is important to remark that, if S consist, in part, of a closed surface, Q, the determination of U within it will be independent of those portions of S, if any, which lie without it; and, *vice versa*, the determination of U through external space will be independent of those portions of S, if any, which lie within Q. Or if S consist, in part, of a surface Q, extending infinitely in all directions, the determination of U through all space on either side of Q, is independent of those portions of S, if any, which lie on the other side. This follows from the preceding investigation, modified by confining the triple integration to one of the two portions of space separated completely from one another by Q.

Isolation of effect by closed portion of surface.

501. Another remark of extreme importance is this:—If $F(E)$ be the potential at E of any distribution, M, of matter, and if S be such as to separate perfectly any portion or portions of space, H, from all of this matter; that is to say, such that it is impossible to pass into H from any part of M without crossing S; then, throughout H, the value of U will be the potential of M.

Green's problem; applied to a given distribution of electricity, M, influencing a conducting surface, S.

For if V denote this potential, we have, throughout H, $\nabla^2 V = 0$; and at every point of the boundary of H, $V = F(E)$. Hence, considering the theorem of Chap. I. App. A. (c), for the space H alone, and its boundary alone, instead of S, we see that, through this space, V satisfies the conditions prescribed for U, and therefore, through this space, $U = V$.

Solved Examples. (1) Let M be a homogeneous solid ellipsoid; and let S be the bounding surface, or any of the external ellipsoidal surfaces confocal with it. The required surface-density is proved in § 494 *g* to be *inversely* proportional to the perpendicular from the centre to the tangent-plane; or, which is the same, directly proportional to the distance between S and another confocal ellipsoid surface infinitely near it. In other words, the attraction of a focaloid (§ 494 *g*, foot-note) of homogeneous matter is, for all points external to it, the same as that of a homogeneous solid of equal mass bounded by any confocal ellipsoid interior to it.

Virtually Maclaurin's theorem, § 494 *i*.

(2) Let M be an elliptic homoeoid (§ 494 *g*, foot-note) of homogeneous matter; and let S be any external confocal ellipsoidal surface. The required surface-density is proved in § 519 below to be *directly* proportional to the perpendicular from the centre to the tangent-plane; and, which is

Elliptic homoeoid, an example belonging to the reducible case, § 505, of Green's problem.

Green's problem.

the same, directly proportional to the distance between S and a similar concentric ellipsoidal surface infinitely near it. In other words, the attractions of confocal infinitely thin elliptic homoeoids of homogeneous matter are the same for all external points, if their masses are equal.

Complex application of § 501.

502. To illustrate more complicated applications of § 501, let S consist of three detached surfaces, S_1, S_2, S_3, as in the diagram, of which S_1, S_2 are closed, and S_3 is an open shell, and if $F(E)$ be the potential due to M, at any point, E, of any of these portions of S; then throughout H_1, and H_2, the spaces within S_1 and without S_2, the value of U is simply the potential of M. The value of U through K, the remainder of space, depends, of course, on the character of the composite surface S, and is a case of the general problem of which the solution was proved to be possible and single in Chap. I. App. A.

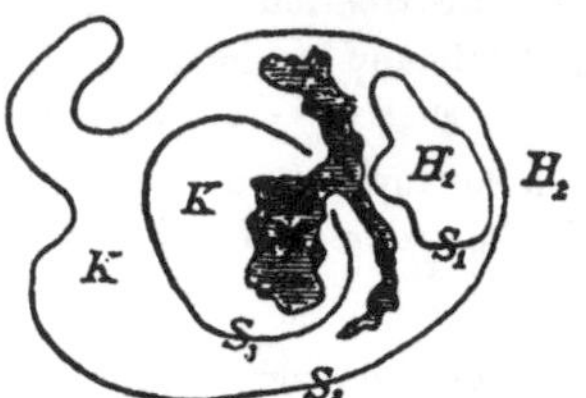

General problem of electric influence possible and determinate.

503. From § 500 follows the grand proposition:—*It is possible to find one, but no other than one, distribution of matter over a surface S which shall produce over S, and throughout all space H separated by S from every part of M, the same potential as any given mass M.*

Thus, in the preceding diagram, it is possible to find one, and but one, distribution of matter over S_1, S_2, S_3 which shall produce over S_3 and through H_1 and H_2 the same potential as M.

The statement of this proposition most commonly made is: *It is possible to distribute matter over any surface, S, completely enclosing a mass M, so as to produce the same potential as M through all space outside S;* which, though seemingly more limited, is, when interpreted with proper mathematical comprehensiveness, equivalent to the foregoing.

Simultaneous electric influences in spaces

504. If S consist of several closed or infinite surfaces, S_1, S_2, S_3, respectively separating certain isolated spaces H_1, H_2, H_3, from

H, the remainder of all space, and if $F(E)$ be the potential of masses m_1, m_2, m_3, lying in the spaces H_1, H_2, H_3; the portions of U due to S_1, S_2, S_3, respectively will throughout H be equal respectively to the potentials of m_1, m_2, m_3, separately. For as we have just seen, it is possible to find one, but only separated by infinitely thin conducting surfaces.

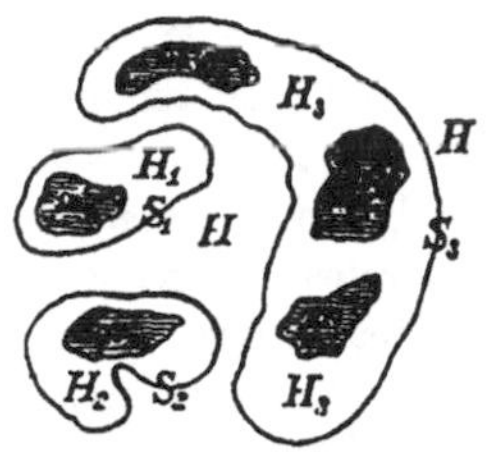

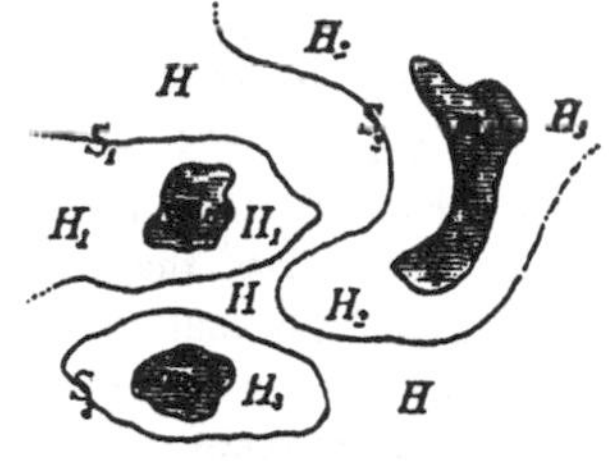

one, distribution of matter over S_1 which shall produce the potential of m_1, throughout all the space H_1, H_2, H_3, etc., and one, but only one, distribution over S_2 which shall produce the potential of m_2 throughout H, H_1, H_3, etc.; and so on. But these distributions on S_1, S_2, etc., jointly constitute a distribution producing the potential $F(E)$ over every part of S, and

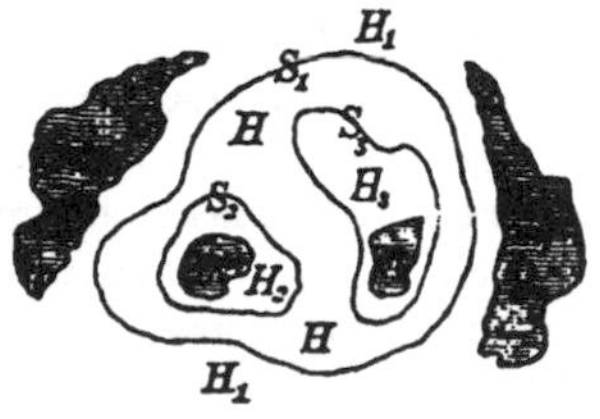

therefore the sum of the potentials due to them all, at any point, fulfils the conditions presented for U. This is therefore (§ 503) *the* solution of the problem.

505. Considering still the case in which $F(E)$ is prescribed to be the potential of a given mass, M: let S be an equipotential surface enclosing M, or a group of isolated surfaces enclosing all the parts of M, and each equipotential for the whole of M. The potential due to the supposed distribution over S will be the same as that of M, through all external space, and will be constant (§ 497) through each enclosed portion of space. Its resultant attraction will therefore be the same as that of M on all external points, and zero on all internal points. Hence we see at once that the density of the matter distributed over it, Reducible case of Green's problem;

Reducible case of Green's problem; to produce $F(E)$, is equal to $\frac{R}{4\pi}$ where R denotes the resultant force of M, at the point E.

We have $[\partial U] = -R$ and $(\partial U) = 0$. Using this in § 500 (2), we find the preceding formula for the required surface-density.

applied to the invention of solved problems of electric influence. **506.** Considering still the case of §§ 501, 505, let S be the equipotential not of M alone, as in § 505, but of M and another mass m completely separated by it from M; so that $V+v=C$ at S, if V and v denote the potentials of M and m respectively.

The potential of the supposed distribution of matter on S, which, (§ 501), is equal to V through all space separated from M by S, is equal to $C-v$ at S, and therefore equal to $C-v$ throughout the space separated from m by S.

Thus, passing from potentials to attractions, we see that the resultant attraction of S alone, on all points on one side of it is the same as that of M; and on the other side is equal and opposite to that of m. The most direct and simple complete statement of this result is as follows:—

If masses m, m', in portions of space, H, H', completely separated from one another by one continuous surface S, whether closed or infinite, are known to produce tangential forces equal and in the same direction at each point of S, one and the same distribution of matter over S will produce the force of m throughout H', and that of m' throughout H. The density of this distribution is equal to $\frac{R}{4\pi}$, if R denote the resultant force due to one of the masses, and the other with its *sign* changed. And it is to be remarked that the direction of this resultant force is, at every point, E, of S, perpendicular to S, since the potential due to one mass, and the other with its sign changed, is constant over the whole of S.

Examples. **507.** Green, in first publishing his discovery of the result stated in § 505, remarked that it shows a way to find an infinite variety of closed surfaces for any one of which we can solve the problem of determining the distribution of matter over it which shall produce a given uniform potential at each point of its surface, and consequently the same also throughout its interior. Thus, an example which Green himself gives, let M be a uniform bar of matter, AA'. The equipotential surfaces round it are, as we have seen above (§ 481 *c*), prolate ellipsoids of revolution, each having A and A' for its foci; and the resultant force at any point P was found to be

Reducible case of Green's problem:—examples.

$$\frac{mp}{l(l^2-a^2)},$$

the whole mass of the bar being denoted by m, and its length by $2a$; $A'P+AP$ by $2l$; and the perpendicular from the centre to the tangent plane at P of the ellipsoid, by p. We conclude that a distribution of matter over the surface of the ellipsoid, having

$$\frac{1}{4\pi}\,\frac{mp}{l(l^2-a^2)}$$

for density at P, produces on all external space the same resultant force as the bar, and zero force or a constant potential through the internal space. This is a particular case of the Example (2) § 501 above, founded on the general result regarding ellipsoidal homoeoids proved below, in §§ 519, 520, 521.

508. As a second example, let M consist of two equal particles, at points I, I'. If we take the mass of each as unity, the potential at P is $\frac{1}{IP}+\frac{1}{I'P}$; and therefore

$$\frac{1}{IP}+\frac{1}{I'P}=C$$

is the equation of an equipotential surface; it being understood that negative values of IP and $I'P$ are inadmissible, and that any constant value, from ∞ to 0, may be given to C. The curves in the annexed diagram have been drawn, from this equation, for the cases of C equal respectively to 10, 9, 8, 7, 6, 5, 4·5, 4·3, 4·2, 4·1, 4, 3·9, 3·8, 3·7, 3·5, 3, 2·5, 2; the value of II' being unity.

The corresponding equipotential surfaces are the surfaces traced by these curves, if the whole diagram is made to rotate round II' as axis. Thus we see that for any values of C less than 4 the equipotential surface is one closed surface. Choosing

Reducible case of Green's problem:—examples.

any one of these surfaces, let R denote the resultant of forces equal to $\frac{1}{IP^2}$ and $\frac{1}{I'P^2}$ in the lines PI and PI'. Then if

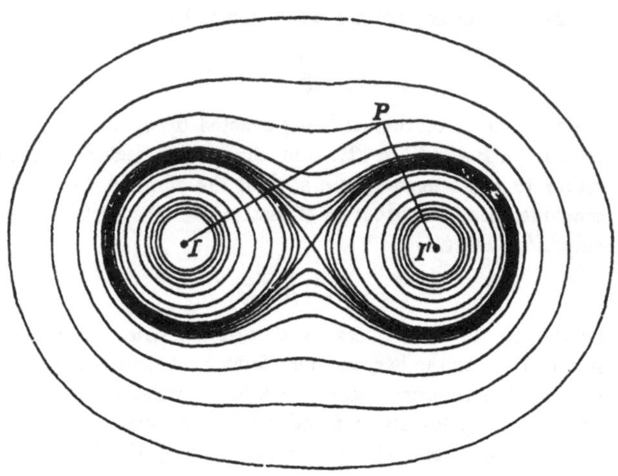

matter be distributed over this surface, with density at P equal to $\frac{R}{4\pi}$, its attraction on any internal point will be zero; and on any external point, will be the same as that of I and I'.

509. For each value of C greater than 4, the equipotential surface consists of two detached ovals approximating (the last three or four in the diagram, very closely) to spherical surfaces, with centres lying between the points I and I', but approximating more and more closely to these points, for larger and larger values of C.

Considering one of these ovals alone, one of the series enclosing I', for instance, and distributing matter over it according to the same law of density, $\frac{R}{4\pi}$, we have a shell of matter which exerts (§ 507) on external points the same force as I'; and on internal points a force equal and opposite to that of I.

510. As an example of exceedingly great importance in the theory of electricity, let M consist of a positive mass, m, concentrated at a point I, and a negative mass, $-m'$, at I'; and let S be a spherical surface cutting II', and II' produced in points A, $A_{,}$, such that $IA : AI' :: IA_{,} : I'A_{,} :: m : m'$. Then, by a well-known geometrical proposition, we shall have $IE : I'E :: m : m'$; and therefore Electric images.

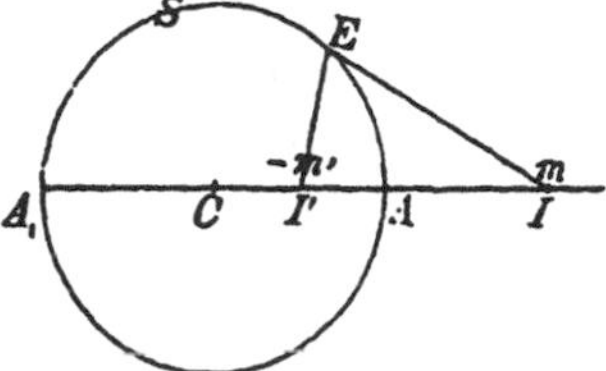

$$\frac{m}{IE} = \frac{m'}{I'E}.$$

Hence, by what we have just seen, one and the same distribution of matter over S will produce the same force as m' through all external space, and the same as m through all the space within S. And, finding the resultant of the forces $\frac{m}{IE^2}$ in EI, and $\frac{m'}{I'E^2}$ in $I'E$ produced, which, as these forces are inversely as IE to $I'E$, is (§ 256) equal to

$$\frac{m}{IE^2 . I'E} II', \text{ or } \frac{m^2 II'}{m'} \frac{1}{IE^3},$$

we conclude that the density in the shell at E is

$$\frac{m^2 II'}{4\pi m'} \cdot \frac{1}{IE^3}.$$

That the shell thus constituted does attract external points as if its mass were collected at I', and internal points as a certain mass collected at I, was proved geometrically in § 474 above.

511. If the spherical surface is given, and one of the points, I, I', for instance I, the other is found by taking $CI' = \frac{CA^2}{CI}$; and for the mass to be placed at it we have

$$m' = m \frac{I'A}{AI} = m \frac{CA}{CI} = m \frac{CI'}{CA}.$$

Hence if we have any number of particles m_1, m_2, etc., at points

Electric images.

I_1, I_2, etc., situated without S, we may find in the same way corresponding internal points I_1', I_2', etc., and masses m_1', m_2', etc.; and, by adding the expressions for the density at E given for each pair by the preceding formula, we get a spherical shell of matter which has the property of acting on all external space with the same force as $-m_1'$, $-m_2'$, etc., and on all internal points with a force equal and opposite to that of m_1, m_2, etc.

512. An infinite number of such particles may be given, constituting a continuous mass M; when of course the corresponding internal particles will constitute a continuous mass, $-M'$, of the opposite kind of matter; and the same conclusion will hold. If S is the surface of a solid or hollow metal ball connected with the earth by a fine wire, and M an external influencing body, the shell of matter we have determined is precisely the distribution of electricity on S called out by the influence of M: and the mass $-M'$, determined as above, is called the *Electric Image* of M in the ball, since the electric action through the whole space external to the ball would be unchanged if the ball were removed and $-M'$ properly placed in the space left vacant. We intend to return to this subject under Electricity.

Transformation by reciprocal radius-vectors.

513. Irrespectively of the special electric application, this method of images gives a remarkable kind of transformation which is often useful. It suggests for mere geometry what has been called the transformation by reciprocal radius-vectors; that is to say, the substitution for any set of points, or for any diagram of lines or surfaces, another obtained by drawing radii to them from a certain fixed point or origin, and measuring off lengths inversely proportional to these radii along their directions. We see in a moment by elementary geometry that any line thus obtained cuts the radius-vector through any point of it at the same angle and in the same plane as the line from which it is derived. Hence any two lines or surfaces that cut one another give two transformed lines or surfaces cutting at the same angle: and infinitely small lengths, areas, and volumes transform into others whose magnitudes are altered respectively in the ratios of the first, second, and third powers of the distances of the latter from the origin, to the same powers of the distances of the former from the same. Hence the lengths, areas, and volumes in the transformed diagram, corresponding to a set of given equal infinitely small lengths, areas, and volumes, however situated, at different distances from the origin, are inversely as the squares, the fourth powers and the sixth powers of these distances. Further, it is easily proved that a straight line and a plane transform into a circle and a spherical surface, each passing through the origin; and that, generally, circles and spheres transform into circles and spheres.

Transformation by reciprocal radius-vectors.

514. In the theory of attraction, the transformation of masses, densities, and potentials has also to be considered. Thus, according to the foundation of the method (§ 512), equal masses, of infinitely small dimensions at different distances from the origin, transform into masses inversely as these distances, or directly as the transformed distances: and, therefore, equal densities of lines, of surfaces, and of solids, given at any stated distances from the origin, transform into densities directly as the first, the third, and the fifth powers of those distances; or inversely as the same powers of the distances, from the origin, of the corresponding points in the transformed system.

515. The statements of the last two sections, so far as proportions alone are concerned, are most conveniently expressed thus:—

General summary of ratios.

Let P be any point whatever of a geometrical diagram, or of a distribution of matter, O one particular point ("the origin"), and a one particular length (the radius of the "reflecting sphere"). In OP take a point P', corresponding to P, and for any mass m, in any infinitely small part of the given distribution, place a mass m'; fulfilling the conditions

$$OP' = \frac{a^2}{OP}, \quad m' = \frac{a}{OP}m = \frac{OP'}{a}m.$$

Then if L, A, V, $\rho(L)$, $\rho(A)$, $\rho(V)$ denote an infinitely small length, area, volume, linear-density, surface-density, volume-density in the given distribution, infinitely near to P, or anywhere at the same distance, r, from O as P, and if the corresponding elements in the transformed diagram or dis-

General summary of ratios. tribution be denoted in the same way with the addition of accents, we have

$$L' = \frac{a^2}{r^2} L = \frac{r'^2}{a^2} L; \; A' = \frac{a^4}{r^4} A = \frac{r'^4}{a^4} A; \; V' = \frac{a^6}{r^6} V = \frac{r'^6}{a^6} V,$$

$$\rho'(L) = \frac{a}{r'} \rho(L) = \frac{r}{a} \rho(L); \; \rho'(A) = \frac{a^3}{r'^3} \rho(A) = \frac{r^3}{a^3} \rho(A);$$

$$\rho'(V) = \frac{a^5}{r'^5} \rho(V) = \frac{r^5}{a^5} \rho(V).$$

The usefulness of this transformation in the theory of electricity, and of attraction in general, depends entirely on the following theorem :—

Application to the potential. **516.** (*Theorem.*)—Let ϕ denote the potential at P due to the given distribution, and ϕ' the potential at P' due to the transformed distribution: then shall

$$\phi' = \frac{r}{a} \phi = \frac{a}{r'} \phi.$$

Let a mass m collected at I be any part of the given distribution, and let m' at I' be the corresponding part in the transformed distribution. We have $a^2 = OI' . OI = OP' . OP$, and therefore $OI : OP :: OP' : OI'$;

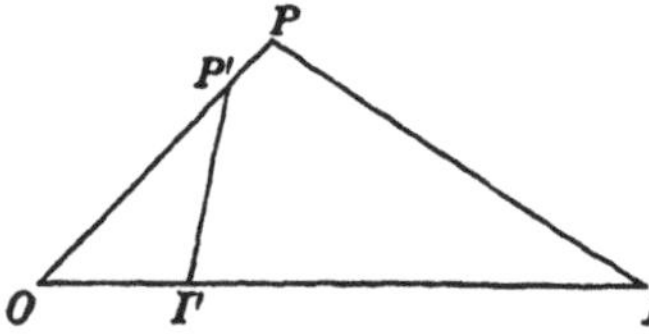

which shows that the triangles IPO, $P'I'O$ are similar, so that

$$IP : P'I' :: \sqrt{OI.OP} : \sqrt{OP'.OI'} :: OI.OP : a^2.$$

We have besides

$$m : m' :: OI : a,$$

and therefore

$$\frac{m}{IP} : \frac{m'}{I'P'} :: a : OP.$$

Hence each term of ϕ bears to the corresponding term of ϕ' the same ratio; and therefore the sum, ϕ, must be to the sum, ϕ', in that ratio, as was to be proved.

517. As an example, let the given distribution be confined to a spherical surface, and let O be its centre and a its radius. The transformed distribution is the same. But the space within it becomes transformed into the space without it. Hence if ϕ be the potential due to any spherical shell at a point P, within it, the potential due to the same shell at the point P' in OP produced till $OP' = \frac{a^2}{OP}$, is equal to $\frac{a}{OP'}\phi$ (which is an elementary proposition in the spherical harmonic treatment of potentials, as we shall see presently). Thus, for instance, let the distribution be uniform. Then, as we know there is no force on an interior point, ϕ must be constant; and therefore the potential at P', any external point, is inversely proportional to its distance from the centre.

Any distribution on a spherical shell.

Or let the given distribution be a uniform shell, S, and let O be any eccentric or any external point. The transformed distribution becomes (§§ 513, 514) a spherical shell, S', with density varying inversely as the cube of the distance from O. If O is within S, it is also enclosed by S', and the whole space within S transforms into the whole space without S'. Hence (§ 516) the potential of S' at any point without it is inversely as the distance from O, and is therefore that of a certain quantity of matter collected at O. Or if O is external to S, and consequently also external to S', the space within S transforms into the space within S'. Hence the potential of S' at any point within it is the same as that of a certain quantity of matter collected at O, which is now a point external to it. Thus, without taking advantage of the general theorems (§§ 499, 506), we fall back on the same results as we inferred from them in § 510, and as we proved synthetically earlier (§§ 471, 474, 475). It may be remarked that those synthetical demonstrations consist merely of transformations of Newton's demonstration, that attractions balance on a point within a uniform shell. Thus the first of them (§ 471) is the image of Newton's in a concentric spherical surface; and the second is its image in a spherical surface having its centre external to the shell, or internal but eccentric, according as the first or the second diagram is used.

Uniform shell eccentrically reflected.

Uniform solid sphere eccentrically reflected.

518. We shall give just one other application of the theorem of § 516 at present, but much use of it will be made later, in the theory of Electricity.

Let the given distribution of matter be a uniform solid sphere, B, and let O be external to it. The transformed system will be a solid sphere, B', with density varying inversely as the fifth power of the distance from O, a point external to it. The potential of B is the same throughout external space as that due to its mass, m, collected at its centre, C. Hence the potential of B' through space external to it is the same as that of the corresponding quantity of matter collected at C', the transformed position of C. This quantity is of course equal to the mass of B'. And it is easily proved that C' is the position of the image of O in the spherical surface of B'. We conclude that a solid sphere with density varying inversely as the fifth power of the distance from an external point, O, attracts any external point as if its mass were condensed at the image of O in its external surface. It is easy to verify this for points of the axis by direct integration, and thence the general conclusion follows according to § 490.

Second investigation of attraction of ellipsoid.

519. One other application of Green's great theorem of § 503, showing us a way to find the potential and the resultant force at any point within or without an elliptic homoeoid, from which we are led to a second very interesting solution of the problem of finding the attraction of an ellipsoid differing greatly from that of § 494, we shall now give.

An elliptic homoeoid exercises no force on internal points.

Elliptic homoeoid exerts zero force on internal point:

To prove this, let the infinitely thin spherical shell of § 462, imagined as bounded by concentric spherical surfaces, be distorted (§§ 158, 160) by simple extensions and compressions in three rectangular directions, so as to become an elliptic homoeoid. In this distorted form, the volumes of all parts are diminished or increased in the proportion of the volume of the ellipsoid to the volume of the sphere; and (§ 158) the ratio of the lines HP, PK is unaltered. Hence the elements IH, KL, still attract P equally; and therefore, as in § 462, we conclude that the resultant force on an internal point is zero.

It follows immediately that the attraction on any point in the hollow space within a homoeoid not infinitely thin is zero. This proposition is due originally to Newton. Theorem due to Newton.

520. In passing it may be remarked that the distribution of electricity on an ellipsoidal conductor, undisturbed by electric influence, is thus proved to be in simple proportion to the thickness of a homoeoid coincident with its surface, and therefore (§ 494, foot-note) directly proportional to the perpendicular from the centre to the tangent plane. Distribution of electricity on ellipsoidal conductor.

521. From § 519 and § 478 it follows that the resultant force on an external point anywhere infinitely near the homoeoid is perpendicular to the surface, and is equal to $4\pi t$, if t denote the thickness of the shell in that neighbourhood (its density being taken as unity). It follows also from § 519 that the potential is constant throughout the interior of the homoeoid and over its surface. Hence the distance from this surface to another equipotential infinitely near it outside is inversely proportional to t: and therefore (§ 494) this second surface is ellipsoidal and confocal with the first. By supposing the proper distribution of matter (§ 505) placed on this second surface to produce over it, and through its interior, its uniform potential, we see in the same way that the third equipotential infinitely near it outside is ellipsoidal and confocal with it; and similarly again that a fourth equipotential is an ellipsoidal surface confocal with the third, and so on. Thus we conclude that the equipotentials external to the original homoeoid are the whole series of external confocal ellipsoidal surfaces. Force external to an elliptic homoeoid found.

522. From this theorem it follows immediately that any two confocal homoeoids of equal masses produce the same attraction on all points external to both. And from this (as pointed out by Chasles, *Journal de l'École Polytechnique*, 25th Cahier, Paris, 1837) follows immediately Maclaurin's theorem thus:—Consider two thick homoeoids having the outer surfaces confocal, and also their inner surfaces confocal. Divide one of them into an infinite number of similar homoeoids; and divide the other in a corresponding manner, so that each of its homoeoidal parts shall be confocal with the corresponding Digression. Second proof of Maclaurin's theorem.

Digression. Second proof of Maclaurin's theorem.

one of the first. These two thick homoeoids produce the same force on any point external to both. Now let the hollow of one of them, and therefore also the hollow of the other, become infinitely small; we have two solid confocal ellipsoids, and it is proved that they exert the same force on all points external to both.

523. A beautiful geometric proof of the theorem of § 521 due to Chasles, is given below, § 532. The proof given in § 521 is from Thomson's "Electrostatics and Magnetism" (§ 812, reprinted from *Camb. Math. Jour.*, Feb. 1842). The theorem itself is due to Poisson, who proved (in the *Connaissance des Temps* for 1837, published in 1834*) that the resultant force of a homoeoid on an external point is in the direction of the interior axis of the tangential elliptic cone through the attracted point circumscribed about the homoeoid; for it is a known geometrical proposition, easily proved, that the three axes of the tangential cone are normal to the three confocal surfaces, ellipsoid, hyperboloid of one sheet, and hyperboloid of two sheets, through its vertex.

524. The magnitude of the resultant force is equal to $4\pi\tau$, where τ denotes the thickness of the confocal homoeoid equal in bulk to the given homoeoid.

Magnitude and direction of attraction of elliptic homoeoid on external point, expressed analytically

To express the magnitude and direction symbolically, let abc be the semi-axes of the given homoeoid, and $\alpha\beta\gamma$ those of the confocal one through P the attracted point; and let p, t and ϖ, τ be the perpendiculars from the centre to the tangent planes, and the thicknesses, at any point of the given homoeoid, and at the point P of the other. The volumes of the two homoeoids are respectively

$$4\pi abc\, t/p, \text{ and } 4\pi \alpha\beta\gamma\tau/\varpi;$$

hence

$$4\pi\tau = 4\pi \frac{abc}{\alpha\beta\gamma}\frac{t}{p}\varpi \quad\ldots\ldots\ldots\ldots(1),$$

and therefore the resultant force is

$$4\pi \frac{abc}{\alpha\beta\gamma}\frac{t}{p}\varpi \quad\ldots\ldots\ldots\ldots(2).$$

* See Todhunter's *History of the Mathematical Theories of Attraction and the Figure of the Earth*, Vol. II. Articles 1391—1415.

Magnitude and direction of attraction of elliptic homoeoid on external point, expressed analytically.

Supposing the rectangular co-ordinates of the attracted point xyz given; to find $\alpha\beta\gamma$ we have

$$\alpha^2 = a^2 + \lambda;\quad \beta^2 = b^2 + \lambda;\quad \gamma^2 = c^2 + \lambda \text{..........}(3),$$

where λ is the positive root of the equation

$$\frac{x^2}{a^2+\lambda} + \frac{y^2}{b^2+\lambda} + \frac{z^2}{c^2+\lambda} = 1 \text{..............}(4),$$

these equations expressing the condition that the two ellipsoidal surfaces are confocal.

To complete the analytical expression remark that

$$\frac{\varpi x}{\alpha^2},\ \frac{\varpi y}{\beta^2},\ \frac{\varpi z}{\gamma^2} \text{.............................}(5)$$

are the direction-cosines of the line of the resultant force.

Potential of an elliptic homoeoid at any point external or internal found.

525. To find the potential at any point remark that the difference of potentials at two of the external equipotential surfaces infinitely little distant from one another is (§ 486) equal to the product of the resultant force at any point into the distance between the two equipotentials in its neighbourhood. Hence, taking the potential as zero at an infinite distance (§ 485), we find by summation (a single integration) the potential at any point external to the given homoeoid. Now let

$$x \pm \tfrac{1}{2}dx,\quad y \pm \tfrac{1}{2}dy,\quad z \pm \tfrac{1}{2}dz$$

be the co-ordinates of the two points infinitely near one another, on two confocal surfaces. The distance between the two surfaces in the neighbourhood of this point is

$$\frac{\varpi x}{a^2+\lambda}dx + \frac{\varpi y}{b^2+\lambda}dy + \frac{\varpi z}{c^2+\lambda}dz \text{.............}(6).$$

Let now the squares of the semi-axes of these surfaces be

$$a^2 + \lambda \pm \tfrac{1}{2}d\lambda;\quad b^2 + \lambda \pm \tfrac{1}{2}d\lambda;\quad c^2 + \lambda \pm \tfrac{1}{2}d\lambda.$$

Now by differentiation of (4) we have

$$2\left(\frac{x\,dx}{a^2+\lambda} + \frac{y\,dy}{b^2+\lambda} + \frac{z\,dz}{c^2+\lambda}\right)$$
$$= \left\{\frac{x^2}{(a^2+\lambda)^2} + \frac{y^2}{(b^2+\lambda)^2} + \frac{z^2}{(c^2+\lambda)^2}\right\} d\lambda = \frac{d\lambda}{\varpi^2} \text{.........}(7).$$

Hence (6) becomes $\dfrac{d\lambda}{2\varpi}$.

Potential of an elliptic homoeoid at any point external or internal found.

Hence, and by § 525 above, and by (2) of § 524 we have

$$dv = -2\pi \frac{abc}{\alpha\beta\gamma}\frac{t}{p}d\lambda \text{.......................(8).}$$

Hence, and by (3) of § 524,

$$v = -2\pi\frac{abct}{p}\int_{\infty}\frac{d\lambda}{(a^2+\lambda)^{\frac{1}{2}}(b^2+\lambda)^{\frac{1}{2}}(c^2+\lambda)^{\frac{1}{2}}} \text{..........(9),}$$

where ∞ denotes that the constant is so assigned as to render the value of the integral zero when $\lambda = \infty$.

Synthesis of concentric homoeoids.

526. Having now found the potential of an elliptic homoeoid, and its resultant force at any point external or internal, we can, by simple integration, find the potential and the resultant force of a homogeneous ellipsoid, or of a heterogeneous ellipsoid with, for its surfaces of equal density, similar concentric ellipsoidal surfaces. To do this we have only to divide the ellipsoid into elliptic homoeoids, and find the potential of each by (9), and the potential of the whole by summation; and again find the rectangular components of the force of each by (2) and (5); and from this by summation* the rectangular components of the required resultant.

Let *abc* be the semi-axes of the whole ellipsoid. Let $\theta a, \theta b, \theta c$, be the semi-axes of the middle surface of one of the interior homoeoids; and

$$(\theta \pm \tfrac{1}{2}d\theta)\,a, \quad (\theta \pm \tfrac{1}{2}d\theta)\,b, \quad (\theta \pm \tfrac{1}{2}d\theta)\,c$$

those of its outer and inner bounding surfaces. From the general definition of a homoeoid, elliptic or not, it follows immediately that $t/p = d\theta/\theta$. Let now ρ, a given function of θ, be the density of the ellipsoid in the homoeoidal stratum corresponding to θ. Hence by (9) remembering that the density there was taken as unity, and putting $\theta a, \theta b, \theta c$ in place of a, b, c, we find for the potential of the homoeoid $\theta \pm \frac{1}{2}d\theta$ the following expression,

$$-2\pi abc\theta^2\rho d\theta\int_{\infty}^{\lambda}\frac{d\zeta}{(\theta^2a^2+\zeta)^{\frac{1}{2}}(\theta^2b^2+\zeta)^{\frac{1}{2}}(\theta^2c^2+\zeta)^{\frac{1}{2}}} \text{........(10),}$$

* Chasles, "Nouvelle solution du problème de l'attraction d'un ellipsoïde hétérogène sur un point extérieur" (Liouville's *Journal*, Dec. 1840). Also W. Thomson, "On the Uniform Motion of Heat in Solid Bodies, and its connection with the Mathematical Theory of Electricity, Electrostatics and Magnetism," § 21—24. (Reprinted from *Cambridge Mathematical Journal*, Feb. 1842.)

where ζ is introduced as the variable of the definite integration, because λ is presently to be made a function of θ. Hence if V denote the potential of the whole ellipsoid, we have

Synthesis of concentric homoeoids.

$$V = -2\pi abc\int_0^1 \theta^2\rho d\theta \int_\infty^\lambda \frac{d\zeta}{(\theta^2a^2+\zeta)^{\frac{1}{2}}(\theta^2b^2+\zeta)^{\frac{1}{2}}(\theta^2c^2+\zeta)^{\frac{1}{2}}} \quad\text{......(11)},$$

where λ is a function of θ given by the equation

$$\frac{x^2}{\theta^2a^2+\lambda}+\frac{y^2}{\theta^2b^2+\lambda}+\frac{z^2}{\theta^2c^2+\lambda}=1 \quad\text{.............(12)}.$$

The expression (11) is simplified by introducing, instead of θ or λ, another variable λ/θ^2. Calling this u, so that

$$\lambda = \theta^2 u \quad\text{..............................(13)},$$

we have by (12)

$$\theta^2 = \frac{x^2}{a^2+u}+\frac{y^2}{b^2+u}+\frac{z^2}{c^2+u} \quad\text{.................(14)}.$$

By differentiation of (12) we have

Potential of heterogeneous ellipsoid.

$$\frac{d\lambda}{d(\theta^2)}\left[\frac{x^2}{(a^2+u)^2}+\frac{y^2}{(b^2+u)^2}+\frac{z^2}{(c^2+u)^2}\right] = -\left[\frac{a^2x^2}{(a^2+u)^2}+\frac{b^2y^2}{(b^2+u)^2}+\frac{c^2z^2}{(c^2+u)^2}\right].$$

And from (13) $du = \frac{1}{\theta^2}\left[\frac{d\lambda}{d(\theta^2)} - u\right] d(\theta^2)$.

Whence, on using (14), we find

$$-2\theta d\theta = \left[\frac{x^2}{(a^2+u)^2}+\frac{y^2}{(b^2+u)^2}+\frac{z^2}{(c^2+u)^2}\right] du.$$

Then changing the variable of integration in the function under the second integral sign in (11) from ζ to $\zeta\theta^2$, and writing u for ζ/θ^2, we find by means of these transformations,

$$V = \pi abc\int_\infty^q \rho du\left\{\frac{x^2}{(a^2+u)^2}+\frac{y^2}{(b^2+u)^2}+\frac{z^2}{(c^2+u)^2}\right\}\int_\infty \frac{du}{(a^2+u)^{\frac{1}{2}}(b^2+u)^{\frac{1}{2}}(c^2+u)^{\frac{1}{2}}}$$
$$\text{......................(15)},$$

where q is the positive root of the equation

$$\frac{x^2}{a^2+q}+\frac{y^2}{b^2+q}+\frac{z^2}{c^2+q}=1 \quad\text{..............(16)}.$$

For the case of uniform density in which we may put $\rho=1$, this becomes simplified by integration by parts, thus:

$$\int_\infty^q du\,\frac{1}{(C+u)^2}\int_\infty f(u)\,du = -\frac{1}{C+q}\int_\infty^q f(u)\,du + \int_\infty^q \frac{du}{C+q}f(u)$$
$$= \frac{1}{C+q}\int_q^\infty f(u)\,du - \int_q^\infty \frac{1}{C+q}f(u).$$

Putting for C successively a^2, b^2, c^2, using the result properly in (15), and taking account of (16), and putting

$$\tfrac{4}{3}\pi\, abc = M \quad \text{.........................(17)},$$

we find

Potential of homogeneous ellipsoid.

$$V = \frac{3M}{4}\int_q^\infty \left(1 - \frac{x^2}{a^2+u} - \frac{y^2}{b^2+u} - \frac{z^2}{c^2+u}\right)\frac{du}{(a^2+u)^{\frac{1}{2}}(b^2+u)^{\frac{1}{2}}(c^2+u)^{\frac{1}{2}}} \quad \text{.........(18)},$$

which agrees with § 494 above.

Attraction of heterogeneous ellipsoid.

Just as we have found (15), we find from (2), (5), (13), and (14), the following expression for the x-components of the resultant force and the symmetricals for the y- and z-components:

$$X = \frac{3Mx}{2}\int_q^\infty \frac{\rho\, du}{(a^2+u)^{\frac{3}{2}}(b^2+u)^{\frac{1}{2}}(c^2+u)^{\frac{1}{2}}} \quad \text{...............(19)},$$

where ρ, a function of θ, is reduced to a function of u by (14).

For the case of a homogeneous ellipsoid ($\rho = 1$), these results become (20) and (21) of § 494. As there they were for external points deduced by aid of Maclaurin's theorem from the attraction of an ellipsoid on a point at its surface, so now when proved otherwise they contain a proof of Maclaurin's theorem. This we see in a moment by putting $u = w + q$ in the integrals, which makes the limits $w = 0$ and $w = \infty$.

527. In the case of a homogeneous ellipsoid of revolution the integrals expressing the potential and the force-components (which for a homogeneous ellipsoid, in general, are elliptic integrals) are reduced to algebraic and trigonometrical forms, thus: let $b = c$ and $z = 0$.

We have

$$V = \frac{3M}{4}\int_q^\infty \frac{du}{(b^2+u)(a^2+u)^{\frac{1}{2}}} - \tfrac{1}{2}(Xx + Yy) \quad \text{.........(20)},$$

$$\left.\begin{aligned} X &= \frac{3M}{2}x\int_q^\infty \frac{du}{(b^2+u)(a^2+u)^{\frac{3}{2}}} \\ Y &= \frac{3M}{2}y\int_q^\infty \frac{du}{(b^2+u)^2(a^2+u)^{\frac{1}{2}}} \end{aligned}\right\} \quad \text{.........................(21)}.$$

To reduce these put

$$b^2 + u = \frac{b^2 - a^2}{\xi^2} \quad \text{.........................(22)}:$$

which reduces the three integrals to $2/(b^2-a^2)^{\frac{1}{2}}.\int d\xi/(1-\xi^2)^{\frac{1}{2}}$, $2/(b^2-a^2)^{\frac{3}{2}}.\int \xi^2 d\xi/(1-\xi^2)^{\frac{3}{2}}$, and $2/(b^2-a^2)^{\frac{3}{2}}.\int \xi^2 d\xi/(1-\xi^2)^{\frac{1}{2}}$; and makes the limits in each of them

$$\xi=0 \text{ to } \xi=\sqrt{\frac{b^2-a^2}{b^2+q}}.$$

We thus find

Potential and attraction of homogeneous ellipsoid of revolution:

$$V=\frac{3M}{2(b^2-a^2)^{\frac{1}{2}}}\tan^{-1}\sqrt{\frac{b^2-a^2}{a^2+q}}-\tfrac{1}{2}(Xx+Yy)\quad\ldots\ldots\ldots\ldots(23).$$

oblate:

$$\left.\begin{aligned}X&=\frac{3Mx}{(b^2-a^2)^{\frac{3}{2}}}\left\{\sqrt{\frac{b^2-a^2}{a^2+q}}-\tan^{-1}\sqrt{\frac{b^2-a^2}{a^2+q}}\right\}\\ Y&=\frac{3My}{2(b^2-a^2)^{\frac{3}{2}}}\left\{\tan^{-1}\sqrt{\frac{b^2-a^2}{a^2+q}}-\frac{(b^2-a^2)^{\frac{1}{2}}(a^2+q)^{\frac{1}{2}}}{b^2+q}\right\}\end{aligned}\right\}\ldots\ldots(24),$$

where, for any external point, q is the positive root of the equation

$$\frac{x^2}{a^2+q}+\frac{y^2}{b^2+q}=1\ldots\ldots\ldots\ldots\ldots\ldots\ldots(25),$$

x and y denoting the co-ordinates of the attracted point respectively along and perpendicular to the axis of revolution, and for any internal point or for points on the surface $q=0$.

Formulas (23) and (24) realized for the case of $a>b$ become

prolate.

$$V=\frac{3M}{2(a^2-b^2)^{\frac{1}{2}}}\log\frac{\sqrt{(a^2-b^2)}+\sqrt{(a^2+q)}}{\sqrt{(b^2+q)}}-\tfrac{1}{2}(Xx+Yy)\ldots\ldots(26),$$

$$\left.\begin{aligned}X&=\frac{3Mx}{(a^2-b^2)^{\frac{3}{2}}}\left\{\log\frac{\sqrt{(a^2-b^2)}+\sqrt{(a^2+q)}}{\sqrt{(b^2+q)}}-\sqrt{\frac{a^2-b^2}{a^2+q}}\right\}\\ Y&=\frac{3My}{2(a^2-b^2)^{\frac{3}{2}}}\left\{\frac{(a^2-b^2)^{\frac{1}{2}}(a^2+q)^{\frac{1}{2}}}{b^2+q}-\log\frac{\sqrt{(a^2-b^2)}+\sqrt{(a^2+q)}}{\sqrt{(b^2+q)}}\right\}\end{aligned}\right\}(27).$$

The structure of these expressions (23), (24), (26), (27), is elucidated, and calculation of results from them is facilitated by taking

$$f=\sqrt{\frac{b^2-a^2}{a^2+q}},\text{ and }\sqrt{(b^2-a^2)}=r\ldots\ldots\ldots\ldots(28),$$

and again $$e=\sqrt{\frac{a^2-b^2}{a^2+q}},\text{ and }\sqrt{(a^2-b^2)}=s\ldots\ldots\ldots\ldots(29);$$

prolate. which reduces them to the following alternative forms:—

$$V = \frac{3M}{2r}\tan^{-1}f - \tfrac{1}{2}(Xx + Yy) = \frac{3M}{2s}\log\sqrt{\frac{1+e}{1-e}} - \tfrac{1}{2}(Xx + Yy)\;\ldots(30),$$

$$\left.\begin{aligned} X &= \frac{3Mx}{r^3}(f - \tan^{-1}f) = \frac{3Mx}{s^3}\left(e - \log\sqrt{\frac{1+e}{1-e}}\right) \\ Y &= \frac{3My}{2r^3}\left(\tan^{-1}f - \frac{f}{1+f^2}\right) = -\frac{3My}{2s^3}\left(\frac{e}{1-e^2} - \log\sqrt{\frac{1+e}{1-e}}\right) \end{aligned}\right\}\ldots\ldots(31).$$

Then, for determining f or e, in the case of an external point, (25) becomes

$$f^2\left(x^2 + \frac{y^2}{1+f^2}\right) = r^2, \text{ and } e^2\left(x^2 + \frac{y^2}{1-e^2}\right) = s^2\ldots\ldots\ldots(32).$$

In the case of an internal point we have

$$f = \sqrt{\frac{b^2 - a^2}{a^2}},\quad e = \sqrt{\frac{a^2 - b^2}{a^2}}\;\ldots\ldots\ldots\ldots\ldots\ldots(33).$$

528. The investigation of the attraction of an ellipsoid which was most popular in England 40 to 50 years ago resembled that of § 494 above, in finding the attraction of an internal point by direct integration, substantially the same as that of § 494, and deducing from the result the attraction of an external point by a special theorem.

Third investigation of the attraction of an ellipsoid.

But the theorem then popularly used for the purpose was not Maclaurin's theorem, which was little known, strange to say, in England at that time; it was Ivory's theorem, much less beautiful and simple and directly suitable for the purpose than Maclaurin's, but still a very remarkable theorem, curiously different from Maclaurin's, and in one respect more important and comprehensive, because, as was shown by Poisson, it is not confined to the Newtonian Law of Attraction, but holds for force varying as any function of the distance. Before enunciating Ivory's theorem, take his following definition:—

Corresponding points on confocal ellipsoids defined.

529. Corresponding points on two confocal ellipsoids are any two points which coincide when either ellipsoid is deformed by a pure strain so as to coincide with the other.

Digression; orthogonal trajectory of confocal

In connection with this definition, it is interesting to remark that each point on the surface of the changing ellipsoid describes an orthogonal trajectory of the intermediate series of confocal ellipsoids if the distortion specified in the definition is produced continuously, in such a manner that the surface of the ellipsoid is always confocal with its original figure.

Ellipsoids traced by any point of a confocally distorted solid ellipsoid:

To prove this proposition, which however is not necessary for our present purpose, let abc be the semi-axes of the ellipsoid in one configuration, and $\sqrt{(a^2+h)}$, $\sqrt{(b^2+h)}$, $\sqrt{(c^2+h)}$ in another. If xyz be the co-ordinates of any point P on the surface in the first configuration, its co-ordinates in the second configuration will be

proof.

$$x\frac{\sqrt{(a^2+h)}}{a},\quad y\frac{\sqrt{(b^2+h)}}{b},\quad z\frac{\sqrt{(c^2+h)}}{c}\;\ldots\ldots\ldots\ldots(32).$$

When h is infinitely small the differences of the co-ordinates of these points are

$$\tfrac{1}{2}h\frac{x}{a^2},\quad \tfrac{1}{2}h\frac{y}{b^2},\quad \tfrac{1}{2}h\frac{z}{c^2}.$$

Hence the direction-cosines of the line joining them are proportional to x/a^2, y/b^2, z/c^2, and therefore it coincides with the normal to the two infinitely nearly coincident surfaces.

530. The property of corresponding points (essential for Ivory's theorem, and for Chasles', § 532 below) is this:—

Ivory's Lemma on corresponding points.

If P, P' be any two points on one ellipsoid, and Q, Q' the corresponding points on any confocal ellipsoid, PQ' is equal to $P'Q$.

To prove this, let xyz be the co-ordinates of P, and $x'y'z'$ those of P'. Taking (32) as the co-ordinates of Q, we find

$$P'Q^2=\left(x'-x\sqrt{\frac{a^2+h}{a^2}}\right)^2+\left(y'-y\sqrt{\frac{b^2+h}{b^2}}\right)^2+\left(z'-z\sqrt{\frac{c^2+h}{c^2}}\right)$$

$$=x'^2-2xx'\sqrt{\frac{a^2+h}{a^2}}+x^2\left(1+\frac{h}{a^2}\right)+\&c.$$

Now because (x, y, z) is on the ellipsoidal surface (a, b, c), we have

$$\frac{x^2}{a^2}+\frac{y^2}{b^2}+\frac{z^2}{c^2}=1.$$

Hence the preceding becomes

$$P'Q^2=x'^2+y'^2+z'^2-2\left(xx'\sqrt{\frac{a^2+h}{a^2}}+yy'\sqrt{\frac{b^2+h}{b^2}}+zz'\sqrt{\frac{c^2+h}{c^2}}\right)+x^2+y^2+z^2+h.$$

This is symmetrical in respect to xyz and $x'y'z'$, and so the proposition is proved.

Ivory's theorem, **531.** The following is Ivory's Theorem :—Let P' and P be corresponding points on the surfaces of two homogeneous confocal ellipsoids (a, b, c) (a', b', c'); the x-component of the attraction of the ellipsoid abc on the point P is to the x-component of the attraction of the ellipsoid $a'b'c'$ on the point P' as bc is to $b'c'$.

proved. Let x, y, z be the co-ordinates of P, the attracted point;

" ξ, η, ζ " co-ordinates of any point of the mass;

" D " distance between the two points;

" $F(D)\, d\xi d\eta d\zeta$ be the attraction of the elemental mass $d\xi d\eta d\zeta$ at (ξ, η, ζ), on (x, y, z);

Let X be the x-component of the attraction of the whole ellipsoid (a, b, c) on (x, y, z).

We have

$$X = \iiint d\xi d\eta d\zeta F(D) \frac{x-\xi}{D} = \iiint d\xi d\eta d\zeta F(D) \times \left(-\frac{dD}{d\xi}\right)$$

$$= \iint d\eta d\zeta \int - F(D)\, dD.$$

Now $F(D)$ being any function of D, let

$$\int F(D)\, dD = -\psi(D);$$

and let E, G be the positive and negative ends of the bar $d\eta d\zeta$ of the ellipsoid, that is to say, the points on the positive and negative sides of the plane yoz in which the surface of the ellipsoid is cut by the line parallel to ox, having $\eta\zeta$ for its other co-ordinates. The proper limits being assigned to the D-integration in the formula for X above being assigned, we find

$$X = \iint d\eta d\zeta \{\psi(EP) - \psi(GP)\}.$$

Now let $E'G'$ be points on a confocal ellipsoidal surface (a', b', c') through P, corresponding to E and G on the surface of the given ellipsoid (a, b, c); and let P' be the point on the first ellipsoidal surface corresponding to P on the second. The y- z- co-ordinates common to $E'G'$ are respectively $b'/b\,.\,\eta$ and $c'/c\,.\,\zeta$;

and by lemma $EP = E'P'$ and $GP = G'P'$. Hence if we change from $\eta\zeta$, as variables for the double integration in the preceding formula for X, to $\eta'\zeta'$, we find

$$X = \frac{bc}{b'c'} \iint d\eta' d\zeta' \{\psi(E'P') - \psi(G'P')\},$$

which is Ivory's theorem.

532. Two confocal homoeoids of equal masses being given, the potential of the first at any point, P, of the surface of the second, is equal to that of the second at the corresponding point, P', on the surface of the first. Chasles' comparison between the potentials of two confocal homoeoids.

Let E be any element of the first and E' the corresponding element of the second. The mass of each element bears to the mass of the whole homoeoid the same ratio as the mass of the corresponding element of a uniform spherical shell, from which either homoeoid may be derived, bears to the whole mass of the spherical shell. Hence the mass of E is equal to the mass of E'; and by Ivory's lemma (§ 530) $PE' = P'E$. Hence the proposition is true for the parts of the potential due to the corresponding elements, and therefore it is due for the entire shells.

This beautiful proposition is due to Chasles. It holds, whatever be the law of force. From it, for the case of the inverse square of the distance, and from Newton's Theorem for this case that the force is zero within an elliptic homoeoid, or, which is the same, that the potential is constant through the interior, it follows that the external equipotential surfaces of an elliptic homoeoid are confocal ellipsoids, and therefore that the attraction on an external point is normal to a confocal ellipsoid passing through the point; which is the same conclusion as that of § 521 above. Proof of Poisson's theorem regarding attraction of elliptic homoeoid.

533. An ingenious application of Ivory's theorem, by Duhamel, must not be omitted here. Concentric spheres are a particular case of confocal ellipsoids, and therefore the attraction of any sphere on a point on the surface of an internal concentric sphere, is to that of the latter upon a point in the surface of the former as the squares of the radii of the spheres. Now *if the law of attraction be such that a homogeneous spherical* Law of attraction when a uniform spherical shell exerts no action on an internal point.

Law of attraction when a uniform spherical shell exerts no action on an internal point. *shell of uniform thickness exerts no attraction on an internal point,* the action of the larger sphere on the internal point is reduced to that of the smaller. Hence the smaller sphere attracts points on its surface and points external to it, with forces inversely as the squares of their distances from its centre. Hence *the law of force is the inverse square of the distance,* as is easily seen by making the smaller sphere less and less till it becomes a mere particle. This theorem is due originally to Cavendish.

Cavendish's theorem.

Centre of gravity.

534. (*Definition.*) If the action of terrestrial or other gravity on a rigid body is reducible to a single force in a line passing always through one point fixed relatively to the body, whatever be its position relatively to the earth or other attracting mass, that point is called its *centre of gravity,* and the body is called a *centrobaric body.*

Centrobaric bodies, proved possible by Green.

One of the most startling results of Green's wonderful theory of the potential is its establishment of the existence of centrobaric bodies; and the discovery of their properties is not the least curious and interesting among its very various applications.

Properties of centrobaric bodies.

534 *a.* If a body (B) is centrobaric relatively to any one attracting mass (A), it is centrobaric relatively to every other: and it attracts all matter external to itself as if its own mass were collected in its centre of gravity *.

Let O be any point so distant from B that a spherical surface described from it as centre, and not containing any part of B, is large enough entirely to contain A. Let A be placed within any such spherical surface and made to rotate about any axis, OK, through O. It will always attract B in a line through G, the centre of gravity of B. Hence if every particle of its mass be uniformly distributed over the circumference of the circle that it describes in this rotation, the mass, thus obtained, will also attract B in a line through G. And this will be the case however this mass is rotated round O; since before obtaining it we might have rotated A and OK in any way round O, hold-

* Thomson, *Proc. R. S. E.*, Feb. 1864.

ing them fixed relatively to one another. We have therefore found a body, A', symmetrical about an axis, OK, relatively to which B is necessarily centrobaric. Now, O being kept fixed, let OK, carrying A' with it, be put successively into an infinite number, n, of positions uniformly distributed round O; that is to say, so that there are equal numbers of positions of OK in all equal solid angles round O: and let $\frac{1}{n}$ part of the mass of A' be left in each of the positions into which it was thus necessarily carried. B will experience from all this distribution of matter, still a resultant force through G. But this distribution, being symmetrical all round O, consists of uniform concentric shells, and (§ 471) the mass of each of these shells might be collected at O without changing its attraction on any particle of B, and therefore without changing its resultant attraction on B. Hence B is centrobaric relatively to a mass collected at O; this being any point whatever not nearer than within a certain limiting distance from B (according to the condition stated above). That is to say, any point placed beyond this distance is attracted by B in a line through G; and hence, beyond this distance, the equipotential surfaces of B are spherical with G for common centre. B therefore attracts points beyond this distance as if its mass were collected at G: and it follows (§ 497) that it does so also through the whole space external to itself. Hence it attracts any group of points, or any mass whatever, external to it, as if its own mass were collected at G.

Properties of centrobaric bodies.

534 b. Hence §§ 497, 492 show that—

(1) *The centre of gravity of a centrobaric body necessarily lies in its interior;* or in other words, *can only be reached from external space by a path cutting through some of its mass.* And

(2) *No centrobaric body can consist of parts isolated from one another, each in space external to all:* in other words, *the outer boundary of every centrobaric body is a single closed surface.*

Thus we see, by (1), that no symmetrical ring, or hollow cylinder with open ends, can have a centre of gravity; for its

Properties of centrobaric bodies.

centre of gravity, if it had one, would be in its axis, and therefore external to its mass.

534 c. *If any mass whatever, M, and any single surface, S, completely enclosing it be given, a distribution of any given amount, M', of matter on this surface may be found which shall make the whole centrobaric with its centre of gravity in any given position (G) within that surface.*

The condition here to be fulfilled is to distribute M' over S, so as by it to produce the potential

$$\frac{M+M'}{EG} - V,$$

any point, E, of S; V denoting the potential of M at this point. The possibility and singleness of the solution of this problem were proved above (§ 499). It is to be remarked, however, that if M' be not given in sufficient amount, an extra quantity must be taken, but neutralized by an equal quantity of negative matter, to constitute the required distribution on S.

The case in which there is no given body M to begin with is important; and yields the following :—

Centrobaric shell.

534 d. *A given quantity of matter may be distributed in one way, but in only one way, over any given closed surface, so as to constitute a centrobaric body with its centre of gravity at any given point within it.*

Thus we have already seen that the condition is fulfilled by making the density inversely as the cube of the distance from the given point, if the surface be spherical. From what was proved in §§ 501, 506 above, it appears also that a centrobaric shell may be made of either half of the lemniscate in the diagram of § 508, or of any of the ovals within it, by distributing matter with density proportional to the resultant force of m at I and m' at I'; and that the one of these points which is within it is its centre of gravity. And generally, by drawing the equipotential surfaces relatively to a mass m collected at a point I, and any other distribution of matter whatever not surrounding this point; and by taking one of these surfaces which encloses I but no other part of the mass, we learn, by

Green's general theorem, and the special proposition of § 506, how to distribute matter over it so as to make it a centrobaric shell with I for centre of gravity. Centrobaric shell.

534 e. Under *hydrokinetics* the same problem will be solved for a cube, or a rectangular parallelepiped in general, in terms of converging series; and under *electricity* (in a subsequent volume) it will be solved in finite algebraic terms for the surface of a lens bounded by two spherical surfaces cutting one another at any sub-multiple of two right angles, and for either part obtained by dividing this surface in two by a third spherical surface cutting each of its sides at right angles.

534 f. *Matter may be distributed in an infinite number of ways throughout a given closed space, to constitute a centrobaric body with its centre of gravity at any given point within it.* Centrobaric solid.

For by an infinite number of surfaces, each enclosing the given point, the whole space between this point and the given closed surface may be divided into infinitely thin shells; and matter may be distributed on each of these so as to make it centrobaric with its centre of gravity at the given point. Both the forms of these shells and the quantities of matter distributed on them, may be arbitrarily varied in an infinite variety of ways.

Thus, for example, if the given closed surface be the pointed oval constituted by either half of the lemniscate of the diagram of § 508, and if the given point be the point I within it, a centrobaric solid may be built up of the interior ovals with matter distributed over them to make them centrobaric shells as above (§ 534*d*). From what was proved in § 518, we see that a solid sphere, with its density varying inversely as the fifth power of the distance from an external point, is centrobaric, and that its centre of gravity is the *image* (§ 512) of this point relatively to its surface. Properties of centrobaric bodies.

534 g. The centre of gravity of a centrobaric body composed of true gravitating matter is its centre of inertia. For a centrobaric body, if attracted only by another infinitely distant body, or by matter so distributed round itself as to produce (§ 499) The centre of gravity (if it exist) is the centre of inertia.

The centre of gravity (if it exist) is the centre of inertia.

uniform force in parallel lines throughout the space occupied by it, experiences (§ 534a) a resultant force always through its centre of gravity. But in this case this force is the resultant of parallel forces on all the particles of the body, which (see *Properties of Matter*, below) are rigorously proportional to their masses: and in § 561 it is proved that the resultant of such a system of parallel forces passes through the point defined in § 230, as the centre of inertia.

A centrobaric body is kinetically symmetrical about its centre of gravity.

535. The moments of inertia of a centrobaric body are equal round all axes through its centre of inertia. In other words (§ 285), all these axes are principal axes, and the body is kinetically symmetrical round its centre of inertia.

Let it be placed with its centre of inertia at a point O (origin of co-ordinates), within a closed surface having matter so distributed over it (§ 499) as to have xyz [which satisfies $\nabla^2(xyz)=0$] for potential at any point (x, y, z) within it. The resultant action on the body is (§ 534a) the same as if it were collected at O; that is to say, zero: or, in other words, the forces on its different parts must balance. Hence (§ 551, I., below) if ρ be the density of the body at (x, y, z)

$$\iiint yz\rho dxdydz = 0, \qquad \iiint zx\rho dxdydz = 0, \qquad \iiint xy\rho dxdydz = 0.$$

Hence OX, OY, OZ are principal axes; and this, however the body is turned, only provided its centre of gravity is kept at O.

To prove this otherwise, let V denote the potential of the given body at (x, y, z); u any function of x, y, z; and ϖ the triple integral

$$\iiint\left(\frac{du}{dx}\frac{dV}{dx}+\frac{du}{dy}\frac{dV}{dy}+\frac{du}{dz}\frac{dV}{dz}\right)dxdydz,$$

extended through the interior of a spherical surface, S, enclosing all of the given body, and having for centre its centre of gravity. Then, as in Chap. I. App. A, we have

$$\varpi = \iint \partial u V d\sigma - \iiint V\nabla^2 u dxdydz$$

$$= \iint \partial V u d\sigma - \iiint u\nabla^2 V dxdydz.$$

But if m be the whole mass of the given body, and a the radius of S, we have, over the whole surface of S,

Properties of centrobaric bodies.

$$V = \frac{m}{a}, \text{ and } \partial V = -\frac{m}{a^2}.$$

Also [§ 491 c] $\nabla^2 V = -4\pi\rho$,

vanishing of course for all points not belonging to the mass of the given body. Hence from the preceding we have

$$4\pi \iiint u\rho dxdydz = \frac{m}{a^2} \iint (a\partial u + u)\, d\sigma - \iiint V\nabla^2 u dxdydz.$$

Let now u be any function fulfilling $\nabla^2 u = 0$ through the whole space within S; so that, by § 492, we have $\iint \partial u d\sigma = 0$, and by § 496, $\iint u\partial\sigma = 4\pi a^2 u_0$, if u_0 denote the value of u at the centre of S. Hence

$$\iiint u\rho dxdydz = mu_0.$$

Let, for instance, $u = yz$. We have $u_0 = 0$, and therefore

$$\iiint yz\rho dxdydz = 0,$$

as we found above. Or let $u = (x^2 + y^2) - (x^2 + z^2)$, which gives $u_0 = 0$; and consequently proves that

$$\iiint (x^2 + z^2)\, \rho dxdydz = \iiint (x^2 + y^2)\, \rho dxdydz,$$

or the moment of inertia round OY is equal to that round OX, verifying the conclusion inferred from the other result.

536. The *spherical harmonic analysis*, which forms the subject of an Appendix to Chapter I., had its origin in the theory of attraction, treated with a view especially to the figure of the earth; having been first invented by Legendre and Laplace for the sake of expressing in converging series the attraction of a body of nearly spherical figure. It is also perfectly appropriate for expressing the potential, or the attraction, of an infinitely thin spherical shell, with matter distributed over it according to any arbitrary law. This we shall take first, being the simpler application.

Origin of spherical harmonic analysis of Legendre and Laplace.

Origin of spherical harmonic analysis of Legendre and Laplace.

Let x, y, z be the co-ordinates of P, the point in question, reckoned from O the centre, as origin of co-ordinates: ρ and ρ' the values of the density of the spherical surface at points E and E', of which the former is the point in which it is cut by OP, or this line produced: $d\sigma'$ an element of the surface at E', a its radius. Then, V being the potential at P, we have

$$V = \iint \frac{\rho' d\sigma'}{E'P} \quad \text{.................................(1).}$$

But, by B (48)

$$\left.\begin{aligned} \frac{1}{E'P} &= \frac{1}{a}\left\{1 + \sum_1^\infty Q_i \left(\frac{r}{a}\right)^i\right\} \text{ when } P \text{ is internal,} \\ \text{and} \quad & \\ &= \frac{1}{r}\left\{1 + \sum_1^\infty Q_i \left(\frac{a}{r}\right)^i\right\} \quad \text{,,} \quad \text{,,} \quad \text{external,} \end{aligned}\right\} \text{..........(2)}$$

where Q_i is the biaxal surface harmonic of (E, E'). Hence, if

$$\rho' = S_0 + S_1 + S_2 + \&c. \text{........................(3)}$$

be the harmonic expansion for ρ, we have, according to B (52),

$$\left.\begin{aligned} V &= 4\pi a \left\{\sum_0^\infty \frac{S_i}{2i+1} \left(\frac{r}{a}\right)^i\right\} \text{ when } P \text{ is internal,} \\ \text{and} \quad & \\ &= \frac{4\pi a^2}{r}\left\{\sum_0^\infty \frac{S_i}{2i+1} \left(\frac{a}{r}\right)^i\right\} \quad \text{,,} \quad \text{,,} \quad \text{external,} \end{aligned}\right\} \text{..........(4)}$$

If, for instance, $\rho = S_i$, we have

$$V = \frac{4\pi r^i}{a^{i-1}} \frac{S_i}{2i+1} \text{ inside,}$$

and

$$V = \frac{4\pi a^{i+2}}{r^{i+1}} \frac{S_i}{2i+1} \text{ outside.}$$

Thus we conclude that

Application of spherical harmonic analysis.

537. A spherical harmonic distribution of density on a spherical surface produces a similar and similarly placed spherical harmonic distribution of potential over every concentric spherical surface through space, external and internal; and so also consequently of radial component force. But the amount of the latter differs, of course (§ 478), by $4\pi\rho$, for points infinitely near one another outside and inside the surface, if ρ

denote the density of the distribution on the surface between them. Application of spherical harmonic analysis.

If R denote the radial component of the force, we have

$$\left.\begin{aligned} R &= -\frac{dV}{dr} = -\frac{4\pi r^{i-1}}{a^{i-1}}\frac{iS_i}{2i+1} \text{ inside,} \\ \text{and} \qquad &= \frac{4\pi a^{i+2}}{r^{i+2}}\frac{(i+1)S_i}{2i+1} \text{ outside,} \end{aligned}\right\} \quad \ldots\ldots\ldots\ldots(5).$$

Hence, if $r = a$, we have

$$R\,(\text{outside}) - R\,(\text{inside}) = 4\pi S_i = 4\pi\rho.$$

538. The potential is of course a solid harmonic through space, both internal and external; and is of positive degree in the internal, and of negative in the external space. The expression for the radial component of the force, in each division of space, is reduced to the same form by multiplying it by the distance from the centre.

539. The harmonic development gives an expression in converging series, for the potential of any distribution of matter through space, which is useful in some applications.

Let x, y, z be the co-ordinates of P, the attracted point, and x', y', z' those of P' any point of the given mass. Then, if ρ' be the density of the matter at P', and V the potential at P, we have

$$V = \iiint \frac{\rho' dx'dy'dz'}{[(x-x')^2 + (y-y')^2 + (z-z')^2]^{\frac{1}{2}}} \ldots\ldots\ldots\ldots\ldots(6).$$

The most convenient view we can take as to the space through which the integration is to be extended is to regard it as infinite in all directions, and to suppose ρ' to be a discontinuous function of x', y', z', vanishing through all space unoccupied by matter.

Now by App. B. (u) we have

$$\left.\begin{aligned} \frac{1}{[(x-x')^2 + (y-y')^2 + (z-z')^2]^{\frac{1}{2}}} &= \frac{1}{r'}\left\{1 + \sum_1^\infty Q_i\left(\frac{r}{r'}\right)^i\right\} \text{ when } r' > r \\ \text{and} \qquad &= \frac{1}{r}\left\{1 + \sum_1^\infty Q_i\left(\frac{r'}{r}\right)^i\right\} \quad ,, \quad r' < r \end{aligned}\right\} \ldots(7).$$

Application of spherical harmonic analysis.

Substituting this in (6) we have

$$V = (\iiint) \frac{\rho' dx' dy' dz'}{r'} + \frac{1}{r} [\iiint] \rho' dx' dy' dz'$$

$$+ \sum_1^{\infty} \left\{ r^i (\iiint) Q_i \frac{\rho' dx' dy' dz'}{r'^{i+1}} + \frac{1}{r^{i+1}} [\iiint] Q_i r'^i \rho' dx' dy' dz' \right\} \ldots (8),$$

where $(\iiint)$ denotes integration through all the space external to the spherical surface of radius r, and $[\iiint]$ integration through the interior space.

Potential of a distant body.

This formula is useful for expressing the attraction of a mass of any figure on a distant point in a single converging series. Thus when OP is greater than the greatest distance of any part of the body from O, the first series disappears, and the expression becomes a single converging series, in ascending powers of $\frac{1}{r}$:—

$$V = \frac{1}{r} \{ \iiint \rho' dx' dy' dz' + \Sigma \frac{1}{r^i} \iiint Q_i r'^i \rho' dx' dy' dz' \} \ldots \ldots \ldots (9).$$

If we use the notation of B. (u) (53), this becomes

$$V = \frac{1}{r} \left\{ \iiint \rho' dx' dy' dz' + \sum_1^{\infty} r^{-2i} \iiint \rho' H_i [(x, y, z), (x', y', z')] dx' dy' dz' \right\} ..(10),$$

and we have, by App. B. (v') and (w),

$$H_i[(x, y, z), (x', y', z')] = \frac{1.3.5\ldots(2i-1)}{1.2.3\ldots i} [\cos^i\theta - \frac{i(i-1)}{2.(2i-1)} \cos^{i-2}\theta + \frac{i(i-1)(i-2)(i-3)}{2.4.(2i-1)(2i-3)} \cos^{i-4}\theta - \text{etc.}] r^i r'^i \ (11)$$

where $$\cos\theta = \frac{xx' + yy' + zz'}{rr'}.$$

From this we find

$$H_1 = xx' + yy' + zz'; \ H_2 = \tfrac{3}{2} [(xx' + yy' + zz')^2 - \tfrac{1}{3}(x^2 + y^2 + z^2)(x'^2 + y'^2 + z'^2)];$$

and so on.

Let now M denote the mass of the body; and let O be taken at its centre of gravity. We shall have

$$\iiint \rho' dx' dy' dz' = M; \text{ and } \iiint \rho' H_1 dx' dy' dz' = 0.$$

Further, let OX, OY, OZ be taken as principal axes (§§ 281, 282), so that $$\iiint \rho' y' z' dx' dy' dz' = 0, \text{ etc.},$$

and let A, B, C be the moments of inertia round these axes. This will give

$$\iiint H_2 \rho' dx' dy' dz' = \tfrac{1}{2} \{ (3x^2 - r^2) \iiint \rho' x'^2 dx' dy' dz' + \text{etc.} \} = \tfrac{1}{2} \{ (3x^2 - r^2) [\tfrac{1}{2}(A + B + C) - A] + \text{etc.} \}$$
$$= \tfrac{1}{2} \{ A(r^2 - 3x^2) + B(r^2 - 3y^2) C + (r^2 - 3z^2) \} = \tfrac{1}{2} \{ (B + C - 2A) x^2 + (C + A - 2B) y^2 + (A + B - 2C) z^2 \}.$$

Hence neglecting terms of the third and higher orders of small quantities $\left(\text{powers of } \frac{r'}{r}\right)$, we have the following approximate expression for the potential:—

Potential a distant body.

$$V = \frac{M}{r} + \frac{1}{2r^3}\{(B+C-2A)x^2+(C+A-2B)y^2+(A+B-2C)z^2\}\ldots(12).$$

As one example of the usefulness of this result, we may mention the investigation of the disturbance in the moon's motion produced by the non-sphericity of the earth, and of the reaction of the same disturbing force on the earth, causing *lunar nutation and precession*, which will be explained later.

Differentiating, and retaining only terms of the first and second degrees of approximation, we have for the components of the mutual force between the body and a unit particle at (x, y, z),

$$\left.\begin{array}{c} X = \frac{Mx}{r^3} - \frac{(B+C-2A)x}{r^5} + \frac{5}{2}\frac{x}{r^7}[(B+C-2A)x^2+(C+A-2B)y^2+(A+B-2C)z^2] \\ Y = \text{etc.}, \quad Z = \text{etc.} \end{array}\right\} (13);$$

whence

$$Zy - Yz = 3\frac{(C-B)yz}{r^5},\ Xz - Zx = 3\frac{(A-C)zx}{r^5},\ Yx - Xy = 3\frac{(B-A)xy}{r^5}\ldots(14).$$

Comparing these with Chap. IX. below, we conclude that

540. The attraction of a distant particle, P, on a rigid body if transferred (according to Poinsot's method explained below, § 555) to the centre of inertia, I, of the latter, gives a couple approximately equal and opposite to that which constitutes the resultant effect of centrifugal force, if the body rotates with a certain angular velocity about IP. The square of this angular velocity is inversely as the cube of the distance of P, irrespectively of its direction; being numerically equal to three times the reciprocal of the cube of this distance, if the unit of mass is such as to exercise the proper kinetic unit (§ 225) force on another equal mass at unit distance. The general tendency of the gravitation couple is to bring the principal axis of least moment of inertia into line with the attracting point. The expressions for its components round the principal axes will be used in Chap. IX. (§ 825) for the investigation of the phenomena of precession and nutation produced, in virtue of

Attraction of a particle on a distant body.

Attraction of a particle on a distant body.

the earth's non-sphericity, by the attractions of the sun and moon. They are available to estimate the retardation produced by tidal friction against the earth's rotation, according to the principle explained above (§ 276).

541. It appears from what we have seen that the amount of the gravitation couple is inversely as the cube of the distance between the centre of inertia and the external attracting point: and therefore that the shortest distance of the line of the resultant force from the centre of inertia varies inversely as the distance of the attracting point. We thus see *how* to a first approximation every rigid body is centrobaric relatively to a distant attracting point.

Principle of the approximation used in the common theory of the centre of gravity.

542. The real meaning and value of the spherical harmonic method for a solid mass will be best understood by considering the following application :—

Let $$\rho = F(r)\, S_i \text{.......................} (15)$$

where $F(r)$ denotes any function of r, and S_i a surface spherical harmonic function of order i, with coefficients independent of r. Substituting accordingly for ρ' in (8), and attending to B. (52) and (16), we find

$$V = \frac{4\pi S_i}{2i+1}\left\{r^i \int_r^\infty r'^{-i+1} F(r')\,dr' + r^{-i-1}\int_0^r r'^{i+2} F(r')\,dr'\right\} \ldots(16).$$

Potential of solid sphere with harmonic distribution of density.

543. As an example, let it be required to find the potential of a solid sphere of radius a, having matter distributed through it according to solid harmonic function V_i.

That is to say, let

$$\rho = V = r^i S_i \text{, when } r < a,$$

and $$\rho' = 0 \quad\quad ,, \quad r > a.$$

Hence in the preceding formula $F(r) = r^i$ from $r = 0$ to $r = a$, and $F(r) = 0$, when $r > a$; and it becomes

$$\left.\begin{aligned} V &= 4\pi V_i \left\{\frac{a^2}{2(2i+1)} - \frac{r^2}{2(2i+3)}\right\} \text{ when } P \text{ is internal,} \\ \text{and} \quad &= \frac{4\pi}{(2i+1)(2i+3)} \frac{a^{2i+3} V_i}{r^{2i+1}} \quad\quad ,, \quad ,, \text{ external.} \end{aligned}\right\} (17).$$

This result may also be obtained by the aid of the algebraical formula B. (12) thus, on the same principle as the potential of a uniform spherical shell was found in § 491 (*d*).

Potential of solid sphere with harmonic distribution of density.

We have by § 491 (*c*)

$$\left.\begin{aligned}\nabla^2 V &= -4\pi V_i, \text{ when } r < a,\\ \text{and}\quad &= 0 \qquad\;\; ,, \quad r > a.\end{aligned}\right\} \qquad (18).$$

But by taking $m = 2$ in B. (12) we have

$$\nabla^2 (r^2 V_i) = 2(2i+3) V_i,$$

and therefore the solution of the equation

$$\nabla^2 V = -4\pi V_i$$

is

$$V = -4\pi \frac{r^2 V_i}{2(2i+3)} + U \qquad (19),$$

where U is any function whatever satisfying the equation

$$\nabla^2 U = 0$$

through the whole interior of the sphere. By choosing U and the external values of V so as to make the values of V equal to one another for points infinitely near one another outside and inside the bounding surface, to fulfil the same condition for $\frac{dV}{dr}$, and to make V vanish when $r = \infty$, and when $r = 0$, we find

$$U = 4\pi V_i \frac{a^2}{2(2i+1)},$$

and obtain the expression of (17) for V external. For in the first place, V external and U must clearly be $A\frac{V_i}{r^{2i+1}}$, and BV_i, where A and B are constants: and the two conditions give the equations to determine them.

544. From App. B. (52) it follows immediately that any function of x, y, z whatever may be expressed, through the whole of space, in a series of surface harmonic functions, each having its coefficients functions of the distance (r) from the origin. Hence (16), with S_i placed under the sign of integration for r', gives the harmonic development of the potential of any mass whatever; being the result of the triple integrations indicated in (8) of § 539, when the mass is specified by means of a harmonic series expressing the density.

Potential of any mass, in harmonic series.

Application to figure of the earth.

545. The most important application of the harmonic development for solid spheres hitherto made is for investigating, in the Theory of the Figure of the Earth, the attraction of a finite mass consisting of approximately spherical layers of matter equally dense through each, but varying in density from layer to layer. The result of the general analytical method explained above, when worked out in detail for this case, is to exhibit the potential as the sum of two parts, of which the first and chief is the potential due to a solid sphere, A, and the second to a spherical shell, B. The sphere, A, is obtained by reducing the given spheroid to a spherical figure by cutting away all the matter lying outside the proper mean spherical surface, and filling the space vacant inside it where the original spheroid lies within it, without altering the density anywhere. The shell, B, is a spherical surface loaded with equal quantities of positive and negative matter, so as to compensate for the transference of matter by which the given spheroid was changed into A. The analytical expression of all this may be written down immediately from the preceding formulæ (§§ 536, 537); but we reserve it until, under hydrostatics and hydrokinetics, we shall be occupied with the theory of the Figure of the Earth, and of the vibrations of liquid globes.

Case of the potential symmetrical about an axis.

546. The analytical method of spherical harmonics is very valuable for several practical problems of electricity, magnetism, and electro-magnetism, in which distributions of force symmetrical round an axis occur: especially in this; that if the force (or potential) at every point through some finite length along the axes be given, it enables us immediately to deduce converging series for calculating the force for points through some finite space not in the axes. (See § 498.)

O being any conveniently chosen point of reference, in the axis of symmetry, let us have, in series converging for a portion AB of the axis,

$$U = a_0 + \frac{b_0}{r} + a_1 r + \frac{b_1}{r^2} + a_2 r^2 + \frac{b_2}{r^3} + \text{etc.} \ldots\ldots\ldots\ldots (a),$$

where U is the potential at a point, Q, in the axis, specified by

$OQ = r$. Then if V be the potential at any point P, specified by $OP = r$ and $QOP = \theta$, and, as in App. B. (47), $Q_1, Q_2, \ldots$ denote the axial surface harmonics of θ, of the successive integral orders, we must have, for all values of r for which the series converges,

Case of the potential symmetrical about an axis.

$$V = a_0 + \frac{b_0}{r} + \left(a_1 r + \frac{b_1}{r^2}\right) Q_1 + \left(a_2 r^2 + \frac{b_2}{r^3}\right) Q_2 + \text{etc.} \ldots\ldots\ldots (b),$$

provided P can be reached from Q and all points of AB within some finite distance from it however small, without passing through any of the matter to which the force in question is due, or any space for which the series does not converge. For throughout this space (§ 498) $V - V'$ must vanish, if V' be the value of the sum of the series; since $V - V'$ is [App. B. (a)] a potential function, and it vanishes for a finite portion of the axis containing Q.

The series (b) is of course convergent for all values of r which make (a) convergent, since the ultimate ratio $Q_{i+1} \div Q_i$ for infinitely great values of i, is unity, as we see from any of the expressions for these functions in App. B.

In general, that is to say unless O be a singular point, the series for U consists, according to Maclaurin's theorem, of ascending integral powers of r only, provided r does not exceed a certain limit. In certain classes of cases there are singular points, such that if O be taken at one of them, U will be expressed in a series of powers of r with fractional indices, convergent and real for all finite positive values of r not exceeding a certain limit. The expression for the potential in the neighbourhood of O in any such case, in terms of solid spherical harmonics relatively to O as centre, will contain harmonics [App. B. (a)] of fractional degrees.

Examples—(I.) The potential of a circular ring of radius a, and linear density ρ, at a point in the axis, distant by r from the centre:—

Examples. (I.) Potential of circular ring;

$$U = \frac{2\pi a\rho}{(a^2 + r^2)^{\frac{1}{2}}} \ldots\ldots\ldots\ldots\ldots\ldots\ldots (1).$$

Hence $$U = 2\pi\rho\left(1 - \tfrac{1}{2}\frac{r^2}{a^2} + \frac{1.3}{2.4}\frac{r^4}{a^4} - \text{etc.}\right) \text{ when } r < a \ldots. (2),$$

and $$U = \frac{2\pi a\rho}{r}\left(1 - \tfrac{1}{2}\frac{a^2}{r^2} + \frac{1.3}{2.4}\frac{a^4}{r^4} - \text{etc.}\right) \text{ when } r > a \ldots (3),$$

Potential symmetrical about an axis.

Potential symmetrical about an axis.

from which we have

$$V = 2\pi\rho\left(1 - \tfrac{1}{2}\frac{r^2}{a^2}Q_2 + \frac{1.3}{2.4}\frac{r^4}{a^4}Q_4 - \text{etc.}\right) \text{ when } r < a \;..(4),$$

and $$V = 2\pi\rho\left(\frac{a}{r} - \tfrac{1}{2}\frac{a^3}{r^3}Q_2 + \frac{1.3}{2.4}\frac{a^5}{r^5}Q_4 - \text{etc.}\right) \text{ when } r > a \;..(5).$$

(II.) of circular disc.

(II.) Multiplying (1) by da, and integrating with reference to a from $a=0$ as lower limit, and now calling U the potential of a circular disc of uniform surface density ρ, and radius a, at a point in its axis, we find

$$U = 2\pi\rho\,\{(a^2+r^2)^{\frac{1}{2}} - r\},$$

r being positive.

Hence, expanding first in ascending, and secondly in descending powers of r, for the cases of $r<a$ and $r>a$, we find

$$V = 2\pi\rho\left\{-rQ_1 + a + \tfrac{1}{2}\frac{r^2}{a}Q_2 - \frac{1.1}{2.4}\frac{r^4}{a^3}Q_4 + \frac{1.1.3}{2.4.6}\frac{r^6}{a^5}Q_6 - \text{etc.}\right\} \text{ when } r<a,$$

and $$V = 2\pi\rho\left\{\tfrac{1}{2}\frac{a^2}{r} - \frac{1.1}{2.4}\frac{a^4}{r^3}Q_2 + \frac{1.1.3}{2.4.6}\frac{a^6}{r^5}Q_4 - \text{etc.}\right\} \text{ when } r>a.$$

It must be remarked that the first of these expressions is only continuous from $\theta=0$ to $\theta=\tfrac{1}{2}\pi$; and that from $\theta=\tfrac{1}{2}\pi$ to $\theta=\pi$ the first term of it must be made

$$+2\pi\rho rQ_1, \text{ instead of } -2\pi\rho rQ_1.$$

(III.) Again, taking $\dfrac{-d}{dr}$ of the expression for U in (II.), and now calling U the potential of a disc of infinitely small thickness c with positive and negative matter of surface density $\dfrac{\rho}{c}$ on its two sides, we have

$$U = 2\pi\rho\left\{1 - \frac{r}{(a^2+r^2)^{\frac{1}{2}}}\right\},$$

[obtainable also from § 477 (e), by integrating with reference to x, putting r for x, and ρ for ρc]. Hence for this case

Potential in the neighbourhood of a circular galvanometer coil.

$$V = 2\pi\rho\left(1 - \frac{r}{a}Q_1 + \tfrac{1}{2}\frac{r^3}{a^3}Q_3 - \frac{1.3}{2.4}\frac{r^5}{a^5}Q_5 + \text{etc.}\right) \text{ when } r<a,$$

and $$V = 2\pi\rho\left(\tfrac{1}{2}\frac{a^2}{r^2}Q_1 - \frac{1.3}{2.4}\frac{a^4}{r^4}Q_3 + \text{etc.}\right) \text{ when } r>a.$$

The first of these expressions also is discontinuous; and when θ

is $> \frac{1}{2}\pi$ and $< \pi$, its first term must be taken as $-2\pi\rho$ instead of $2\pi\rho$.

547. If two systems, or distributions of matter, M and M', given in spaces each finite, but infinitely far asunder, be allowed to approach one another, a certain amount of work is obtained by mutual gravitation: and their mutual potential energy loses, or as we may say *suffers exhaustion*, to this amount: which amount will (§ 486) be the same by whatever paths the changes of position are effected, provided the relative initial positions and the relative final positions of all the particles are given. Hence if m_1, m_2,... be particles of M; m'_1, m'_2,... particles of M'; v'_1, v'_2,... the potentials due to M' at the points occupied by m_1, m_2,...; v_1, v_2,... those due to M at the points occupied by m'_1, m'_2,...; and E the exhaustion of mutual potential energy between the two systems in any actual configurations; we have Exhaustion of potential energy.

$$E = \Sigma mv' = \Sigma m'v.$$

This may be otherwise written, if ρ denote a discontinuous function, expressing the density at any point, (x, y, z) of the mass M, and vanishing at all points not occupied by matter of this distribution, and if ρ' be taken to specify similarly the other mass M'. Thus we have

$$E = \iiint \rho v' dx dy dz = \iiint \rho' v dx dy dz,$$

the integrals being extended through all space. The equality of the second and third members here is verified by remarking that

$$v = \iiint \frac{{}_{,}\rho d_{,}x d_{,}y d_{,}z}{D},$$

if D denote the distance between (x, y, z) and $({}_{,}x, {}_{,}y, {}_{,}z)$, the latter being any point of space, and ${}_{,}\rho$ the value of ρ at it. A corresponding expression of course gives v': and thus we find one sextuple integral to express identically the second and third members, or the value of E, as follows:—

$$E = \iiiint\!\!\iint \frac{{}_{,}\rho\rho' d_{,}x d_{,}y d_{,}z dx dy dz}{D}.$$

548. It is remarkable that it was on the consideration of an analytical formula which, when properly interpreted with reference to two masses, has precisely the same signification as Green's method.

Green's method.

the preceding expressions for E, that Green founded his whole structure of general theorems regarding attraction.

In App. A. (a) let a be constant, and let U, U' be the potentials at (x, y, z) of two finite masses, M, M', finitely distant from one another: so that if ρ and ρ' denote the densities of M and M' respectively at the point (x, y, z), we have [§ 491 (c)]

$$\nabla^2 U = -4\pi\rho, \quad \nabla^2 U' = -4\pi\rho'.$$

It must be remembered that ρ vanishes at every point not forming part of the mass M: and so for ρ' and M'. In the present merely abstract investigation the two masses may, in part or in whole, jointly occupy the same space: or they may be merely imagined subdivisions of the density of one real mass. Then, supposing S to be infinitely distant in all directions, and observing that $U\partial U'$ and $U'\partial U$ are small quantities of the order of the inverse cube of the distance of any point of S from M and M', whereas the whole area of S over which the surface integrals of App. A. (a) (1) are taken as infinitely great, only of the order of the square of the same distance, we have

$$\iint dS U'\partial U = 0, \text{ and } \iint dS U\partial U' = 0.$$

Hence (a) (1) becomes

$$\iiint\left(\frac{dU}{dx}\frac{dU'}{dx}+\frac{dU}{dy}\frac{dU'}{dy}+\frac{dU}{dz}\frac{dU'}{dz}\right)dxdydz = 4\pi\iiint\rho U'dxdydz = 4\pi\iiint\rho' U dxdydz;$$

showing that the first member divided by 4π is equal to the exhaustion of potential energy accompanying the approach of the two masses from an infinite mutual distance to the relative position which they actually occupy.

Without supposing S infinite, we see that the second member of (a) (1), divided by 4π, is the direct expression for the exhaustion of mutual energy between M' and a distribution consisting of the part of M within S and a distribution over S, of density $\frac{1}{4\pi}\partial U'$; and the third member the corresponding expression for M and derivations from M'.

Exhaustion of potential energy, in allowing

549. If, instead of two distributions, M and M', two particles, m_1, m_2 alone be given; the exhaustion of mutual potential energy in allowing them to come together from infinity, to any distance $D(1, 2)$ asunder, is Condensation of diffused matter.

$$\frac{m_1 m_2}{D(1, 2)}.$$

If now a third particle m_3 be allowed to come into their neighbourhood, there is a further exhaustion of potential energy amounting to

$$\frac{m_1 m_3}{D(1, 3)} + \frac{m_2 m_3}{D(2, 3)}.$$

By considering any number of particles coming thus necessarily into position in a group, we find for the whole exhaustion of potential energy

$$E = \Sigma\Sigma \frac{mm'}{D}$$

where m, m' denote the masses of any two of the particles, D the distance between them, and $\Sigma\Sigma$ the sum of the expressions for all the pairs, each pair taken only once. If v denote the potential at the point occupied by m, of all the other masses, the expression becomes a simple sum, with as many terms as there are masses, which we may write thus— Exhaustion of potential energy.

$$E = \tfrac{1}{2} \Sigma mv;$$

the factor $\frac{1}{2}$ being necessary, because Σmv takes each such term as $\frac{m_1 m_2}{D(1, 2)}$ twice over. If the particles form an ultimately continuous mass, with density ρ at any point (x, y, z), we have only to write the sum as an integral; and thus we have

$$E = \tfrac{1}{2} \iiint \rho v \, dx \, dy \, dz$$

as the exhaustion of potential energy of gravitation accompanying the condensation of a quantity of matter from a state of infinite diffusion (that is to say, a state in which the density is everywhere infinitely small) to its actual condition in any finite body.

An important analytical transformation of this expression is suggested by the preceding interpretation of App. A. (*a*); by

Exhaustion of potential energy.

which we find*

$$E = \frac{1}{8\pi}\iiint\left(\frac{dv^2}{dx^2} + \frac{dv^2}{dy^2} + \frac{dv^2}{dz^2}\right)dxdydz,$$

or

$$E = \frac{1}{8\pi}\iiint R^2 dxdydz,$$

if R denote the resultant force at (x, y, z), the integration being extended through all space.

Detailed interpretations in connexion with the theory of energy, of the remainder of App. A., with a constant, and of its more general propositions and formulæ not involving this restriction, especially of the minimum problems with which it deals, are of importance with reference to the dynamics of incompressible fluids, and to the physical theory of the propagation of electric and magnetic force through space occupied by homogeneous or heterogeneous matter; and we intend to return to it when we shall be specially occupied with these subjects.

Gauss's method.

550. The beautiful and instructive manner in which Gauss independently proved Green's theorems is more immediately and easily interpretable in terms of energy, according to the commonly-accepted idea of forces acting simply between particles at a distance without any assistance or influence of interposed matter. Thus, to prove that a given quantity, Q, of matter is distributable in one and only one way over a given single finite surface S (whether a closed or an open shell), so as to produce equal potential over the whole of this surface, he shows (1) that the integral

$$\iiiint\frac{\rho\rho' d\sigma d\sigma'}{PP'}$$

has a minimum value, subject to the condition

$$\iint\rho d\sigma = Q,$$

where ρ is a function of the position of a point, P, on S, ρ' its value at P', and $d\sigma$ and $d\sigma'$ elements of S at these points: and (2) that this minimum is produced by only one determinate distribution of values of ρ. By what we have just seen (§ 549) the first of these integrals is double the potential energy of a

* Nichol's *Encyclopædia*, 2d Ed. 1860. Magnetism, Dynamical Relations of.

distribution over S of an infinite number of infinitely small mutually repelling particles: and hence this minimum problem is (§ 292) merely an analytical statement of the problem to find how these particles must be distributed to be in stable equilibrium.

Gauss's method.

Similarly, Gauss's second minimum problem, of which the preceding is a particular case, and which is, to find ρ so as to make

Equilibrium of repelling particles enclosed in a rigid smooth surface.

$$\iint (\tfrac{1}{2} v - \Omega) \rho \, d\sigma$$

a minimum, subject to

$$\iint \rho \, d\sigma = Q,$$

where Ω is any given arbitrary function of the position of P, and

$$v = \iint \frac{\rho' d\sigma'}{PP'},$$

is merely an analytical statement of the question:—how must a given quantity of repelling particles confined to a surface S be distributed so as to make the whole potential energy due to their mutual forces, and to the forces exerted on them by a given fixed attracting or repelling body (of which Ω is the potential at P), be a minimum? In other words (§ 292), to find how the movable particles will place themselves, under the influence of the acting forces*.

* Gauss's investigations here referred to will be found in Vol. V. of his collected works, p. 197, in a paper entitled "Allgemeine Lehrsätze auf die im verkehrten Verhältnisse des Quadrats der Entfernung wirkenden Anziehungs- und Abstossungs-Kräfte;" originally published in 1839.

CHAPTER VII.

STATICS OF SOLIDS AND FLUIDS.

Rigid body. **551.** We commence with the case of a *rigid body* or system, that is, an ideal substance continuously occupying a given solid figure, admitting no change of shape, but free to move translationally and rotationally. It is sometimes convenient to regard a rigid body as a group of material particles maintained by mutual forces in definite positions relatively to each other, but free to move relatively to other bodies. The condition of perfect rigidity is approximately fulfilled in natural solid bodies, so long as the applied forces are not sufficiently powerful to break them or to distort them, or to condense or rarefy them to a sensible extent. To find the conditions of equilibrium of a rigid body under the influence of any number of forces, we follow the example of Lagrange in using the principle of work (§ 289) and take advantage of our kinematic preliminary (§ 197).

Equilibrium of free rigid body. **552.** First supposing the body to be perfectly free to take any motion possible to a rigid body:—Give it an infinitesimal translation in any direction, and an infinitesimal rotation round any line.

I. In respect to the translational displacement, the work done by the applied forces is equal to the product of the amount of the displacement (being the same for all the points of application) into the algebraic sum of the components of the forces in its direction. Hence for equilibrium (§ 289) the sum of these components must be zero.

II. In respect to rotational displacement the work done by the forces is (§ 240) equal to the product of the infinitesimal angle of rotation into the sum of the moments (§ 231) of the forces round the axis of rotation. Hence for equilibrium (§ 289) the sum of these moments must be zero. Equilibrium of free rigid body.

Since (§ 197) every possible motion of a rigid body may be compounded of infinitesimal translations in any directions, and rotations round any lines, it follows that the conditions necessary and sufficient for equilibrium are that the sum of the components of the forces in any direction whatever must be zero, and the sum of the moments of the forces round any axis whatever must be zero.

Let X_1, Y_1, Z_1 be the components of one of the forces, and x_1, y_1, z_1 the co-ordinates of its point of application relatively to three rectangular axes. Taking successively these axes for directions of the infinitesimal translations, and axes of the infinitesimal rotations, we find, as *necessary* for equilibrium, the following equations:—

$$\Sigma(X_1)=0,\quad \Sigma(Y_1)=0,\quad \Sigma(Z_1)=0 \ldots\ldots\ldots\ldots(1),$$

$$\Sigma(Z_1y_1-Y_1z_1)=0,\quad \Sigma(X_1z_1-Z_1x_1)=0,\quad \Sigma(Y_1x_1-X_1y_1)=0 \ldots(2).$$

Of the latter three equations the first members are respectively the sums of the moments round the three axes of co-ordinates, of the given forces or of the components X_1, Y_1, Z_1, &c., which we take for them.

553. It is interesting and important to remark that the evanescence of the sum of components in any direction whatever is secured if it is ascertained that the sums of the components in the directions of any three lines not in one plane are each nil; and that the evanescence of the sum of moments round any axis whatever is secured if it is ascertained that the sums of the moments round any three axes not in one plane are each nil. Important propositions

Let (l, m, n), (l', m', n'), (l'', m'', n'') be the direction cosines of three lines not in one plane, a condition equivalent to non-evanescence of the determinant $l\,m'\,n''-$&c. Let F, F', F'' be the sums of components of forces along these lines. We have proved.

$$\left.\begin{aligned} F &= l\,\Sigma(X_1)+m\,\Sigma(Y_1)+n\,\Sigma(Z_1)\\ F' &= l'\,\Sigma(X_1)+m'\,\Sigma(Y_1)+n'\,\Sigma(Z_1)\\ F'' &= l''\,\Sigma(X_1)+m''\,\Sigma(Y_1)+n''\,\Sigma(Z_1)\end{aligned}\right\}\ldots\ldots\ldots\ldots(3).$$

Equilibrium of free rigid body.

If each of these is zero, each of the components ΣX, ΣY, ΣZ must be zero, as the determinant is not zero. The corresponding proposition is similarly proved for the moments, because (§ 233) moments of forces round different axes follow the same laws of composition and resolution as forces in different directions.

Equilibrium of constrained rigid body.

554. For equilibrium when the body is subjected to one, two, three, four, or five degrees of constraint, equations to be fulfilled by the applied forces, to ensure equilibrium, correspondingly reduced in number to five, four, three, two or one, are found with the greatest ease by giving direct analytical expression to (§ 289), the principle of work in equilibrium.

Let $\dot{x}$, $\dot{y}$, $\dot{z}$, ϖ, ρ, σ be components of the translational velocity of a point O of the body, and of the angular velocity of the body; and (§ 201) let

$$\left.\begin{array}{l} A\dot{x}+B\dot{y}+C\dot{z}+G\varpi+H\rho+I\sigma=0 \\ A'\dot{x}+B'\dot{y}+C'\dot{z}+G'\varpi+H'\rho+I'\sigma=0 \\ \quad\&c., \qquad \&c., \end{array}\right\} \text{.............(4)},$$

be one, two, three, four, or five equations, representing the constraints. The work done by the applied forces per unit of time is

$$\left.\begin{array}{l} \dot{x}\Sigma(X_1)+\dot{y}\Sigma(Y_1)+\dot{z}\Sigma(Z_1) \\ \quad+\varpi\Sigma(Z_1y_1-Y_1z_1)+\rho\Sigma(X_1z_1-Z_1x_1)+\sigma\Sigma(Y_1x_1-X_1y_1) \end{array}\right\}\text{...(5)},$$

or $$X\dot{x}+Y\dot{y}+Z\dot{z}+L\varpi+M\rho+N\sigma \text{...............(5')},$$

where X, Y, Z, L, M, N denote the sums that appear in (5), that is to say, the sums of the components of the given forces parallel to the axes of co-ordinates, and the sum of their moments round these lines.

This amount of work, (5), must be zero for all values of $\dot{x}$, $\dot{y}$, $\dot{z}$, ϖ, ρ, σ which satisfy equation or equations (4). Hence, by Lagrange's method of indeterminate multipliers, we find

$$\left.\begin{array}{l} \Sigma(X_1)+\lambda A+\lambda' A'+\ldots \qquad =0 \\ \Sigma(Y_1)+\lambda B+\lambda' B'+\ldots \qquad =0 \\ \Sigma(Z_1)+\lambda C+\lambda' C'+\ldots \qquad =0 \\ \Sigma(Z_1y_1-Y_1z_1)+\lambda G+\lambda' G'+\ldots=0 \\ \Sigma(X_1z_1-Z_1x_1)+\lambda H+\lambda' H'+\ldots=0 \\ \Sigma(Y_1x_1-X_1y_1)+\lambda I+\lambda' I'+\ldots=0 \end{array}\right\} \text{............(6)};$$

and the elimination of $\lambda, \lambda', \ldots$ from these six equations gives the correspondingly reduced number of equations of equilibrium among the applied forces.

Equilibrium of constrained rigid body.

To illustrate the use of these equations suppose, for example, the number of constraints to be two, and all except four of the applied forces be given: the six equations (5) determine these four forces, and allow us if we desire it to calculate the two indeterminate multipliers λ, λ'. The use of finding the values of these multipliers is that

Example. Two constraints;—the four equations of equilibrium found;

$$\lambda A,\ \lambda B,\ \lambda C,\ \lambda G,\ \lambda H,\ \lambda I$$

are the components and the moments of the reactions of the first constraining body or system on the given body, and

and the two factors determining the amounts of the constraining forces called into action.

$$\lambda' A',\ \lambda' B',\ \lambda' C',\ \lambda' G',\ \lambda' H',\ \lambda I'$$

are those of the second.

555. When it is desired only to find the equations of equilibrium, not the constraining reactions, the easiest and most direct way to the object is, to first express any possible motion of the body in terms of the five, four, three, two or one freedoms (§§ 197, 200) left to it by the one, two, three, four or five constraints to which it is subjected. The description in § 102 of the most general motion of a rigid body shows that the most general result of five constraints, or the most general way of allowing just one freedom, to a rigid body, is to give it guidance equivalent to that of a nut on a fixed screw shaft. If we unfix this shaft and give it similar guidance to allow it one freedom, the primary rigid body has two freedoms of the most general kind. Its double freedom may be resolved in an infinite number of ways (besides the one way in which it is thus compounded) into two single freedoms. Triple, quadruple, and quintuple freedom may be similarly arranged mechanically.

Equations of equilibrium without expression of constraining reactions.

556. The conditions of equilibrium of a rigid body with single, double, triple, quadruple or quintuple freedom, when each of the constituent freedoms is given in the manner specified in § 555, are found by writing down the equation or equations expressing that the applied forces do no work when the body moves simply according to any one alone of the given freedoms. We shall take first the case of a single freedom of the most general kind.

Equilibrium of forces applied to a nut on a frictionless fixed screw.

Let s* be the axial motion per radian of rotation; so that $q=s\omega$ expresses the relation between axial translational velocity, and angular velocity in the possible motion. Let HK be the axis of the screw, and N_1 the nearest point to it in $L_1 M_1$, the line of P_1, a first of the applied forces. Let i_1 be the inclination of $L_1 M_1$ to HK, and a_1 the distance of N_1 from HK. At any point in $L_1 M_1$, most conveniently at the point N_1, resolve P_1 into two components, $P_1 \cos i_1$, parallel to the axis of freedom, and $P_1 \sin i_1$ perpendicular to it. The former component does work only on the axial component of the motion, the latter on the rotational; and the rate of work done by the two together is

Work done by a single force on a nut, turning on a fixed screw.

$$s\omega\, P_1 \cos i_1 + a\omega\, P_1 \sin i_1 .$$

Hence, if Σ denotes summation for all the given forces, the equation of equilibrium to prevent them from taking advantage of the first freedom is

Equation of equilibrium of forces applied to a nut on a frictionless screw.

$$s\Sigma P_1 \cos i_1 + \Sigma a_1 P_1 \sin i_1 = 0 \ldots\ldots\ldots\ldots\ldots (7);$$

or, in words, *the step of the screw multiplied into the sum of the axial components must be equal to the sum of the moments of the force round the axis of the screw.*

The direction taken as positive for the moments in the preceding statement is the direction opposite to the rotation which the nut would have if it had axial motion in the direction taken as positive for those axial components.

557. The equations of equilibrium when there are two or more freedoms, are merely (7) repeated with accents to denote the elements corresponding to the several guide-screws other than the first. Thus if s, s', s'', &c., denote the screw-steps; a_1, a_1', a_1'', &c., the shortest distances between the axes of the screws and the line of P_1; i_1, i_1', i_1'', &c., the inclinations of this line to the axes; and a_2, a_2', &c., and i_2, i_2', &c., corresponding elements

* The quantity s thus defined we shall, for brevity, henceforth call the screw-step.

for the line of the second force, and so on; we have, for the equations of equilibrium,

$$\left.\begin{array}{l} s\Sigma P_1 \cos i_1 + \Sigma a_1 P_1 \sin i_1 = 0 \\ s'\Sigma P_1 \cos i_1' + \Sigma a_1' P_1 \sin i_1' = 0 \\ s''\Sigma P_1 \cos i_1'' + \Sigma a_1'' P_1 \sin i_1'' = 0 \\ \&c., \quad \&c., \end{array}\right\} \ldots\ldots\ldots\ldots (8).$$

The equations of constraint being, as in § 553, (4),

The same analytically and in terms of rectangular co-ordinates.

$$\left.\begin{array}{l} A\dot{x} + B\dot{y} + C\dot{z} + G\varpi + H\rho + I\sigma = 0 \\ A'\dot{x} + B'\dot{y} + C'\dot{z} + G'\varpi + H'\rho + I'\sigma = 0 \\ \ldots\ldots\ldots\ldots\ldots\ldots\ldots\ldots\ldots\ldots \end{array}\right\} \ldots\ldots\ldots (9),$$

suppose, for example, these equations to be four in number. Take two more equations

$$\left.\begin{array}{l} a\dot{x} + b\dot{y} + c\dot{z} + g\varpi + h\rho + i\sigma = \omega \\ a'\dot{x} + b'\dot{y} + c'\dot{z} + g'\varpi + h'\rho + i'\sigma = \omega' \end{array}\right\} \ldots\ldots (10),$$

where $a, b, \ldots$ and $a', b', \ldots$ are any arbitrarily assumed quantities: and from the six equations (9) and (10) deduce the following:

$$\left.\begin{array}{lll} \dot{x} = \mathfrak{A}\omega + \mathfrak{A}'\omega', & \dot{y} = \mathfrak{B}\omega + \mathfrak{B}'\omega', & \dot{z} = \mathfrak{C}\omega + \mathfrak{C}'\omega', \\ \varpi = \mathfrak{G}\omega + \mathfrak{G}'\omega', & \rho = \mathfrak{P}\omega + \mathfrak{P}'\omega', & \sigma = \mathfrak{I}\omega + \mathfrak{I}'\omega', \end{array}\right\} \ldots\ldots (11);$$

where $\mathfrak{A}, \mathfrak{B}, \ldots$ and $\mathfrak{A}', \mathfrak{B}', \ldots$ are known, being the determinantal ratios found in solving (9) and (10). Thus the *six* rectangular component velocities are expressed in terms of *two* generalized component velocities ω, ω', which, in virtue of the four equations of constraint (9), suffice for the complete specification of whatever motion the constraints leave permissible. In terms of this notation we have, for the rate of working of the applied forces,

Two generalized component velocities corresponding to two freedoms.

$$\left.\begin{array}{l} X\dot{x} + Y\dot{y} + Z\dot{z} + L\varpi + M\rho + N\sigma \\ \quad = (\mathfrak{A}X + \mathfrak{B}Y + \mathfrak{C}Z + \mathfrak{G}L + \mathfrak{P}M + \mathfrak{I}N)\omega \\ \quad + (\mathfrak{A}'X + \mathfrak{B}'Y + \mathfrak{C}'Z + \mathfrak{G}'L + \mathfrak{P}'M + \mathfrak{I}'N)\omega' \end{array}\right\} \ldots\ldots\ldots (12).$$

This must be nil for every permitted motion in order that the forces may balance. Hence the equations of equilibrium are

$$\left.\begin{array}{l} \mathfrak{A}X + \mathfrak{B}Y + \mathfrak{C}Z + \mathfrak{G}L + \mathfrak{P}M + \mathfrak{I}N = 0 \\ \text{and } \mathfrak{A}'X + \mathfrak{B}'Y + \mathfrak{C}'Z + \mathfrak{G}'L + \mathfrak{P}'M + \mathfrak{I}'N = 0 \end{array}\right\} \ldots\ldots\ldots (13).$$

Two generalized component velocities corresponding to two freedoms.

Similarly with one, or two, or three, or five (instead of our example of four) constraining equations (9), we find five, or four, or three, or one equation of equilibrium (13). These equations express obviously the same conditions as those expressed by (8); the first of (13) is identical with the first of (8), the second of (13) with the second of (8), and so on, provided ω, ω',... correspond to the same components of freedom as the several screws of (8) respectively. The equations though identical in substance are very different in form. The purely analytical transformation from either form to the other is a simple enough piece of analytical geometry which may be worked as an exercise by the student, to be done separately for the first of (8) and the first of (13), just as if there were but one freedom.

Equilibrant and resultant.

558. Any system of forces which if applied to a rigid body would balance a given system of forces acting on it, is called an equilibrant of the given system. The system of forces equal and opposite to the equilibrant may be called a resultant of the given system. It is only, however, when the resultant system is less numerous, or in some respect simpler, than the given system that the term resultant is convenient or suitable. It is used with great advantage with respect to the resultant force and couple (§ 559 g, below) to which Poinsot's method leads, or to the two resultant forces which mathematicians before Poinsot had shown to be the simplest system to which any system of forces acting on a rigid body can in general be reduced. It is only when the system is reducible to a single force that the term "resultant" pure and simple is usually applied.

559. As a most useful commentary on and illustration of the general theory of the equilibrium of a rigid body, which we have completed in §§ 552—557, and particularly for the purpose of finding practically convenient resultants in a very simple and clear manner, we may now with advantage introduce the beautiful method of *Couples*, invented by Poinsot.

Couples.

In § 234 we have already defined a couple, and shown that the sum of the moments of its forces is the same about all axes perpendicular to its plane. It may therefore be shifted to any new position in its own plane, or in any parallel plane, without alteration of its effect on the rigid body to which it is applied. Its arm may be turned through any angle in the plane of the forces, and the length of the arm and the magnitudes of the forces may be altered at pleasure, without changing its effect—provided the *moment* remain unchanged. Hence a couple is conveniently specified by the line defined as its "axis" in § 234. According to the convention of § 234 the axis of a couple which tends to produce rotation in the direction contrary to the motion of the hands of a watch, must be drawn through the *front* of the watch and *vice versâ*. This may easily be remembered by the help of a simple diagram such as we give, in which the arrow-heads indicate the directions of rotation, and of the axis, respectively. Couples.

559 *b*. It follows from §§ 233, 234, that couples are to be compounded or resolved by treating their axes by the law of the parallelogram, in a manner identical with that which we have seen must be employed for linear and angular velocities, and forces. Composition of couples.

Hence a couple G, the direction cosines of whose axis are λ, μ, ν, is equivalent to the three couples $G\lambda$, $G\mu$, $G\nu$ about the axes of x, y, z respectively.

559 *c*. If a force, F, act at any point, A, of a body, it may be transferred to any other point, B. Thus: by the principle of superposition of forces, introduce at B, in the line through it parallel to the given force F, a pair of equal and opposite forces F and $-F$. Then F at A, and $-F$ at B, form a couple, and there remains F at B. Force resolved into force and couple.

From this we have, at once, the conditions of equilibrium of a rigid body already investigated in § 552. For, each force may be transferred to any assumed point as origin, if we introduce the corresponding couple. And the forces, which now act at one point, must equilibrate according to the principles of Chap. VI.; while the resultant couple, and therefore its components about any three lines at right angles to each other, must vanish. Application to equilibrium of rigid body.

Forces represented by the sides of a polygon.

559 *d*. Hence forces represented, not merely in magnitude and direction, but in lines of action, by the sides of any closed polygon whether plane or not plane, are equivalent to a single couple. For when transferred to any origin, they equilibrate, by the Polygon of Forces (§§ 27, 256). When the polygon is plane, twice its area is the moment of the couple; when not plane, the component of the couple about any axis is twice the area of the projection on a plane perpendicular to that axis. The resultant couple has its axis perpendicular to the plane (§ 236) on which the projected area is a maximum.

Forces proportional and perpendicular to the sides of a triangle.

559 *e*. Lines, perpendicular to the sides of a triangle, and passing through their middle points, meet; and their mutual inclinations are equal to the changes of direction at the corners, in travelling round the triangle. Hence, if at the middle points of the sides of a triangle, and in its plane, forces be applied all inwards or all outwards; and if their magnitudes be proportional to the sides of the triangle, they are in equilibrium. The same is true of any plane polygon, as we readily see by dividing it into triangles. And if forces equal to the areas of the faces be applied perpendicularly to the faces of any closed polyhedron, at their centres of inertia, all inwards or all outwards, these also will form an equilibrating system; as we see by considering the evanescence of (i) the algebraic sum of the projections of the areas of the faces on any plane, and of (ii) the algebraic sum of the volumes of the rings described by the faces when the solid figure is made to rotate round any axis, these volumes being reckoned by aid of Pappus' theorem (§ 569, below).

Composition of force and couple.

559 *f*. A couple and a force in a given line inclined to its plane may be reduced to a smaller couple in a plane perpendicular to the force, and a force equal and parallel to the given force. For the couple may be resolved into two, one in a plane containing the direction of the force, and the other in a plane perpendicular to the force. The force and the component couple in the same plane with it are equivalent to an equal force acting in a parallel line, according to the converse of § 559 *c*.

559 *g*. We have seen that any set of forces acting on a rigid body may be reduced to a force at any point and a couple. Now (§ 559 *f*) these may be reduced to an equal force acting in a definite line in the body, and a couple whose plane is perpendicular to the force, and which is the least couple which, with a single force, can constitute a resultant of the given set of forces. The definite line thus found for the force is called the *Central Axis*. It is the line about which the sum of the moments of the given forces is least.

Composition of any set of forces acting on a rigid body.

Central axis.

With the notation of §§ 552, 553, let us suppose the origin to be changed to any point x', y', z'. The resultant force has still the components $\Sigma(X)$, $\Sigma(Y)$, $\Sigma(Z)$, or Rl, Rm, Rn, parallel to the axes. But the couples now are

$$\Sigma[Z(y-y')-Y(z-z')],\ \Sigma[X(z-z')-Z(x-x')],\ \Sigma[Y(x-x')-X(y-y')];$$

or

$$G\lambda - R(ny'-mz'),\ G\mu - R(lz'-nx'),\ G\nu - R(mx'-ly').$$

The conditions that the resultant force shall be perpendicular to the plane of the resultant couple are

$$\frac{G\lambda - R(ny'-mz')}{l} = \frac{G\mu - R(lz'-nx')}{m} = \frac{G\nu - R(mx'-ly')}{n}.$$

These two equations among x', y', z' are the equations of the central axis.

We find the same two equations by investigating the conditions that the resultant couple

$$\sqrt{[G\lambda-R(ny'-mz')]^2+[G\mu-R(lz'-nx')]^2+[G\nu-R(mx'-ly')]^2}$$

may be a minimum subject to independent variations of x', y', z'.

560. By combining the resultant force with one of the forces of the resultant couple, we have obviously an infinite number of ways of reducing any set of forces acting on a rigid body to *two* forces whose directions do not meet. But there is one case in which the result is symmetrical, and which is therefore worthy of special notice.

Reduction to two forces.

Supposing the central axis of the system has been found, draw a line, AA', at right angles to it through any point C of

Symmetrical case.

Symmetrical case.

it, and make CA equal to CA'. For R, acting along the central axis, substitute (by § 561) $\frac{1}{2}R$ at each end of AA'. Then, choosing this line AA' as the arm of the couple, and calling it a, we have at one extremity of it, two forces, $\frac{G}{a}$ perpendicular to the central axis, and $\frac{1}{2}R$ parallel to the central axis. Compounding these we get two forces, each equal to $\left(\frac{1}{4}R^2+\frac{G^2}{a^2}\right)^{\frac{1}{2}}$, through A and A' respectively, perpendicular to AA', and inclined to the plane through AA' and the central axis, at angles on the two sides of it each equal to $\tan^{-1}\frac{2G}{Ra}$.

Composition of parallel forces.

561. A very simple, but important, case, is that of any number of *parallel* forces acting at different points of a rigid body.

Here, for equilibrium, obviously it is necessary and sufficient that the algebraic sum of the forces be nil; and that the sum of their moments about any two axes perpendicular to the common direction of the forces be also nil.

This clearly implies (§ 553) that the sum of their moments about any axis whatever is nil.

To express the condition in rectangular coordinates, let P_1, P_2, &c. be the forces; (x_1, y_1, z_1), (x_2, y_2, z_2), &c. points in their lines of action; and l, m, n the direction cosines of a line parallel to them all. The general equations [§ 552 (1), (2)] of equilibrium of a rigid body become in this case,

$$l\Sigma P=0,\quad m\Sigma P=0,\quad n\Sigma P=0;$$

$$n\Sigma Py-m\Sigma Pz=0,\quad l\Sigma Pz-n\Sigma Px=0,\quad m\Sigma Px-l\Sigma Py=0.$$

These equations are equivalent to but three independent equations, which may be written as follows:

$$\Sigma P=0,\quad \frac{\Sigma Px}{l}=\frac{\Sigma Py}{m}=\frac{\Sigma Pz}{n}\quad\ldots\ldots\ldots\ldots\ldots(1).$$

If the given forces are not in equilibrium a single force may be found which shall be their resultant. To prove this let, if possible, a force $-R$, in the direction (l, m, n), at a point

$(\bar{x}, \bar{y}, \bar{z})$ equilibrate the given forces. By (1) we have, for the conditions of equilibrium of $-R, P_1, P_2$, &c., Composition of parallel forces.

$$R = \Sigma P \quad\text{..............................}(2),$$

and

$$\frac{\Sigma Px - R\bar{x}}{l} = \frac{\Sigma Py - R\bar{y}}{m} = \frac{\Sigma Pz - R\bar{z}}{n} \quad\text{.........}(3).$$

Equation (2) determines R, and equations (3) are the equations of a straight line at any point of which a force equal to $-R$, applied in the direction (l, m, n), will balance the given system.

Suppose now the direction (l, m, n) of the given forces to be varied while the magnitude P_1, and one point (x_1, y_1, z_1) in the line of application, of each force is kept unchanged. We see by (3) that one point $(\bar{x}, \bar{y}, \bar{z})$ given by the equations

$$\bar{x} = \frac{\Sigma Px}{R}, \quad \bar{y} = \frac{\Sigma Py}{R}, \quad \bar{z} = \frac{\Sigma Pz}{R} \quad\text{..........}(4),$$

is common to the lines of the resultants.

The point $(\bar{x}, \bar{y}, \bar{z})$ given by equations (4) is what is called the centre of the system of parallel forces P_1 at (x_1, y_1, z_1), P_2 at (x_2, y_2, z_2), &c.: and we have the proposition that a force in the line through this point parallel to the lines of the given forces, equal to their sum, is their resultant. This proposition is easily proved synthetically by taking the forces in any order and finding the resultant of the first two, then the resultant of this and the third, then of this second force, and so on. The line of the first subsidiary resultant, for all varied directions of the given forces, passes through one and the same point (that is the point dividing the line joining the points of application of the first two forces, into parts inversely as their magnitudes). Similarly we see that the second subsidiary resultant passes always through one determinate point: and so for the third, and so on for any number of forces.

562. It is obvious, from the formulas of § 230, that if masses proportional to the forces be placed at the several points of application of these forces, the centre of inertia of these masses will be the same point in the body as the centre of parallel Centre of gravity.

Centre of gravity. forces. Hence the reactions of the different parts of a rigid body against acceleration in parallel lines are rigorously reducible to one force, acting at the centre of inertia. The same is true approximately of the action of gravity on a rigid body of small dimensions relatively to the earth, and hence the centre of inertia is sometimes (§ 230) called the *Centre of Gravity*. But, except on a centrobaric body (§ 534), gravity is in general reducible not to a single force but to a force and couple (§ 559 *g*); and the force does not pass through a point fixed relatively to the body in all the positions for which the couple vanishes.

Parallel forces whose algebraic sum is zero. **563.** In one case the proposition of § 561, that the system has a single resultant force, must be modified: that is the case in which the algebraic sum of the given forces vanishes. In this case the resultant is a couple whose plane is parallel to the common direction of the forces. A good example of this case is furnished by a magnetized mass of steel, of moderate dimensions, subject to the influence of the earth's magnetism. The amounts of the so-called north and south magnetisms in each element of the mass are equal, and are therefore subject to equal and opposite forces, parallel in a rigorously uniform field of force. Thus a compass-needle experiences from the earth's magnetism sensibly a couple (or *directive* action), and is not sensibly attracted or repelled as a whole.

Conditions of equilibrium of three forces. **564.** If three forces, acting on a rigid body, produce equilibrium, their directions must lie in one plane; and must all meet in one point, or be parallel. For the proof we may introduce a consideration which will be very useful to us in investigations connected with the statics of flexible bodies and fluids.

Physical axiom. *If any forces, acting on a solid, or fluid body, produce equilibrium, we may suppose any portions of the body to become fixed, or rigid, or rigid and fixed, without destroying the equilibrium.*

Applying this principle to the case above, suppose any two points of the body, respectively in the lines of action of two of the forces, to be fixed. The third force must have no moment about the line joining these points; in other words, its direction must pass through that line. As any two points in the lines of action may be taken, it follows that the three forces are coplanar. And three forces, in one plane, cannot equilibrate unless their directions are parallel, or pass through a point. Physical axiom.

565. It is easy, and useful, to consider various cases of equilibrium when no forces act on a rigid body but gravity and the pressures, normal or tangential, between it and fixed supports. Thus if one given point only of the body be fixed, it is evident that the centre of inertia must be in the vertical line through this point. For *stable* equilibrium the centre of inertia need not be *below* the point of support (§ 566). Equilibrium under the action of gravity.

566. An interesting case of equilibrium is suggested by what are called Rocking Stones, where, whether by natural or by artificial processes, the lower surface of a loose mass of rock is worn into a convex or concave, or anticlastic form, while the bed of rock on which it rests in equilibrium may be convex or concave, or of an anticlastic form. A loaded sphere resting on a spherical surface is a particular case. Rocking stones.

Let O, O' be the centres of curvature of the fixed, and rocking, bodies respectively, when in the position of equilibrium. Take any two infinitely small, equal arcs PQ, Pp; and at Q make the angle $O'QR$ equal to POp. When, by displacement, Q and p become the points in contact, QR will evidently be vertical; and, if the centre of inertia G, which must be in OPO' when the movable body is in its position of equilibrium, be to the left of QR, the equilibrium will obviously be stable. Hence, if it be below R, the equilibrium is stable, and not unless.

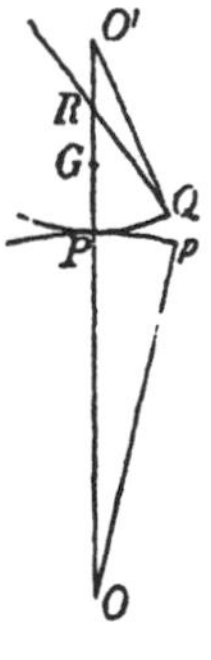

Now if ρ and σ be the radii of curvature OP, $O'P$ of the two surfaces, and θ the angle POp, the angle $QO'R$ will be equal to $\frac{\rho\theta}{\sigma}$; and we have in the triangle $QO'R$ (§ 112)

$$RO' : \sigma :: \sin\theta : \sin\left(\theta + \frac{\rho\theta}{\sigma}\right)$$

$$:: \sigma : \sigma + \rho \text{ (approximately).}$$

Rocking stones.

Hence $$PR = \sigma - \frac{\sigma^2}{\sigma+\rho} = \frac{\rho\sigma}{\rho+\sigma};$$

and therefore, for stable equilibrium,

$$PG < \frac{\rho\sigma}{\rho+\sigma}.$$

If the lower surface be plane, ρ is infinite, and the condition becomes (as in § 291)

$$PG < \sigma.$$

If the lower surface be concave the sign of ρ must be changed, and the condition becomes

$$PG < \frac{\rho\sigma}{\rho-\sigma},$$

which cannot be negative, since ρ *must* be numerically greater than σ in this case.

Equilibrium, about an axis,

567. If two points be fixed, the only motion of which the system is capable is one of rotation about a fixed axis. The centre of inertia must then be in the vertical plane passing through those points. For stability it is necessary (§ 566) that the centre of inertia be *below* the line joining them.

on a fixed surface.

568. If a rigid body rest on a frictional fixed surface there will in general be only *three* points of contact; and the body will be in stable equilibrium if the vertical line drawn from its centre of inertia cuts the plane of these three points *within* the triangle of which they form the corners. For if one of these supports be removed, the body will obviously tend to fall towards that support. Hence each of the three prevents the body from rotating about the line joining the other two. Thus, for instance, a body stands stably on an inclined plane (if the friction be sufficient to prevent it from sliding down) when the vertical line drawn through its centre of inertia falls within the base, or area bounded by the shortest line which can be drawn round the portion in contact with the plane. Hence a body, which cannot stand on a horizontal plane, may stand on an inclined plane.

569. A curious theorem, due to Pappus, but commonly attributed to Guldinus, may be mentioned here, as it is employed with advantage in some cases in finding the centre of gravity (or centre of inertia) of a body. It is obvious from § 230. *If a plane closed curve revolve through any angle about an axis in its plane, the solid content of the surface generated is equal to the product of the area of the curve into the length of the path described by the centre of inertia of the area of the curve; and the area of the curved surface is equal to the product of the length of the curve into the length of the path described by the centre of inertia of the curve.* Pappus' theorem.

570. The general principles upon which forces of constraint and friction are to be treated have been stated above (§§ 293, 329, 452). We add here a few examples for the sake of illustrating the application of these principles to the equilibrium of a rigid body in some of the more important practical cases of constraint.

571. The application of statical principles to the *Mechanical Powers*, or elementary machines, and to their combinations, however complex, requires merely a statement of their kinematical relations (as in §§ 79, 85, 102, &c.) and an immediate translation into Dynamics by Newton's principle (§ 269); or by Lagrange's Virtual Velocities (§§ 289, 290), with special attention to the introduction of forces of friction as in § 452. In no case can this process involve further difficulties than are implied in seeking the geometrical circumstances of any infinitely small disturbance, and in the subsequent solution of the equations to which the translation into dynamics leads us. We will not, therefore, stop to discuss any of these questions; but will take a few examples of no very great difficulty, before quitting for a time this part of the subject. The principles already developed will be of constant use to us in the remainder of the work, which will furnish us with ever-recurring opportunities of exemplifying their use and mode of application. Mechanical powers.

Let us begin with the case of the Balance, of which we promised (§ 431) to give an investigation.

Examples. Balance.

572. *Ex.* I. The centre of gravity of the beam must not coincide with the knife-edge, or else the beam would rest indifferently in any position. We shall suppose, in the first place, that the arms are not of equal length.

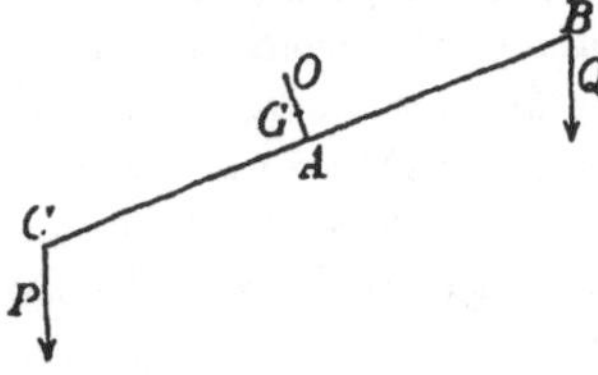

Let O be the fulcrum, G the centre of gravity of the beam, M its mass; and suppose that with loads P and Q in the pans the beam rests (as drawn) in a position making an angle θ with the horizontal line.

Sensibility.

Taking moments about O, and, for convenience (see § 220), using gravitation measurement of the forces, we have

$$Q(AB\cos\theta + OA\sin\theta) + M.\,OG\sin\theta = P(AC\cos\theta - OA\sin\theta).$$

From this we find

$$\tan\theta = \frac{P.\,AC - Q.\,AB}{(P+Q)\,OA + M.\,OG}.$$

If the arms be equal we have

$$\tan\theta = \frac{(P-Q)\,AB}{(P+Q)\,OA + M.\,OG}.$$

Hence the Sensibility (§ 431) is greater, (1) as the arms are longer, (2) as the mass of the beam is less, (3) as the fulcrum is nearer to the line joining the points of attachment of the pans, (4) as the fulcrum is nearer to the centre of gravity of the beam. If the fulcrum be *in* the line joining the points of attachment of the pans, the sensibility is the same for the same *difference* of loads in the pans.

Examples. Rod with frictionless constraint.

Ex. II. Find the position of equilibrium of a rod AB resting on a frictionless horizontal rail D, its lower end pressing against a frictionless vertical wall AC parallel to the rail.

The figure represents a vertical section through the rod, which must evidently be in a plane perpendicular to the wall and rail. The equilibrium is obviously unstable.

The only forces acting are three, R the pressure of the wall on the rod, horizontal; S that of the rail on the rod, perpendicular to the rod; W the weight of the rod, acting vertically downwards at its centre of gravity. If the half-length of the rod be a, and the distance of the rail from the wall b, these are given—and all that is wanted to fix the position of equilibrium is the angle, CAB, which the rod makes with the wall. If we call it θ we have $AD = \frac{b}{\sin\theta}$. Examples. Rod with frictionless constraint.

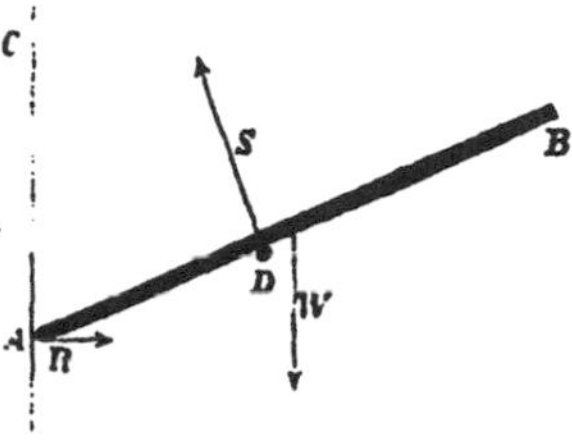

Resolving horizontally, $R - S\cos\theta = 0$ ....(1).

vertically, $W - S\sin\theta = 0$ ....(2).

Taking moments about A

$$S.AD - W.a\sin\theta = 0,$$

or $S.b - W.a\sin^2\theta = 0$ ....(3).

As there are only three unknown quantities R, S, and θ, these three equations contain the complete solution of the problem. By (2) and (3)

$$\sin^3\theta = \frac{b}{a}, \text{ which gives } \theta.$$

And by (2) $$S = \frac{W}{\sin\theta},$$

and by (1) $$R = S\cos\theta = W\cot\theta.$$

Ex. III. As an additional example, suppose the wall and rail to be frictional, and let μ be the coefficient of statical friction for both. If the rod be placed in the position of equilibrium just investigated for the case of no friction, none will be called into play, for there will be no tendency to motion to be overcome. If the end A be brought lower and lower, more Rod constrained by frictional surfaces.

Examples. Rod constrained by frictional surfaces.

and more friction will be called into play to overcome the tendency of the rod to fall between the wall and the rail, until we come to a limiting position in which motion is about to commence. In that position the friction at A is μ times the pressure on the wall, and acts *upwards*. That at D is μ times the pressure on the rod, and acts in the direction DB. Putting $CAD = \theta_1$ in this case, our three equations become

$$R_1 + \mu S_1 \sin\theta_1 - S_1 \cos\theta_1 = 0 \quad\quad (1_1),$$

$$W - \mu R_1 - S_1 \sin\theta_1 - \mu S_1 \cos\theta_1 = 0 \quad\quad (2_1),$$

$$S_1 b - Wa \sin^2\theta_1 = 0 \quad\quad (3_1).$$

The directions of both the friction-forces passing through A, neither appears in (3_1). This is why A is preferable to any other point about which to take moments.

By eliminating R_1 and S_1 from these equations we get

$$1 - \frac{a}{b}\sin^3\theta_1 = \mu\frac{a}{b}\sin^2\theta_1(2\cos\theta_1 - \mu\sin\theta_1) \quad\quad (4_1),$$

from which θ_1 is to be found. Then S_1 is known from (3_1) and R_1 from either of the others.

If the end A be raised above the position of equilibrium without friction, the tendency is for the rod to fall *outside* the rail; more and more friction will be called into play, till the position of the rod (θ_2) is such that the friction reaches its greatest value, μ times the pressure. We may thus find another *limiting* position for stability; and in any position between these the rod is in equilibrium.

It is useful to observe that in this second case the direction of each friction is the opposite to that in the former. Hence equations of the first case, with the sign of μ changed, serve for the second case. Thus for θ_2, by (4_1),

$$1 - \frac{a}{b}\sin^3\theta_2 = -\mu\frac{a}{b}\sin^2\theta_2(2\cos\theta_2 + \mu\sin\theta_2).$$

Ex. IV. A rectangular block lies on a frictional horizontal plane, and is acted on by a horizontal force whose line of action is midway between two of the vertical sides. Find the magnitude of the force when just sufficient to produce motion, and whether the motion will be of the nature of *sliding* or *overturning*. Examples. Block on frictional plane.

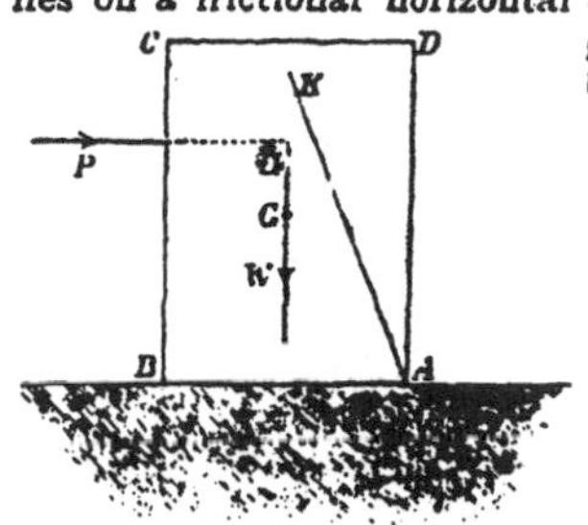

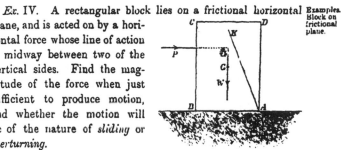

If the force P is on the point of overturning the body, it is evident that it will turn about the edge A, and therefore the pressure, R, of the plane and the friction, S, act at that edge. Our statical conditions are, of course,

$$R = W,$$
$$S = P,$$
$$Wb = Pa,$$

where b is half the length of the solid, and a the distance of P from the plane. From these we have $S = \frac{b}{a} W$.

Now S cannot exceed μR, whence we must not have $\frac{b}{a}$ greater than μ, if it is to be possible to upset the body by a horizontal force in the line given for P.

A simple geometrical construction enables us to solve this and similar problems, and will be seen at once to be merely a graphic representation of the above process. Thus if we produce the directions of the applied force, and of the weight, to meet in H, and make at A the angle BAK whose co-tangent is the coefficient of friction: there will be a tendency to upset, or not, according as H is above, or below, AK.

Ex. V. A mass, such as a gate, is supported by two rings, A and B, which pass loosely round a vertical post. In equilibrium, it is obvious that at A the part of the ring nearest the Mass supported by rings passing round a rough post.

Examples. Mass supported by rings passing round a rough post.

mass, and at B the part farthest from it, will be in contact with the post. The pressures exerted on the rings, R and S, will evidently be in the directions AC, CB, indicated in the diagram, which, if no other force besides gravity act on the mass, must meet in the vertical through its centre of inertia. And it is obvious that, however small be the coefficient of friction, provided there be any force of friction at all, equilibrium is always possible if the distance of the centre of inertia from the post be great enough compared with the distance between the rings.

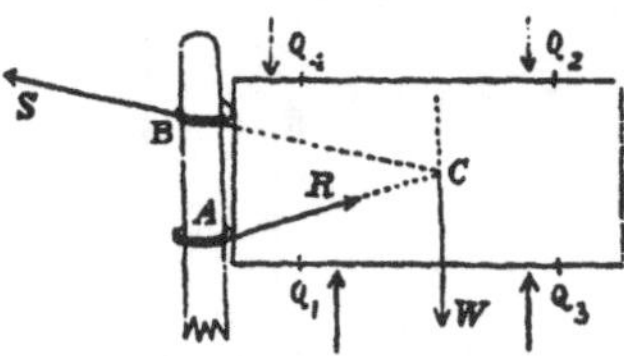

When the mass is just about to slide down, the full amount of friction is called into play, and the angles which R and S make with the horizon are each equal to the sliding angle. If the centre of inertia of the gate be farther from the post than the intersection of two lines drawn from A, B, at the sliding angles, it will hang stably held up by friction; not unless. A force pushing upwards at Q_1, or downwards at Q_2, will remove the tendency to fall; but a force upwards at Q_3, or downwards at Q_4, will produce sliding.

A similar investigation is easily applied to the jamming of a sliding piece or drawer, and to the determination of the proper point of application of a force to move it.

573. Having thus briefly considered the equilibrium of a rigid body, we propose, before entering upon the subject of the deformation of elastic solids, to consider certain intermediate cases, in each of which we make a particular assumption the basis of the investigation, and thereby avoid a very considerable amount of analytical difficulty.

Equilibrium of a flexible and inextensible cord.

574. Very excellent examples of this kind are furnished by the statics of a flexible and inextensible cord or chain, fixed at both ends, and subject to the action of any forces. The curve in which the chain hangs in any case may be called a *Catenary*, although the term is usually restricted to the case of a uniform chain acted on by gravity only.

Catenary.

575. We may consider separately the conditions of equilibrium of each element; or we may apply the general condition (§ 292) that the whole potential energy is a minimum, in the case of any conservative system of forces; or, especially when gravity is the only external force, we may consider the equilibrium of a *finite* portion of the chain treated for the time as a rigid body (§ 564). Three methods of investigation.

576. The first of these methods gives immediately the three following equations of equilibrium, for the catenary in general:— Equations of equilibrium with reference to tangent and osculating plane.

(1) The rate of variation of the tension per unit of length along the cord is equal to the tangential component of the applied force, per unit of length.

(2) The plane of curvature of the cord contains the normal component of the applied force, and the centre of curvature is on the opposite side of the arc from that towards which this force acts.

(3) The amount of the curvature is equal to the normal component of the applied force per unit of length at any point divided by the tension of the cord at the same point.

The first of these is simply the equation of equilibrium of an infinitely small element of the cord relatively to tangential motion. The second and third express that the component of the resultant of the tensions at the two ends of an infinitely small arc, along the normal through its middle point, is directly opposed and is equal to the normal applied force, and is equal to the whole amount of it on the arc. For the plane of the tangent lines in which those tensions act is (§ 8) the plane of curvature. And if θ be the angle between them (or the infinitely small angle by which the angle between their positive directions falls short of π), and T the arithmetical mean of their magnitudes, the component of their resultant along the line bisecting the angle between their positive directions is $2T\sin\frac{1}{2}\theta$, rigorously: or $T\theta$, since θ is infinitely small. Hence $T\theta = N\delta s$, if δs be the length of the arc, and $N\delta s$ the whole

Equations of equilibrium with reference to tangent and osculating plane.

amount of normal force applied to it. But (§ 9) $\theta = \frac{\delta s}{\rho}$ if ρ be the radius of curvature; and therefore

$$\frac{1}{\rho} = \frac{N}{T},$$

which is the equation stated in words (3) above.

Integral for tension.

577. From (1) of § 576, we see that if the applied forces on each particle of the cord constitute a conservative system, and if the cord be homogeneous, the difference of the tensions of the cord at any two points of it when hanging in equilibrium, is equal to the difference of the potential (§ 485) of the forces between the positions occupied by these points. Hence, whatever be the position where the potential is reckoned zero, the tension of the string at any point is equal to the potential at the position occupied by it, with a constant added.

Cartesian equations of equilibrium.

578. Instead of considering forces along and perpendicular to the tangent, we may resolve all parallel to any fixed direction: and we thus see that the component of applied force per unit of length of the chain at any point of it, must be equal to the rate of diminution per unit of length of the cord, of the component of its tension parallel to the fixed line of this component. By choosing any three fixed rectangular directions we thus have the three differential equations convenient for the analytical treatment of catenaries by the method of rectangular co-ordinates.

These equations are

$$\left.\begin{aligned}\frac{d}{ds}\left(T\frac{dx}{ds}\right) &= -\sigma X\\ \frac{d}{ds}\left(T\frac{dy}{ds}\right) &= -\sigma Y\\ \frac{d}{ds}\left(T\frac{dz}{ds}\right) &= -\sigma Z\end{aligned}\right\}\text{.........................(1)},$$

if s denote the length of the cord from any point of it, to a point P; x, y, z the rectangular co-ordinates of P; X, Y, Z the components of the applied forces at P, per unit mass of the cord; σ the mass of the cord per unit length at P; and T its tension at this point.

These equations afford analytical proofs of § 576, (1), (2), and (3) thus:—Multiplying the first by dx, the second by dy, and the third by dz, adding and observing that Cartesian equations of equilibrium.

$$\frac{dx}{ds}d\frac{dx}{ds}+\frac{dy}{ds}d\frac{dy}{ds}+\frac{dz}{ds}d\frac{dz}{ds}=\tfrac{1}{2}d\frac{dx^2+dy^2+dz^2}{ds^2}=0,$$

we have

$$dT=-\sigma(Xdx+Ydy+Zdz)=-\sigma\left(X\frac{dx}{ds}+Y\frac{dy}{ds}+Z\frac{dz}{ds}\right)ds\ldots(2),$$

which is (1) of § 576. Again, eliminating dT and T, we have

$$X\left(\frac{dy}{ds}d\frac{dz}{ds}-\frac{dz}{ds}d\frac{dy}{ds}\right)+Y\left(\frac{dz}{ds}d\frac{dx}{ds}-\frac{dx}{ds}d\frac{dz}{ds}\right)+Z\left(\frac{dx}{ds}d\frac{dy}{ds}-\frac{dy}{ds}d\frac{dx}{ds}\right)=0\ \ldots\ldots\ldots(3),$$

which (§§ 9, 26) shows that the resultant of X, Y, Z is in the osculating plane, and therefore is the analytical expression of § 576 (2). Lastly, multiplying the first by $d\frac{dx}{ds}$, the second by $d\frac{dy}{ds}$, and the third by $d\frac{dz}{ds}$, and adding, we find

$$T=-\sigma\frac{\left(Xd\frac{dx}{ds}+Yd\frac{dy}{ds}+Zd\frac{dz}{ds}\right)ds}{\left(d\frac{dx}{ds}\right)^2+\left(d\frac{dy}{ds}\right)^2+\left(d\frac{dz}{ds}\right)^2}\ \ldots\ldots\ldots\ldots(4),$$

which is the analytical expression of § 576 (3).

579. The same equations of equilibrium may be derived from the energy condition of equilibrium; analytically with ease by the methods of the calculus of variations. Method of energy.

Let V be the potential at (x, y, z) of the applied forces per unit mass of the cord. The potential energy of any given length of the cord, in any actual position between two given fixed points, will be Catenary

$$\int V\sigma ds.$$

This integral, extended through the given length of the cord between the given points, must be a minimum; while the indefinite integral, s, from one end up to the point (x, y, z) remains unchanged by the variations in the positions of this point. Hence, by the calculus of variations,

$$\delta\int V\sigma ds+\int\lambda\delta ds=0,$$

where λ is a function of x, y, z to be eliminated.

Catenary.

Now σ is a function of s, and therefore as s does not vary when x, y, z are changed into $x+\delta x$, $y+\delta y$, $z+\delta z$, the co-ordinates of the same particle of the chain in another position, we have

$$\delta(\sigma V) = \sigma\delta V = -\sigma(X\delta x + Y\delta y + Z\delta z).$$

Using this, and

$$\delta ds = \frac{dx d\delta x + dy d\delta y + dz d\delta z}{ds},$$

in the variational equation; and integrating the last term by parts according to the usual rule; we have

$$\int ds \left\{ \left[\sigma X + \frac{d}{ds}\left(\overline{V\sigma+\lambda}\frac{dx}{ds}\right)\right]\delta x + \left[\sigma Y + \frac{d}{ds}\left(\overline{V\sigma+\lambda}\frac{dy}{ds}\right)\right]\delta y + \left[\sigma Z + \frac{d}{ds}\left(\overline{V\sigma+\lambda}\frac{dz}{ds}\right)\right]\delta z \right\} = 0:$$

Energy equation of equilibrium.

whence finally

$$\frac{d}{ds}\left\{(V\sigma+\lambda)\frac{dx}{ds}\right\} + X\sigma = 0,$$

$$\frac{d}{ds}\left\{(V\sigma+\lambda)\frac{dy}{ds}\right\} + Y\sigma = 0,$$

$$\frac{d}{ds}\left\{(V\sigma+\lambda)\frac{dz}{ds}\right\} + Z\sigma = 0,$$

which, if T be put for $V\sigma+\lambda$, are the same as the equations (1) of § 578.

Common catenary.

580. The form of the common catenary (§ 574) may be of course investigated from the differential equations (§ 578) of the catenary in general. It is convenient and instructive, however, to work it out *ab initio* as an illustration of the third method explained in § 575.

Third method.—The chain being in equilibrium, *any* arc of it may be supposed to become rigid without disturbing the equilibrium. The only forces acting on this rigid body are the tensions at its ends, and its weight. These forces being three in number, must be in one plane (§ 564), and hence, since one of them is vertical, the whole curve lies in a vertical plane. In this plane let $x_0, z_0, s_0, x_1, z_1, s_1$, belong to the two ends of the arc which is supposed rigid, and T_0, T_1, the tensions at those points. Resolving horizontally we have

$$T_0\left(\frac{dx}{ds}\right)_0 = T_1\left(\frac{dx}{ds}\right)_1.$$

Hence $T\frac{dx}{ds}$ is constant throughout the curve. Resolving vertically we have Catenary: common.

$$T_1\left(\frac{dz}{ds}\right)_1 - T_0\left(\frac{dz}{ds}\right)_0 = \sigma(s_1 - s_0),$$

the weight of unit of mass being now taken as the unit of force.

Hence if T_0 be the tension at the lowest point, where $\frac{dz}{ds}=0$, $s=0$, and T the tension at any point (x, z) of the curve, we have

$$T = T_0\frac{ds}{dx} = \sigma s\frac{ds}{dz} \quad\quad (1).$$

Hence

$$T_0\frac{d}{ds}\left(\frac{dz}{dx}\right) = \sigma,$$

or

$$T_0\frac{d^2z}{dx^2} = \sigma\frac{ds}{dx} = \sigma\sqrt{1+\left(\frac{dz}{dx}\right)^2} \quad\quad (2).$$

Integrating we have

$$\log\left\{\frac{dz}{dx} + \sqrt{1+\left(\frac{dz}{dx}\right)^2}\right\} = \frac{\sigma}{T_0}x + C',$$

and the constant is zero if we take the origin so that $x=0$, when $\frac{dz}{dx}=0$, *i.e.*, where the chain is horizontal.

Hence

$$\frac{dz}{dx} + \sqrt{1+\left(\frac{dz}{dx}\right)^2} = \epsilon^{\frac{\sigma}{T_0}x} \quad\quad (3),$$

whence

$$\frac{dz}{dx} = \tfrac{1}{2}\left(\epsilon^{\frac{\sigma}{T_0}x} - \epsilon^{-\frac{\sigma}{T_0}x}\right);$$

and by integrating again

$$z + C'' = \frac{T_0}{2\sigma}\left(\epsilon^{\frac{\sigma}{T_0}x} + \epsilon^{-\frac{\sigma}{T_0}x}\right).$$

This may be written

$$z = \tfrac{1}{2}a\left(\epsilon^{\frac{x}{a}} + \epsilon^{-\frac{x}{a}}\right) \quad\quad (4),$$

the ordinary equation of the catenary, the axis of x being taken at a distance a or $\frac{T_0}{\sigma}$ below the horizontal element of the chain.

Catenary: common.

The co-ordinates of that element are therefore $x = 0$, $z = \frac{T_0}{\sigma} = a$. The latter shows that

$$T_0 = \sigma a,$$

or the tension at the lowest point of the chain (and therefore also the horizontal component of the tension throughout) is the weight of a length a of the chain.

Now, by (1), $T = T_0 \frac{ds}{dx} = \sigma z$, by (4), and therefore

the tension at any point is equal to the weight of a portion of the chain equal to the vertical ordinate at that point.

Relative kinetic problem

Kinetic question relative to catenary.

581. From § 576 it follows immediately that if a material particle of unit mass be carried along any catenary with a velocity, $\dot{s}$, equal to T, the numerical measure of the tension at any point, the force upon it by which this is done is in the same direction as the resultant of the applied force on the catenary at this point, and is equal to the amount of this force per unit of length, multiplied by T. For, denoting by S the tangential and (as before) by N the normal component of the applied force per unit of length at any point P of the catenary, we have, by § 576 (1), S for the rate of variation of $\dot{s}$ per unit length, and therefore $S\dot{s}$ for its variation per unit of time. That is to say,

$$s = S\dot{s} = ST,$$

or (§ 259) the tangential component force on the moving particle is equal to ST. Again, by § 576 (3),

$$NT = \frac{T^2}{\rho} = \frac{\dot{s}^2}{\rho},$$

or the centrifugal force of the moving particle in the circle of curvature of its path, that is to say, the normal component of the force on it, is equal to NT. And lastly, by (2) this force is in the same direction as N. We see therefore that the direction of the whole force on the moving particle is the same as that of the resultant of S and N; and its magnitude is T times the magnitude of this resultant.

Or, by taking

$$\frac{ds}{T} = dt,$$

Kinetic question relative to catenary.

in the differential equation of § 578, we have

$$\frac{d^2x}{dt^2} = -T\sigma X, \quad \frac{d^2y}{dt^2} = -T\sigma Y, \quad \frac{d^2z}{dt^2} = -T\sigma Z,$$

which proves the same conclusion.

When σ is constant, and the forces belong to a conservative system, if V be the potential at any point of the cord, we have, by § 578 (2), $T = \sigma V + C$.

Hence, if $U = \frac{1}{2}(\sigma V + C)^2$, these equations become

$$\frac{d^2x}{dt^2} = -\frac{dU}{dx}, \quad \frac{d^2y}{dt^2} = -\frac{dU}{dy}, \quad \frac{d^2z}{dt^2} = -\frac{dU}{dz}.$$

The integrals of these equations which agree with the catenary, are those only for which the energy constant is such that $\dot{s}^2 = 2U$.

582. Thus we see how, from the more familiar problems of the kinetics of a particle, we may immediately derive curious cases of catenaries. For instance: a particle under the influence of a constant force in parallel lines moves (Chap. VIII.) in a parabola with its axis vertical, with velocity at each point equal to that generated by the force acting through a space equal to its distance from the directrix. Hence, if z denote this distance, and f the constant force, Examples.

$$T = \sqrt{2fz}$$

in the allied parabolic catenary; and the force on the catenary is parallel to the axis, and is equal in amount per unit of length, to

$$\frac{f}{\sqrt{2fz}} \text{ or } \sqrt{\frac{f}{2z}}.$$

Hence if the force on the catenary be that of gravity, it must have its axis vertical (its vertex downwards of course for stable equilibrium) and its mass per unit length at any point must be inversely as the square root of the distance of this point above the directrix. From this it follows that the whole weight of any arc of it is proportional to its horizontal projection. Or,

Examples. again, as will be proved later with reference to the motions of comets, a particle moves in a parabola under the influence of a force towards a fixed point varying inversely as the square of the distance from this point, if its velocity be that due to falling from rest at an infinite distance. This velocity being $\sqrt{\frac{2\mu}{r}}$, at distance r, it follows, according to § 581, that a cord will hang in the same parabola, under the influence of a force towards the same centre, and equal to

$$\frac{\mu}{r^2} \div \sqrt{\frac{2\mu}{r}}, \text{ or } \sqrt{\frac{\mu}{2r^3}}.$$

If, however, the length of the cord be varied between two fixed points, the central force still following the same law, the altered catenary will no longer be parabolic: but it will be the path of a particle under the influence of a central force equal to

$$\left(C + \sqrt{\frac{2\mu}{r}}\right)\sqrt{\frac{\mu}{2r^3}},$$

since (§ 581) we should have,

$$T = \sigma V + C = -\sigma \int \sqrt{\frac{\mu}{2r^3}}\, dr + C = \sigma \sqrt{\frac{2\mu}{r}} + C,$$

instead of $\sqrt{\frac{2\mu}{r}}$.

Catenary. Inverse problem. **583.** Or if the question be, to find what force towards a given fixed point, will cause a cord to hang in any given plane curve with this point in its plane; it may be answered immediately from the solution of the corresponding problem in "central forces."

But the general equations, § 578, are always easily applicable; as, for instance, to the following curious and interesting, but not practically useful, inverse case of the gravitation catenary:—

Catenary of uniform strength. *Find the section, at each point, of a chain of uniform material, so that when its ends are fixed the tension at each point may be proportional to its section at that point. Find also the form of the Curve, called the Catenary of Uniform Strength, in which it will hang.*

Here, as the only external force is gravity, the chain is in a vertical plane—in which we may assume the horizontal axis of x to lie. If μ be the weight of the chain at the point (x, z) reckoned per unit of length; our equations [§ 578 (1)] become

Catenary of uniform strength.

$$\frac{d}{ds}\left(T\frac{dx}{ds}\right) = 0, \quad \frac{d}{ds}\left(T\frac{dz}{ds}\right) = \mu.$$

But, by hypothesis $T \propto \mu$. Let it be $b\mu$. Hence, by the first equation, if μ_0 be the value of μ at the lowest point

$$\mu = \mu_0 \frac{ds}{dx};$$

whence, by the second equation,

$$\frac{d}{ds}\left(\frac{dz}{dx}\right) = \frac{1}{b}\frac{ds}{dx},$$

or

$$\frac{d^2z}{dx^2} = \frac{1}{b}\left[1 + \left(\frac{dz}{dx}\right)^2\right].$$

Integrating we find

$$\tan^{-1}\frac{dz}{dx} = \frac{x}{b},$$

no constant being required if we take the axis of x so as to touch the curve at its lowest point. Integrating again we have

$$\frac{z}{b} = -\log\cos\frac{x}{b},$$

no constant being added, if the origin be taken at the lowest point. We may write the equation in the form

$$\sec\frac{x}{b} = \epsilon^{\frac{z}{b}}.$$

From this form of the equation we see that the curve has vertical asymptotes at a horizontal distance πb from each other. Hence πb is the greatest possible span, if the ends are on the same level, or the horizontal projection of the greatest possible span if they be not on the same level; b denoting the length of a uniform rod or wire of the material equal in weight to the tension of the catenary at any point, and equal in sectional area to the sectional area of the catenary at the same point. The greatest possible value of b is the "length modulus of rupture" (§§ 687, 688 below).

Flexible string on smooth surface.

584. When a perfectly flexible string is stretched over a smooth surface, and acted on by no other force throughout its length than the resistance of this surface, it will, when in stable equilibrium, lie along a line of minimum length on the surface, between any two of its points. For (§ 564) its equilibrium can be neither disturbed nor rendered unstable by placing staples over it, through which it is free to slip, at any two points where it rests on the surface: and for the intermediate part the energy criterion of stable equilibrium is that just stated.

There being no tangential force on the string in this case, and the normal force upon it being along the normal to the surface, its osculating plane (§ 576) must cut the surface everywhere at right angles. These considerations, easily translated into pure geometry, establish the fundamental property of the geodetic lines on any surface. The analytical investigations of §§ 578, 579, when adapted to the case of a chain of *not* given length, stretched between two given points on a given smooth surface, constitute the direct analytical demonstration of this property.

In this case it is obvious that the tension of the string is the same at every point, and the pressure of the surface upon it is [§ 576 (3)] at each point proportional to the curvature of the string.

On rough surface.

585. No real surface being perfectly smooth, a cord or chain may rest upon it when stretched over so great a length of a geodetic on a convex rigid body as to be not of minimum length between its extreme points: but practically, as in tying a cord round a ball, for permanent security it is necessary, by staples or otherwise, to constrain it from lateral slipping at successive points near enough to one another to make each free portion a true minimum on the surface.

Rope coiled about rough cylinder.

586. A very important practical case is supplied by the consideration of a rope wound round a rough cylinder. We may suppose it to lie in a plane perpendicular to the axis, as we thus simplify the question very considerably without sensibly

injuring the utility of the solution. To simplify still further, we shall suppose that no forces act on the rope, but tensions and the reaction of the cylinder. In practice this is equivalent to the supposition that the tensions and reactions are very large compared with the weight of the rope or chain; which, however, is inadmissible in some important cases; especially such as occur in the application of the principle to brakes for laying submarine cables, to dynamometers, and to windlasses (or capstans with horizontal axes). Rope coiled about rough cylinder.

If R be the normal reaction of the cylinder per unit of length of the cord, at any point; T and $T + \delta T$ the tensions at the extremities of an arc δs; $\delta\theta$ the inclination of these lines; we have, as in § 576,

$$T\delta\theta = R\delta s.$$

And the friction called into play is evidently equal to δT. When the rope is about to slip, the friction has its greatest value, and then

$$\delta T = \mu R\delta s = \mu T\delta\theta.$$

This gives, by integration,

$$T = T_0 \epsilon^{\mu\theta},$$

showing that, for equal successive amounts of integral curvature (§ 10), the tension of the rope augments in *geometrical* progression. To give an idea of the magnitudes involved, suppose $\mu = 0{\cdot}25$, $\theta = 2\pi$, then

$$T = T_0 \epsilon^{\cdot 5\pi} = 4{\cdot}81 T_0 \text{ approximately.}$$

Hence if the rope be wound three times round the post or cylinder the ratio of the tensions of its ends, when motion is about to commence, is

$$(4{\cdot}81)^3 : 1 \text{ or about } 111 : 1.$$

Thus we see how, by the aid of friction, one man may easily check the motion of a large ship, by the simple expedient of coiling a rope a few times round a post. This application of friction is of great importance in many other uses, especially for dynamometers.

Rope coiled about rough cylinder.

587. With the aid of the preceding investigations, the student may easily work out for himself the formulæ expressing the solution of the general problem of a cord under the action of any forces, and constrained by a rough surface; they are not of sufficient importance or interest to find a place here.

Elastic wire.

Elastic wire, fibre, bar, rod, lamina, or beam.

588. An elongated body of elastic material, which for brevity we shall generally call a *Wire*, bent or twisted to any degree, subject only to the condition that the radius of curvature and the reciprocal of the twist (§ 119) are everywhere very great in comparison with the greatest transverse dimension, presents a case in which, as we shall see, the solution of the general equations for the equilibrium of an elastic solid is either obtainable in finite terms, or is reducible to comparatively easy questions agreeing in mathematical conditions with some of the most elementary problems of hydrokinetics, electricity, and thermal conduction. And it is only for the determination of certain constants depending on the section of the wire and the elastic quality of its substance, which measure its flexural and torsional rigidity, that the solutions of these problems are required. When the constants of flexure and torsion are known, as we shall now suppose them to be, whether from theoretical calculation or experiment, the investigation of the form and twist of any length of the wire, under the influence of any forces which do not produce a violation of the condition stated above, becomes a subject of mathematical analysis involving only such principles and formulæ as those that constitute the theory of curvature (§§ 5—13) and twist (§§ 119—123) in geometry or kinematics.

589. Before entering on the general theory of elastic solids, we shall therefore, according to the plan proposed in § 573, examine the dynamic properties and investigate the conditions of equilibrium of a perfectly elastic wire, without admitting any other condition or limitation of the circumstances than what is stated in § 588, and without assuming any special quality of isotropy, or of crystalline, fibrous or laminated structure in the substance. The following short geometrical digression is a convenient preliminary:—

590. The geometrical composition of curvatures with one another, or with rates of twist, is obvious from the definition and principles regarding curvature given above in §§ 5—13 and twist in §§ 119—123, and from the composition of angular velocities explained in § 96. Thus if one line, 𝔒𝔗, of a rigid body be always held parallel to the tangent, PT, at a point P moving with unit velocity along a curve, whether plane or tortuous, it will have, round an axis perpendicular to 𝔒𝔗 and to the radius of curvature (that is to say, perpendicular to the osculating plane), an angular velocity numerically equal to the curvature. The body may besides be made to rotate with any angular velocity round 𝔒𝔗. Thus, for instance, if a line of it, 𝔒𝔄, be kept always parallel to a transverse (§ 120) PA, the component angular velocity of the rigid body round 𝔒𝔗 will at every instant be equal to the "rate of twist" (§ 120) of the transverse round the tangent to the curve. Again, the angular velocity round 𝔒𝔄 may be resolved into components round two lines 𝔒𝔎, 𝔒𝔏, perpendicular to one another and to 𝔒𝔗; and the whole curvature of the curve may be resolved accordingly into two component curvatures in planes perpendicular to those two lines respectively. The amounts of these component curvatures are of course equal to the whole curvature multiplied by the cosines of the respective inclinations of the osculating plane to these planes. And it is clear that each component curvature is simply the curvature of the projection of the actual curve on its plane*.

Composition and resolution of curvatures in a curved line.

591. Besides showing how the constants of flexural and torsional rigidity are to be determined theoretically from the form of the transverse section of the wire, and the proper data as to the elastic qualities of its substance, the complete theory simply indicates that, provided the conditional limit (§ 588) of deformation is not exceeded, the following laws will be obeyed by the wire under stress:—

* The curvature of the projection of a curve on a plane inclined at an angle α to the osculating plane, is $(1/\rho)\cos\alpha$ if the plane be parallel to the tangent; and $1/\rho\cos^2\alpha$ if it be parallel to the principal normal (or radius of absolute curvature). There is no difficulty in proving either of these expressions.

Laws of flexure and torsion.

Let the whole mutual action between the parts of the wire on the two sides of the cross section at any point (being of course the action of the matter infinitely near this plane on one side, upon the matter infinitely near it on the other side), be reduced to a single force through any point of the section and a single couple. Then—

I. The twist and curvature of the wire in the neighbourhood of this section are independent of the force, and depend solely on the couple.

II. The curvatures and rates of twist producible by any several couples separately, constitute, if geometrically compounded, the curvature and rate of twist which are actually produced by a mutual action equal to the resultant of those couples.

592. It may be added, although not necessary for our present purpose, that there is one determinate point in the cross section such that if it be chosen as the point to which the forces are transferred, a higher order of approximation is obtained for the fulfilment of these laws than if any other point of the section be taken. That point, which in the case of a wire of substance uniform through its cross section is the centre of inertia of the area of the section, we shall generally call the elastic centre, or the centre of elasticity, of the section. It has also the following important property:—The line of elastic centres, or, as we shall call it, the elastic central line, remains sensibly unchanged in length to whatever stress within our conditional limits (§ 588) the wire be subjected. The elongation or contraction produced by the neglected resultant force, if this is in such a direction as to produce any, will cause the line of *rigorously no elongation* to deviate only infinitesimally from the elastic central line, in any part of the wire finitely curved. It will, however, clearly cause there to be no line of *rigorously unchanged length*, in any straight part of the wire: but as the whole elongation would be infinitesimal in comparision with the effective actions with which we are concerned, this case constitutes no exception to the preceding statement.

593. Considering now a wire of uniform constitution and figure throughout, and naturally straight; let any two planes of reference perpendicular to one another through its elastic central line when straight, cut the normal section through P in the lines PK and PL. These two lines (supposed to belong to the substance, and move with it) will remain infinitely nearly at right angles to one another, and to the tangent, PT, to the central line, however the wire may be bent or twisted within the conditional limits. Let κ and λ be the component curvatures (§ 590) in the two planes perpendicular to PK and PL through PT, and let τ be the twist (§ 120) of the wire at P. We have just seen (§ 590) that if P be moved at a unit rate along the curve, a rigid body with three rectangular axes of reference $\mathfrak{OK}$, $\mathfrak{OL}$, $\mathfrak{OT}$ kept always parallel to PK, PL, PT, will have angular velocities κ, λ, τ round those axes respectively. Hence if the point P and the lines PT, PK, PL be at rest while the wire is bent and twisted from its unstrained to its actual condition, the lines of reference $P'K'$, $P'L'$, $P'T'$ through any point P' infinitely near P, will experience a rotation compounded of $\kappa . PP'$ round $P'K'$, $\lambda . PP'$ round $P'L'$, and $\tau . PP'$ round $P'T'$.

Warping of normal section by torsion and flexure, infinitesimal.

Rotations corresponding to flexure and torsion.

594. Considering now the elastic forces called into action, we see that if these constitute a conservative system, the work required to bend and twist any part of the wire from its unstrained to its actual condition, depends solely on its figure in these two conditions. Hence if $w . PP'$ denote the amount of this work, for the infinitely small length PP' of the rod, w must be a function of κ, λ, τ; and therefore if K, L, T denote the components of the couple-resultant of all the forces which must act on the section through P' to hold the part PP' in its strained state, it follows, from §§ 240, 272, 274, that

Potential energy of elastic force in bent and twisted wire.

$$K\delta\kappa = \delta_\kappa w,\ L\delta\lambda = \delta_\lambda w,\ T\delta\tau = \delta_\tau w \ \ldots\ldots\ldots\ldots(1),$$

where $\delta_\kappa w$, $\delta_\lambda w$, $\delta_\tau w$ denote the augmentations of w due respectively to infinitely small augmentations $\delta\kappa$, $\delta\lambda$, $\delta\tau$, of κ, λ, τ.

595. Now however much the shape of any finite length of the wire may be changed, the condition of § 588 requires

Potential energy of elastic force in bent and twisted wire.

clearly that the changes of shape in each infinitely small part, that is to say, the strain (§ 154) of the substance, shall be everywhere very small (infinitely small in order that the theory may be rigorously applicable). Hence the principle of superposition [§ 591, II.] shows that if κ, λ, τ be each increased or diminished in one ratio, K, L, T will be each increased or diminished in the same ratio: and consequently w in the duplicate ratio, since the angle through which each couple acts is altered in the same ratio as the amount of the couple; or, in algebraic language, w is a homogeneous quadratic function of κ, λ, τ.

Thus if A, B, C, a, b, c denote six constants, we have

$$w = \tfrac{1}{2}(A\kappa^2 + B\lambda^2 + C\tau^2 + 2a\lambda\tau + 2b\tau\kappa + 2c\kappa\lambda) \quad \ldots\ldots\ldots(2).$$

Hence, by § 594 (1),

Components of restituent couple.

$$\left.\begin{aligned} K &= A\kappa + c\lambda + b\tau \\ L &= c\kappa + B\lambda + a\tau \\ T &= b\kappa + a\lambda + C\tau \end{aligned}\right\} \quad \ldots\ldots\ldots\ldots\ldots\ldots(3).$$

By the known reduction of the homogeneous quadratic function, these expressions may of course be reduced to the following simple forms:—

$$\left.\begin{aligned} &w = \tfrac{1}{2}(A_1\vartheta_1^2 + A_2\vartheta_2^2 + A_3\vartheta_3^2) \\ &L_1 = A_1\vartheta_1, \quad L_2 = A_2\vartheta_2, \quad L_3 = A_3\vartheta_3 \end{aligned}\right\} \quad \ldots\ldots\ldots(4),$$

where ϑ_1, ϑ_2, ϑ_3 are linear functions of κ, λ, τ. And if these functions are restricted to being the expressions for the components round three rectangular axes, of the rotations κ, λ, τ viewed as angular velocities round the axes PK, PL, PT, the positions of the new axes, PQ_1, PQ_2, PQ_3, and the values of A_1, A_2, A_3 are determinate; the latter being the roots of the determinant cubic [§ 181 (11)] founded on (A, B, C, a, b, c). Hence we conclude that

Three principal or normal axes of torsion and flexure.

596. There are in general three determinate rectangular directions, PQ_1, PQ_2, PQ_3, through any point P of the middle line of a wire, such that if opposite couples be applied to any two parts of the wire in planes perpendicular to any one of them, every intermediate part will experience rotation in a plane parallel to those of the balanced couples. The moments

Three principal

of the couples required to produce unit rate of rotation round these three axes are called the *principal torsion-flexure* rigidities of the wire. They are the elements denoted by A_1, A_2, A_3 in the preceding analysis.

torsion-flexure rigidities.

597. If the rigid body imagined in § 593 have moments of inertia equal to A_1, A_2, A_3 round three principal axes through ⓟ kept always parallel to the principal torsion-flexure axes through P, while P moves at unit rate along the wire, its moment of momentum round any axis (§§ 281, 236) will be equal to the moment of the component torsion-flexure couple round the parallel axis through P.

598. The form assumed by the wire when balanced under the influence of couples round one of the three principal axes is of course a uniform helix having a line parallel to it for axis, and lying on a cylinder whose radius is determined by the condition that the whole rotation of one end of the wire from its unstrained position, the other end being held fixed, is equal to the amount due to the couple applied.

Three principal or normal spirals.

Let l be the length of the wire from one end, E, held fixed, to the other end, E', where a couple, L, is applied in a plane perpendicular to the principal axis PQ_1 through any point of the wire. The rotation being [§ 595 (4)] at the rate $\frac{L}{A_1}$, per unit of length, amounts on the whole to $l\frac{L}{A_1}$. This therefore is the angular space occupied by the helix on the cylinder on which it lies. Hence if r denote the radius of this cylinder, and i_1 the inclination of the helix to its axis (being the inclination of PQ_1 to the length of the wire), we have

$$r\frac{Ll}{A_1} = l \sin i_1 ;$$

whence $$r = \frac{A_1 \sin i_1}{L} \quad \text{.........................(5).}$$

Case in which elastic central line is a normal axis of torsion.

599. In the most important practical cases, as we shall see later, those namely in which the substance is either "isotropic," as is the case sensibly with common metallic wires, or, as in rods or beams of fibrous or crystalline structure, with an axis of elastic symmetry along the length of the piece, one of the three normal axes of torsion and flexure coincides with the length of the wire, and the two others are perpendicular to it; the first being an axis of pure torsion, and the two others axes of pure flexure. Thus opposing couples round the axis of the wire twist it simply without bending it; and opposing couples in either of the two principal planes of flexure, bend it into a circle. The unbent straight line of the wire, and the circular arcs into which it is bent by couples in the two principal planes of flexure, are what the three principal spirals of the general problem become in this case.

A simple proof that the twist must be uniform (§ 123) is found by supposing the whole wire to turn round its curved axis; and remarking that the work done by a couple at one end must be equal to that undone at the other.

Case of equal flexibility in all directions.

600. In the more particular case in which two principal rigidities against flexure are equal, every plane through the length of the wire is a principal plane of flexure, and the rigidity against flexure is equal in all. This is clearly the case with a common round wire, or rod: or with one of square section. It will be shown later to be the case for a rod of isotropic material and of any form of normal section which is "kinetically symmetrical," § 285, round all axes in its plane through its centre of inertia.

601. In this case, if one end of the rod or wire be held fixed, and a couple be applied in any plane to the other end, a uniform spiral (or helical) form will be produced round an axis perpendicular to the plane of the couple. The lines of the substance parallel to the axis of the spiral are not, however, parallel to their original positions, as (§ 598) in each of the three principal spirals of the general problem: and lines traced along the surface of the wire parallel to its length when straight, become as it were secondary spirals, circling round the main spiral formed by the central line of the deformed wire; instead of being all spirals of equal step, as in each one of the principal spirals of the general problem. Lastly, in the present case, if we suppose the normal section of the wire to be circular, and trace uniform spirals along its surface when deformed in the manner supposed (two of which, for instance, are the lines along which it is touched by the inscribed and the circumscribed cylinder), these lines do not become straight, but become spirals laid on as it were round the wire, when it is allowed to take its natural straight and untwisted condition.

Case of equal flexibility in all directions.

Let, in § 595, PQ_1 coincide with the central line of the wire, and let $A_1 = A$, and $A_2 = A_3 = B$; so that A measures the rigidity of torsion and B that of flexure. One end of the wire being held fixed, let a couple G be applied to the other end, round an axis inclined at an angle θ to the length. The rates of twist and of flexure each per unit of length, according to (4) of § 595, will be

$$\frac{G\cos\theta}{A}, \text{ and } \frac{G\sin\theta}{B},$$

respectively. The latter being (§ 9) the same thing as the curvature, and the inclination of the spiral to its axis being θ, it follows (§ 126, or § 590, footnote) that $\frac{B\sin\theta}{G}$ is the radius of curvature of its projection on a plane perpendicular to this line, that is to say, the radius of the cylinder on which the spiral lies.

602. A wire of equal flexibility in all directions may clearly be held in any specified spiral form, and twisted to any stated degree, by a determinate force and couple applied at one end, the other end being held fixed. The direction of the force must be parallel to the axis of the spiral, and, with the couple, must constitute a system of which this line is (§ 559) the *central axis:* since otherwise there could not be the same system of balancing forces in every normal section of the spiral. All this may be seen clearly by supposing the wire to be first brought by any means to the specified condition of strain; then to have rigid planes rigidly attached to its two ends perpendicular to its axis, and these planes to be rigidly

Wire strained to any given spiral and twist.

Wire strained to any given spiral and twist.

connected by a bar lying in this line. The spiral wire now left to itself cannot but be in equilibrium: although if it be too long (according to its form and degree of twist) the equilibrium may be unstable. The force along the central axis, and the couple, are to be determined by the condition that, when the force is transferred after Poinsot's manner to the elastic centre of any normal section, they give two couples together equivalent to the elastic couples of flexure and torsion.

Let α be the inclination of the spiral to the plane perpendicular to its axis; r the radius of the cylinder on which it lies; τ the rate of twist given to the wire in its spiral form. The curvature is (§ 126) equal to $\frac{\cos^2\alpha}{r}$; and its plane, at any point of the spiral, being the plane of the tangent to the spiral and the diameter of the cylinder through that point, is inclined at the angle α to the plane perpendicular to the axis. Hence the components in this plane, and in the plane through the axis of the cylinder of the flexural couple, are respectively

$$\frac{B\cos^2\alpha}{r}\cos\alpha, \text{ and } \frac{B\cos^2\alpha}{r}\sin\alpha.$$

Also, the components of the torsional couple, in the same planes, are $\quad A\tau\sin\alpha$, and $-A\tau\cos\alpha$.

Hence, for equilibrium,

$$\left.\begin{aligned} G &= \frac{B\cos^2\alpha}{r}\cos\alpha + A\tau\sin\alpha \\ -Rr &= \frac{B\cos^2\alpha}{r}\sin\alpha - A\tau\cos\alpha \end{aligned}\right\}\ldots\ldots\ldots\ldots(6),$$

which give explicitly the values, G and R, of the couple and force required, the latter being reckoned as positive when its direction is such as to pull *out* the spiral, or when the ends of the rigid bar supposed above are pressed *inwards* by the plates attached to the ends of the spiral.

If we make $R = 0$, we fall back on the case considered previously (§ 601). If, on the other hand, we make $G = 0$, we have

$$\tau = -\frac{1}{r}\frac{B}{A}\frac{\cos^3\alpha}{\sin\alpha},$$

and
$$R = -\frac{B}{r^2}\frac{\cos^2\alpha}{\sin\alpha} = \frac{A\tau}{r\cos\alpha},$$

from which we conclude that

603. A wire of equal flexibility in all directions may be held in any stated spiral form by a simple force along its axis between rigid pieces rigidly attached to its two ends, provided that, along with its spiral form, a certain degree of twist be given to it. The force is determined by the condition that its moment round the perpendicular through any point of the spiral to its osculating plane at that point, must be equal and opposite to the elastic unbending couple. The degree of twist is that due (by the simple equation of torsion) to the moment of the force thus determined, round the tangent at any point of the spiral. The direction of the force being, according to the preceding condition, such as to press together the ends of the spiral, the direction of the twist in the wire is opposite to that of the tortuosity (§ 9) of its central curve.

Twist determined for reducing the action to a single force.

604. The principles and formulæ (§§ 598, 603) with which we have just been occupied are immediately applicable to the theory of spiral springs; and we shall therefore make a short digression on this curious and important practical subject before completing our investigation of elastic curves.

Spiral springs.

A common spiral spring consists of a uniform wire shaped permanently to have, when unstrained, the form of a regular helix, with the principal axes of flexure and torsion everywhere similarly situated relatively to the curve. When used in the proper manner, it is acted on, through arms or plates rigidly attached to its ends, by forces such that its form as altered by them is still a regular helix. This condition is obviously fulfilled if (one terminal being held fixed) an infinitely small force and infinitely small couple be applied to the other terminal along the axis and in a plane perpendicular to it, and if the force and couple be increased to any degree, and always kept along and in the plane perpendicular to the axis of the altered spiral. It would, however, introduce useless complication to work out the details of the problem except for the case (§ 599) in which one of the principal axes coincides with the tangent to the central line, and is therefore an axis of pure torsion; as spiral springs in practice always belong to this case. On the other hand, a very interesting complication occurs if we suppose (a thing easily

Spiral springs. realized in practice, though to be avoided if merely a good spring is desired) the normal section of the wire to be of such a figure, and so situated relatively to the spiral, that the planes of greatest and least flexural rigidity are oblique to the tangent plane of the cylinder. Such a spring when acted on in the regular manner at its ends must experience a certain degree of turning through its whole length round its elastic central curve in order that the flexural couple developed may be, as we shall immediately see it must be, precisely in the osculating plane of the altered spiral. But all that is interesting in this very curious effect will be illustrated later (§ 624) in full detail in the case of an open circular arc altered by a couple in its own plane, into a circular arc of greater or less radius; and for brevity and simplicity we shall confine the detailed investigation of spiral springs on which we now enter, to the cases in which either the wire is of equal flexural rigidity in all directions, or the two principal planes of (greatest and least or least and greatest) flexural rigidity coincide respectively with the tangent plane to the cylinder, and the normal plane touching the central curve of the wire, at any point.

605. The axial force, on the moveable terminal of the spring, transferred according to Poinsot's method (§ 555) to any point in the elastic central curve, gives a couple in the plane through that point and the axis of the spiral. The resultant of this and the couple which we suppose applied to the terminal in the plane perpendicular to the axis of the spiral is the effective bending and twisting couple: and as it is in a plane perpendicular to the tangent plane to the cylinder, the component of it to which bending is due must be also perpendicular to this plane, and therefore is in the osculating plane of the spiral. This component couple therefore simply maintains a curvature different from the natural curvature of the wire, and the other, that is, the couple in the plane normal to the central curve, pure torsion. The equations of equilibrium merely express this in mathematical language.

Resolving as before (§ 602) the flexural and the torsional couples each into components in the planes through the axis of

Spiral springs.

the spiral, and perpendicular to it, we have

$$\left.\begin{aligned} G &= B\left(\frac{\cos^2\alpha}{r} - \frac{\cos^2\alpha_0}{r_0}\right)\cos\alpha' + A\tau\sin\alpha', \\ -Rr &= B\left(\frac{\cos^2\alpha}{r} - \frac{\cos^2\alpha_0}{r_0}\right)\sin\alpha' - A\tau\cos\alpha', \\ \text{and, by § 126,}\quad \tau &= \frac{\cos\alpha\sin\alpha}{r} - \frac{\cos\alpha_0\sin\alpha_0}{r_0}, \end{aligned}\right\}\cdots(7),$$

where A denotes the torsional rigidity of the wire, and B its flexural rigidity in the osculating plane of the spiral; α_0 the inclination, and r_0 the radius of the cylinder, of the spiral when unstrained; α and r the same parameters of the spiral when under the influence of the axial force R and couple G; and τ the degree of twist in the change from the unstrained to the strained condition.

These equations give explicitly the force and couple required to produce any stated change in the spiral; or if the force and couple are given they determine α', r' the parameters of the altered curve.

As it is chiefly the external action of the spring that we are concerned with in practical applications, let the parameters α, r of the spiral be eliminated by the following assumptions :—

$$\left.\begin{aligned} x &= l\sin\alpha, \quad \phi = \frac{l\cos\alpha}{r} \\ x_0 &= l\sin\alpha_0, \quad \phi_0 = \frac{l\cos\alpha_0}{r_0} \end{aligned}\right\}\;\ldots\ldots\ldots\ldots\ldots (8),$$

where l denotes the length of the wire, ϕ the angle between planes through the two ends of the spiral, and its axis, and x the distance between planes through the ends and perpendicular to the axis in the strained condition; and, similarly, ϕ_0, x_0 for the unstrained condition; so that we may regard (ϕ, x) and (ϕ_0, x_0) as the co-ordinates of the movable terminal relatively to the fixed in the two conditions of the spring. Thus the preceding equations become

$$\left.\begin{aligned} L &= \frac{B}{l^3}\{\sqrt{(l^2-x^2)}\,\phi - \sqrt{(l^2-x_0^2)}\,\phi_0\}\sqrt{(l^2-x^2)} + \frac{A}{l^3}(x\phi - x_0\phi_0)\,x \\ R &= -\frac{B}{l^3}\{\sqrt{(l^2-x^2)}\,\phi - \sqrt{(l^2-x_0^2)}\,\phi_0\}\frac{x\phi}{\sqrt{(l^2-x^2)}} + \frac{A}{l^3}(x\phi - x_0\phi_0)\,\phi \end{aligned}\right\}(9).$$

Spiral springs.

Here we see that $Ld\phi + Rdx$ is the differential of a function of the two independent variables, x, ϕ. Thus if we denote this function by E, we have

$$\left.\begin{array}{c} E = \frac{1}{2}\frac{B}{l^2}\{\sqrt{(l^2-x^2)}\,\phi - \sqrt{(l^2-x_0^2)}\,\phi_0\}^2 + \frac{1}{2}\frac{A}{l^2}(x\phi - x_0\phi_0)^2 \\ L = \frac{dE}{d\phi}, \quad R = \frac{dE}{dx} \end{array}\right\} \quad (10),$$

a conclusion which might have been inferred at once from the general principle of energy, thus :—

606. The potential energy of the strained spring is easily seen from § 595 (4), above, to be

$$\tfrac{1}{2}[B(\varpi - \varpi_0)^2 + A\tau^2]l,$$

if A denote the torsional rigidity, B the flexural rigidity in the plane of curvature, ϖ and ϖ_0 the strained and unstrained curvatures, and τ the torsion of the wire in the strained condition, the torsion being reckoned as zero in the unstrained condition. The axial force, and the couple, required to hold the spring to any given length reckoned along the axis of the spiral, and to any given angle between planes through its ends and the axes, are of course (§ 272) equal to the rates of variation of the potential energy, per unit of variation of these co-ordinates respectively. It must be carefully remarked, however, that, if the terminal rigidly attached to one end of the spring be held fast so as to fix the tangent at this end, and the motion of the other terminal be so regulated as to keep the figure of the intermediate spring always truly spiral, this motion will be somewhat complicated; as the radius of the cylinder, the inclination of the axis of the spiral to the fixed direction of the tangent at the fixed end, and the position of the point in the axis in which it is cut by the plane perpendicular to it through the fixed end of the spring, all vary as the spring changes in figure. The *effective components* of any infinitely small motion of the moveable terminal are its component translation along, and rotation round, the instantaneous position of the axis of the spiral (two degrees of freedom), along with which it will generally have an infinitely small translation in some direction and rotation round some line, each perpendicular to this axis, to be determined from the two degrees of arbitrary motion, by the condition that the curve remains a true spiral. Spiral springs.

607. In the practical use of spiral springs, this condition is not rigorously fulfilled: but, instead, either of two plans is generally followed:—(1) Force, without any couple, is applied pulling out or pressing together two definite points of the two terminals, each as nearly as may be in the axis of the unstrained spiral; or (2) One terminal being held fixed, the other is allowed to slide, without any turning, in a fixed direction, being as nearly as may be the direction of the axis of the spiral when unstrained. The preceding investigation is applicable to the infinitely small displacement in either case: the couple being put equal to zero for case (1), and the instantaneous rotatory motion round the axis of the spiral equal to zero for case (2).

For infinitely small displacements let $\phi = \phi_0 + \delta\phi$, and $x = x_0 + \delta x$, in (10), so that now

$$L = \frac{dE}{d\delta\phi}, \quad R = \frac{dE}{d\delta x}.$$

Then, retaining only terms of the lowest degree relative to δx and $\delta\phi$ in each formula, and writing x and ϕ instead of x_0 and ϕ_0, we have

$$\left.\begin{aligned} E &= \frac{1}{2l^3}\left\{\left(B\frac{x^2}{l^2-x^2}+A\right)\phi^2\delta x^2+2(A-B)x\phi\delta x\delta\phi+\left[B(l^2-x^2)+Ax^2\right]\delta\phi^2\right\} \\ R &= \frac{1}{l^3}\left\{\left(B\frac{x^2}{l^2-x^2}+A\right)\phi^2\delta x+(A-B)\,x\phi\delta\phi\right\} \\ L &= \frac{1}{l^3}\{(A-B)\,x\phi\delta x+\left[B\,(l^2-x^2)+Ax^2\right]\delta\phi\} \end{aligned}\right\} \quad (11).$$

Example 1.—For a spiral of 45° inclination we have

$$x^2 = \tfrac{1}{2}l^2 \text{ and } \phi^2 = \tfrac{1}{2}\frac{l^2}{r^2}:$$

and the formulæ become

$$\left.\begin{aligned} R &= \tfrac{1}{2}\frac{1}{lr^2}\left[(A+B)\,\delta x+(A-B)\,r\delta\phi\right] \\ L &= \tfrac{1}{2}\frac{1}{lr}\left[(A-B)\,\delta x+(A+B)\,r\delta\phi\right] \end{aligned}\right\} \quad \text{.........} (12).$$

Spiral springs.

A careful study of this case, illustrated if necessary by a model easily made out of ordinary iron or steel wire, will be found very instructive.

Spiral spring of infinitely small inclination:

Example 2.—Let $\frac{x}{l}$ be very small. Neglecting, therefore, its square, we have $\phi = \frac{l}{r}$, and $L = \frac{B}{l}\delta\phi = B\delta\frac{1}{r}$; and $R = \frac{A}{lr^2}\delta x$. The first of these is simply the equation of direct flexure (§ 595). The interpretation of the second is as follows :—

608. In a spiral spring of infinitely small inclination to the plane perpendicular to its axis, the displacement produced in the moveable terminal by a force applied to it in the axis of the spiral is a simple rectilineal translation in the direction of the axis, and is equal to the length of the circular arc through which an equal force carries one end of a rigid arm or crank equal in length to the radius of the cylinder, attached perpendicularly to one end of the wire of the spring supposed straightened and held with the other end absolutely fixed, and the end which bears the crank free to turn in a collar. This statement is due to J. Thomson*, who showed that in pulling out a spiral spring of infinitely small inclination the action exercised and the elastic quality used are the same as in a torsion-balance with the same wire straightened (§ 433). This theory is, as he proved experimentally, sufficiently approximate for most practical applications; spiral springs, as commonly made and used, being of very small inclination. There is no difficulty in finding the requisite correction, for the actual inclination in any case, from the preceding formulæ. The fundamental principle that spiral springs act chiefly by torsion seems to have been first discovered by Binet in 1814†.

virtually a torsion-balance.

Elastic curve transmitting force and couple.

609. In continuation of §§ 590, 593, 597, we now return to the case of a uniform wire straight and untwisted (that is, cylindrical or prismatic) when free from stress. Let us suppose one end to be held fixed in a given direction, and no force from without to influence the wire except that transmitted to it by a rigid frame attached to its other end and acted on by a

* *Camb. and Dub. Math. Jour.* 1848.

† St Venant, *Comptes Rendus*. Sept. 1864.

force, R, in a given line, AB, and a couple, G, in a plane perpendicular to this line. The form and twist it will have when in equilibrium are determined by the condition that the torsion and flexure at any point, P, of its length are those due to the couple G compounded with the couple obtained by bringing R to P. It follows that the rigid body of § 597 will move exactly as there specified if it be set in motion with the proper angular velocity, and, 𝕺 being held fixed, a force equal and parallel to R be applied at a point 𝕽, fixed relatively to the body at unit distance from 𝕺, in the line 𝕺𝕿.

Kirchhoff's kinetic comparison.

This beautiful theorem was discovered by Kirchhoff; to whom also the first thoroughly general investigation of the equations of equilibrium and motion of an elastic wire is due*.

To prove the theorem, it is only necessary to remark that the rate of change of the moment of R round any line through P, kept parallel to itself as P moves along the curve, in the elastic problem, is equal simply to the moment round the parallel line through 𝕺, of R at 𝕽 in the kinetic analogue. It may be added that G of the elastic problem corresponds to the constant moment of momentum round the line through 𝕺 parallel to the constant direction of R in the kinetic analogue.

610. The comparison thus established between the static problem of the bending and twisting of a wire, and the kinetic problem of the rotation of a rigid body, affords highly interesting illustrations, and, as it were, graphic representations, of the circumstances of either by aid of the other; the usefulness of which in promoting a thorough mental appropriation of both must be felt by every student who values rather the physical subject than the mechanical process of working through mathematical expressions, to which so many minds able for better things in science have unhappily been devoted of late years.

When particularly occupied with the kinetic problem in chap. IX., we shall have occasion to examine the rotations corresponding to the spirals of §§ 601—603, and to point out also the general character of the elastic curves corresponding to some of the less simple cases of rotatory motion.

* *Crelle's Journal*, 1859, Ueber das Gleichgewicht und die Bewegung eines unendlich dünnen elastischen Stabes.

Common pendulum and plane elastic curve.

611. For the present we confine ourselves to one example, which, so far as the comparison between the static and kinetic problems is concerned, is the simplest of all—the *Elastic Curve* of James Bernoulli, and the common pendulum. A uniform straight wire, either equally flexible in all planes through its length, or having its directions of maximum and minimum flexural rigidity in two planes through its whole length, is acted on by a force and couple in one of these planes, applied either directly to one end, or by means of an arm rigidly attached to it, the other end being held fast. The force and couple may, of course (§ 558), be reduced to a single force, the extreme case of a couple being mathematically included as an infinitely small force at an infinitely great distance. To avoid any restriction of the problem, we must suppose this force applied to an arm rigidly attached to the wire, although in any case in which the line of the force cuts the wire, the force may be applied directly at the point of intersection, without altering the circumstances of the wire between this point and the fixed end. The wire will, in these circumstances, be bent into a curve lying throughout in the plane through its fixed end and the line of the force, and (§ 599) its curvatures at different points will, as was first shown by James Bernoulli, be simply as their distances from this line. The curve fulfilling this condition has clearly just two independent parameters, of which one is conveniently regarded as the mean proportional, a, between the radius of curvature at any point and its distance from the line of force, and the other, the maximum distance, b, of the wire from the line of force. By choosing any value for each of these parameters it is easy to trace the corresponding curve with a very high approximation to accuracy, by commencing with a small circular arc touching at one extremity a straight line at the given maximum distance from the line of force, and continuing by small circular arcs, with the proper increasing radii, according to the diminishing distances of their middle points from the line of force. The annexed diagrams are, however, not so drawn; but are simply traced from the forms actually assumed by a flat steel spring, of small enough breadth not to be much disturbed by tortuosity in the cases in which different

Graphic construction of elastic curve transmitting force in one plane.

parts of it cross one another. The mode of application of the force is sufficiently explained by the indications in the diagram. Equation of the plane elastic curve.

Let the line of force be the axis of x, and let ρ be the radius of curvature at any point (x, y) of the curve. The dynamical condition stated above becomes

$$\rho y = \frac{B}{T} = a^2 \text{..........................} (1),$$

where B denotes the flexural rigidity, T the tension of the cord, and a a linear parameter of the curve depending on these elements. Hence, by the ordinary formula for ρ^{-1},

$$y = \frac{a^2 \dfrac{d^2y}{dx^2}}{\left(1 + \dfrac{dy^2}{dx^2}\right)^{\frac{3}{2}}} \text{..........................} (2).$$

Multiplying by $2dy$ and integrating, we have

$$y^2 = C - \frac{2a^2}{\left(1 + \dfrac{dy^2}{dx^2}\right)^{\frac{1}{2}}} \text{..........................} (3);$$

and finally,

$$x = \int \frac{(y^2 - C)\, dy}{(4a^4 - C^2 + 2Cy^2 - y^4)^{\frac{1}{2}}} \text{....................} (4),$$

which is the equation of the curve expressed in terms of an elliptic integral.

If, in the first integral, (3), we put $\dfrac{dy}{dx} = 0$, we find

$$y = \pm (C \pm 2a^2)^{\frac{1}{2}} \text{.............................} (5),$$

the upper sign within the bracket giving points of maximum, and the lower, points, if any real, of minimum distance from the axis. Hence there are points of equal maximum distance from the line of force on its two sides, but no real minima when $C < 2a^2$; which therefore comprehends the cases of diagrams 1...5. But there are real minima as well as maxima when $C > 2a^2$, which is therefore the case of diagram 7. In this case it may be remarked that the analytical equations comprehend two equal and similar de-

Equation of the plane elastic curve.

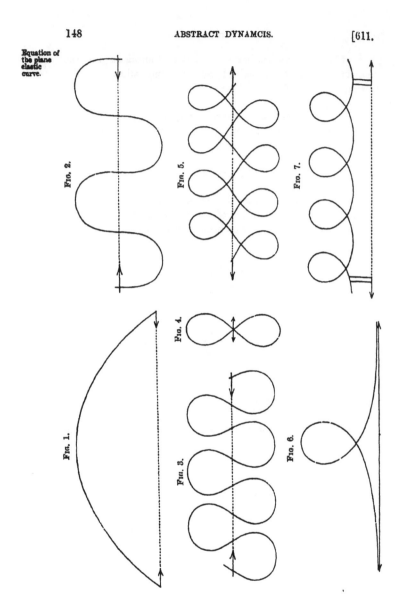

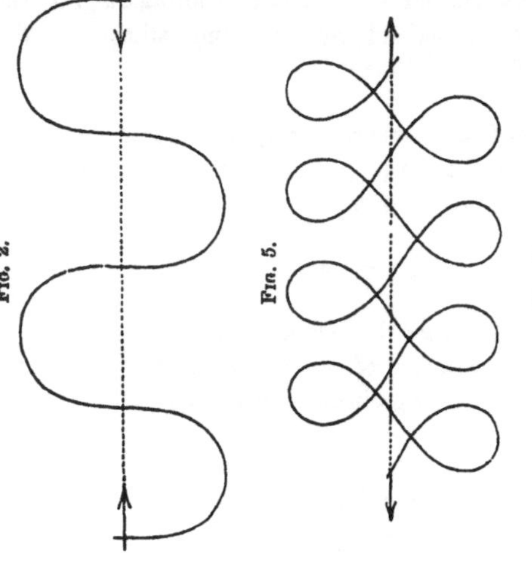

Fig. 2.

Fig. 5.

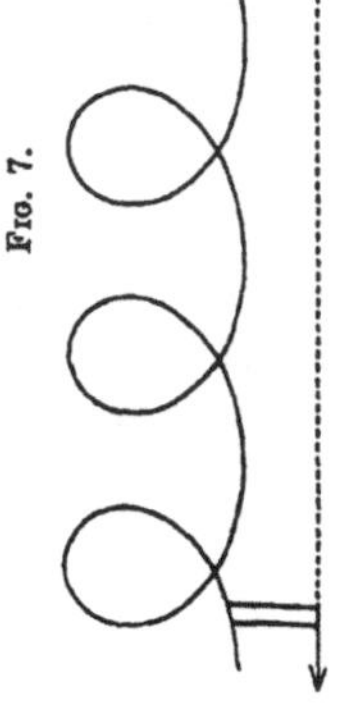

Fig. 7.

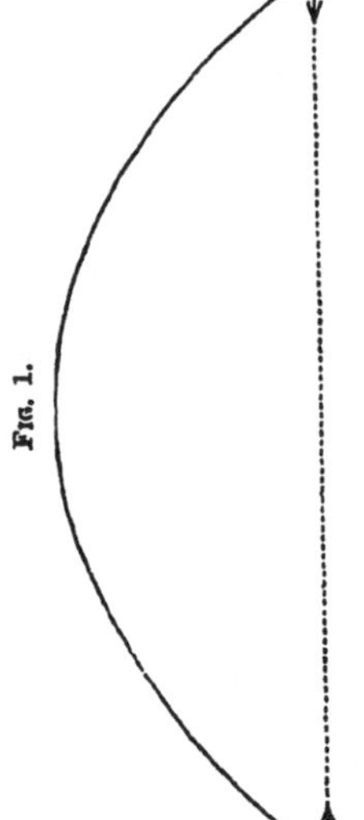

Fig. 1.

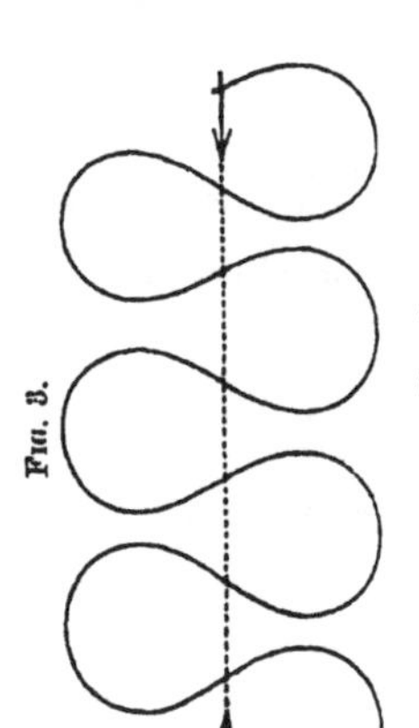

Fig. 4.

Fig. 3.

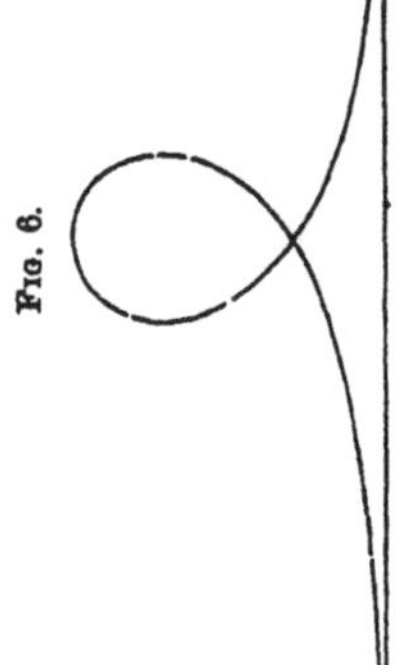

Fig. 6.

tached curves symmetrically situated on the two sides of the line of force; of which one only is shown in the diagram.

Equation of the plane elastic curve.

The intermediate case, $C = 2a^2$, is that of diagram 6. For it the final integral degrades into a logarithmic form, as follows:—

$$x = \int \frac{y\,dy}{(4a^2 - y^2)^{\frac{1}{2}}} - \int \frac{2a^2 dy}{y(4a^2 - y^2)^{\frac{1}{2}}};$$

or, with the integrations effected, and the constant assigned to make the axis of y be that of symmetry,

$$x = -(4a^2 - y^2)^{\frac{1}{2}} + a \log \frac{2a + (4a^2 - y^2)^{\frac{1}{2}}}{y} \quad \text{............(6).}$$

This equation, when the radical is taken with the sign indicated, represents the branch proceeding from the vertex, first to the negative side of the axis of y, crossing it at the double point, and going to infinity towards the positive axis of x as an asymptote. The other branch is represented by the same equation with the sign of the radical reversed in each place.

It may be remarked that in (3) the sign of $\left(1 + \frac{dy^2}{dx^2}\right)^{\frac{1}{2}}$ can only change, for a point moving continuously along the curve, when $\frac{dy}{dx}$ becomes infinite. The interpretation is facilitated by putting

$$\frac{dy}{dx} = \tan\theta, \text{ or } \left(1 + \frac{dy^2}{dx^2}\right)^{\frac{1}{2}} = -\cos\theta,$$

which reduces (3) to

$$y^2 = 2a^2 \cos\theta + C \quad \text{.......................(7).}$$

Here, when $C > 2a^2$ (the case in which, as we have seen above, there are minimum as well as maximum values of y on one side of the line of force), there is no limit to the value of θ. It increases, of course, continuously for a point moving continuously along the curve; the augmentation being 2π for one complete period (diagram 7).

When $C < 2a^2$, θ has equal positive and negative values at the points in which the curve cuts the line of force. These values being given by the equation

$$\cos\theta = -\frac{C}{2a^2} \quad \text{...............................(8),}$$

Equation of the plane elastic curve.

are obtuse when C is positive (diagram 3), and acute when C is negative (diagram 1). The extreme negative value of C is of course $-2a^2$.

If we take $$C = -2a^2 + b^2,$$

$\pm b$ will be the maximum positive or negative value of y, as we see by (7); and if we suppose b to be small in comparison with a, we have the case of a uniform spring bent, as a bow, but slightly, by a string stretched between its ends.

Bow slightly bent.

612. An important particular case is that of figure 1, which corresponds to a bent bow having the same flexural rigidity throughout. If the amount of bending be small, the equation is easily integrated to any requisite degree of approximation. We will merely sketch the process of investigation.

Let e be the maximum distance from the axis, corresponding to $x=0$. Then $y=e$ gives $\frac{dy}{dx}=0$, and (3) becomes

$$e^2 - y^2 = 2a^2\left(1 - \frac{1}{\sqrt{1+\frac{dy^2}{dx^2}}}\right);$$

whence $$\frac{dy}{dx} = \frac{\sqrt{e^2-y^2}\sqrt{4a^2-e^2+y^2}}{2a^2-e^2+y^2} \quad\ldots\ldots\ldots\ldots\ldots\ldots(9).$$

For a first approximation, omit e^2-y^2 in comparison with a^2 where they occur in the same factors, and we have

$$\frac{dy}{dx} = \frac{\sqrt{e^2-y^2}}{a},$$

or, since $y=e$ when $x=0$,

$$y = e\cos\frac{x}{a} \quad\ldots\ldots\ldots\ldots\ldots\ldots\ldots\ldots(10),$$

the harmonic curve, or curve of sines, which is the simplest form assumed by a vibrating cord or pianoforte-wire.

For a closer approximation we may substitute for y, in those factors where it was omitted, the value given by (10); and so on. Thus we have

$$\frac{dy}{dx} = \frac{\sqrt{e^2-y^2}}{a}\left(1 + \frac{3e^2}{8a^2}\sin^2\frac{x}{a}\right), \text{ nearly,}$$

or $$\frac{dy}{\sqrt{e^2-y^2}} = \frac{dx}{a}\left(1+\frac{3e^2}{16a^2}-\frac{3e^2}{16a^2}\cos\frac{2x}{a}\right),$$ Bow slightly bent.

from which, by integration,

$$\cos^{-1}\frac{y}{e} = \frac{x}{a}\left(1+\frac{3e^2}{16a^2}\right) - \frac{3e^2}{32a^2}\sin\frac{2x}{a}$$

and $$y = e\cos\left\{\frac{x}{a}\left(1+\frac{3e^2}{16a^2}\right)\right\} + \frac{3e^3}{32a^2}\sin\frac{x}{a}\sin\frac{2x}{a}.$$

613. As we choose particularly the common pendulum for the corresponding kinetic problem, the force acting on the rigid body in the comparison must be that of gravity in the vertical through its centre of gravity. It is convenient, accordingly, not to take *unity* as the velocity of the point travelling along the bent wire, but the velocity gravity would generate in a body falling through a height equal to half the constant, a, of § 611: and this constant, a, will then be the length of the isochronous simple pendulum. Thus if an elastic curve be held with its line of force vertical, and if a point, P, be moved along it with a constant velocity equal to $\sqrt{ga}$, (a denoting the mean proportional between the radius of curvature at any point and its distance from the line of force,) the tangent at P will keep always parallel to a simple pendulum, of length a, placed at any instant parallel to it, and projected with the same angular velocity. Diagrams 1...5 correspond to *vibrations* of the pendulum. Diagram 6 corresponds to the case in which the pendulum would just reach its position of unstable equilibrium in an infinite time. Diagram 7 corresponds to cases in which the pendulum flies round continuously in one direction, with periodically increasing and diminishing velocity. The extreme case, of the circular elastic curve, corresponds to a pendulum flying round with infinite angular velocity, which of course experiences only infinitely small variation in the course of the revolution. A conclusion worthy of remark is, that the rectification of the elastic curve is the same analytical problem as finding the time occupied by a pendulum in describing any given angle. Plane elastic curve and common pendulum.

Wire of any shape disturbed by forces and couples applied through its length

614. Hitherto we have confined our investigation of the form and twist of a wire under stress to a portion of the whole wire not itself acted on by force from without, but merely engaged in transmitting force between two equilibrating systems applied to the wire beyond this portion; and we have, thus, not included the very important practical cases of a curve deformed by its own weight or centrifugal force, or fulfilling such conditions of equilibrium as we shall have to use afterwards in finding its equations of motion according to D'Alembert's principle. We therefore proceed now to a perfectly general investigation of the equilibrium of a curve, uniform or not uniform throughout its length; either straight, or bent and twisted in any way, when free from stress; and not restricted by any condition as to the positions of the three principal flexure-torsion axes (§ 596); under the influence of any distribution whatever of force and couple through its whole length.

Let α, β, γ be the components of the mutual force, and ξ, η, ζ those of the mutual couple, acting between the matter on the two sides of the normal section through (x, y, z). Those for the normal section through $(x+\delta x,\ y+\delta y,\ z+\delta z)$ will be

$$\alpha+\frac{d\alpha}{ds}\delta s,\ \ \beta+\frac{d\beta}{ds}\delta s,\ \ \gamma+\frac{d\gamma}{ds}\delta s,$$

$$\xi+\frac{d\xi}{ds}\delta s,\ \ \eta+\frac{d\eta}{ds}\delta s,\ \ \zeta+\frac{d\zeta}{ds}\delta s.$$

Hence, if $X\delta s$, $Y\delta s$, $Z\delta s$, and $L\delta s$, $M\delta s$, $N\delta s$ be the components of the applied force, and applied couple, on the portion δs of the wire between those two normal sections, we have (§ 551) for the equilibrium of this part of the wire

$$-X=\frac{d\alpha}{ds},\ \ -Y=\frac{d\beta}{ds},\ \ -Z=\frac{d\gamma}{ds}\ldots\ldots\ldots\ldots\ldots(1),$$

and (neglecting, of course, infinitely small terms of the second order, as $\delta y\delta s$)

$$-L\delta s=\frac{d\xi}{ds}\delta s+\gamma\delta y-\beta\delta z,\ \text{etc.};$$

or

$$-L=\frac{d\xi}{ds}+\gamma\frac{dy}{ds}-\beta\frac{dz}{ds},\ \ -M=\frac{d\eta}{ds}+\alpha\frac{dz}{ds}-\gamma\frac{dx}{ds},\ \ -N=\frac{d\zeta}{ds}+\beta\frac{dx}{ds}-\alpha\frac{dy}{ds}\ldots(2).$$

We may eliminate α, β, γ from these six equations by means of the following convenient assumption—

$$\alpha\frac{dx}{ds}+\beta\frac{dy}{ds}+\gamma\frac{dz}{ds}=T \qquad (3),$$

Longitudinal tension.

T meaning the component of the force acting across the normal section, along the tangent to the middle line. From this, and the second and third of (2), we have

$$\alpha=T\frac{dx}{ds}-\left(M+\frac{d\eta}{ds}\right)\frac{dz}{ds}+\left(N+\frac{d\zeta}{ds}\right)\frac{dy}{ds}.$$

This, and the symmetrical expressions for β and γ, used in (1), give

$$\left.\begin{aligned}X&=-\frac{d}{ds}\left\{T\frac{dx}{ds}-\left(M+\frac{d\eta}{ds}\right)\frac{dz}{ds}+\left(N+\frac{d\zeta}{ds}\right)\frac{dy}{ds}\right\}\\Y&=-\frac{d}{ds}\left\{T\frac{dy}{ds}-\left(N+\frac{d\zeta}{ds}\right)\frac{dx}{ds}+\left(L+\frac{d\xi}{ds}\right)\frac{dz}{ds}\right\}\\Z&=-\frac{d}{ds}\left\{T\frac{dz}{ds}-\left(L+\frac{d\xi}{ds}\right)\frac{dy}{ds}+\left(M+\frac{d\eta}{ds}\right)\frac{dx}{ds}\right\}\end{aligned}\right\} \qquad (4).$$

We have besides, from (2),

$$0=\frac{dx}{ds}\left(L+\frac{d\xi}{ds}\right)+\frac{dy}{ds}\left(M+\frac{d\eta}{ds}\right)+\frac{dz}{ds}\left(N+\frac{d\zeta}{ds}\right) \qquad (5).$$

To complete the mathematical expression of the circumstances, it only remains to introduce the equations of torsion-flexure. For this purpose, let any two lines of reference for the substance of the wire, PK, PL, be chosen at right angles to one another in the normal section through P. Let κ_0, λ_0 be the components of the curvature (§ 589) in the planes perpendicular to these lines, and through the tangent, PT, when the wire is unstrained; and κ, λ what they become under the actual stress. Let τ_0 denote the rate of twist (§ 119) of either line of reference round the tangent from point to point along the wire in the unstrained condition, and τ in the strained, so that $\tau-\tau_0$ is the rate of twist produced at P by the actual stress. Thus [§ 595 (3)] we have

$$\left.\begin{aligned}\xi l+\eta m+\zeta n&=A(\kappa-\kappa_0)+c(\lambda-\lambda_0)+b(\tau-\tau_0)\\\xi l'+\eta m'+\zeta n'&=c(\kappa-\kappa_0)+B(\lambda-\lambda_0)+a(\tau-\tau_0)\\\xi\frac{dx}{ds}+\eta\frac{dy}{ds}+\zeta\frac{dz}{ds}&=b(\kappa-\kappa_0)+a(\lambda-\lambda_0)+C(\tau-\tau_0)\end{aligned}\right\} \qquad (6),$$

Equations of torsion-flexure.

Equations of torsion-flexure.

where (l, m, n), (l', m', n'), $\left(\frac{dx}{ds}, \frac{dy}{ds}, \frac{dz}{ds}\right)$ denote the directions of PK, PL, PT; so that

$$\left.\begin{array}{c} l\frac{dx}{ds}+m\frac{dy}{ds}+n\frac{dz}{ds}=0, \quad l'\frac{dx}{ds}+m'\frac{dy}{ds}+n'\frac{dz}{ds}=0 \\ ll'+mm'+nn'=0 \\ l^2+m^2+n^2=1, \qquad l'^2+m'^2+n'^2=1 \end{array}\right\} \ldots\ldots(7).$$

Now if lines $O_{,}K$, $O_{,}L$, $O_{,}T$, each of unit length, be drawn, as in § 593, always parallel to PK, PL, PT, and if P be carried at unit velocity along the curve, the component velocity of $_{,}L$ parallel to $O_{,}T$, or that of $_{,}T$ parallel to $O_{,}K$ with its sign changed, is (§ 593) equal to κ; and similar statements apply to λ and τ. Hence,

Torsion, and two components of curvature, of wire (or component angular velocities of rotating solid).

$$\left.\begin{array}{l} \kappa=-\left\{l'\frac{d}{ds}\left(\frac{dx}{ds}\right)+m'\frac{d}{ds}\left(\frac{dy}{ds}\right)+n'\frac{d}{ds}\left(\frac{dz}{ds}\right)\right\} \\ \lambda=+\left\{l\frac{d}{ds}\left(\frac{dx}{ds}\right)+m\frac{d}{ds}\left(\frac{dy}{ds}\right)+n\frac{d}{ds}\left(\frac{dz}{ds}\right)\right\} \\ \tau=+\left(l'\frac{dl}{ds}+m'\frac{dm}{ds}+n'\frac{dn}{ds}\right) \end{array}\right\} \ldots(8).$$

Equations (7) reduce (l, m, n), (l', m', n') to one variable element, being the co-ordinate by which the position of the substance of the wire, round the tangent at any point of the central curve, is specified: and (8) express κ, λ, τ in terms of this co-ordinate, and the three Cartesian co-ordinates x, y, z of P. The specification of the unstrained condition of the wire gives κ_0, λ_0, τ_0 as functions of s. Thus (6) gives ξ, η, ζ each in terms of s, and the four co-ordinates, and their differential coefficients relatively to s. Substituting these in (4) and (5) we have four differential equations which, with

$$\frac{dx^2}{ds^2}+\frac{dy^2}{ds^2}+\frac{dz^2}{ds^2}=1 \ldots\ldots\ldots\ldots\ldots\ldots(9),$$

constitute the five equations by which the five unknown functions (the four co-ordinates, and the tension, T) are to be determined in terms of s, or by means of which, with s and T eliminated, the two equations of the curve may be found, and the co-ordinate for the position of the normal section round the tangent determined in terms of x, y, z.

The terminal conditions for any specified circumstances are easily expressed in the proper mathematical terms, by aid of equations (2). Thus, for instance, if a given force and a given couple be directly applied to a free end, or if the problem be limited to a portion of the wire terminated in one direction at a point Q, and if, in virtue of actions on the wire beyond, we have a given force $(\alpha_0, \beta_0, \gamma_0)$ and a given couple (ξ_0, η_0, ζ_0) acting on the normal section through Q of the portion under consideration, and if s_0 is the length of the wire from the zero of reckoning for s up to the point Q, and L_0, M_0, N_0 the values of L, M, N at this point, the equations expressing the terminal conditions will be

Terminal conditions.

$$\left.\begin{aligned} \xi=\xi_0,\ -\frac{d\xi}{ds}&=L_0+\left(\gamma_0\frac{dy}{ds}-\beta_0\frac{dz}{ds}\right) \quad \text{when } s=s_0 \\ \eta=\eta_0,\ -\frac{d\eta}{ds}&=M_0+\left(\alpha_0\frac{dz}{ds}-\gamma_0\frac{dx}{ds}\right) \\ \zeta=\zeta_0,\ -\frac{d\zeta}{ds}&=N_0+\left(\beta_0\frac{dx}{ds}-\alpha_0\frac{dy}{ds}\right) \end{aligned}\right\}\ \ldots\ldots (10).$$

From these we see, by taking $L_0=0$, $M_0=0$, $N_0=0$, $\alpha_0=0$, $\beta_0=0$, $\gamma_0=0$, $\xi_0=0$, $\eta_0=0$, $\zeta_0=0$, that

615. For the simple and important case of a naturally straight wire, acted on by a distribution of force, but not of couple, through its length, the condition fulfilled at a perfectly free end, acted on by neither force nor couple, is that the curvature is zero at the end, and its rate of variation from zero, per unit of length from the end, is, at the end, zero. In other words, the curvatures at points infinitely near the end are as the squares of their distances from the end in general (or, as some higher power of these distances, in singular cases). The same statements hold for the *change* of curvature produced by the stress, if the unstrained wire is not straight, but the other circumstances the same as those just specified.

616. As a very simple example of the equilibrium of a wire subject to forces through its length, let us suppose the natural form to be straight, and the applied forces to be in lines, and the couples to have their axes all perpendicular to its length, and to be not great enough to produce more than an infinitely small deviation from the straight line. Further,

Straight beam infinitely little bent.

Straight beam infinitely little bent.

in order that these forces and couples may produce no twist, let the three flexure-torsion axes be perpendicular to and along the wire. But we shall not limit the problem further by supposing the section of the wire to be uniform, as we should thus exclude some of the most important practical applications, as to beams of balances, levers in machinery, beams in architecture and engineering. It is more instructive to investigate the equations of equilibrium directly for this case than to deduce them from the equations worked out above for the much more comprehensive general problem. The particular principle for the present case is simply that the rate of variation of the rate of variation, per unit of length along the wire, of the bending couple in any plane through the length, is equal, at any point, to the applied force per unit of length, with the simple rate of variation of the applied couple subtracted. This, together with the direct equations (§ 599) between the component bending couples, gives the required equations of equilibrium.

The diagram representing a section of the wire in the plane xy, let $OP=x$, $PP'=\delta x$. Let Y and N be the components

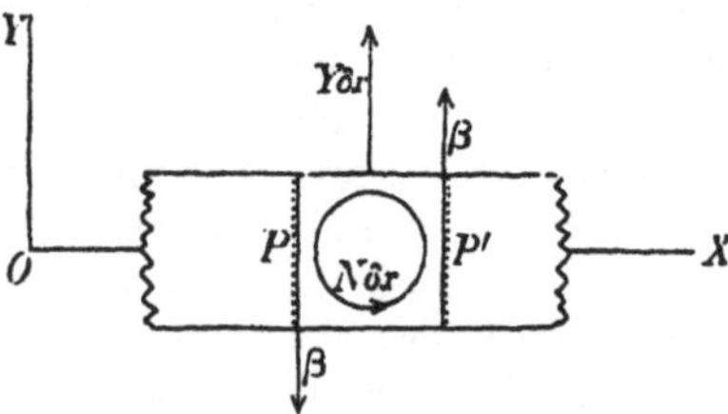

in the plane of the diagram, of the applied force and couple, each reckoned per unit of length of the wire; so that $Y\delta x$ and $N\delta x$ will be the amounts of force and couple in this plane, actually applied to the portions of the wire between P and P'.

Let, as before (§ 614), β and γ denote the components parallel to OY and OZ of the mutual force*, and ζ and η the components

* These forces, being each *in* the plane of section of the solid separating the portions of matter between which they act, are of the kind called *shearing forces*. See below, § 662.

in the plane XOY, XOZ, of the mutual couple, between the portions of matter on the two sides of the normal section through P; and β', γ' and ζ', η' the same for P'. The matter between these two sections is balanced under these actions from the matter contiguous to it beyond them, and the force and couple applied to it from without. These last have, in the plane XOY, components respectively equal to $Y\delta x$ and $N\delta x$: and hence for the equilibrium of the portion PP',

Straight beam infinitely little bent.

$$-\beta + Y\delta x + \beta' = 0, \text{ by forces parallel to } OY,$$

and $$-\zeta + N\delta x + \zeta' + \beta\delta x = 0, \text{ by couples in plane } XOY,$$

the term $\beta\delta x$ in this second equation being the moment of the couple formed by the infinitely nearly equal forces β, β' in the dissimilar parallel directions through P and P'. Now

$$\beta' - \beta = \frac{d\beta}{dx}\delta x, \text{ and } \zeta' - \zeta = \frac{d\zeta}{dx}\delta x.$$

Hence the preceding equations give

$$\left.\begin{aligned} \frac{d\beta}{dx} &= -Y \\ \frac{d\zeta}{dx} &= -N - \beta \end{aligned}\right\} \ldots\ldots\ldots\ldots\ldots\ldots\ldots (1);$$

and these, by the elimination of β,

$$\frac{d^2\zeta}{dx^2} = -\frac{dN}{dx} + Y \ldots\ldots\ldots\ldots\ldots\ldots\ldots\ldots (2).$$

Similarly, by forces and couples in the plane XOZ,

$$\frac{d^2\eta}{dx^2} = -\frac{dM}{dx} + Z \ldots\ldots\ldots\ldots\ldots\ldots\ldots\ldots (3),$$

couples in this plane being reckoned positive when they tend to turn from the direction of OX to that of OZ; which is opposite to the convention (551) generally adopted as being proper when the three axes are dealt with symmetrically.

Since the wire deviates infinitely little from the straight line OX, the component curvatures are

$$\frac{d^2y}{dx^2} \text{ in the plane } XOY,$$

and $$\frac{d^2z}{dx^2} \quad ,, \quad ,, \quad XOZ.$$

Straight beam infinitely little bent.

Hence the equations of flexure are

$$\left.\begin{aligned}\zeta &= B\frac{d^2y}{dx^2} + a\frac{d^2z}{dx^2}\\ \eta &= a\frac{d^2y}{dx^2} + C\frac{d^2z}{dx^2}\end{aligned}\right\} \quad\text{..........................(4)},$$

where B and C are the flexural rigidities (§ 596) in the planes xy and xz, and a the coefficient expressing the couple in either produced by unit curvature in the other; three quantities which are to be regarded, in general, as given functions of x. Substituting these expressions for ζ and η, in (2) and (3), we have the required equations of equilibrium.

Case of independent flexure in two planes.

617. If the directions of maximum and minimum flexural rigidity lie throughout the wire in two planes, the equations of equilibrium become simplified by these planes being chosen as planes of reference, XOY, XOZ. The flexure in either plane then depends simply on the forces in it, and thus the problem divides itself into the two quite independent problems of integrating the equations of flexure in the two principal planes, and so finding the projections of the curve on two fixed planes agreeing with their position when the rod is straight.

In this case, and with XOY, XOZ so chosen, we have $a = 0$. Hence the equations of flexure (4) become simply

$$\zeta = B\frac{d^2y}{dx^2}, \quad \eta = C\frac{d^2z}{dx^2};$$

and the differential equations of the curve, found by using these in (2) and (3),

$$\frac{d^2}{dx^2}\left(B\frac{d^2y}{dx^2}\right) = \mathfrak{Y}, \quad \frac{d^2}{dx^2}\left(C\frac{d^2z}{dx^2}\right) = \mathfrak{Z} \quad\text{............(5)},$$

where $$\mathfrak{Y} = -\frac{dN}{dx} + Y, \quad \mathfrak{Z} = -\frac{dM}{dx} + Z \quad\text{................(6)}.$$

Here $\mathfrak{Y}$ and $\mathfrak{Z}$ are to be generally regarded as known functions of x, given explicitly by (6), being the amounts of component simple forces perpendicular to the wire, reckoned per unit of its length, that would produce the same figure as the distribution of force and couple we have supposed actually applied throughout the length. Later, when occupied with the theory of magnetism, we shall meet with a curious instance of the relation expressed by (6). In the meantime it may be remarked that although the figure of the wire does not sensibly differ when the simple distribution of force is substituted for any given distribution of force and couple, the shearing forces in normal sections become thoroughly altered by this change of circumstances, as is shown by (1). When the wire is uniform, B and C are constant, and the equations of equilibrium become

Case of independent flexure in two planes.

$$\frac{d^4y}{dx^4}=\frac{\mathfrak{Y}}{B},\quad \frac{d^4z}{dx^4}=\frac{\mathfrak{Z}}{C}\;\ldots\ldots\ldots\ldots\ldots\ldots\ldots\ldots(7).$$

The simplest example is obtained by taking $\mathfrak{Y}$ and $\mathfrak{Z}$ each constant, a very interesting and useful case, being that of a uniform beam influenced only by its own weight, except where held or pressed by its supports. Confining our attention to flexure in the one principal plane, XOY, and supposing this to be vertical, so that $\mathfrak{Y}=gw$, if w be the mass per unit of length; we have, for the complete integral, of course

Plank bent by its own weight.

$$y=\frac{gw}{B}(\tfrac{1}{24}x^4+Kx^3+K'x^2+K''x+K''')\ldots\ldots\ldots\ldots(8),$$

where K, K', etc., denote constants of integration. These, four in number, are determined by the terminal conditions; which, for instance, may be that the value of y and of $\frac{dy}{dx}$ is given for each end. Or, as for instance in the case of a plank simply resting with its ends on two edges or trestles, and free to turn round either, the condition may be that the curvature vanishes at each end: so that if OX be taken as the line through the points of support, we have

$$\left.\begin{array}{l} y=0 \\ \dfrac{d^2y}{dx^2}=0 \end{array}\right\}\text{ when } x=0 \text{ and when } x=l,$$

Plank supported by its ends.

l being the length of the plank. The solution then is

$$y=\frac{gw}{B}\cdot\tfrac{1}{24}(x^4-2lx^3+l^3x)\ldots\ldots\ldots\ldots\ldots\ldots\ldots(9).$$

Hence, by putting $x=\frac{1}{2}l$, we find $y=\frac{gw}{B}\cdot\frac{5l^4}{16\times 24}$ for the distance

Plank supported by its ends;

by which the middle point is deflected from the straight line joining the points of support.

Or, as in the case of a plank balanced on a trestle at its middle (taken as zero of x), or hung by a rope tied round it there, we may have

by its middle.

$$\left.\begin{aligned} y&=0 \\ \frac{dy}{dx}&=0 \end{aligned}\right\} \text{ when } x=0,$$

and

$$\left.\begin{aligned} \frac{d^2y}{dx^2}&=0 \\ \frac{d^3y}{dx^3}&=0 \end{aligned}\right\} \text{ when } x=\tfrac{1}{2}l \text{ [see above, § 614 (10)]}.$$

The solution in this case is, for the positive half of the plank,

$$y=\frac{gw}{B}\cdot\tfrac{1}{24}(x^4-2lx^3+\tfrac{3}{2}l^2x^2)\ldots\ldots\ldots\ldots\ldots\ldots(10).$$

By putting $x=\frac{1}{2}l$, we find $y=\dfrac{gw}{B}\cdot\dfrac{3l^4}{16\,.\,24}$. Hence

Droops compared.

618. When a uniform bar, beam, or plank is balanced on a single trestle at its middle, the droop of its ends is only $\frac{3}{8}$ of the droop which its middle has when the bar is supported on trestles at its ends. From this it follows that the former is $\frac{3}{8}$ and the latter $\frac{5}{8}$ of the droop or elevation produced by a force equal to half the weight of the bar, applied vertically downwards or upwards to one end of it, if the middle is held fast in a horizontal position. For let us first suppose the whole to rest on a trestle under its middle, and let two trestles be placed under its ends and gradually raised till the pressure is entirely taken off from the middle. During this operation the middle remains fixed and horizontal, while a force increasing to half the weight, applied vertically upwards on each end, raises it through a height equal to the sum of the droops in the two cases above referred to. This result is of course proved directly by comparing the absolute values of the droop in those two cases as found above, with the deflection from the tangent at the end of the cord in the elastic curve, figure 2, of § 611, which is cut by the cord at right angles. It may be stated otherwise

Plank supported by its ends or middle.

thus: the droop of the middle of a uniform beam resting on trestles at its ends is increased in the ratio of 5 to 13 by laying a mass equal in weight to itself on its middle: and, if the beam is hung by its middle, the droop of the ends is increased in the ratio of 3 to 11 by hanging on each of them a mass equal to half the weight of the beam.

Plank supported by its ends or middle;

619. The important practical problem of finding the distribution of the weight of a solid on points supporting it, when more than two of these are in one vertical plane, or when there are more than three altogether, which (§ 568) is indeterminate* if the solid is perfectly rigid, may be completely solved for a uniform elastic beam, naturally straight, resting on three or more points in rigorously fixed positions all nearly in one horizontal line, by means of the preceding results.

by three or more points.

If there are i points of support, the $i-1$ parts of the rod between them in order and the two end parts will form $i+1$ curves expressed by distinct algebraic equations [§ 617 (8)], each involving four arbitrary constants. For determining these constants we have $4i+4$ equations in all, expressing the following conditions:—

I. The ordinates of the inner ends of the projecting parts of the rod, and of the two ends of each intermediate part, are respectively equal to the given ordinates of the corresponding points of support [$2i$ equations].

II. The curves on the two sides of each support have coincident tangents and equal curvatures at the point of transition from one to the other [$2i$ equations].

III. The curvature and its rate of variation per unit of length along the rod, vanish at each end [4 equations].

Thus the equation of each part of the curve is completely determined: and then, by § 616, we find the shearing force in any normal section. The difference between these in the

* It need scarcely be remarked that indeterminateness does not exist in nature. How it may occur in the problems of abstract dynamics, and is obviated by taking something more of the properties of matter into account, is instructively illustrated by the circumstances referred to in the text.

neighbouring portions of the rod on the two sides of a point of support, is of course equal to the pressure on this point.

Plank supported by its ends and middle.

620. The solution for the case of this problem in which two of the points of support are at the ends, and the third midway between them either exactly in the line joining them, or at any given very small distance above or below it, is found at once, without analytical work, from the particular results stated in § 618. Thus if we suppose the beam, after being first supported wholly by trestles at its ends, to be gradually pressed up by a trestle under its middle, it will bear a force simply proportional to the space through which it is raised from the zero point, until all the weight is taken off the ends, and borne by the middle. The whole distance through which the middle rises during this process is, as we found, $\frac{gw}{B}\cdot\frac{8l^4}{16.24}$; and this whole elevation is $\frac{8}{5}$ of the droop of the middle in the first position. If therefore, for instance, the middle trestle be fixed exactly in the line joining those under the ends, it will bear $\frac{5}{8}$ of the whole weight, and leave $\frac{3}{16}$ to be borne by each end. And if the middle trestle be lowered from the line joining the end ones by $\frac{7}{15}$ of the space through which it would have to be lowered to relieve itself of all pressure, it will bear just $\frac{1}{3}$ of the whole weight, and leave the other two thirds to be equally borne by the two ends.

Rotation of a wire round its elastic central line.

621. A wire of equal flexibility in all directions, and straight when freed from stress, offers, when bent and twisted in any manner whatever, not the slightest resistance to being turned round its elastic central curve, as its conditions of equilibrium are in no way affected by turning the whole wire thus equally throughout its length. The useful application of this principle, to the maintenance of equal angular motion in two bodies rotating round different axes, is rendered somewhat difficult in practice by the necessity of a perfect attachment and adjustment of each end of the wire, so as to have the tangent to its elastic central curve exactly in line with the axis of rotation. But if this condition is rigorously fulfilled, and the wire is of exactly equal flexibility in every direction, and

Elastic universal flexure joint; § 109.

exactly straight when free from stress, it will give, against any constant resistance, an accurately uniform motion from one to another of two bodies rotating round axes which may be inclined to one another at any angle, and need not be in one plane. If they are in one plane, if there is no resistance to the rotatory motion, and if the action of gravity on the wire is insensible, it will take some of the varieties of form (§ 612) of the plane elastic curve of James Bernoulli. But however much it is altered from this; whether by the axes not being in one plane; or by the torsion accompanying the transmission of a couple from one shaft to the other, and necessarily, when the axes are in one plane, twisting the wire out of it; or by gravity; the elastic central curve will remain at rest, the wire in every normal section rotating round it with uniform angular velocity, equal to that of each of the two bodies which it connects. Under Properties of Matter, we shall see, as indeed may be judged at once from the performances of the vibrating spring of a chronometer for twenty years, that imperfection in the elasticity of a metal wire does not exist to any such degree as to prevent the practical application of this principle, even in mechanism required to be durable.

Equable elastic rotating joint.

It is right to remark, however, that if the rotation be too rapid, the equilibrium of the wire rotating round its unchanged elastic central curve may become unstable, as is immediately discovered by experiments (leading to very curious phenomena), when, as is often done in illustrating the kinetics of ordinary rotation, a rigid body is hung by a steel wire, the upper end of which is kept turning rapidly.

Practical inequalities.

622. If the wire is not of rigorously equal flexibility in all directions, there will be a periodic inequality in the communicated angular motion, having for period a half turn of either body: or if the wire, when unstressed, is not exactly straight, there will be a periodic inequality, having the whole turn for its period. In other words, if ϕ and ϕ' be angles simultaneously turned through by the two bodies, with a constant working couple transmitted from one to the other through the wire, $\phi - \phi'$ will not be zero, as in the proper elastic universal

Practical inequalities. flexure joint, but will be a function of $\sin 2\phi$ and $\cos 2\phi$ if the first defect alone exists; or it will be a function of $\sin\phi$ and $\cos\phi$ if there is the second defect whether alone or along with the first. It is probable that, if the bend in the wire when unstressed is not greater than can be easily provided against in actual construction, the inequality of action caused by it may be sufficiently remedied without much difficulty in practice, by setting it at one or at each end, somewhat inclined to the axis of the rotating body to which it is attached. But these considerations lead us to a subject of much greater interest in itself than any it can have from the possibility of usefulness in practical applications. The simple cases we shall choose illustrate three kinds of action which may exist, each either alone or with one or both the others, in the equilibrium of a wire not equally flexible in all directions, and straight when unstressed.

Elastic rotating joint.

Rotation round its elastic central circle of a straight wire made into a hoop.

623. A uniform wire, straight when unstressed, is bent till its two ends meet, which are then attached to one another, with the elastic central curve through each touching one straight line: so that whatever be the form of the normal section, and the quality, crystalline or non-crystalline, of the substance, the whole wire must become, when in equilibrium, an exact circle (gravity being not allowed to produce any disturbance). It is required to find what must be done to turn the whole wire uniformly through any angle round its elastic central circle.

If the wire is of exactly equal flexibility in all directions*, it will, as we have seen (§ 621), offer no resistance at all to this action, except of course by its own inertia; and if it is once set to rotate thus uniformly with any angular velocity, great or small, it would continue so for ever were the elasticity perfect, and were there no resistance from the air or other matter touching the axis.

To avoid restricting the problem by any limitation, we must suppose the wire to be such that, if twisted and bent in any way, the potential energy of the elastic action developed, per

* In this case, clearly it might have been twisted before its ends were put together, without altering the circular form taken when left with its ends joined.

unit of length, is a quadratic function of the twist, and two components of the curvature (§§ 590, 595), with six arbitrarily given coefficients. But as the wire has no twist*, three terms of this function disappear in the case before us, and there remain only three terms,—those involving the squares and the product of the components of curvature in planes perpendicular to two rectangular lines of reference in the normal section through any point. The position of these lines of reference may be conveniently chosen so as to make the product of the components of curvature disappear: and the planes perpendicular to them will then be the planes of maximum and minimum flexural rigidity when the wire is kept free from twist†. There is no difficulty in applying the general equations of § 614 to express these circumstances and answer the proposed question. Leaving this as an analytical exercise to the student, we take a shorter way to the conclusion by a direct application of the principle of energy.

Rotation round its elastic central circle, of a straight wire made into a hoop.

Let the potential energy per unit of length be $\frac{1}{2}(B\kappa^2 + C\lambda^2)$, when κ and λ are the component curvatures in the planes of maximum and minimum flexural rigidity: so that, as in § 617, B and C are the measures of the flexural rigidities in these planes. Now if the wire be held in any way at rest with these planes through each point of it inclined at the angles ϕ and $\frac{1}{2}\pi - \phi$ to the plane of its elastic central circle, the radius of this circle being r, we should have $\kappa = \frac{1}{r}\cos\phi$, $\lambda = \frac{1}{r}\sin\phi$. Hence, since $2\pi r$ is the whole length,

$$E = \pi\left(\frac{B}{r}\cos^2\phi + \frac{C}{r}\sin^2\phi\right) \qquad\qquad (1).$$

* Which we have supposed, in order that it may take a circular form; although in the important case of equal flexibility in all directions this condition would obviously be fulfilled, even with twist.

† When, as in ordinary cases, the wire is either of isotropic material (see § 677 below), or has a normal axis (§ 596) in the direction of its elastic central line, flexure will produce no tendency to twist: in other words, the products of twist into the components of curvature will disappear from the quadratic expressing the potential energy: or the elastic central line is an axis of pure torsion. But, as shown in the text, the case under consideration gains no simplicity from this restriction.

Rotation round its elastic central circle, of a straight wire made into a hoop.

Let us now suppose every infinitely small part of the wire to be acted on by a couple in the normal plane, and let L be the amount of this couple per unit of length, which must be uniform all round the ring in order that the circular form may be retained, and let this couple be varied so that, rotation being once commenced, ϕ may increase at any uniform angular velocity. The equation of work done per unit of time (§§ 240, 287) is

$$2\pi r L\dot{\phi} = \frac{dE}{dt} = \frac{dE}{d\phi}\dot{\phi}.$$

And therefore, by (1),

$$-L = \frac{B-C}{r^2}\sin\phi\cos\phi = \frac{B-C}{2r^2}\sin 2\phi,$$

which shows that the couple required in the normal plane through every point of the ring, to hold it with the planes of greatest flexural rigidity touching a cone inclined at any angle, ϕ, to the plane of the circle, is proportional to $\sin 2\phi$; is in the direction to prevent ϕ from increasing; and when $\phi = \frac{1}{4}\pi$, amounts to $\frac{B-C}{2r^2}$ per unit length of the circumference. From this we see that there are two positions of stable equilibrium,—being those in which the plane of least flexural rigidity lies in the plane of the ring; and two positions of unstable equilibrium,—being those in which the plane of greatest flexural rigidity is in the plane of the ring.

Rotation round its elastic central circle, of a hoop of wire equally flexible in all directions, but circular when unstrained.

624. A wire of uniform flexibility in all directions, so shaped as to be a circular arc of radius a when free from stress, is bent till its ends meet, and these are joined as in § 623, so that the whole becomes a circular ring of radius r. It is required to find the couple which will hold this ring turned round the central curve through any angle ϕ in every normal section, from the position of stable equilibrium (which is of course that in which the naturally concave side of the wire is on the concave side of the ring, the natural curvature being either increased or diminished, but not reversed, when the wire is bent into the ring). Applying the principle of energy exactly as in the preceding section, we find that in this case the couple

is proportional to $\sin\phi$, and that when $\phi = \frac{1}{2}\pi$, its amount per unit of length of the circumference is $\frac{B}{ar}$, if B denote the flexural rigidity.

Rotation round its elastic central circle, of a hoop of wire equally flexible in all directions, but circular when unstrained.

For in this case we have the potential energy

$$E = \pi r B\left\{\left(\frac{1}{a} - \frac{1}{r}\cos\phi\right)^2 + \left(\frac{1}{r}\sin\phi\right)^2\right\} = \pi r B\left(\frac{1}{a^2} - \frac{2}{ar}\cos\phi + \frac{1}{r^2}\right) \text{......(2)},$$

and
$$L = \frac{1}{2\pi r}\frac{dE}{d\phi} = \frac{B}{ar}\sin\phi \text{.....................(3)}.$$

If every part of the ring is turned half round, so as to bring the naturally concave side of the wire to the convex side of the ring, we have of course a position of unstable equilibrium.

625. A wire of unequal flexibility in different directions is formed so that, when free from stress, it constitutes a circular arc of radius a, with the plane of greatest flexural rigidity at each point touching a cone inclined to its plane at an angle α. Its ends are then brought together and joined, as in §§ 623, 624, so that the whole becomes a closed circular ring, of any given radius r. It is required to find the changed inclination, ϕ, to the plane of the ring, which the plane of greatest flexural rigidity assumes, and the couple, G, in the plane of the ring, which acts between the portions of matter on each side of any normal section.

Wire unequally flexible in different directions, and circular when unstrained, bent to another circle by balancing couples applied to its ends.

The two equations between the components of the couple and the components of the curvature in the planes of greatest and least flexural rigidity determine the two unknown quantities of the problem.

These equations are

$$\left.\begin{aligned} B\left(\frac{1}{r}\cos\phi - \frac{1}{a}\cos\alpha\right) &= G\cos\phi \\ C\left(\frac{1}{r}\sin\phi - \frac{1}{a}\sin\alpha\right) &= G\sin\phi \end{aligned}\right\} \text{..............(4)},$$

since $\frac{1}{a}\cos\alpha$ and $\frac{1}{a}\sin\alpha$ are the components of natural curvature in the principal planes, and therefore $\frac{1}{r}\cos\phi - \frac{1}{a}\cos\alpha$, and

Wire unequally flexible in different directions, and circular when unstrained, bent to another circle by balancing couples applied to its ends.

$\frac{1}{r}\sin\phi - \frac{1}{a}\sin\alpha$, are the changes from the natural to the actual curvatures in these planes maintained by the corresponding components $G\cos\phi$ and $G\sin\phi$ of the couple G.

The problem, so far as the position into which the wire turns round its elastic central curve, may be solved by an application of the principle of energy, comprehending those of §§ 623, 624 as particular cases.

Let L be the amount, per unit of length of the ring, of the couple which must be applied from without, in each normal section, to hold it with the plane of maximum flexural rigidity at each point inclined at any given angle, ϕ, to the plane of the ring. We have, as before (§§ 623, 624), for the potential energy of the elastic action in the ring when held so,

$$E = \pi r\left\{B\left(\frac{\cos\phi}{r} - \frac{\cos\alpha}{a}\right)^2 + C\left(\frac{\sin\phi}{r} - \frac{\sin\alpha}{a}\right)^2\right\} \ldots\ldots(5).$$

Hence

$$L = \frac{1}{2\pi r}\frac{dE}{d\phi} = \left\{-B\left(\frac{\cos\phi}{r} - \frac{\cos\alpha}{a}\right)\frac{\sin\phi}{r} + C\left(\frac{\sin\phi}{r} - \frac{\sin\alpha}{a}\right)\frac{\cos\phi}{r}\right\}. \ (6).$$

This equated to zero is the same as (4) with G eliminated, and determines the relation between ϕ and r, in order that the ring when altered to radius r instead of a may be in equilibrium in itself (that is, without any application of couple in the normal section). The present method has the advantage of facilitating the distinction between the solutions, as regards stability or instability of the equilibrium, since (§ 291) for stable equilibrium E is a minimum, and for unstable equilibrium a maximum.

As a particular case, let $C = \infty$, which simplifies the problem very much. The terms involving C as a factor in (5) and (6) become nugatory in this case, and require of course that

$$\frac{\sin\phi}{r} - \frac{\sin\alpha}{a} = 0.$$

But the former method is clearer and better for the present case; as this result is at once given by the second of equations (4); and then the value of G, if required, is found from the first. We conclude what is stated in the following section:—

Conical bendings of developable surface.

626. Let a uniform hoop, possessing flexibility only in one tangent plane to its elastic central line at each point, be given, so shaped that when under no stress (for instance, when cut through in any normal section and uninfluenced by force from other bodies) it rests in the form of a circle of radius a, with its planes of inflexibility all round touching a cone inclined to the plane of this circle. This is very nearly the case with a common hoop of thin sheet-iron fitted upon a conical vat, or on either end of a barrel of ordinary shape. Let such a hoop be shortened (or lengthened), made into a circle of radius a by riveting its ends together (§ 623) in the usual way, and left with no force acting on it from without. It will rest with its plane of inflexibility inclined at the angle $\phi = \sin^{-1}(r \sin a/a)$ to the plane of its circular form, and the elastic couple acting in this plane between the portions of matter on the two sides of any normal section will be

$$G = \frac{B}{\cos\phi}\left(\frac{\cos\phi}{r} - \frac{\cos a}{a}\right).$$

These results we see at once, by remarking that the component curvature in the plane of inflexibility at each point must be invariably of the same value, $\sin a/a$, as in the given unstressed condition of the hoop: and that the component couple, $G\cos\phi$, in the plane perpendicular to that of inflexibility at each point, must be such as to change the component curvature in this plane from $\cos a/a$ to $\cos\phi/r$.

The greatest circle to which such a hoop can be changed is of course that whose radius is $a/\sin a$: and for this $\phi = \frac{1}{2}\pi$, or the surface of inflexibility at each point (the surface of the sheet-metal in the practical case) becomes the plane of the circle: and therefore $G = \infty$, showing that if a hoop approaching infinitely nearly to this condition be made, in the manner explained, the internal couple acting across each normal section will be infinitely great, which is obviously true.

627. Another very important and interesting case readily dealt with by a method similar to that which we have applied to the elastic wire, is the equilibrium of a plane elastic plate

Flexure of a plane elastic plate. bent to a shape differing infinitely little from the plane, by any forces subject to certain conditions stated below (§ 632). Some definitions and preliminary considerations may be conveniently taken first.

Definitions. (1) A *surface of a solid* is a surface passing through always the same particles of the solid, however it is strained.

(2) The middle surface of a plate is the surface passing through all those of its particles which, when it is free from stress, lie in a plane midway between its two plane sides.

(3) A normal section of a plate, or a surface normal to a plate, is a surface which, when the plate is free from stress, cuts its sides and all planes parallel to them at right angles, being therefore, when unstrained, necessarily either a single plane or a cylindrical (or prismatic) surface.

(4) The *deflection* of any point or small part of the plate, is the distance of its middle surface there from the tangent plane to the middle surface at any conveniently chosen point of reference in it.

(5) The *inclination* of the plate, at any point, is the inclination of the tangent plane of the middle surface there to the tangent plane at the point of reference.

(6) The *curvature of a plate* at any point, or in any part, is the curvature of its middle surface there.

(7) In a surface infinitely nearly plane the curvature is said to be *uniform*, if the curvatures in every two parallel normal sections are equal.

(8) Any diameter of a plate, or distance in a plate infinitely nearly plane, is called finite, unless it is an infinitely great multiple of the least radius of curvature multiplied by the greatest inclination.

Geometrical preliminaries. Choosing XOY as the tangent plane at the point of reference, let (x, y, z) be any point of its middle surface, i its inclination there, and $\frac{1}{r}$ its curvature in a normal section through that point, inclined at an angle ϕ to ZOX. We have

Geometrical preliminaries.

$$\tan i = \sqrt{\left(\frac{dz^2}{dx^2} + \frac{dz^2}{dy^2}\right)} \quad\ldots\ldots\ldots\ldots(1),$$

and, if i be infinitely small,

$$\frac{1}{r} = \frac{d^2z}{dx^2}\cos^2\phi + 2\frac{d^2z}{dxdy}\sin\phi\cos\phi + \frac{d^2z}{dy^2}\sin^2\phi \ldots. \quad (2).$$

To prove these, let ξ, η, ζ be the co-ordinates of any point of the surface infinitely near (x, y, z). Then, by the elements of the differential calculus,

$$\zeta = \frac{dz}{dx}\xi + \frac{dz}{dy}\eta + \tfrac{1}{2}\left(\frac{d^2z}{dx^2}\xi^2 + 2\frac{d^2z}{dxdy}\xi\eta + \frac{d^2z}{dy^2}\eta^2\right).$$

Let $\xi = \rho\cos\phi, \quad \eta = \rho\sin\phi,$

so that we have

$$\left.\begin{array}{l} \zeta = A\rho + \tfrac{1}{2}B\rho^2, \text{ where } A = \dfrac{dz}{dx}\cos\phi + \dfrac{dz}{dy}\sin\phi \\ \text{and } B = \dfrac{d^2z}{dx^2}\cos^2\phi + 2\dfrac{d^2z}{dxdy}\sin\phi\cos\phi + \dfrac{d^2z}{dy^2}\sin^2\phi \end{array}\right\} \ldots\ldots(3).$$

Then by the formula for the curvature of a plane curve (§ 9),

$$\frac{1}{r} = \frac{B}{(1+A^2)^{\frac{3}{2}}}, \text{ or, as } A \text{ is infinitely small, } \frac{1}{r} = B,$$

and thus (2) is proved.

It follows that the surface represented by

$$z = \tfrac{1}{2}(Ax^2 + 2cxy + By^2)\ldots\ldots\ldots\ldots(4),$$

is a surface of uniform curvature if A, B, c be constant throughout the admitted range of values of (x, y); these being limited by the condition that $Ax + cy$, and $cx + By$ must be everywhere infinitely small.

628. When a plane surface is bent to any other shape than a developable surface (§ 139), it must experience some degree of stretching or contraction. But an essential condition for the theory of elastic plates on which we are about to enter, is that the amount of the stretching or contraction thus *necessary* in the middle surface is at most incomparably smaller than the stretching and contraction of the two sides (§ 141) due to curvature. It will be shown in § 629 that this condition, if we exclude the case of bending into a surface differing infinitely little from a developable surface, is equivalent to the following:—

Limitation of flexure not to imply a stretching of middle surface comparable with that of either side.

The deflection [§ 627 Def. (4)] *is, at all places finitely* [§ 627 Def. (8)] *distant from the point of reference, incomparably smaller than the thickness.*

And if we extend the signification of "deflection" from that defined in (4) of § 627, to distance from some true developable surface, the excluded case is of course brought under the statement.

Although the truth of this is obvious, it is satisfactory to prove it by investigating the actual degrees of stretching and contraction referred to.

Stretching of a plane by synclastic or anticlastic flexure.

629. Let us suppose a given plane surface to be bent to some curved form without any stretching or contracting of lines radiating from some particular point of it, O; and let it be required to find the stretching or contraction in the circumference of a circle described from O as centre, with any radius a, on the unstrained plane. If the stretching in each part of the circumference, and not merely on the whole, is to be found, something more as to the mode of the bending must be specified; which, for simplicity, in the first place, we shall suppose to be, that any point P of the given surface moves in a plane perpendicular to the tangent plane through O, during the straining.

Let a, θ be polar co-ordinates of P in its primitive position, and r, θ those of the projection on the tangent plane through O, of its position in the bent surface, and let z be the distance of this position from the tangent plane through O. An element, $ad\theta$, of the unstrained circle, becomes

$$(r^2d\theta^2 + dr^2 + dz^2)^{\frac{1}{2}}$$

on the bent surface; and, therefore, for the stretching* of this element we have

$$\epsilon = \left(\frac{r^2}{a^2} + \frac{dr^2}{a^2d\theta^2} + \frac{dz^2}{a^2d\theta^2}\right)^{\frac{1}{2}} - 1 \qquad (1).$$

* Ratio of the elongation to the unstretched length.

Hence if e denote the ratio of the elongation of the whole circumference to its unstretched length, or the mean stretching of the circumference,

Stretching of a plane by synclastic or anticlastic flexure.

$$e=\frac{1}{2\pi}\int_0^{2\pi}d\theta\left\{\left(\frac{r^2}{a^2}+\frac{dr^2}{a^2d\theta^2}+\frac{dz^2}{a^2d\theta^2}\right)^{\frac{1}{2}}-1\right\}\text{ ....(2)},$$

where we must suppose z and r known functions of θ. Confining ourselves now to distances from O within which the curvature of the surface is sensibly uniform, we have

$$z=\frac{a^2}{2\rho},\text{ and } r=\rho\sin\frac{a}{\rho}=a\left(1-\tfrac{1}{6}\frac{a^2}{\rho^2}+\text{etc.}\right)\text{ ....(3)},$$

if ρ be the radius of curvature of the normal section through O and P: and, if we take as the zero line for θ that in which the tangent plane is cut by one of the principal normal planes (§ 130),

$$\frac{1}{\rho}=\frac{1}{\rho_1}\cos^2\theta+\frac{1}{\rho_2}\sin^2\theta=\tfrac{1}{2}\left(\frac{1}{\rho_1}+\frac{1}{\rho_2}\right)+\tfrac{1}{2}\left(\frac{1}{\rho_1}-\frac{1}{\rho_2}\right)\cos 2\theta\text{ ...(4)},$$

where ρ_1, ρ_2 are the principal radii of curvature. Hence the term $dr^2/a^2d\theta^2$ under the radical sign disappears if we include no terms involving higher powers than the first, of the small fraction a^2/ρ^2; and, to this degree of approximation

$$\epsilon=\left\{1-\tfrac{1}{3}\frac{a^2}{\rho^2}+a^2\left(\frac{1}{\rho_2}-\frac{1}{\rho_1}\right)^2\sin^2\theta\cos^2\theta\right\}^{\frac{1}{2}}-1=-\tfrac{1}{6}\frac{a^2}{\rho^2}+\frac{a^2}{2}\left(\frac{1}{\rho_2}-\frac{1}{\rho_1}\right)^2\sin^2\theta\cos^2\theta,$$

or, by (4), and reductions, finally

$$\epsilon=-\tfrac{1}{8}a^2\left\{\left(\frac{1}{\rho_1\rho_2}+\tfrac{1}{2}\left(\frac{1}{\rho_1^2}-\frac{1}{\rho_2^2}\right)\cos 2\theta+\tfrac{1}{2}\left(\frac{1}{\rho_1}-\frac{1}{\rho_2}\right)^2\cos 4\theta\right\}\text{ ...(5)}.$$

Using this in (2) we find

$$e=-\tfrac{1}{6}\frac{a^2}{\rho_1\rho_2}\text{ ....(6)}.$$

The whole amount of stretching thus expressed will, it follows from (5), be distributed uniformly through the circumference, if, instead of compelling each point P to remain in the plane through O, perpendicular to XOY, we allow it to yield in the direction of the circumference through a space equal to

$$\frac{a^3}{24}\left\{\left(\frac{1}{\rho_1^2}-\frac{1}{\rho_2^2}\right)\sin 2\theta+\tfrac{1}{2}\left(\frac{1}{\rho_1}-\frac{1}{\rho_2}\right)^2\sin 4\theta\right\}\text{ ....(7)}.$$

From (6) we conclude that

Stretching of a plane by synclastic or anticlastic flexure.

630. If a plane area be bent to a uniform degree of curvature throughout, without any stretching in any radius through a certain point of it, and with uniform stretching or contraction over the circumference of every circle described from the same point as centre, the amount of this contraction (reckoned negative where the actual effect is stretching) is equal to the ratio of one-sixth of the square of the radius of the circle, to the rectangle under the maximum and minimum radii of curvature of normal sections of the surface; or which is the same thing, the ratio of two-thirds of the rectangle under the maximum and minimum deflections of the circumference from the tangent plane of the surface at the centre, to the square of the radius; or, which is the same, the ratio one-third of the maximum deflection to the maximum radius of curvature.

If the surface thus bent be the middle surface of a plate of uniform thickness, and if each line of particles perpendicular to this surface in the unstrained plate remain perpendicular to it when bent, the stretching on the convex side, and the contraction on the concave side, in any normal section, is obviously equal to the ratio of half the thickness, to the radius of curvature. The comparison of this, with the last form of the preceding statement, proves that the second of the two conditions stated in § 628 secures the fulfilment of the first.

Stretching of a curved surface by flexure not fulfilling Gauss's condition.

631. If a surface already bent as specified, be again bent to a different shape still fulfilling the prescribed conditions, or if a surface given curved be altered to any other shape by bending according to the same conditions, the contraction produced in the circumferences of the concentric circles by this bending, will of course be equal to the increment in the value of the ratio stated in the preceding section. Hence if a curved surface be bent to any other figure, without stretching in any part of it, the rectangle under the two principal radii of curvature at every point remains unchanged. This is Gauss's celebrated theorem regarding the bending of curved surfaces, of which we gave a more elementary demonstration in our introductory Chapter (see §§ 138, 150).

Gauss's theorem regarding flexure.

632. Without further preface we now commence the theory of the flexure of a plane elastic plate with the promised (§ 627) statement of restricting conditions.

Limitations as to the forces and flexures to be admitted in elementary theory of elastic plate.

(1) Of the forces applied from without to any part of the plate, bounded by a normal surface [§ 627 (3)], the components parallel to any line in the plane of the plate are either evanescent or are reducible to *couples*. In other words the algebraic sum of such components, for any part of the plate bounded by a normal surface, is zero.

(2) The principal radii of curvature of the middle surface are everywhere infinitely great multiples of the thickness of the plate.

(3) The deflection is nowhere, within finite distance from the point of reference, more than an infinitely small fraction of the thickness. This condition has a definite meaning for an infinitely large plate, which may be explained thus:—it would be necessary to go to a distance equal to a large multiple of the product of the least radius of curvature into the greatest inclination, to reach a place where the deflection is more than a very small fraction of the thickness of the plate. The consideration of this condition, is of great importance in connection with the theory of the propagation of waves through an infinite plane elastic plate, but scarcely belongs to our present subject.

(4) Neither the thickness of the plate nor the moduluses of elasticity of its substance need be uniform throughout, but if they vary at all they must vary continuously from place to place; and must not any of them be incomparably greater in one place than in another within any finite area of the plate.

633. The general theory of elastic solids investigated later shows that when these conditions are fulfilled the distribution of strain through the plate possesses the following properties, the statement of which at present, although not necessary for the particular problem on which we are entering, will promote a thorough understanding and appreciation of the principles involved.

Results of general theory stated in advance.

Results of general theory stated in advance.

(1) The stretching of any part of the middle surface is infinitely small in comparison with that of either side, in every part of the plate where the curvature is finite.

(2) The particles in any straight line perpendicular to the plate when plane, remain in a straight line perpendicular to the curved surfaces into which its sides, and parallel planes of the substance between them, become distorted when it is bent. And hence the curves in which these surfaces are cut by any plane through that line, have one point in it for centre of curvature of them all.

(3) The whole thickness of the plate remains unchanged, at every point; but the half thickness on one side (which when the curvature is synclastic is the convex side) of the middle surface becomes diminished and on the other side increased, by equal amounts comparable with the elongations and shortenings of lengths equal to the half thickness, measured on the two side surfaces of the plate.

634. The conclusions from the general theory on which we shall found the equations of equilibrium and motion of an elastic plate are as follows:—

Laws for flexure of elastic plate assumed in advance.

Let a naturally plane plate be bent to any surface of uniform curvature [§ 627 (7)] throughout, the applied forces and the extents of displacement fulfilling the conditions and restrictions of § 632: Then—

(1) The force across any section of the plate is, at each point of it, in a line parallel to the tangent plane to the middle surface in the neighbourhood.

(2) The forces across any set of parallel normal sections are equally inclined to the directions of the normal sections at all points (that is to say, are in directions which would be parallel if the plate were bent, and which deviate actually from parallelism only by the infinitely small deviations produced in the normal sections of the flexure).

(3) The amounts of force across one normal section, or any

set of parallel normal sections, on equal infinitely small areas, are simply proportional to the distances of these areas from the middle surface of the plate. Laws for flexure of elastic plate assumed in advance.

(4) The component forces in the tangent planes of the normal sections are equal and in dissimilar directions in sections which are perpendicular to one another. For proof, see § 661. The meaning of "dissimilar directions" in this expression is explained by the diagram; where the arrow-heads indicate the directions in which the portions of matter on the two sides of each normal section would yield if the substance were actually divided, half way through the plate from one side, by each of the normal sections indicated by dotted lines.

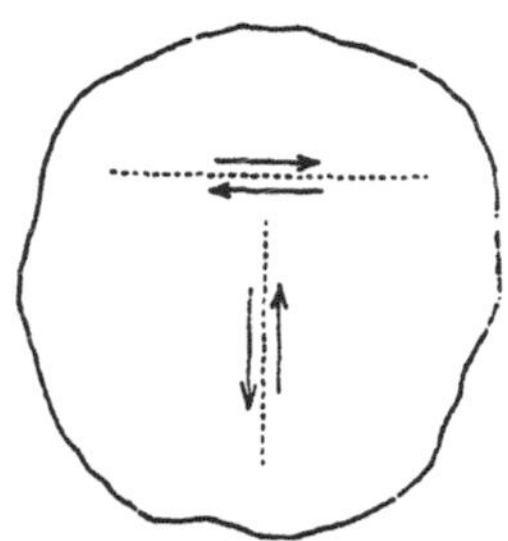

(5) By the law of superposition, we see that if the applied forces be all doubled, or altered in any other ratio, the curvature in every normal section, and all the internal forces specified in (1), (2), (3), (4), are changed in the same ratio; and the potential energy of the internal forces becomes changed according to the square of the same ratio.

635. From § 634 (3) it follows immediately that the forces experienced by any portion of the plate bounded by a normal section through the circumference of a closed polygon or curve of the middle surface, from the action of the contiguous matter of the plate all round it, may be reduced to a set of couples by taking them in groups over infinitely small rectangles into which the bounding normal section may be imagined as divided by normal lines. From § 634 (2) it follows that the distribution of couple thus obtained is uniform along each straight portion, if any there is, of the boundary, and equal per equal lengths in all parallel parts of the boundary. Stress-couple acting across a normal section.

636. From § 634 (4) it follows that the component couples round axes perpendicular to the boundary are equal in parts of the boundary at right angles to one another, and are in Twisting components proved equal round any two perpendicular axes.

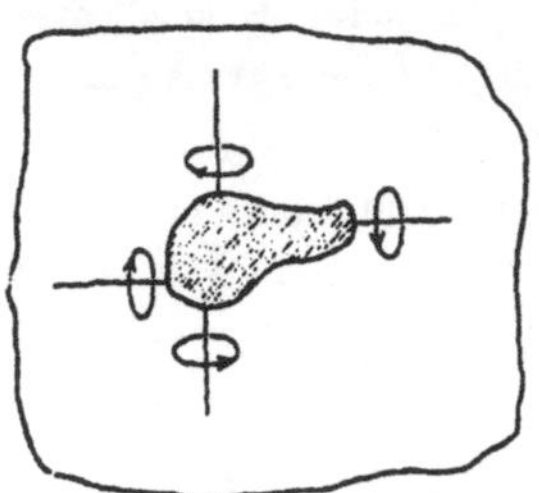

directions related to one another in the manner indicated by the circular arrows in the diagram; that is to say, in such directions that if the axis is, according to the rule of § 234, drawn *outwards* from the portion of the plate under consideration, for one point of the boundary it must be drawn *inwards* for every point where the boundary is perpendicular to its direction at that point.

Principal axes of bending stress.

637. We may now prove that there are two normal sections, at right angles to one another, in which the component couples round axes perpendicular to them vanish, and that in these sections the component couples round axes coincident with the sections are of maximum and minimum values.

Let OAB be a right-angled triangle of the plate. Let Λ and Π be the two component couples acting on the side OA; K and Π those on the side OB; and G and H those on the side AB; the amount of each couple being reckoned per unit of length of the side on which it acts, and the axes and directions of the several couples being as indicated by the circular arrows when each is reckoned as positive.

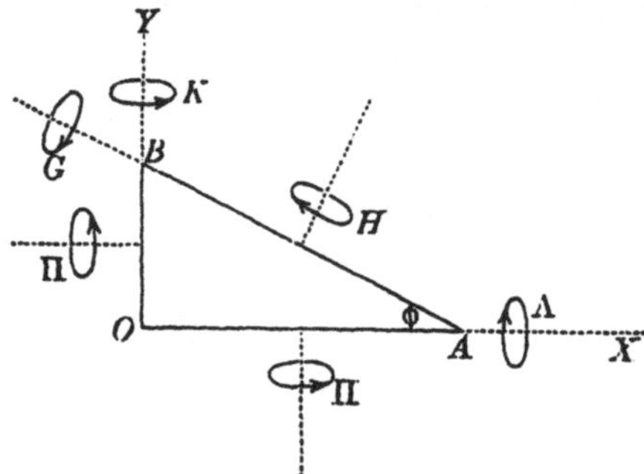

Principal axes of bending stress investigated.

Then, if $AB = a$, and $BAO = \phi$, the whole amounts of the couples on the three sides are respectively

$$\Lambda a \cos\phi, \quad \Pi a \cos\phi,$$
$$\mathrm{K} a \sin\phi, \quad \Pi a \sin\phi,$$
$$Ga, \quad Ha.$$

Resolving the two latter round OX and OY, we have

Principal axes of bending stress investigated.

$$Ga\cos\phi - Ha\sin\phi \text{ round } OX,$$

and $$Ga\sin\phi + Ha\cos\phi \quad ,, \quad OY.$$

But if the portion in question, of the plate, were to become rigid, its equilibrium would not be disturbed (§ 564): and therefore we must have

$$\left.\begin{array}{ll} Ga\cos\phi - Ha\sin\phi = \mathrm{\Lambda}a\cos\phi + \Pi a\sin\phi & \text{by couples round } OX \\ \text{and} & \\ Ga\sin\phi + Ha\cos\phi = \mathrm{K}a\sin\phi + \Pi a\cos\phi & \quad ,, \quad ,, \quad OY \end{array}\right\} \text{(1)}.$$

From these we find immediately

$$\left.\begin{array}{l} G = \mathrm{\Lambda}\cos^2\phi + 2\Pi\sin\phi\cos\phi + \mathrm{K}\sin^2\phi, \\ H = (\mathrm{K} - \mathrm{\Lambda})\sin\phi\cos\phi + \Pi(\cos^2\phi - \sin^2\phi) \end{array}\right\} \text{.........(2)}.$$

Hence the values of ϕ, which make H vanish, give to G its maximum and minimum values, and being determined by the equation

$$\tan 2\phi = -\frac{\Pi}{\frac{1}{2}(\mathrm{K}-\mathrm{\Lambda})} \text{.........................(3)},$$

differ from one another by $\frac{1}{2}\pi$.

A modification of these formulæ, which we shall find valuable, is obtained by putting

$$\Sigma = \tfrac{1}{2}(\mathrm{K}+\mathrm{\Lambda}),\quad \Theta = \tfrac{1}{2}(\mathrm{K}-\mathrm{\Lambda}) \text{..................(4)}.$$

This reduces (2) to

$$\left.\begin{array}{l} G = \Sigma + \Pi\sin 2\phi - \Theta\cos 2\phi \\ H = \qquad \Pi\cos 2\phi + \Theta\sin 2\phi \end{array}\right\} \text{...............(5)},$$

which again become

$$\left.\begin{array}{l} G = \Sigma + \Omega\cos 2(\phi - a) \\ H = -\Omega\sin 2(\phi - a) \end{array}\right\} \text{.......................(6)},$$

where a [being a value of ϕ given by (3)], and Ω are taken so

$$\left.\begin{array}{ll} \text{that} & \Pi = \Omega\sin 2a,\quad \Theta = -\Omega\cos 2a, \\ \text{so that, of course,} & \Omega = (\Pi^2 + \Theta^2)^{\frac{1}{2}} \end{array}\right\} \text{..............(7)}.$$

This analysis demonstrates the following convenient synthesis of the whole system of internal force in question:—

Synclastic and anticlastic stresses defined.

638. The action experienced by each part of the plate, in virtue of the internal forces between it and the surrounding contiguous matter of the plate, being called a *stress* [in accordance with the general use of this term defined below (§ 658)], may be regarded as made up of two distinct elements—(1) a synclastic stress, and (2) an anticlastic stress; as we shall call them.

(1) Synclastic stress consists of equal direct bending action round every straight line in the plane of the plate. Its amount may be conveniently regarded as measured by the amount, Σ, of the mutual couple between the portions of matter on the two sides of any straight normal section of unit length. Its effect would be to produce equal curvature in all normal sections (that is to say, a spherical figure) if the plate were equally flexible in all directions.

Anticlastic stress referred to its principal axes;

(2) Anticlastic stress consists of two simple bending stresses of equal amounts in opposite directions round two sets of parallel straight lines perpendicular to one another in the plane of the plate. Its effect would be uniform anticlastic curvature, with equal convexities and concavities, if the plate were equally flexible in all directions. Its amount is reckoned as the amount, Ω, of the mutual couple between the portions of matter on the two sides of a straight normal section of unit length, parallel to either of these two sets of lines. It gives rise to couples of the same amount, Ω, between the portions of matter on each side of a normal section of unit length parallel to either of the sets of lines bisecting the right angles between those; but the couples now referred to are *in* the plane of the normal section instead of perpendicular to it. This is proved and illustrated by the annexed diagram, representing [a particular case of the diagram and equations (1) of § 637] the equilibrium of an isosceles right-angled triangle under the influence of couples, each equal to $\Omega\sqrt{\frac{1}{2}}$, applied to it round axes coinciding with

referred to axes inclined to them at 45°.

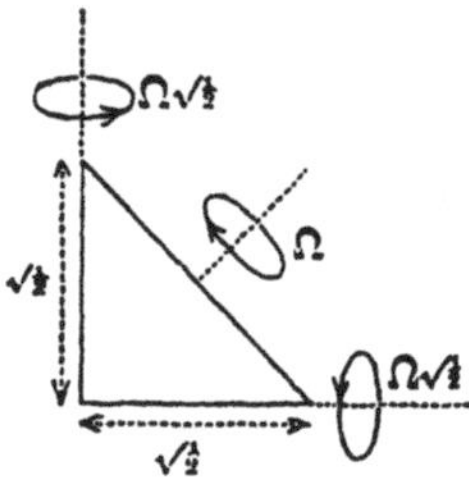

its legs, and a third couple, Ω round an axis perpendicular to its hypotenuse.

Octantal resolution and composition of anticlastic stress.

If two pairs of rectangular axes, each bisecting the right angles formed by the other, be chosen as axes of reference, an anticlastic stress having any third pair of rectangular lines for its axes may, as the preceding formulæ [§ 637 (5)] show, be resolved into two having their axes coincident with the two pairs of axes of reference respectively, by the ordinary cosine formula with each angle doubled. Hence it follows that any two anticlastic stresses may be compounded into one by the same geometrical construction as the parallelogram of forces, made upon lines inclined to one another at an angle equal to twice that between the corresponding axes of the two given stresses; and the position of the axes of the resultant stress will be indicated by the angles of this diagram each halved.

Construction by parallelogram.

Geometrical analogues.

639. Precisely the same set of statements are of course applicable to the curvature of a surface. Thus the proposition proved in § 637 (3) for bending stresses has, for its analogue in curvature, Euler's theorem proved formerly in § 130; and analogues to the series of definitions and propositions founded on it and derived from it may be at once understood without more words or proof.

Two cylindrical curvatures round perpendicular axes, and an anticlastic curvature round axis bisecting their right angles:

Let $$z = \tfrac{1}{2}(\kappa x^2 + 2\varpi xy + \lambda y^2) \qquad \text{.....................(1)}$$

be the equation of a curved surface infinitely near a point O at which it is touched by the plane YOX. Its curvature may be regarded as compounded of a cylindrical curvature, λ, with axis parallel to OX, a cylindrical curvature, κ, with axis parallel to OY, and an anticlastic curvature, ϖ, with axis bisecting the angles XOY, YOX'. Thus, if ϖ and λ each vanished, the surface would be cylindrical, with $1/\kappa$ for radius of curvature and generating lines parallel to OY. Or, if κ and λ each vanished, there would be anticlastic curvature, with sections of equal maximum curvature in the two directions, bisecting the angles XOY and YOX', and radius of curvature in those sections equal to $1/\varpi$.

If now we put

$$\sigma = \tfrac{1}{2}(\kappa + \lambda),\quad \vartheta = \tfrac{1}{2}(\kappa - \lambda) \qquad \text{....................(2),}$$

or a spherical curvature and two anticlastic curvatures;

the equation of the surface becomes

$$z=\tfrac{1}{2}\{\sigma(x^2+y^2)+\vartheta(x^2-y^2)+2\varpi xy\}\text{.................}(3);$$

or, if

$$\left.\begin{aligned}x=r\cos\phi,\ y=r\sin\phi,\\ z=\tfrac{1}{2}\{\sigma+\vartheta\cos 2\phi+\varpi\sin 2\phi\}r^2\end{aligned}\right\}\text{..............}(4);$$

or a spherical and one anticlastic curvature.

or, lastly,

$$\left.\begin{aligned}z=\tfrac{1}{2}\{\sigma+\omega\cos 2(\phi-a)\}r^2,\\ \vartheta=\omega\cos 2a,\ \varpi=\omega\sin 2a\end{aligned}\right\}\text{..............}(5).$$

In these formulæ σ measures the spherical curvature; and ϑ and ϖ two components of anticlastic curvature, referred to the pair of axes $X'X$, $Y'Y$, and the other pair bisecting their angles. The resultant of ϑ and ϖ is an anticlastic curvature ω, with axes inclined, in the angle XOY at angle a to OX, and in YOX' at angle a to OY.

Work done in bending.

640. The notation of §§ 637, 639 being retained, the work done on any area A of the plate experiencing a change of curvature $(\delta\kappa, \delta\lambda, \delta\varpi)$ under the action of a stress $(\mathrm{K}, \Lambda, \Pi)$, is

$$(\mathrm{K}\delta\kappa+\Lambda\delta\lambda+2\Pi\delta\varpi)A\text{........................}(1);$$

or

$$(2\Sigma\delta\sigma+2\Theta\delta\vartheta+2\Pi\delta\varpi)A\text{......................}(2),$$

if, as before,

$$\Sigma=\tfrac{1}{2}(\mathrm{K}+\Lambda),\ \Theta=\tfrac{1}{2}(\mathrm{K}-\Lambda),\ \sigma=\tfrac{1}{2}(\kappa+\lambda),\ \vartheta=\tfrac{1}{2}(\kappa-\lambda)\text{...}(3).$$

Let $PQP'Q'$ be a rectangular portion of the plate with its centre at O, and its sides $Q'P$, $P'Q$ parallel to OX, and $Q'P'$, PQ parallel to OY. If

$$z=\tfrac{1}{2}(\kappa x^2+2\varpi xy+\lambda y^2)$$

be the equation of the curved surface, we have

$$\frac{dz}{dx}=\kappa x+\varpi y,\ \frac{dz}{dy}=\varpi x+\lambda y;$$

and therefore the tangent plane at (x, y) deviates in direction from XOY by an infinitely small rotation

$$\left.\begin{aligned}\kappa x+\varpi y \text{ round } OY\\ \text{and}\qquad \varpi x+\lambda y\ \ \text{,,}\quad OX\end{aligned}\right\}\text{.......................}(4).$$

Hence the rotation from XOY to the mean tangent plane for all points of the side PQ or $Q'P'$ is

$$\mp\tfrac{1}{2}Q'P\,.\,\kappa \text{ round } OY,$$

and

$$\mp\tfrac{1}{2}Q'P\,.\,\varpi\ \ \text{,,}\quad OX.$$

Hence if the tangent plane, XOY, at O remains fixed, while the curvature changes from $(\kappa, \varpi, \lambda)$ to $(\kappa+\delta\kappa, \varpi+\delta\varpi, \lambda+\delta\lambda)$, the work done by the couples $PQ.\mathrm{K}$ round OY, and $PQ.\Pi$ round OX, distributed over the side PQ, will be

Work done in bending.

$$\tfrac{1}{2}\,Q'P.PQ.(\mathrm{K}\delta\kappa+\Pi\delta\varpi),$$

and an equal amount will be done by the equal and opposite couples distributed over the side $Q'P'$ undergoing an equal and opposite rotation. Similarly, we find for the whole work done on the sides $P'Q$ and $Q'P$,

$$PQ.Q'P.(\Pi\delta\varpi+\Lambda\delta\lambda).$$

Hence the whole work done on all the four sides of the rectangle is
$$PQ.Q'P.(\mathrm{K}\delta\kappa+2\Pi\delta\varpi+\Lambda\delta\lambda):$$

whence the proposition to be proved, since any given area of the plate may be conceived as divided into infinitely small rectangles.

It is an instructive exercise to verify the result by beginning with the consideration of a portion of plate bounded by any given curve, and using the expressions (1) of § 637, by which we find, for the couples on any infinitely short portion, ds, of its boundary, specified in position by (x, y),

$$\left.\begin{array}{ll} \left(-\Lambda\dfrac{dx}{ds}+\Pi\dfrac{dy}{ds}\right)ds & \text{round } OX \\ \text{and} \quad \left(\mathrm{K}\dfrac{dy}{ds}-\Pi\dfrac{dx}{ds}\right)ds & \quad ,, \quad OY \end{array}\right\}\ldots\ldots\ldots\ldots\ldots(5).$$

But, as we have just seen in (4), the rotation experienced by the tangent plane to the plate at (x, y), when the curvature changes from $(\kappa, \varpi, \lambda)$ to $(\kappa+\delta\kappa, \varpi+\delta\varpi, \lambda+\delta\lambda)$, is

$$\left.\begin{array}{ll} x\delta\varpi+y\delta\lambda & \text{round } OX \\ \text{and} \quad x\delta\kappa+y\delta\varpi & \quad ,, \quad OY \end{array}\right\}\ldots\ldots\ldots\ldots\ldots\ldots(6),$$

the tangent plane to the plate at O being supposed to remain unchanged in position; and therefore the work done on the portion ds of the edge is

$$\left\{\left(\mathrm{K}\frac{dy}{ds}-\Pi\frac{dx}{ds}\right)(x\delta\kappa+y\delta\varpi)+\left(\Pi\frac{dy}{ds}-\Lambda\frac{dx}{ds}\right)(x\delta\varpi+y\delta\lambda)\right\}ds.$$

The required work, being the integral of this over the whole of the bounding curve, is therefore

$$(\mathrm{K}\delta\kappa+2\Pi\delta\varpi+\Lambda\delta\lambda)\,A;$$

Work done in bending.

since $$\int x\frac{dy}{ds}ds = -\int y\frac{dx}{ds}ds = A,$$

and $$\int x\frac{dx}{ds}ds = 0, \quad \int y\frac{dy}{ds}ds = 0,$$

each integral being round the whole closed curve.

Partial differential equations for work done in bending an elastic plate.

641. Considering now the elastic forces called into action by the flexure $(\kappa, \varpi, \lambda)$ reckoned from the unstressed condition of the plate (plane, or infinitely nearly plane), and denoting by w the whole amount of their potential energy, per unit area of the plate, we have, as in the case of the wire treated in § 594,

$$\mathrm{K}\delta\kappa = \delta_\kappa w, \quad \Lambda\delta\lambda = \delta_\lambda w, \quad 2\Pi\delta\varpi = \delta_\varpi w \quad \text{.........(7)};$$

or, according to the other notation,

$$2\Sigma\delta\sigma = \delta_\sigma w, \quad 2\Theta\delta\vartheta = \delta_\vartheta w, \quad 2\Pi\delta\varpi = \delta_\varpi w \quad \text{........(8)};$$

where, as above explained, K and Λ denote the simple bending stresses (measured by the amount of bending couple, per unit of length) round lines parallel to OY and OX respectively: Π the anticlastic stress with axes at 45° to OX and OY: and Σ and Θ the synclastic stress and the anticlastic stress with OX and OY for axes, together equivalent to K and Λ. Also, as in § 595, we see that whatever be the character, eolotropic or isotropic, § 677, of the substance of the plate, it must be a homogeneous quadratic function of the three components of curvature, whether $(\kappa, \lambda, \varpi)$ or $(\sigma, \vartheta, \varpi)$. From this and (7), or (8), it follows that the coefficients in the linear functions of the three components of curvature which express the components of the stress required to maintain it, must fulfil the ordinary conservative relations of equality in three pairs, reducing the whole number from nine to six.

Potential energy of an elastic plate held bent.

Thus A, B, C, a, b, c denoting six constants depending on the quality of the solid substance and the thickness of the plate, we have $w = \frac{1}{2}(A\kappa^2 + B\lambda^2 + C\varpi^2 + 2a\lambda\varpi + 2b\varpi\kappa + 2c\kappa\lambda)$ ....(9); and hence, by (7),

$$\left.\begin{aligned} \mathrm{K} &= A\kappa + c\lambda + b\varpi \\ \Lambda &= c\kappa + B\lambda + a\varpi \\ 2\Pi &= b\kappa + a\lambda + C\varpi \end{aligned}\right\} \text{.......................(10)}.$$

Transforming these by § 640 (3) we have, in terms of σ, ϑ, ϖ, *Potential energy of an elastic plate held bent.*

$$w = \tfrac{1}{2}\{(A + B + 2c)\sigma^2 + (A + B - 2c)\vartheta^2 + C\varpi^2 + 2(b-a)\vartheta\varpi + 2(b+a)\sigma\varpi + 2(A-B)\sigma\vartheta\} \quad \text{......(11)},$$

and

$$\left.\begin{aligned} 2\Sigma &= (A + B + 2c)\sigma + (A - B)\vartheta + (b + a)\varpi \\ 2\Theta &= (A - B)\sigma + (A + B - 2c)\vartheta + (b - a)\varpi \\ 2\Pi &= (b + a)\sigma + (b - a)\vartheta + C\varpi \end{aligned}\right\} \quad \text{.........(12)}.$$

These second forms are chiefly useful as showing immediately the relations which must be fulfilled among the coefficients for the important case considered in the following section.

642. If the plate be equally flexible in all directions, a synclastic stress must produce spherical curvature: an anticlastic stress having any pair of rectangular lines in the plate for its axes must produce anticlastic curvature having these lines for sections of equal greatest curvature on the opposite sides of the tangent plane: and in either action the amount of the curvature is simply proportional to the amount of the stress. Hence if $\mathfrak{h}$ and $\mathfrak{k}$ denote two coefficients depending on the bulk-modulus and rigidity of the substance if isotropic (see §§ 677, 680, below), and on the thickness of the plate, we have *Case of equal flexibility in all directions.* *Synclastic and anticlastic rigidities of a plate.*

$$\Sigma = \mathfrak{h}\sigma,\ \Theta = \mathfrak{k}\vartheta,\ \pi = \mathfrak{k}\varpi \quad \text{............(13)}.$$

And therefore [§ 640 (2)]

$$w = \mathfrak{h}\sigma^2 + \mathfrak{k}(\vartheta^2 + \varpi^2) \quad \text{..................(14)}.$$

Hence the coefficients in the general expressions of § 641 fulfil, in the case of equal flexibility in all directions, the following conditions :—

$$a = 0,\ b = 0,\ A = B,\ 2(A - c) = C \quad \text{.............(15)};$$

and the newly-introduced coefficients $\mathfrak{h}$ and $\mathfrak{k}$ are related to them thus :—

$$A + c = \mathfrak{h},\ \tfrac{1}{2}C = A - c = \mathfrak{k} \quad \text{..................(16)}.$$

643. Let us now consider the equilibrium of an infinite plate, disturbed from its natural plane by forces applied to it in any way, subject only to the conditions of § 632. The substance may be of any possible quality as regards elasticity in different directions: and the plate itself need not be homogeneous either as to this quality, or as to its thickness, in different parts; provided only that round every point it is in both respects sensibly homogeneous [§ 632 Def. (4)] to distances great in comparison with the thickness at that point. *Plate bent by any forces.*

Plate bent by any forces.

644. Let OX, OY be rectangular axes of reference in the plane of the undisturbed plate; and let z be the infinitely small displacement from this plane, of the point (x, y) of the plate, when disturbed by any forces, specified in their effective components as follows:—Take a portion, E, of the plate bounded by a normal surface cutting the middle surface in a line enclosing an infinitely small area σ in the neighbourhood of the point (x, y), and let $Z\sigma$ denote the sum of the component forces perpendicular to XOY on all the matter of E in the neighbourhood of the point (x, y): and $L\sigma$, $M\sigma$ the component couples round OX and OY obtained by transferring, according to Poinsot, the forces from all points of the portion E, supposed for the moment rigid, to one point of it which it is convenient to take at the centre of inertia of the area, σ, of the part of the middle surface belonging to it. This force and these couples, along with the internal forces of elasticity exerted on the matter of E, across its boundary, by the matter surrounding it, must (§ 564) fulfil the conditions of equilibrium for E treated as a rigid body. And E, being not really rigid, must have the curvature due, according to § 641, to the bending stress constituted by the last-mentioned forces. These conditions expressed mathematically supply five equations from which, four elements specifying the internal forces being eliminated, we have a single partial differential equation for z in terms of x and y, which is the required equation of equilibrium.

Conditions of equilibrium.

Equations of equilibrium of plate bent by any forces, investigated.

Let σ be a rectangle $PQP'Q'$, with sides δx parallel to OX and δy parallel to OY. Let $\alpha\delta y$, $\alpha'\delta y$ be the infinitely nearly equal shearing forces perpendicular to the plate in the normal surfaces through PQ' and QP' respectively: and let β, β' be the corresponding notation for PQ, $P'Q'$.

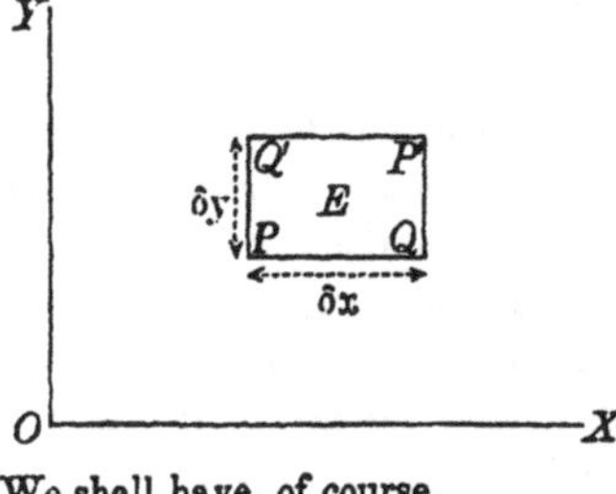

We shall have, of course,

$$\alpha' - \alpha = \frac{d\alpha}{dx}\delta x, \text{ and } \beta' - \beta = \frac{d\beta}{dy}\delta y.$$

The results of these actions on the portion, E, of the plate, considered as rigid, are forces $\alpha'\delta y$, $\beta'\delta x$ through the middle points of QP', $Q'P'$, in the direction of z positive, and forces $\alpha\delta y$, $\beta\delta x$ through the middle points of PQ', PQ, in the direction of z negative. Hence, towards the equilibrium of E as a rigid body, they contribute

Equations of equilibrium of plate bent by any forces, investigated.

$$(\alpha'-\alpha)\,\delta y+(\beta'-\beta)\,\delta x, \text{ or } \left(\frac{d\alpha}{dx}+\frac{d\beta}{dy}\right)\delta x\delta y, \text{ component force parallel to } OZ,$$

$$\alpha\delta y\,.\,\delta x \text{ couple round } OY,$$

and

$$\beta\delta x\,.\,\delta y \quad ,, \quad ,, \quad OX;$$

(in these last two expressions the difference between α and α' and between β and β' are of course neglected). Again, if K, Λ, Π specify, according to the system of § 637, the bending stress at (x, y), we shall have couples infinitely nearly equal and opposite, on the pairs of opposite sides, of which, estimated in components round OX and OY, the differences, representing the residual turning tendencies on E as a rigid body, are as follows:—

round OX, from sides PQ, $Q'P'$, $\dfrac{d\Lambda}{dy}\delta y\,.\,\delta x$,
  ,, ,, PQ', QP', $\dfrac{d\Pi}{dx}\delta x\,.\,\delta y$,

round OY, from sides PQ, $Q'P'$, $\dfrac{d\Pi}{dy}\delta y\,.\,\delta x$,
  ,, ,, PQ', QP', $\dfrac{dK}{dx}\delta x\,.\,\delta y$;

or in all, round OX, $\left(\dfrac{d\Lambda}{dy}+\dfrac{d\Pi}{dx}\right)\delta x\delta y$,

and ,, OY, $\left(\dfrac{d\Pi}{dy}+\dfrac{dK}{dx}\right)\delta x\delta y$.

The equations of equilibrium, therefore, between these and the applied forces on E treated as a rigid body give, if we remove the common factor, $\delta x\delta y$,

Equations of stress in plate bent by any forces.

$$\left.\begin{aligned} Z+\frac{d\alpha}{dx}+\frac{d\beta}{dy}&=0\\ L+\beta+\frac{d\Lambda}{dy}+\frac{d\Pi}{dx}&=0\\ M+\alpha+\frac{d\Pi}{dy}+\frac{dK}{dx}&=0 \end{aligned}\right\}\quad\ldots\ldots\ldots\ldots\ldots\ldots\ldots(1).$$

Equations of stress in plate bent by any forces.

The first of these, with α and β replaced in it by their values from the second and third, becomes

$$\frac{d^2K}{dx^2}+2\frac{d^2\Pi}{dxdy}+\frac{d^2\Lambda}{dy^2}=Z-\frac{dM}{dx}-\frac{dL}{dy}\text{.............. (2).}$$

Now κ, λ, ϖ denoting component curvatures of the plate, according to the system of § 639, we have of course

$$\kappa=\frac{d^2z}{dx^2},\quad \lambda=\frac{d^2z}{dy^2},\quad \varpi=\frac{d^2z}{dxdy}\text{.................. (3),}$$

and hence (10) of § 641 give

Equations connecting stress and curvature.

$$\left.\begin{aligned}K&=A\frac{d^2z}{dx^2}+c\frac{d^2z}{dy^2}+b\frac{d^2z}{dxdy}\\ \Lambda&=c\frac{d^2z}{dx^2}+B\frac{d^2z}{dy^2}+a\frac{d^2z}{dxdy}\\ 2\Pi&=b\frac{d^2z}{dx^2}+a\frac{d^2z}{dy^2}+C\frac{d^2z}{dxdy}\end{aligned}\right\}\text{.................. (4).}$$

Using these in (2) we find the required differential equation of the disturbed surface. On the general supposition (§ 643) we must regard A, B, C, a, b, c as given functions of x and y. In the important practical case of a homogeneous plate they are constants; and the required equation becomes the linear partial differential equation of the fourth order with constant coefficients, as follows:—

$$A\frac{d^4z}{dx^4}+2b\frac{d^4z}{dx^3dy}+(C+2c)\frac{d^4z}{dx^2dy^2}+2a\frac{d^4z}{dxdy^3}+B\frac{d^4z}{dy^4}=Z-\frac{dM}{dx}-\frac{dL}{dy}\text{ (5).}$$

For the case of equal flexibility in all directions, according to § 642 (13), this becomes

Partial differential equation of the bent surface.

$$\left.\begin{aligned}&A\left(\frac{dz^4}{dx^4}+2\frac{d^4z}{dx^2dy^2}+\frac{d^4z}{dy^4}\right)=Z-\frac{dM}{dx}-\frac{dL}{dy}\\ \text{or}\quad &A\left(\frac{d^2}{dx^2}+\frac{d^2}{dy^2}\right)^2z=Z-\frac{dM}{dx}-\frac{dL}{dy}\end{aligned}\right\}\text{.......... (6).}$$

645. To investigate the boundary conditions for a plate of limited dimensions, we may first consider it as forming part of an infinite plate bounded by a normal surface drawn through a closed curve traced on its middle surface. The preceding investigation leads immediately to expressions for the force and couple on any portion of the normal bounding surface. If then

Boundary conditions;

the portion in question be actually cut out from the surrounding sheet, and if a distribution of force and couple identical with that so found be applied to its edge, its elastic condition will remain absolutely unchanged throughout up to the very normal edge. To fulfil this condition requires three equations, expressing (1) that the shearing force applied to the edge (that is, the applied tangential force in the normal surface constituting the edge), which is necessarily in the direction of the normal line to the plate, must be equal to the required amount, and (2) and (3) that the couple applied to any small part of the edge must have components of the proper amounts round any two lines in the plane of the plate. These three equations were given by Poisson as necessary for the full expression of the boundary condition; but Kirchhoff has demonstrated that they express too much, and has shown that two equations suffice. This we shall prove by showing that when a finite plate is given in any condition of stress, or free from stress, we may apply, round axes everywhere perpendicular to its normal surface-edge, any arbitrary distribution of couple without producing any change except at infinitely small distances from the edge, provided a certain distribution of force also, calculated from the distribution of couple, be applied to the edge, perpendicularly to the plate.

Boundary conditions: Poisson's three: two sufficient, proved by Kirchhoff.

Let XY, $=\delta s$, be an infinitely small element at a point (x, y) of a curve traced on the middle surface of an infinite plate; and, PX and PY being parallel to the axes of x and y, let $YXP=\phi$. Then, if $\zeta\delta s$ denote the shearing force in the normal surface to the plate through δs, and, as before (§ 644), $\alpha . PY$ and $\beta . PX$ be those in normal surfaces through PY and PX, we must have, for the equilibrium of the triangle YPX supposed rigid (§ 564),

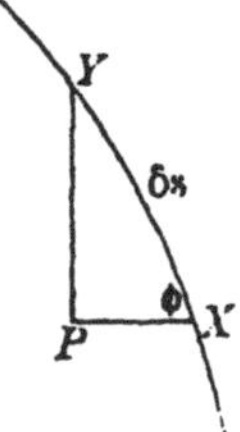

$$\zeta\delta s = \alpha . PY + \beta . PX, \text{ whence } \zeta = \alpha \sin\phi + \beta\cos\phi.$$

Using here for α and β their values by (1) of § 644, we have

$$\zeta = -\left(M + \frac{d\Pi}{dy} + \frac{dK}{dx}\right)\sin\phi - \left(L + \frac{d\Lambda}{dy} + \frac{d\Pi}{dx}\right)\cos\phi \text{......(1)}.$$

Kirchhoff's boundary equations investigated.

Kirchhoff's boundary equations investigated.

Next, if $G\delta s$ and $H\delta s$ denote the components round XY, and round an axis perpendicular to it in the plane of the plate, of the couple acting across the normal surface through δs, we have [(2) of § 637],

$$G = \Lambda \cos^2\phi + 2\Pi \sin\phi\cos\phi + \mathrm{K}\sin^2\phi \text{..............} (2),$$

$$H = (\mathrm{K} - \Lambda)\sin\phi\cos\phi + \Pi(\cos^2\phi - \sin^2\phi) \text{........} (3).$$

If (ζ, G, H) denoted the action experienced by the edge in virtue of applied forces, all the plate outside a closed curve, of which δs is an element, being removed, these three equations would express the same as the three boundary equations given by Poisson. Lastly, let $\mathfrak{Z}\delta s$, $\mathfrak{G}\delta s$, $\mathfrak{H}\delta s$ denote the force perpendicular to the plate, and the components of couple, actually applied at any point (x, y) of a free edge on the length δs of the middle curve. As we shall immediately see (§ 648), if

$$\mathfrak{Z} - \zeta + \frac{d}{ds}(\mathfrak{H} - H) = 0 \text{.......... ....} (4),$$

the plate will be in the same condition of stress throughout, except infinitely near the edge, as with (ζ, G, H) for the action on the edge. Hence, eliminating ζ and H between these four equations, there remain to us (2) unchanged and another, or in all these two—

$$\left.\begin{array}{l} G = \Lambda\cos^2\phi + 2\Pi\sin\phi\cos\phi + \mathrm{K}\sin^2\phi, \text{ and} \\ \mathfrak{Z} + \dfrac{d\mathfrak{H}}{ds} = -\left(M + \dfrac{d\Pi}{dy} + \dfrac{d\mathrm{K}}{dx}\right)\sin\phi - \left(L + \dfrac{d\Lambda}{dy} + \dfrac{d\Pi}{dx}\right)\cos\phi + \dfrac{d}{ds}[(\mathrm{K}-\Lambda)\sin\phi\cos\phi + \Pi(\cos^2\phi - \sin^2\phi)] \end{array}\right\} (5),$$

which are Kirchhoff's boundary equations.

Distribution of shearing force determined, which produces same flexure as a given distribution of couple round axes perpendicular to boundary.

646. The proposition stated at the end of last section is equivalent to this:—That a certain distribution of normal shearing force on the bounding edge of a finite plate may be determined which shall produce the same effect as any given distribution of couple, round axes everywhere perpendicular to the normal surface supposed to constitute the edge. To prove this let equal forces act in opposite directions in lines EF, $E'F'$ on each side of the middle line and parallel to it, constituting the supposed distribution of couple. It must be understood that the forces are actually distributed along their lines of action, and not, as in the abstract dynamics of ideal rigid bodies, applied indifferently at any points of these lines; but the amount of the force per unit of length, though equal in the neighbouring parts of the two lines, must differ from point to point along the edge, to constitute any other than a uniform distribution of couple. Lastly, we may suppose the forces in the opposite directions to be not confined to two lines, as shown in the diagram, but to be diffused over the two halves of the edge on the two sides of its middle line; and further, the amount of them in equal infinitely small breadths at different distances from the middle line must be proportional to these distances, as stated in § 634 (3), if the given distribution of couple is to be thoroughly such as H of § 645.

Distribution of shearing force determined, which produces same flexure as a given distribution of couple round axes perpendicular to boundary.

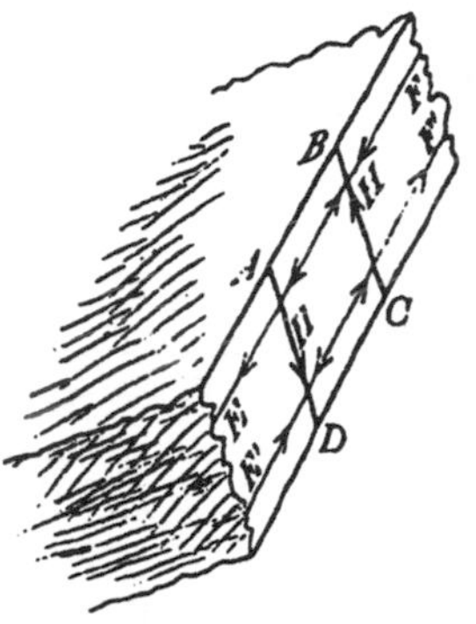

Let now the whole edge be divided into infinitely small rectangles, such as $ABCD$ in the diagram, by lines drawn perpendicularly across it. In one of these rectangles apply a balancing system of couples consisting of a diffused couple equal and opposite to the part of the given distribution of couple belonging to the area of the rectangle, and a couple of single forces in the lines AD, CB, of equal and opposite moment. This balancing system obviously cannot cause any sensible disturbance (stress or strain) in the plate, except within a distance comparable with the sides of the rectangle; and, therefore, when the same thing is done in all the rectangles into which the edge is divided, the plate is only disturbed to an infinitely small distance from the edge inwards all round. But the given distribution of couple is thus removed (being directly balanced by a system of diffused force equal and opposite everywhere to that constituting it), and there remains only the set of forces applied in the cross lines. Of these there are two in each cross line, derived from the operations performed in the two rectangles of which it is a common side, and their difference alone remains effective. Thus we see that if the given distribution of couple be uniform along the edge, it may be removed without disturbing the condition of the plate except infinitely near the edge: in other words,

Uniform distribution of twisting couple produces no flexure.

647. *A uniform distribution of couple along the whole edge of a finite plate, everywhere round axes in the plane of the plate, and perpendicular to the edge, produces distortion, spreading to only infinitely small distances inwards from the edge all round, and no stress or distortion of the plate as a whole.* The truth of this remarkable proposition is also obvious when we consider that the tendency of such a distribution of couple can only be to drag the two sides of the edge infinitesimally in opposite directions round the area of the plate. Later (§ 728) we shall investigate strictly the strain, in the neighbourhood of the edge, produced by it, and we shall find (§ 729) that it diminishes with extreme rapidity inwards from the edge, becoming practically insensible at distances exceeding twice the thickness of the plate.

The distribution of shearing force that produces same flexure as from distribution of twisting couple.

648. *A distribution of couple on the edge of a plate, round axes everywhere in the plane of the plate, and perpendicular to the edge, of any given amount per unit of length of the edge, may be removed, and, instead, a distribution of force perpendicular to the plate, equal in amount per unit length of the edge, to the rate of variation per unit length of the amount of the couple, without altering the flexure of the plate as a whole, or producing any disturbance in its stress or strain except infinitely near the edge.*

In the diagram of § 646 let $AB = \delta s$. Then if H be the amount of the given couple per unit length along the edge, between AD, BC, the amount of it on the rectangle $ABCD$ is $H\delta s$, and therefore H must be the amount of the forces introduced along AD, CB, in order that they may constitute a couple of the requisite moment. Similarly, if $H'\delta s$ denote the amount of the couple in the contiguous rectangle on the other side of BC, the force in BC derived from it will be H' in the direction opposite to H. There remains effective in BC a single force equal to the difference, $H'-H$.

If from A to B be the direction in which we suppose s, a length measured along the edge from any zero point, to increase, we have

$$H' - H = \frac{dH}{ds}\delta s.$$

Thus we are left with single forces, equal to $\frac{dH}{ds}\,\delta s$, applied in lines perpendicularly across the edge, at consecutive distances δs from one another; and for this we may substitute, without causing disturbance except infinitely near the edge, a continuous distribution of transverse force, amounting to dH/ds per unit length; which is the proposition to be proved. The direction of this force, when dH/ds is positive, is that of z negative: whence immediately the form of it expressed in (4) of § 645.

The distribution of shearing force that produces same flexure as from distribution of twisting couple.

649. As a first example of the application of these equations, we shall consider the very simple case of a uniform plate of finite or infinite extent, symmetrically influenced in concentric circles by a load distributed symmetrically, and by proper boundary appliances if required.

Case of circular strain.

Let the origin of co-ordinates be chosen at the centre of symmetry, and let r, θ be polar co-ordinates of any point P, so that

$$x = r\cos\theta, \quad y = r\sin\theta.$$

The second member of (6), § 644, will be a function of r, which for brevity we may now denote simply by Z (being the amount of load per unit area when the applied forces on each small part are reducible to a single normal force through some point of it). Since z is now a function of r, and, as we have seen before [§ 491 (e)],

$$\nabla^2 u = \frac{1}{r}\frac{d}{dr}\left(r\frac{du}{dr}\right)$$

when u is any function of r, equation (6) of § 644 becomes

$$\frac{A}{r}\frac{d}{dr}\left\{r\frac{d}{dr}\left[\frac{1}{r}\frac{d}{dr}\left(r\frac{dz}{dr}\right)\right]\right\} = Z \quad\ldots\ldots\ldots\ldots\ldots(1).$$

Hence

$$z = \frac{1}{A}\int\frac{dr}{r}\int r\int\frac{dr}{r}\int rZdr + \tfrac{1}{4}C(\log r - 1)r^2 + \tfrac{1}{4}C'r^2 + C''\log r + C''' \ldots(2),$$

which is the complete integral, with the four arbitrary constants explicitly shown. The following expressions, founded on intermediate integrals, deserve attention now, as promoting a thorough comprehension of the solution; and some of them will be required later for expressing the boundary conditions. The notation of (7) will be explained in § 650:—

Plate circularly strained.

$$\left.\begin{array}{c}\left(\begin{array}{c}\text{inclination, divided by radius; or curvature in}\\ \text{normal section perpendicular to radius}\end{array}\right)\\ \frac{1}{r}\frac{dz}{dr}=\frac{1}{Ar^2}\int rdr\int\frac{dr}{r}\int rZdr+\tfrac{1}{2}C(\log r-\tfrac{1}{2})+\tfrac{1}{2}C'+\frac{C''}{r^2}\end{array}\right\}\ldots(3),$$

$$\left.\begin{array}{c}\text{(curvature in radial section)}\\ \frac{d^2z}{dr^2}=-\frac{1}{Ar^2}\int rdr\int\frac{dr}{r}\int rZdr+\frac{1}{A}\int\frac{dr}{r}\int rZdr+\tfrac{1}{2}C(\log r+\tfrac{1}{2})+\tfrac{1}{2}C'-\frac{C''}{r^2}\end{array}\right\}\ldots(4),$$

$$\left.\begin{array}{c}\text{(sum of curvatures in rectangular sections)}\\ \nabla^2 z=\frac{1}{A}\int\frac{dr}{r}\int rZdr+C\log r+C'\end{array}\right\}\ldots\ldots(5),$$

$$\left.\begin{array}{l}A\frac{d^2z}{dr^2}+c\frac{dz}{rdr}=G\\ =-\frac{A-c}{Ar^2}\int rdr\int\frac{dr}{r}\int rZdr+\int\frac{dz}{r}\int rZdr+\tfrac{1}{2}C\{(A+c)\log r+\tfrac{1}{2}(A-c)\}\\ \qquad\qquad+\tfrac{1}{2}C'(A+c)-C''(A-c)\frac{1}{r^2}\\ H=0\end{array}\right\}\ldots(6),$$

$$L=c\frac{d^2z}{dr^2}+A\frac{dz}{rdr}\ \ldots\ldots\ldots\ldots\ldots\ldots\ldots(7),$$

$$\left.\begin{array}{l}(A-c)\frac{d}{dr}\left(\frac{1}{r}\frac{dz}{dr}\right)+\frac{dG}{dr}=A\frac{d}{dr}\nabla^2 z=-\zeta\\ =\frac{1}{r}\int rZdr+C\frac{A}{r}\end{array}\right\}\ldots\ldots\ldots(8).$$

Of these (6) and (8) express, according to the notation of § 645, the couple and the shearing force acting on the normal surface cutting the middle surface of the plate in the circle of radius r. They are derivable analytically from our solution (2) by means of (2), (3), and (1) of § 645, with (4) of § 644, and (15) of § 642. The work is of course much shortened by taking $y=0$, and $x=r$, and using (3) and (4) of the present section. The student may go through this process, with or without the abbreviation, as an analytical exercise; but it is more instructive, as well as more direct, to investigate *ab initio* the equilibrium of a plate symmetrically strained in concentric circles, and so, in the course of an independent demonstration of (6) § 644, for this case, or (1) § 649, to find expressions for the flexural and shearing stresses.

650. It is clear that, in every part of the plate, the normal sections (§ 637) of maximum and minimum, or minimum and maximum, bending couples are those through and perpendicular to the radius drawn from O the centre of symmetry. At distance r from O, let L and G be the bending couples in the section through the radius, and in the section perpendicular to it; so that, if λ and κ be the curvatures in these sections, we have, by (10) of § 641 and (15) of § 642,

Independent investigation for circular strain.

$$\left.\begin{aligned} L &= A\lambda + c\kappa \\ G &= c\lambda + A\kappa \end{aligned}\right\} \quad\text{.......................}(9).$$

Let also ζ be the shearing force (§ 616, footnote) in the circular normal section of radius r. The symmetry requires that there be no shearing force in radial normal sections.

Considering now an element, E, bounded by two radii making an infinitely small angle $\delta\theta$ with one another, and two concentric circles of radii $r - \frac{1}{2}\delta r$ and $r + \frac{1}{2}\delta r$; we see that the equal couples $L\delta r$ on its radial normal sections, round axes falling short of direct opposition by the infinitely small angle $\delta\theta$, have a resultant equal to $L\delta r\delta\theta$ round an axis perpendicular to the middle radius, in the negative direction when L is positive; and the infinitely nearly equal couples on its outer and inner circular edges have a resultant round the same axis, equal to $\frac{d}{dr}(Gr\delta\theta)\,\delta r$, being the difference of the values taken by $Gr\delta\theta$ when $r - \frac{1}{2}\delta r$ and $r + \frac{1}{2}\delta r$ are put for r. There is also the couple of the shearing forces on the outer and inner edges, each infinitely nearly equal to $\zeta r\delta\theta$; of which the moment is $\zeta r\delta\theta\delta r$. Hence, for the equilibrium of E under the action of these couples,

$$-L\delta r\delta\theta + \frac{d}{dr}(Gr)\,\delta r\delta\theta + \zeta r\delta\theta\delta r = 0,$$

or

$$-L + \frac{d}{dr}(Gr) + \zeta r = 0 \quad\text{................}(10),$$

if, as we may now conveniently do, we suppose no couples to be applied from without to any part of the plate except its bounding edges. Again, considering normal forces on E, we

Independent investigation for circular strain.

have $\frac{d}{dr}(\zeta r\delta\theta)\,\delta r$ for the sum of those acting on it from the contiguous matter of the plate, and $Zr\delta\theta\delta r$ from external matter if, as above, Z denote the amount of applied normal force per unit area of the plate. Hence, for the equilibrium of these forces,

$$\frac{d}{dr}(\zeta r) + Zr = 0 \qquad\qquad (11).$$

Substituting for ζ in (11) by (10); for L and G in the result by (9); and, in the result of this, for λ and κ their expressions by the differential calculus, which are dz/rdr and d^2z/dr^2, since the plate is a surface of revolution differing infinitely little from a plane perpendicular to the axis, we arrive finally at (1) the differential equation of the problem. Of the other formulæ of § 649, (6), (7), (8) follow immediately from (9) and (10) now proved: except $H=0$, which follows from the fact that the radial and circular normal sections are the sections of maximum and minimum, or minimum and maximum, curvature.

Interpretation of terms in integral.

651. We are now able to perceive the meaning of each of the four arbitrary constants.

(1) C''' is of course merely a displacement of the plate without strain.

(2) $C''\log r$ is a displacement which produces anticlastic curvature throughout, with $\pm C''/r^2$ for the curvatures in the two principal sections: corresponding to which the bending couples, L, G, are equal to $\pm(A-c)\,C''/r^2$. An infinite plane plate, with a circular aperture, and a uniform distribution of bending couple applied to the edge all round, in each part round the tangent as axis, would experience this effect; as we see from the fact that the stress in the plate, due to C'', diminishes according to the inverse square of the distance from the centre of symmetry. It is remarkable that although the absolute value of the deflection, $C''\log r$, is infinite for infinite values of r, the restrictive condition (3) of § 632 is not violated provided C'' is infinitely small in comparison with the thickness: and it may be readily proved that the law (1) of § 633 is, in point of fact, fulfilled by

this deflection, even if the whole displacement has rigorously this value, $C''\log r$, and is precisely in the direction perpendicular to the undisturbed plane. For this case $\zeta=0$, or there is no shear. Interpretation of terms in integral.

(3) $\frac{1}{4}C'r^2$ is a displacement corresponding to spherical curvature: and therefore involving simply a uniform synclastic stress [§ 638 (1)], of which the amount is of course [§ 641 (10) or (11)] equal to $A+c$ divided by the radius of curvature, or $(A+c)\times\frac{1}{2}C'$, agreeing with the equal values given for L and G by (6) and (7) of § 649. In this case also $\zeta=0$, or there is no shearing force. A finite plate of any shape, acted on by a uniform bending couple all round its edge, becomes bent thus spherically.

(4) $\frac{1}{4}C(\log r-1)r^2$ is a deflection involving a shearing force equal to $-AC/r$, and a bending couple,

$$\tfrac{1}{2}C\{(A+c)\log r+\tfrac{1}{2}(A-c)\},$$

in the circle of distance r from the centre of symmetry.

652. It is now a problem of the merest algebra to find the flexure of a flat ring, or portion of plane plate bounded by two concentric circles, when acted on by any given bending couples and transverse forces applied uniformly round its outer and inner edges. For equilibrium, the forces on the outer and inner edges must be in contrary directions, and of equal amounts. Thus we have three arbitrary data: the amounts of the couple applied to the two edges, each reckoned per unit of length, and the whole amount, F, of the force on either edge. By (4), § 651, or (8) of § 649, we see that Symmetrical flexure of flat ring.

$$-C=\frac{F}{2\pi A}\text{......................}(12);$$

and there remain unknown the two constants, C' and C'', to be determined from the two equations given by putting the expression for G [(6) of § 649] equal to the equal values for the values of r at the outer and inner edges respectively.

Example.—A circular table (of isotropic material), with a concentric circular aperture, is supported by its outer edge,

Symmetrical flexure of flat ring.

which rests simply on a horizontal circle; and is deflected by a load uniformly distributed over its inner edge (or *vice versâ*, inner for outer). To find the deflection due to this load (which of course is simply added to the deflection due to the weight, determined below). Here G must vanish at each edge.

The radii of the outer and inner edges being a and a', the equations are

$$\tfrac{1}{2}C\{(A+c)\log a+\tfrac{1}{2}(A-c)\}+\tfrac{1}{2}C'(A+c)-C''(A-c)\frac{1}{a^2}=0,$$

and the same with a' for a. Hence

Flexure of flat ring equilibrated by forces symmetrically distributed over its edges;

$$C''(A-c)\left(\frac{1}{a'^2}-\frac{1}{a^2}\right)=-\tfrac{1}{2}C(A+c)\log\frac{a}{a'},$$

and

$$\tfrac{1}{2}C'(A+c)(a^2-a'^2)=-\tfrac{1}{2}C\left[(A+c)(a^2\log a-a'^2\log a')+\tfrac{1}{2}(A-c)(a^2-a'^2)\right]:$$

and thus, using for C its value (12), we find [(2) § 649]

$$z=\frac{F}{2\pi A}\left[\tfrac{1}{4}\left(-\log r+1+\frac{a^2\log a-a'^2\log a'}{a^2-a'^2}+\tfrac{1}{2}\frac{A-c}{A+c}\right)r^2+\tfrac{1}{2}\frac{A+c}{A-c}\frac{a^2a'^2\log\frac{a}{a'}}{a^2-a'^2}\log r+C'''\right].$$

Putting the factor of r^2 into a more convenient form, and assigning C''' so that the deflection may be reckoned from the level of the inner edge, we have finally

$$z=\frac{F}{2\pi A}\left\{\tfrac{1}{4}\left(-\log\frac{r}{a'}+\frac{a^2}{a^2-a'^2}\log\frac{a}{a'}+\tfrac{1}{2}\frac{3A+c}{A+c}\right)r^2\right.$$

$$\left.+\tfrac{1}{2}\frac{A+c}{A-c}\frac{a^2a'^2\log\frac{a}{a'}}{a^2-a'^2}\log\frac{r}{a'}-\tfrac{1}{4}\frac{a^2a'^2}{a^2-a'^2}\log\frac{a}{a'}-\tfrac{1}{8}\cdot\frac{3A+c}{A+c}a'^2\right\}\ldots(13).$$

Towards showing the distribution of stress through the breadth of the ring, we have from this, by § 649 (6),

$$G=\frac{F}{2\pi a}\cdot\tfrac{1}{2}(A+c)\left(\frac{a^2}{a^2-a'^2}\log\frac{a}{a'}-\log\frac{r}{a'}-\frac{a^2a'^2}{a^2-a'^2}\log\frac{a}{a'}\frac{1}{r^2}\right)\ldots(14),$$

which, as it ought to do, vanishes when $r=a'$, and when $r=a$. Further, by § 649 (8),

$$\zeta=\frac{F}{2\pi r}\ldots\ldots\ldots\ldots\ldots\ldots(15),$$

which shows that, as is obviously true, the whole amount of the transverse force in any concentric circle of the ring is equal to F.

653. The problem of § 652, extended to admit a load distributed in any symmetrical manner over the surface of the ring instead of merely confined to one edge, is solved algebraically in precisely the same manner, when the terms dependent on Z, and exhibited in the several expressions of § 649, are found by integration. One important remark we have to make however: that much needless labour is avoided by treating Z as a discontinuous function in these integrations in cases in which one continuous algebraic or transcendental function does not express the distribution of load over the whole portion of plate considered. Unless this plan were followed, the expression for z, dz/dr, G, and ζ, would have to be worked out separately for each annular portion of plate through which Z is continuous, and their values equated on each side of each separating circle. Hence if there were i annular portions to be thus treated separately there would be $4i$ arbitrary constants, to be determined by the $4(i-1)$ equations so obtained, and the 4 equations expressing that at the outer and inner bounding circular edges G has the prescribed values (whether zero or not) of the applied bending couples, and that z and ζ have each a prescribed value at one or other of these circles. But by the more artful method (due to Fourier and Poisson), the complication of detail required in virtue of the discontinuity of Z is confined to the successive integrations; and the arbitrary constants, of which there are now but four, are determined by the conditions for the two extreme bounding edges. and with load symmetrically spread over its area.

Example.—A circular table (of isotropic material), with a concentric circular aperture, is borne by its outer or inner edge which rests simply on a horizontal circular support, and is loaded by matter uniformly distributed over an annular area of its surface, extending from its inner edge outwards to a concentric circle of given radius, c. It is required to find the flexure.

First, supposing the aperture filled up, and the plate uniform from outer edge to centre, let the whole circle of radius c be uniformly loaded at the rate w, a constant, per unit of its area.

We have

	$Z=$	$\int rZdr=$	$\int\frac{dr}{r}\int rZdr=$	$\int rdr\int\frac{dr}{r}\int rZdr=$	$\int\frac{dr}{r}\int rdr\int\frac{dr}{r}\int rZdr=$
When $r=0$	w	0	0	0	0
" $<c$	w	$\frac{1}{2}wr^2$	$\frac{1}{4}wr^2$	$\frac{1}{16}wr^4$	$\frac{1}{64}wr^4$
" $>c$	0	$\frac{1}{2}wc^2$	$\frac{1}{4}wc^2\left(2\log\frac{r}{c}+1\right)$	$\frac{1}{16}wc^2\left(4r^2\log\frac{r}{c}+c^2\right)$	$\frac{1}{16}wc^2\left(2r^2\log\frac{r}{c}-r^2+c^2\log\frac{r}{c}+\frac{5}{4}c^2\right)$
	I.	II.	III.	IV.	V.

Of these results, V. used in (2) gives the general solution; and IV., III., and II. in (6) and (8) give the corresponding expressions for G and ζ. If, first, we suppose the value of G thus found to have any given value for each of two values, r', r'', of r, and ζ to have a given value for one of these values of r, we have three simple algebraic equations to find C, C', C''; and we solve a more general problem than that proposed; to which we descend by making the prescribed values of G and ζ zero. The power of mathematical expression and analysis in dealing with discontinuous functions, is strikingly exemplified in the applicability of the result not only to the contemplated case, in which c is intermediate between r' and r''; but also to cases in which c is less than either (when we fall back on the previous case, of § 652), or c greater than either (when we have a solution more directly obtainable by taking $Z=w$ for all values of r).

Circular table of isotropic material, supported symmetrically on its edge, and strained only by its own weight.

If the plate is in reality continuous to its centre, and uniformly loaded over the whole area of the circle of radius c, we must have $C=0$ and $C''=0$ to avoid infinite values of ζ and G at the centre: and the equation $G=0$ for the outer boundary of the disc gives C' at once, completing the determination. If, lastly, we suppose c to be not less than the radius of the disc, we have the solution for a uniform circular disc uniformly supported round its edge, and strained only by its own weight.

Reduction of general problem to case of no load over area.

654. If now we consider the general problem,—to determine the flexure of a plate of any form, with an arbitrary distribution of load over it, and with arbitrary boundary appliances, subject of course to the condition that all the applied forces, when the data are entirely of force, must constitute an equilibrating system; we may immediately reduce this problem to the simpler one in which there is no load distributed over the area, but arbitrary boundary appliances only. We shall merely sketch the mathematical investigation.

Reduction of general problem to case of no load over area.

First it is easily proved, as for a corresponding expression for three independent variables in § 491 (c), that

$$\left(\frac{d^2}{dx^2}+\frac{d^2}{dy^2}\right)\iint \rho' \log D\, dx'dy' = 2\pi\rho \ldots\ldots\ldots\ldots\ldots (1),$$

where ρ' is any function of two independent variables, x', y'; ρ the same function of x, y; D denotes $\sqrt{\{(x-x')^2+(y-y')^2\}}$; and $\iint$ denotes integration over an area comprehending all values of x', y', for which ρ' does not vanish. Hence

$$\left(\frac{d^2}{dx^2}+\frac{d^2}{dy^2}\right)^2 u = Z \ldots\ldots\ldots\ldots\ldots\ldots\ldots\ldots (2),$$

if

$$u = \frac{1}{4\pi^2}\iint dx'dy' \log D \iint dx''dy'' Z'' \log D' \ldots\ldots\ldots (3),$$

where $D' = \sqrt{\{(x''-x')^2+(y''-y')^2\}}$; and if Z'' and Z denote the values for (x'', y'') and (x, y) of any arbitrary function of two independent variables. Let this function denote the amount of load per unit of area, which we may suppose to vanish for all values of the co-ordinates not included in the plate; and to avoid trouble regarding limits, let all the integrals be supposed to extend from $-\infty$ to $+\infty$. We thus have, in $z=u$, a solution of our equation (2): and therefore $z-u$ must satisfy the same equation with the second member replaced by zero: or, if $\mathfrak{z}$ denote a general solution of

$$\left(\frac{d^2}{dx^2}+\frac{d^2}{dy^2}\right)^2 \mathfrak{z} = 0 \ldots\ldots\ldots\ldots\ldots\ldots\ldots\ldots (4),$$

then

$$z = u + \mathfrak{z} \ldots\ldots\ldots\ldots\ldots\ldots\ldots\ldots\ldots\ldots (5)$$

is the general solution of (2). The boundary conditions for $\mathfrak{z}$ are of course had by substituting $u+\mathfrak{z}$ for z in the directly prescribed boundary equations, whatever they may be.

655. Mathematicians have not hitherto succeeded in solving this problem with complete generality, for any other form of plate than the circular ring (or circular disc with concentric circular aperture). Having given (§§ 640, 653) a detailed

Flat circular ring the only case hitherto solved.

Flat circular ring the only case hitherto solved.

solution of the problem for this case, subject to the restriction of symmetry, we shall merely indicate the extension of the analysis to include any possible non-symmetrical distribution of strain. The same analysis, under much simpler conditions, will occur to us again and again, and will be on some points more minutely detailed, when we shall be occupied with important practical problems regarding electric influence, fluid motion, and electric and thermal conduction, through cylindrical spaces.

Taking the centre of the circular bounding edges as origin for polar co-ordinates, let

$$x = r\cos\theta,\ y = r\sin\theta.$$

We easily find by transformation

$$\frac{d^2\mathfrak{z}}{dx^2} + \frac{d^2\mathfrak{z}}{dy^2} = \frac{1}{r}\frac{d}{dr}\left(r\frac{d\mathfrak{z}}{dr}\right) + \frac{1}{r^2}\frac{d^2\mathfrak{z}}{d\theta^2} \quad\text{..........} (6).$$

If we put $\log r = \vartheta$, or $r = \epsilon^{\vartheta}$ ....(7),

this becomes $\dfrac{d^2\mathfrak{z}}{dx^2} + \dfrac{d^2\mathfrak{z}}{dy^2} = \epsilon^{-2\vartheta}\left(\dfrac{d^2\mathfrak{z}}{d\vartheta^2} + \dfrac{d^2\mathfrak{z}}{d\theta^2}\right)$ .... (8).

Hence if, as before, ∇^2 denote $\dfrac{d^2}{dx^2} + \dfrac{d^2}{dy^2}$,

$$\nabla^4\mathfrak{z} = \epsilon^{-2\vartheta}\left(\frac{d^2}{d\vartheta^2} + \frac{d^2}{d\theta^2}\right)\epsilon^{-2\vartheta}\left(\frac{d^2}{d\vartheta^2} + \frac{d^2}{d\theta^2}\right)\mathfrak{z} \quad\text{..........(9).}$$

This equated to zero gives

$$\frac{d^2\mathfrak{z}}{d\vartheta^2} + \frac{d^2\mathfrak{z}}{d\theta^2} = \epsilon^{2\vartheta}v \quad\text{..........(10),}$$

if v denote any solution of

$$\frac{d^2v}{d\vartheta^2} + \frac{d^2v}{d\theta^2} = 0 \quad\text{..........(11).}$$

We shall see, when occupied with the electric and other problems referred to above, that a general solution of this equation, appropriate for our present problem as for all involving the expression of arbitrary functions of θ for particular values of ϑ, is

$$v = \sum_0^\infty \{(A_i\cos i\theta + B_i\sin i\theta)\epsilon^{i\vartheta} + (\mathfrak{A}_i\cos i\theta + \mathfrak{B}_i\sin i\theta)\epsilon^{-i\vartheta}\} \text{...(12),}$$

where A_i, B_i, $\mathfrak{A}_i$, $\mathfrak{B}_i$ are constants. That this is a solution, is of course verified in a moment by differentiation. From it we readily find (and the result of course is verified also by differentiation),

Flat circular ring the only case hitherto solved.

$$\mathfrak{z}=\sum_{i=0}^{i=\infty}\left\{\frac{1}{(i+2)^2-i^2}(A_i\cos i\theta+B_i\sin i\theta)\,\epsilon^{(i+2)\vartheta}\right\}$$
$$+\sum_{i=2}^{i=\infty}\left\{\frac{-1}{i^2-(i-2)^2}(\mathfrak{A}_i\cos i\theta+\mathfrak{B}_i\sin i\theta)\,\epsilon^{-(i-2)\vartheta}\right\}-\tfrac{1}{2}(\mathfrak{A}_1\cos\theta+\mathfrak{B}_1\sin\theta)\,\vartheta\epsilon^{\vartheta}+v' \qquad \ldots\ldots\ldots (13),$$

v' being any solution of (11), which may be conveniently taken as given by (12) with accented letters A_i', etc., to denote four new constants. If now the arbitrary periodic functions of θ, with 2π for period, given as the values whether of displacement, or shearing force, or couple, for the outer and inner circular edges, be expressed by Fourier's theorem [§ 77 (14)] in simple harmonic series; the two equations [§ 645 (5)] for each edge, applied separately to the coefficients of $\cos i\theta$ and $\sin i\theta$ in the expressions thus obtained, give eight equations for determining the eight constants A_i, $\mathfrak{A}_i$, B_i, $\mathfrak{B}_i$, A_i', $\mathfrak{A}_i'$, B_i', $\mathfrak{B}_i'$.

656. Although the problem of fulfilling arbitrary boundary conditions has not yet been solved for rectangular plates, there is one remarkable case of it which deserves particular notice; not only as interesting in itself, and important in practical application, but as curiously illustrating one of the most difficult points [§§ 646, 648] of the general theory. A rectangular plate acted on perpendicularly by a balancing system of four equal parallel forces applied at its four corners, becomes strained to a condition of uniform anticlastic curvature throughout, with the sections of no-flexure parallel to its sides, and therefore with sections of equal opposite maximum curvature in the normal planes inclined to the sides at 45°. This follows immediately from § 648, if we suppose the corners rounded off ever so little, and the forces diffused over them.

Rectangular plate, held and loaded by diagonal pairs of corners.

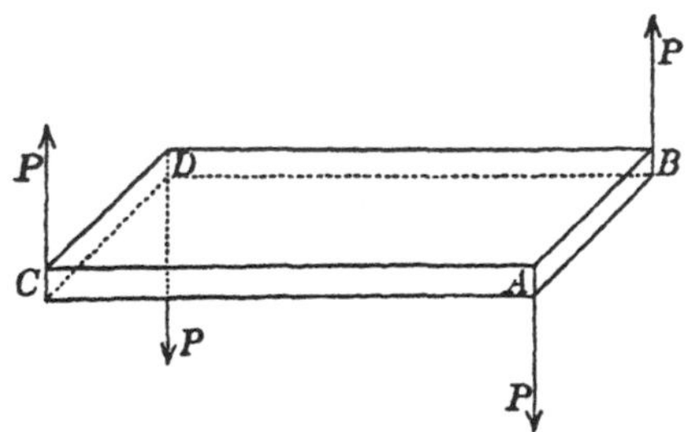

Rectangular plate, held and loaded by diagonal pairs of corners.

Or, in each of an infinite number of normal lines in the edge AB, let a pair of opposite forces each equal to $\frac{1}{2}P$ be applied; which cannot disturb the plate. These, with halves of the single forces P in the dissimilar directions at the corners A and B, constitute a diffused couple over the whole edge AB, amounting in moment per unit of length to $\frac{1}{2}P$, round axes perpendicular to the plane of the edge. Similarly, the other halves of the forces P at the corners A, B, with halves of those at C and D and introduced balancing forces, constitute diffused couples over the edges CA and DB; and the remaining halves of the corner forces at C and D, with introduced balancing forces, constitute a diffused couple over CD; each having $\frac{1}{2}P$ for the amount of moment per unit length of the edge over which it is diffused. Their directions are mutually related in the manner specified in § 638 (2), and thus taken all together, they constitute an anticlastic stress of value $\Omega = \frac{1}{2}P$. Hence (§ 642) the result is uniform anticlastic strain amounting to $\frac{1}{2}P/k$, and having its axes inclined at 45° to the edges; that is to say (§ 639), a flexure with maximum curvatures on the two sides of the tangent plane each equal to $\frac{1}{2}P/k$, and in normal sections in the positions stated.

Transition to finite flexures indicated.

657 Few problems of physical mathematics are more curious than that presented by the transition from this solution, founded on the supposition that the greatest deflection is but a small fraction of the thickness of the plate, to the solution for larger flexures, in which corner portions will bend approximately as developable surfaces (cylindrical, in fact), and a central quadrilateral part will remain infinitely nearly plane; and thence to the extreme case of an infinitely thin perfectly flexible rectangle of inextensible fabric. This extreme case may be easily observed and experimented on by taking a carefully cut rectangle of paper (§ 145), supporting it by fine threads attached to two opposite corners, and kept parallel, while two equal weights are hung by threads from the other corners.

Transmission of force through an elastic solid.

658. The definitions and investigations regarding strain of §§ 154—190 constitute a kinematical introduction to the theory of elastic solids. We must now, in commencing the elementary dynamics of the subject, consider the forces called into play

through the interior of a solid when brought into a condition of strain. We adopt, from Rankine*, the term *stress* to designate such forces, as distinguished from strain defined (§ 154) to express the merely geometrical idea of a change of volume or figure.

Transmission of force through an elastic solid.

659. When through any space in a body under the action of force, the mutual force between the portions of matter on the two sides of any plane area is equal and parallel to the mutual force across any equal, similar, and parallel plane area, the stress is said to be homogeneous through that space. In other words, the stress experienced by the matter is homogeneous through any space if all equal similar and similarly turned portions of matter within this space are similarly and equally influenced by force.

Homogeneous stress.

660. To be able to find the distribution of force over the surface of any portion of matter homogeneously stressed, we must know the direction, and the amount per unit area, of the force across a plane area cutting through it in any direction. Now if we know this for any three planes, in three different directions, we can find it for a plane in any direction, as we see in a moment by considering what is necessary for the equilibrium of a tetrahedron of the substance. The resultant force on one of its faces must be equal and opposite to the resultant of the forces on the three others, which is known if these faces are parallel to the three planes for each of which the force is given.

Force transmitted across any surface in elastic solid.

661. Hence the stress, in a body homogeneously stressed, is completely specified when the direction, and the amount per unit area, of the force on each of three distinct planes is given. It is, in the analytical treatment of the subject, generally convenient to take these planes of reference at right angles to one another. But we should immediately fall into error did we not remark that the specification here indicated consists not of nine but in reality only of six independent elements. For if the equilibrating forces on the six faces of a cube be each resolved into three components parallel to its three edges OX, OY, OZ, we have in all 18 forces; of which each pair acting perpendicularly

Specification of a stress; by six independent elements.

* *Cambridge and Dublin Mathematical Journal*, 1850.

on a pair of opposite faces, being equal and directly opposed, balance one another. The twelve tangential components that remain constitute three pairs of couples having their axes in the direction of the three edges, each of which must separately be in equilibrium. The diagram shows the pair of equilibrating couples having OY for axis; from the consideration of which we infer that the forces on the faces (zy), parallel to OZ, are equal to the forces on the faces (yx), parallel to OX. Similarly, we see that the forces on the faces (yx), parallel to OY, are equal to those of the faces (xz), parallel to OZ; and that the forces on (xz), parallel to OX, are equal to those on (zy), parallel to OY.

Relations between pairs of tangential tractions necessary for equilibrium.

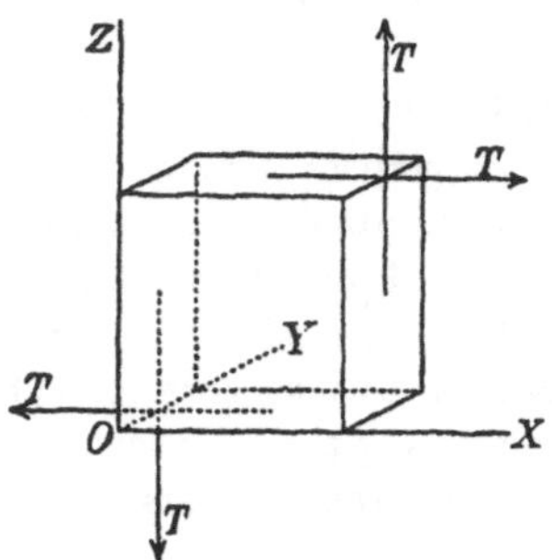

Specification of a stress; by six independent elements: three simple longitudinal stresses, and three simple shearing stresses.

662. Thus, any three rectangular planes of reference being chosen, we may take six elements thus, to specify a stress: P, Q, R the normal components of the forces on these planes; and S, T, U the tangential components, respectively perpendicular to OX, of the forces on the two planes meeting in OX, perpendicular to OY, of the forces on the planes meeting in OY, and perpendicular to OY, of the forces on the planes meeting in OY; each of the six forces being reckoned per unit of area. A normal component will be reckoned as positive when it is a traction tending to separate the portions of matter on the two sides of its plane. P, Q, R are sometimes called longitudinal stresses, sometimes simple normal tractions, and S, T, U shearing stresses.

Simple longitudinal, and shearing, stresses.

Force across any surface in terms of rectangular specification of stress.

From these data, to find in the manner explained in § 660, the force on any plane, specified by l, m, n, the direction-cosines of its normal; let such a plane cut OX, OY, OZ in the three points X, Y, Z. Then, if the area XYZ be denoted for a moment by A, the areas YOZ, ZOX, XOY, being its projections on the three rectangular planes, will be respectively equal to Al, Am, An. Hence, for the equilibrium of the tetrahedron of matter bounded by those four triangles, we have, if F, G, H denote the components of the force experienced by the first of them, XYZ, per unit of its area,

Force across any surface in terms of rectangular specification of stress.

$$F.A = P.lA + U.mA + T.nA,$$

and the two symmetrical equations for the components parallel to OY and OZ. Hence, dividing by A, we conclude

$$\left.\begin{aligned} F &= Pl + Um + Tn \\ G &= Ul + Qm + Sn \\ H &= Tl + Sm + Rn \end{aligned}\right\} \dots\dots\dots\dots\dots\dots\dots(1).$$

These expressions stand in the well-known relation to the ellipsoid

$$Px^2 + Qy^2 + Rz^2 + 2(Syz + Tzx + Uxy) = 1 \dots\dots\dots(2),$$

according to which, if we take

$$x = lr, \quad y = mr, \quad z = nr,$$

and if λ, μ, ν denote the direction-cosines and p the length of the perpendicular from the centre to the tangent plane at (x, y, z) of the ellipsoid, we have

$$F = \frac{\lambda}{pr}, \quad G = \frac{\mu}{pr}, \quad H = \frac{\nu}{pr}.$$

We conclude that

663. For any fully specified state of stress in a solid, a quadric surface may always be determined, which shall represent the stress graphically in the following manner:—

Stress-quadric.

To find the direction, and the amount per unit area, of the force acting across any plane in the solid, draw a radius perpendicular to this plane from the centre of the quadric to its surface. The required force will be equal to the reciprocal of the product of the length of this radius into the perpendicular from the centre to the tangent plane at the extremity of the radius, and will be perpendicular to this tangent plane.

664. From this it follows that for any stress whatever there are three determinate planes at right angles to one another such that the force acting in the solid across each of them is precisely perpendicular to it. These planes are called the principal or normal planes of the stress; the forces upon them, per unit area, —its principal or normal tractions; and the lines perpendicular

Principal planes and axes of a stress.

Principal planes and axes of a stress.

to them,—its principal or normal axes, or simply its axes. The three principal semi-diameters of the quadric surface are equal to the reciprocals of the square roots of the principal tractions. If, however, in any case each of the three principal tractions is negative, it will be convenient to reckon them rather as *pressures;* the reciprocals of the square roots of which will be the semi-axes of a real stress-ellipsoid representing the distribution of force in the manner explained above, with pressure substituted throughout for traction.

Varieties of stress-quadric.

665. When the three principal tractions are all of one sign, the stress-quadric is an ellipsoid; the cases of an ellipsoid of revolution and a sphere being included, as those in which two, or all three, are equal. When one of the three is negative and the two others positive, the surface is a hyperboloid of one sheet. When one of the normal tractions is positive and the two others negative, the surface is a hyperboloid of two sheets.

666. When one of the three principal tractions vanishes, while the other two are finite, the stress-quadric becomes a cylinder, circular, elliptic, or hyperbolic, according as the other two are equal, unequal, of one sign, or of contrary signs. When two of the three vanish, the quadric becomes two planes; and the stress in this case is (§ 662) called a simple longitudinal stress. The theory of principal planes, and principal or normal tractions, just stated (§ 664), is then equivalent to saying that any stress whatever may be regarded as made up of three simple longitudinal stresses in three rectangular directions. The geometrical interpretations are obvious in all these cases.

Composition of stresses.

667. The composition of stresses is of course to be effected by adding the component tractions thus:—If $(P_1, Q_1, R_1, S_1, T_1, U_1)$, $(P_2, Q_2, R_2, S_2, T_2, U_2)$, etc., denote, according to § 662, any given set of stresses acting simultaneously in a substance, their joint effect is the same as that of a single resultant stress of which the specification in corresponding terms is $(\Sigma P, \Sigma Q, \Sigma R, \Sigma S, \Sigma T, \Sigma U)$.

Laws of strain and stress compared.

668. Each of the statements that have now been made (§§ 659, 667) regarding stresses, is applicable to *infinitely small* strains, if for traction perpendicular to any plane, reckoned per unit of its area, we substitute *elongation*, in the lines of the traction, reckoned per unit of length; and for *half the tangential traction* parallel to any direction, *shear* in the same direction reckoned in the manner explained in § 175. The student will find it a useful exercise to study in detail this transference of each one of those statements, and to justify it by modifying in the proper manner the results of §§ 171, 172, 173, 174, 175, 185, to adapt them to infinitely small strains. It must be remarked that the strain-quadric thus formed according to the rule of § 663, which may have any of the varieties of character mentioned in §§ 665, 666, is not the same as the strain-ellipsoid of § 160, which is always essentially an ellipsoid, and which, for an infinitely small strain, differs infinitely little from a sphere.

Laws of strain and stress compared.

The comparison of § 172, with the result of § 661 regarding tangential tractions, is particularly interesting and important.

669. The following schedule of the meaning of the elements constituting the corresponding rectangular specifications of a strain and stress explained in preceding sections, will be found convenient:—

Rectangular elements of strains and stresses.

Components of the strain.	Components of the stress.	Planes; of which relative motion, or across which force is reckoned.	Direction of relative motion or of force.
e	P	yz	x
f	Q	zx	y
g	R	xy	z
a	S	$\begin{cases} yx \\ zx \end{cases}$	y z
b	T	$\begin{cases} zy \\ xy \end{cases}$	z x
c	U	$\begin{cases} xz \\ yz \end{cases}$	x y

670. If a unit cube of matter, given under any stress (P, Q, R, S, T, U), be subjected further to such infinitesimal change of this stress as shall produce an infinitely small simple longitudinal strain e alone, the work done on it will be Pe; since, of

Work done by a stress within a varying solid.

Work done by a stress within a varying solid.

the component forces P, U, T parallel to OX, U and T do no work in virtue of this strain. Similarly Qf, Rg are the works done if, the same stress acting, infinitesimal strains f or g are produced, either of them alone. Again, if the cube experiences a simple shear, a, whether we regard it (§ 172) as a differential sliding of the planes yx, parallel to y, or of the planes zx, parallel to z, we see that the work done is Sa: and similarly, Tb if the strain is simply a shear b, parallel to OZ, of planes zy, or parallel to OX, of planes xy: and Uc if the strain is a shear c, parallel to OX, of planes xz, or parallel to OY, of planes yz. Hence the whole work done by the stress (P, Q, R, S, T, U) on a unit cube taking the additional infinitesimal strain (e, f, g, a, b, c), while the stress varies only infinitesimally, is

Compare § 673, (20).

$$Pe+Qf+Rg+Sa+Tb+Uc \quad\text{.............. (3).}$$

It is to be remarked that, inasmuch as the action called a stress is a system of forces which balance one another if the portion of matter experiencing it is rigid, it cannot (§ 551) do any work when the matter moves in any way without change of shape: and therefore no amount of translation or rotation of the cube taking place along with the strain can render the amount of work done different from that just found.

If the side of the cube be of any length p, instead of unity, each force will be p^2 times, and each relative displacement p times; and therefore the work done p^3 times the respective amounts reckoned above. Hence a body of any shape, and of cubic content C, subjected throughout to a uniform stress (P, Q, R, S, T, U) while taking uniformly throughout an additional strain (e, f, g, a, b, c), experiences an amount of work equal to

$$(Pe+Qf+Rg+Sa+Tb+Uc)C \quad\text{............ (4).}$$

Work done on the surface of a varying solid.

It is to be remarked that this is necessarily equal to the work done on the bounding surface of the body by forces applied to it from without. For the work done on any portion of matter within the body is simply that done on its surface by the matter touching it all round, as no force acts at a distance from without on the interior substance. Hence if we imagine the whole body divided into any number of parts, each of any shape, the sum of the works done on all these parts is, by the disappearance of equal positive and negative terms expressing the portions of the work done on each part by the contiguous parts on all its sides, and spent by these other parts in this action, reduced to the integral amount of work done by force from without, applied all round the outer surface.

Work done on the surface of a varying solid.

The analytical verification of this is instructive with regard to the syntax of the mathematical language in which the theory of the transmission of force is expressed. Let x, y, z be the coordinates of any point within the body; W the whole amount of work done in the circumstances specified above; and $\iiint$ integration extended throughout the space occupied by the body: so that

$$W = \iiint (Pe + Qf + Rg + Sa + Tb + Uc)\, dxdydz \ldots\ldots\ldots (5).$$

If now we denote by α, β, γ the component displacements of any point of the matter infinitely near the point (x, y, z), experienced when the additional strain (e, f, g, a, b, c) takes place, whether non-rotationally (§ 182) and with some point of the body fixed, or with any motion of translation whatever and any infinitely small rotation, by adapting § 181 (5) to infinitely small strains according to our present notation (§ 669), and using in it § 190 (e), we have

Strain-components in terms of displacement.

$$\left.\begin{array}{lll} e = \dfrac{d\alpha}{dx}, & f = \dfrac{d\beta}{dy}, & g = \dfrac{d\gamma}{dz}, \\ a = \dfrac{d\beta}{dz} + \dfrac{d\gamma}{dy}, & b = \dfrac{d\gamma}{dx} + \dfrac{d\alpha}{dz}, & c = \dfrac{d\alpha}{dy} + \dfrac{d\beta}{dx} \end{array}\right\} \ldots\ldots\ldots\ldots (6).$$

With these, (5) becomes

Work done through interior;

$$W = \iiint \left(P\frac{d\alpha}{dx} + U\frac{d\beta}{dx} + T\frac{d\gamma}{dx} + U\frac{d\alpha}{dy} + Q\frac{d\beta}{dy} + S\frac{d\gamma}{dy} + T\frac{d\alpha}{dz} + S\frac{d\beta}{dz} + R\frac{d\gamma}{dz} \right) dxdydz \ldots (7).$$

Hence by integration

$$W = \iint [(P\alpha + U\beta + T\gamma)dydz + (U\alpha + Q\beta + S\gamma)dzdx + (T\alpha + S\beta + R\gamma)dxdy] \ldots\ldots\ldots (8),$$

the limits of the integrations being so taken that, if $d\sigma$ denote an element of the bounding surface, $\iint$ integration all over it, and l, m, n the direction-cosines of the normal at any point of it, the expression means the same as

$$W = \iint \{(P\alpha + U\beta + T\gamma)l + (U\alpha + Q\beta + S\gamma)m + (T\alpha + S\beta + R\gamma)n\} d\sigma \ldots (9);$$

which, with the terms grouped otherwise, becomes

$$W = \iint\{(Pl+Um+Tn)\alpha+(Ul+Qm+Sn)\beta+(Tl+Sm+Rn)\gamma\}d\sigma \ldots (10).$$

agrees with work done on surface.

The second member of this, in virtue of (1), expresses directly the work done by the forces applied from without to the bounding surface.

Differential equation of work done by a stress.

671. If, now, we suppose the body to yield to a stress (P, Q, R, S, T, U), and to oppose this stress only with its innate resistance to change of shape, the differential equation of work done will [by (4) with de, df, etc., substituted for e, f, etc.] be

$$dw = Pde + Qdf + Rdg + Sda + Tdb + Udc \ldots\ldots\ldots (11),$$

if w denote the whole amount of work done per unit of volume in any part of the body while the substance in this part experiences a strain (e, f, g, a, b, c) from some initial state re-

Physical application.

garded as a state of no strain. This equation, as we shall see later, under Properties of Matter, expresses the work done in a natural fluid, by distorting stress (or difference of pressure in different directions) working against its innate viscosity; and w is then, according to Joule's discovery, the dynamic value of the heat generated in the process. The equation may also be applied to express the work done in straining an imperfectly elastic solid, or an elastic solid of which the temperature varies during the process. In all such applications the stress will depend partly on the speed of the straining motion, or on the varying temperature, and not at all, or not solely, on the state of strain at any moment, and the system will not be dynamically conservative.

Perfectly elastic body defined, in abstract dynamics.

672. *Definition.*—A perfectly elastic body is a body which, when brought to any one state of strain, requires at all times the same stress to hold it in this state; however long it be kept strained, or however rapidly its state be altered from any other strain, or from no strain, to the strain in question. Here, according to our plan (§§ 443, 448) for Abstract Dynamics, we ignore variation of temperature in the body. If, however, we add a condition of absolutely no variation of temperature, or of recurrence to one specified temperature after changes of strain, we have a definition of that property of perfect elasticity

towards which highly elastic bodies in nature approximate; and which is rigorously fulfilled by all fluids, and may be so by some real solids, as homogeneous crystals. But inasmuch as the elastic reaction of every kind of body against strain varies with varying temperature, and (a thermodynamic consequence of this, as we shall see later) any increase or diminution of strain in an elastic body is necessarily accompanied by a change of temperature; even a perfectly elastic body could not, in passing through different strains, act as a rigorously conservative system, but, on the contrary, must give rise to dissipation of energy in consequence of the conduction or radiation of heat induced by these changes of temperature.

Its conditional fulfilment in nature.

But by making the changes of strain quickly enough to prevent any sensible equalization of temperature by conduction or radiation (as, for instance, Stokes has shown, is done in sound of musical notes travelling through air); or by making them slowly enough to allow the temperature to be maintained sensibly constant* by proper appliances; any highly elastic, or perfectly elastic body in nature may be got to act very nearly as a conservative system.

673. In nature, therefore, the integral amount, w, of work defined as above, is for a perfectly elastic body, independent (§ 274) of the series of configurations, or states of strain, through which it may have been brought from the first to the second of the specified conditions, provided it has not been allowed to change sensibly in temperature during the process.

Potential energy of an elastic solid held strained.

The analytical statement is that the expression (11) for dw must be the differential of a function of e, f, g, a, b, c, regarded as independent variables; or, which means the same, w is a function of these elements, and

$$\left.\begin{array}{lll} P=\dfrac{dw}{de}, & Q=\dfrac{dw}{df}, & R=\dfrac{dw}{dg}, \\ S=\dfrac{dw}{da}, & T=\dfrac{dw}{db}, & U=\dfrac{dw}{dc}. \end{array}\right\} \quad\ldots\ldots\ldots\ldots(12).$$

* "On the Thermoelastic and Thermomagnetic Properties of Matter" (W. Thomson). *Quarterly Journal of Mathematics.* April, 1855; Mathematical and Physical Papers, Art. XLVIII. Part VII.

Potential energy of an elastic solid held strained.

In Appendix C, we shall return to the comprehensive analytical treatment of this theory, not confining it to infinitely small strains for which alone the notation $(e, f, \ldots)$, as defined in § 669, is convenient. In the meantime, we shall only say that when the whole amount of strain is infinitely small, and the stress-components are therefore all altered in the same ratio as the strain-components if these are altered all in any one ratio; w must be a homogeneous quadratic function of the six variables e, f, g, a, b, c, which, if we denote by (e, e), $(f, f) \ldots (e, f) \ldots$ constants depending on the quality of the substance and on the directions chosen for the axes of co-ordinates, we may write as follows:—

$$\left.\begin{aligned} w = \tfrac{1}{2}\{(e, e)e^2 + (f, f)f^2 + (g, g)g^2 + (a, a)a^2 + (b, b)b^2 + (c, c)c^2 \\ +2(e, f)ef + 2(e, g)eg + 2(e, a)ea + 2(e, b)eb + 2(e, c)ec \\ +2(f, g)fg + 2(f, a)fa + 2(f, b)fb + 2(f, c)fc \\ +2(g, a)ga + 2(g, b)gb + 2(g, c)gc \\ +2(a, b)ab + 2(a, c)ac \\ +2(b, c)bc\} \end{aligned}\right\} \quad (13).$$

The 21 coefficients (e, e), $(f, f) \ldots (b, c)$, in this expression constitute the 21 "coefficients of elasticity," which Green first showed to be proper and essential for a complete theory of the dynamics of an elastic solid subjected to infinitely small strains. The only condition that can be theoretically imposed upon these coefficients is that they must not permit w to become negative for any values, positive or negative, of the strain-components $e, f, \ldots$. Under Properties of Matter, we shall see that an untenable theory (Boscovich's), falsely worked out by mathematicians, has led to relations among the coefficients of elasticity which experiment has proved to be false.

Stress-components expressed in terms of strain.

Eliminating w from (12) by (13) we have

$$\left.\begin{aligned} P &= (e, e)e + (e, f)f + (e, g)g + (e, a)a + (e, b)b + (e, c)c \\ Q &= (e, f)e + (f, f)f + (f, g)g + (f, a)a + (f, b)b + (f, c)c \\ &\quad \text{etc.} \qquad\qquad \text{etc.} \\ &\quad \text{etc.} \qquad\qquad \text{etc.} \end{aligned}\right\} \quad (14).$$

These equations express the six components of stress (P, Q, R, S, T, U) as linear functions of the six components of strain (e, f, g, a, b, c) with 15 equalities [namely $(e, f) = (f, e)$, etc.] among their 36 coefficients, which leave only 21 of them independent. The mere principle of superposition (which we have used above in establishing the quadratic form for w) might have been directly applied to demonstrate linear formulæ for the stress-components. Thus it is that some authors have been led to lay down, as the foundation of the most general possible theory of elasticity, six equations involving 36 coefficients supposed to be independent. But it is only by the principle of energy that, as first discovered by Green, the fifteen pairs of these coefficients are proved to be equal.

Stress-components expressed in terms of strain.

The algebraic transformation of equations (14) to express the strain-components singly, by linear functions of the stress-components, may be directly effected of course by forming the proper determinants from the 36 coefficients, and taking the 36 proper quotients. From a known determinantal theorem, used also above [§ 313 (d)], it follows that there are 15 equalities between pairs of these 36 quotients, because of the 15 equalities in pairs of the coefficients of e, f, etc., in (14). Thus, if we denote by

Strain-components expressed in terms of stress.

$$[P, P], [Q, Q], \ldots [P, Q], \ldots [Q, P] \ldots$$

the set of 36 determinantal quotients found by that process (being, therefore, known algebraic functions of the original coefficients (e, e), (f, f), ... etc.), we have

$$\left.\begin{array}{l} e = [P,P]P+[P,Q]Q+[P,R]R+[P,S]S+[P,T]T+[P,U]U \\ f = [Q,P]P+[Q,Q]Q+[Q,R]R+[Q,S]S+[Q,T]T+[Q,U]U \\ \qquad \text{etc.} \qquad\qquad\qquad \text{etc.} \end{array}\right\} \ldots(16);$$

and these new coefficients satisfy 15 equations

$$[P, Q] = [Q, P], \ [P, R] = [R, P] \ldots\ldots\ldots\ldots\ldots\ldots(17).$$

By what we proved in § 313 (d) when engaged with precisely the same algebraic transformation, we see that $[P, P], [Q, Q], \ldots, [P, Q], \ldots$ are simply the coefficients of $P^2, Q^2, \ldots, 2PQ, \ldots$ in the expression for $2w$ obtained by eliminating $e, f, \ldots$ from (13), so that

$$w = \tfrac{1}{2}\{[P,P]P^2+[Q,Q]Q^2+\ldots+2[P,Q]PQ+2[P,R]PR+\ldots\} \ldots(18);$$

and

$$\left.\begin{array}{l} e = \left[\dfrac{dw}{dP}\right], \ f = \left[\dfrac{dw}{dQ}\right], \ g = \left[\dfrac{dw}{dR}\right], \\ a = \left[\dfrac{dw}{dS}\right], \ b = \left[\dfrac{dw}{dT}\right], \ c = \left[\dfrac{dw}{dU}\right], \end{array}\right\} \ldots\ldots\ldots\ldots(19);$$

Strain-components expressed in terms of stress.

where the brackets [ ] denote the partial differential coefficients taken on the supposition that w is expressed as a function of P, Q, etc., as in (19); to distinguish them from those of equations (12) which were taken on the supposition that w is expressed as a function of e, f, etc., as in (13). We have also, as in § 313 (*d*),

Compare § 670, (3) (4) (5).

$$w = \tfrac{1}{2}(Pe + Qf + Rg + Sa + Tb + Uc) \ldots\ldots\ldots\ldots\ldots (20);$$

which might have been put down in the beginning, as it simply expresses that

Average stress through any changing strain.

674. The average stress, due to elasticity of the solid, when strained from its natural condition to that of strain (e, f, g, a, b, c) is (as from the assumed applicability of the principle of superposition we see it must be) just half the stress required to keep it in this state of strain.

Homogeneousness defined.

675. A body is called homogeneous when any two equal, similar parts of it, with corresponding lines parallel and turned towards the same parts, are undistinguishable from one another by any difference in quality. The perfect fulfilment of this condition without any limit as to the smallness of the parts, though conceivable, is not generally regarded as probable for any of the real solids or fluids known to us, however seemingly homogeneous. It is, we believe, held by all naturalists that there is a *molecular structure*, according to which, in *compound* bodies such as water, ice, rock-crystal, etc., the constituent substances lie side by side, or arranged in groups of finite dimensions, and even in bodies called *simple* (*i.e.*, not known to be chemically resolvable into other substances) there is no ultimate homogeneousness. In other words, the prevailing belief is that every kind of matter with which we are acquainted has a more or less *coarse-grained* texture, whether having visible molecules, as great masses of solid stone- or brick-building, or natural granite or sandstone rocks; or, molecules too small to be visible or directly measureable by us (but *not infinitely small*) * in seemingly homogeneous metals, or continuous crystals, or

Molecular hypothesis

assumes a very fine grained texture in crystals, but no ultimate homogeneousness.

* Probably not *undiscoverably* small, although of dimensions not yet known to us. See Appendix F. on "Size of Atoms."

liquids, or gases. We must of course return to this subject under Properties of Matter; and in the meantime need only say that the definition of *homogeneousness* may be applied practically on a very large scale to masses of building or coarse-grained conglomerate rock, or on a more moderate scale to blocks of common sandstone, or on a very small scale to seemingly homogeneous metals*; or on a scale of extreme, undiscovered fineness, to vitreous bodies, continuous crystals, solidified gums, as India rubber, gum-arabic, etc., and fluids.

Scales of average homogeneousness.

676. The substance of a homogeneous solid is called *isotropic* when a spherical portion of it, tested by any physical agency, exhibits no difference in quality however it is turned. Or, which amounts to the same, a cubical portion cut from any position in an isotropic body exhibits the same qualities relatively to each pair of parallel faces. Or two equal and similar portions cut from *any* positions in the body, not subject to the condition of parallelism (§ 675), are undistinguishable from one another. A substance which is not isotropic, but exhibits differences of quality in different directions, is called *eolotropic*.

Isotropic and eolotropic substances defined.

677. An individual body, or the substance of a homogeneous solid, may be isotropic in one quality or class of qualities, but eolotropic in others.

Isotropy and eolotropy of different sets of properties.

Thus in abstract dynamics a rigid body, or a group of bodies rigidly connected, contained within and rigidly attached to a rigid spherical surface, is kinetically symmetrical (§ 285) if its centre of inertia is at the centre of the sphere, and if its moments of inertia are equal round all diameters. It is also isotropic relatively to gravitation if it is centrobaric (§ 534), so that the centre of a figure is not merely a centre of inertia, but a true centre of gravity. Or a transparent substance may transmit light at different velocities in different directions through it (that is, be *doubly refracting*), and yet a cube of it may (and generally does in natural crystals) absorb the same part of a beam of white light transmitted across it perpendicularly to

* Which, however, we know, as recently proved by Deville and Van Troost, are porous enough at high temperatures to allow very free percolation of gases.

any of its three pairs of faces. Or (as a crystal which exhibits *dichroism*) it may be eolotropic relatively to the latter, or to either optic quality, and yet it may conduct heat equally in all directions.

Practical limitation of isotropy, and homogeneousness of eolotropy, to the average in the aggregate of molecules.

678. The remarks of § 675 relative to homogeneousness in the aggregate, and the supposed ultimately heterogeneous texture of all substances however seemingly homogeneous, indicate corresponding limitations and non-rigorous practical interpretations of isotropy.

Conditions fulfilled in elastic isotropy.

679. To be elastically isotropic, we see first that a spherical or cubical portion of any solid, if subjected to uniform normal pressure (positive or negative) all round, must, in yielding, experience no deformation: and therefore must be equally compressed (or dilated) in all directions. But, further, a cube cut from any position in it, and acted on by *tangential* or shearing stress (§ 662) in planes parallel to two pairs of its sides, must experience simple deformation, or shear (§ 171), in the same direction, unaccompanied by condensation or dilatation*, and the same in amount for all the three ways in which a stress may be thus applied to any one cube, and for different cubes taken from any different positions in the solid.

Measures of resistance to compression and resistance to distortion.

680. Hence the elastic quality of a perfectly elastic, homogeneous, isotropic solid is fully defined by two elements;—its resistance to compression, and its resistance to distortion. The amount of uniform pressure in all directions, per unit area of its surface, required to produce a stated very small compression, measures the first of these, and the amount of the shearing stress required to produce a stated amount of shear measures

* It must be remembered that the changes of figure and volume we are concerned with are so small that the principle of superposition is applicable; so that if any shearing stress produced a condensation, an opposite shearing stress would produce a dilatation, which is a violation of the isotropic condition. But it is possible that a shearing stress may produce, in a truly isotropic solid, condensation or dilatation in proportion to the square of its value: and it is probable that such effects may be sensible in India rubber, or cork, or other bodies susceptible of great deformations or compressions, with persistent elasticity.

the second. The numerical measure of the first is the compressing pressure divided by the diminution of the bulk of a portion of the substance which, when uncompressed, occupies the unit volume. It is sometimes called the *elasticity of volume*, or the *resistance to compression*, or the *bulk-modulus of elasticity* or the *modulus of compression*. Its reciprocal, or the amount of compression on unit of volume divided by the compressing pressure, or, as we may conveniently say, the compression per unit of volume, per unit of compressing pressure, is commonly called the *compressibility*. The second, or resistance to change of shape, is measured by the tangential stress (reckoned as in § 662) divided by the amount of the distortion or shear (§ 175) which it produces, and is called the *modulus of rigidity*, or for brevity *rigidity* of the substance, or its *elasticity of figure*.

Bulk-modulus or modulus of compression.

Compressibility.

Rigidity or elasticity of figure, defined.

681. From § 169 it follows that a strain compounded of a simple extension in one set of parallels, and a simple contraction of equal amount in any other set perpendicular to those, is the same as a simple shear in either of the two sets of planes cutting the two sets of parallels at 45°. And the numerical measure (§ 175) of this shear, or simple distortion, is equal to *double* the amount of the elongation or contraction (each measured, of course, per unit of length). Similarly, we see (§ 668) that a longitudinal traction (or negative pressure) parallel to one line, and an equal longitudinal positive pressure parallel to any line at right angles to it, is equivalent to a shearing stress of tangential tractions (§ 661) parallel to the planes which cut those lines at 45°. And the numerical measure of this shearing stress, being (§ 662) the amount of the tangential traction in either set of planes, is equal to the amount of the positive or negative normal pressure, *not doubled*.

Discrepant reckonings of shear and shearing stress, from the simple longitudinal strains or stresses respectively involved.

682. Since then any stress whatever may be made up of simple longitudinal stresses, it follows that, to find the relation between any stress and the strain produced by it, we have only to find the strain produced by a single longitudinal stress, which we may do at once thus:—A simple longitudinal stress,

Strain produced by a single longitudinal stress.

Strain produced by a single longitudinal stress.

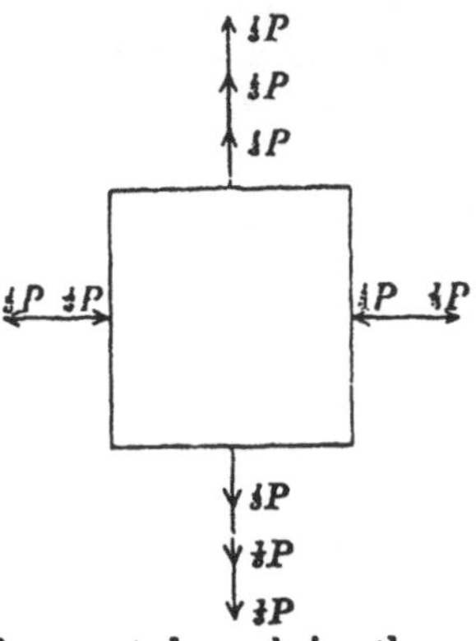

P, is equivalent to a uniform dilating tension $\frac{1}{3}P$ in all directions, compounded with two shearing stresses, each equal to $\frac{1}{3}P$, and having a common axis in the line of the given longitudinal stress, and their other two axes any two lines at right angles to one another and to it. The diagram, drawn in a plane through one of these latter lines, and the former, sufficiently indicates the synthesis; the only forces not shown being those perpendicular to its plane.

Hence if n denote the *rigidity*, and k the *bulk-modulus* [being the same as the reciprocal of the compressibility (§ 680)], the effect will be an equal dilatation in all directions, amounting, per unit of volume, to

$$\frac{\frac{1}{3}P}{k} \qquad\qquad (1),$$

compounded with two equal shears, each amounting to

$$\frac{\frac{1}{3}P}{n} \qquad\qquad (2),$$

and having (§ 679) their axes in the directions just stated as those of the shearing stresses.

683. The dilatation and two shears thus determined may be conveniently reduced to simple longitudinal strains by still following the indications of § 681, thus:

The two shears together constitute an elongation amounting to $\frac{1}{3}P/n$ in the direction of the given force, P, and equal contraction amounting to $\frac{1}{6}P/n$ in all directions perpendicular to it. And the cubic dilatation $\frac{1}{3}P/k$ implies a linear dilatation, equal in all directions, amounting to $\frac{1}{9}P/k$. On the whole, therefore, we have

$$\left.\begin{array}{l}\text{linear elongation} = P\left(\dfrac{1}{3n}+\dfrac{1}{9k}\right)\text{, in the direction of} \\ \qquad\text{the applied stress, and} \\ \text{linear contraction} = P\left(\dfrac{1}{6n}-\dfrac{1}{9k}\right)\text{, in all directions} \\ \qquad\text{perpendicular to the applied stress.}\end{array}\right\}\ldots(3).$$

Hence Young's modulus (§ 686) $=\frac{9nk}{3k+n}$.

684. Hence when the ends of a column, bar, or wire, of isotropic material, are acted on by equal and opposite forces, it experiences a lateral linear contraction, equal to $\frac{3k-2n}{2(3k+n)}$ of the longitudinal dilatation, each reckoned as usual per unit of linear measure. One specimen of the fallacious mathematics above referred to (§ 673), is a celebrated conclusion of Navier's and Poisson's that this ratio is $\frac{1}{4}$, which requires the rigidity to be $\frac{3}{5}$ of the bulk-modulus, for all solids: and which was first shown to be false by Stokes* from many obvious observations, proving enormous discrepancies from it in many well-known bodies, and rendering it most improbable that there is any approach to a constancy of ratio between rigidity and bulk-modulus in any class of solids. Thus clear elastic jellies, and India rubber, present familiar specimens of isotropic homogeneous solids, which, while differing very much from one another in rigidity ("stiffness"), are probably all of very nearly the same compressibility as water. This being $\frac{1}{308000}$ per pound per square inch; the bulk-modulus, measured by its reciprocal, or, as we may read it, "308000 lbs. per square inch," is obviously many hundred times the absolute amount of the rigidity of the stiffest of those substances. A column of any of them, therefore, when pressed together or pulled out, within its limits of elasticity, by balancing forces applied to its ends (or an India-rubber band when pulled out), experiences no sensible change of volume, though very sensible change of length. Hence the proportionate extension or contraction of any transverse diameter must be sensibly equal to $\frac{1}{2}$ the longitudinal contraction or extension:

Ratio of lateral contraction to longitudinal extension

different for different substances from $\frac{1}{2}$ for jelly to 0 for cork.

* On the Friction of Fluids in Motion, and the Equilibrium and Motion of Elastic Solids.—*Trans. Camb. Phil. Jour.*, April, 1845. See also *Camb. and Dub. Math. Jour.*, March, 1848.

different for different substances from ½ for jelly to 0 for cork.

and for all ordinary stresses, such substances may be practically regarded as incompressible elastic solids. Stokes gave reasons for believing that metals also have in general greater resistance to compression, in proportion to their rigidities, than according to the fallacious theory, although for them the discrepancy is very much less than for the gelatinous bodies. This probable conclusion was soon experimentally demonstrated by Wertheim, who found the ratio of lateral to longitudinal change of linear dimensions, in columns acted on solely by longitudinal force, to be about $\frac{1}{3}$ for glass and brass; and by Kirchhoff, who, by a very well-devised experimental method, found ·387 as the value of that ratio for brass, and ·294 for iron. For copper we find that it probably lies between ·226 and ·441, by recent experiments* of our own, measuring the torsional and longitudinal rigidities (§§ 596, 599, 686) of a copper wire.

Supposition of ¼ for ideal perfect solid, groundless.

685. All these results indicate rigidity *less* in proportion to the bulk-modulus than according to Navier's and Poisson's theory. And it has been supposed by many naturalists, who have seen the necessity of abandoning that theory as inapplicable to ordinary solids, that it may be regarded as the proper theory for an ideal *perfect solid*, and as indicating an amount of rigidity not quite reached in any real substance, but approached to in some of the most rigid of natural solids (as, for instance, iron). But it is scarcely possible to hold a piece of cork in the hand without perceiving the fallaciousness of this last attempt to maintain a theory which never had any good foundation. By careful measurements on columns of cork of various forms (among them, cylindrical pieces cut in the ordinary way for bottles) before and after compressing them longitudinally in a Bramah's press, we have found that the change of lateral dimensions is insensible both with small longitudinal contractions and return dilatations, within the limits of elasticity, and with such enormous longitudinal contractions as to $\frac{1}{6}$ or $\frac{1}{3}$ of the original length. It is thus proved decisively that cork is much more rigid, while metals, glass, and gelatinous bodies are

* On the Elasticity and Viscosity of Metals (W. Thomson). *Proc. R. S.*, May, 1865. See Art. 'Elasticity,' *Encyc. Britan.*

all less rigid, in proportion to bulk-modulus than the supposed "perfect solid;" and the utter worthlessness of the theory is experimentally demonstrated.

686. The modulus of elasticity of a bar, wire, fibre, thin filament, band, or cord of any material (of which the substance need not be isotropic, nor even homogeneous within one normal section), as a bar of glass or wood, a metal wire, a natural fibre, an India-rubber band, or a common thread, cord, or tape, is a term introduced by Dr Thomas Young* to designate what we also sometimes call its *longitudinal rigidity:* that is, the quotient obtained by dividing the simple longitudinal force required to produce any infinitesimal elongation or contraction by the amount of this elongation or contraction reckoned as usual per unit of length.

Young's modulus defined.

Same as longitudinal rigidity.

* Extract from *Encycl. Brit.* Art. 'Elasticity,' § 42. "*Young's Modulus*," or *Modulus of Simple Longitudinal Stress.*—Thomas Young called *the modulus of elasticity* of an elastic solid the amount of the end-pull or end-thrust required to produce any infinitesimal elongation or contraction of a wire, or bar, or column of the substance multiplied by the ratio of its length to the elongation or contraction. In this definition the definite article is clearly misapplied. There are, as we have seen, two moduluses of elasticity for an isotropic solid,—one measuring elasticity of bulk, the other measuring elasticity of shape. An interesting and instructive illustration of the confusion of ideas so often rising in physical science from faulty logic is to be found in "An Account of an Experiment on the Elasticity of Ice: By Benjamin Bevan, Esq., in a letter to Dr Thomas Young, Foreign Sec. R. S." and in Young's "Note" upon it, both published in the *Transactions of the Royal Society* for 1826. Bevan gives an interesting account of a well-designed and well-executed experiment on the flexure of a bar, 3·97 inches thick, 10 inches broad, and 100 inches long, of ice on a pond near Leighton Buzzard (the bar remaining attached by one end to the rest of the ice, but being cut free by a saw along its sides and across its other end), by which he obtained a fairly accurate determination of "the modulus of ice" (his result was 21,000,000 feet); and says that he repeated the experiment in various ways on ice bars of various dimensions, some remaining attached by one end, others completely detached, and found results agreeing with the first as nearly "as the admeasurement of the thickness could be ascertained." He then proceeds to compare "the modulus of ice" which he had thus found with "the modulus of water," which he quotes from Young's *Lectures* as deduced from Canton's experiments on the compressibility of water. Young in his "Note" does *not* point out that the two moduluses were essentially different, and that *the modulus of his definition*, the modulus determinable from the flexure of a bar, is essentially zero for every fluid. We now call "Young's modulus" the particular modulus of elasticity defined as above by Young, and so avoid all confusion.

Weight-modulus and length of modulus.

687. Instead of reckoning Young's modulus in units of weight, it is sometimes convenient to express it in terms of the weight of the unit length of the rod, wire, or thread. The modulus thus reckoned, or, as it is called by some writers, the length of the modulus, is of course found by dividing the weight-modulus by the weight of the unit length. It is useful in many applications of the theory of elasticity; as, for instance, in this result, which will be proved later:—the velocity of transmission of longitudinal vibrations (as of sound) along a bar or cord, is equal to the velocity acquired by a body in falling from a height equal to half the length of the modulus*. For other examples see § 791, *a*, below.

Velocity of transmission of a simple longitudinal stress through a rod.

Specific Young's modulus of an isotropic body.

688. The *specific Young's modulus of elasticity of an isotropic substance*, or, as it is most often called, simply the *Young's modulus of the substance*, is the Young's modulus of a bar of it having some definitely specified sectional area. If this be such that the weight of unit length is unity, the Young's *modulus of the substance* will be the same as the length of the modulus of any bar of it: a system of reckoning which, as we have seen, has some advantages in application. It is, however, more usual to choose a common unit of area as the sectional area of the bar referred to in the definition. There must also be a definite understanding as to the unit in terms of which the force is measured, which may be either the *absolute unit* (§ 223): or the gravitation unit for a specified locality; that is (§ 226), the weight in that locality of the unit of mass. Experimenters hitherto have stated their results in terms of the gravitation unit, each for his own locality; the accuracy hitherto attained being scarcely in any cases sufficient to require corrections for the

In terms of the absolute unit; or of the force of gravity on the unit of mass in any particular locality.

* It is to be understood that the vibrations in question are so much spread out through the *length* of the body, that inertia does not sensibly influence the transverse contractions and dilatations which (unless the substance have in this respect the peculiar character presented by cork, § 684) take place along with them. Also, under thermodynamics, we shall see that changes of shape and bulk produced by the varying stresses cause changes of temperature which, in ordinary solids, render the velocity of transmission of longitudinal vibrations sensibly greater than that calculated by the rule stated in the text, if we use the *static modulus* as understood from the definition there given; and we shall learn to take into account the thermal effect by using a definite *static modulus*, or *kinetic modulus*, according to the circumstances of any case that may occur.

different forces of gravity in the different places of observation. Corresponding statements apply to the modulus of rigidity. Young's word "Modulus" is also used conveniently enough in the expression "Modulus of Rupture," which is almost a synonym for "Tenacity." (See table of Moduluses and Strengths, article "Elasticity," *Encyclopædia Britannica*, new edition.) It means the greatest pull that can be applied to a wire, or bar, or rod of the substance without breaking it. It may be reckoned either in units of force per unit of area, of the cross section; or it may be reckoned in terms of the length which the bar must have to be equal in weight to the breaking force, and when so reckoned it is called the "Length-Modulus of Rupture."

In terms of the absolute unit; or of the force of gravity on the unit of mass in any particular locality.

689. The most useful and generally convenient specification of the modulus of elasticity of a substance is in grammes-weight per square centimetre. This has only to be divided by the specific gravity of the substance to give the *length of the modulus*. British measures, however, being still unhappily sometimes used in practical and even in high scientific statements, we may have occasion to refer to reckonings of the modulus in pounds per square inch or per square foot, or to length of the modulus in feet.

690. The reckoning most commonly adopted in British treatises on mechanics and practical statements is pounds per square inch. The modulus thus stated must be divided by the weight of 12 cubic inches of the solid, or by the product of its specific gravity into ·4337*, to find the length of the modulus, in feet.

* This decimal being the weight in lbs. of 12 cubic inches of water. The one great advantage of the French metrical system is, that the mass of the unit volume (1 cubic centimetre) of water at its temperature of maximum density (3°·945) is unity (1 gramme) to a sufficient degree of approximation for almost all practical purposes. Thus, according to this system, the density of a body and its specific gravity mean one and the same thing; whereas on the British no-system the density is expressed by a number found by multiplying the specific gravity by one number or another, according to the choice of a cubic inch, cubic foot, cubic yard, or cubic mile that is made for the unit of volume; and the grain, scruple, gunmaker's drachm, apothecary's drachm, ounce Troy, ounce avoirdupois, pound Troy, pound avoirdupois, stone (Imperial, Ayrshire, Lanarkshire,

To reduce from pounds per square inch to grammes per square centimetre, multiply by 70·31, or divide by ·014223. French engineers generally state their results in kilogrammes per square metre, and so bring them to more convenient numbers, being $\frac{1}{1000000}$ of the inconveniently large numbers expressing moduluses in grammes weight per square centimetre.

Metrical denominations of moduluses of elasticity in general.

691. The same statements as to units, reducing factors, and nominal designations, are applicable to the bulk-modulus of any elastic solid or fluid, and to the rigidity (§ 680) of an isotropic body; or, in general, to any one of the 21 moduluses in the expressions [§ 673. (14)] for stresses in terms of strains, or to the reciprocal of any one of the 21 moduluses in the expressions [§ 673. (16)] for strains in terms of stresses, as well as to the modulus defined by Young.

Practical rules for velocities of waves;

691 *a*. The convenience, for residents on the Earth, of the length-reckoning of moduluses is illustrated by the theorems stated at the end of § 687, and others analogous to it as follows:—

Distortional without change of bulk;

(1) The velocity of propagation of a wave of distortion in an isotropic homogeneous solid is equal to the velocity acquired by a body in falling through a height equal to half the length-modulus of rigidity.

Compressional, in an elastic solid;

(2) The velocity of the other kind of wave possible in an isotropic homogeneous solid, that is to say a wave analogous to that of sound, is equal to the velocity acquired by a body falling through a height equal to half the length-modulus for simple longitudinal strain (compare § 686); just as the Young's modu-

Dumbartonshire), stone for hay, stone for corn, quarter (of a hundredweight), quarter (of corn), hundredweight, or ton, that is chosen for unit of mass. It is a remarkable phenomenon, belonging rather to moral and social than to physical science, that a people tending naturally to be regulated by common sense should voluntarily condemn themselves, as the British have so long done, to unnecessary hard labour in every action of common business or scientific work related to measurement; from which all the other nations of Europe have emancipated themselves. We have been informed, through the kindness of the late Professor W. H. Miller, of Cambridge, that he concludes, from a very trustworthy comparison of standards by Kupffer, of St Petersburgh, that the weight of a cubic decimetre of water at temperature of maximum density is 1000·013 grammes.

lus is reckoned for simple stress. The modulus for simple longitudinal strain may be found by enclosing a rod or bar of the substance in an infinitely rigid, perfectly smooth and frictionless tube fitting it perfectly all round, and then dealing with it as the rod with its sides all free is dealt with for finding the Young's modulus. Of course it is understood that the ideal tube, which gives positive normal pressure when the two ends of the elastic rod within it are pressed together, must be supposed to give the negative normal pressure, or the normal traction, required to prevent lateral shrinkage, when the two ends of the wire are pulled asunder. (Compare § 684 above.)

(3) The velocity of sound in a liquid is the velocity a body would acquire in falling through a height equal to half the length-modulus of compression. **Compressional in liquid;**

(4) The Newtonian velocity of sound (that is to say, the velocity which sound would have in air if the pressure in the course of the vibration varied simply according to Boyle's law without correction for the heat of condensation, and the cold of rarefaction) is equal to the velocity a body would acquire in falling through half the height of the homogeneous atmosphere for the actual temperature of the air whatever it may be. ("The Height of the Homogeneous Atmosphere" is a short expression commonly used to designate the depth that an ideal incompressible liquid of the same density as air must have to give by its weight the same pressure at the bottom as the actual pressure of the air at the supposed temperature and density.) **compressional in gas;**

(5) The velocity of a long wave* in water of uniform depth, supposed incompressible, is the velocity a body would acquire in falling through a height equal to half the depth. **gravitational in liquid;**

(6) The velocity of propagation of a transverse pulse in a stretched cord is equal to the velocity acquired by a body falling through a height equal to half the length of a quantity of cord amounting in weight to the stretching force. **transversal vibration of stretched cord.**

* A "Long wave" is a technical expression in the theory of waves in water used to denote a wave of which the length is a large multiple (20 or 30 or more) of the depth.

Digression on Resilience, from Art. Elasticity *Encyc. Brit.*

691 *b*. "Resilience" is a very useful word, introduced about forty years ago (when the *doctrine of energy* was beginning to become practically appreciated) by Lewis Gordon, first professor of engineering in the university of Glasgow, to denote the quantity of work that a spring (or elastic body) gives back when strained to some stated limit and then allowed to return to the condition in which it rests when free from stress. The word "resilience" used without special qualifications may be understood as meaning *extreme resilience*, or the work given back by the spring after being strained to the extreme limit within which it can be strained again and again without breaking or taking a permanent set. In all cases for which Hooke's law of simple proportionality between stress and strain holds, the resilience is obviously equal to the work done by a constant force of half the amount of the extreme force acting through a space equal to the extreme deflection.

691 *c*. When force is reckoned in "gravitation measure," resilience per unit of the spring's mass is simply the height that the spring itself, or an equal weight, could be lifted against gravity by an amount of work equal to that given back by the spring returning from the stressed condition.

691 *d*. Let the elastic body be a long homogeneous cylinder or prism with flat ends (a bar as we may call it for brevity), and let the stress for which its resilience is reckoned be *positive* normal pressures on its ends. The resilience per unit mass is equal to the greatest height from which the bar can fall with its length vertical, and impinge against a perfectly hard frictionless horizontal plane without suffering stress beyond its limits of elasticity. For in this case (as in the case of the direct impact of two equal and similar bars meeting with equal and opposite velocities, discussed above, §§ 303, 304), the kinetic energy of the translational motion preceding the impact is, during the first half of the collision, wholly converted into potential energy of elastic force, which during the second half of the collision is wholly reconverted into kinetic energy of translational motion in the reverse direction. During the whole time of the collision the stopped end of the bar experiences a constant pressure, and at the middle of the collision the whole substance of the bar is for an instant at rest in the same state of compression as it would have permanently if in equilibrium under the influence of that pressure and an equal and opposite pressure on the other end. From the beginning to the middle of the collision the compression advances at a uniform rate through the bar from the stopped end to the free end. Every particle of the bar which the compression has not reached continues moving uniformly with the velocity of the whole before the collision until the compression reaches it, when it instantaneously comes to rest. The part of the bar which at any instant is all that is compressed remains at rest till the corresponding instant in the second half of the collision.

Digression on Resilience, from Art. Elasticity *Encyc. Brit.*

691 *e.* *From our* preceding view of a bar impinging against an ideal perfectly rigid plane, we see at once all that takes place in the real case of any rigorously direct longitudinal collision between two equal and similar elastic bars with flat ends. In this case the whole of the kinetic energy which the bodies had before collision reappears as purely translational kinetic energy after collision. The same would be approximately true of any two bars, provided the times taken by a pulse of simple longitudinal stress to run through their lengths are equal. Thus if the two bars be of the same substance, or of different substances having the same value for Young's modulus, the lengths must be equal, but the diameters may be unequal. Or if the Young's modulus be different in the two bars, their lengths must be inversely as the square root of its values. To all such cases the laws of "collision between two perfectly elastic bodies," whether of equal or unequal masses, as given in elementary dynamical treatises, are applicable. But in every other case part of the translational energy which the bodies have before collision is left in the shape of vibrations after collision, and the translational energy after collision is accordingly less than before collision. The losses of energy observed in common elementary dynamical experiments on collision between solid globes of the same substance are partly due to this cause. If they were wholly due to it they would be independent of the substance, when two globes of the same substance are used. They would bear the same proportion to

Digression on Resilience, from Art. Elasticity *Encyc. Brit.*

the whole energy in every case of collision between two equal globes, or again, in every case of collision between two globes of any stated proportion of diameters, provided in each case the two which collide are of the same substance; but the proportion of translational energy converted into vibrations would not be the same for two equal globes as for two unequal globes. Hence when differences of proportionate losses of energy are found in experiments on different substances, as in Newton's on globes of glass, iron, or compressed wool, this must be due to imperfect elasticity of the material. It is to be expected that careful experiments upon hard well-polished globes striking one another with such gentle forces as not to produce even at the point of contact any stress approaching to the limit of elasticity, will be found to give results in which the observed loss of translational energy can be almost wholly accounted for by vibrations remaining in the globes after collision.

691 *f. Examples of Resilience.—Example* 1.—In respect to simple longitudinal pull, the extreme resilience of steel pianoforte wire of No. 22 Birmingham wire gauge, of density 7·727, weighing 0·34 grammes per centimetre (calculated by multiplying the breaking weight of 106 kilogrammes into half the elongation produced by it, namely $\frac{1}{86}$) is 6163 metre-grammes (gravitation measure) per ten metres of the wire. Or, whatever the length of the wire, its resilience is equal to the work required to lift its weight through 172 metres.

Example 2.—The torsional resilience of the same wire, twisted in either direction as far as it can be without giving it any notable permanent set, was found to be equal to the work required to lift its weight through 1·3 metres.

Example 3.—The extreme resilience of a vulcanized india-rubber band weighing 12·3 grammes was found to be equal to the work required to lift its weight through 1200 metres. This was found by stretching it by gradations of weights up to the breaking weight, representing the results by aid of a curve, and measuring its area to find the integral work given back by the spring after being stretched by a weight just short of the breaking weight.

Stress required to maintain a simple longitudinal strain.

692. In §§ 681, 682 we examined the effect of a simple longitudinal stress, in producing elongation in its own direction, and contraction in lines perpendicular to it. With stresses substituted for strains, and strains for stresses, we may apply the same process to investigate the longitudinal and lateral tractions required to produce a simple longitudinal strain (that is, an elongation in one direction, with no change of dimensions perpendicular to it) in a rod or solid of any shape.

Thus a simple longitudinal strain e is equivalent to a cubic dilatation e without change of figure (or linear dilatation $\frac{1}{3}e$ equal in all directions), and two shears consisting each of dilatation $\frac{1}{3}e$ in the given direction, and contraction $\frac{1}{3}e$ in each of two directions perpendicular to it and to one another. To produce the cubic dilatation, e, alone requires (§ 680) a normal traction ke equal in all directions. And, to produce either of the shears simply, since the measure (§ 175) of each is $\frac{2}{3}e$, requires a shearing stress equal to $n \times \frac{2}{3}e$, which consists of tangential tractions each equal to this amount, positive (or drawing outwards) in the line of the given elongation, and negative (or pressing inwards) in the perpendicular direction. Thus we have in all

$$\left.\begin{array}{l} \text{normal traction} = (k + \frac{4}{3}n)e, \text{ in the direction of the} \\ \qquad\qquad \text{given strain, and} \\ \text{normal traction} = (k - \frac{2}{3}n)e, \text{ in every direction per-} \\ \qquad \text{pendicular to the given strain.} \end{array}\right\} \ldots(4).$$

Stress-components in terms of strain for isotropic body.

693. If now we suppose any possible infinitely small strain (e, f, g, a, b, c), according to the specification of § 669, to be given to a body, the stress (P, Q, R, S, T, U) required to maintain it will be expressed by the following formulæ, obtained by successive applications of § 692 (4) to the components e, f, g separately, and of § 680 to a, b, c:—

$$\left.\begin{array}{ll} & S = na, \quad T = nb, \quad U = nc, \\ & P = \mathfrak{A}e + \mathfrak{B}(f+g), \\ & Q = \mathfrak{A}f + \mathfrak{B}(g+e), \\ & R = \mathfrak{A}g + \mathfrak{B}(e+f), \\ \text{where} & \mathfrak{A} = k + \dfrac{4}{3}n, \quad \mathfrak{B} = k - \dfrac{2}{3}n, \\ & n = \frac{1}{2}(\mathfrak{A} - \mathfrak{B}) \end{array}\right\} \ldots\ldots\ldots\ldots\ldots(5).$$

694. Similarly, by § 680 and § 682 (3), we have

$$\left.\begin{array}{ll} & a=\frac{1}{n}S,\ b=\frac{1}{n}T,\ c=\frac{1}{n}U, \\ & Me=\{P-\sigma(Q+R)\}, \\ & Mf=\{Q-\sigma(R+P)\}, \\ & Mg=\{R-\sigma(P+Q)\}, \\ \text{where} & M=\frac{9nk}{3k+n}, \\ \text{and} & \sigma=\frac{3k-2n}{2(3k+n)}=\frac{1}{2}\frac{M}{n}-1, \end{array}\right\}\ldots\ldots\ldots\ldots(6),$$

as the formulæ expressing the strain (e, f, g, a, b, c) in terms of the stress (P, Q, R, S, T, U). They are of course merely the algebraic inversions of (5); and (§ 673) they might have been found by solving these for e, f, g, a, b, c, regarded as the unknown quantities. M is here introduced to denote Young's modulus (§ 683).

Equation of energy for the same.

695. To express the equation of energy for an isotropic substance, we may take the general formula, [§ 673 (20)],

$$w=\tfrac{1}{2}(Pe+Qf+Rg+Sa+Tb+Uc)$$

and eliminate from it P, Q, etc., by (5) of § 693, or, again, e, f, etc., by (6) of § 694, we thus find

$$\left.\begin{array}{l} 2w=(k+\frac{4}{3}n)(e^2+f^2+g^2)+2(k-\frac{2}{3}n)(fg+ge+ef)+n(a^2+b^2+c^2) \\ =\frac{1}{3}\left\{\left(\frac{1}{n}+\frac{1}{3k}\right)(P^2+Q^2+R^2)-2\left(\frac{1}{2n}+\frac{1}{3k}\right)(QR+RP+PQ)\right\}+\frac{1}{n}(S^2+T^2+U^2) \end{array}\right\}(7).$$

Fundamental problems of mathematical theory.

696. The mathematical theory of the equilibrium of an elastic solid presents the following general problems:—

A solid of any given shape, when undisturbed, is acted on in its substance by force distributed through it in any given manner, and displacements are arbitrarily produced, or forces arbitrarily applied, over its bounding surface. It is required to find the displacement of every point of its substance.

This problem has been thoroughly solved for a shell of homogeneous isotropic substance bounded by surfaces which, when undisturbed, are spherical and concentric (§ 735); but not hitherto for a body of any other shape. The limitations under which solutions have been obtained for other cases (thin plates, and rods), leading, as we have seen, to important practical results, have been stated above (§§ 588, 632). To demonstrate the laws (§§ 591, 633) which were taken in anticipation will also be one of our applications of the general equations for interior equilibrium of an elastic solid, which we now proceed to investigate.

697. Any portion in the interior of an elastic solid may be regarded as becoming perfectly rigid (§ 564) without disturbing the equilibrium either of itself or of the matter round it. Hence the traction exerted by the matter all round it, regarded as a distribution of force applied to its surface, must, with the applied forces acting on the substance of the portion considered, fulfil the conditions of equilibrium of forces acting on a rigid body. This statement, applied to an infinitely small rectangular parallelepiped of the body, gives the general differential equations of internal equilibrium of an elastic solid. It is to be remarked that *three* equations suffice; the conditions of equilibrium for the *couples* being secured by the relation established above (§ 661) among the six pairs of tangential component tractions on the six faces of the figure.

Conditions of internal equilibrium, expressed by three equations.

Let (x, y, z) be any point within the solid, and δx, δy, δz edges respectively parallel to the rectangular axes of reference, of an infinitely small parallelepiped of the solid having that point for its centre.

If P, Q, R, S, T, U denote (§ 662) the stress at (x, y, z), the average amounts of the component tractions (see table, § 669) on the faces of the parallelepiped will be

$$\text{on the two faces } \delta y\delta z \left\{ \begin{array}{ll} \pm\left(P \pm \dfrac{dP}{dx}\cdot\tfrac{1}{2}\delta x\right)\delta y\delta z, & \text{parallel to } OX, \\ \pm\left(U \pm \dfrac{dU}{dx}\cdot\tfrac{1}{2}\delta x\right)\delta y\delta z, & \quad,, \quad ,, \; OY, \\ \pm\left(T \pm \dfrac{dT}{dx}\cdot\tfrac{1}{2}\delta x\right)\delta y\delta z, & \quad,, \quad ,, \; OZ. \end{array} \right.$$

Taking the symmetrical expressions for the tractions on the two other pairs of faces, and summing for all the faces all the components parallel to the three axes separately, we have

$$\left(\frac{dP}{dx}+\frac{dU}{dy}+\frac{dT}{dz}\right)\delta x\delta y\delta z, \text{ parallel to } OX,$$

$$\left(\frac{dU}{dx}+\frac{dQ}{dy}+\frac{dS}{dz}\right)\delta x\delta y\delta z, \quad ,, \quad ,, \; OY,$$

$$\left(\frac{dT}{dx}+\frac{dS}{dy}+\frac{dR}{dz}\right)\delta x\delta y\delta z, \quad ,, \quad ,, \; OZ.$$

General equations of interior equilibrium.

Let now X, Y, Z denote the components of the applied force on the substance at (x, y, z), reckoned per unit of volume; so that $X\delta x\delta y\delta z$, $Y\delta x\delta y\delta z$, $Z\delta x\delta y\delta z$ will be their amounts on the small portion in question. Adding these to the corresponding components just found for the tractions, equating to zero, and omitting the factor $\delta x\delta y\delta z$, we have

$$\left.\begin{array}{l}\frac{dP}{dx}+\frac{dU}{dy}+\frac{dT}{dz}+X=0\\ \frac{dU}{dx}+\frac{dQ}{dy}+\frac{dS}{dz}+Y=0\\ \frac{dT}{dx}+\frac{dS}{dy}+\frac{dR}{dz}+Z=0\end{array}\right\}\quad\text{..................(2);}$$

which are the general equations of internal stress required for equilibrium.

If for P, Q, R, S, T, U we substitute the linear functions of e, f, g, a, b, c in terms of which they are expressed by (14) of § 673, we have the equations of internal strain. And if we eliminate e, f, g, a, b, c by (6) of § 670 we have, for (α, β, γ) the components of the displacement of any interior point in terms of (x, y, z) its undisplaced position in the solid, three linear partial differential equations of the second degree, which are the equations of internal equilibrium in their ultimate form. It is to be remarked that, by supposing the coefficients (e, e), (e, f), etc., to be not constant, but given functions of (x, y, z), we avoid limiting the investigation to a homogenous body.

Being sufficient, they imply that the forces on any part supposed rigid fulfil the six equations of equilibrium in a rigid body.

698. These equations being sufficient as well as necessary for the equilibrium of the body, they must secure that the condition of § 697 is fulfilled for any and every finite portion of it. This is easily verified.

Let $\iiint$ denote integration throughout any particular part of the solid, $d\sigma$ an element of the surface bounding this part, and $[\iint]$ integration over the whole of this surface. We have

$$\iiint X dxdydz = -\iiint\left(\frac{dP}{dx} + \frac{dU}{dy} + \frac{dT}{dz}\right) dxdydz.$$

Hence, integrating each term once, attending to the limits as in Appendix A., and denoting by l, m, n the direction-cosines of the normal through $d\sigma$,

$$\iiint X dxdydz = -[\iint(Pdydz + Udzdx + Tdxdy)] = -[\iint(Pl + Um + Tn)d\sigma],$$

and therefore [§ 662 (1)]

$$\iiint X dxdydz + [\iint F d\sigma] = 0 \text{ ....(3).}$$

Again we have

$$\iiint\left(yZ - zY\right) dxdydz = -\iiint\left\{y\left(\frac{dT}{dx} + \frac{dS}{dy} + \frac{dR}{dz}\right) - z\left(\frac{dU}{dx} + \frac{dQ}{dy} + \frac{dS}{dz}\right)\right\} dxdydz.$$

Now, integrating by parts, etc., as in Appendix A., we have

$$\iiint y\frac{dS}{dy}\, dxdydz = [\iint ySm d\sigma] - \iiint S dxdydz,$$

and

Verification of equations of equilibrium for any part supposed rigid.

$$\iiint z\frac{dS}{dz}\, dxdydz = [\iint zSn d\sigma] - \iiint S dxdydz.$$

Hence

$$\iiint\left(y\frac{dS}{dy} - z\frac{dS}{dz}\right) dxdydz = [\iint(ySm - zSn) d\sigma].$$

Using this in the preceding expression, integrating the other terms each once simply as before, and using § 662 (1), we find

$$\iiint(yZ - zY)\, dxdydz + [\iint(yH - zG)\, d\sigma] = 0 \text{.........(4).}$$

The six equations of equilibrium being (3), (4), and the symmetrical equations relative to y and z, are thus proved.

Simplified equations for isotropic solid.

For an isotropic solid, the equations (2) become of course much simpler. Thus, using (5) of § 693, eliminating e, f, g, a, b, c by (6) of § 670, grouping conveniently the terms which result, and putting

$$m = (k + \tfrac{1}{3}n) \text{ ....(5),}$$

we find

$$\left.\begin{array}{l} m\dfrac{d}{dx}\left(\dfrac{da}{dx}+\dfrac{d\beta}{dy}+\dfrac{d\gamma}{dz}\right)+n\left(\dfrac{d^2a}{dx^2}+\dfrac{d^2a}{dy^2}+\dfrac{d^2a}{dz^2}\right)+X=0 \\ m\dfrac{d}{dy}\left(\dfrac{da}{dx}+\dfrac{d\beta}{dy}+\dfrac{d\gamma}{dz}\right)+n\left(\dfrac{d^2\beta}{dx^2}+\dfrac{d^2\beta}{dy^2}+\dfrac{d^2\beta}{dz^2}\right)+Y=0 \\ m\dfrac{d}{dz}\left(\dfrac{da}{dx}+\dfrac{d\beta}{dy}+\dfrac{d\gamma}{dz}\right)+n\left(\dfrac{d^2\gamma}{dx^2}+\dfrac{d^2\gamma}{dy^2}+\dfrac{d^2\gamma}{dz^2}\right)+Z=0 \end{array}\right\}\ldots(6),$$

or, as we may write them shortly,

$$m\frac{d\delta}{dx}+n\nabla^2a+X=0,\ m\frac{d\delta}{dy}+n\nabla^2\beta+Y=0,\ m\frac{d\delta}{dz}+n\nabla^2\gamma+Z=0\ldots(7),$$

if we put

$$\frac{da}{dx}+\frac{d\beta}{dy}+\frac{d\gamma}{dz}=\delta\ldots\ldots\ldots\ldots\ldots\ldots\ldots\ldots(8),$$

and

$$\frac{d^2}{dx^2}+\frac{d^2}{dy^2}+\frac{d^2}{dz^2}=\nabla^2\ldots\ldots\ldots\ldots\ldots\ldots\ldots\ldots(9),$$

so that δ shall denote the amount of dilatation in volume experienced by the substance; and ∇^2 the same symbol of operation as formerly [Appendix A. and B., and §§ 491, 492, 499, etc.].

St Venant's application to torsion problems.

699. One of the most beautiful applications of the general equations of internal equilibrium of an elastic solid hitherto made is that of M. de St Venant to "the torsion of prisms.*" To one end of a long straight prismatic rod, wire, or solid or hollow cylinder of any form, a given couple is applied in a plane perpendicular to the length, while the other end is held fast: it is required to find the degree of twist (§ 120) produced, and the distribution of strain and stress throughout the prism. The conditions to be satisfied here are that the resultant action between the substance on the two sides of any normal section is a couple in the normal plane, equal to the given couple. Our work for solving the problem will be much simplified by first establishing the following preliminary propositions:—

Torsion problem stated.

* *Mémoires des Savants Étrangers.* 1855. "De la Torsion des Prismes, avec des considérations sur leur Flexion," etc.

700. Let a solid (whether aeolotropic or isotropic) be so acted on by force applied from without to its boundary, that throughout its interior there is no normal traction on any plane parallel or perpendicular to a given plane, XOY, which implies, of course, that there is no shearing stress with axes in or parallel to this plane, and that the whole stress at any point of the solid is a simple shearing stress of tangential forces in some direction in the plane parallel to XOY, and in the plane perpendicular to this direction. Then— Lemma.

(1.) The interior shearing stress must be equal, and similarly directed, in all parts of the solid lying in any line perpendicular to the plane XOY.

(2.) It being premised that the traction at every point of any surface perpendicular to the plane XOY is, by hypothesis, a distribution of force in lines perpendicular to this plane; the integral amount of it on any closed prismatic or cylindrical surface perpendicular to XOY, and bounded by planes parallel to it, is zero.

(3.) The matter within the prismatic surface and terminal planes of (2.) being supposed for a moment (§ 564) to be rigid, the distribution of tractions referred to in (2.) constitutes a couple whose moment, divided by the distance between those terminal planes, is equal to the resultant force of the tractions on the area of either, and whose plane is parallel to the lines of these resultant forces. In other words, the moment of the distribution of forces over the prismatic surface referred to in (2.) round any line (OY or OX) in the plane XOY, is equal to the sum of the components (T or S), perpendicular to the same line, of the traction in either of the terminal planes multiplied by the distance between these planes.

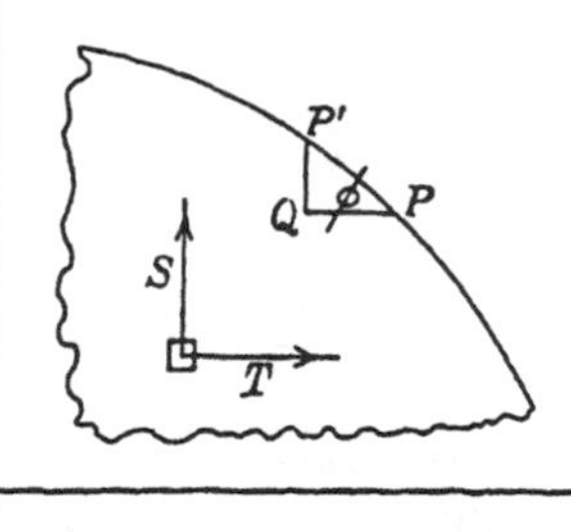

Lemma. To prove (1.) consider for a moment as rigid (§ 564) an infinitesimal prism, AB (of sectional area ω), perpendicular to XOY, and having plane ends, A, B, parallel to it. There being no forces on its sides (or cylindrical boundary) perpendicular to its length, its equilibrium so far as motion in the direction of any line (OX), perpendicular to its length, requires (§ 551, I.) that the components of the tractions on its ends be equal and in opposite directions. Hence, in the notation of § 662, the shearing stress components, T, must be equal at A and B; and so must the stress components S, for the same reason.

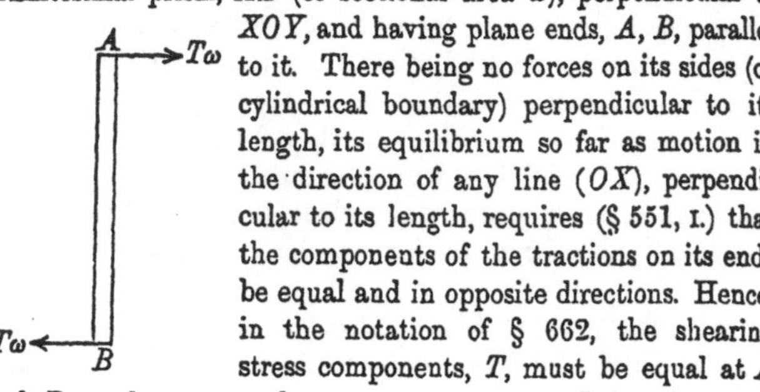

To prove (2.) and (3.) we have only to remark that they are required, according to § 551, I. and II., for the equilibrium of the rigid prism referred to in (3.).

Or, analytically, by the general equations (2) of § 697, since $X=0$, $Y=0$, $Z=0$, $P=0$, $Q=0$, $R=0$, $U=0$, by hypothesis; we have

$$\frac{dT}{dz}=0, \quad \frac{dS}{dz}=0 \quad \ldots\ldots\ldots\ldots\ldots\ldots\ldots (1),$$

and

$$\frac{dT}{dx}+\frac{dS}{dy}=0 \quad \ldots\ldots\ldots\ldots\ldots\ldots\ldots\ldots (2).$$

Of these (1.) prove that S and T are functions of x and y without z, or, in words, (1.) And if $\iint$ denote integration over the whole of any closed area of XOY, we have

$$\iint\left(\frac{dT}{dx}+\frac{dS}{dy}\right)dxdy=\left[\int(Tdy+Sdx)\right],^*$$

of which the second member, when the limits of the effected and indicated integrations are properly assigned, is found to be the same as

$$\int(T\sin\phi+S\cos\phi)\,ds,$$

where $\int$ denotes integration over the whole bounding curve, ds

* The brackets [ ], as here used, denote integrals assigned properly for the bounding curve.

an element of its length, and ϕ the inclination of ds to XO. Lemma.
But, by (1) § 662, with $l=\sin\phi$, $m=\cos\phi$, $n=0$, we have

$$H = T\sin\phi + S\cos\phi \quad\text{.......................}(3),$$

if H denote the traction (parallel to OZ), reckoned as usual per unit of area, experienced by the bounding prismatic surface. Hence

$$\iint\left(\frac{dT}{dx}+\frac{dS}{dy}\right)dxdy = \int Hds \quad\text{...............}(4);$$

and therefore, because of (2),

$$\int Hds = 0 \quad\text{...........................}(5),$$

which is (2.) in symbols. Again we have, by integration by parts, and substitution, (2), of $\frac{dS}{dy}$ for $-\frac{dT}{dx}$,

$$\iint Tdxdy = \left[\int Txdy\right]^* - \iint x\frac{dT}{dx}\,dxdy$$

$$= \left[\int Txdy\right]^* + \iint x\frac{dS}{dy}\,dxdy = \left[\int Txdy\right]^* + \left[\int Sxdx\right]^*$$

$$= \int x(T\sin\phi + S\cos\phi)\,ds = \int xHds \quad\text{..............}(6),$$

which proves (3.)

701. For a solid or hollow circular cylinder, the solution of § 699 (given first, we believe, by Coulomb) obviously is that each circular normal section remains unchanged in its own dimensions, figure, and internal arrangement (so that every straight line of its particles remains a straight line of unchanged length), but is turned round the axis of the cylinder through such an angle as to give a uniform *rate of twist* (§ 120) equal to the applied couple divided by the product of the moment of inertia of the circular area (whether annular or complete to the centre) into the rigidity of the substance.

Torsional rigidity of circular cylinder.

For, if we suppose the distribution of strain thus specified to be actually produced, by whatever application of stress is necessary, we have, in every part of the substance, a simple shear parallel to the normal section, and perpendicular to the radius through it. The elastic reaction against this requires to balance

* The brackets [ ], as here used, denote integrals assigned properly for the bounding curve.

it (§§ 679, 682), a simple distorting stress consisting of forces in the normal section, directed as the shear, and others in planes through the axis, and directed parallel to the axis. The amount of the shear is, for parts of the substance at distance r from the axis, equal obviously to τr, if τ be the rate of twist. Hence the amount of the tangential force in either set of planes is $n\tau r$ per unit of area, if n be the rigidity of the substance. Hence there is no force between parts of the substance lying on the two sides of any element of any circular cylinder coaxal with the bounding cylinder or cylinders; and consequently no force is required on the cylindrical boundary to maintain the supposed state of strain. And the mutual action between the parts of the substance on the two sides of any normal plane section consists of force in this plane, directed perpendicular to the radius through each point, and amounting to $n\tau r$ per unit of area. The moment of this distribution of force round the axis of the cylinder is (if $d\sigma$ denote an element of the area) $n\tau \iint d\sigma r^2$, or the product of $n\tau$ into the moment of inertia of the area round the perpendicular to its plane through its centre, which is therefore equal to the moment of the couple applied at either end.

Prism of any shape constrained to a simple twist,

702. Similarly, we see that if a cylinder or prism of any shape be compelled to take exactly the state of strain above specified (§ 701) with the line through the centres of inertia of the normal sections, taken instead of the axis of the cylinder, the mutual action between the parts of it on the two sides of any normal section will be a couple of which the moment will be expressed by the same formula, that is, the product of the rigidity, into the rate of twist, into the moment of inertia of the section round its centre of inertia.

requires tractions on its sides.

The only additional remark required to prove this is, that if the forces in the normal section be resolved in any two rectangular directions, OX, OY, the sums of the components, being respectively $n\tau \iint x d\sigma$ and $n\tau \iint y d\sigma$, each vanish by the property (§ 230) of the centre of inertia.

Traction on sides of prism constrained to a simple twist.

703. But for any other shape of prism than a solid or symmetrical hollow circular cylinder, the supposed state of strain will require, besides the terminal opposed couples, force parallel to the length of the prism, distributed over the prismatic boundary, in proportion to the distance along the tangent, from each point of the surface, to the point in which this line is cut by a perpendicular to it from the centre of inertia of the normal section. To prove this let a normal section of the prism be represented in the annexed diagram. Let PK, representing the shear at any point, P, close to the prismatic boundary, be resolved into PN and PT respectively along the normal and tangent. The whole shear, PK, being equal to τr, its component, PN, is equal to $\tau r \sin\omega$ or $\tau . PE$. The corresponding component of the required stress is $n\tau . PE$, and involves (§ 661) equal forces in the plane of the diagram, and in the plane through TP perpendicular to it, each amounting to $n\tau . PE$ per unit of area.

Traction on sides of prism constrained to a simple twist.

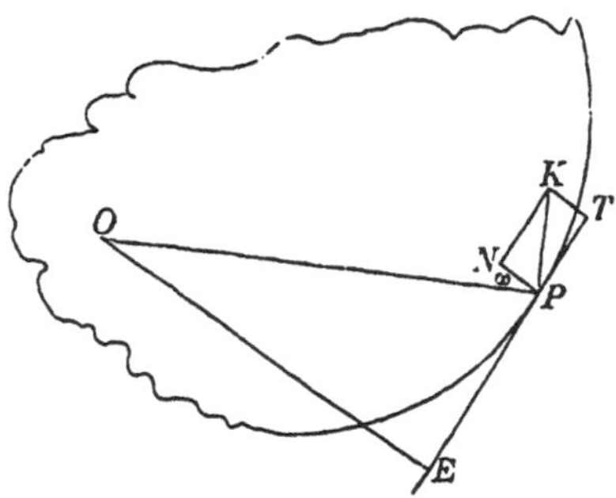

An application of force equal and opposite to the distribution thus found over the prismatic boundary, would of course alone produce in the prism, otherwise free, a state of strain which, compounded with that supposed above, would give the state of strain actually produced by the sole application of balancing couples to the two ends. The result, it is easily seen (and it will be proved below), consists of an increased twist, together with a warping of naturally plane normal sections, by infinitesimal displacements perpendicular to themselves, into certain surfaces of anticlastic curvature, with equal opposite curvatures in the principal sections (§ 130) through every point. This theory is due to St Venant, who not only pointed out the falsity of the supposition admitted by several previous writers, that Coulomb's law holds for other forms of prism than the solid or hollow circular cylinder, but discovered fully the nature of the requisite correction, reduced the determination of it to a problem of pure mathematics, worked out the solution for a great variety of important and curious cases,

St Venant's correction to give the strain produced by mere twisting couples applied to the ends.

compared the results with observation in a manner satisfactory and interesting to the naturalist, and gave conclusions of great value to the practical engineer.

Hydrokinetic analogue to torsion problem.

704. We take advantage of the identity of mathematical conditions in St Venant's torsion problem, and a hydrokinetic problem first solved a few years earlier by Stokes*, to give the following statement, which will be found very useful in estimating deficiencies in torsional rigidity below the amount calculated from the fallacious extension of Coulomb's law:—

705. Conceive a liquid of density n completely filling a closed infinitely light prismatic box of the same shape within as the given elastic prism and of length unity, and let a couple be applied to the box in a plane perpendicular to its length. The *effective* moment of inertia of the liquid† will be equal to the correction by which the torsional rigidity of the elastic prism calculated by the false extension of Coulomb's law must be diminished to give the true torsional rigidity.

Further, the actual *shear* of the solid, in any infinitely thin plate of it between two normal sections, will at each point be, when reckoned as a differential sliding (§ 172) parallel to their planes, equal to and in the same direction as the velocity of the liquid relatively to the containing box.

Solution of torsion problem.

706. To prove these propositions and investigate the mathematical equations of the problem, we first show that the conditions of the case (§ 699) are verified by a state of strain compounded of (1) a simple twist round the line through the centres of inertia, and (2) a distorting of each normal section by infinitesimal displacements perpendicular to its plane: then find the interior and surface equations to determine this warping: and lastly, calculate the actual moment of the couple to which the mutual action between the matter on the two sides of any normal section is equivalent.

Taking OX, OY in any normal section through O any convenient point (not necessarily its centre of inertia), and OZ per-

* "On some cases of Fluid Motion."—*Camb. Phil. Trans.* 1843; or *Mathematical and Physical Papers*, Stokes, Vol. I., page 17.

† That is, the moment of inertia of a rigid solid which, as will be proved in Vol. II., may be fixed within the box, if the liquid be removed, to make its motions the same as they are with the liquid in it.

pendicular to them, let $x+a$, $y+\beta$, $z+\gamma$ be the co-ordinates of the position to which a point (x, y, z) of the unstrained solid is displaced, in virtue of the compound strain just described. Thus γ will be a function of x and y, without z; and, if the twist (1) be denoted by τ according to the simple twist reckoning of § 120, we shall have

$$x+a=x\cos(\tau z)-y\sin(\tau z),\quad y+\beta=x\sin(\tau z)+y\cos(\tau z)\ldots(7).$$

Hence, for infinitely small values of z,

$$a=-\tau yz,\quad \beta=\tau xz\ldots\ldots\ldots\ldots\ldots(8).$$

Adhering to the notation of §§ 670, 693, only changing to Saxon letters, we have

Equations of strain, stress, and internal equilibrium.

$$\mathfrak{e}=0,\ \mathfrak{f}=0,\ \mathfrak{g}=0,\ \mathfrak{a}=\tau x+\frac{d\gamma}{dy},\ \mathfrak{b}=-\tau y+\frac{d\gamma}{dx},\ \mathfrak{c}=0\ \ldots..(9).$$

Hence [§ 693 (5)]

$$P=0,\ Q=0,\ R=0,\ S=n\left(\tau x+\frac{d\gamma}{dy}\right),\ T=n\left(-\tau y+\frac{d\gamma}{dx}\right),\ U=0\ldots(10).$$

And with the notation of § 698, (8) and (9),

$$\delta=0,\ \nabla^2 a=0,\ \nabla^2\beta=0\ \ldots\ldots\ldots\ldots(11).$$

Hence if also

$$\frac{d^2\gamma}{dx^2}+\frac{d^2\gamma}{dy^2}=0\ \ldots\ldots\ldots\ldots\ldots(12),$$

the equations of internal equilibrium [§ 698 (6)] are all satisfied.

For the surface traction, with the notation of §§ 662, 700, we have, by § 662 (1),

$$F=0,\ G=0,\ H=T\sin\phi+S\cos\phi\ \ldots\ldots\ldots\ldots(13);$$

Surface traction to be made zero.

or eliminating T and S by (10), and introducing $d\gamma/dp$ to denote the rate of variation of γ in the direction perpendicular to the prismatic surface, and q (PE of § 703) the distance from the point of the surface for which H is expressed, to the intersection of the tangent plane with a perpendicular from O,

$$\left.\begin{array}{l} H=n\left\{\left(\dfrac{d\gamma}{dy}\cos\phi+\dfrac{d\gamma}{dx}\sin\phi\right)-\tau(y\sin\phi-x\cos\phi)\right\} \\ \text{or } H=n\left(\dfrac{d\gamma}{dp}-\tau q\right). \end{array}\right\}\ldots(14).$$

To find the mutual action between the matter on the two sides of a normal section, we first remark that, inasmuch as each of the two parts of the compound strain considered (the twist and the warping) separately fulfils the conditions of § 700, we must have

Couple resultant of traction in normal section.

$$\iint Tdxdy=\int xHds,\ \text{and}\ \iint Sdxdy=\int yHds\ \ldots\ldots(15).$$

16—2

Hence when the prescribed surface condition $H=0$ is fulfilled, we have $\iint T dxdy = 0,\ \iint S dxdy = 0$ ....(16),

and there remains only a couple

$$N = \iint (Sx - Ty)\,dxdy = n\tau \iint (x^2+y^2)\,dxdy - n\iint \left(y\frac{d\gamma}{dx} - x\frac{d\gamma}{dy}\right) dxdy \ldots (17),$$

in the plane of the normal section. That condition, by (14), gives

$$\frac{d\gamma}{dp} = \tau q, \text{ or } \frac{d\gamma}{dy}\cos\phi + \frac{d\gamma}{dx}\sin\phi = \tau(y\sin\phi - x\cos\phi) \ldots (18),$$

for every point of the prismatic surface.

Hydro-kinetic application of torsional equation. We shall see in Vol. II. that (12) and (18) are differential equations which determine a function, γ, of x, y, such that $d\gamma/dx$ and $d\gamma/dy$ are the components of the velocity of a perfect liquid initially at rest in a prismatic box as described in § 705, and set in motion by communicating to the box an angular velocity, τ, in the direction reckoned negative round OZ: and that the time-integral (§ 297) of the continuous couple by which this is done, however suddenly or gradually, is

$$n\iint \left(x\frac{d\gamma}{dy} - y\frac{d\gamma}{dx}\right) dxdy,$$

which is the excess of $n\tau\iint (x^2+y^2)\,dxdy$ over N. Also, α and b in (9) are the components, parallel to OX and OY, of the velocity of the liquid relatively to the box, since $-\tau y$ and τx are the components of the velocity of a point (x, y) rotating in the positive direction round OZ with the angular velocity τ. Hence the propositions (§ 705) to be proved.

707. M. de St Venant finds solutions of these equations in two ways:—(A.) Taking any solution whatever of (12), he finds a series of curves for each of which (18) is satisfied, and any one of which, therefore, may be taken as the boundary of a prism to which that solution shall be applicable: and (B.) By the purely analytical method of Fourier, he solves (12), subject to the surface equation (18), for the particular case of a rectangular prism.

St Venant's invention of solvable cases. (A.) For this M. de St Venant finds a general integral of the boundary condition, viewed as a differential equation in terms of the two variables x, y, thus:—Multiplying (18) by ds, and replacing $\sin\phi\, ds$ and $\cos\phi\, ds$ by their values dy and $-dx$,

St Venant's invention of solvable cases.

we have $$\frac{d\gamma}{dx}dy - \frac{d\gamma}{dy}dx - \tfrac{1}{2}\tau d(x^2+y^2) = 0 \quad\text{..............(19)}.$$

In this the first two terms constitute a complete differential of a function of x and y, independent variables; because γ satisfies (12). Thus, denoting this function by u, we have

$$\frac{d\gamma}{dx} = \frac{du}{dy}, \text{ and } \frac{d\gamma}{dy} = -\frac{du}{dx} \quad\text{..................(20)},$$

and (19) becomes $du - \tfrac{1}{2}\tau d(x^2+y^2) = 0$,

which requires that $u - \tfrac{1}{2}\tau(x^2+y^2) = C$ .... (21),

for every point in the boundary. It is to be remarked that, because

$$\frac{d}{dx}\frac{d\gamma}{dy} = \frac{d}{dy}\frac{d\gamma}{dx},$$

we have, from (20), $$\frac{d^2u}{dx^2} + \frac{d^2u}{dy^2} = 0 \quad\text{........................(22)};$$

or u also, as γ, fulfils the equation $\nabla^2 u = 0$. A function, algebraically homogeneous as to x, y, which satisfies this equation is [Appendix B. (a)] a spherical harmonic independent of z. Hence a homogeneous solution of integral degree i can only be the part of Appendix B. (39) not containing z. This is

$$C\xi^i + C'\eta^i,$$

where [Appendix B. (26)]

$$\xi = x + vy, \text{ and } \eta = x - vy,$$

v standing for $\sqrt{-1}$;

or, if we change the constants so that the constants may be real,

$$A\{(x+vy)^i + (x-vy)^i\} - vB\{(x+vy)^i - (x-vy)^i\} \quad\text{......(23)},$$

or, in terms of polar co-ordinates,

$$2r^i(A\cos i\theta + B\sin i\theta) \quad\text{...................(24)}.$$

Using this solution for the case $i = 2$ and (without loss of generality) putting $B = 0$, we have

$$u = 2A(x^2 - y^2) \quad\text{.......................(25)};$$

whence by (20) $$\gamma = -4Axy \quad\text{..........................(26)};$$

and the equation (21) of the series of bounding curves to which this solution is applicable is

$$\frac{x^2}{a^2} + \frac{y^2}{b^2} = 1 \quad\text{.......................(27)},$$

if we put, for brevity,

$$\frac{-C}{\frac{1}{2}\tau - 2A} = a^2, \quad \frac{-C}{\frac{1}{2}\tau + 2A} = b^2,$$

which give $$4A = \tau\frac{a^2 - b^2}{a^2 + b^2},$$

so that (26) becomes $$\gamma = -\tau\frac{a^2 - b^2}{a^2 + b^2}xy \text{..................(28)}.$$

Using this in (17) we have

Solution for elliptic cylinder.

$$N = n\tau\left\{\iint(x^2 + y^2)\,dxdy - \frac{a^2 - b^2}{a^2 + b^2}\iint(x^2 - y^2)\,dxdy\right\},$$

or, if I, J denote the moments of inertia of the area of the normal section, round the axes of x and y respectively,

$$N = n\tau\left\{J + I - \frac{a^2 - b^2}{a^2 + b^2}(J - I)\right\} \text{..............(29)};$$

or, lastly, as we have for the elliptic area (27),

$$\left.\begin{aligned} I &= \tfrac{1}{4}\pi ab\,.\,b^2, \quad J = \tfrac{1}{4}\pi ab\,.\,a^2, \\ N &= n\tau(J + I)\left\{1 - \left(\frac{a^2 - b^2}{a^2 + b^2}\right)^2\right\} = n\tau\frac{\pi a^3 b^3}{a^2 + b^2} \end{aligned}\right\} \text{......(30)}.$$

Another very simple but most interesting case investigated by M. de St Venant, is that arrived at by taking a harmonic of the third degree for u. Thus, introducing a factor $\frac{1}{2}\tau/a$ for the sake of homogeneity and subsequent convenience, we have

St Venant's invention of solvable cases.

$$\left.\begin{aligned} &\tfrac{1}{2}\frac{\tau}{a}(x^3 - 3y^2x) - \tfrac{1}{2}\tau(x^2 + y^2) = C, \\ &\text{or in polar co-ordinates,} \\ &\tfrac{1}{2}\frac{\tau}{a}r^3\cos 3\theta - \tfrac{1}{2}\tau r^2 = C, \end{aligned}\right\} \text{............(31)},$$

as an equation giving, by different values of C, a series of bounding lines, for which

Solution for equilateral triangle.

$$\gamma = \tfrac{1}{2}\frac{\tau}{a}(y^3 - 3x^2y) = -\tfrac{1}{2}\frac{\tau}{a}r^3\sin 3\theta \text{.............(32)}$$

is the solution of (12), subject to (18). For the particular value

$$C = -\tfrac{2}{27}a^2\tau$$

(31) gives three straight lines, the sides of an equilateral triangle having a for perpendicular from an angle to the opposite

side, and placed relatively to x and y, as shown in the diagram (§ 708, below). Thus we have the complete solution of the torsion problem for a prism whose normal section is an equilateral triangle. Equation (17) worked out for this area, with (32) for γ, gives

Solution for equilateral triangle.

$$N = n(K - \tfrac{2}{5}K)\tau.$$

But (K being the proper moment of inertia of the triangle, and A its area)

$$K\frac{1}{9\sqrt{3}}a^4 = \frac{1}{9}a^2A = \frac{1}{3\sqrt{3}}A^2;$$

and thus, for the torsional rigidity, we have the several expressions

$$\frac{N}{\tau} = \tfrac{3}{5}nK = \frac{1}{15\sqrt{3}}na^4 = \frac{1}{15}na^2A = \frac{1}{5\sqrt{3}}nA^2 = \frac{1}{45}\frac{nA^4}{K} \ldots\ldots(33).$$

Similarly, taking for u a harmonic of the fourth degree and adjusting the constants to his wants, St Venant finds the equation,

For curvilinear squares.

$$\left.\begin{array}{rl} & x^2 + y^2 - a(x^4 - 6x^2y^2 + y^4) = 1 - a \\ \text{or} & r^2 - ar^4\cos 4\theta = 1 - a \end{array}\right\} \ldots\ldots\ldots\ldots(34)$$

to give, for different values of a, a series of curvilinear squares (see diagram of § 708 (3), below), all having rounded corners, except two similar though differently turned curvilinear squares with concave sides and acute angles corresponding to $a = \cdot 5$, and $a = -\frac{1}{2}(\sqrt{2} - 1)$; for each of which the torsion problem is algebraically solved.

And by taking u the sum of two harmonics, of the fourth and eighth degrees respectively, and properly adjusting the constants, he finds

For star with four rounded points.

$$\left.\begin{array}{l} \dfrac{x^2+y^2}{r_0^2} - \frac{48}{49}\cdot\frac{16}{17}\cdot\dfrac{x^4 - 6x^2y^2 + y^4}{r_0^4} + \frac{12}{49}\cdot\frac{16}{17}\cdot\dfrac{x^8 - 28x^6y^2 + 70x^4y^4 - 28x^2y^6 + y^8}{r_0^8} \\ \qquad\qquad\qquad\qquad = 1 - \frac{36}{49}\cdot\frac{16}{17} \\ \text{or}\quad \dfrac{r^2}{r_0^2} - \frac{48}{49}\cdot\frac{16}{17}\cdot\dfrac{r^4}{r_0^4}\cos 4\theta + \frac{12}{49}\cdot\frac{16}{17}\cdot\dfrac{r^8}{r_0^8}\cos 8\theta = 1 - \frac{36}{49}\cdot\frac{16}{17} \end{array}\right\} \ldots(35),$$

as the equation of the curve shown in § 709, diagram (4), for which therefore the torsion problem is solved.

(B.) The integration (21) of the boundary equation, introduced by St Venant for use in his synthesis, (A.) is also very useful in

St Venant's reduction to Green's problem.

St Venant's reduction to Green's problem.

the analytical investigation, although he has not so applied it. First, we may remark, that the determination of u for a given form of prism is a particular case of "Green's problem" proved possible and determinate in Appendix A. (e); being to find u, a function of x, y which shall satisfy the equation

$$\frac{d^2u}{dx^2}+\frac{d^2u}{dy^2}=0,$$

for every point of the area bounded a certain given closed circuit, subject to the condition,

$$u=\tfrac{1}{2}\tau\,(x^2+y^2)\quad\text{.........................(36)}$$

for every point of the boundary.

When u is found, equations (20) and (17) with (10) complete the solution of the torsion problem.

Solution for rectangular prism,

For the case of a rectangular prism, the solution is much facilitated by taking

$$\left.\begin{array}{l} u=v+A\,(x^2-y^2)+B, \\ \text{which gives}\quad \dfrac{d^2v}{dx^2}+\dfrac{d^2v}{dy^2}=0\,; \\ \text{and for boundary condition,} \\ v=(\tfrac{1}{2}\tau-A)\,x^2+(\tfrac{1}{2}\tau+A)\,y^2-B. \end{array}\right\}\quad\text{...............(37).}$$

If the rectangle be not square, let its longer sides be parallel to OX; and let a, b be the lengths of each of the longer and each of the shorter sides respectively. Take, now,

$$A=\tfrac{1}{2}\tau,\ \text{ and }\ B=\tfrac{1}{4}\tau b^2\quad\text{.......................(38).}$$

The boundary condition becomes

$$\left.\begin{array}{ll} & v=0\ \text{ when }\ y=\pm\tfrac{1}{2}b, \\ \text{and} & v=-\tau\,(\tfrac{1}{4}b^2-y^2)\ \text{ when }\ x=\pm\tfrac{1}{2}a \end{array}\right\}\quad\text{.............(39).}$$

found by Fourier's analysis.

To solve the problem by Fourier's method (compare with the more difficult problem of § 655), the requisite expansion of $\tfrac{1}{4}b^2-y^2$ is clearly*

* Obtainable, as a matter of course, from Fourier's general theorem, but most easily by two successive integrations of the common formula

$$\tfrac{1}{4}\pi=\cos\theta-\tfrac{1}{3}\cos 3\theta+\tfrac{1}{5}\cos 5\theta-\text{etc.}$$

$$\tfrac{1}{4}b^2 - y^2 = \left(\frac{2}{\pi}\right)^3 b^2 \left\{\cos\eta - \frac{1}{3^3}\cos 3\eta + \frac{1}{5^3}\cos 5\eta - \text{etc.}\right\} \ldots (40);$$

where, for brevity $\eta = \pi y/b$.
And, for the same cause, putting $\xi = \pi x/b$ $\Big\}$ ....(41)

we have, for the form of solution,

$$v = \Sigma\{A_{2i+1}\epsilon^{-(2i+1)\xi} + B_{2i+1}\epsilon^{+(2i+1)\xi}\}\cos(2i+1)\eta \ldots\ldots(42),$$

which satisfies (37), and gives $v = 0$ for $y = \pm\frac{1}{2}b$. The residual boundary condition gives, for determining A_{2i+1} and B_{2i+1},

$$\left.\begin{aligned}&\left[A_{2i+1}\epsilon^{-(2i+1)\pi a/2b} + B_{2i+1}\epsilon^{+(2i+1)\pi a/2b}\right]\\ = &\left[A_{2i+1}\epsilon^{+(2i+1)\pi a/2b} + B_{2i+1}\epsilon^{-(2i+1)\pi a/2b}\right] = -\frac{8\tau b^2}{\pi^3}\frac{(-1)^i}{(2i+1)^3}\end{aligned}\right\} (43).$$

These two equations give a common value for the two unknown quantities A_{2i+1}, B_{2i+1}; with which (42) becomes

$$v = -\tau\left(\frac{2}{\pi}\right)^3 b^2 \Sigma \frac{(-1)^i}{(2i+1)^3}\frac{\epsilon^{-(2i+1)\xi} + \epsilon^{+(2i+1)\xi}}{\epsilon^{-(2i+1)\pi a/2b} + \epsilon^{+(2i+1)\pi a/2b}}\cos(2i+1)\eta \ldots (44).$$

From this we find, by (37), (38), and (20),

$$\gamma = -\tau xy + \tau\left(\frac{2}{\pi}\right)^3 b^2 \Sigma \frac{(-1)^i}{(2i+1)^3}\frac{\epsilon^{+(2i+1)\xi} - \epsilon^{-(2i+1)\xi}}{\epsilon^{+(2i+1)\pi a/2b} + \epsilon^{-(2i+1)\pi a/2b}}\sin(2i+1)\eta \ldots (45);$$

and (17) gives, for the torsional rigidity,

$$\frac{N}{\tau} = nab^3\left[\tfrac{1}{3} - \left(\frac{2}{\pi}\right)^5 \frac{b}{a}\Sigma\frac{1}{(2i+1)^5}\frac{1 - \epsilon^{-(2i+1)\pi a/b}}{1 + \epsilon^{-(2i+1)\pi a/b}}\right] \ldots (46).$$

If we had proceeded in all respects as above, only taking $A = -\frac{1}{2}\tau$ instead of $A = \frac{1}{2}\tau$, in (37), we should have obtained expressions for γ and N/τ, seemingly very different, but necessarily giving the same values. These other expressions may be written down immediately by making the interchange x, y, a, b for y, x, b, a in (45) and (46), and changing the sign of each term of (45). They obviously converge less rapidly than (45) and (46) if, as we have supposed, $a > b$, and it is on this account that we proceeded as above rather than in the other way. The comparison of the results gives astonishing theorems of pure mathematics, such as rarely fall to the lot of those mathematicians who confine themselves to pure analysis or geometry, instead of allowing themselves to be led into the rich and beautiful fields of mathematical truth which lie in the way of physical research.

Extension to a class of curvilinear rectangles.

A relation discovered by Stokes* and Lamé† independently [which we have already used in equations (20), (22)] taken in connexion with Lamé's method of curvilinear co-ordinates‡, allows us to extend the Fourier analytical method to a large class of curvilinear rectangles, including the rectlinear rectangle as a particular case, thus:—

Lamé's transformation to plane isothermal co-ordinates

Let ξ be a function of x, y satisfying the equation

$$\frac{d^2\xi}{dx^2} + \frac{d^2\xi}{dy^2} = 0 \quad\text{.......................(47)},$$

and, as this shows that $\frac{d\xi}{dx}dy - \frac{d\xi}{dy}dx$ is a complete differential, let

$$\eta = \int \left(\frac{d\xi}{dx}dy - \frac{d\xi}{dy}dx\right) \quad\text{..... ....(48)};$$

or, which means the same,

$$\frac{d\eta}{dy} = \frac{d\xi}{dx}, \text{ and } \frac{d\eta}{dx} = -\frac{d\xi}{dy} \quad\text{................(49)}.$$

This other function η also, as we see from (49), satisfies the equation

$$\frac{d^2\eta}{dx^2} + \frac{d^2\eta}{dy^2} = 0 \quad\text{.......................(50)}.$$

Theorem of Stokes and Lamé.

And, also because of (49), two intersecting curves, whose equations are

$$\xi = A, \quad \eta = B \quad\text{.......................(51)},$$

cut one another at right angles. Let now, A and B being supposed given, x and y be determined by these two equations. The point whose co-ordinates are x, y may also be regarded as specified by (A, B), or by the values of ξ, η, which give curves

* On the Steady Motion of Incompressible Fluids. *Camb. Phil. Trans.*, 1842; or *Mathematical and Physical Papers*, Stokes, Vol. I., page 1.

† Mémoire sur les lois de l'équilibre du fluide éthéré. *Journal de l'École Polytechnique*, 1834.

‡ See Thomson on the Equations of the Motion of Heat referred to Curvilinear co-ordinates. *Camb. Math. Journal*, 1845; or Reprint of Mathematical and Physical Papers, Art. IX.

intersecting in (x, y). Thus (ξ, η) with any particular values assigned to ξ and η, specifies a point in a plane. Common rectilinear co-ordinates are clearly a particular case (rectilinear orthogonal co-ordinates) of the system of curvilinear orthogonal co-ordinates thus defined. Let now u, any function of x, y, be transformed into terms of ξ, η. We have, by differentiation,

Theorem of Stokes and Lamé.

$$\frac{d^2u}{dx^2}+\frac{d^2u}{dy^2}=\frac{d^2u}{d\xi^2}\left(\frac{d\xi^2}{dx^2}+\frac{d\xi^2}{dy^2}\right)+2\frac{d^2u}{d\xi d\eta}\left(\frac{d\xi}{dx}\frac{d\eta}{dx}+\frac{d\xi}{dy}\frac{d\eta}{dy}\right)$$
$$+\frac{d^2u}{d\eta^2}\left(\frac{d\eta^2}{dx^2}+\frac{d\eta^2}{dy^2}\right)+\frac{du}{d\xi}\left(\frac{d^2\xi}{dx^2}+\frac{d^2\xi}{dy^2}\right)+\frac{du}{d\eta}\left(\frac{d^2\eta}{dx^2}+\frac{d^2\eta}{dy^2}\right)\ldots(52),$$

which is reduced by (49) and (50) to

$$\frac{d^2u}{dx^2}+\frac{d^2u}{dy^2}=\left(\frac{d^2u}{d\xi^2}+\frac{d^2u}{d\eta^2}\right)\left(\frac{d\xi^2}{dx^2}+\frac{d\xi^2}{dy^2}\right)\ldots\ldots\ldots(53).$$

Hence the equation $$\frac{d^2u}{dx^2}+\frac{d^2u}{dy^2}=0$$

transforms into $$\frac{d^2u}{d\xi^2}+\frac{d^2u}{d\eta^2}=0\ldots\ldots\ldots\ldots\ldots\ldots\ldots(54).$$

Also the relations $$\frac{du}{dy}=\frac{d\gamma}{dx},\quad \frac{du}{dx}=-\frac{d\gamma}{dy}$$

transform, in virtue of (49), into

$$\frac{du}{d\eta}=\frac{d\gamma}{d\xi},\quad \frac{du}{d\xi}=-\frac{d\gamma}{d\eta}\ldots\ldots\ldots\ldots\ldots\ldots\ldots(55).$$

Hence the general problem of finding u and γ has precisely the same statement in terms of ξ, η, as that given above, (22), (36), and (20), in terms of x, y, with this exception, that we have not $u=\frac{1}{2}\tau(\xi^2+\eta^2)$, but if $f(\xi, \eta)$ denote the function of ξ, η into which x^2+y^2 transforms,

Solution for rectangle of plane isothermals.

$u=\frac{1}{2}\tau f(\xi, \eta)$ for every point of the boundary......(56).

The solution for the curvilinear rectangle

$$\left.\begin{array}{c|c}\xi=a & \eta=\beta\\ \xi=0 & \eta=0\end{array}\right\}\ldots\ldots\ldots\ldots\ldots\ldots\ldots\ldots(57)$$

is, on Fourier's plan,

$$u=\Sigma\sin\frac{i\pi\xi}{a}\left(A_i\epsilon^{i\pi\eta/a}+A_i'\epsilon^{-i\pi\eta/a}\right)+\Sigma\sin\frac{i\pi\eta}{\beta}\left(B_i\epsilon^{i\pi\xi/\beta}+B_i'\epsilon^{-i\pi\xi/\beta}\right)\ldots(58),$$

where A_i, A_i' are to be determined by two equations, obtained

Solution for rectangle of plane isothermals.

thus:—Equate the coefficient of $\sin i\pi\xi/a$ when $\eta = 0$ and when $\eta = \beta$ respectively to the coefficients of $\sin i\pi\xi/a$ in the expansions of $f(\xi, 0)$ and $f(\xi, \beta)$ in series of the form

$$P_1 \sin\frac{\pi\xi}{a} + P_2 \sin\frac{2\pi\xi}{a} + P_3 \sin\frac{3\pi\xi}{a} + \text{etc.} \quad\text{.........(59)}$$

by Fourier's theorem, § 77. Similarly, B_i, B_i', are determined from the expansions of $f(0, \eta)$ and $f(a, \eta)$, in series of the form

$$Q_1 \sin\frac{\pi\eta}{\beta} + Q_2 \sin\frac{2\pi\eta}{\beta} + Q_3 \sin\frac{3\pi\eta}{\beta} + \text{etc.} \quad\text{.......(60).}$$

Example. Rectangle bounded by two concentric arcs and two radii.

Of one extremely simple example, very interesting in theory and valuable for practical mechanics, we shall indicate the details.

Let
$$\xi = \log\sqrt{\frac{x^2+y^2}{a^2}} \quad\text{....................(61).}$$

This clearly satisfies (47); and it gives, by (48),

$$\eta = \tan^{-1}\frac{y}{x} \quad\text{..........................(62).}$$

The solution may be expressed on the same plan as in (37)... (45) by a series of sines of multiples of $\pi\eta/a$, if we take*

$$u = v + \tfrac{1}{2}\tau a^2 \frac{\epsilon^{2\xi}\cos(\beta - 2\eta)}{\cos\beta} \quad\text{................(63),}$$

which, with (54), gives $\dfrac{d^2v}{d\xi^2} + \dfrac{d^2v}{d\eta^2} = 0$ ....(64),

and leaves, as boundary conditions in the solution for v,

$$\left.\begin{aligned} &v = \tfrac{1}{2}\tau a^2\left\{1 - \frac{\cos(\beta - 2\eta)}{\cos\beta}\right\} \text{ when } \xi = 0,\\ &v = \tfrac{1}{2}\tau a^2\epsilon^{2a}\left\{1 - \frac{\cos(\beta - 2\eta)}{\cos\beta}\right\} \text{ when } \xi = a,\\ \text{and}\quad &v = 0 \text{ when } \eta = 0, \text{ and when } \eta = \beta. \end{aligned}\right\} \quad\text{........(65).}$$

The last condition shows that the B_i and B_i' part of (58) is proper for expressing v, and the first two determine B_i and B_i' as usual.

* It should be noticed that this solution fails for the case of $\beta = \frac{1}{2}(2i+1)\pi$.

Or when it is best to have the result in series of sines of multiples of $\pi\xi/a$, we may take

Rectangle bounded by two concentric arcs and two radii.

$$u = w + \tfrac{1}{2}\tau a^2\left(1 + \frac{\epsilon^{2a}-1}{a}\xi\right) \quad\text{.................(66)},$$

which, with (54), gives $\dfrac{d^2w}{dx^2} + \dfrac{d^2w}{dy^2} = 0$ .... (67),

and leaves, as boundary conditions in the solution for w,

$$\left.\begin{array}{l} w = \tfrac{1}{2}\tau a^2\left\{\epsilon^{2\xi} - 1 - \dfrac{\epsilon^{2a}-1}{a}\xi\right\} \text{ when } \eta = 0, \text{ and when } \eta = \beta, \\ \text{and} \qquad w = 0 \text{ when } \xi = 0, \text{ and when } \xi = a. \end{array}\right\} \text{....(68)}.$$

The last shows that the A_i and A_i' part of (58) is proper for w, and the two first determine A_i, A_i.

708. St Venant's treatise abounds in beautiful and instructive graphical illustrations of his results, from which we select the following:—

(1) *Elliptic cylinder.*—The plain and dotted curvilinear arcs are "contour lines" (*coupes topographiques*) of the section as

Contour lines of normal section of elliptic cylinder, as warped by torsion: equilateral hyperbolas.

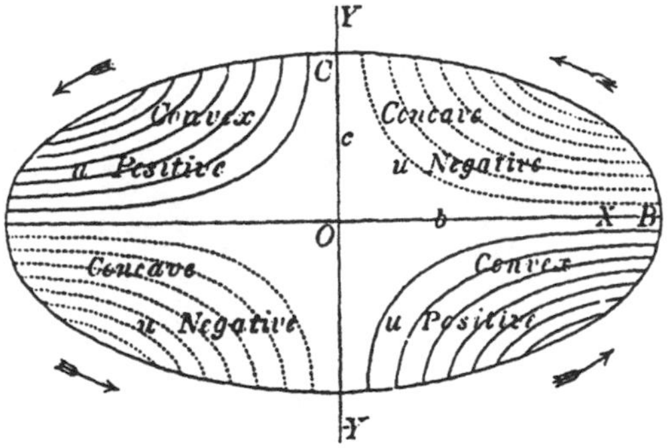

warped by torsion; that is to say, lines in which it is cut by a series of parallel planes, each perpendicular to the axis, or lines for which γ (§ 706) has different constant values. These lines are [§ 707 (28)] equilateral hyperbolas in this case. The arrows indicate the direction of rotation in the part of the prism *above* the plane of the diagram.

Contour lines of normal section of triangular prism, as warped by torsion.

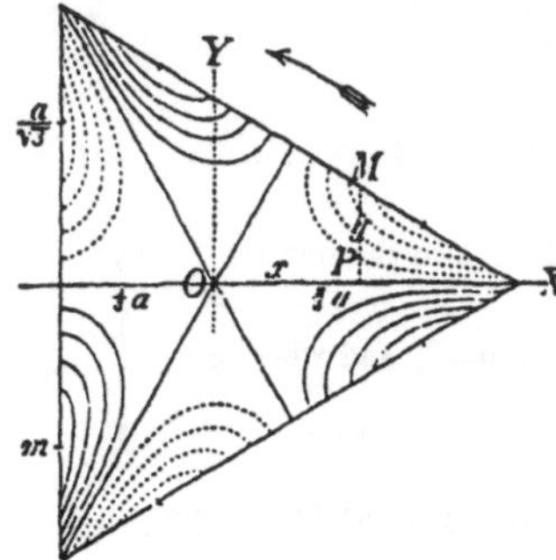

(2) *Equilateral triangular prism.*—The contour lines are shown as in case (1); the dotted curves being those where the warped section falls *below* the plane of the diagram, the direction of rotation of the part of the prism above the plane being indicated by the bent arrow.

Diagram of St Venant's curvilinear squares for which torsion problem is solvable.

(3) This diagram shows the series of lines represented by (34) of § 707, with the indicated values for a. It is remarkable

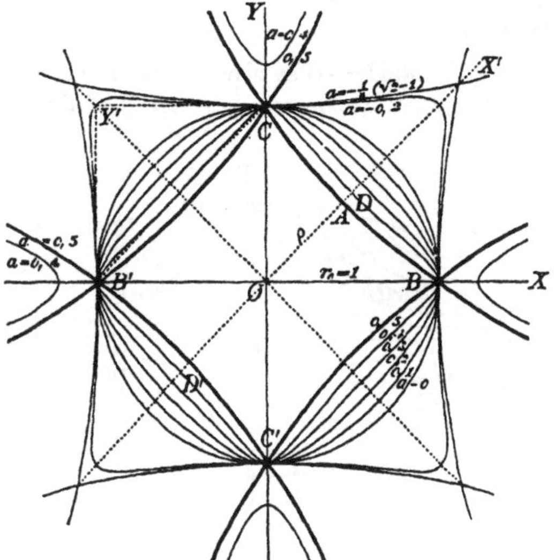

that the values $a=0{\cdot}5$ and $a=-\frac{1}{2}(\sqrt{2}-1)$ give similar but not equal curvilinear squares (hollow sides and acute angles), one of them turned through half a right angle relatively to the other. Everything in the diagram outside the larger of these squares is to be cut away as irrelevant to the physical problem; the series of closed curves remaining exhibits figures of prisms, for any one of which the torsion problem is solved algebraically. These figures vary continuously from a circle, inwards to one of the acute-angled squares, and outwards to the other: each, except these extremes, being a continuous closed curve with no angles. The curves for $a = 0{\cdot}4$ and $a = -0{\cdot}2$ approach remarkably near to the rectilinear squares, partially indicated in the diagram by dotted lines.

(4) This diagram shows the contour lines, in all respects as in the cases (1) and (2), for the case of a prism having for section the figure indicated. The portions of curve outside the continuous closed curve are merely indications of mathematical extensions irrelevant to the physical problem. Contour lines for St Venant's "étoile à quatre points arrondis."

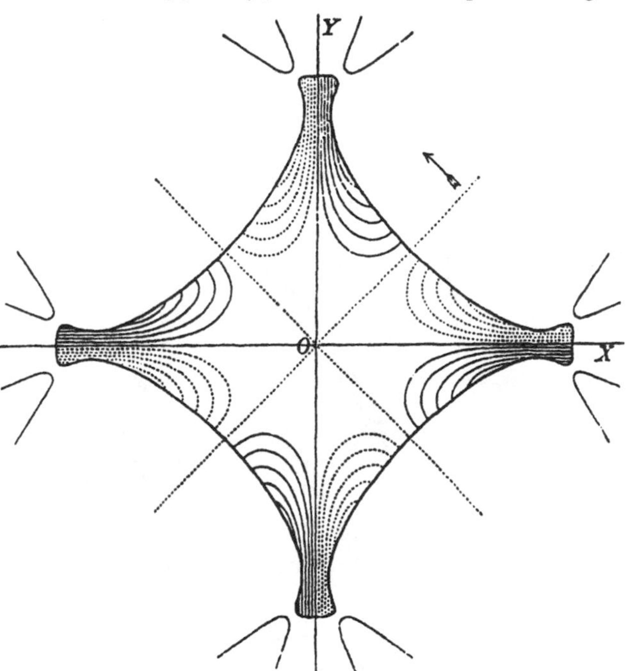

Contour lines of normal section of square prism, as warped by torsion.

(5) This shows as, in the other cases, the contour lines for the warped section of a square prism under torsion.

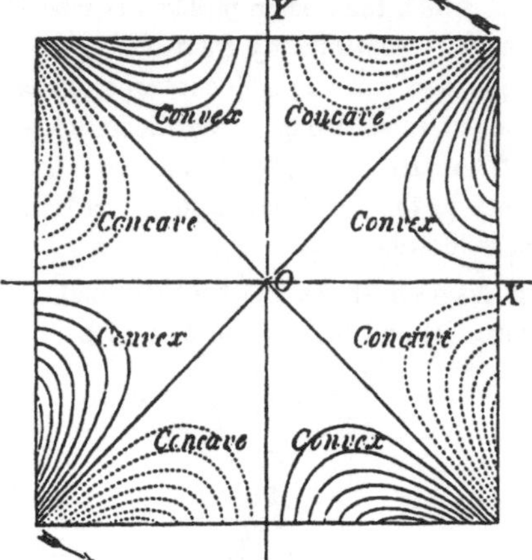

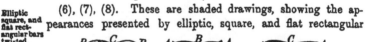

Elliptic square, and flat rectangular bars twisted.

(6), (7), (8). These are shaded drawings, showing the appearances presented by elliptic, square, and flat rectangular

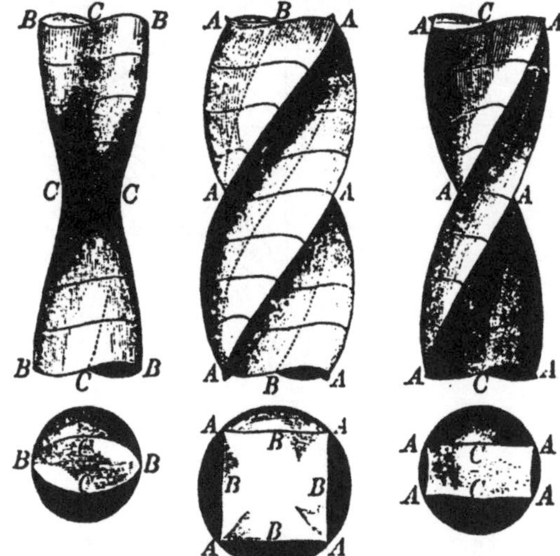

bars under exaggerated torsion, as may be realized with such a substance as India rubber.

709. Inasmuch as the moment of inertia of a plane area about an axis through its centre of inertia perpendicular to its plane is obviously equal to the sum of its moments of inertia round any two axes through the same point, at right angles to one another in its plane, the fallacious extension of Coulomb's law, referred to in § 703, would make the torsional rigidity of a bar of any section equal to n/M (§ 694) multiplied into the sum of its flexural rigidities (see below, § 715) in any two planes at right angles to one another through its length. The true theory, as we have seen (§§ 705, 706), always gives a torsional rigidity less than this. How great the deficiency may be expected to be in cases in which the figure of the section presents projecting angles, or considerable prominences (which may be imagined from the hydrokinetic analogy we have given in § 705), has been pointed out by M. de St Venant, with the important practical application, that strengthening ribs, or projections (see, for instance, the fourth annexed diagram), such as are introduced in engineering to give stiffness to beams, have the reverse of a good effect when *torsional* rigidity or strength is an object, although they are truly of great value in increasing the flexural rigidity, and giving strength to bear ordinary strains, which are always more or less flexural. With remarkable ingenuity and mathematical skill he has drawn beautiful illustrations of this important practical principle from his algebraic and transcendental solutions [§ 707 (32), (34), (35), (45)]. Thus

Torsional rigidity less in proportion to sum of principal flexural rigidities than according to false extension (§ 703) of Coulomb's law.

Ratios of torsional rigidities to those of solid circular rods.

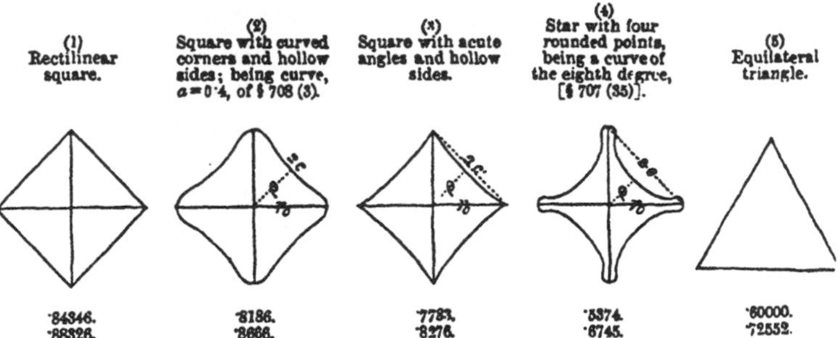

for an equilateral triangle, and for the rectilinear and three curvilinear squares shown in the annexed diagram, he finds for the torsional rigidities the values stated. The number immediately below the diagram indicates in each case the fraction which the true torsional rigidity is of the old fallacious estimate (§ 703); the latter being the product of the rigidity of the substance into the moment of inertia of the cross section round an axis perpendicular to its plane through its centre of inertia. The second number indicates in each case the fraction which the torsional rigidity is of that of a solid circular cylinder of the same sectional area.

(a) of same moment of inertia,

(b) of same quantity of material.

Places of greatest distortion in twisted prisms.

710. M. de St Venant also calls attention to a conclusion from his solutions which to many may be startling, that in his simpler cases the places of greatest distortion are those points of the boundary which are nearest to the axis of the twisted prism in each case, and the places of least distortion those farthest from it. Thus in the elliptic cylinder the substance is most strained at the ends of the smaller principal diameter, and least at the ends of the greater. In the equilateral triangular and square prisms there are longitudinal lines of maximum strain through the middle of the sides. In the oblong rectangular prism there are two lines of greater maximum strain through the middles of the broader pair of sides, and two lines of less maximum strain through the middles of the narrow sides. The strain is, as we may judge from (§ 705) the hydrokinetic analogy, excessively small, but not evanescent, in the projecting ribs of a prism of the figure shown in (4) § 709. It is quite evanescent infinitely near the angle, in the triangular and rectangular prisms, and in each other case as (3) of § 709, in which there is a finite angle, whether acute or obtuse, projecting outwards. This reminds us of a general remark we have to make, although consideration of space may oblige us to leave it without formal proof. A solid of any elastic substance, isotropic or aeolotropic, bounded by any surfaces presenting projecting edges or angles, or re-entrant angles or edges, however obtuse, cannot experience any finite stress or strain in the neighbourhood of a *projecting* angle (trihedral, polyhedral, or conical); in the neighbourhood of an edge, can only experience simple longitudinal stress parallel to the neighbouring part of the edge; and generally

Solid of any shape having edges, or pyramidal or conical angles, under stress.

Strain at projecting angles, evanescent.

experiences infinite stress and strain in the neighbourhood of a *re-entrant* edge or angle; when influenced by any distribution of force, exclusive of surface tractions infinitely near the angles or edges in question. An important application of the last part of this statement is the practical rule, well known in mechanics, that every re-entering edge or angle ought to be rounded to prevent risk of rupture, in solid pieces designed to bear stress. An illustration of these principles is afforded by the concluding example of § 707; in which we have the complete mathematical solution of the torsion problem for prisms of fan-shaped sections, such as the annexed figures. In the cases corresponding to $a=0$, we see, without working out the solution, that the distortion $d\gamma/rd\eta$ vanishes when $r=0$, if β is $<\pi$; becomes infinite when $r=0$, if β is $>\pi$; but is finite and determinate if $\beta=\pi$.

At re-entrant angles infinite.

Liability to cracks proceeding from re-entrant angles, or any places of too sharp concave curvature.

Cases of curvilinear rectangles for which torsion problem has been solved.

The solution indicated above determining v to satisfy (64) and (65) of § 707, if translated into polar co-ordinates r, η, such that $x=r\cos\eta$, and $y=r\sin\eta$, with $\pi/\beta=\nu$, becomes merely this—

$$v=\Sigma\,(B_i r^{i\nu}+B_i' r^{-i\nu})\sin i\nu\eta^{*} \quad\ldots\ldots\ldots\ldots\ldots(69),$$

where B_i, B_i' are to be determined by the equations (65) of § 707, with $r=a$ and $r=a'$ instead of $\xi=0$ and $\xi=a$, and a'^2 instead of $a^2\epsilon^{2a}$ (a and a' denoting the radii of the concave and convex cylindrical surfaces respectively). When $a=0$, these give $B_i=0$; and therefore

Distortion zero at central angle of sector (4). infinite at central angle of sector (6): zero at all the other angles.

$$\left(\frac{dv}{rd\eta}\right)_{r=0} \text{ is zero, or equal to } B_1\cos\eta, \text{ or infinite,}$$

according as $\nu>1$, $=1$, or <1; whence also follow similar results for $\left(\dfrac{d\gamma}{rd\eta}\right)_{r=0}$.

* Compare § 707 (23) (24); by which we see that this solution is merely the general expression in polar co-ordinates for series of spherical harmonics of x, y, with $z=0$, of degrees i, $2i$, $3i$, etc., and $-i$, $-2i$, $-3i$, etc. These are "complete harmonics" when i is unity or any integer.

Problem of flexure.

711. To prove the law of flexure (§§ 591, 592), and to investigate the flexural rigidity (§ 596) of a bar or wire of isotropic substance, we shall first conceive the bar to be bent into a circular arc, and investigate the application of force necessary to do so, subject to the following conditions:—

(1) All lines of it parallel to its length become circular arcs in or parallel to the plane ZOX, with their centres in one line perpendicular to this plane; OZ and all lines parallel to it through OY being bent without change of length.

(2) All normal sections remain plane, and perpendicular to those longitudinal lines, so that their planes come to pass through that line of centres.

(3) No part of any normal section experiences deformation.

Forced condition of no distortion in normal sections.

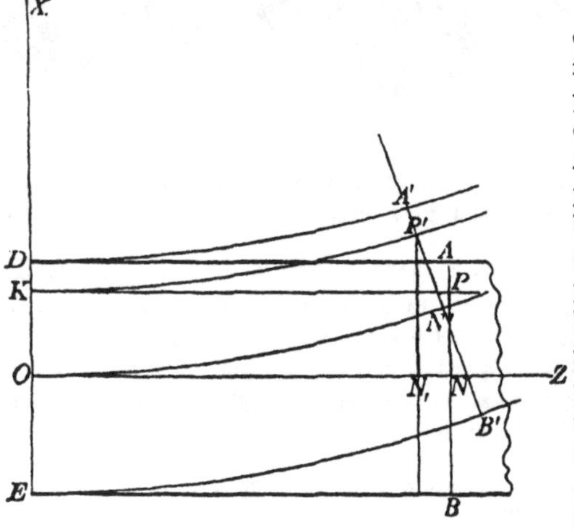

A section DOE of the beam being chosen for plane of reference, XOY, let P, (x, y, z) be any point of the unbent, and P', (x', y', z') the same point of the bent, beam; each seen in projection, on the plane ZOX, in the diagram: and let ρ be the radius of the arc ON', into which the line ON of the straight beam is bent. We have

$$x' = x + (\rho - x)\left(1 - \cos\frac{z}{\rho}\right),\quad y' = y,\quad z' = (\rho - x)\sin\frac{z}{\rho}.$$

But, according to the fundamental limitation (§ 588), x is at most infinitely small in comparison with ρ: and through any length of the bar not exceeding its greatest transverse dimension, z is so also. Hence we neglect higher powers of x/ρ and z/ρ than the second in the preceding expressions; and putting

$$x'-x=\alpha,\ y'-y=\beta,\ z'-z=\gamma,$$

we have $$\alpha=\tfrac{1}{2}\frac{z^2}{\rho},\ \beta=0,\ \gamma=-\frac{xz}{\rho}\ \ldots\ldots\ldots\ldots\ldots\ldots(1).$$

These, substituted in § 693 (5) and § 697 (2), give

Surface traction (P, Q), required to prevent distortion in normal section.

$$\left.\begin{array}{c}P=-(m-n)\frac{x}{\rho},\ Q=-(m-n)\frac{x}{\rho},\ R=-(m+n)\frac{x}{\rho}\\ S=0,\ T=0,\ U=0,\end{array}\right\}\ \ldots(2).$$

$$X=\frac{m-n}{\rho},\ Y=0,\ Z=0\ \ldots\ldots\ldots\ldots\ldots\ldots(3).$$

The interpretation of this result is interesting in itself, but, not requiring it for our present purpose, we leave it as an exercise to the student.

712. The problem of simple flexure supposes that no force is applied from without either as traction on the sides of the bar, or as force acting at a distance on its interior substance, but that, by opposing couples properly applied to its ends, it is kept in a circular form, with strain and stress uniform throughout its length.

Correction to do away with lateral traction, and bodily force.

To the α, β, γ of last section let corrections

$$\alpha'=\tfrac{1}{2}K(x^2-y^2),\ \beta'=Kxy,\ \gamma'=0,$$

be added. This will give, by § 693 (5),

$$P'=Q'=2mKx,\ R'=2(m-n)Kx,\ S'=0,\ T'=0,\ U'=0,$$

and by § 698 (2)

$$X'=-2mK,\ Y'=0,\ Z'=0,$$

to be added to the P, Q...X, Y, Z. Hence if we take

$$K=\frac{m-n}{2m\rho},$$

the surface tractions on the sides of the bar and the bodily forces are reduced to nothing; so that if now

St Venant's solution of flexure problem.

$$\alpha=\frac{1}{2\rho}\left\{z^2+\frac{m-n}{2m}(x^2+y^2)\right\},\ \beta=\frac{1}{\rho}\frac{m-n}{2m}xy,\ \gamma=-\frac{1}{\rho}xz\ldots(1),$$

we have [§ 670 (6) and § 693 (6)]

$$\left.\begin{array}{c}e=\frac{m-n}{2\rho m}x=\frac{\sigma}{\rho}x,\ f=\frac{m-n}{2\rho m}x=\frac{\sigma}{\rho}x,\ g=-\frac{1}{\rho}x,\\ a=b=c=0\end{array}\right\}\ldots.(2),$$

St Venant's solution of flexure problem.

and [§ 693 (5), § 694 (6)]

$$\left.\begin{array}{l} P=0,\ Q=0,\ R=-\dfrac{(3m-n)\,n}{m}\dfrac{x}{\rho}=-M\dfrac{x}{\rho}, \\ X=0,\ Y=0,\ Z=0 \end{array}\right\}\ \ldots\ldots(3).$$

To complete the fulfilment of the conditions, it is only necessary that the traction across each normal section be reducible to a couple. Hence

$$\iint R dxdy = 0,$$

or, by (3),

$$\iint x dxdy = 0;$$

that is to say,

Flexure of a bar.

713. In order that no force, but only a bending couple, may be transmitted along the rod, the centre of inertia of the normal section must be in OY, that line of it in which it is cut by the surface separating longitudinally stretched from longitudinally shortened parts of the substance.

Line through centres of inertia of normal sections remains unchanged in length.

714. In our analytical expressions only an infinitely short part of the beam has been considered; and it has not been necessary to inquire whether the axis of the couple called into play is or is not perpendicular to the plane of flexure. But when so great a length of the beam is concerned, that the change of direction (§ 5) from one end to the other is finite, the couples on the ends could not be directly opposed unless their axes were both perpendicular to the plane of flexure, inasmuch as each axis is in the proper normal section of the rod. For finite flexure in a circular arc, without lateral constraint, we must therefore have

Flexure through finite angle in one plane:

must be in either of two principal planes, if produced simply by balancing couples on the two ends.

$$\iint Rydxdy = 0;\ \text{whence, by (3)},\ \iint xydxdy = 0:$$

that is to say, the plane of flexure must be perpendicular to one of the two principal axes of inertia of the normal section in its own plane. This being the case, the moment of the whole couple acting across each normal section is equal to the product of the curvature, iuto the Young's modulus, into the moment of inertia of the area of the normal section round its principal axis perpendicular to the plane of flexure.

For we have [§ 712 (3)]

$$\iint Rxdxdy = -\frac{M}{\rho}\iint x^2dxdy \qquad (4).$$

715. Hence in a rod of isotropic substance the principal axes of flexure (§ 599) coincide with the principal axes of inertia of the area of the normal section; and the corresponding flexural rigidities [§ 596] are the moments of inertia of this area round these axes multiplied by Young's modulus.

Principal flexural rigidities and axes.

716. The interpretation of the results [§ 712 (2), (3)] to which the analytical investigation has led us is simply that if we imagine the whole rod divided, parallel to its length, into infinitesimal filaments (prisms when the rod is straight), each of these shrinks or swells laterally with sensibly the same freedom as if it were separated from the rest of the substance, and becomes elongated or shortened in a straight line to the same extent as it is really elongated or shortened in the circular arc which it becomes in the bent rod. The distortion of the cross section by which these changes of lateral dimensions are necessarily accompanied is illustrated in the annexed diagram,

Geometrical interpretation of distortion in normal plane.

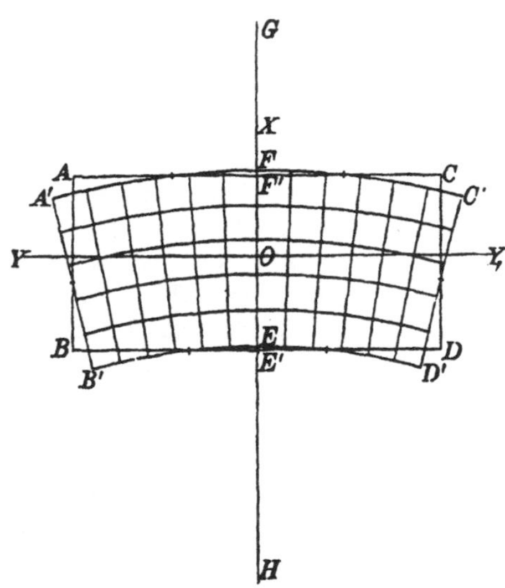

Anticlastic and conical curvatures produced in the four sides of a rectangular prism by flexure in a principal plane.

in which either the whole normal section of a rectangular beam, or a rectangular area in the normal section of a beam of any figure, is represented in its strained and unstrained figures, with the central point O common to the two. The flexure is in planes perpendicular to YOY_1, and concave upwards (or towards X); G the centre of curvature, being in the direction indicated, but too far to be included in the diagram. The straight sides AC, BD, and all straight lines parallel to them, of the unstrained rectangular area become concentric arcs of circles concave in the opposite direction, their centre of curvature, H, being for rods of gelatinous substance, or of glass or metal, from 2 to 4 times as far from O on one side as G is on the other. Thus the originally plane sides AC, BD of a rectangular bar become anticlastic surfaces, of curvatures $1/\rho$ and $-\sigma/\rho$, in the two principal sections. A flat rectangular, or a square, rod of India rubber [for which σ amounts (§ 684) to very nearly $\frac{1}{2}$, and which is susceptible of very great amounts of strain without utter loss of corresponding elastic action], exhibits this phenomenon remarkably well.

Experimental illustration.

717. The conditional limitation (§ 588), that the curvature is to be very small in comparison with that of a circle of radius equal to the greatest diameter of the normal section (not obviously necessary, and indeed not generally known to be necessary, we believe, when the greatest diameter is perpendicular to the plane of curvature), now receives its full explanation. For unless the *breadth*, AC, of the bar (or diameter perpendicular to the plane of flexure) be very small in comparison with the mean proportional between the radius, OH, and the thickness, AB, the distances from OY to the corners A', C' would fall short of the half thickness, OE, and the distances to B', D' would exceed it by differences comparable with its own amount. This would give rise to sensibly less and greater shortenings and stretchings in the filaments towards the corners than those expressed in our formulæ [§ 712 (2)], and so vitiate the solution. Unhappily mathematicians have not hitherto succeeded in solving, possibly not even tried to solve, the beautiful problem thus presented by the flexure of a broad very thin band (such as a watch spring) into a circle of radius

Uncalculated effects of ordinary bendings of a thin flat spring.

comparable with a third proportional to its thickness and its breadth. See § 657.

718. But, provided the radius of curvature of the flexure is not only a large multiple of the *greatest* diameter, but also of a third proportional to the diameters in and perpendicular to the plane of flexure; then however great may be the ratio of the greatest diameter to the least, the preceding solution is applicable; and it is remarkable that the necessary distortion of the normal section (illustrated in the diagram of § 716) does not sensibly impede the free lateral contractions and expansions in the filaments, even in the case of a broad thin lamina (whether of precisely rectangular section, or of unequal thicknesses in different parts).

Hence necessity for stricter limitation, § 628, of curvature than § 588 when a thin flat spring is bent in a plane perpendicular to its breadth.

719. Considering now a uniform thin broad lamina bent in the manner supposed in the preceding solution, we have precisely the case of a plate under the influence of a simple bending stress (§ 638). If the breadth be a, and the thickness b, the moment of inertia of the cross section is $\frac{1}{12}b^2 . ab$, and therefore the flexural rigidity is $\frac{1}{12}Mab^3$, or $\frac{1}{12}Mb^3$ if the breadth be unity. Hence a couple K (§ 637) would bend it to the curvature $12K/Mb^3$ length-wise (or across its length), and (§ 716) would produce the curvature $12\sigma K/Mb^3$ breadth-wise (or across the breadth), but with concavity turned in the contrary direction. Precisely the same solution applies to the effect of a bending stress, consisting of balancing couples applied to the two edges, to bend it across the dimension which hitherto we have been calling its breadth. And by the principle of superposition we may simultaneously apply a pair of balancing couples to each pair of parallel sides of a rectangular plate, without altering by either balancing system the effect of the other; so that the whole effect will be the geometrical resultant of the two effects calculated separately. Thus, a square plate of thickness b, and with each side of length unity, being given, let pairs of balancing couples K on one pair of opposite sides, and Λ on the other pair, be applied, each tending to produce concavity in the same direction when positive. If κ and

Transition to flexure of a plate.

Flexure of a plate: by a single bending stress; by simultaneous bending stresses in two planes at right angles to one another.

λ denote the whole curvatures produced in the planes of these couples, we shall have

$$\kappa = \frac{1}{\frac{1}{12}Mb^3}(\mathrm{K} - \sigma\Lambda) \text{..................} (1),$$

and

$$\lambda = \frac{1}{\frac{1}{12}Mb^3}(\Lambda - \sigma\mathrm{K}) \text{..................} (2).$$

Stress in cylindrical curvature: 720. To find what the couples must be to produce simply cylindrical curvature, κ, let $\lambda = 0$. We have

$$\Lambda = \sigma\mathrm{K}$$

and

$$\mathrm{K} = \frac{1}{12}\frac{Mb^3}{1-\sigma^2}\kappa \text{..................} (3).$$

in spherical curvature: Or to produce spherical curvature, let $\kappa = \lambda$. This gives

$$\mathrm{K} = \Lambda = \frac{1}{12}\frac{Mb^3}{1-\sigma}\kappa \text{..................} (4).$$

in anticlastic curvature. Or lastly, to produce anticlastic curvature, equal in the two directions, let $\kappa = -\lambda$. This gives

$$\mathrm{K} = -\Lambda = \frac{1}{12}\frac{Mb^3}{1+\sigma}\kappa \text{..................} (5).$$

Hence, comparing with § 641 (10) and § 642 (16), we have, for A the cylindrical rigidity, and for $\mathfrak{h}$ and $\mathfrak{k}$ the synclastic and anticlastic rigidities of a uniform plate of isotropic material,

Flexural rigidities of a plate: (A) cylindrical, ($\mathfrak{h}$) synclastic, ($\mathfrak{k}$) anticlastic.

$$\left.\begin{aligned} &A = \frac{1}{12}\frac{Mb^3}{1-\sigma^2}, \\ &\mathfrak{h} = \frac{1}{12}\frac{Mb^3}{1-\sigma}, \quad \mathfrak{k} = \frac{1}{12}\frac{Mb^3}{1+\sigma}, \\ &\text{or [§ 694 (6) and § 698 (5)]} \\ &\mathfrak{h} = \frac{3nkb^3}{2(3k+4n)} = \frac{n(3m-n)b^3}{6(m+n)}, \quad \mathfrak{k} = \tfrac{1}{6}nb^3 \end{aligned}\right\} \text{...} (6).$$

The coefficient A which appears in the equation of equilibrium of a plate urged by any forces [§ 644 (6) and §§ 649...652], and c, which appears in its boundary conditions, are [§ 642 (16)] given in terms of $\mathfrak{h}$ and $\mathfrak{k}$ thus simply :—

$$A = \tfrac{1}{2}(\mathfrak{h} + \mathfrak{k}), \quad c = \tfrac{1}{2}(\mathfrak{h} - \mathfrak{k}) \text{..................} (7).$$

Same result for anticlastic flexure of a plate arrived at also by transition from simple torsion of rectangular prism.

721. It is interesting and instructive to investigate the anticlastic flexure of a plate by viewing it as an extreme case of torsion. Consider first a flat bar of rectangular section uniformly twisted by the proper application of tangential tractions [§ 706 (10)] on its ends. Let now its breadth be comparable with its length; equal, for instance, to its length. We thus have a square plate twisted by opposing couples applied in the planes of two opposite edges, and so distributed over these areas as to cause uniform action in all sections parallel to them when the other two edges are left quite free. If, lastly, we suppose the thickness, b, infinitely small in comparison with the breadth, a, in (46) of § 707, we have

$$N = \tfrac{1}{3} n\tau a b^3 \dots\dots\dots\dots\dots\dots (8).$$

The twist τ per unit of length gives $a\tau$ in the length a, which [§ 640 (4)] is equivalent to an anticlastic curvature ϖ (according to the notation of § 639), equal to τ. And the balancing couple N applied in only one pair of opposite sides of the square is, as we see by § 656, equivalent to an anticlastic stress (according to the notation of § 637) $\Pi = \tfrac{1}{2} N/a$. Hence, for the anticlastic rigidity, according to § 642 (13), we have

$$\mathfrak{k} = \frac{\Pi}{\varpi} = \tfrac{1}{2}\frac{N}{\tau a} = \tfrac{1}{6} n b^3 \dots\dots\dots\dots\dots\dots (9),$$

which agrees with the value (6) otherwise found in § 720, by the composition of flexures.

Analysis of traction in normal section of twisted rectangular prism.

It is most important to remark—(1) That one-half of the part $\tfrac{1}{3} n\tau a b^3$ in the value of N given by the formula (46) of § 707, is derived from α and β as given by (8) of § 706, and the term $-\tau xy$ of γ by (45);—and (2) That if we denote by γ' the transcendental series completing the expression (45) for γ, it is the term $n\iint x\dfrac{d\gamma'}{dy}dxdy$ of § 706 (17), that makes up the other half of the part of N in question, and that it does so as follows, according to the process of integrating by parts, in which it is to be remembered that to change the sign of either x or y, simply changes the sign of γ':—

$$n\int_{-\frac{1}{2}a}^{\frac{1}{2}a}\int_{-\frac{1}{2}b}^{\frac{1}{2}b} x\frac{d\gamma'}{dy}dydx = \int_{-\frac{1}{2}a}^{\frac{1}{2}a} xGdx = a\int_0^{\frac{1}{2}a} Gdx - 2\int_0^{\frac{1}{2}a} dx\int_0^x Gdx \quad (10),$$

Analysis of traction in normal section of twisted rectangular prism.

$$\text{where } G = n\int_{-\frac{1}{2}b}^{\frac{1}{2}b} \frac{d\gamma'}{dy}\,dy = 2n\gamma'_{y=\frac{1}{2}b}$$

$$= 2n\tau\left(\frac{2}{\pi}\right)^3 b^2 \Sigma \frac{1}{(2i+1)^3} \frac{\epsilon^{(2i+1)\pi x/b} - \epsilon^{-(2i+1)\pi x/b}}{\epsilon^{(2i+1)\pi a/2b} + \epsilon^{-(2i+1)\pi a/2b}} \cdots\cdots (11).$$

Thus in N we have a term

$$a\int_0^{\frac{1}{2}a} G dx = \frac{16n\tau a}{\pi^4} b^3 \Sigma \frac{1}{(2i+1)^4}\left\{1 - \frac{2}{\epsilon^{(2i+1)\pi a/2b} + \epsilon^{-(2i+1)\pi a/2b}}\right\},$$

or, because [as we see, by integrating (40) with reference to y, and putting $y = \frac{1}{2}b$],

$$1 + \frac{1}{3^4} + \frac{1}{5^4} + \text{etc.} = \tfrac{1}{6}(\tfrac{1}{2}\pi)^4,$$

$$a\int_0^{\frac{1}{2}a} G dx = \tfrac{1}{6}n\tau ab^3 - \frac{16n\tau}{\pi^4} ab^3 \Sigma \frac{2}{(2i+1)^4\left[\epsilon^{(2i+1)\pi a/2b} + \epsilon^{-(2i+1)\pi a/2b}\right]} \quad (12).$$

The transcendental series constituting the second term of this, together with

$$-2\int_0^{\frac{1}{2}a} dx \int_0^x G dx - n\iint y \frac{d\gamma'}{dx}\,dx dy$$

makes up the transcendental series which appears in the expression (46) for N. This, when a/b is infinite, vanishes in comparison with the first term of (46), as we have seen above § 721 (8). But in examining, as now, the composition of the expression, it is to be remarked that, when a/b is infinite, γ' vanishes except for values of x differing infinitely little from $\pm\frac{1}{2}a$, and therefore we see at once that in this case,

$$n\int_{-\frac{1}{2}a}^{\frac{1}{2}a} dx \int_{-\frac{1}{2}b}^{\frac{1}{2}b} dy \left(x\frac{d\gamma'}{dy} - y\frac{d\gamma'}{dx}\right) = na\int_0^{\frac{1}{2}a} dx \int_{-\frac{1}{2}b}^{\frac{1}{2}b} \frac{d\gamma'}{dy}\,dy = a\int_0^{\frac{1}{2}a} G dx,$$

by which, in connexion with what precedes, we see that

Composition of action in normal section of a long rectangular lamina under torsion.

722. One half of the couple on each of the edges, by which these conditions are fulfilled, consists of two tangential tractions distributed over areas of the edge infinitely near its ends acting perpendicularly to the plate towards opposite parts. The other half consists of forces parallel to the length of the edges, uniformly distributed through the length, and varying across it in simple proportion to the distance, positive or negative, from its middle line.

723. If now we remove the former half, and apply instead, over the edges (BB', AA') hitherto free, a uniform distribution of couple equal and similar to the latter half, and in the proper directions to keep up the same twist through the plate, we have the proper edge tractions to fulfil Poisson's three boundary conditions (§ 645) for the case in question; that is to say, we have such a distribution of tractions on the four edges of a square plate as produces anticlastic stress (§ 638) uniform not only through all of the plate at distances from the edges great in comparison with the thickness, but throughout the plate up to the very edges. The state of strain and stress through the plate is represented by the following formulæ [as we may gather from §§ 706 and 707 (8), (45), (9), (10), (17), and § 722, or, as we see directly, by the verification which the operations now indicated present]:—

Uniform distribution of couple applied to its edges to render the stress uniform from the edges inwards.

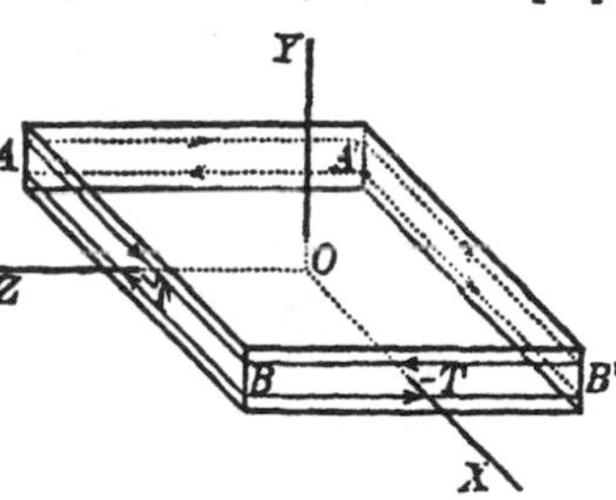

Algebraic solution expressing displacement, strain, and stress, through a plate bent to uniform anticlastic curvature.

$$\left.\begin{aligned} &\alpha=-\tau yz,\ \ \beta=\tau xz,\ \ \gamma=-\tau xy \\ &\mathfrak{e}=\mathfrak{f}=\mathfrak{g}=0,\ \ \mathfrak{a}=0,\ \ \mathfrak{b}=-2\tau y,\ \ \mathfrak{c}=0 \\ &P=Q=R=0,\ \ S=0,\ \ T=-2n\tau y,\ \ U=0 \\ &-L=N=-\int_{-\frac{1}{2}a}^{\frac{1}{2}a}\int_{-\frac{1}{2}b}^{\frac{1}{2}b} Tydydx=\tfrac{1}{6}n\tau ab^3 \end{aligned}\right\}\ldots(13),$$

where L and N denote the moments (with signs reckoned as in § 551) of the whole amounts of couple, applied to the two edges perpendicular to OX and OZ respectively, in the planes of these edges.

By turning the axes OX, OZ through 45° in their own plane, we fall back on the formulæ of flexure as in § 719, for the particular case of equal flexures in the two opposite directions.

724. If, on the other hand, we superimpose on the state of strain investigated in § 721, another produced by applying on

Thin rectangular plate subjected to the edge-traction of § 647.

the pair of edges which it leaves free, precisely the same entire distribution of couple as that described in § 722, but in the direction opposite to the twist which the former gave to the plate (so that now it is not $-L$, but L that is equal to N), we have the square plate precisely in the condition described in § 647, except infinitely near its corners. To find the expressions for the components of displacement, strain, and stress, in this case, we must add to the

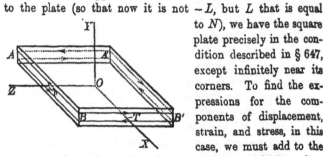

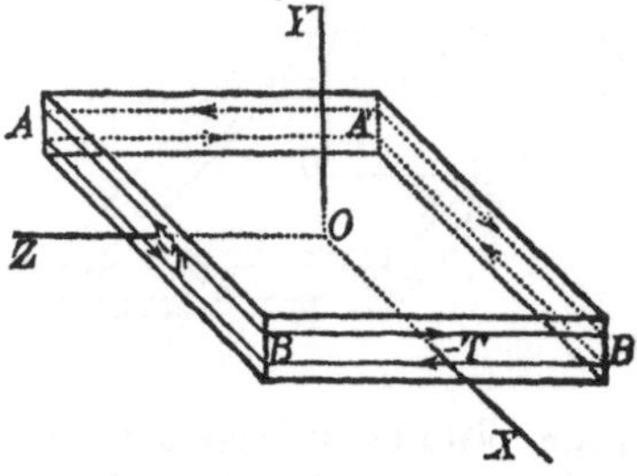

expressions for α, β, γ in (8) of § 706, and (45) of § 707, values obtained by changing the sign of each of these expressions, and interchanging x for z, and α for γ. The consequent values of $\mathfrak{e}$, $\mathfrak{f}$, $\mathfrak{g}$, $\mathfrak{a}$, $\mathfrak{b}$, $\mathfrak{c}$, P, Q, R, S, T, U, are of course obtained in the same way, but need not be written down, as they can be seen in a moment from α, β, γ. Lastly, the strain thus superimposed would, if existing alone, leave the edges parallel to x free from traction, just as the first supposed strain [§ 706 (8)] leaves the edges parallel to z free; and thus, without fresh integration, we see that N has still the value (46), and is the result of the distribution of tractions described in § 722. The parts of the component displacements represented by products of coordinates disappear, and only transcendental series, as follows remain:—

$$\left.\begin{aligned}\alpha &= -\frac{8\tau}{\pi^3}b^2\Sigma\frac{(-1)^i}{(2i+1)^3}\frac{\epsilon^{+(2i+1)\pi z/b}-\epsilon^{-(2i+1)\pi z/b}}{\epsilon^{+(2i+1)\pi a/2b}+\epsilon^{-(2i+1)\pi a/2b}}\sin(2i+1)\frac{\pi y}{b}\\ \gamma &= +\frac{8\tau}{\pi^3}b^2\Sigma\frac{(-1)^i}{(2i+1)^3}\frac{\epsilon^{+(2i+1)\pi x/b}-\epsilon^{-(2i+1)\pi x/b}}{\epsilon^{+(2i+1)\pi a/2b}+\epsilon^{-(2i+1)\pi a/2b}}\sin(2i+1)\frac{\pi y}{b}\end{aligned}\right\}\ (14).$$

725. When a/b is infinite, $\epsilon^{+(2i+1)\pi a/2b}$ becomes infinitely great, and $\epsilon^{-(2i+1)\pi a/2b}$ infinitely small. If then we put

$$\tfrac{1}{2}a - z = z', \text{ and } \tfrac{1}{2}a - x = x',$$

the preceding expressions become

Thin rectangular plate subjected to the edge-traction of § 647.

$$\left.\begin{array}{l}
\alpha = -\dfrac{8\tau}{\pi^3} b^2 \Sigma \dfrac{(-1)^i}{(2i+1)^3} \epsilon^{-(2i+1)\pi x'/b} \sin (2i+1) \dfrac{\pi y}{b} \\
\text{for points not infinitely near the edge } A'B'; \\
\gamma = +\dfrac{8\tau}{\pi^3} b^2 \Sigma \dfrac{(-1)^i}{(2i+1)^3} \epsilon^{-(2i+1)\pi x'/b} \sin (2i+1) \dfrac{\pi y}{b} \\
\text{for points not infinitely near the edge } AA'; \\
\alpha = 0,\ \gamma = 0, \text{ for all points not infinitely near an edge;} \\
\text{and} \qquad \beta = 0 \text{ throughout.} \\
\text{Lastly,} \qquad L = N = \tfrac{1}{3} n\tau a b^3, \\
\text{of each of which one-half is constituted by tractions} \\
\text{uniformly distributed along the corresponding edge,} \\
\text{and proportional to distances from the middle line;} \\
\text{and the other by tractions infinitely near the corners} \\
\text{and perpendicular to the plate.}
\end{array}\right\} \ldots(15).$$

726. It is clear that if the corners were rounded off, or the plate were of any shape without corners, that is to say, with no part of its edge where the radius of curvature is not very great in comparison with the thickness, the effect of applying a distribution of couple all round its edge in the manner defined in § 647 would be expressed by either of these last formulæ for α and γ. Thus the whole displacement of the substance will be parallel to the edge for all points infinitely near it; will vanish for all other points of the plate; and will be equal to the preceding expression (15) for γ if x' denote simply distance from the nearest point of the edge of the plate, and y, as in all these formulæ, distance from the middle surface

Transition to plate without corners subjected to edge-traction of § 647.

727. We may conclude that if a uniform plate, bounded by an edge everywhere perpendicular to its sides, and of thickness a small fraction of the smallest radius of curvature of the edge at any point, be subjected to the action described in § 647, with the more particular condition that the distribution of tangential traction is [as asserted in § 634 (3) for any normal section remote from the boundary of a bent plate] in simple proportion to the distance, positive or negative, from the middle line of the edge; the interior strain and stress will be as specified by the following statement and formulæ :—

Origin shifted from middle plane to one side of plate.

Let O be any point in one corner of the edge: and let OX be perpendicular to the edge inwards, and OY perpendicular to the plane of the plate. The displacement of any particle P, (x, y), at any distance from O not a considerable multiple of the thickness, b, will be perpendicular to the plane YOX, and (denoted by γ) will be given by the formula—

Displacement of substance produced by edge-traction of § 647.

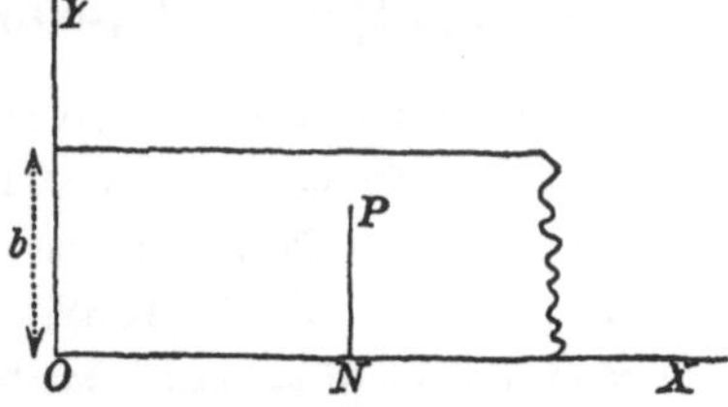

$$\gamma = 6\frac{\Omega}{nb}\left(\frac{2}{\pi}\right)^3\left(\epsilon^{-\frac{\pi x}{b}}\cos\frac{\pi y}{b} + \frac{1}{3^3}\epsilon^{-\frac{3\pi x}{b}}\cos\frac{3\pi y}{b} + \frac{1}{5^3}\epsilon^{-\frac{5\pi x}{b}}\cos\frac{5\pi y}{b} + \text{etc.}\right) \quad (16),$$

where Ω denotes the amount of the couple per unit length of the edge, and n the rigidity (§ 680) of the substance. But the simplest and easiest way of arriving at this result is to solve directly by Fourier's analytical method the following problem, a case of one of the general problems of § 696:—

Case of § 647 independently investigated.

728. A uniform plane plate of thickness b, extending to infinity on one side of a straight edge (or plane perpendicular to its sides) being given,—

It is required to find the displacement, strain, and stress, produced by tangential traction parallel to the edge applied uniformly along the edge, according to a given arbitrary function, $\phi(y)$, of position on its breadth.

Taking co-ordinates as in § 727, we have to solve equations (2) of § 697, with $X = 0$, $Y = 0$, $Z = 0$, for all points of space for which x is positive, and y between 0 and b, subject to the boundary conditions,

See § 661, or § 662 (1); also § 693 (5), and § 670 (6).

$$\left.\begin{array}{l} P=0,\ Q=0,\ R=0,\ S=0,\ T=0,\ U=0, \text{ when } y=0 \text{ or } b: \\ P=0,\ Q=0,\ R=0,\ S=0,\ U=0,\ T=\phi(y), \text{ when } x=0: \\ \text{and } \alpha=0,\ \beta=0,\ \gamma=0, \text{ when } x=\infty. \end{array}\right\} \quad (17).$$

From these, inasmuch as α, β, γ must each be independent of z, we find

Case of § 647 independently investigated.

$$\left.\begin{array}{ll}(a) & \dfrac{d^2\gamma}{dx^2}+\dfrac{d^2\gamma}{dy^2}=0, \text{ throughout the solid;} \\ (b) & \gamma=0 \text{ when } x=\infty\,; \\ (c) & n\dfrac{d\gamma}{dy}=0 \text{ when } y=0 \text{ or } b\,; \\ \text{and } (d) & n\dfrac{d\gamma}{dx}=\phi(y) \text{ when } x=0\,;\end{array}\right\}\text{......(18);}$$

and all the equations, both internal and superficial, involving α and β are satisfied by $\alpha=0$, $\beta=0$, and therefore (App. C.) require $\alpha=0$, $\beta=0$. By means of (a), (b), and (c) the Fourier solution is seen to be of the form

$$\gamma=\Sigma A_i\epsilon^{-\frac{i\pi x}{b}}\cos\frac{i\pi y}{b}\text{ ....(19);}$$

and, because of (d), the coefficients A_i are to be found so as to make

$$-\frac{n\pi}{b}\Sigma iA_i\cos\frac{i\pi y}{b}=\phi(y)\text{ ....(20).}$$

They are therefore [as we see by taking in § 77, (13) and (14), ϕ such that $\phi(p-\xi)=\phi(\xi)$, and putting $p=2b$] as follows:—

$$A_i=-\frac{b}{n\pi}\cdot\frac{1}{i}\cdot\frac{2}{b}\int_0^b\phi(y)\cos\frac{i\pi y}{b}dy\text{ ....(21).}$$

If (for the particular case of § 727) we take

$$\phi(y)=12\frac{\Omega}{b^3}(y-\tfrac{1}{2}b)\text{ ....(22),}$$

we find $\quad A_{2i}=0$, and $A_{2i+1}=6\dfrac{\Omega}{nb}\left(\dfrac{2}{\pi}\right)^3\dfrac{1}{(2i+1)^3}$ ....(23),

and so arrive at the result (16).

729. It is remarkable how very rapidly the whole disturbance represented by this result diminishes inwards from the edge where the disturbing traction is applied (compare § 586): also how very much more rapidly the second term diminishes than the first; and so on.

Rapid decrease of disturbance from edge inwards.

Thus as

$$\epsilon=2{\cdot}71828,\ \epsilon^{\frac{1}{2}\pi}=4{\cdot}801,\ \epsilon^{2{\cdot}303}=10,\ \epsilon^{\pi}=23{\cdot}141,\ \epsilon^{2\pi}=535{\cdot}5,$$

Rapid decrease of disturbance from edge inwards.

we have for

$$x=\frac{1}{3{\cdot}1416}b,\ \gamma=6\frac{\Omega}{nb}\left(\frac{2}{\pi}\right)^3\left(\frac{\cos\pi y/b}{2{\cdot}718}-\frac{\cos 3\pi y/b}{3^3.\,2{\cdot}718^3}+\frac{\cos 5\pi y/b}{5^3.\,2{\cdot}718^5}-\text{etc.}\right)$$

$$x=\tfrac{1}{2}b,\qquad \gamma=6\frac{\Omega}{nb}\left(\frac{2}{\pi}\right)^3\left(\frac{\cos\pi y/b}{4{\cdot}801}-\frac{\cos 3\pi y/b}{3^3.\,4{\cdot}801^3}+\frac{\cos 5\pi y/b}{5^3.\,4{\cdot}801^5}-\text{etc.}\right)$$

$$x=\frac{2{\cdot}303}{\pi}b,\ \gamma=6\frac{\Omega}{nb}\left(\frac{2}{\pi}\right)^3\left(\frac{\cos\pi y/b}{10}-\frac{\cos 3\pi y/b}{3^3.\,10^3}+\frac{\cos 5\pi y/b}{5^3.\,10^5}-\text{etc.}\right)$$

$$x=b,\qquad \gamma=6\frac{\Omega}{nb}\left(\frac{2}{\pi}\right)^3\left(\frac{\cos\pi y/b}{23{\cdot}14}-\frac{\cos 3\pi y/b}{3^3.\,23{\cdot}14^3}+\frac{\cos 5\pi y/b}{5^3.\,23{\cdot}14^5}-\text{etc.}\right)$$

$$x=2b,\qquad \gamma=6\frac{\Omega}{nb}\left(\frac{2}{\pi}\right)^3\left(\frac{\cos\pi y/b}{535{\cdot}5}-\frac{\cos 3\pi y/b}{3^3.\,535{\cdot}5^3}+\frac{\cos 5\pi y/b}{5^3.\,535{\cdot}5^5}-\text{etc.}\right)$$

which proves most strikingly the concluding statement of § 647.

Problems to be solved.

730. We regret that limits of space compel us to leave uninvestigated the torsion-flexure rigidities of a prism and the flexural rigidities of a plate of aeolotropic substance: and to still confine ourselves to isotropic substance when, in conclusion, we proceed to find the complete integrals of the equations [§ 697 (2)] of internal equilibrium for an infinite solid under the influence of any given forces, and the harmonic solutions suitable for problems regarding spheres and spherical shells, and solid and hollow circular cylinders (§ 738) under plane strain. The problem to be solved for the infinite solid is this:

General problem of infinite solid:

Let in (6) *of* § 698, *X, Y, Z be any arbitrary functions whatever of* (x, y, z), *either discontinuous and vanishing in all points outside some finite closed surface, or continuous and vanishing at all infinitely distant points with sufficient convergency to make* RD *converge to* 0 *as* D *increases to* ∞, *if* R *be the resultant of* X, Y, Z *for any point at distance* D *from origin. It is required to find* α, β, γ *satisfying those equations* [(6) *of* § 698], *subject to the condition of each vanishing for infinitely distant points (that is, for infinite values of* x, y, *or* z).

solved for isotropic substance.

(*a*) Taking $\frac{d}{dx}$ of the first of these equations, $\frac{d}{dy}$ of the second, and $\frac{d}{dz}$ of the third, and adding, we have

$$(m+n)\nabla^2\delta+\frac{dX}{dx}+\frac{dY}{dy}+\frac{dZ}{dz}=0\ \ldots\ldots\ldots\ldots\ldots(1).$$

(*b*) This shows that if we imagine a mass distributed through space, with density ρ given by

$$\rho = \frac{1}{4\pi(m+n)}\left(\frac{dX}{dx} + \frac{dY}{dy} + \frac{dZ}{dz}\right) \ldots\ldots\ldots\ldots\ldots(2),$$

δ must be equal to its potential at (x, y, z). For [§ 491 (*c*)] if V be this potential we have

$$\nabla^2 V + 4\pi\rho = 0.$$

Subtracting this from (1) divided by $(m+n)$, we have

$$\nabla^2(\delta - V) = 0 \ldots\ldots\ldots\ldots\ldots\ldots\ldots\ldots(3),$$

for all values of (x, y, z). Now the convergency of XD, YD, ZD to zero when D is infinite, clearly makes $V=0$ for all infinitely distant points. Hence if S be any closed surface round the origin of co-ordinates, everywhere infinitely distant from it, the function $(\delta - V)$ is zero for all points of it, and satisfies (3) for all points within it. Hence [App. A. (*e*)] we must have $\delta = V$. In other words, the fact that (1) holds for all points of space gives determinately

General equations for infinite isotropic solid integrated.

$$\delta = \frac{1}{4\pi(m+n)}\int_{-\infty}^{\infty}\int_{-\infty}^{\infty}\int_{-\infty}^{\infty}\frac{\left(\frac{dX'}{dx'} + \frac{dY'}{dy'} + \frac{dZ'}{dz'}\right)dx'dy'dz'}{\sqrt{[(x-x')^2 + (y-y')^2 + (z-z')^2]}} \ldots(4),$$

where X', Y', Z' denote the values of X, Y, Z for any point (x', y', z').

(*c*) Modifying by integration by parts, and attending to the prescribed condition of convergences, according to which, when x' is infinite,

$$\int_{-\infty}^{\infty}\int_{-\infty}^{\infty}\frac{X'dy'dz'}{\sqrt{[(x-x')^2 + (y-y')^2 + (z-z')^2]}} = 0 \ldots\ldots\ldots(5),$$

we have

$$\delta = \frac{-1}{4\pi(m+n)}\int_{-\infty}^{\infty}\int_{-\infty}^{\infty}\int_{-\infty}^{\infty}\frac{X'(x-x') + Y'(y-y') + Z'(z-z')}{[(x-x')^2 + (y-y')^2 + (z-z')^2]^{\frac{3}{2}}}dx'dy'dz' \;(6),$$

which for most purposes is more convenient than (4).

(*d*) On precisely the same plan as (*b*) we now integrate each of the three equations (6) of § 698 separately for α, β, γ respectively, and find

$$\alpha = u + U, \quad \beta = v + V, \quad \gamma = w + W \ldots\ldots\ldots\ldots\ldots(7)$$

where u, v, w, U, V, W denote the potentials at (x, y, z) of

General equations for infinite isotropic solid integrated.

distributions of matter through all space of densities respectively

$$\frac{m}{4\pi n}\frac{d\delta}{dx},\ \frac{m}{4\pi n}\frac{d\delta}{dy},\ \frac{m}{4\pi n}\frac{d\delta}{dz},\ \frac{X}{4\pi n},\ \frac{Y}{4\pi n},\ \frac{Z}{4\pi n};$$

in other words, such functions that

$$\nabla^2 u + \frac{m}{n}\frac{d\delta}{dx} = 0, \text{ etc., and } \nabla^2 U + \frac{X}{n} = 0, \text{ etc., } \ldots\ldots\ldots(8),$$

each through all space. Thus if δ'', X'', Y'', Z'' denote the values of δ, X, Y, Z for a point (x'', y'', z''), we find, for α,

$$\alpha = \frac{1}{4\pi n}\int_{-\infty}^{\infty}\int_{-\infty}^{\infty}\int_{-\infty}^{\infty}\frac{\left(m\frac{d\delta''}{dx''} + X''\right)dx''dy''dz''}{[(x-x'')^2+(y-y'')^2+(z-z'')^2]^{\frac{1}{2}}}\ldots\ldots(9),$$

if in this we substitute for δ'' its value by (6) we have α expressed by the sum of a sextuple integral and a triple integral, the latter being the U of (7); and similarly for β and γ. These expressions may, however, be greatly simplified, since we shall see presently that each of the sextuple integrals may be reduced to a triple integral.

Force applied uniformly to spherical portion of infinite homogeneous solid.

(e) As a particular case, let X, Y, Z be each constant throughout a spherical space having its centre at the origin and radius a, and zero everywhere else. This by (6) will make $-\delta$ the sum of the products of X, Y, Z respectively into the corresponding component attractions of a uniform distribution of matter of density $1/4\pi(m+n)$ through this space. Hence [§ 491 (b)]

$$\left.\begin{aligned}&\delta = \frac{-1}{3(m+n)}\frac{a^3}{r^3}(Xx+Yy+Zz) \text{ for points outside the spherical space,}\\ &\text{and}\\ &\delta = \frac{-1}{3(m+n)}(Xx+Yy+Zz) \text{ for points within the spherical space.}\end{aligned}\right\}(10).$$

Dilatation produced by it.

Now we may divide u of (8) into two parts, u' and u'', depending on the values of $d\delta/dx$ within and without the spherical space respectively; so that we have,

$$\left.\begin{aligned}&\text{for } r < a, \quad \nabla^2 u' = \frac{mX}{3n(m+n)}, \text{ a constant,}\\ &\text{for } r > a, \quad \nabla^2 u' = 0;\end{aligned}\right\}\ldots\ldots(11);$$

$$\left.\begin{array}{ll}\text{for } r < a, & \nabla^2 u'' = 0, \\ \text{for } r > a, & \nabla^2 u'' = -\frac{m}{n}\frac{d\delta}{dx}, \text{ which is a} \\ \multicolumn{2}{l}{\text{solid spherical harmonic of degree } -3, \text{ because } \delta} \\ \multicolumn{2}{l}{\text{is given by the first of equations (10).}}\end{array}\right\} \quad \ldots\ldots(12).$$

Investigation of displacement.

The solution of (11), being simply the potential due to a uniform sphere of density $-\frac{1}{4\pi}\frac{mX}{3n(m+n)}$, is of course

$$\left.\begin{array}{l} u' = \frac{-mX}{18n(m+n)}(3a^2 - r^2) \text{ for } r < a, \\ u' = \frac{-mX}{9n(m+n)}\frac{a^3}{r} \text{ for } r > a. \end{array}\right\} \quad \ldots\ldots\ldots\ldots(13).$$

Again, if in (12) of App. B. we put $m=2$, $n=-3$, and $V_{-3} = d\delta/dx$, we have

$$\nabla^2\left(r^2\frac{d\delta}{dx}\right) = -6\frac{d\delta}{dx} \text{ for } r > a \quad \ldots\ldots\ldots\ldots(14),$$

since, for $r > a$, $d\delta/dx$ is a spherical harmonic of order -3. And $r^5 d\delta/dx$ is [App. B. (13)] a solid harmonic of degree 2: hence if $[d\delta/dx]$ denote, for any point within the spherical space, the same algebraic expression as $d\delta/dx$ by (10) for the external space, $\frac{r^5}{a^3}\left[\frac{d\delta}{dx}\right]$ is a function which, for all the interior space, satisfies the equation $\nabla^2 u = 0$, and is equal to $r^2 d\delta/dx$ for points infinitely near the surface, outside and inside respectively. Hence $\frac{r^5}{a^3}\left[\frac{d\delta}{dx}\right]$ for interior space, and $r^2 d\delta/dx$ for exterior space, constitute the potential of a distribution of matter of density $\frac{3}{2}d\delta/\pi dx$ outside the spherical space and zero within, and, so far as yet tested, any layer of matter whatever distributed over the separating spherical surface. To find the surface density of this layer we first, for an exterior point infinitely near the surface, take

Force applied uniformly to spherical portion of infinite homogeneous solid.

$$\left(x\frac{d}{dx} + y\frac{d}{dy} + z\frac{d}{dz}\right)\left(r^2\frac{d\delta}{dx}\right), \text{ which may be denoted by } -\{rR\},$$

and, for an interior point infinitely near the surface,

$$\left(x\frac{d}{dx} + y\frac{d}{dy} + z\frac{d}{dz}\right)\left(\frac{r^5}{a^3}\left[\frac{d\delta}{dx}\right]\right), \text{ which may be denoted by } -[rR]$$

Force applied uniformly to spherical portion of infinite homogeneous solid.

Then, remembering that $x\frac{d}{dx}+y\frac{d}{dy}+z\frac{d}{dz}$ is the same as $r\frac{d}{dr}$, according to the notation of App. A. (a); we find [by App. B. (5)]

$$\{R\}=r\frac{d\delta}{dx},\ \text{and}\ [R]=-2\frac{r^4}{a^3}\left[\frac{d\delta}{dx}\right].$$

Therefore, as $r^3\, d\delta/dx$ for external space is independent of r, and as r differs infinitely little from a for each of the two points,

$$\{R\}-[R]=\cdot 3\frac{r^3}{a^3}\frac{d\delta}{dx}.$$

But $\{R\}$ and $[R]$ being the radial components of the force at points infinitely near one another outside and inside, corresponding to the supposed distribution of potential, it follows from § 478 that to produce this distribution there must be a layer of matter on the separating surface, having $\frac{1}{4\pi}(\{R\}-[R])$ for surface density. But, inasmuch as $\{R\}-[R]$ is a surface harmonic of the second order, the potential due to that surface distribution alone is [§ 536 (4)]

$$\tfrac{1}{5}(\{R\}-[R])\frac{r^2}{a}\ \text{through the inner space,}$$

and $$\tfrac{1}{5}(\{R\}-[R])\frac{a^4}{r^3}\ \text{through the outer space;}$$

or, according to the value found above for $\{R\}-[R]$,

$$\tfrac{3}{5}\frac{r^5}{a^3}\left[\frac{d\delta}{dx}\right]\ \text{through the inner space,}$$

and $$\tfrac{3}{5}a^2\frac{d\delta}{dx}\ \text{through the outer space.}$$

Subtracting now this distribution of potential from the whole distribution formerly supposed, we find

$\tfrac{2}{5}\frac{r^5}{a^3}\left[\frac{d\delta}{dx}\right]$ for the inner space, and $(r^2-\tfrac{3}{5}a^2)\frac{d\delta}{dx}$ for the outer, as the distribution of potential due simply to an external distribution of matter, of density $\tfrac{3}{2}d\delta/\pi dx$, with no surface layer. Hence, and by (14), we see that the solution of (12) is

$$\left.\begin{aligned} u''&=\tfrac{1}{6}\cdot\tfrac{2}{5}\frac{m}{n}\frac{r^5}{a^3}\left[\frac{d\delta}{dx}\right]\ \text{for}\ r<a,\\ u''&=\tfrac{1}{6}\frac{m}{n}(r^2-\tfrac{3}{5}a^2)\frac{d\delta}{dx}\ \text{for}\ r>a.\end{aligned}\right\}\dots\dots\dots\dots(15).$$

And [(8) showing that U is the potential of a distribution of matter of density equal to $X/4\pi n$] as X is constant through the spherical space and zero everywhere outside it, we have

$$\left.\begin{aligned} U &= \frac{X}{6n}(3a^2 - r^2) \text{ for } r < a, \\ U &= \frac{X}{3n}\frac{a^3}{r} \text{ for } r > a. \end{aligned}\right\} \quad \text{..............(16).}$$

This, with (13), (15), and (10), gives by (7)

Displacement produced by it.

$$\left.\begin{aligned} &\text{for } r < a, \\ &\alpha = \frac{1}{18n(m+n)}\left\{(2m+3n)X(3a^2-r^2) - \tfrac{2}{5}mr^5\frac{d}{dx}\frac{Xx+Yy+Zz}{r^3}\right\} \\ &\text{and for } r > a, \\ &\alpha = \frac{a^3}{18n(m+n)}\left\{2(2m+3n)\frac{X}{r} - m(r^2 - \tfrac{3}{5}a^2)\frac{d}{dx}\frac{Xx+Yy+Zz}{r^3}\right\} \end{aligned}\right\} \text{(17)},$$

with symmetrical expressions for β and γ.

731. A detailed examination of this result, with graphic illustrations of the displacements, strains, and stresses concerned, is of extreme interest in the theory of the transmission of force through solids; but we reluctantly confine ourselves to the solution of the general problem of § 730.

Displacement produced by a force applied to an infinitely small part of an infinite elastic solid.

To deduce which, we have now only to remark that if a becomes infinitely small, X, Y, Z remaining finite, the expressions for α, β, γ become infinitely small, even within the space of application of the force, and at distances outside it great in comparison with a, they become

$$\left.\begin{aligned} \alpha &= \frac{V}{24\pi n(m+n)}\left\{2(2m+3n)\frac{X}{r} - mr^3\frac{d}{dx}\frac{Xx+Yy+Zz}{r^3}\right\} \\ &\qquad \beta = \text{etc.}, \quad \gamma = \text{etc.} \end{aligned}\right\} \text{..(18)},$$

where V denotes the volume of the sphere. As these depend simply on the whole amount of the force (its components being XV, YV, ZV), and when it is given are independent of the radius of the sphere, the same formulæ express the effect of the same whole amount of force distributed through an infinitely small space of any form not extending in any direction to more than an infinitely small distance from the origin of co-ordinates.

Hence, recurring to the notation of § 730 (*b*), we have for the required general solution

$$\left.\begin{aligned}
\alpha &= \frac{1}{24\pi n(m+n)}\iiint dx'dy'dz'\left\{2(2m+3n)\frac{X'}{D} - mD^2\frac{d}{dx}\frac{X'(x-x')+Y'(y-y')+Z'(z-z')}{D^3}\right\},\\
\beta &= \frac{1}{24\pi n(m+n)}\iiint dx'dy'dz'\left\{2(2m+3n)\frac{Y'}{D} - mD^2\frac{d}{dy}\frac{X'(x-x')+Y'(y-y')+Z'(z-z')}{D^3}\right\},\\
\gamma &= \frac{1}{24\pi n(m+n)}\iiint dx'dy'dz'\left\{2(2m+3n)\frac{Z'}{D} - mD^2\frac{d}{dz}\frac{X'(x-x')+Y'(y-y')+Z'(z-z')}{D^3}\right\},
\end{aligned}\right\} \quad (19).$$

where $D = \sqrt{\{(x-x')^2+(y-y')^2+(z-z')^2\}}$,

$\iiint$ denotes integration through all space, and X', Y', Z' are three arbitrary functions of x', y', z' restricted only by the convergency condition of § 730.

Displacement produced by any distribution of force through an infinite elastic solid.

This solution was first given by Sir William Thomson, though in a somewhat different form, in the *Cambridge and Dublin Mathematical Journal*, 1848, *On the Equations of Equilibrium of an Elastic Solid.* [See *Mathematical and Physical Papers*, Thomson, Vol. I.]

Comparing it with (9), we now see the promised reduction of the sextuple integral involved in that expression to a triple integral.

The process (*e*) by which it is effected consists virtually of the evaluation of a certain triple integral by the proper solution of the partial differential equation $\nabla^2 V + 4\pi\rho = 0$ [like that formerly worked out (§ 649) for the much simpler case of ρ merely a function of r]. Proof of the result by direct integration is a good exercise in the integral calculus.

Application to problem of § 696.

732. In §§ 730, 731 the imagined subject has been a homogeneous elastic solid filling all space, and experiencing the effect of a given distribution of force acting *bodily* on its substance. The solution, besides the interesting application indicated in § 731, is useful for simplifying the practical problem of § 696, by reducing it immediately to the case in which no force acts on the interior substance of the body, thus:—

General problem of § 696 reduced to case of no bodily force.

The equations to be satisfied being (6) of § 698, throughout the portion of space occupied by the body, and certain equations for all points of its boundary expressing that the surface displacements or tractions fulfil the prescribed conditions; let $'\alpha$, $'\beta$, $'\gamma$ be functions of (x, y, z), which satisfy the equations

$$n\nabla^2 {}^{\backprime}\alpha + m\frac{d^{\backprime}\delta}{dx} + X = 0,\ n\nabla^2 {}^{\backprime}\beta + m\frac{d^{\backprime}\delta}{dy} + Y = 0,\ n\nabla^2 {}^{\backprime}\gamma + m\frac{d^{\backprime}\delta}{dz} + Z = 0,$$

where, for brevity,

$${}^{\backprime}\delta = \frac{d^{\backprime}\alpha}{dx} + \frac{d^{\backprime}\beta}{dy} + \frac{d^{\backprime}\gamma}{dz}, \quad (1),$$

through the space occupied by the body. Then, if we put

$$\alpha = {}^{\backprime}\alpha + \alpha_{,},\ \beta = {}^{\backprime}\beta + \beta_{,},\ \gamma = {}^{\backprime}\gamma + \gamma_{,}, \ldots\ldots\ldots\ldots\ldots (2),$$

we see that to complete the solution we have only to find $\alpha_{,}$, $\beta_{,}$, $\gamma_{,}$, as determined by the equations

$$n\nabla^2\alpha_{,} + m\frac{d\delta_{,}}{dx} = 0,\ n\nabla^2\beta_{,} + m\frac{d\delta_{,}}{dy} = 0,\ n\nabla^2\gamma_{,} + m\frac{d\delta_{,}}{dz} = 0,$$

$$\delta_{,} = \frac{d\alpha_{,}}{dx} + \frac{d\beta_{,}}{dy} + \frac{d\gamma_{,}}{dz}, \quad (3)$$

to be fulfilled throughout the space occupied by the body, and certain equations for all points of its boundary, found by subtracting from the prescribed values of the surface displacement or traction, as the case may be, components of displacement or traction calculated from ${}^{\backprime}\alpha$, ${}^{\backprime}\beta$, ${}^{\backprime}\gamma$.

Values for ${}^{\backprime}\alpha$, ${}^{\backprime}\beta$, ${}^{\backprime}\gamma$ may always be found according to §§ 730, 731, by supposing equations (1) § 732 to hold through all space, and X, Y, Z to be discontinuous functions, having the given values for all points of the body, and being each zero for all points of space not belonging to it. But all that is necessary is that (1) be satisfied through the space actually occupied by the body; and in some of the most important practical cases this condition may be more easily fulfilled otherwise than by determining ${}^{\backprime}\alpha$, ${}^{\backprime}\beta$, ${}^{\backprime}\gamma$ in that way with its superadded condition for the rest of space.

733. Thus, for example, let us suppose the forces to be such that $Xdx + Ydy + Zdz$[1] is the differential of a function, W, Important class of cases.

[1] Let m be the mass of any small part of the body, x, y, z its co-ordinates at any time, and Pm, Qm, Rm the components of the force acting on it. If the system be conservative, $Pdx + Qdy + Rdz$ must be the differential of a function of x, y, z. Let, for instance, the forces on all parts of the body be due to attractions or repulsions from fixed matter; and let the particle considered be the matter of the body within an infinitely small volume $\delta x\delta y\delta z$. Then we have $Pm = X\delta x\delta y\delta z$, etc.; and therefore, if ρ be the density of the matter of m, so that $\rho\delta x\delta y\delta z = m$, we have, in the notation of the text, $P\rho = X$, $Q\rho = Y$, $R\rho = Z$; and therefore

Important class of cases.

of x, y, z considered as independent variables. This assumption includes some of the most important and interesting practical applications, among which are—

(1) A homogeneous isotropic body acted on by gravitation sensibly uniform and in parallel lines, as in the case of a body of moderate dimensions under the influence of terrestrial gravity.

(2) A homogeneous isotropic body acted on by any distribution of gravitating matter, and either equilibrated at rest by the aid of surface-tractions if the attracting forces do not of themselves balance on it; or fulfilling the conditions of internal equilibrium by the balancing, according to D'Alembert's principle (§ 264) of the reactions against acceleration of all parts of its mass and the forces of attraction to which it is subjected, when the circumstances are such that no acceleration of rotation has to be taken into account. To this case belongs the problem, solved below, of finding the tidal deformation of the solid Earth, supposed of uniform specific gravity and rigidity throughout, produced by the tide-generating influence of the Moon and Sun.

(3) A uniform body strained by centrifugal force due to uniform rotation round a fixed axis.

But it does not include a solid with any arbitrary non-uniform distribution of specific gravity subjected to any of those influences; nor generally a piece of magnetized steel subjected to magnetic attraction; nor even a uniform body fulfilling the conditions of internal equilibrium under the influence of reactions against acceleration round a fixed axis produced by forces applied to its surface.

$Xdx + Ydy + Zdz$ is or is not a complete differential according as ρ is or is not a function of the potential; that is to say, according as the density of the body is or is not uniform over the equipotential surfaces for the distribution of force to which (P, Q, R) belongs. Thus the condition of the text, if the system of force is conservative, is satisfied when the body is homogeneous. But it is satisfied whether the system be conservative or not if the density is so distributed, that, were the body to lose its rigidity, and become an incompressible liquid held in a closed rigid vessel, it would (§ 755) be in equilibrium.

We have, according to the present assumption, reduced to case of no bodily force.

$$\frac{dW}{dx}=X,\ \frac{dW}{dy}=Y,\ \frac{dW}{dz}=Z\ \ldots\ldots\ldots\ldots\ldots\ldots(4),$$

which give
$$\frac{dX}{dx}+\frac{dY}{dy}+\frac{dZ}{dz}=\nabla^2 W.$$

Hence, for $`\delta$ as in § 730 (a) for δ,

$$(m+n)\,\Delta^2 `\delta+\nabla^2 W=0,$$

which is satisfied by the assumption

$$`\delta=-\frac{W}{m+n}\ \ldots\ldots\ldots\ldots\ldots\ldots\ldots\ldots\ldots(5).$$

Next, introducing these assumptions in (1) of § 732, we see that these equations are finally satisfied by values for $`\alpha$, $`\beta$, $`\gamma$, assumed as follows :—

$$\left.\begin{array}{l} `\alpha=\dfrac{1}{m+n}\dfrac{d\vartheta}{dx},\ `\beta=\dfrac{1}{m+n}\dfrac{d\vartheta}{dy},\ `\gamma=\dfrac{1}{m+n}\dfrac{d\vartheta}{dz} \\ \text{where } \vartheta \text{ is any function satisfying } \nabla^2\vartheta=-W. \end{array}\right\}\ldots\ldots(6).$$

Further, we may remark that if W be a spherical harmonic [App. B. (a)], a supposition including, as we shall see later, the most important applications to natural problems, we have at once, from App. B. (12), an integral of the equation for ϑ, as follows :—

$$\vartheta=-\frac{r^2}{2(2i+3)}W_i\ldots\ldots\ldots\ldots\ldots\ldots\ldots\ldots(7);$$

where the suffix is applied to W to denote that its degree is i.

734. The general problem of § 696 being now reduced to the case in which no force acts on the interior substance, it becomes this, in mathematical language :—To find α, β, γ, three functions of (x, y, z) which satisfy the equations Problem of § 696 with no force except over surface:

$$\left.\begin{array}{l} n\left(\dfrac{d^2\alpha}{dx^2}+\dfrac{d^2\alpha}{dy^2}+\dfrac{d^2\alpha}{dz^2}\right)+m\dfrac{d}{dx}\left(\dfrac{d\alpha}{dx}+\dfrac{d\beta}{dy}+\dfrac{d\gamma}{dz}\right)=0 \\ n\left(\dfrac{d^2\beta}{dx^2}+\dfrac{d^2\beta}{dy^2}+\dfrac{d^2\beta}{dz^2}\right)+m\dfrac{d}{dy}\left(\dfrac{d\alpha}{dx}+\dfrac{d\beta}{dy}+\dfrac{d\gamma}{dz}\right)=0 \\ n\left(\dfrac{d^2\gamma}{dx^2}+\dfrac{d^2\gamma}{dy^2}+\dfrac{d^2\gamma}{dz^2}\right)+m\dfrac{d}{dz}\left(\dfrac{d\alpha}{dx}+\dfrac{d\beta}{dy}+\dfrac{d\gamma}{dz}\right)=0 \end{array}\right\}\ldots\ldots(1)$$

Problem of § 696 with no force except over surface:

for all points of space occupied by the body, and the proper equations for all points of the boundary to express one or other or any sufficient combination of the two surface conditions indicated in § 696. When these conditions are that the surface displacements are given, the equations expressing them are of course merely the assignment of arbitrary values to α, β, γ for every point of the bounding surface. On the other hand, when force is arbitrarily applied in a fully specified manner over the whole surface, subject only to the conditions of equilibrium of forces on the body supposed rigid (§ 564), in its actual strained state, and the problem is to find how the body yields both at its surface and through its interior, the conditions are as follows:—Let $d\Omega$ denote an infinitesimal element of the surface; and F, G, H functions of position on the surface, expressing the components of the applied traction. These functions are quite arbitrary, subject only to the following conditions, being the equations [§ 551 (a), (b)] of equilibrium of a rigid body:—

equations of equilibrium to which the surface-tractions are subject.

$$\left.\begin{array}{c} \iint F d\Omega = 0, \quad \iint G d\Omega = 0, \quad \iint H d\Omega = 0 \\ \iint (Hy - Gz) d\Omega = 0, \ \iint (Fz - Hx) d\Omega = 0, \ \iint (Gx - Fy) d\Omega = 0 \end{array}\right\} \ldots (2);$$

and the strain experienced by the body must be such as to satisfy for every point of the surface the following equations;—

Equations of surface-condition, when tractions are given.

$$\left.\begin{array}{l} \left\{(m+n)\frac{d\alpha}{dx} + (m-n)\left(\frac{d\beta}{dy} + \frac{d\gamma}{dz}\right)\right\} f + n\left(\frac{d\alpha}{dy} + \frac{d\beta}{dx}\right) g + n\left(\frac{d\gamma}{dx} + \frac{d\alpha}{dz}\right) h = F \\ \left\{(m+n)\frac{d\beta}{dy} + (m-n)\left(\frac{d\gamma}{dz} + \frac{d\alpha}{dx}\right)\right\} g + n\left(\frac{d\beta}{dz} + \frac{d\gamma}{dy}\right) h + n\left(\frac{d\alpha}{dy} + \frac{d\beta}{dx}\right) f = G \\ \left\{(m+n)\frac{d\gamma}{dz} + (m-n)\left(\frac{d\alpha}{dx} + \frac{d\beta}{dy}\right)\right\} h + n\left(\frac{d\gamma}{dx} + \frac{d\alpha}{dz}\right) f + n\left(\frac{d\beta}{dz} + \frac{d\gamma}{dy}\right) g = H \end{array}\right\} (3)$$

which we find by (1) of § 662, with (6) of § 670, with (5) of § 693, and (5) of § 698; f, g, h being now taken to denote the direction-cosines of the normal to the bounding surface at (x, y, z).

Problem of § 696 solved for spherical shell.

735. The solution of this problem for the spherical shell (§ 696), found by aid of Laplace's spherical harmonic analysis, was first given by Lamé in a paper published in *Liouville's Journal* for 1854. It becomes much simplified[1] by the plan

[1] "Dynamical Problems regarding Elastic Spheroidal Shells, and Spheroids of Incompressible Liquid." W. Thomson. *Phil. Trans.*, 1862.

we follow of adhering to algebraic notation and symmetrical formulæ [App. B. (1)-(24)], until convenient practical expansions of the harmonic functions, whether in algebraic or trigonometrical forms, are sought [App. B. (25)-(41), (56)-(66)].

Problem of § 698 solved for spherical shell.

(*a*) Using for brevity the same notation δ and ∇^2 as hitherto [§ 698 (8) (9)], we find, from (1) of § 734, by the process (*a*) of § 730, $\nabla^2\delta = 0$.

(*b*) Now let the actual values of δ over any two concentric spherical surfaces of radii a and a' be expanded, by (52) of App. B., in series of surface harmonics, S_0, S_1, S_2, etc., and S'_0, S'_1, S'_2, etc.; so that when

Dilatation proved expressible in convergent series of spherical harmonics.

$$\left.\begin{array}{ll} r = a, & \delta = S_0 + S_1 + S_2 + \ldots S_i + \ldots \\ \text{and} \quad r = a', & \delta = S'_0 + S'_1 + S'_2 + \ldots S'_i + \ldots \end{array}\right\} \ldots\ldots\ldots\ldots(4).$$

Then, throughout the intermediate space, we must have

$$\delta = \sum_0^\infty \frac{(a^{i+1}S_i - a'^{i+1}S'_i)\, r^i - (aa')^{i+1}\,(a'^iS_i - a^iS'_i)\, r^{-i-1}}{a^{2i+1} - a'^{2i+1}} \ldots(5).$$

For (i) this series converges for all values of r intermediate between a and a', as we see by supposing a' to be the less of the two, and writing it thus:—

$$\delta = \sum_0^\infty \delta_i + \sum_0^\infty \delta_{-i-1} \ldots\ldots\ldots\ldots\ldots\ldots\ldots\ldots\ldots(6)$$

where δ_i, δ_{-i-1} are solid harmonics of degrees i and $-i-1$ given by the following:—

$$\delta_i = \frac{S_i - \left(\frac{a'}{a}\right)^{i+1} S'_i}{1 - \left(\frac{a'}{a}\right)^{2i+1}} \left(\frac{r}{a}\right)^i, \text{ and } \delta_{-i-1} = -\frac{\left(\frac{a'}{a}\right) S_i - S'_i}{1 - \left(\frac{a'}{a}\right)^{2i+1}} \left(\frac{a'}{r}\right)^{i+1}.$$

For very great values of i these become sensibly

$$\delta_i = S_i \left(\frac{r}{a}\right)^i, \text{ and } \delta_{-i-1} = S'_i \left(\frac{a'}{r}\right)^{i+1},$$

and therefore, as each of the series (4) is necessarily convergent, the two series into which in (6) the expansion (5) is divided, ultimately converge more rapidly than the geometrical series

$$\left(\frac{r}{a}\right)^i, \left(\frac{r}{a}\right)^{i+1}, \left(\frac{r}{a}\right)^{i+2}, \ldots, \text{ and } \left(\frac{a'}{r}\right)^{i+1}, \left(\frac{a'}{r}\right)^{i+2}, \left(\frac{a'}{r}\right)^{i+3}, \ldots,$$

respectively.

Dilatation proved expressible in convergent series of spherical harmonics.

Again (ii) the expression (5) agrees with (4) at the boundary of the space referred to (the two concentric spherical surfaces).

And (iii) it satisfies $\nabla^2\delta = 0$ throughout the space.

Hence (iv) no function differing in value from that given by (5), for any point of the space between the spherical surfaces, can [App. A. (e)] satisfy the conditions (iii) and (iv) to which δ is subject.

In words, this conclusion is that

General theorem regarding expansibility in solid harmonics.

736. Any function, δ, of x, y, z, which satisfies the equation $\nabla^2\delta = 0$ for any point of the space between two concentric spherical surfaces, may be expanded into the sum of two series of complete spherical harmonics [App. B. (c)] of positive and of negative degrees respectively, which converge for all points of that space.

(c) We may now write (6), for brevity, thus—

$$\delta = \sum_{-\infty}^{\infty} \delta_i \quad \text{.................. . ...(7)},$$

where δ_i, a complete harmonic of any positive or negative degree, i, is to be determined ultimately to fulfil the actual conditions of the problem. But first supposing it known, we find α, β, γ as in § 730 (d), except that now we take advantage of the formulæ appropriate for spherical harmonics instead of proceeding by triple integration. Thus, by (1) and (7), we have

Displacement determined on temporary supposition that dilatation is known.

$$\nabla^2\alpha = -\frac{m}{n}\Sigma\frac{d\delta_i}{dx};$$

and therefore, as $\frac{d\delta_i}{dx}$ is a harmonic of degree $i-1$, by taking, in App. B. (12), $n = i - 1$ and $m = 2$, we see that the complete solution of this equation, regarded as an equation for α, is

$$\alpha = u - \frac{mr^2}{2n}\Sigma\frac{1}{2i+1}\frac{d\delta_i}{dx},$$

where u denotes any solution whatever of the equation $\nabla^2 u = 0$. Similarly, if v and w denote any functions such that $\nabla^2 v = 0$ and $\nabla^2 w = 0$, we have

$$\beta = v - \frac{mr^2}{2n}\Sigma\frac{1}{2i+1}\frac{d\delta_i}{dy}, \text{ and } \gamma = w - \frac{mr^2}{2n}\Sigma\frac{1}{2i+1}\frac{d\delta_i}{dz}.$$

(*d*) Now, in order that (1) may be satisfied, δ_i must be so related to u, v, w that

Complete harmonic solution of equations of interior equilibrium.

$$\frac{d\alpha}{dx}+\frac{d\beta}{dy}+\frac{d\gamma}{dz}=\delta=\Sigma\delta_i.$$

Hence, by differentiating the expressions just found for α, β, γ, and attending to the formula

$$\frac{d}{dx}\left(r^2\frac{d\phi_i}{dx}\right)+\frac{d}{dy}\left(r^2\frac{d\phi_i}{dy}\right)+\frac{d}{dz}\left(r^2\frac{d\phi_i}{dz}\right)=2\left(x\frac{d}{dx}+y\frac{d}{dy}+z\frac{d}{dz}\right)\phi_i+r^2\nabla^2\phi_i$$
$$=2i\phi+r^2\nabla^2\phi_i\ldots(8),$$

ϕ_i being any homogeneous function of degree i, we find

$$\Sigma\delta_i=\frac{du}{dx}+\frac{dv}{dy}+\frac{dw}{dz}-\frac{m}{n}\Sigma\frac{i}{2i+1}\delta_i.$$

This gives

$$\frac{du}{dx}+\frac{dv}{dy}+\frac{dw}{dz}=\Sigma\frac{(2i+1)n+im}{(2i+1)n}\delta_i\ldots\ldots\ldots\ldots(9).$$

If, therefore, Σu_i, Σv_i, Σw_i be the harmonic expansions (§ 736) of u, v, w we must have

$$\delta_i=\frac{(2i+1)n}{(2i+1)n+im}\left(\frac{du_{i+1}}{dx}+\frac{dv_{i+1}}{dy}+\frac{dw_{i+1}}{dz}\right)\ldots\ldots\ldots\ldots(10).$$

Using this, with i changed into $i-1$, in the preceding expressions for α, β, γ, we have finally, as the spherical harmonic solution of (1), § 734,

$$\left.\begin{aligned}
\alpha&=\sum_{i=-\infty}^{i=\infty}\left\{u_i-\tfrac{1}{2}\frac{mr^2}{(2i-1)n+(i-1)m}\frac{d}{dx}\left(\frac{du_i}{dx}+\frac{dv_i}{dy}+\frac{dw_i}{dz}\right)\right\}\\
\beta&=\sum_{i=-\infty}^{i=\infty}\left\{v_i-\tfrac{1}{2}\frac{mr^2}{(2i-1)n+(i-1)m}\frac{d}{dy}\left(\frac{du_i}{dx}+\frac{dv_i}{dy}+\frac{dw_i}{dz}\right)\right\}\\
\gamma&=\sum_{i=-\infty}^{i=\infty}\left\{w_i-\tfrac{1}{2}\frac{mr^2}{(2i-1)n+(i-1)m}\frac{d}{dz}\left(\frac{du_i}{dx}+\frac{dv_i}{dy}+\frac{dw_i}{dz}\right)\right\}
\end{aligned}\right\}\ldots(11),$$

where u_i, v_i, w_i denote any spherical harmonics of degree i.

For the analytical investigations that follow, it is convenient to introduce the following abbreviations:—

$$M_i=\tfrac{1}{2}\frac{m}{(2i-1)n+(i-1)m}\ldots\ldots\ldots\ldots\ldots\ldots(12),$$

and $$\psi_{i-1}=\frac{du_i}{dx}+\frac{dv_i}{dy}+\frac{dw_i}{dz}\ldots\ldots\ldots\ldots\ldots\ldots\ldots(13),$$

Complete harmonic solution of equations of interior equilibrium.

so that (11) becomes

$$\left.\begin{aligned} \alpha &= \sum_{i=-\infty}^{i=\infty}\left(u_i - M_i r^2 \frac{d\psi_{i-1}}{dx}\right) \\ \beta &= \sum_{i=-\infty}^{i=\infty}\left(v_i - M_i r^2 \frac{d\psi_{i-1}}{dy}\right) \\ \gamma &= \sum_{i=-\infty}^{i=\infty}\left(w_i - M_i r^2 \frac{d\psi_{i-1}}{dz}\right) \end{aligned}\right\} \quad \ldots\ldots\ldots\ldots\ldots(14).$$

(*e*) It is important to remark that the addition to u, v, w respectively of terms $d\phi/dx$, $d\phi/dy$, $d\phi/dz$ (ϕ being any function satisfying $\nabla^2\phi = 0$), does not alter the equation (10). This allows us at once to write down as follows the solution of the problem for the solid sphere with surface displacement given.

Solid sphere with surface displacements given

Let a be the radius of the sphere, and let the arbitrarily given values of the three components of displacement for every point of the surface be expressed [App. B. (52)] by series of surface harmonics, ΣA_i, ΣB_i, ΣB_i, respectively. The solution is

$$\left.\begin{aligned} \alpha &= \sum_{i=0}^{i=\infty}\left\{A_i\left(\frac{r}{a}\right)^i + \frac{m(a^2-r^2)}{2a^i[(2i-1)n+(i-1)m]}\frac{d\Theta_{i-1}}{dx}\right\} \\ \beta &= \sum_{i=0}^{i=\infty}\left\{B_i\left(\frac{r}{a}\right)^i + \frac{m(a^2-r^2)}{2a^i[(2i-1)n+(i-1)m]}\frac{d\Theta_{i-1}}{dy}\right\} \\ \gamma &= \sum_{i=0}^{i=\infty}\left\{C_i\left(\frac{r}{a}\right)^i + \frac{m(a^2-r^2)}{2a^i[(2i-1)n+(i-1)m]}\frac{d\Theta_{i-1}}{dz}\right\} \\ \text{where}\quad \Theta_{i-1} &= \frac{d(A_i r^i)}{dx} + \frac{d(B_i r^i)}{dy} + \frac{d(C_i r^i)}{dz} \end{aligned}\right\} \quad (15).$$

For this is what (11) becomes if we take

$$u_i = A_i\left(\frac{r}{a}\right)^i + \frac{m}{2a^i[(2i+3)n+(i+1)m]}\frac{d\Theta_{i+1}}{dx},\ v_i = \text{etc., etc.};$$

and it makes

$$\alpha = \Sigma A_i,\ \beta = \Sigma B_i,\ \gamma = \Sigma C_i,\ \text{when } r = a \ldots\ldots\ldots\ldots\ldots(16).$$

This result might have been obtained, of course, by a purely analytical process; and we shall fall on it again as a particular case of the following:—

Shell with given displacements of its outer and inner surfaces.

(*f*) The problem for a shell with displacements given arbitrarily for all points of each of its concentric spherical bounding surfaces is much more complicated, and we shall find a purely analytical process the most convenient for getting to its solution.

Let a and a' be the radii of the outer and inner spherical surfaces, and let ΣA_i, etc., $\Sigma A'_i$, etc., be the series of surface harmonics expressing [App. B. (52)] the arbitrarily given components of displacement over them; so that our surface conditions are

Shell with given displacements of its outer and inner surfaces.

$$\left.\begin{array}{l}\alpha=\Sigma A_i\\ \beta=\Sigma B_i\\ \gamma=\Sigma C_i\end{array}\right\}\text{ when } r=a;\text{ and }\left.\begin{array}{l}\alpha=\Sigma A'_i\\ \beta=\Sigma B'_i\\ \gamma=\Sigma C'_i\end{array}\right\}\text{ when } r=a' \quad \ldots(17).$$

Using the abbreviated notation (12) and (13), selecting from (14) all terms of α which become surface harmonics of order i for a constant value of r, and equating to the proper harmonic terms of (17), we have

$$u_i+u_{-i-1}-r^2\left(M_{i+2}\frac{d\psi_{i+1}}{dx}+M_{-i+1}\frac{d\psi_{-i}}{dx}\right)\left\{\begin{array}{ll}=A_i \text{ when } & r=a\\ =A'_i \quad ,, & r=a'\end{array}\right\}\ldots(18).$$

Remarking that $r^{-i}u_i$, $r^{i+1}u_{-i-1}$, $r^{-i}d\psi_{i+1}/dx$, and $r^{i+1}d\psi_{-i}/dx$ are each of them independent of r, we have immediately from (18) the following two equations towards determining these four functions:—

$$\left.\begin{array}{l}a^i(r^{-i}u_i)+a^{-i-1}(r^{i+1}u_{-i-1})-a^2\left[M_{i+2}a^i\left(r^{-i}\dfrac{d\psi_{i+1}}{dx}\right)+M_{-i+1}a^{-i-1}\left(r^{i+1}\dfrac{d\psi_{-i}}{dx}\right)\right]=A_i\\ \text{and}\\ a'^i(r^{-i}u_i)+a'^{-i-1}(r^{i+1}u_{-i-1})-a'^2\left[M_{i+2}a'^i\left(r^{-i}\dfrac{d\psi_{i+1}}{dx}\right)+M_{-i+1}a'^{-i-1}\left(r^{i+1}\dfrac{d\psi_{-i}}{dx}\right)\right]=A'_i\end{array}\right\}(19).$$

These, and the symmetrical equations relative to y and z, suffice, with (13), for the determination of u_i, v_i, w_i for every value, positive and negative, of i. The most convenient order of procedure is first to find equations for the determination of the ψ functions by the elimination of the u, v, w, thus:—From (19) we have

$$\left.\begin{array}{l}u_i=\dfrac{(a^{2i+3}-a'^{2i+3})M_{i+2}\dfrac{d\psi_{i+1}}{dx}+(a^2-a'^2)M_{-i+1}r^{2i+1}\dfrac{d\psi_{-i}}{dx}+(a^{i+1}A_i-a'^{i+1}A'_i)r^i}{a^{2i+1}-a'^{2i+1}}\\ u_{-i-1}=\\ \dfrac{-(aa')^{2i+1}(a^2-a'^2)M_{i+2}r^{-2i-1}\dfrac{d\psi_{i+1}}{dx}+(aa')^2(a^{2i-1}-a'^{2i-1})M_{-i+1}\dfrac{d\psi_{-i}}{dx}+(aa')^{i+1}(a^iA'_i-a'^iA_i)r^{-i-1}}{a^{2i+1}-a'^{2i+1}}\end{array}\right\}(20)$$

and symmetrical equations for v and w. Or if, for brevity, we put

$$\mathfrak{A}_i=\frac{a^{i+1}A_i-a'^{i+1}A'_i}{a^{2i+1}-a'^{2i+1}},\quad \mathfrak{A}'_i=\frac{(aa')^{i+1}(a^iA'_i-a'^iA_i)}{a^{2i+1}-a'^{2i+1}}\ \ldots\ldots(21),$$

Shell with given displacements of its outer and inner surfaces.

and

$$\mathfrak{M}_{i+2}=\frac{a^{2i+3}-a'^{2i+3}}{a^{2i+1}-a'^{2i+1}}M_{i+2},\quad \mathfrak{N}_{i+2}=\frac{(aa')^{2i+1}(a^2-a'^2)}{a^{2i+1}-a'^{2i+1}}M_{i+2}\ldots(22),$$

$$\left.\begin{aligned}u_i&=\mathfrak{M}_{i+2}\frac{d\psi_{i+1}}{dx}-\mathfrak{N}_{-i+1}r^{2i+1}\frac{d\psi_{-i}}{dx}+\mathfrak{A}_ir^i\\ u_{-i-1}&=-\mathfrak{N}_{i+2}r^{-2i-1}\frac{d\psi_{i+1}}{dx}+\mathfrak{M}_{-i+1}\frac{d\psi_{-i}}{dx}+\mathfrak{A}'_ir^{-i-1}\\ v_i&=\text{etc.},\ v_{-i-1}=\text{etc.},\ w_i=\text{etc.},\ w_{-i-1}=\text{etc.}\end{aligned}\right\}\ldots\ldots(23).$$

Performing the proper differentiations and summations to eliminate the u, v, w functions between these (23) and (13), and taking advantage of the properties of the ψ functions, that

$$\nabla^2\psi_{i+1}=0,\ \nabla^2\psi_{-i}=0,\ x\frac{d\psi_{i+1}}{dx}+y\frac{d\psi_{i+1}}{dy}+z\frac{d\psi_{i+1}}{dz}=(i+1)\psi_{i+1},$$

$$x\frac{d\psi_{-i}}{dx}+y\frac{d\psi_{-i}}{dy}+z\frac{d\psi_{-i}}{dz}=-i\psi_{-i},$$

we find

$$\left.\begin{aligned}\psi_{i-1}&=(2i+1)\,i\mathfrak{N}_{-i+1}r^{2i-1}\psi_{-i}+\frac{d(\mathfrak{A}_ir^i)}{dx}+\frac{d(\mathfrak{B}_ir^i)}{dy}+\frac{d(\mathfrak{C}_ir^i)}{dz}\\ &\text{and}\\ \psi_{-i-2}&=(2i+1)(i+1)\mathfrak{N}_{i+2}r^{-2i-3}\psi_{i+1}+\frac{d(\mathfrak{A}'_ir^{-i-1})}{dx}+\frac{d(\mathfrak{B}'_ir^{-i-1})}{dy}+\frac{d(\mathfrak{C}'_ir^{-i-1})}{dz}\end{aligned}\right\}(24).$$

Changing i into $i+1$ in the first of these, and into $i-1$ in the second, we have two equations for the two unknown quantities ψ_i and ψ_{-i-1}; which give

$$\left.\begin{aligned}\psi_i&=\frac{\Theta_i+(2i+3)(i+1)\mathfrak{N}_{-i}\Theta'_{-i-1}r^{2i+1}}{1-(2i+3)(2i-1)(i+1)i\mathfrak{N}_{-i}\mathfrak{N}_{i+1}}\\ \psi_{-i-1}&=\frac{(2i-1)\,i\mathfrak{N}_{i+1}\Theta_ir^{-2i-1}+\Theta'_{-i-1}}{1-(2i+3)(2i-1)(i+1)i\mathfrak{N}_{-i}\mathfrak{N}_{i+1}}\end{aligned}\right\}\ldots\ldots..(25),$$

where, for brevity,

$$\left.\begin{aligned}\Theta_i&=\frac{d(\mathfrak{A}_{i+1}r^{i+1})}{dx}+\frac{d(\mathfrak{B}_{i+1}r^{i+1})}{dy}+\frac{d(\mathfrak{C}_{i+1}r^{i+1})}{dz}\\ \text{and}\quad \Theta'_{-i-1}&=\frac{d(\mathfrak{A}'_{-i}r^{-i})}{dx}+\frac{d(\mathfrak{B}'_{-i}r^{-i})}{dx}+\frac{d(\mathfrak{C}'_{-i}r^{-i})}{dx}\end{aligned}\right\}\ldots\ldots(26).$$

The functions ψ_i and ψ_{-i-1} for every value of i being thus given, (23) and (14) complete the solution of the problem.

(*g*) The composition of this solution ought to be carefully studied. Thus separating for simplicity the part due to the terms A_i, etc., A'_i, etc., of the single order i, in the surface data, we see that were there no such terms of other orders, all the ψ functions would vanish except ψ_{i-1}, ψ_{i+1}, ψ_{-i}, ψ_{-i-2}. These would give u_{i-2}, u_i, u_{i+2}, u_{-i+1}, u_{-i-1}, and u_{-i-3}; with symmetrical expressions for the v and w functions; of which the composition will be best studied by first writing them out in full, explicitly in terms of $\mathfrak{A}_i$, $\mathfrak{B}_i$, $\mathfrak{C}_i$, $\mathfrak{A}'_i$, $\mathfrak{B}'_i$, $\mathfrak{C}'_i$, and the derived solid harmonics Θ_{i-1} and Θ'_{-i-2}. Shell with given displacements of its outer and inner surfaces.

737. When, instead of surface displacements, the force applied over the surface is given, the problem, whether for the solid sphere or the shell, is longer because of the preliminary process (*h*) required to express the components of traction on any spherical surface concentric with the given sphere or shell, in proper harmonic forms; and its solution is more complicated, because of the new solid harmonic function ϕ_{i+1} [(32) below] which, besides the function ψ_{i-1} employed above, we are obliged to introduce in this preliminary process. Surface tractions given.

(*h*) Taking F, G, H to denote the components of the traction on the spherical surface of any radius r, having its centre at the origin of co-ordinates, instead of merely for the boundary of the body as supposed formerly in § 734 (3), we have still the same formulæ: but in them we have now to put $f = x/r$, $g = y/r$, $h = z/r$. By grouping their terms conveniently, we may, with the notation (28), put them into the following abbreviated forms:—

$$\left.\begin{aligned} Fr &= (m-n)\,\delta\,.\,x + n\left\{\left(r\frac{d}{dr}-1\right)\alpha + \frac{d\zeta}{dx}\right\} \\ Gr &= (m-n)\,\delta\,.\,y + n\left\{\left(r\frac{d}{dr}-1\right)\beta + \frac{d\zeta}{dy}\right\} \\ Hr &= (m-n)\,\delta\,.\,z + n\left\{\left(r\frac{d}{dr}-1\right)\gamma + \frac{d\zeta}{dz}\right\} \end{aligned}\right\} \quad\ldots\ldots\ldots (27),$$

Component tractions on any spherical surface concentric with origin.

$$\left.\begin{aligned} &\text{where} & \zeta &= \alpha x + \beta y + \gamma z \\ &\text{and} & r\frac{d}{dr} &= x\frac{d}{dx} + y\frac{d}{dy} + z\frac{d}{dz} \end{aligned}\right\}\ldots\ldots\ldots\ldots\ldots\ldots(28),$$

so that ζ/r is the radial component of the displacement at any

Component tractions on any spherical surface concentric with origin.

point, and d/dr prefixed to any function of x, y, z denotes the rate of its variation per unit of length in the radial direction.

It is interesting to remark that if we denote by R the radial component of the traction, we find, from (27) and (28),

$$R = \frac{x}{r}F + \frac{y}{r}G + \frac{z}{r}H = (m-n)\,\delta + \frac{2n}{r}\left(\frac{d\zeta}{dr} - \frac{\zeta}{r}\right) \ldots\ldots (28').$$

(*k*) To reduce these expressions to surface harmonics, let us consider homogeneous terms of degree i of the complete solution (14), which we shall denote* by α_i, β_i, γ_i, and let δ_{i-1}, ζ_{i+1} denote the corresponding terms of the other functions. Thus we have

$$\left.\begin{aligned} Fr &= \Sigma\left\{(m-n)\,\delta_{i-1}x + n(i-1)\,\alpha_i + n\frac{d\zeta_{i+1}}{dx}\right\} \\ Gr &= \Sigma\left\{(m-n)\,\delta_{i-1}y + n(i-1)\,\beta_i + n\frac{d\zeta_{i+1}}{dy}\right\} \\ Hr &= \Sigma\left\{(m-n)\,\delta_{i-1}z + n(i-1)\,\gamma_i + n\frac{d\zeta_{i+1}}{dz}\right\} \end{aligned}\right\} \ldots (29).$$

(*l*) The second of the three terms of order i in these equations, when the general solution of § (*d*) is used, become at the boundary each explicitly the sum of two surface harmonics of orders i and $i-2$ respectively. To bring the other parts of the expressions to similar forms, it is convenient that we should first express ζ_{i+1} in terms of the general solution (14) of § (*d*), by selecting the terms of algebraic degree i. Thus we have

$$\alpha_i = u_i - \frac{mr^2}{2\left[(2i-1)\,n + (i-1)\,m\right]}\frac{d\psi_{i-1}}{dx} \ldots\ldots\ldots\ldots (30),$$

and symmetrical expressions for β_i and γ_i, from which we find

$$\alpha_i x + \beta_i y + \gamma_i z = \zeta_{i+1} = u_i x + v_i y + w_i z - \frac{(i-1)\,mr^2\psi_{i-1}}{2\left[(2i-1)\,n + (i-1)\,m\right]}.$$

Hence, by the proper formulæ [see (36) below] for reduction to harmonics,

$$\zeta_{i+1} = -\frac{1}{2i+1}\left\{\frac{(2i-1)\left[(i-1)\,m - 2n\right]}{2\left[(2i-1)\,n + (i-1)\,m\right]}r^2\psi_{i-1} + \phi_{i+1}\right\} \ldots (31),$$

* The suffixes now introduced have reference solely to the algebraic degree, positive or negative, of the functions, whether harmonic or not, of the symbols to which they are applied.

where

$$\phi_{i+1} = r^{2i+3}\left\{\frac{d(u_i r^{-2i-1})}{dx} + \frac{d(v_i r^{-2i-1})}{dy} + \frac{d(w_i r^{-2i-1})}{dz}\right\} \ldots (32),$$

Component tractions on any spherical surface concentric with origin

and (as before assumed in § 12)

$$\psi_{i-1} = \frac{du_i}{dx} + \frac{dv_i}{dy} + \frac{dw_i}{dz} \ldots\ldots\ldots\ldots\ldots\ldots (33).$$

Also, by (10) of § 736, or directly from (30) by differentiation, we have

$$\delta_{i-1} = \frac{n(2i-1)}{(2i-1)n + (i-1)m}\psi_{i-1} \ldots\ldots\ldots\ldots\ldots\ldots (34).$$

Substituting these expressions for δ_{i-1}, α_i, and ζ_{i+1} in (29), we find

$$Fr = \Sigma\left\{n(i-1)u_i + \frac{n(2i-1)[(i+2)m-(2i-1)n]}{(2i+1)[(2i-1)n+(i-1)m]}x\psi_{i-1}\right.$$

$$\left. - \frac{n[2i(i-1)m-(2i-1)n]}{(2i+1)[(2i-1)n+(i-1)m]}r^2\frac{d\psi_{i-1}}{dx} - \frac{n}{2i+1}\frac{d\phi_{i+1}}{dx}\right\} \ldots (35).$$

harmonically expressed.

This is reduced to the required harmonic form by the obviously proper formula

$$x\psi_{i-1} = \frac{1}{2i-1}\left\{r^2\frac{d\psi_{i-1}}{dx} - r^{2i+1}\frac{d(\psi_{i-1}r^{-2i+1})}{dx}\right\} \ldots\ldots\ldots (36).$$

Thus, and dealing similarly with the expressions for Gr and Hr, we have, finally,

$$\left.\begin{aligned}
Fr &= n\Sigma\left\{(i-1)u_i - 2(i-2)M_i r^2\frac{d\psi_{i-1}}{dx} - E_i r^{2i+1}\frac{d(\psi_{i-1}r^{-2i+1})}{dx} - \frac{1}{2i+1}\frac{d\phi_{i+1}}{dx}\right\}\\
Gr &= n\Sigma\left\{(i-1)v_i - 2(i-2)M_i r^2\frac{d\psi_{i-1}}{dy} - E_i r^{2i+1}\frac{d(\psi_{i-1}r^{-2i+1})}{dx} - \frac{1}{2i+1}\frac{d\phi_{i+1}}{dy}\right\}\\
Hr &= n\Sigma\left\{(i-1)w_i - 2(i-2)M_i r^2\frac{d\psi_{i-1}}{dz} - E_i r^{2i+1}\frac{d(\psi_{i-1}r^{-2i+1})}{dx} - \frac{1}{2i+1}\frac{d\phi_{i+1}}{dz}\right\}
\end{aligned}\right\} (37),$$

$$\left.\begin{aligned}
&\text{where [as above (12)],} \quad M_i = \tfrac{1}{2}\frac{m}{(2i-1)n+(i-1)m}\\
&\text{and now, further,} \quad E_i = \frac{(i+2)m-(2i-1)n}{(2i+1)[(2i-1)n+(i-1)m]}
\end{aligned}\right\} \ldots (38).$$

(m) To express the surface conditions by harmonic equations for the shell bounded by the concentric spherical surfaces, $r = a$, $r = a'$, let us suppose the superficial values of F, G, H to be given as follows:—

Prescribed surface conditions put into harmonics.

$$\left.\begin{aligned}
&\text{when} \quad r = a,\ F = \Sigma A_i,\ G = \Sigma B_i,\ H = \Sigma C_i\\
&\text{and when} \quad r = a',\ F = \Sigma A'_i,\ G = \Sigma B'_i,\ H = \Sigma C'_i
\end{aligned}\right\} \ldots\ldots\ldots (39),$$

Prescribed surface conditions put into harmonics.

where A_i, B_i, C_i, A'_i, B'_i, C'_i denote surface harmonics of order i.

To apply to this harmonic development the conditions § 734 (2) to which the surface traction is subject, let $a^2 d\varpi$ and $a'^2 d\varpi$ be elements of the outer and inner spherical surfaces subtending at the centre (§ 468) a common infinitesimal solid angle $d\varpi$: and let $\iint d\varpi$ denote integration over the whole spherical surface of unit radius. Equations (2) become

Equations of equilibrium to which the surface tractions are subject.

$$\iint d\varpi\Sigma(a^2A_i - a'^2A'_i) = 0, \text{ etc.; and } \iint d\varpi[y\Sigma(a^2C_i - a'^2C'_i) - z\Sigma(a^2B_i - a'^2B'_i)] = 0, \text{ etc. } (40).$$

Now App. B. (16) shows that, of the first three of these, all terms except the first (those in which $i = 0$) vanishes; and that of the second three all the terms except the second (those for which $i = 1$) vanish because x, y, z are harmonics of order 1. Thus the first three become

Limitations imposed on the otherwise arbitrary harmonic data of surface tractions, for their equilibrium.

$$\iint d\varpi\,(a^2A_0 - a'^2A'_0), \text{ etc.};$$

which, as A_0, A'_0, etc., are constants, require simply that

$$a^2A_0 = a'^2A'_0,\ a^2B_0 = a'^2B'_0,\ a^2C_0 = a'^2C'_0 \ \dots\dots\dots (41).$$

The second three are equivalent to

$$r(a^2A_1 - a'^2A'_1) = \frac{dH_2}{dx},\ r(a^2B_1 - a'^2B'_1) = \frac{dH_2}{dy},\ r(a^2C_1 - a'^2C'_1) = \frac{dH_2}{dz} \ (42),$$

where H_2 is a homogeneous function of x, y, z of the second degree. For [App. B. (a)] rA_1, rA'_1, etc., are linear functions of x, y, z. If therefore (A, x), (A, y)...(B, x)...denote nine constants, we have

$$r(a^2A_1 - a'^2A'_1) = (A, x)\,x + (A, y)\,y + (A, z)\,z,$$
$$r(a^2B_1 - a'^2B'_1) = (B, x)\,x + (B, y)\,y + (B, z)\,z,$$
$$r(a^2C_1 - a'^2C'_1) = (C, x)\,x + (C, y)\,y + (C, z)\,z.$$

Using these in the second three of (40) of which, as remarked above, all terms except those for which $i = 1$ disappear, and remarking that yz, zx, xy are harmonics, and therefore (App. B. (16)] $\iint yz\,d\varpi = 0$, $\iint zx\,d\varpi = 0$, $\iint xy\,d\varpi = 0$,

we have $(C, y)\iint y^2 d\varpi - (B, z)\iint z^2 d\varpi = 0$: etc.

From these, because $\iint x^2 d\varpi = \iint y^2 d\varpi = \iint z^2 d\varpi$,

it follows that

$$(C, y) = (B, z),\ (A, z) = (C, x),\ (B, x) = (A, y),$$

which prove (42).

(*n*) The terms of algebraic degree i, exhibited in the preceding expressions (37) for Fr, Gr, Hr, become, at either of the concentric spherical surfaces, sums of surface harmonics of orders i and $i-2$ when i is positive, and of orders $-i-1$ and $-i-3$ when i is negative. Hence, selecting all the terms which lead to surface harmonics of order i, and equating to the proper terms of the data (39), we have

Surface conditions expressed in harmonic equations.

$$\left.\begin{array}{r}\dfrac{n}{r}\left\{\begin{array}{l}(i-1)u_i-(i+2)u_{-i-1}-2iM_{i+2}r^2\dfrac{d\psi_{i+1}}{dx}+2(i+1)M_{-i+1}r^3\dfrac{d\psi_{-i}}{dx}\\ -E_ir^{2i+1}\dfrac{d(\psi_{i-1}r^{-2i+1})}{dx}-E_{-i-1}r^{-2i-1}\dfrac{d(\psi_{-i-2}r^{2i+3})}{dx}-\dfrac{1}{2i+1}\left(\dfrac{d\phi_{i+1}}{dx}-\dfrac{d\phi_{-i}}{dx}\right)\end{array}\right\}\\ =\left\{\begin{array}{l}A_i \text{ when } r=a\\ A'_i \text{ when } r=a'\end{array}\right\}\end{array}\right\}\quad(43)$$

and symmetrical equations relative to y and z.

(*o*) These equations are to be treated precisely on the same plan as formerly were (18). Thus after finding u_i and u_{-i-1}, we perform on u_i, v_i, w_i the operations of (33), and on u_{-i-1}, v_{-i-1}, w_{-i-1} those of (32), and so arrive at two equations which involve as unknown quantities only ψ_{i-1}, ψ_{-i}, ϕ_{-i}, and taking the corresponding expressions for u_{i-2}, u_{-i+1}, and applying (32) to u_{i-2}, v_{i-2}, w_{i-2}, and (33) to u_{-i+1}, v_{-i+1}, w_{-i+1}, we similarly obtain two equations between ϕ_{i-1}, ψ_{i-1}, and ψ_{-i}. Thus we have in all four simple algebraic equations between ψ_{i-1}, ψ_{-i}, ϕ_{i-1}, ϕ_{-i}, by which we find these four unknown functions: and the u, v, w functions having been already explicitly expressed in terms of them, we thus have, in terms of the data of the problem, every unknown function that appears in (14) its solution.

Surface tractions given: general solution for spherical shell;

(*p*) The case of the solid sphere is of course fallen on from the more general problem of the shell, by putting $a'=0$. But if we begin with only contemplating it, we need not introduce any solid harmonics of negative degree (since every harmonic of negative degree becomes infinite at the centre, and therefore is inadmissible in the expression of effects produced throughout a solid sphere by action at its surface); and (43), and all the formulæ described as deducible from it, become much shortened when we thus confine ourselves to this case. Thus, instead of (43), we now have simply

for solid sphere.

$$\left.\dfrac{n}{r}\left\{(i-1)u_i-2iM_{i+2}r^2\dfrac{d\psi_{i+1}}{dx}-E_ir^{2i+1}\dfrac{d}{dx}(\psi_{i-1}r^{-2i+1})-\dfrac{1}{2i+1}\dfrac{d\phi_{i+1}}{dx}\right\}=A_i \atop \text{when } r=a\right\}\quad(44).$$

Surface tractions given: general solution; for solid sphere.

Hence, attending [as formerly in (f)] to the property of a homogeneous function H_j, of any order j, that $r^{-j}H_j$ is independent of r, and depends only on the ratios x/r, y/r, z/r; we have for all values of x, y, z,

$$(i-1)u_i - 2iM_{i+2}a^2\frac{d\psi_{i+1}}{dx} - E_i r^{2i+1}\frac{d}{dx}(\psi_{i-1}r^{-2i+1}) - \frac{1}{2i+1}\frac{d\phi_{i+1}}{dx} = \frac{A_i r^i}{na^{i-1}} \quad (45).$$

From this and the symmetrical equations for v and w, we have by (33),

$$[i-1+(2i+1)iE_i]\psi_{i-1} = \frac{1}{na^{i-1}}\left\{\frac{d(A_i r^i)}{dx} + \frac{d(B_i r^i)}{dy} + \frac{d(C_i r^i)}{dz}\right\} \quad (46);$$

and by (32)

$$2i\phi_{i+1} + 2i(i+1)(2i+1)M_{i+2}a^2\psi_{i+1} = \frac{r^{2i+3}}{na^{i-1}}\left\{\frac{d(A_i r^{-i-1})}{dx} + \frac{d(B_i r^{-i-1})}{dy} + \frac{d(C_i r^{-i-1})}{dz}\right\} \quad (47).$$

Eliminating, by this, ϕ_{i+1} from (45), and introducing the abbreviated notation, Φ_{i+1} [(50) below], we find

$$(i-1)u_i = (i-1)M_{i+2}a^2\frac{d\psi_{i+1}}{dx} + E_i r^{2i+1}\frac{d}{dx}(\psi_{i-1}r^{-2i+1}) + \frac{1}{na^{i-1}}\left[A_i r^i + \frac{1}{2i(2i+1)}\frac{d\Phi_{i+1}}{dx}\right] \quad (48),$$

and (43) gives

$$\psi_{i-1} = \frac{\Psi_{i-1}}{[(i-1)+(2i+1)iE_i]\,na^{i-1}} = \frac{[(i-1)m+(2i-1)n]\Psi_{i-1}}{[(2i^2+1)m-(2i-1)n]\,na^{i-1}} \ldots (49),$$

where
$$\left.\begin{aligned}\Psi_{i-1} &= \frac{d(A_i r^i)}{dx} + \frac{d(B_i r^i)}{dy} + \frac{d(C_i r^i)}{dz}\\ \text{and } \Phi_{i+1} &= r^{2i+3}\left\{\frac{d(A_i r^{-i-1})}{dx} + \frac{d(B_i r^{-i-1})}{dy} + \frac{d(C_i r^{-i-1})}{dz}\right\}\end{aligned}\right\} \quad (50).$$

With these expressions for ψ_i and u_i, (14) is the complete solution of the problem.

(q) The composition and character of this solution are made manifest by writing out in full the terms in it which depend on harmonics of a single order, i, in the surface data. Thus if the components of the surface traction are simply A_i, B_i, C_i, all the Ψ functions except Ψ_{i-1} and all the Φ functions except Φ_{i+1} vanish. Hence (48) shows that all the u functions except u_{i-2} and u_i vanish: and for these it gives

$$\left.\begin{aligned}u_{i-2} &= M_i a^2\frac{d\psi_{i-1}}{dx}\\ u_i &= \frac{1}{i-1}\left\{E_i r^{2i+1}\frac{d}{dx}(\psi_{i-1}r^{-i+1}) + \frac{1}{na^{i-1}}\left[A_i r^i + \frac{1}{2i(2i+1)}\frac{d\Phi_{i+1}}{dx}\right]\right\}\end{aligned}\right\} \quad (51).$$

Using this in (14) and for E_i and M_i substituting their values by (38), we have, explicitly expressed in terms of the data, and the solid harmonics Ψ_{i-1}, Ψ_{i+1}, derived from the data according to the formulæ (50), the final solution of the problem as follows:—

$$\alpha = \frac{1}{na^{i-1}}\left\{\tfrac{1}{2}\frac{m(a^2-r^2)}{(2i^2+1)m-(2i-1)n}\frac{d\Psi_{i-1}}{dx}\right.$$
$$\left.+\frac{1}{i-1}\left[\frac{(i+2)m-(2i-1)n}{(2i^2+1)m-(2i-1)n}\frac{r^{2i+1}d(\Psi_{i-1}r^{-2i+1})}{(2i+1)dx}+\frac{1}{2i(2i+1)}\frac{d\Phi_{i+1}}{dx}+A_i r^i\right]\right\} \quad (52),$$

with symmetrical expressions for β and γ.

(*r*) The case of $i=1$ is interesting, inasmuch as it seems at first sight to make the second part of the expression (52) for α infinite because of the divisor $i-1$. But the terms within the brackets [ ] vanish for $i=1$, owing to the relations (42) proved above, which, for the solid sphere, become

Case of homogeneous strain.

$$rA_1=\frac{dH_2}{dx},\quad rB_1=\frac{dH_2}{dy},\quad rC_1=\frac{dH_2}{dz}\;\ldots\ldots\ldots\ldots(53),$$

H_2 denoting any homogeneous function of x, y, z of the second degree. The verification of this presents no difficulty, and we leave it as an exercise to the student. The true interpretation of the $\frac{0}{0}$ appearing thus in the expressions for α, β, γ is clearly that they are indeterminate: and that they ought to be so, we see by remarking that an infinitesimal rotation round any diameter without strain may be superimposed on any solution without violating the conditions of the problem: in other words (§§ 89, 95),

Indeterminate rotations without strain, necessarily included in general solution for displacement, when the data are merely of force.

$$\omega_2 z-\omega_3 y,\quad \omega_3 x-\omega_1 z,\quad \omega y_1-\omega_2 x$$

may be added to the expressions for α, β, γ in any solution, and the result will still be a solution.

But though α, β, γ are indeterminate, (50) gives ψ_0 and ϕ_2 determinately. The student will find it a good and simple exercise to verify that the determination of ψ_0 and ϕ_2 determines the state of strain [homogeneous (§ 155) of course in this case] actually produced by the given surface traction.

738. A solid is said (§ 730) to experience a plane strain, or to be strained in two dimensions, when it is strained in any manner subject to the condition that the displacements are all in a set of parallel planes, and are equal and parallel for all points in any line perpendicular to these planes: and any one of these planes may be called the plane of the strain. Thus,

Plane strain defined.

Plane strain defined. in plane strain, all cylindrical surfaces perpendicular to the plane of the strain remain cylindrical surfaces perpendicular to the same plane, and nowhere experience stretching along the generating lines.

The condition of plane strain expressed analytically, if we take XOY for the plane, is that γ must vanish, and that α and β must be functions of x and y, without z. Thus we see that

Only two independent variables enter into the analytical expression of plane strain; and thus this case presents a class of problems of peculiar simplicity. For instance, if an infinitely long solid or hollow circular cylinder is the "given solid" of

Problem for cylinders under plane strain, § 696, and if the bodily force (if any) and the surface action consist of forces and tractions everywhere perpendicular to its axis, and equal and parallel at all points of any line parallel to its axis, we have, whether surface displacement or surface traction be given, problems precisely analogous to those of §§ 735, 736, but much simpler, and obviously of very great practical importance in the engineering of long straight tubes under strain.

739. It is interesting to remark, that in these cylindrical problems, instead of surface harmonics of successive orders

solved in terms of "plane harmonics." 1, 2, 3, etc., which are [App. B. (b)] functions of spherical surface co-ordinates (as, for instance, latitude and longitude on a globe), we have simple harmonic functions (§§ 54, 75) of the

Plane harmonic functions defined. same degrees, of the angle between two planes through the axis, and of its successive multiples: and instead of solid harmonic functions [App. B. (a) and (b)], we have what we may call *plane harmonic functions*, being the algebraic functions of two variables (x, y), which we find by expanding $\cos i\theta$ and $\sin i\theta$ in powers of sines or cosines of θ, taking

$$\cos\theta = \frac{x}{\sqrt{(x^2+y^2)}}, \text{ and } \sin\theta = \frac{y}{\sqrt{(x^2+y^2)}},$$

and multiplying the result by $(x^2+y^2)^{\frac{1}{2}i}$.

A plane harmonic function is of course the particular case of a solid harmonic [App. B. (a) and (b)] in which z does not appear;

that is to say, it is any homogeneous function, V, of x and y, which satisfies the equation

Plane harmonic functions defined.

$$\frac{d^2V}{dx^2}+\frac{d^2V}{dy^2}=0, \text{ or, as we may write it for brevity, } \nabla^2V=0.$$

And, as we have seen [§ 707 (23)], the most general expression for a plane harmonic of degree i (positive or negative, integral or fractional) is

$$\left.\begin{array}{l}\tfrac{1}{2}A\{(x+yv)^i+(x-yv)^i\}-\tfrac{1}{2}Bv\{(x+yv)^i-(x-yv)^i\}\\ \text{where } v \text{ stands for } \sqrt{-1}, \text{ or in polar co-ordinates}\\ \qquad (A\cos i\theta+B\sin i\theta)\,r^i\end{array}\right\}\ldots\ldots(1).$$

The equations of internal equilibrium [§ 698 (6)] with no bodily force (that is, $X=0$ and $Y=0$) become, for the case of plane strain,

$$\left.\begin{array}{l}n\left(\dfrac{d^2\alpha}{dx^2}+\dfrac{d^2\alpha}{dy^2}\right)+m\dfrac{d}{dx}\left(\dfrac{d\alpha}{dx}+\dfrac{d\beta}{dy}\right)=0\\ n\left(\dfrac{d^2\beta}{dx^2}+\dfrac{d^2\beta}{dy^2}\right)+m\dfrac{d}{dy}\left(\dfrac{d\alpha}{dx}+\dfrac{d\beta}{dy}\right)=0\end{array}\right\}\ldots\ldots\ldots(2).$$

The plane harmonic solution of these, found by precisely the same process as §§ 735, 736 (a)...(e), but for only two variables instead of three, is

Problem for cylinders under plane strain solved in terms of plane harmonics.

$$\left.\begin{array}{l}\alpha=\Sigma\left[u_i-\dfrac{m}{2(i-1)(2n+m)}r^2\dfrac{d\psi_{i-1}}{dx}\right]\\ \beta=\Sigma\left[v_i-\dfrac{m}{2(i-1)(2n+m)}r^2\dfrac{d\psi_{i-1}}{dy}\right]\\ \text{where}\qquad \psi_{i-1}=\dfrac{du_i}{dx}+\dfrac{dv_i}{dy}\end{array}\right\}\ldots\ldots\ldots(3),$$

and u_i, v_i denote any two plane harmonics of degree i, so that ψ_{i-1} is a plane harmonic of degree $i-1$. Of course i may be positive or negative, integral or fractional.

This solution may be reduced to polar co-ordinates with advantage for many applications, by putting

$$\text{and taking}\quad\left.\begin{array}{l}x=r\cos\theta,\quad y=r\sin\theta,\\ u_i=r^i(A_i\cos i\theta+A'_i\sin i\theta)\\ v_i=r^i(B_i\cos i\theta+B'_i\sin i\theta)\end{array}\right\}\ldots\ldots\ldots\ldots\ldots\ldots(4);$$

which give

$$\frac{2n+m}{2n}\delta_{i-1}=\psi_{i-1}=ir^{i-1}\{(A_i+B'_i)\cos(i-1)\theta+(A'_i-B_i)\sin(i-1)\theta\}\ldots(5),$$

and

$$\left.\begin{array}{l}\alpha=\Sigma r^i\left\{A_i\cos i\theta+A'_i\sin i\theta-\frac{im}{2(2n+m)}[(A_i+B'_i)\cos(i-2)\theta+(A'_i-B_i)\sin(i-2)\theta]\right\}\\ \beta=\Sigma r^i\left\{B_i\cos i\theta+B'_i\sin i\theta-\frac{im}{2(2n+m)}[-(A_i+B'_i)\sin(i-2)\theta+(A'_i-B_i)\cos(i-2)\theta]\right\}\end{array}\right\}\quad(6).$$

Problem for cylinders under plane strain solved in terms of plane harmonics.

The student will find it a good exercise to work out in full, to explicit expressions for the displacement of any point of the solid, in the cylindrical problems corresponding to the spherical problems of § 735 (f), and of § 736 (h)...(r). The process (l) of the latter may be worked through in the symmetrical algebraic form, as an illustration of the plan we have followed in dealing with spherical harmonics; but the result corresponding to (37) of § 737 may be obtained more readily, and in a simpler form, by immediately putting (29) of § 737 into polar co-ordinates, as (4), (5), (6) of § 739. We intend to use, and to illustrate, these solutions under "Properties of Matter."

Small bodies stronger than large ones in proportion to their weights.

740. In our sections on hydrostatics, the problem of finding the deformation produced in a spheroid of incompressible liquid by a given disturbing force will be solved; and then we shall consider the application of the preceding result [§ 736 (51)] for an elastic solid sphere to the theory of the tides and the rigidity of the earth. This proposed application, however, reminds us of a general remark of great practical importance, with which we shall leave elastic solids for the present. Considering different elastic solids of similar substance and similar shapes, we see that if by forces applied to them in any way they are similarly strained, the surface tractions in or across similarly situated elements of surface, whether of their boundaries or of surfaces imagined as cutting through their substances, must be equal, reckoned as usual per unit of area. Hence; the force across, or in, any such surface, being resolved into components parallel to any directions; the whole amounts of each such component for similar surfaces of the different bodies are in proportion to the *squares* of their linear dimensions. Hence, if equilibrated similarly under the action of gravity, or of their kinetic reactions (§ 264) against equal accelerations (§ 28), the greater body would be more strained than the less; as the amounts of gravity or of kinetic reaction of similar portions of them are as the *cubes* of their linear dimensions. Definitively, the strains at similarly situated points of the bodies will be in simple proportion to their linear dimensions, and the displacements will be as the squares of these lines, provided that there is no strain in any part of any of them too great to allow the principle of superposition to hold with sufficient exactness, and that no part is turned through more than a very small angle relatively to any other part. To illustrate by a single example, let us consider a uniform long, thin, round rod held horizontally by its middle. Let its substance be homogeneous, of density ρ, and Young's modulus, M; and let its length, l, be p times its diameter. Then (as the moment of inertia of a circular area of radius r round a diameter is $\frac{1}{4}\pi r^4$) the flexural rigidity of the rod will (§ 715) be $\frac{1}{4}M\pi(l/2p)^4$, which is equal to B/g in the notation of § 610, as B is there reckoned in kinetic or absolute measure (§ 223) instead of the gravitation measure in which we now, according to engineers' usage (§ 220), reckon M. Also $w=\rho\pi(l/2p)^2$, and therefore, for § 617,

Small bodies stronger than large ones in proportion to their weights.

Example: a straight rod held horizontally by its middle.

$$\frac{gw}{B}=\frac{16p^2\rho}{Ml^2}.$$

This, used in § 617 (10), gives us; for the curvature at the middle of the rod; the elongation and contraction where greatest, that is, at the highest and lowest points of the normal section through the middle point; and the droop of the ends; the following expressions,

$$\frac{2p^2\rho}{M};\ \frac{pl\rho}{M};\ \text{and}\ \frac{p^2l^2\rho}{8M}.$$

Thus, for a rod whose length is 200 times its diameter, if its substance be iron or steel, for which $\rho=7{\cdot}75$, and $M=194\times10^7$ grammes per square centimetre, the maximum elongation and contraction (being at the top and bottom of the middle section where it is held) are each equal to $\cdot8\times10^{-6}\times l$, and the droop of its ends to $2\times10^{-5}\times l^2$. Thus a steel or iron wire, ten centimetres long, and half a millimetre in diameter, held horizontally by its middle, would experience only ·000008 as maximum elongation and contraction, and only ·002 of a centimetre of droop in its ends: a round steel rod, of half a centimetre diameter, and one metre long, would experience

Stiffness of uniform steel rods of different dimensions.

Stiffness of uniform steel rods of different dimensions. ·00008 as maximum elongation and contraction, and ·2 of a centimetre of droop: a round steel rod, of ten centimetres diameter, and twenty metres long, need not be of remarkable temper (see Vol. II., Properties of Matter) to bear being held by the middle without taking a very sensible permanent set: and it is probable that any temper of steel or iron except the softest is strong enough in a round shaft forty metres long, if only twenty centimetres in diameter, to allow it to be held by its middle, drooping as it would to the extent of 320 centimetres at its ends, without either bending it beyond elasticity; or breaking it. (See *Encyclopædia Britannica*, Article "Elasticity," § 22.)

Transition to hydrodynamics. **741.** In passing from the dynamics of perfectly elastic solids to abstract hydrodynamics, or the dynamics of perfect fluids, it is convenient and instructive to anticipate slightly some of the views as to intermediate properties observed in real solids and fluids, which, according to the general plan proposed (§ 449) for our work, will be examined with more detail under Properties of Matter.

Imperfectness of elasticity in solids. By induction from a great variety of observed phenomena, we are compelled to conclude that no change of volume or of shape can be produced in any kind of matter without dissipation of energy (§ 275); so that if in any case there is a return to the primitive configuration, some amount (however small) of work is always required to compensate the energy dissipated away, and restore the body to the same physical and the same palpably kinetic condition as that in which it was given. We have seen (§ 672), by anticipating something of thermodynamic principles, how such dissipation is inevitable, even in dealing with the *absolutely perfect* elasticity of volume presented by every fluid, and possibly by some solids, as, for instance, homogeneous crystals. But in metals, glass, porcelain, natural stones, wood, india-rubber, homogeneous jelly, silk fibre, ivory, etc., a distinct *frictional resistance** against every change of shape is, as we shall see in Vol. II., under Properties of Matter, demonstrated by many experiments, and is found to depend on the speed with which the change of

Viscosity of solids.

* See *Proceedings of the Royal Society*, May 1865, "On the Viscosity and Elasticity of Metals" (W. Thomson).

shape is made. A very remarkable and obvious proof of frictional resistance to change of shape in ordinary solids is afforded by the gradual, more or less rapid, subsidence of vibrations of elastic solids; marvellously rapid in india-rubber, and even in homogeneous jelly; less rapid in glass and metal springs, but still demonstrably much more rapid than can be accounted for by the resistance of the air. This molecular friction in elastic solids may be properly called *viscosity of solids*, because, as being an internal resistance to change of shape depending on the rapidity of the change, it must be classed with fluid molecular friction, which by general consent is called *viscosity of fluids*. But, at the same time, we feel bound to remark that the word viscosity, as used hitherto by the best writers, when solids or heterogeneous semisolid-semifluid masses are referred to, has not been distinctly applied to molecular friction, especially not to the molecular friction of a highly elastic solid within its limits of high elasticity, but has rather been employed to designate a property of slow, continual yielding through very great, or altogether unlimited, extent of change of shape, under the action of continued stress. It is in this sense that Forbes, for instance, has used the word in stating that "Viscous Theory of Glacial Motion" which he demonstrated by his grand observations on glaciers. As, however, he, and many other writers after him, have used the words plasticity and plastic, both with reference to homogeneous solids (such as wax or pitch, even though also brittle; soft metals; etc.), and to heterogeneous semisolid-semifluid masses (as mud, moist earth, mortar, glacial ice, etc.), to designate the property*, common to all those cases, of experiencing under continued stress either quite continued and unlimited change of shape, or gradually very great change at a diminishing

Viscosity of solids.

Viscosity of fluids.

Forbes' "Viscous Theory of Glacial Motion."

* Some confusion of ideas might have been avoided on the part of writers who have professedly objected to Forbes' theory while really objecting only (and we believe groundlessly) to his usage of the word viscosity, if they had paused to consider that no one physical explanation can hold for those several cases; and that Forbes' theory is merely the proof by observation that glaciers have the property which mud (heterogeneous), mortar (heterogeneous), pitch (homogeneous), water (homogeneous), all have of changing shape indefinitely and continuously under the action of continued stress.

Plasticity of solids.

(asymptotic) rate through infinite time; and as the use of the term *plasticity* implies no more than does *viscosity* any physical theory or explanation of the property, the word viscosity is without inconvenience left available for the definition we have given of it above.

Perfect and unlimited plasticity unopposed by internal friction, the characteristic of the ideal perfect fluid of abstract hydrodynamics.

742. A *perfect fluid*, or (as we shall call it) a fluid, is an unrealizable conception, like a rigid, or a smooth, body: it is defined as a body incapable of resisting a change of shape: and therefore incapable of experiencing distorting or tangential stress (§ 669). Hence its pressure on any surface, whether of a solid or of a contiguous portion of the fluid, is at every point perpendicular to the surface. In equilibrium, all common liquids and gaseous fluids fulfil the definition. But there is finite resistance, of the nature of friction, opposing change of shape at a finite rate; and therefore, while a fluid is changing shape, it exerts tangential force on every surface other than normal planes of the stress (§ 664) required to keep this change of shape going on. Hence; although the hydrostatical results, to which we immediately proceed, are verified in practice; in treating of hydrokinetics, in a subsequent chapter, we shall be obliged to introduce the consideration of fluid friction, except in cases where the circumstances are such as to render its effects insensible.

Fluid pressure.

743. With reference to a fluid the *pressure at any point in any direction* is an expression used to denote the average pressure per unit of area on a plane surface imagined as containing the point, and perpendicular to the direction in question, when the area of that surface is indefinitely diminished.

744. At any point in a fluid at rest the pressure is the same in all directions: and, if no external forces act, the pressure is the same at every point. For the proof of these and most of the following propositions, we imagine, according to § 564, a definite portion of the fluid to become solid, without changing its mass, form, or dimensions.

Suppose the fluid to be contained in a closed vessel, the pressure within depending on the pressure exerted on it by the vessel, and not on any external force such as gravity.

745. The resultant of the fluid pressures on the elements of any portion of a spherical surface must, like each of its components, pass through the centre of the sphere. Hence, if we suppose (§ 564) a portion of the fluid in the form of a plano-convex lens to be solidified, the resultant pressure on the plane side must pass through the centre of the sphere; and, therefore, being perpendicular to the plane, must pass through the centre of the circular area. From this it is obvious that the pressure is the same at all points of any plane in the fluid. Hence, by § 562, the resultant pressure on any plane surface passes through its centre of inertia.

Fluid pressure proved equal in all directions.

Next, imagine a triangular prism of the fluid, which ends perpendicular to its faces, to be solidified. The resultant pressures on its ends act in the line joining the centres of inertia of their areas, and are equal (§ 552) since the resultant pressures on the sides are in directions perpendicular to this line. Hence the pressure is the same in all parallel planes.

But the centres of inertia of the three faces, and the resultant pressures applied there, lie in a triangular section parallel to the ends. The pressures act at the middle points of the sides of this triangle, and perpendicularly to them, so that their directions meet in a point. And, as they are in equilibrium, they must be, by § 559, *e*, proportional to the respective sides of the triangle; that is, to the breadths, or areas, of the faces of the prism. Thus the resultant pressures on the faces must be proportional to the areas of the faces, and therefore the pressure is equal in any two planes which meet.

Collecting our results, we see that the pressure is the same at all points, and in all directions, throughout the fluid mass.

746. One immediate application of this result gives us a simple though indirect proof of the second theorem in § 559, *e*, for we have only to suppose the polyhedron to be a solidified portion of a mass of fluid in equilibrium under pressures only. The resultant pressure on each side will then be proportional to its area, and, by § 562, will act at its centre of inertia; which, in this case, is the *Centre of Pressure*.

Application to statics of solids.

Centre of pressure.

Application of the principle of energy.

747. Another proof of the equality of pressure throughout a mass of fluid, uninfluenced by other external force than the pressure of the containing vessel, is easily furnished by the energy criterion of equilibrium, § 289; but, to avoid complication, we will consider the fluid to be incompressible. Suppose a number of pistons fitted into cylinders inserted in the sides of the closed vessel containing the fluid. Then, if A be the area of one of these pistons, p the average pressure on it, x the distance through which it is pressed, in or out; the energy criterion is that no work shall be done on the whole, *i.e.* that

Proof by energy of the equality of fluid pressure in all directions.

$$A_1p_1x_1 + A_2p_2x_2 + \ldots = \Sigma\,(Apx) = 0,$$

as much work being restored by the pistons which are forced out, as is done by those forced in. Also, since the fluid is incompressible, it must have gained as much space by forcing out some of the pistons as it lost by the intrusion of the others. This gives

$$A_1x_1 + A_2x_2 + \ldots = \Sigma\,(Ax) = 0.$$

The last is the only condition to which x_1, x_2, etc., in the first equation, are subject; and therefore the first can only be satisfied if

$$p_1 = p_2 = p_3 = \text{etc.},$$

that is, if the pressure be the same on each piston. Upon this property depends the action of Bramah's *Hydrostatic Press.*

If the fluid be compressible, the work expended in compressing it from volume V to $V-\delta V$, at mean pressure p, is $p\delta V$.

If in this case we *assume* the pressure to be the same throughout, we obtain a result consistent with the energy criterion.

The work done on the fluid is $\Sigma\,(Apx)$, that is, in consequence of the assumption, $p\Sigma\,(Ax)$.

But this is equal to $p\delta V$, for, evidently, $\Sigma\,(Ax) = \delta V$.

Fluid pressure depending on external forces.

748. When forces, such as gravity, act from external matter upon the substance of the fluid, either in proportion to the density of its own substance in its different parts, or in proportion to the density of electricity, or of magnetic polarity, or of any other conceivable accidental property of it, the pressure will still be the same in all directions at any one point, but will now vary continuously from point to point. For the preceding demonstration (§ 745) may still be applied by simply taking the dimensions of the prism small enough; since the pressures are as the squares of its linear dimensions, and the effects of the applied forces such as gravity, as the cubes.

Fluid pressure depending on external forces.

749. When forces act on the whole fluid, surfaces of equal pressure, if they exist, must be at every point perpendicular to the direction of the resultant force. For, any prism of the fluid so situated that the whole pressures on its ends are equal must (§ 552) experience from the applied forces no component in the direction of its length; and, therefore, if the prism be so small that from point to point of it the direction of the resultant of the applied forces does not vary sensibly, this direction must be perpendicular to the length of the prism. From this it follows that whatever be the physical origin, and the law, of the system of forces acting on the fluid, and whether it be conservative or non-conservative, the fluid cannot be in equilibrium unless the lines of force possess the geometrical property of being at right angles to a series of surfaces.

Surfaces of equal pressure are perpendicular to the lines of force.

750. Again, considering two surfaces of equal pressure infinitely near one another, let the fluid between them be divided into columns of equal transverse section, and having their lengths perpendicular to the surfaces. The difference of pressures on the two ends being the same for each column, the resultant applied forces on the fluid masses composing them must be equal. Comparing this with § 488, we see that if the applied forces constitute a conservative system, the density of matter, or electricity, or whatever property of the substance they depend on, must be equal throughout the layer under consideration. This is the celebrated hydrostatic proposition that *in a fluid at rest, surfaces of equal pressure are also surfaces of equal density and of equal potential.*

And are surfaces of equal density and of equal potential when the system of force is conservative.

751. Hence, when gravity is the only external force considered, surfaces of equal pressure and equal density are (when of moderate extent) horizontal planes. On this depends the action of levels, syphons, barometers, etc.; also the separation

Gravity the only external force.

Gravity the only external force.

of liquids of different densities (which do not mix or combine chemically) into horizontal strata, etc. etc. The free surface of a liquid is exposed to the pressure of the atmosphere simply; and therefore, when in equilibrium, must be a surface of equal pressure, and consequently level. In extensive sheets of water, such as the American lakes, differences of atmospheric pressure, even in moderately calm weather, often produce considerable deviations from a truly level surface.

Rate of increase of pressure.

752. The rate of increase of pressure per unit of length in the direction of the resultant force, is equal to the intensity of the force reckoned per unit of volume of the fluid. Let F be the resultant force per unit of volume in one of the columns of § 750; p and p' the pressures at the ends of the column, l its length, S its section. We have, for the equilibrium of the column,

$$(p' - p)\, S = SlF.$$

Hence the rate of increase of pressure per unit of length is F.

If the applied forces belong to a conservative system, for which V and V' are the values of the potential at the ends of the column, we have (§ 486)

$$V' - V = -lF\rho,$$

where ρ is the density of the fluid. This gives

$$p' - p = -\rho(V' - V)$$

or

$$dp = -\rho dV.$$

Hence in the case of gravity as the only impressed force the rate of increase of pressure per unit of depth in the fluid is ρ, in gravitation measure (usually employed in hydrostatics). In kinetic or absolute measure (§ 224) it is $g\rho$.

If the fluid be a gas, such as air, and be kept at a constant temperature, we have $\rho = cp$, where c denotes a constant, the reciprocal of H, the "height of the homogeneous atmosphere," defined (§ 753) below. Hence, in a calm atmosphere of uniform temperature we have $dp/p = -cdV$; and from this, by integration, $p = p_0\epsilon^{-cV}$ where p_0 is the pressure at any particular level (the sea-level, for instance) where we choose to reckon the potential as zero.

When the differences of level considered are infinitely small in comparison with the earth's radius, as we may practically regard them, in measuring the height of a mountain, or of a balloon, by the barometer, the force of gravity is constant, and therefore differences of potential (force being reckoned in units of weight) are simply equal to differences of level. Hence if x denote height of the level of pressure p above that of p_0, we have, in the preceding formulæ, $V=x$, and therefore $p=p_0\epsilon^{-cx}$. That is to say— Rate of increase of pressure.

753. If the air be at a constant temperature, the pressure diminishes in geometrical progression as the height increases in arithmetical progression. This theorem is due to Halley. Without formal mathematics we see the truth of it by remarking that differences of pressure are (§ 752) equal to differences of level multiplied by the density of the fluid, or by the proper mean density when the density differs sensibly between the two stations. But the density, when the temperature is constant, varies in simple proportion to the pressure, according to Boyle's and Mariotte's law. Hence differences of pressure between pairs of stations differing equally in level are proportional to the proper mean values of the whole pressure, which is the well-known compound interest law. The rate of diminution of pressure per unit of length upwards in proportion to the whole pressure at any point, is of course equal to the reciprocal of the height above that point that the atmosphere must have, if of constant density, to give that pressure by its weight. The height thus defined is commonly called "the height of the homogeneous atmosphere," a very convenient conventional expression. It is equal to the product of the volume occupied by the unit mass of the gas at any pressure into the value of that pressure reckoned per unit of area, in terms of the weight of the unit of mass. If we denote it by H, the exponential expression of the law is $p=p_0\epsilon^{-x/H}$, which agrees with the final formula of § 752. Pressure in a calm atmosphere of uniform temperature. Height of the homogeneous atmosphere.

The value of H for dry atmospheric air, at the freezing temperature, according to Regnault, is, in the latitude of Paris, 799,020 centimetres, or 26,215 feet. Being inversely as the force of gravity in different latitudes (§ 222), it is 798,533 centimetres, or 26,199 feet, in the latitude of Edinburgh and Glasgow.

Analytical investigation of the preceding theorems.

Let X, Y, Z be the components, parallel to three rectangular axes, of the force acting on the fluid at (x, y, z), reckoned per unit of its mass. Then, inasmuch as the difference of pressures on the two faces $\delta y\delta z$ of a rectangular parallelepiped of the fluid is $\delta y\delta z \frac{dp}{dx}\delta x$, the equilibrium of this portion of the fluid, regarded for a moment (§ 564) as rigid, requires that

$$\delta y\delta z \frac{dp}{dx}\delta x - X\rho\delta x\delta y\delta z = 0.$$

From this and the symmetrical equations relative to y and z we have

$$\frac{dp}{dx} = X\rho, \quad \frac{dp}{dy} = Y\rho, \quad \frac{dp}{dz} = Z\rho \quad \text{..............(1)},$$

which are the conditions necessary and sufficient for the equilibrium of any fluid mass.

From these we have

$$dp = \frac{dp}{dx}dx + \frac{dp}{dy}dy + \frac{dp}{dz}dz = \rho\,(Xdx + Ydy + Zdz) \text{......(2)}.$$

This shows that the expression $Xdx + Ydy + Zdz$ must be the complete differential of a function of three independent variables, or capable of being made so by a factor; that is to say, that a series of surfaces exists which cuts the lines of force at right angles; a conclusion also proved above (§ 749).

When the forces belong to a conservative system no factor is required to make the complete differential; and we have

$$Xdx + Ydy + Zdz = -dV$$

if V denote (§ 485) their potential at (x, y, z): so that (2) becomes

$$dp = -\rho dV \text{..............................(3)}.$$

This shows that p is constant over equipotential surfaces (or is a function of V); and it gives

$$\rho = -\frac{dp}{dV} \text{..............................(4)},$$

showing that ρ also is a function of V; conclusions of which we have had a more elementary proof in § 752. As (4) is an analytical expression equivalent to the three equations (1), for the case of a conservative system of forces, we conclude that

Conditions of equilibrium of

754. It is both necessary and sufficient for the equilibrium of an incompressible fluid completely filling a rigid closed vessel, and influenced only by a conservative system of forces, that its density be uniform over every equipotential surface, that is to say, every surface cutting the lines of force at right angles. If, however, the boundary, or any part of the boundary, of the fluid mass considered, be not rigid; whether it be of flexible solid matter (as a membrane, or a thin sheet of elastic solid), or whether it be a mere geometrical boundary, on the other side of which there is another fluid, or *nothing* [a case which, without believing in vacuum as a reality, we may admit in abstract dynamics (§ 438)], a farther condition is necessary to secure that the pressure from without shall fulfil (4) at every point of the boundary. In the case of a bounding membrane, this condition must be fulfilled either through pressure artificially applied from without, or through the interior elastic forces of the matter of the membrane. In the case of another fluid of different density touching it on the other side of the boundary, all round or over some part of it, with no separating membrane, the condition of equilibrium of a heterogeneous fluid is to be fulfilled relatively to the whole fluid mass made up of the two; which shows that at the boundary the pressure must be constant and equal to that of the fluid on the other side. Thus water, oil, mercury, or any other liquid, in an open vessel, with its free surface exposed to the air, requires for equilibrium simply that this surface be level.

Fluid completely filling a closed vessel.

Free surface in open vessel is level.

755. Recurring to the consideration of a finite mass of fluid completely filling a rigid closed vessel, and supposing that, if the potential of the force-system (as in the case referred to in the sixth and seventh lines of § 758) be a cyclic* function

Fluid, in closed vessel, under a non-conservative system of forces.

* We here introduce term "cyclic function" to designate a function of more than one variable which experiences a constant addition to its value every time the variables are made to vary continuously from a given set of values through some cycle of values back to the same primitive set of values.

Examples (1) $\tan^{-1}(y/x)$. This is the potential of the conservative system referred to in the first clause of the third sentence of § 758.

(2) $f(x^2+y^2)\tan^{-1}(y/x)$. This expresses the fluid pressure in the case of hydrostatic example described in the next to the last sentence of § 758.

(3) The apparent area of a closed curve (plane or not plane) as seen from any point (x, y, z).

Fluid in closed vessel, under a non-conservative system of forces.

tion, the enclosure containing the liquid is singly-continuous, we see, from what precedes, that, if homogeneous and incompressible, the fluid cannot be disturbed from equilibrium by any conservative system of forces; but we do not require the analytical investigation to prove this, as we should have "the perpetual motion" if it were denied, which would violate the hypothesis that the system of forces is conservative. On the other hand, a non-conservative system of forces cannot, under any circumstances, equilibrate a fluid which is either uniform in density throughout, or of homogeneous substance, rendered heterogeneous in density only through difference of pressure. But if the forces, though not conservative, be such that through every point of the space occupied by the fluid a surface can be drawn which shall cut at right angles all the lines of force it meets, a heterogeneous fluid will rest in equilibrium under their influence, provided (§ 750) its density, from point to point of every one of these orthogonal surfaces, varies inversely as the product of the resultant force into the thickness of the infinitely thin layer of space between that surface and another of the orthogonal surfaces infinitely near it on either side. (Compare § 488.)

Fluid under any system of forces.

The same conclusion is proved as a matter of course from (1) since that equation is merely the analytical expression that the force at every point (x, y, z) is along the normal to that surface of the series given by different values of C in $p=C$, which passes through (x, y, z); and that the magnitude of the resultant force is

$$\frac{\sqrt{\left(\frac{dp^2}{dx^2}+\frac{dp^2}{dy^2}+\frac{dp^2}{dz^2}\right)}}{\rho},$$

of which the numerator is equal to $\delta C/\tau$, if τ be the thickness at (x, y, z) of the shell of space between two surfaces $p=C$ and $p=C+\delta C$, infinitely near one another on two sides of (x, y, z).

(4) Functions of any number of variables invented by suggestion from (2).

The designation "many-valued function" which has hitherto been applied to such functions is not satisfactory, if only because it is also applicable to functions of roots of algebraic or transcendental equations.

The analytical expression of the condition which X, Y, Z must fulfil in order that (1) may be possible is found thus; Fluid under any system of forces.

since $$\frac{d}{dz}\frac{dp}{dy} = \frac{d}{dy}\frac{dp}{dz}, \text{ etc.,}$$

we have
$$\left.\begin{aligned} \frac{d}{dz}(\rho Y) &= \frac{d}{dy}(\rho Z) \\ \frac{d}{dx}(\rho Z) &= \frac{d}{dz}(\rho X) \\ \frac{d}{dy}(\rho X) &= \frac{d}{dx}(\rho Y) \end{aligned}\right\} \quad \ldots\ldots\ldots\ldots\ldots\ldots\ldots(5).$$

Performing the differentiations, and multiplying the first of the resulting equations by X, the second by Y, and the third by Z, we have

$$X\left(\frac{dZ}{dy} - \frac{dY}{dz}\right) + Y\left(\frac{dX}{dz} - \frac{dZ}{dx}\right) + Z\left(\frac{dY}{dx} - \frac{dX}{dy}\right) = 0 \ldots(6);$$

which is merely the well-known condition that $Xdx + Ydy + Zdz$ may be capable of being rendered by a factor the complete differential of a function of three independent variables.

Or if we multiply the first of (5) by $d\rho/dx$, the second by $d\rho/dy$, and the third by $d\rho/dz$, and add, we have

$$\frac{d\rho}{dx}\left(\frac{dZ}{dy} - \frac{dY}{dz}\right) + \frac{d\rho}{dy}\left(\frac{dX}{dz} - \frac{dZ}{dx}\right) + \frac{d\rho}{dz}\left(\frac{dY}{dx} - \frac{dX}{dy}\right) = 0 \ldots..(7).$$

This shows that the line whose direction-cosines are proportional to $$\frac{dZ}{dy} - \frac{dY}{dz},\ \frac{dX}{dz} - \frac{dZ}{dx},\ \frac{dY}{dx} - \frac{dX}{dy}$$

is perpendicular to the surface of equal density through (x, y, z); and (6) shows that the same line is perpendicular to the resultant force. It is therefore tangential both to the surface of equal density and to that of equal pressure, and therefore to their curve of intersection. The differential equations of this curve are therefore

$$\frac{dx}{\dfrac{dZ}{dy} - \dfrac{dY}{dz}} = \frac{dy}{\dfrac{dX}{dz} - \dfrac{dZ}{dx}} = \frac{dz}{\dfrac{dY}{dx} - \dfrac{dX}{dy}} \quad \ldots\ldots\ldots\ldots\ldots(8).$$

756. If we imagine all the fluid to become rigid except an infinitely thin closed tubular portion lying in a surface of equal density, and if the fluid in this tubular circuit be moved through any space along the tube and left at rest, it will remain in Equilibrium condition.

Equilibrium condition.

equilibrium in the new position, all positions of it in the tube being indifferent because of its homogeneousness. Hence the work (positive or negative) done by the force (X, Y, Z) on any portion of the fluid in any displacement along the tube is balanced by the work (negative or positive) done on the remainder of the fluid in the tube. Hence a single particle, acted on always by the resultant of X, Y, Z, and kept moving round the circuit, that is to say moving along any closed curve on a surface of equal density, has, at the end of one complete circuit, done just as much work against that resultant force in some parts of its course, as the resultant force does on it in the remainder of the circuit.

An interesting application of (j) § 190 may be made to prove this result analytically. Thus, if we take for α, β, γ our present force-components X, Y, Z; and for the surface there referred to, a surface of equal density in our heterogeneous fluid; the expression

$$\iint dS\left\{l\left(\frac{dZ}{dy}-\frac{dY}{dz}\right)+m\left(\frac{dX}{dz}-\frac{dZ}{dx}\right)+n\frac{dY}{dx}-\frac{dX}{dy}\right)\right\}$$

vanishes because of (7), and we conclude that

$$\int(Xdx+Ydy+Zdz)=0,$$

for any closed circuit on a surface of equal density.

Ideal example of equilibrium under non-conservative forces.

757. The following ideal example, and its realization in a subsequent section (§ 759), show a curiously interesting practical application of the theory of fluid equilibrium under extraordinary circumstances, generally regarded as a merely abstract analytical theory, practically useless and quite unnatural, "because forces in nature follow the conservative law."

758. Let the lines of force be circles, with their centres all in one line, and their planes perpendicular to it. They are cut at right angles by planes through this axis; and therefore a fluid may be in equilibrium under such a system of forces. The system will not be conservative if the intensity of the force be according to any other law than inverse proportionality to distance from this axial line; and the fluid, to be in equilibrium, must be heterogeneous, and be so distributed as to vary

in density from point to point of every plane through the axis, inversely as the product of the force into the distance from the axis. But from one such plane to another it may be either uniform in density, or may vary arbitrarily. To particularize farther, we may suppose the force to be in direct simple proportion to the distance from the axis. Then the fluid will be in equilibrium if its density varies from point to point of every plane through the axis, inversely as the square of that distance. If we still farther particularize by making the force uniform all round each circular line of force, the distribution of force becomes precisely that of the kinetic reactions of the parts of a rigid body against accelerated rotation. The fluid pressure will (§ 749) be equal over each plane through the axis. And in one such plane, which we may imagine carried round the axis in the direction of the force, the fluid pressure will increase in simple proportion to the angle at a rate per unit angle (§ 41) equal to the product of the density at unit distance into the force at unit distance. Hence it must be remarked, that if any closed line (or circuit) can be drawn round the axis, without leaving the fluid, there cannot be equilibrium without a firm partition cutting every such circuit, and maintaining the difference of pressures on the two sides of it, corresponding to the angle 2π. Thus, if the axis pass through the fluid in any part, there must be a partition extending from this part of the axis continuously to the outer bounding surface of the fluid. Or if the bounding surface of the whole fluid be annular (like a hollow anchor-ring, or of any irregular shape), in other words, if the fluid fills a tubular circuit; and the axis (A) pass through the aperture of the ring (without passing into the fluid); there must be a firm partition (CD) extending somewhere continuously across the channel, or passage, or tube, to stop the circulation of the fluid round it; otherwise there could not be equilibrium with the supposed forces in action. If we further suppose the density of the fluid to be uniform round each of the circular lines of force in the

Ideal example of equilibrium under non-conservative forces.

Ideal example of equilibrium under non-conservative forces.

system we have so far considered (so that the density shall be equal over every circular cylinder having the line of their centres for its axis, and shall vary from one such cylindrical surface to another, inversely as the squares of their radii), we may, without disturbing the equilibrium, impose any conservative system of force in lines perpendicular to the axis; that is (§ 488), any system of force in this direction, with intensity varying as some function of the distance. If this function be the simple distance, the superimposed system of force agrees precisely with the reactions against curvature, that is to say, the centrifugal forces, of the parts of a rotating rigid body.

Actual case.

759. Thus we arrive at the remarkable conclusion, that if a rigid closed box be completely filled with incompressible heterogeneous fluid, of density varying inversely as the square of the distance from a certain line, and if the box be moveable round this line as a fixed axis, and be urged in any way by forces applied to its outside, the fluid will remain in equilibrium relatively to the box; that is to say, will move round with the box as if the whole were one rigid body, and will come to rest with the box if the box be brought again to rest: provided always the preceding condition as to partitions be fulfilled if the axis pass through the fluid, or be surrounded by continuous lines of fluid. For, in starting from rest, *if* the fluid moves like a rigid solid, we have reactions against acceleration, tangential to the circles of motion, and equal in amount to $\dot{\omega}r$ per unit of mass of the fluid at distance r from the axis, $\dot{\omega}$ being the rate of acceleration (§ 42) of the angular velocity; and (§ 259) we have, in the direction perpendicular to the axis outwards, reaction against curvature of path, that is to say, "centrifugal force," equal to $\omega^2 r$ per unit of mass of the fluid. Hence the equilibrium which we have demonstrated in the preceding section, for the fluid supposed at rest, and arbitrarily influenced by two systems of force (the circular non-conservative and the radial conservative system) agreeing in law with these forces of kinetic reaction, proves for us now the D'Alembert (§ 264) equilibrium condition for the motion of the whole fluid as of a rigid body experiencing accelerated rotation; that is to say, shows that this kind of motion fulfils

Actual case of fluid equilibrium under non-conservative forces.

for the actual circumstances the laws of motion, and, therefore, that it is *the* motion actually taken by the fluid.

760. If the fluid is of homogeneous substance and uniform temperature throughout, but compressible, as all real fluids are, it can be heterogeneous in density, only because of difference of pressure in different parts; the surfaces of equal density must be also surfaces of equal pressure; and, as we have seen above (§ 753), there can be no equilibrium unless the system of forces be conservative. The function which the density is of the pressure must be supposed known (§ 448), as it depends on physical properties of the fluid. Compare § 752.

Relation between density and potential of applied forces.

Let $$\rho = f(p) \quad\text{.............................(9)}.$$

We have, by § 753 (3), integrated,

$$\int dp/f(p) = C - V \quad\text{.......................(10)},$$

or, if F denote such a function, that

$$F\{\int dp/f(p)\} = p \quad\text{.......................(11)},$$

$$p = F(C - V),$$

and, by (9), $$\rho = f\{F(C - V)\}\quad\text{.......................(12)}.$$

761. In § 746 we considered the resultant pressure on a plane surface, when the pressure is uniform. We may now consider briefly the resultant pressure on a plane area when the pressure varies from point to point, confining our attention to a case of great importance;—that in which gravity is the only applied force, and the fluid is a nearly incompressible liquid such as water. In this case the determination of the position of the Centre of Pressure is very simple; and the whole pressure is the same as if the plane area were turned about its centre of inertia into a horizontal position.

Resultant pressure on a plane area.

The pressure at any point at a depth z in the liquid may be expressed by $$p = \rho z + p_0$$ where ρ is the (constant) density of the liquid, and p_0 the (atmospheric) pressure at the free surface, reckoned in units of weight per unit of area.

Kinetic measure $p = g\rho z + p_0$.

Let the axis of x be taken as the intersection of the plane of the immersed plate with the free surface of the liquid, and that of y perpendicular to it and in the plane of the plate. Let

Resultant pressure on a plane area.

α be the inclination of the plate to the vertical. Let also A be the area of the portion of the plate considered, and $\bar{x}$, $\bar{y}$, the coordinates of its centre of inertia.

Then the whole pressure is

$$\iint p\,dxdy = \iint (p_0 + \rho y \cos\alpha)\,dxdy$$
$$= Ap_0 + A\rho\bar{y}\cos\alpha.$$

The moment of the pressure about the axis of x is

$$\iint py\,dxdy = Ap_0\bar{y} + Ak^2\rho\cos\alpha,$$

k being the radius of gyration of the plane area about the axis of x.

For the moment about y we have

$$\iint px\,dxdy = Ap_0\bar{x} + \rho\cos\alpha \iint xy\,dxdy.$$

The first terms of these three expressions merely give us again the results of § 746; we may therefore omit them. This will be equivalent to introducing a stratum of additional liquid above the free surface such as to produce an equivalent to the atmospheric pressure. If the origin be now shifted to the upper surface of this stratum we have

$$\text{Pressure} = A\rho\bar{y}\cos\alpha,$$
$$\text{Moment about } Ox = Ak^2\rho\cos\alpha,$$
$$\text{Distance of centre of pressure from axis of } x = \frac{k^2}{\bar{y}}.$$

But if k_1 be the radius of gyration of the plane area about a horizontal axis in its plane, and passing through its centre of inertia, we have, by § 283, $k^2 = k_1^2 + \bar{y}^2$.

Hence the distance, measured parallel to the axis of y, of the centre of pressure from the centre of inertia is $k_1^2/\bar{y}$; and, as we might expect, diminishes as the plane area is more and more submerged. If the plane area be turned about the line through its centre of inertia parallel to the axis of x, this distance varies as the cosine of its inclination to the vertical; supposing, of course, that by the rotation neither more nor less of the plane area is submerged.

Loss of apparent weight by immersion in a fluid.

762. A body, wholly or partially immersed in any fluid influenced by gravity, loses, through fluid pressure, in apparent weight an amount equal to the weight of the fluid displaced. For if the body were removed, and its place filled with fluid homogeneous with the surrounding fluid, there would be equi-

librium, even if this fluid be supposed to become rigid. And the resultant of the fluid pressure upon it is therefore a single force equal to its weight, and in the vertical line through its centre of gravity. But the fluid pressure on the originally immersed body was the same all over as on the solidified portion of fluid by which for a moment we have imagined it replaced, and therefore must have the same resultant. This proposition is of great use in Hydrometry, the determination of specific gravity, etc. etc.

Loss of apparent weight by immersion in a fluid.

Analytically, the following demonstration is of interest, especially in its analogies to some preceding theorems, and others which occur in electricity and magnetism.

If V be the potential of the impressed forces, $-dV/dx$ is the force parallel to the axis of x on unit of matter at xyz, and $\rho dxdydz$ is the mass of an element of the fluid, and therefore the whole force parallel to the axis of x on a mass of fluid substituted for the immersed body, is represented by the triple integral $-\iiint\rho\frac{dV}{dx}dxdydz$ taken through the whole space enclosed by the surface. But, by § 752,

$$\frac{dp}{dx}=-\rho\frac{dV}{dx}.$$

Hence the triple integral becomes

$$\iiint\frac{dp}{dx}dxdydz=\iint pdydz$$

extended over the whole surface.

Let dS be an element of any surface at x, y, z; λ, μ, ν the direction-cosines of the normal to the element; p the pressure in the fluid in contact with it. The whole resolved pressure parallel to the axis of x is

$$P_x=\iint\lambda pdS$$
$$=\iint pdydz,$$

the same expression as above.

The couple about the axis of z, due to the applied forces on any fluid mass, is (§ 559) $\Sigma dm(Xy-Yx)$, dm representing the mass of an element of fluid.

This may be written in the form

$$-\iiint\rho dxdydz\left(y\frac{dV}{dx}-x\frac{dV}{dy}\right),$$

the integral being taken throughout the mass.

Loss of apparent weight by immersion in a fluid.

This is evidently equal to

$$\iiint \left(y \frac{dp}{dx} - x \frac{dp}{dy} \right) dxdydz$$
$$= \iint pydydz - \iint pxdzdx$$
$$= \iint p (\lambda y - \mu x) dS,$$

which is the couple due to surface-pressure alone.

Lemma.

763. The following lemma, while in itself interesting, is of great use in enabling us to simplify the succeeding investigations regarding the stability of equilibrium of floating bodies:—

Let a homogeneous solid, the weight of unit of volume of which we suppose to be unity, be cut by a horizontal plane in $XYX'Y'$. Let O be the centre of inertia, and let XX', YY' be the principal axes, of this area.

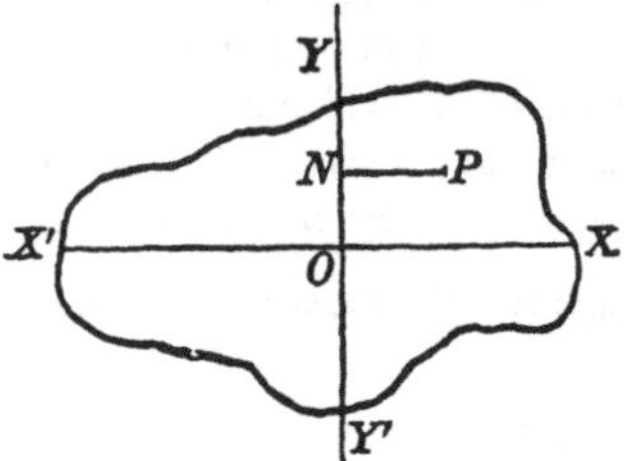

Let there be a second plane section of the solid, through YY', inclined to the first at an infinitely small angle, θ. Then (1) the volumes of the two wedges cut from the solid by these sections are equal; (2) their centres of inertia lie in one plane perpendicular to YY'; and (3) the moment of the weight of each of these, round YY', is equal to the moment of inertia about it of the corresponding portion of the area, multiplied by θ.

Take OX, OY as axes, and let θ be the angle of the wedge: the thickness of the wedge at any point $P(x, y)$ is θx, and the volume of a right prismatic portion whose base is the elementary area $dxdy$ at P is $\theta xdxdy$. Now let [] and () be employed to distinguish integrations extended over the portions of area to the right and left of the axis of y respectively, while integrals over the whole area have no such distinguishing mark. Let a and a' be these areas, v and v' the volumes of the wedges; $(\bar{x}, \bar{y})$, $(\bar{x}', \bar{y}')$ the co-ordinates of their centres of inertia. Then

$$v = \theta \left[\iint xdxdy \right] = a\bar{x}\theta$$
$$-v' = \theta \left(\iint xdxdy \right) = a'\bar{x}'\theta,$$

whence $v - v' = \theta \iint xdxdy = 0$ since O is the centre of inertia.

Hence $v = v'$, which is (1). Lemma.

Again, taking moments about XX',

$$v\bar{y} = \theta\,[\int\int xydxdy],$$

and
$$-v'\bar{y}' = \theta\,(\int\int xydxdy).$$

Hence
$$v\bar{y} - v'\bar{y}' = \theta\int\int xydxdy.$$

But for a principal axis (§ 281) $\Sigma xydm$ vanishes. Hence $v\bar{y} - v'\bar{y}' = 0$, whence, since $v = v'$, we have $\bar{y} = \bar{y}'$, which proves (2). And (3) is merely a statement in words of the obvious equation

$$[\int\int x.x\theta dxdy] = \theta\,[\int\int x^2dxdy].$$

764. If a positive amount of work is required to produce any possible infinitely small displacement of a body from a position of equilibrium, the equilibrium in this position is stable (§ 291). To apply this test to the case of a floating body, we may remark, first, that any possible infinitely small displacement may (§§ 26, 95) be conveniently regarded as compounded of two horizontal displacements in lines at right angles to one another, one vertical displacement, and three rotations round rectangular axes through any chosen point. If one of these axes be vertical, then three of the component displacements, viz. the two horizontal displacements and the rotation about the vertical axis, require no work (positive or negative), and therefore, so far as they are concerned, the equilibrium is essentially neutral. But so far as the other three modes of displacement are concerned, the equilibrium may be stable, or may be unstable, or may be neutral, according to the fulfilment of conditions which we now proceed to investigate. Stability of equilibrium of a floating body.

765. If, first, a simple vertical displacement, downwards let us suppose, be made, the work is done against an increasing resultant of upward fluid pressure, and is of course equal to the mean increase of this force multiplied by the whole space. If this space be denoted by z, the area of the plane of flotation by A, and the weight of unit bulk of the liquid by w, the increased bulk of immersion is clearly Az, and therefore the increase of the resultant of fluid pressure is wAz, and is in a line vertically upward through the centre of gravity of A. The mean force against which the work is done is therefore $\frac{1}{2}wAz$, as this is a case in which work is done against a force Vertical displacements.

Work done in vertical displacement. increasing from zero in simple proportion to the space. Hence the work done is $\frac{1}{2}wAz^2$. We see, therefore, that so far as vertical displacements alone are concerned, the equilibrium is necessarily stable, unless the body is wholly immersed, when the area of the plane of flotation vanishes, and the equilibrium is neutral.

Displacement by rotation about an axis in the plane of flotation. 766. The lemma of § 763 suggests that we should take, as the two horizontal axes of rotation, the principal axes of the plane of flotation. Considering then rotation through an infinitely small angle θ round one of these, let G and E be the

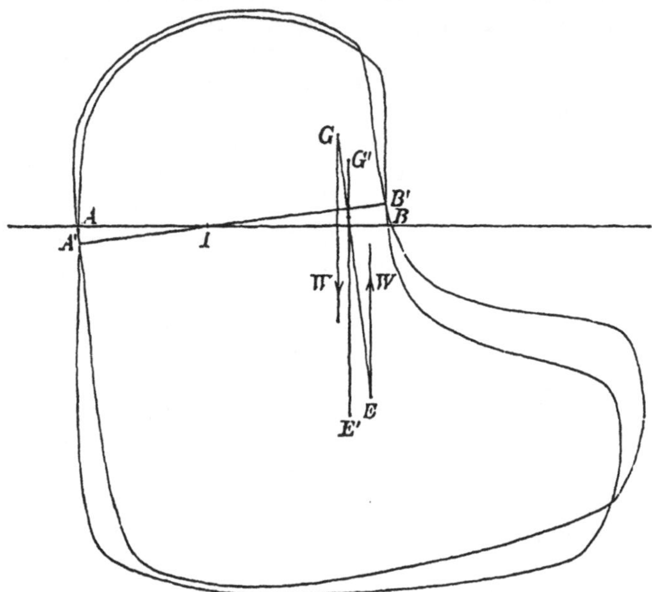

displaced centres of gravity of the solid, and of the portion of its volume which was immersed when it was floating in equilibrium, and G', E' the positions which they then had; all projected on the plane of the diagram which we suppose to be through I the centre of inertia of the plane of flotation. The resultant action of gravity on the displaced body is W, its weight, acting downwards through G; and that of the fluid pressure on it is W upwards through E corrected by the amount (upwards) due to the additional immersion of the wedge AIA', and the amount (downwards) due to the extruded wedge $B'IB$. Hence the whole action of gravity and fluid pressure on the displaced body is the couple of forces up and down in verticals through G and E, and the correction due to the wedges. This correction consists of a force vertically upwards through the centre of gravity of $A'IA$, and downwards through that of BIB'. These forces are equal [§ 763 (1)], and therefore constitute a couple which [§ 763 (2)] has the axis of the displacement for its axis, and which [§ 763 (3)] has its moment equal to θwk^2A, if A be the area of the plane of flotation, and k its radius of gyration (§ 281) round the principal axis in question. But since GE, which was vertical (as shown by $G'E'$) in the position of equilibrium, is inclined at the infinitely small angle θ to the vertical in the displaced body, the couple of forces W in the verticals through G and E has for moment $Wh\theta$, if h denote GE; and is in a plane perpendicular to the axis, and in the direction tending to increase the displacement, when G is above E. Hence the resultant action of gravity and fluid pressure on the displaced body is a couple whose moment is

Displacement by rotation about an axis in the plane of flotation.

$$(wAk^2 - Wh)\,\theta, \text{ or } w\,(Ak^2 - Vh)\,\theta,$$

if V be the volume immersed. It follows that when $Ak^2 > Vh$ the equilibrium is stable, so far as this displacement alone is concerned.

Also, since the couple worked against in producing the displacement increases from zero in simple proportion to the angle of displacement, its mean value is half the above; and therefore the whole amount of work done is equal to

Work done in this displacement.

$$\tfrac{1}{2}w\,(Ak^2 - Vh)\,\theta^2.$$

767. If now we consider a displacement compounded of a vertical (downwards) displacement z, and rotations through infinitely small angles θ, θ' round the two horizontal principal axes of the plane of flotation, we see (§§ 765, 766) that the work required to produce it is equal to

General displacement.

Work required.

$$\tfrac{1}{2}w\,[Az^2 + (Ak^2 - Vh)\,\theta^2 + (Ak'^2 - Vh)\,\theta'^2],$$

Conditions of stability. and we conclude that, for complete stability with reference to all possible displacements of this kind, it is necessary and sufficient that $h < \frac{Ak^2}{V}$, and $< \frac{Ak'^2}{V}$.

The metacentre. Condition of its existence. 768. When the displacement is about any axis through the centre of inertia of the plane of flotation, the resultant of fluid pressure is equal to the weight of the body; but it is only when the axis is a principal axis of the plane of flotation that this resultant is in the plane of displacement. In such a case the point of intersection of the resultant with the line originally vertical, and through the centre of gravity of the body, is called the *Metacentre*. And it is obvious, from the above investigation, that for either of these planes of displacement the condition of stable equilibrium is that the metacentre shall be *above* the centre of gravity.

769. The spheroidal analysis with which we propose to conclude this volume is proper, or practically successful, for hydrodynamic problems only when the deviations from spherical symmetry are infinitely small; or, practically, small enough to allow us to neglect the squares of ellipticities (§ 801); or, which is the same thing, to admit thoroughly the principle of the superposition of disturbing forces, and the deviations produced by them. But we shall first consider a case which admits of very simple synthetical solution, without any restriction to approximate sphericity; and for which the following remarkable theorem was discovered by Newton and Maclaurin:—

A homogeneous ellipsoid is a figure of equilibrium of a rotating liquid mass. 770. An oblate ellipsoid of revolution, of any given eccentricity, is a figure of equilibrium of a mass of homogeneous incompressible fluid, rotating about an axis with determinate angular velocity, and subject to no forces but those of gravitation among its parts.

The angular velocity for a given eccentricity is independent of the bulk of the fluid, and proportional to the square root of its density.

771. The proof of these propositions is easily obtained from

the results already deduced with respect to the attraction of an homogeneous ellipsoid and the properties of the free surface of a fluid as follows:— A homogeneous ellipsoid is a figure of equilibrium of a rotating liquid mass.

We know, from § 522, that if APB be a meridional section of a homogeneous oblate spheroid, OC the polar axis, OA an equatorial radius, and P any point on the surface, the attraction of the spheroid may be resolved into two components; one, Pp, perpendicular to the polar axis, and varying as the ordinate PM; the other, Ps, parallel to the polar axis, and varying as PN. These components are not equal when MP and PN are equal, else the resultant attraction at all points in the surface would pass through O; whereas we know that it is in some such direction as Pf, cutting the radius OA *between* O and A, but at a point nearer to O than n the foot of the normal at P.

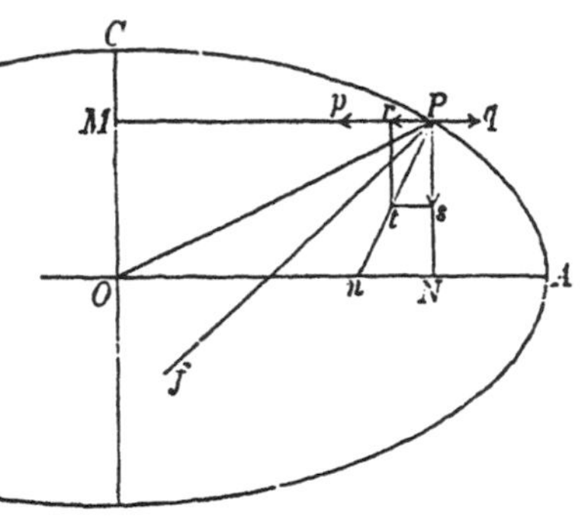

Let then $$Pp = \alpha \, . \, PM,$$

and $$Ps = \gamma \, . \, PN,$$

where α and γ are known constants, depending merely on the density, (ρ), and eccentricity (e), of the spheroid.

Also, we know by geometry that $Nn = (1 - e^2)\, ON$.

Hence; to find the magnitude of a force Pq perpendicular to the axis of the spheroid, which, when compounded with the attraction, will bring the resultant force into the normal Pn: make $pr = Pq$, and we must have

$$\frac{Pr}{Ps} = \frac{Nn}{PN} = (1 - e^2)\frac{ON}{PN} = (1 - e^2)\frac{\gamma \, . \, Pp}{\alpha \, . \, Ps}.$$

Hence $$Pr = (1 - e^2)\frac{\gamma}{\alpha} Pp$$

$$Pp - Pq = (1 - e^2)\frac{\gamma}{\alpha} Pp,$$

A homogeneous ellipsoid is a figure of equilibrium of a rotating liquid mass.

or $$Pq = \left\{1 - (1 - e^2)\frac{\gamma}{\alpha}\right\} Pp$$

$$= \{\alpha - (1 - e^2)\gamma\} PM.$$

Now if the spheroid were to rotate with angular velocity ω about OC, the centrifugal force (§§ 32, 35*a*, 259), would be in the direction Pq, and would amount to $\omega^2 PM$.

Hence, if we make $\omega^2 = \alpha - (1 - e^2)\gamma$......................(1);

the whole force on P, that is, the resultant of the attraction and centrifugal force, will be in the direction of the normal to the surface, which is the condition for the free surface of a mass of fluid in equilibrium.

$$\left.\begin{aligned} \text{Now, § 527 (31)*, } \gamma &= 4\pi\rho\frac{1+f^2}{f^3}(f - \tan^{-1}f) \\ \alpha &= 2\pi\rho\frac{1+f^2}{f^3}\left(\tan^{-1}f - \frac{f}{1+f^2}\right) \end{aligned}\right\} \text{....(2).}$$

Hence by (1) $$\omega^2 = \frac{2\pi\rho}{f^3}\{(3 + f^2)\tan^{-1}f - 3f\} \text{ ....(3).}$$

The square of a requisite angular velocity is as the density of the liquid.

This determines the angular velocity, and proves it to be proportional to $\sqrt{\rho}$.

When e, and therefore also f, is small, this formula is most easily calculated from

$$\frac{\omega^2}{2\pi\rho} = \tfrac{4}{15}f^2 - \tfrac{8}{35}f^4 + \text{etc.................(4),}$$

of which the first term is sufficient when we deal with spheroids so little oblate as the earth.

772. The following table has been calculated by means of these simplified formulæ. The last figure in each of the four last columns is given to the nearest unit. The two last columns will be explained in §§ 775, 776.

From this we see that the value of $\omega^2/2\pi\rho$ increases gradually from zero to a maximum as the eccentricity e rises from zero to

* Remark that the "e" of § 527 is not the eccentricity of the oblate spheroid which we now denote by e, and that with f as there and e as here we have $1 - e^2 = 1/(1 + f^2)$.

about 0 93, and then (more quickly) falls to zero as the eccen- Table of corresponding values of ellipticities and angular velocities.

i.	ii.	iii.	iv. see § 775.	v. see § 776.
eccentricity $e = f/\sqrt{(1+f^2)}$	$f = e/\sqrt{(1-e^2)}$.	$\frac{\omega^2}{2\pi\rho}$	Rotational period, in mean solar seconds, for case of density equal to Earth's mean density.	$\frac{\mu^2}{k} = \frac{(1+f^2)^{\frac{2}{3}}\omega^2}{2\pi\rho}$ where μ is moment of momentum, and k a constant*.
0	0	0	∞	0
0·093	·0934	·0028	86,164	·0023
·1	·1005	·0027	79.966	·0027
·2	·2041	·0107	39,897	·0110
·3	·3145	·0243	26,495	·0258
·4	·4365	·0436	19,780	·0490
·5	·5774	·0690	15,730	·0836
·6	·7502	·1007	13,023	·1356
·7	·9804	·1387	11,096	·2172
·8	1·3333	·1816	9,697	·3588
·8127	1·3946	·1868	9,561	·3838
·9	2·0648	·2203	8,804	·6665
·91	2·1949	·2225	8,759	·7198
·92	2·3474	·2241	8,729	·7813
·93	2·5304	·2247	8,718	·8533
·94	2·7556	·2239	8,732	·9393
·95	3·0423	·2213	8,788	1·045
·96	3·4282	·2160	8,891	1·179
·97	3·9904	·2063	9,098	1·350
·98	4·9261	·1890	9,504	1·627
·99	7·0175	·1551	10,490	2·113
1·00	∞	0·0000	∞	∞

tricity rises from 0·93 to unity. The values of the other quantities corresponding to this maximum are given in the table.

773. If the angular velocity exceed the value calculated from

$$\frac{\omega^2}{2\pi\rho} = 0·2247 \qquad\qquad (5),$$

when for ρ is substituted the density of the liquid, equilibrium is impossible in the form of an ellipsoid of revolution. If the angular velocity fall short of this limit there are always two ellipsoids of revolution which satisfy the conditions of equilibrium. In one of these the eccentricity is greater than 0·93, in the other less.

* Calculated from the mass and density, by the formula

$$k = \frac{3^{\frac{4}{3}}}{5^2}\left(\frac{2}{\pi\rho}\right)^{\frac{1}{3}} M^{\frac{10}{3}}.$$

Mean density of the earth expressed in attraction units.

774. It may be useful, for special applications, to indicate briefly how ρ is measured in these formulæ. In the definitions of §§ 459, 460, on which the attraction formulæ are based, unit mass is defined as exerting unit force on unit mass at unit distance; and unit volume-density is that of a body which has unit mass in unit volume. Hence, with the foot as our linear unit, we have for the earth's attraction on a particle of unit mass at its surface

$$\frac{\frac{4}{3}\pi\sigma R^3}{R^2} = \frac{4}{3}\pi\sigma R = 32 \cdot 2\,;$$

where R is the radius of the earth (supposed spherical) in feet; and σ its mean density, expressed in terms of the unit just defined.

Taking 20,900,000 feet as the value of R, we have

$$\sigma = 0 \cdot 000000368 = 3 \cdot 68 \times 10^{-7} \ldots\ldots\ldots\ldots\ldots (6).$$

As the mean density of the earth is about 5·5 times that of water, § 479, the density of water in terms of our present unit is

$$\frac{3 \cdot 68}{5 \cdot 5} 10^{-7} = 6 \cdot 7 \times 10^{-8}.$$

Time of rotation for spheroid of given eccentricity.

775. The fourth column of the table above gives the time of rotation in seconds, corresponding to each value of the eccentricity, ρ being assumed equal to the mean density of the earth. For a mass of water these numbers must be multiplied by $\sqrt{5 \cdot 5}$, as the time of rotation to give the same figure is inversely as the square root of the density.

For a homogeneous liquid mass, of the earth's mean density, rotating in $23^h\ 56^m\ 4^s$, we find $e = 0 \cdot 093$, which corresponds to an ellipticity of about $\frac{1}{230}$.

Mass and moment of momentum of fluid given.

776. An interesting form of this problem, also discussed by Laplace, is that in which the moment of momentum and the mass of the fluid are given, not the angular velocity; and it is required to find what is the eccentricity of the corresponding

ellipsoid of revolution, the result proving that there can be but one. **Mass and moment of momentum of fluid given.**

Calling M the mass, and μ the moment of momentum, we have

$$M = \tfrac{4}{3}\pi\rho c^3 (1+f^2) \quad\text{.........................(7)},$$

and

$$\mu = \tfrac{2}{5} M c^2 (1+f^2)\,\omega \quad\text{.......................(8)}.$$

These equations, with (3) determine c, f, and ω, for any given values of M and ρ. Eliminating c and ω from (8) by (7) and (3), we find

$$\mu^2 = \frac{3^{\frac{4}{3}}}{5^2}\left(\frac{2}{\pi\rho}\right)^{\frac{2}{3}} M^{\frac{10}{3}} (1+f^2)^{\frac{2}{3}} \left(\frac{3+f^2}{f^3}\tan^{-1} f - \frac{3}{f^2}\right) \quad\text{...... (9)}.$$

It is by this formula, that Col. V. of the table of § 772 has been calculated. The result shows that for any given value of μ, the moment of momentum, there is one and only one value of f.

777. It is evident that a mass of any ordinary liquid (not a *perfect fluid*, § 742), if left to itself in any state of motion, must preserve unchanged its moment of momentum (§ 235). But the viscosity, or internal friction (§ 742), will, if the mass remain continuous, ultimately destroy all relative motion among its parts; so that it will ultimately rotate as a rigid solid. We have seen (§ 776), that if the final form be an ellipsoid of revolution, there is a single definite value of its eccentricity. But, as it has not yet been discovered whether there is any other form consistent with *stable* equilibrium, we do not know that the mass will necessarily assume the form of this particular ellipsoid. Nor in fact do we know* whether even the ellipsoid of rotation may not become an *unstable* form if the moment of momentum exceed some limit depending on the mass of the fluid. We shall return to this subject in Vol. II., as it affords an excellent example of that difficult and delicate question *Kinetic Stability* (§ 346). [See § 778 below.]

* The present tense in this sentence relates to fifteen years ago. We now (Jan. 1882) know that the ellipsoid of revolution is unstable for moment of momentum exceeding some definite multiple of $M^{\frac{5}{3}}/\rho^{\frac{1}{3}}$; or, which comes to the same, the figure is unstable with eccentricity exceeding some definite amount.

Equilibrium ellipsoid of three unequal axes.

778. No one seems yet to have attempted to solve the general problem of finding all the forms of equilibrium which a mass of homogeneous incompressible fluid rotating with uniform angular velocity may assume. Unless the velocity be so small that the figure differs but little from a sphere (a case which will be carefully treated later), the problem presents difficulties of an exceedingly formidable nature. It is therefore of some importance to show by a synthetical process that besides the ellipsoid of revolution, there is an ellipsoid with three unequal axes, which is a figure of equilibrium when the moment of momentum is great enough. This curious theorem was discovered by Jacobi in 1834, and seems, simple as it is, to have been enunciated by him as a challenge to the French mathematicians*. The following proof was given by Archibald Smith in the second number of the *Cambridge Mathematical Journal*†.

The components of the attraction of a homogeneous ellipsoid, whose semi-axes are a, b, c, on a point (x, y, z) at its surface, found in § 526 above, may be written Ax, By, Cz, where

$$A=\tfrac{3}{2}M\int_0^\infty\frac{du}{(a^2+u)D},\quad B=\tfrac{3}{2}M\int_0^\infty\frac{du}{(b^2+u)D},\quad C=\tfrac{3}{2}M\int_0^\infty\frac{du}{(c^2+u)D}\;\ldots(1);$$

where $$D=(a^2+u)^{\frac{1}{2}}(b^2+u)^{\frac{1}{2}}(c^2+u)^{\frac{1}{2}}.$$

If the ellipsoid revolve, with angular velocity ω, about the axis of z, the components of the centrifugal force are ω^2x, ω^2y, 0. Hence the components of the whole resultant of gravity and centrifugal force on a particle at (x, y, z) are

$$(A-\omega^2)\,x,\ (B-\omega^2)\,y,\ Cz.$$

But the direction-cosines of the normal to the surface of the ellipsoid at (x, y, z), are proportional to

$$\frac{x}{a^2},\ \frac{y}{b^2},\ \frac{z}{c^2};$$

and, for equilibrium, the resultant force must be perpendicular to the free surface. Hence

$$a^2(A-\omega^2)=b^2(B-\omega^2)=c^2C.\ \ldots\ldots\ldots\ldots\ldots(2).$$

* See a Paper by Liouville, *Journal de l'École Polytechnique*, cahier XXIII. foot-note to p. 290.

† *Cambridge Math. Journal*, Feb. 1838.

These equations give

Equilibrium ellipsoid of three unequal axes.

$$a^2b^2(A-B)+(a^2-b^2)c^2C=0 \quad\quad (3),$$

and
$$\omega^2=\frac{a^2A-b^2B}{a^2-b^2} \quad\quad (4);$$

which, with A, B, C eliminated by (1), become

$$(a^2-b^2)\int_0^\infty\left\{\frac{c^2}{(c^2+u)}-\frac{a^2}{(a^2+u)}\frac{b^2}{(b^2+u)}\right\}\frac{du}{D}=0 \quad\quad (5),$$

and
$$\omega^2=\tfrac{3}{2}M\int_0^\infty\frac{udu}{(a^2+u)(b^2+u)D} \quad\quad (6).$$

The first factor of (5) equated to zero, gives $a=b$, and (6) gives the angular velocity for any assumed ratio of c to a: thus we fall back on the solution by an ellipsoid of revolution worked out in § 771 above.

Another solution is found by equating the second factor of (5) to zero. This equation which is equivalent to

$$\int_0^\infty\frac{udu}{D^3}\left(\frac{1}{a^2}+\frac{1}{b^2}-\frac{1}{c^2}+\frac{u}{a^2b^2}\right)=0 \quad\quad (7),$$

may be regarded as an equation to determine c^2 for any given values of a and b. It has obviously one and only one real positive root; which is proved by remarking, that while u increases from zero to infinity, u/D^3 decreases continually to zero, and the last factor under the integral sign continuously increases, only reaching a positive value for infinitely great values of u when c is zero, and being positive for all values of u when $1/c^2=$ or $<1/a^2+1/b^2$: and that, for any constant value of u, the last factor increases with increase of c^2. As every element of the integral is positive when $1/c^2=$ or $<1/a^2+1/b^2$ and as we may write this inequality as follows, $c^2=$ or $>b^2/(1+b^2/a^2)$, we see that if $c=$ or $<$ the less of a, or b, every element of the integral is positive, and we infer that the root c is less than the least of a or b.

778′. The solution of (7) for the case of $a=b$ is particularly interesting. It will be interpreted and turned to account in § 778″. It is the case, and obviously the only case, in which (5), regarded as an equation for determining any one of the quantities, a^2, b^2, c^2 in terms of the two others, has equal positive roots. In this case the integral forming the first member of (7) is

Equilibrium ellipsoid of three unequal axes.

reducible from the elliptic function required to express it when a is not equal to b, to a formula involving no other transcendent than an inverse circular function. The reduction is readily performed by aid of the notation of § 527 (22), where however a stood for what we now denote by c. It is to be noted also that the q of § 527 is now zero, because the point we are now considering is on the surface of the ellipsoid. The resulting transcendental equation equivalent to (7) may, if, as in § 527 (28), we put

$$f = \sqrt{\frac{a^2 - c^2}{c^2}} \qquad (8),$$

be written as follows,

$$\frac{\tan^{-1} f}{f} = \frac{1 + \frac{13}{3} f^2}{1 + \frac{14}{3} f^2 + f^4} \qquad (9).$$

When f is increased continuously from zero to infinity the left-hand member of this equation diminishes continuously from unity to zero: the right-hand member diminishes also from unity to zero, but diminishes at first less rapidly and afterwards more rapidly than the other. Thus there is one and only one root, which by trial and error we find to be

$$f = 1{\cdot}39457.$$

Some numerical particulars relating to this case are inserted in the Table of § 772, as amended for the present edition.

General problem of rotating liquid mass.

778″. During the fifteen years which have passed since the publication of our first edition we have never abandoned the problem of the equilibrium of a finite mass of rotating incompressible fluid. Year after year, questions of the multiplicity of possible figures of equilibrium have been almost incessantly before us, and yet it is only now, under the compulsion of finishing this second edition of the second part of our first volume, with hope for a second volume abandoned, that we have succeeded in finding anything approaching to full light on the subject.

Stability and instability of oblate spheroid of revolution.

(*a*) The oblate ellipsoid of revolution is proved by § 776 and by the table of § 772 to be stable, if the condition of being an ellipsoid of revolution be imposed. It is obviously not stable for very great eccentricities without this double condition of being both a figure of revolution and ellipsoidal.

(*b*) If the condition of being a figure of revolution is imposed, without the condition of being an ellipsoid, there is, for large enough moment of momentum, an annular figure of equilibrium which is stable, and an ellipsoidal figure which is unstable. It is probable, that for moment of momentum greater than one definite limit and less than another, there is just one *annular* figure of equilibrium, consisting of a *single ring*. **Annular figures: probably not stable;**

(*c*) For sufficiently large moment of momentum it is certain that the liquid may be in equilibrium in the shape of two, three, four or more separate rings, with its mass distributed among them in arbitrary portions, all rotating with one angular velocity, like parts of a rigid body. It does not seem probable that the kinetic equilibrium in any such case can be stable.

(*d*) The condition of being a figure of equilibrium being still imposed, the single-ring figure, when annular equilibrium is possible at all, is probably stable. It is certainly stable for very large values of the moment of momentum. **unless under constraint to remain symmetrical round an axis.**

(*e*) On the other hand let the condition of being ellipsoidal be imposed, but not the condition of being a figure of revolution. Whatever be the moment of momentum, there is one, and only one revolutional figure of equilibrium, as we have seen in § 776; we now add:

(1) The equilibrium in the revolutional figure is stable, or unstable, according as $f\left(=\frac{\sqrt{a^2-c^2}}{c}\right)$ is < or > 1·39457. **Instability of oblate spheroid and stability of Jacobian figure.**

(2) When the moment of momentum is less than that which makes $f = 1{\cdot}39457$ (or eccentricity $= {\cdot}81266$) for the revolutional figure, this figure is not only stable, but unique.

(3) When the moment of momentum is greater than that which makes $f = 1{\cdot}39457$ for the revolutional figure, there is, besides the unstable revolutional figure, the Jacobian figure (§ 778 above) with three unequal axes, *which is always stable if the condition of being ellipsoidal is imposed.* But, as will be seen in (f) below, the Jacobian figure, without the constraint to ellipsoidal figure, is in some cases certainly unstable, though it seems probable that in other cases it is stable without any constraint.

Unstable Jacobian figures.

(*f*) Looking back now to § 778 and choosing the case of a a great multiple of b, we see obviously that the excess of b above c must in this case be very small in comparison with c. Thus we have a very slender ellipsoid, long in the direction of a, and approximately a prolate figure of revolution relatively to this long a-axis, which, revolving with proper angular velocity round its shortest axis c, is a figure of equilibrium. The motion so constituted, which, without any constraint is, in virtue of § 778 a configuration of minimum energy or of maximum energy, for given moment of momentum, is a configuration of *minimum* energy for given moment of momentum, *subject to the condition that the shape is constrainedly an ellipsoid.* From this proposition, which is easily verified, in the light of § 778, it follows that, with the ellipsoidal constraint, the equilibrium is stable. The revolutional ellipsoid of equilibrium, with the same moment of momentum, is a very flat oblate spheroid; for it the energy is a minimax, because clearly it is the smallest energy that a revolutional ellipsoid with the same moment of momentum can have, but it is greater than the energy of the Jacobian figure with the same moment of momentum.

(*g*) If the condition of being ellipsoidal is removed and the liquid left perfectly free, it is clear that the slender Jacobian ellipsoid of (*f*) is not stable, because a deviation from ellipsoidal figure in the way of thinning it in the middle and thickening it towards its ends, would with the same moment of momentum give less energy. With so great a moment of momentum as to give an exceedingly slender Jacobian ellipsoid, it is clear that another possible figure of equilibrium is, two detached approximately spherical masses, rotating (as if parts of a solid) round an axis through their centre of inertia, and that this figure is stable. It is also clear that there may be an infinite number of such stable figures, with different proportions of the liquid in the two detached masses. With the same moment of momentum there are also configurations of equilibrium with the liquid in divers proportions in more than two detached approximately spherical masses.

Configuration of two detached rotating masses stable.

(*h*) No configuration in more than two detached masses,

has secular stability according to the definition of (k) below, and it is doubtful whether any of them, even if undisturbed by viscous influences, could have true kinetic stability: at all events, unless approaching to the case of the three material points proved stable by Gascheau (see Routh's "Rigid Dynamics," § 475, p. 381).

Configuration of two detached rotating masses stable.

(i) The transition from the stable kinetic equilibrium of a liquid mass in two equal or unequal portions, so far asunder that each is approximately spherical, but disturbed to slightly prolate figures (found by the well-known investigation of equilibrium tides, given in § 804 below), and to the more and more prolate figures which would result from subtraction of energy without change of moment of momentum, carried so far that the prolate figures, now not even approximately elliptic, cease to be stable, is peculiarly interesting. We have a most interesting gap between the unstable Jacobian ellipsoid when too slender for stability, and the case of smallest moment of momentum consistent with stability in two equal detached portions. The consideration of how to fill up this gap with intermediate figures, is a most attractive question, towards answering which we at present offer no contribution.

(j) When the energy with given moment of momentum is either a minimum or a maximum, the kinetic equilibrium is clearly stable, if the liquid is perfectly inviscid. It seems probable that it is essentially unstable, when the energy is a minimax; but we do not know that this proposition has been ever proved.

(k) If there be any viscosity, however slight, in the liquid, or if there be any imperfectly elastic solid, however small, floating on it or sunk within it, the equilibrium in any case of energy either a minimax or a maximum cannot be secularly stable: and the only secularly stable configurations are those in which the energy is a minimum with given moment of momentum. It is not known for certain whether with given moment of momentum there can be more than one secularly stable configuration of equilibrium of a viscous fluid, in one continuous mass, but it seems to us probable that there is only one.

Digression on spherical harmonics.

779. A few words of explanation, and some graphic illustrations, of the character of spherical surface harmonics may promote the clear understanding not only of the potential and hydrostatic applications of Laplace's analysis, which will occupy us presently, but of much more important applications to be made in Vol. II., when waves and vibrations in spherical fluid or elastic solid masses will be treated. To avoid circumlocutions, we shall designate by the term *harmonic spheroid*, or *spherical harmonic undulation*, a surface whose radius to any point differs from that of a sphere by an infinitely small length varying as the value of a surface harmonic function of the position of this point on the spherical surface. The definitions of spherical solid and surface harmonics [App. B. (*a*), (*b*), (*c*)] show that the harmonic spheroid of the second order is a surface of the second degree subject only to the condition of being approximately spherical: that is to say, it may be any elliptic spheroid (or ellipsoid with approximately equal axes). Generally a harmonic spheroid of any order i exceeding 2 is a surface of algebraic degree i, subject to further restrictions than that of merely being approximately spherical.

Harmonic spheroid.

Let S_i be a surface harmonic of the order i with the coefficient of the leading term so chosen as to make the greatest maximum value of the function unity. Then if a be the radius of the mean sphere, and c the greatest deviation from it, the polar equation of a harmonic spheroid of order i will be

$$r = a + cS_i \quad\text{.............................(1)}$$

if S_i is regarded as a function of polar angular co-ordinates, θ, ϕ. Considering that c/a is infinitely small, we may reduce this to an equation in rectangular co-ordinates of degree i, thus:—Squaring each member of (1); and putting cr^i/a^{i+1} for c/a, from which it differs by an infinitely small quantity of the second order, we have

$$r^2 = a^2 + \frac{2c}{a^{i-1}}(r^i S_i) \quad\text{.........................(2).}$$

This, reduced to rectangular co-ordinates, is of algebraic degree i.

Harmonic nodal cone and line.

780. The line of no deviation from the mean spherical surface is called the *nodal line*, or the *nodes* of the harmonic spheroid. It is the line in which the spherical surface is cut

by the *harmonic nodal cone;* a certain cone with vertex at the centre of the sphere, and of algebraic degree equal to the order of the harmonic. An important property of the harmonic nodal line, indicated by an interesting hydrodynamic theorem due to Rankine*, is that when self-cutting at any point or points, the different branches make equal angles with one another round each point of section. Digression on spherical harmonics. Harmonic nodal cone and line.

Denoting r^iS_i of § 779 by V_i, we have

$$V_i = 0 \text{ ....(3)}$$

for the equation of the harmonic nodal cone. As V_i is [App. B. (a)] a homogeneous function of degree i, we may write Theorem regarding nodal cone.

$$V_i = H_0 z^i + H_1 z^{i-1} + H_2 z^{i-2} + H_3 z^{i-3} + \text{etc. ....(4)},$$

where H_0 is a constant, and H_1, H_2, H_3, etc., denote integral homogeneous functions of x, y of degrees 1, 2, 3, etc.; and then the condition $\nabla^2 V_i = 0$ [App. B. (a)] gives

$$\left.\begin{array}{c}\nabla^2 H_2 + i(i-1)H_0 = 0,\ \nabla^2 H_3 + (i-1)(i-2)H_1 = 0\\ \nabla^2 H_s + (i-s+2)(i-s+1)H_{s-2} = 0\end{array}\right\}\text{...(5)},$$

which express all the conditions binding on H_0, H_1, H_2, etc.

Now suppose the nodal cone to be autotomic, and, for brevity and simplicity, take OZ along a line of intersection. Then $z = a$ makes (3) the equation in x, y, of a curve lying in the tangent plane to the spherical surface at a double or multiple point of the nodal line, and touching both or all its branches in this point. The condition that the curve in the tangent plane may have a double or multiple point at the origin of its co-ordinates is, when (4) is put for V_i,

$$H_0 = 0;\ \text{and, for all values of } x,\ y,\ H_1 = 0.$$

Hence (5) gives $\nabla^2 H_2 = 0$,

so that, if $H_2 = Ax^2 + By^2 + 2Cxy$,

we have $A + B = 0$. This shows that the two branches cut one another at right angles.

If the origin be a triple, or n-multiple point, we must have

$$H_0 = 0,\ H_1 = 0, \ldots H_{n-1} = 0,$$

and (5) gives $\nabla^2 H_n = 0$.

* "Summary of the Properties of certain Stream-Lines." *Phil. Mag.*, Oct. 1864.

Digression on spherical harmonics.
Theorem regarding nodal cone.

Hence [§ 707 (23), writing v for $\sqrt{-1}$],

$$H_n = A\{(x+yv)^n + (x-yv)^n\} + Bv\{(x+yv)^n - (x-yv)^n\},$$

or, if $x = \rho\cos\phi$, $y = \rho\sin\phi$,

$$H_n = 2\rho^n (A\cos n\phi + B\sin n\phi),$$

which shows that the n branches cut one another at equal angles round the origin.

Cases of solid harmonics resolvable into factors.
Polar harmonics.

781. The harmonic nodal cone may, in a great variety of cases [V_i resolvable into factors], be composed of others of lower degrees. Thus (the only class of cases yet worked out) each of the $2i+1$ elementary polar harmonics [as we may conveniently call those expressed by (36) or (37) of App. B, with any one alone of the $2i+1$ coefficients A_s, B_s] has for its nodes circles of the spherical surface. These circles, for each such harmonic element, are either (1) all in parallel planes (as circles of latitude on a globe), and cut the spherical surface into zones, in which case the harmonic is called zonal; or (2) they are all in planes through one diameter (as meridians on a globe), and cut the surface into equal sectors, in which case the harmonic is called sectorial; or (3) some of them are in parallel planes, and the others in planes through the diameter perpendicular to those planes, so that they divide the surface into rectangular quadrilaterals, and (next the poles) triangular segments, as areas on a globe bounded by parallels of latitude, and meridians at equal successive differences of longitude.

Zonal and sectorial harmonics defined.

With a given diameter as axis of symmetry there are, for complete harmonics [App. B. (c), (d)], just one zonal harmonic of each order and two sectorial. The zonal harmonic is a function of latitude alone ($\frac{1}{2}\pi - \theta$, according to the notation of App. B.); being the $\Theta_i^{(0)}$ given by putting $s=0$ in App. B. (38). The sectorial harmonics of order i, being given by the same with $s=i$, are

$$\sin^i\theta\cos i\phi, \text{ and } \sin^i\theta\sin i\phi \quad\ldots\ldots\ldots\ldots\ldots\ldots(1).$$

The general polar harmonic element of order i, being the $\Theta_i^{(s)}\cos s\phi$ and $\Theta_i^{(s)}\sin s\phi$ of B. (38), with any value of s from 0 to i, has for its nodes $i-s$ circles in parallel planes, and s great circles intersecting one another at equal angles round their poles; and the variation from maximum to minimum along the equator, or any parallel circle, is according to the simple harmonic law. It is easily proved (as the mathematical student may find for himself) that the law of variation is *approximately* simple harmonic along lengths of each meridian cutting but a small number of the nodal circles of latitude, and not too near either pole, for any polar harmonic element of high order having a large number of such nodes (that is, any one for which $i-s$ is a large number). The law of variation along a meridian in the neighbourhood of either pole, for polar harmonic elements of high orders, will be carefully examined and illustrated in Vol. II., when we shall be occupied with vibrations and waves of water in a circular vessel, and of a circular stretched membrane.

Digression on spherical harmonics.

Tesseral division of surface by nodes of a polar harmonic.

782. The following simple and beautiful investigation of the zonal harmonic due to Murphy* may be acceptable to the analytical student; but (§ 453) we give it as leading to a useful formula, with expansions deduced from it, differing from any of those investigated above in App. B:—

"PROP. I.

"To find a rational and entire function of given dimensions "with respect to any variable, such that when multiplied by "*any* rational and entire function of lower dimensions, the "integral of the product taken between the limits 0 and 1 "shall always vanish.

Murphy's analytical invention of the zonal harmonics.

"Let $f(t)$ be the required function of n dimensions with respect "to the variable t; then the proposed condition will evidently re-"quire the following equations to be separately true; namely,
"(a) ...... $\int f(t)dt=0,\ \int f(t)tdt=0,\ \int f(t)t^2dt=0, \ldots\ldots \int f(t)t^{n-1}dt=0,$
"each integral being taken between the given limits.

"Let the indefinite integral of $f(t)$, commencing when $t=0$, be "represented by $f_1(t)$; the indefinite integral of $f_1(t)$, commencing "also when $t=0$, by $f_2(t)$; and so on, until we arrive at the "function $f_n(t)$, which is evidently of $2n$ dimensions. Then the "method of integrating by parts will give, generally,

$$\text{“}\int f(t)\,t^x dt = t^x f_1(t) - xt^{x-1}f_2(t) + x(x-1)t^{x-2}f_3(t) - \text{etc.}$$

* *Treatise on Electricity.* Cambridge, 1833.

Digression on spherical harmonics. Murphy's analytical invention of the above harmonics.

"Let us now put $t=1$, and substitute for x the values 1, 2, 3, "......$(i-1)$ successively; then in virtue of the equations (a), "we get,

"(b)........$f_1(t)=0,\ f_2(t)=0,\ f_3(t)=0,\ldots\ldots\ldots f_i(t)=0.$

"Hence, the function $f_i(t)$ and its $(i-1)$ successive differential "coefficients vanish, both when $t=0$, and when $t=1$; therefore "t^i and $(1-t)^i$ are each factors of $f_i(t)$; and since this function is "of $2i$ dimensions, it admits of no other factor but a constant c.

"Putting $1-t=t'$, we thus obtain

$$f_i(t)=c\,(tt')^i;$$

"and therefore $$f(t)=c\,\frac{d^i}{dt^i}(tt')^i.$$

"*Corollary.*—If we suppose the first term of $f(t)$, when arranged "according to the powers of t, to be unity, we evidently have "$c=\dfrac{1}{1.2.3\ldots\ldots i}$; on this supposition we shall denote the above "quantity by Q_i.

"Prop. II.

Murphy's analysis.

"The function Q_i which has been investigated in the pre-"ceding proposition, is the same as the coefficient of e^i in the "expansion of the quantity

$$\{1-2e(1-2t)+e^2\}^{-\frac{1}{2}}.$$

"Let u be a quantity which satisfies the equation

$$(c)\ldots\ldots\ldots\ldots u=t+e\,u\,(1-u);$$

"that is, $$u=-\frac{1-e}{2e}+\frac{1}{2e}\{1-2e(1-2t)+e^2\}^{\frac{1}{2}};$$

"therefore $$\frac{du}{dt}=\{1-2e(1-2t)+e^2\}^{-\frac{1}{2}}.$$

"But if, as before, we write t' for $1-t$, we have, by Lagrange's "theorem, applied to the equation (c),

$$u=t+e\,tt'+\frac{e^2}{1.2}\frac{d}{dt}(tt')^2+\frac{e^3}{1.2.3}\frac{d^2}{dt^2}(tt')^3+\text{etc.}$$

"If we differentiate, and put for $\frac{d^i}{dt^i}(tt')^i$ its value $1.2.3\ldots iQ_i$ given "by the former proposition, we get

Digression on spherical harmonics.

Murphy's analysis.

$$\frac{du}{dt} = 1 + Q_1 e + Q_2 e^2 + Q_3 e^3 + \text{etc.}$$

"Comparing this with the above value of $\frac{du}{dt}$ the proposition is "manifest.

"Prop. V.

"To develope the function Q_i.

Expansions of zonal harmonics.

"*First Expansion.*—By Prop. I., we have

$$Q_i = \frac{1}{1.2.3\ldots i}\frac{d^i}{dt^i}(tt')^i.$$

"Hence $Q_i = \frac{1}{1.2.3\ldots i}\frac{d^i}{dt^i}\left\{t^i - it^{i+1} + \frac{i(i-1)}{1.2}t^{i+2} - \text{etc.}\right\}$

$$\text{“}(e)\ldots\ldots = 1 - \frac{i}{1}\frac{(i+1)}{1}t + \frac{i(i-1)}{1.2}\frac{(i+1)(i+2)}{1.2}t^2 - \text{etc.}$$

"*Second Expansion.*—If u and v are functions of any variable t, "then the theorem of Leibnitz gives the identity

$$\frac{d^i}{dt^i}(uv) = v\frac{d^i u}{dt^i} + i\frac{dv}{dt}\frac{d^{i-1}u}{dt^{i-1}} + \frac{i(i-1)}{1.2}\frac{d^2 v}{dt^2}\frac{d^{i-2}v}{dt^{i-2}} + \text{etc.}$$

"Put $u = t^i$ and $v = t'^i$, and dividing by $1.2.3\ldots i$, we have

Formulæ for zonal,

$$\text{“}(f)\ldots\ldots Q_i = t'^i - \left(\frac{i}{1}\right)^2 t'^{i-1}t + \left\{\frac{i(i-1)}{1.2}\right\}^2 t'^{i-2}t^2$$
$$- \left\{\frac{i(i-1)(i-2)}{1.2.3}\right\}^2 t'^{i-3}t^3 + \text{etc.}$$

"*Third Expansion.*—Put $1 - 2t = \mu$, and therefore $tt' = \frac{1-\mu^2}{2^2}$,

"hence $Q_i = \frac{1}{2^i}\frac{1}{1.2.3\ldots i}\frac{d^i}{d\mu^i}(\mu^2 - 1)^i$

$$= \frac{1}{2.4.6\ldots 2i}\frac{d^i}{d\mu^i}\left\{\mu^{2i} - i\mu^{2i-2} + \frac{i(i-1)}{1.2}\mu^{2i-4} - \text{etc.}\right\}$$

$$\text{“}(g)\ldots\ldots\ldots = \frac{1.3.5\ldots(2i-1)}{1.2.3\ldots i}\left\{\mu^i - \frac{i(i-1)}{2(2i-1)}\mu^{i-2}\right.$$
$$\left. + \frac{i(i-1)(i-2)(i-3)}{2.4.(2i-1)(2i-3)}\mu^{i-4} - \text{etc.}\right\}.\text{”}$$

Digression on spherical harmonics.

Formulæ for zonal,

The t, t' and μ of Murphy's notation are related to the θ we have used, thus:—

$$\left.\begin{array}{c} t=(2\sin\tfrac{1}{2}\theta)^2, \quad t'=(2\cos\tfrac{1}{2}\theta)^2 \\ \mu=\cos\theta \end{array}\right\}\ldots\ldots\ldots\ldots\ldots(2).$$

Also it is convenient to recall from App. B. (v'), (38), (40), and (42), that the value of Q_i [or $\vartheta_i^{(0)}$ of App. B. (61)], when $\theta=0$ is unity, and that it is related to the $\Theta_i^{(s)}$, of our notation for polar harmonic elements, thus:—

$$\vartheta_i^{(0)}=Q_i=\frac{1.3.5\ldots(2i-1)}{1.2.3\ldots i}\Theta_i^{(0)}\ldots\ldots\ldots\ldots(3),$$

and tesseral harmonics.

as is proved also by comparing (g) with App. B. (38). We add the following formula, manifest from (38), which shows a derivation of $\Theta_i^{(s)}$ from $\Theta_i^{(0)}$, valuable if only as proving that the $i-s$ roots of $\Theta_i^{(s)}=0$ are all real and unequal, inasmuch as App. B. (p) proves that the i roots of $\Theta_i^{(0)}=0$ are all real and unequal:—

$$\frac{\Theta_i^{(s)}}{\sin\theta}=\frac{1}{i-s+1}\frac{d}{d\mu}\left[\frac{\Theta_i^{(s-1)}}{\sin^{s-1}\theta}\right]\ldots\ldots\ldots\ldots\ldots\ldots(4).$$

From this and (3) we find

$$\Theta_i^{(s)}=\frac{1.2.3\ldots(i-s)}{1.3.5\ldots(2i-1)}\sin^s\theta\frac{d^sQ_i}{d\mu^s}\ldots\ldots\ldots\ldots\ldots(5).$$

And lastly, referring to App. B. (w); let

$$Q'_i \text{ and } Q_i\left[\cos\theta\cos\theta'+\sin\theta\sin\theta'\cos(\phi-\phi')\right]$$

denote respectively what Q_i becomes when $\cos\theta$ is replaced by $\cos\theta'$, and again by $\cos\theta\cos\theta'+\sin\theta\sin\theta'\cos(\phi-\phi')$: and let μ denote $\cos\theta$; and μ', $\cos\theta'$. By what precedes, we may put (61) of App. B into the following much more convenient form, agreeing with that given by Murphy (*Electricity*, p. 24):—

Biaxal harmonic expanded.

$$Q_i\left[\cos\theta\cos\theta'+\sin\theta\sin\theta'\cos(\phi-\phi')\right]$$

$$=Q_iQ_i'+2\left\{\frac{\cos(\phi-\phi')}{i(i+1)}\sin\theta\sin\theta'\frac{dQ_i}{d\mu}\frac{dQ_i'}{d\mu}+\frac{\cos2(\phi-\phi')}{(i-1)i(i+1)(i+2)}\sin^2\theta\sin^2\theta'\frac{d^2Q_i}{d\mu^2}\frac{d^2Q_i'}{d\mu^2}+\text{etc.}\right\}\quad(6).$$

783. Elementary polar harmonics become, in an extreme case of spherical harmonic analysis, the proper harmonics for the treatment, by either polar or rectilinear rectangular co-ordinates, of problems in which we have a plane, or two parallel planes, instead of a spherical surface, or two concentric spherical surfaces, thus:—

Physical problems relative to plane rectangular and circular plates.

First, let S_i be any surface harmonic of order i, and V_i and V_{-i-1} the solid harmonics [App. B. (*b*)] equal to it on the spherical surface of radius a: so that

$$V_i = \left(\frac{r}{a}\right)^i S_i, \text{ and } V_{-i-1} = \left(\frac{a}{r}\right)^{i+1} S_i.$$

Now [compare § 655]

$$\left(\frac{r}{a}\right)^i = \epsilon^{i\log(r/a)};$$

and, therefore, if a be infinite, and $r-a$ a finite quantity denoted by x, which makes $\log(r/a) = x/a$, and if i be infinite, and $a/i = p$, we have

$$\left(\frac{r}{a}\right)^i = \epsilon^{ix/a} = \epsilon^{x/p}, \text{ and similarly } \left(\frac{a}{r}\right)^{i+1} = \epsilon^{-(i+1)x/a} = \epsilon^{-x/p};$$

the solid harmonics then become

$$\epsilon^{x/p} S_i \text{ and } \epsilon^{-x/p} S_i.$$

Supposing now S_i to be a polar harmonic element, and considering, as Green did in his celebrated Essay on Electricity, an area sensibly plane round either pole, or considering any sensibly plane portion far removed from each pole, it is interesting and instructive to examine how the formulæ [App. B. (36)...(40), (61), (65); and § 782, (*e*), (*f*), (*g*)] wear down to the proper plane polar or rectangular formulæ. This we may safely leave to the analytical student. In Vol. II. the plane polar solution will be fully examined. At present we merely remark that, in rectangular surface co-ordinates (y, z) in the spherical surface reduced to a plane, S_i may be any function whatever fulfilling the equation

$$\frac{d^2 S_i}{dy^2} + \frac{d^2 S_i}{dz^2} + \frac{S_i}{p^2} = 0,$$

and that the rectangular solution into which the elementary polar spherical harmonic wears down, for sensibly plane portions of the spherical surface far removed from the poles, is

$$S_i = \cos\frac{y}{q}\cos\frac{z}{q'}$$

where q and q' are two constants such that $q^2 + q'^2 = p^2$.

Examples of polar harmonics.

784. The following tables and graphic representations of all the polar harmonic elements of the 6th and 7th orders may be useful in promoting an intelligent comprehension of the subject.

Sixth order: Zonal,

$$Q_6 = \tfrac{1}{16}(231\mu^6 - 315\mu^4 + 105\mu^2 - 5) \qquad = \tfrac{231}{16}\,\Theta_6^{(0)}.$$

$$\tfrac{1}{21}\cdot\frac{dQ_6}{d\mu} = \tfrac{1}{8}(33\mu^4 - 30\mu^2 + 5)\mu \qquad = \tfrac{33}{8}\,\Theta_6^{(1)}(1-\mu^2)^{-\frac{1}{2}}.$$

Tesseral,

$$\tfrac{1}{210}\cdot\frac{d^2Q_6}{d\mu^2} = \tfrac{1}{16}(33\mu^4 - 18\mu^2 + 1) \qquad = \tfrac{33}{16}\,\Theta_6^{(2)}(1-\mu^2)^{-1}.$$

$$\tfrac{1}{1260}\cdot\frac{d^3Q_6}{d\mu^3} = \tfrac{1}{8}(11\mu^2 - 3)\mu \qquad = \tfrac{11}{8}\,\Theta_6^{(3)}(1-\mu^2)^{-\frac{3}{2}}.$$

$$\tfrac{1}{4725}\cdot\frac{d^4Q_6}{d\mu^4} = \tfrac{1}{10}(11\mu^2 - 1) \qquad = \tfrac{11}{10}\,\Theta_6^{(4)}(1-\mu^2)^{-2}.$$

$$\tfrac{1}{10395}\cdot\frac{d^5Q_6}{d\mu^5} = \mu \qquad = \Theta_6^{(5)}(1-\mu^2)^{-\frac{5}{2}}.$$

Sectorial.

$$\tfrac{1}{10395}\cdot\frac{d^6Q_6}{d\mu^6} = 1 \qquad = \Theta_6^{(6)}(1-\mu^2)^{-3} \text{ not shown.}$$

Seventh order:

$$Q_7 = \tfrac{1}{16}(429\mu^6 - 693\mu^4 + 315\mu^2 - 35)\mu = \tfrac{429}{16}\,\Theta_7^{(0)}.$$

Zonal, Tesseral,

$$\tfrac{1}{28}\cdot\frac{dQ_7}{d\mu} = \tfrac{1}{64}(429\mu^6 - 495\mu^4 + 135\mu^2 - 5) \qquad = \tfrac{429}{64}\,\Theta_7^{(1)}(1-\mu^2)^{-\frac{1}{2}}.$$

$$\tfrac{1}{378}\cdot\frac{d^2Q_7}{d\mu^2} = \tfrac{1}{48}(143\mu^4 - 110\mu^2 + 15)\mu \qquad = \tfrac{143}{48}\,\Theta_7^{(2)}(1-\mu^2)^{-1}.$$

$$\tfrac{1}{3150}\cdot\frac{d^3Q_7}{d\mu^3} = \tfrac{1}{80}(143\mu^4 - 66\mu^2 + 3) \qquad = \tfrac{143}{80}\,\Theta_7^{(3)}(1-\mu^2)^{-\frac{3}{2}}.$$

$$\tfrac{1}{17325}\cdot\frac{d^4Q_7}{d\mu^4} = \tfrac{1}{10}(13\mu^2 - 3)\mu \qquad = \tfrac{13}{10}\,\Theta_7^{(4)}(1-\mu^2)^{-2}.$$

$$\tfrac{1}{62370}\cdot\frac{d^5Q_7}{d\mu^5} = \tfrac{1}{12}(13\mu^2 - 1) \qquad = \tfrac{13}{12}\,\Theta_7^{(5)}(1-\mu^2)^{-\frac{5}{2}}.$$

$$\tfrac{1}{135135}\cdot\frac{d^6Q_7}{d\mu^6} = \mu \qquad = \Theta_7^{(6)}(1-\mu^2)^{-3}.$$

Sectorial.

$$\tfrac{1}{135135}\cdot\frac{d^7Q_7}{d\mu^7} = 1 \qquad = \Theta_7^{(7)}(1-\mu^2)^{-\frac{7}{2}} \text{ not shown.}$$

Polar harmonics of sixth order.

μ.	Q_6.	$\frac{1}{21}\frac{dQ_6}{d\mu}$.	$\frac{33}{8}\Theta_6^{(1)}$.	$\frac{1}{210}\frac{d^2Q_6}{d\mu^2}$.	$\frac{33}{16}\Theta_6^{(2)}$.
·0	− ·3125	·0000	·0000	+ ·0625	+ ·0625
·01	− ·3118	......	......	......	......
·05	− ·2961	+ ·0308	+ ·0307	+ ·0597	+ ·0595
·08	− ·2738	......	......	......	......
·10	− ·2488	+ ·0588	+ ·0585	+ ·0515	+ ·0510
·13	− ·2072	......	......	......	......
·15	− ·1746	+ ·0814	+ ·0805	+ ·0382	+ ·0373
·17	− ·1390	......	......	......	......
·2	− ·0806	+ ·0963	+ ·0944	+ ·0208	+ ·0200
·24	+ ·0029	......	......	......	......
·25	+ ·0243	+ ·1017	+ ·0984	+ ·0002	+ ·0002
·2506	......	......	......	·0000	·0000
·3	+ ·1293	+ ·0966	+ ·0921	− ·0221	− ·0201
·34	+ ·2053	......	......	......	......
·35	+ ·2225	+ ·0796	+ ·0745	− ·0441	− ·0387
·36	+ ·2388	......	......	......	......
·4	+ ·2926	+ ·0522	+ ·0479	− ·0647	− ·0544
·43	+ ·3191	......	......	......	......
·45	......	+ ·0157	+ ·0140	− ·0807	− ·0644
·46	+ ·3314	......	......	......	......
·4688	......	·0000	·0000	......	......
·469	+ ·3321	......	......	......	......
·5	+ ·3233	− ·0273	− ·0237	− ·0898	− ·0674
·54	+ ·2844	......	......	......	......
·55	......	− ·0726	− ·0606	− ·0891	− ·0622
·56	+ ·2546	......	......	......	......
·6	+ ·1721	− ·1142	− ·0914	− ·0752	− ·0481
·63	+ ·0935	......	......	......	......
·65	......	− ·1450	− ·1102	− ·0446	− ·0258
·66	+ ·0038	......	......	......	......
·7	− ·1253	− ·1555	− ·1110	+ ·0064	+ ·0033
·74	− ·2517	......	......	......	......
·75	− ·2808	− ·1344	− ·0889	+ ·0823	+ ·0360
·76	− ·3087	......	......	......	......
·8	− ·3918	− ·0683	− ·0410	+ ·1873	+ ·0674
·82	− ·4119	......	......	......	......
·8302	− ·4147	·0000	·0000	......	......
·84	− ·4119	......	......	......	......
·85	− ·4030	+ ·0586	+ ·0308	+ ·3263	+ ·0905
·87	− ·3638	......	......	......	......
·90	− ·2412	+ ·2645	+ ·1153	+ ·5044	+ ·0958
·92	− ·1084	+ ·1764	+ ·1464	......	......
·93	......	+ ·4346	+ ·1597	......	......
·9325	·0000	......	......	......	......
·94	+ ·0751	+ ·5002	+ ·1706	......	......
·95	......	+ ·5704	+ ·1778	+ ·7271	+ ·0709
·96	+ ·3150	......	......	......	......
·97	......	+ ·7260	+ ·1764	......	......
·98	+ ·6203	+ ·8117	+ ·1615	+ ·8844	+ ·0350
·99	+ ·8003	+ ·9029	+ ·1274	+ ·9411	+ ·0187
1·00	+1·0000	+1·0000	·0000	+1·0000	+ ·0000

Polar harmonics of sixth order.

μ.	$\frac{1}{1260}\frac{d^3Q_6}{d\mu^3}$.	$\frac{3}{8}\Theta_6^{(3)}$.	$\frac{1}{4725}\frac{d^4Q_6}{d\mu^4}$.	$\frac{1}{10}\Theta_6^{(4)}$.	$\Theta_6^{(5)}$.
·0	·0000	·0000	− ·1000	− ·1000	·0000
·05	− ·0186	− ·0185	− ·0975	− ·0970	+ ·0497
·1	− ·0361	− ·0356	− ·0890	− ·0886	+ ·0975
·15	− ·0516	− ·0499	− ·0753	− ·0720	+ ·1417
·2	− ·0640	− ·0602	− ·0560	− ·0516	+ ·1806
·25	− ·0723	− ·0656	− ·0313	− ·0275	+ ·2127
·3	− ·0754	− ·0655	− ·0010	− ·0008	+ ·2370
·35	− ·0767	− ·0630	+ ·0348	+ ·0268	+ ·2524
·4	− ·0620	− ·0477	+ ·0760	+ ·0536	+ ·2586
·45	− ·0435	− ·0310	+ ·1227	+ ·0773	+ ·2555
·5	− ·0156	− ·0101	+ ·1750	+ ·0984	+ ·2436
·55	+ ·0225	+ ·0131	+ ·2327	+ ·1132	+ ·2234
·6	+ ·0720	+ ·0369	+ ·2960	+ ·1211	+ ·1966
·63	......	......	+ ·3366	+ ·1224	......
·65	+ ·1336	+ ·0587	+ ·3647	+ ·1204	+ ·1647
·7	+ ·2091	+ ·0750	+ ·4390	+ ·1139	+ ·1300
·75	+ ·2988	+ ·0865	+ ·5188	+ ·0991	+ ·0949
·8	+ ·4040	+ ·0873	+ ·6040	+ ·0783	+ ·0622
·83	......	......	+ ·6578	+ ·0637	......
·85	+ ·5257	+ ·0768	+ ·6947	+ ·0535	+ ·0344
·87	......	......	+ ·7326	+ ·0433	......
·89	......	......	+ ·7713	+ ·0333	......
·9	+ ·6649	+ ·0551	+ ·7910	+ ·0285	+ ·0150
·92	......	......	......	......	+ ·0085
·93	+ ·7572	+ ·0376	+ ·8514	+ ·0155	......
·95	+ ·8226	+ ·0249	+ ·8928	+ ·0084	+ ·0028
·96	+ ·8565	......	+ ·9138	......	......
·97	+ ·8911	+ ·0128	+ ·9350	+ ·0032	......
·98	+ ·9216	+ ·0073	+ ·9564	+ ·0015	......
·99	+ ·9629	......	+ ·9781	+ ·0004	......
1·00	+ 1·0000	·0000	+ 1·0000	·0000	·0000

Diag. No. 1.

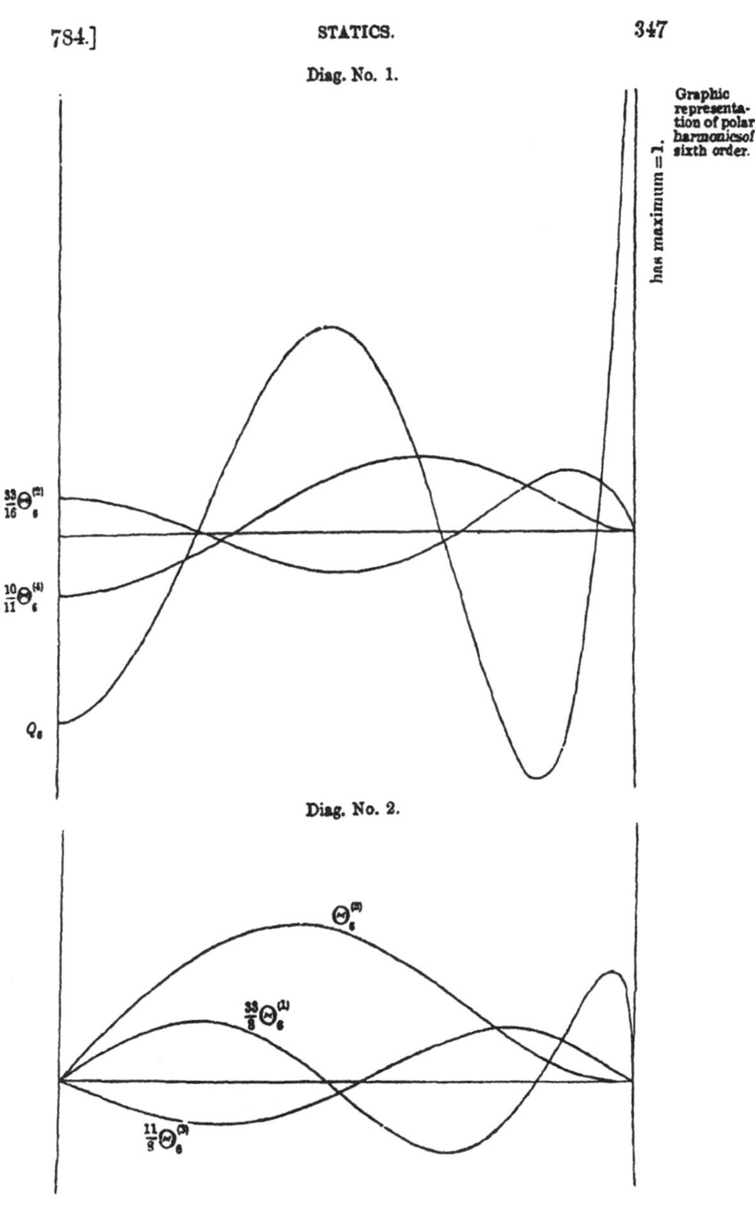

Graphic representation of polar harmonics of sixth order.

Graphic representation of polar harmonics of seventh order.

Diag. No. 3.

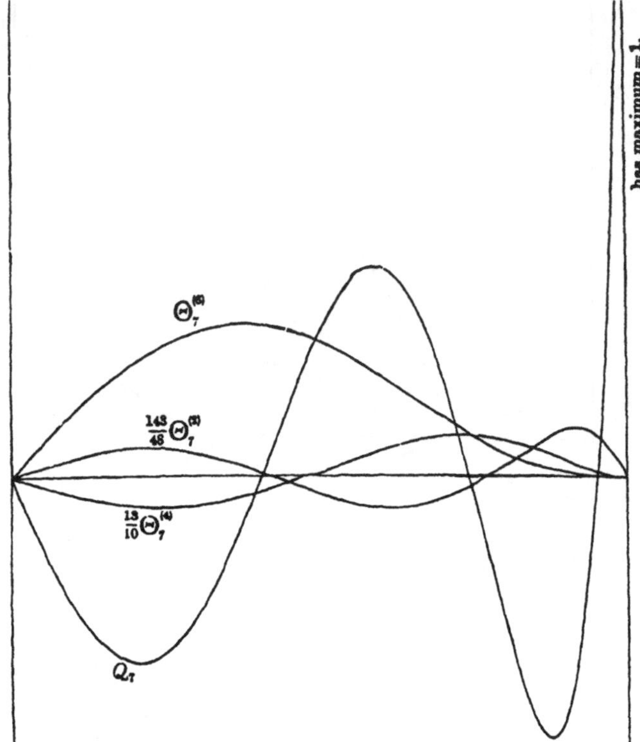

Diag. No. 4.

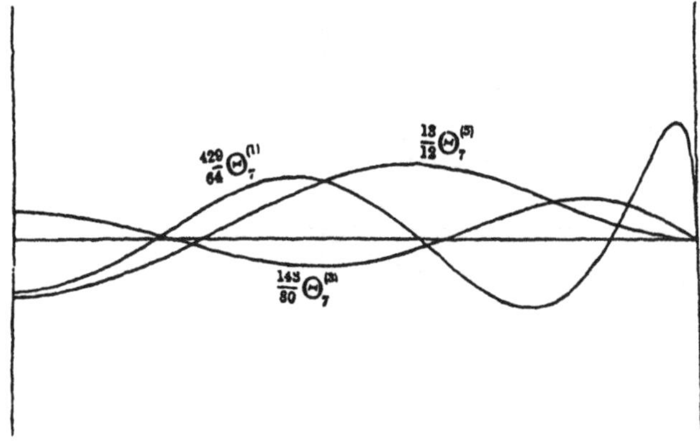

Polar harmonics of seventh order.

μ.	Q_7.	$\frac{1}{28}\frac{dQ_7}{d\mu}$.	$\frac{1}{28}\Theta_7^{(1)}$	$\frac{1}{378}\frac{d^2Q_7}{d\mu^2}$.	$\frac{1}{378}\Theta_7^{(2)}$.
·0	·0000	− ·0781	− ·0781	·0000	·0000
·05	− ·1096	− ·0720	− ·0719	+ ·0153	+ ·0153
·1	− ·1995	− ·0578	− ·0522	+ ·0290	+ ·0287
·15	− ·2649	− ·0345	− ·0341	+ ·0394	+ ·0385
·18	− ·2873	......	......	......	......
·2	− ·2935	− ·0057	− ·0056	+ ·0451	+ ·0433
·2093	......	·0000	......	......	......
·2261	......	......	......	+ ·0459	......
·23	− ·2905	......	......	......	......
·24	......	+ ·0190	+ ·0248	......	......
·25	− ·2799	+ ·0251	+ ·0515	+ ·0452	+ ·0424
·3	− ·2240	+ ·0540	+ ·0717	+ ·0391	+ ·0356
·35	− ·1318	+ ·0765	......	+ ·0268	+ ·0235
·38	− ·0635	......	......	......	......
·4	− ·0865	+ ·0888	+ ·0814	+ ·0084	+ ·0074
·42	+ ·0356	......	......	......	......
·4209	......	+ ·0901	......	·0000	·0000
·45	+ ·1106	+ ·0875	+ ·0782	− ·0132	− ·0105
·5	+ ·2231	+ ·0706	+ ·0611	− ·0371	− ·0278
·53	......	......	......	......	− ·0366
·55	+ ·3007	+ ·0378	+ ·0315	......	− ·0415
·57	+ ·3207	......	......	......	......
·58	......	......	......	......	− ·0415
·5917	+ ·3236	·0000	·0000	......	......
·6	+ ·3226	− ·0115	− ·0092	− ·0758	− ·0485
·62	+ ·3121	......	......	......	......
·6406	......	......	......	− ·0809	......
·65	+ ·2787	− ·0619	− ·0471	− ·0806	− ·0465
·7	+ ·1502	− ·1129	− ·0806	− ·0666	− ·0340
·7415	·0000	......	......	......	......
·75	− ·0842	− ·1458	− ·0964	− ·0254	− ·0111
·7694	......	− ·1490	......	......	......
·7695	......	......	......	·0000	·0000
·8	− ·2307	− ·1390	− ·0834	+ ·0529	+ ·0190
·82	− ·3134	......	......	......	......
·85	− ·3913	− ·0684	− ·0334	+ ·1801	+ ·0500
·86	− ·4054	......	......	......	......
·8717	− ·4117	·0000	·0000	......	......
·88	− ·4082	......	......	......	......
·9	− ·3678	+ ·1183	+ ·0515	+ ·3698	+ ·0723
·92	− ·2718	......	......	......	......
·93	......	......	......	+ ·5276	+ ·0712
·9491	·0000	......	......	......	......
·95	+ ·0112	+ ·4583	+ ·1413	+ ·6373	+ ·0621
·97	+ ·3165	+ ·6421	+ ·1563	+ ·7699	+ ·0455
·98	+ ·5115	+ ·7517	+ ·1458	......	......
·99	+ ·7384	+ ·8706	+ ·1230	+ ·9190	+ ·0184
1·00	+1·0000	+1·0000	·0000	+1·0000	·0000

Polar harmonics of seventh order.

μ.	$\frac{1}{3150}\frac{d^3Q_7}{d\mu^3}$.	$\frac{143}{80}\Theta_7^{(3)}$.	$\frac{1}{17325}\frac{d^4Q_7}{d\mu^4}$.	$\frac{13}{16}\Theta_7^{(4)}$.	$\frac{1}{62370}\frac{d^5Q_7}{d\mu^5}$.	$\frac{13}{12}\Theta_7^{(5)}$.	$\Theta_7^{(6)}$.
·0	+ ·0375	+ ·0375	·0000	·0000	− ·0833	− ·0833	·0000
·05	+ ·0355	·0353	− ·0148	− ·0147	− ·0806	− ·0801	+ ·0496
·1	+ ·0294	·0290	− ·0287	− ·0281	− ·0725	− ·0707	+ ·0970
·15	+ ·0198	·0192	− ·0406	− ·0387	− ·0590	− ·0557	+ ·1401
·2	+ ·0074	·0068	− ·0496	− ·0457	− ·0400	− ·0361	+ ·1769
·2261	·0000	·0000	......	......	......	......	.....
·25	− ·0071	− ·0064	− ·0544	− ·0478	− ·0156	− ·0133	+ ·2059
·2778	......	......	− ·0555	......	·0000	·0000	......
·3	− ·0225	− ·0195	− ·0549	− ·0454	+ ·0142	+ ·0112	+ ·2260
·35	− ·0367	− ·0302	− ·0493	− ·0378	+ ·0494	+ ·0356	+ ·2364
·4	− ·0487	− ·0375	− ·0368	− ·0260	+ ·0900	+ ·0582	+ ·2369
·45	− ·0563	− ·0400	− ·0165	− ·0104	+ ·1361	+ ·0773	+ ·2281
·4804	− ·0577	......	·0000	·0000	......	......	......
·5	− ·0570	− ·0370	+ ·0125	+ ·0070	+ ·1875	+ ·0918	+ ·2110
·55	− ·0485	− ·0282	+ ·0513	+ ·0248	+ ·2564	+ ·1041	+ ·1859
·6	− ·0278	− ·0142	+ ·0708	+ ·0412	+ ·3067	+ ·1004	+ ·1578
·6406	·0000	·0000	......	......	......	......	......
·65	+ ·0080	+ ·0035	+ ·1620	+ ·0540	+ ·3744	+ ·0948	+ ·1252
·7	+ ·0624	+ ·0227	+ ·2359	+ ·0613	+ ·4475	+ ·0831	+ ·0928
·75	+ ·1390	+ ·0401	+ ·3234	+ ·0619	+ ·5260	+ ·0665	+ ·0627
·8	+ ·2417	+ ·0521	+ ·4256	+ ·0551	+ ·6100	+ ·0474	+ ·0373
·85	+ ·3745	+ ·0546	+ ·5434	+ ·0418	+ ·6994	+ ·0288	+ ·0181
·9	+ ·5420	+ ·0448	+ ·6777	+ ·0244	+ ·7942	+ ·0132	+ ·0065
·92	+ ·6197	+ .0378	+ ·7368	+ ·0170	......	......	+ ·0033
·95	+ ·7489	+ ·0227	+ ·8732	+ ·0083	+ ·8944	+ ·0026	+ ·0087
·97	+ ·8062	+ ·0116	+ ·9230	+ ·0032	......	......	+ ·0002
·98	+ ·9564	+ ·0076	......	......	......	......	......
·99	......	......	......	......	......	......	......
1·00	1·0000	·0000	+ 1·0000	·0000	+ 1·0000	·0000	·0000

785. A short digression here on the theory of the potential, and particularly on equipotential surfaces differing little from concentric spheres, will simplify the hydrostatic examples which follow. First we shall take a few cases of purely synthetical investigation, in which, distributions of matter being given, resulting forces and level surfaces (§ 487) are found; and then certain problems of Green's and Gauss's analysis, in which, from data regarding amounts of force or values of potential over individual surfaces, or shapes of individual level surfaces, the distribution of force through continuous void space is to be determined. As it is chiefly for their application to physical geography that we admit these questions at present, we shall occasionally avoid circumlocutions by referring at once to the Earth, when any attracting mass with external equipotential surfaces approximately spherical would answer as well. We shall also sometimes speak of "*the sea level*" (§§ 750, 754) merely as a "level surface," or "surface of equilibrium" (§ 487) just enclosing the solid, or enclosing it with the exception of comparatively small projections, as our dry land. Such a surface will of course be an equipotential surface for mere gravitation, when there is neither rotation nor disturbance due to attractions of other bodies, such as the moon or sun, and due to change of motion produced by these forces on the Earth; but it may be always called an equipotential surface, as we shall see (§ 793) that both centrifugal force and the other disturbances referred to may be represented by potentials.

Digression on theory of potential.

Sea level.

Level surface relatively to gravity and centrifugal force.

786. To estimate how the sea level is influenced, and how much the force of gravity in the neighbourhood is increased or diminished by the existence within a limited volume underground of rocks of density greater or less than the average, let us imagine a mass equal to a very small fraction, $1/n$, of the earth's whole mass to be concentrated in a point somewhere at a depth below the sea level which we shall presently suppose to be small in comparison with the radius, but great in comparison with $1/\sqrt{n}$ of the radius. Immediately over the centre of disturbance, the sea level will be raised in virtue of the disturbing attraction, by a height equal to the same fraction of the radius

Disturbance of sea level by denser than average matter underground.

Disturbance of sea level by a denser than average matter underground. Intensity and direction of gravity altered by underground local excess above average density.

that the distance of the disturbing point from the chief centre is of n times its depth below the sea level as thus disturbed. The augmentation of gravity at this point of the sea level will be the same fraction of the whole force of gravity that n times the square of the depth of the attracting point is of the square of the radius. This fraction, as we desire to limit ourselves to natural circumstances, we must suppose to be very small. The disturbance of *direction* of gravity will, for the sea level, be a maximum at points of a circle described from A as centre, with $D/\sqrt{2}$ as radius; D being the depth of the centre of disturbance. The amount of this maximum deflection will be $\frac{2}{9}\sqrt{3}a^2/nD^2$ of the unit angle of $57°{\cdot}296$ (§ 41), a denoting the earth's radius.

Let C be the centre of the chief attracting mass $(1-n^{-1})$, and B that of the disturbing mass $(1/n)$, the two parts being supposed to act as if collected at these points. Let P be any point on the equipotential surface for which the potential is the same as what it would be over a spherical surface of radius a, and centre C if the whole were collected in C. Then (§ 491)

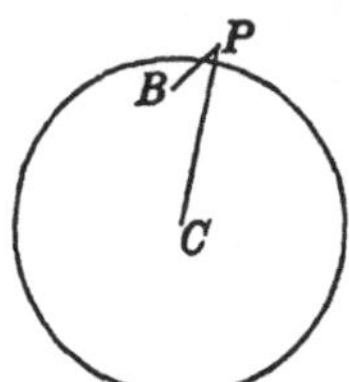

$$\left(1-\frac{1}{n}\right)\cdot\frac{1}{CP}+\frac{1}{n}\cdot\frac{1}{BP}=\frac{1}{a},$$

which is the equation of the equipotential surface in question. It gives

$$CP-a=\frac{a}{nBP}(CP-BP).$$

This expresses rigorously the positive or negative elevation of the disturbed equipotential at any point above the undisturbed surface of the same potential. For the point A, over the centre of disturbance, it gives

$$CA-a=\frac{a}{n\,.\,BA}\,.\,CB,$$

which agrees exactly with the preceding statement: and it proves the approximate truth of that statement as applied to the sea level when we consider that when BP is many times BA, $CP-a$

is many times smaller than its value at A. We leave the proof of the remaining statements of this and the following sections (§§ 787...792) as an exercise for the student.

787. If ρ be the general density of the upper crust, and σ the earth's mean density, and if the disturbance of § 786 be due to there being matter of a different density, ρ', throughout a spherical portion of radius b, with its centre at a depth D below the sea level, the value of n will be $\sigma a^3/(\rho'-\rho)b^3$; and the elevation of the sea level, and the proportionate augmentation of gravity at the point right over it, will be respectively

Effects of local excess above average density on sea level, and on direction and intensity of gravity:

$$\frac{(\rho'-\rho)b^2}{\sigma a D}.\,b, \text{ and } \frac{(\rho'-\rho)b^3}{\sigma a D^2}.$$

The actual value of σ is about double that of ρ. And let us suppose, for example, that $D=b=1000$ feet, or $\frac{1}{21000}$ of the earth's radius, and ρ' to be either equal to 2ρ or to zero. The previous results become

example.

$$\pm\tfrac{1}{42} \text{ of a foot, and } \pm\tfrac{1}{42000} \text{ of gravity,}$$

which are therefore the elevation or depression of sea level, and the augmentation or diminution of gravity, due to there being matter of double or zero density through a spherical space 2000 feet in diameter, with its centre 1000 feet below the surface. The greatest deviation of the plummet is at points of the circle of 707 feet radius round the point; and it amounts to $\frac{1}{109000}$ of the radian, or nearly 2″.

788. It is worthy of remark that, to set off against the increase in the amount of gravity due to the attraction of the disturbing mass, which we have calculated for points of the sea level in its neighbourhood, there is but an insensible deduction on account of the diminution of the attraction of the chief mass, owing to increase of the distance of the sea level from its centre, produced by the disturbing influence. The same remark obviously holds for disturbances in gravity due to isolated mountains, or islands of small dimensions, and it will be proved (§ 794) to hold also for deviations of figure represented by harmonics of high orders. But we shall see (§ 789) that it is otherwise with harmonic deviations of low orders, and conse-

quently with wide-spread disturbances, such as are produced by great tracts of elevated land or deep sea. We intend to return to the subject in Vol. II., under Properties of Matter, when we shall have occasion to examine the phenomenal and experimental foundations of our knowledge of gravity; and we shall then apply §§ 477 (*b*) (*c*) (*d*), 478, 479, and solutions of other allied problems, to investigate the effects on the magnitude and direction of gravity, and on the level surfaces, produced by isolated hills, mountain-chains, large table lands, and by corresponding depressions, as lakes or circumscribed deep places in the sea, great valleys or clefts, large tracts of deep ocean.

Harmonic spheroidal levels.

789. All the level surfaces relative to a harmonic spheroid (§ 779) of homogeneous matter are harmonic spheroids of the same order and type. That one of them, which lies as much inside the solid as outside it, cuts the boundary of the solid in a line (or group of lines)—the mean level line of the surface of the solid. This line lies on the mean spherical surface, and therefore (§ 780) it constitutes the nodes of each of the two harmonic spheroidal surfaces which cut one another in it. If i be the order of the harmonic, the deviation of the level spheroid is (§§ 545, 815) just $3/(2i+1)$ of the deviation of the bounding spheroid, each reckoned from the mean spherical surface.

Thus if $i=1$, the level coincides with the boundary of the solid: the reason of which is apparent when it is considered that any spherical harmonic deviation of the first order from a given spherical surface constitutes an equal spherical surface round a centre at some infinitely small distance from the centre of the given surface.

If $i=2$, the level surface deviates from the mean sphere by $\frac{3}{5}$ of the deviation of the bounding surface. This is the case of an ellipsoidal boundary differing infinitely little from spherical figure. It may be remarked that, as is proved readily from § 522, those of the equipotential surfaces relative to a homogeneous ellipsoid which lie wholly within it are exact ellipsoids, but not so those which cut its boundary or lie wholly without it: these being approximately ellipsoidal only when the deviation from spherical figure is very small.

Harmonic spheroidal levels of high orders.

790. The circumstances for very high orders are sufficiently illustrated if we confine our attention to sectorial harmonics (§ 781). The figure of the line in which a sectorial harmonic spheroid is cut by any plane perpendicular to its polar axis is [§ 781 (1)], as it were, a harmonic curve (§ 62) traced from a circular instead of a straight line of abscissas. Its *wave length* (or double length along the line of abscissas from one zero or nodal point to the next in order) will be $1/i$ of the circumference of the circle. And when i is very great, the factor $\sin\theta$ makes the sectorial harmonic very small, except for values of θ differing little from a right angle, and therefore a sectorial harmonic spheroid of very high order consists of a set of parallel ridges and valleys perpendicular to a great circle of the globe, of nearly simple harmonic form in the section by the plane of this circle (or equator), and diminishing in elevation and depression symmetrically on the two sides of it, so as to be insensible at any considerable angular distance (or latitude) from it on either side. The level surface due to the attraction of a homogeneous solid of this figure is a figure of the same kind, but of much smaller degree of elevations and depressions, that is, as we have seen, only $3/(2i+1)$ of those of the figure: or approximately three times the same fraction of the inequalities of figure that the half-wave length is of the circumference of the globe. It is easily seen that when i is very large the level surface at any place will not be sensibly affected by the inequalities in the distant parts of the figure.

Undulation of level due to parallel mountain-ridges and valleys.

791. Thus we conclude that, if the substance of the earth were homogeneous, a set of several parallel mountain-chains and valleys would produce an approximately corresponding undulation of the level surface in the middle district: the height to which it is raised, under each mountain-crest, or drawn down below the undisturbed level, over the middle of a valley, being three times the same fraction of the height of mountain above or depth of valley below mean level, that the breadth of the mountain or of the valley is of the earth's circumference.

792. If the globe be not homogeneous, the disturbance in magnitude and direction of gravity, due to any inequality in

Practical conclusions as to disturbances of sea-level, and amount and direction of gravity.

the figure of its bounding surface, will (§ 787) be ρ/σ of what it would be if the substance were homogeneous; and further, it may be remarked that, as the disturbances are supposed to be small, we may superimpose such as we have now described, on any other small disturbances, as, for instance, on the general oblateness of the earth's figure, with which we shall be occupied presently.

Practically, then, as the density of the upper crust is somewhere about ½ the earth's mean density, we may say that the effect on the level surface, due to a set of parallel mountain-chains and valleys, is, of the general character explained in § 791, but of half the amounts there stated. Thus, for instance, a set of several broad mountain-chains and valleys twenty nautical miles from crest to crest, or hollow to hollow, and of several times twenty miles extent along the crests and hollows, and 7,200 feet vertical height from hollow to crest, would raise and lower the level by 2½ feet above and below what it would be were the surface levelled by removing the elevated matter and filling the valleys with it.

Determinateness of potential through space from its value over every point of a surface.

793. Green's theorem [App. A. (*e*)]* and Gauss's theorem (§ 497) show that if the potential of any distribution of matter, attracting according to the Newtonian law, be given for every point of a surface completely enclosing this matter, the potential, and therefore also the force, is determined throughout all space external to the bounding surface of the matter, whether this surface consist of any number of isolated closed surfaces, each simply continuous, or of a single one. It need scarcely be said that no general solution of the problem has been obtained. But further, even in cases in which the potential has been fully determined for the space outside the surface over which it is given, mathematical analysis has hitherto failed to determine it through the whole space between this surface and the attracting mass within it. We hope to return, in later

* First apply Green's theorem to the surface over which the potential is given. Then Gauss's theorem shows that there cannot be two distributions of potential agreeing through all space external to this surface, but differing for any part of the space between it and the bounding surface of the matter.

volumes, to the grand problem suggested by Gauss's theorem of § 497. Meantime, we restrict ourselves to questions practically useful for physical geography.

Determination of potential from its value over a spherical surface enclosing the mass.

Example (1)—Let the enclosing surface be spherical, of radius a; and let $F(\theta, \phi)$ be the given potential at any point of it, specified in the usual manner by the polar co-ordinates θ, ϕ. Green's solution [§ 499 (3) and App. B. (46)] of his problem for the spherical surface is immediately applicable to part of our present problem, and gives

$$V=\frac{1}{4\pi a}\int_0^{2\pi}\int_0^{\pi}\frac{(r^2-a^2)F(\theta', \phi')r^2\sin\theta' d\theta' d\phi'}{\{r^2-2ar[\cos\theta\cos\theta'+\sin\theta\sin\theta'\cos(\phi-\phi')]+a^2\}^{\frac{3}{2}}}\ldots(3)$$

for the potential at any point (r, θ, ϕ) external to the spherical surface. But inasmuch as Laplace's equation $\nabla^2 u=0$ is satisfied through the whole internal space as well as the whole external space by the expression (46) of App. B., and in our present problem $\nabla^2 V=0$ is only satisfied [§ 491 (c)] for that part of the internal space which is not occupied by matter, the expression (3) gives the solution for the exterior space only. When $F(\theta, \phi)$ is such that an expression can be found for the definite integral in finite terms, this expression is necessarily the solution of our problem through all space exterior to the actual attracting body. Or when $F(\theta, \phi)$ is such that the definite integral, (3), can be transformed into some definite integral which varies continuously across the whole or across some part of the spherical surface, this other integral will carry the solution through some part of the interior space: that is, through as much of it as can be reached without discontinuity (infinite elements) of the integral, and without meeting any part of the actual attracting mass. To this subject we hope to return later in connexion with Gauss's theorem (§ 497); but for our present purpose it is convenient to expand (3) in ascending powers of a/r, as before in App. B. (s). The result [App. B. (51)] is

$$V=\frac{a}{r}F_0(\theta, \phi)+\left(\frac{a}{r}\right)^2F_1(\theta, \phi)+\left(\frac{a}{r}\right)^3F_2(\theta, \phi)+\text{etc.} \qquad (3 \text{ bis})$$

where $F_0(\theta, \phi)$, $F_1(\theta, \phi)$, etc., are the successive terms of the expansion [App. B. (52)] of $F(\theta, \phi)$ in spherical surface har-

Determination of potential from its value over a spherical surface enclosing the

monics; the general term being given by the formula

$$F_i(\theta, \phi) = \frac{2i+1}{4\pi} \int_0^{2\pi} \int_0^{\pi} Q_i F(\theta', \phi') \sin\theta' d\theta' d\phi' \ldots\ldots(4),$$

where Q_i is the function of (θ, ϕ) (θ', ϕ') expressed by App. B. (61).

In any case in which the actual attracting matter lies all within an interior concentric spherical surface of radius a', the harmonic expansion of $F(\theta, \phi)$ must be at least as convergent as the geometrical series

$$\frac{a'}{a} + \left(\frac{a'}{a}\right)^2 + \left(\frac{a'}{a}\right)^3 + \text{etc.};$$

and therefore (3 bis) will be convergent for every value of r exceeding a', and will consequently continue the solution into the interior at least as far as this second spherical surface.

Determination of potential from the form of an approximately spherical equipotential surface round the mass.

Example (2)—Let the attracting mass be approximately centrobaric (§ 534), and let one equipotential surface completely enclosing it be given. It is required to find the distribution of force and potential through all space external to the smallest spherical surface that can be drawn round it from its centre of gravity as centre. Let a be an approximate or mean radius; and, taking the origin of co-ordinates exactly coincident with the centre of inertia (§ 230), let

$$r = a[1 + F(\theta, \phi)] \ldots\ldots\ldots\ldots\ldots\ldots\ldots(5)$$

be the polar equation of the surface; F being for all values of θ and ϕ so small that we may neglect its square and higher powers. Consider now two proximate points (r, θ, ϕ) (a, θ, ϕ). The distance between them is $aF(\theta, \phi)$ and is in the direction through O, the origin of co-ordinates. And if M be the whole mass, the resultant force at any point of this line is approximately equal to M/a^2 and is along this line. Hence the difference of potentials (§ 486) between them is $MF(\theta, \phi)/a$. And if a be the proper mean radius, the constant value of the potential at the given surface (5) will be precisely M/a. Hence, to a degree of approximation consistent with neglecting squares of $F(\theta, \phi)$, the potential at the point (a, θ, ϕ) will be

$$\frac{M}{a} + \frac{M}{a} F(\theta, \phi) \ldots\ldots\ldots\ldots\ldots\ldots\ldots(6).$$

Hence the problem is reduced to that of the previous example: and remarking that the part of its solution depending on the term M/a of (6) is of course simply M/r, we have, by (3 bis), for the potential now required,

Determination of potential from the form of an approximately spherical equipotential surface round the mass.

$$U = M\left\{\frac{1}{r} + \frac{a}{r^2}F_1(\theta, \phi) + \frac{a^2}{r^3}F_2(\theta, \phi) + \text{etc.}\right\} \dots\dots\dots(7)$$

where F_i is given by (4). F_0 is zero in virtue of a being the proper mean radius; the equation expressing this condition being

$$\iint F(\theta, \phi)\sin\theta d\theta d\phi = 0 \dots\dots\dots\dots\dots\dots\dots(8).$$

If further O be chosen in a proper mean position, that is to say, such that

$$\iint Q_1 F(\theta, \phi)\sin\theta d\theta d\phi = 0 \dots\dots\dots\dots\dots\dots(9)$$

F_1 vanishes and [§ 539 (12)] O is the centre of gravity of the attracting mass; and the harmonic expansion of $F(\theta, \phi)$ becomes

$$F(\theta, \phi) = F_2(\theta, \phi) + F_3(\theta, \phi) + F_4(\theta, \phi) + \text{etc.} \dots\dots\dots(10).$$

If a' be the radius of the smallest spherical surface having O for centre and enclosing the whole of the actual mass, the series (7) necessarily converges for all values of θ and ϕ, at least as rapidly as the geometrical series

$$1 + \frac{a'}{r} + \left(\frac{a'}{r}\right)^2 + \left(\frac{a'}{r}\right)^3 + \text{etc.} \dots\dots\dots\dots\dots\dots(11)$$

for every value of r exceeding a'. Hence (7) expresses the solution of our present particular problem. It may carry it even further inwards; as the given surface (6) may be such that the harmonic expansion (10) converges more rapidly than the series

$$1 + \frac{a'}{a} + \left(\frac{a'}{a}\right)^2 + \left(\frac{a'}{a}\right)^3 + \text{etc.}$$

The direction and magnitude of the resultant force are of course [§§ 486, 491] deducible immediately from (7) throughout the space through which this expression is applicable, that is all space through which it converges that can be reached from the given surface without passing through any part of the actual attracting mass. It is important to remark that as the resultant force deviates from the radial direction by angles of the same order of small quantities as $F(\theta, \phi)$, its magnitude will differ from the radial component by small quantities of the same order as the square of this: and therefore, consistently with our degree

Resultant force.

Resultant force.

of approximation, if R denote the magnitude of the resultant force

$$R = -\frac{dU}{dr} = \frac{M}{r^2}\left\{1 + 3\left(\frac{a}{r}\right)^2 F_2(\theta, \phi) + 4\left(\frac{a}{r}\right)^3 F_3(\theta, \phi) + \text{etc.}\right\} \ldots (12).$$

For the resultant force at any point of the spherical surface agreeing most nearly with the given surface we put in this formula $r = a$, and find

$$\frac{M}{a^2}\{1 + 3F_2(\theta, \phi) + 4F_3(\theta, \phi) + \text{etc.}\} \ldots\ldots\ldots\ldots (13).$$

And at the point (r, θ, ϕ) of the given surface we have $r = a$ nearly enough for our approximation, in all terms except the first, of the series (12): but in the first term, M/r^2, we must put $r = a\{1 + F(\theta, \phi)\}$; so that it becomes

$$\frac{M}{r^2} = \frac{M}{a^2\{1 + F(\theta, \phi)\}^2} = \frac{M}{a^2}\{1 - 2F(\theta, \phi)\} = \frac{M}{a^2}\{1 - 2[F_2(\theta, \phi) + \text{etc.}]\} \ldots (14),$$

Resultant force at any point of approximately spherical level surface, for gravity alone.

and we find for the normal resultant force at the point (θ, ϕ) of the given approximately spherical equipotential surface

$$\frac{M}{a^2}\{1 + F_2(\theta, \phi) + 2F_3(\theta, \phi) + 3F_4(\theta, \phi) + \ldots\} \ldots\ldots (15).$$

Taking for simplicity one term, F_i, alone, in the expansion of F, and considering, by aid of App. B. (38), (40), (p), and §§ 779...784, the character of spherical surface harmonics, we see that the maximum deviation of the normal to the surface

$$r = a\{1 + F_i(\theta, \phi)\} \ldots\ldots\ldots\ldots\ldots\ldots \ldots\ldots (16)$$

from the radial direction is, in circular measure (§ 404), just i times the half range from minimum to maximum in the values of $F_i(\theta, \phi)$ for all harmonics of the second order (case $i = 2$), and for all sectorial harmonics (§ 781) of every order; and that it is approximately so for the equatorial regions of all zonal harmonics of very high order. Also, for harmonics of high order contiguous maxima and minima are approximately equal. We conclude that

794. If a level surface (§ 487), enclosing a mass attracting according to the Newtonian law, deviate from an approximately spherical figure by a pure harmonic undulation (§ 779) of order i; the amount of the force of gravity at any point of it will exceed the mean amount by $i - 1$ times the very small fraction by

which the distance of that point of it from the centre exceeds the mean radius. The maximum inclination of the resultant force to the true radial direction, reckoned in fraction of the unit angle $57^\circ{\cdot}3$ (§ 404) is, for harmonic deviations of the second order, equal to the ratio which the whole range from minimum to maximum bears to the mean magnitude. For the class described above under the designation of *sectorial harmonics*, of whatever order, i, the maximum deviation in direction bears to the proportionate deviation in magnitude from the mean magnitude, exactly the ratio $i/(i-1)$; and approximately the ratio of equality for *zonal harmonics* of high orders.

Example (3).—The attracting mass being still approximately centrobaric, let it rotate with angular velocity ω round OZ, and let one of the level surfaces (§ 487) completely enclosing it be expressed by (5), § 793. The potential of centrifugal force (§§ 800, 813), will be $\frac{1}{2}\omega^2(x^2+y^2)$, or, in solid spherical harmonics, $\frac{1}{3}\omega^2r^2+\frac{1}{6}\omega^2(x^2+y^2-2z^2)$.

Resultant force at any point of approximately spherical level surface for gravity and centrifugal force.

This for any point of the given surface (5) to the degree of approximation to which we are bound, is equal to

$$\frac{1}{3}\omega^2a^2+\frac{1}{2}\omega^2a^2\left(\frac{1}{3}-\cos^2\theta\right);$$

which, added to the gravitation potential at each point of this surface, must make up a constant sum. Hence the gravitation potential at (θ, ϕ) of the given surface (5) is equal to

$$\frac{M}{a}-\frac{1}{2}\omega^2a^2\left(\frac{1}{3}-\cos^2\theta\right);$$

and therefore, all other circumstances and notation being as in Example 2 (§ 793), we now have instead of (6) for gravitation potential at (a, θ, ϕ), the following:

$$\frac{M}{a}+\frac{M}{a}F(\theta, \phi)-\frac{1}{2}\omega^2a^2\left(\frac{1}{3}-\cos^2\theta\right) \;\ldots\ldots\ldots\ldots(16).$$

Hence, choosing the position of O, and the magnitude of a, according to (9) and (8), we now have, instead of (7), for the potential of pure gravitation, at any point (r, θ, ϕ),

$$U=M\left\{\frac{1}{r}+\frac{a^2}{r^3}\left[F_2(\theta, \phi)-\tfrac{1}{2}m\left(\tfrac{1}{3}-\cos^2\theta\right)\right]+\frac{a^3}{r^4}F_3(\theta, \phi)+\frac{a^4}{r^5}F_4(\theta, \phi)+\ldots\right\} \;(17),$$

Resultant force at any point of approximately spherical level surface for gravity and centrifugal force.

where m denotes $\omega^2 a^3/M$, or the ratio of centrifugal force at the equator, to pure gravity at the mean distance a. The force of pure gravity at the point (θ, ϕ) of the given surface (5) is consequently expressed by the following formula instead of (15):—

$$\frac{M}{a^2}\{1 + F_2(\theta, \phi) - 3 \cdot \tfrac{1}{3} m (\tfrac{1}{3} - \cos^2\theta) + 2F_3(\theta, \phi) + 3F_4(\theta, \phi) + \ldots\} \ldots (18).$$

From this must be subtracted the radial component of the centrifugal force, which is (in harmonics)

$$\tfrac{2}{3}\omega^2 a + \omega^2 a (\tfrac{1}{3} - \cos^2\theta),$$

to find the whole amount of the resultant force, g (apparent gravity), normal to the given surface: and therefore

$$g = \frac{M}{a^2}\{1 - \tfrac{2}{3}m + F_2(\theta,\phi) - \tfrac{5}{2}m(\tfrac{1}{3} - \cos^2\theta) + 2F_3(\theta,\phi) + 3F_4(\theta,\phi) + \ldots\} \; (19).$$

If in a particular case we have

$F_i(\theta, \phi) = 0$, except for $i = 2$; and $F_2(\theta, \phi) = e(\tfrac{1}{3} - \cos^2\theta)$:

this becomes

$$g = \frac{M}{a^2}\{1 - \tfrac{2}{3}m - (\tfrac{5}{2}m - e)(\tfrac{1}{3} - \cos^2\theta)\} \ldots\ldots\ldots\ldots (20).$$

795. Hence if outside a rotating solid the lines of resultant force of gravitation and centrifugal force are cut at right angles by an elliptic spheroid* symmetrical round the axis of rotation, the amount of the resultant differs from point to point of this surface as the square of the sine of the latitude: and the excess of the polar resultant above the equatorial bears to the whole amount of either a ratio which added to the ellipticity of the figure is equal to two and a half times the ratio of equatorial centrifugal force to gravity.

Clairaut's theorem.

For the case of a rotating fluid mass, or solid with density distributed as if fluid, these conclusions, of which the second is now generally known as Clairaut's theorem, were first discovered by Clairaut, and published in 1743 in his celebrated treatise *La Figure de la Terre*. Laplace extended them by proving the formula (19) of § 794 for any solid consisting

* Following the best French writers, we use the term spheroid to designate any surface differing very little from spherical figure. The commoner English usage of confining it to an ellipsoid symmetrical round an axis, and of extending it to such figures though not approximately spherical is bad.

Clairaut's theorems.

of approximately spherical layers of equal density. Ultimately Stokes* pointed out that, only provided the surfaces of equilibrium relative to gravitation alone, and relative to the resultant of gravitation and centrifugal force, are approximately spherical; whether the surfaces of equal density are approximately spherical or not, the same expression (19) holds. A most important practical deduction from this conclusion is that, irrespectively of any supposition regarding the distribution of the earth's density, the true figure of the sea level can be determined from pendulum observations alone, without any hypothesis as to the interior condition of the solid.

Figure of the sea level determinable from measurements of gravity;

Let, for brevity,

$$g\{1 + \tfrac{5}{2}m(\tfrac{1}{3} - \cos^2\theta)\} = f(\theta, \phi) \quad \ldots\ldots\ldots\ldots\ldots(21)$$

where m (§ 801) is $\frac{1}{289}$, and g is known by observation in different localities, with reduction to the sea level according to the square of the distance from the earth's centre (not according to Young's rule). Let the expansion of this in spherical surface harmonics be

$$f(\theta, \phi) = f_0 + f_2(\theta, \phi) + f_3(\theta, \phi) + \text{etc.} \quad \ldots\ldots\ldots(22).$$

We have, by (19),

$$F_i(\theta, \phi) = \frac{1}{i}\frac{f_i(\theta, \phi)}{f_0} \quad \ldots\ldots\ldots\ldots\ldots\ldots\ldots\ldots(23),$$

and therefore the equation (5) of the level surface becomes

$$r = a\left\{1 + \frac{1}{f_0}\left[\tfrac{1}{2}f_2(\theta, \phi) + \tfrac{1}{3}f_3(\theta, \phi) + \text{etc.}\right]\right\} \quad \ldots\ldots\ldots(24).$$

Confining our attention for a moment to the first two terms we have for f_2, by App. B. (38), explicitly

$$f_2(\theta, \phi) = A_0(\cos^2\theta - \tfrac{1}{3}) + (A_1\cos\phi + B_1\sin\phi)\sin\theta\cos\theta + (A_2\cos 2\phi + B_2\sin 2\phi)\sin^2\theta \ldots(25).$$

Substituting in (24) squared, putting

$$\cos\theta = \frac{z}{r},\ \sin\theta\cos\phi = \frac{x}{r},\ \sin\theta\sin\phi = \frac{y}{r},$$

and reducing to a convenient form, we find

$$(f_0 + \tfrac{1}{3}A_0 - A_2)x^2 + (f_0 + \tfrac{1}{3}A_0 + A_2)y^2 + (f_0 - \tfrac{2}{3}A_0)z^2 - B_1yz - A_1zx - 2B_2xy = f_0a^2 \ldots(26).$$

* "On the Variation of Gravity at the surface of the Earth."—*Trans. of the Camb. Phil. Soc.*, 1849.

If ellipsoid with three unequal axes must have one of them coincident with axis of revolution.

Now from §§ 539, 534, we see that, if OX, OY, OZ are principal axes of inertia, the terms of f_2 which, expressed in rectangular co-ordinates, involve the products yz, zx, xy must disappear: that is to say, we must have $B_1=0$, $A_1=0$, $B_2=0$. But whether B_2 vanishes or not, if OZ is a principal axis we must have both $A_1=0$ and $B_1=0$; which therefore is the case, to a very minute accuracy, if we choose for OZ the average axis of the earth's rotation, as will be proved in Vol. II., on the assumption rendered probable by the reasons adduced below, that the earth experiences little or no sensible disturbance in its motion from want of perfect rigidity. Hence the expansion (22) is reduced to

$$f(\theta,\phi)=f_0+A_0(\cos^2\theta-\tfrac{1}{3})+(A_2\cos 2\phi+B_2\sin 2\phi)\sin^2\theta+f_3(\theta,\phi)+\text{etc.}\ldots(27).$$

If $f_3(\theta, \phi)$ and higher terms are neglected the sea level is an ellipsoid, of which one axis must coincide with the axis of the earth's rotation. And, denoting by e the mean ellipticity of meridional sections, e' the ellipticity of the equatorial section, and I the inclination of one of its axes to OX, we have

$$e=\tfrac{1}{2}\frac{A_0}{f_0},\quad e'=\frac{\sqrt{(A_2^2+B_2^2)}}{f_0},\quad I=\tfrac{1}{2}\tan^{-1}\frac{B_2}{A_2}.$$

In general, the constants of the expansion (22); f_0 (being the mean force of gravity), A_0, A_2, B_2, the seven coefficients in $f_3(\theta, \phi)$, the nine in $f_4(\theta, \phi)$, and so on; are to be determined from sufficiently numerous and wide-spread observations of the amount of gravity.

Figure of the sea level determinable from measurements of gravity;

796. A first approximate result thus derived from pendulum observations and confirmed by direct geodetic measurements is that the figure of the sea level approximates to an oblate spheroid of revolution of ellipticity about $\frac{1}{295}$. Both methods are largely affected by local irregularities of the solid surface and underground density, to the elimination of which a vast amount of labour and mathematical ability have been applied, with as yet but partial success. Considering the general disposition of the great tracts of land and ocean, we can scarcely doubt that a careful reduction of the numerous accurate pendulum observations that have been made in localities widely spread over the earth* will lead to the determination of an ellipsoid with three unequal axes coinciding more nearly on the whole with the true figure of the sea level than does any spheroid of revolution. Until this has been either accomplished or proved impracticable it would be vain to speculate as to the possibility of obtaining, from attainable data, a yet closer approximation by introducing a harmonic of the third order [$f_3(\theta, \phi)$ in (27)]. But there is little probability that harmonics of the fourth or higher orders will ever be found useful: and local quadratures, after the example first set by Maskelyne in his investigation of the disturbance produced by Schehallien, must be resorted to in order to interpret irregularities in particular districts; whether of the amount of gravity shown by the pendulum; or of its direction, by geodetic observation. We would only remark here, that the problems presented by such local quadratures with reference to the *amount* of gravity seem about as much easier and simpler than those with reference to its direction as pendulum observations are than geodetic measurements: and that we expect much more knowledge regarding the true figure of the sea level from the former than from the latter, although it is to the reduction of the latter that the most laborious efforts have been hitherto applied. We intend to return to this subject in Vol. II. in explaining, under Properties of Matter, the practical foundation of our knowledge of gravity.

Figure of the sea level determinable from measurements of gravity; rendered difficult by local irregularities.

797. Since 1860 geodetic work of extreme importance has been in progress, through the co-operation of the Governments of Prussia, Russia, Belgium, France, and England, in connecting the triangulation of France, Belgium, Russia, and Prussia, which were sufficiently advanced for the purpose in 1860, with the principal triangulation of Great Britain and

Results of geodesy.

* In 1672, a pendulum conveyed by Richer from Paris to Cayenne first proved variation of gravity. Captain Kater and Dr Thomas Young, *Trans. R. S.*, 1819. Biot, Arago, Mathieu, Bouvard, and Chaix; *Base du Système Métrique*, Vol. III., Paris, 1821. Captain Edward Sabine, R.E., "Experiments to determine the Figure of the Earth by means of the Pendulum;" published for the Board of Longitude, London, 1825. Stokes "On the Variation of Gravity at the Surface of the Earth."—*Camb. Phil. Trans.*, 1849.

Results of geodesy.

Ireland, which had been finished in 1851. With reference to this work, General Sir Henry James made the following remarks:—"Before the connexion of the triangulation of the "several countries into one great network of triangles extend-"ing across the entire breadth of Europe, and before the dis-"covery of the Electric Telegraph, and its extension from "Valentia (Ireland) to the Ural mountains, it was not possible "to execute so vast an undertaking as that which is now in "progress. It is, in fact, a work which could not possibly "have been executed at any earlier period in the history "of the world. The exact determination of the Figure and "Dimensions of the Earth has been one great aim of astrono-"mers for upwards of two thousand years; and it is fortunate "that we live in a time when men are so enlightened as to "combine their labours to effect an object which is desired by all, "and at the first moment when it was possible to execute it."

For yet a short time, however, we must be contented with the results derived from the recent British Triangulation, with the separate measurements of arcs of meridians in Peru, France, Prussia, Russia, Cape of Good Hope, and India. The investigation of the ellipsoid of revolution agreeing most nearly with the sea level for the whole Earth, has been carried out with remarkable skill by Captain (now Colonel) A. R. Clarke, R.E., and published in 1858, by order of the Master General and Board of Ordnance (in a volume of 780 pages, quarto, almost every page of which is a record of a vast amount of skilled labour). The following account of conclusions subsequently worked out regarding the ellipsoid of three unequal axes most nearly agreeing with the sea level, is extracted from the preface to another volume recently published as one item of the great work of comparison with the recent triangulations of other countries*:—

"In computing the figures of the meridians and of the

* "Comparisons of the Standards of Length of England, France, Belgium, Prussia, Russia, India, Australia, made at the Ordnance Survey Office, Southampton, by Captain A. R. Clarke, R.E., under the direction of Colonel Sir Henry James, R.E., F.R.S." Published by order of the Secretary of State for War, 1866.

"equator for the several measured arcs of meridian, it is found "that the equator is slightly elliptical, having the longer "diameter of the ellipse in 15° 34′ east longitude. In the "eastern hemisphere the meridian of 15° 34′ passes through "Spitzbergen, a little to the west of Vienna, through the Straits "of Messina, through Lake Chad in North Africa, and along "the west coast of South Africa, nearly corresponding to the "meridian which passes over the greatest quantity of land in "that hemisphere. In the western hemisphere this meridian "passes through Behring's Straits and through the centre of the "Pacific Ocean, nearly corresponding to the meridian which "passes over the greatest quantity of water of that hemi-"sphere.

Results of geodesy.

"The meridian of 105° 34′ passes near North-East Cape, in "the Arctic Sea, through Tonquin and the Straits of Sunda, and "corresponds nearly to the meridian which passes over the "greatest quantity of land in Asia; and in the western hemi-"sphere it passes through Smith's Sound in Behring's Straits, "near Montreal, near New York, between Cuba and St Do-"mingo, and close along the western coast of South America, "corresponding nearly to the meridian passing over the greatest "amount of land in the western hemisphere.

"These meridians, therefore, correspond with the most re-"markable physical features of the globe.

	Feet
"The longest semi-diameter of the equatorial ellipse is	20926350
"And the shortest	20919972
"Giving an ellipticity of the equator equal to	$\frac{1}{3269 \cdot 5}$
"The polar semi-diameter is equal to	20853429
"The maximum and minimum polar compressions "are	$\frac{1}{285 \cdot 97}$ and $\frac{1}{313 \cdot 38}$
"Or a mean compression of very closely	$\frac{1}{300}$."

Fourteen years later Colonel Clarke corrected this result in the following statement*: "But these are affected by the error

* Extracted from pages 308, 309 of "Geodesy," by Col. A. R. Clarke, C.B. Oxford. 1880.

Results of geodesy.

"in the southern half of the old Indian arc. A revision of this "calculation, based on the revision and extension of the Indian "geodetic operations, is to be found in the *Philosophical Maga-* "*zine* for August, 1878, resulting in the following numbers:

"Major semi-axis of equator (long. 8°.15′ W.) $a = 20926629$
"Minor semi-axis ,, (long. 81°.45′ W.) $b = 20925105$
"Polar semi-axis ,, $c = 20854477$

"The meridian of the greater equatorial diameter thus passes "through Ireland and Portugal, cutting off a small bit of the "north-west corner of Africa: in the opposite hemisphere this "meridian cuts off the north-east corner of Asia and passes "through the southern island of New Zealand. The meridian "containing the smaller diameter of the equator passes through "Ceylon on the one side of the earth and bisects North "America on the other. This position of the axes, brought out "by a very lengthened calculation, certainly corresponds very "remarkably with the physical features of the globe—the dis- "tribution of land and water on its surface. On the ellipsoidal "theory of the earth's figure, small as is the difference between "the two diameters of the equator, the Indian longitudes are "much better represented than by a surface of revolution. But "it is nevertheless necessary to guard against an impression "that the figure of the equator is thus definitely fixed, for the "available data are far too slender to warrant such a con- "clusion."

Colonel Clarke had previously found ("Account of Principal Triangulation," 1858) for the spheroid of revolution most nearly representing the same set of observations, the following:—

Equatorial semi-axis $= a = 20926062$ feet,
Polar semi-axis $= c = 20855121$ feet;

$$\text{whence } \frac{c}{a} = \frac{293{\cdot}98}{294{\cdot}98}; \text{ and ellipticity} = \frac{a-c}{a} = \frac{1}{294{\cdot}98}.$$

Colonel Clarke's twenty-two years' labours, from 1858 to 1880, have led him to but very small corrections on these results. In his "Geodesy," page 319, he gives the following as the most probable lengths of the polar semi-axis and of the mean equatorial semi-axis of the terrestrial spheroid so far as all observations and comparisons of standards up to 1880 have allowed him to judge : Results of geodesy.

$$a = 20926202,$$
$$c = 20854895,$$

and their ratio

$$\frac{c}{a} = \frac{292{\cdot}465}{293{\cdot}465}, \text{ and ellipticity } \frac{a-c}{a} = \frac{1}{293{\cdot}465}.$$

798. As an instructive example of the elementary principles of fluid equilibrium, useful also because it includes the celebrated hydrostatic theories of the Tides and of the Figure of the Earth, let us suppose a finite mass of heterogeneous incompressible fluid resting on a rigid spherical shell or solid sphere, under the influence of mutual gravitation between its parts, and of the attraction of the core supposed symmetrical; to be slightly disturbed by any attracting masses fixed either in the core or outside the fluid; or by force fulfilling any imaginable law, subject only to the condition of being a conservative system; or by centrifugal force. Hydrostatic examples resumed.

First we may remark that were there no such disturbance the fluid would come to rest in concentric spherical layers of equal density, the denser towards the centre, this last characteristic being essential for stability, which clearly requires also that the mean density of the nucleus shall be not less than that of the layer of fluid next it; otherwise the nucleus would, as it were, float up from the centre, and either protrude from the fluid at one side, or (if the gradation of density in the fluid permits) rest in an eccentric position completely covered; fulfilling in either case the condition (§ 762) for the equilibrium of floating bodies.

799. The effect of the disturbing force could be at once found without analysis if there were no mutual attraction between parts of the fluid, so that the influence tending to maintain the spherical figure would be simply the symmetrical attraction of the fixed core. For the equipotential surfaces No mutual force between portions of the liquid.

No mutual force between portions of the liquid: would then be known (as directly implied by the data), and the fluid would (§ 750) arrange itself in layers of equal density defined by these surfaces.

Example (1). **800.** *Examples of* § 799.—(1) Let the nucleus act according to the Newtonian law, and be either symmetrical round a point, or (§ 534) of any other centrobaric arrangement; and let the disturbing influence be centrifugal force. In Vol. II. it will appear, as an immediate consequence from the elementary dynamics of circular motion, that kinetic equilibrium under centrifugal force in any case will be the same as the static equilibrium of the imaginary case in which the same material system is at rest, but influenced by repulsion from the axis in simple proportion to distance.

If z be the axis of rotation, and ω the angular velocity, the components of centrifugal force (§§ 32, 35a, 259) are $\omega^2 x$ and $\omega^2 y$. Hence the potential of centrifugal force is

$$\tfrac{1}{2}\omega^2 (x^2 + y^2),$$

reckoned from zero at the axis, and increasing in the direction of the force, to suit the convention (§ 485) adopted for gravitation potentials. The expression for the latter (§§ 491, 534 a.) is

$$\frac{E}{\sqrt{(x^2 + y^2 + z^2)}}$$

where E denotes the mass of the nucleus, and the co-ordinates are reckoned from its centre of gravity (§ 534) as origin. Hence the "level surfaces" (§ 487) external to the nucleus are given by assigning different values to C in the equation

$$\frac{E}{\sqrt{(x^2 + y^2 + z^2)}} + \tfrac{1}{2}\omega^2 (x^2 + y^2) = C \text{ ....(1)},$$

and the fluid when in equilibrium has its layers of equal density and its outer boundary in these surfaces. If ρ be the density and p the pressure of the fluid at any point of one of these surfaces, regarded as functions of C, we have (§ 760)

$$p = \int \rho dC \text{ ....(2)}.$$

Unless the fluid be held in by pressure applied to its bounding surface, the potential must increase from this surface inwards (or the resultant of gravity and centrifugal force, perpendicular as it is to the surface, must be directed inwards), as negative pressure is practically inadmissible. The student will find it an interesting exercise to examine the circumstances under which this condition is satisfied; which may be best done by tracing the meridional curves of the series of surfaces of revolution given by equation (1).

No mutual force between portions of the liquid: Example (1).

Let a and $a(1-e)$ be the equatorial and polar semidiameters of one of these surfaces. We have

$$\frac{E}{a}+\tfrac{1}{2}\omega^2a^2=\frac{E}{a(1-e)},$$

whence

$$e=\frac{\tfrac{1}{2}\omega^2a}{E/a^2+\tfrac{1}{2}\omega^2a}=\frac{m}{2+m}\qquad\text{..................(3)},$$

where m denotes the ratio of centrifugal force at its equator to pure gravity at the same place. (Contrast approximately agreeing definition of m, § 794.) From this, and the form of (1), we infer that

801. In the case of but small deviation from the spherical figure, which alone is interesting with reference to the theory of the earth's figure and internal constitution, the bounding surface and the surfaces of equal density and pressure are very approximately oblate ellipsoids of revolution*; the ellipticity† of each amounting to half the ratio of centrifugal force in its largest circle (or its equator, as we may call this) to gravity at any part of it; and therefore increasing from surface to surface outwards as the cubes of the radii. The earth's equatorial radius is 20,926,000 feet, and its period (the sidereal day) is 86,164 mean solar seconds. Hence in British absolute measure (§ 225) the equatorial centrifugal force is $(2\pi/86164)^2\times 20926000$, or ·11127. This is $\frac{1}{289}$ of 32·158; or very approximately the same fraction of the mean value, 32·14, of apparent gravity over the

* Airy has estimated 24 feet as the greatest deviation of the bounding surface from a true ellipsoid.

† A term used by writers on the figure of the earth to denote the ratio which the difference between the two axes of an ellipse bears to the greater. Thus if ϵ be the ellipticity, and e the eccentricity of an ellipse, we have $e^2=2\epsilon-\epsilon^2$. Hence, when the eccentricity is small, the ellipticity is a small quantity of the same order as its square; and the former is equal approximately to the square root of twice the latter.

No mutual force between portions of the liquid: Example (1). whole sea level, as determined by pendulum observations. It is therefore [§ 794 (20)] $\frac{1}{289\cdot 66}$, or approximately $\frac{1}{290}$, of the mean value of true gravitation. Hence, if the solid earth attracted merely as a point of matter collected at its centre, and there were no mutual attraction between the different parts of the sea, the sea level would be a spheroid of ellipticity $\frac{1}{580}$. In reality, we find by observation that the ellipticity of the spheroid of revolution which most nearly coincides with the sea level is about $\frac{1}{295}$. The difference between these, or $\frac{1}{600}$, must therefore be due to deviation of true terrestrial gravity from spherical symmetry. Thus the whole ellipticity of the actual sea level, $\frac{1}{295}$, may be regarded as made up of two nearly equal parts; of which the greater, $\frac{1}{580}$, is due directly to centrifugal force, and the less, $\frac{1}{600}$, to deviation of solid and fluid attracting mass from any truly centrobaric arrangement (§ 534). A little later (§§ 820, 821) we shall return to this subject.

802. The amount of the resultant force perpendicular to the free surface of the fluid is to be found by compounding the force of gravity towards the centre with the centrifugal force from the axis; and it will be approximately equal to the former diminished by the component of the latter along it, when the deviation from spherical figure is small. And as the former component varies inversely as the square of the distance from the centre, it will be less at the equator than at either pole by an amount which bears to either a ratio equal to twice the ellipticity, and which is therefore (§ 801) equal to the centrifugal force at the equator. Thus in the present case half the difference of apparent gravity between poles and equator is due to centrifugal force, and half to difference of distance from the centre. The gradual increase of apparent gravity in going from the equator towards either pole is readily proved to be as the square of the sine of the latitude; and this not only for the result of the two combined causes of variation, but for each separately. These conclusions needed, however, no fresh proof, as they constitute merely the applications to the present case, of Clairaut's general theorems demonstrated above (§ 795).

Analytically, for the present case, we have

No mutual force between portions of the liquid: Example (1).

$$g = -\frac{dV}{dr}$$

if g denote the magnitude of the resultant of true gravity and centrifugal force; $\frac{d}{dr}$ [as in App. B. (g)] rate of variation per unit of length along the direction of r; and V the first member of (1) § 800. Hence taking $z^2 = r^2 \cos^2\theta$, and $x^2 + y^2 = r^2 \sin^2\theta$ we find

$$g = \frac{E}{r^2} - \omega^2 r \sin^2\theta \text{.........................(4).}$$

On the hypothesis of infinitely small deviation from spherical figure this becomes

$$g = \frac{E}{a^2}(1 - 2u) - \omega^2 a \sin^2\theta \text{..................(5),}$$

if in the small term we put a, a constant, for r, and in the other $r = a(1 + u)$. By (1) we see that E/C is an approximate value for r, and if we take it for a, that equation gives

$$u = \tfrac{1}{2}\frac{\omega^2 a^3}{E} \sin^2\theta \text{........................(6);}$$

and using this in (5) we have

$$g = \frac{E}{a^2}\left(1 - 2\frac{\omega^2 a^3}{E}\sin^2\theta\right) = \frac{E}{a^2}(1 - 2m \sin^2\theta) \text{.........(7),}$$

where, as before, m denotes the ratio of equatorial centrifugal force to gravity.

803. *Examples of* § 799 *continued.*—(2) The nucleus being held fixed, let the fluid on its surface be disturbed by the attraction of a very distant fixed body attracting according to the Newtonian law. Example (2).

Let r, θ be polar co-ordinates referred to the centre of gravity of the nucleus as origin, and the line from it to the disturbing body as axis; let, as before, E be the mass of the nucleus; lastly, let M be the mass of the disturbing body, and D its distance from the centre of the nucleus. The equipotentials have for their equation

$$\frac{E}{r} + \frac{M}{\sqrt{(D^2 - 2rD\cos\theta + r^2)}} = \text{const.} \text{............(8),}$$

No mutual force between portions of the liquid: Example (2).

which, for very small values of r/D, becomes approximately

$$\frac{E}{r}+\frac{M}{D}\left(1+\frac{r}{D}\cos\theta\right)=\text{const.} \quad\text{.................(9)}.$$

And if, as in corresponding cases, we put $r=a(1+u)$ where a is a proper mean value of r, and u is an infinitely small numerical quantity, a function of θ, we have finally

$$u=\frac{Ma^3}{ED^2}\cos\theta \quad\text{.........................(10)}.$$

This is a spherical surface harmonic of the first order, and (§ 789) we conclude that

The fluid will not be disturbed from its spherical figure, but it will be drawn towards the disturbing body, so that its centre will deviate from the centre of the nucleus by a distance amounting to the same fraction of its radius that the attraction of the disturbing body is of the attraction of the nucleus, on a point of the fluid surface. This fraction is about $\frac{1}{300000}$ (being $\frac{1}{83\times60\times60}$) for the earth and moon, as the moon's distance is 60 times the earth's radius, and her mass about $\frac{1}{83}$ of the earth's. Hence if the earth's and moon's centres were both held fixed, there would be a rise of level at the point nearest to the moon, and fall of level at the point farthest from it, each equal to $\frac{1}{300000}$ of the earth's radius, or about 70 feet. Or if we consider the sun's influence under similar unreal circumstances, we should have a tide of 12,500 feet rise on the side next the sun, and the same fall on the remote side; 12,500 feet being (§ 812) $\frac{1}{39\cdot1\times10^6}$ of the sun's distance.

Example for tides.

804. *Examples of* § 799 *continued.*—(3) With other conditions, the same as in Example (2) (§ 803), let one-half of the disturbing body be removed and fixed at an equal distance on the other side.

The equation of the equipotentials, instead of (8), is now

$$\frac{E}{r}+\tfrac{1}{2}M\left[\frac{1}{\sqrt{(D^2-2rD\cos\theta+r^2)}}+\frac{1}{\sqrt{(D^2+2rD\cos\theta+r^2)}}\right]=\text{const.}\ldots(11),$$

and as the first approximation when r/D is treated as very small, instead of (9), we now have

Example for tides:

$$\frac{E}{r}+\frac{M}{D}\left[1+\tfrac{1}{2}\frac{r^2}{D^2}(3\cos^2\theta-1)\right]=\text{const.} \quad \text{........(12)};$$

whence finally, instead of (10), with corresponding notation;

$$u=\tfrac{1}{2}\frac{Ma^3}{ED^3}(3\cos^2\theta-1) \quad \text{....................(13)}.$$

This is a spherical surface harmonic of the second order, and Ma^3/ED^3 is one-quarter of the ratio that the difference between the moon's attraction on the nearest and farthest parts of the earth bears to terrestrial gravity. Hence

The fluid will be disturbed into a prolate ellipsoidal figure, with its long axis in the line joining the two disturbing bodies, and with ellipticity (§ 801) equal to $\frac{3}{4}$ of the ratio which the difference of attractions of one of the disturbing bodies on the nearest and farthest points of the fluid surface bears to the surface value of the attraction of the nucleus. If, for instance, we suppose the moon to be divided into two halves, and these to be fixed on opposite sides of the earth at distances each equal to the true moon's mean distance; the ellipticity of the disturbed terrestrial level would be $\frac{3}{2\times 60\times 300000}$, or $\frac{1}{12.000.000}$; and the whole difference of levels from highest to lowest would be about $1\frac{3}{4}$ feet. We shall have much occasion to use this hypothesis in Vol. II., in investigating the kinetic theory of the tides. We shall see that it (or some equivalent hypothesis) is essential to Laplace's evanescent diurnal tide on a solid spheroid covered with an ocean of equal depth all over; but, on the other hand, we find presently (§ 814) that it agrees very closely with the actual circumstances so far as the foundation of the equilibrium theory is concerned.

result agrees with ordinary equilibrium theory.

The tides: results of ordinary equilibrium theory.

805. The rise and fall of water at any point of the earth's surface we may now imagine to be produced by making these two disturbing bodies (moon and anti-moon, as we may call them for brevity) revolve round the earth's axis once in the lunar twenty-four hours, with the line joining them always inclined to the earth's equator at an angle equal to the moon's declination. If we assume that at each moment the condition of hydrostatic equilibrium is fulfilled; that is, that the free liquid surface is perpendicular to the resultant force, we have what is called the "equilibrium theory of the tides."

Correction of ordinary equilibrium theory.

806. But even on this equilibrium theory, the rise and fall at any place would be most falsely estimated if we were to take it, as we believe it is generally taken, as the rise and fall of the spheroidal surface that would bound the water, if none of the solid were uncovered, that is if there were no dry land. To illustrate this statement, let us imagine the ocean to consist of two circular lakes A and B, with their centres 90° asunder, on the equator, communicating with one another by a narrow channel. In the course of the lunar twelve hours the level of lake A would rise and fall, and that of lake B would simultaneously fall and rise to maximum deviations from the mean level. If the areas of the two lakes were equal, their tides would be equal, and would amount in each to about one foot above and below the mean level; but not so if the areas were unequal. Thus, if the diameter of the greater be but a small part of the earth's quadrant, not more, let us say, than 20°, the amounts of the rise and fall in the two lakes will be inversely as their areas to a close degree of approximation. For instance, if the diameter of B be only $\frac{1}{10}$ of the diameter of A, the rise and fall in A will be scarcely sensible; while the level of B will rise and fall by about two feet above and below its mean; just as the rise and fall of level in the open cistern of an ordinary barometer is but small in comparison with the fall and rise in the tube. Or, if there be two large lakes A, A' at opposite extremities of an equatorial diameter, two small ones B, B' at two ends of the equatorial diameter perpendicular to that one, and two small lakes C, C' at two ends of the polar axis, the largest of these being, however, still supposed to extend over only a small portion of the earth's surface, and if all the six lakes communicate with one another freely by canals, or underground tunnels, there will be no sensible tides in the lakes A and A'; in B and B' there will be high water of two feet above mean level when the moon or anti-moon is in the zenith, and low water of two feet below mean when the moon

is rising or setting; and at C and C' there will be tides rising and falling one foot above and below the mean, the time of low water being when the moon or anti-moon is in the meridian of A, and of high water when they are on the horizon of A. The simplest way of viewing the case for the extreme circumstances we have now supposed is, first, to consider the spheroidal surface that would bound the water at any moment if there were no dry land, and then to imagine this whole surface lowered or elevated all round by the amount required to keep the height at A and A' invariable. Or, if there be a large lake A in any part of the earth, communicating by canals with small lakes over various parts of the surface, having in all but a small area of water in comparison with that of A, the tides in any of these will be found by drawing a spheroidal surface of two feet difference between greatest and least radius, and, without disturbing its centre, adding or subtracting from each radius such a length, the same for all, as shall do away with rise or fall at A.

Correction of ordinary equilibrium theory.

807. It is, however, only on the extreme supposition we have made, of one water area much larger than all the others taken together, but yet itself covering only a small part of the earth's curvature, that the rise and fall can be nearly altogether obliterated in one place, and doubled in another place. Taking the actual figure of the earth's sea-surface, we must subtract a certain positive or negative quantity a from the radius of the spheroid that would bound the water were there no land, a being determined according to the moon's position, to fulfil the condition that the volume of the water remains unchanged, and being the same for all points of the sea, at the same time. Many writers on the tides have overlooked this obvious and essential principle; indeed we know of only one sentence* hitherto published in which any consciousness of it has been indicated.

The tides, mutual attraction of the waters neglected: correction of the ordinary equilibrium theory.

808. The quantity a is a spherical harmonic function of the second order of the moon's declination, and hour-angle from

* "Rigidity of the Earth," § 17, *Phil. Trans.*, 1862.

The tides, mutual attraction of the waters neglected: corrected equilibrium theory.

the meridian of Greenwich, of which the five constant coefficients depend merely on the configuration of land and water, and may be easily estimated by necessarily very laborious quadratures, with data derived from the inspection of good maps.

Let as above

$$r = a(1+u) \text{..........................} (14)$$

be the spheroidal level that would bound the water were the whole solid covered; u being given by (13) of § 804. Thus, if $\iint d\sigma$ denote surface integration over the whole surface of the sea, $a\iint u d\sigma$ expresses the addition (positive or negative as the case may be) to the volume required to let the water stand to this level everywhere. To do away with this change of volume we must suppose the whole surface lowered equally all over by such an amount α (positive or negative) as shall equalize it. Hence if Ω be the whole area of sea, we have

$$\alpha = \frac{a}{\Omega}\iint u d\sigma \text{..........................} (15).$$

And
$$\mathbf{r} = r - \alpha = a\left\{1 + u - \frac{1}{\Omega}\iint u d\sigma\right\} \text{..............} (16),$$

is the corrected equation of the level spheroidal surface of the sea. Hence

$$h = a\left\{u - \frac{1}{\Omega}\iint u d\sigma\right\} \text{......................} (17),$$

where h denotes the height of the surface of the sea at any place, above the level which it would take if the moon were removed.

To work out (15), put first, for brevity,

$$\tau = \tfrac{3}{2}\frac{Ma^3}{ED^3} \text{..........................} (18):$$

and (13) becomes

$$u = \tau(\cos^2\theta - \tfrac{1}{3}) \text{........................} (19).$$

Now let l and λ be the geographical latitude and west longitude of the place, to which u corresponds; and ψ and δ the moon's hour-angle from the meridian of Greenwich, and her declination. As θ is the moon's zenith distance at the place (corrected for parallax), we have by spherical trigonometry

$$\cos\theta = \cos l \cos\delta \cos(\lambda - \psi) + \sin l \sin\delta;$$

which gives

$$3\cos^2\theta - 1 = \tfrac{3}{2}\cos^2 l\cos^2\delta\cos 2(\lambda-\psi) + 6\sin l\cos l\sin\delta\cos\delta\cos(\lambda-\psi) + \tfrac{1}{2}(3\sin^2\delta - 1)(3\sin^2 l - 1) \quad (20).$$

The tides, mutual attraction of the waters neglected: corrected equilibrium theory.

Hence if we take $\mathfrak{A}$, $\mathfrak{B}$, $\mathfrak{C}$, $\mathfrak{D}$, $\mathfrak{E}$ to denote five integrals depending solely on the distribution of laud and water, expressed as follows:

$$\left.\begin{aligned} &\mathfrak{A} = \frac{1}{\Omega}\iint \cos^2 l \cos 2\lambda d\sigma, \quad \mathfrak{B} = \frac{1}{\Omega}\iint \cos^2 l \sin 2\lambda d\sigma, \\ &\mathfrak{C} = \frac{1}{\Omega}\iint \sin l \cos l \cos\lambda d\sigma, \quad \mathfrak{D} = \frac{1}{\Omega}\iint \sin l \cos l \sin\lambda d\sigma, \\ &\mathfrak{E} = \frac{1}{\Omega}\iint (3\sin^2 l - 1)\, d\sigma, \\ &\text{where of course } d\sigma = \cos l\, dl\, d\lambda, \end{aligned}\right\} \quad (21),$$

we have

$$a = \frac{a}{\Omega}\iint u\,d\sigma = \tfrac{1}{3}a\tau\{\tfrac{3}{2}\cos^2\delta(\mathfrak{A}\cos 2\psi + \mathfrak{B}\sin 2\psi) + 6\sin\delta\cos\delta(\mathfrak{C}\cos\psi + \mathfrak{D}\sin\psi) + \tfrac{1}{2}\mathfrak{E}(3\sin^2\delta - 1)\} \quad (22).$$

This, used with (19) and (20) in (17), gives for the full conclusion of the equilibrium theory,

$$\left.\begin{aligned} h = {} & \tfrac{1}{2}a\tau\left[(\cos^2 l\cos 2\lambda - \mathfrak{A})\cos 2\psi + (\cos^2 l \sin 2\lambda - \mathfrak{B})\sin 2\psi\right]\cos^2\delta \\ & + 2a\tau\left[(\sin l\cos l\cos\lambda - \mathfrak{C})\cos\psi + (\sin l\cos l\sin\lambda - \mathfrak{D})\sin\psi\right]\sin\delta\cos\delta \\ & + \tfrac{1}{6}a\tau(3\sin^2 l - 1 - \mathfrak{E})(3\sin^2\delta - 1) \end{aligned}\right\} \quad (23),$$

in which the value of τ may be taken from (18) for either the moon or the sun: and δ and ψ denote the declination and Greenwich hour-angle of one body or the other, as the case may be. In this expression we may of course reduce the semi-diurnal terms to the form $A\cos(2\psi - \epsilon)$, and the diurnal terms to $A'\cos(\psi - \epsilon')$. Interpreting it we have the following conclusions:—

809. In the equilibrium theory, the whole deviation of level at any point of the sea, due to sun and moon acting jointly, is expressed by the sum of six terms, three for each body.

Lunar or solar semi-diurnal tide.

(1) The lunar or solar semi-diurnal tide rises and falls in proportion to a simple harmonic function of the hour-angle from the meridian of Greenwich, having for period 180° of this angle (or in time, half the period of revolution relatively to the earth), with amplitude varying in simple proportion to the square of the cosine of the declination of the sun or moon, as the case may be, and therefore varying but slowly, and through but a small entire range.

Lunar or solar diurnal tide.

(2) The lunar or solar diurnal tide varies as a simple harmonic function of the hour-angle of period 360°, or twenty-four hours, with an amplitude varying always in simple proportion to the sine of twice the declination of the disturbing body, and therefore changing from positive maximum to negative, and back to positive maximum again, in the tropical* period of either body in its orbit.

Lunar fortnightly tide or solar semi-annual tide.

(3) The lunar fortnightly or solar semi-annual tide is a variation on the average height of water for the twenty-four lunar or the twenty-four solar hours, according to which there is on the whole higher water all round the equator and lower water at the poles, when the declination of the disturbing body is zero, than when it has any other value, whether north or south; and maximum height of water at the poles and lowest at the equator, when the declination has a maximum, whether north or south. Gauss's way of stating the circumstances on which "secular" variations in the elements of the solar system depend is convenient for explaining this component of the tides. Let the two parallel circles of the north and south declination of the moon and anti-moon at any time be drawn on a geocentric spherical surface of radius equal to the moon's distance, and let the moon's mass be divided into two halves and distributed over them. As these circles of matter gradually vary each fortnight from the equator to maximum declination and back, the tide produced will be solely and exactly the "fortnightly tide."

Explanation of the lunar fortnightly and solar semi-annual tides.

810. In the equilibrium theory as ordinarily stated there is, at any place, high water of the semi-diurnal tide, *precisely* when the disturbing body, or its opposite, crosses the meridian of the place; and its amount is the same for all places in the same latitude; being as the square of the cosine of the latitude, and therefore, for instance, zero at each pole. In the corrected

* The tropical period is the interval of time between two successive passages of the tide-raising body through the intersection of the orbit of that body with the earth's equator. In the case of the moon this intersection oscillates, with a period of 18½ years, through about 13° on each side of the first point of Aries, as the nodes of the lunar orbit regrede on the ecliptic (see § 848 *a*, *b*). In the case of the sun the intersection is the first point of Aries, which completes its revolution in 26,000 years.

equilibrium theory, high water of the semi-diurnal tides may be either before or after the disturbing body crosses the meridian, and its amount is very different at different places in the same latitude, and is certainly not zero at the poles. In the ordinarily stated equilibrium theory, there is, *precisely* at the time of transit, high water or low water of diurnal tides in the northern hemisphere, according as the declination of the body is north or south; and the amount of the rise and fall is in simple proportion to the sine of twice the latitude, and therefore vanishes both at the equator and at the poles. In the corrected equilibrium theory, the time of high water may be considerably either before or after the time of transit, and its amount is very different for different places in the same latitude, and certainly not zero at either equator or poles. In the ordinary statement there is no lunar fortnightly or solar semi-annual tide in the latitude 35° 16′ (being $\sin^{-1} 1/\sqrt{3}$), and its amount in other latitudes is in proportion to the deviations of the squares of their sines from the value $\frac{1}{3}$. In the corrected equilibrium theory each of these tides is still the same in the same latitude, and vanishes at a certain latitude, and in any other latitudes is in simple proportion to the deviation of the squares of their sines from the square of the sine of that latitude. But the latitude where there is no tide of this class is not $\sin^{-1}(1/\sqrt{3})$, but $\sin^{-1}[\sqrt{\frac{1}{3}(1+\mathfrak{C})}]$, where $\mathfrak{C}$ is the mean value of $3\sin^2 l - 1$, for the whole covered portion of the earth's surface. In § 848 *c* below will be found an approximate evaluation by means of quadratures of the function $\mathfrak{C}$, contributed by Mr G. H. Darwin to our present edition. The uncertainty as to the amount of land in arctic and antarctic regions renders this evaluation to some degree uncertain; but it appears in any case that the distribution of the land is such that the latitude of evanescent fortnightly tide is only removed a little to the southward of 35° 16′. The computations show, in fact, that this latitude is 34°40′ or 34°57′, according to the assumptions made as to the amount of polar land.

The tides, mutual attraction of the waters neglected: practical importance of correction for equilibrium fortnightly and semi-annual tides.

Latitude of evanescent fortnightly tide.

As the fortnightly and semi-annual tides have been supposed by Laplace* to follow in reality very nearly the equilibrium

* In our first edition we undoubtingly accepted this supposition.

law, the determination of the latitude of evanescent tide is a matter of great importance. It is moreover possible that careful determination of the fortnightly and semi-annual tides at various places, by proper reductions of tidal observations, may contribute to geographical knowledge as to the amount of water-surface in the hitherto unexplored districts of the arctic and antarctic regions.

Spring and neap tides: "priming" and "lagging."

811. The superposition of the solar semi-diurnal on the lunar semi-diurnal tide has been investigated above as an example of the composition of simple harmonic motions; and the well-known phenomena of the "spring-tides" and "neap-tides," and of the "priming" and "lagging" have been explained (§ 60). We have now only to add that observation proves the proportionate difference between the heights of spring-tides and neap-tides, and the amount of the priming and lagging to be much less in nearly all places than estimated in § 60 on the equilibrium hypothesis; and to be very different in different places, as we shall see in Vol. II. is to be expected from the kinetic theory.

Discrepancy from observed results, due to inertia of water.

812. The potential expressions used in the preceding investigation are immediately applicable (§§ 802, 804) to the hydrostatic problem. But it is interesting, in connexion with this problem, to know the amount of the disturbing influence on apparent terrestrial gravity at any point of the earth's surface, produced by the lunar or solar influence. We shall therefore —still using the convenient static hypothesis of § 804—determine convenient rectangular components for the resultant of the two approximately equal and approximately opposed disturbing forces assumed in that hypothesis. First, we may remark that these two forces are approximately equivalent to a force equal to their difference in a line parallel to that of the centres of the earth and moon, compounded with another perpendicular to this and equal to twice either, multiplied into the cosine of half the obtuse angle between them.

Lunar and solar influence on apparent terrestrial gravity.

Resolving each of these components along and perpendicular to the earth's radius through the place, we obtain, by a process, the details of which we leave to the student, the following results, which are stated in gravitation measure:—

Lunar and solar influence on apparent terrestrial gravity.

$$\text{Vertical component, upwards} = \frac{Ma^2}{ED^3}(3\cos^2\theta - 1) \;\ldots\; (23').$$

$$\text{Horizontal component} = 3\frac{Ma^2}{ED^3}\sin\theta\cos\theta \ldots\ldots\ldots (23'').$$

The direction of this component is towards the point of the horizon under the moon or anti-moon.

Here, as before, E and M denote the masses of the earth and moon, D the distance between their centres, a the earth's radius, and θ the moon's zenith distance.

Or from the potential expression (12), by taking $\frac{d}{dr}$ and $\frac{d}{rd\theta}$ we find the same expressions.

The vertical component is a maximum upwards, amounting to

$$2\frac{Ma^2}{ED^3},$$

when the moon or anti-moon is overhead; and a maximum downwards of half this amount when the moon is on the horizon. The horizontal component has its maximum value, amounting to

$$\frac{3}{2}\frac{Ma^2}{ED^3},$$

when the moon or anti-moon is 45° above the horizon. Similar statements, of course, apply to the disturbing influence of the sun. For the moon Ma^2/ED^3 is probably equal to about $\frac{1}{83\times(60{\cdot}3)^3}$, or $\frac{1}{18{\cdot}2\times10^6}$: and the corresponding measure of the sun's influence is very approximately $(1+\frac{1}{83})(\frac{27{\cdot}3}{365})^2\frac{1}{(60{\cdot}3)^3}$, or $\frac{1}{39{\cdot}1\times10^6}$. Hence, considering the lunar influence alone, we see that as the moon or anti-moon rises from the horizon to the zenith of any place on the earth's surface, the intensity of apparent gravity is diminished by about one six-millionth part: and the plummet is deflected towards the point of the horizon under either moon or anti-moon, by an amount which reaches its maximum value, $\frac{1}{12\times10^6}$ of the unit angle (57°·3), or 0″·017, when the altitude is 45°. The corresponding effects of solar influence are of nearly half these amounts.

Taking the notation of § 808 above, and using the expansion (20) of that section, we find, from (23′) of the present, the vertical component equal to

Lunar and solar influence on apparent terrestrial gravity.

$$\tfrac{3}{2}\frac{Ma^3}{ED^3}\{\cos^2 l\cos^2\delta\cos 2(\lambda-\psi)+\sin 2l\sin 2\delta\cos(\lambda-\psi)$$
$$+(\tfrac{1}{3}-\sin^2 l)(1-3\sin^2\delta)\}\text{............}(23''').$$

Further remarking that dh/adl and $dh/a\cos l d\lambda$ are respectively the northward and the westward components of the inclination of the apparent level to the undisturbed terrestrial level, we find for the southward and eastward components of the horizontal disturbing force, as given in (23″), the following expressions:

$$\text{Southward component}=\tfrac{3}{4}\frac{Ma^3}{ED^3}\{\sin 2l\cos^2\delta\cos 2(\lambda-\psi)$$
$$-\cos 2l\sin 2\delta\cos(\lambda-\psi)$$
$$+\sin 2l(1-3\sin^2\delta)\}\text{........}(23^{iv});$$

$$\text{Eastward component}=\tfrac{3}{2}\frac{Ma^3}{ED^3}\{\cos l\cos^2\delta\sin 2(\lambda-\psi)$$
$$-\sin l\sin 2\delta\sin(\lambda-\psi)\}\text{......}(23^{v}).$$

These formulas show how in any one place the three components of the lunar disturbing force vary in the course of the 24 hours. They also show how the lunar disturbing force varies in longer periods when we consider them as affected by the monthly and fortnightly variations of δ and D.

Actual tide-generating influence explained by method of centrifugal force.

813. *Examples of* § 799 *continued.*—(4) All other circumstances remaining as in Example (2), let the two bodies be not fixed, but let them revolve in circles round their common centre of inertia, with angular velocity such as to give centrifugal force to each just equal to the force of attraction it experiences from the other.

Let the centre of the earth be origin of rectangular co-ordinates, and OZ perpendicular to the plane of the circular orbits, and let OX revolve so as always to pass through the disturbing body. Then, dealing with centrifugal force by the potential method, as in § 794; for the equation of a series of surfaces cutting everywhere at right angles the resultant of gravity and centrifugal force, we find

$$\frac{E}{\sqrt{(x^2+y^2+z^2)}}+\frac{M}{\sqrt{[(D-x)^2+y^2+z^2]}}+\tfrac{1}{2}\omega^2[(b-x)^2+y^2]=\text{const.}\ldots(24),$$

where ω denotes the angular velocity of revolution of the two bodies round their centre of inertia, and b the distance of this point from the earth's centre:—so that

Actual tide-generating influence explained by method of centrifugal force.

$$M(D-b)\,\omega^2 = Eb\omega^2 = \frac{ME}{D^2} \quad\text{.................}(25).$$

Hence $$\frac{Mx}{D^3} - \omega^2 bx = 0.$$

Using this in (24), expanded and dealt with generally as (12) in Example (3), we see that the first power of x disappears; and, omitting terms of third and higher orders, we have

$$\frac{E}{r} + \frac{M}{D}\left(1 + \frac{1}{2}\,\frac{3x^2 - r^2}{D^2}\right) + \frac{1}{2}\omega^2(x^2 + y^2) = \text{const.} \quad\text{.....}\ (26).$$

To reduce to spherical harmonics we have

$$x^2 + y^2 = \tfrac{2}{3}r^2 - \tfrac{1}{3}(3z^2 - r^2),$$

and therefore, as according to our approximation we may take $\omega^2 a^2$ for $\omega^2 r^2$, we find [with the notation $r = a(1+u)$ as above]

$$u = \frac{1}{2}\,\frac{Ma}{ED}\,\frac{3x^2 - r^2}{D^2} - \frac{1}{6}\omega^2\,\frac{a}{E}(3z^2 - r^2),$$

or in polar co-ordinates

$$u = \frac{1}{2}\,\frac{Ma^2}{ED^3}(3\sin^2\theta\cos^2\phi - 1) - \frac{1}{6}\,\frac{\omega^2 a^3}{E}(3\cos^2\theta - 1) \quad\text{....}(27).$$

This interpreted is as follows:—

The surface of the fluid will be a harmonic spheroid of the second order [that is (§ 799), an ellipsoid differing infinitely little from a sphere], which we may regard as the result of superimposing on the deviation from spherical figure investigated in § 804, another consisting of the oblateness due to rotation with angular velocity ω round the diameter of the earth perpendicular to the plane of the disturbing body's orbit. We may prove this conclusion with less analysis by supposing the purely static system of Example (3), § 804, to rotate, first with any angular velocity ω, about any diameter of the earth perpendicular to the straight line through its centre in which the disturbing bodies are placed; and then supposing this angular velocity to be just such as to balance the earth's attraction on the two disturbing bodies, so that the holdfasts by which they were prevented from falling together may be removed. Then it is easy to prove analytically that the effect of carrying either disturbing body to the other side, and uniting the two, will be a small disturbance in the figure of the fluid amounting to some such fraction of the deviation investigated in Example (3) as the earth's radius is of the distance of the disturber.

814. The purely static system of Example (3), § 804, gives the simplest and most symmetrical foundation for the equilibrium theory of the tides. The kinetic system of Example (4), § 813, is indeed not less purely static in relation to the earth, and is equivalent to an absolutely static ideal system in which repulsion from a fixed line, on parts of a non-rotating system, is substituted for the centrifrugal force of the rotating system. But it is complicated by the oblateness of the fluid surface produced by the centrifugal force or repulsion. This oblateness, as we see from § 801, would amount to as much as $\frac{1}{(27 \cdot 4)^2} \times \frac{1}{580}$, or $\frac{1}{435,000}$, being about 27·8 times the ellipticity of the lunar tide-level for the case of the earth and moon. For the case of the *sun* and earth, the corresponding oblateness amounts only to $\frac{1}{366^2} \times \frac{1}{580}$, or $\frac{1}{77,700,000}$, which is only $\frac{1}{3 \cdot 2}$ of the ellipticity of the solar tide-level.

Augmentation of result by mutual gravitation of the disturbed water.

815. When the attraction of the fluid on itself is sensible, the disturbance in its distribution gives rise to a counter disturbing force, which increases the deviation of the equipotential surfaces from the spherical figure. The general hydrostatic condition (§ 750), that the surfaces of equal density must still coincide with the equipotential surfaces, thus presents an exquisite problem for analysis. It has called forth from Legendre and Laplace an entirely new method in mathematics, commonly referred to by English writers as "Laplace's coefficients" or "Laplace's Functions." The principles have been sketched in the second Appendix to our first Chapter; from which, and the supplementary investigations of §§ 778—784, we have immediately the solution for the case in which the fluid is homogeneous, and the nucleus (being a solid of any shape, and with any internal distribution of density, subject only to the condition that its external equipotential surfaces are approximately spherical) is wholly covered by the fluid.

The conclusion may be expressed thus:—Let ρ be the density of the fluid, and let σ be the mean density of the whole mass, fluid and solid. Let the disturbing influence, whether of external disturbing masses, or of deviation from accurate centrobaric (§ 534) quality in the nucleus, or of centrifugal force due to rotation, be such as to render the level surfaces harmonic spheroids of order i, when the liquid is kept spherical by a rigid envelope in contact with it all round. The tendency of the liquid surface would be to take the figure of that one of these level surfaces which encloses the proper volume. But in changing its figure, if permitted, it would *increase* the deviation of this level surface. The result is, that if the constraint be removed, the level surface of the liquid in equilibrium will be a harmonic spheroid of the same type, but of deviation from sphericity augmented in the ratio of 1 to $1-\frac{3\rho}{(2i+1)\sigma}$.

Augmentation of result by mutual gravitation of the disturbed water.

Let the potential at or infinitely near the bounding surface be

$$\frac{4\pi\sigma a^3}{3r}+S_i \text{ ....} (1)$$

when the liquid is held fixed in shape by a spherical envelope, of radius a. In these circumstances

$$r=a\left(1+\frac{3S_i}{4\pi\sigma a^2}\right) \text{ ....} (2)$$

is the equipotential surface of mean radius a. If now the bounding surface of the liquid be changed into the harmonic spheroid

$$r=a(1+cS_i) \text{ ....} (3),$$

the potential (§ 543) becomes changed from (1) to

$$\frac{4\pi\sigma a^3}{3r}+\left(1+\frac{4\pi\rho ca^2}{2i+1}\right)S_i \text{ ....} (4),$$

and the equipotential surface becomes, instead of (2)

$$r=a\left\{1+\left(1+\frac{4\pi\rho ca^2}{2i+1}\right)\frac{3S_i}{4\pi\sigma a^2}\right\} \text{ ....} (5).$$

Hence that the boundary (3) of the liquid may be an equipotential surface,

$$c=\left(1+\frac{4\pi\rho ca^2}{2i+1}\right)\frac{3}{4\pi\sigma a^2},$$

Augmentation of result by mutual gravitation of the disturbed water.

which gives $$4\pi ca^2 = \frac{1}{\frac{1}{3}\sigma - \frac{\rho}{2i+1}},$$

whence $$1 + \frac{4\pi\rho ca^2}{2i+1} = \frac{1}{1 - \frac{3\rho}{(2i+1)\sigma}} \quad\text{.....................(6).}$$

Using this in (5), and comparing with (2), we prove the proposition.

Stability of the ocean.

816. The instability of the equilibrium in the case in which the density of the liquid is greater than the mean density of the nucleus, already remarked as obvious, is curiously illustrated by the present result, which makes the deviation infinite when $i=1$ and $\sigma=\rho$. But it is to be remarked that it is only when the nucleus is completely covered that the equilibrium would be unstable. However dense the liquid may be, there would be a position of stable equilibrium with the nucleus protruding on one side; and if the bulk of the liquid is either very small or very large in comparison with that of the nucleus, the figure of its surface in stable equilibrium would clearly be approximately spherical. Excluding the case of a very small nucleus of lighter specific gravity (which would become merely a small floating body, not sensibly disturbing the general liquid globe), we have, in the apparently simple question of finding the distribution of a small quantity of liquid on a symmetrical spherical nucleus of less specific gravity, a problem which utterly transcends mathematical skill as hitherto developed.

Augmentations of results by mutual gravitation of water calculated for examples of § 799.

817. The cases of $i=1$ and $i=2$ give the solutions of the several examples of § 799 when the attraction of the liquid on itself is taken into account, provided always that the solid is wholly covered. Thus [§ 799, Example (2)] if the earth and moon were stopped, and each held fixed, the moon's attraction would still not disturb the figure of the liquid surface from true sphericity, but would render it eccentric to a greater degree than that previously estimated, in the ratio of 1 to $1-\rho/\sigma$. For the earth and sea, ρ/σ is about $\frac{2}{11}$, and therefore the spherical liquid surface would be drawn towards the moon

by 86 feet, being $1\frac{2}{9}$ times the amount of 70 feet found above (§ 803). And the tidal and rotational ellipticities estimated in §§ 800, 814, 813 would, on the supposition now made, be augmented each in the ratio of 1 to $1-\frac{3}{5}\sigma/\rho$; or 55 to 49 for the case of earth and sea. The true correction for the attraction of the sea, as altered by tidal disturbance, in the equilibrium theory of the tides, must be less than this, as the liquid does not cover more than about $\frac{2}{3}$ of the surface of the solid. To find the true amount of the correction for the attraction of the water on itself when the whole solid is not covered, even if the arrangement of dry land and sea were quite symmetrical and simple (as, for instance, one circular continent and the rest ocean), belongs to the transcendental problem already referred to (§ 816). It can be practically solved, if necessary, by laborious methods of approximation; but the irregular boundaries of land and sea on the real earth, and the true kinetic circumstances of the tides, are such as to render nugatory any labours of this kind. Happily the error committed in neglecting altogether the correction in question may be safely estimated as less than 10 per cent. ($\frac{6}{49}$ being 12·3 per cent.), and may be neglected in our present uncertainty as to absolute values of causes and effects in the theory of the tides.

Augmentations of results by mutual gravitation of water calculated for examples of § 709.

818. But although the influence on the tides produced by the attraction of the water itself as it rises and falls is not considerable even in any one place; it is a manifest, though not an uncommon, error to suppose that the moon's disturbing influence on terrestrial gravity is everywhere insensible. It was pointed out long ago by Robison* that the great tides of the Bay of Fundy should produce a very sensible deflection on the plummet in the neighbourhood, and that observation of this effect might be turned to account for determining the earth's mean density. But even ordinary tides must produce, at places close to the sea shore, deviations in the plummet considerably exceeding the greatest direct effect of the moon, which, as we have seen (§ 812), amounts to $\frac{1}{12{,}000{,}000}$ of the unit angle (57°·3). Thus, at a point on

Local influence of high water on direction of gravity.

* *Mechanical Philosophy*, 1804. See also Forbes, *Proc. R.S.E.*, April, 1849.

Attraction of high water on a plummet at the sea-side.

or not many feet above the mean sea level, and 100 yards from low-water mark, a deflection, amounting to more than $\frac{1}{8{,}000{,}000}$ of the unit angle on each side of the mean vertical, will be produced by tides of five feet rise and fall on each side of the mean, if the line of coast does not deviate very much from one average direction for 50 miles on either side, and if the rise and fall is approximately simultaneous and equal for 50 miles out to sea. For, a point placed as O in the

Vertical section through O.

sketch will, as the water rises from low tide to high tide, experience the attraction of a plate of water indicated in section by $HKK'L'L$. If we neglect the small part of the whole effect due to the long bar (extending along the coast) shown in section by HKL, we have only to find the attraction of the rectangular plate of water by hypothesis of 50 miles' breadth from KL, 100 miles' length parallel to the coast, and 10 feet thickness (KL). This will not be sensibly altered if O is precisely in the continuation of the middle plane EE' (instead of a few feet above it, as would generally be the case in a convenient sea-side gravitation observatory), and the whole matter of the plate were condensed into its middle plane. But the attraction of a uniform rectangular plate on a point O has, for component parallel to OE,

Horizontal section through O.

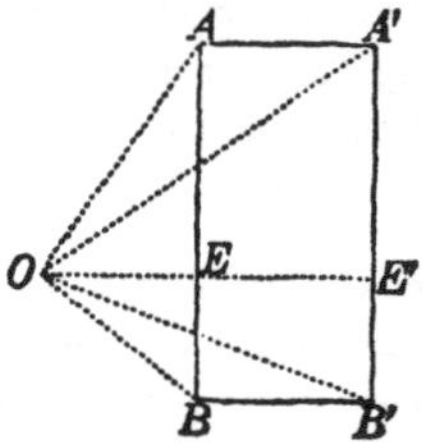

$$\rho t \log\left\{\frac{(OA+AE)(OB+BE)\,OE'^2}{(OA'+A'E')(OB'+B'E')\,OE^2}\right\} \quad\ldots\ldots\ldots\ldots\ldots(7),$$

where ρ denotes the density of the water, and t the thickness of the plate, by hypothesis a small fraction of OE. (We leave the proof as an exercise to the student.) Now, taking the nautical mile as 2000 yards, we have, according to the assumed data, very approximately

$$\frac{OA}{OE} = \frac{AE}{OE} = \frac{OE'}{OE} = 1000, \text{ and } \frac{OA'}{OE} = 1000\sqrt{2}:$$

and B, B' are to be taken as at the same distances on one side of OE' as AA' on the other. Hence the preceding expression becomes

Attraction of high water on a plummet at the sea-side.

$$2\rho t \log \frac{2000}{1+\sqrt{2}}, \text{ which is equal to } 13\cdot 44 \times \rho t.$$

The ratio of this to $\frac{4}{3}\pi\sigma r$, the earth's whole attraction on O, is $3 \times 13\cdot 4\rho t/4\pi\sigma r$: which (as t/r is $\frac{1}{2,100,000}$ by hypothesis, and ρ/σ is about $\frac{2}{11}$) amounts to $\frac{1}{3,580,000}$. The plummet will therefore, at high tide, be disturbed from the position it had at low tide, by a horizontal force of somewhat more than $\frac{1}{4,000,000}$ of the vertical force; and its deviation will of course be this fraction of $57^{\circ}\cdot 3$, the unit angle.

818′. Since the publication of our first edition the British Association has endeavoured to promote the existence of practical gravitational observatories by the appointment of a committee for determining experimentally the lunar disturbance of gravity. The Reports for the years 1881 and 1882 contain accounts of the work which has been done hitherto. In § 818 we did not mean to suggest the seaside as a proper site for a gravitational observatory, and the investigation of that section renders it evident that for the purposes in view of the committee it is essential that the observatory should be remote from the sea-coast.

Gravitational Observatory.

The object of the experimenters for the committee, Mr George and Mr Horace Darwin, being to measure, if possible, the attraction of the moon, and thus to throw light on the elastic yielding of the earth's mass (see § 837 et seq.), care was taken by them to eliminate as far as possible the effects of tremors, either local and seismic. The experiments were, and are still being, carried out at Cambridge, but notwithstanding all the precautions taken to shield the instrument (a pendulum hung in fluid) from disturbance, it was found that the agitation of the soil was incessant. There is strong evidence that this agitation is wholly independent of the tremors produced by traffic in the town, for (amongst other proofs) it appeared that there were periods, lasting during several days, of abnormal

Gravitational Observatory.

agitation and of abnormal quiescence. The experimenters found that superposed on this minute agitation there is a diurnal oscillation of level of some regularity; and that superposed on that again there are continuous changes of level lasting over many weeks. The experiments afford no evidence as to the extent of land over which these changes range; and as the work is still in progress, we should have made no allusion to it here, but that the subject has been attacked from an entirely different point of view, and at earlier dates, by a number of other observers. The general character of the disturbances noted by these other observers agrees in every particular with what is described by the Darwins, and thus we are compelled to believe that none of them were noting a purely local effect. As most of the other experimenters have had in view the observation of minute earthquakes, their instruments have in general been made excessively sensitive to tremor, and the selection of appropriate sites has been rendered very difficult.

We may mention the following instances of observations which agree in character with those of which we have spoken, viz. by D'Abbadie in Brazil and Ethiopia with spirit levels, and on the Pyrenees by reflexion from mercury; by Plantamour at Geneva with spirit-levels; by Zöllner at Leipsig with "a horizontal pendulum"; by Bouquet de la Grye at Campbell Island in the S. Pacific Ocean, with a pendulum. But the observations to which we would especially draw attention are those of the Italians, who have far excelled in zeal all the other nations combined. This has probably been due to the presence in their country of active volcanoes, so that attention has been drawn to the science of earthquakes. In Italy we find Rossi, Bertelli, Palmieri, Mocenigo, Malvasia, Agostini, Galli and many others making continuous observations in many parts of the country for some years past. Their results are being recorded in the *Bulletino del Vulcanismo Italiano**. Milne, Ewing, and Gray have worked in Japan in the same field,—but to note all those who have attended to Seismology would be beyond the scope of our present remarks.

* One of the most interesting points is the use of the microphone for the detection of telluric disturbance.

We here only wish to draw attention to the subject of the slower changes in the direction of gravity relatively to the earth's surface, and to shew that although such results of gravitational observation, as were contemplated by the British Association in the appointment of a Committee, may probably be impossible, yet an important method appears to be initiated for discovery with regard to the mechanical constitution of the upper strata of the earth. For this end it is essential that instruments should be improved, for which there is much scope, and that, following the Italian example, the observations should be simultaneous over large tracts of country.

Gravitational Observatory.

819. Recurring to the case of $\rho=\sigma$, we learn from § 817 that a homogeneous liquid in equilibrium under the influence of centrifugal force, or of tide-generating action, has $2\frac{1}{2}$ times as much ellipticity as it would have if mutual attraction between the parts of the fluid were done away with (§ 800), and gravity were towards a fixed interior centre of force. For a homogeneous liquid of the same mean density as the earth, rotating in a time equal to the sidereal day, the ellipticity is therefore $\frac{1}{232}$, being $2\frac{1}{2}$ times the result, $\frac{1}{580}$, which we found in § 801. This agrees with the conclusion for the case of approximate sphericity, which we derived (§ 775) from the theorem of § 771, regarding the equilibrium of a homogeneous rotating liquid. But even for this case Laplace's spherical harmonic analysis is most important, as proving that the solution is *unique*, when the figure is approximately spherical; so that neither an ellipsoid with three unequal axes, nor any other figure than the oblate elliptic spheroid of revolution, can satisfy the hydrostatic conditions, when the restriction to approximate sphericity is imposed. Our readers will readily appreciate this item of the debt we owe to the great French naturalist, when we tell them that one of us had actually for a time speculated on three unequal axes as a possible figure of terrestrial equilibrium.

Application of § 817 to theory of the earth's figure.

820. As another example of the result of § 817 for the case $i=2$, let us imagine the earth, rotating with the actual angular velocity, to consist of a solid centrobaric nucleus covered with a thin liquid layer of density equal to the true density of the

Application of § 817 to theory of the earth's figure.

upper crust, that is, we may say, half the mean density of the nucleus. The ellipticity of the free surface would be

$$\frac{1}{580} \times \frac{1}{1 - \frac{3}{5} \times \frac{1}{2}}, \text{ or } \frac{1}{406}.$$

Or, lastly, let it be required to find the density of a superficial liquid layer on a centrobaric nucleus which, with the actual angular velocity of rotation, would assume a spheroidal figure with ellipticity equal to $\frac{1}{295}$, the actual ellipticity of the sea level. We should have

$$\frac{1}{1 - \frac{3}{5}\rho/\sigma} = \frac{580}{295},$$

which gives $\rho = {\cdot}819 \times \sigma$.

821. Bringing together the several results of §§ 801, 817, 819, for a centrobaric nucleus revolving with the earth's angular velocity, and covered with a thin layer of liquid of density ρ, the mean density of the whole being σ, we have—

(1) for $\frac{\rho}{\sigma} = 0$, $e = \frac{1}{580}$,

(2) „ $\frac{\rho}{\sigma} = \frac{2}{11}$, $e = \frac{1}{517}$,

(3) „ $\frac{\rho}{\sigma} = \frac{1}{2}$, $e = \frac{1}{406}$,

(4) „ $\frac{\rho}{\sigma} = {\cdot}819$, $e = \frac{1}{295}$,

(5) „ $\frac{\rho}{\sigma} = 1$, $e = \frac{1}{232}$,

where e denotes the ellipticity of the free bounding surface of the liquid. The density of the earth's upper crust may be roughly estimated as $\frac{1}{2}$ the mean density of the entire mass, and is certainly in every part less than ·819 of this mean density. The ellipticity of the sea level does not differ from $\frac{1}{295}$ by more than 2 or 3 per cent., and is therefore decidedly too great to be accounted for by centrifugal force, and ellipticity in the upper crust alone, on the hypothesis that there is a rigid centrobaric nucleus, covered by only a thin upper crust with surface on the whole agreeing in ellipticity with the free liquid surface. It is therefore quite certain that there must be on the whole some degree of oblateness in the lower strata, in the same direction as that which centrifugal force would produce if the mass were fluid. There is, as we shall see in later volumes, a great variety of convincing evidence in support of the common geological hypothesis that the upper crust was at one time all melted by heat. This would account for the general agreement of the boundary of the solid with that of fluid equilibrium, though largely disturbed by upheavals, and shrinkings, in the process of solidification which (App. D.) has probably been going on for a few million years, but is not yet quite complete (witness lava flowing from still active volcanoes). The oblateness of the deeper layers of equal density which we now infer from the figure of the sea level, the observed density of the upper crust, and Cavendish's weighing of the earth as a whole, renders it highly probable that the earth has been at one time melted not merely all round its surface, but either throughout, or to a great depth all round.

Observation shows so great an ellipticity of sea level that there must be oblateness of the solid not only in its bounding surface, but also in interior layers of equal density.

822. We therefore, as our last hydrostatic example, proceed to investigate the conditions of a heterogeneous liquid resting on a rigid spherical centrobaric core or nucleus, and slightly disturbed, as explained in § 815, by attracting masses fixed either externally or in the core (among which, of course, must be included deviations, if any, from a rigorously centrobaric distribution in the matter of the core).

Equilibrium of rotating spheroid of heterogeneous liquid, investigated.

For any point (r, θ, ϕ) in space let

N be the potential due to the core,

V „ „ undisturbed fluid,

Q „ „ disturbing force,

U „ „ disturbance in the distribution of the fluid.

Thus the whole potential at the point in question is $N+V$ when the fluid is undisturbed, and $N+Q+V+U$ when the disturbing force is introduced and equilibrium supervenes. Let also ρ be the density of the undisturbed fluid at (r, θ, ϕ) (which of course would vanish if the point in question were situated in any other

Equilibrium of rotating spheroid of heterogeneous liquid, investigated.

part of space than that occupied by the fluid); and let $\rho + \varpi$ be the altered density at the same point (r, θ, ϕ) when the fluid rests under the disturbing influence. It is to be noticed that N, V, ρ are functions of r alone; while Q, U, ϖ are functions of r, θ, ϕ.

Let now δr be an infinitely small variation of r. The density of the liquid at the point $(r + \delta r, \theta, \phi)$ will be $\rho + \varpi + \frac{d}{dr}(\rho + \varpi)\delta r$, or simply

$$\rho + \varpi + \frac{d\rho}{dr}\delta r,$$

as ϖ is infinitely small by hypothesis. If we equate this to ρ we have

$$\varpi + \frac{d\rho}{dr}\delta r = 0,$$

Spheroidal surface of equal density.

and deduce

$$\delta r = -\frac{\varpi}{d\rho/dr} \quad \text{..........................(1)}$$

for the equation expressing the deviation from the spherical surface of radius r, of the spheroidal surface over which the density in the disturbed liquid is ρ. The liquid being incompressible, the volume enclosed by this spheroidal surface must be equal to that enclosed by the spherical surface, and therefore, if $d\sigma$ denote an element of the spherical surface, and $\iint$ integration over the whole of it,

$$\iint \delta r d\sigma = 0 \quad \text{..........................(2).}$$

Hence, by (1), as $\frac{d\rho}{dr}$ is independent of θ, ϕ,

$$\iint \varpi d\sigma = 0 \quad \text{..........................(3).}$$

Expression of incompressibility.

Now, as before for density, we have for the disturbed potential at $(r + \delta r, \theta, \phi)$

$$N + Q + V + U + \frac{d}{dr}(N + Q + V + U)\delta r,$$

or, because $Q + U$ is infinitely small,

$$N + Q + V + U + \frac{d}{dr}(N + V)\delta r.$$

And, therefore, to express that the spheroidal surface corresponding to (1), with r constant, is an equipotential surface in the disturbed liquid, we have

$$Q + U - \frac{\frac{d}{dr}(N+V)}{\frac{d\rho}{dr}}\varpi + N + V = F(r) \quad\text{.............}(4),$$

Hydrostatic equation.

which (§ 750) is the equation of hydrostatic equilibrium. In this equation we must suppose N and ρ to be functions of r, and Q a function of r, θ, ϕ; all given explicitly: and from ρ we have, by putting $i=0$, in (15) and (16) of § 542,

$$V = 4\pi\left(\int_r^{\mathfrak{r}} r'\rho' dr' + \frac{1}{r}\int_a^r r'^2\rho' dr'\right) \quad\text{................}(5),$$

Equilibrium of rotating spheroid of heterogeneous liquid.

where ρ' is the value of ρ at distance r' from the centre, and $\mathfrak{r}$ the radius of the outer bounding surface of the undisturbed fluid, and a that of the fixed spherical surface of the core on which it rests. To find $V+U$, following strictly the directions of § 545, we add the potential of a distribution of matter with density $\rho+\varpi$ through the space between the spherical surfaces of radii a and $\mathfrak{r}$ to that of the shell B of positive and negative matter there defined. Let the thickness of the latter at the point $(\mathfrak{r}, \theta, \phi)$ be called h, being the value of δr at the surface; and let q denote its density, being the surface value of ρ. Then, subtracting the undisturbed potential V, we have

$$U = \iiint \frac{\varpi' r'^2}{D} d\sigma' dr' + \left[\iint \frac{q' h'}{D} d\sigma'\right] \quad\text{...............}(6),$$

if as usual D denote the distance between the points (r, θ, ϕ), (r', θ', ϕ'), and the accented letters denote the values of the corresponding elements in the latter; and if [ ] denote surface values and integration.

Let us now suppose the required deviation of the surfaces of equal pressure density and potential to be expressed as follows in surface harmonics, of which the term R_0 disappears because of (2):—

Part of the potential, due to oblateness:

$$\left.\begin{array}{l}\text{for the interior of the fluid, } \delta r = R_1 + R_2 + R_3 + \text{etc.,}\\ \text{and for the outer bounding surface, } h = \mathbf{R}_1 + \mathbf{R}_2 + \mathbf{R}_3 + \text{etc.}\end{array}\right\} (7).$$

Hence by (1) $\qquad \varpi = -\frac{d\rho}{dr}(R_1 + R_2 + R_3 + \text{etc.}) \quad\text{...........}(8).$

developed in harmonics.

Using this in (6) according to §§ 544, 542, 536, we have

$$U = -4\pi \sum_1^\infty \frac{1}{2i+1}\left\{r^i \int_r^{\mathfrak{r}} r'^{-i+1}\frac{d\rho'}{dr'}R_i'dr' + r^{-i-1}\int_a^r r'^{i+2}\frac{d\rho'}{dr'}R_i'dr' - q\mathfrak{R}_i\frac{r^i}{\mathfrak{r}^{i-1}}\right\} \text{..(9)},$$

where R_i' denotes the value of R_i from the point (r', θ, ϕ) instead of (r, θ, ϕ).

To complete the expansion of the hydrostatic equation (4) we may suppose the harmonic expression for Q to be either directly given, or be found immediately by Appendix B. (52), or by (8) of § 539, according to the form in which the data are presented. Thus let us have

Harmonic development of disturbing potential.

$$Q = \sum_{i=0}^{i=\infty}\sum_{s=0}^{s=i}(A_i^{(s)}\cos s\phi + B_i^{(s)}\sin s\phi)\,\Theta_i^{(s)} \text{................(10)},$$

according to the notation of App. B. (37) and (38), $A_i^{(s)}$, $B_i^{(s)}$ denoting known functions of r. Using now this and (8) in (4), we have

$$\sum_{i=1}^{i=\infty}\left\{\sum_{s=0}^{s=i}(A_i^{(s)}\cos s\phi + B_i^{(s)}\sin s\phi)\,\Theta_i^{(s)} + R_i\frac{d}{dr}(N+V)\right.$$

$$\left. - \frac{1}{2i+1}\left(r^i\int_r^{\mathfrak{r}} r'^{-i+1}\frac{d\rho'}{dr'}R_i'dr' + r^{-i-1}\int_a^r r'^{i+2}\frac{d\rho'}{dr'}R_i'dr' - \mathfrak{R}_i\frac{r^i}{\mathfrak{r}^{i-1}}\right)\right\}$$

$$+ A_0^{(0)} + N + V = F(r) \text{......................(11)}.$$

Hence: first, for the terms of zero order

$$A_0^{(0)} + N + V = F(r) \text{......... ....(12)},$$

which merely shows the value of $F(r)$, introduced temporarily in (4) and not wanted again: and, by terms of order i,

$$-R_i\frac{d}{dr}(N+V) + \frac{1}{2i+1}\left\{r^i\int_r^{\mathfrak{r}} r'^{-i+1}\frac{d\rho'}{dr'}R_i'dr' + r^{-i-1}\int_a^r r'^{i+2}\frac{d\rho'}{dr'}R_i'dr' - q\mathfrak{R}_i\frac{r^i}{\mathfrak{r}^{i-1}}\right\}$$

$$= \sum_{s=0}^{s=i}(A_i^{(s)}\cos s\phi + B_i^{(s)}\sin s\phi)\,\Theta_i^{(s)} \text{..............(13)}.$$

Equation of equilibrium for general harmonic term:

Lastly, expanding R_i (as above for the i term of Q) by App. B. (37), let us have

$$R_i = \sum_{s=0}^{s=i}(u_i^{(s)}\cos s\phi + v_i^{(s)}\sin s\phi)\,\Theta_i^{(s)} \text{.................(14)},$$

where $u_i^{(s)}$, $v_i^{(s)}$ are functions of r, to the determination of which the problem is reduced. Hence equating separately the coefficients of $\Theta_i^{(s)}\cos s\phi$, etc., on the two sides, and using u_i to denote any

leading to equation for general coefficient u_i regarded as a function of r.

one of the required functions $u_i^{(s)}$, $v_i^{(s)}$, and A_i any of the given functions $A_i^{(s)}$, $B_i^{(s)}$, and u_i', u_i the values of u_i for $r = r'$ and $r = \mathrm{r}$ respectively, we have

$$-u_i \frac{d}{dr}(N+V) + \frac{4\pi}{2i+1}\left\{r^i \int_r^{\mathrm{r}} r'^{-i+1} \frac{d\rho'}{dr'} u_i' dr' + r^{-i-1} \int_a^r r'^{i+2} \frac{d\rho'}{dr'} u_i' dr' - q \frac{\mathrm{u}_i r^i}{\mathrm{r}^{i-1}}\right\} = A_i \quad \ldots (15),$$

or, as it will be convenient sometimes to write it, for brevity, $\sigma_i(u_i) = A_i$

where σ_i denotes a determinate operation, performed on u any function of r, continuous or discontinuous. To reduce (15) to a differential equation, divide by r^i, differentiate, multiply by r^{2i+2}, and differentiate again. If, for brevity, we put

Equilibrium of rotating spheroid of heterogeneous liquid.

$$-\frac{d}{dr}(N+V) = r\psi \quad \ldots\ldots (16),$$

the result is

$$\frac{d}{dr}\left\{r^{2i+2} \frac{d}{dr}(r^{-i+1}\psi u_i)\right\} - 4\pi r^{i+2} \frac{d\rho}{dr} = \frac{d}{dr}\left\{r^{2i+2} \frac{d}{dr}(A_i r^{-i})\right\} \ldots (17),$$

Differential equation to be integrated.

a linear differential equation, of the second order, for u_i, with coefficients and independent terms known functions of r. The general solution, as is known, is of the form

$$u_i = CP + C'P' + \alpha \quad \ldots\ldots (18),$$

where α is a function of r satisfying the integral equation

$$\sigma_i(\alpha) = A_i \quad \ldots\ldots (19) \ [(15) \text{ repeated}];$$

C and C' are two arbitrary constants, and P and P' are two distinct functions of r.

Equation (15) requires that $C = 0$ and $C' = 0$; in other words, u_i, if satisfying it, is fully determinate. This is best shown by remarking that if, instead of (15), we take

$$\sigma_i(u) = A_i + Kr^i + K'r^{-i-1} \quad \ldots\ldots (20)$$

where K, K' are any two constants, these constants disappear in the differentiations, and we have still the same differential equation, (17): and that the two arbitrary constants C and C' of the general solution (18) of this are determined by (20) when any two values are given for K and K'. In fact, the expression (18), used for u_i, reduces (20) to

Determination of constants to complete the required solution.

$$C\sigma_i(P) + C'\sigma_i(P') = Kr^i + K'r^{-i-1} \quad \ldots\ldots (21),$$

which shows that $\sigma_i(P)$ and $\sigma_i(P')$ cannot either of them be

Determination of constants to complete the required solution.

zero, and that they must be distinct linear functions of r^i and r^{-i-1}, and determines C and C'.

Thus we see that whatever be A_i we have, in the integration of the differential equation (19), and the determination of the arbitrary constants to satisfy (15), the complete solution of our problem.

Introduction of the Newtonian law of force.

Unless it is desired, as a matter of analytical curiosity, or for some better reason, to admit the supposition that N is any arbitrary function of r, it is unnecessary to retain both ψ and ρ as two distinct given functions. For the external force of the nucleus, or that part of it of which N is the potential, being by hypothesis symmetrical relatively to the centre, it must in nature vary inversely as the square of the distance from this point; that is to say,

$$-\frac{dN}{dr}=\frac{\mu}{r^2} \quad \text{.........................(22)},$$

μ being a constant, measuring in the usual unit (§ 459) the mass of the nucleus. And by (5)

$$-\frac{dV}{dr}=\frac{4\pi}{r^2}\int_a^r \rho' r'^2 dr' \quad \text{....................(23)}.$$

From this, with (22) and (17), we have

$$\psi=\frac{4\pi}{r^3}\int_a^r \rho' r'^2 dr' + \frac{\mu}{r^3} \quad \text{....................(24)},$$

which gives $4\pi\rho=\dfrac{d(\psi r^3)}{r^2 dr}$ and $4\pi\dfrac{d\rho}{dr}=r\dfrac{d^2\psi}{dr^2}+4\dfrac{d\psi}{dr}$ ......(25).

Simplification by introducing the Newtonian law of force.

Using this last in (17), and reducing by differentiation, we have

$$\frac{d^2u_i}{dr^2}+2\left(\frac{d}{dr}\log\psi+\frac{2}{r}\right)\frac{du_i}{dr}-\frac{(i-1)(i+2)}{r^2}u_i=\frac{1}{r^{2i+3}}\frac{d}{dr}\left\{r^{2i+2}\frac{d}{dr}(r^{-i}A_i)\right\} \text{...(26)}.$$

Another form, convenient for cases in which the disturbing force is due to *external* attracting matter, or to centrifugal force of the fluid itself, if rotating, is got by putting, in (17),

$$r^{-i+1}u_i=e_i \quad \text{.........................(27)},$$

and reducing by differentiation. Thus

$$\frac{d^2e_i}{dr^2}+2\left(\frac{d}{dr}\log\psi+\frac{i+1}{r}\right)\frac{de_i}{dr}+\frac{2(i-1)}{r}e_i\frac{d}{dr}\log\psi=\frac{1}{r^{2i+2}}\frac{d}{dr}\left\{r^{2i+2}\frac{d}{dr}(r^{-i}A_i)\right\} \text{(28)}.$$

With this notation the intermediate integral, obtained from (15) by the first step of the process of differentiating executed in the order specified, gives

Differential equation for proportionate deviation from sphericity.

$$\frac{de_i}{dr} + e_i\frac{d}{dr}\log\psi - r^{-2i-2}\int_a^r\left(r\frac{d^2\psi}{dr^2} + 4\frac{d\psi}{dr}\right)r^{2i+1}e_i\,dr = \frac{d}{dr}(r^{-i}A_i) \;\ldots(29).$$

Equilibrium of rotating spheroid of heterogeneous liquid.

Important conclusions, readily drawn from these forms, are that if Q is a solid harmonic function (as it is when the disturbance is due either to disturbing bodies in the core, or in the space external to the fluid, or to centrifugal force of the fluid rotating as a solid about an axis); then (1) e_i, regarded as positive, and as a function of r, can have no maximum value, although it might have a minimum; and (2) if the disturbance is due to disturbing masses outside, or to any other cause (as centrifugal force) which gives for potential a solid harmonic of order i with only the r^i term, and no term r^{-i-1}, e_i can have no minimum except at the centre, and must increase outwards throughout the fluid.

Layers of greatest and least proportionate deviation from sphericity.

To prove these conclusions, we must first remark that ψ necessarily diminishes outwards. To prove this, let n denote the excess of the mass of the nucleus above that of an equal solid sphere of density s equal to that of the fluid next the nucleus. Then we may put (24) under the form

$$\psi = \tfrac{4}{3}\pi s - \frac{4\pi}{r^3}\int_a^r (s-\rho')\,r'^2dr' + \frac{n}{r^3}\;\ldots\ldots\ldots\ldots(30).$$

For stability it is necessary that n and $s-\rho'$ be each positive; and therefore the last term of the second member is positive, and diminishes as r increases, while the second term of the same is negative, and in absolute magnitude increases, and the first term is constant. Hence ψ diminishes as r increases. Again, when the force is of the kind specified, we must [App. B. (58)] have

$$A_i = Kr^i + K'r^{-i-1}\;\ldots\ldots\ldots\ldots\ldots\ldots\ldots\ldots(31),$$

and therefore the second member of (28) vanishes. Hence if, for any value of r, $de_i/dr = 0$,

for the same, $$\frac{d^2e_i}{dr^2} = -\frac{2(i-1)}{r}e_i\frac{d}{dr}\log\psi,$$

and is therefore positive, which proves (1). Lastly, when the force is such as specified in (2), we have $A_i = Kr^i$ simply, and

Proportionate deviation for case of centrifugal force, or of force from without.

therefore the second member of (29) vanishes. This equation then gives, for values of r exceeding a by infinitely little,

$$\frac{de_i}{dr} = -e_i \frac{d}{dr} \log \psi,$$

which is positive. Hence e_i commences increasing from the nucleus. But it cannot have a minimum (1), and therefore it increases throughout, outwards.

Case of centrifugal force.

823. When the disturbance is that due to rotation of the liquid, the potential of the disturbing force is $\frac{1}{2}\omega^2(x^2+y^2)$, which is equal to a solid harmonic of the second degree with a constant added. From this it follows [§§ 822, 779] that the surfaces of equal density are concentric oblate ellipsoids of revolution, with a common axis, and with ellipticities diminishing from the surface inwards.

We have, in (10) of last section,

$$Q = \tfrac{1}{2}\omega^2(x^2+y^2) = \tfrac{1}{3}\omega^2 r^2 (\Theta_0^{(0)} + \Theta_2^{(0)}).$$

This gives by (7) and (14),

$$\delta r = u_2 \Theta_2^{(0)}.$$

Hence $$r + \delta r = r\left(1 + \frac{u_2}{r}\Theta_2^{(0)}\right) = r\left[1 + \frac{u_2}{r}(\tfrac{1}{3} - \cos^2\theta)\right]$$

$$= r\left(1 - \frac{2u_2}{3r}\right)\left(1 + \frac{u_2}{r}\sin^2\theta\right) \dots\dots\dots\dots\dots\dots (1),$$

neglecting terms of the second order because ω, and therefore also u_2/r, are very small.

Thus the sphere, whose radius was r, has become an oblate ellipsoid of revolution whose ellipticity [§ 822 (27)] is

$$e_2 = \frac{u_2}{r} \dots\dots\dots\dots\dots\dots\dots\dots\dots (2).$$

Its polar diameter is diminished by the fraction $\frac{2}{3}u_2/r$ or $\frac{2}{3}e_2$, and its equatorial diameter is increased by $\frac{1}{3}e_2$; the volume remaining unaltered.

In order to find the value of u_2, we must have data or assumptions which will enable us to integrate equation (15). These may be given in many forms; but one alone, to which we proceed, has been worked out to practical conclusions.

824. To apply the results of the preceding investigation to the determination of the law of ellipticity of the layers of equal density within the earth, on the hypothesis of its original fluidity, it is absolutely essential that we commence with some assumption (in default of information) as to the law which connects the density with the distance from the earth's centre. For we have seen (§ 821) how widely different are the results obtained when we take two extreme suppositions, viz., that the mass is homogeneous; and that the density is infinitely great at the centre. In few measurements hitherto made of the Compressibility of Liquids (see Vol. II., *Properties of Matter*) has the pressure applied been great enough to produce condensation to the extent of one-half per cent. The small condensations thus experimented on have been found, as might be expected, to be very approximately in simple proportion to the pressures in each case; but experiment has not hitherto given any indication of the law of compressibility for any liquid under pressures sufficient to produce considerable condensations. In default of knowledge, Laplace assumed, as an hypothesis, the law of compressibility of the matter of which, before its solidification, the earth consisted, to be that the *increase of the square of the density is proportional to the increase of pressure.* This leads, by the ordinary equation of hydrostatic equilibrium, to a very simple expression for the law of density, which is still further simplified if we assume that the density is everywhere finite.

Laplace's hypothetical law of density within the earth.

Assumed relation between density and pressure.

Neglecting the disturbing forces, we have (§§ 822, 752)

$$dp = \rho d(V+N) \qquad\qquad (1).$$

But, by the hypothesis of Laplace, as above stated, k being some constant

$$dp = k\rho d\rho \qquad\qquad (2).$$

Hence

$$k\rho + C = V + N$$

or, by § 822 (5),

$$= 4\pi \int_r^1 r'\rho' dr' + \frac{4\pi}{r}\int_a^r r'^2 \rho' dr' + \frac{\mu}{r}.$$

Multiplying by r, and differentiating, we get

$$k\frac{d}{dr}(r\rho) + C = 4\pi\int_r^1 r'\rho' dr'$$

Laplace's hypothetical law of density within the earth.

and
$$\frac{d^2}{dr^2}(r\rho) = -\frac{4\pi}{k}r\rho.$$

If we write $4\pi/k = 1/\kappa^2$, the integral may be thus expressed—

$$r\rho = F\sin\left(\frac{r}{\kappa} + G\right).$$

If we suppose the whole mass to be liquid, *i.e.*, if there be no solid core, or, at all events, the same law of density to hold from surface to centre, G must vanish, else the density at the centre would be infinite. Hence, in what follows, we shall take

Law of density.

$$\rho = \frac{F}{r}\sin\frac{r}{\kappa} \qquad\qquad (3).$$

With this value of ρ it is easy to see that

$$\int_0^r r'^2\rho' dr' = -\kappa^2 r^2\frac{d\rho}{dr} \qquad\qquad (4),$$

the common value of these quantities being

$$F\kappa^2\left(\sin\frac{r}{\kappa} - \frac{r}{\kappa}\cos\frac{r}{\kappa}\right).$$

Determination of ellipticities of surfaces of equal density.

We are now prepared to find the value of u_2 in § 823, upon which depends the ellipticity of the strata. For (15) of § 822 becomes, by (23) of that section and the late equation (4),

$$\left(\frac{\mu-\mu'}{r^3} - 4\pi\kappa^2\frac{d\rho}{dr}\right)u_2 + \frac{4\pi}{5}\left[r^2\int_r^{\text{ɪ}}\frac{u_2'}{r'}\frac{d\rho'}{dr'}dr' + r^{-3}\int_a^r r'^4 u_2'\frac{d\rho'}{dr'}dr\right] - \frac{4\pi}{5}q\frac{u_2 r^2}{\text{ɪ}} = \tfrac{1}{3}\omega^2 r^2 \quad ...(5)$$

where μ' is the mass of fluid, following the density law (3), which is displaced by the core μ, and q is the surface density. In the terrestrial problem we may assume $\mu' = \mu$, and of course $a = 0$. For simplicity put

$$r\frac{d\rho}{dr}u_2 = v \qquad\qquad (6),$$

then divide by r^2 and differentiate, and we have

$$\frac{d}{dr}\left(\frac{v}{r^3}\right) + \frac{1}{\kappa^2 r^6}\int_0^r r'^3 v' dr' = 0.$$

Multiply by r^6, and again differentiate; the result is

$$\frac{d^2 v}{dr^2} + \left(\frac{1}{\kappa^2} - \frac{6}{r^2}\right)v = 0 \qquad\qquad (7).$$

The integral of this equation is known to be

Laplace's hypothetical law of density within the earth.

$$v=C\left[\left(\frac{3}{r^2}-\frac{1}{\kappa^2}\right)\sin\left(\frac{r}{\kappa}+C'\right)-\frac{3}{\kappa r}\cos\left(\frac{r}{\kappa}+C'\right)\right]\ldots\ldots(8),$$

so that u_2 is known from (6). Now we have already proved that u_2 increases from the centre outwards, so that we must have $C'=0$, for otherwise u_2 would be infinite at the centre. Thus, dropping the suffix $_2$ to the symbol e for brevity, we have

Consequences as regards ellipticities of surfaces of equal density.

$$e=\frac{u_2}{r}=-\frac{C}{F}\frac{\left(\frac{3}{r^2}-\frac{1}{\kappa^2}\right)\tan\frac{r}{\kappa}-\frac{3}{\kappa r}}{\tan\frac{r}{\kappa}-\frac{r}{\kappa}}\ldots\ldots\ldots\ldots\ldots\ldots(9).$$

Now let $$\vartheta=\frac{r}{\kappa}\ldots\ldots\ldots\ldots\ldots\ldots\ldots\ldots\ldots\ldots(9^{\mathrm{I}}).$$

We may thus write (9) as follows:

$$e=-\frac{C}{F\kappa^2}\left\{\frac{3}{\vartheta^2}-\frac{1}{1-\vartheta\cot\vartheta}\right\}\ldots\ldots\ldots\ldots\ldots\ldots(9^{\mathrm{II}}).$$

The constants are, of course, to be determined by the known values of the ellipticity of the surface and of the angular velocity of the mass.

Now (5) becomes, at the surface,

$$\frac{4\pi}{\mathfrak{r}^2}u_2\int_0^{\mathfrak{r}}\rho r^2dr+\frac{4\pi}{5\mathfrak{r}^3}\int_0^{\mathfrak{r}}r^4\frac{d\rho}{dr}u_2\,dr=\tfrac{1}{3}\mathfrak{r}^2\omega^2+\tfrac{4}{5}\pi q u_2\mathfrak{r}\ldots(10).$$

We may next eliminate ρ, $d\rho/dr$, and q, being the surface value of ρ, by means of (3) (4), (6), and (8), and substitute everywhere re for u_2. Also, if m be the ratio $(\frac{1}{289})$ of centrifugal force to gravity at the equator, ω is to be eliminated by means of the equation

$$m=\frac{\mathfrak{r}\omega^2}{\frac{4\pi}{\mathfrak{r}^2}\int_0^{\mathfrak{r}}\rho r^2dr},$$

from which ρ is to be removed by (3). By the help of these substitutions (10) becomes transformed as follows:—

$$\frac{4\pi F\mathfrak{e}}{\mathfrak{r}}\int_0^{\mathfrak{r}}r\sin\frac{r}{\kappa}dr+\frac{4\pi C}{5\mathfrak{r}^3}\int_0^{\mathfrak{r}}r^3\left[\left(\frac{3}{r^2}-\frac{1}{\kappa^2}\right)\sin\frac{r}{\kappa}-\frac{3}{\kappa r}\cos\frac{r}{\kappa}\right]dr$$
$$=\frac{4\pi mF}{2\mathfrak{r}}\int_0^{\mathfrak{r}}r\sin\frac{r}{\kappa}dr+\frac{4\pi F}{5}\mathfrak{r}\mathfrak{e}\sin\frac{\mathfrak{r}}{\kappa}.$$

If we put $\tan\mathfrak{r}/\kappa=t$, and $\mathfrak{r}/\kappa=\theta$, so that θ is the surface value of ϑ, the integrated expression, divided by $\frac{4}{5}\pi F\mathfrak{e}\kappa^3\cos\theta/\mathfrak{r}$, with

Laplace's hypothetical law of density within the earth.

C eliminated by (9^{ii}), becomes

$$5(t-\theta)-\frac{t-\theta}{(3-\theta^2)t-3\theta}\left[15(t-\theta)+\theta^3-6t\theta^2\right]=\frac{5m}{2e}(t-\theta)+\theta^2t.$$

Ratio of ellipticity of surface to the fraction expressing centrifugal force at equator in terms of surface gravity.

Hence at once

$$\frac{5m}{2e}=\frac{\theta^4+\theta^3t+\theta^4t^2-2t^2\theta^3}{(t-\theta)[(3-\theta^2)t-3\theta]}\dots\dots\dots\dots(11).$$

If we put $1-z$ for $\frac{\theta}{t}$, *i.e.*, for $\frac{\mathfrak{r}/\kappa}{\tan\mathfrak{r}/\kappa}$, this becomes somewhat simpler, and may be written

$$\frac{5m}{2e}=\frac{\theta^4-3z\theta^2+z^2\theta^2}{z(3z-\theta^2)}=\frac{z\theta^2}{3z-\theta^2}-\frac{\theta^2}{z}\dots\dots\dots\dots(12).$$

The mean density of the sphere comprised within the radius r is

$$\frac{\int_0^r\rho r^2dr}{\int_0^r r^2dr}=F\kappa^2\frac{\{\sin(r/\kappa)-(r/\kappa)\cos(r/\kappa)\}}{\frac{1}{3}r^3}=\frac{3F}{\kappa}\left\{\frac{\sin\vartheta-\vartheta\cos\vartheta}{\vartheta^3}\right\}.$$

Let ρ_0 be the mean density of the sphere comprised within this radius r, and ρ, as before, the density at the stratum defined by the radius r. It may be noted in passing that q_0 and q are the values of ρ_0 and ρ corresponding to $r=\mathfrak{r}$.

Then,

$$\rho_0=\frac{3F}{\kappa}\left\{\frac{\sin\vartheta-\vartheta\cos\vartheta}{\vartheta^3}\right\}\dots\dots\dots\dots(12^{i}),$$

$$\rho=\frac{F}{r}\sin\vartheta=\frac{F}{\kappa}\frac{\sin\vartheta}{\vartheta}.$$

If we put f for the ratio of the mean density of this sphere to the density at its bounding surface, we have

$$f=\frac{3}{\vartheta^2}(1-\vartheta\cot\vartheta)\dots\dots\dots\dots(12^{ii}).$$

Ellipticity of internal stratum.

Substituting in (9^{ii})

$$e=-\frac{C}{F\kappa^2}\frac{3}{\vartheta^2}\left(1-\frac{1}{f}\right).$$

Then writing for ϑ its value r/κ, we have

$$e=-\frac{3C}{F}\frac{1}{r^2}\left(1-\frac{1}{f}\right).$$

Since $3C/F$ is constant, it follows that $(er^2)/(1-1/f)$ is the same for all the strata of equal density. If therefore $\mathfrak{f}$ be the surface value of f, that is to say the ratio q_0/q of the mean density to the surface density of the whole earth,—a quantity which may be determined by experiment,

Laplace's hypothetical law of density within the earth.

$$\frac{er^2}{1-1/f} = \frac{\epsilon r^2}{1-1/\mathfrak{f}} \quad \ldots\ldots\ldots\ldots\ldots\ldots(12^{iii}).$$

Ellipticity of internal stratum.

This formula gives the ellipticity of any internal stratum according to the Laplacian theory.

It may be also reduced to another form which is perhaps rather curious than important, as follows :—

Differentiate (12^{i}) logarithmically with regard to ϑ, and we have

$$\frac{d}{d\vartheta}\log\rho_0 = \frac{\vartheta}{1-\vartheta\cot\vartheta} - \frac{3}{\vartheta},$$

Then by (12^{ii})

$$1-\frac{1}{f} = -\tfrac{1}{3}\vartheta\frac{d}{d\vartheta}\log\rho_0.$$

And since

$$\frac{d\vartheta}{\vartheta} = \frac{dr}{r}$$

$$\frac{1}{r^2}\left(1-\frac{1}{f}\right) = -\tfrac{1}{3}\frac{d}{rdr}(\log\rho_0) = -\tfrac{2}{3}\frac{d}{d(r^2)}\log\rho_0.$$

Hence (12^{iii}) shows that e varies as

$$-\frac{d}{d(r^2)}\log\rho_0.$$

Thus we may state verbally that the ellipticity of any internal stratum varies as the rate of decrease, per unit increase of area of the stratum, of the logarithm of the mean density of the sphere comprised within that stratum*.

The formula (12) for $5m/2\epsilon$ may now be more simply expressed. Attributing to f and ϑ their surface values $\mathfrak{f}$ and θ, we have from (12^{ii})

$$\mathfrak{f} = \frac{3}{\theta^2}(1-\theta\cot\theta) = 3\frac{t-\theta}{t\theta^2} = \frac{3z}{\theta^2} \quad \ldots\ldots\ldots\ldots\ldots\ldots(13).$$

Ratio of mean to surface density.

From this equation θ may be found by approximation, and then (12) gives ϵ in terms of known quantities. In fact, it becomes

$$\frac{5m}{2\epsilon} = \frac{\mathfrak{f}\theta^2}{3(\mathfrak{f}-1)} - \frac{3}{\mathfrak{f}} \ldots\ldots\ldots\ldots\ldots\ldots(14).$$

* This and the preceding mode of expressing the ellipticity of an internal stratum are taken (with changed notation) from a paper by Mr G. H. Darwin in the *Messenger of Mathematics* (Vol. VI.), 1877, p. 109.

Laplace's hypothetical law of density within the earth.

From (13) and (14) the numbers in columns iv. and v. of the following table are easily calculated. Column vii. shows the ratio of the moment of inertia about a mean diameter, on the assumed law of density, to what it would be if the earth were homogeneous:—

i.	ii.	iii.	iv.	v.	vi.	vii.
$1-\frac{\theta}{t}$.	$\frac{\theta}{\pi}180^\circ$.	θ.	f.	ϵ.	$\frac{\mathfrak{C}-\mathfrak{A}}{C}$.	$\frac{C}{\frac{2}{3}Mr^2}$.
3·91	140°	2·444	1·966	$\frac{1}{292}$	·00335	·843
4·24	142°·5	2·487	2·057	$\frac{1}{295}$	·00330	·836
4·61	145°	2·531	2·161	$\frac{1}{299}$	·00325	·826
5·04	147°·5	2·574	2·282	$\frac{1}{302\cdot5}$	·00321	·818
5·53	150°	2·618	2·428	$\frac{1}{306\cdot5}$	·00315	·810
6·11	152°·5	2·662	2·589	$\frac{1}{311}$	·00309	·801
6·80	155°	2·705	2·788	$\frac{1}{315}$	·00304	·792

Ellipticity of strata of equal density.

824′*. The table given in § 824 is principally of interest for application to the case of the earth, because it embraces those values of θ which correspond with values of f nearly equal to 2; and experiment has shown that the mean density of the earth is about twice that of superficial rocks. But the march of the functions θ and f, as we pass from the hypothesis of the homogeneity of the planet to that of infinitely small surface density, will afford an interesting illustration of the Laplacian theory, and will besides afford the means of application with some degree of probability, to some of the other planets.

When θ is small we have

$$\left.\begin{aligned} \mathfrak{f} &= 1 + \tfrac{1}{15}\theta^2 \\ \frac{5m}{2\epsilon} &= 2 + \tfrac{8}{15}\theta^2 \end{aligned}\right\} \quad \ldots\ldots\ldots\ldots\ldots\ldots(1),$$

* This section (§ 824′) is derived from a paper by Mr Darwin in the Monthly Notices of the R. Ast. Soc., Dec. 1876.

Ellipticity of strata of equal density.

and when θ is infinitely nearly equal to 180°

$$\left.\begin{aligned} \mathfrak{f} &= \frac{3}{\pi(\pi-\theta)} \\ \frac{5m}{2\mathfrak{e}} &= \tfrac{1}{3}\{\pi^2 - \pi(\pi-\theta)(5-\tfrac{1}{3}\pi^2)\} \end{aligned}\right\} \text{..............(2).}$$

We see from (1) and (2) that as θ ranges from zero to 180°, $\mathfrak{f}$ increases from unity to infinity, and $5m/2\mathfrak{e}$ from 2 to $\frac{1}{3}\pi^2$.

Intermediate values of these functions, computed from the formulæ of § 824, are given in the following table:—

ϑ or θ in degrees.	f or $\mathfrak{f}$.	$\frac{5m}{2\mathfrak{e}}$.
0°	1·0000	2·000
40	1·0341	2·029
50	1·0548	2·046
60	1·0817	2·067
70	1·1161	2·094
80	1·1600	2·126
90	1·2159	2·165
100	1·2879	2·213
110	1·3827	2·270
120	1·5109	2·338
130	1·6922	2·422
140	1·9657	2·525
150	2·4225	2·652
160	3·3363	2·813
170	6·0750	3·019
180	∞	3·290

The numbers here given are applicable in two ways, viz. for determining the ellipticity of any internal stratum of the earth, and for application to the cases of the external figures of the other planets as above stated.

Ellipticity of strata of equal density.

To determine the ellipticity of an internal stratum we write (12''') § 824 in the following form:—

$$\frac{e}{\epsilon}=\left(\frac{\theta}{\vartheta}\right)^2\frac{1-1/f}{1-1/\mathfrak{f}}\text{.........................}(3).$$

We must in (3) take ϑ as the same fraction of θ, as r, the radius of the stratum in question, is of $\mathfrak{r}$ the earth's mean radius. Thus if for example, $r=\frac{5}{12}\mathfrak{r}$, and if (as is probable in the case of the earth) $\mathfrak{f}=2{\cdot}1$, $\theta=144^\circ$, we must take $\vartheta=60^\circ$. The table then shows that $\vartheta=60^\circ$ gives $f=1{\cdot}0817$. By substitution in (3) we get $e=\frac{1}{1{\cdot}204}\epsilon$; which with $\epsilon=\frac{1}{297}$, gives $e=\frac{1}{357}$.

The distribution of density in Jupiter and Saturn.

In the cases of those planets which have satellites, m and $\epsilon-\frac{1}{2}m$ are determinable from observation and from the theory of the satellites; so that $5m/2\epsilon$ is determinable. This function being known, the corresponding value of $\mathfrak{f}$ is determinable from the table, or by direct computation. For example, Mr G. H. Darwin has shown that in the case of Jupiter, where $5m/2\epsilon$ is $3{\cdot}2646$, we must have $\mathfrak{f}=68$, $\theta=179^\circ\ 11'\ 20''$, and $\epsilon=1/16{\cdot}022$*. Different data, perhaps equally probable, give somewhat different results, but in all cases the physical conclusion is that the superficial density of the visible disk of Jupiter is very small compared with the mean density—a conclusion which appears to agree well with the telescopic appearance of that planet. A similar application to the planet Saturn points to a similar result, but the conclusion is less certain on account of the great uncertainty in the data.

Dynamical origin of Precession and Nutation.

825. The phenomena of Precession and Nutation result from the earth's being not centrobaric (§ 534), and therefore attracting the sun and moon, and experiencing reactions from them, in lines which do not pass precisely through the earth's centre of inertia, except when they are in the plane of its equator. The attraction of either body transferred (§ 559, c) from its actual line to a parallel line through the earth's centre of inertia, gives therefore a couple which, if we first assume, for simplicity, gravity to be symmetrical round the polar axis,

* In the *Méc. Cél.* (VIII. vii. § 23) Laplace uses values of m and ϵ which make $5m/2\epsilon$ greater than $\frac{1}{2}\pi^2$. His determination of the Precessional Constant of the planet is thus vitiated.

tends to turn the earth round a diameter of its equator, in the direction bringing the plane of the equator towards the disturbing body. The moment of this couple is [§ 539 (14)] equal to

Dynamical origin of Precession and Nutation.

$$\frac{3S}{D^3}(C-A)\sin\delta\cos\delta \text{ .... } (14),$$

where S denotes the mass of the disturbing body, D its distance, and δ its declination; and C and A the earth's moments of inertia round polar and equatorial diameters respectively. In all probability (§§ 796, 797) there is a sensible difference between the moments of inertia round the two principal axes in the plane (§ 795) of the equator: but it is obvious, and will be proved in Vol. II., that Precession and Nutation are the same as they would be if the earth were symmetrical about an axis, and had for moment of inertia round equatorial diameters, the arithmetical mean between the real greatest and least values. From (12) of § 539 we see that in general the *differences* of the moments of inertia round principal axes, or, in the case of symmetry round an axis, the value of $C-A$, may be determined solely from a knowledge of surface or external gravity, or [§§ 794, 795] from the figure of the sea level, without any data regarding the internal distribution of density.

Equating § 539 (12) to § 794 (17), in which, when the sea level is supposed symmetrical, $F_2(\theta, \phi)$ becomes simply $\epsilon(\frac{1}{3}-\cos^2\theta)$, we find

$$\frac{M\mathfrak{r}^2}{r^3}(\epsilon-\tfrac{1}{2}m)(\tfrac{1}{3}-\cos^2\theta)=\tfrac{3}{2}\frac{C-A}{r^3}(\tfrac{1}{3}-\cos^2\theta),$$

whence
$$C-A=\tfrac{2}{3}M\mathfrak{r}^2(\epsilon-\tfrac{1}{2}m) \text{ .... } (15).$$

Similarly we may prove the same formula to hold for the real case, in which the sea level is an ellipsoid of three unequal axes, one of which coincides with the axis of rotation; provided ϵ denotes the mean of the ellipticities of the two principal sections of this ellipsoid through the axis of rotation, and A the mean of the moments of inertia round the two principal axes in the plane of the equator.

Precession gives information as to the distribution of the earth's mass, while surface-gravity does not.

826. The angular accelerations produced by the disturbing couples are (§ 281) directly as the moments of the couples, and inversely as the earth's moment of inertia round an equatorial diameter. But the integral results, observed in Precession and Nutation, would, if the earth's condition varied, vary directly as $C-A$, and inversely as C. We have seen (§ 794) that if the interior distribution of density were varied in any way subject to the condition of leaving the superficial, and consequently (§ 793) the exterior, gravity unchanged, $C-A$ remains unchanged. But it is not so with C, which will be the less or the greater, according as the mass is more condensed in the central parts, or more nearly homogeneous to within a small distance of the surface: and thus it is that a comparison between dynamical theory and observation of Precession and Nutation gives us information as to the interior distribution of the earth's density (just as from the rate of acceleration of balls or cylinders rolling down an inclined plane we can distinguish between solid brass gilt, and hollow gold, shells of equal weight and equal surface dimensions); while no such information can be had from the figure of the sea level, the surface distribution of gravity, or the disturbance of the moon's motion, without hypothesis as to primitive fluidity or present agreement of surfaces of equal density with the surfaces which would be of equal pressure were the whole deprived of rigidity.

Precession gives information as to the distribution of the earth's mass.

The constant of Precession deduced from Laplace's law.

827. But we shall first find what the magnitude of the terrestrial constant $(C-A)/C$ of *Precession* and *Nutation* would be, if Laplace's were the true law of density in the interior of the earth; and if the layers of equal density were level for the present angular velocity of rotation. Every moment of inertia involving the latter part of this assumption will be denoted by a black-letter capital.

The moment of inertia about the polar axis is, by § 281,

$$\mathfrak{C} = 2\int_0^{\mathrm{r}}\int_0^{\frac{1}{2}\pi}\int_0^{2\pi} \rho \mathrm{r}^2 \sin\theta d\mathrm{r} d\theta d\phi \,.\, \mathrm{r}^2 \sin^2\theta,$$

the first factor under the integral sign being an element of the mass, the second the square of its distance from the axis.

For the moment of inertia about another principal axis (which may be any equatorial radius, but is here taken as that lying in the plane from which ϕ is measured), we have

The constant of Precession deduced from Laplace's law.

$$\mathfrak{A} = 2\int_0^{\mathrm{r}}\int_0^{\frac{1}{2}\pi}\int_0^{2\pi} \rho \mathrm{r}^2 \sin\theta d\mathrm{r} d\theta d\phi \,.\, \mathrm{r}^2 (1 - \sin^2\theta \sin^2\phi).$$

Now, by § 823, we have

$$\mathrm{r} = r\left[1 + e\left(\tfrac{1}{3} - \cos^2\theta\right)\right],$$

where r denotes the mean radius of the surface of equal density passing through r, θ, ϕ; whence

$$\mathrm{r}^4 d\mathrm{r} = \tfrac{1}{5}\frac{d\mathrm{r}^5}{dr}dr = r^4 dr + \left(\tfrac{1}{3} - \cos^2\theta\right)\frac{d}{dr}(r^5 e)\,dr.$$

$$\left.\begin{array}{ll} \text{Let} & \int_0^{\mathrm{r}} \rho r^4 dr = K \\ \text{and} & \int_0^{\mathrm{r}} \rho \frac{d}{dr}(r^5 e)\,dr = K_1 \end{array}\right\} \text{.................(16).}$$

Then $\quad \mathfrak{C} = 2\int_0^{\frac{1}{2}\pi}\int_0^{2\pi} \sin^3\theta d\theta d\phi \left[K + K_1\left(\tfrac{1}{3} - \cos^2\theta\right)\right]$

or $\qquad C = \tfrac{8}{3}\pi K$ nearly.........................(17).

$$\mathfrak{C} - \mathfrak{A} = 2\int_0^{\frac{1}{2}\pi}\int_0^{2\pi} \sin\theta d\theta d\phi \left[K + K_1\left(\tfrac{1}{3} - \cos^2\theta\right)\right]\left(\sin^2\theta - 1 + \sin^2\theta \sin^2\phi\right)$$

$$= \tfrac{8}{15}\pi K_1 \text{ ....(18).}$$

Now we have

$$K = \int_0^{\mathrm{r}} \rho r^4 dr = F\int_0^{\mathrm{r}} r^3 \sin\frac{r}{\kappa}\,dr,$$

or, if we put as before $\theta = \frac{\mathrm{r}}{\kappa}$, $t = \tan\theta$,

$$K = F\kappa^4 \cos\theta\left(-\theta^3 + 3\theta^2 t + 6\theta - 6t\right).$$

Again $\quad K_1 = \int_0^{\mathrm{r}} \rho \frac{d}{dr}(r^5 e)\,dr = \mathrm{r}^5 \epsilon q - \int_0^{\mathrm{r}} r^5 e \frac{d\rho}{dr}\,dr,$

and this, by (10) of last section, becomes

$$K_1 = 5\mathrm{r}^2 \epsilon \int_0^{\mathrm{r}} \rho r^2 dr - \frac{5\mathrm{r}^5 \omega^2}{8\pi} \text{ ....(19).}$$

$$= 5\left(\epsilon - \tfrac{1}{2}m\right) F\kappa^4 \theta^2 (t - \theta)\cos\theta.$$

The constant of Precession deduced from Laplace's Law.

Thus, finally,

$$\frac{\mathfrak{C}-\mathfrak{A}}{C}=\frac{1}{3}\frac{K_1}{K}=(\epsilon-\tfrac{1}{2}m)\frac{\theta^2(t-\theta)}{-\theta^3+3\theta^2t+6\theta-6t}$$

$$=(\epsilon-\tfrac{1}{2}m)\frac{z}{2+(1-6\theta^{-2})z}\quad\ldots\ldots\ldots\ldots\ldots\ldots(20)$$

$$=\frac{\epsilon-\frac{1}{2}m}{1-6(\mathfrak{f}-1)/\mathfrak{f}\theta^2}\quad\ldots\ldots\ldots\ldots\ldots\ldots\ldots\ldots(21).$$

From these formulæ the numbers in Column vi. of the table in § 824 were calculated. By (18) and (19) we see that

$$\mathfrak{C}-\mathfrak{A}=\tfrac{8}{3}\pi\left(\mathrm{r}^2\epsilon\int_0^{\mathrm{r}}\rho r^2dr-\frac{\mathrm{r}^5\omega^2}{8\pi}\right)$$

$$=\tfrac{2}{3}M\mathrm{r}^2(\epsilon-\tfrac{1}{2}m)\quad\ldots\ldots\ldots\ldots\ldots\ldots\ldots\ldots(22),$$

which agrees, as it ought to do, with (15) of § 825.

A comparison of (21) and (22) then shows that

$$C=\tfrac{2}{3}M\mathrm{r}^2\left[1-6\frac{(\mathfrak{f}-1)}{\mathfrak{f}\theta^2}\right]\quad\ldots\ldots\ldots\ldots\ldots\ldots(23).$$

Comparison of Laplace's hypothesis with observation.

828. From the elaborate investigations of Precession and Nutation made by Le Verrier and Serret, it appears that the true value of $(C-A)/C$ is, very approximately, ·00327*. This, according to the table of § 824, agrees with $(\mathfrak{C}-\mathfrak{A})/C$ for $\mathfrak{f}=2{\cdot}1$, which gives $\epsilon=\frac{1}{297}$. These are (§§ 792, 796, 797) about the most probable values which we can assign to these elements by observation. Thus, so far as we have the means of testing it, Laplace's hypothesis is verified.

The compressibility involved in the hypothesis.

829. But, as a further check upon Laplace's assumption, it is necessary to inquire whether the results involve anything inconsistent with experimental knowledge of the compressibility of matter under such pressures as we can employ in the laboratory. For this purpose the first column has been added to the preceding table. From it may be deduced the compressibility of the upper stratum of liquid matter, which composed the crust of the earth, required by the assumed law of density, for the respective values of θ. In fact, the numbers in Col. i. are those by which the earth's radius must be divided to find

* *Annales de l'Observatoire Impérial de Paris*, 1859, p. 324.

the lengths of the modulus of compression (§ 688) of the uppermost layer of fluid, according to the surface value of gravity.

The compressibility involved in the hypothesis.

We have, by § 824 (3),

$$q = \frac{F}{r}\sin\frac{r}{\kappa},\quad \frac{dq}{dr} = -\frac{F}{r}\left\{\frac{\sin(r/\kappa)}{r} - \frac{\cos(r/\kappa)}{\kappa}\right\},$$

whence, at the surface, $\left[-\frac{1}{q}\frac{dq}{dr}\right] = \frac{1}{\mathfrak{r}}\left(1 - \frac{\theta}{t}\right)$.

The corresponding numbers for several different liquid and solid substances are as follows:—

Compressibility of lava required by Laplace's hypothesis, compared with experimental data.

Alcohol	37
Water	29
Mercury	27
Glass	5·0
Copper	8·1
Iron	4·1
Melted Lava, by Laplace's law, with $f = 2{\cdot}1$	4·42

This comparison may be considered as decidedly not adverse to Laplace's law, but actual experiments on the compressibility of melted rock are still a desideratum.

830. In § 276 it was proved that the tides must tend to diminish the angular velocity of the earth's rotation; it may be proved (and it was our intention to do so in a later volume) that this tendency is not counterbalanced to more than a very minute degree by the tendency to acceleration which results from the secular cooling and shrinking of the earth. In observational astronomy the earth's rotation serves as a time-keeper, and thus a retardation of terrestrial rotation will appear astronomically as an acceleration of the motion of the heavenly bodies. It is only in the case of the moon's motion that such an apparent acceleration can be possibly detected. Now, as Laplace first pointed out, there must be a slow variation in the moon's mean motion arising from the secular changes in the eccentricity of the earth's orbit round the sun. At the present time, and for several thousand years in the future, the variation in the moon's motion has been and will be an acceleration. Laplace's theoretical calculation of the amount of that acceleration

Numerical estimates of the amount of tidal friction.

See Appendix G on Tidal Friction.

Numerical estimates of the amount of tidal friction.

See Appendix G on Tidal Friction.

appeared to agree well with the results which were in his day accepted as representing the facts of observations. But in 1853 Adams wrote as follows :—

"In the *Mécanique Céleste*, the approximation to the value "of the acceleration is confined to the principal term, but in the "theories of Damoiseau and Plana the developments are carried "to an immense extent, particularly in the latter, where the mul-"tiplier of the change in the square of the eccentricity of the "earth's orbit, which occurs in the expression of the secular "acceleration, is developed to terms of the seventh order.

"As these theories agree in principle, and only differ slightly "in the numerical value which they assign to the acceleration, "and as they passed under the examination of Laplace, with "especial reference to this subject, it might be supposed that at "most only some small numerical corrections would be required "in order to obtain a very exact determination of the amount of "this acceleration.

"It has therefore not been without some surprise, that I have "lately found that Laplace's explanation of the phenomenon in "question is essentially incomplete, and that the numerical "results of Damoiseau's and Plana's theories, with reference "to it, consequently require to be very sensibly altered *."

Hansen's theory of the secular acceleration is vitiated by an error of principle similar to that which affects the theories of Damoiseau and Plana, but the mathematical process which he followed being different from theirs, he arrived at somewhat different results. From this erroneous theory Hansen found the value 12″·18 for the coefficient of the term in the moon's mean longitude depending on the square of the time, the unit of time being a century; in a later computation given in his *Darlegung*, he found the coefficient to be 12″·56 †.

* "On the Secular Variation of the Moon's Mean Motion," by J. C. Adams. *Phil. Trans.* 1853. Vol. 143, p. 397.

† It appears not unusual for physical astronomers to use an abbreviated phraseology, for specifying accelerations, which needs explanation. Thus when they speak of the secular acceleration being e.g. "12″·56 in a century"; they mean by "acceleration" what is more properly "the effect of the acceleration on the moon's mean longitude." The correct unabbreviated statement is "the acceleration is 25″·12 per century per century." Thus Hansen's result is that

In 1859 Adams communicated to Delaunay his final result, namely that the coefficient of this term appears from a correctly conducted investigation to be 5″·7, so that at the end of a century the moon is 5″·7 before the position it would have had at the same time, if its mean angular velocity had remained the same as at the beginning of the century. Delaunay verified this result, and added some further small terms which increased the coefficient from 5″·7 to 6″·1. **Secular variation of moon's mean motion**

Now, according to Airy, Hansen's value of the "advance" represents very well the circumstances of the eclipses of Agathocles, Larissa and Thales, but is if anything too small. Newcomb on the other hand is inclined from an elaborate discussion of the ancient eclipses to believe Hansen's value to be too large, and gives two competing values, viz. 8″·4 and 10″·9*.

In any case it follows that the value of the advance as theoretically deduced from all the causes, known up to the present time to be operative, is smaller than that which agrees with observation. In what follows 12″ is taken as the observational value of the "advance," and 6″ as the explained part of this phenomenon. About the beginning of 1866 Delaunay suggested that the true explanation of the discrepancy might be a retardation of the earth's rotation by tidal friction. Using this hypothesis, and allowing for the consequent retardation of the moon's mean motion by tidal reaction (§ 276), Adams, in an estimate which he has communicated to us, founded on the rough assumption that the parts of the earth's retardation due **partly explained by tidal friction.**

in each century the mean motion of the moon is augmented by an angular velocity of 25″·12 per century; so that at the end of a century the mean longitude is greater by ½ of 25″·12 than it would have been had the moon's mean motion remained the same as it was at the beginning of the century. Considering how absurd it would be to speak of a falling body as experiencing an acceleration of 16 feet in a second, or of 64 feet in two seconds; and how false and inconvenient it is to speak of a watch being 20 seconds fast when it is 20 seconds in advance of where it ought to be, we venture to suggest that, to attain clearness and correctness without sacrifice of brevity, "advance" be substituted for "acceleration" in the ordinary astronomical phraseology.

* See *Researches on the Motion of the Moon* (Washington, 1878), by Simon Newcomb, Part I. pp. 13 and 280.

Numerical estimate of amount of tidal retardation of earth's rotation.

to solar and lunar tides are as the squares of the respective tide-generating forces, finds 22 sec. as the error by which the earth, regarded as time-keeper, would in a century get behind a perfect clock rated at the beginning of the century. Thus at the end of a century a meridian of the earth is 330″ behind the position in which it would have been, if the earth had continued to rotate with the same angular velocity which it had at the beginning of the century*.

Thermo-dynamic acceleration of earth's rotation.

Besides the secular contraction of the earth in cooling, referred to above, which counteracts the tidal retardation of the earth's rotation to a very minute degree, there exists another counteracting influence, as has been pointed out by Sir William Thomson†, which, though much more considerable, is still but small in the amount of its accelerative effect, compared with the actual retardation as estimated by Adams. It is an observed fact that the barometer indicates variations of pressure during the day and night, and it is found that when these variations are analysed into their diurnal and semi-diurnal harmonic constituents, the semi-diurnal constituent rises to its maximum about 10 a.m. and 10 p.m. The crest of the nearer atmospheric tidal protuberance is thus directed to a point in the heavens westward of the sun, and the solar attraction on these protuberances causes a couple about the earth's axis by which the rotation is accelerated. As the barometric oscillations are due to solar radiation, it follows that the earth and sun together constitute a thermodynamic engine. Sir William Thomson computes, as a rough approximation, that from this cause the earth gains about 2·7 seconds in a century on a perfect chronometer set and rated at the beginning of the century. On the other hand the fall of meteoric dust on to the earth must cause a small retardation of the earth's rotation, although to an amount probably quite insensible in a century.

Retardation from fall of meteoric dust.

* See Appendix G (*a*), where Mr G. H. Darwin verifies Professor Adams's computation, and shows that the combination of Hansen's 12″·56 with Delaunay's 6″·1 would show the earth to be losing 23·4 sec. in the circumstances defined in the text; and that the combination of Newcomb's 8″·4 with Delaunay's 6″·1 would give a result of 8·3 sec. instead of 23·4.

† *Société de Physique*, Sept. 1881; or *Royal Society of Edinburgh*, Session 1881—82, p. 396.

Whatever be the value of the retardation of the earth's rotation, it is necessarily the result of several causes, of which tidal friction is almost certainly preponderant. If we accept Adams's estimate (according to which the earth would in a century get 22 sec. behind a perfect clock rated at the beginning of the century) as applicable to the outcome of the various concurring causes, then if the rate of retardation giving the integral effect were uniform, the earth as a time-keeper would be going slower by 22 of a second per year in the middle, and by ·44 of a second per year at the end, than at the beginning of the century. Causes for retardation preponderant.

The latter is $\frac{1}{71 \cdot 7 \times 10^6}$ of the present angular velocity; and if the rate of retardation had been uniform during ten million centuries past, the earth must have been rotating faster by about one-seventh than at present, and the centrifugal force must have been greater in the proportion of 817^2 to 717^2, or of 67 to 51. If the consolidation took place then or earlier, the ellipticity of the upper layers must have been $\frac{1}{230}$ instead of about $\frac{1}{300}$, as it is at present. It must necessarily remain uncertain whether the earth would from time to time adjust itself completely to a figure of equilibrium adapted to the rotation. But it is clear that a want of complete adjustment would leave traces in a preponderance of land in equatorial regions. The existence of large continents (§ 832'), and the great effective rigidity of the earth's mass (§ 848), render it improbable that the adjustments, if any, to the appropriate figure of equilibrium would be complete. The fact then that the continents are arranged along meridians, rather than in an equatorial belt, affords some degree of proof that the consolidation of the earth took place at a time when the diurnal rotation differed but little from its present value. It is probable therefore that the date of consolidation is considerably more recent than a thousand million years ago. It is proper however to add that Adams lays but little stress on the actual numerical values which have been used in this computation, and is of opinion that the amount of tidal retardation of the earth's rotation is quite uncertain. Date of consolidation of earth.

In Appendix D, § (*j*) it is shown, from the theory of the

Argument from cooling of the earth.

conduction of heat, that the date of consolidation may be about a hundred million years ago; but that in all probability it cannot have been so remote as five hundred million years from the present time.

Abrupt changes of interior density, not improbable.

831. From the known facts regarding compressibilities of terrestrial substances, referred to above (§ 829), it is most probable that even in chemically homogeneous substances there is a continuous increase of density downwards at some rate comparable with that involved in Laplace's law. But it is not improbable that there may be abrupt changes in the quality of the substance, as, for instance, if a large portion of the interior of the earth had at one time consisted of melted metals, now consolidated. We therefore append a solution of the problem of determining the ellipticities of the surfaces of a rotating mass consisting of two non-mixing fluids of different densities, each, however, being supposed incompressible.

Two non-mixing liquids of different densities, each homogeneous.

Let the densities of the two liquids be ρ and $\rho + \rho'$, the latter forming the spheroid

$$r = a'\left[1 + \epsilon'\left(\tfrac{1}{3} - \cos^2\theta\right)\right] \quad\text{......(1)},$$

and the former filling the space between this spheroid and the exterior concentric and coaxal surface

$$r = a\left[1 + \epsilon\left(\tfrac{1}{3} - \cos^2\theta\right)\right] \quad\text{......(2)}.$$

Also let the whole revolve with uniform angular velocity ω. The conditions of equilibrium are that the surface of each spheroid must be an equipotential surface.

Now the potential at a point r, θ, in the outer fluid is

$$\left.\begin{aligned}&\tfrac{4}{3}\pi\rho\left[\tfrac{1}{2}(3a^2 - r^2) + \tfrac{2}{5}r^2\epsilon\left(\tfrac{1}{3} - \cos^2\theta\right)\right]\\ &+ \tfrac{4}{3}\pi\rho'\left[\frac{a'^3}{r} + \tfrac{3}{5}\frac{a'^5}{r^3}\epsilon'\left(\tfrac{1}{3} - \cos^2\theta\right)\right]\\ &+ \tfrac{1}{3}\omega^2 r^2 + \tfrac{1}{2}\omega^2 r^2\left(\tfrac{1}{3} - \cos^2\theta\right)\end{aligned}\right\}\quad\text{......(3)}.$$

The first line is the potential due to a liquid of density ρ filling the larger spheroid, the second that due to a liquid of density ρ' filling the inner spheroid, the third is the potential $(\tfrac{1}{2}\omega^2 r^2 \sin^2\theta)$ of centrifugal force arranged in solid harmonics.

Substituting in (3) the values of r from (1) and (2) successively, neglecting squares, etc., of the ellipticities, and equating to zero the sum of the coefficients of $(\frac{1}{3}-\cos^2\theta)$; we have two equations from which we find

Two non-mixing fluids of different densities, each homogeneous.

$$\epsilon = \frac{\rho + \frac{1}{5}\rho'\left(2 + 3\frac{a'^5}{a^5}\right)}{\left(\rho + \frac{3}{5}\rho'\right)\left(\frac{2}{5}\rho + \frac{a'^3}{a^3}\rho'\right) - \frac{9}{25}\rho\rho'\frac{a'^3}{a^3}} \cdot \frac{3\omega^2}{8\pi} \quad \ldots\ldots\ldots\ldots(4).$$

The corresponding value of ϵ' is to be found from the equation

$$\epsilon\left(\rho + \frac{a'^3}{a^3}\rho'\right) = \epsilon'\left\{\rho + \frac{1}{5}\rho'\left(2 + 3\frac{a'^5}{a^5}\right)\right\}.$$

Expressing ω^2 in terms of the known quantity m we have

Two non-mixing fluids of different densities.

$$\frac{3\omega^2}{8\pi} = \frac{1}{2}m\left(\rho + \frac{a'^3}{a^3}\rho'\right) \quad \ldots\ldots\ldots\ldots\ldots\ldots(5).$$

Also, to a sufficient approximation, we have

$$\left.\begin{aligned} C &= \tfrac{8}{15}\pi a^5\left(\rho + \frac{a'^5}{a^5}\rho'\right) \\ M &= \tfrac{4}{3}\pi a^3\left(\rho + \frac{a'^3}{a^3}\rho'\right) \end{aligned}\right\} \quad \ldots\ldots\ldots\ldots\ldots(6),$$

and the mean density is obviously $\rho + \frac{a'^3}{a^3}\rho'$ ....(7).

The numerical values of the expressions (4) and (7) are approximately known from observation and experiment, so that if we assume a value of a'/a we can at once find ρ and ρ', and, from them, the value of $(C-A)/C$.

From the formulas just given it is easy to show that results closely agreeing with observation as regards precession, ratio of surface density to mean density, and ellipticity of sea level may be obtained without making any inadmissible hypotheses as to the relative volumes and densities of the two assumed liquids. But this must be left as an exercise for the student.

832. These estimates, and all dynamical investigations (whether static or kinetic) of tidal phenomena, and of precession and nutation, hitherto published, with the exceptions referred to below, have assumed that the outer surface of the

Rigidity of the earth:

Rigidity of the earth: solid earth is absolutely unyielding. A few years ago*, for the first time, the question was raised: Does the earth retain its figure with practically perfect rigidity, or does it yield sensibly to the deforming tendency of the moon's and sun's attractions on its upper strata and interior mass? It must yield to *some* extent, as no substance is infinitely rigid: but whether these solid tides are sufficient to be discoverable by any kind of observation, direct or indirect, has not yet been ascertained (see § 847). The negative result of attempts to trace their influence on ocean and lake tides, as hitherto observed, suffices, as we shall see, to disprove the hypothesis, hitherto so prevalent, that we live on a mere thin shell of solid substance, enclosing a fluid mass of melted rocks or metals, and proves, on the contrary, that the earth as a whole is much more rigid than any of the rocks that constitute its upper crust.

great enough to negative the geological hypothesis of a thin solid crust.

The internal stress caused in the earth by the weight of continents.

832′. Since the first edition of this work appeared, certain further investigations have been made, the results of which from a different point of view confirm the conclusion at which we have arrived concerning the solidity of the earth. This subject, forming a point of confluence of the sciences of astronomy and geology, appears of some importance, so that we propose to give a short account of these investigations†.

The mathematical theory of elastic solids imposes no restrictions on the magnitudes of the stresses, except in so far as that mathematical necessity requires the strains to be small enough to admit of the principle of superposition. Nature however does impose a limit on the stresses: if they exceed a limit the elasticity breaks down, and the solid either flows (as in the punching or crushing of metals‡) or ruptures (as when glass or stone breaks under excessive tension). It follows therefore that besides the question of the earth's rigidity, on which depends the

Conditions under which elasticity breaks down and solids rupture.

* "On the Rigidity of the Earth." W. Thomson. *Trans. R. S.*, May 1863, p. 573.

† "On the Stresses caused in the Interior of the Earth by the Weight of Continents and Mountains," by G. H. Darwin. *Phil. Trans.* Vol. 173, Part. I. p. 187. 1882.

‡ See the account of Tresca's most interesting experiments on the flow of solids. *Mémoires Présentés à l'Institut*, Vol. 18. 1868.

amount of straining due to tidal or other stresses, there is an important question as to the strength of the materials of the earth.

Rigidity of the earth. Conditions under which elasticity breaks down and solids rupture.

The theory of elastic solids as developed in §§ 658, 663, &c., shows that when a solid is stressed, the state of stress is completely determined when the amount and direction of the three principal stresses are known, or, speaking geometrically, when the shape, size, and orientation of the stress quadric is given. It is obvious that the tendency of the solid to rupture must be intimately connected with the shape of this quadric.

Provisional reckoning of tendency to rupture by difference between greatest and least principal stresses.

The precise circumstances under which elastic solids break have not hitherto been adequately investigated by experiment. It seems certain that rupture cannot take place without difference of stress in different directions. One essential element therefore is the difference between the greatest and least of the three principal stresses. How much the tendency to break is influenced by the amount of the intermediate principal stress is quite unknown. The difference between the greatest and least stresses may however be taken as the most important datum for estimating tendency to break. This difference has been called by Mr G. H. Darwin (to whom the investigation of which we speak is due) the "stress-difference." It may be proved that the greatest tangential stress at any point is equal to half the stress-difference. In the case of a wire under simple longitudinal stress, "the tenacity" is estimated by the stress per unit area of section under which the wire breaks. In this case two of the principal stresses are zero, and the third is the longitudinal tension; thus tenacity is a word to define "limiting stress-difference" when produced in a special manner. Engineers have made a great many experiments on the strength of materials for sustaining tensional and crushing stresses*, and their experiments afford data for a comparison between the strength which analysis shows that the materials of the earth must possess in the interior, and that of the solids which have been submitted to experiment.

* See, for example, Rankine's *Useful Rules and Tables*. Griffin, London, 1873; and Sir W. Thomson's *Elasticity*. Black, Edinburgh, 1878.

Rigidity of the earth.

We have in § 797 been occupied with the results of observations giving the form of ellipsoid which most nearly satisfies geodetic and gravitational experiments, but the existence of dry land proves that the earth's surface is not a figure of equilibrium appropriate to the diurnal rotation. Hence the interior of the earth must be in a state of stress, and as the land does not sink in, nor the sea-bed rise up, the materials of which the earth is made must be strong enough to bear this stress.

Weight of continents produces internal stress in the earth.

We are thus led to inquire how the stresses are distributed in the earth's mass, and what are magnitudes of the stresses.

Mr Darwin has, by means of the analysis of § 834, solved a problem of the kind indicated for the case of a homogeneous incompressible elastic sphere, and has applied the results to the discussion of the strength of the interior of the earth.

If the earth were formed of a crust with a semi-fluid interior, the stresses in that crust must be greater than if the whole mass be solid, very far greater if the crust be thin; and therefore this investigation cannot give as its result stresses greater than those which exist in reality.

He has only treated the problem for the class of inequalities called zonal harmonics; that is (§ 781) inequalities consisting of a number of undulations running round the globe in parallels of latitude. The number of crests is determined by the order of the harmonic. The second harmonic constitutes simply ellipticity of the spheroid. A harmonic of a high order may be described as a series of mountain chains, with intervening valleys, running round the globe in parallels of latitude, estimated with reference to the chosen equator.

Stress when the ellipticity of the spheroid is not appropriate to the diurnal rotation.

In the case of the second harmonic it is shown by Mr Darwin that the stress-difference rises to a maximum at the centre of the globe, and is constant all over the surface. The central stress-difference is eight times as great as that at the surface.

On evaluating the stress-difference arising from given ellipticity in a rotating spheroid of the size and density of the earth, it appears that if the excess or defect of ellipticity above or below the equilibrium value were $\frac{1}{1000}$, then the stress-difference at the centre would be 12×10^5 grammes weight per square centimetre; and that, if the sphere were made of material as strong as brass, it would be just on the point of rupture. Again, if the homogeneous earth, with ellipticity $\frac{1}{232}$, were to stop rotating, the central stress-difference would be 50×10^5 grammes weight per centimetre, and it would break if made of any material except the finest steel.

Rigidity of the earth.

Stress when the ellipticity of the spheroid is not appropriate to the diurnal rotation.

The stresses produced by harmonic inequalities of high orders are next considered in the paper to which we refer. This is in effect the case of a series of parallel mountains and valleys, corrugating a mean level surface with an infinite series of parallel ridges and furrows.

It is found that the stress-difference depends only on the depth below the mean surface, and is independent of the position of the point considered with regard to ridge and furrow.

Numerical calculation shows that if we take a series of mountains, whose crests are 4,000 metres (or about 13,000 feet) above the intermediate valley bottoms, formed of rock of specific gravity 2·8, then the maximum stress-difference is 4×10^5 grammes weight per square centimetre (about the tenacity of cast tin); also if the mountain chains are 314 kilometres apart, the maximum stress-difference is reached at 50 kilometres below the mean surface.

Stress due to a series of parallel mountain chains.

The solution shows that the stress-difference is *nil* at the surface. It is, however, only an approximate solution, for it will not give the stresses actually in the mountain masses, but it gives correct results at some four or five kilometres below the mean surface.

The cases of the harmonics of the 4th, 6th, 8th, 10th, and 12th orders are then considered; and it is shown that, if we suppose them to exist on a sphere of the mean density and dimensions of the earth, and that the height of the elevation at the equator is in each case 1,500 metres above the mean level of the sphere, then in each case the maximum stress-difference is about 6×10^5 grammes weight per square centimetre. This maximum is reached in the case of the 4th harmonic at 1,840 kilometres, and for the 12th at 560 kilometres, from the earth's surface.

In the second part of the paper it is shown that the great terrestrial inequalities, such as Africa, the Atlantic Ocean, and

Rigidity of the earth.

America, are represented by a harmonic of the 4th order; and that, having regard to the mean density of the earth being about twice that of superficial rocks, the height of the elevation is to be taken as about 1,500 metres.

Six hundred thousand grammes per square centimetre is the crushing stress-difference of average granite, and accordingly it is concluded that at 1,600 kilometres from the earth's surface the materials of the earth must be at least as strong as granite. A very closely analogous result is also found from the discussion of the case in which the continent has not the regular undulating character of the zonal harmonics, but consists of an equatorial elevation with the rest of the spheroid approximately spherical.

Conclusion as to strength of the interior of earth from magnitude of actual continents.

From this we may draw the conclusion, that either the materials of the earth have at least the strength of granite at 1,600 kilometres from the surface, or they must have a much greater strength near to the surface.

For the analysis by which these conclusions are supported we must refer to Mr Darwin's paper.

The subject of this investigation has an important connection with the date of the earth's consolidation as explained in § 830 above.

Tidal influence of sun and moon on the earth.

833. The character of the deforming tidal influence of the sun and moon will be understood readily by considering that if the whole earth were perfectly fluid, its bounding surface would coincide with an equipotential surface relatively to the attraction of its own mass, the centrifugal force of its rotation, and the tide-generating resultant (§ 804) of the moon's and sun's forces, and their kinetic reactions*. Thus

* It was our intention to prove in Vol. II. that the "equilibrium theory" of the tides for an ocean, whether of uniform density or denser in the lower parts, completely covering a solid nucleus, requires correction, on account of the diurnal rotation, but less and less correction the smaller this nucleus is; and that it agrees perfectly with the "kinetic theory" when there is no nucleus, always provided the angular velocity is not too great for the ordinary approximations (§§ 794, 801, 802, 815) which require that there be not, on any account, more than an infinitely small disturbance from the spherical figure. It is interesting to remark that this proposition does not require the tidal deformations to be small in comparison with the 70,000 feet deviation due to centrifugal force of rotation.

(§§ 819, 824) there would be the full equilibrium lunar and solar tides; or $2\frac{1}{2}$ times the amount of the disturbing deviation of level if the fluid were homogeneous, or of nearly twice this amount if it were heterogeneous with Laplace's hypothetical law of increasing density. If now a very thin layer of lighter liquid were added, this layer would rest covering the previous bounding surface to very nearly equal depth all round, and would simply rise and fall with that surface, showing only infinitesimal variations in its own depth, under tidal influences. Hence had the solid part of the earth so little rigidity as to allow it to yield in its own figure very nearly as much as if it were fluid, there would be very nearly nothing of what we call tides—that is to say, rise and fall of the sea relatively to the land; but sea and land together would rise and fall a few feet every twelve lunar hours. This would, as we shall see, be the case if the geological hypothesis of a thin crust were true. The actual phenomena of tides, therefore, give a secure contradiction to that hypothesis. We shall see indeed, presently, (§ 841) that even a continuous solid globe, of the same mass and diameter as the earth, would, if homogeneous and of the same rigidity (§ 680) as glass or as steel, yield in its shape to the tidal influences three-fifths as much, or one-third as much, as a perfectly fluid globe; and further, (§ 842) it will be proved that the effect of such yielding in the solid, according as its supposed rigidity is that of glass or that of steel, would be to reduce the tides to about $\frac{2}{5}$ or $\frac{2}{3}$ of what they would be if the rigidity were infinite.

Rigidity of the earth. Tidal influence of sun and moon on the earth.

834. To prove this, and to illustrate this question of elastic tides in the solid earth, we shall work out explicitly the solution of the general problem of § 696, for the case of a homogeneous elastic solid sphere exposed to no surface traction; but deformed infinitesimally by an equilibrating system of forces acting *bodily* through the interior, which we shall ultimately make to agree with the tide-generating influence of the moon or sun. In the first place, however, we only limit the deforming force by the final assumption of § 733.

Elastic solid tides.

Following the directions of § 732, we are to find, the two constituents $('\alpha, '\beta, '\gamma)$ and $(\alpha_{,}, \beta_{,}, \gamma_{,})$ for the complete solu-

Rigidity of the earth.

tion; of which the first, given by (6) and (7) of § 733, is as follows:—

$$\alpha = -\frac{1}{2(2i+5)(m+n)}\frac{d}{dx}(r^2 W_{i+1})$$
$$= \frac{1}{m+n}\left\{-\frac{r^2}{2(2i+3)}\frac{dW_{i+1}}{dx} + \frac{r^{2i+5}}{(2i+3)(2i+5)}\frac{d}{dx}(W_{i+1}r^{-2i-3})\right\}\ldots(1),$$

with symmetrical formulæ for $'\beta$ and $'\gamma$; which [§ 733 (6)], give

$$\left.\begin{aligned} '\delta &= -\frac{W_i}{m+n} \\ \text{and, [§ 737 (28)],}\quad '\zeta &= -\frac{(i+3)r^2 W_{i+1}}{2(2i+5)(m+n)} \end{aligned}\right\}\ldots\ldots(2).$$

These, used in (29) of § 737 with $i+2$ for i, give

$$-'Fr = \frac{1}{m+n}\left\{(m-n)W_{i+1}x + \frac{(i+2)n}{2i+5}\frac{d}{dx}(r^2 W_{i+1})\right\}\ldots\ldots(3);$$

which, reduced to harmonics by the proper formula [§ 737 (36)], becomes

$$'Fr = \frac{-1}{(2i+3)(m+n)}\left\{[m+(i+1)n]r^2\frac{dW_{i+1}}{dx} - \frac{(2i+5)m-n}{2i+5}r^{2i+5}\frac{d}{dx}(W_{i+1}r^{-2i-3})\right\}(4).$$

This and the symmetrical formulæ for $'Gr$ and $'Hr$, with r taken equal to a, express the components of the force per unit area which would have to be balanced by the application from without of surface traction to the bounding surface of the globe, if the strain through the interior were exactly that expressed by (1). Hence, still according to the directions of § 732, we must now find $(\alpha_{,}, \beta_{,}, \gamma_{,})$ the state of interior strain which with no force from without acting bodily through the interior, would result from surface traction equal and opposite to that (4). Of this part of the problem we have the solution in § 737 (52), the particular data being now

Homogeneous elastic solid globe free at surface; deformed by bodily harmonic force.

$$\frac{A_i}{a^{i+1}} = \frac{m+(i+1)n}{(2i+3)(m+n)}r^{-i}\frac{dW_{i+1}}{dx};\quad \frac{A_{i+2}}{a^{i+1}} = -\frac{(2i+5)m-n}{(2i+3)(2i+5)(m+n)}r^{i+3}\frac{d}{dx}(W_{i+1}r^{i+3})\ldots(5),$$

with symmetrical terms for B_i, C_i, and B_{i+2}, C_{i+2}; but none of other orders than i and $i+2$. Hence for the auxiliary functions of § 737 (50)

$$\left.\begin{aligned} \Psi_{i-1} &= 0,\quad \Phi_{i+1} = -\frac{(i+1)(2i+1)[m+(i+1)n]a^{i+1}}{(2i+3)(m+n)}W_{i+1} \\ \Psi_{i+1} &= \frac{(i+2)[(2i+5)m-n]a^{i+1}}{(2i+3)(m+n)}W_{i+1},\ \text{and}\ \Phi_{i+3} = 0 \end{aligned}\right\}\ldots(6).$$

Now (52), with the proper terms for $i+2$ instead of i added, is to be used to give us $\alpha_{,}$; and through the vanishing of Ψ_{i-1} and Φ_{i+3}, it becomes

Rigidity of the earth.

$$\alpha_{,}=\frac{1}{n}\left\{\frac{1}{i-1}\left[\frac{a^{-i+1}}{2i(2i+1)}\frac{d\Phi_{i+1}}{dx}+\frac{A_i r^i}{a^{i-1}}\right]+\frac{m}{2I}(a^2-r^2)a^{-i-1}\frac{d\Psi_{i+1}}{dx}\right.$$
$$\left.+\frac{a^{-i-1}}{i+1}\left[\frac{(i+4)m-(2i+3)n}{I(2i+5)}r^{2i+5}\frac{d}{dx}(\Psi_{i+1}r^{-2i-3})+A_{i+2}r^{i+2}\right]\right\}\ldots(7),$$

where for brevity we put

$$I=[2(i+2)^2+1]m-(2i+3)n\ldots\ldots\ldots\ldots(7^1).$$

To this we must add $'\alpha$, given by (1), to obtain, according to § 732, the explicit solution, α, of our problem. Thus, after somewhat tedious algebraic reductions in which $m+n$, appearing as a factor in the numerator and denominator of each fraction, is removed, we find a remarkably simple expression for α. This, and the symmetrical formulas for β and γ, are as follows:—

$$\left.\begin{aligned}\alpha&=(\mathfrak{E}a^2-\mathfrak{F}r^2)\frac{dW_{i+1}}{dx}-\mathfrak{G}r^{2i+5}\frac{d}{dx}(W_{i+1}r^{-2i-3})\\ \beta&=(\mathfrak{E}a^2-\mathfrak{F}r^2)\frac{dW_{i+1}}{dy}-\mathfrak{G}r^{2i+5}\frac{d}{dy}(W_{i+1}r^{-2i-3})\\ \gamma&=(\mathfrak{E}a^2-\mathfrak{F}r^2)\frac{dW_{i+1}}{dz}-\mathfrak{G}r^{2i+5}\frac{d}{dz}(W_{i+1}r^{-2i-3})\end{aligned}\right\}\ldots(8),$$

where

$$\left.\begin{aligned}\mathfrak{E}&=\frac{(i+1)[(i+3)m-n]}{2iIn}\\ \mathfrak{F}&=\frac{(i+2)(2i+5)m-(2i+3)n}{2(2i+3)In}\\ \mathfrak{G}&=\frac{(i+1)m}{(2i+3)In}\end{aligned}\right\}\ldots(9).$$

The infinitely great value of $\mathfrak{E}$ for the case $i=0$ depends on the circumstance that the bodily force for this case, being uniform and in parallel lines through the whole mass, is not self equilibrating, and therefore surface stress would be required for equilibrium.

The formulas (8) are susceptible of considerable simplification if we complete the differentiations in their last terms. We shall at the same time separate the formulas into two parts, of which one has for coefficient the bulk-modulus, and the other the rigidity-modulus.

Rigidity of the earth.

If k be the bulk-modulus, or modulus of resistance to compression, we have by § 698 (5),

$$m = k + \tfrac{1}{3}n \text{ ....} (9^{\text{i}});$$

and n is the rigidity modulus.

Thus (7^{i}) becomes

Separation of the two moduluses of elasticity.

$$I = [2(i+2)^2 + 1]k + \tfrac{2}{3}i(i+1)n \text{ ....} (9^{\text{ii}}).$$

Also on completing the differentiation we have

$$\alpha = \{\mathfrak{E}a^2 - (\mathfrak{F} + \mathfrak{G})r^2\}\frac{dW_{i+1}}{dx} + (2i+3)\mathfrak{G}xW_{i+1} \text{....} (9^{\text{iii}}).$$

Then, on substituting in (9) for m from (9^{i}), carrying the results into (9^{iii}) and separating the parts depending on k and n, we have

$$\left.\begin{aligned}(2In)\alpha = k\left\{\left[(i+4)(a^2-r^2) + \frac{3a^2}{i}\right]\frac{dW_{i+1}}{dx} + 2(i+1)xW_{i+1}\right\}\\ + \tfrac{1}{3}n(i+1)\left\{(a^2-r^2)\frac{dW_{i+1}}{dx} + 2xW_{i+1}\right\}\end{aligned}\right\}\text{...} (9^{\text{iv}}).$$

and symmetrical formulæ for β and γ.

In the elastic solids of which we have experimental knowledge [§ 684] the bulk-modulus is larger than the modulus of rigidity, and therefore k is considerably larger than $\tfrac{1}{3}n$; thus the terms written in the first line of (9^{iv}) are practically much more important than those in the second. In the ideal case of an absolutely incompressible elastic solid, the terms in the second line of (9^{iv}) vanish, and I/k becomes simply $2(i+2)^2 + 1$, and thus we have

Case of incompressible elastic solid.

$$\left.\begin{aligned}2n[2(i+2)^2+1]\alpha = \\ \left[(i+4)(a^2-r^2) + \frac{3a^2}{i}\right]\frac{dW_{i+1}}{dx} + 2(i+1)xW_{i+1}\end{aligned}\right\}\text{...} (9^{\text{v}}),$$

and symmetrical formulas for β and γ.

The case of $i=1$ is that with which we are concerned in the tidal problem. In it (7^{i}) and (9^{ii}) give us

$$I = 19m - 5n = 19k + \tfrac{4}{3}n \text{ ....} (10).$$

To prepare for terrestrial applications we may conveniently reduce to polar co-ordinates (distance from the centre, r; latitude, l; longitude, λ) such that

$$x = r\cos l\cos\lambda,\ y = r\cos l\sin\lambda,\ z = r\sin l \text{ ......} (11);$$

and denote by ρ, μ, ν, the corresponding components of displacement. The expressions for these will be precisely the same as those for α, β, γ, except that instead of $\frac{d}{dx}$, as it appears in the expression for α, we have $\frac{d}{dr}$ in the expression for ρ; $\frac{d}{rdl}$ in that for μ, and $\frac{d}{r\cos l d\lambda}$ in that for ν. Also in transforming from α to ρ we must put $x=r$, and in transforming from β and γ to μ and ν, y and z must be put zero. Thus if we put

Rigidity of the earth.

Case of incompressible elastic solid.

$$W_{i+1} = S_{i+1} r^{i+1} \quad \ldots\ldots\ldots\ldots\ldots\ldots\ldots(12),$$

so that S_{i+1} may denote the surface harmonic, or the harmonic function of directional angular co-ordinates l, λ, corresponding to W_{i+1}, we have from (9^{iv})

$$\left.\begin{aligned}(2In)\,\rho &= (i+1)\left[k\left\{\frac{(i+1)(i+3)}{i}a^2-(i+2)r^2\right\}+\tfrac{1}{2}n\,\{(i+1)a^2-(i-1)r^2\}\right]r^i S_{i+1}\\ (2In)\,\mu &= \left[k\left\{\frac{(i+1)(i+3)}{i}a^2-(i+4)r^2\right\}+\tfrac{1}{2}n\,(i+1)\,(a^2-r^2)\right]r^i\frac{dS_{i+1}}{dl}\\ (2In)\,\nu &= \left[k\left\{\frac{(i+1)(i+3)}{i}a^2-(i+4)r^2\right\}+\tfrac{1}{2}n\,(i+1)\,(a^2-r^2)\right]\frac{r^i}{\cos l}\,\frac{dS_{i+1}}{d\lambda}\end{aligned}\right\}\ldots(13).$$

In the case of elastic solids, such as we know them experimentally, the terms in k are much more important than those in n.

Now it is easy to show that, in as far as ρ depends on the term in k, it reaches a maximum value when $r=a\sqrt{1-1/(i+2)^2}$; and in as far as it depends on the term in n it would algebraically reach a maximum when $r=a\sqrt{1+2/\{(i+2)(i-1)\}}$. But this latter point being outside of the sphere it follows that the term in n increases from the centre to the surface. We thus see that ρ increases from zero at the centre, to a maximum value near the surface, and then diminishes again.

In a similar manner it appears that ρ/r reaches a maximum, as far as concerns the term in k, when $r=a\sqrt{1-3/\{i(i+2)\}}$; and as far as concerns the term in n, when $r=a$.

When $i=1$, which corresponds to the case of the tidal problem, we have from (13)

Rigidity of the earth.

$$\left.\begin{aligned}\rho &= \frac{1}{(19k+\frac{4}{3}n)n}\left[(8a^2-3r^2)k+\tfrac{2}{3}a^2n\right]rS_2\\ \mu &= \frac{1}{2(19k+\frac{4}{3}n)n}\left[(8a^2-5r^2)k+\tfrac{2}{3}(a^2-r^2)n\right]r\frac{dS_2}{dl}\\ \nu &= \frac{1}{2(19k+\frac{4}{3}n)n}\left[(8a^2-5r^2)k+\tfrac{2}{3}(a^2-r^2)n\right]\frac{r}{\cos l}\frac{dS_2}{d\lambda}\end{aligned}\right\}\ldots(14).$$

It is obvious that ρ/r diminishes from the centre outwards to the surface; and its extreme values are

$$\left.\begin{aligned}&\text{at the centre } \frac{\rho}{r} = \frac{8k+\frac{2}{3}n}{(19k+\frac{4}{3}n)n}a^2S_2 = \frac{8a^2}{19n}\left(1+\frac{\frac{1}{76}n/k}{1+\frac{4}{57}n/k}\right)S_2\\ &\text{at the surface } \frac{\rho}{r} = \frac{5k+\frac{2}{3}n}{(19k+\frac{4}{3}n)n}a^2S_2 = \frac{5a^2}{19n}\left(1+\frac{\frac{6}{115}n/k}{1+\frac{4}{57}n/k}\right)S_2\end{aligned}\right\}\ldots\ldots(15).$$

Cases:—centrifugal force:—

When the disturbing action is the centrifugal force of uniform rotation with angular velocity ω, we have as found above (§ 794) for the whole potential

$$W_2 = w\{\tfrac{1}{3}\omega^2r^2 + \tfrac{1}{2}\omega^2r^2(\tfrac{1}{3}-\cos^2\theta)\}\ \ldots\ldots\ldots\ldots(16),$$

where w denotes the mass of the solid per unit volume. The effect of the term $\frac{1}{3}w\omega^2r^2$ is merely a drawing outwards of the solid from the centre symmetrically all round; which we may consider in detail later in illustrating properties of matter in our second volume. The remainder of the expression gives us according to our present notation

$$W_2 = \tfrac{1}{3}\tau(x^2+y^2-2z^2);\ \text{or } S_2 = w\tau(\tfrac{1}{3}-\cos^2\theta)\ \ldots\ldots(17),$$

where $$\tau = \tfrac{1}{2}\omega^2\ldots\ldots\ldots\ldots\ldots\ldots(18).$$

tide-generating force.

For tide-generating force the same formulæ (14) and (15) hold if (§§ 804, 808, 813) we take

$$\tau = \tfrac{3}{2}\frac{M}{D^3}\ldots\ldots\ldots\ldots\ldots\ldots(19),$$

and alter signs so as to make the strain-spheroids prolate instead of oblate. The deformed figure of each of the concentric spherical surfaces of the sphere is of course an ellipsoid of revolution; and from (15) we find for the extremes:—

$$\left.\begin{aligned}&\text{ellipticity of central strain spheroids} = \frac{8a^2}{19n}\left(1+\frac{\frac{1}{76}n/k}{1+\frac{4}{57}n/k}\right).w\tau\\ &\quad\text{,, of free surface} = \frac{5a^2}{19n}\left(1+\frac{\frac{6}{115}n/k}{1+\frac{4}{57}n/k}\right).w\tau\end{aligned}\right\}(20).$$

From these results, (8) to (20), we conclude that

Elastic solid tides.

835. The bounding surface and concentric interior spherical surfaces of a homogeneous elastic solid sphere strained slightly by balancing attractions from without, become deformed into harmonic spheroids of the same order and type as the solid harmonic expressing the potential function of these forces, when they are so expressible: and the direction of the component displacement perpendicular to the radius at any point is the same as that of the component of the attracting force perpendicular to the radius. These concentric harmonic spheroids although of the same type are not similar. When they are of the second degree (that is when the force potential is a solid harmonic of the second degree), the proportions of the ellipticities in the three normal sections of each of them are the same in all: but in any one section the ellipticities of the concentric ellipsoids increase from the outermost one inwards to the centre, in the ratio of $5k+\frac{2}{3}n$ to $8k+\frac{2}{3}n$, or

Homogeneous elastic solid globe free at surface; deformed by bodily harmonic force.

$$1-\tfrac{3}{8}\frac{1}{1+\frac{1}{12}n/k} : 1.$$

If $\frac{1}{12}n/k$ be small, as is in general the case, the ratio is approximately $\frac{5}{8}+\frac{3}{32}n/k : 1$.

Case of second degree gives elliptic deformation, diminishing from centre outwards:—higher degrees give greatest proportionate deviation from sphericity neither at centre nor surface.

For harmonic disturbances of higher orders the amount of deviation from sphericity, reckoned of course in proportion to the radius, increases from the surface inwards to a certain distance, and then decreases to the centre. The explanation of this remarkable conclusion is easily given without analysis, but we shall confine ourselves to doing so for the case of ellipsoidal disturbances.

836. Let the bodily disturbing force cease to act, and let the surface be held to the same ellipsoidal shape by such a distribution of surface traction (§§ 693, 662) as shall maintain a homogeneous strain throughout the interior. The interior ellipsoidal surfaces of deformation will now become similar concentric ellipsoids: and the inner ones must clearly be less elliptic than they were when the same figure of outer boundary was maintained by forces acting throughout all the interior;

Synthetic proof of maximum ellipticity at centre, for deformation of second order.

Rigidity of the earth.

and, therefore, they must have been greater for the inner surface. And we may reason similarly for the portion of the whole solid within any one of the ellipsoids of deformation, by supposing all cohesive and tangential force between it and the solid surrounding it to be dissolved; and its ellipsoidal figure to be maintained by proper surface traction to give homogeneous strain throughout the interior when the bodily force ceases to act. We conclude that throughout the solid from surface to centre, when disturbed by bodily force without surface traction, the ellipticities of the concentric ellipsoids increase inwards.

Synthetic proof of maximum ellipticity at centre, for deformation of second order.

837. When the disturbing action is centrifugal force, or tide generating force (as that of the sun or moon on the earth), the potential is, as we have seen, a harmonic of the second degree, symmetrical round an axis. In one case the spheroids of deformation are concentric oblate ellipsoids of revolution; in the other case prolate. In each case the ellipticity increases from the surface inwards, according to the same law [§ 834 (15)] which is, of course, independent of the radius of the sphere. For spheres of different dimensions and similar substances the ellipticities produced by equal angular velocities of rotation are as the squares of the radii. Or, if the equatorial surface velocity (V) be the same in rotating elastic spheres of different dimensions but similar substance, the ellipticities are equal. The values of the surface and central ellipticities are respectively

Oblateness induced in homogeneous elastic solid globe, by rotation.

$$\frac{3}{11}\frac{V^2w}{2n} \text{ and } \frac{14}{33}\frac{V^2w}{2n} \quad\ldots\ldots\ldots\ldots\ldots\ldots(21)$$

for solids fulfilling Poisson's hypothesis (§ 685), according to which $m=2n$, or $k=\frac{5}{3}n$.

If the solid be absolutely incompressible these ellipticities are by § 834 (15)

$$\frac{5}{19}\frac{V^2w}{2n} \text{ and } \frac{8}{19}\frac{V^2w}{2n} \quad\ldots\ldots\ldots\ldots\ldots\ldots(22).$$

Now since $\frac{3}{11}=\cdot 2727$ and $\frac{5}{19}=\cdot 2632$; and $\frac{14}{33}=\cdot 4242$ and $\frac{8}{19}=\cdot 4211$, we see that the compressibility of the elastic solid exercises very little influence on the result.

For steel or iron the values of n and m are respectively 780 × 10⁶ and about 1600 × 10⁶ grammes weight per square centimetre, or 770 × 10⁹ and about 1600 × 10⁹ gramme-centimetre-seconds, absolute units (§ 223), and the specific gravity (w) is about 7·8. Hence a ball of steel of any radius rotating with an equatorial velocity of 10,000 centimetres per second will be flattened to an ellipticity (§ 801) of $\frac{1}{7220}$. For a specimen of flint glass of specific gravity 2·94 Everett finds $n = 244 \times 10^6$ grammes weight per square centimetre and very approximately $m = 2n$. Hence for this substance $n/w = 83 \times 10^6$ [being the length of the modulus of rigidity (§ 678) in centimetres]. But the numbers used above for steel give $n/w = 100 \times 10^6$ centimetres; and therefore (§ 838) the flattening of a glass globe is 1/·83, or $1\frac{1}{5}$ times that of a steel globe with equal velocities.

Elastic solid tides.

Numerical results for iron and glass.

838. For rotating or tidally deformed globes of glass or metals, the amount of deformation is but little influenced by compressibility, as we see from the numerical comparison given in § 837. For any substance for which $3k \gtreqless 5n$ the surface ellipticity is diminished by three per cent. or by less than three per cent., and the centre ellipticity by $\frac{3}{4}$ per cent., or less than $\frac{3}{4}$ per cent. if we suppose the rigidity to remain in any case unchanged, but the substance to become absolutely incompressible. For the surface ellipticity, § 834 (22) gives on this supposition

Rotational or tidal ellipticities but little influenced by compressibility, in globes of metallic, vitreous, or gelatinous elastic solid.

$$e = \frac{5a^2 w}{19n} \tau \quad \text{.........................(23)},$$

or with $n = 770 \times 10^9$ as for steel (§ 837),

$a = 640 \times 10^6$, the earth's radius in centimetres,

and $w = 5·5$, " " mean density,

we have, in anticipation of § 839,

$$e = 77 \times 10^4 . \tau \quad \text{.........................(24)}.$$

839. If now we consider a globe as large as the earth, and of incompressible homogeneous material, of density equal to the earth's mean density, but of the same rigidity as steel or glass; and if, in the first place, we suppose the matter of such a globe to be deprived of the property of mutual gravitation

Value of surface ellipticities for globe same size and mass a earth, of non-gravitating material, homogeneous, incompressible, and same rigidity as steel.

between its parts: the ellipticities induced by rotation, or by tide generating force, will be those given by the preceding formulæ [§ 834 (20)], with the same values of n as before; with $n/k=0$; with 640×10^6 for a, the earth's radius in centimetres; and with 5·5 for w instead of the actual specific gravities of glass and steel.

But without rigidity at all, and mutual gravitation between the parts alone opposing deviation from the spherical figure, we found before (§ 819) for the ellipticity

$$e=\frac{5}{2}\frac{a}{g}\tau=162\times10^4.\tau \qquad (25).$$

Comparison between spheroidal-maintaining powers of gravitation and rigidity, for large homogeneous solid globes.

840. Hence of these two influences which we have considered separately:—on the one hand, elasticity of figure, even with so great a rigidity as that of steel; and, on the other hand, mutual gravitation between the parts: the latter is considerably more powerful than the former, in a globe of such dimensions as the earth. When, as in nature, the two resistances against change of form act jointly, the actual ellipticity of form will be the reciprocal of the sum of the reciprocals of the ellipticities that would be produced in the separate cases of one or other of the resistances acting alone. For we may imagine the disturbing influence divided into two parts: one of which alone would maintain the actual ellipticity of the solid, without mutual gravitation; and the other alone the same ellipticity if the substance had no rigidity but experienced mutual gravitation between its parts. Let τ be the disturbing influence as measured by § 834 (20), (21); and let $\tau/\mathfrak{r}$ and $\tau/\mathfrak{g}$ be the ellipticities of the spheroidal figure into which the globe becomes altered on the two suppositions of rigidity without gravity and gravity without rigidity, respectively. Let e be the actual ellipticity and let τ be divided into τ' and τ'' proportional to the two parts into which we imagine the disturbing influence to be divided in maintaining that ellipticity. We have $\tau=\tau'+\tau''$, and $e=\tau'/\mathfrak{r}=\tau''/\mathfrak{g}$.

$\mathfrak{g}$ $\mathfrak{r}$ denote resistances to deformation due respectively to gravity and to rigidity.

Whence $\frac{\tau}{e}=\mathfrak{r}+\mathfrak{g}$, or $\frac{1}{e}=\frac{\mathfrak{r}}{\tau}+\frac{\mathfrak{g}}{\tau}$, which proves the proposition.

It gives

$$e=\frac{\tau}{\mathfrak{r}+\mathfrak{g}}=\frac{\tau/\mathfrak{g}}{\mathfrak{r}/\mathfrak{g}+1} \qquad (26).$$

By §§ 838, 839 we have $\mathfrak{r} = \frac{19n}{5a^2w}$, and $\mathfrak{g} = \frac{2g}{5a}$ ....(27)

and $\frac{\mathfrak{r}}{\mathfrak{g}} = \frac{19n}{2gaw} = \frac{19n/g}{2aw}$ ....(28)

where n/g is the rigidity in grammes weight per square centimetre. For steel and glass as above (§§ 837, 839) the values of $\mathfrak{r}/\mathfrak{g}$ are respectively 2·1 and ·66. Rigidity of the earth:

840′. Mr G. H. Darwin has shown* how the introduction of the effects of the mutual gravitation of the parts of the spheroid may be also carried out analytically instead of synthetically. The sphere being in a state of strain is distorted into a spheroid (say $r = a + \sigma_i$, where σ_i is a surface harmonic). Then the state of internal stress and strain in the spheroid is due to three causes, (i) the external disturbing potential W_i, (ii) the attraction of the harmonic inequality of which the potential is $3gwr^i\sigma_i/(2i+1)a^i$, (iii) the weight (positive in parts and negative in others) of the inequality σ_i. This last is equivalent to a normal traction per unit area applied to the surface of the sphere equal to $-gw\sigma_i$. It is not possible to arrive at the results due to the last cause without a modification of the analysis of § 834, because we have to introduce the effects of surface tractions. Analytical introduction of effects of gravitation.

But Mr Darwin shows (p. 9 *loc. cit.*) that "if W_i be the potential of the external disturbing influence, the *effective* potential per unit volume at a point within the sphere, now free of surface action and of mutual gravitation, is

$$W_i - 2gw(i-1)\,r^i\sigma_i/(2i+1)\,a^i = r^iT_i \text{ suppose."}$$

The case considered by him is that of an incompressible viscous spheroid, and he goes on to find the height and retardation of tide in such a spheroid. The analysis is, however, almost *literatim* applicable to the case of an elastic incompressible spheroid.

Suppose now that the external disturbing potential is

$$W_2 = w\tau r^2\left(\tfrac{1}{3} - \cos^2\theta\right),$$

* "On the Bodily Tides of Viscous and Semi-elastic Spheroids, &c." *Phil. Trans.* Part I. 1879, p. 1.

Rigidity of the earth. Analytical introduction of effects of gravitation.

and that the sphere consequently becomes distorted into the spheroid whose equation is $r = a[1 + e(\frac{1}{3} - \cos^2\theta)]$, so that $\sigma_2 = ae(\frac{1}{3} - \cos^2\theta)$. Then the effective disturbing potential to produce the same strain in a sphere devoid of gravitation is $(\tau - \mathfrak{g}e)\, wr^2(\frac{1}{3} - \cos^2\theta)$. Such a potential we know by (23) § 838, and (27) § 840, will produce ellipticity e, given by $e = (\tau - \mathfrak{g}e)/\mathfrak{r}$. Whence

$$e = \frac{\tau}{\mathfrak{r} + \mathfrak{g}} \quad\quad (26),$$

which is the result (26) of § 840.

The analytical method has the advantage of showing that we are neglecting as small the tangential action between the inequality σ_i and the true spherical surface, a fact which is not so obvious from the synthetical mode of treatment. In the case of the viscous spheroid considered by Mr Darwin this tangential action (although varying as τ^2) is of much interest, for the sum of the moments of all the tangential actions about the axis of revolution of the spheroid constitutes the tidal frictional retarding couple*.

Hypothesis of imperfect elasticity of the earth.

In the paper to which we refer Mr Darwin has also investigated the consequences which would arise from the hypothesis that the elasticity of the earth is not perfect, but that the stress requisite to maintain a given state of strain diminishes in geometrical progression as the time, measured from the time of straining, increases in arithmetical progression. This hypothesis undoubtedly represents some of the phenomena of the imperfect elasticity of actual solids. He finds, then, that if "the modulus of the time of relaxation of rigidity," being the time in which the stress falls to $1/\epsilon$ or ·378 of its initial value, be about one-third of the period of the tidal disturbance, then the height of the bodily tide scarcely differs sensibly from the height on the hypothesis of perfect elasticity. The phase of tide would still however be considerably affected. The existence of the great continents (§ 832') proves almost conclusively that for

* See "Problems connected with the Tides of a Viscous Spheroid." *Phil. Trans.* Part II. 1879, p. 539.

stresses lasting for a few hours or days the earth has practically perfect elasticity.

841. Reverting now to the results of § 840, it appears that if the rigidity of the earth, on the whole, were only as much as that of steel or iron, the earth as a whole would yield about one-third as much to the tide-generating influences of the sun and moon as it would if it had no rigidity at all; and it would yield by about three-fifths of the fluid yielding, if its rigidity were no more than that of glass.

Rigidity of the earth: unless greater than that of steel would be very imperfectly effective in maintaining figure against tide-generating force.

842. To find the effect of the earth's elastic yielding on the tides, we must recollect (§ 819) that the ellipticity of level due to the disturbing force, and to the gravitation of the undisturbed globe, which [§§ 804, 808, (18), (19)] is $a\tau/g$, will be augmented by $\frac{3}{5}e$ on account of the alteration of the globe into a spheroid of ellipticity e: so that if (§ 799) we neglect the mutual attraction of the waters, we have for the disturbed ellipticity of the sea level (§ 785)

Influence of elastic yielding of the solid earth on the surface-liquid tides.

$$\frac{a}{g}\tau + \tfrac{3}{5}e \quad \text{(29)}.$$

The rise and fall of the water relatively to the solid earth will depend on the excess of this above the ellipticity of the solid. Denoting this excess, or the ellipticity of relative tides, by ϵ, we have

$$\epsilon = \frac{a}{g}\tau - \tfrac{2}{5}e \quad \text{(30)},$$

or by (26) and (27)

$$\epsilon = \frac{a}{g}\tau\frac{\mathfrak{r}}{\mathfrak{r}+\mathfrak{g}} \quad \text{(31)}.$$

Hence the rise and fall of the tides is less than it would be were the earth perfectly rigid, in the proportion that the resistance against tidal deformation of the solid due to its rigidity bears to sum of the resistances due to rigidity of the solid and to mutual gravitation of its parts. By the numbers at the end of § 840 we conclude that if the rigidity were as great as that of steel, the relative rise and fall of the water would be reduced by elastic yelding of the solid to $\frac{2}{3}$, or if the rigidity were only that of glass, the relative rise and fall would be actually reduced to $\frac{2}{5}$, of what it would be were the rigidity perfect.

Rigidity of the earth: probably greater on the whole than that of a solid glass globe.

843. Imperfect as the comparison between theory and observation as to the actual height of the tides has been hitherto, it is scarcely possible to believe that the height is in reality only two-fifths of what it would be if, as has been universally assumed in tidal theories, the earth were perfectly rigid. It seems, therefore, nearly certain, with no other evidence than is afforded by the tides, that the tidal effective rigidity of the earth must be greater than that of glass.

Dynamic theory of tides too imperfect to give any estimate of absolute values for main phenomena: but not so for the fortnightly and semi-annual tides.

844. The actual distribution of land and water, and of depth where there is water, over the globe is so irregular, that we need not expect of even the most powerful mathematical analysis any approach to a direct dynamical estimate of what the ordinary semi-diurnal tides in any one place ought to be if the earth were perfectly rigid. In water 10,000 feet deep (which is considerably less than the general depth of the Atlantic, as demonstrated by the many soundings taken within the last few years, especially those along the whole line of the Atlantic Telegraph Cable, from Valencia to Newfoundland), the velocity of long free waves, as will be proved in Vol. II., is 567 feet per second*. At this rate the time of advancing through 57° (or a distance equal to the earth's radius) would be only ten hours. Hence it may be presumed that, at least at all islands of the Atlantic, any tidal disturbance, whose period amounts to several days or more, ought to give very nearly the true equilibrium tide, not modified sensibly, or little modified, by the inertia of the fluid. Now such tidal disturbances (§ 808) exist in virtue of the moon's and sun's changes of declination, having for their periods the periods of these changes.

Amounts of fortnightly tide estimated on various suppositions as to rigidity.

845. The sum of the rise from lowest to highest at Teneriffe, and simultaneous fall from highest to lowest at Iceland, in the lunar fortnightly tide, would amount to 4·3 inches if the earth were perfectly rigid, or 2·9 inches if the tidal effective rigidity were only that of steel, or 1·7 inches if the tidal effective rigidity were only that of glass. The amounts of the semi-annual tide, whatever be the actual rigidity of the earth, would of course be about half that of the fortnightly tide. The amount

* Airy, *Tides and Waves*, § 170.

of either in any one place would be discoverable with certainty to a small fraction of an inch by a proper application of the method of least squares, such as has hitherto not been made, to the indications of an accurate self-registering tide-gauge. For our present object, the semi-annual tide, though it may have the advantage of being more certainly not appreciably different from the true equilibrium amount, may be sensibly affected by the melting of ice from the arctic and antartic polar regions, and by the fall of rain and drainage of land elsewhere, which will probably be found to give measurable disturbances in the sea level, exhibiting, on the average of many years, an annual and semi-annual harmonic variation. This disturbance will, however, be eliminated for any one fortnight or half-year, by combining observations at well-chosen stations in different latitudes. It seems probable, therefore, that a somewhat accurate determination of the true amount of the earth's elastic yielding to the tide-generating forces of the moon and sun may be deduced from good self-registering tide-gauges maintained for several years at such stations as Iceland, Teneriffe, Cape Verde Islands, Ascension Island, and St Helena. It is probable also that the ratio of the moon's mass to that of the earth may be determined from such observations more accurately than it has yet been. It is to be hoped that these objects may induce the British Government, which has done so much for physical geography in many ways, to establish tide-gauges at proper stations for determining with all possible accuracy the fortnightly and semi-annual tides, and the variations of sea level due to the melting of ice in the polar regions, and the fall of rain and drainage of land over the rest of the world.

Rigidity of the earth: probably to be best learned from observations giving amounts of fortnightly tides.

Tide-gauges to be established at ocean stations.

846. More observation, and more perfect reduction of observations already made, are wanted to give any decided answer to the questions, how much the fortnightly tide and the semi-annual tide really are. "In the *Philosophical Transactions*, "1839, p. 157, Mr Whewell shows that the observations of "high and low water at Plymouth give a mean height of water "increasing as the moon's declination increases, and amounting "to 3 inches when the moon's declination is 25°. This is the

Scantiness of information regarding fortnightly tides, hitherto drawn from observation.

Rigidity of the earth. Scantiness of information regarding fortnightly tides, hitherto drawn from observation.

"same direction as that corresponding in the expression above "to a high latitude. The effect of the sun's declination is not "investigated from the observations. In the *Philosophical* "*Transactions*, p. 163, Mr Whewell has given the observations "of some most extraordinary tides at Petropaulofsk, in Kams- "chatka, and at Novo-Arkhangelsk, in the Island of Sitkhi, on "the west coast of North America.

"From the curves in the *Philosophical Transactions*, as well "as from the remaining curves relating to the same places "(which, by Mr Whewell's kindness, we have inspected), there "appears to be no doubt that the mean level of the water at "Petropaulofsk and Arkhangelsk rises as the moon's declina- "tion increases. We have no further information on this "point."—(Airy's *Tides and Waves*, § 533.)

Advance in knowledge of tides since the first edition.

847. We have left these sections, on the probability of the great effective rigidity of the earth, in the form in which they stood in our first edition in 1867. Since that date great advances have been made in our knowledge of actual tidal phenomena. The Tidal Committee of the British Association "appointed on the motion of Sir William Thomson in 1867, with for one prominent object the evaluation of the long-period tides for the purpose of answering the question of the Earth's rigidity," has done much towards the attainment of a satisfactory knowledge of the tides in the ocean surrounding these islands. But by far the most complete information relates to the Indian Ocean, for in consequence of the exertions of General Walker, Sir William Thomson, General Strachey and others, the Indian Government has taken up the question, and is now issuing, under the direction of General Walker and Major Baird, R.E., tide tables for the principal ports in India. We are thus now able to present the following discussion of the questions raised above, contributed to our present edition by Mr G. H. Darwin.

The theoretical value of the fortnightly and monthly elliptic tides.

848. The expression for a tide should consist of a spherical harmonic function of latitude and longitude of places on the earth's surface multiplied by a simple time-harmonic; but where a correct expansion, rigorously following this definition, would involve some terms of very long period, it is more convenient to regard the spherical harmonic as itself slowly varying between certain limits, and thus to amalgamate a number of terms together. The last term of (23) § 808 will give the theoretical equilibrium values of the fortnightly declinational tide, and of the monthly elliptic tide. The full expansion of this term would involve a certain part going through its period in 19 years, in which time the lunar nodes complete a revolution. This part will, according to Sir William Thomson, be most conveniently included by conceiving the inclination of the lunar orbit to the equator to undergo a slow oscillation in that period. In practice an average value for the inclination, the average being taken over a whole year, is sufficiently exact.

The theoretical value of the fortnightly and monthly elliptic tides.

(*a*) In what follows, the descending node of the equator on the lunar orbit will be called "the intersection." If the lunar orbit were identical with the ecliptic, the intersection would be the vernal equinox or ♈.

The following is a summary of the notation employed below:—

Definition of symbols.

For the moon:

M = mass; D = radius vector; c = mean distance; σ = mean motion; θ = true longitude in the orbit; i = inclination of lunar orbit to the ecliptic: N = longitude of ascending node on the ecliptic; ϖ = longitude of perigee in the orbit; e = eccentricity of orbit; ξ = longitude of "the intersection" in the orbit; ν = right ascension of "the intersection"; δ = declination.

Observe that longitudes "in the orbit" are measured along the ecliptic as far as the lunar node, and thence along the orbit; or are measured altogether in the movable orbit from a point therein, which is at a distance behind the node equal to the distance of the node from ♈.

For the earth:

E = mass; a = mean radius; l, λ = latitude and W. longitude of places on the earth's surface; ω = obliquity of ecliptic; I = inclination of equator to lunar orbit.

For both bodies together, let $\tau = \frac{3}{2} Ma^3/Ec^3$. And let the time t be measured from the instant when the moon's mean longitude vanishes.

The readers of the Tidal Reports of the British Association

Long-period tides.

for 1868, 1870, 1871, 1872, 1876* may find it convenient to note that the symbols employed are frequently the Greek initials of the corresponding words: thus,—γ, σ, η [γῆ, σελήνη, ἥλιος] for the rotation and mean motions of earth, moon, and sun.

We may now write the last term of (23) § 808, thus

$$\left.\begin{aligned} h &= H\frac{c^3}{D^3}(1 - 3\sin^2\delta) \\ \text{where}\qquad H &= \tfrac{1}{2}\tau a\left[\tfrac{1}{3}(1 + \mathfrak{E}) - \sin^2 l\right] \end{aligned}\right\} \quad \text{.....................(1).}$$

It is obvious from the solution of a right-angled spherical triangle that

$$\sin\delta = \sin I \sin(\theta - \xi).$$

Whence

$$1 - 3\sin^2\delta = 1 - \tfrac{3}{2}\sin^2 I + \tfrac{3}{2}\sin^2 I\cos 2(\theta - \xi) \quad \text{.......(2).}$$

By the theory of elliptic motion, we have, on neglecting the solar perturbation of the moon, which causes the 'evection,' the 'variation' and other inequalities,

$$\frac{c}{D}(1 - e^2) = 1 + e\cos(\theta - \varpi) \quad \text{...............(3).}$$

In proceeding to further developments, e and $\sin^2 I$ will be treated as small quantities of the first order, and those of the second order will be neglected. Thus in terms of the first order we have

$$\theta = \sigma t \quad \text{..............................(4).}$$

Then from (3) and (4) we have

$$\frac{c^3}{D^3} = 1 + 3e\cos(\sigma t - \varpi),$$

and from (1), (2), and (4)

$$\frac{h}{H} = \{1 + 3e\cos(\sigma t - \varpi)\}\{1 - \tfrac{3}{2}\sin^2 I + \tfrac{3}{2}\sin^2 I\cos 2(\sigma t - \xi)\}$$

$$= 1 - \tfrac{3}{2}\sin^2 I + 3e(1 - \tfrac{3}{2}\sin^2 I)\cos(\sigma t - \varpi) + \tfrac{3}{2}\sin^2 I\cos 2(\sigma t - \xi) \quad \text{............(5).}$$

In this expression the first term $1 - \frac{3}{2}\sin^2 I$ oscillates with a period of 19 years about the mean value $1 - \frac{3}{2}\sin^2\omega$, the

* Also of papers presented to the British Association by Sir W. Thomson and Capt. Evans, R.N., in 1878 (reprinted in *Nature*, Oct. 24, 1878), and by Mr G. H. Darwin in 1882.

maximum and minimum values of I being $\omega + i$ and $\omega - i$. It represents a small permanent increase to the ellipticity of the oceanic spheroid, on which is superposed a small 19-yearly tide. This part of the expression has no further interest in the present investigation. The last term of (5) goes through a double period in nearly 27·3 m. s. days and constitutes the fortnightly declinational tide. If the approximation were carried to terms of the second order, which may very easily be done, this term would have involved a factor $1 - \frac{5}{2}e^2$. The middle term goes through a single period in something over 27·3 days, the angular motion of the lunar perigee being 40° 40′ per annum. This term as it stands in (5) is complete to the second order. Thus we may write the expressions to the second order of small quantities, for the fortnightly and monthly elliptic tides, thus:— Long-period tides.

$$\left.\begin{aligned} \frac{\phi}{H} &= \tfrac{3}{2}\left(1 - \tfrac{5}{2}e^2\right)\sin^2 I \cos 2(\sigma t - \xi) \\ \frac{\mu}{H} &= 3e\left(1 - \tfrac{3}{2}\sin^2 I\right)\cos(\sigma t - \varpi) \end{aligned}\right\} \quad \ldots\ldots\ldots\ldots(6).$$

(b) We must now show how to compute I and ξ, and it will be expedient (as will appear below) at the same time to compute ν. Formulae for the longitude and R.A. of the intersection.

The accompanying figure exhibits the relation of the three planes to one another.

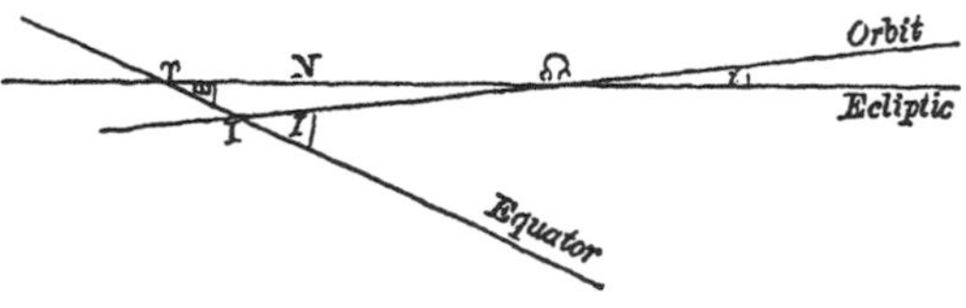

ξ the longitude of I in the orbit is ♈☊ − ☊I, and ν the right ascension of I is ♈I.

Now from the spherical triangle ♈☊I, we have

$$\cot \text{I☊} \sin N = \cos N \cos i + \sin i \cot \omega \quad \ldots\ldots\ldots(7),$$

$$\cot \text{I♈} \sin N = \cos N \cos \omega + \sin \omega \cot i \ldots\ldots\ldots\ldots(8),$$

$$\cos I = \cos i \cos \omega - \sin i \sin \omega \cos N \ldots\ldots\ldots\ldots\ldots(9).$$

Long-period tides.

Also $\tan\xi = \dfrac{(\cot I \, \Omega \sin N - \cos N)\sin N}{\cos N \cot I \quad \sin N + \sin^2 N}$.

Substituting in which from (7), and effecting some reductions in the result, and in (8), we have

$$\left.\begin{aligned}\tan\xi &= \frac{\sin i \cot\omega \sin N\,(1 - \tan\tfrac{1}{2}i \tan\omega \cos N)}{\cos^2\tfrac{1}{2}i + \sin i \cot\omega \cos N - \sin^2\tfrac{1}{2}i \cos 2N}\\ \tan\nu &= \frac{\tan i \operatorname{cosec}\omega \sin N}{1 + \tan i \cot\omega \cos N}\end{aligned}\right\} \ldots(10),$$

Formulas for the longitude and R.A. of the intersection.

These formulas are rigorously true, but since i is small, being about $5^\circ 9'$, we may obtain much simpler approximate formulæ, sufficiently accurate for all practical purposes. Treating then $\sin i$ and $\tan i$ as equal to one another, and to i the circular measure of $5^\circ 9'$, equations (10) become approximately,

$$\left.\begin{aligned}\tan\xi &= i \cot\omega \sin N - \tfrac{1}{2} i^2 \sin 2N \frac{1 - \frac{1}{2}\sin^2\omega}{\sin^2\omega}\\ \tan\nu &= i \operatorname{cosec}\omega \sin N - \tfrac{1}{2} i^2 \sin 2N \frac{\cos\omega}{\sin^2\omega}\end{aligned}\right\} \ldots\ldots\ldots (11).$$

The second terms of these expressions are very nearly equal to one another, because $\cos\omega = 1 - \frac{1}{2}\sin^2\omega$ approximately. And $\nu - \xi$ is a small angle, which is to a close degree of approximation equal to $i \tan\frac{1}{2}\omega \sin N$.

Numerical calculation shows that $i \tan\frac{1}{2}\omega$ is $1^\circ 4'$; hence $\xi = \nu - 1^\circ 4' \sin N$ very nearly.

In the Tidal Report of the British Association for 1876 the treatment of this subject, with notation involving a symbol ☽, is somewhat different from the above, but the result is the same. The symbol ☽ denotes "the equatorial mean moon's" right ascension at the epoch when $t = 0$; which it may be observed is not the same epoch as that chosen here. This fictitious mean moon moves in the equator with an angular velocity equal to the moon's mean motion, and it is at the "intersection" at the instant when the moon's mean longitude is equal to the longitude "in the orbit" of the intersection. In other words, if we take a second fictitious moon moving in the plane of the lunar orbit with an angular velocity equal to the moon's mean motion, and coinciding with the actual moon at the instant when the moon's mean longitude vanishes, then the equatorial mean moon coincides with this orbital mean moon at the intersection.

Long-period tides.

It is obvious then that the right ascension of the equatorial mean moon will always differ from the moon's mean longitude by $\nu - \xi$, and thus

$$☽ = \text{moon's mean longitude at the epoch} + 1°\,4' \sin N.$$

Therefore with the epoch of the Report of 1876 (pp. 299, 302)

$$\sigma t + ☽ - \nu = \text{moon's mean longitude} + 1°\,4' \sin N - \nu$$
$$= \text{moon's mean longitude} - \xi$$

Now according to the Report (p. 305), the fortnightly tide is expressed, (by means of H as defined in (1) above), in the form

Formulas for the longitude and R.A. of the intersection.

$$H\,\tfrac{3}{2}\sin^2 I \cos 2\,(\sigma t + ☽ - \nu).$$

This only differs from (6) in the term $\frac{5}{2}e^2$, which is the correction for the eccentricity of the lunar orbit.

It is to be remarked that in the report $☽ - \varpi'$ is the moon's mean anomaly at the epoch, and therefore ϖ' is equal to the mean longitude of the moon's perigee + $1°\,4' \sin N$, and not simply the mean longitude of the moon's perigee, as defined in the last line of p. 302. Since the moon's mean anomaly is only involved in the arguments of the elliptic tides, which are all small, this correction in ϖ' has no practical importance. It is however important, in regard to clear ideas of the notation and the spherical trigonometry of the subject.

In consequence of not at first apprehending properly the nature of the fictitious "equatorial mean moon," I overlooked the term $1°\,4' \sin N$ in ☽, and in the reductions made below have used ν instead of ξ. Since the difference between ν and ξ is clearly of little importance in respect to the numerical values of the fortnightly tide, I have not repeated the computations with the correct value of ☽, or, in the present notation, with ξ in place of ν.

(c) The factor H or $\frac{1}{2}\tau a\left[\frac{1}{3}(1 + \mathfrak{C}) - \sin^2 l\right]$ involves the function $\mathfrak{C}$, which depends on the distribution of land and water on the earth's surface. By (21) § 808

$$\mathfrak{C} = \frac{1}{\Omega}\iint (3\sin^2 l - 1)\cos l\, dl\, d\lambda$$

where Ω is the total area of ocean, and where the double integral is taken all over the surface of the ocean.

Long-period tides.

The integral of $3\sin^2 l - 1$ taken over the whole sphere vanishes, and therefore the integral taken over the sea is equal to, but opposite in sign to the integral taken over the land. It is more convenient to integrate over the land, because there is less of it, than over the sea.

In order to evaluate this integral, and to determine Ω the total area of sea at the same time, it will be sufficiently accurate if we replace the actual continents and islands of the earth by blocks of land, limited by parallels of latitude and by meridians. The following schedule specifies the blocks which were taken to represent the actual land, together with the names of the land to which they are supposed to correspond.

Since it is impossible that the amount of water, which flows in and out of the Mediterranean Sea in a week or a fortnight, can influence the height of the sea in the open ocean to any sensible extent, that sea has been treated as though it were dry land. The longitudes of the land are given so that any one may verify that the representation of the continents is pretty good; in evaluating the other four functions of (21) § 808 these longitudes would be required; but for $\mathfrak{C}$ we only require the number of degrees of longitude, which are occupied by land, between each pair of parallels of latitude.

Evaluation of the function $\mathfrak{C}$,

As explained above

$$\Omega\mathfrak{C} = -\iint (3\sin^2 l - 1)\cos l\, dl\, d\lambda \;\ldots\ldots\ldots\ldots (12),$$

$$\Omega = 4\pi - \iint \cos l\, dl\, d\lambda \;\ldots\ldots\ldots\ldots\ldots\ldots\ldots\ldots (13),$$

when the integrals are taken all over the land of the globe.

Now $$\int (3\sin^2 l - 1)\cos l\, dl = -\tfrac{1}{4}(\sin l + \sin 3l),$$

and $$\int \cos l\, dl = \sin l.$$

If therefore there be t_1 degrees of land between latitudes l_1 and l_2 of the N. hemisphere, and t_2 degrees of land between the same parallels of the S. hemisphere, it is clear that the contributions to (12) and (13) due to land between these latitudes in both hemispheres, are respectively

$$\frac{\pi}{180}(t_1 + t_2)\,\tfrac{1}{4}\left[\sin l + \sin 3l\right]_{l_2}^{l_1} \text{ and } -\frac{\pi}{180}(t_1 + t_2)\left[\sin l\right]_{l_2}^{l_1}.$$

APPROXIMATE DISTRIBUTION OF LAND ON THE EARTH'S SURFACE.

	N. latitude. W. longitude.	*N. latitude. E. longitude.*
lat. 80° to 90°.	20° to 50°. Arctic land.	
lat. 70° to 80°.	22° to 55°; 85° to 115°. Greenland. Islands.	55° to 60°; 90° to 110°. Nova Zembla. Tunda.
lat. 60° to 70°.	35° to 52°; 65° to 80°; 90° to 165°. Greenland. Baffinland. Brit.N.Am.	10° to 180°. Norway & N. Asia.
lat. 50° to 60°.	0° to 6°; 60° to 78°; 90° to 130°. G. Brit. Canada. Brit. N. Am.	10° to 140°; 155° to 160°. Europe & Asia. Kamschatka.
lat. 40° to 50°.	0° to 5°; 65° to 123°. France & Spain. U. S.	0° to 135°. Asia.
lat. 30° to 40°.	0° to 8°; 78° to 120°. Africa. U. S.	0° to 120°: 135° to 138°. Asia & Medit. Sea. Japan.
lat. 20° to 30°.	0° to 15°; 80° to 82°; 97° to 110°. Africa. Florida & Cuba. Mexico.	0° to 118°. Africa and Asia.
lat. 10° to 20°.	0° to 17°; 87° to 95°. Africa. Mexico.	0° to 50°; 75° to 85°; 95° to 108°; Africa. India. Siam. 122° to 125°. Philip. Isl.
lat. 0° to 10°.	53° to 78°. S. America.	0° to 48°; 98° to 105°; 112° to 117°. Africa. Malayia. Borneo.

	S. latitude. W. longitude.	*S. latitude. E. longitude.*
lat. 0° to 10°.	37° to 80°. S. America.	12° to 40°; 110° to 180°. Africa. Islands.
lat. 10° to 20°.	37° to 74°. S. America.	12° to 38°; 45° to 50°; 120° to 144°. Africa. Madagascar. Australia.
lat. 20° to 30°.	45° to 71°. S. America.	15° to 33°; 115° to 151°. Africa. Australia.
lat. 30° to 40°.	55° to 73°. S. America.	20° to 23°; 132° to 140°. Africa. Australia.
lat. 40° to 50°.	65° to 73°. S. America.	170° to 172°. N. Zealand.
lat. 50° to 60°.	67° to 72°. T. del Fuego.	
lat. 60° to 70°.	55° to 65°. S. Shetland.	120° to 180°. Adelie Land.
lat. 70° to 80°.	about 20° of longitude (Antarctic continent).	
lat. 80° to 90°.	about 180° of longitude (Antarctic continent).	

Now the above table gives t_1 and t_2 for each pair of latitudes 90° to 80°, 80° to 70° &c. in both hemispheres, for example from 20° to 30°, $t_1 + t_2$ is 228; hence it is clear that if Σ denotes sum-

Evaluation of the function $\mathfrak{C}$.

mation for the contributions due to each such pair of latitudes, we have

$$\mathfrak{C} = \frac{\Sigma \frac{1}{4}(t_1 + t_2)\left[\sin l + \sin 3l\right]_{l_2}^{l_1}}{720 - \Sigma (t_1 + t_2)\left[\sin l\right]_{l_2}^{l_1}}.$$

It is only necessary to form tables of $\sin l$ and $\frac{1}{4}(\sin l + \sin 3l)$ for each 10° of latitude from 0° to 90°, and then to form the first differences of these two sets of values, and subsequently to perform a number of multiplications, in order to obtain the required results. As the amount of antarctic land is quite uncertain, two suppositions were taken, namely, first that there is as much antarctic land as is given in the schedule, and secondly, that there is no land between S. latitude 80° and the pole. On the first hypothesis it was found that the fraction of the whole earth's surface which consists of land is $\frac{1}{720}$ of 202·9 = ·283, and in the second that the same proportion is $\frac{1}{720}$ of 200·2 = ·278. Rigaud* has estimated the proportion as ·266; if then it be considered that he too could have no information as to antarctic land, and that the Mediterranean Sea is here treated as solid, it appears that the representation of the continents by square blocks of land has been very satisfactory. The numerator for the expression for $\mathfrak{C}$ was found to be $-7·87$ or $-2·53$ according to the two hypotheses. Hence we have

$$\mathfrak{C} = \frac{-7·87}{517·1} = -·0152, \text{ with antarctic continent}$$

and $$\mathfrak{C} = \frac{-2·53}{519·8} = -·00486, \text{ without antarctic continent}$$

$\frac{1}{3}(1 + \mathfrak{C})$ will be found to be equal $\sin^2 34° 40'$ or $\sin^2 34° 57'$. Since $\frac{1}{3}$ is $\sin^2 35° 16'$, it follows that the latitude of evanescent fortnightly and monthly tides is very little affected by the distribution of land and water on the earth's surface.

In the reductions of the tidal observations I have put

$$\tfrac{1}{3}(1 + \mathfrak{C}) - \sin^2 l = \sin(35° - l)\sin(35° + l).$$

Theoretical expressions for equilibrium value of fortnightly and monthly tides.

Thus from (6) we have

$$\left.\begin{array}{l} \phi = \frac{3}{4}\tau a\left(1 - \frac{5}{2}e^2\right)\sin^2 I \sin(35° - l)\sin(35° + l)\cos 2(\sigma t - \xi) \\ \mu = \frac{3}{2}\tau a e\left(1 - \frac{3}{2}\sin^2 I\right)\sin(35° - l)\sin(35° + l)\cos(\sigma t - \varpi) \end{array}\right\} \ldots (14).$$

* *Trans. Cam. Phil. Soc.* Vol. 6.

Taking $E/M=82$; $c/a=60{\cdot}27$; $a=20{\cdot}9\times10^6$ feet, it will be found that

$$\tfrac{3}{2}\tau a=2{\cdot}6195 \text{ feet.}$$

If we take $\omega=23^\circ 28'$, $i=5^\circ 9'$, the maximum, mean and minimum values of I are $28^\circ 37'$, $23^\circ 28'$, $18^\circ 19'$. Then with $e=\cdot054908$, it will be found that — Maximum, mean and minimum values in British feet.

$$\tfrac{3}{4}\tau a(1-\tfrac{5}{2}e^2)\sin^2 I=\begin{cases}\cdot298, \text{ when } I=28^\circ 37',\\ \cdot206, \text{ when } I=23^\circ 28',\\ \cdot128, \text{ when } I=18^\circ 19';\end{cases}$$

$$\tfrac{3}{2}\tau ae(1-\tfrac{3}{2}\sin^2 I)=\begin{cases}\cdot094, \text{ when } I=28^\circ 37',\\ \cdot109, \text{ when } I=23^\circ 28',\\ \cdot123, \text{ when } I=18^\circ 19'.\end{cases}$$

These numbers are given in feet, and the equatorial semi-ranges of the ϕ and μ tides are (since $\sin^2 35^\circ=\frac{1}{3}$ nearly) about one-third of these numbers. At the time when I is a minimum these two tides have approximately equal ranges; but when I is a maximum the fortnightly is three times as great as the monthly tide.

(*d*) In the Reports of the British Association, and in the "Tide-tables for the Indian Ports"* for 1881 and 1882, the results of the harmonic analysis of the tidal observations are given in the form $R\cos(nt-\epsilon)$, where R, the semi-range of tide, is expressed in British feet, n is the speed of the particular tide in question, and ϵ, the retardation of phase (or shortly the phase), is an angle less than 360°. — Preparation for reduction by least squares.

In the case of the fortnightly and monthly tides n is respectively 2σ and $\sigma-\varpi_1$, where ϖ_1 is the angular velocity of the lunar perigee and therefore $\varpi=\varpi_1 t$. (In the Tidal Report of 1872 that which is here called ϖ_1 is denoted as ϖ.)

Now in order to compare the observed fortnightly tide with its theoretical value, we must write the observation in the form

$$R\cos[2(\sigma t-\xi)-(\epsilon-2\xi)].$$

Or if we put

$$\left.\begin{aligned}R\cos(\epsilon-2\xi)&=\text{A}\\ R\sin(\epsilon-2\xi)&=\text{B}\end{aligned}\right\}\quad\ldots\ldots\ldots\ldots\ldots\ldots(15)$$

* These tables were prepared under the direction of Captain (now Major) A. W. Baird, R.E., and Mr E. Roberts, and are published by "authority of the Secretary of State for India in Council."

Preparation for reduction by least squares.

the observation becomes

$$A\cos 2(\sigma t - \xi) + B\sin 2(\sigma t - \xi) \text{............} (16).$$

In the case of the monthly tide, if we put

$$\left.\begin{aligned} R\cos\epsilon &= C \\ R\sin\epsilon &= D \end{aligned}\right\} \text{......................} (17)$$

the result of observation becomes

$$C\cos(\sigma t - \varpi) + D\sin(\sigma t - \varpi) \text{............} (18).$$

The expressions for the theoretical equilibrium fortnightly and monthly tides are given in (14). If however the solid earth yields tidally, either as an elastic body, or as a viscous one, the height of the tide will fall below its equilibrium value. Moreover on the hypothesis of viscosity the phase of the tide will be affected; a result which would also follow from the effects of fluid friction.

Theoretical expression for the tides when the earth yields bodily, and when there is friction.

Thus the actual fortnightly and monthly tides must be expressed in the forms

$$\left.\begin{aligned} \phi &= \tfrac{3}{4}\tau a\left(1 - \tfrac{5}{2}e^2\right)\sin^2 I \sin(35^\circ - l)\sin(35^\circ + l)\,\{x\cos 2(\sigma t - \xi) + y\sin 2(\sigma t - \xi)\} \\ \mu &= \tfrac{3}{2}\tau a e\left(1 - \tfrac{3}{2}\sin^2 I\right)\sin(35^\circ - l)\sin(35^\circ + l)\,\{u\cos(\sigma t - \varpi) + v\sin(\sigma t - \varpi)\} \end{aligned}\right\} 19,$$

where x, y, u, v are numerical coefficients. If the equilibrium theory be nearly true (compare § 808 above) for the fortnightly and monthly tides, y and v will be small; and x and u will be fractions approaching unity, in proportion as the rigidity of the earth's mass approaches infinity.

If we now put

$$\left.\begin{aligned} a &= \tfrac{3}{4}\tau a\left(1 - \tfrac{5}{2}e^2\right)\sin^2 I \sin(35^\circ - l)\sin(35^\circ + l) \\ c &= \tfrac{3}{2}\tau a e\left(1 - \tfrac{3}{2}\sin^2 I\right)\sin(35^\circ - l)\sin(35^\circ + l) \end{aligned}\right\} \text{....} (20),$$

Equations for reduction by least squares.

then for the fortnightly tide

$$\left.\begin{aligned} ax &= A \\ ay &= B \end{aligned}\right\} \text{.........................} (21),$$

and for the monthly tide

$$\left.\begin{aligned} cu &= C \\ cv &= D \end{aligned}\right\} \text{.........................} (22).$$

Every set of tidal observations will give equations for x, y, u, v; and the most probable values of these quantities must be determined by the method of least squares.

Equations for reduction by least squares.

For places north of 35° N. lat., or south of 35° S. lat. the coefficients a and c become negative. This would be inconvenient for the arithmetical operations of reduction, and therefore for such places it is convenient to subtract 180° from the phases $\epsilon - 2\xi$, and ϵ which occur in the expressions for A, B, C, D; after doing this the coefficients a and c may in all cases be treated as positive, for we may suppose $(l - 35°)$ to be taken for places in the northern hemisphere North of 35°, and $35° - l$ for places in the same hemisphere to the South of 35°; and similarly for the southern hemisphere.

Numerical results of harmonic analysis of tidal observations.

(*e*) In collecting the results of tidal observation I have to thank Sir William Thomson, General Strachey, and Major Baird for placing all the materials in my hands, and for giving me every facility. As above stated the observations are to be found in the British Association Reports for 1872 and 1876, and in the Tide-tables of the Indian Government.

The results of the harmonic analysis of the tidal observations are given altogether for 22 different ports, but of these only 14 are here used. The following are the reasons for rejecting those made at 8 out of the 22 ports.

One of these stations is Cat Island in the Gulf of Mexico; this place, in latitude 30° 14′ N., lies so near to the critical latitude of evanescent fortnightly and monthly tides, that considering the uncertainty in the exact value of that latitude, it is impossible to determine the proper weight which should be assigned to the observation. The result only refers to a single year, viz. 1848, and as its weight must in any case be very small, the omission can exercise scarcely any effect on the result.

Another omitted station is Toulon; this being in the Mediterranean Sea cannot exhibit the true tide of the open ocean.

Another is Hanstal in the Gulf of Cutch. The result is given in an Indian Blue Book. I do not know the latitude, and General Strachey informs me that he believes the observations were only made during a few months for the purpose of determining the mean level of the sea, for the levelling operations of the great survey of India.

The other omitted stations are Diamond Harbour, Fort Gloster and Kidderpore in the Hooghly estuary, and Rangoon,

Numerical results of harmonic analysis of tidal observations.

and Moulmein. All these are river stations, and they all exhibit long period tides of such abnormal height as to make it nearly certain that the shallowness of the water has exercised a large influence on the results. The observations higher up the Hooghly seem more abnormal than those lower down. I also learn that the tidal predictions are not found to be satisfactory at these stations.

The following tables exhibit the results for the 14 remaining ports. The rows R and ϵ are extracted from the printed tidal results, and the rest of the values are the reductions effected in accordance with the investigations of the preceding sections. It has already been explained why, in the case of the fortnightly tide, $\epsilon - 2\nu$ is given in place of the more correct $\epsilon - 2\xi$. It must also be added that in many cases there is no information as to the days on which the observations began and ended; it was thus impossible to use the rigorously correct value for ν, namely that corresponding to the middle day of the period embraced by the observations. These details might no doubt have been obtained by means of correspondence with various persons in India; but considering the uncertainty in the tabular results it did not seem worth while to incur this delay.

Sir William Thomson placed in my hands a table of the values of I and ν corresponding to the 1st of July of each year. Accordingly when the observations are stated to be, for example, for 1874—5, I assume that the observations began early in 1874, and the values for I and ν for July 1, 1874, are used. In several cases it appears that the observations began in March, and here but little error has been incurred. In the few cases in which only a single year is named (*e.g.* Ramsgate), it is assumed that values for July 1 will be proper.

No attempt has been made to assign weight to each year's observations according to the exact number of months over which the tidal records extend. The data for such weighting are in many cases wanting. In computing the value for a the factor $1 - \frac{3}{2}e^2$ was omitted, but it has been introduced finally as explained below.

BRITISH AND FRENCH PORTS, NORTH OF LATITUDE 35°.

[*Tidal Reports of Brit. Assoc.* 1872 *and* 1876.]

	PLACE......... N. Latitude..	RAMSGATE. 51° 21′	LIVERPOOL. 53° 40′				WEST HARTLEPOOL. 54° 41′			BREST. 48° 23′
		1.	2.	3.	4.	5.	6.	7.	8.	9.
	YEAR	1864.	1857-8.	1858–9.	1859-60.	1866-7.	1858-9.	1859-60.	1860-1.	1875.
Fortnightly Tide.	*R.*	·0331	·093	·087	·024	·086	·052	·058	·078	·099
	ϵ.	268°·29	170°·7	148°·8	72°·9	340°·6	190°·84	222°·34	158°·62	80°·65
	$\epsilon - 2\nu - \pi$.	$\pi - 70^\circ\,17'$	$-9^\circ\,42'$	$-24^\circ\,26'$	$+86^\circ\,24' - \pi$	$\pi - 15^\circ\,20'$	$+17^\circ\,6'$	$+55^\circ\,50'$	$-2^\circ\,7'$	$+75^\circ\,47' - \pi$
	A.	− ·0112	+ ·0917	+ ·0337	− ·0015	− ·0346	+ ·0497	+ ·0297	+ ·0729	− ·0244
	B.	+ ·0311	− ·0156	− ·0153	− ·0240	+ ·0095	+ ·0153	+ ·0488	− ·0027	− ·0959
	a.	·0439	·0055	·0946	·0905	·0416	·0996	·0952	·0884	·0684
Monthly Tide.	*R.*	·0316	·046	·198	·152	·072	·075	·185	·189	·0331
	ϵ.	45°·09	289°·4	81°·6	172°·8	259°·8	23°·53	175°·75	79°·19	327°·51
	$\epsilon - \pi$.	$+45^\circ\,5' - \pi$	$\pi - 70^\circ\,36'$	$+31^\circ\,30' - \pi$	$-7^\circ\,12'$	$+79^\circ\,48'$	$+23^\circ\,32' - \pi$	$-4^\circ\,15'$	$+79^\circ\,11' - \pi$	$\pi - 32^\circ\,29'$
	C.	− ·0223	− ·0153	− ·1687	+ ·1508	+ ·0127	− ·0688	+ ·1847	− ·0261	− ·0279
	D.	− ·0224	+ ·0484	− ·1088	− ·0190	+ ·0708	− ·0290	− ·0100	− ·1865	+ ·0178
	c.	·0332	·0302	·0304	·0312	·0392	·0320	·0328	·0339	·0218

INDIAN PORTS.

[*Indian Tide Tables for 1881.*]

	PLACE ..... *Latitude*..	ADEN. 12° 47′		KURRACHEE. 24° 47′							
		10.	11.	12.	13.	14.	15.	16.	17.	18.	19.
	YEAR	1877-8.	1879-80.	1868-9.	1869-70.	1870-1.	1873-4.	1874-5.	1875-6.	1876-7.	1877-8.
Fortnightly Tide.	*R*.	·062	·062	·038	·064	·035	·016	·054	·014	·046	·065
	ϵ.	2°·80	354°·47	335°·40	333°·91	283°·22	237°·40	49°·18	25°·97	356°·22	19°·22
	$\epsilon - 2\nu$.	+12° 12′	+15° 40′	−40° 56′	−48° 53′	+77°29′−π	+89°32′−π	+37° 26′	+21° 7′	−1°·26′	+38° 37′
	A.	+·0606	+·0597	+·0287	+·0421	−·0076	−·0123	+·0429	+·0131	+·0460	+·0508
	B.	+·0131	+·0167	−·0249	−·0482	−·0342	−·0102	+·0328	+·0050	−·0012	+·0406
	a.	·0818	·0706	·0218	·0248	·0287	·0411	·0440	·0456	·0459	·0448
Monthly Tide.	*R*.	·025	·033	·076	·043	·032	·050	·057	·058	·085	·110
	ϵ.	320°·76	4°·62	247°·73	175°·27	115°·90	55°·83	23°·66	103°·22	42°·11	49°·03
	ϵ.	−39° 14′	+4° 37′	π+67° 44′	π−4° 44′	π−64° 6′	+55° 50′	+23° 39′	π−76° 47′	+42° 7′	+49° 2′
	C.	+·0194	+·0329	−·0288	−·0429	−·0140	+·0281	+·0522	−·0133	+·0631	+·0722
	D.	−·0158	+·0027	−·0703	+·0035	+·0288	+·0414	+·0229	+·0565	+·0570	+·0831
	c.	·0268	·0286	·0185	·0180	·0178	·0153	·0148	·0145	·0145	·0147

INDIAN PORTS (*continued*).

[*Indian Tide Tables for 1881-2.*]

	PLACE	OKHA POINT AND BEYT HARBOUR.	BOMBAY—APOLLO BUNDER.			KÁRWÁR.		BEYPORE.		PAUMBEN-PASS, Island of Rameswerann.	
	Latitude..	22° 28′	18° 55′			14° 48′		11° 10′		9° 16′	
		20.	**21.**	**22.**	**23.**	**24.**	**25.**	**26.**	**27.**	**28.**	**29.**
	YEAR	1874-5.	1876-7.	1878-9.	1879-80.	1878-9.	1879-80.	1878-9.	1879-80.	1878-9.	1879-80.
Fortnightly Tide.	*R*.	·070	·071	·091	·066	·067	·070	·106	·095	·056	·045
	ϵ.	52°·73	7°·51	331°·73	347°·49	329°·90	340°·88	355°·69	356°·03	343°·41	333°·31
	$\epsilon - 2\nu$.	+40° 59′	+9° 51′	−12° 40′	+8° 41′	−14° 30′	+2° 5′	+11° 17′	+17° 14′	−0° 59′	−5° 29′
	A.	+·0529	+·0699	+·0888	+·0653	+·0649	+·0699	+·1040	+·0907	·0560	·0448
	B.	+·0459	+·0121	−·0199	+·0100	−·0168	+·0025	+·0208	+·0281	−·0010	−·0043
	a.	·0525	·0671	·0617	·0565	·0726	·0665	·0803	·0735	·0835	·0764
Monthly Tide.	R.	·050	·027	·053	·046	·042	·058	·069	·071	·059	·052
	ϵ.	311°·98	100°·24	314°·52	355°·46	350°·58	14°·28	5°·05	85°·42	348°·98	57°·52
	ϵ.	−48° 37′	π − 79° 46′	−45° 29′	−4° 32′	−9° 25′	+14° 17′	+5° 57′	+85° 25′	−11° 1′	+57° 31′
	C.	+·0331	−·0048	+·0372	+·0459	+·0415	+·0562	+·0687	+·0057	+·0579	+·0279
	D.	−·0375	+·0266	−·0378	−·0036	−·0069	+·0143	+·0072	+·0708	−·0113	+·0439
	c.	·0177	·0212	·0220	·0229	·0260	·0270	·0287	·0298	·0298	·0310

INDIAN PORTS (*continued*).

[*Indian Tide Tables*, 1881-2.]

	PLACE	VIZAGAPATAM.		MADRAS.	PORT BLAIR, ROSS ISLAND.
	Latitude..	17° 41′		13° 4′	11° 40½′
		30.	31.	32.	33.
	YEAR	1879-80.	1880-1.	1880-1.	1880-1.
Fortnightly Tide.	*R*.	·036	·055	·032	·059
	ϵ.	3°·07	317°·54	341°·95	332°·33
	$\epsilon-2\nu$.	+24° 16′	−17° 43′	+6° 41′	−2° 56′
	A.	+·0328	+·0524	+·0318	+·0589
	B.	+·0148	−·0167	+·0037	−·0029
	a.	·0597	·0534	·0626	·0649
Monthly Tide.	*R*.	·021	·077	·040	·020
	ϵ.	22°·48	52°·54	40°·65	12°·95
	ϵ.	+22° 29′	+52° 33′	+40° 39′	+12° 57′
	C.	+·0194	+·0468	+·0304	+·0195
	D.	+·0080	+·0611	+·0260	+·0045
	c.	·0242	·0253	·0296	·0307

Gauss's notation is adopted for the reductions*. That is to say, [AA] denotes the sum of the squares of the A's, and [Aa] the sum of the products of each A into its corresponding a.

In computing the value of a for the fortnightly tide the factor $(1-\frac{5}{2}e^2)$ which occurs therein was treated as being equal to unity; since $\frac{5}{2}e^2=\cdot00754$, it follows that the [aa], which would be found from the numbers given in the table, must be multiplied by (1−·01508), and the [Aa] and [Ba] by (1−·00754). After introducing these correcting factors the following results were found;

Results of reduction.

[aa]=·14573, [AA]=·09831, [BB]=·02576, [Aa]=·09836, [Ba]=·00291

[cc]=·02253, [CC]=·11588, [DD]=·07552, [Cc]=·01533, [Dc]=·00202.

* See Gauss's works, or the Appendix to Chauvenet's *Astronomy*.

Then according to the method of least squares, the following are the most probable values of x, y, u, v. Results of reduction.

$$x=\frac{[Aa]}{[aa]}, \quad y=\frac{[Ba]}{[aa]}, \quad u=\frac{[Cc]}{[cc]}, \quad v=\frac{[Dc]}{[cc]}.$$

And if m be the number of observations (which in the present case is 33) the mean errors of x, y, u, v are respectively

$$\frac{1}{[aa]}\sqrt{\frac{[AA][aa]-[Aa]^2}{m-1}}, \quad \frac{1}{[aa]}\sqrt{\frac{[BB][aa]-[Ba]^2}{m-1}},$$

$$\frac{1}{[cc]}\sqrt{\frac{[CC][cc]-[Cc]^2}{m-1}}, \quad \frac{1}{[cc]}\sqrt{\frac{[DD][cc]-[Dc]^2}{m-1}}.$$

The probable errors are found from the mean errors by multiplying by ·6745.

I thus find that

$$x=\cdot675\pm\cdot056, \; y=\cdot020\pm\cdot055, \; u=\cdot680\pm\cdot258, \; v=\cdot090\pm\cdot218.$$

The smallness of the values of y and v is satisfactory; for, as stated above (§ 848 (d)), if the equilibrium theory were true for the two tides under discussion, they should vanish. Moreover the signs are in agreement with what they should be, if friction be a sensible cause of tidal retardation. But considering the magnitude of the probable errors, it is of course rather more likely that the non-evanescence of y and v is due to errors of observation*.

If the solid earth does not yield tidally, and if the equilibrium theory is fulfilled, x and u should each be approximately

* Shortly after these computations were completed Professor Adams happened to observe a misprint in the Tidal Report for 1872. This Report gives the method employed in the reduction by harmonic analysis of the tidal observations, and the erroneous formula relates to the reduction of the tides of long period. On inquiring of Mr Roberts, who has superintended the harmonic analysis, it appears that the erroneous formula has been throughout used in the reductions. A discussion of this mistake and of its effects will be found in a paper communicated to the British Association by me in 1882. It appears that the values of the fortnightly tide are not seriously vitiated, but the monthly elliptic tide will have suffered much more. This will probably account for the large probable error which I have found for the value of the monthly tide. If a recomputation of all the long-period tides should be carried out, I think there is good hope that the probable error of the value of the fortnightly tide may also be reduced.

It appears from a communication from Major Baird, R.E., that the erroneous formula referred to has *not* been used in the reduction of the Indian Tidal Observations (March 20, 1883).

unity, and if it yields tidally they should have equal values. The very close agreement between them is probably somewhat due to chance. From this point of view it seems reasonable to combine all the observations, resulting from 66 years of observation, for both sorts of tides together.

Then writing X and Y for the numerical factors by which the equilibrium values of the two components of either tide are to be multiplied in order to give the actual results, I find

$$X = \cdot 676 \pm \cdot 076, \quad Y = \cdot 029 \pm \cdot 065.$$

Tidal yielding of the earth's mass. Rigidity about equal to or greater than that of steel.

These results really seem to present evidence of a tidal yielding of the earth's mass, showing that it has an effective rigidity about equal to that of steel *.

But this result is open to some doubt for the following reason:

Taking only the Indian results (48 years in all), which are much more consistent than the English ones, I find

$$X = \cdot 931 \pm \cdot 056, \quad Y = \cdot 155 \pm \cdot 068.$$

We thus see that the more consistent observations seem to bring out the tides more nearly to their theoretical equilibrium-values with no elastic yielding of the solid.

It is to be observed however that the Indian results being confined within a narrow range of latitude give (especially when we consider the absence of minute accuracy in the evaluation of $\mathfrak{C}$ in § 848 (c)) a less searching test for the elastic yielding, than a combination of results from all latitudes.

On the whole we may fairly conclude that, whilst there is some evidence of a tidal yielding of the earth's mass, that yielding is certainly small, and that the effective rigidity is at least as great as that of steel.

* It is remarkable that elastic yielding of the upper strata of the earth, in the case where the sea does not cover the whole surface, may lead to an apparent augmentation of oceanic tides at some places, situated on the coasts of continents. This subject is investigated in the Report for 1882 of the Committee of the British Association on "The Lunar Disturbance of Gravity." It is there, however, erroneously implied that this kind of elastic yielding would cause an apparent augmentation of tide at all stations of observation.

APPENDIX TO CHAPTER VII.

The following Appendices are reprints of papers published at various times. Excepting where it is expressly so stated, or where it is obvious from the context, they speak as from the date of publication. The marginal notes however to the appendices which appeared in the first edition speak as at the date of issue of that edition, viz. 1867; in the new appendices the marginal notes are now added for the first time.

(C.)—Equations of Equilibrium of an Elastic Solid deduced from the Principle of Energy*.

(*a*) Let a solid composed of matter fulfilling no condition of isotropy in any part, and not homogeneous from part to part, be given of any shape, unstrained, and let every point of its surface be altered in position to a given distance in a given direction. It is required to find the displacement of every point of its substance, in equilibrium. Let x, y, z be the co-ordinates of any particle, P, of the substance in its undisturbed position, and $x+\alpha$, $y+\beta$, $z+\gamma$ its co-ordinates when displaced in the manner specified: that is to say, let α, β, γ be the components of the required displacement. Then, if for brevity we put

Strain of any magnitude specified by six elements.

$$\left.\begin{aligned}
A &= \left(\frac{d\alpha}{dx}+1\right)^2 + \left(\frac{d\beta}{dx}\right)^2 + \left(\frac{d\gamma}{dx}\right)^2\\
B &= \left(\frac{d\alpha}{dy}\right)^2 + \left(\frac{d\beta}{dy}+1\right)^2 + \left(\frac{d\gamma}{dy}\right)^2\\
C &= \left(\frac{d\alpha}{dz}\right)^2 + \left(\frac{d\beta}{dz}\right)^2 + \left(\frac{d\gamma}{dz}+1\right)^2\\
a &= \frac{d\alpha}{dy}\frac{d\alpha}{dz} + \left(\frac{d\beta}{dy}+1\right)\frac{d\beta}{dz} + \frac{d\gamma}{dy}\left(\frac{d\gamma}{dz}+1\right)\\
b &= \frac{d\alpha}{dz}\left(\frac{d\alpha}{dx}+1\right) + \frac{d\beta}{dz}\frac{d\beta}{dx} + \left(\frac{d\gamma}{dz}+1\right)\frac{d\gamma}{dx}\\
c &= \left(\frac{d\alpha}{dx}+1\right)\frac{d\alpha}{dy} + \frac{d\beta}{dx}\left(\frac{d\beta}{dy}+1\right) + \frac{d\gamma}{dx}\frac{d\gamma}{dy}
\end{aligned}\right\}\quad\text{..........(1);}$$

these six quantities A, B, C, a, b, c are proved [§ 190 (*e*) and § 181 (5)] to thoroughly determine the strain experienced by the

* Appendix to a paper by Sir W. Thomson on "Dynamical problems regarding Elastic Spheroidal Shells and Spheroids of incompressible liquid." *Phil. Trans.* 1863, Vol. 153, p. 610.

Strain specified by six elements.

substance infinitely near the particle P (irrespectively of any rotation it may experience), in the following manner:

(*b*.) Let ξ, η, ζ be the undisturbed co-ordinates of a particle infinitely near P, relatively to axes through P parallel to those of x, y, z respectively; and let $\xi_{,}, \eta_{,}, \zeta_{,}$ be the co-ordinates relative still to axes through P, when the solid is in its strained condition. Then

$$\xi_{,}^2 + \eta_{,}^2 + \zeta_{,}^2 = A\xi^2 + B\eta^2 + C\zeta^2 + 2a\eta\zeta + 2b\zeta\xi + 2c\xi\eta \;\ldots\ldots(2);$$

and therefore all particles which in the strained state lie on a spherical surface

$$\xi_{,}^2 + \eta_{,}^2 + \zeta_{,}^2 = r_{,}^2,$$

are in the unstrained state, on the ellipsoidal surface,

$$A\xi^2 + B\eta^2 + C\zeta^2 + 2a\eta\zeta + 2b\zeta\xi + 2c\xi\eta = r_{,}^2.$$

This (§§ 155—165) completely defines the homogeneous strain of the matter in the neighbourhood of P.

Anticipatory application of the Carnot and Clausius thermodynamic law:

(*c*.) Hence, the thermodynamic principles by which, in a paper on the "Thermo-elastic Properties of Matter*," Green's dynamical theory of elastic solids was demonstrated as part of the modern dynamical theory of heat, show that if $wdxdydz$ denote the work required to alter an infinitely small undisturbed volume, $dxdydz$, of the solid, into its disturbed condition, when its temperature is kept constant, we must have

$$w = f(A, B, C, a, b, c) \;\ldots\ldots\ldots\ldots\ldots\ldots (3)$$

its combination with Joule's law expressed analytically for elastic solid.

where f denotes a positive function of the six elements, which vanishes when $A-1$, $B-1$, $C-1$, a, b, c each vanish. And if W denote the whole work required to produce the change actually experienced by the whole solid, we have

Potential energy of deformation;

$$W = \iiint wdxdydz \;\ldots\ldots\ldots\ldots\ldots\ldots\ldots\ldots(4)$$

where the triple integral is extended through the space occupied by the undisturbed solid.

a minimum for stable equilibrium.

(*d*.) The position assumed by every particle in the interior of the solid will be such as to make this a minimum subject to the condition that every particle of the surface takes the position given to it; this being the elementary condition of stable equilibrium. Hence, by the method of variations

$$\delta W = \iiint \delta wdxdydz = 0 \ldots\ldots\ldots\ldots\ldots\ldots\ldots(5).$$

* *Quarterly Journ. of Math.*, April, 1855, or *Mathematical and Physical Papers* by Sir W. Thomson, 1882, Art. XLVIII. Part VII.

But, exhibiting only terms depending on δa, we have

Potential energy of deformation; a minimum for stable equilibrium.

$$\delta w = \left\{2\frac{dw}{dA}\left(\frac{da}{dx}+1\right)+\frac{dw}{db}\frac{da}{dz}+\frac{dw}{dc}\frac{da}{dy}\right\}\frac{d\delta a}{dx}$$
$$+\left\{2\frac{dw}{dB}\frac{da}{dy}+\frac{dw}{da}\frac{da}{dz}+\frac{dw}{dc}\left(\frac{da}{dx}+1\right)\right\}\frac{d\delta a}{dy}$$
$$+\left\{2\frac{dw}{dC}\frac{da}{dz}+\frac{dw}{da}\frac{da}{dy}+\frac{dw}{db}\left(\frac{da}{dx}+1\right)\right\}\frac{d\delta a}{dz}$$
$$+\text{ etc.}$$

Hence, integrating by parts, and observing that δa, $\delta\beta$, $\delta\gamma$ vanish at the limiting surface, we have

$$\delta W = -\iiint dxdydz\left\{\left(\frac{dP}{dx}+\frac{dQ}{dy}+\frac{dR}{dz}\right)\delta a+\text{etc.}\right\}\ \ldots\ldots(6)$$

where for brevity P, Q, R denote the multipliers of $\frac{d\delta a}{dx}$, $\frac{d\delta a}{dy}$, $\frac{d\delta a}{dz}$ respectively, in the preceding expression. In order that δW may vanish, the multipliers of δa, $\delta\beta$, $\delta\gamma$, in the expression now found for it, must each vanish, and hence we have, as the equations of equilibrium

Equations of internal equilibrium of an elastic solid experiencing no bodily force.

$$\left.\begin{array}{l}\frac{d}{dx}\left\{2\frac{dw}{dA}\left(\frac{da}{dx}+1\right)+\frac{dw}{db}\frac{da}{dz}+\frac{dw}{dc}\frac{da}{dy}\right\}\\ +\frac{d}{dy}\left\{2\frac{dw}{dB}\frac{da}{dy}+\frac{dw}{da}\frac{da}{dz}+\frac{dw}{dc}\left(\frac{da}{dx}+1\right)\right\}\\ +\frac{d}{dz}\left\{2\frac{dw}{dC}\frac{da}{dz}+\frac{dw}{da}\frac{da}{dy}+\frac{dw}{db}\left(\frac{da}{dx}+1\right)\right\}=0\\ \qquad\text{etc.}\quad\text{etc.}\end{array}\right\}\ \ldots\ldots(7),$$

of which the second and third, not exhibited, may be written down merely by attending to the symmetry.

(*e.*) From the property of w that it is necessarily positive when there is any strain, it follows that there must be some distribution of strain through the interior which shall make $\iiint w dxdydz$ *the least possible*, subject to the prescribed surface condition; and therefore that the solution of equations (7) subject to this condition, is possible. If, whatever be the nature of the solid as to difference of elasticity in different directions, in any part, and as to heterogeneity from part to part, and whatever be the extent of the change of form and dimensions to which it is subjected, there cannot be any internal configuration of unstable

Their solution proved possible and unique when surface displacement is given, unless there can be unstable equilibrium:

equilibrium, nor consequently any but one of stable equilibrium, with the prescribed surface displacement, and no disturbing force on the interior; then, besides being always positive, w must be such a function of A, B, etc., that there can be only one solution of the equations. This is obviously the case when the unstrained solid is homogeneous.

hence necessarily unique for a homogeneous solid.

(*f.*) It is easy to include, in a general investigation similar to the preceding, the effects of any force on the interior substance, such as we have considered particularly for a spherical shell, of homogeneous isotropic matter, in §§ 730...737 above. It is also easy to adapt the general investigation to superficial data of *force*, instead of displacement.

Extension of the analysis to include bodily force, and data of surface easy.

(*g.*) Whatever be the general form of the function f for any part of the substance, since it is always positive it cannot change in sign when $A-1, B-1, C-1, a, b, c$, have their signs changed; and therefore for infinitely small values of these quantities it must be a homogeneous quadratic function of them with constant coefficients. (And it may be useful to observe that for all values of the variables A, B, etc., it must therefore be expressible in the same form, with varying coefficients, each of which is always finite, for all values of the variables.) Thus, for infinitely small strains we have Green's theory of elastic solids, founded on a homogeneous quadratic function of the components of strain, expressing the work required to produce it. Thus, putting

Transition to case of infinitely small strains.

Green's theory:

$$A-1=2e,\quad B-1=2f,\quad C-1=2g \quad\text{............(8)}$$

and denoting by $\frac{1}{2}(e,e)$, $\frac{1}{2}(f,f)$,...(e,f),...(e,a),... the coefficients, we have, as above (§ 673),

$$\left.\begin{aligned} w=\tfrac{1}{2}\{(e,e)\,e^2+(f,f)f^2+(g,g)\,g^2+(a,a)\,a^2+(b,b)\,b^2+(c,c)\,c^2\}& \\ +(e,f)\,ef+(e,g)\,eg+(e,a)\,ea+(e,b)\,eb+(e,c)\,ec& \\ +(f,g)fg+(f,a)fa+(f,b)fb+(f,c)fc& \\ +(g,a)\,ga+(g,b)\,gb+(g,c)\,gc& \\ +(a,b)\,ab+(a,c)\,ac& \\ +(b,c)\,bc& \end{aligned}\right\}\text{(9).}$$

(*h.*) When the strains are infinitely small the products $\frac{dw}{dA}\frac{da}{dx}$, $\frac{dw}{db}\frac{da}{dz}$, etc., are each infinitely small, of the second order. We therefore omit them; and then attending to (8), we reduce (7) to

Case of infinitely small strains:— dynamic equations of internal equilibrium:

$$\left.\begin{aligned}\frac{d}{dx}\frac{dw}{de}+\frac{d}{dy}\frac{dw}{dc}+\frac{d}{dz}\frac{dw}{db}=0\\ \frac{d}{dx}\frac{dw}{dc}+\frac{d}{dy}\frac{dw}{df}+\frac{d}{dz}\frac{dw}{da}=0\\ \frac{d}{dx}\frac{dw}{db}+\frac{d}{dy}\frac{dw}{da}+\frac{d}{dz}\frac{dw}{dg}=0\end{aligned}\right\}\quad\ldots\ldots\ldots\ldots(10),$$

which are the equations of interior equilibrium. Attending to (9) we see that $\frac{dw}{de}\ldots\frac{dw}{da}\ldots$ are linear functions of e, f, g, a, b, c the components of strain. Writing out one of them as an example we have

$$\frac{dw}{de}=(e,e)\,e+(e,f)\,f+(e,g)\,g+(e,a)\,a+(e,b)\,b+(e,c)\,c\ldots(11).$$

And, α, β, γ denoting, as before, the component displacements of any interior particle, P, from its undisturbed position (x, y, z) we have, by (8) and (1)

and relative kinematic equations.

$$\left.\begin{aligned}e=\frac{d\alpha}{dx},\quad f=\frac{d\beta}{dy},\quad g=\frac{d\gamma}{dz}\\ a=\frac{d\beta}{dz}+\frac{d\gamma}{dy},\quad b=\frac{d\gamma}{dx}+\frac{d\alpha}{dz},\quad c=\frac{d\alpha}{dy}+\frac{d\beta}{dx}\end{aligned}\right\}\quad\ldots\ldots\ldots(12).$$

It is to be observed that the coefficients (e, e), (e, f), etc., will be in general functions of (x, y, z), but will be each constant when the unstrained solid is homogeneous.

Solution proved unique whether the body is homogeneous or heterogeneous, when surface displacement is given:

(*i.*) It is now easy to prove directly, for the case of infinitely small strains, that the solution of the equations of interior equilibrium, whether for a heterogeneous or a homogeneous solid, subject to the prescribed surface condition, is unique. For, let α, β, γ be components of displacement fulfilling the equations, and let α', β', γ' denote any other functions of x, y, z, having the same surface values as α, β, γ, and let $e', f', \ldots, w'$ denote functions depending on them in the same way as $e, f, \ldots, w$ depend on α, β, γ. Thus by Taylor's theorem,

$$w'-w=\frac{dw}{de}(e'-e)+\frac{dw}{df}(f'-f)+\frac{dw}{dg}(g'-g)+\frac{dw}{da}(a'-a)+\frac{dw}{db}(b'-b)+\frac{dw}{dc}(c'-c)+H,$$

where H denotes the same homogeneous quadratic function of

Case of infinitely small strains:—

$e'-e$, etc., that w is of e, etc. If for $e'-e$, etc., we substitute their values by (12), this becomes

$$w'-w=\frac{dw}{de}\frac{d(\alpha'-\alpha)}{dx}+\frac{dw}{db}\frac{d(\alpha'-\alpha)}{dz}+\frac{dw}{dc}\frac{d(\alpha'-\alpha)}{dy}+\text{etc.}+H.$$

Multiplying by $dxdydz$, integrating by parts, observing that $\alpha'-\alpha$, $\beta'-\beta$, $\gamma'-\gamma$ vanish at the bounding surface, and taking account (10), we find simply

$$\iiint(w'-w)\,dxdydz=\iiint H dxdydz\dots\dots\dots\dots\dots(13).$$

But H is essentially positive. Therefore every other interior condition than that specified by α, β, γ, provided only it has the same bounding surface, requires a greater amount of work than w to produce it: and the excess is equal to the work that would be required to produce, from a state of no displacement, such a displacement as superimposed on α, β, γ, would produce the other. And inasmuch as α, β, γ, fulfil only the conditions of satisfying (11) and having the given surface values, it follows that no other than one solution can fulfil these conditions.

solution not necessarily unique, when the surface data are of force.

(*j*.) But (as has been pointed out to us by Stokes) when the surface data are of force, not of displacement, or when force acts from without, on the interior substance of the body, the solution is not in general unique, and there may be configurations of unstable equilibrium even with infinitely small displacement. For instance, let part of the body be composed of a steel-bar magnet; and let a magnet be held outside in the same line, and with a pole of the same name in its end nearest to one end of the inner magnet. The equilibrium will be unstable, and there will be positions of stable equilibrium with the inner bar slightly inclined to the line of the outer bar, unless the rigidity of the rest of the body exceed a certain limit.

Condition that substance may be isotropic, without limitation to infinitely small strains:

(*k*.) Recurring to the general problem, in which the strains are not supposed infinitely small; we see that if the solid is isotropic in every part, the function of A, B, C, a, b, c which expresses w, must be merely a function of the roots of the equation [§ 181 (11)]

$$(A-\zeta^2)(B-\zeta^2)(C-\zeta^2)-a^2(A-\zeta^2)-b^2(B-\zeta^2)-c^2(C-\zeta^2)+2abc=0\dots(14)$$

which (that is the positive values of ζ) are the ratios of elongation along the principal axes of the strain-ellipsoid. It is unnecessary here to enter on the analytical expression of this condition. For in the case of $A-1$, $B-1$, $C-1$, a, b, c each infinitely small, it obviously requires that

$$\left.\begin{array}{l}(e,e)=(f,f)=(g,g);\ (f,g)=(g,e)=(e,f);\ (a,a)=(b,b)=(c,c);\\ (e,a)=(f,b)=(g,c)=0;\ (b,c)=(c,a)=(a,b)=0;\ \text{and}\\ (e,b)=(e,c)=(f,c)=(f,a)=(g,a)=(g,b)=0.\end{array}\right\}\ldots(15).$$

Thus the 21 coefficients are reduced to three—

(e, e) which we may denote by the single letter $\mathfrak{A}$,
(f, g) " " " " $\mathfrak{B}$,
(a, a) " " " " n.

It is clear that this is necessary and sufficient for insuring *cubic isotropy;* that is to say, perfect equality of elastic properties with reference to the three rectangular directions OX, OY, OZ. But for *spherical isotropy*, or complete isotropy with reference to all directions through the substance, it is further necessary that

expressed in equations among the moduli of elasticity for case of infinitely small strains.

$$\mathfrak{A}-\mathfrak{B}=2n\ldots\ldots(16);$$

as is easily proved analytically by turning two of the axes of co-ordinates in their own plane through 45°; or geometrically by examining the nature of the strain represented by any one of the elements a, b, c (a simple shear) and comparing it with the resultant of c, and $f=-e$ (which is also a simple shear). It is convenient now to put

$$\mathfrak{A}+\mathfrak{B}=2m;\ \text{so that}\ \mathfrak{A}=m+n,\ \mathfrak{B}=m-n\ldots\ldots(17);$$

and thus the expression for the potential energy per unit of volume becomes

$$2w=m(e+f+g)^2+n(e^2+f^2+g^2-2fg-2ge-2ef+a^2+b^2+c^2)\ldots(18).$$

Potential energy of infinitely small strain in isotropic solid.

Using this in (9), and substituting for e, f, g, a, b, c their values by (12), we find immediately the equations of internal equilibrium, which are the same as (6) of § 698.

(*l*.) To find the mutual force exerted across any surface within the solid, as expressed by (1) of § 662, we have clearly, by considering the work done respectively by P, Q, R, S, T, U (§ 662) on any infinitely small change of figure or dimensions in the solid,

Components of stress required for infinitely small strain.

$$P=\frac{dw}{de},\quad Q=\frac{dw}{df},\quad R=\frac{dw}{dg},\quad S=\frac{dw}{da},\quad T=\frac{dw}{db},\quad U=\frac{dw}{dc}\ldots(19).$$

Hence, for an isotropic solid, (18) gives the expressions which we have used above, (12) of § 673.

Moduli of resistance to compression and of rigidity.

(*m*.) To interpret the coefficients m and n in connexion with elementary ideas as to the elasticity of the solid; first let $a=b=c=0$, and $e=f=g=\frac{1}{3}\delta$: in other words, let the substance experience a uniform dilatation, in all directions, producing an expansion of volume from 1 to $1+\delta$. In this case (18) becomes

$$w=\tfrac{1}{2}(m-\tfrac{1}{3}n)\delta^2;$$

and we have

$$\frac{dw}{d\delta}=(m-\tfrac{1}{3}n)\delta.$$

Hence $(m-\frac{1}{3}n)\delta$ is the normal force per unit area of its surface required to keep any portion of the solid expanded to the amount specified by δ. Thus $m-\frac{1}{3}n$ measures the elastic force called out by, or the elastic resistance against, change of volume: and viewed as a *modulus of elasticity*, it may be called the bulk-modulus. [Compare §§ 692, 693, 694, 688, 682, and 680.] What is commonly called the "compressibility" is measured by $1/(m-\frac{1}{3}n)$.

And let next $e=f=g=b=c=0$; which gives

$$w=\tfrac{1}{2}na^2; \text{ and, by (19), } S=na.$$

This shows that the tangential force per unit area required to produce an infinitely small shear (§ 171), amounting to a, is na. Hence n measures the innate power of the body to resist change of shape, and return to its original shape when force has been applied to change it: that is to say, it measures *the rigidity* of the substance.

(D).—On the Secular Cooling of the Earth*.

Appendix D. Dissipation of energy disregarded by many followers of Hutton.

(*a*.) For eighteen years it has pressed on my mind, that essential principles of Thermo-dynamics have been overlooked by those geologists who uncompromisingly oppose all paroxysmal hypotheses, and maintain not only that we have examples now before us, on the earth, of all the different actions by which its crust has been modified in geological history, but that these actions have never, or have not on the whole, been more violent in past time than they are at present.

* *Transactions of the Royal Society of Edinburgh*, 1862 (W. Thomson).

Dissipation of energy from the solar system.

(*b.*) It is quite certain the solar system cannot have gone on, even as at present, for a few hundred thousand or a few million years, without the irrevocable loss (by dissipation, not by *annihilation*) of a very considerable proportion of the entire energy initially in store for sun heat, and for Plutonic action. It is quite certain that the whole store of energy in the solar system has been greater in all past time than at present; but it is conceivable that the rate at which it has been drawn upon and dissipated, whether by solar radiation, or by volcanic action in the earth or other dark bodies of the system, may have been nearly equable, or may even have been less rapid, in certain periods of the past. But it is far more probable that the secular rate of dissipation has been in some direct proportion to the total amount of energy in store, at any time after the commencement of the present order of things, and has been therefore very slowly diminishing from age to age.

Terrestrial climate influenced by the probably hotter sun of a few million years ago.

(*c.*) I have endeavoured to prove this for the sun's heat, in an article recently published in *Macmillan's Magazine* (March 1862)*, where I have shown that most probably the sun was sensibly hotter a million years ago than he is now. Hence, geological speculations assuming somewhat greater extremes of heat, more violent storms and floods, more luxuriant vegetation, and hardier and coarser grained plants and animals, in remote antiquity, are more probable than those of the extreme quietist, or "uniformitarian" school. A middle path, not generally safest in scientific speculation, seems to be so in this case. It is probable that hypotheses of grand catastrophes destroying all life from the earth, and ruining its whole surface at once, are greatly in error; it is impossible that hypotheses assuming an equability of sun and storms for 1,000,000 years, can be wholly true.

(*d.*) Fourier's mathematical theory of the conduction of heat is a beautiful working out of a particular case belonging to the general doctrine of the "Dissipation of Energy †." A characteristic of the practical solutions it presents is, that in each case a

* Reprinted as Appendix E, below.

† *Proceedings of Royal Soc. Edin.*, Feb. 1852. "On a universal Tendency in Nature to the Dissipation of Mechanical Energy," *Mathematical and Physical Papers*, by Sir W. Thomson, 1882, Art. LIX. Also, "On the Restoration of Energy in an unequally Heated Space," *Phil. Mag.*, 1853, first half year, *Mathematical and Physical Papers*, by Sir W. Thomson, 1882, Art. LXII.

distribution of temperature, becoming gradually equalized through an unlimited future, is expressed as a function of the time, which is infinitely divergent for all times longer past than a definite determinable epoch. The distribution of heat at such an epoch is essentially *initial*—that is to say, it cannot result from any previous condition of matter by natural processes. It is, then, well called an "*arbitrary* initial distribution of heat," in Fourier's great mathematical poem, because that which is rigorously expressed by the mathematical formula could only be realized by action of a power able to modify the laws of dead matter. In an article published about nineteen years ago in the *Cambridge Mathematical Journal**, I gave the mathematical criterion for an essentially initial distribution; and in an inaugural essay, "De Motu Caloris per Terræ Corpus," read before the Faculty of the University of Glasgow in 1846, I suggested, as an application of these principles, that a perfectly complete geothermic survey would give us data for determining an initial epoch in the problem of terrestrial conduction. At the meeting of the British Association in Glasgow in 1855, I urged that special geothermic surveys should be made for the purpose of estimating absolute dates in geology, and I pointed out some cases, especially that of the salt-spring borings at Creuznach, in Rhenish Prussia, in which eruptions of basaltic rock seem to leave traces of their igneous origin in residual heat†. I hope this suggestion may yet be taken up, and may prove to some extent useful; but the disturbing influences affecting underground temperature, as Professor Phillips has well shown in a recent inaugural address to the Geological Society, are too great to allow us to expect any very precise or satisfactory results‡.

Mathematicians' use of word "arbitrary" metaphysically significant.

Criterion of an essentially "initial" distribution of heat in a solid:

now applied to estimate date of earth's consolidation, from data of present underground temperature.

Value of local geothermic surveys, for estimation of absolute dates in geology.

(*e*.) The chief object of the present communication is to estimate from the known general increase of temperature in the earth downwards, the date of the first establishment of that *consistentior status*, which, according to Leibnitz's theory, is the initial date of all geological history.

* Feb. 1844.—"Note on Certain Points in the Theory of Heat," *Mathematical and Physical Papers*, by Sir W. Thomson, 1882, Vol. I. Art. X.

† See British Association Report of 1855 (Glasgow) Meeting.

‡ Much work in the direction suggested above has been already carried out by the Committee of the British Association, on Underground Temperature.

(*f*.) In all parts of the world in which the earth's crust has been examined, at sufficiently great depths to escape large influence of the irregular and of the annual variations of the superficial temperature, a gradually increasing temperature has been found in going deeper. The rate of augmentation (estimated at only $\frac{1}{110}$th of a degree, Fahr., in some localities, and as much as $\frac{1}{15}$th of a degree in other, per foot of descent) has not been observed in a sufficient number of places to establish any fair average estimate for the upper crust of the whole earth. But $\frac{1}{50}$th is commonly accepted as a rough mean; or, in other words, it is assumed as a result of observation, that there is, on the whole, about 1° Fahr. of elevation of temperature per 50 British feet of descent.

Increase of temperature downwards in earth's crust; but very imperfectly observed hitherto.

(*g*.) The fact that the temperature increases with the depth implies a continual loss of heat from the interior, by conduction outwards through or into the upper crust. Hence, since the upper crust does not become hotter from year to year, there must be a secular loss of heat from the whole earth. It is possible that no cooling may result from this loss of heat, but only an exhaustion of potential energy, which in this case could scarcely be other than chemical affinity between substances forming part of the earth's mass. But it is certain that either the earth is becoming on the whole cooler from age to age, or the heat conducted out is generated in the interior by temporary dynamical (that is, in this case, chemical) action*. To suppose, as Lyell, adopting the chemical hypothesis, has done†, that the substances, combining together, may be again separated electrolytically by thermo-electric currents, due to the heat generated by their combination, and thus the chemical action and its heat continued in an endless cycle, violates the principles of natural philosophy in exactly the same manner, and to the same degree, as to believe that a clock constructed with a self-winding movement may fulfil the expectations of its ingenious inventor by going for ever.

Secular loss of heat out of the earth demonstrated:

but not so any present or past secular cooling, however probable.

Fallacy of a thermo-electric perpetual motion.

* Another kind of dynamical action, capable of generating heat in the interior of the earth, is the friction which would impede tidal oscillations, if the earth were partially or wholly constituted of viscous matter. See a paper by Mr G. H. Darwin, "On problems connected with the tides of a viscous spheroid." *Phil. Trans.* Part II. 1879.

† *Principles of Geology*, chap. xxxi. ed. 1853.

Exception to the soundness of arguments adduced in the promulgation and prosecution of the Huttonian reform.

(*h*.) It must indeed be admitted that many geological writers of the "Uniformitarian" school, who in other respects have taken a profoundly philosophical view of their subject, have argued in a most fallacious manner against hypotheses of violent action in past ages. If they had contented themselves with showing that many existing appearances, although suggestive of extreme violence and sudden change, may have been brought about by long-continued action, or by paroxysms not more intense than some of which we have experience within the periods of human history, their position might have been unassailable; and certainly could not have been assailed except by a detailed discussion of their facts. It would be a very wonderful, but not an absolutely incredible result, that volcanic action has never been more violent on the whole than during the last two or three centuries; but it is as certain that there is now less volcanic energy in the whole earth than there was a thousand years ago, as it is that there is less gunpowder in a "Monitor" after she has been seen to discharge shot and shell, whether at a nearly equable rate or not, for five hours without receiving fresh supplies, than there was at the beginning of the action. Yet this truth has been ignored or denied by many of the leading geologists of the present day *, because they believe that the facts within their province do not demonstrate greater violence in ancient changes of the earth's surface, or do demonstrate a nearly equable action in all periods.

Secular diminution of whole amount of volcanic energy quite certain: but not in 1862 admitted by some of the chief geologists.

(*i*) The chemical hypothesis to account for underground heat might be regarded as not improbable, if it was only in isolated localities that the temperature was found to increase with the depth; and, indeed, it can scarcely be doubted that chemical action exercises an appreciable influence (possibly negative, however) on the action of volcanoes; but that there is slow uniform "combustion," *eremacausis*, or chemical combination of any kind going on, at some great unknown depth under the surface everywhere, and creeping inwards gradually as the chemical affinities in layer after layer are successively saturated, seems extremely improbable, although it cannot be pronounced to be absolutely impossible, or contrary to all analogies in nature. The less

Chemical hypothesis to account for ordinary underground heat not impossible, but very improbable.

* It must be borne in mind that this was written in 1862. The opposite statement concerning the beliefs of geologists would probably be now nearer the truth.

hypothetical view, however, that the earth is merely a warm chemically inert body cooling, is clearly to be preferred in the present state of science.

Poisson's hypothesis to account for ordinary underground heat proved impossible without destruction of life.

(*j.*) Poisson's celebrated hypothesis, that the present underground heat is due to a passage, at some former period, of the solar system through hotter stellar regions, cannot provide the circumstances required for a palæontology continuous through that epoch of external heat. For from a mean of values of the conductivity, in terms of the thermal capacity of unit volume, of the earth's crust, in three different localities near Edinburgh, deduced from the observations on underground temperature instituted by Principal Forbes there, I find that if the supposed transit through a hotter region of space took place between 1250 and 5000 years ago, the temperature of that supposed region must have been from 25° to 50° Fahr. above the present mean temperature of the earth's surface, to account for the present general rate of underground increase of temperature, taken as 1° Fahr. in 50 feet downwards. Human history negatives this supposition. Again, geologists and astronomers will, I presume, admit that the earth cannot, 20,000 years ago, have been in a region of space 100° Fahr. warmer than its present surface. But if the transition from a hot region to a cool region supposed by Poisson took place more than 20,000 years ago, the excess of temperature must have been more than 100° Fahr., and must therefore have destroyed animal and vegetable life. Hence, the further back and the hotter we can suppose Poisson's hot region, the better for the geologists who require the longest periods; but the best for their view is Leibnitz's theory, which simply supposes the earth to have been at one time an incandescent liquid, without explaining how it got into that state. If we suppose the temperature of melting rock to be about 10,000° Fahr. (an extremely high estimate), the consolidation may have taken place 200,000,000 years ago. Or, if we suppose the temperature of melting rock to be 7000° Fahr. (which is more nearly what it is generally assumed to be), we may suppose the consolidation to have taken place 98,000,000 years ago.

Poisson's hypothesis disproved as any acceptable mitigation of Leibnitz's theory.

Probable limits of uncertainty as to thermal con-

(*k.*) These estimates are founded on the Fourier solution demonstrated below. The greatest variation we have to make in them, to take into account the differences in the ratios of conductivities and capacities of surface rocks.

ductivities to specific heats of the three Edinburgh rocks, is to reduce them to nearly half, or to increase them by rather more than half. A reduction of the Greenwich underground observations recently communicated to me by Professor Everett of Windsor, Nova Scotia, gives for the Greenwich rocks a quality intermediate between those of the Edinburgh rocks. But we are very ignorant as to the effects of high temperatures in altering the conductivities and specific heats of rocks, and as to their latent heat of fusion. We must, therefore, allow very wide limits in such an estimate as I have attempted to make; but I think we may with much probability say that the consolidation cannot have taken place less than 20,000,000 years ago, or we should have more underground heat than we actually have, nor more than 400,000,000 years ago, or we should not have so much as the least observed underground increment of temperature. That is to say, I conclude that Leibnitz's epoch of emergence of the *consistentior status* was probably between those dates.

Extreme admissible limits of date of earth's consolidation.

(*l*.) The mathematical theory on which these estimates are founded is very simple, being, in fact, merely an application of one of Fourier's elementary solutions to the problem of finding at any time the rate of variation of temperature from point to point, and the actual temperature at any point, in a solid extending to infinity in all directions, on the supposition that at an initial epoch the temperature has had two different constant values on the two sides of a certain infinite plane. The solution for the two required elements is as follows:—

Mathematical expression for interior temperature near the surface of a hot solid commencing to cool:

$$\frac{dv}{dx} = \frac{V}{\sqrt{\pi\kappa t}}\epsilon^{-x^2/4\kappa t}$$

$$v = v_0 + \frac{2V}{\sqrt{\pi}}\int_0^{x/2\sqrt{\kappa t}} dz\epsilon^{-z^2}$$

where κ denotes the conductivity of the solid, measured in terms of the thermal capacity of the unity of bulk;

V, half the difference of the two initial temperatures;

v_0, their arithmetical mean;

t, the time;

x, the distance of any point from the middle plane;

v, the temperature of the point x and t;

and, consequently (according to the notation of the differential calculus), dv/dx the rate of variation of the temperature per unit of length perpendicular to the isothermal planes.

(*m.*) To demonstrate this solution, it is sufficient to verify—

(1.) That the expression for v satisfies Fourier's equation for the linear conduction of heat, viz.:

$$\frac{dv}{dt} = \kappa \frac{d^2v}{dx^2}.$$

(2.) That when $t = 0$, the expression for v becomes $v_0 + V$ for all positive, and $v_0 - V$ for all negative values of x; and (3.) That the expression for dv/dx is the differential coefficient of the expression for v with reference to x. The propositions (1.) and (3.) are proved directly by differentiation. To prove (2.) we have, when $t = 0$, and x positive, Proof.

$$v = v_0 + \frac{2V}{\sqrt{\pi}} \int_0^\infty dz \epsilon^{-z^2}$$

or according to the known value, $\frac{1}{2}\sqrt{\pi}$, of the definite integral

$\int_0^\infty dz \epsilon^{-z^2}$, $\qquad v = v_0 + V$;

and for all values of t, the second term has equal positive and negative values for equal positive and negative values of x, so that when $t = 0$ and x negative,

$$v = v_0 - V.$$

The admirable analysis by which Fourier arrived at solutions including this, forms a most interesting and important mathematical study. It is to be found in his *Théorie Analytique de la Chaleur.* Paris, 1822.

(*n.*) The accompanying diagram (page 477) represents, by two curves, the preceding expressions for dv/dx and v respectively.

(*o.*) The solution thus expressed and illustrated applies, for a certain time, without sensible error, to the case of a solid sphere, primitively heated to a uniform temperature, and suddenly exposed to any superficial action, which for ever after keeps the surface at some other constant temperature. If, for instance, the case considered is that of a globe 8000 miles diameter of solid rock, the solution will apply with scarcely sensible error for more than 1000 millions of years. For, if the rock be of a certain average quality as to conductivity and specific heat, the value of κ, as found in a previous communication to the Royal Expression for interior temperature near surface of a hot body commencing to cool:

proved to be practically approximate for the earth for 100 million years.

Society,* will be 400, for unit of length a British foot and unit of time a year; and the equation expressing the solution becomes

$$\frac{dv}{dx} = \frac{1}{35\cdot4}\frac{V}{\sqrt{t}}\epsilon^{-x^2/1600t};$$

and if we give t the value 1,000,000,000, or anything less, the exponential factor becomes less than $\epsilon^{-5\cdot6}$ (which being equal to about $\frac{1}{270}$, may be regarded as insensible), when x exceeds 3,000,000 feet, or 568 miles. That is to say, during the first 1000 million years the variation of temperature does not become sensible at depths exceeding 568 miles, and is therefore confined to so thin a crust, that the influence of curvature may be neglected.

(*p*.) If, now, we suppose the time to be 100 million years from the commencement of the variation, the equation becomes

$$\frac{dv}{dx} = \frac{1}{3\cdot54\times10^5}V\epsilon^{-x^2/1600\times10^8}.$$

Distribution of temperature 100 million years after commencement of cooling of a great enough mass of average rock.

The diagram, therefore, shows the variation of temperature which would now exist in the earth if, its whole mass being first solid and at one temperature 100 million years ago, the temperature of its surface had been everywhere suddenly lowered by V degrees, and kept permanently at this lower temperature: the scales used being as follows:—

(1) For depth below the surface,—scale along OX, length a, represents 400,000 feet.

(2) For rate of increase of temperature per foot of depth,—scale of ordinates parallel to OY, length b, represents $\frac{1}{354000}$ of V per foot. If, for example, $V = 7000°$ Fahr. this scale will be such that b represents $\frac{1}{50\cdot6}$ of a degree Fahr. per foot.

(3) For excess of temperature,—scale of ordinates parallel to OY, length b, represents $V/\frac{1}{2}\sqrt{\pi}$, or 7900°, if $V = 7000°$ Fahr.

Thus the rate of increase of temperature from the surface downwards would be sensibly $\frac{1}{51}$ of a degree per foot for the first 100,000 feet or so. Below that depth the rate of increase per foot would begin to diminish sensibly. At 400,000 feet it would have diminished to about $\frac{1}{141}$ of a degree per foot. At

* "On the Periodical Variations of Underground Temperature." *Trans. Roy. Soc. Edin.*, March 1860.

INCREASE OF TEMPERATURE DOWNWARDS IN THE EARTH.

Distribution of temperature 100 million years after commencement of cooling of a great enough mass of average rock:

$ON = x.$

$NP' = b\epsilon^{-x^2/a^2} = y'.$

$NP = \text{area } ONP'A \div a = \dfrac{1}{a}\displaystyle\int_0^x y'dx.$

$a = 2\sqrt{\kappa t}.$

$\dfrac{dv}{dx} = \dfrac{V}{a} \cdot \dfrac{NP}{b\frac{1}{2}\sqrt{\pi}}.$

$v - v_0 = V \cdot \dfrac{NP}{b \cdot \frac{1}{2}\sqrt{\pi}}.$

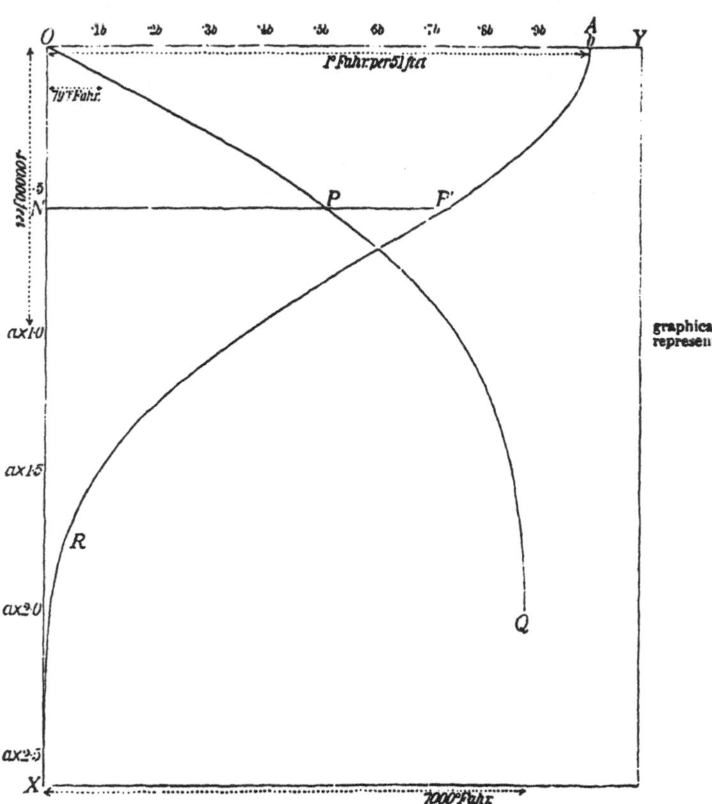

graphically represented.

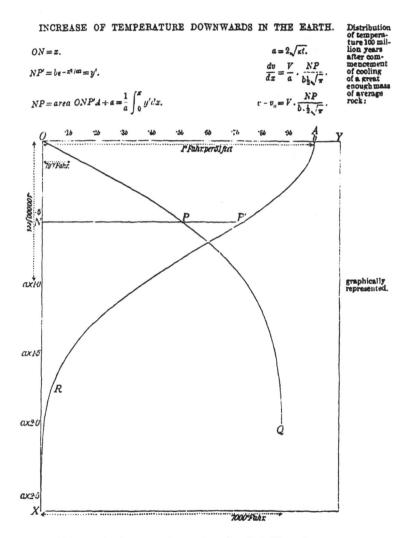

OPQ **curve showing excess of temperature above that of the surface.**

AP'R **curve showing rate of augmentation of temperature downwards.**

800,000 feet it would have diminished to less than $\frac{1}{50}$ of its initial value,—that is to say, to less than $\frac{1}{2550}$ of a degree per foot; and so on, rapidly diminishing, as shown in the curve. Such is, on the whole, the most probable representation of the earth's present temperature, at depths of from 100 feet, where the annual variations cease to be sensible, to 100 miles; below which the whole mass, or all except a nucleus cool from the beginning, is (whether liquid or solid), probably at, or very nearly at, the proper melting temperature for the pressure at each depth.

Terrestrial climate not sensibly influenced by underground heat.

(*q.*) The theory indicated above throws light on the question so often discussed, as to whether terrestrial heat can have influenced climate through long geological periods, and allows us to answer it very decidedly in the negative. There would be an increment of temperature at the rate of 2° Fahr. per foot downwards near the surface 10,000 years after the beginning of the cooling, in the case we have supposed. The radiation from earth and atmosphere into space (of which we have yet no satisfactory absolute measurement) would almost certainly be so rapid in the earth's actual circumstances, as not to allow a rate of increase of 2° Fahr. per foot underground to augment the temperature of the surface by much more than about 1°; and hence I infer that the general climate cannot be sensibly affected by conducted heat at any time more than 10,000 years after the commencement of superficial solidification. No doubt, however, in particular places there might be an elevation of temperature by thermal springs, or by eruptions of melted lava, and everywhere vegetation would, for the first three or four million years, if it existed so soon after the epoch of consolidation, be influenced by the sensibly higher temperature met with by roots extending a foot or more below the surface.

Rates of increase of temperature inwards in a great enough mass of average rock, at various times after commencement of cooling from a primitive temperature of 7000° Fahr.

(*r.*) Whatever the amount of such effects is at any one time, it would go on diminishing according to the inverse proportion of the square roots of the times from the initial epoch. Thus, if at 10,000 years we have 2° per foot of increment below ground,

At	40,000	years we should have		1° per foot.	
„	160,000	„	„	$\frac{1}{2}$°	„
„	4,000,000	„	„	$\frac{1}{10}$°	„
„	100,000,000	„	„	$\frac{1}{50}$°	„

It is therefore probable that for the last 96,000,000 years the rate of increase of temperature under ground has gradually diminished from about $\frac{1}{10}$th to about $\frac{1}{50}$th of a degree Fahrenheit per foot, and that the thickness of the crust through which any stated degree of cooling has been experienced has in that period gradually increased up to its present thickness from $\frac{1}{5}$th of that thickness. Is not this, on the whole, in harmony with geological evidence, rightly interpreted? Do not the vast masses of basalt, the general appearances of mountain-ranges, the violent distortions and fractures of strata, *the great prevalence of metamorphic action* (which must have taken place at depths of not many miles, if so much), all agree in demonstrating that the rate of increase of temperature downwards must have been much more rapid, and in rendering it probable that volcanic energy, earthquake shocks, and every kind of so-called plutonic action, have been, on the whole, more abundantly and violently operative in geological antiquity than in the present age?

(*s.*) But it may be objected to this application of mathematical theory—(1), That the earth was once all melted, or at least melted all round its surface, and cannot possibly, or rather cannot with any probability, be supposed to have been ever a uniformly heated solid, 7000° Fahr. warmer than our present surface temperature, as assumed in the mathematical problem; and (2) No natural action could possibly produce at one instant, and maintain for ever after, a seven thousand degrees' lowering of the surface temperature. Taking the second objection first, I answer it by saying, what I think cannot be denied, that a large mass of melted rock, exposed freely to our air and sky, will, after it once becomes crusted over, present in a few hours, or a few days, or at the most a few weeks, a surface so cool that it can be walked over with impunity. Hence, after 10,000 years, or, indeed, I may say after a single year, its condition will be sensibly the same as if the actual lowering of temperature experienced by the surface had been produced in an instant, and maintained constant ever after. I answer the first objection by saying, that if experimenters will find the latent heat of fusion, and the variations of conductivity and specific heat of the earth's crust up to its melting point, it will be easy to modify the solution given above, so as to make it applicable to the case of a liquid globe gradually solidifying from without inwards, in consequence of

Objections to terrestrial application raised and removed.

Objections to terrestrial application raised and removed.

heat conducted through the solid crust to a cold external medium. In the meantime, we can see that this modification will not make any considerable change in the resulting temperature of any point in the crust, unless the latent heat parted with on solidification proves, contrary to what we may expect from analogy, to be considerable in comparison with the heat that an equal mass of the solid yields in cooling from the temperature of solidification to the superficial temperature. But, what is more to the purpose, it is to be remarked that the objection, plausible as it appears, is altogether fallacious, and that the problem solved above corresponds much more closely, in all probability, with the actual history of the earth, than does the modified problem suggested by the objection. The earth, although once all melted, or melted all round its surface, did, in all probability, really become a solid at its melting temperature all through, or all through the outer layer, which had been melted; and not until the solidification was thus complete, or nearly so, did the surface begin to cool. That this is the true view can scarcely be doubted, when the following arguments are considered.

Present underground temperature probably due to heat generated in the original building of the earth.

(*t.*) In the first place, we shall assume that at one time the earth consisted of a solid nucleus, covered all round with a very deep ocean of melted rocks, and left to cool by radiation into space. This is the condition that would supervene, on a cold body much smaller than the present earth meeting a great number of cool bodies still smaller than itself, and is therefore in accordance with what we may regard as a probable hypothesis regarding the earth's antecedents. It includes, as a particular case, the commoner supposition, that the earth was once melted throughout, a condition which might result from the collision of two nearly equal masses. But the evidence which has convinced most geologists that the earth had a fiery beginning, goes but a very small depth below the surface, and affords us absolutely no means of distinguishing between the actual phenomena, and those which would have resulted from either an entire globe of liquid rock, or a cool solid nucleus covered with liquid to any depth exceeding 50 or 100 miles. Hence, irrespectively of any hypothesis as to antecedents from which the earth's initial fiery condition may have followed by natural causes, and simply assuming, as rendered probable by geological evidence, that there was at one time melted rock all over the surface, we need not assume the

Primitive heating may have been throughout, or merely through a superficial layer of no greater depth than $\frac{1}{40}$ of the radius.

depth of this lava ocean to have been more than 50 or 100 miles; although we need not exclude the supposition of any greater depth, or of an entire globe of liquid.

(*u.*) In the process of refrigeration, the fluid must [as I have remarked regarding the sun, in a recent article in *Macmillan's Magazine* (March, 1862)*, and regarding the earth's atmosphere, in a communication to the Literary and Philosophical Society of Manchester †] be brought by convection, to fulfil a definite law of distribution of temperature which I have called "convective equilibrium of temperature." That is to say, the temperatures at different parts in the interior must [in any great fluid mass which is kept well stirred] differ according to the different pressures by the difference of temperatures which any one portion of the liquid would present, if given at the temperature and pressure of any part, and then subjected to variation of pressure, but prevented from losing or gaining heat. The reason for this is the extreme slowness of true thermal conduction; and the consequently preponderating influence of great currents throughout a continuous fluid mass, in determining the distribution of temperature through the whole.

"Convective equilibrium of temperature" defined: must have been approximately fulfilled until solidification commenced.

(*v.*) The thermo-dynamic law connecting temperature and pressure in a fluid mass, not allowed to lose or gain heat, investigated theoretically, and experimentally verified in the cases of air and water, by Dr Joule and myself‡, shows, therefore, that the temperature in the liquid will increase from the surface downwards, if, as is most probably the case, the liquid contracts in cooling. On the other hand, if the liquid, like water near its

* See Appendix E, below.

† *Proceedings*, Jan. 1862. "On the Convective Equilibrium of Temperature in the Atmosphere."

‡ Joule, "On the Changes of Temperature produced by the Rarefaction and Condensation of Air," *Phil. Mag.* 1845. Thomson, "On a Method for Determining Experimentally the Heat evolved by the Compression of Air;" Dynamical Theory of Heat, Part IV., *Trans. R. S. E.*, Session 1850-51; and reprinted *Phil. Mag.* Joule and Thomson, "On the Thermal Effects of Fluids in Motion," *Trans. R. S. Lond.*, June 1853 and June 1854. Joule and Thomson, "On the Alterations of Temperature accompanying Changes of Pressure in Fluids," *Proceedings R. S. Lond.*, June 1857. These articles, except the first by Joule, are all now republished in Vol. I. Arts. XLVIII. and XLIX. of *Mathematical and Physical Papers*, by Sir W. Thomson.

Alternative cases as to distribution of temperature before solidification.

freezing-point, expands in cooling, the temperature, according to the convective and thermo-dynamic laws just stated (§§ *u*, *v*), would actually be lower at great depths than near the surface, even although the liquid is cooling from the surface; but there would be a very thin superficial layer of lighter and cooler liquid, losing heat by true conduction, until solidification at the surface would commence.

Effect of pressure on the temperature of solidification.

(*w*.) Again, according to the thermo-dynamic law of freezing, investigated by my brother*, Professor James Thomson, and verified by myself experimentally for water†, the temperature of solidification will, at great depths, because of the great pressure, be higher there than at the surface if the fluid contracts, or lower than at the surface if it expands, in becoming solid.

(*x*.) How the temperature of solidification, for any pressure, may be related to the corresponding temperature of fluid convective equilibrium, it is impossible to say, without knowledge, which we do not yet possess, regarding the expansion with heat, and the specific heat of the fluid, and the change of volume, and the latent heat developed in the transition from fluid to solid.

Question whether solidification commenced at surface or centre or bottom.

(*y*.) For instance, supposing, as is most probably true, both that the liquid contracts in cooling towards its freezing-point, and that it contracts in freezing, we cannot tell, without definite numerical data regarding those elements, whether the elevation of the temperature of solidification, or of the actual temperature of a portion of the fluid given just above its freezing-point, produced by a given application of pressure is the greater. If the former is greater than the latter, solidification would commence at the bottom, or at the centre, if there is no solid nucleus to begin with, and would proceed outwards; and there could be no complete permanent incrustation all round the surface till the whole globe is solid, with, possibly, the exception of irregular, comparatively small spaces of liquid.

(*z*.) If, on the contrary, the elevation of temperature, produced

* "Theoretical Considerations regarding the Effect of Pressure in lowering the Freezing-point of Water," *Trans. R. S. E.*, Jan. 1849. Republished by permission of the author, in Vol. I. (pp. 156—164) of *Mathematical and Physical Papers*, by Sir W. Thomson, 1882.

† *Proceedings R. S. E.*, Session 1849-50. *Mathematical and Physical Papers*, by Sir W. Thomson, 1882, p. 165.

by an application of pressure to a given portion of the fluid, is greater than the elevation of the freezing temperature produced by the same amount of pressure, the superficial layer of the fluid would be the first to reach its freezing-point, and the first actually to freeze.

(*aa.*) But if, according to the second supposition of § *v*, the liquid expanded in cooling near its freezing-point, the solid would probably likewise be of less specific gravity than the liquid at its freezing-point. Hence the surface would crust over permanently with a crust of solid, constantly increasing inwards by the freezing of the interior fluid in consequence of heat conducted out through the crust. The condition most commonly assumed by geologists would thus be produced.

Importance of experimental investigation of contraction or expansion of melted rocks in solidification.

(*bb.*) But Bischof's experiments, upon the validity of which, as far as I am aware, no doubt has ever been thrown, show that melted granite, slate, and trachyte, all contract by something about 20 per cent. in freezing. We ought, indeed, to have more experiments on this most important point, both to verify Bischof's results on rocks, and to learn how the case is with iron and other unoxydised metals. In the meantime we must consider it as probable that the melted substance of the earth did really contract by a very considerable amount in becoming solid.

Bischof's experiments proving contraction make it probable that the surface was never allowed to cool till solidification was very nearly complete through the interior.

(*cc.*) Hence if, according to any relations whatever among the complicated physical circumstances concerned, freezing did really commence at the surface, either all round or in any part, before the whole globe had become solid, the solidified superficial layer must have broken up and sunk to the bottom, or to the centre, before it could have attained a sufficient thickness to rest stably on the lighter liquid below. It is quite clear, indeed, that if at any time the earth were in the condition of a thin solid shell of, let us suppose 50 feet or 100 feet thick of granite, enclosing a continuous melted mass of 20 per cent. less specific gravity in its upper parts, where the pressure is small, this condition cannot have lasted many minutes. The rigidity of a solid shell of superficial extent so vast in comparison with its thickness, must be as nothing, and the slightest disturbance would cause some part to bend down, crack, and allow the liquid to run out over the whole solid. The crust itself would in consequence become shattered into fragments, which must all sink to the bottom, or meet in

31—2

the centre and form a nucleus there if there is none to begin with.

(*dd*.) It is, however, scarcely possible, that any such continuous crust can ever have formed all over the melted surface at one time, and afterwards have fallen in. The mode of solidification conjectured in § *y*, seems on the whole the most consistent with what we know of the physical properties of the matter concerned. So far as regards the result, it agrees, I believe, with the view adopted as the most probable by Mr Hopkins*. But whether from the condition being rather that described in § *z*, which seems also possible, for the whole or for some parts of the heterogeneous substance of the earth, or from the viscidity as of mortar, which necessarily supervenes in a melted fluid, composed of ingredients becoming, as the whole cools, separated by crystallizing at different temperatures before the solidification is perfect, and which we actually see in lava from modern volcanoes; it is probable that when the whole globe, or some very thick superficial layer of it, still liquid or viscid, has cooled down to near its temperature of perfect solidification, incrustation at the surface must commence.

(*ee*.) It is probable that crust may thus form over wide extents of surface, and may be temporarily buoyed up by the vesicular character it may have retained from the ebullition of the liquid in some places, or, at all events, it may be held up by the viscidity of the liquid; until it has acquired some considerable thickness sufficient to allow gravity to manifest its claim, and sink the heavier solid below the lighter liquid. This process must go on until the sunk portions of crust build up from the bottom a sufficiently close ribbed solid skeleton or frame, to allow fresh incrustations to remain bridging across the now small areas of lava pools or lakes.

Probable cause of volcano and earthquakes.

(*ff*.) In the honey-combed solid and liquid mass thus formed, there must be a continual tendency for the liquid, in consequence of its less specific gravity, to work its way up; whether by masses of solid falling from the roofs of vesicles or tunnels, and causing earthquake shocks, or by the roof breaking quite through when very thin, so as to cause two such hollows to unite, or the liquid of

* See his report on "Earthquakes and Volcanic Action." British Association Report for 1847.

any of them to flow out freely over the outer surface of the earth; or by gradual subsidence of the solid, owing to the thermodynamic melting, which portions of it, under intense stress, must experience, according to views recently published by Professor James Thomson*. The results which must follow from this tendency seem sufficiently great and various to account for all that we see at present, and all that we learn from geological investigation, of earthquakes, of upheavals, and subsidences of solid, and of eruptions of melted rock.

(*gg.*) These conclusions, drawn solely from a consideration of the necessary order of cooling and consolidation, according to Bischof's result as to the relative specific gravities of solid and of melted rock, are in perfect accordance with §§ 832...848, regarding the present condition of the earth's interior,—that it is not, as commonly supposed, all liquid within a thin solid crust of from 30 to 100 miles thick, but that it is on the whole more rigid certainly than a continuous solid globe of glass of the same diameter, and probably than one of steel.

(E.) On the Age of the Sun's Heat†.

The second great law of Thermodynamics involves a certain principle of *irreversible action in nature*. It is thus shown that, although mechanical energy is *indestructible*, there is a universal tendency to its dissipation, which produces gradual augmentation and diffusion of heat, cessation of motion, and exhaustion of potential energy through the material universe‡. The result would inevitably be a state of universal rest and death, if the universe were finite and left to obey existing laws. But it is impossible to conceive a limit to the extent of matter in the universe; and therefore science points rather to an endless progress, through an endless space, of action involving the trans-

Dissipation of Energy.

* *Proceedings of the Royal Society of London*, 1861, "On Crystallization and Liquefaction as influenced by Stresses tending to Change of Form in Crystals."

† From *Macmillan's Magazine*, March 1862.

‡ See *Proceedings R.S.E.* Feb. 1852, or *Phil. Mag.* 1853, first half year, "On a Universal Tendency in Nature to the Dissipation of Mechanical Energy." *Math. and Phys. Papers*, by Sir W. Thomson, 1882, Art. LIX.

formation of potential energy into palpable motion and thence into heat, than to a single finite mechanism, running down like a clock, and stopping for ever. It is also impossible to conceive either the beginning or the continuance of life, without an overruling creative power; and, therefore, no conclusions of dynamical science regarding the future condition of the earth, can be held to give dispiriting views as to the destiny of the race of intelligent beings by which it is at present inhabited.

The object proposed in the present article is an application of these general principles to the discovery of probable limits to the periods of time, past and future, during which the sun can be reckoned on as a source of heat and light. The subject will be discussed under three heads:—

I. The secular cooling of the sun.

II. The present temperature of the sun.

III. The origin and total amount of the sun's heat.

PART I.

ON THE SECULAR COOLING OF THE SUN.

Rate of cooling of sun unknown.

How much the sun is actually cooled from year to year, if at all, we have no means of ascertaining, or scarcely even of estimating in the roughest manner. In the first place we do not know that he is losing heat at all. For it is quite certain that *some heat* is generated in his atmosphere by the influx of meteoric matter; and it is possible that the *amount* of heat so generated from year to year is sufficient to compensate the loss by radiation. It is, however, also possible that the sun is now an incandescent liquid mass, radiating away heat, either primitively created in his substance, or, what seems far more probable, generated by the falling in of meteors in past times, with no sensible compensation by a continuance of meteoric action.

Heat generated by fall of meteors into the sun

It has been shown* that, if the former supposition were true, the meteors by which the sun's heat would have been produced during the last 2,000 or 3,000 years must have been during all

* "On the Mechanical Energies of the Solar System." *Transactions of the Royal Society of Edinburgh*, 1854, and *Phil. Mag.* 1854, second half-year. *Math. and Phys. Papers*, by Sir W. Thomson (Art. LXVI. of Vol. II. now in the press).

that time much within the earth's distance from the sun, and must therefore have approached the central body in very gradual spirals; because, if enough of matter to produce the supposed thermal effect fell in from space outside the earth's orbit, the length of the year would have been very sensibly shortened by the additions to the sun's mass which must have been made. The quantity of matter annually falling in must, on that supposition, have amounted to $\frac{1}{47}$ of the earth's mass, or to $\frac{1}{17,000,000}$ of the sun's; and therefore it would be necessary to suppose the zodiacal light to amount to at least $\frac{1}{5000}$ of the sun's mass, to account in the same way for a future supply of 3,000 years' sun-heat. When these conclusions were first published it was pointed out that "disturbances in the motions of visible planets" should be looked for, as affording us means for estimating the possible amount of matter in the zodiacal light; and it was conjectured that it could not be nearly enough to give a supply of 300,000 years' heat at the present rate. These anticipations have been to some extent fulfilled in Le Verrier's great researches on the motion of the planet Mercury, which have recently given evidence of a sensible influence attributable to matter circulating as a great number of small planets within his orbit round the sun. But the amount of matter thus indicated is very small; and, therefore, if the meteoric influx taking place at present is enough to produce any appreciable portion of the heat radiated away, it must be supposed to be from matter circulating round the sun, within very short distances of his surface. The density of this meteoric cloud would have to be supposed so great that comets could scarcely have escaped, as comets actually have escaped, showing no discoverable effects of resistance, after passing his surface within a distance equal to $\frac{1}{8}$ of his radius. All things considered, there seems little probability in the hypothesis that solar radiation is compensated, to any appreciable degree, by heat generated by meteors falling in, at present; and, as it can be shown that no chemical theory is tenable*, it must be concluded as most probable that the sun is at present merely an incandescent liquid mass cooling.

Insufficient to give heat supply.

because the matter in zodiacal light and intra-mercurial planets is certainly small.

The sun an incandescent cooling mass.

How much he cools from year to year, becomes therefore a

* "Mechanical Energies," &c. referred to above.

question of very serious import, but it is one which we are at present quite unable to answer. It is true we have data on which we might plausibly found a probable estimate, and from which we might deduce, with at first sight seemingly well founded confidence, limits, not very wide, within which the present true rate of the sun's cooling must lie. For we know, from the independent but concordant investigations of Herschel and Pouillet, that the sun radiates every year from his whole surface about 6×10^{30} (six million million million million million) times as much heat as is sufficient to raise the temperature of 1 lb. of water by 1° Cent. We also have excellent reason for believing that the sun's substance is very much like the earth's. Stokes's principles of solar and stellar chemistry have been for many years explained in the University of Glasgow, and it has been taught as a first result that sodium does certainly exist in the sun's atmosphere, and in the atmospheres of many of the stars, but that it is not discoverable in others. The recent application of these principles in the splendid researches of Bunsen and Kirchhof (who made an independent discovery of Stokes's theory) has demonstrated with equal certainty that there are iron and manganese, and several of our other known metals, in the sun. The specific heat of each of these substances is less than the specific heat of water, which indeed exceeds that of every other known terrestrial body, solid or liquid. It might, therefore, at first sight seem probable that the mean specific heat* of the sun's whole substance is less, and very certain that it cannot be much greater, than that of water. If it were equal to the specific heat of water we should only have to divide the preceding number (6×10^{30}), derived from Herschel's and Pouillet's observations, by the number of pounds ($4{\cdot}23 \times 10^{30}$) in the sun's mass, to find 1°·4 Cent. for the present annual rate of

Pouillet's and Herschel's estimates of solar radiation.

Largeness of specific heat of sun

* The "specific heat" of a homogeneous body is the quantity of heat that a unit of its substance must acquire or must part with, to rise or to fall by 1° in temperature. The mean specific heat of a heterogeneous mass, or of a mass of homogeneous substance, under different pressures in different parts, is the quantity of heat which the whole body takes or gives in rising or in falling 1° in temperature, divided by the number of units in its mass. The expression, "mean specific heat" of the sun, in the text, signifies the total amount of heat actually radiated away from the sun, divided by his mass, during any time in which the average temperature of his mass sinks by 1°, whatever physical or chemical changes any part of his substance may experience.

cooling. It might therefore seem probable that the sun cools more, and almost certain that he does not cool less, than a centigrade degree and four-tenths annually. But, if this estimate were well founded, it would be equally just to assume that the sun's expansibility* with heat does not differ greatly from that of some average terrestrial body. If, for instance, it were the same as that of solid glass, which is about $\frac{1}{40000}$ of bulk, or $\frac{1}{120000}$ of diameter, per 1° Cent. (and for most terrestrial liquids, especially at high temperatures, the expansibility is much more), and if the specific heat were the same as that of liquid water, there would be in 860 years a contraction of one per cent. on the sun's diameter, which could scarcely have escaped detection by astronomical observation. There is, however, a far stronger reason than this for believing that no such amount of contraction can have taken place, and therefore for suspecting that the physical circumstances of the sun's mass render the condition of the substances of which it is composed, as to expansibility and specific heat, very different from that of the same substances when experimented on in our terrestrial laboratories.

and smallness of expansibility

rendered probable by absence of sensible contraction in solar diameter.

Mutual gravitation between the different parts of the sun's contracting mass must do an amount of work, which cannot be calculated with certainty, only because the law of the sun's interior density is not known. The amount of work performed during a contraction of one-tenth per cent. of the diameter, if the density remained uniform through the interior, would, as Helmholtz showed, be equal to 20,000 times the mechanical equivalent of the amount of heat which Pouillet estimated to be radiated from the sun in a year. But in reality the sun's density must increase very much towards his centre, and probably in varying proportions, as the temperature becomes lower and the whole mass contracts. We cannot, therefore, say whether the work actually done by mutual gravitation during a contraction of one-tenth per cent. of the diameter, would be

Work done in contraction of solar diameter by $\frac{1}{1000}$ may give heat supply for perhaps 20,000 years.

* The "expansibility in volume," or the "cubical expansibility," of a body, is an expression technically used to denote the proportion which the increase or diminution of its bulk, accompanying a rise or fall of 1° in its temperature, bears to its whole bulk at some stated temperature. The expression, "the sun's expansibility," used in the text, may be taken as signifying the ratio which the actual contraction, during a lowering of his mean temperature by 1° Cent., bears to his present volume.

more or less than the equivalent of 20,000 years' heat; but we may regard it as most probably not many times more or less than this amount. Now, it is in the highest degree improbable that mechanical energy can in any case increase in a body contracting in virtue of cooling. It is certain that it really does diminish very notably in every case hitherto experimented on. It must be supposed, therefore, that the sun always radiates away in heat something more than the Joule-equivalent of the work done on his contracting mass, by mutual gravitation of its parts. Hence, in contracting by one-tenth per cent. in his diameter, or three-tenths per cent. in his bulk, the sun must give out something either more, or not greatly less, than 20,000 years' heat; and thus, even without historical evidence as to the constancy of his diameter, it seems safe to conclude that no such contraction as that calculated above one per cent. in 860 years can have taken place in reality. It seems, on the contrary, probable that, at the present rate of radiation, a contraction of one-tenth per cent. in the sun's diameter could not take place in much less than 20,000 years, and scarcely possible that it could take place in less than 8,600 years. If, then, the mean specific heat of the sun's mass, in its actual condition, is not more than ten times that of water, the expansibility in volume must be less than $\frac{1}{4000}$ per 100° Cent., (that is to say, less than $\frac{1}{10}$ of that of solid glass,) which seems improbable. But although from this consideration we are led to regard it as probable that the sun's specific heat is considerably more than ten times that of water (and, therefore, that his mass cools considerably less than 100° in 700 years, a conclusion which, indeed, we could scarcely avoid on simply geological grounds), the physical principles we now rest on fail to give us any reason for supposing that the sun's specific heat is more than 10,000 times that of water, because we cannot say that his expansibility in volume is probably more than $\frac{1}{400}$ per 1° Cent. And there is, on other grounds, very strong reason for believing that the specific heat is really much less than 10,000. For it is almost certain that the sun's mean temperature* is even now as high as 14,000°

* [Rosetti (*Phil. Mag.* 1879, 2nd half year) estimates the effective radiational temperature of the sun as "not much less than ten thousand degrees Centigrade:" (9965° is the number expressing the results of his measurements). On the other hand, C. W. Siemens estimates it at as low as 3000° Cent. The mean tem-

Cent. ; and the greatest quantity of heat that we can explain, with any probability, to have been by natural causes ever acquired by the sun (as we shall see in the third part of this article), could not have raised his mass at any time to this temperature, unless his specific heat were less than 10,000 times that of water.

Sun's specific heat probably between 10 and 10,000 times that of water; and fall of temperature 100° Cent. in from 700 to 700,000 years.

We may therefore consider it as rendered highly probable that the sun's specific heat is more than ten times, and less than 10,000 times, that of liquid water. From this it would follow with certainty that his temperature sinks 100° Cent. in some time from 700 years to 700,000 years.

PART II.

ON THE SUN'S PRESENT TEMPERATURE.

At his surface the sun's temperature cannot, as we have many reasons for believing, be incomparably higher than temperatures attainable artificially in our terrestrial laboratories.

Sun's superficial temperature comparable with what may be artificially produced.

Among other reasons it may be mentioned that the sun radiates heat, from every square foot of his surface, at only about 7,000 horse power*. Coal, burning at a rate of a little less than a pound per two seconds, would generate the same amount; and it is estimated (Rankine, 'Prime Movers,' p. 285, Ed. 1859) that, in the furnaces of locomotive engines, coal burns at from one pound in thirty seconds to one pound in ninety seconds, per square foot of grate-bars. Hence heat is radiated from the sun at a rate not more than from fifteen to forty-five times as high as that at which heat is generated on the grate-bars of a locomotive furnace, per equal areas.

perature of the whole sun's mass must (Part II. below) be much higher than the "surface temperature," or "effective radiational temperature."—W. T. Nov. 9, 1882.]

* One horse power in mechanics is a technical expression (following Watt's estimate), used to denote a rate of working in which energy is evolved at the rate of 33,000 foot pounds per minute. This, according to Joule's determination of the dynamical value of heat, would, if spent wholly in heat, be sufficient to raise the temperature of 23¾ lbs. of water by 1° Cent. per minute.

[Note of Nov. 11, 1882. This is sixty-seven times the rate per unit of radiant surface at which energy is emitted from the incandescent filament of the Swan electric lamp when at the temperature which gives about 240 candles per horse power.]

Interior temperature probably far higher.

The interior temperature of the sun is probably far higher than that at his surface, because direct conduction can play no sensible part in the transference of heat between the inner and outer portions of his mass, and there must in virtue of the prodigious convective currents due to cooling of the outermost portions by radiation into space, be an approximate *convective* equilibrium of heat throughout the whole, if the whole is fluid. That is to say, the temperatures, at different distances from the centre, must be approximately those which any portion of the substance, if carried from the centre to the surface, would acquire by expansion without loss or gain of heat.

Law of temperature probably roughly that of convective equilibrium.

PART III.

ON THE ORIGIN AND TOTAL AMOUNT OF THE SUN'S HEAT.

The sun being, for reasons referred to above, assumed to be an incandescent liquid now losing heat, the question naturally occurs, How did this heat originate? It is certain that it cannot have existed in the sun through an infinity of past time, since, as long as it has so existed, it must have been suffering dissipation, and the finiteness of the sun precludes the supposition of an infinite primitive store of heat in his body.

The sun must, therefore, either have been created an active source of heat at some time of not immeasurable antiquity, by an over-ruling decree; or the heat which he has already radiated away, and that which he still possesses, must have been acquired by a natural process, following permanently established laws. Without pronouncing the former supposition to be essentially incredible, we may safely say that it is in the highest degree improbable, if we can show the latter to be not contradictory to known physical laws. And we do show this and more, by merely pointing to certain actions, going on before us at present, which, if sufficiently abundant at some past time, must have given the sun heat enough to account for all we know of his past radiation and present temperature.

Solar heat must arise from conversion of kinetic and potential energy.

It is not necessary at present to enter at length on details regarding the meteoric theory, which appears to have been first proposed in a definite form by Mayer, and afterwards independently by Waterston; or regarding the modified hypothesis of meteoric vortices, which the writer of the present article showed to be necessary, in order that the length of the year, as known for the last 2,000 years, may not have been sensibly disturbed by the accessions which the sun's mass must have had during that period, if the heat radiated away has been always compensated by heat generated by meteoric influx.

For the reasons mentioned in the first part of the present article, we may now believe that all theories of complete, or nearly complete, contemporaneous meteoric compensation, must be rejected; but we may still hold that—

"*Meteoric action.... is.... not only proved to exist as a cause of solar heat, but it is the only one of all conceivable causes which we know to exist from independent evidence**."

The form of meteoric theory which now seems most probable, and which was first discussed on true thermodynamic principles by Helmholtz†, consists in supposing the sun and his heat to have originated in a coalition of smaller bodies, falling together by mutual gravitation, and generating, as they must do according to the great law demonstrated by Joule, an exact equivalent of heat for the motion lost in collision.

That some form of the meteoric theory is certainly the true and complete explanation of solar heat can scarcely be doubted, when the following reasons are considered:

Chemical action insufficient, but meteoric theory may easily explain heat for 20 million years.

(1) No other natural explanation, except by chemical action, can be conceived.

(2) The chemical theory is quite insufficient, because the most energetic chemical action we know, taking place between substances amounting to the whole sun's mass, would only generate about 3,000 years' heat‡.

(3) There is no difficulty in accounting for 20,000,000 years' heat by the meteoric theory.

* "Mechanical Energies of the Solar System," referred to above.

† Popular lecture delivered on the 7th February, 1854, at Königsberg, on the occasion of the Kant commemoration.

‡ "Mechanical Energies of the Solar System."

It would extend this article to too great a length, and would require something of mathematical calculation, to explain fully the principles on which this last estimate is founded. It is enough to say that bodies, all much smaller than the sun, falling together from a state of relative rest, at mutual distances all large in comparison with their diameters, and forming a globe of uniform density equal in mass and diameter to the sun, would generate an amount of heat which, accurately calculated according to Joule's principles and experimental results, is found to be just 20,000,000 times Pouillet's estimate of the annual amount of solar radiation. The sun's density must, in all probability, increase very much towards his centre, and therefore a considerably greater amount of heat than that must be supposed to have been generated if his whole mass was formed by the coalition of comparatively small bodies. On the other hand, we do not know how much heat may have been dissipated by resistance and minor impacts before the final conglomeration; but there is reason to believe that even the most rapid conglomeration that we can conceive to have probably taken place could only leave the finished globe with about half the entire heat due to the amount of potential energy of mutual gravitation exhausted. We may, therefore, accept, as a lowest estimate for the sun's initial heat, 10,000,000 times a year's supply at present rate, but 50,000,000 or 100,000,000 as possible, in consequence of the sun's greater density in his central parts.

Only about half the heat due to energy of matter concentrating in sun available for explaining solar temperature.

The considerations adduced above, in this paper, regarding the sun's possible specific heat, rate of cooling, and superficial temperature, render it probable that he must have been very sensibly warmer one million years ago than now; and, consequently, that if he has existed as a luminary for ten or twenty million years, he must have radiated away considerably more than ten or twenty million times the present yearly amount of loss.

The sun has probably not lighted the earth for 100 million years.

It seems, therefore, on the whole most probable that the sun has not illuminated the earth for 100,000,000 years, and almost certain that he has not done so for 500,000,000 years. As for the future, we may say, with equal certainty, that inhabitants of the earth cannot continue to enjoy the light and heat essential to their life, for many million years longer, unless sources now unknown to us are prepared in the great storehouse of creation.

(F.)—On the Size of Atoms*.

The idea of an atom has been so constantly associated with incredible assumptions of infinite strength, absolute rigidity, mystical actions at a distance, and indivisibility, that chemists and many other reasonable naturalists of modern times, losing all patience with it, have dismissed it to the realms of metaphysics, and made it smaller than "anything we can conceive." But if atoms are inconceivably small, why are not all chemical actions infinitely swift? Chemistry is powerless to deal with this question, and many others of paramount importance, if barred by the hardness of its fundamental assumptions, from contemplating the atom as a real portion of matter occupying a finite space, and forming a not immeasurably small constituent of any palpable body.

More than thirty years ago naturalists were scared by a wild proposition of Cauchy's, that the familiar prismatic colours proved the "sphere of sensible molecular action" in transparent liquids and solids to be comparable with the wave-length of light. The thirty years which have intervened have only confirmed that proposition. They have produced a large number of capable judges; and it is only incapacity to judge in dynamical questions that can admit a doubt of the substantial correctness of Cauchy's conclusion. But the "sphere of molecular action" conveys no very clear idea to the non-mathematical mind. The idea which it conveys to the mathematical mind is, in my opinion, irredeemably false. For I have no faith whatever in attractions and repulsions acting at a distance between centres of force according to various laws. What Cauchy's mathematics really proves is this: that in palpably homogeneous bodies such as glass or water, contiguous portions are not similar when their dimensions are moderately small fractions of the wave-length. Thus in water contiguous cubes, each of one one-thousandth of a centimetre breadth are sensibly similar. But contiguous cubes of one ten-millionth of a centimetre must be very sensibly different. So in a solid mass of brickwork, two adjacent lengths of 20,000 centimetres each, may contain, one of them nine hundred and ninety-nine bricks and two half bricks, and the

Meaning of sphere of molecular action.

Meaning of homogeneity.

* *Nature*, March 1870.

other one thousand bricks: thus two contiguous cubes of 20,000 centimetres breadth may be considered as sensibly similar. But two adjacent lengths of forty centimetres each might contain one of them, one brick, and two half bricks, and the other two whole bricks; and contiguous cubes of forty centimetres would be very sensibly dissimilar. In short, optical dynamics leaves no alternative but to admit that the diameter of a molecule, or the distance from the centre of a molecule to the centre of a contiguous molecule in glass, water, or any other of our transparent liquids and solids, exceeds a ten-thousandth of the wave-length, or a two-hundred-millionth of a centimetre.

Contact electricity of metals.

By experiments on the contact electricity of metals made in the year 1862, and described in a letter to Dr Joule*, which was published in the proceedings of the Literary and Philosophical Society of Manchester [Jan. 1862], I found that plates of zinc and copper connected with one another by a fine wire attract one another, as would similar pieces of one metal connected with the two plates of a galvanic element, having about three-quarters of the electro-motive force of a Daniel's element.

Energy of electric attraction between plates of different metals in metallic contact.

Measurements published in the Proceedings of the Royal Society for 1860 showed that the attraction between parallel plates of one metal held at a distance apart small in comparison with their diameters, and kept connected with such a galvanic element, would experience an attraction amounting to two ten-thousand-millionths of a gramme weight per area of the opposed surfaces equal to the square of the distance between them. Let a plate of zinc and a plate of copper, each a centimetre square and a hundred-thousandth of a centimetre thick, be placed with a corner of each touching a metal globe of a hundred-thousandth of a centimetre diameter. Let the plates, kept thus in metallic communication with one another be at first wide apart, except at the corners touching the little globe, and let them then be gradually turned round till they are parallel and at a distance of a hundred-thousandth of a centimetre asunder. In this position they will attract one another with a force equal in all to two grammes weight. By abstract dynamics and the theory of energy, it is readily proved that the work done by the changing force of attraction during the motion by which we have supposed

* [Now published as Art. XXII. in a "Reprint of Papers on Electrostatics and Magnetism" by Sir William Thomson. New edition, 1883.]

this position to be reached, is equal to that of a constant force of two grammes weight acting through a space of a hundred-thousandth of a centimetre; that is to say, to two hundred-thousandths of a centimetre-gramme. Now let a second plate of zinc be brought by a similar process to the other side of the plate of copper; a second plate of copper to the remote side of this second plate of zinc, and so on till a pile is formed consisting of 50,001 plates of zinc and 50,000 plates of copper, separated by 100,000 spaces, each plate and each space one hundred-thousandth of a centimetre thick. The whole work done by electric attraction in the formation of this pile is two centimetre-grammes.

Work done in forming pile of zinc and copper plates.

The whole mass of metal is eight grammes. Hence the amount of work is a quarter of a centimetre-gramme per gramme of metal. Now 4,030 centimetre-grammes of work, according to Joule's dynamical equivalent of heat, is the amount required to warm a gramme of zinc or copper by one degree Centigrade. Hence the work done by the electric attraction could warm the substance by only $\frac{1}{16120}$ of a degree. But now let the thickness of each piece of metal and of each intervening space be a hundred-millionth of a centimetre instead of a hundred thousandth. The work would be increased a million-fold unless a hundred-millionth of a centimetre approaches the smallness of a molecule. The heat equivalent would therefore be enough to raise the temperature of the material by 62°. This is barely, if at all, admissible, according to our present knowledge, or, rather, want of knowledge, regarding the heat of combination of zinc and copper. But suppose the metal plates and intervening spaces to be made yet four times thinner, that is to say, the thickness of each to be a four hundred-millionth of a centimetre. The work and its heat equivalent will be increased sixteen-fold. It would therefore be 990 times as much as that required to warm the mass by 1° cent., which is very much more than can possibly be produced by zinc and copper in entering into molecular combination. Were there in reality anything like so much heat of combination as this, a mixture of zinc and copper powders would, if melted in any one spot, run together, generating more than heat enough to melt each throughout; just as a large quantity of gunpowder if ignited in any one spot burns throughout without fresh application of heat. Hence plates of zinc and copper of a

The heat of combination of zinc and copper shows that molecules probably are at least 10^{-8} cm. and certainly more than $\frac{1}{2}\times10^{-8}$ cm in diameter.

three hundred-millionth of a centimetre thick, placed close together alternately, form a near approximation to a chemical combination, if indeed such thin plates could be made without splitting atoms.

Work done in stretching fluid film against surface tension.

The theory of capillary attraction shows that when a bubble—a soap-bubble for instance—is blown larger and larger, work is done by the stretching of a film which resists extension as if it were an elastic membrane with a constant contractile force. This contractile force is to be reckoned as a certain number of units of force per unit of breadth. Observation of the ascent of water in capillary tubes shows that the contractile force of a thin film of water is about sixteen milligrammes weight per millimetre of breadth. Hence the work done in stretching a water film to any degree of thinness, reckoned in millimetre-milligrammes, is equal to sixteen times the number of square millimetres by which the area is augmented, provided the film is not made so thin that there is any sensible diminution of its contractile force. In an article "On the Thermal effect of drawing out a Film of Liquid," published in the Proceedings of the Royal Society for April 1858, I have proved from the second law of thermodynamics that about half as much more energy, in the shape of heat, must be given to the film to prevent it from sinking in temperature while it is being drawn out. Hence the intrinsic energy of a mass of water in the shape of a film kept at constant temperature increases by twenty-four milligramme-millimetres for every square millimetre added to its area.

Intrinsic energy of a mass of water estimated from the heat required to prevent film from cooling as it extends.

Suppose then a film to be given with a thickness of a millimetre, and suppose its area to be augmented ten thousand and one fold: the work done per square millimetre of the original film, that is to say per milligramme of the mass would be 240,000 millimetre-milligrammes. The heat equivalent of this is more than half a degree centigrade of elevation of temperature of the substance. The thickness to which the film is reduced on this supposition is very approximately a ten-thousandth of millimetre. The commonest observation on the soap-bubble (which in contractile force differs no doubt very little from pure water) shows that there is no sensible diminution of contractile force by reduction of the thickness to the ten-thousandth of a millimetre; inasmuch as the thickness which gives the first maximum brightness round the black spot seen where the bubble is thinnest, is only about an eight-thousandth of a millimetre.

Surface tension falls much before the film is reduced to $\frac{1}{2}\times10^{-8}$ cm., and there are probably few molecules in that thickness.

The very moderate amount of work shown in the preceding estimates is quite consistent with this deduction. But suppose now the film to be farther stretched until its thickness is reduced to a twenty-millionth of a millimetre. The work spent in doing this is two-thousand times more than that which we have just calculated. The heat equivalent is 1,130 times the quantity required to raise the temperature of the liquid by one degree centigrade. This is far more than we can admit as a possible amount of work done in the extension of a liquid film. A smaller amount of work spent on the liquid would convert it into vapour at ordinary atmospheric pressure. The conclusion is unavoidable, that a water-film falls off greatly in its contractile force before it is reduced to a thickness of a twenty-millionth of a millimetre. It is scarcely possible, upon any conceivable molecular theory, that there can be any considerable falling off in the contractile force as long as there are several molecules in the thickness. It is therefore probable that there are not several molecules in a thickness of a twenty-millionth of a millimetre of water.

Kinetic theory of gases.

The kinetic theory of gases suggested a hundred years ago by Daniel Bernoulli has, during the last quarter of a century, been worked out by Herapath, Joule, Clausius, and Maxwell, to so great perfection that we now find in it satisfactory explanations of all non-chemical properties of gases. However difficult it may be to even imagine what kind of thing the molecule is, we may regard it as an established truth of science that a gas consists of moving molecules disturbed from rectilinear paths and constant velocities by collisions or mutual influences, so rare that the mean length of nearly rectilinear portions of the path of each molecule is many times greater than the average distance from the centre of each molecule to the centre of the molecule nearest it at any time. If, for a moment, we suppose the molecules to be hard elastic globes all of one size, influencing one another only through actual contact, we have for each molecule simply a zigzag path composed of rectilinear portions, with abrupt changes of direction. On this supposition Clausius proves, by a simple application of the calculus of pro-

Meaning of molecule, free path and collision.

Average length of free path estimated by Clausius.

Kinetic theory of gases. Meaning of molecule, free path and collision.

babilities, that the average length of the free path of a particle from collision to collision bears to the diameter of each globe, the ratio of the whole space in which the globes move, to eight times the sum of the volumes of the globes. It follows that the number of the globes in unit volume is equal to the square of this ratio divided by the volume of a sphere whose radius is equal to that average length of free path. But we cannot believe that the individual molecules of gases in general, or even of any one gas, are hard elastic globes. Any two of the moving particles or molecules must act upon one another somehow, so that when they pass very near one another they shall produce considerable deflexion of the path and change in the velocity of each. This mutual action (called force) is different at different distances, and must vary, according to variations of the distance so as to fulfil some definite law. If the particles were hard elastic globes acting upon one another only by contact, the law of force would be—zero force when the distance from centre to centre exceeds the sum of the radii, and infinite repulsion for any distance less than the sum of the radii. This hypothesis, with its "hard and fast" demarcation between no force and infinite force, seems to require mitigation. Without entering on the theory of vortex atoms at present, I may at least say that soft elastic solids, not necessarily globular, are more promising than infinitely hard elastic globes. And, happily, we are not left merely to our fancy as to what we are to accept as probable in respect to the law of force. If the particles were hard elastic globes the average time from collision to collision would be inversely as the average velocity of the particles. But Maxwell's experiments on the variation of the viscosities of gases with change of temperature prove that the mean time from collision to collision is independent of the velocity if we give the name collision to those mutual actions only which produce something more than a certain specified degree of deflection of the line of motion. This law could be fulfilled by soft elastic particles (globular or not globular); but, as we have seen, not by hard elastic globes. Such details, however, are beyond the scope of our present argument. What we want now are rough approximations to absolute values, whether of time or space or mass—not delicate differential results. From Joule, Maxwell, and Clausius we know that the average velocity of the molecules of oxygen or nitrogen or common air, at ordinary atmospheric temperature and pressure, is about 50,000 centimetres per second, and the average time from collision to collision a five-thousand-millionth of a second. Hence the average length of path of each molecule between collisions is about $\frac{1}{100000}$ of a centimetre. Now, having left the idea of hard globes, according to which the dimensions of a molecule and the distinction between collision and no collision are perfectly sharp, something of circumlocution must take the place of these simple terms.

Kinetic theory of gases.

Average free path 10^{-5} cm.

First, it is to be remarked that two molecules in collision will exercise a mutual repulsion in virtue of which the distance between their centres, after being diminished to a minimum, will begin to increase as the molecules leave one another. This minimum distance would be equal to the sum of the radii, if the molecules were infinitely hard elastic spheres; but in reality we must suppose it to be very different in different collisions. Considering only the case of equal molecules, we might, then, define the radius of a molecule as half the average shortest distance reached in a vast number of collisions. The definition I adopt for the present is not precisely this, but is chosen so as to make as simple as possible the statement I have to make of a combination of the results of Clausius and Maxwell. Having defined the radius of a gaseous molecule, I call the double of the radius the diameter; and the volume of a globe of the same radius or diameter I call the volume of the molecule.

Meaning of collision and diameter of molecule.

The experiments of Cagniard de la Tour, Faraday, Regnault, and Andrews, on the condensation of gases do not allow us to believe that any of the ordinary gases could be made forty thousand times denser than at ordinary atmosphere pressure and temperature, without reducing the whole volume to something less than the sum of the volume of the gaseous molecules, as now defined. Hence, according to the grand theorem of Clausius quoted above, the average length of path from collision to collision cannot be more than five thousand times the diameter of the gaseous molecule; and the number of molecules in unit of volume cannot exceed 25,000,000 divided by the volume of a globe whose radius is that average length of path. Taking now the preceding estimate, $\frac{1}{100000}$ of a centimetre, for the average length of path from collision to collision we conclude that the

Free path cannot be more than 5000 times diameter of molecule;

and diameter cannot be less than 2×10^{-9} cm.

diameter of the gaseous molecule cannot be less than $\frac{1}{500\,000{,}000}$ of a centimetre; nor the number of molecules in a cubic centimetre of the gas (at ordinary density) greater than 6×10^{21} (or six thousand million million million).

Average distance from centre to centre of molecules in solids and liquids between 7×10^{-9} and 2×10^{-9} cm.

The densities of known liquids and solids are from five hundred to sixteen thousand times that of atmospheric air at ordinary pressure and temperature; and, therefore, the number of molecules in a cubic centimetre may be from 3×10^{24} to 10^{26} (that is, from three million million million million to a hundred million million million million). From this (if we assume for a moment a cubic arrangement of molecules), the distance from centre to nearest centre in solids and liquids may be estimated at from $\frac{1}{140{,}000{,}000}$ to $\frac{1}{460{,}000{,}000}$ of a centimetre.

The four lines of argument which I have now indicated, lead all to substantially the same estimate of the dimensions of molecular structure. Jointly they establish with what we cannot but regard as a very high degree of probability the conclusion that, in any ordinary liquid, transparent solid, or seemingly opaque solid, the mean distance between the centres of contiguous molecules is less than the hundred-millionth, and greater than the two thousand-millionth of a centimetre*.

Illustration of size of molecules.

To form some conception of the degree of coarse-grainedness indicated by this conclusion, imagine a rain drop, or a globe of glass as large as a pea, to be magnified up to the size of the earth, each constituent molecule being magnified in the same proportion. The magnified structure would be more coarse grained than a heap of small shot, but probably less coarse grained than a heap of cricket-balls.

* I find that M. Loschmidt had preceded me in the fourth of the preceding methods of estimating the size of atoms [Sitzungsberichte of the Vienna Acad., 12 Oct., 1865, p. 395]. He finds the diameter of a molecule of common air to be about a ten-millionth of a centimetre. M. Lippmann has also given a remarkably interesting and original investigation relating to the size of atoms *Comptes Rendus*, Oct. 16th, 1882, basing his argument on the variations of capillarity under electrification. He finds that the thickness of the double electric layer, according to Helmholtz's theory, is about a 35-millionth of a centimetre. W. T., Dec. 13, 1882.

(G.)—On Tidal Friction, by G. H. Darwin, F.R.S.

(a.) *The retardation of the earth's rotation, as deduced from the secular acceleration of the Moon's mean motion.*

Retardation of earth's rotation. Numerical estimates.

In my paper on the precession of a viscous spheroid [*Phil. Trans.* Pt. II., 1879], all the data are given which are requisite for making the calculations for Professor Adams' result in § 830, viz.: that if there is an unexplained part in the coefficient of the secular acceleration of the moon's mean motion amounting to 6″, and if this be due to tidal friction, then in a century the earth gets 22 seconds behind time, when compared with an ideal clock, going perfectly for a century, and perfectly rated at the beginning of the century. In the paper referred to however the earth is treated as homogeneous, and the tides are supposed to consist in a bodily deformation of the mass. The numerical results there given require some modification on this account.

If E, E', E'' be the heights of the semidiurnal, diurnal and fortnightly tides, expressed as fractions of the equilibrium tides of the same denominations; and if $\epsilon, \epsilon', \epsilon''$ be the corresponding retardations of phase of these tides due to friction; it is shown on p. 476 and in equation (48), that in consequence of lunar and solar tides, at the end of a century, the earth, as a time-keeper, is behind the time indicated by the ideal perfect clock

$$1900{\cdot}27\, E \sin 2\epsilon + 423{\cdot}49\, E' \sin \epsilon' \text{ seconds of time} \ldots\ldots(a),$$

and that if the motion of the moon were unaffected by the tides, an observer, taking the earth as his clock, would note that at the end of the century the moon was in advance of her place in her orbit by

$$1043''{\cdot}28\, E \sin 2\epsilon + 232''{\cdot}50\, E' \sin \epsilon' \ldots\ldots\ldots\ldots\ldots\ldots\ldots\ldots(b).$$

This is of course merely the expression of the same fact as (*a*), in a different form.

Lastly it is shown in equation (60) that from these causes in a century, the moon actually lags behind her place

$$630''{\cdot}7\, E \sin 2\epsilon + 108''{\cdot}6\, E' \sin \epsilon' - 7''{\cdot}042\, E'' \sin 2\epsilon'' \ldots\ldots(c).$$

In adapting these results to the hypothesis of oceanic tides on a heterogeneous earth, we observe in the first place that, if the

Retardation of earth's rotation.

fluid tides are inverted, that is to say if for example it is low water under the moon, then friction advances the fluid tides*, and therefore in that case the ϵ's are to be interpreted as advancements of phase; and secondly that the E's are to be multiplied by $\frac{2}{11}$, which is the ratio of the density of water to the mean density of the earth. Next the earth's moment of inertia (as we learn from col. vii. of the table in § 824) is about ·83 of its amount on the hypothesis of homogeneity, and therefore the results (a) and (b) have both to be multiplied by 1/·83 or 1·2; the result (c) remains unaffected except as to the factor $\frac{2}{11}$.

Thus subtracting (c) from (b) as amended, we find that to an observer, taking the earth as a true time-keeper, the moon is, at the end of the century, in advance of her place by

$$\tfrac{2}{11}\{(1\cdot 2 \times 1043''\cdot 28 - 630''\cdot 7)\,E \sin 2\epsilon \\ + (1\cdot 2 \times 232''\cdot 50 - 108''\cdot 6)\,E' \sin \epsilon' + 7''\cdot 042\, E'' \sin 2\epsilon''\},$$

which is equal to

$$\tfrac{2}{11}\{621''\cdot 24\, E \sin 2\epsilon + 170''\cdot 40\, E' \sin \epsilon' + 7''\cdot 04\, E'' \sin 2\epsilon''\} \ldots (d)$$

and from (a) as amended that the earth, as a time-keeper, is behind the time indicated by the ideal clock, perfectly rated at the beginning of the century, by

$$\tfrac{2}{11}\{2280\cdot 32\, E \sin 2\epsilon + 508\cdot 19\, E' \sin \epsilon'\} \text{ seconds of time } \ldots\ldots\ldots (e).$$

Now if we suppose that the tides have their equilibrium height, so that the E's are each unity; and that ϵ' is one half of ϵ (which must roughly correspond to the state of the case), and that ϵ'' is insensible, and ϵ small, (d) becomes

$$\tfrac{4}{11}\{621''\cdot 24 + \tfrac{1}{4} \times 170''\cdot 40\}\,\epsilon \ldots\ldots\ldots\ldots\ldots (f)$$

and (e) becomes

$$\tfrac{4}{11}\{2280\cdot 32 + \tfrac{1}{4} \times 508\cdot 19\}\,\epsilon \text{ seconds of time } \ldots\ldots\ldots (g).$$

If (f) were equal to 1″, then (g) would clearly be

$$\frac{2280\cdot 32 + \frac{1}{4} \times 508\cdot 19}{621\cdot 24 + \frac{1}{4} \times 170\cdot 40} \text{ seconds of time } \ldots\ldots\ldots (h).$$

The second term, both in the numerator and denominator of (h), depends on the diurnal tide, which only exists when the ecliptic

* That this is true may be seen from considerations of energy. If it were approximately low water under the moon, the earth's rotation would be accelerated by tidal friction, if the tides of short period lagged; and this would violate the principles of energy.

is oblique. Now Adams' result was obtained on the hypothesis that the obliquity of the ecliptic was nil, therefore according to his assumption, 1″ in the coefficient of lunar acceleration means that the earth, as compared with a perfect clock rated at the beginning of the century, is behind time **Adams' result.**

$$\frac{2280{\cdot}32}{621{\cdot}24} = 3\tfrac{2}{3} \text{ seconds at the end of a century.}$$

Accordingly 6″ in the coefficient gives 22 secs. at the end of a century, which is his result given in § 830. If however we include the obliquity of the ecliptic and the diurnal tide, we find that 1″ in the coefficient means that the earth, as compared with the perfect clock, is behind time

$$\frac{2407{\cdot}37}{663{\cdot}80} = 3{\cdot}6274 \text{ seconds at the end of a century.}$$

Thus taking Hansen's 12″·56 with Delaunay's 6″·1, we have the earth behind $6{\cdot}46 \times 3{\cdot}6274 = 23{\cdot}4$ sec., and taking Newcomb's 8″·4 with Delaunay's 6″·1, we have the earth behind $2{\cdot}3 \times 3{\cdot}6274 = 8{\cdot}3$ sec. **Other results.**

It is worthy of notice that this result would be only very slightly vitiated by the incorrectness of the hypothesis made above as to the values of the E's and ϵ's; for $E \sin 2\epsilon$ occurs in the important term both in the numerator and denominator of the result for the earth's defect as a time-keeper, and thus the hypothesis only enters in determining the part played by the diurnal tide. Hence the result is not sensibly affected by some inexactness in this hypothesis, nor by the fact that the oceans in reality only cover a portion of the earth's surface.

(*b*.) *The Determination of the Secular Effects of Tidal Friction by a Graphical Method.* (Portion of a paper published in the *Proc. Roy. Soc.* No. 197, 1879, but with alterations and additions.)

Suppose an attractive particle or satellite of mass m to be moving in a circular orbit, with an angular velocity Ω, round a planet of mass M, and suppose the planet to be rotating about an axis perpendicular to the plane of the orbit, with an angular velocity n; suppose, also, the mass of the planet to be partially or wholly imperfectly elastic or viscous, or that there are oceans **General problem of tidal friction.**

on the surface of the planet; then the attraction of the satellite must produce a relative motion in the parts of the planet, and that motion must be subject to friction, or, in other words, there must be frictional tides of some sort or other. The system must accordingly be losing energy by friction, and its configuration must change in such a way that its whole energy diminishes.

Such a system does not differ much from those of actual planets and satellites, and, therefore, the results deduced in this hypothetical case must agree pretty closely with the actual course of evolution, provided that time enough has been and will be given for such changes.

Let C be the moment of inertia of the planet about its axis of rotation;

r the distance of the satellite from the centre of the planet;

h the resultant moment of momentum of the whole system;

e the whole energy, both kinetic and potential of the system.

It will be supposed that the figure of the planet and the distribution of its internal density are such that the attraction of the satellite causes no couple about any axis perpendicular to that of rotation.

Special units.

I shall now adopt a special system of units of mass, length, and time such that the analytical results are reduced to their simplest forms.

Let the unit of mass be $Mm/(M+m)$.

Let the unit of length γ be such a distance, that the moment of inertia of the planet about its axis of rotation may be equal to the moment of inertia of the planet and satellite, treated as particles, about their centre of inertia, when distant γ apart from one another. This condition gives

$$M\left(\frac{m\gamma}{M+m}\right)^2 + m\left(\frac{M\gamma}{M+m}\right)^2 = C$$

whence $$\gamma = \left\{\frac{C(M+m)}{Mm}\right\}^{\frac{1}{2}}.$$

Let the unit of time τ be the time in which the satellite revolves through 57°·3 about the planet, when the satellite's radius vector is equal to γ. In this case $1/\tau$ is the satellite's orbital angular velocity, and by the law of periodic times we have

$$\tau^{-2}\gamma^3 = \mu(M+m)$$

where μ is the attraction between unit masses at unit distance. Then by substitution for γ

$$\tau = \left\{\frac{C^3(M+m)}{\mu^2(Mm)^3}\right\}^{\frac{1}{2}}.$$

This system of units will be found to make the three following functions each equal to unity, viz. $\mu^{\frac{2}{3}}Mm(M+m)^{-\frac{1}{3}}$, μMm, and C. The units are in fact derived from the consideration that these functions are each to be unity.

In the case of the earth and moon, if we take the moon's mass as $\frac{1}{82}$nd of the earth's, and the earth's moment of inertia as $\frac{1}{3}Ma^2$ [see § 824], it may easily be shown that the unit of mass is $\frac{1}{83}$ of the earth's mass, the unit of length is 5·26 earth's radii or 33,506 kilometres, and the unit of time is 2 hrs. 41 minutes. In these units the present angular velocity of the earth's diurnal rotation is expressed by ·7044, and the moon's present radius vector by 11·454. **Numerical values of the units for earth and moon.**

The two bodies being supposed to revolve in circles about their common centre of inertia with an angular velocity Ω, the moment of momentum of orbital motion is **Moment of momentum and energy of system.**

$$M\left(\frac{mr}{M+m}\right)^2\Omega + m\left(\frac{Mr}{M+m}\right)^2\Omega = \frac{Mm}{M+m}r^2\Omega.$$

Then, by the law of periodic times, in a circular orbit,

$$\Omega^2r^3 = \mu(M+m)$$

$$\text{whence } \Omega r^2 = \mu^{\frac{1}{2}}(M+m)^{\frac{1}{2}}r^{\frac{1}{2}}.$$

And the moment of momentum of orbital motion

$$= \mu^{\frac{1}{2}}Mm(M+m)^{-\frac{1}{2}}r^{\frac{1}{2}},$$

and in the special units this is equal to $r^{\frac{1}{2}}$.

The moment of momentum of the planet's rotation is Cn, and $C = 1$, in the special units.

Therefore $$h = n + r^{\frac{1}{2}} \quad\text{..........................(1).}$$

Again, the kinetic energy of orbital motion is

$$\tfrac{1}{2}M\left(\frac{mr}{M+m}\right)^2\Omega^2 + \tfrac{1}{2}m\left(\frac{Mr}{M+m}\right)^2\Omega^2 = \tfrac{1}{2}\frac{Mm}{M+m}r^2\Omega^2 = \tfrac{1}{2}\frac{\mu Mm}{r}.$$

Moment of momentum and energy of system.

The kinetic energy of the planet's rotation is $\frac{1}{2}Cn^2$.

The potential energy of the system is $-\mu Mm/r$.

Adding the three energies together, and transforming into the special units, we have

$$2e = n^2 - \frac{1}{r} \quad\text{..........................(2).}$$

Since the moon's present radius vector is 11·454, it follows that the orbital momentum of the moon is 3·384. Adding to this the rotational momentum of the earth which is ·704, we obtain 4·088 for the total moment of momentum of the moon and earth. The ratio of the orbital to the rotational momentum is 4·80, so that the total moment of momentum of the system would, but for the obliquity of the ecliptic, be 5·80 times that of the earth's rotation. In § 276, where the obliquity is taken into consideration, the number is given as 5·38.

Now let $\quad x = r^{\frac{1}{2}}, \quad y = n, \quad Y = 2e.$

It will be noticed that x, the moment of momentum of orbital motion, is equal to the square root of the satellite's distance from the planet.

Then the equations (1) and (2) become

$$h = y + x \quad\text{..............................(3).}$$

$$Y = y^2 - \frac{1}{x^2} = (h - x)^2 - \frac{1}{x^2} \quad\text{..................(4).}$$

(3) is the equation of conservation of moment of momentum, or shortly, the equation of momentum; (4) is the equation of energy.

Two configurations of maximum and minimum energy for given momentum, determined by quartic equation.

Now, consider a system started with given positive (or say clockwise*) moment of momentum h; we have all sorts of ways in which it may be started. If the two rotations be of opposite kinds, it is clear that we may start the system with any amount of energy however great, but the true maxima and minima of energy compatible with the given moment of momentum are given by $dY/dx = 0$,

or $$x - h + \frac{1}{x^3} = 0,$$

that is to say, $$x^4 - hx^3 + 1 = 0 \quad\text{........................(5).}$$

* This is contrary to the ordinary convention, but I leave this passage as it stood originally.

We shall presently see that this quartic has either two real roots and two imaginary, or all imaginary roots*.

This quartic may be derived from quite a different consideration, viz., by finding the condition under which the satellite may move round the planet, so that the planet shall always show the same face to the satellite, in fact, so that they move as parts of one rigid body.

The condition is simply that the satellite's orbital angular velocity $\Omega = n$ the planet's angular velocity of rotation; or since $n = y$ and $r^{\frac{1}{2}} = \Omega^{-\frac{1}{3}} = x$, therefore $y = 1/x^3$.

In these configurations the satellite moves as though rigidly connected with the planet.

By substituting this value of y in the equation of momentum (3), we get as before

$$x^4 - hx^3 + 1 = 0 \quad \text{.........................} \quad (5).$$

In my paper on the "Precession of a Viscous Spheroid †," I obtained the quartic equation from this last point of view only, and considered analytically and numerically its bearings on the history of the earth.

Sir William Thomson, having read the paper, told me that he thought that much light might be thrown on the general physical meaning of the equation, by a comparison of the equation of conservation of moment of momentum with the energy of the system for various configurations, and he suggested the appropriateness of geometrical illustration for the purpose of this comparison. The method which is worked out below is the result of the suggestions given me by him in conversation.

The simplicity with which complicated mechanical interactions may be thus traced out geometrically to their results appears truly remarkable.

At present we have only obtained one result, viz.: that if with given moment of momentum it is possible to set the satellite and planet moving as a rigid body, then it is possible to do so in two ways, and one of these ways requires a maximum amount of energy and the other a minimum; from which it is clear that one must be a rapid rotation with the satellite near the planet, and the other a slow one with the satellite remote from the planet.

* I have elsewhere shown that when it has real roots, one is greater and the other less than $\frac{3}{4}h$. *Proc. Roy. Soc.* No. 202, 1880.

† *Trans. Roy. Soc.* Part I. 1879.

In these configurations the satellite moves as though rigidly connected with the planet.

Now, consider the three equations,

$$h = y + x \text{ ....(6)},$$

$$Y = (h - x)^2 - \frac{1}{x^2} \text{ ....(7)},$$

$$x^3 y = 1 \text{ ....(8)}.$$

(6) is the equation of momentum; (7) that of energy; and (8) we may call the equation of rigidity, since it indicates that the two bodies move as though parts of one rigid body.

Graphical solution.

Now, if we wish to illustrate these equations geometrically, we may take as abscissa x, which is the moment of momentum of orbital motion; so that the axis of x may be called the axis of orbital momentum. Also, for equations (6) and (8) we may take as ordinate y, which is the moment of momentum of the planet's rotation; so that the axis of y may be called the axis of rotational momentum. For (7) we may take as ordinate Y, which is twice the energy of the system; so that the axis of Y may be called the axis of energy. Then, as it will be convenient to exhibit all three curves in the same figure, with a parallel axis of x, we must have the axis of energy identical with that of rotational momentum.

It will not be necessary to consider the case where the resultant moment of momentum h is negative, because this would only be equivalent to reversing all the rotations; thus h is to be taken as essentially positive.

Then the line of momentum, whose equation is (6), is a straight line inclined at 45° to either axis, having positive intercepts on both axes.

The curve of rigidity, whose equation is (8), is clearly of the same nature as a rectangular hyperbola, but having a much more rapid rate of approach to the axis of orbital momentum than to that of rotational momentum.

The intersections (if any) of the curve of rigidity with the line of momentum have abscissæ which are the two roots of the quartic $x^4 - hx^3 + 1 = 0$. The quartic has, therefore, two real roots or all imaginary roots. Then, since $x = \sqrt{r}$, the intersection which is more remote from the origin, indicates a configuration where the satellite is remote from the planet; the other gives the configuration where the satellite is closer to the planet. We have already learnt that these two correspond respectively to minimum and maximum energy. Graphical solution.

When x is very large, the equation to the curve of energy is $Y=(h-x)^2$, which is the equation to a parabola, with a vertical axis parallel to Y and distant h from the origin, so that the axis of the parabola passes through the intersection of the line of momentum with the axis of orbital momentum.

When x is very small the equation becomes $Y=-1/x^2$.

Fig. 1.

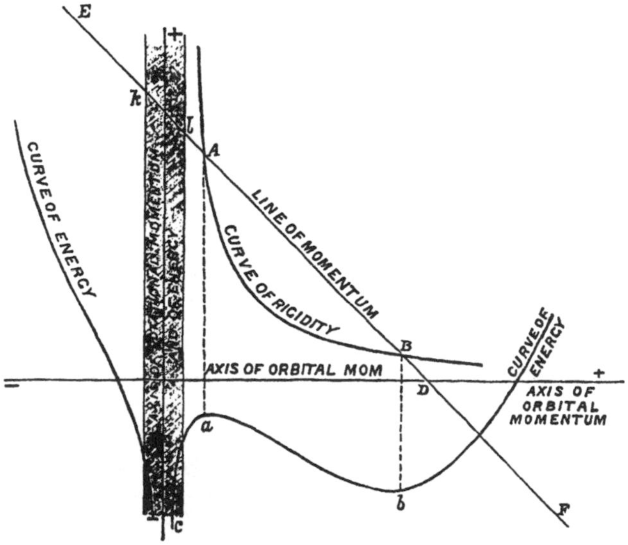

Hence, the axis of Y is asymptotic on both sides to the curve of energy.

Then, if the line of momentum intersects the curve of rigidity, the curve of energy has a maximum vertically underneath the point of intersection nearer the origin, and a minimum underneath the point more remote. But if there are no intersections, it has no maximum or minimum.

It is not easy to exhibit these curves well if they are drawn to scale, without making a figure larger than it would be

Graphical solution.

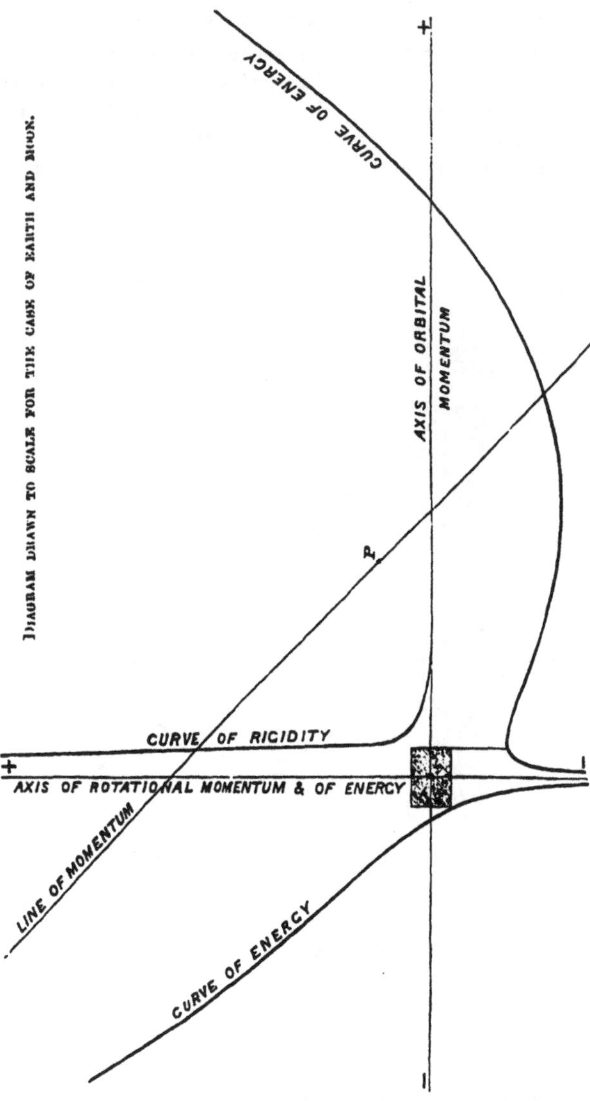

Fig. 2.

convenient to print, and accordingly fig. 1 gives them as drawn with the free hand. As the zero of energy is quite arbitrary, the origin for the energy curve is displaced downwards, and this prevents the two curves from crossing one another in a confusing manner. The same remark applies also to figs. 2 and 3. Graphical solution.

Fig. 1 is erroneous principally in that the curve of rigidity ought to approach its horizontal asymptote much more rapidly, so that it would be difficult in a drawing to scale to distinguish the points of intersection B and D.

Fig. 2 exhibits the same curves, but drawn to scale, and designed to be applicable to the case of the earth and moon, that is to say, when $h = 4$ nearly.

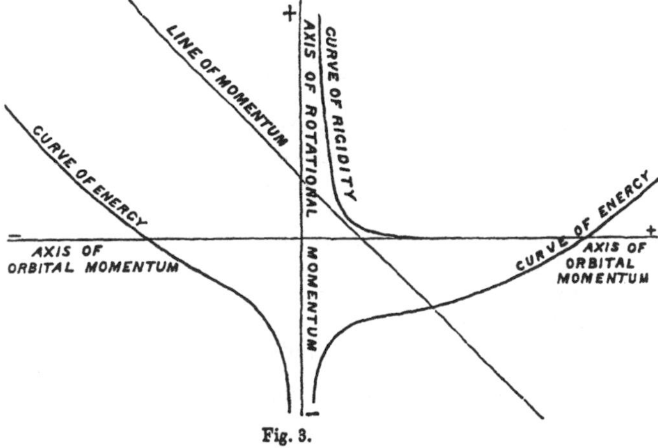

Fig. 3.

Fig. 3 shows the curves when $h = 1$, and when the line of momentum does not intersect the curve of rigidity; and here there is no maximum or minimum in the curve of energy.

These figures exhibit all the possible methods in which the bodies may move with given moment of momentum, and they differ in the fact that in figs. 1 and 2 the quartic (5) has real roots, but in the case of fig. 3 this is not so. Every point of the line of momentum gives by its abscissa and ordinate the square root of the satellite's distance and the rotation of

Graphical solution.

the planet, and the ordinate of the energy curve gives the energy corresponding to each distance of the satellite.

Parts of these figures have no physical meaning, for it is impossible for the satellite to move round the planet at a distance which is less than the sum of the radii of the planet and satellite. Accordingly in fig. 1 a strip is marked off and shaded on each side of the vertical axis, within which the figure has no physical meaning.

Since the moon's diameter is about 2,200 miles, and the earth's about 8,000, therefore the moon's distance cannot be less than 5,100 miles; and in fig. 2, which is intended to apply to the earth and moon and is drawn to scale, the base of the strip is only shaded, so as not to render the figure confused.

The point P in fig. 2 indicates the present configuration of the earth and moon.

The curve of rigidity $x^3y=1$ is the same for all values of h, and by moving the line of momentum parallel to itself nearer or further from the origin, we may represent all possible moments of momentum of the whole system.

Critical value of moment of momentum.

The smallest amount of moment of momentum with which it is possible to set the system moving as a rigid body, with centrifugal force enough to balance the mutual attraction, is when the line of momentum touches the curve of rigidity. The condition for this is clearly that the equation $x^4-hx^3+1=0$ should have equal roots. If it has equal roots, each root must be $\frac{3}{4}h$, and therefore

$$(\tfrac{3}{4}h)^4-h(\tfrac{3}{4}h)^3+1=0,$$

whence $h^4=4^4/3^3$ or $h=4/3^{\frac{3}{4}}=1{\cdot}75$.

The actual value of h for the moon and earth is about 4, and hence if the moon-earth system were started with less than $\frac{7}{16}$ of its actual moment of momentum, it would not be possible for the two bodies to move so that the earth should always show the same face to the moon.

Again if we travel along the line of momentum there must be some point for which yx^3 is a maximum, and since $yx^3=n/\Omega$ there must be some point for which the number of planetary rotations is greatest during one revolution of the satellite, or shortly there must be some configuration for which there is a maximum number of days in the month.

Now yx^3 is equal to $x^3(h-x)$, and this is a maximum when $x=\frac{3}{4}h$ and the maximum number of days in the month is $(\frac{3}{4}h)^3(h-\frac{3}{4}h)$ or $3^3h^4/4^4$; if h is equal to 4, as is nearly the case for the earth and moon, this becomes 27. **Maximum number of days in the month.**

Hence it follows that we now have very nearly the maximum number of days in the month. A more accurate investigation in my paper on the "Precession of a Viscous Spheroid," showed that taking account of solar tidal friction and of the obliquity to the ecliptic the maximum number of days is about 29, and that we have already passed through the phase of maximum.

We will now consider the physical meaning of the several parts of the figures.

It will be supposed that the resultant moment of momentum of the whole system corresponds to a clockwise rotation.

Now imagine two points with the same abscissa, one on the momentum line and the other on the energy curve, and suppose the one on the energy curve to guide that on the momentum line.

Then since we are supposing frictional tides to be raised on the planet, therefore the energy must degrade, and however the two points are set initially, the point on the energy curve must always slide down a slope carrying with it the other point.

Now looking at fig. 1 or 2, we see that there are four slopes in the energy curve, two running down to the planet, and two others which run down to the minimum. In fig. 3 on the other hand there are only two slopes, both of which run down to the planet. **Various modes of degradation according to initial circumstances.**

In the first case there are four ways in which the system may degrade, according to the way it was started; in the second only two ways.

i. Then in fig. 1, for all points of the line of momentum from C through E to infinity, x is negative and y is positive; therefore this indicates an anti-clockwise revolution of the satellite, and a clockwise rotation of the planet, but the moment of momentum of planetary rotation is greater than that of the orbital motion. The corresponding part of the curve of energy slopes uniformly down, hence however the system be started, for this part of the line of momentum, the satellite must approach the planet, and will fall into it when its distance is given by the point k.

Various modes of degradation according to initial circumstances.

ii. For all points of the line of momentum from D through F to infinity, x is positive and y is negative; therefore the motion of the satellite is clockwise, and that of the planetary rotation anti-clockwise, but the moment of momentum of the orbital motion is greater than that of the planetary rotation. The corresponding part of the energy curve slopes down to the minimum b. Hence the satellite must approach the planet until it reaches a certain distance where the two will move round as a rigid body. It will be noticed that as the system passes through the configuration corresponding to D, the planetary rotation is zero, and from D to B the rotation of the planet becomes clockwise.

If the total moment of momentum had been as shown in fig. 3, then the satellite would have fallen into the planet, because the energy curve would have no minimum.

From i and ii we learn that if the planet and satellite are set in motion with opposite rotations, the satellite will fall into the planet, if the moment of momentum of orbital motion be less than or equal to or only greater by a certain critical amount (viz. $4/3^{\frac{3}{4}}$, in our special units), than the moment of momentum of planetary rotation, but if it be greater by more than a certain critical amount the satellite will approach the planet, the rotation of the planet will stop and reverse, and finally the system will come to equilibrium when the two bodies move round as a rigid body, with a long periodic time.

Case of Martian Satellite.

iii. We now come to the part of the figure between C and D. For the parts AC and BD of the line AB in fig. 1, the planetary rotation is slower than that of the satellite's revolution, or the month is shorter than day, as in one of the satellites of Mars. In fig. 3 these parts together embrace the whole. In all cases the satellite approaches the planet. In the case of fig. 3, the satellite must ultimately fall into the planet; in the case of figs. 1 and 2 the satellite will fall in if its distance from the planet is small, or move round along with the planet as a rigid body if its distance be large.

General case for satellites of solar system.

For the part of the line of momentum AB, the month is longer than the day, and this is the case of all known satellites except the nearer one of Mars. As this part of the line is non-existent in fig. 3, we see that the case of all existing satellites (except the Martian one) is comprised within this part of figs. 1 and 2. Now if a satellite be placed in the condition A, that is to say, moving rapidly round a planet, which always shows the same face to the satellite, the condition is clearly dynamically unstable, for the least disturbance will determine whether the system shall degrade down the slopes *ac* or *ab*, that is to say, whether it falls into or recedes from the planet. If the equilibrium breaks down by the satellite receding, the recession will go on until the system has reached the state corresponding to B. Compare § 778" (g).

The point P, in fig. 2, shows approximately the present state of the earth and moon, viz., when $x = 3{\cdot}2$, $y = {\cdot}8$ *.

It is clear that, if the point *l*, which indicates that the satellite is just touching the planet, be identical with the point A, then the two bodies are in effect parts of a single body in an unstable configuration. If, therefore, the moon was originally part of the earth, we should expect to find A and *l* identical. The figure 2, which is drawn to represent the earth and moon, shows that there is so close an approach between the edge of the shaded band and the intersection of the line of momentum and curve of rigidity, that it would be scarcely possible to distinguish them on the figure. Hence, there seems a considerable probability that the two bodies once formed parts of a single one, which broke up in consequence of some kind of instability. This view is confirmed by the more detailed consideration of the case in the paper on the "Precession of a Viscous Spheroid," and subsequent papers, which have appeared in the Philosophical Transactions of the Royal Society. Suggested origin of the moon. Compare § 778" (i).

The remainder of the paper, of which this Appendix forms a part, is occupied with a similar graphical treatment of the problem involved in the case of a planet and satellite or a system of two stars, each raising frictional tides in the other, and revolving round one another orbitally. This problem involves the construction of a surface of energy. Double-star system.

* The proper values for the present configuration of the earth and moon are $x = 3{\cdot}4$, $y = {\cdot}7$. Figure (2) was drawn for the paper as originally presented to the Royal Society, and is now merely reproduced.

INDEX.

The arabic numbers refer to sections. The numbers I. *and* II. *preceding them indicate the volume (parts of the original first volume) in which they will be found. The capital letters refer to the Appendixes. In the case of the Appendixes A_0, A, B, references are given to the pages of* I.; *the other Appendixes are at the ends of their respective volumes.*

CPSIA information can be obtained
at www.ICGtesting.com
Printed in the USA
BVHW04s2241080418
512812BV00009B/61/P